QUANTUM THEORY OF THE ELECTRON LIQUID

Modern electronic devices and novel materials often derive their extraordinary properties from the intriguing, complex behavior of large numbers of electrons forming what is known as an electron liquid. This book provides an in-depth introduction to the physics of the interacting electron liquid in a broad variety of systems, including metals, semiconductors, artificial nano-structures, atoms, and molecules.

One-, two- and three-dimensional systems are treated separately and in parallel. Different phases of the electron liquid, from the Landau Fermi liquid to the Wigner crystal, from the Luttinger liquid to the quantum Hall liquid, are extensively discussed. Both static and time-dependent density functional theory are presented in detail. Although the emphasis is on the development of the basic physical ideas and on a critical discussion of the most useful approximations, the formal derivation of the results is highly detailed and based on the simplest, most direct methods. A self-contained, comprehensive presentation of the necessary techniques, from second quantization to canonical transformations to both zero and finite temperature Green's functions is provided.

This comprehensive text will be of value to graduate students in physics, electrical engineering and quantum chemistry, as well as practicing researchers in those areas.

GABRIELE F. GIULIANI is a Professor of Physics at Purdue University, Indiana, USA. Since gaining his PhD at the Scuola Normale Superiore, Pisa, Italy in 1983, he has held postdoctoral positions at Purdue University and Brown University. He has also conducted research at the University of Pisa, the University of Rome, and the International Centre for Theoretical Physics, Trieste, Italy. He joined the faculty of Purdue University in 1984 and has been a Professor of Physics since 1995. Professor Giuliani's main areas of research are many-body theory, transport in low-dimensional electronic systems, and superconductivity. He has published more than 60 papers.

GIOVANNI VIGNALE is the Millsap Professor of Physics at the University of Missouri-Columbia. After graduating from the Scuola Normale Superiore in Pisa in 1979 and gaining his PhD at Northwestern University in 1984, he has carried out research at the Max-Planck-Institute for Solid State Research in Stuttgart, Germany and Oak Ridge National Laboratory. He was made a Fellow of the American Physical Society in 1997. Professor Vignale's main areas of research are many-body theory and density functional theory, and he has over 100 papers in print.

QUANTUM THEORY OF THE ELECTRON LIQUID

Gabriele Giuliani
Purdue University

Giovanni Vignale
University of Missouri

CAMBRIDGE UNIVERSITY PRESS
Cambridge, New York, Melbourne, Madrid, Cape Town, Singapore, São Paulo

Cambridge University Press
The Edinburgh Building, Cambridge CB2 8RU, UK

Published in the United States of America by Cambridge University Press, New York

www.cambridge.org
Information on this title: www.cambridge.org/9780521821124

First published 2005
This digitally printed version 2008

A catalogue record for this publication is available from the British Library

ISBN 978-0-521-82112-4 hardback
ISBN 978-0-521-52796-5 paperback

To my parents Ada and Federico
GV

To Pamela, Daniela, Adriana and Giuseppe
GFG

Contents

Appendices

Preface

The electron liquid paradigm is at the basis of most of our current understanding of the physical properties of electronic systems. Quite remarkably, the latter are nowadays at the intersection of the most exciting areas of science: materials science, quantum chemistry, nano-electronics, biology, and quantum computation. Accordingly, its importance can hardly be overestimated. The field is particularly attractive not only for the simplicity of its classic formulation, but also because, by its very nature, it is still possible for individual researchers, armed with thoughtfulness and dedication, and surrounded by a small group of collaborators, to make deep contributions, in the best tradition of "small science".

When we began to write this book, more than five years ago, our goal was to bring up to date the masterly treatise of the 1960s by Pines and Noziéres on quantum liquids – the very same book on which we had first studied the subject. There were good reasons for wanting to do this. During the past 40 years the field has witnessed momentous developments. Advances in semiconductor technology have allowed the realizations of ultra-pure electron liquids whose density, unlike that of the ones spontaneously occurring in nature, can be tuned by electrical means, allowing a systematic exploration of both strongly and weakly correlated regimes. Most of these system are two- or even one-dimensional, and can be coupled together in the form of multi-layers or multi-wires, opening observational possibilities that were undreamed of in the 1960s. On the theoretical side, quantum Monte Carlo methods, implemented on powerful computers, have allowed an essentially exact determination of the ground-state energy of the electron liquid, and have provided partial answers to the still open question of the structure of its phase diagram. The Landau theory of the Fermi liquid, which in the 1960s was in its infancy, has been fully vindicated by detailed and often painstaking microscopic calculations. The emergence of density functional theory as the standard tool for the calculation of the electronic structure of matter has anointed the electron liquid as the holder of the prototypical correlations in electronic systems.

Starting from the 1980s some truly revolutionary concepts have emerged, which we wanted to be well represented in our book: for example, the notion of fractionally charged excitations in one-dimensional systems and in the quantum Hall liquid, the Luttinger liquid model for one-dimensional systems and for the edges of a quantum Hall liquid, and the beautiful composite-fermion theory of the quantum Hall effect. These concepts transcend the traditional Landau picture of the interacting electron liquid as the "continuation" of the noninteracting one. What makes these developments particularly significant is the fact that the new scenarios have been found to emerge in the low-energy and low-temperature limit, subverting a traditional wisdom which saw in the high-energy limit the true frontier of physics.

As we advanced in the project, the natural desire to make the book truly accessible to graduate students and as self-contained as possible, and the explicit design to discuss openly and critically the approximations on which the theory is based, caused the length of the manuscript to grow beyond our original intentions. We hope that the reader will find the length of the treatise justified by a corresponding increase in clarity and readability. In the end, however, we just had to throw in the towel, and accept to live with many imperfections we were not able to get rid of. To assuage this problem we point the reader to the book web site **http://www.missouri.edu/~physvign/qtel.htm** where we will post the corrections and clarifications that will undoubtedly prove necessary a few seconds after publication. We apologize in advance to all the authors whose important work has not been properly referenced.

A few words concerning our choice of topics are now in order. As a rule, we have refrained from treating in any depth a topic when we had nothing to add to treatments already in print. Examples of such reasoned omissions are the electron–phonon interaction, superconductivity, weak localization theory, the renormalization group, classical plasma analogies, and lattice models of strong correlation. For most of these topics, we have limited ourselves to broad-brush discussions, summarizing the main results of more technical treatments. On the other hand, the reader will find in this book several in-depth discussions of topics never presented before in a pedagogical form, such as the time-dependent current density functional theory, the visco-elastic description of the collective dynamics of the electron liquid, with and without a magnetic field, and the renormalized hamiltonian approach to Fermi liquid theory.

Many people from around the world have in a variety of ways helped us to complete this work. Our special thanks go to David Ceperley, Bahman Davoudi, Paola Gori-Giorgi, Jainendra Jain, Albert Overhauser, Marco Polini, George Simion, and Carsten Ullrich. It is also a pleasure to thank Klaus Capelle, Stefano Chesi, Irene D'Amico, Roberto D'Agosta, Maurizio Ferconi, Michael Geller, Matt Grayson, Catalina Marinescu, Gerardo Ortiz, Vincenzo Piazza, Vittorio Pellegrini, Zhixin Qian, Roberto Raimondi, Stefano Roddaro, Gaetano Senatore, Carlos Wexler and, of course, the Purdue and UMC graduate students who for the last few years have had to put up with lectures based on early, unpolished drafts of this book. GV also thanks the National Science Foundation for providing continuous support during the completion of this work.

As in any endeavor of this magnitude motivations must exist that come from the depths of one's soul. In our case love for the still intriguing field of interacting electrons and inspiration for this work have sprang from our fortunate and early interaction with our mentors and electron gas theory pioneers Franco Bassani, Mario P. Tosi, Albert W. Overhauser, and our beloved Kundan S. Singwi who is no longer with us to see this.

We finally must also express our gratitude to and hope for forgiveness from our families, especially the children who have endured for much too long a time high doses of paternal absenteeism.

PS: Due to life's serendipitous nature, this book has already met with a great deal of success, having afforded one of us (GFG) the possibility of remaining in touch with a professional endeavor during particularly challenging times. In this respect GFG must also heartily thank Geoffrey B. Thompson, John H. Edmonson and Leonard L. Gunderson for having given him, through their singular abilities a chance of seeing the completion of this work.

Gabriele F. Giuliani and Giovanni Vignale
West Lafayette (IN) and Columbia (MO), May 2004

1

Introduction to the electron liquid

1.1 A tale of many electrons

The twentieth century has witnessed some of the greatest revolutions in the history of science: in physics we have gone from classical to quantum mechanics, our views of space and time have forever been changed by the theory of relativity, a comprehensive microscopic theory of matter has emerged, the invention of solid state electronics has ushered in the information age. Through all these changes one thing has always kept the center stage: the electron. Discovered by J. J. Thomson in 1897, the electron was the first elementary particle to be clearly identified. Although the discovery took place in an entirely classical context,[1] further investigation of the properties of the electron soon led the way into the new world of quantum mechanics. In particular, the Davisson–Germer electron diffraction experiment (1927) established the wave-like properties of matter particles; the discovery of the half-integer spin (Goudsmit and Uhlenbeck, 1925) and the related statistical properties of the electron (Pauli, 1925; Fermi, 1925; Dirac, 1929) laid the foundation for the understanding of the atomic structure; Dirac's relativistic treatment of the electron (Dirac, 1928) created a new branch of theoretical physics: quantum field theory.

It was clear from the beginning that electrons are a pervasive component of matter.[2] An electrical current in a metal wire is nothing but a flow of many electrons: this is the basic assumption of the classic Drude–Lorentz model of electrical conduction in metals, which was proposed as early as 1900. In 1911, however, Kamerlingh Onnes astonished the world (and himself) by showing that a metal (Hg) cooled below liquid helium temperature lost all electrical resistance and became a perfect conductor. The phenomenon of superconductivity, which was completely inexplicable within the framework of the Drude–Lorentz model (or, for that matter, of its subsequent quantum mechanical versions), was perhaps the first demonstration that the properties of a many-electron system can be dramatically different from the properties of the individual electrons that constitute it. In this case, it was felt that

---

[1] For a detailed account of the discovery of the electron see the excellent book by Steven Weinberg *The discovery of subatomic particles* (Freeman, New York, 1990). Thomson referred to electrons as *cathode rays*: the modern name was apparently introduced by the physicist and astronomer G. J. Stoney (1826–1911).

[2] In the words of J. J. Thomson (1897): *"... we have in the cathode rays matter in a new state, a state in which the subdivision of matter is carried very much further than in the ordinary gaseous state: a state in which all matter – that is, matter derived from different sources such as hydrogen, oxygen, etc. – is of one and the same kind; this matter being the substance from which the chemical elements are built up."*

(a)   (b)

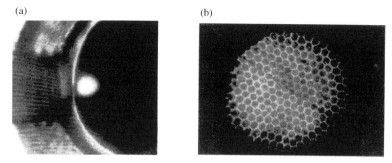

Fig. 1.1. Two extreme regimes of a many-electron system, namely (a) a liquid droplet of nearly free electrons and holes in Ge, visualized through the electron–hole recombination radiation (photograph by Wolfe *et al.*, 1975), (b) a crystalline arrangement of electrons on the surface of ^{4}He, visualized through the dimples it produces on the latter (each dimple accommodates several thousands of electrons) (courtesy of P. Leiderer, 1997).

some kind of subtle cooperative effect was responsible for the vanishing of the resistance: the effect was, in fact, so subtle that it had to wait until 1957 for a convincing theoretical explanation (Bardeen, Cooper and Schrieffer, 1957). On the other hand, several properties of electrons in metals, for example the diamagnetism and the low temperature specific heat, could be successfully explained, as early as 1928, in terms of a quantum mechanical model (Pauli, 1927; Sommerfeld, 1928) which treated the electrons as independent particles in a box, subject only to the *Pauli exclusion principle* (no two electrons can be in the same quantum state). In a way, this is no less amazing than superconductivity: how can the electrons be considered independent, when we know that they strongly interact via the coulomb force?[3] Here the puzzle is not the presence, but the *absence* of strong collective effects.

This book is largely about the competition between individualistic and collective behavior of electrons in metals, or in systems that are conceptually equivalent to metals, even though they are realized in non-metallic hosts. The individualistic behavior is best exemplified by the Pauli exclusion principle, which leads electrons to avoid each other while still roaming more or less freely in space (see Fig. 1.1(a)). This "Pauli repulsion" is ultimately responsible for the structure of the atom and the stability of ordinary matter. The collective behavior, on the other hand, is manifested in the emergence of spatially ordered structures, such as the Wigner crystal first predicted by Eugene Wigner in 1934 (a closely related structure is shown in Fig. 1.1(b)), and magnetically ordered structures, such as the ferromagnetic electron liquid predicted by Bloch in 1929. Additional collective phases are characterized by anomalous electrical conduction properties, such as superconductivity and fractionally quantized Hall conductivity. All these phases, with the exception of the last one, are broken-symmetry phases, that is, the symmetry of the state is lower than that of the environment in which the electrons live.

---

[3] A similar but less striking situation occurs in atomic physics where a *shell model* based on independent electrons and the Pauli exclusion principle leads to an excellent qualitative description of the energy levels.

This book is above all about the role of electron–electron interactions, which are always assumed to be relevant, unless proved otherwise.[4] For this reason, we choose from the beginning to focus on a model translationally invariant system (the so-called *jellium model*), which allows us to study the interaction effects without the distracting and often inessential influence of the periodic crystal potential.

The deepest insight about the nature of an interacting electron system at low temperature[5] is provided by the character of the *elementary excitations* the system supports in a given phase. This point of view, originally introduced by Landau, was forcefully promoted during the 1950s by pioneers like Bohm, Pines, Noziéres, and Anderson.[6] Elementary excitations are approximate, long-lived eigenstates of a many-body system, which evolve essentially independently of each other and are characterized (in a translationally invariant system) by a definite energy–momentum relation. Landau, in particular, noticed that a system of particles satisfying Fermi statistics (not necessarily electrons!) admits a phase that is qualitatively similar to an independent particle model, in the sense that there is a one-to-one correspondence between the low-energy elementary excitations of the interacting system and those of the noninteracting model. These elementary excitations are called *quasiparticles*, and while they may have effective masses orders of magnitude larger than the bare particle masses, they still exhibit the phenomenology of a noninteracting system, thus explaining the success of the Pauli–Sommerfeld theory of electrons in metals.

Different phases will have different elementary excitations, of course. For example, the low energy excitations of a Wigner crystal are quantized lattice vibrations (phonons), similar to those of an ordinary crystal. In a ferromagnetic liquid, on the other hand, the low energy excitations are spin waves. Both are examples of *collective excitations*. Even the basic *Landau Fermi liquid phase* – as the independent-particle-like phase introduced by Landau is usually called – supports, in addition to individual quasiparticles, collective density oscillations, which are known as *plasmons* in electron liquids and zero-sound waves in charge-neutral Fermi liquids (e.g. ^{3}He); but these are relatively high in energy and do not contribute much to the low-temperature properties of the system.

The appearance of the BCS theory of superconductivity in 1957 unsettled the potentially dull scenario of the Landau theory, showing that an interacting electron liquid can support homogeneous, nonmagnetic states, that are qualitatively different from any noninteracting system. In fact the elementary excitations of a superconductor arise when one breaks the "bond" that ties two electrons together in the ground-state, which is assumed to consist of an infinite sea of electron pairs known as *Cooper pairs*.[7] Because the creation of a quasiparticle in a superconductor disrupts the ground-state by breaking a bond between two electrons, we see that there must be a finite energy cost – the so-called energy *gap* – associated with the creation of such a quasiparticle. This is a qualitative difference with a noninteracting

----

[4] To emphasize the relevance of the interactions we normally describe our system as an *electron liquid* as opposed to an electron gas.

[5] The sense in which the temperature must be low will be explained in the next section.

[6] Many of the pioneering papers on the electron liquid are collected in the book by D. Pines (Pines, 1962).

[7] The "bond" that holds together the electron in a Cooper pair should not be regarded as a physical bond in real space: rather, it is a correlation that forces one member of the pair to have momentum $-\vec{p}$ whenever the other has momentum $\vec{p}$.

system, and shows that a superconducting state cannot be obtained as a smooth continuation of the latter.

The BCS theory marked the beginning of the contemporary era in which the interest of condensed matter physicists is focused on identifying new models of correlated electronic behavior. One of the first suggestions in this direction was Overhauser's proposal of spin and charge density waves, which will be discussed in greater detail later in this and the next chapter. A nice description of the work on strong electronic correlations during the 1960s and 1970s can be found in the review article by Singwi and Tosi (1981). It was during this period that the description of the electron liquid in terms of effective local potentials came to full maturity, and a powerful new tool for the study of inhomogeneous systems – density functional theory – made its first appearance on the scene. These efforts culminated with the first essentially exact determination of the ground-state energy of selected phases of the electron liquid by the numerical quantum Monte Carlo method (Ceperley and Alder, 1980).

In the meanwhile, major technological advances in the processing of semiconductor materials allowed the realization of systems in which both the density and the effective dimensionality of the electron liquid could be artificially controlled. Both innovations offered good prospects of breaking away from the Landau Fermi liquid paradigm; the first by exploring a regime in which the interaction potential energy becomes gradually more important than the kinetic energy, and the second by effectively reducing the kinetic energy through geometrical constraints on the motion of the electrons.

It turns out that the very existence of the Landau Fermi liquid phase depends crucially on dimensionality, and becomes more problematic as the latter is decreased. Already in two dimensions the Fermi liquid is found to be marginally stable in the presence of disorder. In one dimension, the Landau theory fails completely, and one must resort to a novel paradigm which, in recent years, has come to be known as the *Luttinger liquid*. The elementary excitations of a Luttinger liquid are collective density waves (very much like plasmons in two and three dimensions): however, the injection of an additional electron in this system disrupts pre-existing correlations, causing the appearance of a *pseudogap* in the one-particle density of states.[8]

The quest for novel non-Fermi liquid phases in two-dimensional systems has been long and frustrating. A refreshing exception has come from the discovery, during the 1980s, that the electronic state responsible for the fractionally quantized Hall effect in two-dimensional electron layers at high magnetic field is a new type of incompressible, homogeneous quantum liquid. This state can be beautifully described in terms of a new type of particle – the *composite fermion* – which effectively carries a fractional electric charge. The special feature of this state is that the kinetic energy is completely suppressed by the strong magnetic field: in this case the noninteracting model would have an infinite degeneracy, and perturbation theory is not just incorrect, but impossible.

---

[8] The difference between a strict gap and a pseudogap is that in the latter case the extra electron can be accommodated at an arbitrarily small energy cost, but the probability of succeeding to do so tends to zero as the energy cost tends to zero.

A very recent, surprising twist in the theory in the quantum Hall liquid is the realization that the *edge* of this system – a kind of one-dimensional system in its own right – is a realization of the *chiral* Luttinger liquid, a special type of Luttinger liquid in which the collective excitations travel only in one direction, determined by the gradient of the density and by the direction of the magnetic field.

This completes our bird's eye view of the tale we are going to tell, sometimes in excruciating detail, in the next few hundred pages. A chronology of significant events in the history of the electron liquid is presented in Table 1.1.

Ironically, we end up saying very little about superconductivity, partly because it is a lattice-related effect, and mostly because we have nothing to add to what is already said in many excellent textbooks. Other painful omissions concern the physics of the strongly correlated systems (metal oxides, heavy-fermion materials, high-temperature superconductors) for which we feel a coherent theoretical picture has not yet emerged. Disorder effects (including electron–phonon interactions) are only touched at the most elementary level. On the other hand, Chapter 7 presents a rather detailed treatment of the *density functional theory* – a powerful method through which the information gained from the study of the homogeneous electron liquid can be put to work in realistic situations including systems of interest in chemistry and biology.

1.2 Where the electrons roam: physical realizations of the electron liquid

Let us now examine the systems in which the physics of the electron liquid can be studied most clearly, i.e., with the least possible interference from foreign effects. The most important attribute of such systems is a clean and smooth microscopic environment in which the electrons can roam essentially freely, interacting only among themselves. The number of such systems has considerably increased in the past few decades, and is still steadily growing as new ways are found to control the size and shape of semiconductor devices. We find it to convenient to classify these systems, according to their effective dimensionality, as three-, two-, and one-dimensional systems.

1.2.1 Three dimensions

Natural realizations of a three-dimensional electron liquid (3-DEL) are provided by most elemental metals. The main requirement is that the Fermi surface of the conducting electrons be nearly spherical, as this indicates that the electrons move as approximately free particles, i.e., with a quadratic relation between their kinetic energy and their momentum. The very nature of the alkali, the so-called "simple" metals, i.e. Li, Na, K, Rb, and Cs makes them the obvious 3-DEL prototypes.[9]

---

[9] The Brillouin zone of an alkali metal has 12 congruent faces, each perpendicular to a [110] lattice vector. The distance of each face from $\vec{k} = 0$ is 14% larger than k_F. The energy gaps at the zone surfaces, caused by the ionic potential, are about 0.4 eV in K, and leave the Fermi surface undistorted from its spherical shape. See the textbook by Ashcroft and Mermin (1976) for the definition of Brillouin zone and Fermi surface in a crystal.

Table 1.1. *A chronology of significant events in the history of the electron liquid.*

Year	Event
1897	Discovery of the electron and measure of e/m
1900	Drude–Lorentz model of electrical conduction
1902	Discovery of the photoelectric effect
1905	Einstein's interpretation of the photoelectric effect
1911	Measure of the electron charge
	Discovery of superconductivity
1924	De Broglie hypothesis
1925	Uhlenbeck and Goudsmit's proposal of the electron spin
	Fermi statistics
	Pauli exclusion principle
1926	Schrödinger equation
1927	Davisson–Germer experiment (electron diffraction)
	Thomas–Fermi theory of the atom
	Slater exchange
1928	Dirac's relativistic theory of the electron
	Bloch's theory of electrons in solids
	Pauli–Sommerfeld free-electron theory of metals
1929	Bloch's theory of exchange and ferromagnetic instability
1934	Wigner's proposal of the Wigner crystal
1947	Invention of the transistor
1956	Landau theory of Fermi liquids
1957	BCS theory of superconductivity
	Random phase approximation (RPA)
	Hubbard local field factor
1962	Overhauser's proposal of spin and charge density waves
1964	Hohenberg–Kohn–Sham density functional theory
1968	Singwi–Tosi–Land–Sjölander (STLS) generalization of the RPA
1974	Solution of the Luttinger model by Mattis and Lieb
1980	Discovery of the integral quantum Hall effect
	Ceperley–Alder Quantum Monte Carlo
1981	Haldane's definition of the "Luttinger liquid" paradigm
1982	Discovery of the fractional quantum Hall effect
1983	Laughlin's theory of the quantum Hall liquid and fractionally charged excitations
	Discovery of heavy fermion materials
1986	Discovery of high-temperature superconductors
1989	Jain's composite fermions theory of the quantum Hall effect
1990	Wen's definition of the chiral Luttinger liquid
1996	Observation of the chiral Luttinger liquid

(a)   (b)

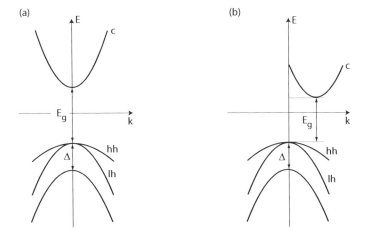

Fig. 1.2. (a) Schematic band diagram of a typical direct gap semiconductor, such as GaAs, featuring one conduction band (c) and three valence bands known as the "heavy hole" (hh), the "light hole" (lh), and the spin–orbit split-off band. The different curvatures of these bands near $k = 0$ define the effective band masses of electrons, heavy holes, and light holes. (b) Same as (a) for a typical indirect gap semiconductor, such as Ge or Si. In this case there are several equivalent conduction bands (4 in Ge and 6 in Si) with minima at wavevectors $\vec{k} \neq 0$ (only one is shown in the figure). The number of equivalent conduction bands is determined by symmetry.

Each metal comes with its own fixed electron density. The available values of the *dimensionless Wigner–Seitz radius* r_s, i.e., the radius of the sphere that encloses, on the average, one electron, expressed in units of the Bohr radius $a_B \equiv \frac{\hbar^2}{me^2} \simeq 0.529$ Å, range from 3.25 for Li to 5.62 for Cs (see Table 1.2 later in this chapter). This constitutes the standard metallic density range. The remaining elemental metals have r_s values between 1.87 (Be) and 3.71 (Ba).[10] A remarkable property of the alkali metals is the relative softness of their ionic lattice as compared to that of other elemental metals. In particular their measured *bulk moduli* range from $1.43 \times 10^{10} \frac{erg}{cm^3}$ for Cs to $11.5 \times 10^{10} \frac{erg}{cm^3}$ of Li, while the values for the noble metals Cu, Ag, and Al are of the order of $10^{12} \frac{erg}{cm^3}$. This feature will allow us to describe the alkali metals in terms of a *deformable jellium* model, while the *rigid jellium* model is more suitable for the noble metals.

Doped semiconductors, such as Ge, Si, and GaAs can also host a 3-DEL, with the advantage that the electron density is, to a certain extent, controllable via the dopant concentration. A fundamental property of doped semiconductor systems is that on the scale of the typical interelectron separation, ~ 100 Å, the crystal lattice can be treated as a smooth background of positive charge, whose only role is to set the effective mass, m_b, of the electrons (determined by the curvature of the conduction bands near their minimum – see Fig. 1.2), and the effective coulomb interaction, $\frac{e^2}{\epsilon_b r}$, between them – ϵ_b being the static dielectric constant

---

[10] The necessary density data can be found in the classic resource by Wyckoff (1963). Useful tables can also be found in the textbook by Ashcroft and Mermin (1976).

of the semiconductor.[11] Unfortunately, this nice feature is offset by the fact that the dopant atoms usually introduce a considerable amount of disorder, which may be more disruptive than a periodic crystal potential.[12]

A nice system, which does not suffer from this problem, is the *electron–hole liquid* in pure semiconductors. A pure semiconductor, at the absolute zero of temperature, has a completely full valence band and an empty conduction band (again a consequence of the Pauli exclusion principle!) separated by an energy gap of the order of 1 eV. If we shine laser light of the appropriate frequency on the system, we can excite a large number of electrons in the conduction band, while leaving an equal number of "holes" in the valence band. Clearly, this is not an equilibrium situation, and the electron–hole liquid will quickly disappear by recombination if the laser is turned off. However, the recombination time is very long ($\sim 10^{-6}$ s) in comparison with the thermal equilibration time ($\sim 10^{-12}$ s), and this leaves ample time for the system to be studied in a condition of quasi-equilibrium. The photograph shown in Fig. 1.1(a) was taken precisely during this time. Notice that, in contrast with conventional one-component systems, whose density is fixed either by the lattice parameter or by the dopant's concentration, the density of an electron–hole liquid is intrinsically determined by the minimum of its own free energy: as such, its equilibrium density does not depend on the laser's intensity, but only on the temperature.

Finally, we mention a recently developed system, the *wide parabolic quantum well* which is, in a sense, the closest realization of the idea of free electrons on a uniform background of positive charge (Rimberg and Westervelt, 1989). Recall that according to Gauss' theorem a uniform slab of charge of infinite area and finite thickness perpendicular to the z-axis produces in its interior an electrostatic potential $\varphi(z) = 2\pi\rho z^2$, where ρ is the charge density and z is zero at the center of the slab. Therefore, by engineering the conduction band edge of a semiconductor to smoothly follow a parabolic profile proportional to z^2 one can simulate the electrostatic potential of a uniform charge distribution.[13] The well can then be filled with electrons donated by remote dopants: these electrons will automatically adjust their density so as to neutralize the effective "charge density" of the well, which is set by the curvature of the parabolic potential (see Fig. 1.3 and Exercise 1). In this manner, uniform electron densities of the order of 10^{15} cm^{-3}, or $r_s \sim 6 - 7$ have been realized.

1.2.2 Two dimensions

It has been known for a long time that electrons localized at the surface of elemental metals can provide a natural realization of the two-dimensional electron liquid (Tamm, 1932). For instance, the Cu [111] surface accommodates electronic states that are exponentially

---

[11] The fundamental reason for this, a priori surprising, behavior, lies in the validity of the *effective mass approximation*, whereby Bloch states near the bottom of, say, a conduction band, can be put in one-to-one correspondence with the states of a free particle (Luttinger and Kohn, 1957).

[12] Another peculiarity of semiconductor systems is that, due to the presence of several equivalent conduction bands (4 in Ge and 6 in Si) electrons are endowed with an additional discrete quantum number – the *valley index*. This can have interesting consequences.

[13] This is accomplished by *alloying* GaAs with Al$_x$Ga$_{1-x}$As with an Al concentration x that varies quadratically as a function of z.

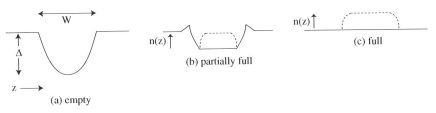

Fig. 1.3. A wide parabolic well is a nearly ideal realization of the 3-D jellium model. Here the energy of the bottom of the conduction band (solid line), and the electron density profile (dashed line) are plotted as functions of z – the coordinate perpendicular to the plane of the well – in the following three cases (a) empty well, (b) partially full well, and (c) full well (the width W of the empty well is typically a few thousands Å and its depth Δ is 10–100 meV).

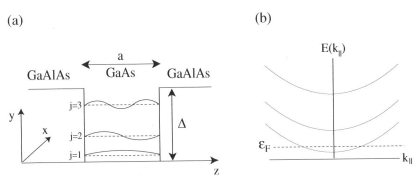

Fig. 1.4. Schematic diagram of the electron subbands in a two-dimensional GaAs-GaAlAs quantum well. (a) Quantized energy levels and wave functions in the direction z perpendicular to the plane of the well. (b) Electron energies vs wave vector in the plane of the quantum well. Only the lowest subband is occupied in this example.

attenuated in the direction perpendicular to the surface, while completely delocalized in the plane of the surface. Recently, this has been vividly visualized by means of low temperature scanning tunneling microscopy techniques (Crommie *et al.*, 1993).

A two-dimensional electron gas can also be realized on the free surface of liquid He. Electrons spilled onto such a surface, are bound to it by the image potential. The electrons are otherwise free to move on the surface as a handy potential barrier of the order of 1 eV prevents them from diffusing into the bulk. In this manner it is possible to create 2D electronic systems with an areal charge density of the order of 10^9 cm^{-2}. These two examples share with the 3D elemental metals the drawback of not having a tunable electron density.

The development of semiconductor electronics in the second half of the 1900s has had a major impact on our ability to study the physics of low-dimensional electron liquids. It is now possible to fabricate high quality Si- and GaAs-based quantum wells in which the carriers motion in one of the space directions is quantized. The system is schematically shown in Fig. 1.4. In the simplest case, a single electron (or hole) behaves as a free quantum

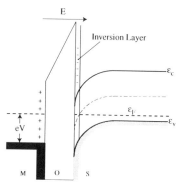

Fig. 1.5. Schematic band diagram of an MOS inversion layer structure. The 2DEL is formed in the semiconductor, near the interface with the oxide.

particle as far as its motion in the plane of the quantum well (x–y) is concerned. Thus, its energy levels are described by a wavevector $\vec{k}_\parallel$ in the x–y plane, and by a discrete index $j = 1, 2 ..$ in the z-direction: the complete energy spectrum, shown in Fig. 1.4, consists of a set of two-dimensional *subbands*, one for each value of j.[14]

According to elementary quantum mechanics, the bottoms of the subbands occur at energies $\varepsilon_j = \frac{\hbar^2}{2m_b a^2} j^2$ where a is the width of the well. By making a suitably small, it is possible to create a situation in which the energy separation between the two lowest subbands is much larger than the characteristic kinetic energy of the in-plane motion.[15] In such a situation all the electrons reside in the lowest subband and their kinetic energies depend on just the two-dimensional wave vector $\vec{k}_\parallel$: as long as there is not enough energy to populate higher subbands we have effectively realized a two-dimensional electron liquid.

For our purposes the most interesting feature of these systems is represented by the possibility of varying the electronic density over an unprecedentedly broad range, thereby allowing the study of electron–electron interactions effects in a variety of regimes.[16] Even more important is the fact that control over the density is achieved by electrical means, that is, by changing the voltage of certain control electrodes (known as gates), as opposed to the traditional doping method which, as we saw, introduces disorder in the system.

Fig. 1.5 shows a typical realization of electrical control of the density, namely a Metal–Oxide–Semiconductor (MOS) structure grown on the surface of p-doped Si. In the bulk

[14] In reality, deviations from the ideal free-particle picture arise from the periodicity of the crystal potential in the plane of the well: this affects the holes in particular, leading to additional band structure, and to non-parabolicity and warping of the hole bands. In a typical direct gap semiconductor the confinement potential lifts the degeneracy of heavy and light hole bands at $k = 0$ and leads to an interesting reversal of roles: the formerly heavy hole band actually has the smaller effective mass in the plane of the well, while the formerly light hole band becomes the heavier in the plane (see Chow, Koch, and Sargent, 1994).

[15] The precise condition for this to happen is $na^3 \ll 1$, where n is the average three-dimensional electron density in the well.

[16] In dealing with two-dimensional systems we will adhere to the standard practice of calling "density" what is really the *sheet density*, i.e., the ordinary three-dimensional density integrated over the transverse direction (perpendicular to the plane of the well). The sheet density of 2DEL systems typically ranges from 10^{12} cm^{-2} to 10^9 cm^{-2} and the corresponding r_s values range from ~1 to ~20.

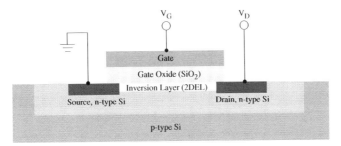

Fig. 1.6. Schematic diagram of a Metal–Oxide–Semiconductor Field Effect Transistor (MOSFET). An electric potential applied to the gate changes the density of carriers in the underlying inversion layer, and thus controls the current that flows between the source and the drain electrodes.

of silicon the majority carriers are holes, which means that the Fermi level is close to the edge of the valence band. As one approaches the interface with the oxide (SiO_2), both the conduction and the valence band edges begin to bend downward, due to the electrostatic potential created by the charges on the metal gate. Eventually, the conduction band gets closer to the Fermi level than the valence band: within a thin layer next to the oxide the sytem is effectively n-doped, and electrons accumulate. Because the carriers at the surface and those in the bulk have opposite polarities, this system is often called an *inversion layer*. Clearly, the density of electrons in the inversion layer can be controlled by changing the voltage on the gate. In addition, the original dopants are far removed from the layer and have therefore limited influence on the motion of the electrons.

Quantum well technology has had a tremendous impact on fundamental research on the electron liquid, as one can see from the following examples:

- The discovery of the fractional quantum Hall effect would have been impossible without the availability of very clean low-density electron liquids in GaAs.[17]
- The *scaling theory of localization* (Abrahams *et al.*, 1979), according to which a disordered 2DEL always behaves as an insulator at sufficiently low temperature and large length scales, was created in response to the development of semiconductor inversion layers.[18]
- Strong evidence for the existence of a Wigner crystal has been found in low-density electron and hole liquids in Si-MOS and GaAs quantum wells.
- The possibility of fabricating multiple quantum wells close enough to interact and/or exchange electrons has led to the study of the *layered electron liquid* and the exploration of new concepts such as *coulomb drag, interlayer coherence*, and *charge-transfer instability*.

The technological applications of 2DELs have been equally impressive. Just to limit ourselves to a single example, in Fig. 1.6 we show the schematics of an MOS-based field-effect transistor (MOSFET) which replaces the time-honored bipolar transistor of

---

[17] The standard measure of cleanness of an electronic system is the *electron mobility* μ, which gives the average drift velocity of the electrons in response to an applied electric field E: $v_d = \mu E$. In a perfectly pure system μ would be infinite. The electron liquids in which the fractional quantum Hall effect was first discovered had $\mu \sim 10^5$ cm^2 V/s. Nowadays the purest electron liquids in GaAs quantum wells have mobilities of the order of 10^7 cm^2 V/s.
[18] In clean two-dimensional devices the localization length exceeds the size of the system. It is for this reason that the system behaves as a metal. Metallic behavior in 2D can also arise from electron–electron interations.

(a) Quantum Wire   (b) Carbon Nanotube

Fig. 1.7. Some modern marvels in one-dimension: a quantum wire on GaAs (from Roukes *et al.*, 1987) and a carbon nanotube (courtesy of Richard Smalley).

Bardeen, Brattain, and Shockley in contemporary computer chips. The caption gives a brief description of the operation of this device.[19]

1.2.3 One dimension

Experimental studies of quasi-one dimensional conductors began during the 1970s with the development of the first organic conductor, the charge-transfer salt TTF-TCNQ (*tetrathiafulvalene-tetracyanoquinodimethane*). This is a highly anisotropic material comprised of chains of molecules with a large room temperature conductivity along the chain direction. The material stays metallic down to a temperature of 53 K where it undergoes a charge-density-wave transition (see Section 1.7.4) to an insulating state.

TTF-TCNQ, and the related *Bechgaard salts* $(TMTSF)_2X$ and $(TMTTF)_2X^{[20]}$ serve as a prime examples of one-dimensional conductors and, at least before turning into charge or spin density waves, or superconductors, show some of the characteristic features of a Luttinger liquid, like a strong suppression of the density of one-electron states at the Fermi level.

Unfortunately, the band structure of these materials is far too complex to permit a clear disentanglement of electron–electron interaction effects from one-electron effects. More importantly, although the systems are strongly anisotropic, they do not satisfy the technical condition for quasi-one dimensionality, namely that all the electrons should reside in a single one-dimensional subband.

True one-dimensional systems can be fabricated by etching nanowires on the surface of semiconductors (see Fig. 1.7(a)). In fact, the existence of nanowires supporting multiple one-dimensional channels was demonstrated in a landmark experiments (van Wees *et al.*, 1980), which proved the quantization of the conductance of clean quantum wires in integer multiples of e^2/h – each e^2/h coming from a 1D channel and one spin orientation. It is probably just a matter of time before single channel wires will be realized, that are clean

---

[19] In modelling low-dimensional *gated devices*, such as MOSFETs, one must keep in mind that the coulomb interaction between the electrons is screened by image charges induced on the gates: this leads to an effective interaction potential that decreases faster than $\frac{1}{r}$ at large distances, and can lead to modifications in the phase diagram and the excitation spectrum of the system.

[20] TMTSF stands for *tetramethyl tetraselena fulvalene* and X stands for the ions PF_6, AsF_6, $Cl\,O_4$.

enough to permit a quantitative investigation of Luttinger liquid effects. To date, however, the best confirmations of Luttinger liquid theory have come from experiments involving tunneling of electrons into the edges of a quantum Hall liquid in GaAs quantum wells, as will be discussed in Chapter 10.

Finally, we note that a very promising class of quasi-1D systems is provided by carbon nanotubes of which one example is shown in Fig. 1.7(b) (Iijima, 1991). In contrast to organic polymers these systems have a relatively simple, one dimensional band structure (Dresselhaus *et al.*, 1996) and are promising candidates for studies of many-body effects in low-dimensional electron liquids.

1.3 The model hamiltonian

1.3.1 Jellium model

Unlike other popular models in condensed matter physics, the electron liquid is described by a truly fundamental hamiltonian, $\hat{H}$, representing N electrons that interact with one another and with a uniform background of positive charge:

$$\hat{H} = \sum_i \frac{\hat{p}_i^2}{2m} + \frac{1}{2} \sum_{i \neq j} \frac{e^2}{|\hat{\vec{r}}_i - \hat{\vec{r}}_j|} + \hat{H}_{e-b} + \hat{H}_{b-b} . \tag{1.1}$$

Here $\hat{\vec{r}}_i$ and $\hat{\vec{p}}_i$ are the quantum mechanical operators of the coordinate and the momentum of the i-th electron, m is the electron mass, and e is the magnitude of the electron charge. Notice that operators are always distinguished by a "hat" accent. The electrons are assumed to be enclosed in a d-dimensional cubic box of volume L^d. In the *thermodynamic limit* both N and L tend to infinity in such a way that the average electronic density $n \equiv \frac{N}{L^d}$ remains constant.

The last two terms in $\hat{H}$ represent the electron–background and background–background interactions and are given by:

$$\hat{H}_{e-b} = -e^2 \int d\vec{r} \int d\vec{r}\,' \frac{\hat{n}(\vec{r})n_b(\vec{r}\,')}{|\vec{r} - \vec{r}\,'|} , \tag{1.2}$$

and

$$\hat{H}_{b-b} = \frac{e^2}{2} \int d\vec{r} \int d\vec{r}\,' \frac{n_b(\vec{r})n_b(\vec{r}\,')}{|\vec{r} - \vec{r}\,'|} , \tag{1.3}$$

where

$$\hat{n}(\vec{r}) = \sum_{i=1}^{N} \delta \left(\vec{r} - \hat{\vec{r}}_i \right) \tag{1.4}$$

is the electron number density operator, and

$$en_b(\vec{r}) = en \tag{1.5}$$

is the (uniform) charge density of the background.

In writing Eq. (1.5) we have replaced the actual structure of the background on which the electrons roam (e.g., the atomic lattice) by a homogeneous jelly-like continuum of positive charge: hence the name of "jellium model".[21] The main idea is to get rid of the complications associated with the structure of the host material, while focusing on the distinctive effects of the electron–electron interaction. Although this simplification is probably too drastic in traditional metallic systems, it is quite realistic in the semiconductor systems described in Section 1.2.[22]

In spite of its apparent simplicity, and even after decades of intense theoretical work, the jellium model presents an unsolved, challenging problem. This chapter surveys some of the "partial truths" that have been established so far, highlights the main issues, and sets the stage for the specific discussions provided in the remaining chapters of the book.

1.3.2 Coulomb interaction regularization

Before proceeding we must clear up a disturbing feature of the jellium model, namely the fact that, due to the long range of the coulomb interaction, the integrals in Eqs. (1.3) and (1.2) fail to converge in the thermodynamic limit $L \to \infty$. In brief, this difficulty is solved by showing that the infinities in Eqs. (1.3) and (1.2) are cancelled by an equivalent infinity of opposite sign in the electron–electron part of the hamiltonian. This cancellation would not take place without the external background, which is thus seen to be essential to the existence of the thermodynamic limit.

To explicitly show the cancellation of the infinities in $\hat{H}$ we resort to a mathematical procedure known as *coulomb interaction regularization*. The idea is to replace the coulomb interaction with one with a very long, yet finite range, and to study the system in the limit in which this range is much larger than any other physical length scale except L. A convenient (but by no means necessary) choice is the Yukawa interaction

$$v(r, \kappa) = e^2 \frac{e^{-\kappa r}}{r} , \tag{1.6}$$

which has a characteristic range κ^{-1}. The physical hamiltonian is recovered by letting $\kappa \to 0$ *after* going to the thermodynamic limit $L \to \infty$. The procedure is justified *a posteriori* since the physical quantities obtained by taking the limits in this order are finite and independent of κ.

We begin by rewriting the interaction potential in the form

$$v(r, \kappa) = \frac{1}{L^d} \sum_{\vec{q}} v_q(\kappa) e^{i\vec{q}\cdot\vec{r}} , \tag{1.7}$$

---

[21] Strictly speaking Eq. (1.1) describes an infinitely *rigid* jellium model. As anticipated in Section 1.2.1 we will also have some use for a *deformable* jellium model, in which the charge density of the background can be locally changed at a finite energy cost.

[22] Another limitation of the jellium hamiltonian is that relativistic effects, such as retarded and/or velocity-dependent electron–electron interactions, and spin–orbit coupling, are neglected. These effects are negligible when the average speed of the electrons is much smaller than the speed of light – a condition that is well satisfied in the systems under study.

where

$$v_q(\kappa) \equiv \int v(r, \kappa) \, e^{-i\vec{q}\cdot\vec{r}} \, d\vec{r} \tag{1.8}$$

is the Fourier transform of the interaction potential.[23] Notice that the presence of the fi-nite length scale κ^{-1} ensures the existence of the the $q \to 0$ limit of $v_q(\kappa)$. It is now a simple exercise to show that the electron–electron interaction term in Eq. (1.1) can be rewritten as

$$\begin{aligned}
\hat{H}_{e-e} &\equiv \frac{1}{2} \sum_{i \neq j} \frac{e^2}{|\hat{\vec{r}}_i - \hat{\vec{r}}_j|} \\
&= \frac{1}{2L^d} \sum_{\vec{q}} v_q(\kappa) \left[\hat{n}_{-\vec{q}} \hat{n}_{\vec{q}} - \hat{N} \right] ,
\end{aligned} \tag{1.10}$$

where $\hat{n}_{\vec{q}}$ is the Fourier transform of the electronic density operator:

$$\hat{n}_{\vec{q}} = \sum_{i=1}^{N} e^{-i\vec{q}\cdot\hat{\vec{r}}_i} , \tag{1.11}$$

and $\hat{n}_{\vec{q}=0} = \hat{N}$ is just the number operator $\hat{N}$. Notice that the term proportional to $\hat{N}$ in $\hat{H}_{e-e}$ is needed to disallow the interaction of an electron with itself, i.e., to implement the restriction $i \neq j$ in the sum of Eq. (1.1).

In a similar way, one can rewrite $\hat{H}_{e-b}$ and $\hat{H}_{b-b}$ in terms of $\hat{N}$, so the complete hamiltonian takes the form

$$\begin{aligned}
\hat{H} &= \sum_i \frac{\hat{p}_i^2}{2m} + \frac{1}{2L^d} \sum_{\vec{q}} v_q(\kappa) \left[\hat{n}_{-\vec{q}} \hat{n}_{\vec{q}} - \hat{N} \right] \\
&\quad - v_{\vec{q}=0}(\kappa) n_B \hat{N} + \frac{1}{2} v_{\vec{q}=0}(\kappa) n_B^2 L^d ,
\end{aligned} \tag{1.12}$$

where the last two terms on the right-hand side describe the electron–background and the background–background interactions respectively.

We now observe that the $\vec{q} = 0$ contribution to the electron–electron part of the hamiltonian is:

$$(\hat{H}_{e-e})_{\vec{q}=0} = \frac{v_{\vec{q}=0}(\kappa)}{2L^d} \left(\hat{N}^2 - \hat{N} \right) . \tag{1.13}$$

Combining this with the electron–background and background–background terms, and

---

[23] We follow the common practice of imposing *periodic boundary conditions*, i.e., the coordinates $\vec{r}$ and $\vec{r} + L\vec{e}_\alpha$ (where $\vec{e}_\alpha$ is a unit vector in the x, y, or z direction) are considered identical. Accordingly, the sum in Eq. (1.7) runs over a discrete set of wave vectors, such that each component of $\vec{q}$ is a multiple of $\frac{2\pi}{L}$. In the $L \to \infty$ limit the discrete sum can be replaced by an integral in the following manner:

$$\frac{1}{L^d} \sum_{\vec{q}} \cdots \stackrel{L\to\infty}{\to} \int \frac{d\vec{q}}{(2\pi)^d} \cdots \tag{1.9}$$

making use of the fact that in the thermodynamic limit $\hat{N} = nL^d = n_B L^d$ we find that

$$\frac{(\hat{H}_{e-b})_{\vec{q}=0} + (\hat{H}_{b-b})_{\vec{q}=0} + (\hat{H}_{e-e})_{\vec{q}=0}}{L^d} = -\frac{nv_{\vec{q}=0}(\kappa)}{2L^d} \,, \tag{1.14}$$

which tends to zero when L tends to infinity for $\kappa > 0$. Thus, in the thermodynamic limit, all the terms that depend on $v_{\vec{q}=0}(\kappa)$ cancel out. It is only at this point that we are allowed to take the limit $\kappa \to 0$, which is no longer singular, since $\vec{q}$ is never equal to zero. The final expression for the regularized jellium hamiltonian is therefore

$$\hat{H} = \sum_i \frac{\hat{p}_i^2}{2m} + \frac{1}{2L^d} \sum_{\vec{q} \neq 0} v_q \left[\hat{n}_{-\vec{q}} \hat{n}_{\vec{q}} - \hat{N} \right] \,, \tag{1.15}$$

where

$$v_q = \lim_{\kappa \to 0} v_q(\kappa) \tag{1.16}$$

is our operational definition of the Fourier transform of the coulomb interaction. The entire regularization procedure amounts in the end to dropping the $\vec{q} = 0$ component of the interaction, as well as the electron–background and background–background terms. This hamiltonian is perfectly well defined in the thermodynamic limit. For simplicity, in what follows we will indicate with $\hat{H}_{e-e}$ not the original expression (1.10), but the second term of Eq. (1.15).

Notice that, up to this point, we have not made any use of the explicit form (1.6) of the interaction. It turns out that Fourier transforming Eq. (1.6) is in fact one of the most efficient ways to calculate v_q. It is, in fact an easy exercise to show that

$$v_q(\kappa) = \begin{cases} \frac{4\pi e^2}{q^2 + \kappa^2} \,, & \text{3-D}, \\ \frac{2\pi e^2}{\sqrt{q^2 + \kappa^2}}, & \text{2-D} \,. \end{cases} \tag{1.17}$$

The one-dimensional case is more tricky, as one also needs a short-distance cutoff a to avoid the non-integrable divergence of the interaction at $r = 0$. This can be done by modelling the 1D system as an infinitely long 3D cylinder of radius a, with a so small that the transverse motion of the electrons is effectively "frozen" in the lowest energy state described by a gaussian wave function (see Appendix 1 for a detailed discussion of this procedure). The resulting expression for the Fourier transform of Eq. (1.6) is

$$v_q(\kappa, a) = -e^2 e^{(q^2 + \kappa^2)a^2} \,\text{Ei}[-(q^2 + \kappa^2)a^2] \,, \quad \text{1-D} \,, \tag{1.18}$$

where $\text{Ei}(x)$ is the exponential-integral function (Gradshteyn and Rhyzhik, 8.21). Taking the limit $\kappa \to 0$ in the above expressions we finally get

$$v_q = \begin{cases} \frac{4\pi e^2}{q^2}, & 3D, \\ \frac{2\pi e^2}{q}, & 2D, \end{cases} \tag{1.19}$$

and

$$v_q(a) = -e^2 e^{q^2 a^2} \operatorname{Ei}\left(-q^2 a^2\right) , \quad 1D .$$ (1.20)

Notice that in one dimension $v_q(a) \simeq -2e^2 \ln qa$ for $qa \ll 1$ while $v_q(a) \simeq \frac{e^2}{q^2 a^2}$ for $qa \gg 1$.

1.3.3 The electronic density as the fundamental parameter

The jellium model hamiltonian (1.15) appears to contain three parameters, namely the electron mass m, the electron charge e, and, implicity the electron density n. Actually, we now show that in three and two dimensions there is only one fundamental parameter, the *dimensionless electronic density*, $\bar{n} = n a_B^d$, where d (2 or 3) is the number of spatial dimensions and a_B is the *Bohr radius*

$$a_B = \frac{\hbar^2}{me^2} .$$ (1.21)

To prove this, we express the electronic coordinates in terms of the "natural" length scale $r_s a_B$ – the "typical" distance between two electrons. It is customary to define this typical distance as the radius of the d-dimensional sphere that encloses, on the average, exactly one electron. This gives

$$\frac{1}{n} = \begin{cases} \frac{4\pi}{3}(r_s a_B)^3 , & 3D, \\ \pi (r_s a_B)^2 , & 2D, \\ 2r_s a_B , & 1D, \end{cases}$$ (1.22)

and, accordingly,

$$r_s a_B = \begin{cases} \left(\frac{3}{4\pi n}\right)^{\frac{1}{3}} , & 3D, \\ \left(\frac{1}{\pi n}\right)^{\frac{1}{2}} , & 2D, \\ \frac{1}{2n} , & 1D. \end{cases}$$ (1.23)

Let us now define the rescaled variables

$$\hat{\vec{r}} = \frac{\hat{\vec{r}}}{r_s a_B} , \quad \hat{\vec{p}} = \frac{r_s a_B}{\hbar} \hat{\vec{p}} , \quad \tilde{\vec{q}} = r_s a_B \vec{q} , \quad \tilde{L} = \frac{L}{r_s a_B} :$$ (1.24)

It is straightforward to verify that the hamiltonian, written in terms of these variables, takes the form

$$\hat{H} = \left(\frac{1}{r_s^2} \sum_i \hat{\vec{p}}_i^2 + \frac{1}{r_s \tilde{L}^d} \sum_{\tilde{\vec{q}} \neq 0} v_{\tilde{q}} \left[\hat{n}_{-\tilde{\vec{q}}} \hat{n}_{\tilde{\vec{q}}} - \hat{N} \right] \right) Ry ,$$ (1.25)

where the *Rydberg*

$$Ry = \frac{e^2}{2a_B} ,$$ (1.26)

is the global energy scale. Notice that the rescaled hamiltonian (i.e., the quantity in the brackets of Eq. (1.25)) does not contain any longer the physical constants e and m. Moreover, the quantity $\frac{N}{L^d}$ can be identified, in the thermodynamic limit, with $n(r_s a_B)^d$, which is just a numerical constant. Thus, the physical properties of the hamiltonian depend only on the value of r_s. This result depends crucially on the homogeneous form of the coulomb interaction (i.e., on the fact that $v_q \propto q^{2-d}$) in two and three dimension, and would not hold true for different forms of the interaction such as the one dimensional "coulomb interaction" $v_q(a)$ of Eq. (1.20).

The form of Eq. (1.25) clearly shows that the kinetic energy and the coulomb interaction scale with different powers of r_s, and that r_s represents, in a sense, the ratio of the average potential energy to the average kinetic energy of the system.[24] This scaling property allows us to clearly identify two quite different limiting behaviors of the electron liquid. For small r_s the kinetic energy rules the physics of the problem and the system behaves as a non-interacting Fermi gas. In the opposite limit, the interaction term prevails and the system behaves as a classical assembly of charged particles. It is widely believed that in this case the electrons form a crystalline structure, that is generally referred to as a *Wigner crystal*.

As it is clear from the two asymptotic scenarios, as r_s varies from zero (high density limit) to infinity (low density limit) the ground-state energy of the system (alongside its associated many-body wave function) varies in a nontrivial way as the electron liquid may go through several phase transformations in which its symmetry properties change. It is a challenge for the modern theory of condensed matter to establish the correct picture, i.e. the *phase diagram* of the electron liquid.

Before closing this section we note that in many realistic systems, the material serving as background for the electron liquid sets the values of the "effective mass" m_b and the "effective coulomb interaction" $\frac{e^2}{\epsilon_b r}$ to be used in the jellium model hamiltonian. For example, in GaAs, a typical host of high-purity two-dimensional electron liquids, the effective mass of electrons near the bottom of the conduction band is $m_b \simeq 0.067\, m$. At the same time, this material has a static dielectric constant $\epsilon_b \sim 12$, which reduces the strength of the coulomb interaction as if the electron had a reduced charge $\frac{e}{\sqrt{\epsilon_b}}$. In the light of the results obtained in this section, these background effects can be taken into account by redefining r_s in terms of an *effective Bohr radius*

$$a_B^* = \frac{\hbar^2 \epsilon_b}{m_b e^2} \,, \tag{1.27}$$

which is typically much larger than the standard 0.5294 Å value. The energy scale is also modified to an *effective Rydberg* $Ry^* = \frac{e^2}{2\epsilon_b a_B^*}$, which is typically much smaller than the standard value. Table 1.2 helps to quickly determine r_s in cases of practical interest.

---

[24] In a classical electron liquid the quantum scales $r_s a_B$ and Ry are replaced by $n^{-1/d}$ and $k_B T$ respectively, where k_B is the Boltzmann constant and T is the temperature. Accordingly, the role of r_s is played by $\Gamma = \frac{e^2 n^{1/d}}{k_B T}$.

Table 1.2. *Quick conversion table between the electron density and* r_s. *Here* m_b *is the band effective mass in units of the bare electron mass and* ϵ_b *is the static dielectric constant of the background material.*

d	$\bar{n} = n a_B^d$	r_s
3	$\bar{n} = \dfrac{n}{10^{22} cm^{-3}}$	$r_s \simeq \dfrac{5.45}{\bar{n}^{\frac{1}{3}}} \dfrac{m_b}{\epsilon_b}$
2	$\bar{n} = \dfrac{n}{10^{11} cm^{-2}}$	$r_s \simeq \dfrac{337}{\bar{n}^{\frac{1}{2}}} \dfrac{m_b}{\epsilon_b}$
1	$\bar{n} = \dfrac{n}{10^{7} cm^{-1}}$	$r_s \simeq \dfrac{9.45}{\bar{n}} \dfrac{m_b}{\epsilon_b}$

1.4 Second quantization

Although the real space representation, Eq. (1.1) is useful in some situations, more often a considerable formal simplification can be achieved by rewriting the hamiltonian (1.1) in the *occupation number representation* – a procedure that is also known as *second quantization*.

Forced by necessity in quantum field theory, where the number of particles is a dynamical variable, second quantization is widely used in nonrelativistic many-body theory (where the particle number is conserved) as a powerful tool to take advantage of the indistinguishability of the particles.

The reader not familiar with this technique will find in the next few pages a self-contained description of the method. The "experts" can skip to the next section in which we work out the second-quantized representation of the electron–gas hamiltonian. In Appendix 2 second-quantized expressions for a number of useful operators are provided. The more advanced topics of normal-ordering and Wick's theorem are tackled in Appendix 3.

1.4.1 Fock space and the occupation number representation

Second quantization is made possible by the fact that identical particles are indistinguishable: as we will see, the great advantage of this representation is to eliminate any reference to the individual electrons. Let $|\phi_\alpha\rangle$ be a complete orthonormal set of *single particle* states, where α is a short-hand for the whole set of quantum numbers, (e.g., momentum, spin, band index, etc.) needed to specify each state.[25] The wave functions of these states will be denoted by $\phi_\alpha(\vec{r}, s)$ where $\vec{r}$ and s are the position and the spin of the particle. Indistinguishability implies that one can specify the state of an N-particle system simply by saying how many particles are in each of the single-particle states, without specifying "who" those particles

---

[25] The choice of this set is suggested by practical considerations: for example, in a translationally invariant system, plane wave states are the natural choice.

are. We thus denote by

$$|N_1, N_2, \ldots, N_\alpha, \ldots\rangle \tag{1.28}$$

the state of an N-particle system in which there are *definitely* N_1 particles in state $|\phi_1\rangle$, N_2 particles in state $|\phi_2\rangle$, and so on. The N_α's are called *occupation numbers*, and their knowledge uniquely specifies the state. States of this form will be referred to as *Fock states*.

There are, of course, some restrictions on the admissible values of the occupation numbers. First of all they must all be non negative integers, and secondly they must add up to the total number of particles N in the system – if the latter has a definite value. Additional restrictions, may arise from the *statistics* of the particles under consideration.

It turns out that there are two types of particles in nature: bosons and fermions.

A system of N identical bosons has the property that its wave function must be *symmetric* with respect to the interchange of any two particles. Since the state $\phi_\alpha(\vec{r}_1, s_1)\phi_\alpha(\vec{r}_2, s_2)\ldots\phi_\alpha(\vec{r}_N, s_N)$ satisfies this property for any N, we see that in a bosonic system the occupation number of a single-particle state can be arbitrarily large.

In contrast to this, a system of N identical fermions has the property that its wave function must be *antisymmetric* with respect to the interchange of any two particles. Since the state $\phi_\alpha(\vec{r}_1, s_1)\phi_\alpha(\vec{r}_2, s_2)$ does not survive antisymmetrization, this property implies that each single particle state cannot be occupied by more than one electron. The occupation numbers of a fermionic system can only take the values 0 and 1. This extremely important property is known as the *Pauli exclusion principle*.

Thus, the Fock states of an N-fermion system are described by a set of N occupied states $\{\alpha_1, \alpha_2, \ldots, \alpha_N\}$ such that $N_\alpha = 1$ when α belongs to the set, and $N_\alpha = 0$ otherwise. We shall make use of the symbol

$$|\alpha_1, \alpha_2, \ldots, \alpha_N\rangle \tag{1.29}$$

to denote such states. In the ordinary Schrödinger representation, these states are represented by an antisymmetrized product of single-particle wave functions in the following manner:

$$\psi_{\alpha_1, \alpha_2, \ldots, \alpha_N}(\vec{r}_1, s_1; \vec{r}_2, s_2; \ldots; \vec{r}_N, s_N)$$
$$= \frac{1}{\sqrt{N!}} \sum_{P\{1\ldots N\}} S_P \phi_{\alpha_1}(\vec{r}_{P1}, s_{P1})\phi_{\alpha_2}(\vec{r}_{P2}, s_{P2})\ldots\ldots\phi_{\alpha_N}(\vec{r}_{PN}, s_{PN}), \tag{1.30}$$

where the sum runs over the permutations P of the set $\{1\ldots N\}$, Pi denotes the image of $i, (1 \leq i \leq N)$ under the permutation P, S_P is the "signature" of the permutation, which equals 1 if the permutation can be realized by an *even* number of interchanges, and -1 otherwise, and $1/\sqrt{N!}$ is a normalization factor originating from the fact that Eq. (1.30) is the sum of $N!$ mutually orthogonal terms. Notice the ordering of the state labels on the right-hand side, which matches the ordering on the left-hand side.

Eq. (1.30) can also be written more compactly as the *determinant* of the $N \times N$ matrix $\phi_{\alpha_i}(\vec{r}_j, s_j)$ $(1 \leq i, j \leq N)$

$$\psi_{\alpha_1, \alpha_2, \ldots, \alpha_N}(\vec{r}_1, s_1; \vec{r}_2, s_2; \ldots; \vec{r}_N, s_N) = \frac{1}{\sqrt{N!}} \det[\phi_{\alpha_i}(\vec{r}_j, s_j)], \tag{1.31}$$

where the row index runs over the labels of the occupied states, $(i = 1, \ldots, N)$ and the column index over the particles labels $(j = 1, \ldots, N)$: wave functions of this type are known as *Slater determinants*. It is evident from this representation that any attempt to put more than one electron in the same state will give the null wave function – the determinant of a matrix with two identical rows being zero. Formulas analogous to Eq. (1.30) can be written for N-boson wave functions, but we shall not pursue the subject here.

It is important to appreciate that *not all* the states in the Hilbert space of an N-Fermion system have the form of Eq. (1.29). It is possible, for example, to superimpose two different Fock states to obtain a new state in which the *average occupation number* n_α is neither definitely zero nor definitely 1. What makes the Fock states (1.29) special is that they are the *eigenstates* of the observables $\hat{N}_1, \hat{N}_2, \ldots, \hat{N}_\alpha$, which measure the number of particles in states $1, 2, \ldots, \alpha$. It is evident, on physical grounds, that these observables exist (the occupation numbers can, in principle, be measured) and are described by mutually commuting operators (measurements of different occupation numbers can be done independently). It follows therefore from the general principles of quantum mechanics that their simultaneous eigenstates form a *complete* set of states in the Hilbert space.

We now wish to choose Fock states as the basis for an *occupation number representation* in the N-particle Hilbert space. There is, however, a small technical complication. Although the states (1.28) are physically well defined, their overall sign is ambiguous. This can be seen most clearly from the determinantal representation: any interchange in the ordering of two state labels amounts to an interchange of two rows in the Slater determinant, and therefore flips the sign of the wave function. In order to unambiguosly determine the sign of the basis states we introduce, once and for all, a *conventional ordering* of the single particle state labels, and stipulate that our basis set will be constituted by Fock states (Eq. (1.28)) with the labels α_i arranged according to the conventional ordering, i.e., $\alpha_1 < \alpha_2 < \cdots < \alpha_N$. At this point any physical state in the Hilbert space can be written as a linear superposition of the basis states, and the amplitudes of the superposition constitute its "wave function" in the occupation number representation.

1.4.2 Representation of observables

Let us now consider the problem of expressing *symmetric operators*[26] in the occupation number representation. When such an operator acts on a Fock state it produces, in general, a linear superposition of Fock states that differ from the initial one in the values of some of the occupation numbers. Each of these states can be obtained through the successive application of elementary *creation* and *destruction operators*, which increase or decrease by *one* a given occupation number. Thus every operator, no matter how complicated, can ultimately be decomposed into a sum of products of creation and destruction operators that carry out in a step-by-step fashion the required changes in the values of the occupation numbers.

We are now in a position to clearly state the program of second quantization. It consists of two steps: in the first we define the elementary creation and destruction operators

---

[26] "Symmetric operators" are, by definition, invariant under permutations of the particles. These are the only operators that always preserve the symmetry of the wave function and hence respect the principle of indistinguishability of the particles.

and work out their algebra; in the second we discover the rule for expressing a general symmetric operator as a sum of products of creation and destruction operators. We discuss the case of fermions in detail. The modifications needed for bosons are briefly described in the footnotes.

Step 1: Definition of creation and destruction operators

We define the fermion creation operator $\hat{a}_\alpha^\dagger$ for the single-particle state $|\phi_\alpha\rangle$ by specifying its action on the basis states (1.28). The creation operator must satisfy the following requirements:

(i) $\hat{a}_\alpha^\dagger$ must increase the occupation number N_α by one if $N_\alpha = 0$, and destroy the state (i.e., give zero) if $N_\alpha = 1$ (Pauli exclusion principle);

(ii) The states $|\alpha_1, \alpha_2, \ldots, \alpha_N\rangle$ defined in Eq. (1.30) must be obtainable from the "vacuum state";

$$|0\rangle = |0, 0, \ldots, 0\rangle, \tag{1.32}$$

i.e. the state with no particles at all, through successive applications of creation operators:

$$|\alpha_1, \alpha_2, \ldots, \alpha_N\rangle = \hat{a}_{\alpha_1}^\dagger \hat{a}_{\alpha_2}^\dagger \ldots \hat{a}_{\alpha_N}^\dagger |0\rangle, \tag{1.33}$$

for any ordering of the α's.

Since the state vectors $|\alpha_1, \alpha_2, \ldots, \alpha_N\rangle$ are antisymmetric with respect to the interchange of any two occupied state labels (row indices in the Slater determinant) we see that condition (ii) can only be satisfied if the creation operators *anticommute* with one another, that is, if

$$\hat{a}_\alpha^\dagger \hat{a}_\beta^\dagger = -\hat{a}_\beta^\dagger \hat{a}_\alpha^\dagger \tag{1.34}$$

for any α and β. This, of course, implies $(\hat{a}_\alpha^\dagger)^2 = 0$ – a requirement already evident from (i).

All of the above requirements are satisfied by the following definition of the creation operator:

$$\hat{a}_\alpha^\dagger |N_1, N_2, \ldots, N_\alpha, \ldots\rangle = \delta_{N_\alpha 0}(-1)^{S_\alpha} |N_1, N_2, \ldots, N_\alpha + 1, \ldots\rangle, \tag{1.35}$$

where the multiplying sign $(-1)^{S_\alpha}$ is determined by the parity of the number S_α of occupied states that *precede* α in the conventional ordering of the single-particle states:

$$S_\alpha \equiv \sum_{\gamma < \alpha} N_\gamma \tag{1.36}$$

and $\delta_{ij} = 1$ if $i = j$ and 0 otherwise.

Let us check that the operators defined in this way do what they are supposed to do. Condition (i) is obviously satisfied. Condition (ii) is satisfied when the α_i's are in the conventional order because then the phase factors $(-1)^{S_\alpha}$ are all equal to $+1$: at the time that $\hat{a}_\alpha^\dagger$ operates the occupation numbers of the states $|\phi_\gamma\rangle$ with $\gamma < \alpha$ are still zero. But, if (ii) holds for one particular ordering of the α_i's, it will necessarily hold for any other ordering, provided the creation operators satisfy the anticommutation relations (1.34). The latter can easily be verified by direct calculation. Assuming, for definiteness $\alpha < \beta$ and making use of Eq. (1.35) we obtain

$$\hat{a}_\alpha^\dagger \hat{a}_\beta^\dagger |N_1, \ldots, N_\alpha, \ldots, N_\beta, ..\rangle = \delta_{N_\alpha 0}\delta_{N_\beta 0}(-1)^{S_\alpha + S_\beta}|N_1, \ldots, N_\alpha + 1, \ldots, N_\beta + 1, \ldots\rangle.$$
(1.37)

Here the phase factor $(-1)^{S_\alpha}$ is not affected by the previous action of $\hat{a}_\beta^\dagger$, because $\alpha < \beta$. On the other hand, after interchanging the order of the two operators we get

$$\hat{a}_\beta^\dagger \hat{a}_\alpha^\dagger |N_1, \ldots N_\alpha, ..N_\beta, ..\rangle = \delta_{N_\beta 0}\delta_{N_\alpha 0}(-1)^{S_\beta + 1}(-1)^{S_\alpha}|N_1, \ldots N_\alpha + 1, .. N_\beta + 1, ..\rangle.$$
(1.38)

The phase factor associated with $\hat{a}_\beta^\dagger$ is now changed to $(-1)^{S_\beta + 1}$ because the previous action of $\hat{a}_\alpha^\dagger$ has increased by one the number of occupied states preceding β. Thus

$$\hat{a}_\beta^\dagger \hat{a}_\alpha^\dagger |N_1, N_2, \ldots N_\alpha, .. N_\beta, ..\rangle = -\hat{a}_\alpha^\dagger \hat{a}_\beta^\dagger |N_1, N_2, \ldots N_\alpha, .. N_\beta..\rangle$$
(1.39)

for *any* basis state, and therefore Eq. (1.34) holds as an operator identity.

Once the creation operators have been constructed, we see that their hermitian conjugates $\hat{a}_\alpha$ decrease the occupation number N_α by one if $N_\alpha = 1$, or destroy the state if $N_\alpha = 0$. This can easily be verified by transposing the matrix elements of $\hat{a}_\alpha^\dagger$ in the occupation number representation. It follows that $\hat{a}_\alpha$ *is* the destruction operator, and its action on the basis states is described by

$$\hat{a}_\alpha |N_1, N_2, \ldots N_\alpha \ldots\rangle = \delta_{N_\alpha 1}(-1)^{S_\alpha}|N_1, N_2, \ldots N_\alpha - 1 \ldots\rangle .$$
(1.40)

Taking the hermitian conjugate of Eq. (1.34) we see that the destruction operators also anticommute with one another and, in particular, $(\hat{a}_\alpha)^2 = 0$.

Let us now consider the action of the operator $\hat{a}_\alpha^\dagger \hat{a}_\alpha$ on one of the basis states. Making use of Eqs. (1.40) and (1.35) (in this order!) we obtain

$$\hat{a}_\alpha^\dagger \hat{a}_\alpha |N_1, N_2 \ldots N_\alpha \ldots\rangle = N_\alpha |N_1, N_2 \ldots N_\alpha \ldots\rangle .$$
(1.41)

This is the same as one would obtain by operating with the occupation number *operator* $\hat{N}_\alpha$ on the state in question. We conclude that

$$\hat{a}_\alpha^\dagger \hat{a}_\alpha = \hat{N}_\alpha.$$
(1.42)

In a similar manner we prove that

$$\hat{a}_\alpha \hat{a}_\alpha^\dagger = 1 - \hat{N}_\alpha.$$
(1.43)

Summing Eqs. (1.42) and (1.43) we obtain the important result

$$\hat{a}_\alpha^\dagger \hat{a}_\alpha + \hat{a}_\alpha \hat{a}_\alpha^\dagger = 1.$$
(1.44)

On the other hand, if $\alpha \neq \beta$, it is not difficult to verify from the definitions (1.40) and (1.35) that

$$\hat{a}_\alpha^\dagger \hat{a}_\beta = -\hat{a}_\beta \hat{a}_\alpha^\dagger , \quad (\alpha \neq \beta) .$$
(1.45)

Thus, in summary, we have shown that the creation and destruction operators obey the following *anti-commutation rules*:

$$\{\hat{a}_\alpha, \hat{a}_\beta\} = \{\hat{a}_\alpha^\dagger, \hat{a}_\beta^\dagger\} = 0 , \quad \{\hat{a}_\alpha, \hat{a}_\beta^\dagger\} = \delta_{\alpha,\beta} \tag{1.46}$$

where $\{\hat{A}, \hat{B}\} = \hat{A}\hat{B} + \hat{B}\hat{A}$ is the "anticommutator".[27]

The field operators

Creation and destruction operators can be introduced in conjunction with *any* complete set of single particle states. In particular, we can define the operators $\hat{\Psi}_\sigma(\vec{r})$ which destroy a particle in the eigenstate of position and spin with eigenvalues $\vec{r}$ and σ respectively. This set of operators is collectively described as an *electron field operator*. Thus our study of many-particle systems has smoothly led us into quantum field theory!

The relation between the field operator and any other set of creation operators is given by

$$\hat{\Psi}_\sigma(\vec{r}) = \sum_\alpha \phi_\alpha(\vec{r}, \sigma) \, \hat{a}_\alpha , \tag{1.47}$$

and similarly for the hermitian conjugate. It is a simple matter to prove that the following relations hold:

$$\{\hat{\Psi}_\sigma(\vec{r}), \hat{\Psi}_{\sigma'}(\vec{r}')\} = \{\hat{\Psi}_\sigma^\dagger(\vec{r}), \hat{\Psi}_{\sigma'}^\dagger(\vec{r}')\} = 0 , \tag{1.48}$$

and

$$\{\hat{\Psi}_\sigma(\vec{r}), \hat{\Psi}_{\sigma'}^\dagger(\vec{r}')\} = \delta_{\sigma\sigma'}\delta(\vec{r} - \vec{r}') . \tag{1.49}$$

Furthermore, it is clear from Eq. (1.42) that

$$\hat{n}_\sigma(\vec{r}) = \hat{\Psi}_\sigma^\dagger(\vec{r})\hat{\Psi}_\sigma(\vec{r}) , \tag{1.50}$$

is the operator of the density of σ-spin electrons.

Step 2: Representation of observables

We now turn to the problem of expressing physical observables in terms of creation and destruction operators.

One-particle operators

The simplest case is that of observables that are represented by symmetric "one-particle" operators. A one-particle operator $\hat{V}^{(1)}$, consists of a sum of N identical operators $\hat{V}_i$ acting only on the Hilbert space of the i-th electron

$$\hat{V}^{(1)} = \sum_{i=1,\dots,N} \hat{V}_i . \tag{1.51}$$

---

[27] For particles obeying Bose statistics, the definitions (1.35) and (1.40) are modified by dropping the sign and Pauli restriction factors and multiplying the kets on the right by $\sqrt{N_\alpha + 1}$ and $\sqrt{N_\alpha}$ respectively. In Eq. (1.46) the anti-commutators are simply replaced by commutators.

Typical examples are the particle density operator $\hat{n}(\vec{r}) = \sum_i \delta(\vec{r} - \hat{\vec{r}}_i)$ and the kinetic energy operator, both of which will be considered in detail in the next section.

When $\hat{V}_i$ operates on a one-electron wave function $\phi_\alpha(\vec{r}_i, \sigma_i)$ it produces a superposition of one-electron wave functions

$$\hat{V}_i \phi_\alpha(\vec{r}_i, s_i) = \sum_\beta V_{\beta\alpha} \phi_\beta(\vec{r}_i, s_i) \,, \tag{1.52}$$

with the amplitudes obtained from the equation

$$V_{\beta\alpha} \equiv \langle i, \beta | \hat{V}_i | i, \alpha \rangle$$
$$= \sum_{s,t} \int d\vec{r}_i \phi_\beta^*(\vec{r}_i, t) [\hat{V}_i]_{t,s} \phi_\alpha(\vec{r}_i, s) \,, \tag{1.53}$$

where we have explicitly introduced the spin indices of the operator $\hat{V}_i$. Notice that these amplitudes do not depend on i.

Making use of Eq. (1.52), it is not hard to see that the *complete* operator $\hat{V}^{(1)}$ acting on one of the basis states (1.33) generates a superposition of states in which each of the occupied one-particle states α_i is replaced in turn by β with an amplitude $V_{\beta\alpha_i}$:

$$\hat{V}^{(1)} |\alpha_1, \dots, \alpha_i, \dots, \alpha_N\rangle = \sum_\beta \sum_{i=1}^N V_{\beta\alpha_i} |\alpha_1, \dots, \alpha_i \to \beta, \dots, \alpha_N\rangle \,. \tag{1.54}$$

Here the ket $|\alpha_1, \dots, \alpha_i \to \beta, \dots, \alpha_N\rangle$ denotes the state obtained from $|\alpha_1, \dots, \alpha_i, \dots, \alpha_N\rangle$ upon replacing ϕ_{α_i} by ϕ_β.

Now observe that $|\alpha_1, \dots, \alpha_i \to \beta, \dots, \alpha_N\rangle$ is obtained from $|\alpha_1, \dots, \alpha_i, \dots, \alpha_N\rangle$ through the application of the operators $\hat{a}_\beta^\dagger$ and $\hat{a}_{\alpha_i}$ in the following manner:

$$|\alpha_1, \dots, \alpha_i \to \beta, \dots, \alpha_N\rangle = \hat{a}_\beta^\dagger \hat{a}_{\alpha_i} |\alpha_1, \dots, \alpha_i, \dots, \alpha_N\rangle \,. \tag{1.55}$$

This can be easily verified starting from the representation (1.33) of the basis states, and going through the following chain of transformations:

$$\hat{a}_\beta^\dagger \hat{a}_{\alpha_i} (\hat{a}_{\alpha_1}^\dagger \hat{a}_{\alpha_2}^\dagger \dots \hat{a}_{\alpha_i}^\dagger \dots \hat{a}_{\alpha_N}^\dagger) |0\rangle$$
$$= \hat{a}_{\alpha_1}^\dagger \hat{a}_{\alpha_2}^\dagger \dots \hat{a}_\beta^\dagger \hat{a}_{\alpha_i} \hat{a}_{\alpha_i}^\dagger \dots \hat{a}_{\alpha_N}^\dagger |0\rangle$$
$$= \hat{a}_{\alpha_1}^\dagger \hat{a}_{\alpha_2}^\dagger \dots \hat{a}_\beta^\dagger [1 - \hat{a}_{\alpha_i}^\dagger \hat{a}_{\alpha_i}] \dots \hat{a}_{\alpha_N}^\dagger |0\rangle$$
$$= \hat{a}_{\alpha_1}^\dagger \hat{a}_{\alpha_2}^\dagger \dots \hat{a}_\beta^\dagger \dots \hat{a}_{\alpha_N}^\dagger |0\rangle = |\alpha_1, \alpha_2, \dots, \alpha_i \to \beta, \dots, \alpha_N\rangle \,. \tag{1.56}$$

The first equality follows from the fact that $\hat{a}_\beta^\dagger \hat{a}_{\alpha_i}$ commutes with all the $\hat{a}_\alpha^\dagger$'s with $\alpha < \alpha_i$. In the second equality we make use of the anticommutation rule (1.45), and, in the third one, of the fact that $\hat{a}_{\alpha_i}$ anticommutes with all the $\hat{a}_\alpha^\dagger$'s with $\alpha > \alpha_i$, and thus annihilates the vacuum.

Returning to Eq. (1.54) we see that the action of the operator $\hat{V}^{(1)}$ is identical with that of the operator $\sum_\beta \sum_{i=1}^N V_{\beta\alpha_i} \hat{a}_\beta^\dagger \hat{a}_{\alpha_i}$. We can in fact extend the sum over occupied states

α_i to a sum over *all* states, since the destruction operators of the empty states give a null contribution. We thus arrive at the important connection formula for one-particle operators:

$$\hat{V}^{(1)} = \sum_{i=1}^{N} \hat{V}_i = \sum_{\alpha,\beta} V_{\beta\alpha} \hat{a}_\beta^\dagger \hat{a}_\alpha , \qquad (1.57)$$

where $V_{\beta\alpha}$ is given by Eq. (1.53).

Two-particle operators

The next case, in order of increasing complexity, is that of a two-particle operator $\hat{V}^{(2)}$ consisting of a sum of identical operators $\hat{V}_{ij}$ which concomitantly act on the Hilbert spaces of two electrons:

$$\hat{V}^{(2)} = \frac{1}{2} \sum_{i \neq j} \hat{V}_{ij} . \qquad (1.58)$$

Notice the important condition $i \neq j$, which excludes what would be effectively a one-electron operator. The coulomb interaction term in (1.1) is a typical example of such a two-particle operator, and the condition $i \neq j$ means, in that case, that an electron does not interact with itself.

When $\hat{V}_{ij}$ operates on a two-electron wave function $\phi_\alpha(\vec{r}_j, s_j)\phi_\beta(\vec{r}_i, s_i)$ it produces a superposition of such two-electron wave functions

$$\hat{V}_{ij}\phi_\alpha(\vec{r}_j, s_j)\phi_\beta(\vec{r}_i, s_i) = \sum_{\gamma\delta} V_{\gamma\delta\alpha\beta}\phi_\gamma(\vec{r}_i, s_i)\phi_\delta(\vec{r}_j, s_j) , \qquad (1.59)$$

where the amplitudes $V_{\gamma\delta\alpha\beta}$ are obtained as follows:

$$\begin{aligned} V_{\gamma\delta\alpha\beta} &\equiv ((\langle i, \gamma | \langle j, \delta |) \hat{V}_{ij} (| j, \alpha \rangle | i, \beta \rangle)) \\ &= \sum_{s,s',t,t'} \int d\vec{r}_i \int d\vec{r}_j \phi_\gamma^*(\vec{r}_i, s')\phi_\delta^*(\vec{r}_j, t') \\ &\quad \times [\hat{V}_{ij}]_{t't,s's}\phi_\alpha(\vec{r}_j, t)\phi_\beta(\vec{r}_i, s) , \end{aligned} \qquad (1.60)$$

where again the spin indices have been written explicitly. Notice that these amplitudes do not depend on i and j. When the complete operator $\hat{V}^{(2)}$ acts on one of the basis states (1.33) it generates a superposition of states in which each pair of occupied one-particle states α_i, α_j is replaced in turn by δ, γ (respectively) with an amplitude $V_{\gamma\delta\alpha_i\alpha_j}$:

$$\hat{V}^{(2)}|\alpha_1, ..\alpha_i, ..\alpha_j, ..\alpha_N\rangle = \frac{1}{2} \sum_{\gamma\delta} \sum_{i \neq j} V_{\gamma\delta\alpha_i\alpha_j}|\alpha_1, ..\alpha_i \to \delta, ..\alpha_j \to \gamma, ..\alpha_N\rangle. \qquad (1.61)$$

Now observe that the state $|\alpha_1, \ldots, \alpha_i \to \delta, \ldots, \alpha_j \to \gamma, \ldots, \alpha_N\rangle$ in which ϕ_δ and ϕ_γ replace ϕ_{α_i} and ϕ_{α_j} is obtained from $|\alpha_1, \ldots, \alpha_i, \ldots, \alpha_j, \ldots, \alpha_N\rangle$ in the following manner

$$|\alpha_1, \ldots, \alpha_i \to \delta, \ldots, \alpha_j \to \gamma, \ldots, \alpha_N\rangle = \hat{a}_\gamma^\dagger \hat{a}_\delta^\dagger \hat{a}_{\alpha_i} \hat{a}_{\alpha_j} |\alpha_1, \ldots, \alpha_i, \ldots, \alpha_j, \ldots, \alpha_N\rangle. \qquad (1.62)$$

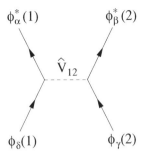

Fig. 1.8. Visual guide to the calculation of the matrix element $V_{\alpha\beta\gamma\delta}$.

The proof of this lemma is quite simple. If the pairs γ, δ and α_i, α_j have no element in common then we can change the order of the four operators on the right-hand side of Eq. (1.62) to $(\hat{a}_{\gamma}^{\dagger}\hat{a}_{\alpha_j})(\hat{a}_{\delta}^{\dagger}\hat{a}_{\alpha_i})$. According to Eq. (1.55), the two pairs of operators perform the replacements $\phi_{\alpha_j} \rightarrow \phi_{\gamma}$ and $\phi_{\alpha_i} \rightarrow \phi_{\delta}$ in the rows of the Slater determinant, as desired. A similar analysis can be carried out for all the remaining cases. For example, if $\gamma = \alpha_i$ and $\delta = \alpha_j$ (which amounts to interchanging the i-th and the j-th row in the initial Slater determinant which therefore changes sign), then the four operators on the right-hand side of Eq. (1.62) can be rearranged as $-(\hat{a}_{\alpha_i}^{\dagger}\hat{a}_{\alpha_i})(\hat{a}_{\alpha_j}^{\dagger}\hat{a}_{\alpha_j})$ which correctly multiplies the ket on its right by a factor -1.

Returning to Eq. (1.61), we now see that the action of the operator $\hat{V}^{(2)}$ is identical with that of the operator $\frac{1}{2}\sum_{\gamma\delta}\sum_{i\neq j}V_{\gamma\delta\alpha_i\alpha_j}\hat{a}_{\gamma}^{\dagger}\hat{a}_{\delta}^{\dagger}\hat{a}_{\alpha_i}\hat{a}_{\alpha_j}$. We can extend the sum over occupied states α_i, α_j to a sum over *all* states, since the additional terms give zero. We thus obtain the formula expressing a two-particle operator as a sum of products of four creation and destruction operators:

$$\hat{V}^{(2)} = \frac{1}{2}\sum_{i\neq j}\hat{V}_{ij} = \frac{1}{2}\sum_{\alpha,\beta,\gamma,\delta}V_{\alpha\beta\gamma\delta}\hat{a}_{\alpha}^{\dagger}\hat{a}_{\beta}^{\dagger}\hat{a}_{\gamma}\hat{a}_{\delta} . \tag{1.63}$$

The schematic diagram of Fig. 1.8 provides a helpful visual guide to the construction of the matrix elements in Eq. (1.63). The rule is that the wave functions corresponding to the states γ and β are attributed to the same particle, and so are the ones corresponding to the states α and δ. This "natural" way of grouping the labels $\alpha, \beta, \gamma, \delta$ is commonly referred to as *direct pairing*.

More complicated observables – acting on groups of three or more particles at a time – can also be handled in a similar way. However, due to the limited practical relevance of such cases, we conclude here our presentation of second quantization.[28]

1.4.3 Construction of the second-quantized hamiltonian

Let us apply the formalism of second quantization to the construction of the jellium model hamiltonian of Eq. (1.15). As a basis in the single particle space we choose the eigenstates

[28] The rules (1.57) and (1.63) for the construction of one and two particle operators remain valid in the case of bosons.

$|\vec{k}\sigma\rangle$ of the momentum operator $\hat{\vec{p}}$ with eigenvalue $\hbar\vec{k}$ and of the z-component of the spin, with eigenvalue σ (in units of $\hbar/2$). The corresponding normalized wave functions are the plane waves

$$\phi_{\vec{k}\sigma}(\vec{r}, s) = \frac{1}{\sqrt{L^d}} e^{i\vec{k}\cdot\vec{r}} \delta_{s\sigma} . \qquad (1.64)$$

The creation and destruction operators for this state will be denoted by $\hat{a}_{\vec{k}\sigma}^{\dagger}$ and $\hat{a}_{\vec{k}\sigma}$, respectively.

The kinetic energy, being of the form (1.57), is represented as

$$\hat{T} \equiv \sum_i \frac{\hat{p}_i^2}{2m} = \sum_{\vec{k}\sigma, \vec{k}'\sigma'} \langle \vec{k}\sigma | \frac{\hat{p}^2}{2m} | \vec{k}'\sigma' \rangle \hat{a}_{\vec{k}\sigma}^{\dagger} \hat{a}_{\vec{k}'\sigma'}$$

$$= \sum_{\vec{k}\sigma} \frac{\hbar^2 k^2}{2m} \hat{a}_{\vec{k}\sigma}^{\dagger} \hat{a}_{\vec{k}\sigma} . \qquad (1.65)$$

To construct the remaining coulomb interaction term, we first work out the second-quantized representation of the electron density operator (1.11). This is still of the form (1.57), so we immediately get

$$\hat{n}_{\vec{q}} \equiv \sum_i e^{-i\vec{q}\cdot\hat{\vec{r}}_i} = \sum_{\vec{k}\sigma, \vec{k}'\sigma'} \langle \vec{k}\sigma | e^{-i\vec{q}\cdot\vec{r}} | \vec{k}'\sigma' \rangle \hat{a}_{\vec{k}\sigma}^{\dagger} \hat{a}_{\vec{k}'\sigma'}$$

$$= \sum_{\vec{k}\sigma} \hat{a}_{\vec{k}-\vec{q}\,\sigma}^{\dagger} \hat{a}_{\vec{k}\sigma} . \qquad (1.66)$$

Substituting this expression in the interaction term of Eq. (1.15), and making use of the anticommutation rules to move all the creation operators to the left, we easily arrive at

$$\sum_{\vec{q}\neq 0} v_q \left[\hat{n}_{-\vec{q}}\hat{n}_{\vec{q}} - \hat{N} \right] = \sum_{\vec{q}\neq 0} v_q \sum_{\vec{k}_1\sigma_1, \vec{k}_2\sigma_2} \hat{a}_{\vec{k}_1+\vec{q}\,\sigma_1}^{\dagger} \hat{a}_{\vec{k}_2-\vec{q}\,\sigma_2}^{\dagger} \hat{a}_{\vec{k}_2\sigma_2} \hat{a}_{\vec{k}_1\sigma_1} . \qquad (1.67)$$

Notice how efficiently the second quantization formalism takes care of the self-interaction-removing term $-\hat{N} = -\sum_{\vec{k}\sigma} \hat{a}_{\vec{k}\sigma}^{\dagger} \hat{a}_{\vec{k}\sigma}$ through the appropriate ordering of the creation operators.

Combining Eqs. (1.65) and (1.67) we finally obtain the *second-quantized form of the jellium model hamiltonian*:

$$\hat{H} = \sum_{\vec{k}\sigma} \frac{\hbar^2 k^2}{2m} \hat{a}_{\vec{k}\sigma}^{\dagger} \hat{a}_{\vec{k}\sigma} + \frac{1}{2L^d} \sum_{\vec{q}\neq 0} v_q \sum_{\vec{k}_1\sigma_1, \vec{k}_2\sigma_2} \hat{a}_{\vec{k}_1+\vec{q}\,\sigma_1}^{\dagger} \hat{a}_{\vec{k}_2-\vec{q}\,\sigma_2}^{\dagger} \hat{a}_{\vec{k}_2\sigma_2} \hat{a}_{\vec{k}_1\sigma_1} . \qquad (1.68)$$

Notice that $\hat{H}$ can be also readily expressed in terms of the field operators defined in (1.47). In particular the interaction term reads:

$$\hat{H}_{e-e} = \frac{1}{2L^d} \sum_{\vec{q}\neq 0, \sigma_1, \sigma_2} v_q \int d\vec{r}_1 \int d\vec{r}_2 \, e^{i\vec{q}\cdot(\vec{r}_1-\vec{r}_2)} \hat{\Psi}_{\sigma_1}^{\dagger}(\vec{r}_1) \hat{\Psi}_{\sigma_2}^{\dagger}(\vec{r}_2) \hat{\Psi}_{\sigma_2}(\vec{r}_2) \hat{\Psi}_{\sigma_1}(\vec{r}_1) . \qquad (1.69)$$

1.5 The weak coupling regime

As anticipated in Section 1.3.3 there are two quite different regimes of the electron liquid: one is the weak coupling regime (small r_s) in which the kinetic energy rules the physics and the system remains qualitatively similar to a noninteracting gas; the other is the strong coupling regime, in which the potential energy dominates and the liquid displays the collective behavior characteristic of a crystal. We begin with a brief survey of the weak coupling regime.

1.5.1 The noninteracting electron gas

When the potential energy is completely neglected, we are left with a noninteracting electron gas described by the hamiltonian

$$\hat{H}_0 = \sum_{\vec{k}\sigma} \frac{\hbar^2 k^2}{2m} \hat{a}_{\vec{k}\sigma}^{\dagger} \hat{a}_{\vec{k}\sigma} \ . \tag{1.70}$$

As established in Section 1.3.3, this hamiltonian describes the electron gas in the limit of high density, or $r_s \to 0$.

For N electrons confined to a volume L^d, the ground-state of $\hat{H}_0$ is obtained by singly occupying plane wave states of wave vector $\vec{k}$ and spin projection $\sigma = \uparrow$ or $\downarrow$ in order of increasing energy, i.e. starting from $k = 0$ all the way up to a maximum value k_F, known as the *Fermi wave vector*. All the plane wave states with $|\vec{k}| = k_F$ have the same (highest) energy $\epsilon_F = \frac{\hbar^2 k_F^2}{2m}$, known as the *Fermi energy*. The surface that delimits the region of occupied states in $\vec{k}$-space is called the *Fermi surface*: it is a spherical surface in 3D, the circumference of a circle in 2D, and just two points in 1D.

A formal representation of the noninteracting ground-state $|\Psi_0\rangle$ is

$$|\Psi_0\rangle = \prod_{|\vec{k}| \le k_F, \sigma} \hat{a}_{\vec{k}\sigma}^{\dagger} |0\rangle \ , \tag{1.71}$$

so that

$$\langle \Psi_0 | \hat{a}_{\vec{k}\sigma}^{\dagger} \hat{a}_{\vec{k}\sigma} | \Psi_0 \rangle = \Theta(k_F - |\vec{k}|) \ , \tag{1.72}$$

where $\Theta(x)$ is the *step function* equal to 1 for $x > 0$, and 0 otherwise. Accordingly k_F is determined by the condition

$$\sum_{\vec{k}\sigma} \Theta(k_F - |\vec{k}|) = N \ , \tag{1.73}$$

This amounts to requiring that the number of states within the Fermi surface equal $\frac{N}{2}$, i.e.,

$$\frac{V_{FS}^{(d)}}{\frac{(2\pi)^d}{L^d}} = \frac{N}{2} \ , \tag{1.74}$$

where $V_{FS}^{(d)}$, the volume of the generalized Fermi sphere, is given by

$$V_{FS}^{(d)} = \int_{|\vec{k}| \leq k_F} d^d k = \frac{\Omega_d k_F^d}{d} \ , \tag{1.75}$$

and Ω_d is the solid angle in d dimensions (4π in 3D, 2π in 2D, and 2 in 1D).

Equation (1.74) leads to the following fundamental relationship between the Fermi wave vector and the density:

$$k_F^d = \frac{(2\pi)^d d}{2\Omega_d} n \ , \tag{1.76}$$

or, more explicitly,

$$k_F = \begin{cases} (3\pi^2 n)^{\frac{1}{3}}, & 3D, \\ (2\pi n)^{\frac{1}{2}}, & 2D, \\ \frac{\pi}{2} n, & 1D. \end{cases} \tag{1.77}$$

It is often useful to express k_F in terms of the parameter r_s. From an elementary dimensional argument one can see that k_F is in general inversely proportional to r_s and can be therefore written as

$$k_F = \frac{1}{\alpha_d r_s a_B} \ . \tag{1.78}$$

In three, two and one dimensions the constants α_d are specified by

$$\alpha_d = \begin{cases} \left(\frac{4}{9\pi}\right)^{1/3}, & 3D, \\ \frac{1}{\sqrt{2}}, & 2D, \\ \frac{4}{\pi}, & 1D. \end{cases} \tag{1.79}$$

By making use of Eqs. (1.78) and (1.79) the Fermi energy can then be expressed in Rydberg units as:

$$\epsilon_F = \frac{1}{\alpha_d^2 r_s^2} Ry \ . \tag{1.80}$$

The total energy for this state is readily evaluated from the relation:

$$E_0 = \sum_{|\vec{k}| \leq k_F, \sigma} \frac{\hbar^2 k^2}{2m} = \frac{2\Omega_d L^d}{(2\pi)^d} \int_0^{k_F} dk k^{d-1} \frac{\hbar^2 k^2}{2m} \ , \tag{1.81}$$

where Ω_d was defined above. Performing the trivial integration and making use of (1.76) and of the definition of the Fermi energy, we arrive at the following expression for the energy per electron:

$$\epsilon_0 = \frac{E_0}{N} = \frac{d}{d+2} \epsilon_F. \tag{1.82}$$

We can then write:

$$\epsilon_0(r_s) = \begin{cases} \frac{3}{5}\epsilon_F \simeq \frac{2.21}{r_s^2}\ \text{Ry}, & 3D, \\ \frac{1}{2}\epsilon_F = \frac{1}{r_s^2}\ \text{Ry}, & 2D, \\ \frac{1}{3}\epsilon_F \simeq \frac{0.205}{r_s^2}\ \text{Ry}, & 1D. \end{cases} \qquad (1.83)$$

This state of the electron gas, which is *spatially homogeneous* and has equal numbers of up- and down-spin electrons, is commonly referred to as the *paramagnetic Fermi sea.*

1.5.2 Noninteracting spin polarized states

Since the hamiltonian $\hat{H}_0$ (Eq. (1.70)) is spin-independent, all its eigenstates can be classified according to the eigenvalue, S_z, of the total spin angular momentum along the z-axis. Admissible values of S_z (in units of $\frac{\hbar}{2}$) range from 0, for the spin-compensated state considered in the previous section, to $\pm N$ for the state in which all the spins point in the same direction. A partially spin-polarized state will have $N_\uparrow$ electrons pointing up and $N_\downarrow$ electrons pointing down, so that $N_\uparrow + N_\downarrow = N$ and $N_\uparrow - N_\downarrow = S_z$. The situation is best described by the *fractional spin polarization* $p = \frac{S_z}{N}$ via

$$N_{\uparrow(\downarrow)} = \frac{N}{2}(1 \pm p), \qquad (1.84)$$

where the upper sign refers to $\uparrow$ spins. The unpolarized state corresponds to $p = 0$, while the fully polarized state has $p = 1$.

We have seen in the previous section that the ground-state of $\hat{H}_0$ has $p = 0$. What if $p \neq 0$? All such states are necessarily excited, but we may still ask: what is the minimum energy state for a given value of $p \neq 0$? This is, in a sense, a generalization of the problem of finding the absolute ground-state: namely, we are now trying to find a "constrained ground-state" within a restricted sector of the Hilbert space, characterized by a spin polarization p. The question is interesting and important because it might well happen that the true ground-state of the interacting system has $p \neq 0$, in which case the constrained ground-state of the noninteracting system with the same value of p would provide the natural starting point for a perturbative treatment of the interaction.

A little thought should be sufficient to see that the "constrained ground-state" for $p \neq 0$ is obtained by filling two unequal Fermi surfaces in $\vec{k}$ space, one with radius $k_{F\uparrow}$ for up-spin electrons, and one with radius $k_{F\downarrow}$ for down-spin electrons. The up-spin Fermi surface encloses $N_\uparrow$ states while the down-spin Fermi surface encloses the remaining $N_\downarrow$ states. With this picture in mind it is easy to see that the spin-dependent Fermi wavevectors are given by

$$k_{F\uparrow(\downarrow)} = k_F(1 \pm p)^{\frac{1}{d}}, \qquad (1.85)$$

where k_F is the Fermi wave vector of the unpolarized state, given by Eq. (1.77). This allows us to generalize the concept of Fermi energy to account for a different numbers of $\uparrow$ and $\downarrow$

electrons. We have:

$$\epsilon_{F\uparrow(\downarrow)} = \frac{\hbar^2 k_{F\uparrow(\downarrow)}^2}{2m} = \epsilon_F (1 \pm p)^{\frac{2}{d}} . \tag{1.86}$$

The kinetic energy of the up- (down-) spin electrons is then given by

$$E_{0\uparrow(\downarrow)} = \frac{d}{d+2}\epsilon_{F\uparrow(\downarrow)} N_{\uparrow(\downarrow)} = \frac{d}{d+2}\epsilon_F (1 \pm p)^{\frac{d+2}{d}} \frac{N}{2} = \frac{E_0}{2}(1 \pm p)^{\frac{d+2}{d}} , \tag{1.87}$$

where E_0, the energy of the unpolarized state, given in Eq. (1.82).

Accordingly the energy per particle in a noninteracting spin polarized "ground-state" is

$$\epsilon_0(r_s, p) = \frac{E_{0\uparrow} + E_{0\downarrow}}{N} = \epsilon_0(r_s) \frac{(1+p)^{\frac{d+2}{d}} + (1-p)^{\frac{d+2}{d}}}{2} , \tag{1.88}$$

where ϵ_0 is given in Eq. (1.83). Notice that these spin-polarized states are spatially homogeneous but obviously not isotropic in spin space and are therefore examples of *broken symmetry states*.

1.5.3 The exchange energy

The first step toward including the coulomb interaction in the treatment of the electron liquid is to do perturbation theory to first order in the interaction hamiltonian $\hat{H}_{e-e}$.[29] According to standard quantum mechanical perturbation theory (see Landau-Lifshitz, *Course of Theoretical Physics*, Vol. 3) the first-order interaction correction to the energy of an eigenstate of $\hat{H}_0$ is given by the expectation value of $\hat{H}_{e-e}$ in that eigenstate. Accordingly, the first order approximation to the ground-state energy of the jellium model is

$$E \simeq E_0 + E_1 = E_0 + \langle \Psi_0 | \hat{H}_{e-e} | \Psi_0 \rangle , \tag{1.89}$$

where $|\Psi_0\rangle$ is the ground-state of $\hat{H}_0$, explicitly given by Eq. (1.71). The calculation of E_1 involves evaluating the expectation value of the product $\hat{a}^\dagger_{\vec{k}_1+\vec{q},\sigma_1} \hat{a}^\dagger_{\vec{k}_2-\vec{q},\sigma_2} \hat{a}_{\vec{k}_2\sigma_2} \hat{a}_{\vec{k}_1\sigma_1}$ in the noninteracting ground-state. This can be done most swiftly with the help of *Wick's theorem*, described in Appendix 3. However, in this introductory chapter, we adopt a more pedagogical procedure that illustrates in a step-by-step fashion the use of creation and destruction operators.

Since one must return to the non-interacting ground-state (a single Slater determinant) after having applied all the operators to it, this calculation can be done by first deciding which creation operator is to make up for the action of which destruction operator. Accordingly each creation operator is formally *paired up* with a destruction operator. In this case there

---

[29] The validity of the perturbative approach depends on the assumption of continuity, i.e. on the possibility that the unperturbed ground-state and the fully interacting one can indeed be connected by adiabatically switching on the interaction term in the hamiltonian. This can be taken as a somewhat safe assumption only when the chosen unperturbed ground-state and the fully interacting one have the same symmetry. That this is the case is by no means necessary and, as it turns out, it is the most subtle and therefore difficult property of an interacting system to ascertain. It is common to refer to the many-body state of a system of fermions obtained through a perturbative procedure as a *Fermi liquid*. This concept will be discussed in detail in Chapter 8.

are of course only two possibilities. The first one, referred to as the *direct pairing*, pairs $\hat{a}^{\dagger}_{\vec{k}_1+\vec{q}\ \sigma_1}$ with $\hat{a}_{\vec{k}_1\sigma_1}$, and $\hat{a}^{\dagger}_{\vec{k}_2-\vec{q}\ \sigma_2}$ with $\hat{a}_{\vec{k}_2\sigma_2}$, while the second one, referred to as the *exchange pairing*, pairs $\hat{a}^{\dagger}_{\vec{k}_1+\vec{q}\ \sigma_1}$ with $\hat{a}_{\vec{k}_2\sigma_2}$, and $\hat{a}^{\dagger}_{\vec{k}_2-\vec{q}\ \sigma_2}$ with $\hat{a}_{\vec{k}_1\sigma_1}$.

Since in the jellium model only $q \neq 0$ is allowed, the direct pairing is impossible: $\hat{a}^{\dagger}_{\vec{k}_1+\vec{q}\ \sigma_1}$ can never make up for what $\hat{a}_{\vec{k}_1\sigma_1}$ has done. We are then left with the sole exchange pairing which leads to the conditions: $\vec{k}_1 + \vec{q} = \vec{k}_2$ and $\sigma_1 = \sigma_2$. By implementing these conditions on the quantum numbers and noticing that the two operators in the middle can be simply anticommuted (as for $\vec{q} \neq 0$ they pertain to different wave vectors) we can write:

$$
\begin{aligned}
\left\langle \hat{a}^{\dagger}_{\vec{k}_1+\vec{q}\ \sigma_1} \hat{a}^{\dagger}_{\vec{k}_2-\vec{q}\ \sigma_2} \hat{a}_{\vec{k}_2\sigma_2} \hat{a}_{\vec{k}_1\sigma_1} \right\rangle_0 &= -\left\langle \hat{a}^{\dagger}_{\vec{k}_1+\vec{q}\ \sigma_1} \hat{a}_{\vec{k}_2\sigma_2} \hat{a}^{\dagger}_{\vec{k}_2-\vec{q}\ \sigma_2} \hat{a}_{\vec{k}_1\sigma_1} \right\rangle_0 \\
&= -\delta_{\vec{k}_1+\vec{q},\vec{k}_2}\delta_{\sigma_1\sigma_2} \left\langle \hat{N}_{\vec{k}_1+\vec{q}\ \sigma_1} \hat{N}_{\vec{k}_1\sigma_1} \right\rangle_0 ,
\end{aligned}
\tag{1.90}
$$

where $\langle \hat{A} \rangle_0$ is a shorthand for the expectation value of an operator $\hat{A}$ in the filled Fermi sea $|\Psi_0\rangle$. Since $|\Psi_0\rangle$ is manifestly an eigenstate of the operator $\hat{N}_{\vec{k}\sigma}(= \hat{a}^{\dagger}_{\vec{k}\sigma}\hat{a}_{\vec{k}\sigma})$ with eigenvalue $n_{\vec{k}\sigma}(= \Theta(k_F - k))$, the last expectation value in (1.90) can be immediately evaluated, yielding

$$
-\delta_{\vec{k}_1+\vec{q},\vec{k}_2}\delta_{\sigma_1\sigma_2}n_{\vec{k}_1+\vec{q}\ \sigma_1}n_{\vec{k}_1\sigma_1} .
\tag{1.91}
$$

Multiplying the above expression by v_q and summing over $\vec{k}_1, \vec{k}_2, \sigma_1, \sigma_2$, and $\vec{q}$ we finally get

$$
E_1 \equiv E_x = -\frac{1}{2L^d} \sum_{\vec{q}\neq 0} v_q \sum_{\vec{k}\sigma} n_{\vec{k}+\vec{q}\ \sigma}n_{\vec{k}\sigma} .
\tag{1.92}
$$

This (always negative) contribution to the total energy of the electron liquid is known as the *exchange energy* (E_x) since, as explained in Section 1.5.5, it arises entirely from the antisymmetry of the N-electron wave function under exchange of two electrons.

The exchange energy can be easily calculated from Eq. (1.92). As a first step we calculate the sum over $\vec{k}$ and σ, which is found to be given by

$$
\frac{1}{N} \sum_{\vec{k}\sigma} n_{\vec{k}+\vec{q}\sigma} n_{\vec{k}\sigma} =
\begin{cases}
1 - \frac{3}{2}\frac{q}{2k_F} + \frac{1}{2}\left(\frac{q}{2k_F}\right)^3 , & 3D , \\[2mm]
1 - \frac{2}{\pi}\left[\sin^{-1}\frac{q}{2k_F} - \frac{q}{2k_F}\sqrt{1-\left(\frac{q}{2k_F}\right)^2}\right] , & 2D , \\[2mm]
1 - \frac{q}{2k_F} , & 1D ,
\end{cases}
\tag{1.93}
$$

for $|q| < 2k_F$ and 0 otherwise. We then substitute these expressions in Eq. (1.92) and perform an elementary integral over $\vec{q}$ to find the exchange energy per electron. The results in three and two dimensions are

$$
\epsilon_x = \frac{E_x}{N} =
\begin{cases}
-\frac{3}{4}\frac{e^2 k_F}{\pi} , & 3D , \\[2mm]
-\frac{4}{3}\frac{e^2 k_F}{\pi} , & 2D ,
\end{cases}
\tag{1.94}
$$

or, in terms of r_s,

$$\epsilon_x(r_s) = \begin{cases} -\frac{3}{2\pi\alpha_3 r_s} \text{ Ry } \simeq -\frac{0.916}{r_s} \text{ Ry, } & 3D \text{ ,} \\ -\frac{8\sqrt{2}}{3\pi r_s} \text{ Ry } \simeq -\frac{1.200}{r_s} \text{ Ry, } & 2D \text{ .} \end{cases} \tag{1.95}$$

In one dimension the calculation is slightly more complicated due to the non-homogeneous form of $v_q(a)$ (see Eq. (1.20)), which spoils the scaling of the energy with r_s. We recover r_s-scaling if we use, for illustration purposes, a short-ranged interaction potential of the form

$$v(x - x') = \gamma e^2 \delta(x - x') \text{ ,} \tag{1.96}$$

where γ is a dimensionless constant. The Fourier transform of this interaction is $v_q = \gamma e^2$, which, substituted in Eq. (1.92), gives

$$\epsilon_x(r_s) = -\frac{\gamma}{4r_s} \text{ Ry, } \quad 1D \text{ (short-range interaction).} \tag{1.97}$$

1.5.4 Exchange energy in spin polarized states

The formulas obtained in the previous section for the exchange energy per electron in the paramagnetic ground-state ($S_z = 0$) can be easily generalized to the spin-polarized states introduced in Section 1.5.2. The key point is that the full Fermi spheres of $\uparrow$ and $\downarrow$ spins contribute independently to the total exchange energy: there is no exchange between electrons of opposite spins. The contribution of each spin component can be calculated from Eq. (1.94) where the Fermi wave vector k_F is appropriately replaced by $k_{F\sigma}$, given by Eq. (1.78). It is then an easy exercise to show that

$$\epsilon_x(r_s, p) = \epsilon_x(r_s) \frac{(1 + p)^{\frac{d+1}{d}} + (1 - p)^{\frac{d+1}{d}}}{2} \text{ ,} \tag{1.98}$$

where p is the fractional polarization. Notice that the exchange energy decreases, at fixed density, with increasing spin polarization. Since at low densities the exchange energy dominates, we can immediately guess that the ferromagnetic state ($p = 1$) will be lower in energy than the paramagnetic state ($p = 0$) above a certain critical value of r_s (~ 5.45 in 3D). Thus, the first-order approximation predicts a ferromagnetic ground-state at sufficiently low density – the so-called *Bloch transition* (Bloch, 1929). It turns out that this prediction is *quantitatively* incorrect: nevertheless, more accurate calculations confirm the prediction of a ferromagnetic transition, even though at a much lower density than in the first-order theory.

1.5.5 Exchange and the pair correlation function

The connection between the exchange energy and the symmetry of the wave function can be clearly seen from the following argument. If the electrons were distinguishable then in the noninteracting ground-state there could be no correlation between the positions of any two electrons. As a result the average interaction energy of an electron with all the other electrons would be exactly cancelled by its interaction with the positively charged background. In reality, even in the absence of interactions, the antisymmetry of the wave function implies a high degree of correlation between the positions of two electrons with the same spin orientation: for example, two such electrons can never be found at the same point in space. As a result, each electron is surrounded by an *"exchange hole"* – a region in which the density of same-spin electrons is smaller than average, and in which, therefore, the positive background charge is not exactly cancelled. It is the interaction of each electron with the positive charge of the exchange hole that gives rise to the exchange energy – hence the negative sign.

This physical picture can be formalized with the introduction of the *pair correlation function* $g(\vec{r}_2, \vec{r}_1)$ which is defined as the normalized probability of finding an electron at position $\vec{r}_1$ given that, at the same time, there is another electron at position $\vec{r}_2$, i.e.

$$g(\vec{r}_2, \vec{r}_1) \equiv \frac{1}{n(\vec{r}_2)n(\vec{r}_1)} \left\langle \sum_{i \neq j} \delta(\vec{r}_1 - \vec{r}_i)\delta(\vec{r}_2 - \vec{r}_j) \right\rangle , \qquad (1.99)$$

where the angular brackets denote here the average in the ground-state. While a more detailed discussion of the physical relevance of the pair correlation function and the related structure factor is provided in Appendix 4, it will suffice to notice here that according to the general rules of second quantization $g(\vec{r}_2, \vec{r}_1)$ can be also expressed in terms of the field operators $\hat{\Psi}_\sigma(\vec{r})$ (defined in Section 1.4.2) in the following manner:

$$g(\vec{r}_2, \vec{r}_1) = \sum_{\sigma_1 \sigma_2} \frac{\left\langle \hat{\Psi}_{\sigma_2}^\dagger(\vec{r}_2)\hat{\Psi}_{\sigma_1}^\dagger(\vec{r}_1)\hat{\Psi}_{\sigma_1}(\vec{r}_1)\hat{\Psi}_{\sigma_2}(\vec{r}_2) \right\rangle}{n(\vec{r}_2)n(\vec{r}_1)} . \qquad (1.100)$$

A comparison with Eq. (1.69) shows that $g(\vec{r}_2, \vec{r}_1)$ is directly related to the expectation value of the electron–electron interaction. Furthermore, in the jellium model the total potential energy can be expressed as

$$\frac{U}{N} = \frac{n}{2} \int d\vec{r} \, v(r)[g(r) - 1] , \qquad (1.101)$$

where the "−1" stems from the interaction between the electrons and the background (or, equivalently, from the removal of the $q = 0$ component of the electron–electron interaction). Of course, we have used the fact that in the jellium model $g(\vec{r}_2, \vec{r}_1)$ only depends on the magnitude of the difference $\vec{r}_2 - \vec{r}_1$ and can be therefore simply represented as a function $g(|\vec{r}_2 - \vec{r}_1|)$ of this variable.

It can now be seen that the exchange energy of the uniform electron liquid, defined in Eq. (1.89) as the expectation value of the potential energy in the non-interacting ground-state, is given by Eq. (1.101), provided we use for $g(r)$ the pair correlation function of the non-interacting ground-state. This function, denoted by $g^{(0)}(r)$, is easily calculated (see the analytical expression in Appendix 4), and its form is shown in Fig. 1.9.

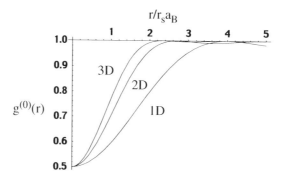

Fig. 1.9. The noninteracting pair correlation function in 3, 2, and 1 dimensions. Notice the "exchange hole" around the origin, becoming more extended in lower spatial dimension. Analytic expressions for $g^{(0)}(r)$ are given in Appendix 4.

We can now return to the physical argument presented at the beginning of this section and restate it in terms of the properties of $g^{(0)}(r)$. It is evident that in a non-interacting system only pairs of electrons with the same spin orientation are affected by the antisymmetry of the wave function: hence the pair correlation function for opposite-spin electrons, $g^{(0)}_{\uparrow\downarrow}(r)$, equals 1 for all distances, while the pair correlation function for same-spin electrons, $g^{(0)}_{\uparrow\uparrow}(r)$ vanishes for $r \to 0$ and tends to 1 for $r \to \infty$. It follows that the full pair correlation function, $g^{(0)}(r) = \frac{g^{(0)}_{\uparrow\uparrow}(r)+g^{(0)}_{\uparrow\downarrow}(r)}{2}$, being the average of the two possible relative spin orientations, is reduced to $\frac{1}{2}$ for $r \to 0$ and tends to 1 for $r \to \infty$. The depletion region at small separations is precisely the exchange hole introduced at the beginning of this section, and its existence leads to the negative exchange energy via Eq. (1.101).

1.5.6 All-orders perturbation theory: the RPA

In principle one could think of improving on the first-order result discussed above by evaluating higher-order corrections. Standard time-independent perturbation theory allows one to write:

$$E = E_0 + \langle\Psi_0|\hat{H}_{e-e}|\Psi_0\rangle - \sum_{n\neq 0}\frac{\langle\Psi_0|\hat{H}_{e-e}|\Psi_n\rangle\langle\Psi_n|\hat{H}_{e-e}|\Psi_0\rangle}{E_n - E_0} + \cdots, \qquad (1.102)$$

where $|\Psi_n\rangle$ are the excited eigenstates of $\hat{H}_0$ (i.e., Fermi seas of plane wave states with a number of electron–hole pairs present) and E_n are their unperturbed energies. But here comes a surprise: in three dimensions already the second order term in the above expansion is divergent, and higher order terms diverge even more strongly.

To understand how this comes about observe that the intermediate states $|\Psi_n\rangle$ in the second order term of Eq. (1.102) contain two holes at wave vectors and spins $\vec{k}_1\sigma_1$ and $\vec{k}_2\sigma_2$ within the Fermi sphere and two electrons at wavectors and spins $\vec{k}_1 + \vec{q}\ \sigma_1$ and

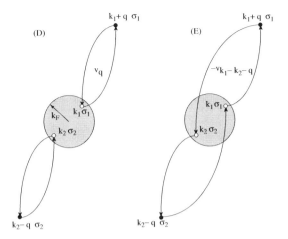

Fig. 1.10. Schematic illustration of the direct and exchange processes that contribute to the second-order correction to the ground-state energy.

$\vec{k}_2 - \vec{q}\,\sigma_2$ outside the Fermi sphere (see Fig. 1.10). Their energy, relative to the ground-state is therefore

$$E_n - E_0 = \varepsilon_{\vec{k}_1+\vec{q}} - \varepsilon_{\vec{k}_1} + \varepsilon_{\vec{k}_2-\vec{q}} - \varepsilon_{\vec{k}_2}$$
$$= \frac{\hbar^2}{m}\left[q^2 + (\vec{k}_1 - \vec{k}_2)\cdot\vec{q}\right], \qquad (1.103)$$

where we have introduced the *free particle energy* $\varepsilon_{\vec{k}} \equiv \frac{\hbar^2 k^2}{2m}$.

These excited states are connected to the ground-state via an amplitude v_q. Now, in order to have a finite result, the electron–electron interaction, acting on $|\Psi_n\rangle$ must de-excite the electrons and reconstitute the original Fermi sphere. This can be accomplished via two essentially distinct routes, as illustrated in Fig. 1.10. In the first route, known as *direct*, the electron at $\vec{k}_1 + \vec{q}\,\sigma_1$ recombines with the hole at $\vec{k}_1, \sigma_1$ with amplitude v_q. In the second route, known as *exchange*, the same electron recombines with the hole at $\vec{k}_2\sigma_2$. Obviously, the exchange process can occur only if $\sigma_1 = \sigma_2$, and its amplitude is $-v_{|\vec{k}_1-\vec{k}_2-\vec{q}|}\delta_{\sigma_1\sigma_2}$ (notice the minus sign arising from the antisymmetry under exchange of the two electrons).

Thus, keeping a careful record of the multiplicities of the various terms, we arrive at the following expression for the second order correction to the ground-state energy:

$$\frac{E^{(2)}}{N} = -\frac{1}{2nL^{2d}}\sum_{\vec{k}_1\sigma_1,\vec{k}_2\sigma_2}\int\frac{d^d q}{(2\pi)^d}v_q^2\frac{n_{\vec{k}_1\sigma_1}(1-n_{\vec{k}_1+\vec{q}\,\sigma_1})n_{\vec{k}_2\sigma_2}(1-n_{\vec{k}_2-\vec{q}\,\sigma_2})}{\frac{\hbar^2 q^2}{m}+\frac{\hbar^2(\vec{k}_1-\vec{k}_2)\cdot\vec{q}}{m}}$$

$$+\frac{1}{2nL^{2d}}\sum_{\vec{k}_1,\vec{k}_2,\sigma}\int\frac{d^d q}{(2\pi)^d}v_q v_{|\vec{k}_1-\vec{k}_2-\vec{q}|}\frac{n_{\vec{k}_1\sigma}(1-n_{\vec{k}_1+\vec{q}\,\sigma})n_{\vec{k}_2\sigma}(1-n_{\vec{k}_2-\vec{q}\,\sigma})}{\frac{\hbar^2 q^2}{m}+\frac{\hbar^2(\vec{k}_1-\vec{k}_2)\cdot\vec{q}}{m}}. \qquad (1.104)$$

In three dimensions, the evaluation of this expression proceeds are follows. The exchange contribution (second term) corresponds to a well behaved, yet very complicated multidimensional integral, which can in fact be evaluated exactly (Onsager, Mittag, and Stephen, 1966).[30] The result is:

$$\frac{\epsilon_x^{(2)}}{\text{Ry}} = \frac{1}{3}\ln 2 - \frac{3}{2\pi^2}\zeta(3) \approx +0.0484 , \qquad (1.105)$$

where $\zeta(n) = \sum_{k=1}^{\infty} k^{-n}$ is the Riemann zeta function.

The direct term has a simpler structure, but presents us with the problem of an *infrared* divergence. In fact, a careful inspection of the integrand shows that, except for unimportant constants, the integral diverges like

$$\int_0^{q_c} \frac{dq}{q} , \qquad (1.106)$$

i.e., logarithmically in the small q region.[31] The physical origin of this divergence lies in the long range of the coulomb interaction $v_q \sim \frac{1}{q^2}$ in 3D.[32]

Higher order terms of the perturbation expansion can be systematically examined[33] and are found to diverge more and more strongly, making the perturbation expansion apparently worthless.

From the mathematical point of view, the appearance of a divergence in the perturbative expansion of the energy as a function of r_s simply means that this function is nonanalytic at $r_s = 0$. It is a perfectly smooth and well behaved function, but the perturbative approach assumes it to be expandable in powers of r_s when, in fact, it does not admit such an expansion!

This serious difficulty can happily be circumvented in the limit of small r_s, i.e. at high electronic densities. In a now classic paper that pioneered the application of quantum field theoretical methods to the electron gas problem, Gell-Mann and Brueckner (1957) discovered that the perturbative series can be actually resummed to infinite order *before doing the momentum integrals*. The physics that is captured in this resummation is that the coulomb interaction is effectively "screened" at large distances, or, equivalently, below a characteristic wave vector $\kappa \sim k_F \sqrt{r_s}$ (the physics of the screening will be discussed in detail in Chapter 5).[34] Convergence is achieved by realizing that in the high-density limit the contribution to the energy arising from the small q part of the interaction ($q < q_c$, where q_c is an arbitrary cutoff of the order of the screening wave vector κ) can be calculated *exactly*, i.e., the perturbative series can be summed to infinite order. The contribution from the $q > q_c$

---

[30] This is the famed *Onsager's exchange integral*. The rather herculean task of its analytical evaluation was dedicated by Lars Onsager to Erich Hueckel in occasion of his 70th birthday.

[31] In counting powers of q it should be noted that the products $n_{\vec{k}\sigma}[1 - n_{\vec{k}+\vec{q}\,\sigma}]$ in Eq. (1.104) vanish as q for $q \to 0$.

[32] In the two dimensional case the divergence first appears in the third order term of the perturbative expansion: the physical picture remains the same.

[33] A way to expedite this process is to make use of diagrammatic methods. In this particular case the rules involved are rather simple and the corresponding diagrams are referred to as *Goldstone diagrams*. An introduction to this approach can be found in Feynman's book *Statistical Mechanics* (Chapter 9), or can be inferred from the paper of Gell-Mann and Brueckner (Gell-Mann and Brueckner, 1957).

[34] This point was first realized by Macke (1950), who also determined the leading logarithmic correction to the energy. Macke's theory is discussed in Exercise 9 in Chapter 5.

components, on the other hand, can be expanded in powers of r_s without encountering divergences. When the two contributions are added together the dependence on q_c cancels out, and a finite result is obtained. The complete procedure, nowadays known, for historical reasons, as the *random phase approximation* (RPA), will be described in detail in Chapter 5.

Following this procedure, it has been possible to derive an asymptotic expansion of the electron gas energy for $r_s \to 0$. The result for the jellium model, expressed in Rydberg per electron, is as follows:

$$\epsilon(r_s) \simeq \begin{cases} \left(\frac{2.210}{r_s^2} - \frac{0.916}{r_s} + 0.062 \ln r_s - 0.093 + \mathcal{O}(r_s \ln r_s) \right) \text{ Ry}, & (3D), \\ \left(\frac{1}{r_s^2} - \frac{1.20}{r_s} - (0.38 \pm 0.04) - 0.1726\, r_s \ln r_s + \mathcal{O}(r_s) \right) \text{ Ry}, & (2D). \end{cases} \quad (1.107)$$

where we have used Eqs. (1.83) and (1.95) respectively for the zeroth and first order terms.[35] The appearance of powers of $\ln r_s$ in this formula shows that the energy is a nonanalytic function of r_s for $r_s \to 0$ and explains the failure of the naive perturbative approach. The coefficients of the leading logarithmic terms in Eqs. (1.107) are in fact known exactly, being equal to $\frac{2}{\pi^2}(1 - \ln 2)$ and $\frac{2^{3/2}}{3\pi}(10 - 3\pi)$ in three and two dimensions respectively, and will be discussed in Chapter 5. The constant term is the sum of the second order Onsager's exchange integral and a numerical constant stemming (among other terms) from the sum over the diverging contributions. In practice, the asymptotic formulas are accurate only for very small values of r_s (see Figs. 1.11 and 1.15).

The difference between the exact energy and the sum of the noninteracting kinetic energy and the exchange energy is commonly referred to as the *correlation energy*. Accordingly a small correlation energy indicates that the description of the ground-state as a single Slater determinant of plane waves (in which there are no correlations other than the ones implied by the antisymmetry of the wave function) is somewhat accurate. As it will be made clear in Chapter 2 however, it is rather difficult to find a physically compelling definition of the correlation energy in general.[36]

1.6 The Wigner crystal

Having (for the time being) exhausted the discussion of the high-density behavior of the ground-state energy of the jellium model, we now ask ourselves what can be learned from consideration of the low-density limit. At very low density a system of electrically neutral particles becomes effectively noninteracting because the average interparticle separation is much larger than the (finite) range of the interaction. A system of electrons is anomalous in this respect because, as the average distance between the electrons increases, the coulomb potential energy decreases only as $\frac{1}{r_s}$ and thus eventually becomes much larger than the

---

[35] The result in the two-dimensional case has been worked by Rajagopal and Kimball (1977). A calculation of the next order term has been performed (Isihara and Toyoda, 1978). The result appears to be $(0.865 \pm 0.009)\, r_s$.

[36] It is perhaps for this reason that Richard Feynman rather colorfully referred to this quantity as *the stupidity energy* (Feynman, 1972).

average kinetic energy, which, as we will show later, scales as $\frac{1}{r_s^{3/2}}$ for large r_s.[37] In this limit it is possible, as a starting point, to neglect the kinetic energy part of the hamiltonian $\hat{H}_{jellium}$: the ground-state of the system is therefore close to the equilibrium state of a system of N classical point charges distributed on a uniform background of density $n = \frac{N}{L^d}$. Such a state is believed to be crystalline. This implies that, in the low-density limit, the ground-state wave function of the jellium model should reduce to an antisymmetrized product of δ-functions centered at the positions of a regular lattice. Spin and statistics are largely irrelevant to the energetics of such a state. The possibility of a crystalline state of the electron gas was first pointed out by Wigner (Wigner, 1934), and, quite fittingly, the state takes the name of "Wigner crystal".[38]

The structure of the Wigner crystal cannot be determined *a priori*.[39] In principle, one should try all the structures, and choose the one with the lowest energy. The only obvious constraint is that the unit cell must be charge neutral, to ensure global charge neutrality. This determines the volume of the unit cell Ω via the relation $n\Omega = Z$, where Z is the number of electrons per unit cell. For simple Bravais lattices $Z = 1$. If the lattice is characterized by a single parameter (as is the case for *cubic*, *square*, and *hexagonal* lattices) then this parameter is completely determined by the background density. We now proceed to the calculation of the energy.

1.6.1 Classical electrostatic energy

The energy of a classical electron crystal is calculated as the sum of electron–electron, electron–background and background–background potential energies, according to the formula

$$U = \frac{1}{2}\sum_{i \neq j}\frac{e^2}{|\vec{R}_i - \vec{R}_j|} - \sum_i \int d\vec{r}\,\frac{ne^2}{|\vec{r} - \vec{R}_i|} + \frac{1}{2}\int d\vec{r}\int d\vec{r}'\,\frac{n^2e^2}{|\vec{r} - \vec{r}'|}, \quad (1.108)$$

where $\vec{R}_i$ are the lattice vectors. The numerical evaluation of this formula is tricky, because each term is divergent, yet the divergences cancel out in the sum, the basic physical problem being the same as the one faced in Section 1.3.2 on the regularization of the coulomb interaction.

If a high degree of precision is not required, a quick estimate of the Wigner crystal energy can be obtained as follows. Treat the crystal as a collection of electrically neutral unit cells – each cell containing one electron and a total charge $+1$ from the background. In a Bravais lattice, the unit cells can be chosen to have no electric dipole moment. Because the electrostatic interaction between two unit cells decreases rapidly with increasing distance, we see that each cell interacts effectively only with a finite number of neighboring cells,

---

[37] Of course this is true only at very low temperatures, such that $k_B T \ll \frac{e^2}{r_s a_B}$.

[38] Although it is physically reasonable, appealing, and probably true, at the moment we know of no rigorous proof of the statement that the ground-state of the jellium model in the low-density limit is a Wigner crystal. We will therefore take this as a rather safe working hypothesis.

[39] We will however provide a simple argument based on the expression for the total energy in wave vector space which leads to the correct answer.

independent of the total number of cells N: it is for this reason that the total energy scales as N and not N^2. To estimate the energy, approximate the unit cell as a sphere of radius $r_s a_B$ with the electron at the center. What is lost in this approximation is the true shape of the unit cell, while the correct volume is preserved. In three dimensions, these neutral spherical cells do not interact with each other (Gauss' theorem): therefore, in this approximation, the energy of the crystal is simply

$$U = N U_{cell} \tag{1.109}$$

where U_{cell} is the electrostatic energy of a single spherical unit cell. The calculation of U_c is a simple exercise in electrostatics. By making use of Gauss' theorem, the potential generated by the positive charge within each sphere works out to be

$$V_{sphere}(r) = \frac{e}{2 r_s a_B} \left[3 - \left(\frac{r}{r_s a_B} \right)^2 \right], \tag{1.110}$$

where $r < r_s a_B$ is the distance from the center. Then, the energy of the unit cell is calculated as the energy of the electron in the potential (1.110), plus the self-energy of the sphere

$$U_{sphere} = \frac{1}{2} \int_{r \le r_s a_B} en V_{sphere}(r) d\vec{r} = \frac{3}{5} \frac{e^2}{r_s a_B}. \tag{1.111}$$

We therefore get

$$\frac{U}{N} = U_{cell} = -e V_{sphere}(0) + U_{sphere} = -\frac{1.8}{r_s} \text{Ry.} \tag{1.112}$$

This apparently rough estimate gives a surprisingly good approximation to the accurate results of Table 1.3 later in this chapter. In addition, it can be shown to be a *rigorous lower bound* on the electrostatic energy of *any* classical configuration (crystalline or not!) of N point charges on a uniform neutralizing background (Lieb and Narnhofer, 1975). The proof of this fact is presented in Appendix 6.

The same rough calculation can be done in two dimensions, but here we find a basic difficulty, namely, Gauss' theorem does not hold for *disks* of charge, and non-overlapping neutral disks do interact with each other. Ignoring this complication and proceeding as in the three-dimensional case, we calculate the potential from the positive charge within the disk to be

$$V_{disk}(r) = \frac{4e}{\pi r_s a_B} \mathbf{E} \left(\frac{r}{r_s a_B} \right), \tag{1.113}$$

$(r < r_s a_B)$ where $\mathbf{E}(x)$ is the *complete elliptic integral of the second kind* (see Gradshteyn and Ryzhik, 8.11), and the self-energy of the disk is

$$U_{disk} = \frac{en}{2} \int_{r \le r_s a_B} d\vec{r} \, V_{disk}(r) = \frac{8e^2}{3\pi r_s a_B}. \tag{1.114}$$

Thus, the approximate energy of the two-dimensional Wigner crystal is

$$\frac{U}{N} = -eV_{disk}(0) + U_{disk} \simeq -\frac{2.3}{r_s}\text{Ry} . \qquad (1.115)$$

Comparison with the data of Table 1.3 shows that this is still, perhaps surprisingly, quite a reasonable estimate.

A simple yet powerful argument can also be constructed to predict the actual crystalline structure of lowest electrostatic energy.[40] It is readily seen that if classical particles are located at the lattice sites $\vec{R}$ of a simple *Bravais lattice*, then their density can be expressed as

$$n(\vec{r}) = \sum_{\vec{R}} \delta(\vec{r} - \vec{R}) = n\sum_{\vec{G}} e^{i\vec{G}\cdot\vec{r}} , \qquad (1.116)$$

where the sums run over real lattice vectors $\vec{R}$, and corresponding reciprocal lattice vectors $\vec{G}$. By making use of this formula in Eq. (1.108) one can derive (see Exercise 7) the following alternative expression for the electrostatic energy per particle of a Wigner crystal:

$$\frac{U}{N} = \frac{n}{2}\sum_{\vec{G}\neq 0} v_{|\vec{G}|} - \frac{1}{2L^d}\sum_{\vec{q}} v_q , \qquad (1.117)$$

While this formula is not useful for quantitative studies (both terms are formally divergent), what is noteworthy here is that only the first term contains the effects of the actual crystal structure. Then, in view of the wave vector dependence of v_q (see Eq. (1.19)), it is reasonable to expect that the crystal structure with the lowest energy will be the one that, for a given volume of the primitive unit cell, has the first reciprocal lattice vector of the largest magnitude.[41] In two and three dimensions, the crystal structures with this property are hexagonal and body-centered-cubic (bcc) respectively: hence, one is led to predict that the Wigner crystal will be hexagonal in two dimensions and bcc in three. As we shall see, these predictions are again accurate (see Table 1.3).

Although these simple physical arguments provide satisfactory answers, ingenious mathematical transformations have been devised to carry out the calculation of the energy *exactly*. A particularly convenient expression for U has been obtained by Bonsall and Maradudin (1977) and is based on the original method of Ewald (1921). Its derivation is presented in Appendix 5. For a Bravais lattice in two or three dimensions one has

$$\frac{U}{N} = -\frac{ne^2}{2}\left(\frac{\pi}{\eta}\right)^{\frac{d-1}{2}}\left[\frac{2}{d-1} - \sum_{\vec{G}\neq 0}\phi_{\frac{d-3}{2}}\left(\frac{G^2}{4\eta}\right)\right]$$

$$- \frac{e^2}{2}\left(\frac{\eta}{\pi}\right)^{\frac{1}{2}}\left[2 - \sum_{\vec{R}\neq 0}\phi_{-\frac{1}{2}}(\eta R^2)\right] , \qquad (1.118)$$

---

[40] This argument is due to A. W. Overhauser.

[41] The first reciprocal lattice vector, $\vec{G}_1$, is the reciprocal lattice vector closest to the origin of reciprocal space. For example, in a 3D bcc lattice of density n one has $|\vec{G}_1| = n^{1/3}2^{7/6}\pi$ while in a fcc lattice of the same density $|\vec{G}_1| = n^{1/3}2^{1/3}\sqrt{3}\pi$.

Table 1.3. *Energies per electron of the
classical Wigner crystal for different Bravais
lattices in three and two dimensions.*

d	Lattice	$\frac{U}{N}\left(\frac{Ry}{r_s}\right)$
3	Simple Cubic	−1.760
3	Face Centered Cubic	−1.79175
3	Body Centered Cubic	−1.79186
3	Hexagonal Close Packed	−1.79168
2	Square	−2.2
2	Hexagonal	−2.212

where the *Misra functions* $\phi_\nu(z)$ are defined as

$$\phi_\nu(z) \equiv \int_1^\infty dt\, e^{-zt} t^\nu . \tag{1.119}$$

The parameter η is arbitrary: a typical choice is $\eta = \pi n$ in two dimensions, and $\eta = \pi n^{\frac{2}{3}}$ in three dimensions. Because the $\vec{R}$ and $\vec{G}$ sums in Eq. (1.118) are rapidly converging, the energy can be evaluated with any desired degree of accuracy.

Table 1.3 summarizes the results obtained for the classical energy of several Bravais lattices in three and two dimensions (Scholl, 1967; Bonsall and Maradudin, 1977). From these calculations one can conclude that, as anticipated, *among the Bravais lattices, the body-centered-cubic is favored in three dimensions and the hexagonal in two dimensions.*[42]

1.6.2 Zero-point motion

In order to refine the calculation of the ground-state energy of the low-density electron system, we now examine the leading quantum mechanical corrections to the classical electrostatic energy. The main correction arises from the fact that, as a consequence of quantum mechanical uncertainty, electrons localized near a lattice site perform a *zero-point motion*, which rises the energy relative to the classical value.

To address this question in detail, let us expand the potential energy of the classical Wigner crystal up to second order in the displacements $\vec{r}_i - \vec{R}_i$ of the *i-th* electron from its equilibrium position, which is taken to coincide with the *i-th* site of the classical Wigner lattice.[43] The first order terms in the expansion vanish, as the Wigner crystal is a local

---

[42] See Eq. (1.131) for a useful interpolation formula for the total energy of a Wigner crystal.
[43] Notice that this classical description treats the electrons as *distinguishable* particles, by associating them with different lattice sites. In the low density limit, however, this approximation has a negligible effect on the energy.

minimum of the potential energy. Thus, the hamiltonian takes the form:

$$\hat{H} \simeq \sum_i \frac{\hat{p}_i^2}{2m} + \frac{1}{2} \sum_{ij,\alpha\beta} V_{ij,\alpha\beta}(\hat{\vec{r}}_i - \vec{R}_i)_\alpha (\hat{\vec{r}}_j - \vec{R}_j)_\beta + U(\vec{R}_1 \ldots \vec{R}_n), \quad (1.120)$$

where α, β are cartesian labels, and $V_{ij,\alpha\beta}$ represents the second derivative of the potential energy with respect to $r_{i\alpha}$ and $r_{j\beta}$, calculated at the equilibrium positions $\vec{r}_i = \vec{R}_i$. $U(\vec{R}_1, \ldots, \vec{R}_n)$ is the energy of the classical Wigner crystal. Note that the sum over i, j now includes the case $i = j$. Because the momenta $\hat{p}_i$ are canonically conjugated to the displacements $\hat{\vec{r}}_i - \vec{R}_i$, this quadratic hamiltonian can be solved by a standard transformation to normal modes. The transformed hamiltonian represents a set of Nd *independent* harmonic oscillators of mass m and frequency $\omega_{\vec{q}\lambda}$, where $\vec{q}$ is a vector in the first Brillouin zone (BZ) of the Wigner crystal, and $\lambda = 1, \ldots, d$ is a polarization index. Making use of the well-known results for the ground-state energy of the quantum harmonic oscillator, we see that the ground-state energy of this hamiltonian is

$$E = U(\vec{R}_1, \ldots, \vec{R}_n) + \sum_{\vec{q} \in BZ, \lambda} \frac{\hbar \omega_{\vec{q}\lambda}}{2}, \quad (1.121)$$

where the last term is the sum of the zero point energies of the independent harmonic oscillators.

For the present discussion we only need to know how the oscillator frequencies scale with r_s. From the homogeneity of the coulomb interaction we immediately see that the second derivatives of the potential energy $V_{ij,\alpha\beta}$ scale as $\frac{e^2}{r_s^3 a_B^3}$ and, therefore, the average phonon energy, $\hbar\bar{\omega}$, scales as $\hbar\sqrt{\frac{V_{ij,\alpha\beta}}{m}} \sim \frac{Ry}{r_s^{3/2}}$. Thus, the quantum mechanical correction to the classical Wigner crystal energy scales as $\sim \frac{Ry}{r_s^{3/2}}$, which, for large r_s, is much smaller than the classical potential energy, as anticipated. Notice that half of this correction is kinetic energy, and half is potential. Also, the characteristic amplitude of the oscillations of the electrons about their equilibrium positions is $l \sim \frac{\hbar}{\sqrt{m\bar{\omega}}} \sim r_s^{\frac{3}{4}} a_B$, which is much smaller than the average inter-electron distance, $r_s a_B$, proving, *a posteriori*, the consistency of the Wigner crystal picture.

The calculation of the characteristic frequencies $\omega_{\vec{q}\lambda}$ and of the zero point energy has been carried out numerically by several authors (Coldwell, Horsfall, and Maradudin, 1960; Carr, 1961; Bonsall and Maradudin, 1977). Two representative results for the leading correction to the classical energy are

$$\frac{\Delta U}{N} = \frac{1}{N} \sum_{\vec{q} \in BZ, \lambda} \frac{\hbar \omega_{\vec{q}\lambda}}{2} \approx \begin{cases} \frac{2.66}{r_s^{\frac{3}{2}}} Ry, & (3D - bcc), \\ \frac{1.59}{r_s^{\frac{3}{2}}} Ry, & (2D - hexagonal). \end{cases} \quad (1.122)$$

In addition to phonons, the Wigner crystal also supports collective spin excitations, which are governed by the weak dependence of the energy on the relative orientation of the spin on different lattice sites. The energy scale of these excitations is much lower than the energy scale for phonons. A useful formula for the total energy of the Wigner crystal (Eq. (1.131)) is discussed in the next section.

1.7 Phase diagram of the electron liquid

The Wigner crystal is a spontaneously broken symmetry state since the translational invariance of the jellium hamiltonian is broken. We have just argued that, for very large r_s, it is the ground-state of the electron gas. On the other hand, for sufficiently small r_s the ground-state is certainly translationally invariant. This leads to a fascinating question: At what value of r_s does the transition between the uniform liquid and the Wigner crystal take place? This is a hard question because, near the transition, the system is very far from both the noninteracting uniform liquid and the classical Wigner crystal. Specifically, anharmonicity and exchange effects, which are neglected in the elementary treatment of the Wigner crystal, play an important role near the transition. Furthermore, it may well be the case that the system is already in some kind of broken symmetry state (e.g., a spin-polarized liquid) when the transition to the Wigner crystal occurs. It is therefore clear that the perturbation theory outlined in Section 1.5.6 is not applicable at the characteristically large r_s of this transition, while unaided mean field approximations, based on the uncontrolled neglect of fluctuations, can be dangerously misleading. On the other hand, an exact solution of the quantum mechanical problem for systems of reasonable size is still beyond the power of even the fastest computers.

A major technical breakthrough occurred in the early 1980s, with the introduction of numerical *Quantum Monte Carlo* (QMC) methods which, for the first time, yielded highly accurate values for the ground-state energies of the paramagnetic and spin polarized phases of the uniform electron liquid (Ceperley and Alder, 1980), as well as useful information about the pair correlation functions and the static response functions of these phases (see Chapter 5).[44]

Virtually, all we know today about the phase diagram of the electron liquid is based on the ground-state energies obtained from such calculations (Ceperley and Alder, 1980; Tanatar and Ceperley, 1989; Rapisarda and Senatore, 1996). The results of these studies will be presented in the next section. Unfortunately, even the most sophisticated numerical methods do not *determine* the symmetry of the ground-state, but rather *assume it* from the start. The phase diagram is then constructed by comparing the energies of states of different symmetry. The obvious weakness of this approach is that the true ground-state might have a symmetry that was not considered, either because it was too difficult to implement, or just for lack of imagination.

1.7.1 The Quantum Monte Carlo approach

Under the generic name of Quantum Monte Carlo (QMC) one groups several different computational methods of varying degree of sophistication, all of which have in common the use of random walks to sample the configuration space of the system. The three basic flavors of QMC are (i) the Variational Monte Carlo (VMC), (ii) the Diffusion Monte Carlo (DMC), and (iii) the Path Integral Monte Carlo (PIMC). In this section we very briefly

[44] A basic review of these methods can be found in the paper by Foulkes *et al.* (2001) or, on a more elementary level, in the book by Hammond, Lester and Reynolds (1994).

describe the main characteristics, strengths and limitations of each. Our discussion closely follows the recent review by Ceperley (2004).

Variational Monte Carlo. In this approach the Monte Carlo method is used to evaluate the expectation value of the hamiltonian in a trial wave function $\Psi_t(R)$ (R is here a short-hand for the spatial and spin coordinates of the N electrons), i.e.,

$$E_V = \frac{\int \Psi_t^*(R)\hat{H}\Psi_t(R)\,dR}{\int \Psi_t^*(R)\Psi_t(R)\,dR} \,. \tag{1.123}$$

According to the Rayleigh–Ritz variational principle of quantum mechanics, E_V is an upper bound to the ground-state energy, provided Ψ_t has the proper symmetry and the correct continuity and differentiability properties. To evaluate the frightening expression (1.123) one uses a random walk to generate a set of points in configuration space that are distributed according to the probability distribution

$$\Pi(R) = \frac{|\Psi_t(R)|^2}{\int \Psi_t^*(R)\Psi_t(R)\,dR} \,. \tag{1.124}$$

The ground-state energy is then computed as

$$E_V = \int \Pi(R)E_L(R)\,dR \,, \tag{1.125}$$

where

$$E_L(R) \equiv \Re e\,\frac{\hat{H}\Psi_t(R)}{\Psi_t(R)} \tag{1.126}$$

is the local energy.[45]

The main strength of the VMC method (besides its conceptual simplicity) lies in the fact that, through the choice of the trial wave function, it allows us to exert ingenuity and physical insight (we shall see in Chapter 10 examples in which an inspired choice of the trial wave function has given truly spectacular results). This is also, in a sense, the main weakness of the method, since it is clear that a VMC calculation will only be as good as the trial wave function one chooses. For example, in performing comparisons between the energies of different states, the VMC usually favors the "simpler" state, namely the one for which we can more easily guess a good trial wave function (e.g. the Wigner crystal state is favored upon the liquid state). In addition, the variational energy is insensitive to long range order, such as the long-range correlations that are responsible for superconductivity. This is because the energy is dominated by short-range correlations. Therefore one can get very good values of the energy even with a fundamentally incorrect wave function.

---

[45] The desired probability distribution is achieved with the help of the *Metropolis algorithm*, a simple set of rules for constructing a random walk in which a trial step from R_a to R_b is rejected or accepted with probabilities $P_r = \min\{1, |\Psi_t(R_b)|^2/|\Psi_t(R_a)|^2\}$ and $1 - P_r$ respectively.

Diffusion Monte Carlo. The DMC is a more powerful Monte Carlo method which can, in principle, *determine* the ground-state wave function. The basic idea is to let the initial wave function $\Psi_t(R)$ evolve in imaginary time according to the rule

$$\Psi_t(R, \tau) = e^{-(\hat{H}-E_t)\tau/\hbar}\Psi_t(R) . \tag{1.127}$$

Unless the initial wave function Ψ_t is chosen with a deliberate lack of ingenuity (i.e., orthogonal to the ground-state) $\Psi_t(R, \tau)$ will converge, for large τ, to the true ground-state wave function. The constant E_t will be adjusted to keep the normalization constant: when this is accomplished E_t equals the ground-state energy.

The time evolution (1.127) is numerically implemented with the help of random walks. Consider first the case of a system of N bosons. The ground-state wave function of such a system is completely symmetric under exchange of any two particles, and this, combined with the variational principle for the energy, implies that it has to be real and *nodeless*, i.e. its sign is everywhere the same, say positive. One can then interpret the wave function itself, rather than its square, as a probability distribution in configuration space. The time-evolution of this probability distribution can be simulated as a diffusion, drifting, and branching process, where the three terms refer respectively to the spreading of the probability distribution, the drifting of its center of gravity, and the generation and death of members of the distribution. In this way one arrives at an "equilibrium" distribution that reflects the magnitude of the ground-state wave function.

Applying this strategy to a system of N fermions, which is the case of interest here, presents a major difficulty, namely, the fact that the ground-state wave function is antisymmetric under interchange of two particles and therefore must necessarily change sign across the so-called nodal surfaces, where it vanishes. This fact prevents a direct interpretation of the wave function as a probability distribution. The difficulty could be avoided if one knew *a priori* the location of the nodal surfaces, for, in that case, one could solve the problem within a *nodal region* – the region of configuration space bounded by a nodal surface – in which the sign of wave function is constant, and then reconstruct the complete solution through the use of antisymmetry.

Unfortunately, the nodal surfaces are generally unknown, except in the special case of one dimensional systems, where they are completely determined by the conditions $x_i = x_j$, where x_i and x_j are the coordinates of any two parallel-spin electrons (Pauli exclusion principle).[46]

A solution to the fermion problem can be achieved by means of a two-step procedure (Ceperley and Alder, 1980). In the first step, the nodal surfaces are assumed to coincide with those of the trial wave function chosen as the starting point of the calculation. Within a nodal region, the problem is equivalent to a bosonic problem and can be solved with QMC. The energy calculated with this *fixed node* procedure provides an upper bound for the exact ground-state energy, and this is already much lower than the energy obtained in a

---

[46] The nodal surfaces are hypersurfaces of dimension $Nd - 1$ in the Nd-dimensional configuration space. In more than one spatial dimension, the condition $\vec{r}_i = \vec{r}_j$ determines a surface of dimension $Nd - d$, which is *lower* than the dimension of the nodal surfaces.

typical variational approach. In the second step, called *nodal relaxation*, the random walk is allowed to cross the nodal surfaces, but a minus sign is attached to a walker each time it does cross a nodal surface. It can be shown that the antisymmetric ground-state wave function is, in principle, proportional to the *difference* between the density of positive and negative walkers. Unfortunately, the loss of accuracy inherent in taking the difference of two large numbers limits in general the effectiveness of this method (this is the so called *fermion sign problem*). The method only works when the convergence to the equilibrium distribution, starting from the trial wave function, occurs rapidly on a time scale set by the inverse of the energy difference between the fermionic and bosonic ground-state. This means, in practice, that the nodal surfaces of the trial wave function must already be a good approximation to the true ones, in order to achieve convergence to the exact distribution. Ceperley and Alder determined that this condition is satisfied in the case of the electron liquid when the nodes of the noninteracting ground-state wave function Ψ_0 are used to start the procedure. Subsequent calculations verified this with better initial guesses for the nodal surfaces.

Path Integral Monte Carlo. While VMC and DMC are intrinsically zero temperature methods, the PIMC method focuses on thermodynamic quantities which are obtained from the N-particle density matrix $\rho(R, R', T) = \langle R|e^{-\hat{H}/k_B T}|R' \rangle$ – again, a kind of evolution operator in imaginary time. One of the characteristic features of this approach is that it does not rely on a trial wave function: the density matrix is calculated as a sum of probability amplitudes associated with paths (in configuration space) that start at R at $\tau = 0$ and return to PR (P being a permutation of N things) at time $\hbar/k_B T$ (see Feynman, 1972). Because an initial guess of the solution is not required, the PIMC method has great potential as a general-purpose solver of complex quantum mechanical system. In electron liquids, this method has been successfully employed to date in the calculation of properties of the Wigner crystal, such as: the energies of defects (interstitial and vacancies), the melting temperature, and the effective interactions between spins (Ceperley, 2003).

A serious challenge in the implementation of all Monte Carlo methods is the limit to the size of the systems that can be computed in practice (currently up to a thousand electrons or so). For the results to be meaningful, they must be interpreted via suitable size-scaling procedures which ultimately add to the uncertainty of the computed numbers. The finite size of the system also leads to a certain fuzziness in the definition of wave vectors: accordingly it is difficult to detect any small, yet potentially crucial, distortion of the Fermi surface.[47] The above limitations add to the challenge of deciding upon the relative stability of competing electronic states which happen to be very close in energy.

1.7.2 The ground-state energy

The ground-state energy of the electron liquid has been calculated by both Variational Monte Carlo and Diffusion Monte Carlo methods in two and in three dimensions. Technical

---

[47] This difficulty currently hinders the study of charge or spin density wave states.

advances in the last few years have greatly improved the accuracy of these calculations. The results are most conveniently presented in terms of analytical fitting formulas.

Three dimensions. In three dimensions, the best known representations of the energy have been proposed by Perdew and Wang (Perdew and Wang, 1992), by Vosko, Wilk and Nusair (Vosko, Wilk, and Nusair, 1980) and by Perdew and Zunger (Perdew and Zunger 1980).[48] All these formulas reduce to the Macke–Gell-Mann and Brückner high-density limit (see Eq. (1.107)) for small r_s and to the Ceperley and Alder results for large $r_s (\leq 100)$. In the limit of large r_s they can be written as power series of the form $g_1/r_s + g_{3/2}/r_s^{3/2} + g_2/r_s^2 + \cdots$, which is reminiscent of the expression for the energy of a Wigner crystal (see Section 1.6 and Eq. (1.131)). Here we quote the formula proposed by Perdew and Wang for the correlation energy of the three dimensional electron gas (in Rydberg per electron) based on the original QMC calculations of Ceperley and Alder:

$$\epsilon_c^{(3D)}(r_s, p) = \epsilon_c^{(3D)}(r_s) + \alpha_c(r_s) \frac{f(p)}{f''(0)}(1 - p^4)$$
$$+ [\epsilon_c^{(3D)}(r_s, 1) - \epsilon_c^{(3D)}(r_s)] f(p) p^4 \qquad (1.128)$$

where the interpolation function $f(p)$ has the form:

$$f(p) = \frac{(1+p)^{\frac{4}{3}} + (1-p)^{\frac{4}{3}} - 2}{(2^{\frac{4}{3}} - 2)}, \qquad (1.129)$$

so that $f(0) = 0$, $f(1) = 1$, and $f''(0) = 1.709921$. The three functions $\epsilon_c^{(3D)}(r_s)$ (the correlation energy at $p = 0$), $\alpha_c(r_s)$ (the spin stiffness, defined in Section 1.8) and $\epsilon_c^{(3D)}(r_s, 1)$ (the correlation energy at $p = 1$) can all be represented by means of the general expression

$$G(r_s, A, B, C, D, E, F) = -4A (1 + B r_s)$$
$$\times \ln \left[1 + \frac{1}{2A \left(C r_s^{1/2} + D r_s + E r_s^{3/2} + F r_s^2 \right)} \right] \text{Ry},$$
$$(1.130)$$

where the constants A–F for each case are given in Table 1.4.

It should be noted that the spin-polarization dependence of the ground-state energy implied by this formula is not very accurate. The calculation of the p-dependence of the energy is difficult because, due to the very small energy difference between states with different polarizations, systematic errors greatly affect the QMC results. Only very recently, technical advances, such as the use of twist-averaged boundary conditions (as opposed to standard periodic ones) have allowed reasonably accurate calculations of the p-dependent energy, resulting in the curves that are plotted in Fig. 1.13. Figs. 1.11 and 1.12 show the total ground-state energy of the uniform liquid phase in three dimensions at $p = 0$ (paramagnetic phase) and $p = 1$ (ferromagnetic phase).

---

[48] QMC studies of the three dimensional electron gas have been carried out more recently by Ortiz and Ballone, 1994; Ortiz, Harris, and Ballone, 1999; and Zong, Lin, and Ceperley, 2002.

Table 1.4. *Parameters for the 3D correlation*
energy, Eqs. (1.128)–(1.130).

	$\epsilon_c(r_s)$	$\epsilon_c(r_s, 1)$	$-\alpha_c(r_s)$
A	0.031091	0.015545	0.016887
B	0.21370	0.20548	0.11125
C	7.5957	14.1189	10.357
D	3.5876	6.1977	3.6231
E	1.6382	3.3662	0.88026
F	0.49294	0.62517	0.49671

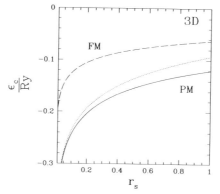

Fig. 1.11. Comparison of the correlation energy per particle (in Rydbergs) for the three dimensional electron gas for small r_s. The curves labeled PM and FM correspond to the paramagnetic and ferromagnetic states and have been obtained via the parametrizations of the QMC results given in the text. The dotted line represents the asymptotic expression of Eq. (1.107) for the paramagnetic state.

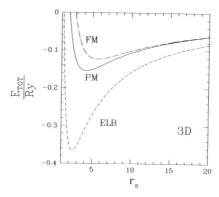

Fig. 1.12. Comparison of the total energy per particle for the paramagnetic (PM) and the ferromagnetic (FM) phases of the electron gas in three dimensions for intermediate density values. The curve labeled ELB represents the exact lower bound of Eq. (A6.1).

Table 1.5. *Parameters for the Wigner crystal energy, Eq. (1.131). (From Drummond et al. (2004) for the bcc lattice in 3D and from Rapisarda and Senatore (1996) for the hexagonal lattice in 2D.)*

	2D	3D
c_1	−2.20943	−1.79186
$c_{3/2}$	1.58948	2.6758
c_2	0.146762	−1.1054

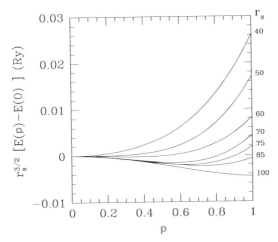

Fig. 1.13. The spin polarization dependence of the ground-state energy of a uniform 3D electron liquid. (Adapted from Zong *et al.*, 2002.)

At large values of r_s these energies should be compared with the energy of the Wigner crystal, which is also calculated by QMC and can be represented by the expression

$$\epsilon_{WC}(r_s) \approx \left(\frac{c_1}{r_s} + \frac{c_{3/2}}{r_s^{3/2}} + \frac{c_2}{r_s^2} \right) \text{Ry} , \qquad (1.131)$$

with parameters given in Table 1.5.

From this comparison one arrives at the conclusion that in three dimensions, among the three possible states of the electron gas considered here (paramagnetic liquid, ferromagnetic liquid, and Wigner crystal), the paramagnetic liquid has the lowest energy for $r_s < 75$, while the ferromagnetic liquid has the lowest energy in the range $75 < r_s < 100$. Moreover for $r_s > 100$ it is the Wigner crystal to have the lowest energy.

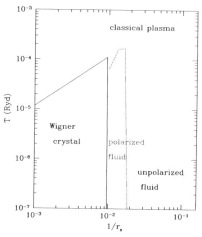

Fig. 1.14. Tentative QMC phase diagram for a three dimensional electron gas. The magnetic phase boundaries at finite temperature are estimated with the help of the *Stoner model*, discussed in Exercise 8. (Ceperley, 2004); used with permission.

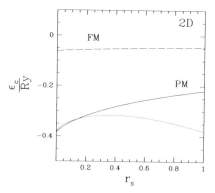

Fig. 1.15. Same as in Fig. 1.11 but for the case of a two dimensional electron gas. The dotted line represents the asymptotic expression of Eq. (1.107) for the paramagnetic state.

The dependence of the energy on spin polarization is shown in Fig. 1.13. A polarization transition is evident. At $r_s = 40$, the system is still paramagnetic, with the unpolarized phase stable. As the density decreases, at $r_s \approx 50$, the liquid becomes unstable with respect to a partially spin polarized state. As the electronic density continues to decrease, the stable state becomes more and more polarized, approaching full polarization ($p = 1$) at the freezing transition $r_s \approx 100$.

Fig. 1.14 summarizes the discussion by presenting a tentative phase diagram based on comparison of the energies of only three different phases of the jellium model, namely the uniform paramagnetic liquid, the uniform (partially) spin-polarized liquid, and the Wigner crystal.

Table 1.6. *Parameters for the 2D correlation energy, Eqs. (1.132)–(1.134).*

	$i = 0$	$i = 1$	$i = 2$
A_i	-0.1925	0.117331	0.0234188
B_i	0.0863136	-3.394×10^{-2}	-0.037093
C_i	0.0572384	-7.66765×10^{-3}	0.0163618
E_i	1.0022	0.4133	1.424301
F_i	-0.02069	0	0
G_i	0.33997	6.68467×10^{-2}	0
H_i	1.747×10^{-2}	7.799×10^{-4}	1.163099
$D_i \equiv -A_i H_i$	3.36298×10^{-3}	-9.15064×10^{-5}	-0.0272384

Two dimensions. The ground-state energy of the two dimensional electron liquid has been calculated by Tanatar and Ceperley (Tanatar and Ceperley, 1989), by Rapisarda and Senatore (Rapisarda and Senatore, 1996), and, very recently, by Attaccalite *et al.* (Attaccalite *et al.*, 2002). We quote the parametrization proposed by the latter, which reads

$$\epsilon_c^{(2D)}(r_s, p) = \left(e^{-\beta r_s} - 1\right) \epsilon_x^{(6)}(r_s, p) + \alpha_0(r_s) + \alpha_1(r_s)p^2 + \alpha_2(r_s)p^4 , \quad (1.132)$$

where $\beta = 1.3886$,

$$\epsilon_x^{(6)}(r_s, p) = \epsilon_x(r_s, p) - \epsilon_x(r_s)\left(1 + \frac{3}{8}p^2 + \frac{3}{128}p^4\right) \quad (1.133)$$

is the Taylor expansion of $\epsilon_x(r_s, p)$ beyond the fourth order in p, and the functions $\alpha_i(r_s)$, with $i = 0, 1, 2$, have the form

$$\alpha_i(r_s) = 2\left[A_i + \left(B_i r_s + C_i r_s^2 + D_i r_s^3\right)\right.$$
$$\left. \times \ln\left(1 + \frac{1}{E_i r_s + F_i r_s^{3/2} + G_i r_s^2 + H_i r_s^3}\right)\right] \text{Ry}. \quad (1.134)$$

The parameters of the formula are given in Table 1.6.

Fig. 1.16 shows the total ground-state energy of the two-dimensional uniform liquid phase at $p = 0$ and $p = 1$, while Fig. 1.17 shows the dependence of the energy on spin polarization, according to Eq. (1.132). A comparison between these energies and those of the Wigner crystal at low density (see Eq. (1.131)) shows that the paramagnetic liquid is stable for $r_s \le 20$, while the Wigner crystal has lower energy for $r_s > 34$. At variance with the three dimensional situation, in the intermediate range $20 < r_s < 34$ it is very hard to decide which between the paramagnetic and the ferromagnetic liquid has the lower energy. The problem is to date unsettled. Fig. 1.18 provides a schematic QMC phase diagram in two dimensions.

We conclude this section by noting that, although the Monte Carlo method has provided valuable information about the relative stability of the simplest phases of the jellium model,

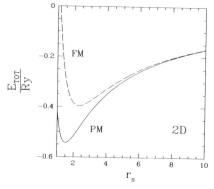

Fig. 1.16. Comparison of the total energy per particle for the paramagnetic (PM) and the ferromagnetic (FM) phases of the electron gas in two dimensions for intermediate density values.

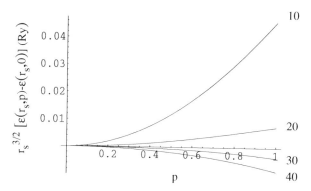

Fig. 1.17. The spin polarization dependence of the ground-state energy of a uniform 2D electron liquid, calculated from Eq. (1.132).

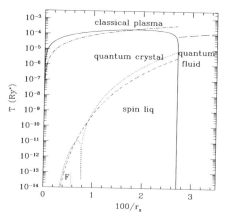

Fig. 1.18. QMC phase diagram for a two dimensional electron gas From Bernu *et al.* (2001); used with permission.

one clearly cannot claim to have achieved a complete understanding of the phase diagram of the electron liquid. The occurrence of transitions between states of different symmetry indicates that the *ground-state energy* of the system is a nonanalytic function of r_s, and therefore it is conceptually incorrect, albeit at times practically useful, to approximate it with an analytic function.

1.7.3 Experimental observation of the electron gas phases

We have thus far examined three possible phases of the electron liquid: the paramagnetic Fermi liquid, the spin-polarized Fermi liquid, and the Wigner crystal. The character of the first two phases is defined by the quantum kinetic energy, with interactions playing a subordinate, if quantitatively important, role.[49] The crystal phase, on the other hand, is classical in nature, with quantum effects becoming important only near the liquid–solid transition. The QMC method has given us valuable predictions about the relative stability of these three phases: it is now natural to ask what actual experimental observations tell us.

It turns out that most realizations of the electron gas so far explored fall squarely under the paradigm of the paramagnetic Fermi liquid. The reason for this can be readily seen from the QMC phase diagram. The magnetic and the crystal phase are predicted to be stable only at extremely low densities, densities that have long been beyond the reach of experiment and are just now beginning to be explored.[50] One of the main problems, at such low densities, is that even a small concentration of impurities can completely destroy the intrinsic behavior of the electron gas. Thus, for example, a system of electrons bound to randomly distributed impurities might be mistaken for a Wigner crystal. Sample growers must go through great pain to ensure the necessary degree of purity.

Electrons in real metals, of course, do undergo phase transitions, of which the best known are superconductivity and magnetism. Both these transitions involve the ionic lattice in essential ways: in the case of superconductivity it is the electron–phonon coupling that normally drives the transition,[51] while magnetism is typically observed in *narrow band systems* (metal oxides) or metals with complex Fermi surfaces: these systems are not adequately described by the jellium model.[52] Examples of strongly interacting Fermi liquid at high density are provided by heavy-fermion materials such as UPt_3, which exhibit a rich phase diagram including normal, superconducting, and magnetic phases. This and similar systems are usually modelled in terms of localized atomic f-orbitals immersed in a continuum of extended d-states – the so-called *Anderson model*. Again, this can hardly be considered a realization of the jellium model.

---

[49] The precise definition of the "Fermi liquid" phases will be given in Chapter 8.
[50] Recent work reports a study of the two-dimensional electron gas at the *unprecedented low density* of $r_s = 13.4$ (Zhu *et al.*, 2003).
[51] A very interesting and still unresolved question is whether superconductivity can arise in the rigid jellium model purely from electron–electron interactions, and, if so, at what densities. Existing estimates of the transition temperature (Kohn and Luttinger, 1965; Grabowski and Sham, 1984; Takada, 1993) differ literally by orders of magnitude.
[52] A canonical model that plays the role of the jellium model in such systems is the *Hubbard model*, in which electrons hop from one lattice site to the next experiencing a large coulomb repulsion only when two of them, with opposite spin orientations, happen to be on the same site.

In view of these difficulties, several ingenious attempts have been made to create situations in which the effective coupling constant r_s is larger than the density would imply. A good example is provided by the *hole liquid* in p-doped GaAs. Due to the large effective mass of holes in GaAs ($m_h \simeq 0.34\, m$, to be compared with $m_b \simeq 0.067\, m$ for electrons) the r_s value for holes is about four times larger than for electrons at the same density. Indeed evidence has been found for the formation of a Wigner crystal of holes in this system at a value of $r_s \simeq 35$, which closely matches the QMC prediction (Yoon *et al.*, 1999). Unfortunately, holes in GaAs present a number of problems related to the non-parabolicity of the valence band and strong spin–orbit coupling, so the confirmation of the theory cannot be considered definitive.[53]

Magnetic fields can also facilitate the observation of different phases of the electron liquid. In two dimensions, in particular, the application of a magnetic field to a 2DEL causes a severe suppression of the kinetic energy (see Chapter 10), which, in turn, favors the emergence of broken symmetry phases at relatively small values of r_s. For example, transport experiments in the 2DEL in n-doped GaAs at magnetic fields of the order of 10 T indicate the existence of a Wigner crystal of electrons at $r_s \simeq 2.5$, which is very much smaller than the critical value predicted by QMC at zero magnetic field (Goldman, 1990).

Efforts to observe the ferromagnetic phase predicted by Bloch back in 1929 have likewise been frustrated by the difficulty of achieving large values of r_s. Again, the closest thing to an experimental observation of this transition has come so far from experiments in the 2DEL at high magnetic field. Under appropriate conditions, which will be fully described in Chapter 10, the magnetic field suppresses not only the kinetic energy, but also the correlation energy. This leaves the exchange energy master of the field, and leads to a ferromagnetic transition of the type envisaged by Bloch (Giuliani and Quinn, 1985; Daneshvar *et al.*, 1997).

1.7.4 Exotic phases of the electron liquid

In addition to the phases discussed in the previous sections a few exotic phases have been proposed and studied over the years.

In three dimensions, the most remarkable are Overhauser's charge density wave (CDW) and the closely related spin density wave (SDW). These are non uniform states in which the the charge density or the spin density (or both) exhibit spatial oscillations at a wave vector $\vec{Q}$ close in magnitude to $2k_F$. Thus, in a typical CDW/SDW state one has

$$n_\uparrow(\vec{r}) = \frac{n}{2} + A\cos\left(\vec{Q}\cdot\vec{r} + \frac{\phi}{2}\right)$$

$$n_\downarrow(\vec{r}) = \frac{n}{2} + A\cos\left(\vec{Q}\cdot\vec{r} - \frac{\phi}{2}\right), \qquad (1.135)$$

---

[53] Notice that the Wigner crystal phase being sought here is quite different from the classical crystal of electrons that is observed on the surface of helium. The latter occurs in a temperature regime in which the electrons are still classical, i.e., the thermal wavelength $\lambda_T \equiv \dfrac{\hbar}{\sqrt{2mk_B T}}$ is smaller than the average distance between the electrons.

where A is the amplitude of the wave, and ϕ is the *relative phase* of the up-spin and down-spin density oscillations. For $\phi = 0$ Eq. (1.135) describes a pure CDW with uniform spin density, while for $\phi = \pi$ it describes a pure SDW with uniform charge density. Intermediate values of ϕ describe mixed CDW-SDW states (Overhauser, 1960, 1961).

The possibility of CDW/SDW phases was initially suggested by a Hartree–Fock analysis (see Chapter 2) which took into account exchange, but not correlation. The physical idea is as follows. We have seen that the exchange energy always favors the development of a spin polarization, while the kinetic energy opposes it. The kinetic energy cost arises from the fact that the *whole* Fermi surface of each of the spin species must be modified in order to generate even a small fractional polarization. Now, as will be shown in detail in Chapter 2, a spin density wave state is constructed in such a way as to involve the modification of only a *small portion* of the original (paramagnetic) Fermi surface. If one is careful enough to chose the distortion in such a way as to involve only regions of the Fermi surface for which the corresponding gain in exchange energy is optimal, then it is conceivable (although by no means obvious) that the distorted state could be energetically favored for densities considerably higher than that of the Bloch transition.

Indeed, in the early 1960s, Overhauser proved an important theorem (of which a proof is given in Chapter 2) according to which the paramagnetic state of the jellium model is *always* unstable, within the Hartree–Fock approximation, with respect to a pure SDW state. Interestingly, the kinetic and the exchange energies of the pure SDW and CDW states (or any combination of the two with $\phi \neq 0, \pi$) are identical.[54] Thus, at the Hartree–Fock level, the only difference between the CDW and the SDW comes from the electrostatic energy, which is obviously positive in the CDW while vanishing in the SDW state. This difference can be considerably reduced if one allows the positive background to be deformable so as to neutralize, at least in part, the charge density of the CDW. In physical terms, this is equivalent to taking into account the finite stiffness of the lattice. In the limit that the background is deformable at zero energy cost (the so-called *deformable jellium model*), the energy difference between the CDW and the SDW disappears in the Hartree–Fock approximation. Overhauser went on to argue that correlations will likely decrease the energy of the CDW relative to that of the SDW and of the homogeneous paramagnetic state (Overhauser, 1968). A hint of this effect can be seen in the plots of the compressibility and the spin susceptibility presented in the next section. Both quantities are enhanced by the exchange energy term in precisely the same manner, but correlations affects them in opposite ways, pushing the compressibility up, and the spin susceptibility down (see Fig. 1.21 later). In other words, correlations favor density modulations over spin modulations. If the net gain in exchange and correlation energy overcomes the inevitable electrostatic energy cost, then the CDW is favored. According to Overhauser (1985), several simple metals, and potassium (K) in particular, satisfy this requirement, since their lattice is soft enough to effectively screen the charge of the CDW at a modest energy cost. He has been able to produce an impressive

---

[54] Exchange only couples electrons of the same spin orientation, and will therefore be insensitive to the relative phase of the up- and down-spin modulations.

list of experimental facts that suggest the existence of a CDW ground-state in potassium (some of the evidence is presented below).

To fully appreciate the depth of Overhauser's idea it must be kept in mind that his CDW state would occur in a system that has an essentially spherical Fermi surface to begin with. While spin and charge density waves are commonplace in strongly anisotropic systems with "nested" Fermi surfaces (for example, in chromium),[55] and in quasi one-dimensional systems with strong electron–lattice coupling (Peierls, 1955), and while they can be easily induced by the application of magnetic fields (Celli and Mermin, 1965) or even by random impurities (Overhauser, 1960; Lamelas *et al.*, 1995), the Overhauser CDW stands alone in its stark simplicity, being driven only by isotropic interactions in an isotropic system – a perfect example of *spontaneously broken symmetry*.

Short of directly observing the Bragg's reflection peaks associated with the CDW in X-ray or neutron scattering experiments (a daunting task that has not yet been accomplished at this time) the best evidence for a CDW might come from the observation of gaps on the otherwise gapless and nearly spherical Fermi surface of K. Strong indication that such gaps exist is provided by the observation of an anisotropic optical absorption peak at about 0.8 eV in potassium well below the threshold of ordinary interband transitions which occur at 1.3 eV (Mayer and El Naby, 1963). Another important piece of evidence comes from the anomalous dependence of the resistance on magnetic field. According to standard transport theory (see Ashcroft and Mermin, 1976) the resistance of a metal with a spherical Fermi surface should be independent of magnetic field at sufficiently high field. But, if the Fermi surface has gaps, then some of the classical orbits of the electron on the Fermi surface may fail to close: the presence of such *open orbits* leads to a dramatic increase of the resistance when the magnetic field is perpendicular to the plane of the open orbit. Such anomalous magnetoresistance has indeed been measured via the *induced-torque method* (Coulter and Datars, 1980).[56] The puzzle remains as to why effects of the CDW structure do not appear in what is supposed to be the most direct probe of the topology of the Fermi surface – the measurement of the de Haas–van Alphen effect. This contradiction being still clearly unresolved and the CDW Bragg spots having to date eluded discovery, makes the ground-state properties of the simple metals still a hotly debated subject.

As anticipated in the introduction, additional exotic phases have been found in low-dimensional electronic systems. In two dimensions, at very high magnetic field, the electron gas forms the so called *quantum Hall liquid* – an incompressible quantum liquid with fractionally charged excitations (Laughlin, 1983) – which is characterized by a vanishing longitudinal resistivity and a perfectly quantized Hall conductivity that depends only on the fundamental constants e and $\hbar$. The single-electron properties of this system are as far from those of a normal Fermi liquid as one can imagine (for example, there is a gap for the injection of an electron into the liquid), and yet the system is not a Wigner crystal.

---

[55] A Fermi surface is said to be "nested" if it contains two parallel or nearly parallel sections. It is worthwhile mentioning that an ellipsoid is not a nested surface (see Exercise 3).

[56] In the induced torque method a single crystal sample is supported by a vertical rod in a horizontal magnetic field. As the field is slowly rotated in the horizontal plane, induced currents in the sample create a magnetic moment perpendicular to the field. This causes a torque (on the rod), which is monitored vs. field and angle. The method allows a probeless measure of the conductivity tensor.

Other more or less hypothetical phases of this system are the *Hall crystal* (Tesanovic, Axel, and Halperin, 1988) – i.e., a weak CDW coexisting with a quantum Hall liquid – and, at somewhat weaker magnetic fields, the so-called *stripe phase* (Koulakov, Fogler, and Shklovskii, 1996) – essentially a CDW in the distribution of the centers of the quasi-classical electron orbits (some details will be supplied in Chapter 10).

In quasi-one-dimensional systems the normal Fermi liquid phase does not exist: the basic paradigm for the liquid phase is provided in this case, by the *Luttinger liquid* (LL) (Haldane, 1983) – a highly collective state that can be viewed as an incipient CDW quenched by strong quantum fluctuations. Like a crystal, this system supports only collective excitations, and exhibits a pseudogap in the single-particle spectrum. Its incipient order is manifested in the slow power-law decay of its correlation functions, whose exponents depend on the strength of the interaction.

An intriguing variation on the ordinary Luttinger liquid is the *chiral Luttinger liquid* (χLL) (Wen, 1990) – a Luttinger liquid in which the excitations travel only in one direction. Such a system is realized at the edge of a fractional quantum Hall liquid and is responsible for a number of spectacular properties that are just beginning to be experimentally observed. Ordinary and chiral Luttinger liquids will be studied in Chapters 9 and 10 respectively.

We conclude here our brief excursion through the largely uncharted world of the possible phases of the electron liquid and return to the study of its basic equilibrium properties.

1.8 Equilibrium properties of the electron liquid

The ground-state energy is by far the most studied and best understood property of the electron liquid. Unfortunately, there is no experiment that directly measures this quantity. On the other hand, several in principle observable properties can be calculated from the *derivatives* of the energy with respect to the density and the spin polarization. In this section, we take advantage of the availability of accurate expressions for the ground-state energy per electron to calculate some of these observable properties.

1.8.1 Pressure, compressibility, and spin susceptibility

Starting from the expression for the energy as a function of r_s,

$$E = N\epsilon(r_s) , \qquad (1.136)$$

the *pressure*, P, the *chemical potential*, μ, and the *compressibility*, K, can be readily calculated as follows:

$$\frac{P}{n} = -\frac{1}{n}\left(\frac{\partial E}{\partial V}\right)_N = n\frac{\partial \epsilon(n)}{\partial n} = -\frac{r_s}{d}\epsilon'(r_s) , \qquad (1.137)$$

$$\mu = \left(\frac{\partial E}{\partial N}\right)_V = \frac{\partial [n\epsilon(n)]}{\partial n} = \epsilon(r_s) - \frac{r_s}{d}\epsilon'(r_s) , \qquad (1.138)$$

$$\frac{1}{nK} = -\frac{V}{n}\left(\frac{\partial P}{\partial V}\right)_N = \frac{\partial P}{\partial n} = m\frac{\partial \mu}{\partial n} = \frac{r_s}{d}\left[\frac{1-d}{d}\epsilon'(r_s) + \frac{r_s}{d}\epsilon''(r_s)\right] , \quad (1.139)$$

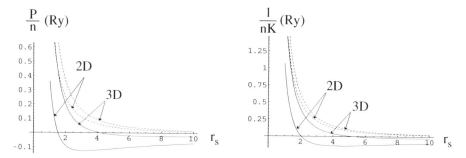

Fig. 1.19. Pressure and bulk modulus of the electron liquid vs r_s. The dashed lines represent the noninteracting results. Notice that both quantities become negative at low density, the pressure for $r_s > 1.51$ in 2D and $r_s > 4.18$ in 3D, the compressibility for $r_s > 2.03$ in 2D and $r_s > 5.25$ in 3D.

where $\epsilon'(r_s)$ and $\epsilon''(r_s)$ are, respectively, the first and second derivative of $\epsilon(r_s)$ with respect to r_s, and V is the volume of the d-dimensional system. The inverse of the compressibility $\frac{1}{K}$ is also known as the *bulk modulus*, denoted by $\mathcal{B}$, and is related to the chemical potential by the useful equation

$$\mathcal{B} = \frac{1}{K} = n^2 \frac{\partial \mu}{\partial n} . \qquad (1.140)$$

Fig. 1.19 shows the pressure and the bulk modulus calculated from the formulas of Section 1.7.2 for the ground-state energy. A puzzling feature jumps immediately to the eye, namely both P and $\mathcal{B}$ become negative at low density. Since the positivity of the bulk modulus is normally considered a necessary condition for the stability of a thermodynamic system, we clearly have some explaining to do.

The paradox is resolved by noting that what we are calculating here is the pressure and compressibility of a fictitious charge-neutral system in which the positive background of charge automatically adjusts itself to neutralize the electronic charge *at no energy cost*. In what follows, these two quantities will be referred to as the *proper pressure* and the *proper compressibility* of the electron liquid. Because we are thus (unphysically) neglecting the stiffness of the background, it is not surprising that the proper compressibility can be negative. In a physical electronic system (of which the jellium model is an idealization) the negative proper compressibility of the electrons would be more than compensated by the positive compressibility of the background. Similarly, a negative *proper pressure* simply means that the positive background, besides neutralizing the coulomb interaction, must also be exerting an outward force to keep itself from collapsing (together with the electrons).[57]

A completely parallel treatment can be developed for the derivatives of the energy with respect to spin polarization. The role of the pressure is now played by the magnetic field

---

[57] Since the proper compressibility has been defined in an apparently unphysical manner, one might ask whether there is any hope to experimentally *measure* this quantity. In Chapter 5 we will answer this question in the affirmative.

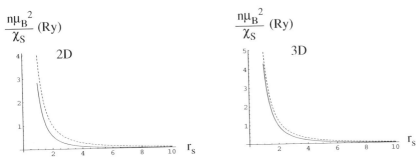

Fig. 1.20. The spin stiffness of the electron liquid as a function of r_s. The dashed lines represent the noninteracting results. Notice that the spin stiffness is still positive at $r_s = 10$.

$B(r_s, p)$ which produces a given polarization p in the ground-state. We have

$$B(r_s, p) = \frac{\partial E(M)}{\partial M} = -\frac{2}{g\mu_B}\frac{\partial \epsilon(r_s, p)}{\partial p}, \qquad (1.141)$$

where $M = -\frac{g\mu_B}{2}Np$ is the magnetization, $\mu_B \equiv \frac{e\hbar}{2mc}$ is the Bohr magneton, and $-\frac{g\mu_B}{2}$ is the magnetic moment per electron.

The analogue of the compressibility is the spin susceptibility per unit volume, χ_S, which is defined as

$$\frac{1}{\chi_S} \equiv V\frac{\partial B}{\partial M} = \frac{4}{n(g\mu_B)^2}\frac{\partial^2\epsilon(r_s, p)}{\partial p^2}. \qquad (1.142)$$

A plot of the spin stiffness (the inverse of the spin susceptibility) vs r_s at $p = 0$ is presented in Fig. 1.20. Notice that $\frac{1}{\chi_S}$ is always *smaller* than in the noninteracting electron gas, yet it remains positive up to large values of r_s ($r_s < 25$ in 2D and $r_s < 75$ in 3D).

We can present the results for the spin stiffness and the bulk modulus in a more incisive form by plotting the ratios $\frac{K_0}{K}$ and $\frac{\chi_P}{\chi_S}$, which describe the many-body enhancement of the compressibility K and spin susceptibility χ_S relative to noninteracting values K_0 and χ_P. Writing the energy as $\epsilon = \epsilon_0 + \epsilon_x + \epsilon_c$, where ϵ_0 and ϵ_x are given by Eqs. (1.83)–(1.88) and (1.95)–(1.98) respectively, it is easy to show that

$$\begin{cases} \frac{K_0}{K} = 1 + \frac{d+1}{d+2}\frac{\epsilon_x(r_s)}{2\epsilon_0(r_s)} - \frac{r_s}{2\epsilon_0(r_s)(d+2)}\left[(d-1)\epsilon_c'(r_s) - r_s\epsilon_c''(r_s)\right] \\[2mm] \frac{\chi_P}{\chi_S} = 1 + \frac{d+1}{d+2}\frac{\epsilon_x(r_s)}{2\epsilon_0(r_s)} + \frac{d^2}{2\epsilon_0(r_s)(d+2)}\frac{\partial^2\epsilon_c(r_s,p)}{\partial p^2}\Big|_{p=0}. \end{cases} \qquad (1.143)$$

These two ratios are plotted in Fig. 1.21 in three and two dimensions. Notice that the compressibility and the spin susceptibility are enhanced in exactly the same manner by the exchange energy term (dashed line in Fig. 1.21). However, the correlation energy pushes the two ratios in opposite directions: it further enhances the compressibility while reducing the spin susceptibility.

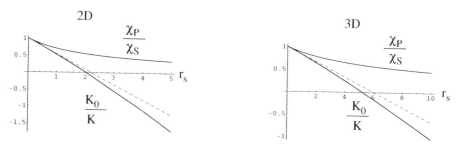

Fig. 1.21. The inverse many-body enhancement factors $\frac{K_0}{K}$ and $\frac{\chi_P}{\chi_S}$ for the compressibility and the spin susceptibility in two and in three dimensions. The two ratios coincide in the first-order approximation (exchange only) and are shown, in that case, by the dashed line.

1.8.2 The virial theorem

The total energy of the electron liquid can be decomposed into its kinetic and potential energy components with the help of the *virial theorem* (see for instance, Landau-Lifshitz, *Course of Theoretical Physics*, Vol. 5, Part 1), which is a direct consequence of the homogenous coordinate scaling of the potential energy. The theorem states that

$$2t + u = d\,\frac{P}{n} = -r_s\epsilon'(r_s)\,,\tag{1.144}$$

where t and u are, respectively, the kinetic and potential energy per particle, and we have made use of Eq. (1.137) for $\frac{P}{n}$. The proof of the theorem will be presented at the end of the next section. Since, by definition

$$t + u = \epsilon\,,\tag{1.145}$$

we see that the virial theorem is equivalent to the following expressions for the kinetic and the potential energies:

$$t(r_s) = -\left[\epsilon(r_s) + r_s\epsilon'(r_s)\right]\,,$$
$$u(r_s) = 2\epsilon(r_s) + r_s\epsilon'(r_s)\,.\tag{1.146}$$

It is important to appreciate that the average kinetic energy of the interacting electron liquid, denoted by $t(r_s)$, is not the same as the average non-interacting kinetic energy, which coincides with the non-interacting ground-state energy $\epsilon_0(r_s)$. In fact, it is not difficult to see that

$$t(r_s) \geq \epsilon_0(r_s)\,.\tag{1.147}$$

since the kinetic energy is at its minimum in the noninteracting ground-state. Notice that the difference $t_c(r_s) \equiv t(r_s) - \epsilon_0(r_s)$ vanishes to first order in perturbation theory due to the fact that $\epsilon_x(r_s) + r_s\epsilon'_x(r_s) = 0$. This quantity is therefore called the *correlation kinetic*

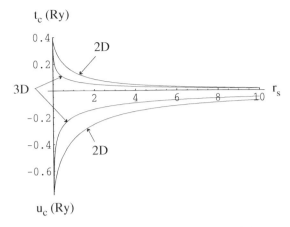

Fig. 1.22. The correlation kinetic energy t_c and the correlation potential energy u_c vs r_s in two and three dimensions.

energy. In Fig. 1.22 we plot the correlation components of the kinetic and the potential energy, namely, $t_c(r_s)$, defined above, and $u_c(r_s) \equiv u(r_s) - \epsilon_x(r_s)$. The two contributions have comparable magnitudes and opposite signs.

1.8.3 The ground-state energy theorem

A very useful theorem enables us to calculate the ground-state energy $\epsilon(r_s)$ from the integral of the potential energy $u(r'_s)$ with respect to r'_s from 0 to r_s. To see how this can be done consider a rescaled jellium model in which the electrons have charge $-e\sqrt{\lambda}$ ($0 \le \lambda \le 1$) and the background has charge density $ne\sqrt{\lambda}$ to keep the system neutral. The hamiltonian has the form

$$\hat{H}(\lambda) = \hat{T} + \lambda\hat{U}, \qquad (1.148)$$

where $\hat{T}$ and $\hat{U}$ are respectively the kinetic and potential energy operators of the standard jellium model hamiltonian. It is evident that $\hat{H}(\lambda)$ describes a noninteracting electron gas for $\lambda = 0$ and a standard jellium model for $\lambda = 1$. Let $|\Psi_0(\lambda)\rangle$ be the ground-state of $\hat{H}(\lambda)$ and $E(\lambda)$ the corresponding ground-state energy, so that

$$\hat{H}(\lambda)|\Psi_0(\lambda)\rangle = E(\lambda)|\Psi_0(\lambda)\rangle . \qquad (1.149)$$

Clearly $E(1) \equiv N\epsilon(r_s)$ is the ground-state energy of the standard jellium, while $E(0) \equiv N\epsilon_0(r_s)$ is that of the noninteracting electron gas. Making use of the fact that the normalization of the ground-state, $\langle\Psi_0(\lambda)|\Psi_0(\lambda)\rangle = 1$, is independent of λ, one can easily verify the following chain of identities known as the *Hellman–Feynman theorem*

$$\frac{dE(\lambda)}{d\lambda} = \frac{d}{d\lambda} \left\langle \Psi_0(\lambda) \left| \hat{H}(\lambda) \right| \Psi_0(\lambda) \right\rangle$$

$$= \left\langle \Psi_0(\lambda) \left| \frac{\partial \hat{H}(\lambda)}{\partial \lambda} \right| \Psi_0(\lambda) \right\rangle = \frac{U(\lambda)}{\lambda}, \qquad (1.150)$$

where $U(\lambda)$ is the potential energy of the scaled system at charge $e\sqrt{\lambda}$. Integrating Eq. (1.150) with respect to λ we obtain

$$E(1) = E(0) + \int_0^1 d\lambda \frac{U(\lambda)}{\lambda}. \qquad (1.151)$$

This result, commonly referred to as the *integration over the coupling constant algorithm*, is not specific to the jellium model, as it applies as well to any many-body system described by a hamiltonian of the form (1.148). In the jellium model one can go further and note that, because of the universal dependence of the energy on r_s, we also have:

$$E(\lambda) = N \lambda^2 \epsilon(\lambda r_s), \qquad (1.152)$$

and

$$U(\lambda) = N \lambda^2 u(\lambda r_s), \qquad (1.153)$$

where $\epsilon(r_s)$ and $u(r_s)$ are the energy and the potential energy of the standard jellium model expressed in Rydbergs. The form of Eqs. (1.152) and (1.153) arises as follows: λr_s is the effective r_s value of the rescaled jellium model with Bohr radius $a_B(\lambda) \equiv \frac{a_B}{\lambda}$, and the prefactor λ^2 arises from the effective Rydberg, $Ry(\lambda) = \frac{\lambda e^2}{2a_B(\lambda)}$, of the rescaled jellium model. Substituting these formulas in Eq. (1.151) and changing the integration variable from λ to $r_s' = \lambda r_s$, we obtain the following useful and exact expression for the energy per particle:

$$\epsilon(r_s) = \epsilon_0(r_s) + \frac{1}{r_s^2} \int_0^{r_s} dr_s' r_s' u(r_s'). \qquad (1.154)$$

Notice that the second term on the right-hand side of this equation is the sum of the exchange energy and the correlation energy defined in Section 1.5.6. For this reason, it is usually called the *exchange-correlation energy* $\epsilon_{xc}(r_s)$. Thus, we have

$$\epsilon_{xc}(r_s) \equiv \epsilon(r_s) - \epsilon_0(r_s) = \frac{1}{r_s^2} \int_0^{r_s} dr_s' r_s' u(r_s'). \qquad (1.155)$$

It is easy to verify that the potential and the kinetic energies are expressed in terms of the exchange-correlation energy as follows:

$$u(r_s) = 2\epsilon_{xc}(r_s) + r_s \epsilon_{xc}'(r_s), \qquad (1.156)$$

and

$$t(r_s) = \epsilon(r_s) - u(r_s)$$
$$= \epsilon_0(r_s) - \epsilon_{xc}(r_s) - r_s \epsilon'_{xc}(r_s) . \qquad (1.157)$$

The virial theorem, Eq. (1.144), follows immediately from the above expressions.

Exercises

1.1 **Wide parabolic quantum well.** The bottom of the conduction band in a wide parabolic quantum well of width W perpendicular to the z-axis has the spatial dependence $E_b^{(0)}(z) = E_0 + \Delta \left(\frac{z}{W}\right)^2$ for $|z| \le W$, and $E_b^{(0)}(z) = E_0 + \Delta$ for $|z| > W$. Use simple stability considerations and Gauss' theorem to show that electrons placed in such a well must form an electron liquid slab of width $w \le W$ and approximately uniform density $n = \frac{\Delta}{2\pi e^2 W^2}$. Also calculate the renormalized position of the bottom of the band, $E_b(z) = E_b^0(z) - e\varphi(z)$, where $\varphi(z)$ is the potential created by the electrons. (Note: the potential created by the remote donors can be neglected if the donors are placed symmetrically on the two sides of the well.)

1.2 **Density of states of free fermions.** Calculate the density of states at the Fermi level for free fermions of density n in 3, 2, and 1 dimensions.

1.3 **Heat capacity of free fermions.** Calculate the heat capacity of a noninteracting electron gas in 3, 2, and 1 dimensions. Show that in all cases it is linear in T with a coefficient related to the density of states at the Fermi level.

1.4 **Hamiltonian of a bilayer system.** Write the second-quantized hamiltonian for a *bilayer system* consisting of two parallel two-dimensional electron liquids separated by a distance d. Assume, at first, that no tunneling is allowed between the layers. Generalize then to the case in which electrons can tunnel from a layer to the the other with an amplitude t, conserving the momentum parallel to the layers.

1.5 **Multiplicity of terms in the perturbative expansion for the energy.** With reference to the discussion of Section 1.5.6, show that the multiplicity of each distinct contribution to the energy in the n-th order of the perturbative expansion for the ground-state energy of a many-particle system is $2^{n-1}(n-1)!$. [Hint: see the discussion of perturbation theory in Feynman's book (Feynman, 1972), Sections 9.8 and 9.9.]

1.6 **First-order calculation of the spin susceptibility.** Calculate the spin suceptibility of the electron gas to first-order in the coulomb interaction by making use of Eq. (1.142), together with the expressions for the non-interacting kinetic energy (1.88) and the exchange energy (1.98) of a spin-polarized electron gas in two and three dimensions.

1.7 **Electrostatic energy of a Wigner crystal.** By making use of the ideas developed in Section 1.3.2 derive Eq. (1.117) for the total electrostatic energy of a Wigner crystal. [Hint: calculate the classical electrostatic energy of the net charge distribution $\rho(\vec{r}) = -e[n(\vec{r}) - n]$ and subtract the unphysical interaction of each electron with itself.]

1.8 **Stoner model of the electron liquid.** In the *Stoner model*, the electrons are assumed
to interact via the contact interaction

$$\frac{\gamma r_s}{2} e^2 a_B^{d-1} \delta(\vec{r}) \,,$$

where γ is a dimensionless constant (notice that, due to Pauli's exclusion principle,
only electrons of opposite spin orientations interact).

(a) Show that the first-order ground-state energy per electron can be written as

$$\epsilon(r_s, p) = \left[\epsilon_0(r_s, p) + \frac{3\gamma}{16\pi r_s^2}(1 - p^2) \right] \text{Ry}$$

where the noninteracting kinetic energy $\epsilon_0(r_s, p)$ is given by Eq. (1.88).

(b) Determine the value of γ in such a way that a partially spin-polarized ground-
state become stable at $r_s \approx 50$ in 3D. At what value of r_s does one then predict a
fully spin-polarized ground-state? (Answer: at $r_s \simeq 54$, with $\gamma \simeq 0.0082$.)

(c) Using the value of γ obtained in (b), determine the region of stability of the
spin polarized phase in the $r_s - T$ plane. [Hint: at finite temperature one must study
the evolution of the minimum in the *free energy* of the model. The noninteracting
free energy of each spin species is given by the formula

$$F_{0\sigma} = N_\sigma \mu_\sigma - k_B T \sum_{\vec{k}} \ln \left(1 + e^{-(\varepsilon_k - \mu_\sigma)/k_B T} \right),$$

where $\varepsilon_k = \frac{\hbar^2 k^2}{2m}$ and the chemical potential μ_σ is determined by the number of
electrons with spin σ. The interaction contribution to the free energy per particle is
still assumed to be given by $\frac{3\gamma}{16\pi r_s^2}(1 - p^2)$Ry. This is how the phase diagram of the
spin polarized liquid has been estimated in Fig. 1.14.]

1.9 **Ferrell's inequality.** Let ΔE be the change in the ground-state energy of the electron
liquid caused by an infinitesimal variation of the electronic charge such that e^2 is
replaced by $e^2(1 + \delta)$, with $\delta \ll 1$.

(a) Make use of the scaling properties of the ground-state energy to show that

$$\Delta E = [(1 + \delta)^2 \epsilon(r_s + \delta r_s) - \epsilon(r_s)] N \text{Ry} \,. \tag{E1.1}$$

(b) Prove *Ferrell's inequality* (Ferrell, 1958):

$$\frac{d^2}{dr_s^2}[r_s^2 \epsilon_{xc}(r_s)] \leq 0 \,, \tag{E1.2}$$

for all r_s.

[Hint: make use of the well known theorem according to which the second-order
change in the ground-state energy caused by a perturbation is always negative. In
this case the "perturbation" is of first order in δ. Hence, the $\mathcal{O}(\delta^2)$ contribution to
ΔE must be negative.]

(c) Verify that the expressions for the energy in Eq. (1.107) satisfy Ferrell's inequality.

1.10 **Density scaling of equilibrium properties.** Show that any term δE appearing in the total energy and scaling as n^α with the particle number density n contributes the following corresponding terms to the chemical potential, the pressure and the inverse compressibility at zero temperature:

$$(\mu)_{\delta E} = (1+\alpha)\frac{\delta E}{N}, \quad (P)_{\delta E} = \alpha\frac{\delta E}{V}, \quad \left(\frac{1}{K}\right)_{\delta E} = (\alpha+\alpha^2)\frac{\delta E}{V}.$$

1.11 **First-order calculation of equilibrium properties.** Use the formulas derived in the previous exercise to evaluate μ, P and K at zero temperature to first order in the coulomb interaction. Carry out the calculations for three and two dimensions.

1.12 **Exchange energy for screened interaction potential.** Calculate the exchange energy for a system of particles interacting via the screened interaction potential $v_q(\kappa)$ of Eq. (1.17) (Answer: $\epsilon_x = -\frac{3e^2 k_F}{4\pi}\left[4 + \frac{\lambda^2}{k_F^2}\ln\left(1 + \frac{4k_F^2}{\lambda^2}\right) - \frac{4\lambda}{k_F}\tan^{-1}\frac{2k_F}{\lambda}\right]$ in 3D). Discuss how the compressibility and the spin susceptibility of this system depend on the screening wave vector, to first order in the interaction.

1.13 **Correlation contribution to equilibrium properties.** Study the effect of correlation on the chemical potential, the pressure, and the proper compressibility at high density by repeating the calculations of the previous exercise with the energy expressions given by Eq. (1.107).

1.14 **High density expansion for the kinetic energy.** Making use of Eqs. (1.107) and (1.146) calculate the kinetic energy per particle in the interacting electron liquid up to terms of order r_s^0 in 3D and $r_s \ln r_s$ in 2D.

1.15 **Pair correlation function of non-interacting fermions.** Make use of Wick's theorem to express the pair correlation function of a non-interacting Fermi system in terms of the square of the one-particle density matrix, as in Eq. (1.100).

1.16 **Pauli principle for the pair correlation function.** Make use of Eq. (A4.8) to prove that $g_{\sigma\sigma}(\vec{r}, \vec{r}) = 0$.

1.17 **Cusp conditions for the pair correlation function.** Derive the cusp conditions (A4.12) and (A4.13) for the pair correlation function from the short-distance behavior of the two-electron scattering wave function.

1.18 **Matrix elements of the current operator.** Show that the matrix elements of the paramagnetic current operator between plane wave states are given by Eq. (A2.18).

1.19 **Continuity equation.** Derive the continuity equation (A2.22) by calculating the commutator of the density operator with the hamiltonian. Notice that the form of this equation is not affected by interactions.

1.20 **Large-q behavior of the static structure factor.** To derive the large-q behavior of the static structure factor, Eq. (A4.19), begin by showing that the difference $g(r) - g(0)$ can be expressed as an integral over q in the following manner:

$$g(r) - g(0) = \frac{1}{2\pi^{d-1}n}\int_0^\infty dq\, q^{d-1}[S(q) - 1] \times \begin{cases} \frac{\sin qr}{qr} - 1 & (3D) \\ J_0(qr) - 1 & (2D) \end{cases}.$$

[Hint: make use of Eq. (A4.22).]

Now observe that part of the integral, from $q = 0$ to a finite but arbitrarily large cutoff q_c, is proportional to r^2 for $r \to 0$: it follows that the cusp in $g(r)$ (i.e. the linear dependence of $g(r)$ on r for small r) must arise entirely from the large-q part of the integral, i.e., from the region $q > q_c$. Changing integration variable in this part of the integral from q to $x = qr$ and making use of the cusp conditions (A4.12) and (A4.13), one arrives at Eq. (A4.19).

2

The Hartree–Fock approximation

2.1 Introduction

The Hartree–Fock (HF) approximation has historically provided the starting point for many-body theories of the electron liquid. Although an effective implementation of this approximation is, in general, far from trivial, its basic idea can be simply described as an attempt to approximate the ground-state wavefunction of the system by that of an *effective hamiltonian*, which is quadratic in the electron creation and destruction operators and can therefore be readily diagonalized.

The oldest and most common version of this procedure is based on number-conserving quadratic hamiltonians and the concept of Slater determinant. In this version the HF theory provides the answer to the question: what is the best independent-electron approximation to an interacting many-electron system?

On the other hand, quadratic hamiltonians that *do not* conserve particle number can also be contemplated (as exemplified by the BCS theory of superconductivity) and will be briefly discussed later in the chapter. Most of our discussion, however, will focus on the number-conserving version of the HF theory.

It should be clear from the discussion in Section 1.4 that any N-electron wave function can be written as a superposition of Slater determinants $\psi_{\alpha_1,\alpha_2,\ldots\alpha_N}(\vec{r}_1, s_1; \vec{r}_2, s_2; \ldots; \vec{r}_N, s_N)$ (defined in Eq. (1.31)) in the following manner:

$$\sum_{\alpha_1,\alpha_2,\ldots\alpha_N} C_{\alpha_1,\alpha_2,\ldots,\alpha_N} \psi_{\alpha_1,\alpha_2,\ldots,\alpha_N}(\vec{r}_1, s_1; \vec{r}_2, s_2; \ldots; \vec{r}_N, s_N), \tag{2.1}$$

where $C_{\alpha_1,\ldots,\alpha_N}$ are complex coefficients, and the sum runs over all possible subsets, $\{\alpha_1, \ldots, \alpha_N\}$, of N single particle states. In the (number-conserving) HF approximation one first approximates the true ground-state of the system by limiting the above sum to a single term, i.e.

$$\Psi_{HF}(\vec{r}_1, s_1; \vec{r}_2, s_2; \ldots; \vec{r}_N, s_N) \approx \psi_{\alpha_1,\alpha_2,\ldots,\alpha_N}(\vec{r}_1, s_1; \vec{r}_2, s_2; \ldots; \vec{r}_N, s_N), \tag{2.2}$$

and then determines the single particle states labelled by $\alpha_1, \alpha_2, \ldots, \alpha_N$, by requiring that the expectation value of the hamiltonian

$$\langle \psi_{\alpha_1,\alpha_2,\ldots,\alpha_N} | \hat{H} | \psi_{\alpha_1,\alpha_2,\ldots,\alpha_N} \rangle \tag{2.3}$$

69

be a minimum. Although this idea may appear to be rather simple, it must be kept in mind that, to date, even the HF ground-state of the jellium model is unknown. In view of the infinite possibilities, one is forced to guess the symmetries of the many-body wave function and then to restrict the search for the minimum energy to single particle states which manifestly guarantee such symmetries.[1]

In practice what one does is to first require that the set of single particle states give an *extremum* for the expectation value of the energy. The extremum condition leads to a set of coupled nonlinear integro-differential equations for the N single particle wave functions that make up the Slater determinant: in the general case, these are rather difficult to solve even with the help of a powerful computer. The stability of the extremum must then be tested with respect to local (small) variations of the single-particle wave functions, as well as global changes that may completely alter the symmetry of the solution. As one can easily guess, there is no known algorithm to efficiently carry out this program, or even just to verify that a given putative HF ground-state is actually such! It is on the other hand a very simple matter to verify (see Exercise 1) that a set of plane waves, which represents the noninteracting ground-state of the electron gas, provides an extremum solution for the HF equations.[2] The properties of this solution are very useful for an understanding of the electron liquid and will be therefore discussed at length in this chapter.

The approximate nature of the HF ground-states naturally leads to the concept of *correlation*. Correlation effects are, by definition, all the effects that stem from the fact that the true ground-state wave function is *not* a single Slater determinant.[3] In particular, as already briefly mentioned in Section 1.5.6, the difference between the *exact ground-state energy* and the HF ground-state energy defines the *correlation energy*.

Although physically appealing, this definition is, in a sense, empty as long as we do not know the energy of the *true* HF ground-state. As a practical solution, it is customary to define the correlation energy with reference to a particular HF state. For the jellium model, this reference state is chosen to be the plane waves state, i.e. the ground-state of the non interacting system.

A central concept in the HF theory is that of *exchange potential* – a nonlocal potential that is naturally associated to the exchange hole, much as the electrostatic potential is associated to the charge density. The concept of exchange potential stems from the Schrödinger-like form of the HF equations and is what characterizes the HF theory as the most elementary form of an *effective mean field theory*. Since we believe this aspect to be of fundamental importance, we have chosen to discuss the HF approximation from this point of view.

---

[1] Restricted HF calculations play an important role in condensed matter physics, mostly as a source of *suggestions* for novel broken-symmetry phases. It is often quite easy to find broken-symmetry solutions of the HF equations (we will see a few examples in this chapter), and this may be taken as a hint that the same symmetry may be broken in the exact ground-state too. However, experience shows that quantum fluctuations, which are not included in the HF approximation, tend to restore the full symmetry. Therefore, suggestions of broken-symmetry states based on HF calculations cannot be accepted without validation from more accurate studies.

[2] Unfortunately, as proved in Section 2.6.3, this solution does not correspond to an HF energy minimum – it merely represents a saddle point – and therefore does not constitute an acceptable solution of the HF energy minimization problem.

[3] Strictly speaking even a single Slater determinant is a correlated wave function due to its antisymmetric structure. However, the word correlation is normally reserved only for those correlations that go beyond the inevitable antisymmetry-related ones. Keeping more than one Slater determinant in Eq. (2.2) is commonly described as allowing for *configuration interaction*. We will later present a criterion for deciding whether an antisymmetric wave function is a single Slater determinant or not.

2.2 Formulation of the Hartree–Fock theory

2.2.1 The Hartree–Fock effective hamiltonian

The first step is to find an *effective single-particle* hamiltonian which can be diagonalized exactly and whose ground-state will be used as an approximation to the true ground-state. Such a hamiltonian has quite generally the form[4]

$$\hat{H}_{HF} = \sum_{\vec{k}\sigma} \frac{\hbar^2 k^2}{2m} \hat{a}^\dagger_{\vec{k}\sigma} \hat{a}_{\vec{k}\sigma} + \sum_{\vec{k}\sigma} \sum_{\vec{k}'\sigma'} V^{HF}_{\vec{k}\sigma;\vec{k}'\sigma'} \hat{a}^\dagger_{\vec{k}\sigma} \hat{a}_{\vec{k}'\sigma'} , \tag{2.4}$$

where the coefficients $V^{HF}_{\vec{k}\sigma;\vec{k}'\sigma'}$ $(= [V^{HF}_{\vec{k}'\sigma';\vec{k}\sigma}]^*$ by virtue of hermiticity) are as yet unspecified. Unlike the standard potential energy, which may or may not be present in the original hamiltonian, the HF potential energy term

$$\hat{V}^{HF} = \sum_{\vec{k}\sigma} \sum_{\vec{k}'\sigma'} V^{HF}_{\vec{k}\sigma;\vec{k}'\sigma'} \hat{a}^\dagger_{\vec{k}\sigma} \hat{a}_{\vec{k}'\sigma'} , \tag{2.5}$$

is generally *nonlocal* in space. This can be seen most clearly in the first quantization representation, where $\hat{V}^{HF} = \sum_i \hat{V}^{HF}_i$ and $\hat{V}^{HF}_i$ is an integral operator, which acts on a wave function $\psi(\vec{r}_i, s_i)$ of the i-th electron in the following manner:

$$\hat{V}^{HF}_i \psi(\vec{r}_i, s_i) = \sum_{s'_i} \int d\vec{r}'_i \, V^{HF}_{s_i s'_i}(\vec{r}_i, \vec{r}'_i) \psi(\vec{r}'_i, s'_i) , \tag{2.6}$$

where

$$V^{HF}_{ss'}(\vec{r}, \vec{r}') \equiv \frac{1}{L^d} \sum_{\vec{k},\vec{k}'} V^{HF}_{\vec{k}s;\vec{k}'s'} e^{i\vec{k}\cdot\vec{r}} e^{-i\vec{k}'\cdot\vec{r}'} . \tag{2.7}$$

By contrast, a local potential would simply multiply $\psi(\vec{r}_i, s_i)$ by a function of $\vec{r}_i$ and s_i: this situation corresponds to the special case that $V^{HF}_{\vec{k}s;\vec{k}'s'}$ is a function of the difference $\vec{k} - \vec{k}'$.

The effective hamiltonian (2.4) describes a system of independent electrons moving under the influence of an effective nonlocal potential. In reality each electron experiences the fluctuating coulomb potential created by all other electrons at a given point. This potential is in fact local but varies in time in an unpredictable manner. The HF approximation attempts to replace the fluctuating potential by a static one that gives a good description on the average. In this sense it is a *mean field theory*. Nonlocality is a small price to pay for this simplification.

2.2.2 The Hartree–Fock equations

The quadratic effective hamiltonian (2.4) is readily diagonalized by a linear transformation to a suitable new basis set. In practice this amounts to changing the basis states in Eq. (2.4)

---

[4] This is the general form of a quadratic *number-conserving* hamiltonian. As mentioned in Section 2.1 an even more general choice would include the number-non-conserving terms $\hat{a}^\dagger \hat{a}^\dagger$ and $\hat{a}\hat{a}$. This generalized form is at the basis of the BCS theory of superconductivity (Section 2.7).

from plane waves to eigenfunctions of the single-particle Schrödinger equation

$$-\frac{\hbar^2}{2m}\nabla^2\phi_\alpha(\vec{r},s) + \sum_{s'}\int V_{ss'}^{HF}(\vec{r},\vec{r}')\phi_\alpha(\vec{r}',s')d\vec{r}' = \varepsilon_\alpha\phi_\alpha(\vec{r},s), \qquad (2.8)$$

where $\alpha = 1, 2, \ldots$ labels the eigenvalues ε_α in increasing order. Let $\hat{a}_\alpha^\dagger$ be the creation operators for this basis set. Then, the effective hamiltonian takes the simple form

$$\hat{H}_{HF} = \sum_\alpha \varepsilon_\alpha \hat{a}_\alpha^\dagger \hat{a}_\alpha , \qquad (2.9)$$

and its eigenfunctions are single Slater determinants of N different states from the basis set. In particular, the ground-state of $\hat{H}_{HF}$ is given by

$$|\Psi_{HF}\rangle = \prod_{\alpha=1}^{N} \hat{a}_\alpha^\dagger|0\rangle , \qquad (2.10)$$

where $|0\rangle$ is the vacuum and $\alpha \leq N$ labels the N lowest lying states.

We now use the ground-state of $\hat{H}_{HF}$ as a variational approximation to the true ground-state.[5] The simplest way to calculate the expectation value of the exact hamiltonian $\hat{H}$ in the approximate ground-state $|\Psi_{HF}\rangle$ is to rewrite $\hat{H}$ in the basis of the HF eigenfunctions $|\phi_\alpha\rangle$. Following the standard procedure described in Chapter 1, we obtain:

$$\hat{H} = \sum_{\alpha\beta} (T_{\alpha\beta} + V_{\alpha\beta}^{ext})\,\hat{a}_\alpha^\dagger\hat{a}_\beta + \frac{1}{2}\sum_{\alpha\beta\gamma\delta} v_{\alpha\beta\gamma\delta}\,\hat{a}_\alpha^\dagger\hat{a}_\beta^\dagger\hat{a}_\gamma\hat{a}_\delta , \qquad (2.11)$$

where $v_{\alpha\beta\gamma\delta}$ – the matrix elements of the coulomb interaction in the chosen basis – are explicitly given by Eq. (1.60) of Section 1.4. $T_{\alpha\beta}$ and $V_{\alpha\beta}^{ext}$ are, respectively, the matrix elements of the kinetic energy and of any external static potential, which we have included here for greater generality.

The expectation value of $\hat{H}$ in $|\Psi_{HF}\rangle$ can be calculated by the elementary method presented in Section 1.5.3 or, more formally, by making use of Wick's theorem (see Appendix 3). Taking the first path, one just notices that the electrons destroyed by $\hat{a}_\delta$ and $\hat{a}_\gamma$ must be reinstated by $\hat{a}_\alpha^\dagger$ and $\hat{a}_\beta^\dagger$ either in the same order (direct term, positive sign) or in reverse order (exchange-term, negative sign). This gives

$$\langle\Psi_{HF}|\hat{H}|\Psi_{HF}\rangle = \sum_\alpha n_\alpha(T_{\alpha\alpha} + V_{\alpha\alpha}^{ext}) + \frac{1}{2}\sum_{\alpha\beta} n_\alpha n_\beta(v_{\alpha\beta\beta\alpha} - v_{\alpha\beta\alpha\beta}) , \qquad (2.12)$$

where $n_\alpha = 1$ for the occupied states ($\alpha \leq N$) and $n_\alpha = 0$ for the empty ones ($\alpha > N$). This energy is a functional of the (as yet unknown) HF potential $\hat{V}^{HF}$, and we now attempt to determine $\hat{V}^{HF}$ from the condition that the energy be as low as possible.

A little thought shows that this determination cannot be unambiguous. If we add to $\hat{V}^{HF}$ a single particle potential that has zero matrix elements between occupied and unoccupied states, the variational wave function does not change. This is because an additional potential

---

[5] The choice of the ground-state, as opposed to any other eigenstate of $\hat{H}_{HF}$, may at this point seem arbitrary but we will see later that it is the correct one.

of this kind is diagonalized by means of a unitary transformation of the N occupied states among themselves: any Slater determinant constructed from the occupied states does not change under such a transformation. Let us then consider small changes in $\hat{V}^{HF}$ that *do* cause correspondingly small changes in the variational wave function:

$$|\delta\Psi_{HF}\rangle = \sum_{\alpha>N,\beta\leq N} \eta_{\beta\alpha}\hat{a}_\alpha^\dagger\hat{a}_\beta|\Psi_{HF}\rangle, \tag{2.13}$$

where $\eta_{\beta\alpha}$ are small arbitrary parameters, and the restrictions on α and β exclude terms that would give zero when acting on $|\Psi_{HF}\rangle$. To *first order* in $\eta_{\beta\alpha}$ the new wave function is still a single Slater determinant formed from the N wave functions $\phi_\beta(\vec{r},s)+\sum_{\alpha>N}\eta_{\alpha\beta}\phi_\alpha(\vec{r},s)$ ($\beta=1,\ldots,N$) (see Exercise 4). The corresponding first-order variation in the expectation value of the energy is easily worked out to be

$$\delta\langle\Psi_{HF}|\hat{H}|\Psi_{HF}\rangle = \langle\delta\Psi_{HF}|\hat{H}|\Psi_{HF}\rangle + \langle\Psi_{HF}|\hat{H}|\delta\Psi_{HF}\rangle$$
$$= \sum_{\alpha\beta}\left[T_{\alpha\beta}+V_{\alpha\beta}^{ext}+\sum_\gamma n_\gamma(v_{\alpha\gamma\gamma\beta}-v_{\alpha\gamma\beta\gamma})\right]\eta_{\alpha\beta}. \tag{2.14}$$

A necessary (but certainly not sufficient) condition for the existence of a local minimum in the energy is that the variation of $\langle\Psi_{HF}|\hat{H}|\Psi_{HF}\rangle$ vanish to first order in $\eta_{\alpha\beta}$. This immediately gives us the condition

$$T_{\alpha\beta}+V_{\alpha\beta}^{ext}+\sum_\gamma n_\gamma(v_{\alpha\gamma\gamma\beta}-v_{\alpha\gamma\beta\gamma})=0, \quad (\alpha>N,\beta\leq N). \tag{2.15}$$

At the same time, in view of the fact that the basis wave functions ϕ_α are eigenfunctions of $\hat{H}_{HF}$, we also know that

$$T_{\alpha\beta}+V_{\alpha\beta}^{HF}=0 \quad for \quad \alpha\neq\beta: \tag{2.16}$$

the HF hamiltonian is certainly diagonal in the basis of its own eigenstates! Thus, subtracting (2.16) from (2.15) we see that the matrix elements of the kinetic energy cancel out, and we obtain our extremum condition in the form

$$V_{\alpha\beta}^{HF}=V_{\alpha\beta}^{ext}+\sum_\gamma n_\gamma(v_{\alpha\gamma\gamma\beta}-v_{\alpha\gamma\beta\gamma}), \tag{2.17}$$

which is an implicit equation for the matrix elements of $\hat{V}^{HF}$ with $\alpha>N$ and $\beta\leq N$.

In the light of our earlier comments it should not come as a surprise that the extremum condition has determined only the matrix elements of $\hat{V}^{HF}$ between occupied and unoccupied states. The effect of the remaining matrix elements is simply to unitarily transform the occupied (or unoccupied) states among themselves, with no effect on the determinantal wave function. Those matrix elements are therefore arbitrary, and we take advantage of their arbitrariness to *define* the HF potential as the operator whose matrix elements are given by Eq. (2.17) *for all values of α and β*. This is a self-consistent definition, because the basis states ϕ_α are themselves determined by the HF potential.[6]

---

[6] It should be noted that the definition of the matrix elements of $\hat{V}^{HF}$ involves a sum – or trace – over the occupied states, which makes it invariant under a unitary transformation of the occupied states among themselves.

At this point the HF eigenvalues ε_α can be written down explicitly as they simply represent the diagonal matrix elements of the HF effective hamiltonian in the $|\phi_\alpha\rangle$ basis. We find

$$\varepsilon_\alpha = T_{\alpha\alpha} + V_{\alpha\alpha}^{HF} = T_{\alpha\alpha} + V_{\alpha\alpha}^{ext} + \sum_\gamma n_\gamma (v_{\alpha\gamma\gamma\alpha} - v_{\alpha\gamma\alpha\gamma}) \,. \tag{2.18}$$

Since the self-consistent orbitals are determined by the single-particle Schrödinger equation (2.8) it is very useful to work out the explicit expression for $V_{ss'}^{HF}(\vec{r}, \vec{r}')$ in the first quantized representation. This is done with the help of the transformation

$$V_{ss'}^{HF}(\vec{r}, \vec{r}') = \sum_{\alpha\beta} V_{\alpha\beta}^{HF} \phi_\alpha(\vec{r}, s) \phi_\beta^*(\vec{r}', s') \,, \tag{2.19}$$

where the sum runs over the whole complete set of self-consistent states. Substituting Eq. (2.17) and making use of the completeness relation

$$\sum_\alpha \phi_\alpha(\vec{r}, s) \phi_\alpha^*(\vec{r}', s') = \delta(\vec{r} - \vec{r}')\delta_{ss'} \,, \tag{2.20}$$

we obtain, after straightforward manipulations

$$V_{ss'}^{HF}(\vec{r}, \vec{r}') = \delta(\vec{r} - \vec{r}')\delta_{ss'} V^{ext}(\vec{r})$$

$$+ \delta(\vec{r} - \vec{r}')\delta_{ss'} \int d\vec{r}_2 \frac{e^2}{|\vec{r} - \vec{r}_2|} \sum_\beta n_\beta \sum_{s_2} |\phi_\beta(\vec{r}_2, s_2)|^2$$

$$- \frac{e^2}{|\vec{r} - \vec{r}'|} \sum_\beta n_\beta \phi_\beta(\vec{r}, s) \phi_\beta^*(\vec{r}', s') \,. \tag{2.21}$$

The second term on the right-hand side of this expression is known as the *Hartree potential* and is local in space: it represents the average electrostatic potential created by the electronic charge density. The last term is the *exchange potential*: it has no classical analogue (its origin being the antisymmetry of the wave function) and is nonlocal in space. A more compact expression for the exchange potential can be given in terms of the *one-particle density matrix*,

$$\rho_{ss'}(\vec{r}, \vec{r}') \equiv \sum_\beta n_\beta \phi_\beta(\vec{r}, s) \phi_\beta^*(\vec{r}', s') \,, \tag{2.22}$$

a very useful quantity about which more will be said in Section 2.5. This expression has the form

$$V_{x,ss'}(\vec{r}, \vec{r}') = -\frac{e^2}{|\vec{r} - \vec{r}'|} \rho_{ss'}(\vec{r}, \vec{r}') \,, \tag{2.23}$$

which shows $V_{x,ss'}(\vec{r}, \vec{r}')$ in a new light, namely, as the electrostatic energy of an electron in the field of the "charge distribution" $e\rho_{ss'}(\vec{r}, \vec{r}')$.[7]

The set of equations (2.8) and (2.21) constitute the standard self-consistent HF equations. If a minimum of the energy exists in the class of single Slater determinants, then the orbitals

---

[7] The language is purely evocative, since the density matrix is, in general, a complex quantity.

in this determinant must necessarily satisfy the HF equations. On the other hand, satisfaction of the HF equations does not guarantee that one has really found a minimum of the energy. In fact, the solution could correspond to a maximum, or, more generally, to a saddle point of the energy. The problem of the stability of the HF solutions will be addressed in Section 2.5.

2.2.3 Ground-state and excitation energies

Once the HF orbitals and eigenvalues are known it is an easy matter to express the (approximate) ground-state energy in terms of them. All we have to do is return to Eq. (2.12) and make use of Eqs. (2.17) and (2.18) to eliminate the matrix elements of the coulomb interaction. This yields

$$E_{HF} = \frac{1}{2} \sum_\alpha n_\alpha \left(T_{\alpha\alpha} + V_{\alpha\alpha}^{ext} + \varepsilon_\alpha \right)$$

$$= \sum_\alpha n_\alpha \varepsilon_\alpha - \frac{1}{2} \sum_\alpha n_\alpha \left(V_{\alpha\alpha}^{HF} - V_{\alpha\alpha}^{ext} \right) . \qquad (2.24)$$

Notice that the HF ground-state energy is *not* the sum of the HF eigenvalues. In particular the last expression makes it clear that the sum of the eigenvalues ε_α would double-count the interaction energy.

We consider next the energy of the state $|\Psi_{HF}^{(e,\gamma)}\rangle$, which is obtained from the HF ground-state by adding an electron into the otherwise unoccupied orbital $\gamma > N$.[8] The second-quantized representation of this state is

$$|\Psi_{HF}^{(e,\gamma)}\rangle = \hat{a}_\gamma^\dagger |\Psi_{HF}\rangle , \qquad \gamma > N. \qquad (2.25)$$

The expectation value of $\hat{H}$ in $|\Psi_{HF}^{(e,\gamma)}\rangle$ is easily calculated from Eq. (2.12), with the proviso that the occupation number n_γ, which was zero in $|\Psi_{HF}\rangle$, is equal to 1 in $|\Psi_{HF}^{(e,\gamma)}\rangle$, while the other occupation numbers are unchanged. This gives

$$E_{HF}^{(e,\gamma)} = \langle \Psi_{HF}^{(e,\gamma)} | \hat{H} | \Psi_{HF}^{(e,\gamma)} \rangle = E_{HF} + \varepsilon_\gamma, \qquad (2.26)$$

where E_{HF} is the energy of the N-electron HF ground-state. Thus, for $\gamma > N$, the HF eigenvalues ε_γ can be interpreted as the the energy required to add an electron into the empty state γ, without allowing for a possible "reconstruction" of the HF orbitals.

In a similar way we compute the energy of the state $|\Psi_{HF}^{(h,\delta)}\rangle$ obtained by *removing* an electron from an occupied orbital $\delta \leq N$. The second-quantized representation of this state is

$$|\Psi_{HF}^{(h,\delta)}\rangle = \hat{a}_\delta |\Psi_{HF}\rangle , \qquad \delta \leq N, \qquad (2.27)$$

and its energy is given by

$$E_{HF}^{(h,\delta)} = \langle \Psi_{HF}^{(h,\delta)} | \hat{H} | \Psi_{HF}^{(h,\delta)} \rangle = E_{HF} - \varepsilon_\delta . \qquad (2.28)$$

---

[8] It must be borne in mind that $|\Psi_{HF}^{(e,\gamma)}\rangle$ is *not* an eigenstate of the HF hamiltonian for $N + 1$ electrons, because it is constructed from orbitals that are solutions of the N-electron HF equation.

Thus, the HF eigenvalues of an occupied state can be interpreted as the the negative of the energy required to remove an electron from that state (i.e., to create a hole), without allowing for a possible "reconstruction" of the HF orbitals.

Finally, let us consider the energy of the state obtained by promoting an electron from an occupied state $\delta \leq N$ to an unoccupied one $\gamma > N$. The second-quantized representation of this state is

$$|\Psi_{HF}^{(e,\gamma;h,\delta)}\rangle = \hat{a}_\gamma^\dagger \hat{a}_\delta |\Psi_{HF}\rangle , \qquad \gamma > N , \; \delta \leq N. \tag{2.29}$$

Calculating the expectation value of $\hat{H}$ in this state gives the important result:

$$E_{HF}^{(e,\gamma;h,\delta)} = \langle \Psi_{HF}^{(e,\gamma;h\delta)} | \hat{H} | \Psi_{HF}^{(e,\gamma;h,\delta)} \rangle = E_{HF} + \varepsilon_\gamma - \varepsilon_\delta - \Delta_{\gamma\delta} , \tag{2.30}$$

where we have defined

$$\Delta_{\gamma\delta} = v_{\gamma\delta\delta\gamma} - v_{\gamma\delta\gamma\delta} . \tag{2.31}$$

Clearly, the HF energy of the electron–hole pair is *not* just the sum of the HF energies of the electron and the hole: the correction $\Delta_{\gamma\delta}$ takes into account the interaction energy between the two excitations. More explicitly in terms of HF orbitals one has

$$\Delta_{\gamma\delta} = \frac{1}{2} \sum_{s_1 s_2} \int d\vec{r}_1 \int d\vec{r}_2 \frac{e^2}{|\vec{r}_1 - \vec{r}_2|} |\phi_\gamma(\vec{r}_1, s_1)\phi_\delta(\vec{r}_2, s_2) - \phi_\delta(\vec{r}_1, s_1)\phi_\gamma(\vec{r}_2, s_2)|^2, \tag{2.32}$$

which is *always positive*, i.e., the electron and the hole always *attract* each other.

2.2.4 Two stability theorems and the coulomb gap

The positivity of $\Delta_{\gamma\delta}$ has several interesting consequences. These all stem from the fact that for a stable HF solution the inequality

$$\varepsilon_\gamma - \varepsilon_\delta - \Delta_{\gamma\delta} > 0 \tag{2.33}$$

must be strictly satisfied for any pair of HF states γ and δ such that δ is occupied and γ is empty. If this condition were violated we could *lower* the energy of the given HF solution by promoting an electron from δ to γ. This, in turn, would mean that the given solution is *not* a minimum of the energy.

An immediate consequence of Eq. (2.33) *and* the positivity of $\Delta_{\gamma\delta}$ is that the eigenvalues of all the occupied HF orbitals (δ) must be strictly less than the eigenvalues of all the unoccupied HF orbitals (γ). This proves formally the intuitively evident fact that any stable solution of the HF equations must necessarily be the *ground-state* of the self-consistent HF hamiltonian.

A second consequence of Eq. (2.33) is that there cannot be both occupied and un-occupied states of the same spin orientation at the Fermi level: there must necessarily be a "gap" separating the lowest unoccupied orbital from the highest occupied orbital of a given spin. The gap may tend to zero in extended systems (because the matrix element of the coulomb interaction between extended states scales as the inverse of the volume of the system), but it is finite in systems with localized states, such as atoms and molecules. Thus an "open shell" situation (that is, a situation with orbital degeneracy at the Fermi level) can never occur in an *unrestricted* HF theory (Bach *et al.*, 1994).

A vivid demonstration of the usefulness of Eq. (2.33) is provided by the classic argument (Efros and Shklovskii, 1975) for the existence of a *coulomb gap* in the distribution of energy levels of localized electrons in strongly disordered semiconductors. In this system the HF states are localized in the neighborhood of randomly distributed impurities. Each state has a localization center $\vec{R}$ and its amplitude decreases as $e^{-|\vec{r}-\vec{R}|/\xi}$ (with a localization length ξ) as one moves away from $\vec{R}$. A Fermi energy ϵ_F separates the occupied states ($\varepsilon < \epsilon_F$) from the empty ones ($\varepsilon > \epsilon_F$). Consider a pair of states γ, δ respectively above and below the Fermi energy, with localization centers at $\vec{R}_\gamma$ and $\vec{R}_\delta$. Eq. (2.33) implies that

$$\varepsilon_\gamma - \varepsilon_\delta > \Delta_{\gamma\delta} \simeq \frac{e^2}{\epsilon_b |\vec{R}_\gamma - \vec{R}_\delta|} , \tag{2.34}$$

where ϵ_b is the static dielectric constant of the semiconductor.

The approximate expression for $\Delta_{\gamma\delta}$ is valid only when the separation between $\vec{R}_\gamma$ and $\vec{R}_\delta$ exceeds the localization length, but Eq. (2.34) shows that this must necessarily happen when ε_γ and ε_δ are sufficiently close to the Fermi energy. In fact, states on opposite sides of the Fermi level that differ in energy by less than a small value η must be separated in space by a distance *larger* than $e^2/\epsilon_b\eta$. Hence, the spatial density of such states must vanish at least as fast as $(\epsilon_b\eta/e^2)^d$, where d is the dimensionality of the system.

Let us denote by $N(\epsilon)$ the density of single-particle states – that is, the number of states per unit volume per unit energy range about ϵ. Then the spatial density of states within a band of width η around the Fermi level will be given by

$$\int_{\epsilon_F - \frac{\eta}{2}}^{\epsilon_F + \frac{\eta}{2}} N(\epsilon)d\epsilon \tag{2.35}$$

and the above arguments imply

$$N(\epsilon) < 2^{d-1} d \left(\frac{\epsilon_b}{e^2}\right)^d (\epsilon - \epsilon_F)^{d-1} , \tag{2.36}$$

for $\epsilon \to \epsilon_F$. This leads us to the striking conclusion that the density of HF eigenvalues must vanish at least as fast as $(\epsilon - \epsilon_F)^{d-1}$ as ϵ tends to the Fermi energy:[9] this effect has been experimentally verified (Shklovskii and Efros, 1980).[10]

---

[9] A more detailed argument, due to Efros and Shklovskii, shows that the the density of states scales *precisely* as $(\epsilon - \epsilon_F)^{d-1}$.
[10] We emphasize that the appearance of the coulomb gap is due to the localization of the electronic states. In systems in which the electronic states are extended $\Delta_{\gamma\delta}$ vanishes as the inverse of the volume, and Eq. (2.33) reduces to the statement that the all the occupied states lie below all the empty states.

2.3 Hartree–Fock factorization and mean field theory

The HF theory can also be described as an approximate *factorization* of the interaction hamiltonian: this point of view emphasizes the key role played by the neglect of fluctuations in the development of the theory, and thus reveals its deeper significance as a *mean field theory*.

The main idea, in this approach, is to extract the "one-body part" of the two-body interaction. By this, we mean the part of the interaction which, when acting on the (yet to be determined) HF ground-state, $|\Psi_{HF}\rangle$, promotes just one electron from an occupied state to an empty state, as a single-particle potential would. This one-body part of the interaction, combined with the kinetic energy and the external potential, will constitute our Hartree–Fock hamiltonian $\hat{H}_{HF}$. The remaining part of the interaction, denoted by $\hat{H}_{e-e}^{(2)}$, is a genuinely two-body operator, which, acting on $|\Psi_{HF}\rangle$, promotes *two* electrons from occupied to empty states. This operator has two important properties: first, its expectation value $\langle \Psi_{HF}|\hat{H}_{e-e}^{(2)}|\Psi_{HF}\rangle$ vanishes, because a state with two electron–hole excitations is necessarily orthogonal to the HF ground-state; second, and most important, the variation $\langle \delta\Psi_{HF}|\hat{H}_{e-e}^{(2)}|\Psi_{HF}\rangle$ also vanishes because, by virtue of Eq. (2.13), a small variation of the HF wave function (preserving its single-determinantal character) introduces at most *one* electron–hole pair. It follows that we can neglect $\hat{H}_{e-e}^{(2)}$ without changing either the expectation value of the energy, or its first variation. The expectation value of the full hamiltonian in $|\Psi_{HF}\rangle$ will then be stationary with respect to small variations of $|\Psi_{HF}\rangle$ provided the expectation value of $\hat{H}_{HF}$ is stationary: this leads us to define $|\Psi_{HF}\rangle$ as the ground-state of $\hat{H}_{HF}$ and closes the self-consistency loop.

The normal-ordering, described in Appendix 3, offers an elegant way to accomplish the separation of the interaction operator into one- and two-body parts. Recall that the standard second-quantized form of the many-body hamiltonian is naturally normal-ordered with respect to the vacuum, i.e., all the creation operators are on the left of the destruction operators. In Chapter 1, Section 1.4, we pointed out that this particular ordering automatically excludes the unphysical interaction of a particle with itself. Similarly, by normal-ordering the interaction with respect to the HF ground-state (yet to be determined) we are able to isolate the truly two-body part of the interaction. The remainder is a one-body operator that describes the interaction of an electron with itself, mediated by other electrons. This, now physical, self-interaction (also known as *self-energy*), becomes part of the effective HF hamiltonian.

Let us work out the idea in detail.[11] We start from the second-quantized form of the electron-electron interaction, written on the basis of the (yet to be determined) HF single particle orbitals

$$\hat{H}_{e-e} = \frac{1}{2}\sum_{\alpha\beta\gamma\delta} v_{\alpha\beta\gamma\delta}\, \hat{a}_\alpha^\dagger \hat{a}_\beta^\dagger \hat{a}_\gamma \hat{a}_\delta \,, \qquad (2.37)$$

and make use of Wick's theorem (stated and proved in Appendix 3) to expand the product

---

[11] The reader who is not familiar with normal ordering should study Appendix 3 before proceeding in this section.

of four creation and destruction operators as a sum of terms that are *normal ordered* relative to the HF ground-state. The result of this rewriting is (see Eq. (A3.12))

$$\hat{a}_\alpha^\dagger \hat{a}_\beta^\dagger \hat{a}_\gamma \hat{a}_\delta = \langle \hat{a}_\alpha^\dagger \hat{a}_\delta \rangle \langle \hat{a}_\beta^\dagger \hat{a}_\gamma \rangle - \langle \hat{a}_\alpha^\dagger \hat{a}_\gamma \rangle \langle \hat{a}_\beta^\dagger \hat{a}_\delta \rangle$$
$$+ \langle \hat{a}_\alpha^\dagger \hat{a}_\delta \rangle : \hat{a}_\beta^\dagger \hat{a}_\gamma : + \langle \hat{a}_\beta^\dagger \hat{a}_\gamma \rangle : \hat{a}_\alpha^\dagger \hat{a}_\delta :$$
$$- \langle \hat{a}_\alpha^\dagger \hat{a}_\gamma \rangle : \hat{a}_\beta^\dagger \hat{a}_\delta : - \langle \hat{a}_\beta^\dagger \hat{a}_\delta \rangle : \hat{a}_\alpha^\dagger \hat{a}_\gamma :$$
$$+ : \hat{a}_\alpha^\dagger \hat{a}_\beta^\dagger \hat{a}_\gamma \hat{a}_\delta : , \tag{2.38}$$

where the angular brackets denote the average in the HF ground-state. The first line of this expression is just a number – the average of the interaction hamiltonian in the HF ground-state. The one-body operators on the second and the third line contain the HF self-energy described in the previous paragraph. The quartic term on the last line is the normal-ordered interaction, now free of self-energy effects. It is evident that this term, acting on $|\Psi_{HF}\rangle$ will give zero unless all the operators are of the creation type, i.e., unless two electron–hole pairs are created: hence this term must be identified with $\hat{H}_{e-e}^{(2)}$ of the previous discussion.

Up to this point everything is exact. We now introduce the key approximation, namely, we discard the normal-ordered interaction. This term is purely "fluctuational" since its average value and its first variation both vanish by the basic properties of the normal-ordered product. In this sense our approximation can be described as a neglect of quantum fluctuations. The result of the approximation is that we are left with a quadratic interaction operator, which looks as if we had *factored* the original interaction into a product of simpler one-body terms. More precisely, we have

$$\hat{H}_{e-e} \simeq \sum_{\alpha\beta\gamma\delta} (v_{\alpha\beta\gamma\delta} - v_{\alpha\beta\delta\gamma}) \langle \hat{a}_\alpha^\dagger \hat{a}_\delta \rangle : \hat{a}_\beta^\dagger \hat{a}_\gamma : , \tag{2.39}$$

where we have dropped the constant terms, and made use of the symmetry $v_{\alpha\beta\gamma\delta} = v_{\beta\alpha\delta\gamma}$.

To check that this is indeed equivalent to the HF potential derived in the previous section we notice that the ground-state of the quadratic hamiltonian with interaction given by (2.39) has occupation numbers $n_\alpha = 1$ for occupied states and $n_\alpha = 0$ for unoccupied states. Thus we have

$$\langle \hat{a}_\alpha^\dagger \hat{a}_\beta \rangle = n_\alpha \delta_{\alpha\beta} . \tag{2.40}$$

Substituting this in Eq. (2.39), making use of the fact that $\hat{a}_\alpha^\dagger \hat{a}_\beta = \langle \hat{a}_\alpha^\dagger \hat{a}_\beta \rangle + : \hat{a}_\alpha^\dagger \hat{a}_\beta :$, and discarding another constant, we arrive at

$$\hat{H}_{e-e} \simeq \sum_{\alpha,\beta} \sum_{\gamma \leq N} (v_{\alpha\gamma\gamma\beta} - v_{\alpha\gamma\beta\gamma}) \hat{a}_\alpha^\dagger \hat{a}_\beta , \tag{2.41}$$

which indeed coincides with the interaction part of the HF potential (2.17).

2.4 Application to the uniform electron gas

As a first simple application of the HF theory we consider here the case of the uniform electron liquid. In this case the simplest, although not necessarily the correct or best choice, is to assume a translationally invariant HF potential of the form $V_s^{HF}(\vec{r} - \vec{r}')\delta_{ss'}$ that depends only on the difference $\vec{r} - \vec{r}'$ and does not mix the two components of the spin. Such a potential is automatically diagonal in the plane wave representation and therefore provides a self-consistent solution of the HF equations which we will refer to as the HF plane wave state. Even though this state does not represent the actual HF ground-state of the jellium model (see Section 2.6.3), it nevertheless provides the correct reference state for the construction of the Fermi liquid theory (see Chapter 8): for this reason we now examine its properties in detail.

The second-quantized expression for the self consistent HF potential in the plane wave state is given by

$$\hat{V}^{HF} = \sum_{\vec{k}\sigma} V_{\vec{k}\sigma,\vec{k}\sigma}^{HF} \hat{a}_{\vec{k}\sigma}^{\dagger} \hat{a}_{\vec{k}\sigma} \ . \tag{2.42}$$

The numerical coefficients in this expression are immediately obtained from Eq. (2.17), keeping in mind that $V^{ext} = 0$ and that the eigenfunctions of the HF hamiltonian are just plane waves:

$$V_{\vec{k}\sigma;\vec{k}\sigma}^{HF} = \sum_{\vec{k}'\sigma'} n_{\vec{k}'\sigma'} (v_{\vec{k}\sigma \ \vec{k}'\sigma' \ \vec{k}'\sigma' \ \vec{k}\sigma} - v_{\vec{k}\sigma \ \vec{k}'\sigma' \ \vec{k}\sigma \ \vec{k}'\sigma'}) \ . \tag{2.43}$$

We notice that the Hartree term $\sum_{\vec{k}'\sigma'} n_{\vec{k}'\sigma'} v_{\vec{k}\sigma \ \vec{k}'\sigma' \ \vec{k}'\sigma' \ \vec{k}\sigma} = n v_{\vec{q}=0}$ is identically zero in view of the fact that $v_{q=0} = 0$ in our definition of the jellium model (see Eq. (1.15)). By substituting in the exchange term the expression

$$v_{\vec{k}\sigma \ \vec{k}'\sigma' \ \vec{k}\sigma \ \vec{k}'\sigma'} = \frac{1}{L^d} v_{\vec{k}-\vec{k}'} \delta_{\sigma\sigma'} \ , \tag{2.44}$$

we obtain

$$V_{\vec{k}\sigma;\vec{k}\sigma}^{HF} \equiv \varepsilon_{\vec{k}\sigma}^{(x)} = -\frac{1}{L^d} \sum_{\vec{k}'} n_{\vec{k}'\sigma} v_{\vec{k}-\vec{k}'} \ . \tag{2.45}$$

The HF eigenvalues are therefore given by

$$\varepsilon_{\vec{k}\sigma} = \frac{\hbar^2 k^2}{2m} + \varepsilon_{\vec{k}\sigma}^{(x)} \ , \tag{2.46}$$

an expression that includes the free electron energy term and the exchange contribution $\varepsilon_{\vec{k}\sigma}^{(x)}$ defined by Eq. (2.45).

Because, by assumption, the HF potential does not mix the two components of the spin, it follows that the HF ground-state is given by the product of two independent full Fermi spheres, one for each spin component:

$$|\Psi_p\rangle = \left(\prod_{|\vec{k}| \leq k_{F\uparrow}} \hat{a}_{\vec{k},\uparrow}^{\dagger} \right) \left(\prod_{|\vec{k}| \leq k_{F\downarrow}} \hat{a}_{\vec{k},\downarrow}^{\dagger} \right) |0\rangle \,. \qquad (2.47)$$

The radii $k_{F\uparrow}$ and $k_{F\downarrow}$ are related to the numbers $N_\uparrow$ and $N_\downarrow$ (with $N_\uparrow + N_\downarrow = N$) by Eq. (1.84). In the above equation the sub index p for the wave function refers to the fractional polarization $p = \frac{N_\uparrow - N_\downarrow}{N}$.

2.4.1 The exchange energy

The HF ground-state energy can be readily evaluated from the general expression (2.24). Setting $\alpha = \vec{k}\sigma$ we immediately get

$$E_{HF} = \sum_{\vec{k}\sigma} n_{\vec{k}\sigma} \left(\frac{\hbar^2 k^2}{2m} + \frac{1}{2} \varepsilon_{\vec{k}\sigma}^{(x)} \right) \,. \qquad (2.48)$$

Making use of Eq. (2.45) for $\varepsilon_{\vec{k}\sigma}^{(x)}$ we see that the above expression coincides with the energy calculated to first order in the coulomb interaction in Chapter 1, i.e.,

$$\epsilon_{HF}(r_s) = \frac{E_{HF}}{N} = \epsilon_0(r_s) + \epsilon_x(r_s) \,. \qquad (2.49)$$

The coincidence of the HF energy and the first-order energy is a peculiarity of the jellium model, stemming from the fact that the plane wave HF state coincides with the non interacting ground-state. Of course, this would *not* be true in the case of a non-homogeneous system.

Let us now examine the HF eigenvalues in greater detail. The calculation of $\varepsilon_{\vec{k}\sigma}^{(x)}$ is straightforward at zero temperature. In three dimensions we have

$$\varepsilon_{\vec{k}\sigma}^{(x)} = -\int_{|\mathbf{q}| \leq k_{F\sigma}} \frac{d^3 q}{(2\pi)^3} \frac{4\pi e^2}{|\mathbf{k} - \mathbf{q}|^2}$$

$$= -\frac{e^2 k_{F\sigma}}{\pi} \int_0^1 dx\, x^2 \int_{-1}^1 \frac{d\mu}{x^2 + y^2 - 2xy\mu} \,, \qquad (2.50)$$

where we have set $x = \frac{q}{k_{F\sigma}}$, $y = \frac{k}{k_{F\sigma}}$, and μ is the cosine of the angle between $\vec{k}$ and $\vec{q}$. The integration is elementary and gives:

$$\varepsilon_{\vec{k}\sigma}^{(x)} = -\frac{2e^2 k_{F\sigma}}{\pi} f_{3D} \left(\frac{k}{k_{F\sigma}} \right) \,, \qquad (2.51)$$

where the function f_{3D} is given by

$$f_{3D}(y) = \frac{1}{2} + \frac{1 - y^2}{4y} \ln \left| \frac{1 + y}{1 - y} \right| \,, \qquad (2.52)$$

with limits $f(0) = 1$, $f(1) = \frac{1}{2}$ and $f(\infty) = 0$. A plot of the single particle energy $\epsilon_{\vec{k}\sigma}^{HF}$ in the three-dimensional case is provided in Fig. 2.1 for the case of $r_s = 4$. With the same

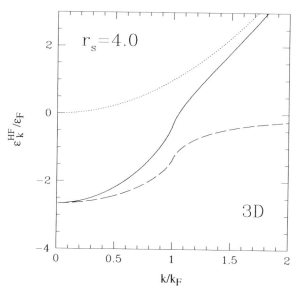

Fig. 2.1. Momentum dependence of the single particle energy $\varepsilon_{\vec{k}}^{HF}$ in HF theory for a homogeneous paramagnetic three dimensional electron gas (full line). The dashed line is the exchange contribution. The dotted line represents the non-interacting kinetic energy. Here $r_s = 4$.

notation, in the two dimensional case we have instead (Chaplik, 1971)

$$\varepsilon_{\vec{k}\sigma}^{(x)} = -\int_{|\mathbf{q}|\leq k_{F\sigma}} \frac{d^2q}{(2\pi)^2} \frac{2\pi e^2}{|\mathbf{k} - \mathbf{q}|}$$

$$= -\frac{e^2 k_{F\sigma}}{2\pi} \int_0^1 dx\, x \int_0^{2\pi} \frac{d\varphi}{\sqrt{x^2 + y^2 - 2xy\cos\varphi}}$$

$$= -\frac{2e^2 k_{F\sigma}}{\pi} f_{2D}\left(\frac{k}{k_{F\sigma}}\right), \tag{2.53}$$

where the function f_{2D} is given by

$$f_{2D}(y) = \begin{cases} \mathbf{E}(y), & y \leq 1, \\ y\left[\mathbf{E}\left(\frac{1}{y}\right) - \left(1 - \frac{1}{y^2}\right)\mathbf{K}\left(\frac{1}{y}\right)\right], & y \geq 1. \end{cases} \tag{2.54}$$

Here $\mathbf{K}(y)$ and $\mathbf{E}(y)$ are the complete elliptic integrals respectively of the first and second kind as defined in Gradshteyn and Ryzhik (1965). A plot of the single particle energy $\epsilon_{\vec{k}\sigma}^{HF}$ in two dimensions is shown in Fig. 2.2 for $r_s = 1$.

Inspecting Figs. 2.1 and 2.2 we notice that a key feature of the HF theory for the uniform electron liquid is a sizable increase of the single-particle bandwidth $\epsilon_B = \varepsilon_{k_F} - \varepsilon_0$. In the non interacting electron gas the bandwidth is simply the Fermi energy ϵ_F. In HF on the

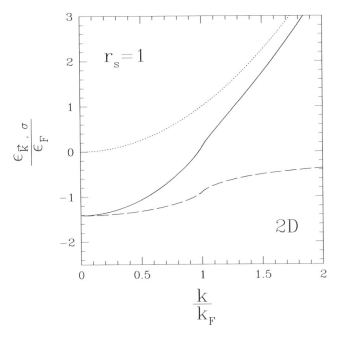

Fig. 2.2. Same as Fig. 2.1 but for the two dimensional case. Here $r_s = 1$.

other hand, Eq. (2.46) gives

$$\epsilon_B = \epsilon_F + \varepsilon_{k_F}^{(x)} - \varepsilon_0^{(x)} . \tag{2.55}$$

Because $\varepsilon_k^{(x)}$ is a negative and monotonically increasing function of k we see that the energy of the state at the bottom of the band ($k = 0$) is lowered by a larger amount (in fact two times larger) than the energy of an electron at the Fermi surface ($k = k_F$). Consequently, the bandwidth is predicted to be *larger* than in the free electron case. This prediction of the Hartree–Fock theory turns out to be in serious disagreement with experiments performed in simple metals (Jensen and Plummer, 1985; Lyo and Plummer, 1988), where, in fact, a small *reduction* of the bandwidth has been measured. A partial resolution of this difficulty will be discussed in Chapter 8.

Another difficulty of the HF theory is that, because of the logarithmically divergent derivative of $\varepsilon^{(x)}(k)$ with respect to k at $k = k_F$, the density of single-electron states vanishes (as $|\ln |\epsilon - \epsilon_F||^{-1}$) near the Fermi level. Such a vanishing density of states, if real, would cause the low-temperature specific heat to behave "sublinearly" as a function of temperature, i.e., $c_V(T) \sim \frac{T}{|\ln T|}$. Again, this prediction is incorrect: a linear temperature dependence with a prefactor very near to the free-electron value is measured instead.

Both difficulties can ultimately be traced to the fact that the HF approximation fails to take into account the phenomenon of the *screening*, that is, the readjustment of the electronic density in response to the motion of one electron. This will be examined in Chapter 5. On

Table 2.1. *Kinetic and exchange energy coefficients in 1, 2 and 3 dimensional jellium. For the one dimensional case we have given here the exchange energy coefficient appropriate to the contact interaction discussed at the end of Section 1.5.3.*

d	A_d	B_d
1	$\dfrac{\pi^2}{48}$	$\dfrac{\gamma}{4}$
2	1	$\dfrac{8\sqrt{2}}{3\pi}$
3	$\dfrac{3}{5}\left(\dfrac{9\pi}{4}\right)^{\frac{2}{3}}$	$\dfrac{3}{2\pi}\left(\dfrac{9\pi}{4}\right)^{\frac{1}{3}}$

the positive side, the large HF exchange energy gives a good account of the lowering of the electron energy, which is responsible for the cohesion of simple metals.

2.4.2 Polarized versus unpolarized states

Let us now examine the dependence of the HF energy on the degree of spin polarization. An interesting question is which value of p yields the lowest HF energy. Will the lowest energy state be paramagnetic ($p = 0$), ferromagnetic ($p = 1$), or partially spin polarized ($0 < p < 1$)? This question is readily settled by inspecting the expressions for the total HF energy as a function of r_s and p that were already obtained in Chapter 1, i.e. Eqs. (1.88) and (1.98). The general expression, valid in any number of dimensions, is

$$\epsilon_{HF}(r_s, p) = \left(\frac{[(1+p)^{\frac{d+2}{d}} + (1-p)^{\frac{d+2}{d}}] A_d}{2} \frac{}{r_s^2} \right.$$
$$\left. - \frac{[(1+p)^{\frac{d+1}{d}} + (1-p)^{\frac{d+1}{d}}] B_d}{2} \frac{}{r_s} \right) \text{Ry} , \qquad (2.56)$$

where the coefficients A_d and B_d are given in Table 2.1.

Clearly the answer to our question depends on r_s, i.e. on the electronic density. The reason is that the physics of the situation is ruled by the interplay between the increase of kinetic energy (a positive contribution to the energy) and the gain in exchange energy (a negative contribution to the energy) associated with the existence of a finite polarization. In view of the different density dependence of the two terms one expects the state of lowest energy to be paramagnetic at high densities and ferromagnetic at low densities. A direct inspection of Eq. (2.56) confirms this scenario (see Fig. 2.3) and also shows that the partially polarized

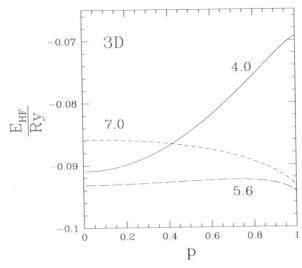

Fig. 2.3. Polarization dependence of the total HF energy of a homogeneous three dimensional electron gas for various representative values of r_s.

state is *never* the state of lowest energy. In this respect the HF phase diagram of the system has only two phases separated by a critical value r_{sB} which marks what is referred to as the *Bloch transition* (Bloch, 1929). Except for the peculiar case of a one dimensional system, the Bloch transition is a *sudden phase transition* from the paramagnetic state to the fully spin polarized state. Accordingly the value of r_{sB} is determined by the condition that the two states are degenerate. From Eq. (2.56) one finds:

$$r_{sB} = \frac{(2^{\frac{2}{d}} - 1)A_d}{(2^{\frac{1}{d}} - 1)B_d} \simeq \begin{cases} 5.45 & , \quad 3D, \\ 2.01 & , \quad 2D, \\ \frac{\pi^2}{4\gamma} & , \quad 1D. \end{cases} \tag{2.57}$$

We hasten to say that this phase diagram is quite far from reality, as the Monte Carlo calculations reviewed in Chapter 1 clearly indicate. The fact that a ferromagnetic transition occurs in reality at much lower densities indicates that some important physics is missing in the HF description of the electron liquid. We will see in Chapter 5 that the missing physics is the existence of correlations between electrons of opposite spin orientations.

2.4.3 *Compressibility and spin susceptibility*

It is straightforward to calculate the compressibility K and the spin susceptibility χ_S of the homogeneous paramagnetic HF state. This is most simply accomplished by means of the thermodynamic relationships of Section 1.8 (see Eqs. (1.137) and (1.142)). One readily finds the following interesting result:

$$\left(\frac{K}{K_0}\right)_{HF} = \left(\frac{\chi_S}{\chi_P}\right)_{HF} = \frac{1}{1 - \frac{r_s}{r_s^*}}, \tag{2.58}$$

where K_0 and χ_P are the non-interacting values of these quantities and, with reference to Table 2.1, we have defined

$$r_s^{\star} = \frac{2(d+2)A_d}{(d+1)B_d} \simeq \begin{cases} 6.03 \ , & 3D \, , \\ 2.22 \ , & 2D \, , \\ \frac{\pi^2}{4\gamma} \ , & 1D \, . \end{cases} \tag{2.59}$$

The fact that these two ratios coincide is a peculiarity of the HF theory. In general one expects $\frac{K}{K_0}$ and $\frac{\chi_S}{\chi_P}$ to differ due to the presence of correlation, as discussed in Chapter 1 (the inverse ratios $\frac{K_0}{K}$ and $\frac{\chi_P}{\chi_S}$ are plotted in Fig. 1.22). Eq. (2.58) shows that within HF both K and χ_S diverge as r_s approaches $r_s^{\star}$ from below. Clearly the divergence signals a potential instability of the paramagnetic state. The problem of stability is studied in the following section.

2.5 Stability of Hartree–Fock states

2.5.1 Basic definitions: local versus global stability

We now turn to the interesting and difficult problem of the *stability* of the HF solutions. We shall entertain two basic notions of stability: *local stability* and *global stability*.

If a solution of the HF equations is at hand, we shall refer to it as a *locally* or *differentially stable solution* if it represents a local minimum of the HF energy. The solution will be called *globally stable* if it corresponds to the absolute minimum of the energy. It is evident that the energy of a locally stable solution can and will in general lie above that of other self-consistent minima of the energy somewhere else in the space of single Slater determinants. In this case we shall say that the original solution is locally stable but globally unstable. The Bloch ferromagnetic transition introduced in the previous section offers the simplest demonstration of this scenario (see Section 2.5.3 below).

It should immediately be said that, to date, there is no known practical algorithm to assess the global stability of a solution and one is limited to a case-by-case comparison of the total energies of different locally stable solutions. In other words one is never guaranteed that a given solution is globally stable.[12] In practice, one often restricts at the outset the space of the allowed orbitals by imposing a certain type of symmetry: these *restricted* HF schemes usually lead to much more tractable "energy landscapes".

2.5.2 Local stability theory

A well defined procedure can be formally established to determine if a given solution of the HF equations corresponds to a true local minimum and not, as in the most common case, to a saddle point. The idea is to compare the energy of the putative HF ground-state, $|\Psi_{HF}\rangle$, with that of other single-determinant states that are close to it. To this end we need

---

[12] The complexity of the HF equations can be better put in perspective by realizing that the problem at hand is essentially the minimization of the energy functional in a space of infinite dimensions.

a simple criterion to distinguish between states that can be represented as single Slater determinants and states that cannot be so represented. Such a criterion is provided by the *one-particle density matrix*, which, for an arbitrary many-body state state $|\Psi\rangle$, is defined by the expectation value

$$\rho_{\alpha\beta} \equiv \langle\Psi|\hat{a}_{\alpha}^{\dagger}\hat{a}_{\beta}|\Psi\rangle \qquad (2.60)$$

(here α and β are labels for the wavefunctions of a complete but otherwise arbitrary one-electron basis set). In particular, the density matrix of the state $|\Psi_{HF}\rangle$, written in the basis of the HF orbitals, has the form $\rho_{\alpha\beta} = n_{\alpha}\delta_{\alpha\beta}$ with $n_{\alpha} = 1$ for $\alpha \leq N$ and $n_{\alpha} = 0$ otherwise. The same matrix, written in a different basis, would no longer be diagonal, but its eigenvalues would still be 1s and 0s. It is quite evident that this property must hold in any many-body state that can be represented as a Slater determinant of N single-particle orbitals. Conversely, the fact that the density matrix of a given state has only 0s and 1s for eigenvalues implies that the state can be represented as a single Slater determinant. We can express these facts in a compact and elegant form by saying that the many-body state $|\Psi\rangle$ can be represented as a single Slater determinant if and only if its one-electron density matrix $\hat{\rho}$ satisfies the *idempotency condition*

$$\hat{\rho}^{2} = \hat{\rho} \qquad (2.61)$$

(see Exercise 6).

Another useful feature of the density matrix is that it allows us to express the expectation value of the many-body hamiltonian *in any single-determinant state* $|\Psi_{S}\rangle$ in the following compact form:

$$\langle\Psi_{S}|\hat{H}|\Psi_{S}\rangle = \sum_{\alpha\beta}(T_{\alpha\beta} + V_{\alpha\beta}^{ext})\rho_{\alpha\beta} + \frac{1}{2}\sum_{\alpha\beta\gamma\delta}(v_{\alpha\beta\gamma\delta} - v_{\alpha\beta\delta\gamma})\rho_{\alpha\delta}\rho_{\beta\gamma}, \qquad (2.62)$$

which follows immediately from the application of Wick's theorem (see Appendix 3 and Exercise 3). Notice that his expression reduces to Eq. (2.12) when $|\Psi_{S}\rangle$ is the HF ground-state.

Let us now consider a Slater determinant Ψ_{S} which is very close to Ψ_{HF} in the sense that its density matrix (in the basis of the HF orbitals) has the form

$$\rho_{\alpha\beta} = n_{\alpha}\delta_{\alpha\beta} + \delta\rho_{\alpha\beta}, \qquad (2.63)$$

where $\delta\rho_{\alpha\beta}$ are infinitesimal increments. The key question is whether this density matrix, when substituted in Eq. (2.62) gives an energy higher or lower than the energy of the putative HF ground-state. From Eq. (2.14) we know that the first-order variation of the energy will be zero by virtue of $|\Psi_{HF}\rangle$ being a solution of the HF equation. We must therefore expand the energy to second order in $\delta\rho$.

In order to do this expansion properly it is essential to realize that the increments $\delta\rho_{\alpha\beta}$ are not independent, because the density matrix of a single Slater determinant must satisfy the condition (2.61), or, in components,

$$\sum_{\gamma} \rho_{\alpha\gamma}\rho_{\gamma\beta} = \rho_{\alpha\beta}. \tag{2.64}$$

Substituting Eq. (2.63) into (2.64) we obtain a relation between the increments of the density matrix

$$(1 - n_{\alpha} - n_{\beta})\delta\rho_{\alpha\beta} = \sum_{\gamma} \delta\rho_{\alpha\gamma}\delta\rho_{\gamma\beta}. \tag{2.65}$$

This equation implies that:

$$\delta\rho_{\alpha\beta} = \pm \sum_{\gamma} \delta\rho_{\alpha\gamma}\delta\rho_{\gamma\beta}, \tag{2.66}$$

where the positive sign applies when both α and β are unoccupied states (i.e. α *and* $\beta > N$), while the negative sign applies when both α and β are occupied states (i.e. α *and* $\beta \leq N$). Notice that the left-hand side of Eq. (2.65) vanishes if α and β are on opposite sides of the Fermi level, i.e. if $\alpha \leq N$ and $\beta > N$ *or* $\beta \leq N$ and $\alpha > N$. Thus, only the increments $\delta\rho_{\alpha\beta}$ with α and β on opposite sides of the Fermi level can be regarded as independent first order increments, while $\delta\rho_{\alpha\beta}$ with α and β on the same side of the Fermi level are dependent second-order quantities, determined by Eqs. (2.66). It is not hard to show that the independent increments $\delta\rho_{\alpha\beta}$, with α and β on opposite sides of the Fermi level, determine, up to corrections of second order in $\delta\rho$, the form of the determinantal wave function $|\Psi\rangle$ as

$$|\Psi\rangle \propto \prod_{\alpha \leq N} \left(\hat{a}_{\alpha}^{\dagger} + \sum_{\beta > N} \delta\rho_{\alpha\beta}\hat{a}_{\beta}^{\dagger} \right) |0\rangle$$

$$= \prod_{\alpha \leq N} \left(1 + \sum_{\beta > N} \delta\rho_{\alpha\beta}\hat{a}_{\beta}^{\dagger}\hat{a}_{\alpha} \right) |\Psi_{HF}\rangle. \tag{2.67}$$

The proof of this statement is left as an exercise for the reader.

We now return to Eq. (2.62) for the energy of the state $|\Psi\rangle$ and substitute into it Eq. (2.63) for the density matrix. The zero-th order term is, of course, the HF energy of the state Ψ_{HF}. The first order variation vanishes, due to the extremal property of the HF solution. Let us therefore focus on the *second order variation* of the energy. Making use of the self-consistency condition $T_{\alpha\beta} + V_{\alpha\beta}^{HF} = \varepsilon_{\alpha}\delta_{\alpha\beta}$ (see Eqs. (2.16) and (2.18)), and carefully keeping track of the conditions (2.66), the reader should be able to verify that

$$\Delta^{(2)}E_{HF} = \frac{1}{2}\sum_{(\alpha\beta)} \delta\rho_{\alpha\beta}\frac{\varepsilon_{\beta} - \varepsilon_{\alpha}}{n_{\alpha} - n_{\beta}}\delta\rho_{\beta\alpha} + \frac{1}{2}\sum_{(\alpha\delta)}\sum_{(\beta\gamma)}(v_{\alpha\beta\gamma\delta} - v_{\alpha\beta\delta\gamma})\delta\rho_{\alpha\delta}\delta\rho_{\beta\gamma}, \tag{2.68}$$

where the subscript $(\alpha\beta)$ means that the sum is restricted to pairs of states α and β such that ε_{α} and ε_{β} lie on opposite sides of the Fermi level.

This expression can be cast in the form

$$\Delta^{(2)}E_{HF} = -\frac{1}{2} \sum_{(\alpha\delta),(\beta\gamma)} [\chi_{HF}]^{-1}_{(\alpha\beta);(\gamma\delta)} \delta\rho_{\alpha\beta}\delta\rho_{\gamma\delta} \,, \tag{2.69}$$

where we have defined the symmetric matrix

$$[\chi_{HF}]^{-1}_{(\alpha\beta);(\gamma\delta)} \equiv -\frac{\varepsilon_\beta - \varepsilon_\alpha}{n_\alpha - n_\beta}\delta_{\alpha\delta}\delta_{\beta\gamma} - (v_{\alpha\gamma\delta\beta} - v_{\alpha\gamma\beta\delta}) \,. \tag{2.70}$$

Now, for the given HF solution to be stable, this quadratic form in $\delta\rho_{(\alpha\beta)}$ must be positive for any choice of the $\delta\rho$'s. Accordingly, the condition of stability is that every eigenvalue of the matrix $[\chi_{HF}]^{-1}_{(\alpha\beta);(\gamma\delta)}$ be positive.

We shall see later that $[\chi_{HF}]_{\alpha\delta;\beta\gamma}$ has a simple physical interpretation: it is the linear response of the density-matrix element $\rho_{\alpha\beta}$ to a static perturbing potential that couples linearly to the density matrix element $\rho_{\gamma\delta}$.

2.5.3 Local and global stability for a uniformly polarized electron gas

The case of a homogeneous, uniformly polarized electron gas discussed in Section 2.4, offers a simple yet instructive stability scenario. As we have seen, if we limit our considerations to the states described by Eq. (2.47), i.e. homogeneous HF states with a fractional polarization p, we find that the paramagnetic state is globally stable for r_s smaller than the Bloch critical value r_{sB}, whereas for r_s exceeding this value it is the ferromagnetic state that becomes globally stable. What about the local stability of these two states? In order to study this, we simply evaluate the change in the energy as p deviates slightly from its putative equilibrium values, i.e. $p = 0$ and $p = 1$.[13]

Expanding for small p about the paramagnetic state Eq. (2.56) gives:[14]

$$\epsilon_{HF}(r_s, p) \simeq \epsilon_{HF}(r_s, 0) + \left(\frac{2+d}{d^2}\frac{A_d}{r_s^2} - \frac{1+d}{2d^2}\frac{B_d}{r_s}\right)p^2 + \mathcal{O}(p^4), \tag{2.71}$$

where $\epsilon_{HF}(r_s, 0)$ is the energy of the paramagnetic state at the given r_s. Clearly the state is locally stable if the coefficient of the p^2 term is positive. This is the case for r_s smaller than the critical value $r_s^\star$, defined in Eq. (2.59). This conclusion simply reflects the divergence of the spin susceptibility χ_S at $r_s = r_s^\star$ displayed in Eq. (2.58). Notice that in one dimension the critical density value for the Bloch transition does coincide with that of the differential instability.

By comparison with the values of r_{sB} of Eq. (2.57) we infer the following scenario for the two and three dimensional cases: the paramagnetic state is differentially (i.e. locally) stable for all $r_s < r_s^\star$; at $r_s^\star$ the systems can then in principle undergo a *continuous phase transition* into a polarized state. This transition is preempted by the sudden Bloch transition which occurs for $r_{sB} < r_s^\star$.

[13] As shown in Chapters 3 and 5, this is equivalent to calculating the long wavelength limit of the static spin susceptibility.
[14] This equation is in fact exact in the one dimensional case with delta function interactions for which the energy is given by the simple quadratic expression of Eq. (1.97).

For the fully spin polarized state the situationt is quite different. Expanding Eq. (2.56) about $p = 1$, one quickly realizes that the leading correction to the ferromagnetic energy is linear in $1 - p$, i.e.

$$\epsilon_{HF}(r_s, p) \simeq \epsilon_{HF}(r_s, 1) + \left(\frac{1+d}{2^{\frac{d-1}{d}} d} \frac{B_d}{r_s} - \frac{2+d}{2^{\frac{d-2}{d}} d} \frac{A_d}{r_s^2} \right) (1 - p)$$

$$+ \mathcal{O}((1 - p)^{\frac{d+1}{d}}), \tag{2.72}$$

where now $\epsilon_{HF}(r_s, 1)$ is the energy of the ferromagnetic state at the given r_s. Clearly the ferromagnetic state is never stationary as a function of the fractional polarization, even though it is a minimum for $r_s > r_{sB}$. The linear term in Eq. (2.72) is negative for r_s smaller than

$$\tilde{r}_s = \frac{r_s^\star}{2^{\frac{d-1}{d}}}, \tag{2.73}$$

and becomes then positive at lower densities. In three and in two dimensions this occurs respectively for $r_s \simeq 3.80$ and $r_s = \frac{\pi}{2} \simeq 1.57$. Accordingly, in the regime in which the ferromagnetic state corresponds to the lowest energy, deviations from total polarization lead to an increase of the energy, a fact that stabilizes the state. In the one dimensional case $\tilde{r}_s$ again coincides with r_{sB}.

We finally turn to the more general question: does the homogeneous paramagnetic HF state correspond to a genuine minimum of the HF energy for r_s smaller than the Bloch transition critical value r_{sB}? The answer, as we show in the next section, is a resounding, and perhaps surprising, "never". The uniform state of the electron gas corresponds in fact to a *saddle point* of the HF energy *for all densities*. To date there are no known states corresponding to genuine minima of the HF energy.

2.6 Spin density wave and charge density wave Hartree–Fock states

Between 1960 and 1962, shortly after the appearance of the BCS theory of superconductivity (about which more will be said later), nontrivial spontaneously broken symmetry HF states of the electron liquid were discovered by A. W. Overhauser (Overhauser, 1960, 1962). Overhauser proved an *instability theorem*, which states that:

Within the Hartree–Fock approximation, the homogeneous paramagnetic states of the three-dimensional electron gas is unstable with respect to the formation of spin- or charge-modulated states.

The theorem holds for all electron densities.

Spin- and charge-density wave states (SDW and CDW respectively) were briefly described in Section 1.7.4, and a qualitative reason was given there to explain why such states could be energetically advantageous in comparison with more conventional phases. In the next section we will concentrate on a special type of SDW, known as *spiral spin density wave* (SSDW) in which the spin polarization varies in direction but remains constant in magnitude. The local spin density vector $\vec{S}(\vec{r})$ of a generic SSDW rotates about the spin quantization axis ($\hat{z}$) with an angle $\vec{Q} \cdot \vec{r}$ that grows linearly along the direction of a wave vector $\vec{Q}$:[15]

$$\vec{S}(\vec{r}) = A \left[\vec{e}_x \cos \vec{Q} \cdot \vec{r} + \vec{e}_y \sin \vec{Q} \cdot \vec{r} \right]. \qquad (2.74)$$

These states lead, perhaps against one's naive intuition, to a relatively tractable mathematical problem from which a non-trivial HF solution can be obtained. We will then proceed to a simple, yet rigorous proof of Overhauser's instability theorem.

2.6.1 Hartree–Fock theory of spiral spin density waves

It must be immediately said that, in view of the geometry of the Fermi surface, the optimal magnitude of the wave vector $\vec{Q}$ of a generic SDW will have to be quite close to the diameter $2k_F$ of the Fermi surface in the paramagnetic state. It is only at about this wavevector that the HF potential can effectively mix occupied and unoccupied states of the original (paramagnetic) Fermi sea, with minimal disturbance of the Fermi surface. The presence of scattering between plane wave states with different wave vectors leads to a much more difficult problem than the homogeneous cases considered so far. One might think to simplify the problem by choosing for the matrix elements of the HF potential in Eq. (2.5) the simpler form

$$V^{HF}_{\vec{k}\sigma;\vec{k}'\sigma'} = V_{0,\sigma\sigma'}(\vec{k})\delta_{\vec{k},\vec{k}'} + \frac{1}{2} \left[g_{\sigma\sigma'}(\vec{k}')\delta_{\vec{k},\vec{k}'+\vec{Q}} + g_{\sigma'\sigma}(\vec{k})\delta_{\vec{k}',\vec{k}+\vec{Q}} \right] \qquad (2.75)$$

in which the second term mixes a plane wave of wave vector $\vec{k}$ with plane waves of wave vectors $\vec{k} \pm \vec{Q}$. While this Ansatz is admissible for generating a variational wave function, it is evident that it cannot be in general a self-consistent solution of the HF equations. For, upon diagonalization of the HF hamiltonian, the ground-state density will be found, in general, to contain higher Fourier components with wave vectors $2\vec{Q}, 3\vec{Q}$, etc.: these higher harmonics will in turn give rise to a potential that scatters from $\vec{k}$ to $\vec{k} \pm 2\vec{Q}, \vec{k} \pm 3\vec{Q}$, etc., in contradiction with the simple Ansatz of Eq. (2.75). The spiral SDW state introduced in the previous section provides a rather unique case in which a self-consistent solution of the HF equations can be found, characterized by a *single* wave vector $\vec{Q}$. In a SSDW state the HF potential scatters an up-spin electron with wave vector $\vec{k}$ to a down-spin state with wave vector $\vec{k} + \vec{Q}$ and vice versa. Mathematically we have

$$V^{HF}_{\vec{k}\sigma;\vec{k}'\sigma'} = V_{0\sigma}(\vec{k})\delta_{\vec{k},\vec{k}'}\delta_{\sigma\sigma'} + \frac{1}{2} \left[g_{\vec{k}'} \, \delta_{\vec{k},\vec{k}'+\vec{Q}}\hat{\sigma}_- + g_{\vec{k}} \, \delta_{\vec{k}',\vec{k}+\vec{Q}}\hat{\sigma}_+ \right]_{\sigma\sigma'}, \qquad (2.76)$$

where $\hat{\sigma}_\pm = \hat{\sigma}_x \pm i\hat{\sigma}_y$ are spin rising and lowering operators relative to the axis of quantization (z). The reader can easily check that this is a special instance of the general form (2.75). With a potential of this form the HF eigenvalue problem can be solved exactly and the resulting spin density is given by Eq. (2.74). This spin density in turn generates an exchange potential having precisely the form of Eq. (2.76): thus, the Ansatz is self-consistent.

---

[15] $\vec{Q}$ and $\hat{z}$ need not be parallel.

To see how it all comes together, let us rewrite the complete HF hamiltonian

$$\hat{H}_{HF} = \sum_{\vec{k}\sigma} \left[\frac{\hbar^2 k^2}{2m} + V_{0\sigma}(\vec{k}) \right] \hat{a}_{\vec{k}\sigma}^\dagger \hat{a}_{\vec{k}\sigma}$$

$$+ \sum_{\vec{k}} g_{\vec{k}} \left[\hat{a}_{\vec{k}+\vec{Q}\downarrow}^\dagger \hat{a}_{\vec{k}\uparrow} + \hat{a}_{\vec{k}\uparrow}^\dagger \hat{a}_{\vec{k}+\vec{Q}\downarrow} \right] \tag{2.77}$$

in the following form:

$$\hat{H}_{HF} = \sum_{\vec{k}} \varepsilon_{\vec{k}+\vec{Q}\downarrow} \hat{a}_{\vec{k}+\vec{Q}\downarrow}^\dagger \hat{a}_{\vec{k}+\vec{Q}\downarrow} + \sum_{\vec{k}} \varepsilon_{\vec{k}\uparrow} \hat{a}_{\vec{k}\uparrow}^\dagger \hat{a}_{\vec{k}\uparrow}$$

$$+ \sum_{\vec{k}} g_{\vec{k}} [\hat{a}_{\vec{k}+\vec{Q}\downarrow}^\dagger \hat{a}_{\vec{k}\uparrow} + \hat{a}_{\vec{k}\uparrow}^\dagger \hat{a}_{\vec{k}+\vec{Q}\downarrow}] , \tag{2.78}$$

where

$$\varepsilon_{\vec{k}\sigma} \equiv \frac{\hbar^2 k^2}{2m} + V_{0\sigma}(\vec{k}) . \tag{2.79}$$

This hamiltonian is easily diagonalized by means of a unitary transformation to new fermion operators $\alpha_{\vec{k}}$ and $\beta_{\vec{k}}$ such that

$$\hat{a}_{\vec{k}\uparrow} = \hat{\alpha}_{\vec{k}} \cos \theta_{\vec{k}} - \hat{\beta}_{\vec{k}} \sin \theta_{\vec{k}} ,$$

$$\hat{a}_{\vec{k}+\vec{Q}\downarrow} = \hat{\alpha}_{\vec{k}} \sin \theta_{\vec{k}} + \hat{\beta}_{\vec{k}} \cos \theta_{\vec{k}} . \tag{2.80}$$

The reader can readily verify that $\hat{\alpha}_{\vec{k}}$ and $\hat{\beta}_{\vec{k}}$ satisfy the canonical anticommutation relations for fermions. The "rotation angle" $\theta_{\vec{k}}$ is chosen so as to suppress "off-diagonal" terms of the form $\hat{\alpha}_{\vec{k}}^\dagger \hat{\beta}_{\vec{k}}$ in the transformed hamiltonian. This is accomplished by the choice

$$\tan 2\theta_{\vec{k}} = \frac{2 g_{\vec{k}}}{\varepsilon_{\vec{k}\uparrow} - \varepsilon_{\vec{k}+\vec{Q}\downarrow}} , \tag{2.81}$$

which gives two possible values for $\theta_{\vec{k}}$ differing by $\frac{\pi}{2}$, and hence two solutions for each $\vec{k}$. Thus, up to a trivial additive constant, the transformed hamiltonian takes the form

$$\hat{H}_{HF} = \sum_{\vec{k}} \left(\varepsilon_{\vec{k},-} \hat{\alpha}_{\vec{k}}^\dagger \hat{\alpha}_{\vec{k}} + \varepsilon_{\vec{k},+} \hat{\beta}_{\vec{k}}^\dagger \hat{\beta}_{\vec{k}} \right) , \tag{2.82}$$

where

$$\varepsilon_{\vec{k},\pm} = \frac{\varepsilon_{\vec{k}\uparrow} + \varepsilon_{\vec{k}+\vec{Q}\downarrow}}{2} \pm \sqrt{\left(\frac{\varepsilon_{\vec{k}\uparrow} - \varepsilon_{\vec{k}+\vec{Q}\downarrow}}{2} \right)^2 + g_{\vec{k}}^2} \tag{2.83}$$

are the two branches of the HF eigenvalue spectrum shown in Fig. 2.4. The corresponding wave functions are spinors with components

$$\phi_{\vec{k},-}(\vec{r}, s) = \frac{1}{L^{\frac{d}{2}}} \left[\cos \theta_{\vec{k}} e^{i\vec{k}\cdot\vec{r}} \delta_{s\uparrow} + \sin \theta_{\vec{k}} e^{i(\vec{k}+\vec{Q})\cdot\vec{r}} \delta_{s\downarrow} \right] ,$$

$$\phi_{\vec{k},+}(\vec{r}, s) = \frac{1}{L^{\frac{d}{2}}} \left[-\sin \theta_{\vec{k}} e^{i\vec{k}\cdot\vec{r}} \delta_{s\uparrow} + \cos \theta_{\vec{k}} e^{i(\vec{k}+\vec{Q})\cdot\vec{r}} \delta_{s\downarrow} \right] , \tag{2.84}$$

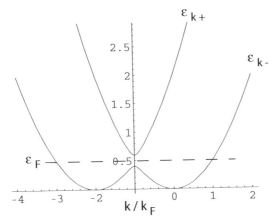

Fig. 2.4. HF eigenvalue spectra $\varepsilon_{\vec{k},\pm}$ (in units of $2\epsilon_F$) vs. k/k_F for the SDW state with $Q = 2k_F$ and $g_{\vec{k}} = 0.2\epsilon_F$. The eigenvalues are plotted for $\vec{k}$ parallel to $\vec{Q}$. The Fermi energy is such that only the "$-$" band is occupied. The diagonal part of the exchange energy, $V_{0\sigma}(\vec{k})$, has been assumed to be independent of $\vec{k}$ and has been subtracted for clarity.

where $\theta_{\vec{k}}$ is the solution of Eq. (2.81) that lies in the interval $-\frac{\pi}{2} \le \theta_{\vec{k}} < 0$ (see Exercise 9). A more compact way of writing these wave functions is

$$\phi_{\vec{k}b}(\vec{r}, s) = \frac{1}{L^{\frac{d}{2}}} \left[\cos\theta_{\vec{k}b} e^{i\vec{k}\cdot\vec{r}} \delta_{s\uparrow} + \sin\theta_{\vec{k}b} e^{i(\vec{k}+\vec{Q})\cdot\vec{r}} \delta_{s\downarrow} \right] , \qquad (2.85)$$

where $b = +, -$ is the branch index, and

$$\theta_{\vec{k}-} = \theta_{\vec{k}} , \qquad (2.86)$$
$$\theta_{\vec{k}+} = \theta_{\vec{k}} + \frac{\pi}{2} .$$

One can easily verify that, when the interaction energy is much smaller than the Fermi energy, the states of the lower branch are basically up-spin states for $k_\parallel \gg -Q/2$, and down-spin states for $k_\parallel \ll -Q/2$ (where $k_\parallel$ is the projection of $\vec{k}$ along $\vec{Q}$) but become strong mixtures of the two spin orientations in the wave vector region $k_\parallel \simeq -Q/2$.

The states (2.85) must be populated in order of increasing energy, until all the N electrons are accommodated. The particular occupation of the available states depends crucially on the magnitude of $\vec{Q}$: for $Q = 2k_F$ all the electrons can be accommodated in the lower band, $b = -$; for smaller values of Q the upper band will also be occupied.

It is straightforward to verify that the spin density of the single-determinant state constructed from the wave functions (2.85) has the form of Eq. (2.74) with an amplitude A given by (in units of $\frac{\hbar}{2}$):

$$A = \frac{1}{L^d} \sum_{\vec{k}b} n_{\vec{k}b} \sin 2\theta_{\vec{k}b} , \qquad (2.87)$$

where $n_{\vec{k}b}$ is 1 for the occupied states and 0 otherwise (see Exercise 10).

The final and crucial step in the implementation of the HF theory is the derivation of the equations that self-consistently determine the diagonal and off-diagonal parts of the HF potential, i.e., the functions $V_{0\sigma}(\vec{k})$ and $g_{\vec{k}}$. To accomplish this, we return to the general expression (2.17) for the HF potential. Setting $V^{ext} = 0$, $\alpha = \vec{k} + \vec{Q} \downarrow$ and $\beta = \vec{k} \uparrow$, and comparing with Eq. (2.76) we obtain the self-consistent equations

$$g_{\vec{k}} = V^{HF}_{\vec{k}+\vec{Q}\downarrow;\vec{k}\uparrow} = \sum_{\gamma \leq N}[v_{\vec{k}+\vec{Q}\downarrow\ \gamma\ \gamma\ \vec{k}\uparrow} - v_{\vec{k}+\vec{Q}\downarrow\ \gamma\ \vec{k}\uparrow\ \gamma}]\,, \tag{2.88}$$

and

$$V_{0\sigma}(\vec{k}) = V^{HF}_{\vec{k}\sigma;\vec{k}\sigma} = \sum_{\gamma \leq N}[v_{\vec{k}\sigma\ \gamma\ \gamma\ \vec{k}\sigma} - v_{\vec{k}\sigma\ \gamma\ \vec{k}\sigma\ \gamma}]\,, \tag{2.89}$$

where γ labels the HF eigenvalues in increasing order. We now substitute the explicit forms of the HF eigenfunctions (2.84) characterized by a wavevector $\vec{k}'$, and a branch index b, and evaluate the matrix elements of the coulomb interaction. The Hartree terms vanish by spin orthogonality, consistent with the fact that there is no net charge density in this state. Let us illustrate the calculation of the exchange terms. Setting $\gamma = \vec{k}'b$, and making use of the definition of the coulomb matrix elements we write

$$v_{\vec{k}+\vec{Q}\downarrow\ \vec{k}'b\ \vec{k}\uparrow\ \vec{k}'b} = \frac{1}{L^{2d}} \sum_{s_1,s_2} \int d\vec{r}_1 \int d\vec{r}_2 e^{-i(\vec{k}+\vec{Q})\cdot\vec{r}_1} \delta_{s_1\downarrow} \phi^*_{\vec{k}'b}(\vec{r}_2, s_2)$$

$$\times \frac{e^2}{|\vec{r}_1 - \vec{r}_2|} e^{i\vec{k}\cdot\vec{r}_2} \delta_{s_2\uparrow} \phi_{\vec{k}'b}(\vec{r}_1, s_1)$$

$$= \frac{1}{L^d} v_{\vec{k}-\vec{k}'} \sin\theta_{\vec{k}'b} \cos\theta_{\vec{k}'b}\,. \tag{2.90}$$

Substituting this into Eqs. (2.88) and (2.89) we obtain the self-consistency equations

$$g_{\vec{k}} = -\frac{1}{2L^d} \sum_{\vec{k}',b} n_{\vec{k}'b} v_{\vec{k}-\vec{k}'} \sin 2\theta_{\vec{k}'b}\,, \tag{2.91}$$

and

$$V_{0\uparrow}(\vec{k}) = -\frac{1}{L^d} \sum_{\vec{k}'b} n_{\vec{k}'b} v_{\vec{k}-\vec{k}'} \cos^2\theta_{\vec{k}'b}\,,$$

$$V_{0\downarrow}(\vec{k}+\vec{Q}) = -\frac{1}{L^d} \sum_{\vec{k}'b} n_{\vec{k}'b} v_{\vec{k}-\vec{k}'} \sin^2\theta_{\vec{k}'b}\,. \tag{2.92}$$

Equations (2.91) and (2.81) form a set of coupled nonlinear equations for the mixing angle $\theta_{\vec{k}b}$ *and* for the occupied region in momentum space. They can be solved numerically on a computer, even though, to the best of our knowledge, the results of such a calculation have not been published to date. An analytic solution can be obtained only in the special case of a one dimensional electron gas, which will be briefly considered in the next section.

Once $\theta_{\vec{k}b}$ is known, it is an easy matter to calculate the HF ground-state energy according to the general formula (2.24). The result is (see Exercise 11)

$$
E = \frac{\hbar^2}{2m} \sum_{\vec{k}b} n_{\vec{k}b} \left[k^2 \cos^2 \theta_{\vec{k}b} + (\vec{k} + \vec{Q})^2 \sin^2 \theta_{\vec{k}b} \right]
$$

$$
- \frac{1}{2L^d} \sum_{\vec{k},\vec{k}',b} n_{\vec{k}b} n_{\vec{k}'b} v_{\vec{k}-\vec{k}'} \cos^2(\theta_{\vec{k}b} - \theta_{\vec{k}'b}) . \tag{2.93}
$$

2.6.2 Spin density wave instability with contact interactions in one dimension

An analytic solution of the HF equations for a spiral spin density wave can be obtained in the case of a one dimensional system of electrons interacting through the contact interaction

$$
\hat{H}_{e-e} = \gamma e^2 \sum_{i>j} \delta(x_i - x_j) . \tag{2.94}
$$

Because the Fourier transform of the interaction $v_q = \gamma e^2$ is independent of momentum transfer, a quick inspection of Eqs. (2.91) and (2.92) reveals that in this case both $g_{\vec{k}}$ and $V_{0\sigma}(\vec{k})$ are independent of $\vec{k}$ and σ. This simplifies the problem enormously.

For simplicity we specialize our analysis to a SSDW with a predetermined value of the magnitude of the wave vector, i.e. $Q = 2k_F$. In this case only the lower band with $b = -$ will be occupied, with wave vectors ranging from $-3k_F$ to $+k_F$ (this is the situation depicted in Fig. 2.4).

Of particular interest is the self-consistent value of $g_{\vec{k}} \equiv g$, which controls the splitting of the bands at $k = -k_F$. This quantity can be directly obtained from Eq. (2.91). To this end we find useful to first derive from Eq. (2.81) the following expression for $\sin 2\theta_{\vec{k}-}$:

$$
\sin 2\theta_{\vec{k}-} = -\frac{g}{\sqrt{g^2 + (\mathcal{E}_{\vec{k}} - \mathcal{E}_{\vec{k}+\vec{Q}})^2}} . \tag{2.95}
$$

At this point Eq. (2.91) can be immediately cast in the form:

$$
1 = \frac{\gamma e^2}{4\pi} \int_{-3k_F}^{k_F} \frac{dk}{\sqrt{g^2 + \left(\frac{\hbar^2 k^2}{2m} - \frac{\hbar^2 (k+2k_F)^2}{2m} \right)^2}} , \tag{2.96}
$$

where we have made use of the fact that, in this case, $V_{0\sigma}(\vec{k})$ being independent of $\vec{k}$ and σ, the difference $\mathcal{E}_{\vec{k}} - \mathcal{E}_{\vec{k}+\vec{Q}}$ reduces to the difference between the corresponding free-particle energies.

The integral in Eq. (2.96) is elementary and a few straightforward manipulations lead to the expression:

$$
g = \frac{8\epsilon_F}{\sinh \frac{16\epsilon_F}{n\gamma e^2}} , \tag{2.97}
$$

where we have made use of the relation $k_F = \frac{\pi}{2}n$.

We immediately notice that the solution exists for arbitrarily small values of the interaction strength γ. For small values of γ the expression for g reduces to

$$g \simeq 16\epsilon_F \, e^{-\frac{16\epsilon_F}{n\gamma e^2}} . \tag{2.98}$$

This expression cannot be expanded in a power series of γ: hence the SSDW state cannot be obtained perturbatively from the non interacting one.

2.6.3 Proof of Overhauser's instability theorem

We now turn to an explicit proof of the SDW instability, i.e. the statement that *for all values of r_s, one can construct a Slater determinant of SDW symmetry, such that its energy is lower than the energy of a Slater determinant made up of plane waves with wave vector $\vec{k}$, with $k \le k_F$*. This implies that the homogeneous paramagnetic state of the electron gas is not a locally stable HF state.

The original proof was given by A. W. Overhauser in a classic 1962 paper (Overhauser, 1962). Here we present what we believe is a simpler proof along the lines suggested by Martin and Fedders (Martin and Fedders, 1966). The overall strategy is simple: following Overhauser's original idea we construct a Slater determinant out of N single particle wave functions, which are admixtures of a plane wave of wave vector $\vec{k}$ with $k < k_F$ and spin projection σ (along an arbitrary quantization axis) with plane waves of wave vector $\vec{k} \pm \vec{Q}$ of magnitude greater than k_F and the same spin orientation. These wave functions have the form

$$\phi_{\vec{k}}(\vec{r}, \sigma) \propto e^{i\vec{k}\cdot\vec{r}} + A_{\vec{k}+\vec{Q},\vec{k},\sigma} e^{i(\vec{k}+\vec{Q})\cdot\vec{r}} + A_{\vec{k}-\vec{Q},\vec{k},\sigma} e^{i(\vec{k}-\vec{Q})\cdot\vec{r}}, \tag{2.99}$$

where $A_{\vec{k}\pm\vec{Q},\vec{k},\sigma}$ is the mixing amplitude of the plane wave state of wave vector $\vec{k} \pm \vec{Q}$ into the plane wave state of wave vector $\vec{k}$. Here $\vec{Q}$ represents a fixed suitable wave vector whose optimal magnitude must eventually be determined by minimizing the total energy of the system.

To compare the energy of the Slater determinant constructed from the wavefunctions (2.99) to that of a Slater determinant of simple plane waves, we make use of Eq. (2.68), which expresses the change in energy in terms of the change in the density matrix. The latter is given by

$$\delta\rho_{\vec{k}\pm\vec{Q}\sigma,\vec{k}\sigma} = \delta\rho_{\vec{k}\sigma,\vec{k}\pm\vec{Q}\sigma} = A_{\vec{k}\pm\vec{Q},\vec{k}\sigma} = A_{\vec{k}\sigma,\vec{k}\pm\vec{Q}\sigma}, \tag{2.100}$$

where $k < k_F$ and $|\vec{k} \pm \vec{Q}| > k_F$. However $\delta\rho_{\alpha\beta} = 0$ for any other pair of states α and β which, while being on opposite sides of the Fermi surface, do not satisfy the condition that their wave vectors differ by $\pm\vec{Q}$. As for the matrix elements between states which lie on the *same side* of the Fermi surface, they are given by Eqs. (2.66) (where γ is on the opposite side of α and β) and are therefore of second order in $A_{\vec{k}\sigma}$. The presence of these matrix elements assures that the distorted state is still a single Slater determinant, but does not affect the calculation of the energy to second order in $A_{\vec{k}\sigma}$ since the latter depends only on matrix elements between states on opposite sides of the Fermi surface (see Eq. (2.68)).

Physical considerations presented in the previous section led us to the conclusion that the magnitude of $\vec{Q}$ must be close to $2k_F$ as this choice allows for the admixture of plane wave states with nearly equal unperturbed energy. In the following we choose $\vec{Q} = 2k_F\vec{e}_z$ so that the condition $|\vec{k} \pm \vec{Q}| > k_F$ is automatically satisfied when $k < k_F$. With this choice of $\vec{Q}$ it is easy to verify that the only nonvanishing elements of $\delta\rho$ have the form

$$\delta\rho_{\vec{k}+\vec{Q}/2\sigma,\vec{k}-\vec{Q}/2\sigma} = \delta\rho_{\vec{k}-\vec{Q}/2\sigma,\vec{k}+\vec{Q}/2\sigma} = (n_{\vec{k}+\vec{Q}/2} + n_{\vec{k}-\vec{Q}/2})A_{\vec{k}\sigma} , \qquad (2.101)$$

where $n_{\vec{k}} = \Theta(k_F - k)$ are the occupation numbers of the homogeneous paramagnetic state, and we have defined

$$A_{\vec{k}\sigma} \equiv A_{\vec{k}+\vec{Q}/2,\vec{k}-\vec{Q}/2,\sigma} . \qquad (2.102)$$

The density modulation implied by Eq. (2.101) is, to first order in $\delta\rho$, given by

$$\delta n(\vec{r}) = 2 \sum_{\vec{k}\sigma} \delta\rho_{\vec{k}+\vec{Q}/2\sigma,\vec{k}-\vec{Q}/2\sigma} \cos \vec{Q} \cdot \vec{r} , \qquad (2.103)$$

while the modulation in the z-component of the spin density (expressed in units of $\hbar/2$) is

$$\delta S_z(\vec{r}) = 2 \sum_{k<k_F,\sigma} \sigma \, \delta\rho_{\vec{k}+\vec{Q}/2\sigma,\vec{k}-\vec{Q}/2\sigma} \cos \vec{Q} \cdot \vec{r} . \qquad (2.104)$$

Thus, we see that Eq. (2.100) truly describes a CDW or a SDW of vector $\vec{Q}$ depending on whether $A_{\vec{k}\uparrow} = A_{\vec{k}\downarrow}$ (CDW) or $A_{\vec{k}\uparrow} = -A_{\vec{k}\downarrow}$ (SDW).

Now to calculate the energy change we substitute Eq. (2.101) into Eq. (2.68), which we rewrite here for convenience with labels $\alpha = \vec{k}\sigma$, $\beta = \vec{p}\tau$, $\gamma = \vec{p}'\tau'$, and $\delta = \vec{k}'\sigma'$:

$$\Delta^{(2)}E_{HF} = \frac{1}{2} \sum_{(\vec{k}\sigma,\vec{p}\tau)} \frac{\varepsilon_{\vec{p}} - \varepsilon_{\vec{k}}}{n_{\vec{k}} - n_{\vec{p}}} \delta\rho_{\vec{k}\sigma,\vec{p}\tau} \delta\rho_{\vec{p}\tau,\vec{k}\sigma}$$

$$+ \frac{1}{2} \sum_{(\vec{k}\sigma,\vec{k}'\sigma')} \sum_{(\vec{p}\tau,\vec{p}'\tau')} (v_{\vec{k}\sigma\ \vec{p}\tau\ \vec{p}'\tau'\ \vec{k}'\sigma'} - v_{\vec{k}\sigma\ \vec{p}\tau\ \vec{k}'\sigma'\ \vec{p}'\tau'}) \delta\rho_{\vec{k}\sigma,\vec{k}'\sigma'} \delta\rho_{\vec{p}\tau,\vec{p}'\tau'} , \tag{2.105}$$

where $\varepsilon_{\vec{k}}$ are the HF eigenvalues *of the uniform paramagnetic phase*. Substituting in the above equation the following expressions for the matrix elements of the coulomb interaction

$$v_{\vec{k}-\vec{Q}/2\sigma\ \vec{p}+\vec{Q}/2\tau\ \vec{p}-\vec{Q}/2\tau\ \vec{k}+\vec{Q}/2\sigma} = \frac{1}{L^d} v_{\vec{Q}} , \qquad (2.106)$$

and

$$v_{\vec{k}-\vec{Q}/2\sigma\ \vec{p}+\vec{Q}/2\tau\ \vec{k}+\vec{Q}/2\sigma\ \vec{p}-\vec{Q}/2\tau} = \frac{1}{L^d} v_{\vec{k}-\vec{p}}\delta_{\sigma\tau} , \qquad (2.107)$$

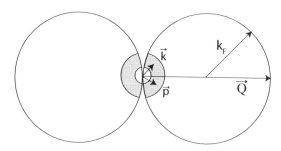

Fig. 2.5. Integration region for the calculation of the SDW exchange energy. $\vec{k}$ and $\vec{p}$ are confined to the shaded region where $\mu k_F < k, p < \mu e k_F$.

we obtain

$$\Delta^{(2)}E_{HF} = \sum_{\vec{k}\sigma}[n_{\vec{k}+\vec{Q}/2} + n_{\vec{k}-\vec{Q}/2}]\frac{\varepsilon_{\vec{k}-\vec{Q}/2} - \varepsilon_{\vec{k}+\vec{Q}/2}}{n_{\vec{k}+\vec{Q}/2} - n_{\vec{k}-\vec{Q}/2}}A_{\vec{k}\sigma}^2$$

$$+ \frac{2}{L^d}\sum_{\vec{k}\sigma, \vec{p}\tau}\left(v_{\vec{Q}} - v_{\vec{k}-\vec{p}}\delta_{\sigma\tau}\right)[n_{\vec{k}+\vec{Q}/2} + n_{\vec{k}-\vec{Q}/2}][n_{\vec{p}+\vec{Q}/2} + n_{\vec{p}-\vec{Q}/2}]A_{\vec{k}\sigma}A_{\vec{p}\tau}\,.$$

$$(2.108)$$

Notice that the factors $[n_{\vec{k}+\vec{Q}/2} + n_{\vec{k}-\vec{Q}/2}]$ and $[n_{\vec{p}+\vec{Q}/2} + n_{\vec{p}-\vec{Q}/2}]$ in Eq. (2.108) restrict the $\vec{k}$ and $\vec{p}$ integrals to the region of momentum space shown in Fig. 2.5, i.e. to two spheres of radius k_F touching at the origin. The question is whether the energy change (2.108) can be made negative by a suitable choice of the amplitudes $A_{\vec{k}\sigma}$. First of all, we notice that by choosing a pure spin–density modulation, i.e.,

$$A_{\vec{k}\uparrow} = -A_{\vec{k}\downarrow} \equiv A_{\vec{k}} \qquad (2.109)$$

we eliminate the Hartree term – the term proportional to $v_{\vec{Q}}$ in Eq. (2.108) – which is positive and opposes the instability. While the exchange contribution is the same in the CDW and in the SDW, the Hartree contribution is present only in the CDW. It is therefore evident that the SDW instability is much more likely to occur than the CDW instability.[16]

We now explicitly construct a pure SDW distortion which, *in three dimensions*, makes $\Delta^{(2)}E_{HF}$ negative. Try

$$A_{\vec{k}} = \begin{cases} \frac{(bk_F)^2}{\ln\frac{2}{b}}\left|\frac{n_{\vec{k}+\vec{Q}/2} - n_{\vec{k}-\vec{Q}/2}}{k^2\cos\theta}\right| & bk_F < k < bek_F, \\ 0 & \text{otherwise}, \end{cases} \qquad (2.110)$$

where θ is the angle between $\vec{k}$ and $\vec{Q}$, b is an arbitrary positive number, which we take to be $\ll 1$, and e is the base of natural logarithms.

---

[16] This argument is valid for the standard jellium model. In a *perfectly deformable* jellium model, in which the density of the positively charged background can neutralize the electronic charge density at no energy cost, the CDW and SDW states have exactly the same energy in the HF approximation.

The region of integration in $\vec{k}$-space is shown in Fig. 2.5. The integrand is nonzero only when $\vec{k} + \frac{\vec{Q}}{2}$ and $\vec{k} - \frac{\vec{Q}}{2}$ are on opposite sides of the Fermi surface and the condition $bk_F < k < bek_F$ is satisfied. From the geometry of Fig. 2.5 we see that the angle θ between $\vec{k}$ and $\vec{Q}$ is thus restricted to the region $|\cos\theta| > k/2k_F$. In the limit of small b, this can safely be replaced by the simpler condition

$$|\cos\theta| > \frac{b}{2} , \tag{2.111}$$

since the error caused by the replacement is of higher order in b. For the same reason, we can approximate

$$\left|\varepsilon_{\vec{k}-\vec{Q}/2} - \varepsilon_{\vec{k}+\vec{Q}/2}\right| \simeq \frac{2\hbar^2 k_F k |\cos\theta|}{m}\left(1 + \frac{\alpha r_s}{2\pi}\ln\frac{2k_F}{k|\cos\theta|}\right) , \tag{2.112}$$

where $\alpha = \left(\frac{4}{9\pi}\right)^{\frac{1}{3}} \simeq 0.52$. (The second term in the brackets accounts for the logarithmic divergence in the slope of the exchange energy at $k = k_F$)

Substituting Eqs. (2.110) and (2.112) into the first term on the right-hand side of Eq. (2.108) and making use of Eq. (2.111), we obtain

$$12N\epsilon_F \frac{b^4}{\left[\ln\frac{2}{b}\right]^2}\int_b^{be}\frac{dx}{x}\int_{b/2}^1\frac{dy}{y}\left(1 + \frac{\alpha r_s}{2\pi}\ln\frac{2}{xy}\right) , \tag{2.113}$$

where $x \equiv k/k_F$ and $y \equiv \cos\theta$. These integrals can be evaluated analytically. To leading order in b the y-integral yields $\frac{\alpha}{2}\left[\ln\left(\frac{2}{b}\right)\right]^2$ and the complete result is

$$\frac{6N\epsilon_F\alpha r_s}{2\pi}b^4 + \mathcal{O}\left(\frac{b^4}{|\ln b|}\right) . \tag{2.114}$$

Now we show that the second term on the right-hand side of Eq. (2.108) is proportional to $b^4\ln\left(\frac{1}{b}\right)$ and therefore dominates in the limit $b \to 0$, leading to the SDW instability. We begin by writing the Fourier transform of the coulomb interaction in the form

$$v_{\vec{k}-\vec{p}} = \frac{4\pi e^2}{k_F^2(x^2 + x'^2 - 2xx'\cos\psi)} , \tag{2.115}$$

where $x = k/k_F$, $x' = p/k_F$, and ψ is the angle between $\vec{k}$ and $\vec{p}$. The latter is given by

$$\cos\psi = yy' + \sqrt{(1 - y^2)(1 - y'^2)}\cos(\varphi - \varphi') , \tag{2.116}$$

where $y = \cos\theta$ and φ is the azimuthal angle of $\vec{k}$ relative to an arbitrary reference plane. The primed variables are similarly defined in terms of $\vec{p}$.

Because the trial amplitude (2.110) does not depend on the azimuthal angle, the coulomb interaction in the second term of Eq. (2.108) is automatically averaged over φ and φ'. This azimuthal averaging is easily carried out with the help of the integral

$$\frac{1}{2\pi}\int_0^{2\pi}\frac{d\varphi}{a + b\cos\varphi} = \frac{1}{\sqrt{a^2 - b^2}} \quad , \quad |b| \leq |a| . \tag{2.117}$$

We obtain

$$
\begin{aligned}
\left\langle v_{\vec{k}-\vec{p}}\right\rangle &\equiv \int_0^{2\pi}\frac{d\varphi}{2\pi}\int_0^{2\pi}\frac{d\varphi'}{2\pi}\,v_{\vec{k}-\vec{p}} \\
&= \frac{4\pi e^2}{k_F^2\sqrt{(x^2-x'^2)^2+4xx'[xx'(y^2+y'^2)-yy'(x^2+x'^2)]}}\,.
\end{aligned}
\tag{2.118}
$$

It is evident that we *underestimate* the exchange energy by restricting the integral (2.108) to regions in which y and y' have the same sign (i.e., $\vec{k}$ and $\vec{p}$ both lie within the right or the left sphere in Fig. 2.5). In this region the coulomb interaction is underestimated by neglecting the last term in the denominator of Eq. (2.118), which is negative:

$$
\left\langle v_{\vec{k}-\vec{p}}\right\rangle \geq \frac{4\pi e^2}{k_F^2\sqrt{(x^2-x'^2)^2+4x^2x'^2(y^2+y'^2)}}\,.
\tag{2.119}
$$

Substituting this in Eq. (2.108) we see that the absolute value of the exchange energy is certainly *larger* than

$$
\begin{aligned}
&\frac{12N\epsilon_F\alpha r_s}{\pi}\frac{b^4}{\left[\ln\frac{2}{b}\right]^2}\int_b^{be}dx\int_b^{be}dx'\int_{b/2}^1\frac{dy}{y}\int_{b/2}^1\frac{dy'}{y'} \\
&\times\frac{1}{\sqrt{(x^2-x'^2)^2+4x^2x'^2(y^2+y'^2)}}\,.
\end{aligned}
\tag{2.120}
$$

It is not difficult to see that this expression diverges as $\left[\ln\frac{1}{b}\right]^3$ for $b\to 0$. Two logarithms arise from the lower limits of the y and y' integrations; the third and decisive one arises from the region where $x\simeq x'$. For the purpose of evaluating the integral to leading order in b it is legitimate to put $y=y'=\frac{b}{2}$ and $x^2-x'^2\simeq 2x(x-x')$ in the square root of Eq. (2.120). In addition we note that $\int_b^{be}dx\int_b^{be}dx'\ldots=2\int_b^{be}dx\int_b^x dx'\ldots$. Thus Eq. (2.120) becomes

$$
\frac{12N\epsilon_F\alpha r_s}{\pi}\frac{b^4}{\left[\ln\frac{2}{b}\right]^2}\int_b^{be}\frac{dx}{x}\int_b^x\frac{dx'}{\sqrt{(x-x')^2+\frac{x^2b^2}{2}}}\int_{b/2}^1\frac{dy}{y}\int_{b/2}^1\frac{dy'}{y'}\,,
\tag{2.121}
$$

an expression that can be evaluated analytically. To leading order in b the result is readily found to be

$$
\frac{12N\epsilon_F\alpha r_s}{\pi}b^4\ln\frac{\sqrt{2}}{b}\,.
\tag{2.122}
$$

Due to the additional logarithmic factor the exchange energy gain dominates the energy of the SDW state for sufficiently small b. Since higher order contributions to the energy are negligible in the $b\to 0$ limit, the SDW instability is proved. The instability occurs for arbitrary values of the density. Note, however, that the long range of the coulomb interaction, and specifically the fact that $v_k\propto\frac{1}{k^2}$ for small k, is an essential ingredient of the proof.

One should also appreciate the fact that the trial state that we have constructed in this section is unrelated, in fact essentially orthogonal to, the state one would obtain by perturbing the paramagnetic state with a periodic magnetic field of wave vector $\vec{Q}$. The non-perturbative character of the trial state is evident from the fact that the energy lowering is a nonanalytic function of the SSDW amplitude A. The latter is proportional to b^3 (as can be seen from Eq. (2.104)), and therefore $\Delta E \propto -A^{4/3} \ln A$ in contrast to the standard A^2 behavior one would expect for a perturbative state.

2.7 BCS non number-conserving mean field theory

In order to explain the superconductivity of metals Bardeen, Cooper, and Schrieffer (BCS, 1957) introduced what was at that time a revolutionary form of mean field theory, based on the quadratic hamiltonian

$$\hat{H}_{BCS} = \sum_{\vec{k}} \left[\xi_{\vec{k}} \left(\hat{a}^{\dagger}_{\vec{k}\uparrow} \hat{a}_{\vec{k}\uparrow} + \hat{a}^{\dagger}_{-\vec{k}\downarrow} \hat{a}_{-\vec{k}\downarrow} \right) + \Delta_{\vec{k}} \left(\hat{a}^{\dagger}_{\vec{k}\uparrow} \hat{a}^{\dagger}_{-\vec{k}\downarrow} + \hat{a}_{-\vec{k}\uparrow} \hat{a}_{\vec{k}\uparrow} \right) \right], \quad (2.123)$$

where $\xi_k = \frac{\hbar^2 k^2}{2m} - \mu$ is the kinetic energy of an electron relative to the chemical potential μ. The most striking feature of this hamiltonian is that it does not conserve particle number (hence the need for introducing the chemical potential). The unusual term $\sum_{\vec{k}} \Delta_{\vec{k}} \hat{a}^{\dagger}_{\vec{k}\uparrow} \hat{a}^{\dagger}_{-\vec{k}\downarrow}$ creates a pair of electrons with opposite momenta and opposite spin orientations, rather than transferring a single electron from one state to another.[17] For this reason, this term is usually referred to as a *pairing potential*.

The ground-state of $\hat{H}_{BCS}$ consists of a coherent superposition of Slater determinants in which the two partner states of each pair $(\vec{k} \uparrow, -\vec{k} \downarrow)$ are either simultaneously occupied or simultaneously empty. It is in this way that one takes maximal advantage of the pairing potential term. Excitations are created by breaking pairs, i.e., by introducing configurations in which one partner state is occupied while the other is empty. Such excitations are separated from the ground-state by a finite energy gap. Although neither the ground-state nor the excited states have a definite particle number, it turns out that the fluctuations about the average value N are of order $\sqrt{N}$ and therefore negligible in the thermodynamic limit. The average value of N is fixed by appropriately choosing the chemical potential.

To prove the assertions we have just made we need first of all to diagonalize $\hat{H}_{BCS}$. This is best done with the help of a unitary transformation to a new set of fermion operators $\hat{\gamma}_{\vec{k}\uparrow}$, $\hat{\gamma}_{\vec{k}\downarrow}$ such that

$$\hat{a}_{\vec{k}\uparrow} = u_{\vec{k}} \hat{\gamma}_{\vec{k}\uparrow} + v_{\vec{k}} \hat{\gamma}^{\dagger}_{-\vec{k}\downarrow}$$

$$\hat{a}_{-\vec{k}\downarrow} = u_{\vec{k}} \hat{\gamma}_{-\vec{k}\downarrow} - v_{\vec{k}} \hat{\gamma}^{\dagger}_{\vec{k}\uparrow}, \quad (2.124)$$

---

[17] When acting on the vacuum this term creates a pair of electrons described by the wave function

$$\Delta(\vec{r}) = \sum_{\vec{k}} \Delta_{\vec{k}} e^{i\vec{k}\cdot\vec{r}},$$

where $\vec{r}$ is the relative coordinate of the two electrons. The center of mass of the pair is uniformly distributed in space.

with $u_{\vec{k}}^2 + v_{\vec{k}}^2 = 1$. After expressing $\hat{H}_{BCS}$ in terms of the $\hat{\gamma}$'s we choose the coefficients $u_{\vec{k}}$ and $v_{\vec{k}}$ in such a way as to eliminate terms of the form $\hat{\gamma}^\dagger \hat{\gamma}^\dagger$ and $\hat{\gamma}\hat{\gamma}$: this yields

$$\hat{H}_{BCS} = \sum_{\vec{k}} \varepsilon_{\vec{k}} \left(\hat{\gamma}_{k\uparrow}^\dagger \hat{\gamma}_{k\uparrow} + \hat{\gamma}_{-k\downarrow}^\dagger \hat{\gamma}_{-k\downarrow} \right) + \text{constant} , \qquad (2.125)$$

where

$$\varepsilon_{\vec{k}} = \sqrt{\xi_{\vec{k}}^2 + \Delta_{\vec{k}}^2} ,$$

$$u_{\vec{k}} = \sqrt{\frac{1}{2}\left(1 + \frac{\xi_{\vec{k}}}{\varepsilon_{\vec{k}}}\right)} ,$$

$$v_{\vec{k}} = \sqrt{\frac{1}{2}\left(1 - \frac{\xi_{\vec{k}}}{\varepsilon_{\vec{k}}}\right)} . \qquad (2.126)$$

It is then evident that the ground-state of $\hat{H}_{BCS}$ is the state that is annihilated by all the $\hat{\gamma}_{k\sigma}$. This state has the form

$$|\psi_{BCS}\rangle = \prod_{\vec{k}} \left(u_{\vec{k}} + v_{\vec{k}}\hat{a}_{\vec{k}\uparrow}^\dagger \hat{a}_{-\vec{k}\downarrow}^\dagger \right) |0\rangle , \qquad (2.127)$$

where $|0\rangle$ is the vacuum, $v_{\vec{k}}$ is the probability amplitude of finding the states $(\vec{k}\uparrow, -\vec{k}\downarrow)$ simultaneously occupied, and $u_{\vec{k}}$ is the amplitude of finding them simultaneously empty. As we have already remarked, this many-body state cannot be written as a single Slater determinant. Rather, one can show that its projection onto the N-electron subspace yields an antisymmetrized product of pair wavefunctions (N even):

$$\psi_{BCS}^{(N)}(\vec{r}_1 s_1 \ldots \vec{r}_N s_N) = \mathcal{A}\left[\Delta(\vec{r}_1 - \vec{r}_2)\chi(s_1, s_2)\ldots\Delta(\vec{r}_{N-1} - \vec{r}_N)\chi(s_{N-1}, s_N)\right] , \quad (2.128)$$

where χ is the singlet spin wave function, $\Delta(\vec{r})$ is the Fourier transform of $\Delta_{\vec{k}}$ (see footnote[17]), and the antisymmetrization operator $\mathcal{A}$ sums over all the permutations of the coordinates and the spin with the appropriate signs. Excited states are obtained by applying the rising operators $\hat{\gamma}_{k\uparrow}^\dagger$ and $\hat{\gamma}_{-\vec{k}\downarrow}^\dagger$ to the ground-state (2.128). This results in breaking a pair (leaving an unpaired electron either in $\vec{k}\uparrow$ or in $-\vec{k}\downarrow$) and increasing the energy by $\varepsilon_{\vec{k}} \geq |\Delta_{\vec{k}}|$. Thus, the excitation spectrum has a gap.

Now that we have the mean field ground-state we can calculate the expectation value of the many-electron hamiltonian $\hat{H}$, and minimize it with respect to $\Delta_{\vec{k}}$. Following tradition we shall consider a hamiltonian in which the electrons interact only via an effective short-range attractive interaction, which arises from the ionic screening of the coulomb interaction. It is a simple exercise in second quantization to show that

$$\langle\psi_{BCS}|\hat{H}|\psi_{BCS}\rangle = 2\sum_{\vec{k}} \xi_{\vec{k}} v_{\vec{k}}^2 + \frac{1}{2}\sum_{\vec{k}\vec{k}'} \left[V_{\vec{0}} - V_{\vec{k}-\vec{k}'}\right] v_{\vec{k}}^2 v_{\vec{k}'}^2$$

$$+ \frac{1}{2}\sum_{\vec{k}\vec{k}'} V_{\vec{k}-\vec{k}'} u_{\vec{k}} v_{\vec{k}} u_{\vec{k}'} v_{\vec{k}'} , \qquad (2.129)$$

where $V_{\vec{q}}$ is the Fourier transform of the effective electron-electron interaction, here denoted by a capital V to avoid confusion with the amplitude $v_{\vec{k}}$. It is customary to ignore the ordinary Hartree–Fock term (the second term on the right-hand side) on the grounds that the effective interaction is short-ranged and therefore its Fourier transform is weakly dependent on momentum transfer. On the other hand, the pairing energy is negative for an attractive interaction and can overcome the increase in kinetic energy. Indeed, by minimizing the energy with respect to $\Delta_{\vec{k}}$ we find the *gap equation*

$$\Delta_{\vec{k}} = -\sum_{\vec{k}'} V_{\vec{k}-\vec{k}'} \frac{\Delta_{\vec{k}'}}{2\varepsilon_{\vec{k}'}} , \qquad (2.130)$$

which can be shown to have a solution for any attractive $V_{\vec{k}}$, no matter how small. A straigthforward calculation of the energy shows that this solution has indeed lower energy than the ordinary HF solution. At finite temperature, however, one must minimize the free energy rather than the energy: then the BCS paired state is found to be advantageous only up to a critical temperature T_c.

The BCS state has many striking properties, the most important one being superconductivity itself. A current-carrying state can be constructed by simply shifting the center of mass momentum of all the pairs so that the paired states are $(\vec{k} + \vec{q}/2 \uparrow, -\vec{k} + \vec{q}/2 \downarrow)$ rather than $(\vec{k} \uparrow, -\vec{k} \downarrow)$ (try this!). Once the system is locked in this state it is virtually impossible for impurities, phonons etc. acting on individual electrons to change the total momentum of the system. In fact, scattering processes cannot alter the equilibrium distribution of the quasiparticles compatible with a given value of the total momentum. Thus, although the current-carrying state is not the absolute equilibrium state, its lifetime is infinite in the thermodynamic limit.

Another defining property of the superconducting state is the *Meissner effect*, i.e., the exclusion of the magnetic flux from the interior of the sample. This follows from the fact that the response of the paired state to a magnetic field (represented by a transverse vector potential) is purely diamagnetic (see Section 3.4 for an accurate definition of paramagnetic and diamagnetic responses of an electronic system). The large diamagnetic currents induced at the surface of the sample effectively screen the magnetic field in its interior and produce the Meissner effect.

2.8 Local approximations to the exchange

The exchange potential in HF theory is a *nonlocal potential* because it acts on one-electron wave functions as an integral operator, rather than a simple multiplicative operator. A question of both fundamental and practical interest is, to what extent can the physics of the exact nonlocal exchange be captured by an approximate *local* exchange potential? We will see that local approximations to the exchange often work surprisingly well, yielding energies almost as good as HF, without suffering from some of the characteristic drawbacks of that theory.[18]

---

[18] For example, the local exchange potential in a homogeneous electron liquid is just a constant, which uniformly shifts the single-particle eigenvalues, without affecting the bandwidth or the heat capacity (see discussion at the end of Section 2.4.1).

The fundamental reason for this striking fact will be discussed in Chapter 8 (Section 8.8), where we show how the inclusion of correlation effects suppresses the momentum dependence of the self-energy, making it effectively equivalent to a local potential.[19]

Attempts to derive *local approximations* to the exchange potential date back to the early 1950s, but it is only with the advent of the density functional theory, which proposes to derive the exact ground-state density of the system from a local exchange-correlation potential, that the fundamental importance of this question has come clearly into focus.

In what follows we describe the two main approaches to "local exchange", leading to the exchange potential (Slater, 1951) and to the *optimized effective potential* (Sharp and Horton, 1953), respectively. The full significance of these approximations will be better appreciated in Chapter 7, where the problem is revisited in the context of the density functional theory.

2.8.1 Slater's local exchange potential

To understand Slater's approach consider first a system of *noninteracting* electrons in a local external potential $V_{ext}(\vec{r})$. To avoid unnecessary complication in the notation we ignore in this section the spin degree of freedom. Let $\phi_\alpha(\vec{r})$ be the one-electron wave functions (also referred to as *orbitals*) for this system. The potential energy is given by

$$E_{pot} = \sum_\alpha n_\alpha \int V_{ext}(\vec{r})\phi_\alpha^*(\vec{r})\phi_\alpha(\vec{r})d\vec{r}, \qquad (2.131)$$

where $n_\alpha = 1$ for occupied orbitals and $n_\alpha = 0$ for empty ones. Notice that the external potential can be expressed as the functional derivative of the potential energy with respect to the occupied orbitals in the following manner:

$$V_{ext}(\vec{r}) = \frac{1}{\phi_\alpha^*(\vec{r})} \frac{\delta E_{pot}}{\delta\phi_\alpha(\vec{r})}, \qquad (2.132)$$

where the right-hand side is independent of α for $\alpha \leq N$. In analogy to this, one might think to define a local exchange potential as the functional derivative of the exchange energy

$$E_x = -\frac{1}{2}\sum_{\alpha\beta} n_\alpha n_\beta \int d\vec{r} \int d\vec{r}' \phi_\beta^*(\vec{r})\phi_\alpha^*(\vec{r}')\frac{e^2}{|\vec{r}-\vec{r}'|}\phi_\beta(\vec{r}')\phi_\alpha(\vec{r}) \qquad (2.133)$$

with respect to the occupied orbitals. Unfortunately, this gives

$$\begin{aligned}
V_{x,\alpha}(\vec{r}) &\equiv \frac{1}{\phi_\alpha^*(\vec{r})} \frac{\delta E_x}{\delta\phi_\alpha(\vec{r})} \\
&= -\frac{1}{\phi_\alpha^*(\vec{r})}\sum_\beta n_\beta \int d\vec{r}' \phi_\beta^*(\vec{r})\phi_\alpha^*(\vec{r}')\frac{e^2}{|\vec{r}-\vec{r}'|}\phi_\beta(\vec{r}'),
\end{aligned} \qquad (2.134)$$

---

[19] On the other hand, one must never forget that the possibility of certain exotic phases of the electron liquid (e.g., Overhauser's spin density wave – see Section 2.6.3) would not have been discovered without taking into account the nonlocality of the exchange potential.

which still depends on the state label α ($\leq N$) and therefore is not a local potential. Notice, however, that in the limit of $\vec{r} \to \infty$ (we assume a finite, if possibly very large system), this potential reduces to

$$V_{x,\alpha}(\vec{r} \to \infty) = \begin{cases} -\frac{e^2}{r}, & (\alpha \text{ occupied}), \\ 0, & (\alpha \text{ empty}). \end{cases} \tag{2.135}$$

In order to obtain a local potential, Slater suggested to take the *average* of $V_{x,\alpha}(\vec{r})$ over the occupied states: this yields the *Slater exchange potential*

$$V_x^S(\vec{r}) = \frac{\sum_\alpha n_\alpha |\phi_\alpha(\vec{r})|^2 V_{x,\alpha}(\vec{r})}{\sum_\alpha n_\alpha |\phi_\alpha(\vec{r})|^2}. \tag{2.136}$$

The Slater potential has an elegant physical interpretation in terms of the *pair correlation function* $g(\vec{r}, \vec{r}')$ defined in Eqs. (1.99) and (1.100). To arrive at it, we begin with a very useful observation, namely, we notice that the pair correlation function of any state described by a single Slater determinant can be expressed in terms of its density matrix, $\rho(\vec{r}, \vec{r}') = \sum_\alpha n_\alpha \phi_\alpha^*(\vec{r})\phi_\alpha(\vec{r}')$, in the following manner:

$$g(\vec{r}, \vec{r}') = 1 - \frac{|\rho(\vec{r}, \vec{r}')|^2}{n(\vec{r})n(\vec{r}')}, \tag{2.137}$$

where $n(\vec{r}) \equiv \rho(\vec{r}, \vec{r})$ is the electronic density (see Exercise 12). Armed with this result, it is not difficult to verify that (Exercise 13)

$$V_x^S(\vec{r}) = \int \frac{e^2}{|\vec{r} - \vec{r}'|} n(\vec{r}')[g(\vec{r}, \vec{r}') - 1] d\vec{r}'. \tag{2.138}$$

Thus, the Slater exchange potential is just the electrostatic potential created by the exchange-hole density. Notice that, because of Eq. (2.135), it vanishes as $-\frac{e^2}{r}$ for $\vec{r} \to \infty$.

Consider, for example, a large but finite electron liquid confined to a box of linear size L. The Slater potential for such a system is constant within the box, where its value is equal to the average of the homogeneous exchange energies $\varepsilon_k^{(x)}$ over the Fermi sphere. This is just twice the exchange energy per electron per spin. Thus we have

$$V_x^S(\vec{r}) = 2\frac{E_x}{N} = -e^2 n^{1/d} \begin{cases} \frac{3^{4/3}}{(4\pi)^{1/3}}, & 3D, \\ \frac{16}{3\sqrt{\pi}}, & 2D. \end{cases} \tag{2.139}$$

It turns out that this value is *larger* (by a factor 3/2 in 3D) than the correct value obtained from the density functional theory of Kohn and Sham (see Section 7.3.2). The latter is given by $\varepsilon_{k_F}^{(x)}$, that is, the exchange energy of an electron *at the top* of the Fermi distribution. Use of the Slater approximation therefore tends to overestimate the magnitude of exchange effects.

2.8.2 The optimized effective potential

A systematic approach to the problem of finding a local approximation to the exchange potential was proposed by Sharp and Horton (1953). They posed the following question: which *local* potential, acting on a system of N noninteracting electrons, yields a ground-state that minimizes the expectation value of the interacting hamiltonian $\hat{H}$? The requirement of locality is of essence here: without it, the answer to the question would be "the Hartree–Fock potential". By insisting on locality we actually define a new variational object, not quite as good as the Hartree–Fock potential from the point of view of the energy, but definitely easier to use in the Schrödinger equation. To this new object we give the name of *optimized effective potential* (OEP), and denote it by $V^O(\vec{r})$.

To derive a formal equation for the OEP, notice that a necessary condition is that the expectation value of $\hat{H}$ be stationary with respect to a small variation of the OEP, that is

$$\sum_\alpha n_\alpha \int \frac{\delta\langle\hat{H}\rangle_O}{\delta\phi_\alpha^O(\vec{r}')}\frac{\delta\phi_\alpha^O(\vec{r}')}{\delta V^O(\vec{r})}d\vec{r}' = 0, \qquad (2.140)$$

where $\phi_\alpha^O(\vec{r})$ are the eigenfunctions of the noninteracting hamiltonian $\hat{H}^O = \hat{T} + \sum_{i=1}^N V^O(\hat{r}_i, \hat{\sigma}_{i,z})$, and $\langle\hat{H}\rangle_O$ denotes the expectation value of $\hat{H}$ in the ground-state of $\hat{H}^O$. It is evident from these definitions that the first variational derivative $\frac{\delta\langle\hat{H}^O\rangle_O}{\delta\phi_\alpha^O(\vec{r}',s)}$ is zero. Therefore, we can write

$$\frac{\delta\langle\hat{H}\rangle_O}{\delta\phi_\alpha^O(\vec{r}')} = \frac{\delta\langle\hat{H}-\hat{H}^O\rangle_O}{\delta\phi_\alpha^O(\vec{r}')}$$
$$= [V_{x,\alpha}(\vec{r}') - V_x^O(\vec{r}')]\phi_\alpha^{O*}(\vec{r}'), \qquad (2.141)$$

where $V_{x,\alpha}(\vec{r}')$ is calculated from Eq. (2.134) with the optimized orbitals and $V_x^O(\vec{r}')$ is the exchange component of the OEP. As for the functional derivative of the optimized orbitals with respect to the optimized potential, standard first order perturbation theory tells us that

$$\frac{\delta\phi_\alpha^O(\vec{r}')}{\delta V^O(\vec{r})} = -G_\alpha(\vec{r}',\vec{r})\phi_\alpha^O(\vec{r}), \qquad (2.142)$$

where

$$G_\alpha(\vec{r}',\vec{r}) = \sum_{\beta,\beta\neq\alpha} \frac{\phi_\beta^O(\vec{r}')\phi_\beta^{O*}(\vec{r})}{\varepsilon_\beta^O - \varepsilon_\alpha^O}, \qquad (2.143)$$

and ε_α^O are the eigenvalues of $\hat{H}^O$ associated with the orbitals ϕ_α^O. Substituting Eqs. (2.141) and (2.142) into Eq. (2.140) we obtain the OEP equation

$$\sum_\alpha n_\alpha \int d\vec{r}'[V_{x,\alpha}(\vec{r}') - V_x^O(\vec{r}')]G_\alpha(\vec{r}',\vec{r})\phi_\alpha^{O*}(\vec{r}')\phi_\alpha^O(\vec{r}) = 0. \qquad (2.144)$$

Notice that $V_x^O(\vec{r})$ appears not only explicitly but also implicitly, that is, via the eigenfunctions ϕ_α^O and the eigenvalues ε_α^O.

An exact analytic solution of Eq. (2.144) is not available, but some insight into the behavior of the exact OEP can be obtained from a careful analysis of this equation. First of all, we notice that if $V_x^O(\vec{r})$ is a solution of Eq. (2.144) then $V_x^O(\vec{r}) + C$, where C is an arbitrary constant, also is a solution. The constant can be chosen in such a way that, like the Slater exchange potential, the optimized exchange potential vanishes as $-\frac{e^2}{r}$ for $r \to \infty$. In order to prove this assertion, we observe that, in the limit of $r \to \infty$ all the orbitals decrease exponentially with a characteristic length scale, which increases with increasing energy. The most energetic orbital is also the most extended in space, and thus dominates the sum on the right-hand side of Eq. (2.144). For large r, we thus obtain the simpler equation

$$\int d\vec{r}' [V_{x,m}(\vec{r}') - V_x^O(\vec{r}')] G_m(\vec{r}', \vec{r}) \phi_m^{O*}(\vec{r}') \phi_m^O(\vec{r}) = 0 , \qquad (2.145)$$

where m is the label of the most energetic orbital. Operating on Eq. (2.145) with $\varepsilon_m^O - \hat{H}_O$, and making use of the equation of motion

$$[\varepsilon_m^O - \hat{H}_O] G_m(\vec{r}', \vec{r}) = - \sum_{\beta \neq m} \phi_\beta^O(\vec{r}') \phi_\beta^{O*}(\vec{r})$$

$$= -[\delta(\vec{r} - \vec{r}') - \phi_m^O(\vec{r}') \phi_m^{O*}(\vec{r})] , \qquad (2.146)$$

we obtain, *for large r*,

$$V_x^O(\vec{r}) \stackrel{\vec{r} \to \infty}{=} V_{x,m}(\vec{r}) + (\bar{V}_{x,m} - \bar{V}_{x,m}^O) , \qquad (2.147)$$

where the constants $\bar{V}_{x,m}$ and $\bar{V}_{x,m}^O$ are, respectively, the expectation values of the exact exchange potential and of the optimized exchange potential in the most energetic orbital m, i.e., for instance

$$\bar{V}_{x,m} \equiv \int V_{x,m}(\vec{r}) |\phi_m^O(\vec{r})|^2 d\vec{r}, \qquad (2.148)$$

and similarly for $\bar{V}_{x,m}^O$. This shows that the behavior of the OEP for large r coincides, up to a constant, with that of $V_{x,m}(\vec{r})$, which is known from Eq. (2.135) to be $-e^2/r$. Choosing the arbitrary constant in $V_x^O(\vec{r})$ such that its asymptotic behavior is *exactly* $-e^2/r$ we see that

$$\bar{V}_{x,m} = \bar{V}_{x,m}^O , \qquad (2.149)$$

i.e., V_x and V_x^O have the same expectation value in the most energetic orbital.

Let us now return to the special case discussed at the end of the previous section, namely that of a homogeneous electron liquid confined to a box of large size L. In this case we expect $V_x^O(\vec{r})$ to have a constant value within the box, and this constant value must perforce coincide with $\bar{V}_{x,m}^O$ – the expectation value of the potential in the most energetic orbital. Then Eq. (2.149) implies that the optimized exchange potential of a uniform electron liquid coincides with the exchange energy of an electron *at the top* of the Fermi distribution, in agreement with the local density approximation of Kohn and Sham (see Section 7.3.2), and at variance with Slater's surmise.

Table 2.2. *Comparison of HF energies (in mRy) with energies obtained from local approximations to the exchange in closed shell atoms. (Adapted from Krieger, Li, and Iafrate, 1992.)*

Atom	$-E_{HF}(mRy)$	$E_O - E_{HF}$	$E_{KLI} - E_O$	$E_S - E_O$	$E_{LSDA} - E_O$
Be	29146.0	1.1	0.3	22.1	8.7
Ne	257094.2	3.4	1.1	89.4	35.7
Mg	399229.2	6.0	1.8	157.2	28.5
Ar	1053635.0	10.6	3.4	218.3	34.4
Ca	1353516.4	12.6	4.4	291.5	32.2
Zn	3555696.2	27.5	7.3	516.1	101.6
Kr	5504110.0	24.1	6.3	574.0	64.3
Sr	6263091.4	24.5	7.1	648.5	58.5
Cd	10930266.2	37.4	12.0	837.8	88.7
Xe	14464276.8	35.0	12.1	897.5	68.0

The OEP equations can and have been solved numerically for a variety of atomic and molecular systems. A very successful approximation, which preserves the main features of the exact solution while tremendously reducing the numerical effort was proposed by Krieger, Li, and Iafrate (KLI) (1992). In this approximate treatment of the OEP equation one replaces the denominator of Eq. (2.143) by a constant $\Delta\varepsilon$ independent of α and β. Then the function $G_\alpha(\vec{r}, \vec{r}')$ takes the form

$$G_\alpha(\vec{r}, \vec{r}') = (\Delta\varepsilon)^{-1}[\delta(\vec{r} - \vec{r}') - \phi_\alpha^O(\vec{r}')\phi^{O*}_\alpha(\vec{r})] . \qquad (2.150)$$

Substituting this into Eq. (2.144) leads to the KLI form of the optimized exchange potential

$$V_x^{KLI}(\vec{r}) = V_x^S(\vec{r}) + \sum_\alpha \frac{n_\alpha(\vec{r})}{n(\vec{r})} \left(\bar{V}_{x,\alpha} - \bar{V}_{x,\alpha}^{KLI} \right) , \qquad (2.151)$$

where $n_\alpha(\vec{r}) \equiv n_\alpha|\phi_\alpha^O(\vec{r})|^2$ and $n(\vec{r}) \equiv \sum_\alpha n_\alpha(\vec{r})$. This must be interpreted as an integral equation, in which $\bar{V}_{x,\alpha}^{KLI}$ is the average of the KLI potential in the state ϕ_α^O which, in turn is determined by the KLI potential. We note that this result has the form of a correction to the Slater potential $V_x^S(\vec{r})$ introduced in the previous section.

Table 2.2 presents a comparison of ground-state energies of closed-shell atoms calculated in the full HF approximation and in the various local approximations described above. We see that the exact OEP energy E_O lies above the HF energy, as it should, since the set of the Slater determinants that arise from a local potential is a subset of all the Slater determinants. The energy calculated in the KLI approximation, E_{KLI}, always lies above E_O, but significantly below the energy given by the Slater approximation, E_S. The *local spin density approximation* (LSDA), whose results are presented on the rightmost column of Table 2.2, will be discussed in detail in Chapter 7.

2.9 Real-world Hartree–Fock systems

Once an interesting HF state has been discovered the question arises of its actual physical reality, i.e. will the electron liquid in fact find itself in that particular state, or at least in one whose features (for instance the symmetry) are very close to it? The obvious answer to this question is that if the correlation corrections to the energy of the state are negligible or otherwise irrelevant the HF state will be in fact be realized.

An example of this situation is provided by the two dimensional electron gas at high magnetic field – a system we will discuss in detail in Chapter 10. Due to the quantization of the orbital motion, an interesting scenario can be realized, in which the ground-state of the liquid remains paramagnetic up to high values of the magnetic field and can be approximated, with excellent accuracy, as a single Slater determinant of one-electron orbitals. Thus we have a physical system for which the HF theory gives essentially the exact answer! The reason for this surprising occurrence is that the magnetic field creates a gap in the single particle excitation spectrum, and this gap, in turn, suppresses the mixing of excited configurations into the HF ground-state.

Another possible HF state, in this regime, is the ferromagnetic state – also a single Slater determinant, but one in which all the electrons have the same spin orientation. The kinetic energy of this state is higher than that of the paramagnetic state, while its exchange energy is lower. By tilting the magnetic field away from the normal to the plane of the electron liquid it is possible to gradually reduce the energy difference between the ferromagnetic and the paramagnetic state until, at a critical value of the tilt angle, a transition occurs between these two HF states: the gain in exchange energy offsets the cost in kinetic energy and the system turns ferromagnetic. This phenomenon, first predicted theoretically (Giuliani and Quinn, 1985) and recently verified experimentally (Daneshvar *et al.*, 1997), represents, in a sense, the high magnetic field version of the Bloch transition discussed in this chapter.

The lesson to be learned is that HF states, as well as transitions between them, can actually be realized in physical systems provided that some mechanism exists to suppress correlations.

Exercises

2.1 **Traditional derivation of the Hartree–Fock equations.** Derive the Hartree–Fock equations (2.8) and (2.21) from the requirement that the energy of a single Slater determinant, given by Eq. (2.12), be minimum with respect to the single-particle orbitals. [Hint: at a minimum the energy must be stationary with respect to small variations of the single-particle orbitals.]

2.2 **Electron density in a Slater determinant wave function.** Show that the expectation value of the density operator $\hat{n}(\vec{r})$ in a single Slater determinant made out of orthogonal functions $\phi_\alpha(\vec{r})$ is given by the sum $\sum_\alpha n_\alpha |\phi_\alpha(\vec{r})|^2$, where n_α limits the sum to the occupied states.

2.3 **Energy in a Slater determinant wave function.** Making use of Wick's theorem, show that the expectation value of the many-electron hamiltonian in any single Slater determinant is given by Eq. (2.62).

2.4 **Variation of a Slater determinant I.** Let $|\Psi_{HF}\rangle$ be the Slater determinant formed from the N lowest-lying eigenfunctions, $\phi_\beta(\vec{r}, s)$ ($\beta = 1, \ldots, N$), of the HF equations. Show that the Slater determinant formed from the N modified wave functions $\phi_\beta(\vec{r}, s) + \sum_{\alpha>N} \eta_{\alpha\beta}\phi_\alpha(\vec{r}, s)$ can be written in the form of Eq. (2.13), to first order in $\eta_{\alpha\beta}$.

2.5 **Variation of a Slater determinant energy I.** Verify that the variation in the energy of a single Slater determinant is given by Eq. (2.14). [Hint: this can be done most efficiently using the expression (2.62), for the energy of a single Slater determinant.]

2.6 **Criterion for telling if a many-body wave function can be written as a single Slater determinant.** Show that the *idempotency relation* $\hat{\rho}^2 = \hat{\rho}$ implies the existence of a basis set in which the one-particle density matrix $\hat{\rho}$ is diagonal and has eigenvalues 0 (for empty states) and 1 (for occupied states). Because $Tr(\hat{\rho}) = N$, there are exactly N occupied states. Conclude that a many-body wave function can be represented as a single Slater determinant if and only if $\hat{\rho}^2 = \hat{\rho}$.

2.7 **Variation of a Slater determinant II.** Show the equivalence of the two forms given in Eq. (2.67) for the variation of a Slater determinant; then calculate the density matrix associated with it, and show that it is given by Eq. (2.63), up to corrections of second order in $\delta\rho_{\alpha\beta}$.

2.8 **Variation of a Slater determinant energy II.** Show that Eq. (2.68) is the correct expression for the change in energy due to a small variation of the determinantal wave function about a self-consistent Hartree–Fock solution.

2.9 **Diagonalization of the HF hamiltonian for spiral spin density waves (SSDW).** By explicitly diagonalizing the SSDW mean-field hamiltonian (2.78) show that the eigenfunctions associated with the $+$ and $-$ branches of eigenvalues are given by Eqs. (2.84), where $\theta_{\vec{k}}^-$ is the negative solution of Eq. (2.81).

2.10 **Amplitude of the spin modulation in a SSDW state.** Calculate the amplitude of the spin modulation in a SSDW state obtained by occupying states in the single-particle bands of Fig. 2.4 with occupation numbers $n_{\vec{k}b}$. Compare the result with Eq. (2.87).

2.11 **Energy of a SSDW state.** Calculate the energy of the spiral-spin–density wave state described in the previous exercise. Verify that it is given by Eq. (2.93).

2.12 **Pair correlation function in a Slater determinant wave function.** Show that the pair correlation function of a state described by a single Slater determinant is related to the one-particle density matrix by Eq. (2.137). [Hint: make use of the definition (A4.8) and evaluate the average with the help of Wick's theorem.]

2.13 **Slater exchange potential.** Derive the expression (2.138) for the Slater local exchange potential in terms of the exchange hole.

3

Linear response theory

3.1 Introduction

There are countless situations in physics when one is interested in calculating the response of a system to a small time-dependent perturbation acting on it. With some luck the response can be expanded in a power series of the strength of the perturbation, so that, to first order, it is a *linear* function of the latter. To compute this function is the objective of the *linear response theory* (LRT).

Linear response theory has many important applications to the study of electronic matter. Virtually all interactions of electrons with experimental probes (electromagnetic fields, beams of particles) can be regarded as small perturbations to the system: if they were not, one would not be probing the system, but the system modified by the probe. Consequently, the results of these experiments can be expressed in terms of linear response functions, which are properties of the unperturbed system. In particular it will turn out that the analytic structure of these functions is entirely determined by the eigenvalues and eigenfunctions of the unperturbed system. Conversely, a measure of the linear response as a function of the frequency of the perturbation enables us to determine the excitation energies of the system.

Beside being a cornerstone for the theory of single-particle properties to be developed in Chapter 8, the linear response functions also provide invaluable information in their own right. For example, as we will discuss in Chapter 5, the extent to which an external electrostatic potential is reduced by screening is controlled by the dynamical dielectric function which, in turn, is determined by the density–density response function. Moreover, knowledge of the dielectric function allows one to calculate the ground-state energy and the spectrum of collective excitations, and explains electron-energy-loss experiments.

We begin this chapter by first introducing the basic ideas of the LRT method in an intuitive, and algebraically simplified manner, while the formal development of the theory and some of the most relevant results and applications are provided in the following sections. Excellent discussions of linear response theory can be found in the classic book by Pines and Nozières (1966), and in the seminal paper by Kubo (1957).

As an introductory example consider a system in the ground-state. By definition, if the average value of a physical quantity $\hat{B}$ is made to deviate from its ground-state value by an amount b, then the dependence of the internal energy on this parameter will be of the type:

$$E(b) \approx E_0 + \frac{\alpha_b}{2}b^2 , \qquad (3.1)$$

where E_0 is the ground-state energy, and the quantity α_b, formally given by $\frac{\partial^2 E(b)}{\partial b^2}\Big|_{b=0}$, depends on the properties of the system in its ground-state. This parameter is a measure of the *stiffness* offered by the system to small modifications of the average value of the quantity $\hat{B}$.[1] The second term in Eq. (3.1) will be referred to as *stiffness energy*.

We now apply to the system an external field linearly coupled to the quantity $\hat{B}$ and such that the coupling energy is of the type ϵb, where ϵ is positive and can be made arbitrarily small. In this situation the energy acquires the form

$$E_\epsilon(b) \approx E_0 + \frac{\alpha_b}{2}b^2 + \epsilon b . \qquad (3.2)$$

Minimization of this relationship leads immediately to the following finite equilibrium value for b, i.e.

$$b_{eq} = -\frac{\epsilon}{\alpha_b} , \qquad (3.3)$$

which corresponds to the new total ground-state energy

$$E_\epsilon(b_{eq}) \approx E_0 - \frac{\epsilon^2}{2\alpha_b} . \qquad (3.4)$$

The physical situation is quite clear: the linear coupling to the external perturbation leads to a new ground-state of lower total energy, in which the average value of the quantity $\hat{B}$ deviates from its original value by an amount directly proportional to the strength of the coupling (a linear response) and inversely proportional to the corresponding stiffness α_b.

The ratio of b_{eq} to the strength of the coupling ϵ, in the limit in which ϵ becomes infinitesimally small, is well defined, and allows us to define the *linear response function* χ_{bb} (the reason for the double subscript will become clear in the following sections) which here is simply given by

$$\chi_{bb} = \lim_{\epsilon \to 0} \frac{b_{eq}}{\epsilon} = -\frac{1}{\alpha_b} , \qquad (3.5)$$

a relationship which shows the simple connection between χ_{bb} and the stiffness α_b. At the same time the change in internal energy (i.e., the stiffness energy) can also be expressed quite generally as

$$\delta E = -\frac{b_{eq}^2}{2\chi_{bb}} , \qquad (3.6)$$

a relationship of considerable fundamental and practical relevance.

This scenario is generic to all linear response problems. In order to exemplify it, we discuss next, anticipating a few of the basic results, the case of a non-interacting electron gas

---

[1] The concept of stiffness will be discussed more rigorously in Section 3.2.9.

in its homogeneous, paramagnetic ground-state as perturbed by an external scalar electric potential – a probe that directly couples to the electron density.[2]

We begin by writing the external potential energy associated with a static sinusoidal electric field as follows:

$$\hat{V} = \epsilon \sum_{i=1}^{N} \cos \vec{q} \cdot \hat{\vec{r}}_i , \tag{3.7}$$

where the vector $\vec{q}$ dictates the direction of the field and the wavelength of the perturbation. The quantity $\left\langle \sum_{i=1}^{N} \cos \vec{q} \cdot \hat{\vec{r}}_i \right\rangle = \frac{1}{2} \left(n_{\vec{q}} + n_{-\vec{q}} \right)$, where $n_{\vec{q}}$ is the average value of the Fourier component of the electronic density at wave vector $\vec{q}$, plays the role of the quantity b of the previous discussion.

What we will calculate is the response of the electronic density to this perturbation. In this simple case the average electronic density can be obtained directly from the expression for the single particle wave functions $\phi_{\vec{k}}(\vec{r})$ via the formula (see Exercise 2)

$$n(\vec{r}) = \sum_{|\vec{k}| \le k_F, \sigma} |\phi_{\vec{k}}(\vec{r})|^2 . \tag{3.8}$$

Here, the functions $\phi_{\vec{k}}(\vec{r})$ must be calculated in the presence of the perturbation. To linear order in ϵ we have:

$$\phi_{\vec{k}}(\vec{r}) = \phi_{\vec{k}}^{(0)}(\vec{r}) + \delta\phi_{\vec{k}}(\vec{r}) , \tag{3.9}$$

where $\phi_{\vec{k}}^{(0)}(\vec{r}) = \frac{e^{i\vec{k}\vec{r}}}{\sqrt{L^d}}$ is the unperturbed wave function and $\delta\phi_{\vec{k}}(\vec{r})$ is given by elementary first-order perturbation theory in the form

$$\delta\phi_{\vec{k}}(\vec{r}) = \sum_{\vec{p}} \frac{\langle \phi_{\vec{p}}^{(0)} | \epsilon \cos \vec{q} \cdot \vec{r} | \phi_{\vec{k}}^{(0)} \rangle}{\varepsilon_{\vec{k}} - \varepsilon_{\vec{p}}} \phi_{\vec{p}}^{(0)}(\vec{r})$$

$$= \frac{\epsilon}{2} \left(\frac{\phi_{\vec{k}+\vec{q}}^{(0)}(\vec{r})}{\varepsilon_{\vec{k}} - \varepsilon_{\vec{k}+\vec{q}}} + \frac{\phi_{\vec{k}-\vec{q}}^{(0)}(\vec{r})}{\varepsilon_{\vec{k}} - \varepsilon_{\vec{k}-\vec{q}}} \right) , \tag{3.10}$$

where $\varepsilon_{\vec{k}} = \frac{\hbar^2 k^2}{2m}$.

Since we are only interested in the linear response, the expression for the electronic density can be rewritten as:

$$n(\vec{r}) = n^{(0)} + 2\mathcal{R}e \sum_{|\vec{k}| \le k_F, \sigma} \phi_{\vec{k}}^{(0)}(\vec{r}) \delta\phi_{\vec{k}}^{\star}(\vec{r}) + \mathcal{O}(\epsilon^2) , \tag{3.11}$$

where $n^{(0)}$ is the value of the unperturbed density. Now, by making use of Eq. (3.10), we see that, to linear order in the perturbation, the deviation of the electron density from $n^{(0)}$ is

---

[2] A complete treatment of the linear response of noninteracting electrons is provided in Chapter 4.

given by

$$n_1(\vec{r}) = \epsilon \left(\frac{2}{L^d} \sum_{|\vec{k}| \le k_F, \sigma} \frac{1}{\varepsilon_{\vec{k}} - \varepsilon_{\vec{k}+\vec{q}}} \right) \cos \vec{q} \cdot \vec{r} , \tag{3.12}$$

from which one obtains the perturbed equilibrium value of b:

$$b_{eq} = \frac{1}{2} \left(n_{\vec{q}} + n_{-\vec{q}} \right) = \epsilon \sum_{|\vec{k}| \le k_F, \sigma} \frac{1}{\varepsilon_{\vec{k}} - \varepsilon_{\vec{k}+\vec{q}}} . \tag{3.13}$$

At this point we can calculate the response function χ_{bb} directly from its definition (3.5), namely, as the ratio of b_{eq} to ϵ:

$$\chi_{bb} = \lim_{\epsilon \to 0} \frac{b_{eq}}{\epsilon} = \sum_{|\vec{k}| \le k_F, \sigma} \frac{1}{\varepsilon_{\vec{k}} - \varepsilon_{\vec{k}+\vec{q}}} . \tag{3.14}$$

This allows us to express the induced density as

$$n_1(\vec{r}) = \epsilon \frac{2\chi_{bb}}{L^d} \cos \vec{q} \cdot \vec{r} . \tag{3.15}$$

Clearly in this non-interacting case the function $\chi_{nn}(\vec{q}) = \frac{2\chi_{bb}}{L^d}$ measures the ability of the system to *polarize* as a consequence of the application of a perturbation coupled to the electronic density. Quite generally in a homogeneous system $\chi_{nn}(\vec{q})$ relates the corresponding Fourier components of the external field and the induced linear density modulation, via the relation $n_{\vec{q}} \equiv \chi_{nn}(\vec{q}) \frac{\epsilon L^d}{2}$, and is known as the *density–density response function*.

To complete the present discussion we now relate the response function χ_{bb} to the stiffness energy for density modulations of the system at hand. It is manifest that for a non-interacting electron gas the only source of "resistance" to such a perturbation is provided by the kinetic energy. It is a rather simple and instructive exercise (see Exercise 1) to show that, in this situation, the (positive) change in kinetic energy to second order in ϵ is given by:

$$\delta T = -\frac{\epsilon^2}{2} \sum_{|\vec{k}| \le k_F, \sigma} \frac{1}{\varepsilon_{\vec{k}} - \varepsilon_{\vec{k}+\vec{q}}} . \tag{3.16}$$

(Notice that this is the opposite of the change in the total ground-state energy.) By making use of Eqs. (3.13) and (3.14) we rewrite this as

$$\delta T = -\frac{b_{eq}^2}{2\chi_{bb}} , \tag{3.17}$$

and comparing with Eq. (3.6) we see that χ_{bb} is indeed related to the stiffness energy by Eq. (3.5).

The above discussion is limited to the case of *static* perturbations. What happens if the perturbation depends on time? In this case a new physical phenomenon appears. In order to understand the problem we consider the linear response of a non-interacting electron gas to a particularly simple quasi-periodic perturbation of the form

$$\hat{V}(t) = \epsilon \, \cos \omega t \, e^{\eta t} \sum_{i=1}^{N} \cos \vec{q} \cdot \hat{\vec{r}}_i , \qquad (3.18)$$

where ω is the frequency of the external field and η is a positive infinitesimal whose only purpose is to make sure the system starts from its equilibrium state at $t = -\infty$.[3] The analysis is basically a repeat of that for the static situation, the only difference being that now one must make use of first order *time-dependent* perturbation theory. Although cumbersome, the calculation represents a straightforward and highly recommended exercise. The result for the time-dependent density modulation can be cast in the form:

$$n_1(\vec{r}, t) = \epsilon \left[\Re e \chi_{nn}(\vec{q}, \omega) \cos \omega t \, + \, \Im m \chi_{nn}(\vec{q}, \omega) \sin \omega t \right] \cos \vec{q} \cdot \vec{r} , \qquad (3.19)$$

where $\Re e \chi_{nn}(\vec{q}, \omega)$ and $\Im m \chi_{nn}(\vec{q}, \omega)$ are the real and the imaginary part of a complex density–density response function $\chi_{nn}(\vec{q}, \omega)$.[4] We see therefore that a time-dependent perturbation leads in general to an *in-phase* and a $90°$ *out-of-phase* component of the response. While the in-phase component is just a generalization to a time-dependent situation of the response function of Eq. (3.14), i.e. it describes a dynamical polarization phenomenon, the out-of-phase component will be shown to be associated with *energy absorption* (see Section 3.2.5) and therefore describes dissipation.[5]

Among all conceivable response functions, those associated with density and current play a central role in the physics of the electron liquid. The density response function, for instance, can be shown to determine both the ground-state energy and the spectrum of the collective excitations. The behavior of the current response function, in turn, determines the important distinction between conductors and insulators.

In most situations an exact calculation of the response functions is out of the question. One must therefore recur to reasonable approximations. While a discussion of modern approximate theories of the response of an interacting electron system is provided in Chapter 5, we present in this chapter a number of exact relationships, notably symmetries and sum rules which can be used to devise and check useful approximation schemes.

3.2 General theory of linear response

3.2.1 Response functions

To set up the problem in the most general terms we consider a system described by a time-independent hamiltonian $\hat{H}$, which is acted upon by an external time-dependent field $F(t)$ that couples linearly to an observable $\hat{B}$ of the system. The complete time-dependent

---

[3] The precise meaning of this quantity is discussed in Section 3.2.2.
[4] An alternative, general approach, as well explicit expressions for these functions are provided in Sections 4.2 and 4.3.
[5] The reason for the difference between the time-dependent response and the static one lies entirely in the innocent-looking switching-on factor $e^{\eta t}$ (see Eq. (3.18)), which introduces an asymmetry between past and future times. Nothing comparable exists in space.

hamiltonian is therefore

$$\hat{H}_F(t) = \hat{H} + F(t)\hat{B} , \tag{3.20}$$

and we shall assume, for the time being, that the external field $F(t)$ vanishes for t earlier than a certain initial time t_0. For example, the coupling of the electron gas to a time-dependent external potential $V_{ext}(\vec{r}, t)$ can be written in the form

$$\int V_{ext}(\vec{r}, t)\hat{n}(\vec{r})d\vec{r} , \tag{3.21}$$

where $\hat{n}(\vec{r}) = \sum_i \delta(\vec{r} - \hat{\vec{r}}_i)$ is the density operator. Eq. (3.21) is a sum of perturbing terms of the form (3.20), where $F(t) \rightarrow V^{ext}(\vec{r}, t)$ and $\hat{B} \rightarrow \hat{n}(\vec{r})$. Since in the linear response approximation the responses to different perturbing terms add up independently, there is no loss of generality in considering the response to only one of them, as done in Eq. (3.20).

For $t \leq t_0$ the system is assumed to be in the ground-state, or, more generally, in thermal equilibrium with a large "thermal reservoir" at temperature T. This implies that the n-th eigenstate $|\psi_n\rangle$ of the unperturbed hamiltonian $\hat{H}$, with eigenvalue E_n, is populated with probability

$$P_n = \frac{e^{-\beta E_n}}{\mathcal{Z}} , \tag{3.22}$$

where $\beta = \frac{1}{k_B T}$ and $\mathcal{Z} = \sum_n e^{-\beta E_n}$ is the canonical partition function. For $t \leq t_0$, the time evolution of this state amounts to a changing phase factor, which does not affect the physical properties of the state.

At time $t = t_0$ the external field is turned on, and the system begins to evolve under its influence. A key assumption is that the weak interactions between the system and the thermal reservoir, which are needed to establish the initial thermal equilibrium, do not significantly affect the evolution of the ensemble during the time scale of our observations. In particular, the occupation probabilities P_n stay constant: no transitions are allowed between orthogonal states.[6] Thus, the time-evolution of the system, in the Schrödinger picture, is completely determined by the Schrödinger equation

$$i\hbar \frac{\partial}{\partial t}|\psi_n(t)\rangle = \hat{H}_F(t)|\psi_n(t)\rangle \tag{3.23}$$

with the initial condition

$$|\psi_n(t_0)\rangle = |\psi_n\rangle. \tag{3.24}$$

The solution of this linear equation can be written in the form

$$|\psi_n(t)\rangle = \hat{U}(t, t_0)|\psi_n(t_0)\rangle \tag{3.25}$$

---

[6] This feature is characteristic of *adiabatic* processes. The assumption is justified only for time evolutions that are *fast* on the scale of the thermal equilibration time. When the time evolution is slow on the scale of the equilibration time, it becomes possible for the system to exchange energy (heat) with the reservoir, causing *repopulation* of the energy levels.

where $\hat{U}(t, t_0)$ is the unitary time-evolution operator, which relates the state at time t to the state at time t_0.

In the absence of the perturbation, we would have

$$\hat{U}(t, t_0) = e^{-\frac{i}{\hbar}\hat{H}(t-t_0)} . \qquad (3.26)$$

In order to set up a perturbative expansion of $\hat{U}$ in powers of $F(t)$ it is convenient to write

$$\hat{U}(t, t_0) = e^{-\frac{i}{\hbar}\hat{H}(t-t_0)}\hat{U}_F(t, t_0) \qquad (3.27)$$

where $\hat{U}_F(t, t_0)$ – the part of the time-evolution operator that is due to the external field F – obeys the equation of motion

$$i\hbar\frac{\partial}{\partial t}\hat{U}_F(t, t_0) = F(t)\hat{B}(t - t_0)\hat{U}_F(t, t_0), \qquad (3.28)$$

with initial condition $\hat{U}_F(t_0, t_0) = \hat{1}$. The time-dependent operator

$$\hat{B}(t) \equiv e^{\frac{i}{\hbar}\hat{H}t}\hat{B}e^{-\frac{i}{\hbar}\hat{H}t} \qquad (3.29)$$

is the Heisenberg picture version of the Schrödinger operator $\hat{B}$, and coincides with the latter only at $t = 0$.

The usefulness of Eq. (3.28) stems from the fact that the time dependence of $\hat{U}_F$ is entirely due to the perturbation. In particular, substituting the zero-order approximation $\hat{U}_{F,0}(t, t_0) = \hat{1}$ on the right-hand side of Eq. (3.28) and integrating with respect to time we obtain the first order approximation to $\hat{U}_F$ in the following form.

$$\hat{U}_{F,1}(t, t_0) = \left[\hat{1} - \frac{i}{\hbar}\int_{t_0}^t \hat{B}(t' - t_0)F(t')dt'\right]. \qquad (3.30)$$

Consequently, the complete time evolution operator, *to first order in F*, takes the form

$$\hat{U}_1(t, t_0) = e^{-\frac{i}{\hbar}\hat{H}(t-t_0)}\left[\hat{1} - \frac{i}{\hbar}\int_{t_0}^t \hat{B}(t' - t_0)F(t')dt'\right]: \qquad (3.31)$$

this result is the cornerstone of the linear response theory.

Let us now consider a second observable $\hat{A}$ which, up to the time t_0, had the average equilibrium value

$$\langle\hat{A}\rangle_0 = \sum_n P_n\langle\psi_n|\hat{A}|\psi_n\rangle , \quad t \le t_0 . \qquad (3.32)$$

Our goal is to calculate the expectation value of $\hat{A}$ at times later than t_0, under the influence of the perturbation. Formally, this is given by

$$\langle\hat{A}\rangle_F(t) = \sum_n P_n\langle\psi_n(t)|\hat{A}|\psi_n(t)\rangle, \qquad (3.33)$$

where the $|\psi_n(t)\rangle$'s are given by Eq. (3.25). Since we are interested only in the linear response to the perturbing field $F(t)$ we can make use of the linearized form of the time-evolution operator Eq. (3.31) and its hermitian conjugate. After some simple manipulations we obtain the fundamental result

$$\langle \hat{A} \rangle_F(t) - \langle \hat{A} \rangle_0 = -\frac{i}{\hbar} \int_{t_0}^{t} \langle [\hat{A}(t), \hat{B}(t')] \rangle_0 F(t') dt' , \qquad (3.34)$$

where both $\hat{A}(t)$ and $\hat{B}(t)$ are calculated via Eq. (3.29), $[\hat{A}, \hat{B}]$ is the commutator of two operators $\hat{A}$ and $\hat{B}$, and $\langle \ldots \rangle_0$ denotes the average in the thermal equilibrium ensemble.[7]

Using again the time independence of the unperturbed $\hat{H}$ we can write

$$\langle [\hat{A}(t), \hat{B}(t')] \rangle_0 = \langle [\hat{A}(\tau), \hat{B}] \rangle_0 , \qquad (3.35)$$

where $\tau \equiv t - t' > 0$, and define the *retarded* linear response function $\chi_{AB}(\tau)$ as follows:

$$\chi_{AB}(\tau) \equiv -\frac{i}{\hbar} \Theta(\tau) \langle [\hat{A}(\tau), \hat{B}] \rangle_0 , \qquad (3.36)$$

where the step-function $\Theta(\tau)$ vanishes for $\tau < 0$ and equals 1 for $\tau > 0$.

After making the change of variables $t' = t - \tau$ in Eq. (3.34) we finally write the linear response for the observable $\hat{A}$ in the form

$$\langle \hat{A} \rangle_1(t) \equiv \langle \hat{A} \rangle_F(t) - \langle \hat{A} \rangle_0$$
$$= \int_0^{t-t_0} \chi_{AB}(\tau) F(t - \tau) d\tau . \qquad (3.37)$$

It is evident that $\chi_{AB}(\tau)$ describes the response of the observable $\hat{A}$ at time t to an impulse that coupled to the observable $\hat{B}$ at an *earlier* time $t - \tau$. Because it describes the *after-effect* of a perturbation it is appropriately called *retarded*, or *causal* response function. Notice that the formula for the response function makes no reference to the initial "switching-on" time. We can therefore let t_0 tend to $-\infty$, and all the formulas will still be valid provided that the perturbing field tends to zero for $t \to -\infty$ in such a way that the system can be assumed to have been in the unperturbed equilibrium state in the far past:

$$\langle \hat{A} \rangle_1(t) = \int_0^{\infty} \chi_{AB}(\tau) F(t - \tau) d\tau. \qquad (3.38)$$

Notice that, if $\hat{A}$ and $\hat{B}$ are hermitian operators, as we have implicitly assumed so far, then $\chi_{AB}(\tau)$ is also real as it connects two real quantities.[8] However, Eq. (3.36) will be taken as to define more generally the linear response function for any pair of operators, regardless of their being hermitian or not.[9]

---

[7] An important step in the derivation of Eq. (3.34) is the realization that for any eigenstate $|\psi_n\rangle$ of the unperturbed hamiltonian $\hat{H}$, we have $\langle \psi_n | [\hat{A}(t), \hat{B}(t')] | \psi_n \rangle = \langle \psi_n | [\hat{A}(t - t_0), \hat{B}(t' - t_0)] | \psi_n \rangle$.

[8] The commutator of two hermitian operators $\hat{A}(\tau)$ and $\hat{B}$ is *anti-hermitian*, i.e. it changes sign upon hermitian conjugation. Its expectation value is a purely imaginary number, which combines with the factor $-i$ in the definition (3.36) to give a real response function.

[9] Any operator $\hat{B}$ can be expressed as $\hat{B}' + i\hat{B}''$ where $\hat{B}'$ and $\hat{B}''$ are hermitian. We define the response to the external perturbation $F(t)\hat{B}$ as the response to $F(t)\hat{B}'$ plus i times the response to $F(t)\hat{B}''$.

3.2.2 Periodic perturbations

In practical applications of the linear response formalism a central role is played by the response to a periodic perturbation of the type

$$F(t) = F_\omega e^{-i\omega t} + c.c. , \tag{3.39}$$

where F_ω is a complex amplitude and $c.c$ stands for *complex conjugate*. The importance of such perturbations stems from the fact that every "well behaved" function of time[10] can be written as a superposition of periodic functions according to the well known formula

$$F(t) = \int_{-\infty}^{\infty} \tilde{F}(\omega) e^{-i\omega t} \frac{d\omega}{2\pi} , \tag{3.40}$$

where

$$\tilde{F}(\omega) = \int_{-\infty}^{\infty} F(t) e^{i\omega t} \, dt . \tag{3.41}$$

Knowledge of the linear response to the first term of Eq. (3.39) will therefore suffice to construct the response to a general well behaved $F(t)$.

There is, however, a difficulty: the periodic potential of Eq. (3.39) does not vanish for $t \to -\infty$, and, for this reason, it seems that we are *not* entitled to assume that the system was in equilibrium in the far past. The standard trick by which one avoids this difficulty is the "*adiabatic switching-on*" of the perturbation: we assume that the amplitude of the periodic potential is slowly turned on according to the law $e^{\eta t}$, where η is positive and η^{-1} represents a time scale much longer than the period of the perturbation. Mathematically, this amounts to considering a perturbing potential of the form

$$F(t) = F_\omega e^{-i(\omega + i\eta)t} + c.c. \tag{3.42}$$

As long as the convergence factor η is present, the linear response formalism is applicable in the form described in the previous section. We can then take the limit $\eta \to 0^+$ at the end of the calculation. If this limit exists, we expect it to describe the physical response of the system to a steady periodic field, i.e., a periodic field that has been going on long enough to erase any memory of the artificial switching-on process.[11]

Insertion of Eq. (3.42) into Eq. (3.38) yields

$$\langle \hat{A} \rangle_1(t) = \langle \hat{A} \rangle_1(\omega) e^{-i\omega t} + c.c. , \tag{3.43}$$

where

$$\langle \hat{A} \rangle_1(\omega) = \chi_{AB}(\omega) F_\omega , \tag{3.44}$$

---

[10] In mathematical language $f(t)$ should be an L^2 function, that is a function for which the integral $\int_{-\infty}^{\infty} |f(t)|^2 dt$ exists.

[11] Strictly speaking, if the population factors P_n are to be treated as constants, then η^{-1} must remain shorter than the thermal equilibration time. This means that η cannot become smaller than γ where γ is of the order of the inverse of the thermal equilibration time and is also proportional to the magnitude of the interaction between the system and its environment. Thus, the adiabatic linear response theory is consistent only for frequencies $\omega \gg \gamma$.

and

$$\chi_{AB}(\omega) = -\frac{i}{\hbar} \lim_{\eta \to 0} \int_0^\infty \langle [\hat{A}(\tau), \hat{B}] \rangle_0 e^{i(\omega + i\eta)\tau} d\tau \qquad (3.45)$$

is the Fourier transform of the response function. Eq. (3.45) defines a frequency dependent response function for *any* pair of (not necessarily hermitian) operators $\hat{A}$ and $\hat{B}$.

3.2.3 Exact eigenstates and spectral representations

We gain insight into the structure of the causal response function by expanding the commutator in Eq. (3.36) in a complete set of exact eigenstates $|\psi_n\rangle$ of $\hat{H}$:

$$\langle [\hat{A}(\tau), \hat{B}] \rangle_0 = \sum_{m,n} P_m (e^{i\omega_{mn}\tau} A_{mn} B_{nm} - e^{i\omega_{nm}\tau} B_{mn} A_{nm})$$

$$= \sum_{m,n} (P_m - P_n) e^{i\omega_{mn}\tau} A_{mn} B_{nm}, \qquad (3.46)$$

where we have introduced the notation $O_{nm} \equiv \langle \psi_n | \hat{O} | \psi_m \rangle = [O_{mn}^\dagger]^*$, to denote the matrix elements of an operator $\hat{O}$, and $\omega_{nm} = \frac{E_n - E_m}{\hbar} = -\omega_{mn}$ are the excitation frequencies of the system. Inserting Eq. (3.46) into Eq. (3.45) and performing the elementary required time integration, we obtain the *exact eigenstates representation* – also known as *Lehmann representation* – of the response function:

$$\chi_{AB}(\omega) = \frac{1}{\hbar} \sum_{nm} \frac{P_m - P_n}{\omega - \omega_{nm} + i\eta} A_{mn} B_{nm} , \qquad (3.47)$$

where it is understood that $\eta \to 0^+$. Notice that $\chi_{AB}(\omega)$ is analytic in the upper half of the complex plane and has only simple poles in the lower half. The zero-temperature limit of this formula is obtained by setting $P_n = 1$ for the ground-state ($n = 0$) and $P_n = 0$ for all the other states.[12]

We can go further, and separate the real and imaginary parts of χ_{AB} with the help of the useful formula

$$\lim_{\eta \to 0} \frac{1}{\omega - y + i\eta} = \mathcal{P} \frac{1}{\omega - y} - i\pi \delta(\omega - y) , \qquad (3.48)$$

where the first term is the Cauchy–Hadamard *principal value* distribution defined through the integral relation

$$\mathcal{P} \int_{-\infty}^\infty \frac{f(\omega)}{\omega - y} d\omega = \lim_{\eta \to 0^+} \left(\int_{-\infty}^{y-\eta} \frac{f(\omega)}{\omega - y} d\omega + \int_{y+\eta}^\infty \frac{f(\omega)}{\omega - y} d\omega \right) \qquad (3.49)$$

---

[12] An interesting property of Eq. (3.47) is that the terms with $n = m$ and, more generally, those with $E_n = E_m$, do not contribute to the sum, since $P_n = P_m$ in such cases. This implies that, if we denote by $\hat{A}_0$ the diagonal part of an operator $\hat{A}$ in the representation in which $\hat{H}$ is diagonal, and by $\tilde{A} = \hat{A} - \hat{A}_0$ the off-diagonal part of the same operator in that representation, then $\chi_{AB} = \chi_{\tilde{A}\tilde{B}}$. We conclude that the response function vanishes if either $\tilde{A}$ or $\tilde{B}$ are constants of the unperturbed motion (i.e., commute with $\hat{H}$) since either $\tilde{A}$ or $\tilde{B}$ is 0 in that case.

for any test function $f(\omega)$. The separation of real and imaginary parts takes a particularly simple form in the important case $\hat{B} = \hat{A}^\dagger$. Then $A_{mn}B_{nm}$ is the real quantity $|A_{mn}|^2$ and we can write

$$\Re \chi_{AA^\dagger}(\omega) = \frac{1}{\hbar}\mathcal{P}\sum_{nm}\frac{|A_{mn}|^2}{\omega - \omega_{nm}}(P_m - P_n) , \tag{3.50}$$

and

$$\Im \chi_{AA^\dagger}(\omega) = -\frac{\pi}{\hbar}\sum_{nm}(P_m - P_n)|A_{mn}|^2\delta(\omega - \omega_{nm})$$

$$= -\frac{\pi}{\hbar}\sum_{nm}P_m[|A_{mn}|^2\delta(\omega - \omega_{nm}) - |A^\dagger_{mn}|^2\delta(\omega + \omega_{nm})] . \tag{3.51}$$

We immediately notice an important fact: $\Im \chi_{AA^\dagger}(\omega)$ is always negative for positive ω and positive for negative ω.

It is often useful to extend the definition of the response function $\chi_{AA^\dagger}$ to complex values of the frequency. To this end we define the function

$$\tilde{\chi}_{AA^\dagger}(z) = \frac{1}{\hbar}\sum_{nm}\frac{|A_{mn}|^2}{z - \omega_{nm}}(P_m - P_n) , \tag{3.52}$$

where z is a complex frequency. It must be noted that $\tilde{\chi}_{AA^\dagger}(\omega + i\eta) = \chi_{AA^\dagger}(\omega)$, but $\tilde{\chi}_{AA^\dagger}(\omega - i\eta) = \chi^*_{AA^\dagger}(\omega)$ for infinitesimal η. Thus, $\tilde{\chi}_{AA^\dagger}(z)$ is discontinuous across the real axis and the magnitude of the discontinuity is $2i\Im \chi_{AA^\dagger}(\omega)$.

A crucial tool in linear response theory is the *spectral representation*, which allows us to express $\tilde{\chi}_{AA^\dagger}(z)$ in the entire complex plane in terms of the imaginary part of $\chi_{AA^\dagger}(\omega)$ along the real axis. The spectral representation has the form

$$\tilde{\chi}_{AA^\dagger}(z) = -\frac{1}{\pi}\int_{-\infty}^{\infty}\frac{\Im \chi_{AA^\dagger}(\omega)}{z - \omega}d\omega , \tag{3.53}$$

as one can easily verify by substituting Eq. (3.51) on the right-hand side, and showing that the result is identical with Eq. (3.52).

3.2.4 Symmetry and reciprocity relations

The very structure of Eqs. (3.45) and (3.47) allows us to derive a number of useful and exact relationships satisfied by the linear response functions.

Let us first consider the static case, $\omega = 0$. Starting from Eq. (3.47) we see that at $\omega = 0$ only the principal part term in the expansion (3.48) of $\frac{1}{\omega - \omega_{nm} + i\eta}$ survives, because $(P_n - P_m)\delta(\omega_{nm}) = 0$. This immediately implies that $\chi_{AA^\dagger}(0)$ is *real* and *negative*,

$$\chi_{AA^\dagger}(0) \leq 0 , \tag{3.54}$$

a fact that will be seen to be crucial to the stability of the system (see Section 3.2.9). Another interesting property of static linear response functions is the reciprocity relationship

$$\chi_{AB}(0) = \chi_{BA}(0) , \qquad (3.55)$$

which is proved by interchanging the indices n and m in Eq. (3.47).

In the general case of finite frequency, a direct inspection of Eq. (3.47) leads to the conclusion that for real ω, $\chi_{AB}(\omega)$ satisfies the symmetry relation

$$\chi_{AB}(-\omega) = [\chi_{A^\dagger B^\dagger}(\omega)]^* . \qquad (3.56)$$

This implies that the real and imaginary parts of $\chi_{AB}(\omega) \equiv \Re e \chi_{AB}(\omega) + i \Im m \chi_{AB}(\omega)$ satisfy the relations

$$\Re e \chi_{AB}(\omega) = \Re e \chi_{A^\dagger B^\dagger}(-\omega) ,$$
$$\Im m \chi_{AB}(\omega) = -\Im m \chi_{A^\dagger B^\dagger}(-\omega) . \qquad (3.57)$$

These formulas are particularly useful in the special case in which both $\hat{A}$ and $\hat{B}$ are hermitian as we then have:

$$\Re e \chi_{AB}(-\omega) = \Re e \chi_{AB}(\omega) ,$$
$$\Im m \chi_{AB}(-\omega) = -\Im m \chi_{AB}(\omega) ; \qquad (3.58)$$

i.e. $\Re e \chi_{AB}(\omega)$ is an even function of frequency while $\Im m \chi_{AB}(\omega)$ is odd.[13]

The linear response functions of a time-reversal invariant system, such as the electron liquid in the absence of a magnetic field, satisfy the following *reciprocity relations*:

$$\chi_{AB}(\omega) = \chi_{{}^t B^t A}(\omega), \qquad (3.59)$$

where ${}^t \hat{A} = (\hat{A}^\dagger)^*$ is the *transpose* of $\hat{A}$. For instance, ${}^t \hat{n}(\vec{r}) = \hat{n}(\vec{r})$, but for the paramagnetic current density $\hat{j}_p(\hat{r})$ we have ${}^t \hat{j}_p(\vec{r}) = -\hat{j}_p(\vec{r})$ due to the i in the definition of the momentum $\hat{\vec{p}} = -i\hbar\vec{\nabla}$ (see Eq. A2.15).

The proof is quite simple. From the hypothesis of time-reversal invariance it follows that all the wave functions of the system can be chosen to be real. Then the matrix element A_{mn} in Eq. (3.47) can be rewritten as

$$A_{mn} = \langle \psi_m | \hat{A} | \psi_n \rangle = \left(\langle \psi_n | \hat{A}^\dagger | \psi_m \rangle \right)^*$$
$$= \langle \psi_n | (\hat{A}^\dagger)^* | \psi_m \rangle = {}^t A_{nm} , \qquad (3.60)$$

where in the last line we have made use of the reality of the ψ_n's. Putting Eq. (3.60) and the corresponding equation for B_{nm} in Eq. (3.47) we obtain

$$\chi_{AB}(\omega) = \frac{1}{\hbar} \sum_{nm} \frac{P_m - P_n}{\omega - \omega_{nm} + i\eta} [{}^t \hat{B}]_{mn} [{}^t \hat{A}]_{nm} = \chi_{{}^t B^t A}(\omega) . \qquad (3.61)$$

Let us consider next the electron liquid in the presence of a magnetic field $\vec{B}$, which breaks time-reversal invariance. Now the eigenfunctions are no longer real but, if $\psi_n(\vec{B})$ is

---

[13] For hermitian operators this property follows directly from the reality of $\chi_{AB}(\tau)$.

an eigenfunction of energy E_n, then $\psi_n^*(\vec{B})$ is an eigenfunction of the same energy in a magnetic field $-\vec{B}$. Thus, the derivation proceeds as in the time-reversal-invariant case, except that in the last step we must write $\psi_n^*(\vec{B}) = \psi_n(-\vec{B})$. Since the energies are invariant under reversal of the magnetic field we conclude that, in general

$$\chi_{AB}(\omega, \vec{B}) = \chi_{B'A'}(\omega, -\vec{B}) , \tag{3.62}$$

where $\chi_{AB}(\omega, \vec{B})$ is the response function in the presence of a magnetic field $\vec{B}$.

The reciprocity relations are quite remarkable, because they establish a mathematical link between physical processes that are quite different. For example the density induced by a magnetic field parallel to the z-axis in a spin polarized system is related to the spin density induced by an electric potential in the same system: $\chi_{nS_z}(\omega) = \chi_{S_zn}(\omega)$.

We conclude this section by mentioning that many of the remarkable properties of the linear response functions associated with hermitian operators, like for instance Eq. (3.58), also hold for a particularly important class of non-hermitian operators in homogeneous systems with inversion symmetry. These are the "Fourier transforms" $\hat{A}_{\vec{q}} = \frac{1}{2} \sum_{i=1}^{N} \left[\hat{A}_i e^{-i\vec{q}\cdot\hat{\vec{r}}_i} + e^{-i\vec{q}\cdot\hat{\vec{r}}_i} \hat{A}_i \right]$ of hermitian operators of the form $\hat{A} = \sum_{i=1}^{N} \hat{A}_i$, where $\hat{A}_i$ is a hermitian operator acting on the i-th particle. Since $\hat{A}_{\vec{q}}^\dagger = \hat{A}_{-\vec{q}}$, it follows from inversion symmetry that $\chi_{A_{\vec{q}}A_{-\vec{q}}}(\omega)$ enjoys the property (3.58).

3.2.5 Origin of dissipation

Consider a system governed by a hamiltonian of the form

$$\hat{H}_F(t) = \hat{H} + F_\omega e^{-i(\omega+i\eta)t} \hat{A}^\dagger + F_\omega^* e^{i(\omega-i\eta)t} \hat{A} . \tag{3.63}$$

The average power delivered by the external field to the system during one period of oscillation (T) is

$$W = \frac{1}{T} \int_0^T \frac{\partial}{\partial t} \langle \psi(t) | \hat{H}_F(t) | \psi(t) \rangle dt , \tag{3.64}$$

that is, the change in energy divided by the time over which it takes place. Evidently, this would vanish for a strictly periodic time evolution: hence the crucial importance of the infinitesimal η.

The time derivative can be carried inside the thermal average thanks to the identity (reminiscent of the Hellmann–Feynman theorem)

$$\frac{\partial}{\partial t} \langle \psi(t) | \hat{H}_F(t) | \psi(t) \rangle = \langle \psi(t) | \frac{\partial \hat{H}_F(t)}{\partial t} - \frac{i}{\hbar} [\hat{H}_F(t), \hat{H}_F(t)] | \psi(t) \rangle$$

$$= \langle \psi(t) | \frac{\partial \hat{H}_F(t)}{\partial t} | \psi(t) \rangle , \tag{3.65}$$

which holds whenever $|\psi(t)\rangle$ satisfies the Schrödinger equation with the hamiltonian $\hat{H}_F(t)$. Substituting Eq. (3.63) in this expression, and neglecting terms that vanish for $\eta \to 0$, we obtain

$$W = \frac{1}{T} \int_0^T [-i\omega F_\omega e^{-i\omega t} \langle \hat{A}^\dagger \rangle_1(t) + i\omega F_\omega^* e^{i\omega t} \langle \hat{A} \rangle_1(t)] dt \; . \qquad (3.66)$$

The expectation values of $\hat{A}$ and $\hat{A}^\dagger$ are now evaluated using linear response theory. It is evident that only the component of $\langle \hat{A} \rangle_1(t)$ proportional to $e^{-i\omega t}$ and the component of $\langle \hat{A}^\dagger \rangle_1(t)$ proportional to $e^{i\omega t}$ give nonvanishing contributions to the integral over time. For the purpose of evaluating the absorption rate we can therefore write

$$\langle \hat{A} \rangle_1(t) \sim \Re e \chi_{AA^\dagger}(\omega) F_\omega e^{-i\omega t} + i \Im m \chi_{AA^\dagger}(\omega) F_\omega e^{-i\omega t} \; , \qquad (3.67)$$

and, similarly,

$$\langle \hat{A}^\dagger \rangle_1(t) \sim \Re e \chi_{A^\dagger A}(-\omega) F_\omega^* e^{i\omega t} + i \Im m \chi_{A^\dagger A}(-\omega) F_\omega^* e^{i\omega t} \; . \qquad (3.68)$$

These responses contain a component that is in phase with the driving field, and one that is 90° out-of-phase with it. The in-phase component describes a *polarization* process in which the wave function is modified in a periodic manner, but no energy is absorbed or released (on average) by the field. The out-of-phase component, on the other hand, gives rise to *energy absorption*. To see this, substitute Eqs. (3.67) and (3.68) in Eq. (3.66), integrate over time, and make use of the symmetry (3.56) $\chi_{AA^\dagger}(\omega) = \chi_{A^\dagger A}^*(-\omega)$. The result is simply

$$W = -2\omega \Im m \chi_{AA^\dagger}(\omega) |F_\omega|^2. \qquad (3.69)$$

Because, as previously noted, $\Im m \chi_{AA^\dagger}(\omega)$ is negative for positive ω and positive for negative ω, we see that the average rate of energy absorption from the external field is always *positive*, in accord with the second law of thermodynamics.[14]

At this point the reader may be left with the uneasy feeling of being the victim of some kind of mathematical trick. After all, we have already pointed out that, in the $\eta \to 0$ limit, Eq. (3.64) presents the integral over one period of the derivative of a periodic function of time: how can such an integral fail to be zero? From a physical point of view, one may note that while the system oscillates in time, its average energy ought to remain constant. This implies that any work done by the external field on the system is somehow dissipated in the environment rather than being fed into the system. The problem is that the hamiltonian description of the time evolution does not seem to offer any mechanism for this to happen.

The resolution of these difficulties requires a different discussion depending on whether one considers a finite system with a discrete spectrum of energy levels (for example, an atom), or an infinite system with a continuum spectrum of energy levels (for example, the electron liquid in the thermodynamic limit).

In a finite system, Eq. (3.51) predicts that there is no energy absorption unless the frequency of the field exactly matches one of the transition frequencies ω_{n0}. At $\omega = \omega_{n0}$ the

[14] In the general case that there are multiple fields F_i with conjugate variables A_i the formula (3.69) for the absorption rate becomes

$$W = -2\omega \sum_{ij} \Im m \chi_{A_i A_j^\dagger}(\omega) F_{i\omega}^* F_{j\omega} \; .$$

linear response theory breaks down: a careful study of the singularity in χ (similar to what one does in the theory of resonance in the classical forced harmonic oscillator) reveals that in this case the response of the system is no longer periodic – rather, the amplitude of the oscillations grows linearly in time no matter how small the external field is. This explains the non vanishing value of Eq. (3.64) and shows that the work done by the external field is stored in the growing amplitude of the oscillations. Paradoxically, the linear response theory allows us to calculate the energy absorption rate by failing to converge to a steady solution.

The convergence factor η regularizes the linear response theory by spreading the resonance over a finite range of frequency and thus leading to a steady periodic response of finite amplitude. Now the response is a periodic function of time, but the presence of η breaks the perfect periodicity of $H_F(t)$, allowing the integral in Eq. (3.64) to differ from zero. With a *finite* η the system does reach a steadily oscillating regime, but the energy supplied by the external field is *dissipated* through the coupling of the system to its *environment*.[15] This environment (which for a single atom might consist simply of the ever-present electromagnetic field) is not explicitly included in $\hat{H}$, but its existence is implied by the presence of a finite η. The role of η is thus seen to be analogous to that of the damping constant in the classical theory of the forced harmonic oscillator: it describes the coupling between the system and its environment.

Let us now consider an extended system in the thermodynamic limit. The novel feature is that dissipation now occurs over a continuous range of frequencies even in the limit $\eta \to 0$, provided the thermodynamic limit is taken first. This happens because in the thermodynamic limit the energy spectrum of the system is continuous: an infinite number of transition frequencies fall within a range of width η, no matter how small η is. Thus, in this sense, dissipation is an intrinsic feature of extended systems: it does not require the explicit presence of an "environment". One may say that an extended system is its own heat bath since it can carry away to infinity (in practice, out of any finite "observation box") the energy supplied by the driving field.

3.2.6 Time-dependent correlations and the fluctuation–dissipation theorem

We have seen that $\Im m \chi_{AA^\dagger}(\omega)$ describes the rate of energy absorption from a field that couples linearly to $\hat{A}^\dagger$. There exists a deep connection, known as the *fluctuation-dissipation theorem*, between this absorption rate and the spectrum of time-dependent fluctuations of $\hat{A}$ (Callen *et al.*, 1951). The latter are described by the so-called *dynamical structure factor*

$$S_{AA^\dagger}(\omega) = \frac{1}{2\pi} \int_{-\infty}^{\infty} \langle \hat{A}(t)\hat{A}^\dagger \rangle_0 e^{i\omega t} \, dt \,, \tag{3.70}$$

which can also be written as

$$S_{AA^\dagger}(\omega) = \sum_{nm} P_m |A_{mn}|^2 \delta(\omega - \omega_{nm}) \,, \tag{3.71}$$

---

[15] The environment of a system is sometimes described by the fancy words *heat bath*. This language emphasizes the randomness of many minute external influences acting on the system of interest.

a manifestly real and positive quantity. Notice that by setting $\hat{A} = \langle A \rangle_0 + \delta\hat{A}$, where $\langle A \rangle_0$ is the equilibrium average of $\hat{A}$ and $\delta\hat{A}$ is its fluctuation, one can rewrite Eq. (3.70) as

$$S_{AA^\dagger}(\omega) = |\langle\hat{A}\rangle_0|^2\delta(\omega) + \frac{1}{2\pi}\int_{-\infty}^{\infty}\langle\delta\hat{A}(t)\delta\hat{A}^\dagger\rangle_0 e^{i\omega t}dt \ . \tag{3.72}$$

Thus, we see that the behavior of $S_{AA^\dagger}(\omega)$ for $\omega \neq 0$, is indeed controlled by the dynamics of fluctuations.

The physical interpretation of $S_{AA^\dagger}(\omega)$ depends on whether the frequency is positive or negative. For $\omega > 0$, $S_{AA^\dagger}(\omega)$ gives the *absorption spectrum* from a field of frequency ω that couples linearly to $\hat{A}$, while, for $\omega < 0$, it gives the *stimulated emission spectrum* in the same field. Interchanging n and m and changing $\omega \to -\omega$ in Eq. (3.71) we see that the absorption and emission spectra are related by

$$S_{AA^\dagger}(-\omega) = e^{-\beta\hbar\omega}S_{A^\dagger A}(\omega) \ . \tag{3.73}$$

This relation implies that the emission spectrum vanishes identically at the absolute zero of temperature.

The dual meaning of $S_{AA^\dagger}(\omega)$ as absorption/emission spectrum and fluctuation autocorrelation spectrum is the mathematical basis of the fluctuation-dissipation theorem. This theorem asserts that

$$\Im m\chi_{AA^\dagger}(\omega) = -\frac{\pi}{\hbar}(1 - e^{-\beta\hbar\omega})S_{AA^\dagger}(\omega) \ . \tag{3.74}$$

Its proof follows immediately from Eq. (3.51), combined with the definition (3.71) and the relation (3.73). An equivalent but more symmetric form of the theorem is obtained by combining Eqs. (3.74) and (3.73):

$$\Im m\chi_{AA^\dagger}(\omega) = -\frac{\pi}{\hbar}\coth\left(\frac{\beta\hbar\omega}{2}\right)[S_{AA^\dagger}(\omega) + S_{A^\dagger A}(-\omega)] \ . \tag{3.75}$$

At $T = 0$ only the term with a positive argument survives, and the hyperbolic cotangent reduces to sign(ω). The opposite limit of high temperature (i.e., the classical limit) is worked out in Exercise 4.

From a physical point of view the existence of a connection between fluctuations and dissipation is required by the actual maintenance of thermal equilibrium. For, while dissipation causes an irreversible decay of every observable towards its equilibrium value, random fluctuations continuously occur as to preserve the correct statistical distribution of the values of the same observable.

The integral over all frequencies of the dynamical structure factor equals the average value of $\hat{A}\hat{A}^\dagger$ in the equilibrium state, i.e.,

$$\int_{-\infty}^{\infty}S_{AA^\dagger}(\omega)d\omega = \langle\hat{A}\hat{A}^\dagger\rangle_0$$

$$= |\langle\hat{A}\rangle_0|^2 + \langle\delta\hat{A}\delta\hat{A}^\dagger\rangle_0 \ . \tag{3.76}$$

This quantity is commonly referred to as the *static structure factor* even though "instantaneous structure factor" would be more accurate, since it depends on the *instantaneous* correlation between fluctuations of $\hat{A}$.

3.2.7 Analytic properties and collective modes

Because all the response functions vanish for negative times (a property directly associated with causality), it is intuitively clear what the *Titchmarsh lemma* rigorously states, namely that the Fourier transform $\chi_{AB}(\omega)$, regarded as a function of a complex frequency ω, is an analytic function of ω in the upper half of the complex plane (UCP). This is because a positive imaginary part in ω makes the factor $e^{i\omega t}$ decrease exponentially at large positive times, assuring the convergence of the integral in Eq. (3.45). The argument fails for ω in the lower half plane, or for a function that does not vanish at times earlier than some "initial time".

The analyticity property is confirmed by the more explicit Lehmann representation. In a *finite* system this representation shows that all the singularities of a linear response function are simple poles, located infinitesimally *below* the real axis in correspondence to the transition frequencies of the system.

In extended systems, the operation of taking the thermodynamic limit, introduces qualitatively new features. First of all, the response function that results from the sum over a continuous excitation spectrum has a *branch cut* just infinitesimally below the real axis: the cut arises from the merging of infinitely many poles of vanishing strength. Second, and most important, the analytic continuation of the response function from the upper half plane across the branch cut may have an isolated pole of finite strength in the *lower half* of the complex plane. Such a pole, occurring at a complex frequency $\Re e\,\omega + i\Im m\,\omega$, with $\Im m\,\omega < 0$, signals the existence of a collective resonant state – also known as a *collective mode* – with an energy approximately equal to $\hbar\Re e\,\omega$ and a lifetime proportional to $|\Im m\,\omega|^{-1}$. This description makes intuitive sense because an infinite susceptibility implies that the system can exhibit a finite dynamical "response" even in the absence of an external perturbation.[16] The presence of a pole near the real axis usually shows up as a sharp peak in the absorption spectrum $\Im m\,\chi(\omega)$: such a peak is readily identifiable above the structureless continuum of the exact eigenstates. The occurrence of sharply defined collective excitations – which are conceptually distinct from exact eigenstates – is a distinctive feature of systems with infinitely many degrees of freedom. The collective pole appears only after taking the thermodynamic limit, that is, after the sum over n has been replaced by a continuum integral. It should be realized that care must be exercised here, since this is quite different from what one would get by doing the analytic continuation *before* taking the thermodynamic limit.

Returning to our discussion of formal properties we note that the analyticity of χ_{AB} in the UCP makes the following contour integral vanish

$$\int_\Gamma \frac{\chi_{AB}(z)}{z - \omega} \frac{dz}{2\pi i} = 0 \,, \qquad (3.77)$$

---

[16] The pole must lie in the lower complex plane for the simple reason that only in such a case will the amplitude of the mode decay rather than grow out of bound with time.

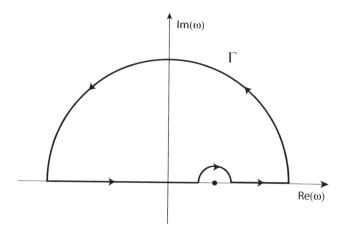

Fig. 3.1. Contour of integration for Eq. (3.77).

where the closed contour Γ is shown in Fig. 3.1, z is a complex variable, and ω a real frequency. By separating the contour integral into its contributions we obtain

$$\mathcal{P}\int_{-\infty}^{\infty}\frac{\chi_{AB}(\nu)}{\nu-\omega}\frac{d\nu}{2\pi i}-\frac{1}{2}\chi_{AB}(\omega)=0\,,\tag{3.78}$$

where the first term on the left-hand side comes from the part of the contour that lies along the real axis, and the second from the semicircle about $z=\omega$. As usual, $\mathcal{P}$ represents the Cauchy–Hadamard principal part operator, defined in Eq. (3.49). The contribution from the semicircle at infinity vanishes because $\chi_{AB}(z)\to 0$ for $|z|\to\infty$.

A slight rearrangement of Eq. (3.78) yields the *Kramers–Krönig dispersion relation*:

$$\chi_{AB}(\omega)=-i\,\mathcal{P}\int_{-\infty}^{\infty}\frac{\chi_{AB}(\nu)}{\nu-\omega}\frac{d\nu}{\pi}\,.\tag{3.79}$$

A more familiar form of this important result, which separates the real and imaginary parts of χ_{AB} and makes use of the parity relations (3.58) (and therefore is valid only when they hold) is provided by the formulas

$$\mathfrak{Re}\,\chi_{AB}(\omega)=\frac{2}{\pi}\mathcal{P}\int_{0}^{\infty}\frac{\nu\,\mathfrak{Im}\,\chi_{AB}(\nu)}{\nu^2-\omega^2}d\nu\tag{3.80}$$

and

$$\mathfrak{Im}\,\chi_{AB}(\omega)=-\frac{2\omega}{\pi}\mathcal{P}\int_{0}^{\infty}\frac{\mathfrak{Re}\,\chi_{AB}(\nu)}{\nu^2-\omega^2}d\nu\,.\tag{3.81}$$

The dispersion relations apply to all causal response functions. They establish a non-trivial relationship between polarizability and absorption and have quite a variety of useful applications. In typical cases, one first calculates the imaginary part, which is simpler, and then obtains the real part from the one-dimensional integral. They also allow us to confirm that $\chi_{AA'}(0)\le 0$ – a fact already noted in Section 3.2.4.

We now turn to more specific properties of the response function $\chi_{AA^\dagger}(\omega)$ again in the case in which the parity relations of Eq. (3.58) hold. As inferred from Eq. (3.51) and discussed in Section 3.2.5 in the context of dissipation, we have $\omega\Im m\,\chi_{AA^\dagger}(\omega) \le 0$. From this fact, we can deduce an interesting property: $\chi_{AA^\dagger}(\omega)$ not only is analytic, but also *has no zeroes* on the UCP except, possibly, on the real axis itself. This means that its *inverse* $\chi_{AA^\dagger}^{-1}$ is also an analytic function of ω in the UCP, except for possible divergencies at infinity and on the real axis.

The proof is as follows. The response function $\chi_{AA^\dagger}(z)$ is analytic when z is in the UCP and therefore coincides, in this domain, with the extended function $\tilde{\chi}_{AA^\dagger}(z)$ defined in Eq. (3.52). Setting $z = z_1 + iz_2$, where z_1 is real and z_2 is real and positive, and making use of the spectral representation (3.53) and of the parity relations (3.58) we find

$$\Im m\,\chi_{AA^\dagger}(z) = \frac{4z_1z_2}{\pi}\int_0^\infty \frac{\nu\,\Im m\,\chi_{AA^\dagger}(\nu)}{\left|\nu^2 - z^2\right|^2}d\nu . \tag{3.82}$$

Because $\Im m\,\chi_{AA^\dagger}(\omega)\omega$ is always negative, we see that $\Im m\,\chi_{AA^\dagger}(z)$ is zero along the imaginary axis ($z_1 = 0$) and may also vanish along portions of the real axis, but nowhere else. Consequently, the zeroes of the full response function $\chi_{AA^\dagger}(z) = \Re e\,\chi_{AA^\dagger}(z) + i\Im m\,\chi_{AA^\dagger}(z)$, can only occur either on the imaginary axis or on the real axis. The first possibility is ruled out by noting that the real part of $\chi_{AA^\dagger}(z)$ along the imaginary axis $z = iz_2$ is given by Eq. (3.53) as

$$\Re e\,\chi_{AA^\dagger}(iz_2) = \frac{2}{\pi}\int_0^\infty \frac{\nu\,\Im m\,\chi_{AA^\dagger}(\nu)}{(\nu)^2 + z_2^2}d\nu , \tag{3.83}$$

which is a *negative, monotonically increasing function of z_2*. We conclude that $\chi_{AA^\dagger}(z)$ has no zeroes in the upper half of the complex plane (excluding the real axis).

3.2.8 Sum rules

One of the best known and useful applications of the dispersion relations is the derivation of *sum rules* – identities that connect the moments of the absorption spectrum to ground-state (equilibrium) averages of observables.[17]

In order to derive general sum rules, we start from the dispersion relation (3.80), and note that the spectral function $\Im m\,\chi_{AB}(\omega)$ quickly falls to zero for $\omega > \omega_{max}$, where ω_{max} is a characteristic cutoff frequency of the excitation spectrum. For frequencies $\omega \gg \omega_{max}$, we can therefore expand the denominator in the integral of Eq. (3.80) as follows:

$$\frac{1}{\omega' - \omega} = -\sum_{k=0}^\infty \frac{\omega'^k}{\omega^{k+1}} . \tag{3.84}$$

In the case that the relations (3.58) are satisfied, the integrals containing even powers of ω' vanish by symmetry and we obtain the following high-frequency expansion for the real

---

[17] While our treatment concerns hermitian operators, the results have a wider range of application (see the end of Section 3.2.4).

part of $\chi_{AB}(\omega)$:

$$\Re\chi_{AB}(\omega) = \sum_{k=0}^{\infty} \frac{M_{AB}^{(2k+1)}}{\omega^{2k+2}}, \qquad (3.85)$$

where the quantities

$$M_{AB}^{(2k+1)} \equiv -\frac{2}{\pi} \int_0^{\infty} \omega^{2k+1} \Im\chi_{AB}(\omega) d\omega \qquad (3.86)$$

are the *odd moments* of the spectral function $-\pi^{-1}\Im\chi_{AB}(\omega)$.

Of course the series (3.85) has a purely formal meaning because, unless the excitation spectrum happens to vanish exponentially for large ω, the integral (3.86) diverges for k larger than a certain minimum value, and the corresponding moment does not exist. However, the first few moments are in fact usually finite, and they control the strength of the leading terms in the high-frequency expansion. More importantly, Eq. (3.85) enables us to make a connection between the spectral moments and the short-time behavior of the response function.

To see this, let us go back to Eq. (3.36) and expand $\chi_{AB}(\tau)$ in a Taylor series of τ:

$$\chi_{AB}(\tau) = -\frac{i}{\hbar} \sum_{l=0}^{\infty} \frac{\tau^l}{l!} \langle [\hat{A}^{(l)}, \hat{B}] \rangle_0 , \qquad (3.87)$$

where we have defined

$$\hat{A}^{(l)} = \left(\frac{d^l \hat{A}(\tau)}{d\tau^l} \right)_{\tau=0}$$

$$= \left(\frac{i}{\hbar} \right)^l [\hat{H}, [\hat{H}, \ldots, [\hat{H}, \hat{A}] \ldots]] , \qquad (3.88)$$

an expression that determines the *l*-th time derivative of $\hat{A}(\tau)$ at $\tau = 0$ in terms of l nested commutators.[18] Substituting this series into Eq. (3.45) for the frequency-dependent susceptibility, and performing the time integral, we obtain an alternative form of the high-frequency expansion

$$\chi_{AB}(\omega) = \frac{1}{\hbar} \sum_{l=0}^{\infty} \frac{i^l}{\omega^{l+1}} \langle [\hat{A}^{(l)}, \hat{B}] \rangle_0 . \qquad (3.89)$$

Unlike Eq. (3.85) this expansion applies to both the real and the imaginary parts of χ_{AB}. We then extract the real part by noting that the commutator of two hermitian operators is anti-hermitian and therefore all its expectation values are purely imaginary. Then, setting $l = 2k + 1$ we obtain

$$\Re\chi_{AB}(\omega) = \frac{i}{\hbar} \sum_{k=0}^{\infty} (-1)^k \frac{\langle [\hat{A}^{(2k+1)}, \hat{B}] \rangle_0}{\omega^{2k+2}}. \qquad (3.90)$$

---

[18] By definition, for $l = 0$, $\hat{A}^{(0)}$ coincides with $\hat{A}$.

Comparing Eq. (3.90) and Eq. (3.85) we arrive at the sum rules

$$M_{AB}^{(2k+1)} \equiv -\frac{2}{\pi}\int_0^\infty \omega^{2k+1}\Im m\,\chi_{AB}(\omega)d\omega = \frac{i}{\hbar}(-1)^k\langle[\hat{A}^{(2k+1)},\hat{B}]\rangle_0\,, \qquad (3.91)$$

which connect any *odd* moment of the spectrum to the equilibrium expectation value of an equal-time commutator (assuming, of course, that the integral in Eq. (3.91) exists).

What about the *even* moments of the spectrum? The dispersion relations do not tell us anything about them, but they can be evaluated directly from Eq. (3.51). Thus, extending in an obvious way the definition of Eq. (3.86) we find:

$$M_{AB}^{(2k)} = \frac{2}{\hbar}\sum_{nm} P_m[A_{mn}B_{nm}\Theta(\omega_{nm}) - B_{mn}A_{nm}\Theta(\omega_{mn})]\omega_{nm}^{2k}\,. \qquad (3.92)$$

This cannot be reduced to an equilibrium average, *except in the special case of zero temperature*, in which $P_m = \delta_{m0}$ and all the ω_{nm} are positive, so that (3.92) simplifies to

$$M_{AB}^{(2k)}(T=0) = \frac{2}{\hbar}(-1)^k\langle\hat{A}^{(k)}\hat{B}^{(k)}\rangle_0\bigg|_{T=0}. \qquad (3.93)$$

Applications of the moment sum rules will be presented in Section 3.3.

3.2.9 The stiffness theorem

As we have seen in Sections 3.1 and 3.2.4, the inverse of the static response function $\chi_{AA}(0)$ is a measure of the stiffness of the system against a field that attempts to change the value of $\langle\hat{A}\rangle$. Here we provide a formal proof of this assertion.

We begin by giving a precise definition of the notion of stiffness for a single observable $\hat{A}$ such that $\hat{A}^\dagger = \hat{A}$. For simplicity, we work at $T = 0$, but all that follows is generalizable to finite temperature ensembles. We can assume, without loss of generality, that the expectation value of $\hat{A}$ in the ground-state is zero.[19] Let $E(A)$ be the lowest possible energy of the system subject to the constraint that the expectation value of $\hat{A}$ has a given value $A \neq 0$:

$$E(A) \equiv \min_{\psi \to A}\langle\psi|\hat{H}|\psi\rangle\,, \qquad (3.94)$$

where the subscript $\psi \to A$ indicates that the minimum must be sought among the normalized antisymmetric wave functions which yield expectation value $\langle\psi|\hat{A}|\psi\rangle = A$. Hence $E(0)$ is the ground-state energy and $E(A) \geq E(0)$ for any $A \neq 0$: for small A, this implies that $E(A) - E(0) = \alpha A^2/2$, with $\alpha > 0$, if $E(A)$ is an analytic function of A. The constant α is by definition, the *stiffness* of the system with respect to a change in A. We are now in a position to prove the *stiffness theorem* which, for the case of just one variable, takes the form

$$\alpha = -\frac{1}{\chi_{AA}(0)}\,. \qquad (3.95)$$

---

[19] If $\hat{A}$ has a nonvanishing ground-state expectation value A_0, all the following considerations apply to the operator $\hat{A} - A_0$.

The proof is quite simple. First note that, in the limit of $A \to 0$, the solution of the minimization problem posed by Eq. (3.94) is the ground-state $|\psi_A\rangle$ of the hamiltonian

$$\hat{H}_A = \hat{H} + F_A \hat{A} \tag{3.96}$$

where F_A is given by

$$F_A = \frac{A}{\chi_{AA}(0)} . \tag{3.97}$$

That the expectation value of $\hat{A}$ in $|\psi_A\rangle$ has the desired value A follows from the very definition of $\chi_{AA}(0)$. Furthermore, it is easy to see that

$$E(A) = \langle \psi_A | \hat{H} | \psi_A \rangle , \tag{3.98}$$

for, if there existed another state $|\psi'_A\rangle$ with the same expectation value for $\hat{A}$, but with a lower expectation value for $\hat{H}$, then we would have

$$\langle \psi'_A | \hat{H}_A | \psi'_A \rangle < \langle \psi_A | \hat{H}_A | \psi_A \rangle , \tag{3.99}$$

which is impossible, since $|\psi_A\rangle$ is the ground-state of $\hat{H}_A$.

Let us then calculate

$$E(A) = \langle \psi_A | \hat{H}_A | \psi_A \rangle - A F_A . \tag{3.100}$$

To calculate the first term on the right-hand side, we introduce the hamiltonian

$$\hat{H}_A(\lambda) = \hat{H} + \lambda F_A \hat{A} \tag{3.101}$$

where $0 < \lambda < 1$, and we denote by $|\psi_A(\lambda)\rangle$ its ground-state, and by $E_A(\lambda)$ its ground-state energy. To first order in F_A, the expectation value of $\hat{A}$ in $|\psi_A(\lambda)\rangle$ is

$$A(\lambda) = \langle \psi_A(\lambda) | \hat{A} | \psi_A(\lambda) \rangle \approx \lambda \chi_{AA}(0) F_A . \tag{3.102}$$

Making use of the Hellman–Feynman theorem (see Chapter 1) we can write

$$\begin{aligned} E_A(1) = \langle \psi_A | \hat{H}_A | \psi_A \rangle &= E(0) + \int_0^1 d\lambda \langle \psi_A(\lambda) | \frac{\partial \hat{H}_A(\lambda)}{\partial \lambda} | \psi_A(\lambda) \rangle \\ &\approx E(0) + \int_0^1 d\lambda \, \lambda \chi_{AA}(0) F_A^2 \\ &= E(0) + \frac{1}{2} \chi_{AA}(0) F_A^2 . \end{aligned} \tag{3.103}$$

Substituting this for the first term on the right-hand side of Eq. (3.100) and making use of Eq. (3.97) for the field F_A in the second term, we finally obtain

$$E(A) \approx E(0) - \frac{1}{2} \frac{A^2}{\chi_{AA}(0)} , \tag{3.104}$$

which is what we wanted to prove.[20]

---

[20] It must be firmly kept in mind that the change in the internal energy $E(A) - E(0)$ – a positive quantity – is the opposite of the change in the total energy $E_A(1) - E_A(0)$.

The formulation of the stiffness theorem can be readily generalized to the case of n variables $\hat{A}_i$, $i = 1, \ldots, n$. Accordingly for the coefficients α_{ij} of the terms $A_i A_j$ in the expansion of the energy one finds the expression

$$\alpha_{ij} = -\chi^{-1}_{A_i A_j}(0) , \tag{3.105}$$

where $\chi^{-1}_{A_i A_j}$ is the matrix inverse of $\chi_{A_i A_j}$. This formula will be extremely useful in later applications (see, e.g., Section 3.3.4).

As a final point we notice that, beside being necessarily negative for stability, the static response function at finite temperature must also satisfy the inequality

$$|\chi_{AA}(0)| < \frac{\langle \hat{A}^2 \rangle_0}{k_B T} \tag{3.106}$$

(see Exercise 6). It is not difficult to prove that the expression on the right-hand side is just the classical static susceptibility. Thus, Eq. (3.106) implies that the quantum static susceptibility is always smaller than its classical counterpart.

3.2.10 Bogoliubov inequality

Another interesting and general property of response functions is the *Bogoliubov inequality*. This asserts that for arbitrary operators $\hat{A}$ and $\hat{C}$ the static susceptibility $\chi_{A^\dagger A}(0)$ satisfies the inequality

$$\chi_{A^\dagger A}(0) \langle [[\hat{H}, \hat{C}], \hat{C}^\dagger]] \rangle_0 \geq |\langle [\hat{A}, \hat{C}] \rangle_0|^2. \tag{3.107}$$

One way to prove this result is to notice that the static response function can be used to define a "scalar product" of operators according to the definition

$$(\hat{A}, \hat{B}) \equiv -\chi_{A^\dagger B}(0) . \tag{3.108}$$

It is easy to verify that this product has all the formal properties of a scalar product, namely, it is linear in $\hat{B}$ and anti-linear in $\hat{A}$, it satisfies the symmetry $(\hat{A}, \hat{B}) = (\hat{B}, \hat{A})^*$, and it is positive semidefinite because $\chi_{A^\dagger A}(0) \leq 0$ if the ground-state is stable (see Eq. (3.54)). Given these properties, the Schwartz inequality

$$(\hat{A}, \hat{A})(\hat{B}, \hat{B}) \geq |(\hat{A}, \hat{B})|^2 \tag{3.109}$$

follows automatically. Setting $\hat{B} = [\hat{H}, \hat{C}^\dagger]$ we may write

$$(\hat{A}, \hat{B}) = \sum_{mn}(P_m - P_n)A^\dagger_{mn}C^\dagger_{nm} = \langle [\hat{A}^\dagger, \hat{C}^\dagger] \rangle_0 , \tag{3.110}$$

and

$$(\hat{B}, \hat{B}) = \sum_{mn}(P_m - P_n)\omega_{nm}C_{mn}C^\dagger_{nm} = -\langle [[\hat{H}, \hat{C}], \hat{C}^\dagger] \rangle_0 . \tag{3.111}$$

The inequality (3.107) results from inserting Eqs. (3.110) and (3.111) in Eq. (3.109).[21]

---

[21] A physically transparent derivation of the Bogoliubov inequality for two hermitian operators $\hat{A}$ and $\hat{C}$ at $T = 0$ can be given as follows. Consider the "distorted ground-state" obtained by applying the infinitesimal unitary transformation $\exp[i\lambda\hat{C}]$

Over the years, there have been several interesting applications of the Bogoliubov inequality. The most famous is perhaps the proof of the Mermin–Wagner theorem (Mermin and Wagner, 1966), which asserts the absence of long-range order in a class of two-dimensional statistical mechanical models.

For an application of Eq. (3.107) to the electron liquid see Exercise 9, where we show how the third and the first moments of the density–density response function and the density stiffness are constrained by the Bogoliubov inequality.

3.2.11 Adiabatic versus isothermal response

Up to this point we have uncritically assumed that the static response of the system can be obtained simply by setting $\omega = 0$ in the expression for the frequency-dependent response. We now show that this is a somewhat dangerous assumption. It often works, but sometimes it fails in a spectacular way (see for instance the examples discussed in Appendix 8). This section is devoted to the clarification of the conditions under which the static response and the $\omega = 0$ limit of the dynamical response coincide.

The time-dependent LRT is based on the assumption that each state of the initial equilibrium ensemble undergoes a unitary time evolution with the hamiltonian $\hat{H}_F(t)$. Such a unitary evolution does not allow for discontinuous transitions (also known as *level crossings*) between orthogonal states:[22] the occupation probabilities of these states (the P_n's in Eq. (3.46)) therefore remain constant in time.

This assumption is correct only if the system is kept (after the initial preparation) strictly isolated from the rest of the world. More realistically, the measurements must be performed under *adiabatic* conditions, that is, the heat exchange with the environment must be kept at a negligible level. In this sense, the $\omega \to 0$ limit of the frequency-dependent susceptibility of Eq. (3.47) yields the *adiabatic static susceptibility* of the system at hand.

Now it is known even to theorists that thermal insulation becomes more and more difficult to maintain as the time-scale of the measurement becomes longer and longer. Therefore, as the frequency of the perturbing field tends to zero, thermal interactions with the external environment become more and more important, and finally dominate the time evolution. These interactions can and do cause transitions between different orthogonal states, so that the correct equilibrium populations may be established. Eventually, one reaches a state of quasi-equilibrium in which the density matrix has the canonical form

$$\hat{\rho}_F(t) = \frac{e^{-\beta \hat{H}_F(t)}}{\mathcal{Z}_F(t)} , \qquad (3.112)$$

(λ infinitesimal) to the ground-state. The energy of the distorted ground-state, to second order in λ, is $-\frac{\lambda^2}{2}\langle[[\hat{H},\hat{C}],\hat{C}]\rangle_0$ above the energy of the undistorted ground-state. On the other hand, the change in the expectation value of $\hat{A}$ due to this transformation is given by $i\lambda\langle[\hat{C},\hat{A}]\rangle_0$. According to the stiffness theorem, the minimum energy required to effect such a change in the expectation value of $\hat{A}$ is $-\frac{1}{2}\lambda^2 \chi_{AA}^{-1}(0)|\langle[\hat{C},\hat{A}]\rangle_0|^2$. Requiring that the excess energy of the distorted ground-state be greater than this minimum energy yields the Bogoliubov inequality.

[22] Recall that, under a unitary transformation, states that are initially orthogonal, remain orthogonal at all times.

where $\mathcal{Z}_F(t) = Tr[e^{-\beta \hat{H}_F(t)}]$ is the quasi-equilibrium partition function in the presence of the field $F(t)$.

The quasi-equilibrium time evolution described by Eq. (3.112) is conceptually different from the unitary time evolution described by Eq. (3.25), as it allows for "level crossings". Correspondingly, the *isothermal static susceptibility* defined as

$$\chi_{AB}^{iso} = \lim_{F \to 0} \frac{Tr[\hat{A}\hat{\rho}_F] - Tr[\hat{A}\hat{\rho}_{F=0}]}{F} \qquad (3.113)$$

differs, in general, from the adiabatic static susceptibility.

In order to see more clearly the origin of the difference, let us rewrite Eq. (3.113) in a way that closely resembles Eq. (3.47) for the adiabatic susceptibility. We expand the density matrix, $\hat{\rho}_F$, to first order in F with the help of the useful formula[23]

$$e^{-\beta \hat{H}_F(t)} = e^{-\beta \hat{H}} \left(1 - F \int_0^\beta d\beta' e^{\beta' \hat{H}} \hat{B} e^{-\beta' \hat{H}} \right) + \mathcal{O}(F^2). \qquad (3.114)$$

Substitution in Eq. (3.113) and a little manipulation leads to

$$\chi_{AB}^{iso} = - \int_0^\beta d\beta' \langle e^{\beta' \hat{H}} \hat{B} e^{-\beta' \hat{H}} \hat{A} \rangle_0 - \beta \langle \hat{A} \rangle_0 \langle \hat{B} \rangle_0. \qquad (3.115)$$

The thermal average under the integral sign can be rewritten as

$$\langle e^{\beta' \hat{H}} \hat{B} e^{-\beta' \hat{H}} \hat{A} \rangle_0 = \sum_{nm} P_m B_{mn} A_{nm} e^{\beta' \hbar \omega_{mn}}, \qquad (3.116)$$

and the integral over β' can be done analytically, yielding

$$\int_0^\beta d\beta' e^{\beta' \hbar \omega_{mn}} = \frac{e^{\beta \hbar \omega_{mn}} - 1}{\hbar \omega_{mn}} (1 - \delta_{mn}) + \beta \delta_{mn}. \qquad (3.117)$$

Putting all the pieces together, we obtain the Lehmann representation for the isothermal susceptibility

$$\chi_{AB}^{iso} = \frac{1}{\hbar} \sum_{n \neq m} \frac{P_m - P_n}{\omega_{mn}} A_{mn} B_{nm} - \beta \left(\sum_n P_n A_{nn} B_{nn} - \langle \hat{A} \rangle_0 \langle \hat{B} \rangle_0 \right). \qquad (3.118)$$

The first term on the right-hand side of this expression is easily seen to coincide with the $\omega \to 0$ limit of the adiabatic response function (3.47). The last two terms can be written in terms of the conserved parts of the observables $\hat{A}$ and $\hat{B}$, i.e., in a representation in which $\hat{H}$ is diagonal, the operators whose matrix elements are $(\hat{A}_0)_{nm} = \hat{A}_{nn} \delta_{nm}$ and $(\hat{B}_0)_{nm} = \hat{B}_{nn} \delta_{nm}$. Thus, in summary we find the relationship

$$\chi_{AB}^{iso} = \lim_{\omega \to 0} \chi_{AB}(\omega) - \beta[\langle \hat{A}_0 \hat{B}_0 \rangle_0 - \langle \hat{A}_0 \rangle_0 \langle \hat{B}_0 \rangle_0]. \qquad (3.119)$$

The last term, which manifestly represents a correlation between the diagonal matrix elements of $\hat{A}$ and $\hat{B}$, does not necessarily vanish, so the isothermal and adiabatic responses

---

[23] A little thought shows that this the same as Eq. (3.30) written in imaginary time.

can differ. Two important examples, in which the isothermal and adiabatic susceptibilities differ dramatically are discussed in Section 3.4.4 and in Appendix 8.

It is indeed fortunate that in most cases of interest what one needs to calculate is the response of operators whose conserved part is zero. A typical example is the Fourier component of a density operator at finite wave vector $\vec{q}$: its diagonal elements in the energy representation vanish by translational invariance. This is why a macroscopic isothermal susceptibility can be safely calculated as the $q \to 0$ limit of a static response function $\chi(q, 0)$, whereas the $\omega \to 0$ limit of $\chi(0, \omega)$ would yield the wrong result.

3.3 Density response

To illustrate the general theory, in this section we begin the study of specific response functions of the electron liquid. We shall only be concerned with general properties: approximations will be presented in Chapters 4–6.

3.3.1 The density–density response function

The *density–density response function* $\chi_{nn}(\vec{r}, \vec{r}', t)$ describes the response of the expectation value of the number density operator

$$\hat{n}(\vec{r}) = \sum_i \delta(\vec{r} - \hat{\vec{r}}_i) , \qquad (3.120)$$

at point $\vec{r}$ to a scalar potential $V_{ext}(\vec{r}', t)$ that couples linearly to the density $\hat{n}(\vec{r}')$ at the point $\vec{r}'$. Thus, in this case, the perturbed hamiltonian is

$$\hat{H}_V(t) = \hat{H} + \int V_{ext}(\vec{r}', t)\hat{n}(\vec{r}')d\vec{r}' . \qquad (3.121)$$

Accordingly the induced time-dependent density $n_1(\vec{r}, t)$ is given by (see Eq. (3.38))

$$n_1(\vec{r}, t) = \int_0^\infty d\tau \int d\vec{r}' \, \chi_{nn}(\vec{r}, \vec{r}', \tau)V_{ext}(\vec{r}', t - \tau) , \qquad (3.122)$$

where we have introduced the notation

$$\chi_{nn}(\vec{r}, \vec{r}', t) \equiv \chi_{n(\vec{r})n(r')}(t) = -\frac{i}{\hbar}\Theta(t)\langle[\hat{n}(\vec{r}, t), \hat{n}(\vec{r}')]\rangle_0 . \qquad (3.123)$$

We are now in a position to calculate the linear response to an external potential of arbitrary shape, both in space and in time. The most important case is, of course, that of a periodic potential

$$V_{ext}(\vec{r}', t) = \frac{1}{L^d} V_{ext}(\vec{q}', \omega)e^{i(\vec{q}'\cdot\vec{r}' - \omega t)} + c.c. , \qquad (3.124)$$

the general situation being the result of a linear superposition of potentials of this form. Substituting Eq. (3.124) in (3.122) we readily obtain

$$n_1(\vec{r}, t) = \frac{1}{L^d} \sum_{\vec{q}} n_1(\vec{q}, \omega) e^{i(\vec{q} \cdot \vec{r} - \omega t)} + c.c. , \qquad (3.125)$$

where

$$n_1(\vec{q}, \omega) = \chi_{nn}(\vec{q}, \vec{q}', \omega) V_{ext}(\vec{q}', \omega) , \qquad (3.126)$$

and we have defined

$$\chi_{nn}(\vec{q}, \vec{q}', \omega) \equiv \frac{1}{L^d} \int d^d r \, e^{-i\vec{q} \cdot \vec{r}} \int d^d r' e^{i\vec{q}' \cdot \vec{r}'} \int_0^\infty dt \, \chi_{nn}(\vec{r}, \vec{r}', t) e^{i\omega t} . \qquad (3.127)$$

Combining Eqs. (3.127) and (3.123) one can readily see that:

$$\chi_{nn}(\vec{q}, \vec{q}', \omega) = \frac{1}{L^d} \chi_{n_{\vec{q}} n_{-\vec{q}'}}(\omega) , \qquad (3.128)$$

where $\hat{n}_{\vec{q}} = \sum_i e^{-i\vec{q} \cdot \hat{r}_i}$ is the density fluctuation operator at wave vector $\vec{q}$.

Notice that in general a perturbation with wave vector $\vec{q}'$ induces a density modulation at all the wave vectors $\vec{q}$ for which the response function $\chi_{nn}(\vec{q}, \vec{q}', \omega)$ differs from zero. A major simplification occurs in translationally invariant systems, where $\chi_{nn}(\vec{r}, \vec{r}', \tau)$ only depends on the difference $\vec{r} - \vec{r}'$.[24] Then one of the spatial integrals in Eq. (3.127) vanishes unless $\vec{q} = \vec{q}'$ and we are left with the simpler form

$$\chi_{nn}(\vec{q}, \vec{q}', \omega) = \chi_{nn}(\vec{q}, \omega) \delta_{\vec{q}, \vec{q}'} , \qquad (3.129)$$

where, according to Eqs. (3.47) and (3.128), the explicit expression for $\chi_{nn}(\vec{q}, \omega)$ is

$$\chi_{nn}(\vec{q}, \omega) = \frac{1}{L^d} \chi_{n_{\vec{q}} n_{-\vec{q}}}(\omega) = \frac{1}{\hbar L^d} \sum_{nm} \frac{P_m - P_n}{\omega - \omega_{nm} + i\eta} |(\hat{n}_{\vec{q}})_{nm}|^2 . \qquad (3.130)$$

In particular for zero temperature this gives

$$\Re e \, \chi_{nn}(\vec{q}, \omega) = \frac{1}{\hbar L^d} \sum_n \left(\frac{|(\hat{n}_{\vec{q}})_{n0}|^2}{\omega - \omega_{n0}} - \frac{|(\hat{n}_{-\vec{q}})_{n0}|^2}{\omega + \omega_{n0}} \right) , \qquad (3.131)$$

and

$$\Im m \, \chi_{nn}(\vec{q}, \omega) = -\frac{\pi}{\hbar L^d} \sum_n \left[|(\hat{n}_{\vec{q}})_{n0}|^2 \delta(\omega - \omega_{n0}) - |(\hat{n}_{-\vec{q}})_{n0}|^2 \delta(\omega + \omega_{n0}) \right] , \qquad (3.132)$$

where we have used the relation $n_{\vec{q}}^\dagger = \hat{n}_{-\vec{q}}$.

From the above formulas one sees that $\chi_{nn}(q, 0)$ satisfies the stability condition

$$\chi_{nn}(\vec{q}, 0) \leq 0 , \qquad (3.133)$$

---

[24] This relation also applies to what are referred to as self-averaging response functions in a disordered system. The idea of self-averaging quantities is again touched upon in Section 3.4 and revisited in Section 8.9.

as expected from the general theory of Section 3.2.4: this result will be used in Section 5.2 to determine the physically admissible range of values of the static dielectric constant of matter. Furthermore, by making use of inversion symmetry and of Eq. (3.57), we find that $\chi_{nn}(\vec{q}, \omega)$ satisfies the symmetry relations[25]

$$\chi_{nn}(-\vec{q}, \omega) = \chi_{nn}(\vec{q}, \omega) ,$$
$$\Re e\, \chi_{nn}(\vec{q}, -\omega) = \Re e\, \chi_{nn}(\vec{q}, \omega) ,$$
$$\Im m\, \chi_{nn}(\vec{q}, -\omega) = -\Im m\, \chi_{nn}(\vec{q}, \omega) . \tag{3.134}$$

Another case of great interest is that of *spatially periodic structures*, in which case the function $\chi_{nn}(\vec{q}, \vec{q}', \omega)$ is finite only when the vectors $\vec{q}$ and $\vec{q}'$ differ by a reciprocal lattice vector. Therefore $\chi_{nn}(\vec{q}, \vec{q}', \omega)$ has the structure $\chi_{nn}(\vec{k} + \vec{G}, \vec{k} + \vec{G}', \omega)$ where $\vec{G}$ and $\vec{G}'$ are reciprocal lattice vectors and the vector $\vec{k}$ lies within the first Brillouin zone for the system at hand. The density response of a crystal is discussed in Appendix 7.

3.3.2 The density structure factor

The density structure factor (or structure factor *tout court*) plays an important role in the theory of the electron liquid. Its dynamic version is defined via Eq. (3.71) as follows:

$$S(\vec{q}, \omega) = S_{n_{\vec{q}} n_{-\vec{q}}}(\omega) = \sum_{nm} P_m |(\hat{n}_{\vec{q}})_{nm}|^2 \delta(\omega - \omega_{nm}) . \tag{3.135}$$

Inversion symmetry immediately gives $S(-\vec{q}, \omega) = S(\vec{q}, \omega)$, a result which, together with the relationships of Section 3.2.6, leads us to the formula[26]

$$S(\vec{q}, -\omega) = e^{-\beta\hbar\omega} S(\vec{q}, \omega) , \tag{3.136}$$

and to the density version of the fluctuation-dissipation theorem:

$$\Im m\, \chi_{nn}(\vec{q}, \omega) = -\frac{\pi}{\hbar L^d}(1 - e^{-\beta\hbar\omega})S(\vec{q}, \omega) . \tag{3.137}$$

By integrating $S(\vec{q}, \omega)$ over the frequency and dividing by the total number of electrons we then obtain the static structure factor

$$S(q) = \frac{1}{N} \int_{-\infty}^{\infty} S(\vec{q}, \omega) . \tag{3.138}$$

From Eq. (3.76) we see that $S(q)$ can be expressed as

$$S(q) = \frac{\langle \hat{n}_{-\vec{q}} \hat{n}_{\vec{q}} \rangle_0}{N} , \tag{3.139}$$

an equation that was derived in Appendix 4 in connection with our discussion of the static pair correlation function $g(r)$.

---

[25] See also the discussion at the end of Section 3.2.4.
[26] That $S(-\vec{q}, \omega) = S(\vec{q}, \omega)$ also follows from the assumption of time reversal and is in this case valid also for non homogeneous systems.

3.3.3 High-frequency behavior and sum rules

In view of the parity relations of Eq. (3.134), and the simple form of the operators involved, both the expansion (3.85) and the formula (3.91) for the odd moments apply to $\chi_{nn}(\vec{q}, \omega)$. Accordingly the high-frequency behavior of this function is given by

$$\Re\chi_{nn}(\vec{q}, \omega) \stackrel{\omega \to \infty}{\to} \frac{M_{nn}^{(1)}(\vec{q})}{\omega^2} + \frac{M_{nn}^{(3)}(\vec{q})}{\omega^4} , \tag{3.140}$$

where $M_{nn}^{(1)}(\vec{q})$ and $M_{nn}^{(3)}(\vec{q})$ are the first and third moments of the density–density fluctuation spectrum and are calculated directly from Eq. (3.91).[27] For ease of reference we first state the results and then sketch the calculations.

The f-sum rule:

$$M_{nn}^{(1)}(\vec{q}) = -\frac{2}{\pi}\int_0^{\infty}\omega\Im\chi_{nn}(\vec{q}, \omega)d\omega = \frac{nq^2}{m} . \tag{3.141}$$

The third-moment sum rule:

$$M_{nn}^{(3)}(\vec{q}) = -\frac{2}{\pi}\int_0^{\infty}\omega^3\Im\chi_{nn}(q, \omega)d\omega$$

$$= \frac{nq^2}{m}\left\{\left(\frac{\hbar q^2}{2m}\right)^2 + \frac{nq^2v_q}{m} + \frac{6q^2}{md}t + \frac{1}{mL^d}\sum_{\vec{k}\neq\vec{q}}v_k\left(\frac{\vec{k}\cdot\vec{q}}{q}\right)^2[S(\vec{q}-\vec{k})-S(\vec{k})]\right\}, \tag{3.142}$$

where d is the number of dimensions, t is the average kinetic energy per particle in the ground-state, and $S(\vec{k})$ is the static structure factor. Notice that the sum over $\vec{k}$ in the last equation *excludes* the singular term $\vec{k} = \vec{q}$ as well as the $\vec{k} = 0$ term, so the argument of the structure factor is never equal to zero.

Eq. (3.141) is called the *f-sum rule* from an equivalent sum rule for dipole oscillator strengths in atomic physics. It can be directly obtained from the observation that at very high frequency the electrons must respond to the external field as free independent particles. To formally derive Eq. (3.141) recall that, according to Eq. (3.91) and the definition Eq. (3.128), the left-hand side of Eq. (3.141) is given by

$$\frac{i}{\hbar L^d}\left\langle\left[\frac{\partial\hat{n}_{\vec{q}}}{\partial t}, \hat{n}_{-\vec{q}}\right]\right\rangle_0 . \tag{3.143}$$

Making use of the continuity equation (see Appendix 2, Eq. (A2.24)) we express the time derivative of $\hat{n}_{\vec{q}}$ in terms of the current operator

$$\hat{\vec{j}}_{p\vec{q}} = \frac{1}{m}\sum_i\left(\hat{\vec{p}}_i + \frac{\hbar\vec{q}}{2}\right)e^{-i\vec{q}\cdot\hat{\vec{r}}_i} . \tag{3.144}$$

---

[27] The high frequency expansion of the *imaginary* part of χ_{nn} poses a more difficult problem. It will be shown in Chapter 5 that $\Im\chi_{nn}(q, \omega) \sim \omega^{-4-\frac{d}{2}}$ in d dimensions.

We thus have

$$M_{nn}^{(1)}(\vec{q}) = \frac{\vec{q} \cdot \langle [\hat{\vec{j}}_{p\vec{q}}, \hat{n}_{-\vec{q}}] \rangle_0}{\hbar m L^d} . \tag{3.145}$$

The commutator between the current and the density fluctuation operators is an important ingredient of many calculations. It can be easily evaluated from the first quantization expressions of $\hat{n}_{\vec{q}}$ and $\hat{\vec{j}}_{\vec{q}}$, or from the corresponding second-quantized representation (see Appendix 2). The result is

$$[\hat{\vec{j}}_{\vec{q}}, \hat{n}_{-\vec{q}'}] = \frac{\hbar \vec{q}' \hat{n}_{\vec{q}-\vec{q}'}}{m} . \tag{3.146}$$

Substituting this in Eq. (3.145), and making use of $\langle \hat{n}_{\vec{q}} \rangle_0 = N \delta_{\vec{q},\vec{0}}$, we obtain the right-hand side of Eq. (3.141).

Although more complicated, the calculation of the third moment proceeds straightforwardly once it is recognized that, by virtue of the continuity equation and translational invariance in time, one has with the same notation as in Eq. (3.88)

$$M_{nn}^{(3)}(q) = -\frac{i}{\hbar L^d} \langle [\hat{n}_{\vec{q}}^{(3)}, \hat{n}_{-\vec{q}}] \rangle_0 = \frac{iq^2}{\hbar L^d} \langle [\hat{j}_{L,\vec{q}}^{(1)}, \hat{j}_{L,-\vec{q}}] \rangle_0 , \tag{3.147}$$

where $\hat{j}_{L,\vec{q}} = \frac{\vec{q} \cdot \hat{\vec{j}}_{\vec{q}}}{q}$ is the longitudinal component of the current operator. Thus, in view of Eq. (3.91), the third moment of the density fluctuation spectrum is q^2 times the first moment of the longitudinal current fluctuation spectrum. We shall present the calculation of this moment in Section 3.4.1 where the current–current response function is studied in detail.

3.3.4 The compressibility sum rule

A fundamental relationship can be established between the proper compressibility K, introduced in Section 1.81, and the long-wavelength limit of the static density–density response function $\chi_{nn}(\vec{q}, 0)$ of the homogeneous electron liquid. This goes under the name of *compressibility sum rule* because $\chi_{nn}(\vec{q}, 0)$ is formally associated (through Eq. (3.80)) with the first negative moment of $\Im m \chi_{nn}(q, \omega)$.

Consider a slightly inhomogeneous electron liquid with a density of the form

$$n(\vec{r}) = n \left(1 + \gamma \cos \vec{q} \cdot \vec{r}\right) , \tag{3.148}$$

where n is the uniform background density, and $\gamma \ll 1$. According to the stiffness theorem (see also Eq. (3.128)), the excess energy due to the inhomogeneity can be written as

$$\delta E_{\vec{q}} = -\frac{1}{2} \frac{|n_{\vec{q}}|^2}{\chi_{n_{\vec{q}} n_{-\vec{q}}}(\omega = 0)} = -\frac{1}{2} \frac{|n_{\vec{q}}|^2}{L^d \chi_{nn}(q, 0)} . \tag{3.149}$$

Since here the only finite Fourier components are $n_{\pm\vec{q}} = \frac{\gamma N}{2}$ we readily obtain

$$\delta E = -N\frac{n}{4\chi_{nn}(q,0)}\gamma^2 \ . \tag{3.150}$$

On the other hand, in the limit $q \to 0$ the excess energy can also be calculated directly from physical considerations. It consists of two parts: (i) the electrostatic energy, which is given (for any $\vec{q}$) by

$$\delta E_{el} = \frac{n^2\gamma^2}{2}\int d\vec{r}d\vec{r}'\cos\vec{q}\cdot\vec{r}\frac{e^2}{|\vec{r}-\vec{r}'|}\cos\vec{q}\cdot\vec{r}'$$

$$= N\frac{nv_q}{4}\gamma^2 \ , \tag{3.151}$$

and (ii) the kinetic plus exchange-correlation energy, which can be obtained, in the $q \to 0$ limit, from a spatial integral of $\epsilon(n(\vec{r}))$ – the energy of the homogeneous jellium model evaluated at the local density – as follows:[28]

$$\delta E_{loc} = \int d\vec{r}\left[n(\vec{r})\epsilon(n(\vec{r})) - n\epsilon(n)\right] \ . \tag{3.152}$$

We now Taylor-expand the energy to second order in γ and, noting that the integral of $\cos\vec{q}\cdot\vec{r}$ is zero for any finite $\vec{q}$, we get

$$\delta E_{loc} \approx N\frac{n\gamma^2}{4}\frac{\partial^2[n\epsilon(n)]}{\partial n^2} \ . \tag{3.153}$$

Comparing Eq. (3.150) with the sum of Eqs. (3.151) and (3.153) we arrive at the desired result

$$\lim_{q\to 0}\left[\frac{1}{\chi_{nn}(q,0)} + v_q\right] = -\frac{\partial^2[n\epsilon(n)]}{\partial n^2} \ . \tag{3.154}$$

This equation tells us that the density stiffness $-1/\chi_{nn}(q,0)$ consists of two parts: (i) an electrostatic part v_q which tends to infinity for $q \to 0$ due to the infinite coulomb energy cost associated with a uniform change in the electronic density (the background charge remaining constant) and (ii) a finite additional contribution known as "proper stiffness" arising from the change in kinetic and exchange-correlation energies in the assumedly neutral jellium model.

At this point, recalling the discussion in Section 1.8.1, we write

$$\lim_{q\to 0}\frac{\chi_{nn}(\vec{q},0)}{1 + v_q\chi_{nn}(\vec{q},0)} = -\frac{1}{\frac{\partial^2[n\epsilon(n)]}{\partial n^2}} = -n^2K \ . \tag{3.155}$$

Note that, due to the diverging electrostatic energy, $\chi_{nn}(q,0)$ vanishes in the long wave-length limit. By contrast, the combination appearing on the left side of Eq. (3.154) has a finite limit, which is controlled by the proper compressibility of the jellium model. This

---

[28] We hold this *locality property* of the kinetic+exchange-correlation energy as a self-evident fact.

combination is therefore known as the *proper* density–density response function and its physical significance will be discussed further in Chapter 5.

Eq. (3.155) provides an important consistency check for any theory of the electron gas in that it allows one to compare the value obtained for the compressibility from a calculation of the total energy as a function of the density (see Section 3.3.5) to that inferred from the long wavelength limit of $\chi_{nn}(\vec{q}, 0)$.

3.3.5 Total energy and density response

The density–density response function, in conjunction with the fluctuation-dissipation theorem (3.74), has an interesting application in the calculation of the ground-state energy of the electron liquid.

Recall that in Chapter 1, the ground-state energy was expressed as a coupling-constant integral of the potential energy per particle. The latter, in turn, can be obtained from the static structure factor $S(q)$ of Eqs. (3.138) and (3.139), via (compare with Eq. (1.15))

$$\frac{U}{N} \equiv u(r_s) = \frac{1}{2L^d} \sum_{\vec{q} \neq 0} v_q [S(q) - 1] . \tag{3.156}$$

Then, by making use of the fluctuation-dissipation theorem in the limit of zero temperature, we express the potential energy in terms of the density–density response function as follows:

$$U = \frac{n}{2} \sum_{\vec{q} \neq 0} v_q \left(-\frac{\hbar}{n\pi} \int_0^\infty \Im m \, \chi_{nn}(q, \omega) d\omega - 1 \right) . \tag{3.157}$$

The ground-state energy can be calculated from the potential energy by making use of the coupling constant integration algorithm described in Section 1.8.3:

$$E = E_0 + \frac{n}{2} \int_0^1 d\lambda \sum_{\vec{q} \neq 0} v_q \left(-\frac{\hbar}{n\pi} \int_0^\infty \Im m \, \chi_{nn}(q, \omega; \lambda) d\omega - 1 \right) , \tag{3.158}$$

where the spectral function $\Im m \, \chi_{nn}(q, \omega; \lambda)$ is calculated for a fictitious electron gas in which the electrons interact via the rescaled interaction λv_q, with $0 < \lambda < 1$. As it turns out, in order to carry out this calculation in a most efficient way, it is convenient to first change the path of the frequency integral in Eq. (3.158) from the real to the imaginary axis of the complex frequency plane. This can be done with impunity because the density–density response function is analytic in the upper half-plane and vanishes at infinity faster than ω^{-1}. It is then straightforward to verify that

$$\int_0^\infty \Im m \, \chi_{nn}(q, \omega) d\omega = \Im m \int_0^\infty \chi_{nn}(q, \omega) d\omega$$

$$= \Im m \int_0^\infty \chi_{nn}(q, i\omega) d(i\omega)$$

$$= \int_0^\infty \chi_{nn}(q, i\omega) d\omega, \tag{3.159}$$

where in the last step we have exploited the fact that here $\chi_{nn}(q, i\omega)$ is real.

Thus we arrive at the equivalent formula for the ground-state energy:

$$E = E_0 + \frac{n}{2} \sum_{\vec{q} \neq 0} v_q \left(-\frac{\hbar}{n\pi} \int_0^1 d\lambda \int_0^\infty \chi_{nn}(q, i\omega; \lambda) \, d\omega - 1 \right). \tag{3.160}$$

As we will see in Chapter 5, Eq. (3.160) is often applied to the calculation of the ground-state energy of the electron gas starting from various approximate forms of the dielectric function.

3.4 Current response

3.4.1 The current–current response function

We now turn to another very important application of the formalism: the calculation of the current and density response of a uniform electron liquid subjected to a *vector potential* $\vec{A}(\vec{r}, t)$.

It is well known from classical electrodynamics that an electromagnetic field can be represented in terms of a scalar potential ϕ and a vector potential $\vec{A}$ according to the formulas

$$\vec{E}(\vec{r}, t) = -\vec{\nabla}\phi(\vec{r}, t) - \frac{1}{c} \frac{\partial \vec{A}(\vec{r}, t)}{\partial t},$$

$$\vec{B}(\vec{r}, t) = \vec{\nabla} \times \vec{A}(\vec{r}, t). \tag{3.161}$$

These formulas are invariant under the *gauge transformation*

$$\phi(\vec{r}, t) \to \phi(\vec{r}, t) - \frac{1}{c} \frac{\partial \Lambda(\vec{r}, t)}{\partial t}$$

$$\vec{A}(\vec{r}, t) \to \vec{A}(\vec{r}, t) + \vec{\nabla}\Lambda(\vec{r}, t), \tag{3.162}$$

where $\Lambda(\vec{r}, t)$ is a differentiable, but otherwise arbitrary function of $\vec{r}$ and t. One can often exploit the freedom in the choice of Λ to map a given problem to an equivalent one that is easier to solve, or offers different insights. For example, the scalar potential $V(\vec{r}, t) = -e\phi(\vec{r}, t)$ considered in the previous Sections can always be eliminated by an appropriate choice of $\Lambda(\vec{r}, t)$. This transformation results in the appearance of a vector potential

$$\vec{A}(\vec{r}, t) = -\frac{c}{e} \int_0^t \vec{\nabla} V(\vec{r}, t') dt'. \tag{3.163}$$

The problem of calculating the response of the system to a scalar potential is thus tranformed into the problem of calculating the response of the same system to a special type of vector potential, namely, one that can be written as the gradient of a scalar function. Such a vector field is called *longitudinal*, because its Fourier transform, $\vec{A}(\vec{q})$, is parallel to $\vec{q}$ for every $\vec{q}$. There are also *transverse* vector potentials (such as the ones used to describe a static magnetic field), whose Fourier transform is perpendicular to $\vec{q}$ for every $\vec{q}$.[29] The most general vector field is neither longitudinal nor transverse, but can be written as a superposition of the two kinds. In this section we study the response of the electron liquid to such a general field.

The perturbed hamiltonian is

$$\hat{H}_{\vec{A}}(t) = \frac{1}{2m} \sum_i \left(\hat{p}_i + \frac{e}{c} \vec{A}(\hat{\vec{r}}_i, t) \right)^2 + U(\hat{\vec{r}}_1, \dots \hat{\vec{r}}_N) , \qquad (3.164)$$

where U is the total potential energy function. Upon linearization with respect to $\vec{A}$, this takes the standard form

$$\hat{H}_{\vec{A}}(t) = \hat{H} + \frac{e}{c} \int \hat{\vec{j}}_p(\vec{r}) \cdot \vec{A}(\vec{r}, t) d\vec{r}, \qquad (3.165)$$

where

$$\hat{\vec{j}}_p(\vec{r}) = \frac{1}{2m} \sum_i [\hat{\vec{p}}_i \delta(\vec{r} - \hat{\vec{r}}_i) + \delta(\vec{r} - \hat{\vec{r}}_i) \hat{\vec{p}}_i] \qquad (3.166)$$

is the *paramagnetic* component of the current density as defined in Appendix 2.

Recall that it is the physical current $\hat{\vec{j}}$ (as defined for instance in Eq. (A2.14)), not $\hat{\vec{j}}_p$, that satisfies the continuity equation and is invariant under gauge transformations. As for $\vec{j}_p$, it has none of these nice properties: its crucial advantage is that it does not depend on $\vec{A}$, and therefore is the natural "intrinsic" quantity coupled to $\vec{A}$. It is for this precise reason that we define the linear response function $\chi_{j_{p\alpha} j_{p\beta}}(\vec{q}, \omega)$ as the linear response function of the paramagnetic current density to the external vector potential, i.e.

$$\langle \hat{j}_{p\alpha} \rangle(\vec{q}, \omega) = \frac{e}{c} \sum_\beta \sum_{\vec{q}'} \chi_{j_{p\alpha} j_{p\beta}}(\vec{q}, \vec{q}', \omega) A_\beta(\vec{q}', \omega) , \qquad (3.167)$$

where α and β are labels for the cartesian components of $\hat{\vec{j}}_p$ and $\vec{A}$. Here $\chi_{j_{p\alpha} j_{p\beta}}(\vec{q}, \vec{q}', \omega)$ is the Fourier transform of the response function

$$\chi_{j_{p\alpha} j_{p\beta}}(\vec{r}, \vec{r}', t) = -\frac{i}{\hbar} \Theta(t) \langle [\hat{j}_{p\alpha}(\vec{r}, t), \hat{j}_{p\beta}(\vec{r}')] \rangle_0 , \qquad (3.168)$$

defined as in Eq. (3.127). The response of the physical current is then obtained by adding to Eq. (3.167) the expectation value of the diamagnetic current (see Eq. (A2.14)), which is obviously already linear in the perturbation:

$$\chi^J_{\alpha\beta}(\vec{q}, \vec{q}', \omega) = -\frac{n}{m} \delta_{\alpha\beta} + \chi_{j_{p\alpha} j_{p\beta}}(\vec{q}, \vec{q}', \omega) . \qquad (3.169)$$

Notice that we also have the general relation:

$$\chi_{j_{p\alpha} j_{p\beta}}(\vec{q}, \vec{q}', \omega) = \frac{1}{L^d} \chi_{j_{p\vec{q}\alpha} j_{p-\vec{q}'\beta}}(\omega) . \qquad (3.170)$$

---

[29] It is evident from these definitions that a longitudinal vector field has finite divergence and zero curl, while a transverse one has zero divergence and finite curl.

Of particular relevance is the case of translationally invariant (homogeneous) systems, or even disordered, but "self-averaging" systems.[30] In such cases the paramagnetic current–current response function, as well as the full one, becomes a function of a single wave vector, i.e. we have

$$\chi_{j_{p\alpha} j_{p\beta}}(\vec{q}, \vec{q}', \omega) = \chi_{j_{p\alpha} j_{p\beta}}(\vec{q}, \omega)\delta_{\vec{q},\vec{q}'} , \qquad (3.171)$$

and

$$\chi_{\alpha\beta}^{J}(\vec{q}, \omega) = \frac{n}{m}\delta_{\alpha\beta} + \chi_{j_{p\alpha} j_{p\beta}}(\vec{q}, \omega) . \qquad (3.172)$$

Some interesting properties of the current response function of an electron liquid are immediately evident from symmetry considerations. First of all, the homogeneity and isotropy of the system ensure that the tensor $\chi_{\alpha\beta}^{J}(\vec{q}, \omega)$ can be decomposed into longitudinal and transverse components, relative to the direction of $\vec{q}$, each depending only on the magnitude of $\vec{q}$. In other words, the current induced by a purely longitudinal vector potential is purely longitudinal, and that induced by a purely transverse vector potential is purely transverse. More formally, one can write

$$\chi_{\alpha\beta}^{J}(\vec{q}, \omega) = \chi_L(q, \omega)\frac{q_\alpha q_\beta}{q^2} + \chi_T(q, \omega)\left(\delta_{\alpha\beta} - \frac{q_\alpha q_\beta}{q^2}\right) , \qquad (3.173)$$

where χ_L and χ_T are the current–current response functions in the longitudinal and transverse channels respectively, i.e.,

$$\vec{j}_{L(T)}(q, \omega) = \frac{e}{c}\chi_{L(T)}(q, \omega)\vec{A}_{L(T)}(q, \omega) , \qquad (3.174)$$

where $\vec{j}_{L(T)}$ and $\vec{A}_{L(T)}$ are the longitudinal (transverse) components of the current density and vector potential.[31]

It should be evident from the above discussion that the problem of calculating the current response to a vector potential includes, as a special case, the problem of calculating the density response to a scalar potential. The vector potential formulation, however, is far more general, as it also includes the calculation of the response to transverse electromagnetic waves, and, in the static limit, the orbital magnetization response to a magnetic field.

---

[30] Self-averaging refers to the property of some of the macroscopic physical quantities of a disordered system of having a value that only depends on the density of the impurities and not on their specific positions. This amounts to state that the value of these quantities can be obtained by either averaging them over the impurity positions of a given system or, equivalently, over an ensemble of replicas of the original system only differing for the particular realization of the impurity distribution (the average impurity density being identical). This situation obtains when each electron is unable to propagate in a coherent fashion (i.e. without suffering phase destroying scattering events) through the entire system, but only through finite – yet still macroscopic – subsystems. Since these subsystems differ for their specific impurity distribution, it ensues that by summing over the electrons one is left with the average effect of the disorder. Since inelastic scattering is phase destroying it follows that self-averaging typically occurs in situations in which the temperature is not too low (a precise condition can be obtained once the inelastic lifetime has been determined). Dephasing is also caused by spin–orbit scattering. Conversely deviations from self-averaging occur when all the electrons can propagate coherently throughout the whole system. In this case even the displacement of one impurity can lead to finite and observable corrections. It is then clear that if a response function is self-averaging then it has the same translational invariance properties as the same response function in the corresponding clean (i.e. homogeneous) system.

[31] Notice that we omit the superscript J from the longitudinal and transverse components of the current response function, which are thus denoted simply by χ_L and χ_T.

3.4.2 Gauge invariance

Some properties of the current–current response function follow immediately from the requirements of gauge invariance. According to Eq. (3.163) a longitudinal vector potential $\vec{A}_L(q, \omega) = \frac{cq}{e\omega} V(q, \omega)$ should be equivalent, in its physical effects, to the scalar potential $V(q, \omega)$. The former induces a longitudinal current density $\vec{j}_L(q, \omega) = \frac{q}{\omega} \chi_L(q, \omega) V(q, \omega)$, and therefore (via the continuity equation) a density $n_1(q, \omega) = \frac{q}{\omega} j_L(q, \omega) = \frac{q^2}{\omega^2} \chi_L(q, \omega) V(q, \omega)$. This chain of transformations leads us to the conclusion that

$$\chi_{nn}(q, \omega) = \frac{q^2}{\omega^2} \chi_L(q, \omega) . \tag{3.175}$$

As a special case, this result implies

$$\chi_L(q, 0) = 0 \tag{3.176}$$

for any finite q.[32] This is, of course, nothing but the familiar statement that a purely longitudinal *and static* vector potential cannot induce any physical current.

As for the *transverse* current–current response function, *in the absence of long-range order*, one expects[33]

$$\lim_{q \to 0} \chi_T(q, 0) = \lim_{q \to 0} \chi_L(q, 0) = 0 . \tag{3.177}$$

This result is known as the *diamagnetic sum rule* and some of its implications will be discussed in the next section.

3.4.3 The orbital magnetic susceptibility

A physical problem in which the transverse current response is needed is that of calculating the *orbital magnetization* induced in the electron liquid by a static, sinusoidal magnetic field.

To be specific we assume that the field is oriented along the z axis and oscillates with wave vector $\vec{q}$ along the y axis: $\vec{B}(y) = \vec{e}_z B_q \cos qy$. We describe this field by means of a vector potential $\vec{A}$ oriented along the x axis, $\vec{A}(y) = -\vec{e}_x \frac{B_q \sin qy}{q}$, such that $\vec{B} = \vec{\nabla} \times \vec{A}$. The corresponding linear response current is then directed along the $\hat{x}$ axis and, like the vector potential, is purely transverse. Accordingly, by making use of the results and the notation established in the previous section, we can write

$$\vec{j}(y) = -\vec{e}_x \frac{e}{c} \chi_T(q, 0) \frac{B_q}{q} \sin qy . \tag{3.178}$$

---

[32] The case of $q = 0$ is trickier, and is discussed at the end of Section 3.4.4.
[33] A notable exception to this relation occurs in superconductors, where $\lim_{q \to 0} \chi_T(q, 0) \neq \lim_{q \to 0} \chi_L(q, 0)$ due to the existence of long range order and the consequent singularity of the current–current response functions for $q \to 0$.

We now recall that in electrodynamics the *charge* current density $-e\vec{j}(\vec{r})$ is related to the *orbital magnetization density* $\vec{M}_{orb}(\vec{r})$ by the relation

$$-e\vec{j}(\vec{r}) = c\vec{\nabla} \times \vec{M}_{orb}(\vec{r}) . \tag{3.179}$$

Since, in the present case, $\vec{M}_{orb}$ is along the $\hat{z}$ axis, the above equation takes the form

$$\frac{\partial M_{orb}(y)}{\partial y} = \frac{e}{c} j(y) . \tag{3.180}$$

Substituting Eq. (3.178) and integrating over y we get

$$\vec{M}_{orb}(y) = -\vec{e}_z \frac{e^2}{c^2} \frac{\chi_T(q,0)}{q^2} B_q \cos qy . \tag{3.181}$$

Then, taking the $q \to 0$ limit, we obtain

$$M_{orb} = \chi_{orb} B , \tag{3.182}$$

where M_{orb} and B are the macroscopic components of the corresponding fields, and

$$\chi_{orb} = -\frac{e^2}{c^2} \lim_{q \to 0} \frac{\chi_T(q,0)}{q^2} \tag{3.183}$$

is the *orbital magnetic susceptibility*. Thus, we see that a necessary condition for the existence of a finite orbital magnetic susceptibility is that $\chi_T(q,0)$ vanish as q^2 for $q \to 0$, thus satisfying the diamagnetic sum rule of Eq. (3.177). As discussed in footnote[33], this condition is satisfied in ordinary electronic systems but not in superconductors. Superconducting systems have, therefore, an infinite orbital susceptibility, which is manifested in the perfect screening of an applied magnetic field (the Meissner effect).

3.4.4 Electrical conductivity: conductors versus insulators

Let us now consider the response of an electronic system to a *uniform* electric field, $\vec{E}(t)$, which can be represented as the time-derivative of a uniform vector potential $\vec{A}(t) = -c \int_0^t \vec{E}(t')dt'$. The Fourier transform of $\vec{A}(t)$ is $\vec{A}(\omega) = -\frac{ic}{\omega}\vec{E}(\omega)$. According to Eqs. (3.174) and (3.172) the $q = 0$ component of the induced charge current is given by

$$-ej_\alpha(0,\omega) = \frac{ie^2}{\omega} \sum_\beta \chi_{\alpha\beta}^J(0,0,\omega)E_\beta(\omega) , \tag{3.184}$$

where the notation reflects the fact that, as already noted, in a generic inhomogeneous system $\chi_{\alpha\beta}^J$ and $\chi_{j_{p\alpha} j_{p\beta}}$ are functions of two wave vectors and one frequency. The factor multiplying $E_\beta(\omega)$ in the above equations is the *macroscopic electrical conductivity*

$$\sigma_{\alpha\beta}(\omega) = \frac{ie^2}{\omega} \chi_{\alpha\beta}^J(0,0,\omega) = \frac{ie^2}{\omega} \left[\frac{n}{m} \delta_{\alpha\beta} + \chi_{j_{p\alpha} j_{p\beta}}(0,0,\omega) \right] . \tag{3.185}$$

This relation is usually referred to as the *Kubo formula* for the electrical conductivity.[34]

Also, in view of Eq. (3.170) in the long wavelength limit we have

$$\chi_{j_{P\alpha}j_{P\beta}}(0, 0, \omega) = \frac{1}{m^2 L^d} \chi_{P_\alpha P_\beta}(\omega) \,, \tag{3.186}$$

where $\hat{P}_\alpha$ is one of the components of the *total momentum*. Accordingly the Kubo formula can be rewritten as:

$$\sigma_{\alpha\beta}(\omega) = \frac{ine^2}{m\omega} \left[\delta_{\alpha\beta} + \frac{\chi_{P_\alpha P_\beta}(\omega)}{nmL^d} \right] \,. \tag{3.187}$$

Let us now examine the low frequency behavior of $\sigma_{\alpha\beta}(\omega)$. From the fact that a uniform and static vector potential can be removed by a gauge transformation one might expect $\chi^J_{\alpha\beta}(0, 0, \omega)$ to vanish for $\omega \to 0$. Actually this is a non-trivial expectation because the wave vectors tend to zero before the frequency, when the vector potential is still time-dependent. The underlying issue is whether the order of the $q \to 0$ and $\omega \to 0$ limits can be interchanged or not. We know that such an interchange in the order of the limits is *not* allowed in the case of a strictly homogeneous electron liquid (see discussion below) or in the case of superconductors, which exhibit long-range order. In any case, the crucial question is whether $\chi^J_{\alpha\beta}(0, 0, \omega)$ tends to zero faster than ω, as ω, or more slowly than ω – the last case including the possibility that it may not vanish at all. Corresponding to these three possibilities, we are led to the following classification of the possible types of electrical response of an electronic system:

- *Insulators.* In this case $\chi^J_{\alpha\beta}(0, 0, \omega)$ tends to zero faster than ω, leading to a vanishing conductivity for $\omega \to 0$. This situation typically occurs when the system has a *gap* in its excitation spectrum, so that the current response function is purely real for small ω. The simplest analytic behavior of $\Re e \chi^J_{\alpha\beta}(0, 0, \omega)$ that is consistent with the known symmetry of the real part of a response function is proportionality to ω^2, which implies $\sigma_{\alpha\beta} \propto i\omega$ for $\omega \to 0$.[35]
- *Ordinary conductors.* In these systems, due to the presence of low energy excitations, the imaginary part of the current response function remains finite as the frequency goes to zero. The simplest analytic behavior of $\Im m \chi^J_{\alpha\beta}(0, 0, \omega)$ that is consistent with symmetry requirements at low frequency is linearity in ω. This leads to a finite real conductivity at zero frequency.
- *Superconductors.* In superconductors, the assumption of regularity of the current response function breaks down and it is no longer true that $\chi^J_{\alpha\beta}(0, 0, \omega)$ tends to zero for $\omega \to 0$. In the simplest case the second term in the square bracket of Eq. (3.184) vanishes (as it would in a perfectly homogeneous electron liquid) due to a peculiar rigidity of the ground-state wave function. As a result, the conductivity is purely imaginary (there is no dissipation) and diverges as ω^{-1}.

In addition to the three physical cases listed above, it is worthwhile considering the ideal case of a perfectly homogeneous electron liquid. Because in this case the total momentum is a constant of motion, the second term in the square bracket of Eq. (3.184) vanishes (as in

---

[34] Strictly speaking, the current–current response function to be used in Eqs. (3.184) and (3.185) is the *proper* current–current response function, which will be discussed in detail in Chapter 5, Sections 5.2.2 and 5.4.6.

[35] This behavior is characteristic of systems that exhibit a finite static dielectric polarizability – see Exercise 11.

a superconductor) and we are left with the simple result

$$\sigma_{\alpha\beta}(\omega) = \frac{ine^2}{m\omega}\delta_{\alpha\beta} \tag{3.188}$$

which demonstrates the infinite conductivity of this model.[36] What is interesting here is the difference between $\lim_{\omega\to0}\lim_{q\to0}\chi^J_{\alpha\beta}(q,\omega)$, which is finite, and $\lim_{q\to0}\lim_{\omega\to0}\chi^J_{\alpha\beta}(q,\omega)$, which vanishes in accordance with the diamagnetic sum rule.

3.4.5 The third moment sum rule

We now calculate the high-frequency behavior of the current–current response functions. According to the definition (3.172) and the theory of Section (3.2.8) we have

$$\Re\chi_\mu(q,\omega) \overset{\omega\to\infty}{\to} \frac{n}{m} + \frac{M_\mu^{J(1)}(q)}{\omega^2}, \tag{3.189}$$

where $M_\mu^{J(1)}(q)$ is the first moment of the longitudinal or transverse current excitation spectrum, depending on whether $\mu = L$ or T, and

$$M_\mu^{J(1)}(q) = -\frac{i}{\hbar L^d}\langle[\hat{j}^{(1)}_{\mu\vec{q}}, \hat{j}_{\mu-\vec{q}}]\rangle_0, \tag{3.190}$$

with the notation $\hat{j}^{(1)}$ established by Eq. (3.88). Making use of Eq. (3.175) we see that, as anticipated in Section (3.3.3), $M_L^{J(1)}(q) = M^{(3)}(q)/q^2$ where $M^{(3)}(q)$ is the *third moment* of the density–density response function. For ease of reference we begin by stating the sum rule

$$
\begin{aligned}
M_\mu^{J(1)}(\vec{q}) &= -\frac{2}{\pi}\int_0^\infty \omega\,\Im\chi_{j_{p\mu},j_{p\mu}}(q,\omega)\,d\omega \\
&= \frac{n}{m}\left\{\left(\frac{\hbar qq_\mu}{2m}\right)^2 + \frac{nq_\mu^2 v_q}{m} + \frac{2q^2+4q_\mu^2}{md}t + \frac{1}{mL^d}\sum_{\vec{k}\neq\vec{q}}v_k k_\mu^2[S(\vec{q}-\vec{k})-S(\vec{k})]\right\},
\end{aligned}
\tag{3.191}
$$

where d is the number of dimensions, t the kinetic energy per particle, $S(\vec{k})$ the static structure factor, and $\mu = L$ or T. Clearly $q_L = q$ and $q_T = 0$.

Let us describe in some detail the proof of this result for the longitudinal case[37] (Puff, 1965). The transverse case is left as an exercise. The time derivative of $\hat{j}_{\vec{q}\mu}$ is easily computed in first quantization. We find

$$\hat{j}^{(1)}_{\mu\vec{q}} = \frac{1}{m}\sum_i\left[\hat{F}_{i\mu} - \frac{i}{m}\left(\hat{p}_{i\mu}+\frac{\hbar q_\mu}{2}\right)\vec{q}\cdot\left(\hat{p}_i+\frac{\hbar\vec{q}}{2}\right)\right]e^{-i\vec{q}\cdot\hat{r}_i}, \tag{3.192}$$

---

[36] Of course, in reality, the electrical response is limited by impurity scattering.
[37] Interestingly, this calculation was first done for a bosonic system: it turns out that the only difference between bosons and fermions is the expression of the kinetic energy in terms of the density. Eq. (3.191) is valid in both cases.

where

$$\hat{F}_{i\mu} = -\nabla_{i\mu} \sum_{j\neq i} \frac{e^2}{|\hat{r}_i - \hat{r}_j|} = -\frac{1}{L^d} \sum_k k_\mu v_k [\hat{n}_{\vec{k}} e^{i\vec{k}\cdot\hat{\vec{r}}_i} - 1] \tag{3.193}$$

is the time derivative of the i-th momentum operator, that is, the force acting on the i-th particle. The second term in the square brackets of Eq. (3.192) arises from the time derivative of $e^{-i\vec{q}\cdot\hat{\vec{r}}_i}$, which is not affected by the interactions and is therefore equal to minus the divergence of the single particle current, that is $-i\vec{q} \cdot \sum_i \left(\frac{\hat{\vec{p}}_i}{2m} e^{-i\vec{q}\cdot\hat{\vec{r}}_i} + e^{-i\vec{q}\cdot\hat{\vec{r}}_i} \frac{\hat{\vec{p}}_i}{2m} \right)$. The -1 in the square brackets of Eq. (3.193) gives a vanishing contribution to the wave vector sum and can be dropped.

Putting all this back in Eq. (3.192) we obtain

$$\hat{j}_{\mu\vec{q}}^{(1)} = -\frac{i}{mL^d} \sum_k k_\mu v_k \hat{n}_{\vec{k}} \hat{n}_{\vec{q}-\vec{k}} - \frac{i}{m^2} \sum_i \left(\hat{p}_{i\mu} + \frac{\hbar q_\mu}{2} \right) \vec{q} \cdot \left(\hat{\vec{p}}_i + \frac{\hbar\vec{q}}{2} \right) e^{-i\vec{q}\cdot\hat{\vec{r}}_i} . \tag{3.194}$$

We are now ready to take the commutator of this expression with $\hat{j}_{\mu-\vec{q}}$. The coulomb term is straightforwardly handled by applying the commutator (3.146) to each of the two density operators:

$$[\hat{n}_{\vec{k}} \hat{n}_{\vec{q}-\vec{k}}, \hat{j}_{\mu-\vec{q}}] = -\frac{\hbar}{m}[k_\mu \hat{n}_{\vec{k}-\vec{q}} \hat{n}_{\vec{q}-\vec{k}} + (q_\mu - k_\mu)\hat{n}_{\vec{k}} \hat{n}_{-\vec{k}}] . \tag{3.195}$$

Taking the ground-state average and using the standard definition of the static structure factor yields the second and the last term in the curly brackets of Eq. (3.191).

The kinetic term can be easily evaluated with the help of the identity

$$\left(\hat{p}_{i\mu} + \frac{\hbar q_\mu}{2} \right) e^{-i\vec{q}\cdot\hat{\vec{r}}_i} = e^{-i\vec{q}\cdot\hat{\vec{r}}_i} \left(\hat{p}_{i\mu} - \frac{\hbar q_\mu}{2} \right) . \tag{3.196}$$

This leads to:

$$\sum_i \left[\left(\hat{p}_{i\mu} + \frac{\hbar q_\mu}{2} \right) \vec{q} \cdot \left(\hat{\vec{p}}_i + \frac{\hbar\vec{q}}{2} \right) e^{-i\vec{q}\cdot\hat{\vec{r}}_i} , \hat{j}_{\mu-\vec{q}} \right]$$

$$= \sum_i \left[\left(\hat{p}_{i\mu} + \frac{\hbar q_\mu}{2} \right)^2 \vec{q} \cdot \left(\hat{\vec{p}}_i + \frac{\hbar\vec{q}}{2} \right) - \left(\hat{p}_{i\mu} - \frac{\hbar q_\mu}{2} \right)^2 \vec{q} \cdot \left(\hat{\vec{p}}_i - \frac{\hbar\vec{q}}{2} \right) \right] . \tag{3.197}$$

The terms containing one or three momentum operators vanish upon averaging in the ground-state. The term containing no momentum operators yields the first term in the square bracket of Eq. (3.191). Finally, the terms containing two momentum operators can be simplified and collected together with the help of the identity

$$\left\langle \sum_i \frac{p_{i\mu} p_{i\beta}}{m} \right\rangle_0 = \frac{2\delta_{\mu\beta}}{d} \langle \hat{T} \rangle_0 , \tag{3.198}$$

that follows from the isotropy of the ground-state. This gives the third term within the square bracket of Eq. (3.191), and completes the proof.

3.5 Spin response

3.5.1 Density and longitudinal spin response

Let us finally consider the linear response of the σ-spin component of the density $n_{1\sigma}(\vec{r}, t)$ to an external scalar potential $V_{ext,\sigma'}(\vec{r}, t)$ that couples only to the density of σ'-spin electrons.[38] In this context the relevant quantities are the spin-resolved number density operator $\hat{n}_\sigma(\vec{r})$ and the spin–density operator $\hat{S}_z(\vec{r})$ (in units of $\frac{\hbar}{2}$) which were defined in Appendix 2.

The linear response formalism gives

$$n_{1\sigma}(\vec{r}, t) = \int_0^\infty d\tau \int d\vec{r}' \chi_{\sigma\sigma'}(\vec{r}, \vec{r}', \tau) V_{ext,\sigma'}(\vec{r}', t - \tau), \qquad (3.199)$$

where $\chi_{\sigma\sigma'}$ is the spin-resolved density–density response function

$$\chi_{\sigma\sigma'}(\vec{r}, \vec{r}', t) \equiv \chi_{n_\sigma(\vec{r})n_{\sigma'}(r')}(t) = -\frac{i}{\hbar}\Theta(t)\langle[\hat{n}_\sigma(\vec{r}, t), \hat{n}_{\sigma'}(\vec{r}')]\rangle_0 . \qquad (3.200)$$

In a non-interacting electron gas the two spin components respond independently, so $\chi_{\sigma\sigma'}$ is diagonal.[39] This is no longer true in the presence of interactions, and we can only say that $\chi_{\uparrow\downarrow} = \chi_{\downarrow\uparrow}$ by virtue of the reciprocity relations.

Knowing the spin-resolved density–density response function $\chi_{\sigma\sigma'}$ it is easy to calculate the perhaps more immediately useful response functions χ_{nn}, $\chi_{S_z S_z}$ and $\chi_{nS_z} = \chi_{S_z n}$.

The first is the already familiar density–density response function, which is seen to have the expression

$$\chi_{nn} = \sum_{\sigma\sigma'} \chi_{\sigma\sigma'} . \qquad (3.201)$$

The second is the *longitudinal* spin–spin response function, which we denote with χ_S, and gives the response of the z-component of the spin density to a space and time-dependent magnetic field parallel to the z-axis. Such a magnetic field enters the hamiltonian with an interaction of the form

$$\frac{g\mu_B}{2} \int d\vec{r}' \hat{S}_z(\vec{r}') B_z(\vec{r}', t) , \qquad (3.202)$$

(where μ_B is the Bohr magneton) which can be viewed as the interaction with two scalar potentials of opposite sign coupling to up-spins and down-spins respectively. Hence it is immediate to see that

$$\chi_S = \chi_{S_z S_z} = \sum_{\sigma\sigma'} \sigma\sigma' \chi_{\sigma\sigma'} , \qquad (3.203)$$

where we assign the value $\sigma = 1$ to up-spins and $\sigma = -1$ to down-spins.

---

[38] The direction of the spin quantization axis, denoted by $\vec{e}_z$, is arbitrary in the paramagnetic state, or it coincides with the direction of the static spin polarization in the spin-polarized case.

[39] This is also true in Hartree–Fock theory, since electrons of opposite spin orientations are not coupled by the exchange field.

Similar reasoning shows that χ_{nS_z}, the response function of the density to a longitudinal magnetic field, is given by

$$\chi_{nS_z} = \sum_{\sigma\sigma'} \sigma' \chi_{\sigma\sigma'} = \chi_{S_z n} \,. \tag{3.204}$$

A conclusion that one can draw immediately is that the spin–density response functions vanish in the paramagnetic state due to the obvious additional symmetry $\chi_{\uparrow\uparrow} = \chi_{\downarrow\downarrow}$. Thus, *spin and density responses are perfectly decoupled in a paramagnetic electron liquid*: this is an exact result.

3.5.2 High-frequency expansion

The high-frequency expansion of the spin-resolved density–density response functions is remarkable because correlation effects appear already at order q^2 rather than at order q^4. The calculation is similar to the one outlined in Section 3.4.1, with spin-resolved current densities replacing the ordinary current density. The final result is (Goodman *et al.*, 1973)

$$\Re e \chi_{\sigma\sigma'}(q, \omega) \xrightarrow{\omega \to \infty} \frac{n_\sigma q^2}{m\omega^2}\delta_{\sigma\sigma'} + \frac{M^{(3)}_{\sigma\sigma'}(q)}{\omega^4} \,, \tag{3.205}$$

where

$$M^{(3)}_{\sigma\sigma'}(\vec{q}) = \frac{q^2}{m} \left\{ \left[\left(\frac{\hbar q^2}{2m}\right)^2 + \frac{6q^2}{md}t_\sigma \right] n_\sigma \delta_{\sigma\sigma'} + \frac{n_\sigma n_{\sigma'} q^2 v_q}{m} \right.$$
$$\left. + \frac{n}{mL^d} \sum_{\vec{k} \neq \vec{q}} v_k \left(\frac{\vec{k}\cdot\vec{q}}{q}\right)^2 \left[S_{\sigma\sigma'}(\vec{q} - \vec{k}) - \delta_{\sigma\sigma'} \sum_\tau S_{\sigma\tau}(\vec{k}) \right] \right\} \,, \tag{3.206}$$

where t_σ is the average kinetic energy of σ-spin electrons, and $S_{\sigma\sigma'}(k)$ is the spin-resolved structure factor defined in Appendix 4.

Compare this with the third moment of the full density–density spectral function, given by Eq. (3.142). The last term in the curly brackets of Eq. (3.206) has a non-zero limit for $q \to 0$, at variance with the corresponding term of Eq. (3.142), which vanishes as q^2 for $q \to 0$. This implies that interaction effects, beyond the "trivial" Hartree term (the term proportional to v_q), make a contribution of order q^2 to the spin-resolved third moment, but only of order q^4 to the full third moment. Of course, summing Eq. (3.206) over the spin indices gives back Eq. (3.142). We will see in Chapter 5 that the physical reason for the different behavior of the post-Hartree term in the spin-resolved case is that the net force exerted by the down-spins on the up-spins does not vanish, whereas the net force exerted by the electron liquid on itself does vanish.

The small-q behavior of the spin-resolved third moment can be calculated analytically, with the following result:

$$M^{(3)}_{\sigma\sigma'}(\vec{q}) \overset{q\to 0}{\to} \frac{nq^2}{m^2 L^d} \lim_{q\to 0} \sum_{\vec{k}\neq\vec{q}} v_k \left(\frac{\vec{k}\cdot\vec{q}}{q}\right)^2 \left[S_{\sigma\sigma'}(\vec{q}-\vec{k}) - \delta_{\sigma\sigma'}\sum_\tau S_{\sigma\tau}(\vec{k})\right]$$

$$= -\sigma\sigma' \frac{nq^2}{m^2 L^d} \sum_{\vec{k}\neq 0} \left(\frac{v_k k^2}{d}\right) S_{\uparrow\downarrow}(k) . \tag{3.207}$$

In three dimensions $v_k k^2$ is independent of k and the above expression can be further simplified, making use of Eq. (A4.22), to

$$M^{(3)}_{\sigma\sigma'}(\vec{q}) \overset{q\to 0}{\to} = -\sigma\sigma' \frac{4\pi e^2 q^2}{3m^2} n_\uparrow n_\downarrow [g_{\uparrow\downarrow}(0) - 1] , \quad (3D) , \tag{3.208}$$

where $g_{\uparrow\downarrow}(0)$ is given by Eq. (A4.11). No such simplification is possible in two dimensions: in fact, the large-k behavior of $S_{\uparrow\downarrow}(k) \sim \frac{1}{k^3}$ in 2D (see Exercise 20 in Chapter 1) causes the wave vector sum in Eq. (3.207) to diverge logarithmically (Qian, 2004).

3.5.3 Transverse spin response

It is often necessary to consider the response of the electron liquid to a magnetic field that is perpendicular to the direction of the static spin polarization. The perturbation has the form

$$\frac{g\mu_B}{4} \int d\vec{r}' \left[\hat{S}_+(\vec{r}')B_-(\vec{r}', t) + \hat{S}_-(\vec{r}')B_+(\vec{r}', t)\right] , \tag{3.209}$$

where $B_\pm = B_x \pm i B_y$ and $\hat{S}_\pm = \hat{S}_x \pm i\hat{S}_y$.[40]

Because the system is symmetric under rotations about the z-axis, it is evident that the response functions $\chi_{n S_\pm}$ and $\chi_{S_z S_\pm}$ vanish identically. This means that *in a system with axial rotational symmetry the transverse spin response is completely decoupled from the density and longitudinal spin responses.* For the same reason (axial rotational symmetry) $\chi_{S_x S_x} = \chi_{S_y S_y}$. In addition, the reciprocity relation 3.59, combined with the property ${}^t\hat{S}_y = -\hat{S}_y$, yields $\chi_{S_x S_y} = -\chi_{S_y S_x}$. The above relations imply that $\chi_{S_+ S_+} = \chi_{S_- S_-} = 0$. Thus, finally, the only two nonvanishing response functions in the transverse spin sector are

$$\chi_{S_+ S_-} = 2[\chi_{S_x S_x} + i\chi_{S_x S_y}]$$
$$\chi_{S_- S_+} = 2[\chi_{S_x S_x} - i\chi_{S_x S_y}] . \tag{3.210}$$

It is interesting to notice that, by virtue of the standard relations $\chi_{S_x S_x}(q, \omega) = \chi^*_{S_x S_x}(q, -\omega)$ and $\chi_{S_x S_y}(q, \omega) = \chi^*_{S_x S_y}(q, -\omega)$ the transverse spin response functions have the property

$$\chi_{S_+ S_-}(q, -\omega) = \chi^*_{S_- S_+}(q, \omega) . \tag{3.211}$$

---

[40] The second quantized representation of $\hat{\vec{S}}(\vec{r})$ is given in Appendix 2.

The two functions transform into each other upon reversing the direction of the static spin polarization.

Exercises

3.1 **Kinetic energy stiffness.** Making use of second-order static perturbation theory derive the expression (3.16) for the increase of the kinetic energy of a non-interacting electron gas in the presence of the static periodic potential (3.7).

3.2 **First-order time-dependent density response.** Derive the linear density response formula (3.19) from first-order time-dependent perturbation theory.

3.3 **Equations of motion for linear response functions.** Let $\dot{\hat{O}} \equiv \frac{i}{\hbar}[\hat{H}, \hat{O}]$ be the time derivative of an operator $\hat{O}$ in the Heisenberg picture. Use the general properties of linear response functions *vis à vis* time homogeneity to prove the following useful relationships:

$$\chi_{\dot{A}B}(t) = -\chi_{A\dot{B}}(t),$$

$$\frac{\partial \chi_{AB}(t)}{\partial t} = -\frac{i}{\hbar}\delta(t)\langle[\hat{A}, \hat{B}]\rangle_0 + \chi_{A\dot{B}}(t),$$

and

$$\frac{\partial^2 \chi_{AA}(t)}{\partial t^2} = -\frac{i}{\hbar}\delta(t)\langle[\dot{\hat{A}}, \hat{A}]\rangle_0 - \chi_{\dot{A}\dot{A}}(t).$$

3.4 **Classical fluctuation-dissipation theorem.** Obtain the classical form of the fluctuation-dissipation theorem from the high-temperature limit of Eq. (3.74). Combining this result with the Kramers–Krönig dispersion relations show that in the classical theory the static density–density response function $\chi_{A^\dagger A}(0)$ (which is real) is related to the static structure factor $S_{A^\dagger A} = \langle \hat{A}^\dagger \hat{A}\rangle$ by

$$\chi_{A^\dagger A}(0) = -\frac{S_{A^\dagger A}}{k_B T}. \tag{E3.1}$$

3.5 **Stability condition for a thermal ensemble.** Show that the stability condition (3.54) at finite temperature follows from the exact eigenstates representation of the response function, Eq. (3.47), and the fact that the canonical distribution $e^{-\beta E}$ is a monotonically decreasing function of energy.

3.6 **Bound for static response functions.** Prove the inequality (3.106) for the static response function $\chi_{AA}(0)$ starting from Eq. (3.47) and noting that

$$0 \geq \frac{P_n - P_m}{E_n - E_m} > -\frac{P_n + P_m}{2k_B T}.$$

3.7 **Berry curvature and linear response functions.** Let $\hat{A}_i$ $(i = 1, \ldots, n))$ be a set of hermitian observables and F_i a corresponding set of external fields that are linearly coupled to $\hat{A}_i$. Starting from the ground-state at $F_i = 0$ imagine a process in which the F_i's are slowly changed (always remaining small) around the closed loop $0 \rightarrow$

$\delta F_i^{(1)} \to \delta F_i^{(1)} + \delta F_i^{(2)} \to \delta F_i^{(2)} \to 0$, where $\delta F_i^{(1)}$ and $\delta F_i^{(2)}$ are two sets of small increments. At the end of the cycle the system will return to the ground-state having acquired a *Berry phase* $e^{i\gamma}$ where

$$\gamma = \sum_{ij} \delta F_i^{(1)} \Omega_{ij} \delta F_j^{(2)}$$

and

$$\Omega_{ij} \equiv 2\,\Im m \left\langle \frac{\partial \psi}{\partial F_i} \middle| \frac{\partial \psi}{\partial F_j} \right\rangle \qquad (\text{E3.2})$$

is the *Berry curvature*. Here $\left| \frac{\partial \psi}{\partial F_j} \right\rangle = \frac{\partial}{\partial F_j} |\psi\rangle$, where $|\psi\rangle$ is the state of the system regarded as a functional of the F_i, and $\left\langle \frac{\partial \psi}{\partial F_i} \right|$ is the bra conjugate of $\frac{\partial}{\partial F_i} |\psi\rangle$. By evaluating $\frac{\partial}{\partial F_j} |\psi_0\rangle$ by first-order perturbation theory and comparing the resulting expression for Ω_{ij} with the Lehmann representation (3.47) for the linear response function $\chi_{A_i A_j}(\omega)$ (at $T = 0$) show that the Berry curvature can be expressed in the following manner:

$$\Omega_{ij} = \frac{i}{2} \lim_{\omega \to 0} \frac{\partial}{\partial \omega} [\chi_{A_i A_j}(\omega) - \chi_{A_j A_i}(\omega)] . \qquad (\text{E3.3})$$

3.8 **Physical derivation of the Bogoliubov inequality.** Work out in detail the derivation of the Bogoliubov inequality outlined in footnote[21].

3.9 **Bound on the density stiffness of the electron liquid.** The Bogoliubov inequality leads to an interesting upper bound for the density stiffness of the electron liquid. Putting $\hat{A} = \hat{n}_{\vec{q}}$ and $\hat{B} = \hat{j}_{L,-\vec{q}}$ in Eq. (3.107) and making use of Eqs. (3.147) and (3.146) show that

$$\frac{n^2}{|\chi_{nn}(q, 0)|} \leq \frac{m^2 M_{nn}^{(3)}(q)}{q^4} .$$

Combine this with the compressibility sum rule to get an upper bound for the inverse of the proper compressibility:

$$\frac{1}{K} \leq \lim_{q \to 0} \frac{m^2 \tilde{M}_{nn}^{(3)}(q)}{q^4} , \qquad (\text{E3.4})$$

where $\tilde{M}_{nn}^{(3)}(q)$ is given by Eq. (3.142) *without* the term $\frac{nq^2 v_q}{m}$.

3.10 **Physical derivation of the f-sum rule.** Derive the high-frequency behavior of the density–density response function (and hence the f-sum rule) from the physical assumption that, at very high frequency, the electrons respond to the external potential as free independent particles. [Hint: use the Kramer–Krönig dispersion relations.]

3.11 **Low-frequency conductivity of insulators.** Use the constitutive relations of classical electromagnetism in macroscopic media to argue that the dynamical conductivity of an isotropic insulating material must vanish, for $\omega \to 0$, as $\sigma(\omega) \sim 4\pi i \omega \epsilon$ where ϵ is the static dielectric tensor of the material.

3.12 **An orgy of spectral moments.** Calculate
      (a) The first-moment of the transverse current excitation spectrum, Eq. (3.191);
      (b) The third moment of the spin dependent excitation spectrum, Eq. (3.206);
      (c) The first moment of the transverse spin current spectrum.

3.13 **Symmetries of spin–spin and spin–density response functions.** Show that in a uniformly spin-polarized electron liquid, where z is the polarization axis, the response functions $\chi_{nS_\pm}(q, \omega)$ and $\chi_{S_z S_\pm}(q, \omega)$ vanish for all values of q and ω due to rotational symmetry in the $x - y$ plane. Then make use of Eqs. (3.59) and (3.210) to prove that $\chi_{S_x S_y}(q, 0) = 0$.

3.14 **Macroscopic dielectric constant of a crystal.** The macroscopic dielectric constant, ϵ_m, of an insulating crystal of cubic symmetry is defined as $\epsilon_m = \frac{E_{\vec{0}} + 4\pi P_{\vec{0}}}{E_{\vec{0}}}$, where the polarization vector $\vec{P}$ is related to the induced electron density by $-en_1 = -\vec{\nabla} \cdot \vec{P}$, and the sub-index indicates the $\vec{q} = 0$ components of the fields. Make use of the formalism presented in Appendix 7 to show that $\epsilon_m = \left(\left[\epsilon^{-1} \right]_{\vec{0},\vec{0}} (0, 0) \right)^{-1}$.

4

Linear response of independent electrons

4.1 Introduction

The calculation of the linear response functions of an electron liquid is a very important, but obviously extremely difficult task. Even after many years of study, and in spite of much progress, a complete solution of the problem is still lacking. In preparation to the study of this difficult problem, we consider in this chapter the main response functions of the *non-interacting* electron gas, namely, the density–density, spin–spin, and current–current response functions, all of which can be calculated analytically. It turns out that understanding the response of the non-interacting electron gas is an essential prerequisite for understanding the richer and more complex response of the interacting liquid. Indeed, one of the most fruitful ideas in many-body physics is that the response of an interacting system can be pictured as the response of a non-interacting system to an effective self-consistent field, which depends on global properties such as the particle density, the density matrix, the current–density etc. . . This is true, in particular, for every *dynamical mean field theory*, which can be derived from the corresponding static mean field theory (such as the HF theory) with the help of the techniques introduced in Section 4.7.

In the first part of this chapter we present a detailed study of the response functions of the homogeneous non-interacting electron gas in three, two, and one dimension. Electron–impurity scattering is included only at the most elementary level, and only to demonstrate its main effect, namely, the emergence of *diffusion* in the dynamics of density fluctuations. In the next chapter we shall present approximations for the effective self-consistent field. By putting together the two concepts – non-interacting response and effective field – we will be able to construct approximations for the response functions of the interacting electron liquid.

4.2 Linear response formalism for non-interacting electrons

Although in the case of non-interacting electrons the general result of Eq. (3.47) can be evaluated directly and without great difficulty for any pair of operators (see Exercise 1), we find convenient to follow here a somewhat different path which leads to a very simple and useful formula for the response associated with single-particle operators.

157

Let us consider, for definiteness, two operators of the form

$$\hat{A} = \sum_{i=1}^{N} \hat{A}_i = \sum_{\alpha\beta} A_{\alpha\beta} \hat{a}_\alpha^\dagger \hat{a}_\beta \ ,$$

$$\hat{B} = \sum_{i=1}^{N} \hat{B}_i = \sum_{\alpha\beta} B_{\alpha\beta} \hat{a}_\alpha^\dagger \hat{a}_\beta \ , \tag{4.1}$$

where $\hat{a}_\alpha^\dagger$ are the creation operators of one-electron states that diagonalize the independent electron hamiltonian

$$\hat{H}_0 = \sum_{\alpha} \varepsilon_\alpha \hat{a}_\alpha^\dagger \hat{a}_\alpha \ , \tag{4.2}$$

and ε_α are their energies. It should be noted that, as from our discussion of the second quantization formalism, $A_{\alpha\beta}$ and $B_{\alpha\beta}$ are the matrix elements of the single particle operators $\hat{A}$ and $\hat{B}$, e.g. $\langle i, \alpha | \hat{A}_i | i, \beta \rangle$.

Our problem is to calculate the time-dependence of the expectation value of $\hat{A}$ due to a perturbation that couples linearly to $\hat{B}$, i.e., for a time-dependent hamiltonian of the form

$$\hat{H}_{0F}(t) = \hat{H}_0 + F(t)\hat{B} \ . \tag{4.3}$$

It is evident from the form of $\hat{H}_0$ that the time-dependence of the operators $\hat{a}_\alpha$ under the unperturbed hamiltonian amounts to multiplication by a phase factor

$$\hat{a}_\alpha(t) \equiv e^{i\hat{H}_0 t/\hbar} \hat{a}_\alpha e^{-i\hat{H}_0 t/\hbar} = e^{-i\varepsilon_\alpha t/\hbar} \hat{a}_\alpha \ ,$$

$$\hat{a}_\alpha^\dagger(t) \equiv e^{i\hat{H}_0 t/\hbar} \hat{a}_\alpha^\dagger e^{-i\hat{H}_0 t/\hbar} = e^{i\varepsilon_\alpha t/\hbar} \hat{a}_\alpha^\dagger \ . \tag{4.4}$$

Making use of this result, and inserting the single particle operators $\hat{A}$ and $\hat{B}$ from Eq. (4.1) into Eq. (3.36) for the linear response function we obtain

$$\chi_{AB}^{(0)}(t) = -\frac{i}{\hbar}\Theta(t) \sum_{\alpha\beta\gamma\delta} A_{\alpha\beta} B_{\gamma\delta} e^{i(\varepsilon_\alpha - \varepsilon_\beta)t/\hbar} \left\langle [\hat{a}_\alpha^\dagger \hat{a}_\beta, \hat{a}_\gamma^\dagger \hat{a}_\delta] \right\rangle_0 \ . \tag{4.5}$$

The equilibrium average in this equation can be performed with the help of Wick's theorem. Alternatively, one can work out the commutator

$$[\hat{a}_\alpha^\dagger \hat{a}_\beta, \hat{a}_\gamma^\dagger \hat{a}_\delta] = \delta_{\beta\gamma} \hat{a}_\alpha^\dagger \hat{a}_\delta - \delta_{\alpha\delta} \hat{a}_\gamma^\dagger \hat{a}_\beta \tag{4.6}$$

and subsequently calculate its average in the equilibrium ensemble of $\hat{H}_0$. This gives

$$\left\langle [\hat{a}_\alpha^\dagger \hat{a}_\beta, \hat{a}_\gamma^\dagger \hat{a}_\delta] \right\rangle_0 = \delta_{\alpha\delta} \delta_{\beta\gamma} (n_\alpha - n_\beta) \ , \tag{4.7}$$

where $n_\alpha = \frac{1}{e^{\beta(\varepsilon_\alpha - \mu)}+1}$ is the Fermi–Dirac average occupation of state α at temperature T and chemical potential μ. Substituting Eq. (4.7) in Eq. (4.5), we obtain the following result for the non-interacting linear response function.

$$\chi_{AB}^{(0)}(t) = -\frac{i}{\hbar}\Theta(t) \sum_{\alpha\beta} A_{\alpha\beta} B_{\beta\alpha} e^{i(\varepsilon_\alpha - \varepsilon_\beta)t/\hbar} (n_\alpha - n_\beta) \ . \tag{4.8}$$

Finally, taking the Fourier transform of this expression with respect to time, according to the prescription of Eq. (3.45), we obtain

$$\chi_{AB}^{(0)}(\omega) = \sum_{\alpha\beta} \frac{n_\alpha - n_\beta}{\hbar\omega + \varepsilon_\alpha - \varepsilon_\beta + i\hbar\eta} A_{\alpha\beta} B_{\beta\alpha} . \tag{4.9}$$

Notice that this expression, at variance with the general exact eigenstates representation of Eq. (3.47), contains the matrix elements of $\hat{A}$ and $\hat{B}$ between *single particle states*, and the denominator likewise contains the differences of *single particle energies*.

4.3 Density and spin response functions

The calculation of the density–density response function $\chi_{nn}^{(0)}(q, \omega)$ for a homogeneous non-interacting electron gas was discussed by means of a method based on elementary perturbation theory in the introduction to Chapter 3. Although that analysis is certainly valid, we find useful to apply the formalism developed in the previous section to this simple but fundamental calculation.

According to Eq. (3.130), what we need is the second quantized representation of the density fluctuation operators $\hat{n}_{\vec{q}}$ $(=\hat{A})$ and $\hat{n}_{-\vec{q}}$ $(=\hat{B})$ in a basis of single particle states, α, that are characterized by a wave vector $\vec{k}$ and a spin orientation σ. This representation should by now be familiar (see Eq. (1.66)) and it implies that $(n_{\vec{q}})_{\vec{k}\sigma,\vec{k}'\sigma'} = \delta_{\vec{k},\vec{k}'-\vec{q}}\delta_{\sigma\sigma'} = (n_{-\vec{q}})_{\vec{k}'\sigma',\vec{k}\sigma}$. Substituting in Eq. (4.9) we immediately get

$$\chi_{nn}^{(0)}(q, \omega) \equiv \chi_0(q, \omega) = \frac{1}{L^d} \sum_{\vec{k}\sigma} \frac{n_{\vec{k}\sigma} - n_{\vec{k}+\vec{q}\sigma}}{\hbar\omega + \varepsilon_{\vec{k}\sigma} - \varepsilon_{\vec{k}+\vec{q}\sigma} + i\hbar\eta} , \tag{4.10}$$

which is a function of $q = |\vec{q}|$ only. Explicitly,

$$\Re e \chi_0(q, \omega) = \frac{1}{L^d} \mathcal{P} \sum_{\vec{k}\sigma} \frac{n_{\vec{k}\sigma} - n_{\vec{k}+\vec{q}\sigma}}{\hbar\omega + \varepsilon_{\vec{k}\sigma} - \varepsilon_{\vec{k}+\vec{q}\sigma}} , \tag{4.11}$$

$$\Im m \chi_0(q, \omega) = -\frac{\pi}{L^d} \sum_{\vec{k}\sigma} (n_{\vec{k}\sigma} - n_{\vec{k}+\vec{q}\sigma}) \, \delta(\hbar\omega + \varepsilon_{\vec{k}\sigma} - \varepsilon_{\vec{k}+\vec{q}\sigma}) , \tag{4.12}$$

two formulas to which it proves often useful to return. This response function is known as the *Lindhard function* (Lindhard, 1954) and will henceforth be denoted simply as $\chi_0(q, \omega)$.

A direct inspection of the formulas of Section 3.5 of Chapter 3 shows that in a non-interacting electron gas the longitudinal spin–spin response function coincides with the density–density response function, i.e.

$$\chi_{S_z S_z}^{(0)}(q, \omega) = \chi_0(q, \omega) . \tag{4.13}$$

This is a direct consequence of the absence of correlation between up and down spin electrons. Finally, the transverse spin–spin response function $\chi_{S_+ S_-}^{(0)}(q, \omega)$ introduced in

Section 3.5.3 is readily seen to be given by

$$\chi_{S_+ S_-}^{(0)}(q,\omega) = \frac{2}{L^d} \sum_{\vec{k}} \frac{n_{\vec{k}\uparrow} - n_{\vec{k}+\vec{q}\downarrow}}{\hbar\omega + \varepsilon_{\vec{k}\uparrow} - \varepsilon_{\vec{k}+\vec{q}\downarrow} + i\hbar\eta} , \tag{4.14}$$

which is closely related to (but not identical with) the Lindhard function. Because of its importance in the problem of the electron gas, the whole next section is devoted to a study of the Lindhard function.

4.4 The Lindhard function

The calculation of the seemingly complicated sum in Eq. (4.10) can be done analytically at zero temperature, and constitutes a mandatory exercise for any student of electron gas theory. Notice that the complete result can be written as the sum of two terms, each associated with one spin orientation. Thus we write

$$\chi_0(q,\omega) = \chi_{0\uparrow}(q,\omega) + \chi_{0\downarrow}(q,\omega) , \tag{4.15}$$

and focus henceforth on the calculation of the one-spin Lindhard function

$$\chi_{0\sigma}(q,\omega) \equiv \frac{1}{L^d} \sum_{\vec{k}} \frac{n_{\vec{k}\sigma} - n_{\vec{k}+\vec{q}\sigma}}{\hbar\omega + \varepsilon_{\vec{k}\sigma} - \varepsilon_{\vec{k}+\vec{q}\sigma} + i\hbar\eta} . \tag{4.16}$$

A careful study of this function and of its graphical representation is necessary to achieve a real understanding of its physical content. The reader who is already familiar with the material, or is not interested in the details of the derivations, will find the main results summarized in Tables 4.1 and 4.2 and Figs. 4.4–4.8.

We begin by noting that, with a change of variable $\vec{k} \to -\vec{k} - \vec{q}$, Eq. (4.16) can be rewritten as a sum over occupied states:

$$\chi_{0\sigma}(q,\omega) = \frac{1}{L^d} \sum_{\vec{k}} \frac{n_{\vec{k}\sigma}}{\hbar\omega + \varepsilon_{\vec{k}\sigma} - \varepsilon_{\vec{k}+\vec{q}\sigma} + i\hbar\eta}$$
$$+ \frac{1}{L^d} \sum_{\vec{k}} \frac{n_{\vec{k}\sigma}}{-\hbar\omega + \varepsilon_{\vec{k}\sigma} - \varepsilon_{\vec{k}+\vec{q}\sigma} - i\hbar\eta} . \tag{4.17}$$

The second term on the right-hand side of this equation is obtained from the first with the replacement $\omega + i\eta \to -\omega - i\eta$, which will be indicated below as $(\omega^+ \to -\omega^+)$. After inserting the expressions for the free electron energies, Eq. (4.17) can be rewritten as

$$\chi_{0\sigma}(q,\omega) = \frac{m}{\hbar^2 k_{F\sigma} q L^d} \sum_{\vec{k}} \frac{n_{\vec{k}\sigma}}{\frac{\omega}{q v_{F\sigma}} - \frac{q}{2k_{F\sigma}} - \frac{k}{k_{F\sigma}}\cos\theta + i\eta}$$
$$+ (\omega^+ \to -\omega^+) , \tag{4.18}$$

where θ is the angle between $\vec{k}$ and $\vec{q}$, $k_{F\sigma}$ is the Fermi wave vector for spin σ, and $v_{F\sigma} = \frac{\hbar k_{F\sigma}}{m}$ is the corresponding Fermi velocity.

Explicit calculations can be readily carried out by converting the sum into an integral over dimensionless variables. First Eq. (4.18) is cast in the form

$$\chi_{0\sigma}(q,\omega) = \sum_{\sigma} \frac{mk_{F\sigma}^{d-1}}{(2\pi)^d \hbar^2 q} \int_0^1 dx x^{d-1} \int \frac{d\Omega_d}{\frac{\omega}{q v_{F\sigma}} - \frac{q}{2k_{F\sigma}} - x\cos\theta + i\eta}$$
$$+ (\omega^+ \to -\omega^+),$$
(4.19)

where $x = \frac{k}{k_{F\sigma}}$ and $d\Omega_d$ is the element of solid angle in d dimensions. At this point, in order to further simplify the notation, we define a complex function $\Psi_d(z)$ of complex argument z in the following manner:

$$\Psi_d(z) \equiv \int_0^1 dx x^{d-1} \int \frac{d\Omega_d}{\Omega_d} \frac{1}{z - x\cos\theta},$$
(4.20)

where $\Omega_d (= \int d\Omega_d) = 4\pi, 2\pi, 2$ in $d = 3, 2, 1$ respectively. Notice that Ψ_d is an antisymmetric function of z, i.e., $\Psi_d(z) = -\Psi_d(-z)$. Then, for any number of spatial dimensions, the one-spin Lindhard function is given by

$$\chi_{0\sigma}(q,\omega) = N_\sigma(0)\frac{k_{F\sigma}}{q}\left[\Psi_d\left(\frac{\omega+i\eta}{q v_{F\sigma}} - \frac{q}{2k_{F\sigma}}\right) - \Psi_d\left(\frac{\omega+i\eta}{q v_{F\sigma}} + \frac{q}{2k_{F\sigma}}\right)\right],$$
(4.21)

where

$$N_\sigma(0) = \frac{dn_\sigma}{2\epsilon_{F\sigma}} = \begin{cases} \frac{mk_{F\sigma}}{2\pi^2\hbar^2} & , \ 3D, \\ \frac{m}{2\pi\hbar^2} & , \ 2D, \\ \frac{m}{\pi\hbar^2 k_{F\sigma}} & , \ 1D, \end{cases}$$
(4.22)

is the density of states of spin σ per unit volume at the Fermi surface.

In order to complete the calculation we must evaluate the double integral (4.20) in three, two, and one dimension. This can be done analytically, and the resulting formulas for $\Psi_d(z)$ are conveniently presented in the last three lines of Table 4.1. Since z is a complex number, for these formulas to be correctly applied it is essential that the branch cut of the complex logarithm ($\ln z = \ln|z| + i\,\mathrm{Arg}(z)$) and the complex square root ($\sqrt{z} = \sqrt{|z|}\exp\left[\frac{i}{2}\mathrm{Arg}(z)\right]$) lie along the negative real axis, so that $\mathrm{Arg}(z)$ takes values between $-\pi$ and π (this is the standard definition adopted by computer libraries). Also, sign(x) denotes the sign function, defined in the caption of Table 4.1.

We are now in a position to explicitly write down the real and the imaginary parts of the one-spin Lindhard function in three dimensions (Lindhard, 1954), two dimensions (Stern, 1967), and one dimension (Stern, 1967). The final formulas are best written in terms of the dimensionless variables

$$\bar{q}_\sigma \equiv \frac{q}{k_{F\sigma}},$$

$$\nu_{\pm\sigma} \equiv \frac{\omega}{q v_{F\sigma}} \pm \frac{q}{2k_{F\sigma}},$$
(4.23)

Table 4.1. *Computationally useful expressions for the σ-spin component of the frequency dependent complex Lindhard function $\chi_{0\sigma}(q,\omega)$ at zero temperature. The formula for the paramagnetic case (summed over spin) is immediately obtained by dropping the label σ above. The complex logarithm and the complex square root are defined with branch cuts along the negative real axis. The sign function is defined as $\mathrm{sign}(x) = 1, 0, -1$ for $x > 0, x = 0,$ and $x < 0$ respectively. The expressions on the right are the leading terms of the expansion in powers of $\frac{1}{z}$ for $|z| \to \infty$.*

d	$\chi_{0\sigma}(q,\omega) = N_\sigma(0)\frac{k_{F\sigma}}{q}\left[\Psi_d\left(\frac{\omega+i\eta}{qv_{F\sigma}} - \frac{q}{2k_{F\sigma}}\right) - \Psi_d\left(\frac{\omega+i\eta}{qv_{F\sigma}} + \frac{q}{2k_{F\sigma}}\right)\right]$	
3	$\Psi_3(z) = \frac{z}{2} + \frac{1-z^2}{4}\ln\frac{z+1}{z-1}$,	$\left(\to \frac{1}{3z} + \frac{1}{15z^3}\right)$
2	$\Psi_2(z) = z - \mathrm{sign}(\Re e z)\sqrt{z^2 - 1}$,	$\left(\to \frac{1}{2z} + \frac{1}{8z^3}\right)$
1	$\Psi_1(z) = \frac{1}{2}\ln\frac{z+1}{z-1}$,	$\left(\to \frac{1}{z} + \frac{1}{3z^3}\right)$

and have the following form:

$$\frac{\Re e\,\chi_{0\sigma}(q,\omega)}{N_\sigma(0)} = -\begin{cases} \frac{1}{2} - \frac{1-v_{-\sigma}^2}{4\bar{q}_\sigma}\ln\left|\frac{v_{-\sigma}+1}{v_{-\sigma}-1}\right| + \frac{1-v_{+\sigma}^2}{4\bar{q}_\sigma}\ln\left|\frac{v_{+\sigma}+1}{v_{+\sigma}-1}\right|, & 3D \\[2mm] 1 + \frac{1}{\bar{q}_\sigma}\left[\mathrm{sign}(v_{-\sigma})\Theta(v_{-\sigma}^2 - 1)\sqrt{v_{-\sigma}^2 - 1}\right. \\ \qquad \left. - \mathrm{sign}(v_{+\sigma})\Theta(v_{+\sigma}^2 - 1)\sqrt{v_{+\sigma}^2 - 1}\right], & 2D \\[2mm] \frac{1}{2\bar{q}_\sigma}\ln\left|\frac{v_{-\sigma}-1}{v_{-\sigma}+1}\right| - \frac{1}{2\bar{q}_\sigma}\ln\left|\frac{v_{+\sigma}-1}{v_{+\sigma}+1}\right|, & 1D \end{cases} \qquad (4.24)$$

and

$$\frac{\Im m\,\chi_{0\sigma}(q,\omega)}{N_\sigma(0)} = -\begin{cases} \frac{\pi}{4\bar{q}_\sigma}\left[\Theta(1 - v_{-\sigma}^2)(1 - v_{-\sigma}^2) - \Theta(1 - v_{+\sigma}^2)(1 - v_{+\sigma}^2)\right], & 3D \\[2mm] \frac{1}{\bar{q}_\sigma}\left[\Theta(1 - v_{-\sigma}^2)\sqrt{1 - v_{-\sigma}^2} - \Theta(1 - v_{+\sigma}^2)\sqrt{1 - v_{+\sigma}^2}\right], & 2D \\[2mm] \frac{\pi}{2\bar{q}_\sigma}\left[\Theta(1 - v_{-\sigma}^2) - \Theta(1 - v_{+\sigma}^2)\right]. & 1D \end{cases} \qquad (4.25)$$

The regions of the (q,ω) plane in which $|v_{-\sigma}|$ and $|v_{+\sigma}|$ are smaller or larger than 1 are shown in Fig. 4.2. Plots of $\Re e\,\chi_{0\sigma}(q,\omega)$ and $\Im m\,\chi_{0\sigma}(q,\omega)$ vs ω for different values of q are shown in Figs. 4.4–4.8. The special case of the static Lindhard function $\chi_{0\sigma}(q,0)$ is discussed in the next section.

4.4.1 The static limit

In the static limit ($\omega = 0$) the Lindhard function is purely real and is given by

$$\chi_{0\sigma}(q,0) = \frac{1}{L^d}\sum_{\vec{k}}\frac{n_{\vec{k}\sigma} - n_{\vec{k}+\vec{q}\sigma}}{\varepsilon_{\vec{k}\sigma} - \varepsilon_{\vec{k}+\vec{q}\sigma}} . \qquad (4.26)$$

Table 4.2. *Explicit expressions for* $\chi_{0\sigma}(q,0)$ – *the static limit of the one-spin Lindhard function. Here* $\bar{q}_\sigma = \frac{q}{k_{F\sigma}}$. *The corresponding Fourier transforms are given in Table 4.3.*

d	$\chi_{0\sigma}(q,0)$
3	$-N_\sigma(0)\left[\frac{1}{2} + \frac{\bar{q}_\sigma^2 - 4}{8\bar{q}_\sigma} \ln\left\|\frac{\bar{q}_\sigma - 2}{\bar{q}_\sigma + 2}\right\|\right]$
2	$-N_\sigma(0)\left[1 - \Theta(\bar{q}_\sigma - 2)\frac{\sqrt{\bar{q}_\sigma^2 - 4}}{\bar{q}_\sigma}\right]$
1	$-N_\sigma(0)\left[\frac{1}{\bar{q}_\sigma} \ln\left\|\frac{\bar{q}_\sigma + 2}{\bar{q}_\sigma - 2}\right\|\right]$

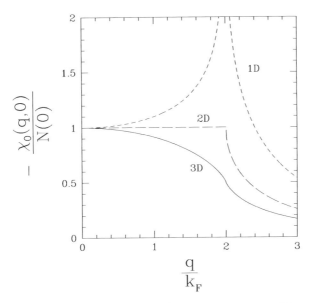

Fig. 4.1. The behavior of the function $\chi_0(q,0) = \sum_\sigma \chi_{0\sigma}(q,0)$ as obtained from Table 4.2 in the paramagnetic state for $d = 1, 2, 3$.

Notice that the principal part prescription, apparently required by the presence of the infinitesimal η in the denominator of Eq. (4.16), is actually not needed here, since the denominator of this expression can only vanish simultaneously with the numerator.[1] At $T = 0$ the static Lindhard function can either be obtained analytically from a direct evaluation of Eq. (4.26), or deduced from the $\omega \to 0$ limit of Eq. (4.24). The resulting analytic form is given in Table 4.2 and plotted in Fig. 4.1.

---

[1] Eq. (4.26) is mathematically equivalent to Eq. (3.14), which was obtained in Chapter 3 by more elementary means.

The function $\chi_{0\sigma}(q,0)$ displays a number of noteworthy features. We begin by noticing that the long wavelength limit is readily evaluated with the result

$$\lim_{q\to 0} \chi_{0\sigma}(q,0) = -N_\sigma(0) , \qquad (4.27)$$

where $N_\sigma(0)$ is given by Eq. (4.22).

The fact that $\lim_{q\to 0}\chi_0(q,0)$ is finite (see Fig. 4.1) and equal to the density of states at the Fermi energy, simply follows from the chain of identities

$$\lim_{q\to 0} \frac{1}{L^d}\sum_{\vec{k}} \frac{n_{\vec{k}\sigma} - n_{\vec{k}+\vec{q}\sigma}}{\varepsilon_{\vec{k}\sigma} - \varepsilon_{\vec{k}+\vec{q}\sigma}} = \frac{1}{L^d}\sum_{\vec{k}} \frac{\partial n_{\vec{k}\sigma}}{\partial\varepsilon_{\vec{k}\sigma}}$$

$$\overset{(T=0)}{=} -\frac{1}{L^d}\sum_{\vec{k}} \delta(\epsilon_{F\sigma} - \varepsilon_{\vec{k}\sigma})$$

$$= -N_\sigma(0) . \qquad (4.28)$$

Thus, we see that the $q \to 0$ limit of the static Lindhard function is a measure of the number of excited states available to the system for vanishing excitation energy. Therefore $\lim_{q\to 0}\chi_0(q,0)$ vanishes in systems that have a gap in the excitation spectrum, e.g. a two-dimensional electron gas in the presence of a perpendicular magnetic field. A direct comparison between Fig. 4.1 and Fig. 10.9 of Chapter 10 vividly illustrates the point.

An interesting implication of Eq. (4.27) is that, in any number of dimensions, the static homogeneous spin susceptibility per unit volume is given by (see Eq. (4.13))

$$\chi_P = -\left(\frac{g\mu_B}{2}\right)^2 \lim_{q\to 0}\chi_{S_zS_z}^{(0)}(q,0) = \left(\frac{g\mu_B}{2}\right)^2 N(0) , \qquad (4.29)$$

at $T=0$, and remains essentially constant as long as $k_BT \ll \epsilon_F$.[2] This behavior, commonly known as *Pauli paramagnetism* (Pauli, 1927), is in stark contrast with the susceptibility of an assembly of non-interacting classical magnetic moments of magnitude $-\frac{g\mu_B}{2}$, which is given by the Curie law $\chi_C = \frac{n(g\mu_B)^2}{4k_BT}$ and diverges for $T \to 0$. The reason for the difference is that in the ground-state of the non-interacting electron gas Pauli's exclusion principle effectively prevents most electrons from flipping their spin in response to the magnetic field – the spin-flipped state being already occupied by another electron. This general phenomenon is appropriately referred to as *Pauli blocking*. In such a situation, only the electrons in partially occupied states within an energy k_BT of the Fermi level can flip their spins and contribute to the susceptibility. Since there are approximately $N(0)k_BT$ such electrons per unit volume, each contributing a Curie-like susceptibility $\frac{(g\mu_B)^2}{4k_BT}$, we see that the temperature dependence drops out, leaving the temperature-independent result of Eq. (4.29).

A second feature implied by the formulae of Table 4.2 and clearly displayed in Fig. 4.1 is that for $q=2k_F$ the Lindhard function is singular. While for three and two dimensions the singularity only appears in the derivatives of $\chi_0(q,0)$, in one dimension the

[2] Rather conveniently, the spin susceptibility is a dimensionless number in 3D.

Table 4.3. *Explicit expressions for $\chi_{0\sigma}(r,0)$, the spatial*
Fourier transform of $\chi_{0\sigma}(q,0)$. These expressions are
obtained from the integral in Eq. (4.30).

d	$\chi_{0\sigma}(r,0)$
3	$12\pi n_\sigma N_\sigma(0)\frac{\sin 2k_{F\sigma}r - 2k_{F\sigma}r\cos 2k_{F\sigma}r}{(2k_{F\sigma}r)^4}$
2	$2\pi n_\sigma N_\sigma(0)\,[J_0(k_{F\sigma}r)N_0(k_{F\sigma}r) + J_1(k_{F\sigma}r)N_1(k_{F\sigma}r)]$
1	$\pi n_\sigma N_\sigma(0)\,\mathrm{si}(2k_{F\sigma}x)$

function itself has a logarithmic divergence. This singular behavior is responsible for several interesting phenomena. Specific examples are the *Friedel oscillations* and the associated *RKKY interaction*, which will be discussed in the next chapter. Moreover, in one-dimensional systems the divergence of $\chi_0(q,0)$ at $q = 2k_F$ leads to the so called *Peierls instability* (Peierls, 1954), i.e., the spontaneous formation of a density wave at wave vector $2k_F$.[3]

We conclude this section by noting that the static Lindhard function $\chi_{0\sigma}(\vec{q},0)$ can be Fourier transformed to obtain $\chi_{0\sigma}(r,0)$, which physically represents the response of the density (or spin density) of the non-interacting electron gas to a potential (or magnetic field) that is localized at the origin of the coordinate system.

The calculation of the Fourier transform can be done analytically with the help of the formulas of Table 4.2 for the static Lindhard function:

$$\chi_{0\sigma}(r,0) = \int_{-\infty}^{\infty} \frac{d\vec{q}}{(2\pi)^d} \chi_{0\sigma}(\vec{q},0)e^{i\vec{q}\cdot\vec{r}}. \tag{4.30}$$

The final result is expressed in terms of trigonometric functions in 3D, Bessel functions of the first and second kind ($J_n(x)$ and $N_n(x)$ respectively) in 2D, and the sine-integral function

$$\mathrm{si}(x) = \int_0^x \frac{\sin t}{t}dt - \frac{\pi}{2}, \tag{4.31}$$

in 1D. The definitions and the properties of these functions can be found in Gradshteyn and Ryzhik (1965). The results of the calculations are summarized in Table 4.3.

Of particular interest is the behavior of $\chi_{0\sigma}(r,0)$ for large distances, $r \gg k_{F\sigma}^{-1}$. It is readily found from the formulae of Table 4.3 that in this limit $\chi_{0\sigma}(r,0)$ oscillates with periodicity $\frac{1}{2k_{F\sigma}}$ and decays as $\frac{1}{r^d}$.[4] This result implies that a weak impurity placed at the origin of the electron gas will produce long-range oscillations in the (spin) density very far away. These

---

[3] At variance with the Overhauser charge density wave discussed in Chapters 1 and 2, the Peierls instability is not driven by the electron-electron interaction, but by the electron–lattice coupling.
[4] Notice that in the one dimensional case one can directly use the asymptotic formula $\mathrm{si}(x) \simeq -\frac{\cos x}{x}$. In the two dimensional case, on the other hand, at least the first two terms of the asymptotic expansions of the Bessel functions are needed to determine the behavior of the expression.

are known as *Friedel oscillations* and are a direct consequence of the existence of the Fermi surface. A more complete discussion of Friedel oscillations in the interacting electron liquid will be presented in Section 5.4.3.

4.4.2 The electron–hole continuum

One of the most relevant features of the Lindhard function is the structure of its imaginary part, a quantity that, as we have seen in Chapter 3, is related to the dynamical structure factor by Eq. (3.74). For simplicity's sake, our discussion will now be concerned with the paramagnetic electron gas only.

At zero temperature $\Im m \chi_0(q, \omega)$ is in general given by (see Eq. (4.12)):

$$\Im m \chi_0(q, \omega) = -\frac{\pi}{L^d} \sum_{\vec{k}, \sigma} \left[\Theta(\varepsilon_{\vec{k}\sigma} - \epsilon_F) - \Theta(\varepsilon_{\vec{k}+\vec{q}\sigma} - \epsilon_F) \right] \delta(\hbar\omega + \varepsilon_{\vec{k}\sigma} - \varepsilon_{\vec{k}+\vec{q}\sigma}) . \quad (4.32)$$

For a given q this expression differs from zero only in a well defined *range* of frequencies. This range is determined by the geometry of the Fermi surface and is directly related to the possible excitation energies of *electron–hole pairs*, i.e. the states that are obtained from the ground-state of the non-interacting electron gas by promoting an electron from an occupied state of wave vector $\vec{k}$ (which then remains empty and is referred to as the *hole*) to an empty one of wave vector $\vec{k} + \vec{q}$. The region of the q, ω plane in which $\Im m \chi_0$ differs from zero is accordingly referred to as the *electron–hole continuum*.[5]

As it turns out, there is a qualitative difference between the structure of the electron–hole continuum of three and two dimensional systems and that of a one dimensional one. In three and two dimensions the electron–hole continuum is determined by the inequalities

$$\max(0, \omega_-(q)) \leq |\omega| \leq \omega_+(q) , \qquad 3D, 2D , \qquad (4.33)$$

where

$$\omega_\pm(q) = \frac{\hbar q^2}{2m} \pm v_F q . \qquad (4.34)$$

This is represented by the shaded region in Fig. 4.2. The lower and the upper limits on $|\omega|$ represent respectively the minimum and the maximum electron–hole excitation energy for given q. It is evident that, for any $q \leq 2k_F$, the minimum electron–hole excitation energy vanishes because one can move an electron from a state $\vec{k}$ infinitesimally below the Fermi surface to a state $\vec{k} + \vec{q}$ infinitesimally above it (see Fig. 4.3). The maximum excitation energy, on the other hand, is obtained when an electron of wave vector $\vec{p}$ at the Fermi surface is displaced radially outward by a wave vector of magnitude q (see Fig. 4.3): this gives $\hbar\omega_+(q) = \hbar v_F q + \frac{\hbar^2 q^2}{2m}$.

---

[5] Because of the general symmetry $\Im m \chi_0(q, -\omega) = -\Im m \chi_0(q, \omega)$ it is evident that the electron–hole continuum is symmetric about the $\omega = 0$ axis.

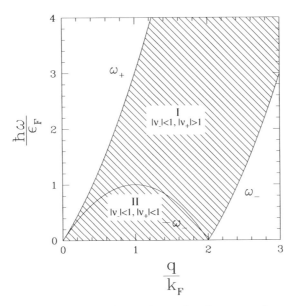

Fig. 4.2. The electron–hole continuum in the non-interacting electron gas. In one dimension, region II is absent.

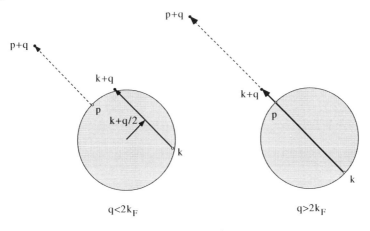

Fig. 4.3. electron–hole pairs of minimum energy ($\vec{k} \to \vec{k} + \vec{q}$ – solid line) and maximum energy ($\vec{p} \to \vec{p} + \vec{q}$ – dashed line) for a given wave vector $\vec{q}$. Notice that for zero energy excitations the sum of the initial and final wave vectors, that is, $2\left(\vec{k} + \frac{\vec{q}}{2}\right)$, is orthogonal to $\vec{q}$.

For $q \geq 2k_F$, the minimum excitation energy is finite because it is no longer possible to move an electron from one point of the Fermi surface to another, also on the Fermi surface. In this case the minimum excitation energy is obtained when an electron at the Fermi surface is displaced radially inward by a wave vector of magnitude q (see Fig. 4.3): this gives $\hbar\omega_-(q) = \frac{\hbar^2 q^2}{2m} - \hbar v_F q$.

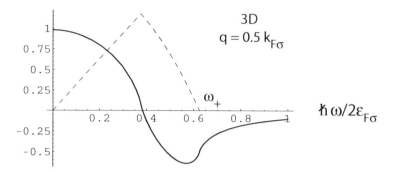

Fig. 4.4. The real part (solid line) and the imaginary part (dashed line) of the dimensionless Lindhard function $-\frac{\chi_{0\sigma}(q,\omega)}{N_\sigma(0)}$ in three dimensions for $q = 0.5k_{F\sigma}$. ω_+ represents the upper edge of the electron–hole continuum. The cusp in the imaginary part occurs at the boundary between regions I and II of Fig. 4.2, i.e. at $\omega = |\omega_-(q)|$.

Notice that in the region $|\omega_-(q)| \leq |\omega| \leq \omega_+(q)$ (labeled as I in Fig. 4.2) one has $|v_-| < 1$ and $|v_+| < 1$, so both terms in Eq. (4.25) contribute to the imaginary part of χ_0, while in the region $0 \leq |\omega| \leq |\omega_-(q)|$ (labeled as II in Fig. 4.2) one has $|v_-| < 1$ and $|v_+| > 1$ so only the first term in Eq. (4.25) contributes.

The details of the behavior of $\Im m \chi_0(q, \omega)$ depend on dimensionality. For $q \leq 2k_F$, in three dimensions, we see from Eq. (4.25) that $\Im m \chi_{0\sigma}(q, \omega)$ is a *linear* function of frequency

$$\Im m \chi_{0\sigma}(q, \omega) = -\frac{\pi}{2} N_\sigma(0) \frac{\omega}{q v_{F\sigma}} , \quad 3D , \tag{4.35}$$

for $0 \leq \omega \leq |\omega_-(q)|$ (i.e., in region II), and then becomes an arc of parabola for $|\omega_-(q)| \leq \omega \leq \omega_+(q)$ (region I). For $q > 2k_F$ only the parabola is present. This behavior is displayed in Figs. 4.4 and 4.5.

In the two dimensional case the situation is only slightly different and the low frequency expansion for $\Im m \chi_0(q, \omega)$ in region II, for $\omega \ll |\omega_-(q)|$ is

$$\Im m \chi_{0\sigma}(q, \omega) \simeq -\frac{N_\sigma(0)}{\sqrt{1 - \left(\frac{q_-}{2k_{F\sigma}}\right)^2}} \frac{\omega}{q v_{F\sigma}} , \quad 2D . \tag{4.36}$$

The behavior of $\Im m \chi_0$ is displayed in Figs. 4.6 and 4.7.

Notice that the sharp cutoffs in the imaginary part of $\chi_0(\vec{q}, \omega)$ as a function of frequency are intimately related to the rapid swing in the real part of this function: at small wave vector $\Re e \chi_0(\vec{q}, \omega)$ changes sign from negative to positive, as ω sweeps across the electron–hole continuum.[6] For $\omega \ll q v_F$ on the other hand, the real part of $\chi_0(q, \omega)$ has a parabolic shape $\chi_0(q, \omega) \approx \chi_0(q, 0) + \mathcal{O}(\omega^2)$.

In one dimension, the structure of the electron–hole continuum is different. Because the "Fermi surface" in this case consists of just the two points $\pm k_F$, excitations of vanishing

---

[6] This behavior is a direct consequence of the fact that $\Re e \chi_0(\vec{q}, \omega)$ and $\Im m \chi_0(\vec{q}, \omega)$ are related via the Kramers–Krönig relations (see Eq. (3.80)).

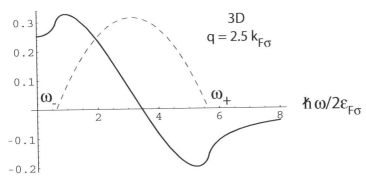

Fig. 4.5. Same as Fig. 4.4 for $q = 2.5 k_{F\sigma}$. ω_- and ω_+ represent the lower and upper edges of the electron–hole continuum.

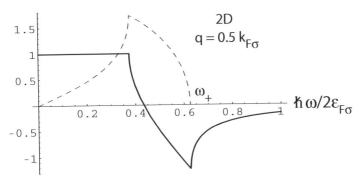

Fig. 4.6. Characteristic *shark fin* shape of the imaginary part of the dimensionless Lindhard function $-\frac{\chi_{0\sigma}(q,\omega)}{N_\sigma(0)}$ (dashed line) in two dimensions for wave vectors less than $2k_F$. Here $q = 0.5 k_{F\sigma}$. ω_+ represents the upper edge of the electron–hole continuum. The cusp occurs at $\omega = |\omega_-(q)|$. The solid line represent the real part of the same function, which is constant for $\omega < |\omega_-(q)|$.

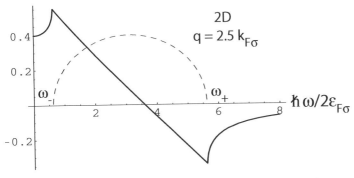

Fig. 4.7. Same as Fig. 4.6 for $q = 2.5 k_{F\sigma}$. ω_- and ω_+ represent the lower and upper edges of the electron–hole continuum.

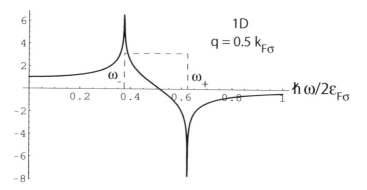

Fig. 4.8. The real part (solid line) and the imaginary part (dashed line) of the dimensionless Lindhard function $-\frac{\chi_{0\sigma}(q,\omega)}{N_\sigma(0)}$ in one dimension for $q = 0.5 k_{F\sigma}$. $|\omega_-|$ and ω_+ represent the lower and upper edges of the electron–hole continuum.

energy are impossible except at wave vectors 0 and $2k_F$. Accordingly, for all values of q the electron–hole continuum is defined by the inequalities

$$|\omega_-(q)| \leq |\omega| \leq \omega_+(q), \qquad 1D. \tag{4.37}$$

This means that the 1D electron–hole continuum consists only of the region labeled I in Fig. 4.2: region II is absent. Furthermore, within region I, $\Im m\, \chi_0(q, \omega)$ is independent of frequency and given by $-\frac{m}{\hbar^2 q}$ sign ω. This situation is depicted in Fig. 4.8.

4.4.3 The nature of the singularity at small q and ω

The presence of low-energy excitations across the Fermi surface causes the Lindhard function to be a singular function of wave vector and frequency in the key region in which both these quantities are small. The existence of a mathematical singularity is seen from the fact that the limit of $\chi_{0\sigma}(q, \omega)$ for $q \to 0$ and $\omega \to 0$ depends on the ratio $\nu = \frac{\omega}{v_{F\sigma} q}$, that is, on the *direction* along which the origin of the (q, ω) plane is approached: different limits are obtained for different values of ν.

For example if the origin is approached along the q axis at $\omega = 0$ (corresponding to $\nu = 0$), we are in the so called *static limit* and, as discussed above (see Eq. (4.27)), the Lindhard function tends to a finite limit – the negative of the total density of states at the Fermi surface.

If, on the other hand, the origin is approached along the ω axis at $q = 0$, i.e. at $\nu = \infty$, a case referred to as the *dynamical long-wavelength limit*, then the Lindhard function can be readily shown to have the limiting form

$$\chi_{0\sigma}(q, \omega) \overset{q\to 0}{\approx} -\frac{n_\sigma q^2}{m\omega^2}, \tag{4.38}$$

a result that, incidentally, holds true even if ω is not small.

Table 4.4. *Analytical
expressions for* $\Psi'_d(z)$.

$$\Psi'_3(z) = 1 - \frac{z}{2}\ln\frac{z+1}{z-1}$$

$$\Psi'_2(z) = 1 - \text{sign}\,(\Re ez)\,\frac{z}{\sqrt{z^2-1}}$$

$$\Psi'_1(z) = \frac{1}{1-z^2}$$

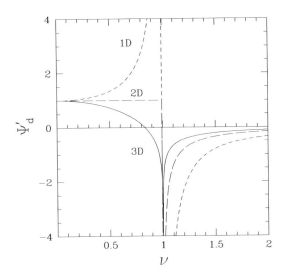

Fig. 4.9. The behavior of the real part of the functions $\Psi'_d(\nu)$.

From a physical point of view, the dramatic difference between the static and the dynamic limits (finite in one case, zero in the other) is due to the presence of low energy electron–hole excitations which are able to "follow" the external potential in the first case, but not in the second.[7]

In general, the ν-dependence of the small q limit of the Lindhard function is given by

$$\lim_{q\to 0}\chi_{0\sigma}(q,qv_{F\sigma}\nu) = -N_\sigma(0)\Psi'_d(\nu+i\eta)\,, \tag{4.39}$$

where $\Psi'_d(z)$ is the derivative of the function $\Psi_d(z)$ defined in Eq. (4.20) with respect to z. The functions $\Psi'_d(z)$ are listed in Table 4.4 and plotted in Fig. 4.9. Note their singular behavior when $z \to \pm 1$.

The knowledge of the appropriate limiting behavior of the Lindhard function will be crucial in analyzing the excitation spectrum of systems that support different types

---

[7] In order to "follow" the evolution of a potential with wave vector q and frequency ω an excitation must be able to travel with a speed larger than $\frac{\omega}{q}$. For electron–hole pairs, this is only possible when $\frac{\omega}{q}$ is less than the Fermi velocity v_F.

of collective excitations (see Section 5.3.3). This is a situation in which ν remains finite as $q \to 0$ and therefore particular care must be exercised (Santoro and Giuliani, 1988).

4.4.4 The Lindhard function at finite temperature

At finite temperature, T, the Lindhard function is still given by Eq. (4.10), but the occupation numbers are given by the Fermi–Dirac distribution $n_{\vec{k}\sigma}(T, \mu_\sigma) = \frac{1}{e^{\beta(\epsilon_{\vec{k}\sigma} - \mu_\sigma)}+1}$, where, as usual, $\beta = \frac{1}{k_B T}$. The chemical potentials μ_σ are determined by the conditions

$$\frac{1}{L^d} \sum_{\vec{k}} n_{\vec{k}\sigma} = n_\sigma , \qquad (4.40)$$

so the μ_σ are ultimately functions of T, and will be denoted by $\mu_\sigma(T)$ in the rest of this section.

Unlike its zero-temperature counterpart, the finite-temperature Lindhard function cannot be expressed entirely in terms of elementary functions, but typically requires the numerical evaluation of a one-dimensional integral. A particularly elegant approach (Maldague, 1978) makes use of the identity

$$\frac{1}{e^x+1} = \int_{-\infty}^{\infty} \frac{\Theta(y-x)}{4\cosh^2\left(\frac{y}{2}\right)} \, dy \qquad (4.41)$$

to express the finite-temperature Lindhard function as an integral of the zero-temperature Lindhard function over the Fermi energy:

$$\chi_\sigma^{(0)}(q, \omega, T) = \int_0^{\infty} dE \, \frac{[\chi_\sigma^{(0)}(q, \omega)]_{T=0,\epsilon_F=E}}{4k_B T \cosh^2\left(\frac{E-\mu(T)}{2k_B T}\right)} \qquad (4.42)$$

(see Exercise 7).

As it turns out, in a few special cases the calculation can be carried out analytically leading to closed form expressions. Making use of the same notation as in Section 4.4, and introducing the function

$$F_\sigma(x, T) = \frac{x}{e^{\beta[x^2 \epsilon_{F\sigma} - \mu_\sigma(T)]} + 1} \qquad (4.43)$$

we can summarize the results as follows:

$$\frac{\Re e \chi_{0\sigma}(q, \omega)}{N_\sigma(0)} = -\int_0^{\infty} dx \, \frac{F_\sigma(x, T)}{\bar{q}_\sigma} \times \begin{cases} \frac{1}{2}\left(\ln\left|\frac{x-\nu_{-\sigma}}{x+\nu_{-\sigma}}\right| - \ln\left|\frac{x-\nu_{+\sigma}}{x+\nu_{+\sigma}}\right|\right), & 3D \\[2ex] \frac{\nu_{-\sigma}}{|\nu_{-\sigma}|}\frac{\Theta(|\nu_{-\sigma}|-x)}{\sqrt{\nu_{-\sigma}^2-x^2}} - \frac{\nu_{+\sigma}}{|\nu_{+\sigma}|}\frac{\Theta(|\nu_{+\sigma}|-x)}{\sqrt{\nu_{+\sigma}^2-x^2}}, & 2D \\[2ex] \frac{1}{x}\left(\frac{\nu_{-\sigma}}{\nu_{-\sigma}^2-x^2} - \frac{\nu_{+\sigma}}{\nu_{+\sigma}^2-x^2}\right), & 1D \end{cases} \qquad (4.44)$$

and

$$
\frac{\Im m \chi_{0\sigma}(q,\omega)}{N_\sigma(0)} = -\pi \begin{cases} \dfrac{\omega}{v_{F\sigma}q} + \dfrac{k_B T}{\hbar v_{F\sigma}q} \ln \dfrac{1+e^{\beta\left[v_{-\sigma}^2 \epsilon_{F\sigma} - \mu_\sigma(T)\right]}}{1+e^{\beta\left[v_{+\sigma}^2 \epsilon_{F\sigma} - \mu_\sigma(T)\right]}}, & 3D \\[3mm] 2\displaystyle\int_0^\infty dx \, \dfrac{F_\sigma(x,T)}{\bar{q}_\sigma} \left(\dfrac{\Theta(x-|v_{+\sigma}|)}{\sqrt{x^2-v_{+\sigma}^2}} - \dfrac{\Theta(x-|v_{-\sigma}|)}{\sqrt{x^2-v_{-\sigma}^2}} \right), & 2D \\[3mm] \dfrac{1}{2\bar{q}_\sigma} \left(\dfrac{1}{e^{\beta\left[v_{-\sigma}^2 \epsilon_{F\sigma} - \mu_\sigma(T)\right]}+1} - \dfrac{1}{e^{\beta\left[v_{+\sigma}^2 \epsilon_{F\sigma} - \mu_\sigma(T)\right]}+1} \right). & 1D \end{cases}
\tag{4.45}
$$

4.5 Transverse current response and Landau diamagnetism

The *longitudinal* current response of a non-interacting electron gas can be either directly calculated (see Exercise 8) or immediately inferred from the density response according to the exact relation Eq. (3.175) of Chapter 3. In contrast to this, the transverse current response must be calculated independently.

As we saw in Chapter 3 the transverse current response is needed to calculate the orbital magnetization induced in the electron gas by a weak magnetic field. We will now show that the orbital magnetic response of the electron gas is *diamagnetic*, that is, the induced orbital magnetization tends to *reduce* the applied field. This behavior of the induced orbital magnetism is opposite to the paramagnetic behavior of the induced spin magnetism, which, as we saw in Section 4.4.1, tends to reinforce the external field ($\chi_P > 0$).

To calculate $\chi_T(q,\omega)$ for the non-interacting electron gas we first assume that $\vec{q}$ lies along the y-axis and then recall the general relation valid for homogeneous systems

$$
\chi_T(q,\omega) = \frac{n}{m} + \chi_{j_{pT} j_{pT}}(q,\omega),
\tag{4.46}
$$

where in the present geometry we have $\chi_{j_{pT} j_{pT}}(q,\omega) = \chi_{j_{px} j_{px}}(\vec{q},\omega)$. The essential ingredient for the calculation of $\chi^{(0)}_{j_{px} j_{px}}(\vec{q},\omega)$ is the matrix element of the paramagnetic current operator $\hat{j}_{px}$ between two plane wave states of wave vectors $\vec{k}$ and $\vec{k}+\vec{q}$. As shown in Appendix 2, this is given by $\frac{\hbar}{m}\left(k_x + \frac{q_x}{2}\right)$, which reduces to $\frac{\hbar k_x}{m}$ under the present assumption of $q_x = 0$. Substituting this in Eq. (4.9) we arrive at the formula

$$
\chi^{(0)}_{j_{px} j_{px}}(\vec{q},\omega) = \frac{\hbar^2}{m^2 L^d} \sum_{\vec{k}\sigma} k_x^2 \frac{n_{\vec{k}\sigma} - n_{\vec{k}+\vec{q}\sigma}}{\hbar\omega + \varepsilon_{\vec{k}\sigma} - \varepsilon_{\vec{k}+\vec{q}\sigma} + i\hbar\eta}.
\tag{4.47}
$$

The calculation of this expression is very similar to the calculation of the ordinary Lindhard function, described in Section 4.4. Taking the same steps one easily arrives at the following expression for the transverse current response function of a non-interacting electron gas

$$
\chi_T^{(0)}(q,\omega) = \frac{n}{m} + \sum_\sigma \frac{\hbar^2 k_{F\sigma}^3 N_\sigma(0)}{m^2 q} \left[\Psi_d^{(T)}(v_{-\sigma}) - \Psi_d^{(T)}(v_{+\sigma}) \right],
\tag{4.48}
$$

Table 4.5. *Computationally useful expressions for the spin σ component of the transverse current–current response function $\chi^{(0)}_{T\sigma}(q,\omega)$ at zero temperature.*

d	$\chi^{(0)}_{T\sigma}(q,\omega) = \frac{n_\sigma}{m} + N_\sigma(0)\frac{\hbar^2 k_{F\sigma}^3}{m^2 q}\left[\Psi_d^{(T)}\left(\frac{\omega+i\eta}{qv_{F\sigma}} - \frac{q}{2k_{F\sigma}}\right) - \Psi_d^{(T)}\left(\frac{\omega+i\eta}{qv_{F\sigma}} + \frac{q}{2k_{F\sigma}}\right)\right]$
3	$\Psi_3^T(z) = \frac{z(5-3z^2)}{24} + \left(\frac{1-z^2}{4}\right)^2 \ln\frac{z+1}{z-1}$
2	$\Psi_2^T(z) = \frac{z(3-2z^2)}{6} - \text{sign}\,(\Re ez)\,\frac{1-z^2}{3}\sqrt{z^2-1}$

where

$$\Psi_d^{(T)}(z) = \int_0^1 dx\, x^{d+1} \int \frac{d\Omega_d}{\Omega_d} \frac{n_x^2}{z - x\cos\theta}\,, \qquad (4.49)$$

θ is the angle between $\vec{k}$ and $\vec{q}$, and n_x is the x-component of the unit vector along the radial direction, i.e., $\sin\theta\cos\varphi$ in three dimensions and $\sin\theta$ in two. The functions $\Psi_d^{(T)}(z)$ can be calculated explicitly in three and two dimensions (there is no "transverse" direction in the strictly 1D case!) and are listed in Table 4.5.

Let us now focus on the calculation of $\lim_{q\to 0} \frac{\chi_T^{(0)}(q,0)}{q^2}$, which, as discussed in Section 3.4.3, determines the orbital magnetic susceptibility. A necessary condition for the existence of this limit is that the first term on the right-hand side of Eq. (4.48) exactly cancel the $q \to 0$ limit of the second term, so that only terms of order q^2 survive. A direct calculation confirms that the "diamagnetic sum rule" is satisfied in the non-interacting electron gas. Making use of Eq. (4.48) one can show that, in both three and two dimensions,

$$\chi_{orb}^{(0)} = -\frac{e^2}{c^2}\lim_{q\to 0}\frac{\chi_T^{(0)}(q,0)}{q^2} = -\frac{1}{3}\left(\frac{g\mu_B}{2}\right)^2 N(0)\,, \qquad (4.50)$$

where $\mu_B = \frac{e\hbar}{2mc}$ is the Bohr magneton. This result was first obtained by Landau via energy considerations (Landau, 1930). Notice that the orbital susceptibility is *diamagnetic*, i.e., the induced moment is opposite to the applied magnetic field, and its numerical value is one third of the Pauli paramagnetic spin susceptibility.[8]

The reader may wonder whether a diamagnetic susceptibility violates the stiffness theorem of Section 3.2.9, according to which a static response function must always be negative. The reason why the orbital susceptibility is anomalous in this respect is that the vector potential enters the hamiltonian with both a linear and a quadratic term. Taking this into account one sees that the stiffness theorem only requires the paramagnetic current response to be negative but allows $\chi_T(q,0)$ to be positive.

We finally remark that the formal coincidence of the diamagnetic susceptibilities in two and three dimensions is not an accident: its deeper reason is explained in Exercise 10.

---

[8] Curiously, the Landau diamagnetism is referred to as "paramagnetism" in Landau's original paper (Landau, 1930).

4.6 Elementary theory of impurity effects

In this section we present the basic theory of impurity effects, mainly to illustrate the effect of electron–impurity scattering on the linear response functions of the non-interacting electron liquid.

At the classical level, the presence of impurities implies that each electron performs a classical random walk with a velocity of order v_F, changing direction randomly at each collision with an impurity.[9] Thus, during a time interval t that is much larger than a characteristic time interval τ between collisions, the electrons experiences $\frac{t}{\tau}$ collisions with impurities and travels a typical distance $\langle r^2 \rangle = (v_F \tau)^2 \frac{t}{\tau}$. The mean square displacement in any direction is then given by $\langle x^2 \rangle = \frac{\langle r^2 \rangle}{d}$, and the average distance travelled in any direction grows with the square root of time according to the law $\sqrt{\langle x^2 \rangle} = \sqrt{Dt}$, where $D = \frac{v_F^2 \tau}{d}$ is the *diffusion constant* in d dimensions.

To understand the impact of the diffusive dynamics on the density–density response function, consider the evolution of a local macroscopic density disturbance, e.g., a local excess density distributed on a region that is small compared to the system size, yet large compared to any microscopic length scale. If the positive background remained uniform (as in the rigid jellium model) such a density disturbance would spread out in a very short time under the action of the electric field it produces (this is further explored in Exercise 13). Let us instead assume that the excess electronic density is perfectly neutralized by a corresponding deformation of the positive background, and that this neutrality condition is preserved throughout the evolution of the disturbance.[10] How will such a disturbance evolve in time? The answer is that the excess density will slowly relax to zero according to the *diffusion equation*

$$\frac{\partial n(\vec{r}, t)}{\partial t} = D \nabla^2 n(\vec{r}, t), \qquad (4.51)$$

which is simply obtained from the continuity equation and the assumption that $\vec{j}(\vec{r}, t) = -D\vec{\nabla} n(\vec{r}, t)$. This result will be derived in Section 4.6.3. The dynamics generated by the diffusion equation can be best exemplified by noting that in a d-dimensional system an initial density distribution $n(\vec{r}, 0) = \gamma \delta(\vec{r})$ will evolve as follows (see Exercise 11)

$$n(\vec{r}, t) = \gamma \frac{e^{-\frac{r^2}{4Dt}}}{(4\pi Dt)^{\frac{d}{2}}}. \qquad (4.52)$$

Corresponding to this form of the time evolution, we will show in this section that the small q and ω behavior of the Lindhard function (now averaged over many random impurity

---

[9] We assume, in this qualitative discussion, that the electron–impurity interaction is short-ranged in space.
[10] It turns out that the deformable jellium description is actually closer to physical reality, as any excess electronic density in a physical system is usually compensated, within a very short time, by a corresponding excess of oppositely charged carriers, e.g., holes in a semiconductor. The deformable jellium model adopted here is just an idealization, in which the compensating carriers are assumed to have infinite mobility. The more realistic case of electrons and holes with finite mobilities will be examined in Exercise 14.

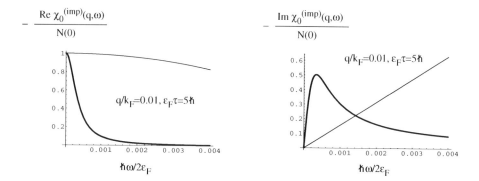

Fig. 4.10. The real and the imaginary part of the Lindhard function in the presence of impurities, calculated from Eq. (4.67) for $q = 0.01k_F$ and $\epsilon_F \tau/\hbar = 5$ in 3D (thick lines). For comparison, the real and imaginary parts of the pure Lindhard function are also plotted (thin lines). Notice the effect of the diffusion pole on $\Im m \chi_0^{imp}(q, \omega)$ at low frequency.

configurations) takes the form

$$\chi_0^{imp}(q, \omega) \approx -N(0)\frac{Dq^2}{-i\omega + Dq^2}, \quad (v_F q\tau \ll 1, \omega\tau \ll 1). \tag{4.53}$$

The main difference between this and the pure Lindhard function arises from the presence of the *diffusion pole* at $\omega = -i Dq^2$. Because of this pole, the density fluctuation spectrum $|\Im m \chi_0^{imp}(q, \omega)|$ exhibits a pronounced peak at low frequency, in sharp contrast with the linear behavior of $|\Im m \chi_0(q, \omega)|$ (see Fig. 4.10). In Chapter 8, we will see that this impurity-induced enhancement of the low-frequency part of the density fluctuation spectrum has a deep influence on the properties of quasiparticles in a disordered electron liquid.

The above description of diffusion is, of course, purely classical, being based on the assumption that the mean free path $v_F \tau$ between collisions is much larger than the typical electron wavelength $\lambda \approx k_F^{-1}$. It is truly amazing that such a simple-minded description gives essentially exact results. In fact, the main thing we are going to learn from the detailed microscopic analysis presented below is that τ is not the average time between collisions, but rather the *momentum relaxation time*. We will also show how τ can be calculated from first principles. When λ is comparable to $v_F \tau$, however, two consecutive electron-impurities can no longer be considered independent, and quantum interference effects can arise. The study of the quantum corrections to diffusion to leading order in $\frac{\lambda}{v_F \tau} = \frac{\hbar}{2\epsilon_F \tau}$ constitutes the field of "*weak localization*" a fascinating topic which, unfortunately, will not be seriously addressed in this book.

Quantum mechanical effects can also lead to *localization* of electronic states below a critical energy known as *mobility edge*, so that a *metal-insulator transition* occurs when the Fermi level of an initially metallic system drops below the mobility edge. An even richer, and to date not fully understood scenario, opens up when electron-electron interactions are taken into account, particularly in reduced dimensionality. It is not difficult to envision a whole book dedicated just to this subject!

4.6.1 Derivation of the Drude conductivity

To begin with, let us show how the classical formula for the Drude conductivity $\sigma = \frac{ne^2\tau}{m}$ (Ashcroft and Mermin, 1976) is derived from the current–current response function i.e., from Eq. (3.187). The derivation is quite subtle and gives us an explicit microscopic expression for $\frac{1}{\tau}$ to second-order in the electron–impurity interaction.

We begin by introducing the model hamiltonian

$$\hat{H} = \hat{H}_0 + \frac{1}{L^d} \sum_{\vec{k}} U(\vec{k}) n_{\vec{k}}^{(i)} \hat{n}_{-\vec{k}} , \qquad (4.54)$$

where $U(\vec{k})$ is the Fourier transform of the electron–impurity potential and $n_{\vec{k}}^{(i)} = \sum_i e^{-i\vec{k}\cdot\vec{R}_i}$ is the Fourier transform of the impurity density, with $\vec{R}_i$ denoting the positions of the impurities. Assuming that there are N_i randomly distributed impurities we have $|n_{\vec{k}}^{(i)}|^2 = N_i$ for all $\vec{k}$'s, with negligible error in the thermodynamic limit.

In the absence of impurities, $\hat{P}_x$ is a constant of the motion, so $\chi_{P_x P_x} = 0$, and $\sigma(\omega)$ reduces to the free-electron value $\frac{ine^2}{m\omega}$. To proceed, observe that we can, without loss of generality, represent the conductivity in the form

$$\sigma(\omega) = \frac{\frac{ine^2}{m}}{\omega + \frac{i}{\tau(\omega)}} , \qquad (4.55)$$

where $\frac{1}{\tau(\omega)}$ is a complex function of frequency, which describes the effect of the impurities. Formally assuming that this effect is small, we can write an asymptotic expansion to leading order in the strength of disorder as follows:

$$\sigma(\omega) = \frac{ine^2}{m\omega} + \frac{ne^2}{m\omega^2} \frac{1}{\tau(\omega)} + \cdots. \qquad (4.56)$$

Comparing this expansion with Eq. (3.187) we obtain

$$\frac{1}{\tau(\omega)} = \frac{i\omega}{nmL^d} \chi_{P_x P_x}(\omega) . \qquad (4.57)$$

The approximate character of this result lies in the fact that the expansion of Eq. (4.56) fails to converge for $\omega \ll \frac{1}{\tau(\omega)}$ and, therefore, Eq. (4.57) is correct only for frequencies larger than $\frac{1}{\tau(\omega)}$. Ignoring this restriction is the key, and, as we shall see, the only approximation needed for the derivation of the classical conductivity. The justification of this procedure rests on the fact that $\frac{1}{\tau(\omega)}$, as calculated from Eq. (4.57), is essentially independent of ω on the scale of $\frac{1}{\tau(0)}$ itself.

To calculate Eq. (4.57) we make use of the exact identity

$$\chi_{P_x P_x}(\omega) = \frac{1}{\omega^2} \left\{ \frac{i}{\hbar} \langle [\hat{F}_x, \hat{P}_x] \rangle + \chi_{F_x F_x}(\omega) \right\}, \tag{4.58}$$

where $\hat{F}_x = \dot{\hat{P}}_x = \frac{i}{\hbar}[\hat{H}, \hat{P}_x]$ is the time derivative of the x component of the momentum, i.e. the x component of the force operator

$$\hat{F}_x = \frac{i}{L^d} \sum_{\vec{k}} k_x U(\vec{k}) n_{\vec{k}}^{(i)} \hat{n}_{-\vec{k}}, \tag{4.59}$$

and $\chi_{F_x F_x}$ is the force–force response function. This identity can be easily derived from the formulas of Exercise 3. The two terms on the right-hand side of Eq. (4.58) are now calculated to second order in the electron–impurity potential. For the first term we get

$$\frac{i}{\hbar} \langle [\hat{F}_x, \hat{P}_x] \rangle = -\frac{1}{L^d} \sum_{\vec{k}} k_x^2 n_{\vec{k}}^{(i)} U(\vec{k}) \langle \hat{n}_{-\vec{k}} \rangle$$

$$\simeq -\frac{n_i}{d} \sum_{\vec{k}} k^2 |U(k)|^2 \chi_0(k, 0), \tag{4.60}$$

where we have used the fact that, to first order in $U(\vec{k})$, the ground-state density is given by $\langle \hat{n}_{-\vec{k}} \rangle = \chi_0(k, 0) n_{-\vec{k}}^{(i)} U(-\vec{k})$. As for the force-force response function, making use of Eq. (4.59) we get,

$$\chi_{F_x F_x}(\omega) = -\frac{1}{L^d} \sum_{\vec{k}, \vec{k}'} k_x k_x' n_{\vec{k}}^{(i)} n_{-\vec{k}'}^{(i)} U(\vec{k}) U(-\vec{k}') \chi_{nn}(\vec{k}, \vec{k}'; \omega)$$

$$\simeq \frac{n_i}{d} \sum_{\vec{k}} k^2 |U(k)|^2 \chi_0(k, \omega), \tag{4.61}$$

where $n_i = \frac{N_i}{L^d}$ is the average density of impurities. To obtain the last line of Eq. (4.61) we have approximated $\chi_{nn}(\vec{k}, \vec{k}'; \omega)$ with $\chi_0(k, \omega) \delta_{\vec{k}, \vec{k}'}$ – a legitimate step, since two powers of $U(\vec{q})$ are already explicitly present. We have also replaced $\frac{|n_{\vec{k}}^{(i)}|^2}{L^d}$ by its average value n_i.

By substituting the above expressions in Eqs. (4.57) and (4.56) we obtain our main result:

$$\frac{1}{\tau(\omega)} = -\frac{n_i}{mnd L^d} \sum_{\vec{k}} k^2 |U(k)|^2 \frac{\chi_0(k, \omega) - \chi_0(k, 0)}{i\omega}. \tag{4.62}$$

Several observations follow from this interesting formula.

- First of all, at low frequency the imaginary part of $\chi_0(k, \omega) - \chi_0(k, 0)$, which is linear in ω, dominates the real part, which goes as ω^2. Hence we see that the zero-frequency limit of $\frac{1}{\tau(\omega)}$ is purely real, as expected:

$$\frac{1}{\tau(0)} = -\frac{n_i}{mnd} \frac{1}{L^d} \sum_{\vec{k}} k^2 |U(k)|^2 \lim_{\omega \to 0} \frac{\Im m \chi_0(k, \omega)}{\omega}. \tag{4.63}$$

We can write this result in a perhaps more appealing form by making use of the fact that

$$\lim_{\omega \to 0} \frac{\Im m \, \chi_0(k, \omega)}{\omega} = -\frac{\pi \hbar}{L^d} \lim_{\omega \to 0} \sum_{\vec{p}, \sigma} \frac{n_{\vec{p}\sigma} - n_{\vec{p}+\vec{k}\sigma}}{\omega} \delta(\hbar\omega + \epsilon_{\vec{p}\sigma} - \epsilon_{\vec{p}+\vec{k}\sigma})$$

$$= -\frac{\pi \hbar}{L^d} \sum_{\vec{p}, \sigma} \delta(\epsilon_{\vec{p}\sigma} - \epsilon_F) \delta(\epsilon_{\vec{p}+\vec{k}\sigma} - \epsilon_F) . \tag{4.64}$$

Substituting Eq. (4.64) into Eq. (4.63) we arrive, after simple manipulations, at the standard expression for the Drude relaxation time of an electron at the Fermi surface

$$\frac{1}{\tau(0)} = \frac{2\pi}{\hbar} \frac{n_i}{L^d} \sum_{\vec{q}} (1 - \cos\theta) |U(k_F \hat{n} - \vec{q})|^2 \delta(\epsilon_{\vec{q}\sigma} - \epsilon_F) , \tag{4.65}$$

where θ is the angle between $\vec{q}$ and any arbitrary direction $\hat{n}$.

- Second, because the frequency dependence of $\chi_0(q, \omega)$, for the most important range of wave vectors $q \sim k_F$, is on the scale of the Fermi energy, we see that the frequency dependence of $\frac{1}{\tau}$ can indeed be neglected on the scale of $\frac{1}{\tau(0)}$. This guarantees the consistency, if not the correctness, of the calculation.
- Third, we see that the real part of $\chi_0(q, \omega)$, varying as ω^2 for $\omega \to 0$, leads to an imaginary part of $\frac{1}{\tau(\omega)}$ that is linear in ω at low frequency. From the form of Eq. (4.56) we see that this imaginary scattering time is equivalent to a renormalization of the electron mass.

All the above results are valid in any number of dimensions and the zero-frequency scattering time can be easily calculated with the help of the formulas derived earlier in this chapter.

4.6.2 The density–density response function in the presence of impurities

A rigorous calculation of the density–density response function of the non-interacting electron gas in the presence of impurities is not easy. The approximate approach presented here is based on the notion that the electronic current responds with the Drude conductivity to the gradient of a local *electrochemical potential*, defined precisely below. This is the first of a series of approximate local field theories, which we shall explore mainly in the next chapter.

At long wavelengths and small frequencies (more precisely, for $q \ll k_F$ and $\omega \ll \frac{\epsilon_F}{\hbar}$) the final result can be written in the following appealing form:

$$\frac{1}{\chi_0^{imp}(q, \omega)} = \frac{\omega}{\omega + \frac{i}{\tau}} \frac{1}{\chi_0\left(q, \omega + \frac{i}{\tau}\right)} + \frac{\frac{i}{\tau}}{\omega + \frac{i}{\tau}} \frac{1}{\chi_0(q, 0)}, \tag{4.66}$$

where $\tau = \tau(0)$ is the zero-frequency limit of Eq. (4.62), and χ_0 is the Lindhard function. Note that this approximation yields $\chi_0^{imp}(q, 0) = \chi_0(q, 0)$, so that impurities have no effect on the static response function. In reality $\chi^{imp}(q, 0)$ and $\chi_0(q, 0)$ differ by corrections of the of order of $\frac{\hbar}{\epsilon_F \tau} \ll 1$, which are assumed to be negligible in the weak-disorder approximation.

Eq. (4.66) can also be cast in the equivalent form

$$\chi_0^{imp}(q, \omega) = \frac{\chi_0\left(q, \omega + \frac{i}{\tau}\right)}{1 + (1 - i\omega\tau)^{-1} \left[\frac{\chi_0(q, \omega + \frac{i}{\tau})}{\chi_0(q, 0)} - 1\right]}, \tag{4.67}$$

which is the expression originally derived from the solution of the quantum Boltzmann equation in the relaxation time approximation (Green *et al.*, 1969; Mermin, 1970). The real and imaginary parts of $\chi_0^{imp}(q, \omega)$ in 3D are plotted vs frequency in Fig. 4.10 for $q = 0.01 k_F$ and $\frac{\epsilon_F \tau}{\hbar} = 5$, and compared there with the corresponding graphs for the Lindhard function.

It should be noticed that the general formula (4.67) is correct in the weak-disorder limit, as one can verify by evaluating $\chi_0^{imp}(q, \omega)$ by means of diagrammatic techniques (Abrikosov, Gorkov, and Dzyaloshinski, 1963). As mentioned above however, coherence effects that lead to localization are not accounted for in this elementary theory. Remarkably, Eq. (4.67) allows one to also derive formulas that span the whole range between the clean and the diffusive regime. An example is given by the following useful expression for the two dimensional case which is easily derived (see Exercise 15):

$$\chi_0^{imp}(q, \omega) \simeq -N(0)\left(1 - \frac{\omega}{\sqrt{\omega^2 - \frac{1}{\tau^2} + \frac{2}{\tau}(i\omega - Dq^2)} - \frac{i}{\tau}}\right), \tag{4.68}$$

where here $D = \frac{v_F^2 \tau}{2}$ (see also Section 4.6.3).

To derive Eq. (4.66) we reason as follows. In an inhomogeneous system the effective electric field, which drives the current, is the gradient of the external potential V plus the gradient of the local chemical potential associated with the inhomogeneous density.[11] For a disturbance of wave vector $\vec{q}$ and frequency ω, while $\delta n(\vec{q}, \omega) = \chi_0^{imp}(q, \omega)V(\vec{q}, \omega)$, the Fourier amplitude of the local chemical potential is given by

$$\mu(\vec{q}, \omega) = -\frac{\delta n(\vec{q}, \omega)}{\chi_0^{imp}(q, 0)}, \tag{4.69}$$

implying that $-\mu(\vec{q}, t)$ is the potential that would yield the density $\delta n(\vec{q}, t)$ *if one allowed the system to reach equilibrium* in the absence of driving terms.[12] The full effective electric field (also referred to as the electrochemical field) is then given by

$$-e\vec{E}_{eff}(\vec{q}, \omega) = -i\vec{q}\left[V(\vec{q}, \omega) + \mu(\vec{q}, \omega)\right]. \tag{4.70}$$

Notice that the effective field vanishes at $\omega = 0$, because $\delta n(\vec{q}, 0)$ is given by $\chi_0^{imp}(q, 0)V(\vec{q}, 0)$: this fact is crucial, as it guarantees that no current flows in equilibrium.

We now formulate our linear response theory in terms of $\vec{E}_{eff}$. We write

$$-e\vec{j}(\vec{q}, \omega) = \tilde{\sigma}(q, \omega)\vec{E}_{eff}(\vec{q}, \omega), \tag{4.71}$$

where $\tilde{\sigma}(q, \omega)$ is a "proper" conductivity, which is, in fact, *defined* by Eq. (4.71).

Then, combining Eqs. (4.69), (4.70) and (4.71), and making use of the continuity equation $\delta n(\vec{q}, \omega) = \frac{\vec{q} \cdot \vec{j}(\vec{q}, \omega)}{\omega}$, it is straightforward to arrive at the following expression for the

---

[11] In keeping with the conventions adopted so far, the external potential V is related to the usual electric potential ϕ by $V = -e\phi$ where e is the electron charge.

[12] Of course, the notion of a local chemical potential is valid only if the space and time variations of the density are sufficiently slow on microscopic scales: hence the need for the restrictions $q \ll k_F$ and $\omega \ll \epsilon_F/\hbar$.

density–density response function in terms of $\tilde{\sigma}(q, \omega)$:

$$\frac{1}{\chi_0^{imp}(q,\omega)} = \frac{1}{\chi_0^{imp}(q,0)} + \frac{ie^2\omega}{q^2\tilde{\sigma}(q,\omega)} . \qquad (4.72)$$

It is clear that $\tilde{\sigma}(q, \omega)$ is not the ordinary conductivity defined in Section 3.4.4. However, and this is essential to our derivation, we know that it must coincide with $\sigma(\vec{q}, \omega)$ in the homogeneous limit ($q = 0$), since, in that case, there is no induced density, and thus no variation in the chemical potential.

This observation, combined with the form of Eq. (4.55), suggests that we write[13]

$$\tilde{\sigma}(q, \omega) \approx \frac{\frac{ine^2}{m}}{\omega + \frac{i}{\tau(q,\omega)}} = \sigma_0\left(\omega + \frac{i}{\tau(q,\omega)}\right) , \qquad (4.73)$$

where $\sigma_0(\omega) = \frac{ine^2}{m\omega}$ is the conductivity of the clean homogeneous electron gas, and $\tau(q, \omega)$ is a wave vector and frequency-dependent relaxation time, which, for $q \to 0$, must reduce to the Drude relaxation time $\tau(\omega)$ of Eq. (4.62).

We now recall that the proper energy scale for variations of $\tau(q, \omega)$ is the Fermi energy, and, similarly, the proper scale for its wave-vector dependence is the Fermi wave vector. Thus, as long as q and ω are sufficiently small on the scale of k_F and $\frac{\epsilon_F}{\hbar}$, we are justified in approximating $\tau(q, \omega)$ by $\tau(0, 0)$, whose microscopic expression is known from Eq. (4.63).

With these approximations, Eq. (4.72) can be rewritten as

$$\frac{1}{\chi_0^{imp}(q,\omega)} = \frac{1}{\chi_0^{imp}(q,0)} + \frac{\omega}{\omega + \frac{i}{\tau}}\frac{ie^2\left(\omega + \frac{i}{\tau}\right)}{q^2\sigma_0(\omega + \frac{i}{\tau})} , \qquad (4.74)$$

where we have set $\tau(0, 0) = \tau$. This very same formula, applied to the clean system ($\tau \to \infty$), gives

$$\frac{1}{\chi_0(q,\omega)} = \frac{1}{\chi_0(q,0)} + \frac{ie^2\omega}{q^2\sigma_0(\omega)} . \qquad (4.75)$$

Comparing Eqs. (4.74) and (4.75), and ignoring the difference between $\chi_0^{imp}(q,0)$ and $\chi_0(q, 0)$, we arrive at Eq. (4.66).

4.6.3 The diffusion pole

The formulas derived in the previous section for $\chi_0^{imp}(q, \omega)$ assumed only $q \ll k_F$ and $\omega \ll \frac{\epsilon_F}{\hbar}$. Let us now examine what happens at frequencies and wave vectors that satisfy the more stringent conditions $\omega \ll \frac{1}{\tau}$ and $q \ll 1/v_F\tau$.

As anticipated in Section 4.6, we find that a new structure, known as the "*diffusion pole*", makes its appearance. The small q and ω behavior of χ_0^{imp} is most easily obtained from

---

[13] The temptation to write from the outset the seemingly reasonable relation $\chi_0^{imp}(q, \omega) = \chi_0(q, \omega + \frac{i}{\tau})$ must be resisted for, as our discussion shows, this leads to an unacceptable non number conserving approximation (see Exercise 8 as well Mermin, 1970). The problem arises from the fact that while in the presence of impurities the total momentum of the electrons is not conserved, their local number still is and particular care must be exercised in formulating the theory.

Eq. (4.72), making use of the limiting forms $\tilde{\sigma}(0,0) = \frac{ne^2\tau}{m}$, and $\chi_0^{imp}(0,0) = -N(0)$ (see also Exercise 16).[14] Simple algebra then leads to the expression anticipated in Eq. (4.53):

$$\chi_0^{imp}(q,\omega) \approx -N(0)\frac{Dq^2}{-i\omega + Dq^2}, \quad (qv_F\tau \ll 1, \omega\tau \ll 1), \tag{4.76}$$

where the *diffusion constant* D is given by

$$D = \frac{\sigma}{e^2 N(0)} = \frac{v_F^2\tau}{d}. \tag{4.77}$$

The above equation, connecting the diffusion constant, D, to the static conductivity, σ, is known as *Einstein relation*.

Notice that the diffusion pole occurs in the lower half of the complex plane, in agreement with the general causality requirements discussed in Chapter 3. As shown in Fig. 4.10, the spectral density $|\Im m \chi_0^{imp}(q,\omega)|$ in the presence of impurities has a pronounced peak at low frequency, which completely swamps the electron–hole continuum. Also the real part of $\chi_0^{imp}(q,\omega)$ is found to decrease to zero very rapidly, and the characteristic "swing" associated with the electron–hole continuum (see Figs. 4.4 and 4.5) is absent. The mathematical form of Eq. (4.53) implies that a charge-neutral density fluctuation relaxes according to the standard diffusion equation (4.51).

Many interesting many-body effects are directly associated with the diffusion pole. For example, one can show that the slowing down of density fluctuations is responsible for the non-analytic behavior of the density of states, and a substantially reduced lifetime of electron quasiparticles. These effects will be discussed in Section 8.9.

4.7 Mean field theory of linear response

We conclude this chapter by presenting a method based on the independent electron picture but ultimately meant to describe the generally quite complex response of an interacting electron gas. We shall describe how a static mean-field approximation for the ground-state properties of an interacting many-body system can be turned into a dynamical mean field theory for the calculation of the linear response of that system to a *time-dependent* external field. Applications of this theory will be encountered in Chapters 5, 7, and 10.

We consider, as in Chapter 3, a system governed by a hamiltonian $\hat{H}$ subjected to an external time-dependent field $F(t)$ that couples linearly to an observable $\hat{B}$.

$$\hat{H}(t) = \hat{H} + F(t)\hat{B}. \tag{4.78}$$

Let us assume that the ground-state properties of $\hat{H}$ have been approximated by a single-particle mean-field hamiltonian

$$\hat{H}^{MF} = \hat{H}_0 + \hat{V}^{MF}, \tag{4.79}$$

---

[14] Unlike $\frac{1}{\tau(q,\omega)}$, $\tilde{\sigma}_0(q,\omega)$ does depend on ω and q on the smaller scales of $\frac{1}{\tau}$ and $\frac{1}{v_F\tau}$ – hence the need for the more stringent restrictions on q and ω.

where $\hat{H}_0$ is the non-interacting part of $\hat{H}$ (generally including a static external potential) and $\hat{V}^{MF}$ is the (generally nonlocal) mean-field potential. For example $\hat{V}^{MF}$ could represent the Hartree plus exchange part of the ground-state Hartree–Fock potential described in Chapter 2. Simpler approximations such as $\hat{V}^{MF} = \hat{V}^{H}$, where $\hat{V}^{H}$ is the classical Hartree potential, or $\hat{V}^{MF} = \hat{V}^{H} + \hat{V}_x^S$, where $\hat{V}_x^S$ is a Slater-type local approximation to the exchange potential, would also be admissible. Yet another possibility would be to use $\hat{V}^{MF} = \hat{V}^{H} + \hat{V}_{xc}$ where $\hat{V}_{xc}$ is an approximation to the local exchange-correlation potential of the static density functional theory.[15]

In all these examples $\hat{V}^{MF}$ is a one-particle operator that can be represented in the form

$$\hat{V}^{MF} = \sum_{\alpha\beta} V^{MF}_{\alpha\beta} \hat{a}^\dagger_\alpha \hat{a}_\beta. \tag{4.80}$$

Furthermore the matrix elements $V^{MF}_{\alpha\beta}$ have self-consistent expressions in terms of the density matrix $\rho_{\gamma\delta} = \langle \hat{a}^\dagger_\gamma \hat{a}_\delta \rangle$

$$V^{MF}_{\alpha\beta} = V^{MF}_{\alpha\beta}[\rho_{\gamma\delta}]. \tag{4.81}$$

The explicit form of these expression depends, of course, on the nature of the mean field theory: the case of the Hartree–Fock theory will be considered in some detail in Section 5.7.1.

Let us denote by ε_α the eigenvalues of $\hat{H}^{MF}$ and by n_α the occupation numbers of the corresponding eigenstates $|\alpha\rangle$. The linear response functions of the non-interacting system described by $\hat{H}^{MF}$ can be expressed in terms of $|\alpha\rangle$ and ε_α, according to the formulas derived in Section 4.2. For any two single-particle operators $\hat{A}$ and $\hat{B}$, the linear response of $\langle \hat{A} \rangle$ to a field that couples linearly to $\hat{B}$ is given by the function

$$\chi^f_{AB}(\omega) = \sum_{\alpha\beta} \frac{n_\alpha - n_\beta}{\hbar\omega + \varepsilon_\alpha - \varepsilon_\beta + i\hbar\eta} A_{\alpha\beta} B_{\beta\alpha} , \tag{4.82}$$

where the superscript "f" reminds us that this is the response function of the *fictitious* non-interacting system governed by $\hat{H}^{MF}$.

It would be a serious error to think that χ^f_{AB} is the correct mean-field approximation to the response function χ_{AB} of the physical system. The problem is that when a time-dependent external field is applied the mean-field potential itself becomes a time-dependent quantity:

$$\hat{V}^{MF}(t) = \hat{V}^{MF}_0 + \hat{V}^{MF}_1(t) , \tag{4.83}$$

where the first term is the ground-state mean-field potential, and the second term is the change induced by the perturbation. However, the fictitious response function of Eq. (4.82) is based on the unphysical assumption that the mean field is "locked" to its ground-state value.

Fortunately, it turns out that the self-consistent mean-field response functions can still be expressed in terms of the fictitious ones (χ^f) through a rather simple and general construction. The central idea is that the system responds with the fictitious response function

---

[15] The exact density functional theory (see Chapter 7) *is not* a mean-field approximation. In practice, however, the formalism presented in this section applies to that theory as well.

χ^f to the sum of two fields: the external perturbation $F(t)\hat{B}$, and the additional mean field $\hat{V}_1^{MF}(t)$ induced by this perturbation. The latter can be written as

$$\hat{V}_1^{MF}(t) = \sum_{\alpha\beta} V_{1,\alpha\beta}^{MF}(t)\hat{a}_\alpha^\dagger\hat{a}_\beta \, , \qquad (4.84)$$

with

$$V_{1,\alpha\beta}^{MF}(t) = \sum_{\gamma\delta} \left(\frac{\delta V_{\alpha\beta}^{MF}[\rho_{\gamma\delta}]}{\delta\rho_{\gamma\delta}}\right)_0 \rho_{1,\gamma\delta}(t) \, , \qquad (4.85)$$

where the derivative is calculated from the expression of the mean field potential in terms of the density matrix evaluated at the ground-state (hence the "0" subscript). The first-order variation of the density matrix, $\rho_{1,\gamma\delta}(t)$, must be calculated self-consistently from

$$\rho_{1,\gamma\delta}(\omega) = \chi_{\rho_{\gamma\delta}B}^{MF}(\omega)F(\omega) \, , \qquad (4.86)$$

where $\chi_{\rho_{\gamma\delta}B}^{MF}(\omega)$ is the mean-field linear response function of $\hat{\rho}_{\gamma\delta} \equiv \hat{a}_\gamma^\dagger\hat{a}_\delta$ to a perturbation linear in $\hat{B}$ – a quantity that we are going to determine momentarily.

According to our discussion, the induced change in the expectation value of an observable $\hat{A}$ due to a perturbation linear in $\hat{B}$ is given by

$$\langle\hat{A}\rangle_1(\omega) \equiv \chi_{AB}^{MF}(\omega)F(\omega)$$

$$= \chi_{AB}^f(\omega)F(\omega) + \sum_{\alpha\beta\gamma\delta} \chi_{A\rho_{\alpha\beta}}^f \left(\frac{\delta V_{\alpha\beta}^{MF}[\rho_{\gamma\delta}]}{\delta\rho_{\gamma\delta}}\right)_0 \chi_{\rho_{\gamma\delta}B}^{MF}F(\omega) \, . \qquad (4.87)$$

It is convenient at this point to define the interaction kernel

$$I_{\alpha\beta,\gamma\delta} \equiv \left(\frac{\delta V_{\alpha\beta}^{MF}[\rho_{\gamma\delta}]}{\delta\rho_{\gamma\delta}}\right)_0 \, . \qquad (4.88)$$

Then putting $\hat{A} = \hat{\rho}_{\alpha\beta}$ in Eq. (4.87) we obtain a closed set of equations for $\chi_{\rho_{\alpha\beta}B}^{MF}(\omega)$:

$$\chi_{\rho_{\alpha\beta}B}^{MF} = \chi_{\rho_{\alpha\beta}B}^f + \sum_{\alpha'\beta'\gamma\delta} \chi_{\rho_{\alpha\beta},\rho_{\alpha'\beta'}}^f I_{\alpha'\beta',\gamma\delta}\chi_{\rho_{\gamma\delta}B}^{MF} \, . \qquad (4.89)$$

The formal solution of this equation is easily written down even though the actual calculation may be frightening:

$$\chi_{\rho_{\alpha\beta}B}^{MF} = \sum_{\gamma\delta}[\underline{1} - \underline{\chi}^f \cdot \underline{I}]_{\alpha\beta,\gamma\delta}^{-1}\chi_{\rho_{\gamma\delta}B}^f \, . \qquad (4.90)$$

Here the underlined quantities represent matrices in the composite indices $\alpha\beta$ and $\gamma\delta$, $\underline{1}$ is the unit matrix

$$\underline{1}_{\alpha\beta,\gamma\delta} = \delta_{\alpha\beta}\delta_{\gamma\delta} \, , \qquad (4.91)$$

and all the indicated operations (multiplication, inversion, etc.) must be carried out according to the usual rules of matrix algebra.

With this solution we have accomplished our goal of expressing the self-consistent mean field response function $\chi_{\rho_{\alpha\beta}B}^{MF}$ – and therefore any linear response function involving only single-particle operators – in terms of the response functions χ^f of a fictitious non-interacting system. Although the resulting formulae appear rather complicated, the reader should concentrate on the central idea of the dynamical mean field approximation, which is quite simple. In the following chapters will see several instances in which the present formulation is useful.

Exercises

4.1 **Alternative derivation of the Lindhard function.** Derive the expression (4.10) for the Lindhard function by applying the general formula (3.47) to the density–density response function of non-interacting electrons.

4.2 **A useful integral.** Calculate $\psi_2(z)$ from the double integral (4.20). An intermediate result that is useful to arrive at the result listed in Table 4.1 is

$$\int_0^{2\pi} \frac{d\varphi}{z - \cos\varphi} = \frac{z + z^*}{|z + z^*|} \frac{2\pi}{\sqrt{z^2 - 1}} \ .$$

4.3 **Lindhard function for an ellipsoidal band.** Calculate the Lindhard function for free electrons in an ellipsoidal band with dispersion

$$\varepsilon_{\vec{k}} = \frac{\hbar^2}{2} \left(\frac{k_x^2}{m_x} + \frac{k_y^2}{m_y} + \frac{k_z^2}{m_z} \right) \ .$$

[Hint: make the scale transformation $k_i \to k_i \sqrt{\frac{m_i}{m_d}}$ ($i = x, y$, or z) to reduce the calculation to that of the isotropic Lindhard function. The *density of states mass* $m_d = (m_x m_y m_z)^{1/3}$ is so chosen to ensure that the transformation preserve the volume element in momentum space.]

4.4 **Density–density response function for a non-interacting inhomogeneous Fermi system.** Write the expression for the density–density response function $\chi_{nn}^{(0)}(\vec{r}, \vec{r}', \omega)$ of a non-interacting inhomogeneous electron gas in terms of one-electron orbitals $\varphi_\alpha(\vec{r})$ and their eigenvalues.

4.5 **Static limit of the Lindhard function at finite temperature.** Equation (4.28) is valid only at $T = 0$. Make use of the compressibility sum rule to show that *at any temperature* one has

$$\lim_{q \to 0} \chi_0(\vec{q}, 0) = -\frac{\partial n(\mu)}{\partial \mu} = -n^2 K_0 \ ,$$

where $n(\mu)$ is the electronic density of the non-interacting electron gas as a function of the chemical potential, and K_0 is the compressibility.

4.6 **High-frequency limit of the Lindhard function.** Starting from the general formulas for the Lindhard function verify the high-frequency limiting form (4.38), and show that it describes the small-q behavior of the Lindhard function for any $\omega \neq 0$.

4.7 **Lindhard function at finite temperature.** Show that the Fermi–Dirac distribution $n_F(E,T) = \frac{1}{e^{E/k_B T}+1}$ at energy E, temperature T and chemical potential 0 satisfies the identity

$$n_F(E,T) = \int_{-\infty}^{\infty} dE' \, \frac{n_F(E',0)}{4k_B T \cosh^2 \frac{E'-E}{2k_B T}} .$$

Make use of this identity to prove Eq. (4.42) for the Lindhard function at finite temperature.

4.8 **Longitudinal current–current response function of the non-interacting electron gas.** Show by direct calculation that the longitudinal current–current response function of a non-interacting electron gas, defined by means of the general formulas (3.174) and (3.172), satisfies Eq. (3.175). This amounts to proving that the Lindhard function is a number-conserving response function.

4.9 **Transverse spin–spin response functions of the non-interacting electron gas.** Explicitly evaluate Eq. (4.14) for the transverse spin–spin response function of the non-interacting electron gas. The result is

$$\chi_{S_+ S_-}^{(0)}(q,\omega) = 2N_\uparrow(0)\frac{k_{F\uparrow}}{q}\Psi_d\left(\frac{\omega-\Delta+i\eta}{qv_{F\uparrow}} - \frac{q}{2k_{F\uparrow}}\right)$$
$$- 2N_\downarrow(0)\frac{k_{F\downarrow}}{q}\Psi_d\left(\frac{\omega-\Delta+i\eta}{qv_{F\downarrow}} + \frac{q}{2k_{F\downarrow}}\right), \qquad \text{(E4.1)}$$

where $\Delta = g\mu_B B$ is the energy splitting between the up and down spin bands. Notice that in the paramagnetic state this reduces to twice the longitudinal spin–spin response function.

4.10 **Orbital magnetic susceptibility in two and three dimensions.** In the presence of a small magnetic field a three-dimensional electron gas can be described as a stack of two-dimensional gases obtained by slicing the three-dimensional Fermi sphere with planes perpendicular to the direction of the field. Because the two-dimensional density of states does not depend on Fermi wave vector all the two-dimensional slices give equal contributions to the total density of states and to the orbital magnetization. Use this observation to explain why the Landau result (4.50) for the orbital magnetic susceptibility is valid in both two and three dimensions.

4.11 **Evolution of a density packet in the deformable jellium model.** Obtain Eq. (4.52) for the evolution of a density packet in the deformable jellium model with random impurities from the solution of the diffusion equation (4.51). [Hint: make good use of Laplace and Fourier transforms.]

4.12 **Diffusion pole and diffusion equation.** Show that the form (4.76) for the density–density response function of the electron liquid implies that neutral density fluctuations evolve according to the diffusion equation (4.51).

4.13 **Evolution of a density packet in the rigid jellium model.** Calculate the time evolution of the density packet considered in Exercise 11, but now for a *rigid* jellium model with random impurities. [Hint: combine the continuity equation with the Poisson equation for the electric field produced by the excess charge density and use the Drude formula to connect the current density to the electric field. Notice that the diffusion equation (4.51) does not apply to this situation.]

4.14 **Evolution of a neutral electron–hole packet.** Same as the previous exercise for the case of a charge-neutral packet of electrons and holes, with the electrons and the holes having different conductivities and diffusion constants.

4.15 **Small q and ω form of Mermin's response function in 2D.** Derive Eq. (4.68) for the density–density response function of a disordered 2D electron gas at small q and ω.

4.16 **Diffusion pole in Mermin's density–density response function.** By explicitly evaluating the small q and ω expression for $\chi_0(q, \omega + \frac{i}{\tau})$, show that Mermin's density–density response function (4.67) has a diffusion pole in the form of Eq. (4.76).

5

Linear response of an interacting electron liquid

5.1 Introduction and guide to the chapter

Linear response functions contain a wealth of information about the physical properties of a many-body system. In the case of the electron liquid, for example, the density–density response function provides a unified framework for the understanding of different phenomena such as static screening, effective interaction, collective modes, electron energy loss spectra (inelastic scattering of electrons), and Raman spectra (inelastic scattering of photons). The spin–spin response function provides the corresponding information for the spin density fluctuations as probed, for example, by cross-polarized Raman scattering experiments, in which the incident and scattered photon have perpendicular polarizations, or by spin-flip electron energy loss spectroscopy, in which the incoming and outgoing electrons have opposite spin orientations.[1]

As we have seen, the Hartree–Fock approximation describes each electron as an independent particle moving in a self-consistent field generated by all the other electrons. There is no correlation between this self-consistent field and the instantaneous position of the electron. Reality is however quite different: whenever an electron moves, it acts on the surrounding electrons, causing a collective disturbance which eventually feeds back on its own motion. In order to study these effects we need, first of all, to learn how the electron liquid as a whole responds to disturbances caused by the charge, the spin, and the current of a single electron. The linear response theory of Chapter 3 provides the natural framework for this description. Strictly speaking, the linear response functions describe the readjustments of the electronic density, spin, or current, in response to *externally controlled* fields. As we will see in this chapter, however, the theory of linear response can be generalized to also describe the self-consistent adjustment of the electronic density to *internal fields* created by electrons within the system, an idea that provides the physical basis of the time-dependent mean field theory developed in Section 4.7.

There are two different but closely related lines of attack to the problem of calculating the linear response functions of an interacting many-body system. The first one is based on the notion that many-body effects can be in principle exactly described in terms of an effective

---

[1] The coupling between the photon and the spin in cross-polarized Raman scattering is mediated by the spin-orbit interaction. Spin-flip electron scattering, on the other hand, can occur even in the absence of spin–orbit interaction, since, for example, an up-spin electron can kick out a down-spin electron, and take its place within the liquid.

local potential acting on each particle independently. This will be the main topic of this chapter. The second approach is based on the diagrammatic expansion of the perturbation series for the response function, and will be presented in Chapter 6.

The best known and most successful approximation in the first class is the *time-dependent Hartree approximation*, universally known (for historical reasons) as the *random phase approximation (RPA)*. In this theory one assumes that the electrons respond to the external field plus the time-dependent Hartree potential. To understand in simple terms some of the physics behind the RPA, it is useful at first to revisit the simple exercise worked out in Section 3.1 by generalizing it to the case of interacting electrons. The problem of calculating the density–density response function, can also in this case be accomplished in a straightforward manner by making use of the concept of stiffness. As we have seen, the application of the static external potential of Eq. (3.7) to a homogeneous system leads to the onset of density fluctuations of identical amplitude $n_{\vec{q}}$ and $n_{-\vec{q}}$, which in turn determine the density–density response function $\chi_{nn}(\vec{q}, 0)$ via the exact relation

$$\delta E = -\frac{n_{\vec{q}}^2}{\chi_{nn}(q, 0)L^d} \,, \tag{5.1}$$

where δE is the stiffness energy, i.e., the change in the *internal* energy upon application of the external potential. The crux of the matter is finding a satisfactory approximation for δE. A bold and often physically reasonable point of view is to postulate that δE is simply given by the sum of the electrostatic (Hartree) energy associated with the density modulation

$$\delta E_C = \frac{v_q n_{\vec{q}}^2}{L^d} \,, \tag{5.2}$$

and the kinetic stiffness energy of a system of non-interacting electrons in the presence of the same density modulations

$$\delta E_T = -\frac{n_{\vec{q}}^2}{L^d \chi_0(q, 0)} \,, \tag{5.3}$$

where $\chi_0(q, 0)$ is the static Lindhard function. This approximation, which obviously neglects important exchange and correlation effects, immediately leads via (5.1) to the famous result[2]

$$\chi_{nn}^{RPA}(q, 0) = \frac{\chi_0(q, 0)}{1 - v_q \chi_0(q, 0)} \,. \tag{5.4}$$

We will see that this result readily generalizes to the dynamic case.

The next logical step beyond the RPA would appear to be the time-dependent Hartree–Fock (TDHF) approximation. Unfortunately, the nonlocal nature of the exchange potential makes the TDHF equations very complex. To avoid this difficulty, it is highly desirable to develop a mathematical scheme that goes beyond the RPA without sacrificing its simplicity. We shall show in detail in this chapter that local approximations to the exchange-correlation potential (of which Slater's approximation described in Chapter 2 is the prototype), when

---

[2] See also Exercise 1.

generalized to the time-dependent context, lead precisely to such schemes. A local potential is one that simply multiplies the wave function $\psi(\vec{r}_1, \ldots, \vec{r}_N; t)$ by a sum of one-body terms $V(\vec{r}_1, t) + \cdots + V(\vec{r}_2, t)$. According to the basic theorems of density functional theory (see Chapter 7) such a potential can be found under broad hypotheses to produce the exact density response of a general inhomogeneous system. In this chapter we will discuss the construction of such an effective potential in the important case of the homogeneous electron liquid.

A particularly simple route to the introduction of exchange and correlation corrections begins by postulating the existence of an additional term of the type

$$\delta E_{xc} = -\frac{v_q G(q) n_{\vec{q}}^2}{L^d}, \tag{5.5}$$

in the energy stiffness. Here the function $G(q)$ is meant to represent the fractional modification of the coulomb energy associated with the exchange-correlation hole as discussed in Chapter 1.[3] The inclusion of this term into the left-hand side of Eq. (5.1) leads to the expression

$$\chi_{nn}(q, 0) \simeq \frac{\chi_0(q, 0)}{1 - v_q(1 - G(q))\chi_0(q, 0)}, \tag{5.6}$$

a result originally due to Hubbard (1957).

This simple picture can be generalized to the dynamic case, where

$$-v_q G(q, \omega) n_{\vec{q}}(\omega)$$

is easily seen to be the Fourier transform of the exchange-correlation potential created by a density fluctuation of amplitude $n_{\vec{q}}(\omega)$, which is periodic in space and time. Hence $G(q, \omega)$ is commonly referred to as a *many-body local field factor*. Spin effects are handled by introducing a whole family of dynamic many-body local field factors $G_{\sigma\sigma'}(q, \omega)$, which describe the local environment of an electron. Within this scenario the RPA simply amounts to setting $G_{\uparrow\uparrow} = G_{\uparrow\downarrow} = 0$, thus ignoring the correlation between an electron and its surrounding medium.

Of particular interest are the spin symmetric and spin anti-symmetric combinations

$$G_{\pm}(q, \omega) \equiv \frac{G_{\uparrow\uparrow}(q, \omega) \pm G_{\uparrow\downarrow}(q, \omega)}{2}. \tag{5.7}$$

which will be shown to control, respectively, the density–density and spin–spin response function of a paramagnetic electron liquid. It is in fact G_+ which properly enters Eq. (5.6) above.

The local field factors are clearly the central quantities in this class of mean-field-like theories. Inclusion of local field corrections even at the most elementary level leads to a considerably improved description of the physics of the exchange and correlation hole, better values of the correlation energy, and a much more accurate description of the linear

---

[3] Note that by modifying χ_{nn} the inclusion of this term will lead to a change in both the Hartree and the kinetic energy stiffness via a modification of the magnitude of the induced density modulations $n_{\pm q}$.

response. The quest for increasingly accurate local field factors has recently led to an essentially exact determination of the *zero-frequency limit*, $G_{\sigma\sigma'}(q,0)$, in the homogeneous, paramagnetic phase. In Sections 5.6–5.8 we will present exact results and approximate computational schemes for $G_{\sigma\sigma'}(q,\omega)$ and will discuss the outcome of these calculations in some detail.

In the concluding section of this chapter, we will show that the dynamics of the electron liquid in the *collective regime* ($q \to 0, \omega$ finite) is equivalent to that of a visco-elastic medium with frequency-dependent visco-elastic "constants", which can in turn be expressed in terms of the exact local field factors.

5.2 Screened potential and dielectric function

5.2.1 The scalar dielectric function

When an electron gas is perturbed by a time-dependent scalar potential $V_{ext}(\vec{r},t)$ – for example, the coulomb potential created by a fast moving external electron in an electron-energy-loss experiment – its density deviates from the equilibrium value $n(\vec{r})$. The change in density creates an additional coulomb field $V_{ind}(\vec{r},t)$ which is superimposed to the external field. The total resulting potential seen by a *test charge* is referred to as the *screened scalar potential* and is given by[4]

$$V_{sc}(\vec{r},t) = V_{ext}(\vec{r},t) + V_{ind}(\vec{r},t), \tag{5.8}$$

where

$$V_{ind}(\vec{r},t) = \int d\vec{r}' \frac{e^2 n_1(\vec{r}',t)}{|\vec{r}-\vec{r}'|}, \tag{5.9}$$

where $n_1(\vec{r},t) \equiv n(\vec{r},t) - n(\vec{r})$ is the induced density.[5] In the limit that the perturbation is weak, one expects a linear relationship between $n_1(\vec{r},t)$ and the external field. Accordingly, making use of the linear response theory of Chapter 3, we write

$$n_1(\vec{r},\omega) = \int_{-\infty}^{\infty} dt\, n_1(\vec{r},t)e^{i\omega t}$$
$$= \int d\vec{r}' \chi_{nn}(\vec{r},\vec{r}';\omega)V_{ext}(\vec{r}',\omega), \tag{5.10}$$

where χ_{nn} is the density–density response function. Substituting this in Eq. (5.9) we obtain for the Fourier transform of the screened potential

$$V_{sc}(\vec{r},\omega) = V_{ext}(\vec{r},\omega) + V_{ind}(\vec{r},\omega)$$
$$\equiv \int d\vec{r}' \epsilon^{-1}(\vec{r},\vec{r}';\omega)V_{ext}(\vec{r}',\omega), \tag{5.11}$$

---

[4] A *test charge* can be defined as a classical particle of vanishingly small charge. Its main characteristic is that it does not disturb the system in which it is embedded. See also Sections 5.5.1 and 5.5.2.
[5] We neglect retardation effects that are due to the finite speed of propagation of electromagnetic fields. This approximation is fully justified when dealing with nonrelativistic particles whose speed is much smaller than the speed of light. The vector potential created by the induced electronic current is neglected for the same reason.

where

$$\epsilon^{-1}(\vec{r}, \vec{r}'; \omega) \equiv \delta(\vec{r} - \vec{r}') + \int d\vec{r}'' \frac{e^2}{|\vec{r} - \vec{r}''|} \chi_{nn}(\vec{r}'', \vec{r}'; \omega) \tag{5.12}$$

is (by definition) the *inverse* of the *scalar dielectric function* $\epsilon(\vec{r}, \vec{r}', \omega)$.[6]

The above equations take a much simpler form in the case of a *homogeneous* electron gas, since $\chi_{nn}(\vec{r}, \vec{r}'; \omega)$ depends only on the distance $|\vec{r} - \vec{r}'|$: one can then Fourier-transform Eqs. (5.11) and (5.12) with respect to $\vec{r} - \vec{r}'$ to get the algebraic relation

$$V_{sc}(\vec{q}, \omega) = \frac{V_{ext}(\vec{q}, \omega)}{\epsilon(q, \omega)}, \tag{5.13}$$

where

$$\frac{1}{\epsilon(q, \omega)} = 1 + v_q \chi_{nn}(q, \omega). \tag{5.14}$$

Thus, in the homogeneous case, $\epsilon(q, \omega)$ is simply the ratio between the Fourier components of V_{ext} and V_{sc}.

As discussed in Section 3.2.4, the stability of the ground-state, combined with the stiffness theorem, requires $\chi_{nn}(q, 0) < 0$. This immediately implies (Dolgov *et al.*, 1981)

$$\frac{1}{\epsilon(q, 0)} < 1, \tag{5.15}$$

an inequality that only allows for values of $\epsilon(q, 0)$ either greater than 1 *or* less than 0. Notice that, as far as the energetic stability of the ground-state is concerned, negative values of $\epsilon(q, 0)$ are allowed.

5.2.2 Proper versus full density response and the compressibility sum rule

Since, as we will soon see, the screened effective potential experienced by a test-charge is often quite different from the bare external potential, it makes sense to define, along with $\chi_{nn}(\vec{r}, \vec{r}'; \omega)$, a *proper density–density response function*, denoted by $\tilde{\chi}_{nn}(\vec{r}, \vec{r}'; \omega)$, which gives the response of the density $n_1(\vec{r}, \omega)$ to the screened potential $V_{sc}(\vec{r}', \omega)$:

$$n_1(\vec{r}, \omega) = \int d\vec{r}' \tilde{\chi}_{nn}(\vec{r}, \vec{r}', \omega) V_{sc}(\vec{r}', \omega). \tag{5.16}$$

Making use of Eq. (5.9) for the induced potential, it is easy to see that the relation between the proper density–density response function and the full one can be cast in the form

$$\tilde{\chi}_{nn}^{-1}(\vec{r}, \vec{r}', \omega) = \chi_{nn}^{-1}(\vec{r}, \vec{r}', \omega) + \frac{e^2}{|\vec{r} - \vec{r}'|}, \tag{5.17}$$

where χ_{nn}^{-1} is the matrix inverse of χ_{nn} ($\vec{r}$ and $\vec{r}'$ are the "indices" of this matrix).

---

[6] This must be understood as the inverse of a matrix whose "indices" are $\vec{r}$ and $\vec{r}'$. See also the discussions of the dielectric function in Appendix 7.

We now have two expressions connecting the screened potential to the external potential: one is Eq. (5.11), which defines the dielectric function, and the other can be inferred from the relation

$$V_{sc}(\vec{r}, \omega) = V_{ext}(\vec{r}, \omega) + \int d\vec{r}' \int d\vec{r}'' \frac{e^2}{|\vec{r} - \vec{r}'|} \tilde{\chi}_{nn}(\vec{r}', \vec{r}''; \omega) V_{sc}(\vec{r}'', \omega) . \quad (5.18)$$

Comparing these two expressions we get a direct relationship between the dielectric function and the proper density–density response function:

$$\epsilon(\vec{r}, \vec{r}'; \omega) = \delta(\vec{r} - \vec{r}') - \int d\vec{r}'' \frac{e^2}{|\vec{r} - \vec{r}''|} \tilde{\chi}_{nn}(\vec{r}'', \vec{r}'; \omega) . \quad (5.19)$$

In the case of a homogeneous electron liquid, the above formulas reduce to

$$\frac{1}{\tilde{\chi}_{nn}(q, \omega)} = \frac{1}{\chi_{nn}(q, \omega)} + v_q \quad (5.20)$$

and

$$\epsilon(q, \omega) = 1 - v_q \tilde{\chi}_{nn}(q, \omega) , \quad (5.21)$$

so that

$$\chi_{nn}(q, \omega) = \frac{\tilde{\chi}_{nn}(q, \omega)}{\epsilon(q, \omega)} . \quad (5.22)$$

It must be borne in mind that, in spite of its formal similarity to the full density–density response function, $\tilde{\chi}_{nn}(q, \omega)$, and hence $\epsilon(q, \omega)$, is not, in general, a causal response function. The reason is that the screened potential is not an externally controlled probe, but is itself a function of the density response. Then, for example, the Kramers–Krönig dispersion relations apply to $\frac{1}{\epsilon(q,\omega)}$ but not to $\epsilon(q, \omega)$.[7]

A good case in point is provided by the static dielectric function $\epsilon(q, 0)$. By making use of Eq. (5.20) and recalling the discussion of Section 3.3.4 (in particular Eq. (3.155)) we see that the compressibility sum rule can be written in the form

$$\lim_{q \to 0} \tilde{\chi}_{nn}(q, 0) = -\frac{\partial n}{\partial \mu} = -n^2 K , \quad (5.23)$$

where $\mu(n)$ and K are, respectively, the chemical potential, and the compressibility of the *locally neutral* jellium model at density n (i.e., what we have called the *proper compressibility* in Chapter 1). This result can be combined with Eq. (5.21) to yield

$$\lim_{q \to 0} \epsilon(q, 0) = 1 + v_q \frac{\partial n}{\partial \mu}$$

$$= 1 + v_q N(0) \frac{K}{K_0} , \quad (5.24)$$

where $K_0 = \frac{N(0)}{n^2}$ is the non-interacting compressibility.

---

[7] The $q = 0$ case is somewhat anomalous, since, in this case, one *can* control the screened potential by placing the system between the plates of a capacitor connected to a battery (see Dolgov *et al.* (1981)).

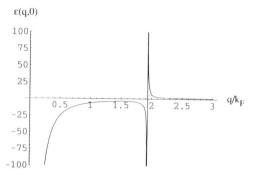

Fig. 5.1. The static dielectric function of a 3D electron liquid at $r_s = 10$. For all values of q, $\epsilon(q, 0)$ lies outside the forbidden interval $[0, 1]$, as required by Eq. (5.15).

In Chapter 1 we saw that the proper compressibility K changes sign from positive at high density to negative at low density. We also pointed out that the negative proper compressibility does not indicate an instability of the system because the full stiffness includes also an electrostatic contribution, which is positive and more than compensates the negative value of K^{-1}.

Eq. (5.24) shows that, when $K < 0$ the static dielectric function is large and negative at small q. This is not in contradiction with the stability condition $1/\epsilon(q, 0) < 1$ derived in Chapter 3, and, again, at the end of the preceding section. For large q, however, $\epsilon(q, 0)$ must tend to 1 from above. The only way this can happen without violating Eq. (5.15) is that $\epsilon(q, 0)$ tend to $\pm\infty$ at some finite $q = q_0$. This behavior is exemplified in Fig. 5.1. The pole in $\epsilon(q, 0)$ is admissible because $\epsilon(q, \omega)$ is *not* a causal reponse function at finite q. The physical significance of the pole is that the screening field exactly cancels the external field at $q = q_0$.

5.2.3 Compressibility from capacitance

The compressibility of a two-dimensional electron liquid has an interesting interpretation in terms of *capacitance*. Let us consider the capacitor shown in Fig. 5.2, in which one plate is an ordinary three-dimensional metal sheet of macroscopic thickness (the "gate"), and the other is a very thin layer that accommodates a two-dimensional electron liquid. A potential difference V is applied between the gate and the two dimensional liquid. Our goal is to calculate the charge ρ per unit area of the gate, i.e., the capacitance per unit area.

A purely classical calculation would begin with the observation that $V = Ed$, where E is the electric field between the plates and d the distance between them. By making use of Gauss' theorem $E = 4\pi \frac{\rho}{\epsilon_b}$ (ϵ_b being here the dielectric constant of the embedding medium) one obtains the classical result $\frac{C}{A} = \frac{\epsilon_b}{4\pi d}$ for the capacitance per unit area. Corrections to the classical result arise from the quantum mechanical nature of the two-dimensional electron liquid. To see how this comes about, consider a small change δV in the gate potential. This

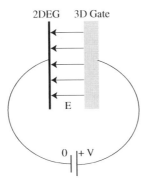

Fig. 5.2. The measurement of the capacity between a three dimensional gate and a two dimensional electron gas (2DEG) allows, in principle, a determination of the proper compressibility K of the 2DEG.

produces a change δE in the electric field between the capacitor plates, but also a small change $\delta \mu$ in the chemical potential of the two dimensional electron liquid. The latter is given by $\delta \mu = \frac{\partial \mu}{\partial n} \delta n$, where $\delta n = \frac{\delta \rho}{e}$ is the change in the density of the electron liquid.[8] Thus we now have a potential difference (including both the electric and the "chemical" contribution) $\delta V = \delta E d + \frac{\partial \mu}{\partial n} \frac{\delta \rho}{e^2}$. Introducing the *screening length* $\lambda = \frac{\epsilon_b}{4\pi e^2} \frac{\partial \mu}{\partial n}$ and using Gauss' theorem $\delta E = 4\pi \delta \rho / \epsilon_b$, we see that the differential capacitance $C/A = \delta \rho / \delta V$ is given by

$$\frac{C}{A} = \frac{\epsilon_b}{4\pi(d + \lambda)} . \tag{5.25}$$

Thus, the capacitance is smaller or larger than the classical value depending on whether the screening length, which is inversely proportional to the compressibility, is positive or negative. Eq. (5.25) can be used, in principle, to extract the value of $\frac{\partial \mu}{\partial n}$ from a measurement of the capacitance between the two dimensional electron liquid and the gate.

In practice, such a measurement would be extremely difficult because the separation d is macroscopic while the screening length λ is of the order of the Bohr radius. A much more sensitive measurement can be done with the ingenious set-up devised by Eisenstein *et al.* (1992) and shown in Fig. 5.3. They use *two* two-dimensional electron layers (denoted by "2DEL" in the figure) that are kept *at the same electro-chemical potential* via ohmic contacts. The idea is to measure the change in the charge density of the bottom layer caused by a small *ac* modulation of the gate voltage. Classically, this would be zero, but quantum mechanics allows the penetration of a small electric field E_p in the region between the 2D layers. A simple way to understand the effect is to notice that any change in the chemical potential of the top 2D layer will cause a flow of charge between the top and the bottom layer, and hence a change in the electric field in the region between the layers. In Exercise 3 the reader is encouraged to discover that a change δE_0 in the electric field between the top

---

[8] The corresponding change in the chemical potential of the thick metal plate is negligible due to the large density of states.

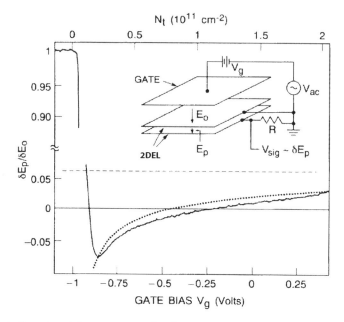

Fig. 5.3. Inset: Experimental arrangement for the measurement of the compressibility. Graph: $\delta E_p/\delta E_0$ vs. gate bias V_g at $T = 1.2K$. Note broken vertical axis. N_t is the density of electrons in the top layer. Dotted curve calculated from Eq. (5.26) using Tanatar and Ceperley's compressibility; the dashed line is the non-interacting result, which does not depend on density. (Adapted from Eisenstein *et al.* (1992)).

2D layer and the gate causes E_p to change by

$$\delta E_p = \frac{\lambda_t}{a + \lambda_t + \lambda_b} \delta E_0 \qquad (5.26)$$

where $\lambda_{t(b)}$ is the screening length of the top (bottom) layer, and a is the distance between them. Because $a \gg \lambda_{t(b)}$, δE_p is approximately proportional to λ_t and a measurement of the former allows an accurate determination of the latter. The results of the measurement are shown in the lower part of Fig. 5.3. The change in sign of λ_t, and hence of the proper compressibility, with decreasing density is consistent with the prediction of the jellium model for the two-dimensional electron liquid.

5.3 The random phase approximation

Perhaps the most popular and historically significant theory that attempts to go beyond the Hartree–Fock approximation in the electron gas problem is the *random phase approximation* (RPA) originally introduced, via two seemingly different approaches, by Gell-Mann and Brückner (Gell-Mann and Brückner, 1957) and by Bohm and Pines (Bohm and Pines, 1957). Although the RPA can be defined in a quite general way and therefore is applicable also to the one dimensional electron liquid, the detailed applications that will be discussed in this chapter will deal with three and two dimensional systems. The study of the one-dimensional case is deferred to Chapter 9.

5.3.1 The RPA as time-dependent Hartree theory

The RPA can be operationally defined in a number of ways.[9] A physically appealing possibility is to develop the approximation from the mean field theory of linear response (as discussed in Section 4.7), choosing the Hartree field as the static mean field. Such a theory could well be called *time-dependent Hartree approximation*, except that the latter is more general, being not necessarily restricted to the linear response regime. In the following, we first present the RPA for a general inhomogeneous electron liquid, and then specialize to the case of the homogeneous liquid.

5.3.1.1 Inhomogeneous electron liquid

For a system of electrons in a static potential $V(\vec{r})$ the static Hartree mean field approximation assumes that each electron is independently subjected to the self-consistent potential

$$V_H(\vec{r}) = V(\vec{r}) + \int \frac{e^2 n(\vec{r}')}{|\vec{r} - \vec{r}'|} d\vec{r}' , \qquad (5.27)$$

where $n(\vec{r})$ is the ground-state density. This independent electron system will be referred to as the *Hartree system*, and we denote by $\chi_{nn}^H(\vec{r}, \vec{r}'; \omega)$ its density–density response function. According to the basic mean-field Ansatz described in Section 4.7 we now assume that the physical system responds as the Hartree system would to the sum of the perturbing potential $V_1(\vec{r}, t)$ and the electrostatic field $\int \frac{e^2 n_1(\vec{r}', t)}{|\vec{r} - \vec{r}'|} d\vec{r}'$ created by the induced density $n_1(\vec{r}, t)$. Thus we have

$$n_1(\vec{r}, \omega) = \int d\vec{r}' \chi_{nn}^H(\vec{r}, \vec{r}', \omega) V_{sc}(\vec{r}', \omega) \qquad (5.28)$$

where

$$V_{sc}(\vec{r}, \omega) = V_{ext}(\vec{r}, \omega) + \int d\vec{r}' \frac{e^2}{|\vec{r} - \vec{r}'|} \chi_{nn}^{RPA}(\vec{r}', \vec{r}'', \omega) V_{ext}(\vec{r}'', \omega) , \qquad (5.29)$$

is the screened potential introduced in Section 5.2.

Comparing the above equation (5.28) with the definition (5.16) of the proper response function we immediately see that *the RPA is the approximation in which the proper density–density response function is replaced by the Hartree response function*:

$$\tilde{\chi}_{nn}^{RPA}(\vec{r}, \vec{r}', \omega) = \chi_{nn}^H(\vec{r}, \vec{r}', \omega) . \qquad (5.30)$$

---

[9] The most common, quick definition for the homogeneous case can be found in Eqs. (5.34) and (5.33). A definition based on the diagrammatic formalism is discussed in Chapter 6.

Then, according to Eq. (5.17) we have

$$[\chi_{nn}^{RPA}]^{-1}(\vec{r}, \vec{r}', \omega) = [\chi_{nn}^{H}]^{-1}(\vec{r}, \vec{r}', \omega) - \frac{e^2}{|\vec{r} - \vec{r}'|} . \tag{5.31}$$

5.3.1.2 Homogeneous electron liquid

For a homogeneous electron liquid the Hartree system coincides with the non-interacting electron gas and the RPA therefore amounts to approximating the proper response function with the Lindhard function, i.e.

$$\tilde{\chi}_{nn}^{RPA}(q, \omega) = \chi_0(q, \omega) . \tag{5.32}$$

According to Eq. (5.22) we immediately have:

$$\chi_{nn}^{RPA}(q, \omega) = \frac{\chi_0(q, \omega)}{\epsilon^{RPA}(q, \omega)} = \frac{\chi_0(q, \omega)}{1 - v_q \chi_0(q, \omega)}, \tag{5.33}$$

where

$$\epsilon^{RPA}(q, \omega) = 1 - v_q \chi_0(q, \omega) \tag{5.34}$$

is the dynamical RPA dielectric function. Explicit expressions for $\chi_{nn}^{RPA}(q, \omega)$ and $\epsilon^{RPA}(q, \omega)$ can be readily obtained by making use of the results of Section 4.4.

The density–density response function in RPA is essentially different from the Lindhard function. For example in Figs. 5.4 and 5.5 the full line represents the behavior of the static limit $\chi_{nn}^{RPA}(q, 0)$ respectively in three and two dimensions. The difference with the Lindhard function is especially noticeable in the small-q limit, where the diverging dielectric function makes $\chi_{nn}(q, 0)$ vanish as q^d in d dimensions. (In both figures the long-dashed curves represent the exact result as obtained by means of the numerical interpolation of the appropriate many-body local fields as obtained by QMC methods which we provide in Appendix 11. This improvement over the RPA will be discussed in Section 5.4.)

Plots of $\epsilon_1 \equiv \Re e \epsilon^{RPA}$ and $\epsilon_2 \equiv \Im m \epsilon^{RPA}$ as functions of frequency are presented in Figs. 5.6 and 5.7 for fixed values of q. Although the behavior exemplified in these figures is generic, the actual numerical values of ϵ_1 and ϵ_2 depend strongly on the value of r_s. Notice that there is a value of the frequency (denoted in the plots by ω_p) for which both and ϵ_1 and ϵ_2 vanish. As we will see in Section 5.3.3, this corresponds to a long lived collective mode of the system: the plasmon mode.

5.3.2 Static screening

The static screening properties of the electron gas in the RPA are controlled by the static dielectric function $\epsilon^{RPA}(q, 0)$, which is plotted in Fig. 5.1. Because of the long range of the coulomb interaction the most relevant limit is that of long wavelengths, i.e. $q \ll k_F$.

In this limit we see from Eqs. (5.34) and (4.27) that

$$\epsilon^{RPA}(\vec{q}, 0) \simeq 1 + \frac{\kappa_3^2}{q^2}, \quad q \ll k_F, \quad 3D, \tag{5.35}$$

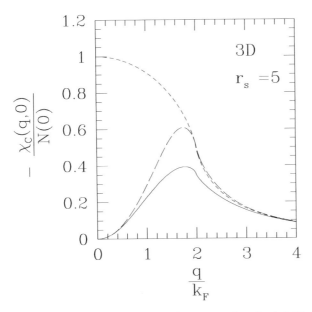

Fig. 5.4. Behavior of the three dimensional static density response function in RPA for a three dimensional electron gas (full line). The short dashed line reproduces the corresponding non-interacting (Lindhard) function. The long dashed curve shows the sizable effect of the appropriate static many-body local field correction beyond RPA as discussed in Section 5.4.1 and in Appendix 11, and agrees with the result of the Monte Carlo calculation.

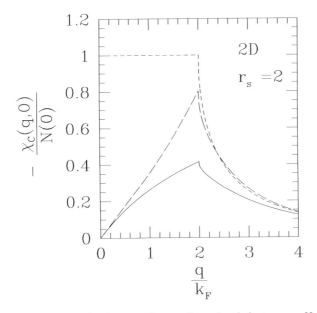

Fig. 5.5. Same as in Fig. 5.4 for the case of a two-dimensional electron gas. Here $r_s = 2.0$.

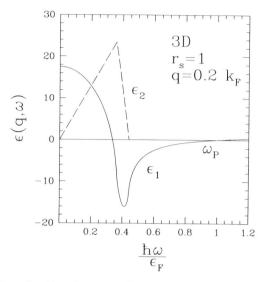

Fig. 5.6. Behavior of the real and imaginary part of the RPA dielectric function for a three dimensional electron gas. The label ω_P indicates the value of the frequency where both functions vanish.

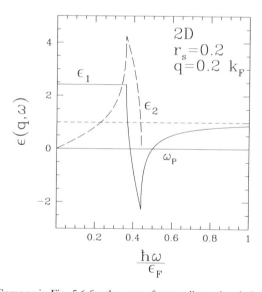

Fig. 5.7. Same as in Fig. 5.6 for the case of a two dimensional electron gas.

where we have defined the three dimensional *Thomas–Fermi screening wave vector*:

$$\kappa_3 = \sqrt{4\pi e^2 N(0)} = \frac{2}{\sqrt{\pi\alpha_3 r_s}}\frac{1}{a_B}, \quad 3D, \tag{5.36}$$

where $\alpha_3 = (4/9\pi)^{1/3}$. Notice that the divergence of the dielectric function at small q is consistent with the well known *perfect screening* property of classical metallic systems.

The corresponding expression in two dimensions is

$$\epsilon^{RPA}(\vec{q}, 0) \simeq 1 + \frac{\kappa_2}{q}, \quad q \ll k_F, \quad 2D, \tag{5.37}$$

where the two-dimensional Thomas–Fermi wave vector is given by

$$\kappa_2 = 2\pi e^2 N(0) = \frac{2}{a_B}, \quad 2D, \tag{5.38}$$

and is notably a mere constant, independent of the density.[10]

The physical significance of the Thomas–Fermi wave vector is better appreciated if one considers the screened electrostatic potential produced by a point impurity charge e embedded in the electron gas. Since in this case we have simply $V_{ext}(\vec{q}) = v_q$, this quantity is approximately given by:

$$V_{sc}(\vec{r}) = \int \frac{d^3q}{(2\pi)^3} \frac{v_{\vec{q}}}{\epsilon(\vec{q}, 0)} e^{i\vec{q}\cdot\vec{r}}. \tag{5.39}$$

Making use of the small-q approximation (5.35) for the static dielectric function we obtain

$$V_{sc}^{TF}(\vec{r}) = \int \frac{d^3q}{(2\pi)^3} \frac{4\pi e^2}{q^2 + \kappa_3^2} e^{i\vec{q}\cdot\vec{r}} = e^2 \frac{e^{-\kappa_3 r}}{r}, \quad 3D, \tag{5.40}$$

which has precisely the form of the Yukawa screened potential introduced in Chapter 1 to regularize the coulomb interaction. Clearly within this approximation the impurity potential is completely "screened out" within a distance of the order of the inverse of the Thomas–Fermi wave vector. Notice that in the high density limit (i.e. for $r_s \to 0$) the Thomas–Fermi wave vector ($\kappa_3 \sim r_s^{-1/2}$) is much smaller than the Fermi wave vector ($k_F \sim r_s^{-1}$): it is precisely this circumstance that leads to the logarithmic divergence of the correlation energy, as discussed in Chapter 1 (see Eq. (1.107)).

Proceeding in a similar manner in two dimensions we obtain

$$V_{sc}^{TF}(\vec{r}) = \int \frac{d^2q}{(2\pi)^2} \frac{2\pi e^2}{q + \kappa_2} e^{i\vec{q}\cdot\vec{r}} = e^2 \left(\frac{1}{r} - \frac{\pi\kappa_2}{2} [\mathbf{H}_0(\kappa_2 r) - N_0(\kappa_2 r)] \right), \quad 2D, \tag{5.41}$$

where $\mathbf{H}_0(x)$ is a Struve function and $N_0(x)$ a Bessel function of the second kind (Gradshteyn and Ryzhik, 1965). This rather unwieldy expression recovers the usual $\frac{e^2}{r}$ law at small distances and, at variance with the three dimensional case, decreases only as r^{-3} for $r \gg \kappa_2^{-1}$. The inverse of the Thomas–Fermi wave vector κ_2 clearly plays the role of crossover length. The power law behavior at large r arises from the fact that the charge distribution comprised of the external and the induced screening charges is confined to a plane and therefore has a finite quadrupole moment in the physical three dimensional space.

We emphasize that the Thomas–Fermi screened potential is not the same thing as the true RPA screened potential. The latter exhibits long-range Friedel oscillations, associated

---

[10] This peculiarity is the reason for the much milder $r_s \ln r_s$ behavior at high density of the correlation energy in two dimensions (see Eq. (1.107)) with respect to the $\ln r_s$ of the three dimensional case.

with the existence of a singularity of $\epsilon(q, 0)$ at $q = 2k_F$ – see Section 5.4.3 and in particular Fig. 5.14.

5.3.3 Plasmons

We now turn to the study of what is perhaps the most distinctive property of the RPA, namely, the fact that the spectrum of the density fluctuations as given by $-\pi^{-1}\Im m \chi_{nn}^{RPA}(\vec{q}, \omega)$ is completely dominated, at long wavelength, by a collective excitation known as the *plasmon*.

Plasmons and electron–hole pairs are the two basic density excitations of the electron liquid. They can be thought of as liquid-state versions of longitudinal and transverse phonons[11] respectively. The RPA is just the simplest theory that treats them both on equal footing. In this section we concentrate on plasmons, and in the next we discuss electron–hole pairs.

Before embarking in calculations, let us try and understand why a collective density excitation should be expected on purely physical grounds.

Classical approach

The idea that ordinary gases and liquids support longitudinal density oscillations (sound waves) is rather familiar. The restoring force responsible for such oscillations is a pressure gradient that becomes weaker and weaker as the wavelength increases: therefore the frequency of the wave vanishes in the limit $q \to 0$. An elementary theory of sound waves starts with the classical Euler equation of motion for the current density, i.e.

$$m\frac{\partial \vec{j}(\vec{r}, t)}{\partial t} = -\vec{\nabla} P(\vec{r}, t) , \tag{5.42}$$

where $P(\vec{r}, t)$ is the pressure field. The essential assumption underlying the use of Eq. (5.42) is that the oscillations are sufficiently slow for the system to remain in thermodynamic equilibrium at all times. In other words, if τ is the time needed for the establishment of thermodynamic equilibrium, then the period of the wave must be *long* compared to τ: i.e. $\omega\tau \ll 1$. This regime is commonly referred to as the *collisional* or *hydrodynamic regime*.

In order to solve Eq. (5.42) we write the density as

$$n(\vec{r}, t) = n + n_1(\vec{r}, t) , \tag{5.43}$$

where n is uniform and time-independent, and $n_1(\vec{r}, t) \ll n$, and make the linear approximation

$$\vec{\nabla} P(\vec{r}, t) \simeq \frac{\partial P(n)}{\partial n} \vec{\nabla} n_1(\vec{r}, t) , \tag{5.44}$$

---

[11] The essentially transverse nature of low-energy electron-hole pair excitations can be understood from Fig. 4.3 in Chapter 4. Evaluating the expectation value of the current operator $\vec{j}_{\vec{q}}$ in the state $\hat{a}_{\vec{k}+\vec{q}}^{\dagger}\hat{a}_{\vec{k}}|\Psi_0\rangle$ ($|\Psi_0\rangle$ being the non-interacting Fermi sea) one finds it to be equal to $\frac{\hbar}{m}\left(\vec{k} + \frac{\vec{q}}{2}\right)$, where $\vec{k} + \frac{\vec{q}}{2}$ is essentially orthogonal to $\vec{q}$. See Sections 5.6.2.6 and 5.9 for additional discussion of this point.

where $P(n)$ is the equation of state relating the pressure to the density at equilibrium. Taking the divergence of both sides of Eq. (5.42) and making use of the continuity equation we obtain the wave equation

$$\frac{\partial^2 n_1(\vec{r}, t)}{\partial t^2} - \frac{1}{m} \frac{\partial P(n)}{\partial n} \nabla^2 n_1(\vec{r}, t) = 0 . \tag{5.45}$$

This implies the existence of a density wave with frequency $\omega = \sqrt{\frac{1}{m} \frac{\partial P(n)}{\partial n}} q$.

Plasmons in the electron liquid are superficially similar to sound waves, but there is one crucial difference: the frequency of the oscillations is much larger than the inverse of the equilibration time and therefore the system cannot be assumed to be in local thermodynamic equilibrium. To see how this comes about consider that in the electron liquid the force attempting to restore equilibrium is provided primarily by a long-range electrostatic field. At long wavelength this force is much larger than the gradient of the local pressure and leads to a frequency that remains finite in three dimensions, or vanishes as $\sqrt{q}$ in two dimensions. Because this frequency is usually much larger than the frequency of equilibrium-restoring collisions (i.e., $\omega \tau \gg 1$) the latter play no significant role, and the system is said to be in the *collisionless regime*.[12] This circumstance makes plasmons much more similar to longitudinal phonons in a solid than to ordinary sound waves in a liquids. The Euler equation of motion for the current density now takes the form

$$m \frac{\partial \vec{j}(\vec{r}, t)}{\partial t} = -n \vec{\nabla} \int d\vec{r}' \frac{e^2}{|\vec{r} - \vec{r}'|} n_1(\vec{r}', t) . \tag{5.46}$$

Again by taking the divergence, using the continuity equation, and Fourier-transforming both sides of this equation with respect to $\vec{r}$ and t, we obtain

$$\left(\omega^2 - \frac{nq^2 v_q}{m} \right) n_1(\vec{q}, \omega) = 0 . \tag{5.47}$$

This implies the existence of density waves[13] with frequency

$$\omega_p(q) = \sqrt{\frac{nq^2 v_q}{m}} = \begin{cases} \sqrt{\frac{4\pi n e^2}{m}} , & 3D , \\ \sqrt{\frac{2\pi n e^2 q}{m}} , & 2D . \end{cases} \tag{5.48}$$

Notice that in both cases the plasmon frequency lies well above the maximum electron–hole excitation frequency (of order of $v_F q$): this fact will turn out crucial to the stability of the collective mode since it implies that plasmons cannot decay into electron–hole pairs via a momentum and energy-conserving process.

---

[12] In fact, even neutral systems, such as ^{3}He, can support a collisionless sound mode, known as *zero sound*, since the strong interactions between the atoms push the frequency of such a mode well above the typical low-temperature relaxation rates.

[13] These density waves should not be confused with the global oscillation of the center of mass that is often observed in *finite* electronic systems in parabolic quantum wells.

Quantum mechanical calculation

We now turn to the quantum mechanical calculation of the plasmon frequency. As discussed in Section 3.2.7, the frequencies of collective modes in a many-body systems are determined by the poles of the density–density response function in the lower half of the complex frequency plane. These poles give rise to sharp peaks (resonances) in the spectral function, provided they occur sufficiently close to the real frequency axis.

Let us then search for the poles of the RPA density–density response function, Eq. (5.33). The Lindhard function $\chi_0(q, \omega)$ has no poles, but only a branch cut along the real axis in the frequency range of electron–hole pair excitations. The poles of χ_{nn}^{RPA} must therefore arise from the vanishing of the denominator in Eq. (5.33). This leads to the condition[14]

$$\epsilon^{RPA}(q, \omega) = 1 - v_q \chi_0(q, \omega) = 0 . \tag{5.49}$$

Inspecting Figs. 5.6 and 5.7 we see that the real part of $\epsilon^{RPA}(q, \omega)$ at long wavelength has zeroes for two values of the frequency. The first one falls well within the electron–hole continuum (where $\Im m\epsilon \neq 0$) and is therefore not a solution of Eq. (5.49). The second one, however, lies above the electron–hole continuum, and does represent a solution of Eq. (5.49). To obtain explicit formulas for the plasmon dispersion at long wavelength one makes use of the expansion of the Lindhard function for $q \to 0$ and $\omega \gg q v_F$, which is easily obtained from Eq. (4.39) and Table 4.4 of Chapter 4. One readily finds

$$\chi_0(q, \omega) \simeq \frac{nq^2}{m\omega^2}\left[1 + a_d \frac{q^2 v_F^2}{\omega^2}\right] , \tag{5.50}$$

where $a_3 = \frac{3}{5}$ in three dimensions and $a_2 = \frac{3}{4}$ in two dimensions. Substituting in Eq. (5.49) and solving for ω up to terms of order q^2, we have:

$$\Omega_p^2(q) \simeq \omega_p^2(q) + a_d q^2 v_F^2 + \cdots , \tag{5.51}$$

with $\omega_p(q)$ given by Eq. (5.48). This can also be rewritten in terms of the three dimensional Thomas–Fermi screening wave vectors $\kappa_3^2 = \frac{3\omega_p^2}{v_F^2}$ and $\kappa_2 = \frac{2}{a_B}$ as follows:

$$\Omega_p(q) \simeq \omega_p(q)\left[1 + \frac{9q^2}{10\kappa_3^2} + \cdots\right] , \quad 3D \tag{5.52}$$

and

$$\Omega_p(q) \simeq \sqrt{\frac{2\pi n e^2 q}{m}}\left[1 + \frac{3q}{4\kappa_2} + \cdots\right] , \quad 2D . \tag{5.53}$$

While the zero-order terms in these expansions agree with the corresponding results obtained from the classical treatment, the leading order corrections for small q are of quantum mechanical origin and will be discussed in the last part of this section.

[14] From the point of view of macroscopic electrodynamics, $\epsilon = 0$ means that one can have a finite electric field E even though there are no external charges, i.e., $D = \epsilon E = 0$: the electric field is entirely sustained by electronic density fluctuations within the material.

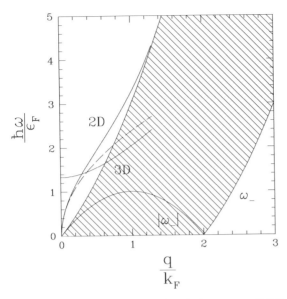

Fig. 5.8. Plasmon dispersion in three and two dimensional electron gas (full lines) for $r_s = 2$. The dashed line represents the approximate two dimensional dispersion given by Eq. (5.53).

As it turns out, in the two dimensional case Eq. (5.49) can be solved analytically at all wave vectors (Czachor *et al.*, 1982). The result (see Exercise 5) can be cast in the form:

$$\Omega_p(q) = \sqrt{\frac{2\pi n e^2 q}{m} \frac{\left(1 + \frac{q}{\kappa_2}\right)^2 \left(1 + \frac{q^3}{4\pi n \kappa_2} + \frac{q^4}{8\pi n \kappa_2^2}\right)}{1 + \frac{q}{2\kappa_2}}} , \quad 2D. \qquad (5.54)$$

For the three dimensional case a numerical solution of Eq. (5.49) proves necessary. Fig. 5.8 provides a plot of the RPA plasmon dispersion in two and three dimensions. Notice that in all cases, the condition $\omega \gg q v_F$ is satisfied.

Both in three and two dimensions a solution is found only up to a *critical wave vector* q_c where the plasmon dispersion impinges onto the electron–hole continuum. In three dimensions, for wave vectors exceeding this threshold, the plasma mode is heavily damped and in practice ceases to exist. In the two dimensional case, on the other hand, the plasmon dispersion eventually touches the upper edge of the electron–hole continuum remaining parallel to it and then also ceases to exist (see discussion below). It is a straightforward exercise to derive the equations determining the density dependent plasmon critical wave vector within the RPA (see Exercises 4 and 5).

Plasmon oscillator strength

We now examine the strength of the plasmon contribution to the density fluctuation spectrum

$$\Im m \chi_{nn}^{RPA}(q, \omega) = \frac{\Im m \chi_0(q, \omega)}{[1 - v_q \Re e \chi_0(q, \omega)]^2 + [v_q \Im m \chi_0(q, \omega)]^2} . \qquad (5.55)$$

From the fact that $\Im m \chi_0(q, \omega_p) = 0$ one might be tempted to conclude that also $\Im m \chi_{nn}^{RPA}(q, \omega_p) = 0$, but this is wrong. Remember that the plasmon pole must be located in the *lower* half of the complex frequency plane: in the RPA this means that the pole really occurs at $\omega = \Omega_p - i\eta$, i.e., infinitesimally below the real axis. At this frequency $\Im m \chi_0(q, \omega)$ is of the order of η: this is infinitesimally small, yet it cannot be neglected when $\omega = \Omega_p$, since in that case $1 - v_q \Re e \chi_0(q, \Omega_p) = 0$ making $\Im m \chi_{nn}^{RPA}(q, \omega_p)$ proportional to η^{-1}, which tends to ∞ for $\eta \to 0$. One therefore sees that $\Im m \chi_{nn}^{RPA}(q, \omega_p)$ has a δ-function peak at $\omega = \Omega_p(q)$.

To determine the strength of the plasmon peak let us consider the behavior of $\chi_{nn}^{RPA}(q, \omega)$ in the vicinity of the plasmon frequency. Expanding the denominator of Eq. (5.33) to first order in $\omega - \Omega_p(q)$, and using the fact that $\epsilon(q, \Omega_p(q)) = 0$ we obtain

$$\chi_{nn}^{RPA}(q, \omega) \simeq \frac{\chi_0\left(q, \Omega_p(q)\right)}{\left(\frac{\partial \Re e \epsilon(q, \omega)}{\partial \omega}\right)_{\omega=\Omega_p(q)} \left[\omega - \Omega_p(q) + i\eta\right]} . \tag{5.56}$$

In the limit $\eta \to 0$ this gives us

$$-\frac{1}{\pi} \Im m \chi_{nn}^{RPA}(q, \omega)\bigg|_{plasmon} = \frac{\delta\left(\omega - \Omega_p(q)\right)}{v_q \left|\frac{\partial \Re e \epsilon(q, \omega)}{\partial \omega}\right|_{\omega=\Omega_p(q)}} . \tag{5.57}$$

In the small-q limit, by making use of the approximate formula for $\chi_0(q, \omega)$ given in Eq. (5.50), we find that $\left(\frac{\partial \Re e \epsilon(q, \omega)}{\partial \omega}\right)_{\omega=\Omega_p(q)} \simeq \frac{2}{\Omega_p(q)}$ so that

$$-\frac{1}{\pi} \Im m \chi_{nn}^{RPA}(q, \omega)\bigg|_{plasmon} \simeq \frac{\Omega_p(q)}{2v_q} \delta\left(\omega - \Omega_p(q)\right) . \tag{5.58}$$

Eq. (5.58) shows that the strength of the plasmon pole at long wavelength behaves as q^2 in three dimensions and as $q^{\frac{3}{2}}$ in two dimensions. Moreover the plasmon peak exhausts the f-sum rule to order q^2, as well as the third-moment sum rule deduced from the high frequency expansion of the RPA response function (see Exercise 7). This implies that, in the three-dimensional RPA, the plasmon peak accounts for the entire spectral strength up to order q^4. Contributions from single electron–hole pair excitations, which are present for frequencies $\omega \le qv_F$, begin only at order q^6 (see Section 5.3.4).

Eq. (5.57) can also be used to show that the the plasmon oscillator strength vanishes as its dispersion approaches the upper edge of the electron–hole continuum. This follows from the fact that $\left|\frac{\partial \epsilon^{RPA}(q, \omega)}{\partial \omega}\right|_{\omega=\Omega_p(q)}$ tends to infinity for $q \to q_c$. In Exercise 6 the reader is invited to show that the oscillator strength vanishes as $-\frac{1}{\ln|q-q_c|}$ in three dimensions and $\sqrt{|q - q_c|}$ in two dimensions.

Plots of the electron–hole contribution to $\Im m \chi_{nn}^{RPA}(q, \omega)$ are presented in Figs. 5.9 and 5.10. It is evident that the magnitude of $\Im m \chi_{nn}^{RPA}(q, \omega)$ is greatly reduced with respect to the non-interacting case (notice the factors of 10 and 50 that are needed to restore the actual

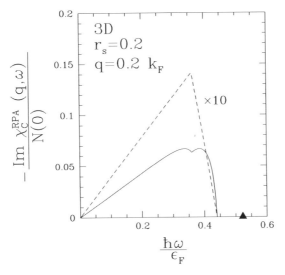

Fig. 5.9. Plot of $-\Im m\,\chi_{nn}^{RPA}(\vec{q},\omega)$ versus frequency at a fixed wave vector q (full line) for a three dimensional electron gas. The dashed curve represents the rescaled values (by a factor of 10^{-1}) of the corresponding non-interacting quantity $-\Im m\,\chi_0(\vec{q},\omega)$. The solid triangle on the frequency axis marks the position of the delta-function representing the plasmon contribution.

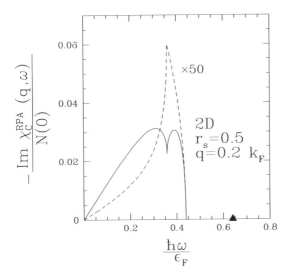

Fig. 5.10. Same as in Fig. 5.9 but for a two dimensional electron gas.

values of the Lindhard function respectively in two and three dimensions) as the oscillator strength is transferred to the plasma mode.

It is reasonable to inquire whether these are genuine properties of the interacting electron liquid, or artifacts of the RPA. It turns out that, to order q^2, the RPA is qualitatively correct:

the plasmon peak does absorb all the spectral strength to this order. At order q^4 however, the situation is more delicate: while it is true that single electron–hole pairs do not contribute to the spectral strength at this order, there is a finite contribution from *double* electron–hole pairs. These two-electron–two-hole excitations take some spectral strength away from the plasmon and, in fact, cause the latter to be lightly damped, with a decay rate proportional to q^2. Of course, these effects lie beyond the power of the RPA, and will be discussed in detail in Section 5.6.2.

The failure of hydrodynamics

Before ending this section we wish to comment on the nature of the leading corrections (q^2 in three dimensions and q in two) to the long-wavelength plasmon frequency. It is tempting to explain these contributions in terms of additional short-range forces that are not included in Eq. (5.46). For a liquid in local equilibrium the only additional force is the gradient of the local pressure

$$\vec{F} = -\vec{\nabla} P(\vec{r}, t) = -\frac{\partial P(n)}{\partial n} \vec{\nabla} n_1(\vec{r}, t) \,. \tag{5.59}$$

The pressure of a noninteracting electron gas at zero temperature is given by[15]

$$P(n) = \frac{2}{2+d} n\epsilon_F(n) \,. \tag{5.60}$$

Solving the equation of motion (5.46), with the pressure term (5.59) added on the right-hand side, we obtain the following result for the plasmon dispersion in three dimensions:

$$\Omega_p^2(q) \simeq \omega_p^2 + \frac{1}{3} q^2 v_F^2 \,. \tag{5.61}$$

This is very similar to the RPA result, Eq. (5.51), but notice that the coefficient of q^2 is wrong: $\frac{1}{3}$ instead of $\frac{3}{5}$. The reason for this incorrect result is that the assumption of local equilibrium, on which Eq. (5.59) rests, fails for plasmon waves, because electron-electron collisions are unable to establish equilibrium at the high frequency of these waves. In a plasmon oscillation, both the volume and the shape of the Fermi surface change in time, the latter becoming alternately oblate and prolate with respect to the direction of the wave vector $\vec{q}$. By contrast, in a local equilibrium situation, such as occurs in an ordinary sound wave, the Fermi surface retains its spherical shape, merely expanding and contracting its radius to accommodate the varying density. Since the change in shape of the Fermi surface costs additional kinetic energy (remember that the spherical shape is the one that minimizes the kinetic energy for a given volume) the plasmon frequency turns out larger than predicted on the basis of the local equilibrium assumption. We shall see in Section 5.9 that the correct plasmon dispersion can be obtained from a macroscopic equation of motion including an elastic *shear force* in addition to the pressure force.

---

[15] This expression for $P(n)$, also known as the *Fermi pressure*, is of purely quantum mechanical origin, and equals $\frac{2}{d}$ of the kinetic energy density as one can readily infer from Eq. (1.82).

5.3.4 The electron–hole continuum in RPA

The electron–hole continuum in the RPA occurs within the same kinematic boundaries as in the noninteracting gas case discussed in Section 4.4.2. This is displayed in Figs. 5.9 and 5.10. The essential difference is that the existence at long wavelength of the plasmon reduces the overall strength of the single electron–hole pairs by a factor q^{2d-2} in the long wavelength limit.

To see how this comes about, let us rewrite Eq. (5.55) as

$$-\frac{1}{\pi}\Im m\,\chi_{nn}^{RPA}(\vec{q},\omega) = -\frac{1}{\pi}\frac{\Im m\,\chi_0(\vec{q},\omega)}{\left|\epsilon^{RPA}(\vec{q},\omega)\right|^2} \tag{5.62}$$

and specialize to the long wavelength limit $q \ll k_F$. The numerator differs from zero only in the region $|\omega| < qv_F$. In this region the RPA dielectric function can be approximated (see Eq. (4.39)) as

$$\epsilon^{RPA}(q,\omega) \simeq 1 + \left(\frac{\kappa_d}{q}\right)^{d-1}\Psi'_d\left(\frac{\omega}{qv_F}\right), \tag{5.63}$$

where the function $\Psi'_d(\nu)$ is given in Table 4.4.

Denoting by C_d the minimum of $|\Psi'_d(\nu)|$ in the interval $0 \le \nu \le 1$ (see Fig. 4.9) we see that the electron–hole part of the spectral function is bounded by

$$\left|\frac{1}{\pi}\Im m\,\chi_{nn}^{RPA}(\vec{q},\omega)\right|_{e-h} \le \left|\left(\frac{q}{\kappa_d}\right)^{2d-2}\frac{\Im m\,\chi_0(\vec{q},\omega)}{\pi C_d^2}\right|, \qquad q \ll k_F, \tag{5.64}$$

and therefore vanishes as q^4 in three dimensions and q^2 in two for small q. Its contribution to the f-sum rule vanishes as q^{2d}, consistent with the fact, noted above, that this sum rule is entirely exhausted, to order q^2, by the plasmon.

5.3.5 The static structure factor and the pair correlation function

The static structure factor can be calculated from the imaginary part of the density–density response function according to the general formulas derived in Chapter 3. At zero temperature, the relationship is

$$S(q) = -\frac{\hbar}{\pi n}\int_0^\infty \Im m\,\chi_{nn}(q,\omega)d\omega. \tag{5.65}$$

We have seen that, at long wavelength, the RPA density excitation spectrum (as well as the exact spectrum) is completely dominated by the plasmon. Substituting Eq. (5.57) in Eq. (5.65) and performing the trivial frequency integration we see that in this regime, for both three and two dimensional systems, the RPA structure factor can be cast in the form

$$S^{RPA}(q) = \frac{\hbar q^2}{2m\omega_p(q)}, \qquad q \ll k_F, \tag{5.66}$$

where $\omega_p(q)$ is given by Eq. (5.48). This expression differs markedly from the noninteracting result, $S^{(0)}(q) \sim q$, that was derived in Chapter 1. Due to the long-range of the coulomb interaction, long wavelength density fluctuations are severely suppressed (their energy cost tends to infinity for $q \to 0$) and therefore $S(q)$ vanishes as q^2 in three dimensions and as $q^{\frac{3}{2}}$ in two.

Eq. (5.66) is a special case of the so-called *Bijl–Feynman formula*, which follows from the assumption that the density fluctuation spectrum is exhausted by a single collective excitation of frequency $\Omega_p(q)$. Then $S(q, \omega)$ must be of the form $S(q)\delta(\omega - \Omega_p(q))$ where the strength of the δ-function is given by the static structure factor. Requiring that this expression satisfy the f-sum rule yields the Bijl–Feynman expression

$$S(q) = \frac{\hbar q^2}{2m\Omega_p(q)} \,, \tag{5.67}$$

of which Eq. (5.66) is a special case. Notably, since the RPA spectral function is exact to order q^2, Eq. (5.66) is in fact an exact long wavelength property of the interacting electron gas.

The situation is not nearly as rosy as far as the calculation of the pair correlation function is concerned. According to Eq. (A4.17) the calculation of $g(r)$ requires a knowledge of the structure factor for *all* wave vectors. In particular, the short range behavior of $g(r)$ is controlled by the behavior of $S(q)$ at large wave vectors, which is *not* well approximated by the RPA.

Fig. 5.16 later shows the RPA pair correlation function at $r_s = 3$ and 4, i.e., in the middle of the metallic density range. It is evident that the RPA grossly overestimates the short-range correlations to the point of giving unphysical negative values for this positive-definite quantity. This rather disappointing failure provided much of the initial impetus for developing better approximations. The pair correlation functions obtained from two of these "better approximations" are also shown in Fig. 5.16.

5.3.6 The RPA ground-state energy

In this section we present the calculation of the ground-state energy based on Eq. (3.160) of Chapter 3 and on the RPA expression for the density–density response function. From the introductory discussion of Section 1.5.6 we already know the ground-state energy to first order in the coulomb interaction. This is the sum of the kinetic energy of the non-interacting gas and the exchange energy and can be written as

$$\frac{E^{(1)}}{N} = \frac{E_0}{N} + \frac{1}{2L^d} \sum_{\vec{q}} v_q \left(-\frac{\hbar}{n\pi} \int_0^1 d\lambda \int_0^\infty \chi_0(q, i\omega)d\omega - 1 \right). \tag{5.68}$$

The residual energy (i.e., the correlation energy) is then given by

$$\frac{E^{RPA} - E^{(1)}}{N} = \frac{1}{2L^d} \sum_{\vec{q}} \left(-\frac{\hbar}{n\pi} \int_0^1 d\lambda \int_0^\infty \frac{\lambda v_q^2 \chi_0^2(q, i\omega)}{1 - \lambda v_q \chi_0(q, i\omega)} d\omega \right). \tag{5.69}$$

The integration over the coupling constant λ is easily done, and we arrive at

$$\frac{E^{RPA} - E_1}{N} = \frac{\hbar}{2\pi N} \sum_{\vec{q}} \int_0^\infty \{v_q \chi_0(q, i\omega) + \ln[1 - v_q \chi_0(q, i\omega)]\} d\omega . \quad (5.70)$$

It is convenient, at this point to introduce dimensionless wave vectors and frequencies $y = \frac{q}{k_F}$ and $v = \frac{\omega}{q v_F}$. Making use of the representation (4.21) for the Lindhard function (and the fact that $\Psi_d(z) = -\Psi_d(-z)$) we can write

$$v_q \chi_0(q, i\omega) = - \left(\frac{\kappa_d}{k_F}\right)^{d-1} \frac{f_d(y, v)}{y^{d-1}} , \quad (5.71)$$

where we have defined the function

$$f_d(y, v) = \frac{2\Re e \Psi_d(\frac{y}{2} + iv)}{y} , \quad (5.72)$$

and κ_d is the Thomas–Fermi wave vector in d dimensions.

Notice that here the role of the *dimensionless coupling constant* is played by the quantity

$$\gamma_d \equiv \left(\frac{\kappa_d}{k_F}\right)^{d-1} = \begin{cases} \frac{4}{\pi} \frac{1}{k_F a_B} = \frac{4}{\pi} \alpha_3 r_s = \frac{4}{\pi} \left(\frac{4}{9\pi}\right)^{\frac{1}{3}} r_s, & 3D, \\ \frac{2}{k_F a_B} = 2\alpha_2 r_s = \sqrt{2} r_s, & 2D . \end{cases} \quad (5.73)$$

The complete expression for the RPA correlation energy per particle in d dimensions therefore becomes

$$\frac{\epsilon_c^{RPA}}{Ry} = -\frac{d}{2\pi \alpha_d^2 r_s^2} \int_0^\infty dy \, y^d \int_0^\infty dv \left\{ \frac{\gamma_d}{y^{d-1}} f(y, v) - \ln\left[1 + \frac{\gamma_d}{y^{d-1}} f(y, v)\right] \right\} , \quad (5.74)$$

where the constants α_d can be obtained from Eq. (1.79).

Two important things should be noted about this expression. The first is that the integral exists and is easily computable. The second is that, if one attempts to expand the integrand in powers of γ_d (alias r_s) one encounters divergencies in the individual terms of the expansion – a phenomenon already noted in Chapter 1. For example, to order γ_d^2 in three dimensions and to order γ_d^3 in two, the integrand diverges as y^{-1} at long wavelength. Care must therefore be exercised in the perturbative treatment of this expression.

5.3.6.1 Asymptotic high density form of ϵ_c^{RPA}

In order to extract the leading terms in the high density (small r_s) expansion of Eq. (5.74) an elegant procedure, reminiscent of the Hadamard definition of the principal part of an improper integral, is employed. In order to be explicit, the calculation will be carried out for the three dimensional electron gas. For the two dimensional case the method is the same and the calculation is left as an exercise to the reader.

We begin by splitting the troublesome momentum integral into two parts, from 0 to y_c and from y_c to infinity, where y_c is a small dimensionless wave vector which, however, we assume to be much larger than the coupling constant γ_d. The final result of the calculation will be independent of y_c.

We notice next that in the small-y part of the integral ($y < y_c$) expanding with respect to the ratio $\frac{\gamma_d}{y^2}$ is inappropriate. On the other hand the smallness of y justifies the use of the limiting form

$$\lim_{y \to 0} f_d(y, v) = \Psi'_d(iv), \tag{5.75}$$

where $\Psi'_d(z)$ is the derivative of $\Psi_d(z)$ with respect to its argument. This substitution leads to an integrand that is simple enough for the y integration to be done analytically. Only then are we entitled to expand in powers of $\frac{\gamma_d}{y_c^2}$ and readily obtain

$$\int_0^{y_c} dy\, y^3 \int_0^\infty dv \left\{ \frac{\gamma_3}{y^2} \Psi'_3(iv) - \ln\left[1 + \frac{\gamma_3}{y^2} \Psi'_3(iv) \right] \right\}$$

$$\simeq \frac{\gamma_3^2}{4} \int_0^\infty dv \left(\frac{1}{2} - \ln \gamma_3 - \ln \Psi'_3(iv) + 2 \ln y_c \right) [\Psi'_3(iv)]^2 . \tag{5.76}$$

There still remains however a term that is logarithmically divergent in the limit $y_c \to 0$. This problem is taken care by the result of the remaining integral from y_c to infinity. In the latter, since the singular point $y = 0$ is excluded, one can expand the integrand to order γ_3^2 and integrate by parts with respect to y getting

$$\frac{\gamma_3^2}{2} \int_{y_c}^\infty dy \int_0^\infty dv\, \frac{f_3^2(y, v)}{2y}$$

$$= \frac{\gamma_3^2}{2} \int_0^\infty dv \left(\int_{y_c}^\infty dy\, \ln y \frac{\partial f_3^2(y, v)}{\partial y} - f_3^2(y_c, v) \ln y_c \right) . \tag{5.77}$$

Combining the small and large momentum contributions we see that the singular dependence $\ln y_c$ on the cutoff cancels out. Only at this point are we allowed to take the limit $y_c \to 0$: this leaves us with

$$\frac{\epsilon_c^{RPA}}{\mathrm{Ry}} \simeq -\frac{6}{\pi^3} \int_0^\infty dv \left\{ \left(\frac{1}{2} - \ln \gamma_3 - \ln \Psi'_3(iv) \right) [\Psi'_3(iv)]^2 \right.$$

$$\left. + 2 \int_0^\infty dy\, \frac{\partial f_3^2(y, v)}{\partial y} \ln y \right\} , \tag{5.78}$$

where we have made use of Eq. (5.73). Then, with the help of the explicit formula

$$\lim_{y \to 0} f_3(y, v) = \Psi'_3(iv) = 1 - v \tan^{-1} \frac{1}{v} , \tag{5.79}$$

which is obtained by substituting the imaginary argument iv in the formulas of Table 4.4 for $\Psi'_3(z)$, we can write the leading ($\ln r_s$) term of the correlation energy in terms of an integral

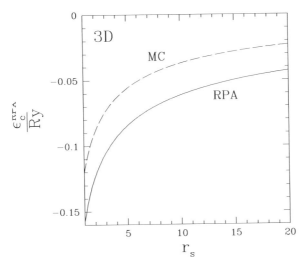

Fig. 5.11. RPA correlation energy for a three dimensional electron gas. The dashed line corresponds to the result obtained via the Monte Carlo method (see Section 1.7).

that can be evaluated analytically:

$$\frac{6}{\pi^3} \int_0^\infty dv \left(1 - v \tan^{-1} \frac{1}{v} \right)^2 \ln r_s = \frac{2}{\pi^2}(1 - \ln 2) \ln r_s \simeq 0.062 \ln r_s \ . \tag{5.80}$$

This result was originally obtained by W. Macke (Macke, 1950). The remaining integrations can be carried out numerically so that one obtains

$$\epsilon_c^{RPA} \simeq (0.062 \ln r_s - 0.141) \text{ Ry} \ . \tag{5.81}$$

Finally, by adding to this the Onsager result of Eq. (1.105) for the second-order exchange contribution we obtain the classic expression for the high density expansion of the correlation energy quoted in Chapter 1 (Eq. (1.107)). The corresponding result in two dimensions is given in Eq. (1.107).

5.3.6.2 Numerical values of ϵ_c^{RPA} for all densities

The calculation of the RPA correlation energy can readily be carried out for all r_s by numerical means. One simply employs Eq. (5.74) to obtain the results displayed in Figs. 5.11 and 5.12 where the RPA correlation energy is displayed for a range of r_s values in which the electrons are expected to be in an unpolarized liquid state.

In these figures we also plot for comparison the correlation energies obtained from much more accurate Monte Carlo calculations, described in Section 1.7. It is clear that except for very small values of r_s the RPA leads in general to rather inaccurate results for the total energy.[16]

---

[16] The fact that the magnitude of the RPA correlation energy calculated from Eq. (5.74) is sensibly larger than that obtained with the Monte Carlo method simply reflects the non-variational nature of the RPA.

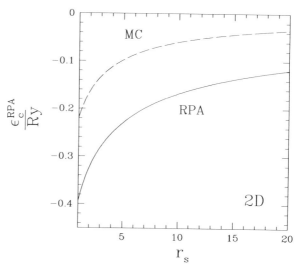

Fig. 5.12. Same as in Fig. 5.11 but for a two dimensional electron gas.

5.3.6.3 Asymptotic low-density form of ϵ_c^{RPA}

It is instructive, even though apparently unphysical, to examine the behavior of the RPA energy in the low density limit of large r_s. We begin by noticing that for $r_s \to \infty$ the radius of the Fermi surface shrinks to zero and, as we shall presently see, the leading contributions to the energy comes from wave vectors and frequencies that are much larger than k_F and $\frac{\epsilon_F}{\hbar}$ respectively. We are therefore entitled to use in Eq. (3.160) the large q and large ω form of the density response function. In this limit the Lindhard function for imaginary frequency becomes

$$\chi_0(q, i\omega) \simeq -\frac{\frac{nq^2}{m}}{\omega^2 + \left(\frac{\hbar q^2}{2m}\right)^2} , \tag{5.82}$$

so that the RPA response function, evaluated for the scaled interaction λv_q, has the form

$$\chi_\lambda^{RPA}(q, i\omega) \simeq -\frac{\frac{nq^2}{m}}{\omega^2 + \lambda\omega_p^2 + \left(\frac{\hbar q^2}{2m}\right)^2} . \tag{5.83}$$

Integrating this expression over frequency (see Eq. (5.65)) yields the static structure factor

$$S_\lambda(q) = -\frac{\frac{\hbar q^2}{2m}}{\left(\frac{\hbar q^2}{2m}\right)^2 + \lambda\omega_p^2} , \tag{5.84}$$

which can then be substituted in Eq. (3.160) to calculate the total energy per particle. In three dimensions this is given by

$$\epsilon^{RPA} \simeq \frac{1}{\pi \alpha_3 r_s} \int_0^1 d\lambda \int_0^\infty dy \left(\frac{y^2}{\sqrt{y^4 + \frac{16\alpha_3 \lambda r_s}{3\pi}}} - 1 \right) \text{Ry}, \qquad (5.85)$$

where we have neglected the rapidly vanishing ($\sim r_s^{-2}$) contribution of the non-interacting kinetic energy. Notice that here for large r_s the main contribution to the integral comes from values of $y \sim r_s^{\frac{1}{4}} \gg 1$, consistent with our initial assumption.

With the change of variable $y \to y(\lambda r_s)^{\frac{1}{4}}$ the y integral becomes independent of the coupling constant, and we get simply

$$\epsilon^{RPA} \simeq \frac{4}{5\pi \alpha_3 r_s^{\frac{3}{4}}} \int_0^\infty dy \left(\frac{y^2}{\sqrt{y^4 + \frac{16\alpha_3}{3\pi}}} - 1 \right) \text{Ry} \simeq -\frac{0.712}{r_s^{\frac{3}{4}}} \text{Ry}. \qquad (5.86)$$

A similar result holds in two dimensions where an expression proportional to $r_s^{-\frac{2}{3}}$ is found instead.

Although not physically meaningful for the one-component electron liquid, the low density limit of the RPA energy is actually relevant to the case of multicomponent systems like multivalley doped semiconductors where the Fermi energy and the Fermi wave vector in each valley can be small even as the total electronic density is high. As it turns out, the RPA becomes exact in the limit in which the number of electronic species tends to infinity at fixed total density.

5.3.7 Critique of the RPA

The main advantage and appeal of the RPA is its simplicity. As we have seen, this approximation leads to a basically exact expression for the total energy of the electron gas in the high density limit. Moreover it provides a simple and versatile framework to study collective plasmon oscillations as well as, to some extent, the single electron properties of the system – a problem that will be tackled in Chapter 8. While appreciating the main benefits of this approach it is also important to be aware of its, at time striking, failures.

As we have seen, the RPA is very accurate in describing the long wavelength longitudinal spectrum of the electron gas. On the other hand, one must keep in mind that in this approximation the imaginary part of the response function $\chi_{nn}(q, \omega)$ is finite only for values of q and ω that fall within the region of the non-interacting electron–hole continuum. This feature results in the lack of a damping mechanism for the plasmon excitations at small wave vectors. In addition, while the description of the plasmon dispersion is accurate at long wavelengths, the same statement cannot be made at finite q.

One must also notice that, unfortunately, many of the the RPA-based results have very limited applicability in the standard metallic density range. The main problem was already pointed out at the end of Sections 5.3.3 and 5.3.6: while yielding an essentially exact description of long wavelength density fluctuations, the RPA grossly overestimates the strength of the exchange-correlation hole, on which the energy critically depends. Even worse, the pair correlation function at small separations becomes unphysically negative at metallic densities.

It is quite clear that the problems with the RPA are bound to become more obvious at low densities. This fact is exemplified by the behavior of the RPA ground-state energy in the large r_s limit, which was studied in the previous section. The $r_s^{-\frac{3}{4}}$ behavior of the RPA ground-state energy at low density in three dimensions is in violation of the rigorous lower bound proportional to r_s^{-1}, which was derived in Chapter 1. Of course this state of affairs should not come as a surprise as we know that the energy of the electron gas for small densities must be that of a basically classical crystal, a physical scenario rather remote from the liquid state described by the RPA.

Another major problem with the RPA is represented by a serious violation of the *compressibility sum rule*. This is simply established by noticing that while the compressibility obtained from the density dependence of the RPA energy behaves reasonably, the limit $\lim_{\vec{q}\to 0} \tilde{\chi}_{nn}^{RPA}(\vec{q}, 0)$ leads instead to the compressibility of a non-interacting electron gas in clear violation of the sum rule.

5.4 The many-body local field factors

It quickly becomes very difficult to go beyond the simple RPA. Even what would seem to be the obvious next step, namely the time-dependent Hartree–Fock approximation (see Section 5.7.1), turns out to be mathematically involved, and physically unsatisfactory. Fortunately, it is possible to develop a systematic and elegant description of the the linear response of an electron liquid, which preserves the simple mathematical structure of the RPA while going far beyond the RPA in the treatment of exchange and correlation effects.

At the heart of this generalization of the RPA lies the concept of *local effective potential*. In the RPA the effective field, to which an electron responds, is the field that would be seen by a classical test charge embedded in the electron gas. This approximate picture fails to account for the correlation that exists between the electron under scrutiny and the other electrons in the system. Here are a few examples that should clarify the nature of the missing correlation

- The electrostatic field seen by a real electron must not include the contribution from that very same electron.
- Because of the antisymmetry of the N-electron wave function, the presence of an electron at any given point excludes the presence of a second electron with the same spin orientation at that point, and, more generally, reduces the probability of finding such an electron at any finite

distance from the first. This so-called "exchange-hole" is present even in a non-interacting electron gas.

- The probability of finding an electron within a finite distance from the one under scrutiny is further altered by the coulomb repulsion: this gives rise to what is usually called (with somewhat improper terminology) a "correlation-hole", even though it only redistributes the electronic density.

To overcome this limitation it is quite natural to try and replace the average electrostatic potential by a *local effective potential* $V_{eff,\sigma}(\vec{r}, t)$ which is postulated to be the effective potential to which a real electron of spin projection σ, as opposed to a mere test charge, is subject in an average sense. The power of this idea squarely rests on the assumption that the effective potential is *local*, i.e., it acts on the wave function as a simple multiplicative operator.[17] Thus, $V_{eff,\sigma}$ is formally simpler than, say, the Fock nonlocal exchange potential, and yet it includes much more physics, albeit in an approximate fashion.

In what follows we develop the theory of the effective potential in a homogeneous electron liquid in the *linear response* regime. In this regime the Fourier amplitude $V_{eff,\sigma}(\vec{q}, \omega)$ is a linear function of the Fourier amplitudes, $n_{\sigma'}(\vec{q}, \omega)$, of the $\uparrow$- and $\downarrow$- spin densities. Thus, we can write

$$V_{eff,\sigma}(\vec{q}, \omega) = V_{ext,\sigma}(\vec{q}, \omega) + \sum_{\sigma'} v_q n_{\sigma'}(\vec{q}, \omega) - \sum_{\sigma'} v_q G_{\sigma\sigma'}(q, \omega) n_{\sigma'}(\vec{q}, \omega) , \quad (5.87)$$

where on the right-hand side the first term is an external spin-dependent potential, the second the electrostatic potential seen by a test charge (the term retained by the RPA), and the last, featuring the *local field factors* $G_{\sigma\sigma'}(q, \omega)$, contains the corrections to the effective potential stemming from the exchange interaction between electrons of the same spin, plus any correlation effects. Notice that the G's satisfy the symmetry relation

$$G_{\uparrow\downarrow} = G_{\downarrow\uparrow} , \quad (5.88)$$

while the relation $G_{\uparrow\uparrow} = G_{\downarrow\downarrow}$ is satisfied only in a paramagnetic system.[18]

Undoubtedly, Eq. (5.87) is a rather bold attempt to take into account the physics of the exchange and correlation hole, i.e. the "local environment" in which a physical electron lives. Its practical usefulness depends on how credibly we can connect the unknown functions $G_{\uparrow\uparrow}$ and $G_{\uparrow\downarrow}$ to the exchange-correlation hole.

The first attempt in this direction was J. Hubbard's suggestion of a frequency-independent approximation of the form

$$G_{\uparrow\uparrow}(q, \omega) \approx G_{\uparrow\uparrow}^H(q) = \frac{q^2}{q^2 + k_F^2} , \qquad G_{\uparrow\downarrow}(q, \omega) \approx 0 , \quad (5.89)$$

---

[17] This notion of locality is not to be confused with the *local density approximation*, now widely used in density functional theory (see Chapter 7), in which the effective potential $V_{eff,\sigma}(\vec{r}, t)$ is assumed to be a *function* of the density $n(\vec{r}, t)$ at the same point in space. The effective potential discussed in this section while a "local" quantum mechanical operator, may be a highly nonlocal *functional* of the density.

[18] The exchange (x) contribution to the local field factor can be separated from the correlation (c) contribution by writing $G_{\uparrow\uparrow} = G_{\uparrow\uparrow}^x + G_{\uparrow\uparrow}^c$ and $G_{\uparrow\downarrow} = G_{\uparrow\downarrow}^c$. Notice that there is no exchange contribution to the local field factor between electrons with opposite spin orientation.

in three dimensions.[19] Hubbard's original physical explanation of the form of Eq. (5.89) was that (i) the primary source of correlation between electrons in the weak coupling limit is the Pauli principle, which affects only pairs of electrons with parallel spin orientation: hence $G_{\uparrow\downarrow} \approx 0$; (ii) because the exchange hole is short-ranged in space the exchange part of the effective potential should be short-ranged too: hence $G_{\uparrow\uparrow}$ must vanish for $q \to 0$ as q^2, to cancel the $\frac{1}{q^2}$ divergence of v_q; (iii) at very short range, the interaction between two electrons with the same spin is effectively suppressed by the zero of the antisymmetric wave function: therefore, $G_{\uparrow\uparrow}$ should tend to 1 for $q \to \infty$. Eq. (5.89) is the simplest interpolation between the limits (ii) and (iii), with a scale of variation set by k_F, the only available scale in the non-interacting electron gas.

The soundness of Hubbard's physical argument is confirmed by the following analysis. Let us return to Slater's local approximation for the exchange potential, described in Section 2.8, i.e.

$$V_\sigma^x(\vec{r}) = \int \frac{e^2}{|\vec{r} - \vec{r}'|} n_\sigma(\vec{r}')[g_0(\vec{r}, \vec{r}') - 1]d\vec{r}' , \tag{5.90}$$

where g_0 is the pair correlation function of the non-interacting electron gas. For small deviations from the homogeneous state, i.e., for densities of the form $n_\sigma(\vec{r}, t) = n/2 + n_{1\sigma}(\vec{r}, t)$ with $n_{1\sigma} \ll n$ we may expand the exchange potential to first order in $n_{1\sigma}$, obtaining (up to an irrelevant constant)

$$V_{1\sigma}^x(\vec{r}, \omega) \simeq \int \left. \frac{\delta V_\sigma^x(\vec{r})}{\delta n_\sigma(\vec{r}')} \right|_{n_\sigma(\vec{r}')=n/2} n_{1\sigma}(\vec{r}', t)d\vec{r}' . \tag{5.91}$$

It is actually possible to calculate the functional derivative exactly but we will not pursue this possibility here. After all, the Slater potential itself is the result of an approximation in which the density dependence of the pair correlation function is neglected (see discussion in Section 2.8). Continuing to ignore the density dependence of g_0 and making use of the translational invariance of the ground-state we are therefore led to the approximation

$$V_{1\sigma}^x(\vec{r}, t) \simeq \int \frac{e^2}{|\vec{r} - \vec{r}'|} [g_0(\vec{r} - \vec{r}') - 1]n_{1\sigma}(\vec{r}', t)d\vec{r}' . \tag{5.92}$$

Taking the Fourier transform with respect to $\vec{r}$ and t we obtain

$$V_{1\sigma}^x(q, \omega) \simeq \frac{1}{n} \int v_{\vec{q}-\vec{q}'}[S_0(q') - 1]\frac{d\vec{q}'}{(2\pi)^d} n_{1\sigma}(q, \omega) , \tag{5.93}$$

where $S_0(q)$ is the static structure factor of the non-interacting electron gas. Comparing this

[19] In two dimensions, the Hubbard approximation takes the form

$$G_{\uparrow\uparrow}(q, \omega) \approx G_{\uparrow\uparrow}^H(q) = \frac{q}{\sqrt{q^2 + k_F^2}} , \qquad G_{\uparrow\downarrow}(q, \omega) \approx 0 .$$

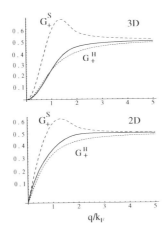

Fig. 5.13. Approximate forms for the static local field factor $G_+(q) = \frac{G_{\uparrow\uparrow}(q) + G_{\uparrow\downarrow}(q)}{2}$ in three and two dimensions: the Hubbard approximation and the "Slater" approximation are plotted with short and long dashes, respectively. The solid line is the "zero-order" STLS approximation (discussed in Section 5.4.4). Notice that, in these approximations, the local field factors do not depend on r_s when plotted as functions of q/k_F.

with Eq. (5.87), we see that the "Slater" local field factor is given by

$$G^S_{\uparrow\uparrow}(q) = -\frac{2}{nv_{\vec{q}}} \int v_{\vec{q}-\vec{q}'}[S_0(q') - 1]\frac{d\vec{q}'}{(2\pi)^d}$$

$$G^S_{\uparrow\downarrow}(q) = 0 \, . \tag{5.94}$$

How does this compare to the local field factor proposed by Hubbard? The wave vector integral in Eq. (5.94) can be done analytically, and the result is plotted in Fig. 5.13 vis-à-vis Hubbard's proposal. First of all, we notice that the two limits for $q \to 0$ and $q \to \infty$ coincide with the limits of the Hubbard local field factor.[20] The only differences occur at intermediate wave vectors, where $G^S_+(q)$ has a small hump, which is absent in $G^H(q)$.

Of course, both Eqs. (5.94) and (5.89) are rather crude approximations to the "true" local field factor. Here is a list of their main weaknesses.

- Neither (5.94) nor (5.89) take into account the correlation hole. This is deadly for $G_{\uparrow\downarrow}$ which is thereby set to zero (see footnote [18]). Among the many unphysical consequences of this approximation we mention only the most serious one, namely the spin susceptibility and the proper compressibility come out to have identical values, in conflict with theory and experiment.
- In the limit $q \to 0$ the compressibility sum rule is violated even in an "exchange-only" approximation. In three dimensions, for example, the compressibility sum rule calculated from the exchange energy would demand

$$\lim_{\vec{q}\to 0} G_{\uparrow\uparrow}(q, 0) = \frac{q^2}{2k_F^2} \, . \tag{5.95}$$

[20] The $q \to \infty$ limit can be seen without calculating the integral using the fact that $n^{-1}\int[S_0(q') - 1]\frac{d\vec{q}'}{(2\pi)^d} = 1 - g_0(0) = \frac{1}{2}$.

Thus, we see that both (5.94) and (5.89) severely overestimate (by a factor of two or so) the strength of the local field factor at small wave vectors.

- The $q \to \infty$ limit of $G_{\uparrow\uparrow}(q, 0)$, in three dimensions, is incorrect even at the exchange-only level. The correct limit (deduced from the first row of Table 5.1 in the exchange-only approximation) is $\frac{2}{3}$ in three dimensions and 1 in two dimensions.
- The frequency dependence is absent.

Setting aside, for the time being, all these difficulties, in the following sections we shall see how the local field factors enable us to express in a simple form key physical quantities such as the linear response functions and the effective electron-electron interaction. The connection with the density–density and spin–spin response functions, in particular, has allowed a virtually exact determination of most of the *static* local field factors $G_{\sigma\sigma'}(q, 0)$ via the quantum Monte Carlo method (see Section 5.6.1). No comparable success can be reported, at this time, for the problem of the frequency dependence of the local field factors. The status of this outstanding problem will be discussed in Section 5.6.2.

5.4.1 Local field factors and response functions

In order to calculate the linear response functions we assume that the system responds to the effective local potential as a non-interacting electron gas would, that is, all the effects of the coulomb interaction are by definition embedded in $V_{eff,\sigma}$. Thus, we have

$$n_{1\sigma}(q, \omega) = \chi_{0\sigma}(q, \omega) V_{eff,\sigma}(q, \omega) , \qquad (5.96)$$

where $\chi_{0\sigma}$ is the Lindhard function for spin projection σ, as defined in Eq. (4.21).

Substitution of this expression in Eq. (5.87) leads to the linearly coupled equations

$$V_{eff,\sigma} = V_{ext,\sigma} + \sum_{\sigma'} v_q (1 - G_{\sigma\sigma'}) \chi_{0\sigma'} V_{eff,\sigma'} , \qquad (5.97)$$

which can immediately be solved for $V_{eff,\uparrow}$ and $V_{eff,\downarrow}$.

Before examining the solution in general, let us focus on the case of a paramagnetic system ($n_\uparrow = n_\downarrow$), in which we evidently have $\chi_{0\uparrow} = \chi_{0\downarrow} = \frac{\chi_0}{2}$, $G_{\uparrow\uparrow} = G_{\downarrow\downarrow}$. In this case the solution of Eq. (5.97) is

$$V_{eff,\uparrow(\downarrow)} = \frac{[1 - \frac{v_q}{2}(1 - G_+ - G_-)\chi_0] V_{ext,\uparrow(\downarrow)} + \frac{v_q}{2}(1 - G_+ + G_-)\chi_0 V_{ext,\downarrow(\uparrow)}}{[1 - v_q(1 - G_+)\chi_0][1 + v_q G_- \chi_0]} , \qquad (5.98)$$

where we have introduced the following *symmetric* and *antisymmetric* combinations of the many-body local field factors:

$$G_+(q, \omega) = \frac{G_{\uparrow\uparrow} + G_{\uparrow\downarrow}}{2} , \qquad (5.99)$$

and

$$G_-(q, \omega) = \frac{G_{\uparrow\uparrow} - G_{\uparrow\downarrow}}{2} . \qquad (5.100)$$

It must be realized that even in the paramagnetic state, the existence of two distinct many-body local fields is crucial. This has often been not appreciated and in many studies the antiparallel spin fields have been neglected altogether ($G_{\uparrow\downarrow} = 0$). The neglect of the antiparallel spin contributions leads to unphysical results for, as we shall see below and in Section 5.4.2, $G_+(q, \omega)$ and $G_-(q, \omega)$ do differ in substantial ways.

We now proceed to express some of the most important response functions of the interacting electron liquid in terms of local field factors.

5.4.1.1 Density response (paramagnetic case)

In this case $V_{ext\uparrow} = V_{ext\downarrow} = V_{ext}$ so that χ_{nn} can be obtained from the relationship

$$\chi_{nn}(\vec{q}, \omega) = \frac{n_\uparrow(\vec{q}, \omega) + n_\downarrow(\vec{q}, \omega)}{V_{ext}(\vec{q}, \omega)} = \frac{\chi_0(\vec{q}, \omega)}{2} \frac{V_{eff,\uparrow}(\vec{q}, \omega) + V_{eff,\downarrow}(\vec{q}, \omega)}{V_{ext}(\vec{q}, \omega)}. \quad (5.101)$$

The sum $V_{eff,\uparrow} + V_{eff,\downarrow}$ is readily obtained from Eq. (5.98) and one immediately arrives at the important result

$$\chi_{nn}(\vec{q}, \omega) = \frac{\chi_0(\vec{q}, \omega)}{1 - v_q[1 - G_+(q, \omega)]\chi_0(\vec{q}, \omega)}. \quad (5.102)$$

Thus only the symmetric many-body local field factor $G_+(q, \omega)$ enters the density response problem *in the paramagnetic state*. This expression generalizes the original result (5.6) due to Hubbard. As anticipated in the introduction one sees immediately that the RPA amounts to neglecting this many-body correction to the density response.

The corresponding expression for the proper response function (Eq. (5.20)) is

$$\tilde{\chi}_{nn}(q, \omega) = \frac{\chi_0(q, \omega)}{1 + v_q G_+(q, \omega)\chi_0(q, \omega)}. \quad (5.103)$$

From the definition of dielectric function (Eq. (5.21)) it then follows immediately that we have

$$\epsilon(\vec{q}, \omega) = 1 - \frac{v_q \chi_0(\vec{q}, \omega)}{1 + v_q G_+(q, \omega)\chi_0(\vec{q}, \omega)}. \quad (5.104)$$

Equation (5.103) can also be cast in the form

$$v_{\vec{q}} G_+(q, \omega) = \frac{1}{\tilde{\chi}_{nn}(q, \omega)} - \frac{1}{\chi_0(q, \omega)}, \quad (5.105)$$

which can in fact be taken as a formal definition of $G_+(q, \omega)$. At a first glance, Eqs. (5.102) and (5.105) appear to add no new insight or offer obvious advantages since the unknown function $\chi_{nn}(\vec{q}, \omega)$ is merely replaced by the also unknown function $G_+(q, \omega)$. We notice however that our derivation of this expression and the very structure of Eq. (5.102) reveal the underlying physics as well the origin of the terminology used in this section. It is in fact clear that in the present context the quantity $G_+(q, \omega)$ plays a role very similar to that of the classical molecular local field, a concept introduced in the middle of the 19th century by O. F. Mossotti (1791–1863)[21] and R. J. E. Clausius (1822–1888). It is therefore only within this physical context that the otherwise formally trivial introduction of the local fields acquires real meaning.

In practice, the definition (5.105) is useful provided that $G_+(q, \omega)$ can be approximated more easily than $\tilde{\chi}_{nn}(q, \omega)$ itself. If, as one hopes, the excitation spectra of χ_0 and $\tilde{\chi}_{nn}$ are qualitatively similar, then their frequency dependences should tend to cancel out. Under this scenario a frequency-independent approximation to the dynamical local field factor is expected to work very well. We shall see later that this is precisely what happens in many cases.

The compressibility sum rule leads to an explicit connection between the long wavelength limit of $G_+(q, 0)$ and the enhancement of the proper compressibility K of the interacting electron liquid as compared to that of the non-interacting electron gas. Specifically by making use of Eqs. (3.155) and (5.102) we can write

$$\frac{K}{K_0} = \lim_{q \to 0} \frac{1}{1 - v_q G_+(q, 0)N(0)} . \tag{5.106}$$

The introduction of the many-body local field factors allows for the possibility of an instability of the electron liquid. This is suggested by the structure of the denominator of Eq. (5.102) for, if the condition

$$1 - v_q [1 - G_+(q, 0)] \chi_0(\vec{q}, 0) = 0 , \tag{5.107}$$

is satisfied for some particular value of q, then the system is expected to be unstable with respect to charge modulations of wavelength $\frac{2\pi}{q}$.[22] Since Eq. (5.107) implies

$$G_+(q, 0) = 1 - \frac{1}{v_q \chi_0(\vec{q}, 0)} , \tag{5.108}$$

and $\chi_0(q, 0)$ is negative, we see that for any number of spatial dimensions the occurrence of the instability requires $G_+(q, 0)$ to be larger than one at the wave vector of the unstable density fluctuation.

5.4.1.2 Spin response (paramagnetic case)

We arrive at a similarly structured expression for the spin–spin response function $\chi_{S_z S_z}(\vec{q}, \omega)$ by imposing an external field of the form $V_{ext,\uparrow} = B_{ext}$ and $V_{ext,\downarrow} = -B_{ext}$, so that

$$\chi_{S_z S_z}(\vec{q}, \omega) = \frac{n_\uparrow(\vec{q}, \omega) - n_\downarrow(\vec{q}, \omega)}{B_{ext}(\vec{q}, \omega)} = \frac{\chi_0(\vec{q}, \omega)}{2} \frac{V_{eff,\uparrow}(\vec{q}, \omega) - V_{eff,\downarrow}(\vec{q}, \omega)}{B_{ext}(\vec{q}, \omega)} . \tag{5.109}$$

---

[21] Alongside his pioneering research and teaching activity Ottaviano Fabrizio Mossotti, mostly known in his time for his work in astronomy, also distinguished himself by valiantly leading on May 29, 1848, a battalion drawn from the student body of the University of Pisa in the battle of Curtatone and Montanara fought between the forces of the Grand-duchy of Tuscany and the Austrian army during the first Italian independence war. The anniversary of this event is still celebrated in Pisa.
[22] Within the RPA ($G_+ = 0$) this condition can never be satisfied since $\chi_0(\vec{q}, 0)$ is negative for all q.

In this case, in view of the absence of net charge modulation, the Hartree terms cancel out when taking the difference of the effective potentials in Eq. (5.98), and we are left with:

$$\chi_{S_z S_z}(\vec{q}, \omega) = \frac{\chi_0(\vec{q}, \omega)}{1 + v_q G_-(q, \omega)\chi_0(q, \omega)} . \qquad (5.110)$$

This establishes that only the antisymmetric many-body local field factor $G_-(q, \omega)$ determines the spin response of the paramagnetic electron liquid. In particular the long wavelength limit of $G_-(q, 0)$ determines the so called *many-body enhancement* of the spin susceptibility, i.e. the ratio

$$\frac{\chi_S}{\chi_P} = -\left(\frac{g\mu_B}{2}\right)^2 \lim_{q\to 0} \frac{\chi_{S_z S_z}(\vec{q}, 0)}{\chi_P} = \lim_{q\to 0} \frac{1}{1 - v_q G_-(q, 0)N(0)} , \qquad (5.111)$$

where χ_P is the Pauli susceptibility discussed in Section 4.4.1. Notice that the above equation predicts a ferromagnetic instability at the value of r_s for which $N(0)\lim_{q\to 0} v_q G_-(q, 0) = 1$.

5.4.2 Many-body enhancement of the compressibility and the spin susceptibility

For historical, physical and practical reasons, the compressibility and the many-body spin enhancement play a particularly important role in the theory of the electron liquid. Both can be obtained from the behavior of the total energy, from linear response theory, and from a microscopic approach to Fermi liquid theory.[23] Accordingly they offer the possibility of checking the internal consistency of any approximate calculation.

As we have seen, both the compressibility K and the spin susceptibility χ_S can be either obtained as long wavelength limits of the corresponding response functions or from the density and polarization dependence of the energy per particle via the thermodynamic relationships of Section 1.8. The latter procedure is certainly more practical. For the homogeneous paramagnetic state Eqs. (1.143) and (2.58) together yield

$$\frac{K_0}{K} = 1 - \frac{r_s}{r_s^\star} - \frac{r_s}{2\epsilon_F}\left(\frac{d-1}{d}\epsilon_c'(r_s) - \frac{r_s}{d}\epsilon_c''(r_s)\right) , \qquad (5.112)$$

and

$$\frac{\chi_P}{\chi_S} = 1 - \frac{r_s}{r_s^\star} + \frac{d}{2\epsilon_F}\frac{\partial^2 \epsilon_c}{\partial p^2} , \qquad (5.113)$$

where $r_s^\star$ is defined in Eq. (2.59).

The structure of these expressions can be readily understood in terms of the stiffness concept. The right-hand sides of these formulas contain the sum of three stiffnesses: the non-interacting electron gas stiffness, the exchange stiffness and, lastly, the correlation stiffness.

As we have already noted in Chapter 2, within the standard HF theory of the homogeneous electron gas, $\frac{K}{K_0}$ and $\frac{\chi_S}{\chi_P}$ coincide. Inclusion of correlation effects leads to a striking removal

[23] This will be discussed in Chapter 8.

of this degeneracy, as can be clearly seen in Fig. 1.22. The representation of the response in terms of local field factors gives us insight into the origin of this effect.

First of all we notice that the HF result, when expressed in terms of local field factors, amounts to the relation

$$\lim_{q \to 0} G_+^{HF}(q, 0) = \lim_{q \to 0} G_-^{HF}(q, 0) = \frac{r_s}{r_s^\star} \frac{q^{d-1}}{\kappa_d^{d-1}}. \tag{5.114}$$

This is a special instance of a more general fact: within HF theory the ↑↓ local field factor vanishes (and hence $G_+ = G_-$) because there is no correlation between electrons of opposite spin orientation. Any approximate theory (e.g., the Hubbard–Slater theory) in which one explicitly (or tacitly) sets $G_{\uparrow\downarrow} = 0$ will give exactly the same enhancement factor for the proper compressibility and for the spin susceptibility. It is evident that such an approximation implies an unwarranted neglect of antiparallel spin correlations. Thus, *the difference between the compressibility enhancement and the spin susceptibility enhancement provides a quantitative measure of the relevance of correlations between up- and down-spin electrons.*

The repulsive character of the correlation hole between electrons of opposite spin orientation implies $G_{\uparrow\downarrow}(q, 0) > 0$ and hence, $G_+(q, 0) > G_-(q, 0)$. This explains why correlations increase the value of the proper compressibility enhancement, while markedly reducing the spin susceptibility enhancement. As a consequence, the ferromagnetic instability is pushed to high values of r_s, while the proper compressibility diverges at relatively small values of r_s (see Chapter 1).

5.4.3 Static response and Friedel oscillations

The proper compressibility and the spin susceptibility are experimentally accessible quantities at $q = 0$. One may ask whether the wave vector dependence of the static density–density and spin–spin response functions can also be subjected to a direct experimental test. Indeed, a measurement of, say, $\chi_{nn}(q, 0)$ vs q would amount to a determination of the static local field factor $G_+(q, 0)$. At the time of writing there are several sophisticated calculations of $\chi_{nn}(q, 0)$ and $\chi_{S_z S_z}(q, 0)$ based on the QMC method, but no direct experimental check of these results. Scanning tunneling microscopy is a most promising technique, in this sense, since, by measuring the local density of states, it allows, in principle, the reconstruction of the density profile induced by an impurity in the electron liquid.

Consider, for simplicity, a single impurity embedded in an otherwise homogeneous paramagnetic electron liquid. The impurity potential is assumed to have the form $V_{ei}(\vec{r}) = C\delta(\vec{r})$, coupling to the electronic density and/or the spin density. We will focus on the density case, as the spin case is completely analogous. Even though $V_{ei}(\vec{r})$ is extremely short-ranged, the density response is not short-ranged at all. At large distances from the impurity this response can be calculated by linear response theory and is given by

$$n_1(\vec{r}) = \frac{C}{L^d} \sum_{\vec{q}} \chi_{nn}(\vec{q}, 0)e^{i\vec{q}\cdot\vec{r}}. \tag{5.115}$$

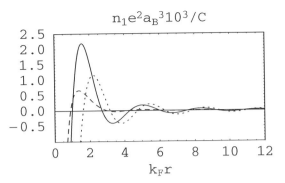

Fig. 5.14. Charge density modulation of a two dimensional electron liquid due to an impurity with a short range potential for $r_s = 4.0$. Results are shown for the noninteracting case (dotted line), RPA (dashed line) and the fully interacting case (solid line).

As briefly noted in Chapter 4, the asymptotic behavior of $\delta n(\vec{r})$ for $r \to \infty$ contains characteristic oscillations at wave vector $2k_F$, known as Friedel oscillations. The origin of these oscillations lies in the non-analyticity of $\chi_{nn}(q, 0)$ at $q = 2k_F$, which in turn, reflects the existence of a sharp Fermi surface. The effect is already evident in the static Lindhard function, which exhibits a logarithmic divergence in the derivative in three dimensions, a discontinuity in the same derivative in two dimensions, and a logarithmic divergence in one dimension. The Fourier transform of $\chi_0(q, 0)$ was shown in Chapter 4 to behave as $\frac{\sin 2k_F r}{(k_F r)^d}$ for $r \to \infty$ in d dimensions. We will now show that the amplitude of the Friedel oscillations allows, in principle, an exact determination of the local field factor $G_+(2k_F, 0)$. Our key assumption is that $G(q, 0)$ is an analytic function of q, so that the singular behavior of $\chi_{nn}(q, 0)$ has qualitatively the same form as that of $\chi_0(q, 0)$.[24] Then, one can show (Giuliani and Simion, 2005) that at large distance from the impurity, the induced density has the form

$$n_1(r) \approx \frac{C}{(\pi e r_s)^2 a_B^3 \left[1 + \frac{r_s}{\sqrt{2}}(1 - G_+(2k_F, 0))\right]^2} \frac{\sin(2k_F r + \delta)}{(k_F r)^2}, \quad (2D), \quad (5.116)$$

where δ is a phase. A similar result is obtained in three dimensions. In one dimension, on the other hand, the amplitude of the Friedel oscillations decays with a non universal exponent that depends on the interaction strength and reduces to $d = 1$ only in the non-interacting limit (see Section 9.10).[25]

Notice that the factor that multiplies the oscillating term depends strongly on the local field factor $G_+(2k_F, 0)$. Fig. 5.14 shows the difference between the oscillations calculated in the non-interacting electron gas, in the RPA, and, finally, with the inclusion of G_+. These differences could be used, in principle, to determine the magnitude of $G_+(2k_F, 0)$.

---

[24] This may be regarded as a property of the "normal Fermi liquid state" – see Chapter 8.
[25] In this case there is no sharp Fermi surface and the oscillations originate from the fact that, as we will see, in a sense the ground-state of the system contains an incipient charge density wave.

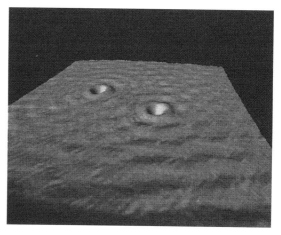

Fig. 5.15. Details of the Friedel-like oscillations of the local density of states of the two dimensional electron gas thriving on the [111] surface of a Cu sample induced by the presence of two impurities. (Courtesy of IBM Research.)

Scanning tunneling microscopy provides vivid pictures of the Friedel oscillations of the local density of states in the proximity of an impurity in the two dimensional electron gas that exists on the surface of elemental metals in suitably prepared quantum wells (see Crommie *et al.* (1993) and Fig. 5.15).

Consider now *two* impurities embedded in an electron gas. One impurity is at the origin while the other sits in the position $\vec{r}$. Each impurity interacts (via a short-range interaction) with the density induced by the other. Thus, according to our previous discussion, the interaction energy between the two impurities is seen to be directly proportional to $\chi_{nn}(r, 0)$. For two impurities that are magnetically coupled to the spin–density of the electron gas this interaction is universally known as the Ruderman–Kittel–Kasuya–Yosida (RKKY) interaction, from the name of the authors who first published the result (Rudermann and Kittel, 1954; Kasuya, 1956; Yosida, 1957).

Because of its oscillatory character this interaction may promote parallel or antiparallel alignment of two impurity spins, depending on their separation. This may lead to what is commonly referred to as *frustration*, and *spin-glass* behavior, when many randomly distributed impurities are present.

5.4.4 The STLS scheme

We have seen how the local field factors enable us to write the response functions of an electron liquid in an intuitively appealing form. But, conversely, once the linear response functions are known, they can be used to calculate the exchange-correlation hole, and hence feed back on the local field factor. This is the central idea of the ingenious approach developed by Singwi, Tosi, Land, and Sjölander (1968), henceforth referred to as "STLS" – one of the most successful schemes ever developed (before the advent of quantum Monte Carlo) for the calculation of the local field factor.

Eq. (5.94) suggests a natural way to achieve a consistency between the local field factor and the density–density response function that derives from it. We replace the non-interacting structure factor $S_0(q)$ in that equation by the interacting structure factor $S(q)$ determined from the fluctuation-dissipation theorem. Thus we have

$$G_+(q) = -\frac{1}{n v_{\vec{q}}} \int v_{\vec{q}-\vec{q}'}[S(q')-1]\frac{d\vec{q}'}{(2\pi)^d} , \qquad (5.117)$$

where

$$S(q) = -\frac{\hbar}{\pi n}\int \Im m\, \chi_{nn}(q,\omega)d\omega . \qquad (5.118)$$

By the use of the interacting $S(q)$, not only do we take into account some correlation effects that are completely ignored in the Slater approximation, but, more importantly, we obtain a closed set of nonlinear equations, (5.117) and (5.102), from which $G_+(q)$, $S(q)$, and $\chi_{nn}(q,\omega)$ can be calculated.

However, the STLS expression for the local field factor is not exactly given by Eq. (5.117). Because STLS derive their local field factor from a classical analogy (see Appendix 9) the central quantity in their approach is not the effective potential, but the effective *force*, i.e., the gradient of the effective potential. The effective force is computed as the convolution of the classical electrostatic force with the pair correlation function. Thus, the "exchange-correlation force" is given by

$$\vec{\nabla} V_{xc}^{STLS}(\vec{r},t) = \int [g(\vec{r}-\vec{r}')-1]\vec{\nabla}_{\vec{r}}\frac{e^2}{|\vec{r}-\vec{r}'|}n_1(\vec{r}',t)d\vec{r}'. \qquad (5.119)$$

Taking the Fourier transform of this expression with respect to $\vec{r}$ and t, and setting $V_{xc}^{STLS}(q,\omega) = -v_{\vec{q}}G_+^{STLS}(q)n_1(q,\omega)$ we obtain, in analogy to Eq. (5.94),

$$G_+^{STLS}(q) = -\frac{1}{n}\int \frac{\vec{q}\cdot\vec{q}'}{q^2}\frac{v_{q'}}{v_q}\left[S(\vec{q}-\vec{q}')-1\right]\frac{d\vec{q}'}{(2\pi)^d}. \qquad (5.120)$$

The zero-th order approximation to this expression (obtained by substituting $S(q) \rightarrow S_0(q)$) is plotted as a solid line in Fig. 5.13.

Eqs. (5.120), (5.118), and (5.102) constitute the complete STLS scheme for the calculation of the local field factor. Notice that there are no adjustable parameters in this theory.

How well does the STLS scheme perform in the electron liquid at metallic densities? First of all, the pair correlation function is greatly improved in comparison with the RPA: in three dimensions it remains positive up to $r_s = 4$ and it becomes only very slightly negative for larger values of r_s. Accordingly, the correlation energy is in excellent agreement with Monte Carlo outputs and exhibits the expected r_s^{-1} behavior (rather than the unphysical $r_s^{-\frac{3}{4}}$ of the RPA) at large values of r_s. Other physical properties, such as the plasmon dispersion, obtained from the poles of the density–density response function, are also in reasonably good agreement with experiment: the improved treatment of the exchange-correlation hole leads to a slightly negative plasmon dispersion, starting at $r_s \simeq 6$ (see Fig. 5.28 and discussion in Section 5.8.1). The compressibility sum rule, however, is violated: in other words, the

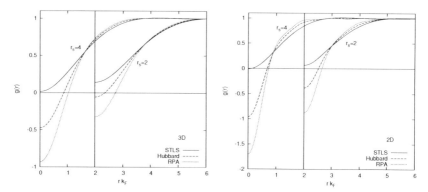

Fig. 5.16. Comparison of the pair correlation function $g(r)$ in RPA, Hubbard approximation, and STLS, in three and two dimensions.

compressibility calculated from the long wavelength limit of the proper density–density response function is found to be significantly smaller than the compressibility calculated from the second derivative of the ground-state energy with respect to the density. This means that the small-q behavior of the local field factor continues to be quite unreliable in STLS, even though this inaccuracy has little impact on the ground-state energy. A variation of the STLS scheme that does not suffer from this drawback was developed by Vashishta and Singwi (1972) and is outlined in Exercise 16.

The large-q limit of the local field factor is also incorrect, being equal to

$$\lim_{q \to \infty} G_{STLS}(q) = 1 - g_{STLS}(0) , \tag{5.121}$$

where $g_{STLS}(0)$ is the value of the STLS pair correlation function at zero separation. As we will show later in Section 5.6.1, the exact $G_+(q)$ diverges as q^{d-1} for large q.

Finally, the STLS approximation fails to account for the frequency dependence of the local field factor. This problem is, in part, related to the fact that the original derivation of the STLS scheme was based on the decoupling of the classical equations of motion for the Boltzmann distribution function (see Appendix 10). Several attempts have been made to convert the STLS scheme into a fully quantum mechanical theory starting from the equations of motion for the Wigner distribution function (Hasegawa and Shimizu, 1975; Holas and Rahman, 1987). We will not review these efforts here, but simply mention that, while the quantum STLS approach does indeed produce a frequency-dependent local field factor (the high-frequency limit of which is, in fact, the classical STLS result), it also suffers from some inconsistencies (e.g., wrong sign of the imaginary part in certain frequency intervals) which have discouraged its application so far.

5.4.5 *Multicomponent and spin-polarized systems*

The present formulation of the response in terms of effective potentials can readily be generalized to the case of a multicomponent system. Interesting examples of multicomponent electronic systems are provided by the electron–hole liquid (Vignale and Singwi, 1984, 1985) in semiconductors such as Ge and Si, and by the quasi two dimensional electron gas in Si inversion layers. In the latter the various electronic components are associated with the possible valley degeneracy of the band structure (Ando et al., 1982). The partially spin-polarized electron liquid, with $n_\uparrow \neq n_\downarrow$ is another system that can be most effectively treated with the multicomponent formalism.

To streamline the calculations we introduce the following notation: $V_{ext,i}$ is an external field acting only on component i ($i = 1, \ldots, N$), $\chi_{0i}(q, \omega)$ is the Lindhard function for component i, $v_{ij}(q)$ is the Fourier transform of the bare interaction between component i and component j, and $G_{ij}(q, \omega)$ is the total local field correction for the $i - j$ interaction, including both exchange and correlation. Notice that in this formalism the label i may denote a spin component of a given type of electron or hole carrier. The effective field acting on component i can be written as

$$V_{eff,i} = V_{ext,i} + \sum_{j=1}^{N} v_{ij}[1 - G_{ij}]n_{1j}, \tag{5.122}$$

where n_{1i} is the density deviation from equilibrium induced in the i-th component. In analogy to Eq. (5.96) we then write

$$n_{1i} = \chi_{0i} V_{eff,i}$$

$$= \chi_{0i} \left[V_{ext,i} + \sum_{j=1}^{N} v_{ij}[1 - G_{ij}]n_{1j} \right]. \tag{5.123}$$

Solving these linear equations for n_{1i} we obtain

$$n_{1i} = \sum_{j=1}^{N} \chi_{ij} V_{ext,j}, \tag{5.124}$$

where the multicomponent density–density response function χ_{ij} is related to the Lindhard functions and the local field factors in the following manner:

$$[\chi]_{ij}^{-1} = [\chi_0]_i^{-1}\delta_{ij} - v_{ij}[1 - G_{ij}]. \tag{5.125}$$

The quantity on the left-hand side of this equation is the *matrix inverse* of χ_{ij}.

It is easy to see that all the response functions of interest, for the total density, spin, etc., can be constructed in terms of χ_{ij}. As a simple application, let us calculate the density–density, longitudinal spin–longitudinal spin, and density–longitudinal spin response functions of a partially polarized homogeneous electron liquid. These are given respectively by

$$\chi_{nn} = \chi_{\uparrow\uparrow} + \chi_{\downarrow\downarrow} + 2\chi_{\uparrow\downarrow}$$

$$\chi_{S_z S_z} = \chi_{\uparrow\uparrow} + \chi_{\downarrow\downarrow} - 2\chi_{\uparrow\downarrow}$$

$$\chi_{nS_z} = \chi_{\uparrow\uparrow} - \chi_{\downarrow\downarrow}, \tag{5.126}$$

where we have used the symmetry relations derived in Chapter 3: $\chi_{\uparrow\downarrow} = \chi_{\downarrow\uparrow}$ and $\chi_{nS_z} =$

$\chi_{S_z n}$. Here the z-axis is along the direction of the spin polarization in the ground-state. The partial response functions $\chi_{\sigma\sigma'}$ are obtained from Eq. (5.125), where the subscripts 1 and 2 are replaced by $\uparrow$ and $\downarrow$ respectively, and all the interactions v_{ij} are set equal to v_q. The calculation requires only the inversion of a 2×2 matrix and the result is

$$\chi_{\uparrow\uparrow} = \frac{\chi_{0\uparrow}[1 - v_q(1 - G_{\downarrow\downarrow})\chi_{0\downarrow}]}{\Delta},$$

$$\chi_{\downarrow\downarrow} = \frac{\chi_{0\downarrow}[1 - v_q(1 - G_{\uparrow\uparrow})\chi_{0\uparrow}]}{\Delta},$$

$$\chi_{\uparrow\downarrow} = \frac{v_q(1 - G_{\uparrow\downarrow})\chi_{0\uparrow}\chi_{0\downarrow}}{\Delta}, \qquad (5.127)$$

where

$$\Delta = [1 - v_q(1 - G_{\uparrow\uparrow})\chi_{0\uparrow}][1 - v_q(1 - G_{\downarrow\downarrow})\chi_{0\downarrow}] - [v_q(1 - G_{\uparrow\downarrow})]^2 \chi_{0\uparrow}\chi_{0\downarrow} . \quad (5.128)$$

In the paramagnetic case one has $\chi_{0\uparrow} = \chi_{0\downarrow}$ and $G_{\uparrow\uparrow} = G_{\downarrow\downarrow}$ and the above formulas recover Eqs. (5.102) and (5.110). In particular, the density–spin response function vanishes.

5.4.6 *Current and transverse spin response*

We have already seen in Chapter 3 that the current response function of an isotropic electron liquid has two independent components: the longitudinal one χ_L, parallel to $\vec{q}$, and the transverse one, χ_T perpendicular to $\vec{q}$.[26]

The longitudinal component is related to the density–density response function by the exact relation $\chi_L = (\omega^2/q^2)\chi_{nn}$ and vanishes identically for $\omega = 0$. Making use of Eq. (5.102) we can then write (in the paramagnetic case, for simplicity)

$$\chi_L(\vec{q}, \omega) = \frac{\chi_L^{(0)}(\vec{q}, \omega)}{1 - v_q[1 - G_+(q, \omega)](q^2/\omega^2)\chi_L^{(0)}(\vec{q}, \omega)}, \qquad (5.129)$$

where $\chi_{L,T}^{(0)}$ are the current response functions of the non-interacting electron gas (see Chapter 4). This equation can be interpreted as follows: the electron liquid responds as a non-interacting electron gas would to an effective longitudinal vector potential that is given by

$$A_{eff,L} = A_{ext,L} + \frac{q^2}{\omega^2}[v_q(1 - G_+)]j_{1L}. \qquad (5.130)$$

The term $(q^2/\omega^2)v_q j_{1L}$ is readily seen to be the Hartree potential $v_q n_1$ expressed in terms of the current, and gauge-transformed into a longitudinal vector potential. The remaining term $-(q^2/\omega^2)v_q G_+ j_{1L}$ is the exchange-correlation contribution to the longitudinal effective potential.

---

[26] To avoid misunderstandings we underline that the current–current response function considered in this section is the full one, including the diamagnetic term.

It is often useful to introduce the *proper* current response function $\tilde{\chi}_L(\vec{q}, \omega)$, which determines the current response to a self-consistent vector potential that includes the Hartree potential. This is important, for example, in calculating the electrical conductivity according to the Kubo formula, Eq. (3.184), since what one wants, in that case, is the current that flows in response to the actual electric potential experienced by the electrons (see footnote [34]). It is evident from the above discussion that the form of the proper response function is

$$\tilde{\chi}_L(\vec{q}, \omega) = \frac{\chi_L^{(0)}(\vec{q}, \omega)}{1 + v_q G_+(q, \omega)(q^2/\omega^2)\chi_L^{(0)}(\vec{q}, \omega)} \, . \qquad (5.131)$$

The transverse current response is constructed in a similar way. One writes the effective transverse vector potential as

$$A_{eff,T} = A_{ext,T} - \frac{q^2}{\omega^2} v_q G_{T+} j_{1T}, \qquad (5.132)$$

where G_{T+} is the *transverse local field factor*, which is different from the longitudinal one $G_{L+} \equiv G_+$. In writing (5.132) we have taken into account the fact that there is no Hartree potential associated with a transverse current in an isotropic system. Assuming, as usual, that the electron liquid responds to $A_{eff,T}$ as a non-interacting gas would, the following expression for the transverse current–current response function is obtained

$$\chi_T(\vec{q}, \omega) = \tilde{\chi}_T(\vec{q}, \omega) = \frac{\chi_T^{(0)}(\vec{q}, \omega)}{1 + v_q G_{T+}(q, \omega)(q^2/\omega^2)\chi_T^{(0)}(\vec{q}, \omega)} \, . \qquad (5.133)$$

Notice that the proper response function $\tilde{\chi}_T(\vec{q}, \omega)$ coincides with $\chi_T(\vec{q}, \omega)$.

Yet another local field factor arises when we consider the response of a spin-polarized electron liquid to a magnetic field $\vec{B}_{ext}$ that is directed perpendicular to the polarization axis (z). By isotropy the induced spin polarization must lie in the $x - y$ plane. As discussed in Chapter 3, the transverse spin response decouples into two independent channels: the response function $\chi_{S_+ S_-}(q, \omega)$ describes the response of $\langle \hat{S}_+ \rangle$ to the right-handed component $\frac{B_{ext,+}}{2}$ of the external field, while $\chi_{S_- S_+}(q, \omega)$ describes the response of $\langle \hat{S}_- \rangle$ to the left-handed component $\frac{B_{ext,-}}{2}$. The two functions are connected by the reciprocity relation $\chi_{S_+ S_-}(q, -\omega) = \chi_{S_- S_+}^*(q, \omega)$. For the effective field we now write

$$B_{eff,\pm} = B_{ext,\pm} - v_q G_{S\pm} \langle \hat{S}_\pm \rangle \, , \qquad (5.134)$$

where G_{S+} and G_{S-} are the local field factors for right and left handed fields respectively. The transverse spin response function $\chi_{S_+ S_-}$ is then found to be given by

$$\chi_{S_+ S_-}(\vec{q}, \omega) = \frac{\chi_{S_+ S_-}^{(0)}(\vec{q}, \omega)}{1 + \frac{1}{2} v_q G_{S+}(q, \omega)\chi_{S_+ S_-}^{(0)}(\vec{q}, \omega)} \, , \qquad (5.135)$$

where the reciprocity relation is satisfied if and only if $G_{S+}(q, -\omega) = G_{S-}^*(q, \omega)$.

In the paramagnetic limit $\chi_{S_+ S_-}$ and $\chi_{S_- S_+}$ are both equal to $2\chi_{S_z S_z}$. Therefore, according to Eq. (5.110), both G_{S+} and G_{S-} reduce to the longitudinal spin local field factor G_-.

Thus we have shown that the key concept of local effective field allows one to express a broad variety of linear response functions in terms of non-interacting response functions and local field factors. Another important application of the local field concept is the calculation of the effective interaction between charged particles in the electron liquid: we turn to this topic next.

5.5 Effective interactions in the electron liquid

The theory of screening based on linear response and the idea of many-body local fields allow us to discuss in simple terms the problem of the "effective interaction" in an electron liquid. In the simplest case, the effective interaction between two particles can be operationally defined as follows. One begins by introducing in the system a perturbing particle whose charge distribution has Fourier transform $\rho(q, \omega)$. As discussed in Section 5.4 any other particle in the system will experience an effective potential as a consequence of this perturbation. The effective interaction is then simply defined as the ratio of this potential to $\rho(q, \omega)$. As we shall discuss below however the calculation of the effective potential is quite delicate and depends crucially on whether the particles in questions are "test charges" or physical electrons. As it turns out there are three relevant effective interactions and we will treat them separately.

5.5.1 Test charge–test charge interaction

The first and simplest example is provided by the effective interaction of two test charges embedded in the electron liquid. A "test charge" in this context is a foreign charged particle, for example an ion, that is distinguishable from the electrons and is fully disentangled from the wave function of the electron liquid. This case is pertinent to the study of the ion–ion interaction, and provides a basis for the study of lattice dynamics in metals. The effective potential felt by a spectator test charge due to the introduction of a perturbing test charge is given by

$$V_{eff} = v_q \rho + v_q n_1 , \qquad (5.136)$$

where n_1 is the electronic density modulation induced by the external potential $v_q \rho$ (we omit for simplicity the arguments q and ω). Substituting for n_1 the linear response expression $\chi_{nn} v_q \rho$ we obtain

$$V_{eff} = v_q (1 + v_q \chi_{nn}) \rho , \qquad (5.137)$$

so that the test charge–test charge effective interaction is found to be given by:

$$W_{tt}(q, \omega) = \frac{v_q}{\epsilon(q, \omega)} , \qquad (5.138)$$

where $\epsilon(q, \omega)$ is the dielectric function of the electron gas as given for instance by Eq. (5.104). We reach therefore the not so surprising conclusion that the effective interaction between two test charges embedded in an electron liquid, is obtained by dividing the bare coulomb interaction by the dielectric function.

5.5.2 Electron–test charge interaction

We consider next the effective interaction between a perturbing test charge and an electron. This case is pertinent to the study of the electron–ion interaction, and provides a basis for the study of the electron–phonon interaction in metals. The effective potential experienced by the electron is given by Eq. (5.98) with $V_{ext,\uparrow} = V_{ext,\downarrow} = v_q \rho$. We therefore readily obtain:

$$V_{eff} = \frac{v_q \rho}{1 - v_q(1 - G_+)\chi_0} .$$ (5.139)

This has the same form as Eq. (5.138) except for the fact that the standard dielectric function must be replaced by an *electron–test charge dielectric function*, $\epsilon_{et}(q, \omega)$, defined as

$$\epsilon_{et}(q, \omega) \equiv 1 - v_q[1 - G_+(q, \omega)]\chi_0(q, \omega) .$$ (5.140)

Notice that the form of ϵ_{et} is such that

$$W_{et}(q, \omega) = v_q + v_q \chi_{nn}(q, \omega)v_q[1 - G_+(q, \omega)] .$$ (5.141)

This formula has a simple physical interpretation. The electron interacts with the test particle in two different ways: (i) directly via the coulomb interaction in vacuum (first term) and (ii) indirectly via the density induced in the surrounding electron liquid (second term). The key point is that the electron interacts with the induced density with the "corrected" interaction $v_q[1 - G_+]$, where the local field factor attempts to take into account the exchange-correlation hole; whereas the test particle, carrying no exchange-correlation hole, interacts with the induced density with the bare interaction v_q (see Fig. 5.17(a)).

Another suggestive way of writing the electron–test charge effective interaction is

$$W_{et}(q, \omega) = \frac{v_q \Lambda(q, \omega)}{\epsilon(q, \omega)} ,$$ (5.142)

where

$$\Lambda(q, \omega) = \frac{1}{1 + v_q G_+(q, \omega)\chi_0(q, \omega)}$$ (5.143)

is a *vertex correction* to the plain screened interaction (see Chapter 6).

It is interesting to notice that in the small-q limit Eq. (5.106) yields $\lim_{q\to 0} \Lambda(q, 0) = K/K_0$. In this limit, therefore, the vertex correction exactly cancels against the local field correction contained in the dielectric function (recall that $\epsilon(q, 0)/\epsilon^{RPA}(q, 0) \to K/K_0$): hence in this limit the electron–test charge interaction does not depend on G_+ and can be calculated in RPA.

(a) Test charge–electron interaction

(b) Electron–electron interaction

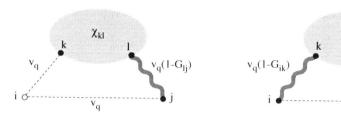

Fig. 5.17. Schematic diagrams for the effective test charge–electron and electron–electron interactions. The indices i, j, k, l run over different electron "flavors", such as, for example, up- or down-spin. In both cases the interaction consists of a direct term, which is invariably the bare coulomb interaction, and an indirect term, which is mediated by fluctuations in the densities of the various electron flavors. A test charge (open circle) perturbs the system via the bare coulomb potential, while a physical electron (solid circle) perturbs it via the local field-corrected coulomb potential. $\chi_{kl}(q, \omega)$ is the flavor-resolved density–density response function of the electron liquid.

5.5.3 *Electron-electron interaction*

We turn next to what is perhaps the most important of the effective interactions, namely the one between two physical electrons. We will see in Chapter 8 that this interaction emerges naturally as the central quantity in microscopic calculations of the electronic self-energy, quasiparticle properties, etc. Furthermore, an attractive electron-electron interaction can drive the system to a superconducting BCS state.

The determination of the effective electron-electron interaction is complicated by the fact that the putative "external potential" is not really external since, if one electron is picked as the "perturbing electron", it will still be *indistinguishable* from any other electron in the system. Accordingly, depending on the relative spin orientation, any other spectator electron will experience direct exchange and correlation with the perturbing electron. As a consequence, the effective interaction can no longer be obtained simply as the ratio between the effective potential and the perturbing density: getting the correct answer requires a sound dose of creativity and a firm mastery of quantum mechanics. In this introductory treatment we closely follow the original derivation of Kukkonen and Overhauser (1979) after whom the resulting interaction is named. Because of its central importance in the theory of the electron liquid, the effective interaction will also be derived by two different procedures in Chapters 6 and 8.

5.5.3.1 *The Kukkonen–Overhauser interaction (paramagnetic case)*

We begin by characterizing the "perturbing electron" by a charge density distribution with Fourier transform $\rho_\uparrow(q, \omega)$ where explicit reference is made to the (up) spin of the electron. The effective potential experienced by a spectator electron with spin $\uparrow$ or $\downarrow$ can still be obtained by means of Eq. (5.98), but now particular care must be exercised in the choice of the "external potential". In order to account for exchange and correlation effects we take for the "external potential" seen by an electron with spin ↑ the following expression:

$$V_{ext,\uparrow} = v_q \rho_\uparrow - v_q G_{\uparrow\uparrow} \rho_\uparrow = v_q \rho_\uparrow (1 - G_+ - G_-), \qquad (5.144)$$

where the new terms include the effect of direct exchange and correlation between the perturbing and the spectator electron. In the same way, the corresponding expression for the "external potential" seen by an electron with spin ↓ is taken to be:

$$V_{ext,\downarrow} = v_q \rho_\uparrow - v_q G_{\downarrow\uparrow} \rho_\uparrow = v_q \rho_\uparrow (1 - G_+ + G_-). \qquad (5.145)$$

By substituting these expressions in Eq. (5.98), the effective potential $V_{eff,\downarrow\uparrow}$ experienced by an electron with spin ↓ due to the presence of the ↑-spin electron is readily found to be given by:

$$V_{eff,\downarrow\uparrow} = \frac{1}{\chi_0} \left\{ \frac{1}{1 - v_q(1 - G_+)\chi_0} - \frac{1}{1 + v_q G_- \chi_0} \right\} \rho_\uparrow, \qquad (5.146)$$

while the potential felt by an ↑ spin electron under the same circumstances is

$$V_{eff,\uparrow\uparrow} = V_{eff,\downarrow\uparrow} - \frac{2v_q G_-}{1 + v_q G_- \chi_0} \rho_\uparrow. \qquad (5.147)$$

As we have already anticipated, in this case the effective interactions cannot be simply obtained by dividing the effective potentials by $\rho_\uparrow$. The origin of the difficulty lies in the fact that *the effective interaction must not include the direct exchange and correlation between the two electrons.* One can grasp this fact by reasoning that, in the solution of any genuine two-body system (i.e., two particles in a vacuum), what we have called "direct exchange and correlation" is not contained in the interaction between the particles but arises from the nature of the two-body wave function which is (i) antisymmetric under interchange of the particles and (ii) non representable as a single determinant of one-particle orbitals. By the same reasoning, we arrive at the conclusion that the effective interaction between two particles in the presence of a many-body medium should not include the direct exchange-correlation terms. Notice however, and this is crucial, that we must still include exchange and correlation between each electron and the many-body medium. Therefore, the correct prescription for deriving the effective interaction is to first subtract from the potentials (5.146) and (5.147) the direct exchange-correlation contributions $-v_q G_{\uparrow\uparrow}$ and $-v_q G_{\downarrow\uparrow}$ respectively, and then divide what is left by $\rho_\uparrow$. This leads us, after some simple algebra, to the following expressions for the ↑↑ and ↑↓ components of the electron-electron interaction (not to be confused with the corresponding scattering amplitudes – see Eqs. (5.153) and (5.154) below):

$$W_{\uparrow\uparrow} = v_q + \left[v_q(1 - G_+) \right]^2 \chi_{nn} + \left[v_q G_- \right]^2 \chi_{S_z S_z}, \qquad (5.148)$$

and

$$W_{\downarrow\uparrow} = v_q + \left[v_q(1 - G_+) \right]^2 \chi_{nn} - \left[v_q G_- \right]^2 \chi_{S_z S_z}. \qquad (5.149)$$

Of course, $W_{\uparrow\uparrow}$ and $W_{\downarrow\uparrow}$ are functions of $\vec{q}$ and ω, or, after an inverse Fourier transformation, of the separation $\vec{r}$ between the particles and the time difference t between the instant at which the disturbance is produced by one electron and the instant at which it produces its effect on the second electron.

The spin dependence of the effective interaction can be explicitly displayed by writing the Fourier transform of the interaction as a two-body operator in spin space:[27]

$$\hat{W} = W_+ + \hat{\vec{\sigma}}_1 \cdot \hat{\vec{\sigma}}_2 W_- , \qquad (5.150)$$

where $\hat{\vec{\sigma}}_1$ and $\hat{\vec{\sigma}}_2$ are spin-operators acting on the first and the second electron respectively, and

$$W_\pm = \frac{W_{\uparrow\uparrow} \pm W_{\downarrow\uparrow}}{2} . \qquad (5.151)$$

By means of the explicit formulas (5.148) and (5.149) the above Eq. (5.150) can be cast in the elegant form

$$\hat{W}_{\vec{\sigma}_1, \vec{\sigma}_2}(q, \omega) = v_q + \left\{ v_q [1 - G_+(q, \omega)] \right\}^2 \chi_{nn}(q, \omega)$$
$$+ \hat{\vec{\sigma}}_1 \cdot \hat{\vec{\sigma}}_2 \left\{ v_q G_-(q, \omega) \right\}^2 \chi_{S_z S_z}(q, \omega) . \qquad (5.152)$$

This formula is known as the *Kukkonen–Overhauser (KO) electron-electron effective interaction*.

The physical content of the KO interaction is illustrated in Fig. 5.17(b). The first term in the KO expression is just the interaction one would have if the two particles were in a vacuum. The second and the third term are interactions mediated by charge fluctuations and spin density fluctuations respectively. It should be noted that the latter include both longitudinal and transverse spin fluctuations, with respect to the (arbitrary) direction of the spin quantization axis.[28] The essential point is that the two electrons under study interact with all other electrons via the "vertex corrected" spin-dependent interaction $v_q[1 - G_{\sigma\sigma'}]$, but, with each other, only via the bare coulomb interaction.

We now calculate the *scattering amplitude* for a process in which two electrons with momenta and spins $\vec{p}_1, \sigma_1$ and $\vec{p}_2, \sigma_2$ scatter to final states $\vec{p}_1 - \vec{q}, \sigma_1$ and $\vec{p}_2 + \vec{q}, \sigma_2$. In the Born approximation, this is given by the matrix element of the Kukkonen–Overhauser interaction between the properly antisymmetrized initial and final two-electron states. For parallel spin electrons, the scattering amplitude is thus given by

$$A_{\uparrow\uparrow} = W_{\uparrow\uparrow}(q, \epsilon_{\vec{p}_1} - \epsilon_{\vec{p}_1 - \vec{q}}) - W_{\uparrow\uparrow}(\vec{p}_1 - \vec{p}_2 - \vec{q}, \epsilon_{\vec{p}_1} - \epsilon_{\vec{p}_2 - \vec{q}}), \qquad (5.153)$$

where the two terms correspond to the direct and exchange scattering processes, as illustrated in upper half of Fig. 5.18. Similarly, for electrons of opposite spins, we have

$$A_{\uparrow\downarrow} = W_{\uparrow\downarrow}(q, \epsilon_{\vec{p}_1} - \epsilon_{\vec{p}_1 - \vec{q}}) - 2W_-(\vec{p}_1 - \vec{p}_2 - \vec{q}, \epsilon_{\vec{p}_1} - \epsilon_{\vec{p}_2 - \vec{q}}). \qquad (5.154)$$

The curious exchange-like term in this expression originates from the transverse spin part of the KO interaction, which couples the "direct" component of the incoming state

---

[27] The form of this equation reflects the isotropy of the paramagnetic system. Spin polarized systems will be considered in the next section.

[28] The scalar product $\hat{\vec{\sigma}}_1 \cdot \hat{\vec{\sigma}}_2$ can be rewritten as $\hat{\sigma}_{1z}\hat{\sigma}_{2z} + \frac{1}{2}(\hat{\sigma}_{1+}\hat{\sigma}_{2-} + \hat{\sigma}_{1-}\hat{\sigma}_{2+})$, where $\hat{\sigma}_{iz}$ are the longitudinal spin operators and $\hat{\sigma}_{i\pm} = \hat{\sigma}_{ix} \pm \hat{\sigma}_{iy}$ are the transverse ones.

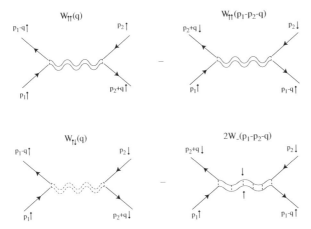

Fig. 5.18. Contributions to the scattering amplitude for parallel spin electrons (upper half) and antiparallel spin electrons (lower half). Notice, in particular, the exchange-like contribution to the scattering amplitude for antiparallel spin. This process is mediated by transverse spin fluctuations and its magnitude is controlled by $2W_- = W_{\uparrow\uparrow} - W_{\uparrow\downarrow}$.

$e^{i\vec{p}_1\cdot\vec{r}_1}|\uparrow\rangle_1 e^{i\vec{p}_2\cdot\vec{r}_2}|\downarrow\rangle_2$ with the "exchange" component of the outgoing state $-e^{i(\vec{p}_2+\vec{q})\cdot\vec{r}_1}|\downarrow\rangle_1$ $e^{i(\vec{p}_1-\vec{q})\cdot\vec{r}_2}|\uparrow\rangle_2$. The underlying physical process is illustrated in the lower half of Fig. 5.18: the incoming electron with wave vector $\vec{p}_1$ and spin $\uparrow$ flips its spin and emits a transverse spin wave of wave vector $\vec{p}_1 - \vec{p}_2 - \vec{q}$. This wave is absorbed by the second electron, which thereby flips its spin from $\downarrow$ to $\uparrow$, and emerges with momentum $\vec{p}_1 - \vec{q}$. The factor 2 arises from the fact that there are two transverse spin orientations.

The KO effective interaction provides a powerful tool for the study of many-body effects in metals. Explicit examples will be discussed in Chapter 8.

5.5.3.2 Spin polarized and multicomponent systems

The construction of the KO interaction operator $\hat{W}$ from its components $W_{\uparrow\uparrow}$ (5.148) and $W_{\downarrow\uparrow}$ (5.149) relies on the spin isotropy of the paramagnetic electron liquid. The expression for the effective interaction becomes considerably more complicated in partially spin-polarized electron liquids, or, more generally, in multicomponent systems. It turns out that it is far easier, in such cases, to obtain expressions for the scattering amplitudes rather than for the effective interaction operator. The derivation is most efficiently done with the help of the diagrammatic formalism developed in Chapter 6. Here we preview the results, which are simple and intuitively appealing. Following the notation of Section 5.4.5, we use latin indices $i, j \ldots$ to label the N different components of the system. We consider the scattering process $(\vec{p}_1\, i, \vec{p}_2\, j \to \vec{p}_1 - \vec{q}\, i, \vec{p}_2 + \vec{q}\, j)$ in which two particles of species i and j with initial momenta $\vec{p}_1$ and $\vec{p}_2$ collide and emerge with final momenta $\vec{p}_1 - \vec{q}$ and $\vec{p}_2 + \vec{q}$. For two particles of the same species ($i = j$) the probability amplitude of this process is found to be given by

$$A_{ii} = W_{ii}(q, \epsilon_{\vec{p}_1\, i} - \epsilon_{\vec{p}_1-\vec{q}\, i}) - W_{ii}(\vec{p}_1 - \vec{p}_2 - \vec{q}, \epsilon_{\vec{p}_1\, i} - \epsilon_{\vec{p}_2+\vec{q}\, i}), \qquad (5.155)$$

where

$$W_{ii}(q,\omega) = v_{ii}(q) + \sum_{k,l=1}^{N} v_{ik}(q)[1 - G_{ik}(q,\omega)]\,\chi_{kl}(q,\omega)v_{li}(q)[1 - G_{li}(q,\omega)] \ .$$

(5.156)

The first term on the right-hand side of Eq. (5.156) is the direct interaction between the particles. The second term is an interaction mediated by the following mechanism: the first electron, acting on electrons of the species k, induces a density change in the species l, which finally acts on the second electron. The first and last interaction in this chain are mediated by the "pseudopotentials" $v_{ik}(q)[1 - G_{ik}]$ and $v_{li}(q)[1 - G_{li}]$, while the intermediate link is provided by the density–density response function χ_{kl}. The two terms on the right side of Eq. (5.155) are the direct and exchange contributions to the scattering amplitude.

For the scattering amplitude between different species ($i \neq j$) one finds

$$A_{ij} = W_{ij}(q, \epsilon_{\vec{p}_1\,i} - \epsilon_{\vec{p}_1 - \vec{q}\,j}) - W_{ij,ij}(\vec{p}_1 - \vec{p}_2 - \vec{q}, \epsilon_{\vec{p}_1\,i} - \epsilon_{\vec{p}_2 + \vec{q}\,j}), \quad (5.157)$$

where

$$W_{ij}(q,\omega) = v_{ij}(q) + \sum_{k,l=1}^{N} v_{ik}(q)[1 - G_{ik}(q,\omega)]\chi_{kl}(q,\omega)v_{lj}(q)\Big[1 - G_{lj}(q,\omega)\Big],$$

(5.158)

and

$$W_{ij,ij}(q,\omega) = \Big[v_{ij}(q)G_{ij,ij}(q,\omega)\Big]^2 \chi_{ij,ij}(q,\omega)\ .$$

(5.159)

The first term in Eq. (5.158) is completely analogous to the first term of Eq. (5.156), but the second term contains two new objects, which are met here for the first time. They are

- The response function $\chi_{ij,ij}$ of the generalized density operator $\hat{\rho}_{ij}(\vec{q}) \equiv \sum_k \hat{a}^\dagger_{\vec{k}-\vec{q}\,i} \hat{a}_{\vec{k}\,j}$. This is defined as

$$\chi_{ij,ij}(q,\omega) \equiv \frac{1}{L^d}\chi_{\rho_{ij}(\vec{q})\rho_{ji}(-\vec{q})}(\omega)\ .$$

(5.160)

- The local field factor $G_{ij,ij}(q,\omega)$ associated with the response function $\chi_{ij,ij}$. This is defined by

$$v_{ij}G_{ij,ij}(q,\omega) = \frac{1}{\chi_{ij,ij}(q,\omega)} - \frac{1}{\chi^{(0)}_{ij,ij}(q,\omega)}\ .$$

(5.161)

In the special case that the two species i and j are completely equivalent in mass, density, and charge, one can show that

$$G_{ij,ij} = G_{ii} - G_{jj}\ .$$

(5.162)

For orientation, consider the paramagnetic case. Because the up- and down-spin components are completely equivalent we have $G_{\uparrow\uparrow} = G_{\downarrow\downarrow}$ and $\chi_{\uparrow\uparrow} = \chi_{\downarrow\downarrow}$. Furthermore, Eq. (5.162) tells us that $G_{\uparrow\downarrow,\uparrow\downarrow} = G_{\uparrow\uparrow} - G_{\downarrow\downarrow} = 2G_-$. Finally, $\hat{\rho}_{\uparrow\downarrow}(\vec{q}) = \frac{1}{2}\hat{S}_+(\vec{q})$ so that $\chi_{\uparrow\downarrow,\uparrow\downarrow} = \frac{1}{4}\chi_{S_+S_-} = \frac{1}{2}\chi_{S_zS_z}$. It is easy then to verify that Eqs. (5.155) and (5.157) coincide with the scattering amplitudes calculated from the Kukkonen–Overhauser interaction (Eqs. (5.153) and (5.154)).

The calculation of the scattering amplitudes can be straightforwardly extended to the partially spin polarized electron liquid, bearing in mind that (i) $\chi_{\sigma\sigma'}$ has the form worked out in Section 5.4.5, (ii) $\chi_{\uparrow\downarrow,\uparrow\downarrow} = \frac{1}{4}\chi_{S_+S_-}$, and (iii) according to Eqs. (5.161) and (5.135) $G_{\uparrow\downarrow,\uparrow\downarrow} = 2G_{S+}$, where G_{S+} is the transverse spin local field factor introduced and discussed in Section 5.4.6.

Another interesting application of Eqs. (5.155) and (5.157) is the calculation of the scattering amplitude for two electrons in an *electron–hole liquid*. The amplitude can be separated into two contributions, one that is formally mediated by electrons only, and is given by the Kukkonen–Overhauser formula, and one that is mediated by the holes, in close analogy to a phonon-mediated interaction. The derivation of this useful interaction is the object of Exercise 15.

5.5.3.3 Inclusion of lattice screening

We end this section by providing a very useful generalization of the Kukkonen–Overhauser interaction designed to also include the effects of lattice screening. More precisely we will consider the effective interaction between two electrons in the presence of a deformable positive ionic background. The additional "lattice-mediated" interaction can be quite relevant, leading, for example, to the onset of superconductivity in metals.

The complete analysis is reported in Appendix 9: we only present and discuss the main results here. The lattice of ions (nuclei plus core electrons) gives an additional contribution to the the spin-independent part of the effective interaction, Eq. (5.152). This contribution has the form

$$\delta W_+^{ph}(q,\omega) = \frac{2\omega_q}{\omega^2 - \omega_q^2} \left(\frac{q \sqrt{\frac{n_{ion}}{2M\omega_q}} V_{e-i}(q)}{\epsilon_{et}(q,\omega)} \right)^2 , \tag{5.163}$$

where ω_q is the phonon frequency (i.e., the frequency of the low-lying collective modes of the electron–ion system), M is the mass of the ions, n_{ion} the ionic density, $V_{e-i}(q)$ the bare electron–ion interaction, and $\epsilon_t(q,\omega)$ the electron-test charge dielectric function defined in Eq. (5.140).

The first factor is known in the literature as a *phonon propagator*, while the second one is the square of the appropriate *electron–phonon vertex*. The latter is given by the product of the amplitude of the ionic displacement associated with the particular phonon (the quantity inside the square root), the wave vector q (stemming from having taken in real space the divergence of the ionic displacement field), and the screened electron–ion interaction. As it should be clear by now, since the ionic potential is a truly external potential for the electrons, the screening is provided here by the electron-test charge dielectric function we have discussed in Section 5.5.1. Assuming that the electron–ion interaction is purely coulomb (i.e., setting $V_{e-i}(q) = v_q$ in Eq. (5.163)) we obtain the simpler form

$$\delta W_+^{ph} = \frac{n_{ion}q^2}{M\left(\omega^2 - \omega_q^2\right)} \left(\frac{v_q}{\epsilon_{et}} \right)^2 . \tag{5.164}$$

If, furthermore, the many-body local field factor is neglected (i.e. for $G_+ = 0$) this expression recovers what is referred to as the *Fröhlich interaction* (Bardeen, 1951 and Fröhlich, 1952).[29]

For ease of reference, the full expression of the Kukkonen–Overhauser interaction in the presence of lattice screening is provided below

$$\hat{W}^{KO} = v_q + v_q^2(1 - G_+)^2 \chi_{nn} + \frac{n_{ion}q^2}{M\left(\omega^2 - \omega_q^2\right)}\left(\frac{V_{e-i}(q)}{\epsilon_{et}(q,\omega)}\right)^2$$

$$+ \hat{\sigma}_1 \cdot \hat{\sigma}_2 \left(v_q G_-\right)^2 \chi_{S_z S_z} . \tag{5.165}$$

5.6 Exact properties of the many-body local field factors

Although a complete determination of the many-body local field factors is still beyond the reach of present-day theoretical techniques, the limiting behaviors of these functions are well established, and will be discussed in the next two sections. We will first consider the limiting cases of the wave vector dependence of $G_\pm(q, \omega)$ for different fixed frequencies; then, in Section 5.6.2 we will examine the limiting cases of the frequency dependence for different wave vectors. As shown in Appendix 11, these results turn out to be very useful in constructing practical interpolation formulas for the local field factors.

5.6.1 Wave vector dependence

5.6.1.1 Limit of $q \ll k_F$, $\omega \ll q v_F$

In this regime both q and ω are small, but the $\omega \to 0$ limit is taken before the $q \to 0$ limit: thus, we are in the long-wavelength static regime. Consider first the static spin-symmetric local field factor $G_+(q, 0)$. Its long wavelength limit is immediately established by means of the compressibility sum rule. By making use of Eqs. (5.106) we obtain

$$\lim_{q \to 0} \frac{G_+(q, 0)}{q^{d-1}} = \frac{1 - \frac{K_0}{K}}{\kappa_d^{d-1}}, \tag{5.166}$$

where κ_d is the Thomas-Fermi wave vector in $d = 2, 3$ dimensions as defined in Eqs. (5.36) and (5.38). Notice that $G_+(q, 0)$ vanishes as q^{d-1} for $q \to 0$.

In a completely analogous way the long wavelength limit of the static spin-antisymmetric field $G_-(q, 0)$ is instead related to the many-body enhancement of the static spin susceptibility (Eq. 5.111)). Making use of Eq. (5.111) we obtain

$$\lim_{q \to 0} \frac{G_-(q, 0)}{q^{d-1}} = \frac{1 - \frac{\chi_P}{\chi_S}}{\kappa_d^{d-1}}, \tag{5.167}$$

showing that the qualitative behaviors of $G_+(q, 0)$ and $G_-(q, 0)$ for $q \to 0$ are similar.

---

[29] The latter is derived in Exercise 20.

Table 5.1. *Explicit expressions for the constant term in Eq. (5.170) for* $q \to \infty$ *and finite* ω. *The results are due to: (a) Niklasson (1974); (b) Zhu and Overhauser (1984); (c) Santoro and Giuliani (1988).*

3D		2D	
$\bar{G}_+^{(\infty)}$	$\bar{G}_-^{(\infty)}$	$\bar{G}_+^{(\infty)}$	$\bar{G}_-^{(\infty)}$
$\frac{2}{3}[1 - g(0)]^{(a)}$	$\frac{1}{3}[4g(0) - 1]^{(b)}$	$1 - g(0)^{(c)}$	$g(0)^{(c)}$

5.6.1.2 Limit of $q \gg k_F$, finite ω

The limit of $q \to \infty$ and finite frequency can be obtained from the following chain of arguments. We begin by introducing the following modified many-body local fields $\bar{G}_\pm(\vec{q}, \omega)$ through the equations

$$\chi_{nn}(\vec{q}, \omega) = \frac{\bar{\chi}_0(\vec{q}, \omega)}{1 - v_q[1 - \bar{G}_+(q, \omega)]\bar{\chi}_0(\vec{q}, \omega)} , \qquad (5.168)$$

and

$$\chi_{S_z S_z}(\vec{q}, \omega) = \frac{\bar{\chi}_0(\vec{q}, \omega)}{1 + v_q \bar{G}_-(q, \omega)\bar{\chi}_0(\vec{q}, \omega)} , \qquad (5.169)$$

where $\bar{\chi}_0(\vec{q}, \omega)$ is obtained from the general expression (4.10) using the exact occupation numbers $n_{\vec{k}}$ instead of the non-interacting ones. By making use of these definitions a cumbersome but straightforward analysis of the equation of motion of the two-particle distribution function allows one to establish that the quantities $\bar{G}_\pm(\vec{q}, \omega)$ tend to a finite limit for $q \to \infty$ (Niklasson, 1974):

$$\lim_{q \to \infty} \bar{G}_\pm(q, \omega) = \bar{G}_\pm^{(\infty)}, \qquad \omega \ll \frac{\hbar q^2}{2m} . \qquad (5.170)$$

The explicit results for two and three dimensions are presented in Table 5.1.[30]

At this point in order to make the connection between $\bar{G}_\pm(q, \omega)$ and the original objects of our analysis, i.e. $G_\pm(q, \omega)$, we introduce a local field factor $G_n(q, \omega)$ through the expression

$$\bar{\chi}_0(q, \omega) = \frac{\chi_0(\vec{q}, \omega)}{1 + v_q G_n(\vec{q}, \omega)\chi_0(\vec{q}, \omega)} , \qquad (5.171)$$

which relates the modified and the usual Lindhard functions, i.e. contains the effect of the exact occupation numbers on the Lindhard function (Richardson and Ashcroft, 1994).

---

[30] Equation (5.170) must be modified in the case that ω also tends to infinity, in such a way that the ratio $x = \frac{\omega}{\hbar q^2/2m}$ remains constant. In three dimensions one finds (Niklasson, 1974)

$$\lim_{q \to \infty} \bar{G}_+ \left(q, x\frac{\hbar q^2}{2m} \right) = \left\{ 1 - \frac{1}{6}\left[\left(\frac{x+1}{x-1}\right)^2 + \left(\frac{x-1}{x+1}\right)^2 \right] \right\} [1 - g(0)], \quad (x \neq 1).$$

No change is needed in two dimensions. In Exercise 18 the reader is invited to show that the above formula is consistent with the exact large-q behavior of the static structure factor.

Substituting this equation in (5.168) and (5.169) one obtains the relation

$$G_\pm(\vec{q}, \omega) = \bar{G}_\pm(\vec{q}, \omega) + G_n(\vec{q}, \omega) . \tag{5.172}$$

The next task is then to find the relevant limiting behavior of $G_n(\vec{q}, \omega)$. This is easily accomplished. We first notice that $\chi_0(\vec{q}, \omega)$ has the large-q expansion

$$\chi_0(\vec{q}, 0) \simeq -\frac{4nm}{\hbar^2 q^2}\left(1 + \frac{4\langle(\vec{k}\cdot\hat{q})^2\rangle_0}{q^2} + \frac{16\langle(\vec{k}\cdot\hat{q})^4\rangle_0}{q^4} + \mathcal{O}(q^{-6})\right), \tag{5.173}$$

where we have used the notation $\langle\ldots\rangle_0 = \frac{1}{N}\sum_{\vec{k},\sigma}(\ldots)n_{\vec{k}}$ to indicate an average over the non-interacting occupation numbers. In particular we have

$$\langle(\vec{k}\cdot\hat{q})^2\rangle_0 = \frac{2m\epsilon_0}{d\hbar^2} , \tag{5.174}$$

where ϵ_0 is the energy per particle in the non-interacting system. On the other hand, by definition, $\bar{\chi}_0(q, \omega)$ admits an expansion formally identical to that of Eq. (5.173) with the only difference that in this case the averages must be evaluated with the *exact* occupation numbers of the interacting system. This implies that the leading correction to $\bar{\chi}_0$ is proportional to

$$\langle(\vec{k}\cdot\hat{q})^2\rangle = \frac{2mt}{d\hbar^2} , \tag{5.175}$$

where t is the *exact kinetic energy per particle* of the interacting system. Then by making use of Eq. (5.173) in (5.171) and comparing term by term the resulting expansion with that of $\bar{\chi}_0(\vec{q}, \omega)$, one arrives at the conclusion that in the large-q limit $G_n(q, \omega)$ must have an expansion of the type

$$G_n(q, \omega) \approx \beta q^{d-1} + \alpha , \quad q \gg k_F, \ \omega \ll \frac{\hbar q^2}{2m} . \tag{5.176}$$

The constants α and β depend on dimensionality in an interesting way. In three dimensions one finds (Holas, 1992):

$$G_n(q, \omega) \approx \frac{\pi}{3\alpha_3 r_s}(9\delta_4 - \delta_2) + \frac{\pi\alpha_3 r_s}{4}\frac{t - \epsilon_0}{Ry}\frac{q^2}{k_F^2}, \quad 3D, \tag{5.177}$$

where $\alpha_3 = \left(\frac{4}{9\pi}\right)^{1/3}$, and

$$\delta_2 = \frac{\langle k^2\rangle^2 - \langle k^2\rangle_0^2}{k_F^4},$$

$$\delta_4 = \frac{\langle(\vec{k}\cdot\hat{q})^4\rangle - \langle(\vec{k}\cdot\hat{q})^4\rangle_0}{k_F^4} . \tag{5.178}$$

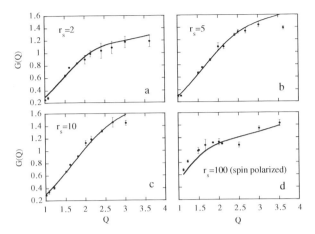

Fig. 5.19. The density local field factor $G_+(q, 0)$ in three dimensions calculated by the quantum Monte Carlo technique for different values of r_s (dots with error bars). The solid line is the fit proposed by Corradini *et al.* (1998) (see Eq. (A11.5)).

In two dimensions, on the other hand, we get the simpler expression

$$G_n(q, \omega) \approx \frac{r_s}{2\sqrt{2}} \frac{t - \epsilon_0}{\text{Ry}} \frac{q}{k_F}, \qquad 2D. \qquad (5.179)$$

Clearly in two dimensions the knowledge of t and that of the asymptotic values of Table 5.1 are sufficient to determine the large-q limit of the local field factors.

Finally the sought expansion for $G_\pm(q, \omega)$ is obtained by means of Eqs. (5.172), (5.177), (5.179), and the formulas of Table 5.1. Thus, the static many-body local fields tend to infinity at large wave vectors. This divergence, however, is rather weak since it is determined by the change in the kinetic energy that is brought about the coulomb interaction.[31]

These results have been explicitly verified in two and three dimensions by the quantum Monte Carlo technique (Moroni, Ceperley, and Senatore, 1992 and 1995). The behavior of the static many-body local field $G_+(q, 0)$ for the three and two dimensional electron gas obtained via this technique is presented in Figs. 5.19 and 5.20.

It is important to remark that the quantum Monte Carlo results for the static many-body local field $G_\pm(q, 0)$ do much more than simply confirm the asymptotic values. A careful inspection of Figs. 5.19 and 5.20 actually indicates that for most of the relevant range, the wave vector dependence of $G_\pm(q, 0)$ is determined by the small q limit. This provides the basis for the establishment of practical interpolation formulas which are presented in Appendix 11.

---

[31] In the HF approximation the kinetic energy of the homogeneous electron liquid coincides with that of the noninteracting gas. Therefore, in this approximation the divergence disappears and $G_\pm(q, \omega)$ tends to a finite value. Moreover, since in Hartree–Fock we also have $\delta_2 = \delta_4 = 0$, this value coincides with $\bar{G}_\pm^{(\infty)}$ (see Table 5.1) with the proviso that we take $g(0) = \frac{1}{2}$. We therefore arrive at the values of $\frac{1}{3}$ in three dimensions and $\frac{1}{2}$ in two.

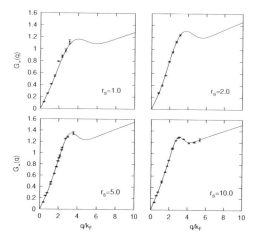

Fig. 5.20. Same as Fig. 5.19 for the two-dimensional electron gas. The solid line is the fit proposed by Davoudi *et al.* (from Davoudi *et al.* (2001)).

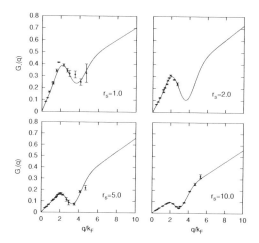

Fig. 5.21. The spin local field factor $G_-(q, 0)$ in two dimensions calculated by the quantum Monte Carlo technique for different values of r_s (dots with error bars). The solid line is the fit proposed by Davoudi *et al.* (2001).

The behavior of the static spin local field factor $G_-(q, 0)$ calculated by the QMC in the two-dimensional case is shown in Fig. 5.21. The corresponding results in three dimensions are not available to date.

5.6.1.3 The $2k_F$ "hump" puzzle

Although the behavior of the static local field $G_+(q, 0)$ is well known for small and large q, an important unsolved question is posed by the behavior of this function for $q \simeq 2k_F$.

Specifically it would be very important to determine if, as suggested by a number of approximate calculations, $G_+(q, 0)$ does have a peak in this region.[32] If indeed a peak exists at $q = q_{peak}$ one should next compare the value of $G_+(q_{peak}, 0)$ with the critical value determined for $q = q_{peak}$ by the instability condition (5.108).

At his time there is no consensus among theorists on this question. That $G_+(q, 0)$ could display an interesting structure in the vicinity of $2k_F$ is suggested by the fact that it is in principle possible, although by no means necessary, that a vestige of the CDW/SDW instability discussed in Chapter 2 could still be present in the static response of the system. A $2k_F$-peak is in fact obtained within theories based on the time-dependent HF approximation (Brosens *et al.*, 1980). It is by no means clear however that these results are valid in view of the neglect of important correlation effects. Other theories occasionally display just a small hump, or none at all, or a rather suspect peak that is unphysically located at wave vectors much larger than $2k_F$. The basically exact Monte Carlo calculations have led to smooth curves with no visible hump (see Fig. 5.20).[33] This however does not resolve the quandary for, due to the intrinsic uncertainty in the determination of the wave vector in this finite-size method, the calculations that have been performed to date do not exclude the possibility of a narrow peak occurring at about $2k_F$. Interestingly, a $2k_F$-hump does appear in the *spin* local field factor in two dimensions (see Fig. 5.21). The puzzle here is that, from general arguments, one would expect a $2k_F$-peak to be stronger in the density channel than in the spin channel (this is what led Overhauser to predict the CDW as more likely than the SDW in the electron gas at metallic density). This feature of the QMC results is intriguing and deserves further investigation.

5.6.1.4 High-frequency limit

The high-frequency limit of the local field factor is easily obtained, for any value of q, from the sum rules derived in Section 3.4.5. For example, starting from Eq. (5.102), and expanding the Lindhard function according to the general high-frequency expansion (3.140) we obtain

$$\chi_{nn}(q, \omega) \rightarrow \frac{nq^2}{m\omega^2} + \left(\frac{nq^2}{m\omega^2}\right)^2 v_q [1 - G_+(q, \infty)] + \frac{[M_{nn}^{(3)}]_0(q)}{\omega^4}, \qquad (5.180)$$

where the third moment of the non-interacting density fluctuation spectrum, $[M_{nn}^{(3)}]_0$, is given by Eq. (3.142) with $v_q = 0$ and all other quantities calculated in the non-interacting system. Comparing the above expansion with Eq. (3.140) we see that

$$\lim_{\omega \to \infty} v_q [1 - G_+(q, \omega)] = \left(\frac{m}{nq^2}\right)^2 \{M_{nn}^{(3)}(q) - [M_{nn}^{(3)}]_0(q)\}, \qquad (5.181)$$

where $M_{nn}^{(3)}(q)$ is the third moment of the interacting density fluctuation spectrum. The

---

[32] We are here assuming that the density is not so low as to be near to the Wigner crystal transition where a complicated structure for $G_+(q, 0)$ is obviously expected.

[33] Actually a small hump develops about $q = 4k_F$ in the two dimensional case at very low density ($r_s > 6$). The reason for this behavior is not understood to date.

generalization of this result to the spin-dependent local field factors $G_{\sigma\sigma'}$ reads

$$\lim_{\omega\to\infty} v_q \left[1 - G_{\sigma\sigma'}(q,\omega)\right] = \left(\frac{m}{nq^2}\right)^2 \frac{n}{n_\sigma} \left\{M_{\sigma\sigma'}^{(3)}(q) - \left[M_{\sigma\sigma'}^{(3)}\right]_0 (q)\right\} \frac{n}{n_{\sigma'}}, \quad (5.182)$$

where $M_{\sigma\sigma'}^{(3)}(q)$ and its non-interacting counterpart $\left[M_{\sigma\sigma'}^{(3)}\right]_0 (q)$ are both obtained from Eq. (3.206). The above results indicate that the high frequency limit of the local field factor is a purely real quantity, which can be expressed in terms of the static structure factor.

5.6.2 Frequency dependence

In spite of considerable effort over the past thirty years, not much is known about the frequency dependence of the local field factors. The reason for this state of affairs is that none of the known many-body techniques is sufficiently powerful to tackle the problem properly, i.e., in a controlled and affordable way. Moreover the dynamics of the local fields cannot be studied with present-day Monte Carlo techniques. As a consequence only a hint of the actual physical behavior has been to date garnered by theoretical efforts. There is however a small number of exact results, which are extremely useful both in checking approximate calculations and in suggesting the form of empirical interpolation functions for the frequency dependence. We now present and discuss these results.

5.6.2.1 Dispersion relation

Because the linear response functions are not only analytic but also free of zeroes in the in the upper half of the complex plane, it turns out the the local field factors, which are defined in terms of inverses of linear response functions (see Eq. (5.105)), are analytic functions of ω in the upper half of the complex plane. It follows then from the general theory of Chapter 3 that they satisfy Kramers–Krönig dispersion relations, i.e., for example

$$\Re eG_+(q,\omega) = \Re eG_+(q,\infty) + \mathcal{P} \int_{-\infty}^{\infty} \frac{d\omega'}{\pi} \frac{\Im mG_+(q,\omega')}{\omega' - \omega}. \quad (5.183)$$

Analogous relations hold for G_-, $G_{\sigma\sigma'}$, G_T, etc. Thus, the full local field factor can always be computed from its imaginary part together with the high-frequency limit of its real part.

5.6.2.2 Limit of $q \to 0$ and finite frequency

The behavior of the local field factor for $q \to 0$ and finite frequency is described by the relation

$$-\lim_{q\to 0} v_q G_+(q,\omega) \equiv f_{xcL}(\omega), \quad (5.184)$$

where $f_{xcL}(\omega)$ is known as the *longitudinal exchange-correlation kernel*.[34] Together with its transversal counterpart $f_{xcT}(\omega)$, defined in Eq. (5.188) below, this function will play a central role in the development of time-dependent density functional theory in Chapter 7.

---

[34] The reason why this is "longitudinal" is that it relates to the *density* local field factor, where the density is determined by the longitudinal component of the current density.

The existence of the limit (5.184) is, in essence, a consequence of the translational invariance of the system. To understand why this is so, recall the relation (3.169) for the longitudinal current response function

$$\chi_L(q, \omega) = \frac{n}{m} + \chi_{j_{pL} j_{pL}}(q, \omega) , \qquad (5.185)$$

where the second term on the right-hand side is the response function of the paramagnetic current density. In a translationally invariant system one would expect $\chi_{j_{pL} j_{pL}}(q, \omega)$ to vanish for $q \to 0$ because in this limit $\hat{j}_p(q)$ reduces to $\hat{P}/m$, where $\hat{P}$, the total canonical momentum, is a constant of motion, and therefore has a null response function (see footnote [12] in Chapter 3).

Actually, this is not quite right, because the stability of the coulomb system requires the presence of a neutralizing background of positive charge, which inevitably breaks translational invariance at the boundary of the system. As a result $\hat{P}$ is not a constant, but obeys the equation of motion $\ddot{\hat{P}} = -\omega_p^2 \hat{P}$ where ω_p is the $q = 0$ plasmon frequency. Hence $\chi_{j_{pL} j_{pL}}(q, \omega)$ does not vanish for $q \to 0$, but tends to $\frac{n}{m} \frac{\omega_p^2}{\omega^2 - \omega_p^2}$.[35]

We can make more direct use of translational invariance by working with the *proper* current–current response function, which was defined in Section 5.4.6 (see Eq. (5.131)) as the response function of the current to a *screened* vector potential. Because the electric field due to charges at the boundary of the system is now treated as a part of the external potential, it is no longer necessary to worry about the loss of translational invariance at the system boundary. We can therefore correctly conclude that

$$\tilde{\chi}_L(q, \omega) \overset{q \to 0}{\to} \frac{n}{m} + \mathcal{O}(q^2) , \qquad (5.186)$$

where the $\mathcal{O}(q^2)$ term arises from the fact that $\hat{j}_p(q)$ differs from $\hat{P}$ by a quantity that is linear in q.

The small q expansion of $\chi_L(q, \omega)$ can be straightforwardly worked out from the definition (5.131), with the following result:

$$\tilde{\chi}_L(q, \omega) \overset{q \to 0}{\to} \frac{n}{m} - \left(\frac{n}{m}\right)^2 \frac{q^2}{\omega^2} \left[v_q G_+(q, \omega) - \frac{6\epsilon_F}{(d + 2)n} \right] + \mathcal{O}\left(\frac{q^4}{\omega^4}\right) . \qquad (5.187)$$

This expression is consistent with Eq. (5.186) if and only if $\lim_{q \to 0} v_q G_+(q, \omega)$ is a finite quantity, which is what we set out to prove.

It is evident that the above analysis also applies to the transverse current response function $\tilde{\chi}_T(q, \omega)$, which must therefore tend to $\frac{n}{m} + \mathcal{O}(q^2)$ for $q \to 0$. Accordingly, we conclude that

$$- \lim_{q \to 0} v_q G_{T+}(q, \omega) \equiv f_{xcT}(\omega) , \qquad (5.188)$$

where $f_{xcT}(\omega)$ is a finite function of frequency.

---

[35] This problem appears only in 3D, since, in 2D, $\omega_p = 0$.

We emphasize that the above results hold only for the spin-symmetric local field factors, $G_{L(T)+}(q, \omega)$. The spin-antisymmetric local field factor $G_-(q, \omega)$ is qualitatively different because $\lim_{q \to 0} \left[\hat{j}_{p\uparrow}(q) - \hat{j}_{p\downarrow}(q) \right] = \frac{\vec{P}_\uparrow - \vec{P}_\downarrow}{m}$ is not a constant of the motion, and therefore its response function does not vanish.

Physically, the interaction between up- and down-spin electrons causes a "spin drag" effect (D'Amico and Vignale, 2000), whereby the relative momentum $\vec{P}_\uparrow - \vec{P}_\downarrow$, if initially different from zero, decays to zero with a characteristic relaxation time τ_{sd} ("sd" for "spin drag"). The small-q behavior of $v_q G_-(q, \omega)$ can be deduced from the equation of motion for the *average spin current* $\vec{j}_s \equiv \vec{j}_\uparrow - \vec{j}_\downarrow$ that follows from the above physical discussion. For small q and ω, such that $q \ll \frac{\omega}{v_F} \ll \frac{\epsilon_F}{\hbar}$ we expect:

$$\frac{\partial \vec{j}_s}{\partial t} = -\frac{\vec{j}_s}{\tau_{sd}} - \frac{ne}{m} \vec{E}_s, \tag{5.189}$$

where $\vec{E}_s = \vec{E}_\uparrow - \vec{E}_\downarrow$ is the difference of two electric fields that act only on up- and down-spins respectively. Since $\vec{E}_\sigma = -\frac{1}{c} \frac{\partial \vec{A}_\sigma}{\partial t}$, this immediately yields the small-q limits of the spin current–spin current and spin–spin response functions as follows:

$$\lim_{q \to 0} \chi_L^{J_s}(q, \omega) = \lim_{q \to 0} \frac{\omega^2}{q^2} \chi_{S_z S_z}(q, \omega) = \frac{n}{m} \frac{1}{1 + \frac{i}{\omega \tau_{sd}}}. \tag{5.190}$$

Comparison with the small q expansion of Eq. (5.110) reveals that for small q we must have

$$v_q G_-(q, \omega) \overset{q \to 0}{\approx} \frac{im\omega}{nq^2 \tau_{sd}}. \tag{5.191}$$

Thus, we see that $v_q G_-(q, \omega)$ *diverges* for small q and finite ω, since the spin–drag relaxation time is finite.[36] This is completely different from the behavior we found in Section 5.6.1 for small q and $\omega \ll qv_F$, and shows once again how cautious one must be in dealing with the order of the small q and ω limits.[37]

In summary, the small q-finite ω limit of the spin-resolved local field factors $G_{L(T),\sigma\sigma'}(q, \omega)$ has the form

$$-v_q G_{L(T),\sigma\sigma'}(q, \omega) \overset{q \to 0}{\approx} \frac{A(\omega)}{q^2} \frac{\sigma\sigma' n^2}{4n_\sigma n_{\sigma'}} + B_{L(T),\sigma\sigma'}(\omega) + O(q^2), \tag{5.192}$$

where $A(\omega)$ and $B_{\sigma\sigma'}(\omega)$ are finite complex functions of frequency, n_σ is the density of σ-spin electrons, and $\sigma, \sigma' = \pm 1$. Notice that there is a single function, $A(\omega)$, for both the longitudinal and the transverse channel.

---

[36] In a normal Fermi liquid at finite temperature $1/\tau_{sd}$ scales as T^2 in three dimensions and $T^2 \ln T$ in two dimensions, while at $T = 0$ it goes as ω^2 and $\omega^2 \ln \omega$ respectively. In a more rigorous treatment the mass m in Eq. (5.189) is replaced by the "spin mass", $m_S > m$, defined and discussed at the end of Section 8.3.6. Then the right-hand side of Eq. (5.191) acquires an additional real part $\frac{(m_S - m)\omega^2}{nq^2}$.

[37] The $\frac{1}{q^2}$ divergence of $v_q G_-(q, \omega)$ is also visible in the high-frequency limit of Eq. (5.182), since the spin-dependent third moments $M_{\sigma\sigma'}^{(3)}(q)$ are proportional to q^2 for $q \to 0$ (Goodman and Sjölander, 1973).

5.6.2.3 Limit of $q \to 0$ and low frequency

The general discussion of the preceding section can be specialized to the limiting cases of low and high frequency, where some exact results can be established. In the low frequency limit $qv_F < \omega \ll \min(\epsilon_F/\hbar, \bar{\omega}_p)$, where $\bar{\omega}_p$ is the characteristic plasmon frequency ($\sqrt{2\pi n e^2 k_F/m}$ in 2D) one can write

$$f_{xcL(T)}(\omega) \equiv - \lim_{q \to 0} v_q G_{+L(T)}(q, \omega) \overset{\omega \to 0}{\approx} f_{xcL(T)}(0) - i\omega f'_{xcL(T)}(0) , \qquad (5.193)$$

where $f_{xcL(T)}(0)$ and $f'_{xcL(T)}(0)$ – the latter being the derivative of $f_{xcL(T)}(\omega)$ with respect to ω at $\omega = 0$ – are real quantities.[38] The physical significance of these quantities is best appreciated within the context of the generalized elasticity theory for the electron liquid, which will be presented in Section 5.9. There we show that $f_{xcL}(0)$ and $f_{xcT}(0)$ determine the exchange-correlation contributions to the *bulk modulus* and the *shear modulus* of the electron liquid.[39] Getting a bit ahead of ourselves we anticipate here for ease of reference the basic results

$$f_{xcL}(0) - \left(2 - \frac{2}{d}\right) f_{xcT}(0) = \frac{d^2[n\epsilon_{xc}(n)]}{dn^2}$$

$$= \frac{\epsilon_F}{n} \frac{1 + F_0^s}{1 + F_1^s} \times \begin{cases} \frac{2}{3} & 3D, \\ \frac{1}{2} & 2D , \end{cases} \qquad (5.194)$$

and

$$f_{xcT}(0) = \frac{\epsilon_F}{n} \left(1 - \frac{1}{d}\right) \frac{F_2^s - F_1^s}{1 + F_1^s} . \qquad (5.195)$$

In these formulas F_1^s and F_2^s are *Landau parameters*, which will be defined and discussed at length in Chapter 8. The values of these parameters can be obtained, in principle, from microscopic calculations. The first equality in Eq. (5.194) will be proved in Section 5.9. There, we will also show that the initial slopes of $\Im m f_{xcL}(\omega)$ and $\Im m f_{xcT}(\omega)$, denoted by $f'_{xcL}(0)$ and $f'_{xcT}(0)$ in Eq. (5.193), also have a simple physical interpretation as bulk and shear *viscosity coefficients* of the electron liquid.

A microscopic analysis, presented in the next section, reveals that

$$f'_{xcT}(0) = \frac{d}{2(d-1)} f'_{xcL}(0). \qquad (5.196)$$

Furthermore, a perturbative calculation to leading order in the strength of the coulomb interaction yields (Qian and Vignale, 2002)

$$f'_{xcL}(0) = \begin{cases} \dfrac{\hbar}{90n^2} \left(\dfrac{k_F}{\pi a_B}\right)^{\frac{3}{2}} & 3D \\[4mm] \dfrac{\hbar}{6\pi^2 n^2} \dfrac{\ln(k_F a_B)}{a_B^2} & 2D . \end{cases} \qquad (5.197)$$

---

[38] This follows from the fact that the real and imaginary parts of $f_{xcL}(\omega)$ are, respectively, even and odd functions of frequency.

[39] For a discussion of the somewhat surprising appearance of a nonvanishing shear modulus at zero frequency see Section 5.9.

All of the above results apply to the spin-symmetric combination of the exchange-correlation kernel in the paramagnetic state. Similar results can be established for the function $A(\omega)$ that controls the singular behavior of the spin-resolved exchange-correlation kernel according to Eq. (5.192). For example, one has

$$\lim_{\omega \to 0} \frac{\Re e \, A(\omega)}{\omega^2} = \frac{m}{n} \frac{F_1^a - F_1^s}{1 + F_1^a} \qquad (5.198)$$

in any number of dimensions and, to leading order in the coulomb interaction,

$$\lim_{\omega \to 0} \frac{\Im m \, A(\omega)}{\omega^3} \approx -\frac{m^4 e^4}{36\pi^2 \hbar^5 n^2 \kappa_3}, \quad 3D . \qquad (5.199)$$

Notice that the real part of $A(\omega)$ at zero temperature vanishes as ω^2, while the imaginary part of $A(\omega)$ varies as ω^3 in 3D ($\omega^3 |\ln \omega|$ in 2D).

This concludes our brief summary of exact low-frequency properties of $f_{xcL,T}(\omega)$. For a more complete discussion we refer the reader to the recent literature (Qian and Vignale, 2002, 2003).

5.6.2.4 *Limit of $q \to 0$, $\omega \to \infty$*

We have already seen that the high-frequency limit of the local field factors can be expressed in terms of the third moment of the density fluctuation spectrum (see Eqs. (5.181) and (5.182)). In the $q \to 0$ limit the third moment can in turn be expressed in terms of the exchange-correlation energy, so that the high-frequency limit of $f_{xcL(T)}(\omega)$ can be written in a simple and useful form:

$$f_{xcL}(\infty) = \frac{1}{2n}\left[-(1+3\beta_d)n^{1+2/d}\frac{d}{dn}\left(\frac{\epsilon_{xc}}{n^{2/d}}\right) + 12n^{1+1/d}\frac{d}{dn}\left(\frac{\epsilon_{xc}}{n^{1/d}}\right)\right],$$

$$f_{xcT}(\infty) = \frac{1}{2n}\left[-(\beta_d-1)n^{1+2/d}\frac{d}{dn}\left(\frac{\epsilon_{xc}}{n^{2/d}}\right) + 4n^{1+1/d}\frac{d}{dn}\left(\frac{\epsilon_{xc}}{n^{1/d}}\right)\right], \qquad (5.200)$$

where $\beta_3 = 1/5$ and $\beta_2 = 1/2$.

The imaginary part of $f_{xcL,T}(\omega)$ vanishes at high frequency with a power-law behavior, which can be calculated by perturbation theory. This calculation was first done by Glick and Long (1971) for the longitudinal local field factor G_+ in three dimensions, and subsequently extended by other authors to different local field factors in three and two dimensions. The result for $\Im m f_{xcL(T)}(\omega)$ is

$$\Im m f_{xcL(T)}(\omega) \overset{\omega \to \infty}{\to} -c_{d(L)T}\frac{\pi^{4-d}2^{1+d/2}}{(\hbar\omega/\mathrm{Ry})^{d/2}} \, \mathrm{Ry} \, a_B^d, \qquad (5.201)$$

where $c_{3L} = 23/15$, $c_{3T} = 16/15$ in three dimensions, and $c_{2L} = 11/32$, $c_{2T} = 9/32$ in two dimensions. The derivation of this formula will be presented in the next section.

Recently, the analysis has been extended to include the spin-resolved exchange-correlation kernels $f_{xc,\sigma\sigma'}(q,\omega) = -v_q G_{\sigma\sigma'}(q,\omega)$. As discussed at the end of Section 5.6.2.2 these kernels diverge for $q \to 0$, and the strength of their divergence is characterized by the function $A(\omega)$, defined by Eq. (5.192). The limit of $\Re e\, A(\omega)$ for $\omega \to \infty$ can easily be inferred from Eqs. (5.182) and (3.207). The result is

$$\lim_{\omega \to \infty} \Re e\, A(\omega) = -\frac{4}{nL^d} \sum_{\vec{k} \neq 0} \frac{v_k k^2}{d} S_{\uparrow\downarrow}(k),\tag{5.202}$$

where the right-hand side equals $-\frac{4\pi e^2}{3}[g_{\uparrow\downarrow}(0) - 1]$ in 3D and diverges logarithmically in 2D (see discussion after Eq. (3.207)).[40] As for the imaginary part, it can be shown that

$$\Im m\, A(\omega) \overset{\omega \to \infty}{\longrightarrow} -c_d \frac{4n_\uparrow n_\downarrow}{n^2} \frac{1}{(\hbar\omega/\text{Ry})^{(d-2)/2}} \, \text{Ry} \cdot a_B^{d-2},\tag{5.203}$$

where $c_3 = 8\pi/3$ and $c_2 = \pi^2/2$.

5.6.2.5 Calculation of the high frequency limit of $\Im m f_{xcL(T)}(\omega)$

The original derivation of Glick and Long, based on diagrammatic perturbation theory, is quite difficult. Here we outline a more recent derivation (Hasegawa and Watabe (1969), Nifosí, Conti, and Tosi (1998)) which is simpler and provides a good starting point for microscopic calculations of $f_{xcL}(\omega)$ at finite frequency. The reader, however, will still have to work pretty hard to verify all the steps of the derivation.

We start from the identity

$$\Im m f_{xc\alpha}(\omega) = \lim_{q \to 0} \frac{m^2 \omega^2}{n^2 q^2} \Im m \chi_{j_{p\alpha} j_{p\alpha}}(q, \omega),\tag{5.204}$$

where $\alpha = L$ or T and $\omega \neq \omega_p$. (This is best proved in two steps, namely, by first showing that the identity holds when $\Im m \chi_{j_{p\alpha} j_{p\alpha}}(q, \omega)$ is replaced by $\Im m \tilde{\chi}_{j_{p\alpha} j_{p\alpha}}(q, \omega)$, and then showing that the imaginary parts of the full and the proper response functions coincide for $q \to 0$ and $\omega \neq \omega_p$.) The right-hand side is then transformed with the help of the identity

$$\omega^2 \Im m \chi_{AA^\dagger}(\omega) = \Im m \chi_{\dot{A}\dot{A}^\dagger}(\omega),\tag{5.205}$$

(see Exercise 3 in Chapter 3) where $\dot{A} = -i\hbar^{-1}[\hat{A}, \hat{H}]$. This identity holds for any operator $\hat{A}$ (in this case, $\hat{A} = \hat{j}_{p\alpha\vec{q}}$) such that $[\hat{A}, \hat{A}^\dagger] = 0$. (Note: the expectation value $\langle [[\hat{H}, \hat{A}], \hat{A}^\dagger]\rangle$ is purely real.)

After lengthy calculations involving the commutators of $\hat{j}_p$ and $\hat{n}$ with the hamiltonian, we arrive at

$$\Im m f_{xc\alpha}(\omega) = \frac{1}{n^2 q^2 \omega^2 L^{2d}} \sum_{\beta,\gamma,\vec{k},\vec{k}'} \Gamma_{\alpha\beta}(\vec{k}, \vec{q}) \Gamma_{\alpha\gamma}(-\vec{k}', q)$$

$$\times \Im m \chi_{j_{p\beta\vec{k}} n_{-\vec{k}}, j_{p\gamma\vec{k}'} n_{-\vec{k}'}}(\omega),\tag{5.206}$$

---

[40] In view of the Kramers–Krönig relations (5.183), the fact that the $\omega \to \infty$ limit of $\Re e\, A(\omega)$ diverges logarithmically in 2D is perfectly consistent with the failure of $\Im m\, A(\omega)$ to vanish in the same limit.

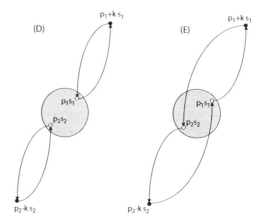

Fig. 5.22. Double electron–hole pair excitations contributing to $\Im m f_{xcL}(\omega)$ in the high-frequency limit. (D) – direct process, (E) – exchange process.

where

$$\Gamma_{\alpha\beta}(\vec{k},\vec{q}) = \frac{dv_k}{dk}\vec{q}\cdot\vec{e}_{\vec{k}}k_{\alpha}k_{\beta} + v_k\left[k_{\beta}q_{\alpha} - q_{\beta}k_{\alpha} - \delta_{\alpha\beta}\vec{k}\cdot\vec{q}\right]. \qquad (5.207)$$

Apparently, we have only succeeded in expressing the spectrum of the current response function at small q in terms of the even more complicated spectrum

$$\Im m \chi_{j_{p\beta\vec{k}}n_{-\vec{k}}, j_{p\gamma\vec{k}'}n_{-\vec{k}'}}(\omega) = -\pi\sum_{n}\langle 0|\hat{j}_{p\beta\vec{k}}\hat{n}_{-\vec{k}}|n\rangle\langle n|\hat{j}_{p\gamma\vec{k}'}\hat{n}_{-\vec{k}'}|0\rangle\delta(\omega - \omega_{n0}), \qquad (5.208)$$

where $|0\rangle$ is the ground-state, and the sum runs over the excited states $|n\rangle$ of the system. The real advantage of this representation lies in the fact that the high-frequency limit of Eq. (5.208) can be safely evaluated in the non-interacting approximation, since interaction corrections are expected to become unimportant at very high frequency, and since two interaction factors (Γ) are already in evidence in Eq. (5.206).

The basic idea of the calculation is illustrated in Fig. 5.22. The operators $\hat{n}_{-\vec{k}}$ and $\hat{j}_{p\gamma\vec{k}}$ acting on the non-interacting ground-state (the full Fermi sphere) create two electron hole pairs with opposite momenta: $(\vec{p}_1\sigma_1, \vec{p}_1 + \vec{k}\,\sigma_1)$ and $(\vec{p}_2\sigma_2, \vec{p}_2 - \vec{k}\,\sigma_2)$. Since ω is much larger than ϵ_F, k is necessarily much larger than k_F and the excitation energy ω_{n0} of the two-pair state can be approximated as $\hbar^2 k^2/m$, independent of the momenta $\vec{p}_1$ and $\vec{p}_2$ of the two holes. The two electron–hole pairs must now be annihilated by the application of $\hat{n}_{-\vec{k}'}$ and $\hat{j}_{p\gamma\vec{k}'}$. This can be done in two ways: in a "direct" process the electron at $\vec{p}_1 + \vec{k}\,\sigma_1$ recombines with the hole at $\vec{p}_1\sigma_1$ as shown in Fig. 5.22 (D); in an "exchange process" the electron at $\vec{p}_1 + \vec{k}\,\sigma_1$ recombines with the hole at $\vec{p}_2\sigma_1$, as shown in Fig. 5.22 (E).

The direct process can occur either for $\vec{k}' = \vec{k}$ or for $\vec{k}' = -\vec{k}$. In both cases, the product of the matrix elements of the two current operators gives a factor $k'_{\beta}k_{\gamma}/4m^2$, and the sum over the independent spin orientations σ_1, σ_2 gives a factor 4.

The exchange processes, instead, can only occur for $\sigma_1 = \sigma_2$ and either $\vec{k}' = \vec{p}_2 - \vec{p}_1 - \vec{k} \approx -\vec{k}$ or $\vec{k}' = -\vec{p}_2 + \vec{p}_1 + \vec{k} \approx \vec{k}$ (we are using the fact that $p_1, p_2 < k_F$ while $k \gg k_F$). The product of the matrix elements of the current operators still yields $k'_\beta k_\gamma / 4m^2$, but the spin multiplicity is now reduced to 2 due to the condition $\sigma_1 = \sigma_2$, and the overall sign of the amplitude is reversed.

Combining the direct and the exchange terms we obtain

$$\Im m \chi^{(0)}_{j_{p\beta\vec{k}}n_{-\vec{k}}, j_{p\gamma\vec{k}'}n_{-\vec{k}'}}(\omega) \overset{\omega\to\infty}{\simeq} -\frac{\pi n^2}{8m^2} k_\beta k'_\gamma (\delta_{\vec{k},\vec{k}'} + \delta_{\vec{k},-\vec{k}'})\delta\left(\omega - \frac{\hbar k^2}{m}\right), \quad (5.209)$$

(the (0) stands for "non-interacting") and substituting this in Eq. (5.206) we arrive at the high-frequency behavior of Eq. (5.201).

5.6.2.6 Relation between $f'_{xcL}(0)$ and $f'_{xcT}(0)$

Another interesting application of Eq. (5.206) is the derivation of the relation (5.196) between the longitudinal and transverse spectra at low frequency. We shall see in Section 5.9 that this relation is equivalent to the statement that the *bulk viscosity* of the electron liquid vanishes at zero frequency. The physical reason for this result is that the current operator, $\vec{j}_{p\vec{k}} = \sum_{\vec{p}} (\vec{p} - \vec{k}/2)\hat{a}^\dagger_{\vec{p}-\vec{k}}\hat{a}_{\vec{p}}$, is essentially *transverse* in the subspace of low-energy electron–hole excitations. This can be easily seen from Fig. 4.3: the scarcity of longitudinal excitations is what causes the bulk viscosity to vanish (see also footnote [11]). The transversality condition $\vec{k} \cdot \vec{j}_{p\vec{k}} \sim 0$ allows us to drop the terms proportional to k_β in $\Gamma_{\alpha\beta}$ (see Eq. (5.206)): this leads to the simpler expression

$$\Gamma_{\alpha\beta}(\vec{k}, \vec{q}) \approx -v_k \left[k_\alpha q_\beta + \delta_{\alpha\beta}\vec{k} \cdot \vec{q}\right]. \quad (5.210)$$

Next we notice that, because of isotropy, Eq. (5.206) can be averaged with respect to the direction of $\vec{q}$. Calculating in this way both the longitudinal component f_{xcL} and the trace $\sum_\alpha f_{xc\alpha} = f_{xcL} + (d-1)f_{xcT}$ we finally find that

$$\lim_{\omega\to 0} \frac{\Im m f_{xcL}(\omega)}{\Im m f_{xcL}(\omega) + (d-1)\Im m f_{xcT}(\omega)} = \frac{2}{d+2}, \quad (5.211)$$

from which Eq. (5.196) follows. The details of the derivation are left as an exercise.

5.7 Theories of the dynamical local field factor

The calculation of the dynamical local field factor is perhaps the most important challenge in the modern theory of the electron liquid. As explained earlier, the quantum Monte Carlo method is still far from being able to adequately treat the dynamics of many-body systems, so one is forced to rely on analytical approximation methods. These methods fall roughly into two classes, namely: (1) approximate decoupling schemes, of which the time-dependent Hartree–Fock theory and the quantum STLS scheme are prime examples, and (2) perturbative calculations based on diagrammatic techniques. None of these methods has been particularly successful to date. So, with the exception of the exact limiting forms presented earlier in this chapter, the frequency dependence of the local field factors is still largely unknown. In this section, we review some of the most significant attempts that have been made to solve this remarkably difficult problem.

5.7.1 The time-dependent Hartree–Fock approximation

The time-dependent Hartree–Fock approximation (TDHFA) bears a peculiar relation to the other microscopic theories described in this chapter. On one hand, it is the natural "next step" beyond the time-dependent Hartree approximation, i.e., the RPA. On the other hand, this theory is not based on the notion of local field factor (in fact, the Hartree–Fock effective potential is nonlocal!), but can be used to compute a frequency-dependent local field factor *a posteriori*, that is, by making use of Eq. (5.105). For this reason the TDHFA turns out to be considerably more complex than any of the theories considered so far, even the ones that are arguably superior, by virtue of their taking into account correlation effects. It is nevertheless useful to study the TDHFA for the following two reasons: (i) the theory is exact to first order in the coulomb interaction, (ii) the local field factors calculated in this approximation provide some insight into the general frequency dependence of these quantities.

The TDHFA is derived following the general procedure outlined in Section 4.7. Recall that the matrix elements of the HF potential in the HF ground-state are given by Eq. (2.17). For a state that is not the HF ground-state, but still a single Slater determinant, Eq. (2.17) generalizes to

$$V_{\alpha\beta}^{HF} = V_{\alpha\beta}^{ext} + \sum_{\gamma\delta} \rho_{\gamma\delta}(v_{\alpha\gamma\delta\beta} - v_{\alpha\gamma\beta\delta}) , \qquad (5.212)$$

where $\rho_{\gamma\delta}$ is the one-particle density matrix of the state under consideration, and $V_{\alpha\beta}^{ext}$ are the matrix elements of the *static* external potential.[41]

Let us now apply a small *time-dependent* external potential $V_1(t)$, which causes the HF single-determinant state to oscillate about the ground-state solution, and the density matrix to become a time-dependent object of the form

$$\rho_{\gamma\delta} = n_\gamma \delta_{\gamma\delta} + \rho_{\gamma\delta}^{(1)}(t). \qquad (5.213)$$

According to Eq. (5.212), the Hartree–Fock potential also acquires a time-dependent component

$$V_{1,\alpha\beta}^{HF}(t) = V_{1,\alpha\beta}(t) + \sum_{\gamma\delta} \left(v_{\alpha\gamma\delta\beta} - v_{\alpha\gamma\beta\delta} \right) \rho_{\gamma\delta}^{(1)}(t) , \qquad (5.214)$$

which adds to the ground-state potential. The fundamental Ansatz of the TDHFA is that the system responds to $V_{1,\alpha\beta}^{HF}(t)$ as if it were a non-interacting system governed by the self-consistent HF hamiltonian for the ground-state.

Let us denote by $\chi_{\alpha\beta,\gamma\delta}^f$ the response function that gives the change in $\rho_{\alpha\beta}$ due to a potential that couples linearly to $\rho_{\gamma\delta}$ in the system governed by the ground-state HF hamiltonian.

---

[41] Eq. (5.212) follows from the derivative of the energy of the Slater determinant (Eq. (2.62)) with respect to $\rho_{\alpha\beta}$.

According to the general formalism of Chapter 3 (see in particular Eq. (4.82)), this is given by

$$\chi^f_{\alpha\beta,\gamma\delta}(\omega) = \frac{n_\alpha - n_\beta}{\hbar\omega + \varepsilon_\alpha - \varepsilon_\beta + i\hbar\eta} \delta_{\alpha\delta}\delta_{\beta\gamma} , \qquad (5.215)$$

where ε_α denotes the HF eigenvalues and n_α the corresponding occupation numbers. Then, according to Eq. (4.89), the density matrix – density matrix response function in TDHFA will be given by the solution of the linear equation

$$\chi^{HF}_{\alpha\beta,\gamma\delta}(\omega) = \chi^f_{\alpha\beta,\gamma\delta}(\omega) + \sum_{\alpha'\beta'\gamma'\delta'} \chi^f_{\alpha\beta,\alpha'\beta'}(\omega) I_{\alpha'\beta',\gamma'\delta'} \chi^{HF}_{\gamma'\delta',\gamma\delta}(\omega) , \qquad (5.216)$$

where the interaction kernel $I_{\alpha\beta,\gamma\delta} = \frac{\delta V^{HF}_{1,\alpha\beta}}{\delta\rho^{(1)}_{\gamma\delta}}$ can be directly derived from Eq. (5.214):

$$I_{\alpha\beta,\gamma\delta} = v_{\alpha\gamma\delta\beta} - v_{\alpha\gamma\beta\delta}. \qquad (5.217)$$

The linear equations (5.216) are easily "solved" (in a formal sense) yielding the inverse of the response function

$$[\chi^{HF}]^{-1}_{\alpha\beta,\gamma\delta}(\omega) = [\chi^f]^{-1}_{\alpha\beta,\gamma\delta}(\omega) - v_{\alpha\gamma\delta\beta} + v_{\alpha\gamma\beta\delta}$$
$$= \frac{\hbar\omega + \varepsilon_\alpha - \varepsilon_\beta}{n_\alpha - n_\beta}\delta_{\alpha\delta}\delta_{\beta\gamma} - v_{\alpha\gamma\delta\beta} + v_{\alpha\gamma\beta\delta} . \qquad (5.218)$$

All the quantities that appear in this equation must be interpreted as matrices in two composite indices $\alpha\beta$ and $\gamma\delta$, and their inverses must be computed accordingly. In the limit $\omega \to 0$, Eq. (5.218) yields the static susceptibility first encountered in Eq. (2.70) of Chapter 2. In that chapter we derived, with considerable effort, an expression for the second-order change of the HF energy due to a small change in the density matrix. In view of the physical interpretation of the inverse static susceptibility as *stiffness* (see Section 3.2.9), that result appears now almost trivial.

Knowledge of $\chi^{HF}_{\alpha\beta,\gamma\delta}$ allows one, in principle, to calculate any linear response function involving single-particle operators. In practice, the inversion of Eq. (5.218) poses a difficult mathematical problem leading, in general, to an integral equation. Let us see, for example, what happens in the simple case of the homogeneous electron gas. Assume the solutions of the static HF equation to be plane waves of wave vector $\vec{k}$ and spin projection σ. With the introduction of the symmetric labelling $\alpha = \vec{k} + \frac{\vec{q}}{2}\sigma$, $\beta = \vec{k} - \frac{\vec{q}}{2}\sigma$, $\gamma = \vec{k}' - \frac{\vec{q}}{2}\sigma'$, and $\delta = \vec{k}' + \frac{\vec{q}}{2}\sigma'$, the response functions of interest take the form $\chi^{HF}_{\vec{k}\sigma,\vec{k}'\sigma'}(\vec{q},\omega)$ and $\chi^f_{\vec{k}\sigma,\vec{k}'\sigma'}(\vec{q},\omega)$. These functions describe the response of the density matrix element $\rho_{\vec{k}+\frac{\vec{q}}{2}\sigma,\vec{k}-\frac{\vec{q}}{2}\sigma}$ to a potential that couples linearly to $\rho_{\vec{k}'-\frac{\vec{q}}{2}\sigma',\vec{k}'+\frac{\vec{q}}{2}\sigma'}$. The reference response function χ^f and the interaction kernel I are given by Eqs. (5.215) and (5.217) respectively as

$$\chi^f_{\vec{k}\sigma,\vec{k}'\sigma'}(\vec{q},\omega) = \frac{n_{\vec{k}+\frac{\vec{q}}{2}\sigma} - n_{\vec{k}-\frac{\vec{q}}{2}\sigma}}{\hbar\omega + \varepsilon_{\vec{k}+\frac{\vec{q}}{2}\sigma} - \varepsilon_{\vec{k}-\frac{\vec{q}}{2}\sigma} + i\hbar\eta}\delta_{\vec{k}\vec{k}'}\delta_{\sigma\sigma'} , \qquad (5.219)$$

and

$$I_{\vec{k}+\frac{\vec{q}}{2}\sigma,\vec{k}-\frac{\vec{q}}{2}\sigma,\vec{k}''-\frac{\vec{q}}{2}\sigma'',\vec{k}''+\frac{\vec{q}}{2}\sigma''} = \frac{1}{L^d}\left(v_{\vec{q}} - v_{\vec{k}-\vec{k}''}\right). \tag{5.220}$$

Thus, Eq. (5.216) takes the form

$$\chi^{HF}_{\vec{k}\sigma,\vec{k}'\sigma'}(\vec{q},\omega) = \chi^{f}_{\vec{k}\sigma,\vec{k}'\sigma'}(\vec{q},\omega) + \frac{1}{L^d}\sum_{\vec{k}''\sigma''} \chi^{f}_{\vec{k}\sigma,\vec{k}\sigma}(\vec{q},\omega)[v_{\vec{q}} - v_{\vec{k}-\vec{k}''}\delta_{\sigma\sigma''}]\chi^{HF}_{\vec{k}''\sigma'',\vec{k}'\sigma'}. \tag{5.221}$$

Once $\chi^{HF}_{\vec{k}\sigma,\vec{k}'\sigma'}(\vec{q},\omega)$ is known, one can easily calculate the density–density response function

$$\chi^{HF}_{nn}(q,\omega) = \sum_{\vec{k}\sigma,\vec{k}'\sigma'} \chi^{HF}_{\vec{k}\sigma,\vec{k}'\sigma'}(\vec{q},\omega), \tag{5.222}$$

the spin density–spin density response function

$$\chi^{HF}_{S_z S_z}(q,\omega) = \sum_{\vec{k}\sigma,\vec{k}'\sigma'} \sigma\sigma' \chi^{HF}_{\vec{k}\sigma,\vec{k}'\sigma'}(\vec{q},\omega), \tag{5.223}$$

and the (paramagnetic) current–current response function

$$\chi^{HF}_{j_{pi}j_{pj}}(q,\omega) = \sum_{\vec{k}\sigma,\vec{k}'\sigma'} \frac{\hbar k_i}{m} \chi^{HF}_{\vec{k}\sigma,\vec{k}'\sigma'}(\vec{q},\omega)\frac{\hbar k_j}{m}, \tag{5.224}$$

as well as any other response function involving single-particle observables.

The solution of the linear integral equation (5.221) is however a nontrivial task. If the exchange term is neglected, both in the interaction kernel *and* in the HF eigenvalues (which thus reduce to free-particle energies) one immediately recovers the RPA. In fact, it is not difficult to verify that the *proper* density–density response function is given by

$$\tilde{\chi}^{HF}_{nn}(q,\omega) = \sum_{\vec{k},\vec{k}',\sigma} \tilde{\chi}^{HF}_{\vec{k}\sigma,\vec{k}'\sigma}(\vec{q},\omega), \tag{5.225}$$

where $\tilde{\chi}^{HF}_{\vec{k}\sigma,\vec{k}'\sigma'}(\vec{q},\omega)$ is diagonal with respect to the spin index and satisfies an integral equation similar to Eq. (5.221), but *without* the Hartree term $v_{\vec{q}}$:

$$\tilde{\chi}^{HF}_{\vec{k}\sigma,\vec{k}'\sigma}(\vec{q},\omega) = \chi^{f}_{\vec{k}\sigma,\vec{k}'\sigma}(\vec{q},\omega) - \frac{1}{L^d}\sum_{\vec{k}''} \chi^{f}_{\vec{k}\sigma,\vec{k}\sigma}(\vec{q},\omega)v_{\vec{k}-\vec{k}''}\tilde{\chi}^{HF}_{\vec{k}''\sigma,\vec{k}'\sigma}. \tag{5.226}$$

The derivation of this result is left as an exercise. Notice that, because $\tilde{\chi}^{HF}_{\vec{k}\sigma,\vec{k}'\sigma'}(\vec{q},\omega)$ is diagonal with respect to the spin index, *the proper density–density response function and the longitudinal spin–spin response function coincide in the TDHFA.*

The numerical solution of Eq. (5.226) has been attempted by several authors (Hamann and Overhauser (1966); Brosens *et al.* (1980); Hameeuw *et al.* (2003)) but not without resorting to additional approximations.[42] In the end, the results of the approximate solution

---

[42] An exception is the calculation of the static limit, which can be done analytically for $q \to 0$ (Yarlagadda and Giuliani, 1989): the reason for this will become clear in Chapter 8.

turn out to be qualitatively similar to those of the first-order perturbation theory, which we now describe.

5.7.2 First order perturbation theory and beyond

The perturbation theoretical formalism for the calculation of linear response functions will be presented in detail in Chapter 6. A calculation to first order in the coulomb interaction, however, hardly necessitates such a complex formalism, as the relevant correction to the proper density–density response function can be obtained simply by expanding Eq. (5.226) to first order in v_q. The result of this expansion is

$$\tilde{\chi}^{(1)}_{\vec{k}\sigma,\vec{k}'\sigma'}(\vec{q},\omega) = \chi^{f(1)}_{\vec{k}\sigma,\vec{k}'\sigma'}(\vec{q},\omega) - \chi^{(0)}_{\vec{k}\sigma,\vec{k}\sigma}(\vec{q},\omega)v_{\vec{k}-\vec{k}'}\chi^{(0)}_{\vec{k}'\sigma,\vec{k}'\sigma}\delta_{\sigma\sigma'} , \qquad (5.227)$$

where $\chi^{(0)}_{\vec{k}\sigma,\vec{k}'\sigma'}(\vec{q},\omega)$ is the *non-interacting* density matrix–density matrix response function, given by Eq. (5.219) with free particle energies $\varepsilon^{(0)}_{\vec{k}}$ replacing the HF eigenvalues, and $\chi^{f(1)}_{\vec{k}\sigma,\vec{k}'\sigma'}(\vec{q},\omega)$ is the first order term in the expansion of Eq. (5.219) in powers of the interaction strength. Summing over $\vec{k}\sigma$ and $\vec{k}'\sigma'$ and making use of Eq. (5.219) we obtain

$$\tilde{\chi}^{(1)}(q,\omega) = \sum_{\vec{k},\vec{k}'\sigma} v_{\vec{k}-\vec{k}'} \frac{\left(n_{\vec{k}+\frac{\vec{q}}{2}\sigma} - n_{\vec{k}-\frac{\vec{q}}{2}\sigma}\right)\left(n_{\vec{k}'+\frac{\vec{q}}{2}\sigma} - n_{\vec{k}'-\frac{\vec{q}}{2}\sigma}\right)}{\left(\hbar\omega + \varepsilon^{(0)}_{\vec{k}+\frac{\vec{q}}{2}} - \varepsilon^{(0)}_{\vec{k}-\frac{\vec{q}}{2}} + i\eta\right)^2}$$

$$- \sum_{\vec{k},\vec{k}'\sigma} v_{\vec{k}-\vec{k}'} \frac{\left(n_{\vec{k}+\frac{\vec{q}}{2}\sigma} - n_{\vec{k}-\frac{\vec{q}}{2}\sigma}\right)\left(n_{\vec{k}'+\frac{\vec{q}}{2}\sigma} - n_{\vec{k}'-\frac{\vec{q}}{2}\sigma}\right)}{\left(\hbar\omega + \varepsilon^{(0)}_{\vec{k}+\frac{\vec{q}}{2}} - \varepsilon^{(0)}_{\vec{k}-\frac{\vec{q}}{2}} + i\eta\right)\left(\hbar\omega + \varepsilon^{(0)}_{\vec{k}'+\frac{\vec{q}}{2}} - \varepsilon^{(0)}_{\vec{k}'-\frac{\vec{q}}{2}} + i\eta\right)} .$$

$$(5.228)$$

The above expression has been calculated by several authors in different regimes, namely (i) as a function of frequency, ω, at constant q (Holas *et al.*, 1979, 1982); (ii) as a function of wave vector, q, at $\omega = 0$ (Engel and Vosko, 1990); and (iii) as a function of imaginary frequency (Richardson and Ashcroft, 1994; Atwal *et al.*, 2003). The local field factor at this order of approximation is given by

$$v_q G^{(1)}_\pm(q,\omega) = -\frac{\tilde{\chi}^{(1)}(q,\omega)}{\chi_0^2(q,\omega)} , \qquad (5.229)$$

as one can easily verify by expanding Eq. (5.105) to first order in the interaction. Representative results for the frequency dependence of the first order local field factor are presented in Fig. 5.23.

Several problems are immediately evident. First of all, the singularities that appear in the real and in the imaginary parts of $G(q,\omega)$ at the two limits of the electron–hole continuum indicate a breakdown of the first-order approximation at those frequencies (more will be said about this in Chapter 6). Furthermore, there is a small region of frequencies, near

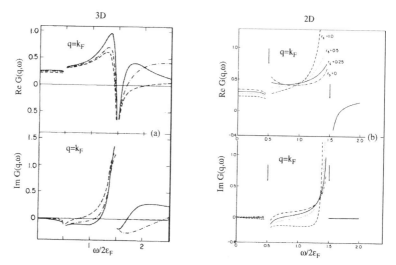

Fig. 5.23. Real and imaginary parts of the dynamic local field factor $G(q, \omega)$ vs ω at $q = k_F$ in first order perturbation theory (recall that $G_+ = G_- = G$ in this approximation). (a) 3D case: $r_s = 0.5$ (solid and short-dashed lines), and $r_s = 2$ (long-dashed line). The two curves at $r_s = 0.5$ correspond to two slightly different ways of extracting $G(q, \omega)$ from the first order response function, with the solid line coming from Eq. (5.229) (adapted from Holas *et al.* (1979)). (b) 2D case: $r_s = 0, 0.25, 0.5$ and 1, as indicated (adapted from Czachor *et al.* (1982)). The arrows point to the singularities of $G(q, \omega)$.

the limit of the electron hole continuum, in which the imaginary part of $G(q, \omega)$ turns negative, violating general requirements of causality and positivity of dissipation. It is because of these pathological features, in addition to other obvious limitations, that the first order perturbation theory has never really "caught up" as a viable approximation for $G(q, \omega)$. A similar fate has befallen the quantum mechanical version of the STLS scheme, briefly mentioned at the end of Section 5.4.4, as that scheme was found to violate causality within certain frequency ranges (Holas and Rahman, 1987). More recently Richardson and Ashcroft (1994) and Atwal, Khalil, and Ashcroft (2003), have performed non-perturbative calculations of the local field factors $G_{\pm}(q, i\omega)$ *for imaginary frequency* $i\omega$, in three and two dimensions respectively (see also Lein *et al.*, 2000). The resulting local field factors satisfy the compressibility and the third moment sum rules, and have the right asymptotic behavior for large q and $i\omega$ (explicit expressions are provided in the cited references). The local field factors for imaginary frequency are very useful in the calculation of equilibrium properties of the electron liquid, where an integral over real frequencies can be converted in an integral along the imaginary frequency axis (see Section 3.3.5). Unfortunately, the restriction to imaginary frequencies prevents the application of these local field factors to genuinely dynamical problems. This restriction has been partially lifted in a recent paper by Sturm and Gusarov (2000), in which a selected group of diagrams of first and second-order in the screened coulomb interaction is evaluated to yield the imaginary part of the dielectric function (and hence the local field factor) outside the single electron–hole pair continuum. The reader is referred to the original paper for a detailed description of this difficult calculation.

5.7.3 The mode-decoupling approximation

An interesting approach to the calculation of the dynamical local field factors at small wave vector has been recently proposed by Nifosí, Conti, and Tosi (NCT) (1998) (see also Böhm, Conti, and Tosi, 1996), building on earlier mode-decoupling ideas by several authors (e.g. Hasegawa and Watabe, 1969).

Starting from the exact expression (5.206) NCT propose the following decoupling of the imaginary part of the four-point response function:

$$\Im m\,\chi_{j_{p\beta\vec{k}}n_{-\vec{k}},\,j_{p\gamma\vec{k}'}n_{-\vec{k}'}}(\omega) \approx$$
$$-\int_0^\omega \frac{d\omega'}{\pi}\left[\Im m\,\chi_{j_{p\beta\vec{k}}j_{p\gamma\vec{k}'}}(\omega')\,\Im m\,\chi_{n_{-\vec{k}}n_{-\vec{k}'}}(\omega-\omega')\right.$$
$$\left.+\,\Im m\,\chi_{j_{p\beta\vec{k}}n_{-\vec{k}'}}(\omega')\,\Im m\,\chi_{n_{-\vec{k}}j_{p\gamma\vec{k}'}}(\omega-\omega')\right]. \qquad (5.230)$$

The two-point response functions on the right-hand side of this equation are immediately recognized as the familiar density–density, density-current, and current–current response functions: in a homogeneous liquid they differ from zero only if $\vec{k}' = -\vec{k}$ in the first term and $\vec{k}' = \vec{k}$ in the second. Thus, Eqs. (5.206) and (5.230) express $\Im m f_{xcL(T)}(\omega)$ as a *convolution of two spectral functions*, which can in turn be approximated in terms of static local field factors. Although the decoupling approximation is not exact in any limit, the underlying physical idea, namely that the density fluctuation spectrum should be understandable in terms of *pairs* of elementary excitations, such as two electron–hole pairs, two plasmons, or an electron–hole pair and a plasmon, is both reasonable and appealing. The very same idea has provided valuable guidance in constructing simple interpolation formulas for $\Im m f_{xcL(T)}(\omega)$, which are presented in Appendix 11.

Fig. 5.24 shows representative results of the NCT calculation (short-dashed line) for the important quantity $f_{xcL}(\omega) = -\lim_{q\to 0} v_q G_+(q,\omega)$ in three dimensions, along with two interpolation formulas (GK and QV) which will be discussed in Appendix 11. The most striking feature of the NCT result for $\Im m f_{xcL}(\omega)$ is its sharp rise just above twice the plasmon frequency ω_p. This arises naturally from the convolution of the two spectral functions on the right-hand side of Eq. (5.230), since each of them has a sharp peak at the plasmon frequency.[43] However, the strength of the peak is overestimated by NCT since the approximate spectral functions used in that calculation do not include the broadening of the plasmon due to its interaction with multiple electron–hole pairs. This unphysical feature is remedied by the empirical QV interpolation formula (see Appendix 11).

Fig. 5.25 shows the result of the NCT theory for $f_{xcL}(\omega)$ in a two-dimensional electron liquid. Due to the different form of the plasmon dispersion the two-plasmon peak in

---

[43] The curve shown in Fig. 5.24 was obtained using the RPA for the response functions in Eq. (5.230).

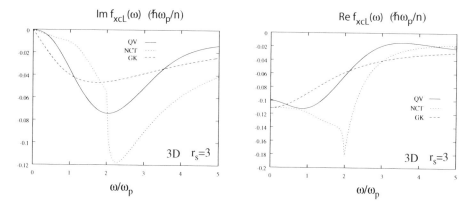

Fig. 5.24. The imaginary and the real parts of $f_{xcL}(\omega)$ (in units of $\hbar\omega_p/n$) in the three-dimensional electron liquid at $r_s = 3$. The short-dashed line (NCT) is the result of the mode-decoupling calculation. The long-dashed line (GK) and the solid line (QV) are interpolation formulas, which will be described in Appendix 11.

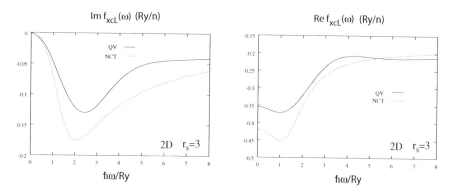

Fig. 5.25. Imaginary and the real parts of $f_{xcL}(\omega)$ (in units of Ry/n) in the two-dimensional electron liquid at $r_s = 3$. The short-dashed line (NCT) is the result of the mode-decoupling calculation, and the solid line (QV) is the interpolation formula described in Appendix 11.

$\Im m f_{xcL}(\omega)$ is replaced by a smooth maximum. Qualitatively similar results are obtained for the transverse exchange-correlation kernel, $f_{xcT}(\omega)$.

5.8 Calculation of observable properties

Let us briefly discuss how the many-body local field factors introduced in this chapter can be used to calculate quantities of experimental interest. Experiments performed on the best available realizations of the homogeneous electron liquid can be grouped in three classes:

(1) Static measurements at $q = 0$, e.g. measurements of the compressibility and susceptibility enhancement, and dc conductivities;

(2) Static measurements at finite q, e.g. scanning tunneling microscopy measurements of Friedel oscillations;
(3) Dynamic measurements, e.g. electron-energy-loss, Raman, and X-ray scattering.

We have already touched on (1) and (2) earlier in this chapter, and more will be said on the experimental determination of the spin susceptibility in Chapter 8, where the topic is put in the broader context of the Landau theory of Fermi liquids. Hence in this section we focus on item (3), in particular on the experimental determination of (i) the plasmon dispersion, (ii) the plasmon damping, and (iii) the dynamical structure factor.

5.8.1 Plasmon dispersion and damping

To determine the complex plasmon dispersion, $\Omega_p(q)$, we simply seek the zero of the denominator of Eq. (5.102) as a function of q. At long wavelength, making use of the limiting forms of χ_0 (Eq. (5.50)) and $v_q G_+(q, \omega)$ (Eq. (5.184)) we obtain

$$\Omega_p^2(q) = \omega_p^2(q) + \left[a_d v_F^2 + \frac{n}{m} f_{xcL}(\omega_p)\right] q^2 + \mathcal{O}(q^4), \qquad (5.231)$$

where the constants a_d are given immediately after Eq. (5.50), and $\omega_p(q)$ is specified in Eq. (5.48). Comparing this with Eq. (5.51), we see that exchange and correlation effects modify the RPA result in two important ways.

First, because the real part of $f_{xcL}(\omega_p)$ is negative (see Figs. 5.24 and 5.25), the plasmon frequency is *lower* than in RPA. The dispersion may even bend downward at sufficiently low density: this effect is known as *negative dispersion*. A slight negative dispersion is indeed visible in the plots of Fig. 5.26, but only at a very large value of r_s.

Second, because $f_{xcL}(\omega_p)$ has an imaginary part, the plasmon resonance peak – which is observed, for example, in electron-energy-loss experiments – acquires a finite linewidth.

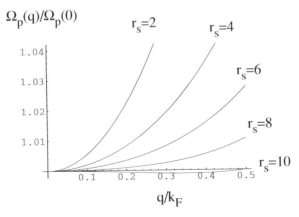

Fig. 5.26. Theoretical results for the plasmon dispersions in three dimensions. The dispersion is calculated with the local field factor $G_+(q, 0)$ of Eq. (A11.5). Notice the negative dispersion at $r_s = 10$.

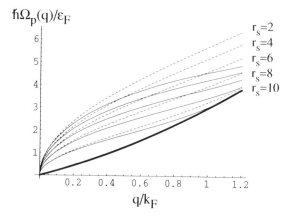

Fig. 5.27. Theoretical results for the plasmon dispersions in two dimensions. The dispersion is calculated in RPA (dashed line) and with the local field factor $G_+(q, 0)$ of Eq. (A11.8) (solid line). The thick solid line marks the boundary of the electron–hole continuum.

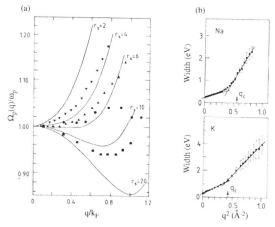

Fig. 5.28. (a) Plasmon dispersion normalized to the plasmon frequency at $q = 0$. The symbols represent the experimental data for Na ($r_s = 3.93$, downward triangles), K ($r_s = 4.86$, upward triangles), Rb ($r_s = 5.20$, dots) and Cs ($r_s = 5.62$, squares). The solid lines give the theoretical plasmon dispersions obtained with the use of the STLS static local field factor. (b) q-dependence of the plasmon half-width in Na and K. The straight lines are fits to the experimental data. q_c is the critical wave vector at which the plasmon enters the electron–hole continuum. Notice that the half-width does not vanish at $q = 0$ due to the presence of the crystal potential. Adapted from vom Felde *et al.*, 1989.

This effect is caused by the interaction of the plasmon with multiple electron–hole pair excitations. The linewidth is proportional to q^2, and is therefore much smaller than the plasmon frequency itself for $q \to 0$.

A reduction of the plasmon frequency with decreasing density is clearly seen in experiments. Fig. 5.28(a) shows the experimental plasmon dispersion for several elemental metals, *vis-à-vis* the theoretical dispersion obtained with the use of the STLS local field factor, given by Eq. (5.120). Notice that the STLS dispersion (Fig. 5.28) lies well below the nominally more accurate dispersion calculated with the QMC local field factor at the same r_s (Fig. 5.26). In spite of this, even the STLS is unable to fit quantitatively the experimental dispersion, unless one resorts to values of r_s much larger than the physical ones.

Consider now the determination of the plasmon linewidth, i.e., the imaginary part of the plasmon frequency. From Eq. (5.231) we immediately see that the plasmon linewidth at long wavelength is given by

$$\Im m \Omega_p(q) \approx \frac{nq^2}{2m\omega_p(q)} \Im m f_{xcL}(\omega_p(q)),$$
(5.232)

where the exchange-correlation kernel $f_{xcL}(\omega)$ is discussed in Section 5.6.2.2 and its parametrization is provided in Eqs. (A11.13)–(A11.14) of Appendix 11. Thus, the linewidth is predicted to vary as q^2 both in three and in two dimensions.[44]

Experimental results for the plasmon linewidth in elemental metals are shown in Fig. 5.28 (panel (b)). The lack of translational invariance explains why the line-width does not vanish for $q \to 0$, but, obviously, complicates the comparison between theory and experiment. After subtracting the $q = 0$ contribution, one sees that the q^2 scaling is well satisfied, but the coefficient is considerably larger than implied by Eq. (5.232) above.

Of course the interpretation of experiments on "simple" metals is seriously complicated by band structure effects. Because the plasmon energy is larger than the inter-band gap, electron–hole transitions between different bands cannot be ignored: in fact, the presence of inter-band excitations has been found to lower the plasmon frequency in aluminum by as much as 4 eV (Quong and Eguiluz, 1993)! The influence of the crystal lattice on the plasmon damping mechanism in potassium has recently been studied in detail by Ku and Eguiluz (1999) and the role of inter-band transition has been found to be paramount.

Plasmons have also been studied in two-dimensional systems, notably in Si inversion layers (Allen, Tsui, and Logan, 1977; Theis, 1980), in a metallic surface state band on an Si surface (Nagao *et al.*, 2001), and more recently in low density GaAs quantum wells (Hirjibehedin *et al.*, 2002).

5.8.2 Dynamical structure factor

The dynamical structure factor $S(q, \omega)$ of electrons in Al has been measured by several independent groups (Platzman *et al.*, 1992; Schülke *et al.*, 1993; Larson *et al.*, 1996) by means of inelastic X-ray scattering. In comparison with the RPA, local field factors tend to strengthen the electron–hole contribution at low frequency. The most interesting observations, however, were done in the large q regime ($q \sim 2k_F$). Here one sees a puzzling peak-and-shoulder structure (see Fig. 5.29) which is not obtained in any of the available local field factor-based theories. During the 1970s and 1980s, this observation led many

---

[44] In two dimensions the q^2 scaling of the linewidth follows from the fact that $\Im m f_{xcL}(\omega_p(q)) \propto \omega_p(q) \propto q^{1/2}$.

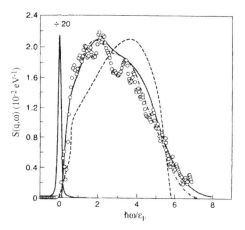

Fig. 5.29. The dynamical structure factor of Al at $q = 1.7k_F$. Notice the peak-and-shoulder structure. Also included are the predictions of the damped electron–hole pair theory by Mukhopadhyay *et al.* (1975) (solid line), and the RPA (dashed line). Adapted from Platzman *et al.*, 1992.

theorists (e.g. Mukhopadhyay *et al.*, 1975) to speculate about the role of single particle lifetimes, which are completely ignored in the local field factor theories discussed so far (single particle lifetimes will be discussed in detail in Chapter 8). While several explanations were proposed, some of which met with a certain degree of success, it is now clear that band structure effects play a crucial role in the explanation of this observation.

Another interesting question is the extent to which the two-plasmon peak discussed in Section 5.7.3 is visible in the high-frequency part of the spectrum. A tiny structure in the measured spectrum of Al at $q = 0.66k_F$ at an energy of 31eV (Schülke *et al.*, 1993) has recently been interpreted as a manifestation of the two-plasmon peak (Sturm and Gusarov, 2000): more detailed work is clearly needed to establish the correctness of this interpretation.

As pointed out in the introduction, although photons do not directly interact with the spin of the electrons, the presence of spin-orbit interaction establishes an indirect coupling. The spin density fluctuation spectrum can then be observed in experiments in which the incident light and the scattered light are polarized at right angles with each other (Pinczuk *et al.*, 1989). It is interesting to notice that, at small to moderate wave vector, the spin fluctuation spectrum is actually much stronger than the density fluctuation spectrum, since the latter is screened by the large dielectric fucnction, while the former is not. This is clearly visible in the different scale of the two panels of Fig. 5.30. Local field factors further enhance the strength of the low-frequency portion of the spectrum relative to the RPA, but these advantages are partially offset by the weakness of the spin-orbit coupling.

5.9 Generalized elasticity theory

We conclude this chapter by showing that, in the limit $q \to 0$ with ω finite, the linear response theory for the current density is equivalent to a description of the electron liquid as a *visco-elastic medium* whose visco-elastic constants are actually functions of frequency,

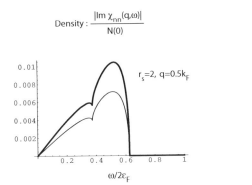
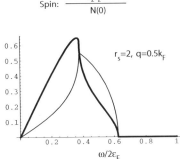

Fig. 5.30. Density fluctuation spectrum (left) and spin–density fluctuation spectrum (right) of a 2D electron liquid at $r_s = 2$ and $q = 0.5k_F$. The solid lines respresent the spectra calculated by using the local field factors $G_+(q)$ and $G_-(q)$ of Davoudi *et al.* for the density and spin channels respectively. The thinner lines show the results of the RPA (which, in the spin case, coincides with the non-interacting result). Notice the completely different scales of the left and right graphs, reflecting the presence of screening in the density case, and the absence of it in the spin case.

which can be expressed in terms of the exchange-correlation kernels $f_{xcL}(\omega)$ and $f_{xcT}(\omega)$. This formal development will play a crucial role in the formulation of the time-dependent current–density functional theory in Chapter 7.

5.9.1 Elasticity and hydrodynamics

We begin with a very brief review of the relevant basic notions of elasticity and hydrodynamics. In classical elasticity theory the state of a continuous medium is described by a displacement field, $\vec{u}(\vec{r}, t)$, which measures the displacement of a small volume element from its equilibrium position $\vec{r}$ at time t. In a homogeneous and isotropic medium the equation of motion for $\vec{u}(\vec{r}, t)$ has the form [Landau and Lifshitz, *Course of Theoretical Physics*, Vol. 7]

$$ mn\frac{\partial^2 \vec{u}}{\partial t^2} = \left[\mathcal{B} + \left(1 - \frac{2}{d}\right)\mathcal{S} \right]\vec{\nabla}(\vec{\nabla} \cdot \vec{u}) + \mathcal{S}\nabla^2\vec{u} + \vec{F}(\vec{r}, t), \qquad (5.233) $$

where $\mathcal{B}$ and $\mathcal{S}$ are two constants, the *bulk modulus* and the *shear modulus* respectively, d is the number of spatial dimensions, $\vec{F}(\vec{r}, t)$ is an externally applied volume force density, and mn is the equilibrium mass density of the medium.[45] We consider periodic force densities of the form

$$ \vec{F}(\vec{r}, t) = \vec{F}(\vec{q}, \omega)e^{i(\vec{q}\cdot\vec{r}-\omega t)} + c.c., \qquad (5.234) $$

which induce periodic displacements

$$ \vec{u}(\vec{r}, t) = \vec{u}(\vec{q}, \omega)e^{i(\vec{q}\cdot\vec{r}-\omega t)} + c.c. \qquad (5.235) $$

[45] The electron mass m is introduced to facilitate the comparison with the microscopic theory of the electron liquid.

In order to make contact with the microscopic theory of the electron liquid we write the force density as the time derivative of a vector potential

$$\vec{F}(\vec{r}, t) = n \frac{e}{c} \frac{\partial \vec{A}(\vec{r}, t)}{\partial t} \tag{5.236}$$

and introduce the current density

$$\vec{j}(\vec{r}, t) = n \frac{\partial \vec{u}(\vec{r}, t)}{\partial t} \tag{5.237}$$

as its conjugate field. The equation of motion (5.233), when written in terms of the Fourier transform of the current density, takes the form

$$-i\omega m \vec{j}(\vec{q}, \omega) = \left[\frac{B}{n} + \left(1 - \frac{2}{d} \right) \frac{S}{n} \right] \frac{\vec{q}[\vec{q} \cdot \vec{j}(\vec{q}, \omega)]}{i\omega}$$

$$+ \frac{S}{n} \frac{q^2}{i\omega} \vec{j}(\vec{q}, \omega) - i\omega n \frac{e}{c} \vec{A}(\vec{q}, \omega). \tag{5.238}$$

Both the current and the vector potential can be split into longitudinal and transverse components (parallel and perpendicular to $\vec{q}$ respectively) $\vec{j} = \vec{j}_L + \vec{j}_T$, $\vec{A} = \vec{A}_L + \vec{A}_T$, and the two components satisfy independent equations of motion. The solution of Eq. (5.238) therefore is

$$\vec{j}_L(\vec{q}, \omega) = \frac{n/m}{1 - \left[\frac{B}{n^2} + 2 \left(1 - \frac{1}{d} \right) \frac{S}{n^2} \right] \frac{nq^2}{m\omega^2}} \frac{e}{c} \vec{A}_L(\vec{q}, \omega) \tag{5.239}$$

and

$$\vec{j}_T(\vec{q}, \omega) = \frac{n/m}{1 - \frac{S}{n^2} \frac{nq^2}{m\omega^2}} \frac{e}{c} \vec{A}_T(\vec{q}, \omega) . \tag{5.240}$$

In writing these equations we have assumed that the longitudinal external field $\vec{A}_L$ (which is completely equivalent, up to a gauge transformation, to a scalar potential) includes the contribution of the Hartree field generated by the induced density n_1. We also recall that $\vec{j}_L(\vec{q}, \omega)$ is related to the induced density change $n_1(\vec{q}, \omega)$ by the continuity equation

$$n_1(\vec{q}, \omega) = \frac{\vec{q}}{\omega} \cdot \vec{j}_L(\vec{q}, \omega) . \tag{5.241}$$

Next we consider the classical hydrodynamical equation of motion for the current density in a liquid – the so-called *Navier–Stokes equation* [Landau and Lifshitz, *Course of Theoretical Physics*, Vol. 6]. Assuming that the fluid remains close to a homogeneous equilibrium state, the linearized form of this equation is

$$m \frac{\partial \vec{j}}{\partial t} = -\vec{\nabla} P + \left[\frac{\eta}{n} + \left(1 - \frac{2}{d} \right) \frac{\zeta}{n} \right] \vec{\nabla}(\vec{\nabla} \cdot \vec{j}) + \frac{\eta}{n} \nabla^2 \vec{j} + \vec{F}(\vec{r}, t) , \tag{5.242}$$

where P is the local equilibrium pressure (a function of density and temperature), and η and ζ are two constants, the shear and the bulk viscosity respectively. The gradient of the pressure is directly proportional (neglecting temperature variations) to the gradient of the density and can be written as follows:

$$\vec{\nabla} P = \frac{\partial P(n)}{\partial n} \vec{\nabla} n = \frac{\mathcal{B}(n)}{n} \vec{\nabla} n , \qquad (5.243)$$

where $\mathcal{B}(n) = n \frac{\partial P(n)}{\partial n}$ is the bulk modulus of the homogeneous liquid. Finally, Fourier transforming Eq. (5.242) exactly as we did earlier for the equation of elasticity, and making use of the continuity equation $\frac{\partial n}{\partial t} = -\vec{\nabla} \cdot \vec{j}$, it is easy to arrive at

$$-i\omega m \vec{j}(\vec{q}, \omega) = \left[\frac{\mathcal{B}}{n} - \frac{i\omega\zeta}{n} - \left(1 - \frac{2}{d}\right) \frac{i\omega\eta}{n} \right] \frac{\vec{q}[\vec{q} \cdot \vec{j}(\vec{q}, \omega)]}{i\omega}$$
$$+ \frac{i\omega\eta}{n} \frac{q^2}{i\omega} \vec{j}_1(\vec{q}, \omega) - i\omega n \frac{e}{c} \vec{A}(\vec{q}, \omega) . \qquad (5.244)$$

Eqs. (5.244) and (5.238) are obviously very similar; they only differ as follows:

1. The shear modulus S of Eq. (5.238) is replaced by the imaginary quantity $-i\omega\eta$ in Eq. (5.244). This quantity vanishes at $\omega = 0$ in agreement with the intuitive notion that a liquid has no resistance to shear.
2. The bulk modulus $\mathcal{B}$ of Eq. (5.238) acquires an imaginary part $-i\omega\zeta$ in Eq. (5.244).

These observations suggest the use of a single language – that of elasticity theory – to describe both the liquid and the solid. In this unified language the equation of motion takes the form

$$-i\omega m \vec{j}(\vec{q}, \omega) = \left[\frac{\tilde{\mathcal{B}}(\omega)}{n} + \left(1 - \frac{2}{d}\right) \frac{\tilde{S}(\omega)}{n} \right] \frac{\vec{q}[\vec{q} \cdot \vec{j}(\vec{q}, \omega)]}{i\omega}$$
$$+ \frac{\tilde{S}(\omega)}{n} \frac{q^2}{i\omega} \vec{j}(\vec{q}, \omega) - i\omega n \frac{e}{c} \vec{A}(\vec{q}, \omega), \qquad (5.245)$$

where

$$\tilde{S}(\omega) = S(\omega) - i\omega\eta(\omega) \qquad (5.246)$$

and

$$\tilde{\mathcal{B}}(\omega) = \mathcal{B}(\omega) - i\omega\zeta(\omega) \qquad (5.247)$$

are *complex visco-elastic "constants"*. Notice that we are now allowing the real quantities S, η, $\mathcal{B}$, and ζ to be functions of frequency. The solution of Eq. (5.245) is

$$\vec{j}_L(\vec{q}, \omega) = \frac{n/m}{1 - \left[\frac{\tilde{\mathcal{B}}(\omega)}{n^2} + 2 \left(1 - \frac{1}{d}\right) \frac{\tilde{S}(\omega)}{n^2} \right] \frac{nq^2}{m\omega^2}} \frac{e}{c} \vec{A}_L(\vec{q}, \omega) \qquad (5.248)$$

and

$$\vec{j}_T(\vec{q}, \omega) = \frac{n/m}{1 - \frac{\tilde{S}(\omega)}{n^2} \frac{nq^2}{m\omega^2}} \frac{e}{c} \vec{A}_T(\vec{q}, \omega) . \qquad (5.249)$$

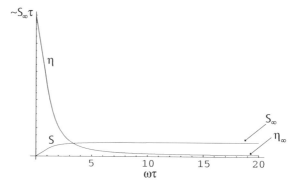

Fig. 5.31. Qualitative evolution of the shear modulus S and the shear viscosity η vs frequency. The frequency scale is set by the relaxation rate $\frac{1}{\tau}$ which, in the electron liquid, varies as T^2 (see Chapter 8). $S(\omega)$ goes from ~ 0 for $\omega \ll \frac{1}{\tau}$ to S_∞ for $\frac{1}{\tau} \ll \omega \ll \frac{\epsilon_F}{\hbar}$. Over the same frequency range η goes from $\sim S_\infty \tau$ to η_∞. The quantities S_∞ and η_∞ coincide with $S(0)$ and $\eta(0)$ of the zero-temperature theory.

The key difference between a solid and a liquid is that the solid has an essentially real $\tilde{S}$ (S finite, $\eta \sim 0$), whereas the liquid has an essentially imaginary $\tilde{S}$ (η finite, $S \sim 0$). Of course, whether liquid-like or solid-like behavior prevails in a given material is ultimately a matter of time scales: Fig. 5.31 shows the crossover from the former to the latter as the frequency of the disturbance increases from below to above the inverse of the appropriate equilibration time τ. On the other hand, the bulk modulus does not show a significant frequency dependence on the scale of $\frac{1}{\tau}$. The bulk viscosity is generally of the same order of magnitude as the shear viscosity, but, in the case of the electron liquid, it will been shown to vanish for ω tending to 0.

5.9.2 Visco-elastic constants of the electron liquid

To identify the visco-elastic constants of the electron liquid all we need to do is to compare Eqs. (5.248) and (5.249) with the expressions obtained from the microscopic linear response theory in the limit of small wave vector. We start from the defining relations

$$\vec{j}_{L(T)}(\vec{q}, \omega) = \chi_{L(T)}(q, \omega)\vec{A}_{L(T)}(\vec{q}, \omega), \tag{5.250}$$

and notice that the small-q expansion of the proper current–current response functions (5.131) and (5.133) at finite frequency has the form

$$\tilde{\chi}_{L(T)}(q, \omega) \simeq \frac{n}{m}\left[1 + \alpha_{L(T)}^{(d)}\frac{\epsilon_F}{n}\frac{nq^2}{m\omega^2} + f_{xcL(T)}(\omega)\frac{nq^2}{m\omega^2}\right], \tag{5.251}$$

where $\alpha_L^{(3)} = 6/5, \alpha_L^{(2)} = 3/2, \alpha_T^{(3)} = 2/5, \alpha_T^{(2)} = 1/2$.

Comparing Eq. (5.251) with the macroscopic response functions (5.248) and (5.249) we arrive at the following identifications:

$$\tilde{S}(\omega) = \alpha_T^{(d)}n\epsilon_F + n^2 f_{xcT}(\omega) \tag{5.252}$$

and

$$\tilde{\mathcal{B}}(\omega) = \left[\alpha_L^{(d)} - \left(2 - \frac{2}{d}\right)\alpha_T^{(d)}\right] n\epsilon_F$$
$$+ n^2\left[f_{xcL}(\omega) - \left(2 - \frac{2}{d}\right)f_{xcT}(\omega)\right]. \qquad (5.253)$$

Separating the real and imaginary parts of these equations we arrive at the promised expressions for the visco-elastic "constants" in terms of the long wavelength limit of the local field factors:

$$\mathcal{S}(\omega) = \alpha_T^{(d)} n\epsilon_F + n^2 f_{xcT}(\omega), \qquad (5.254)$$

$$\mathcal{B}(\omega) = \left[\alpha_L^{(d)} - \left(2 - \frac{2}{d}\right)\alpha_T^{(d)}\right] n\epsilon_F$$
$$+ n^2\left[f_{xcL}(\omega) - \left(2 - \frac{2}{d}\right)f_{xcT}(\omega)\right], \qquad (5.255)$$

$$\eta(\omega) = -n^2\frac{\Im m f_{xcT}(\omega)}{\omega}, \qquad (5.256)$$

$$\zeta(\omega) = -n^2\left[\frac{\Im m f_{xcL}(\omega)}{\omega} - \left(2 - \frac{2}{d}\right)\frac{\Im m f_{xcT}(\omega)}{\omega}\right] = 0. \qquad (5.257)$$

Thus, we see that the quantity $n^2\left[f_{xcL}(0) - \left(2 - \frac{2}{d}\right)f_{xcT}(0)\right]$ is indeed the exchange-correlation contribution to the static bulk modulus, and should therefore be identified with $n^2\frac{d^2[n\epsilon_{xc}(n)]}{dn^2}$ as anticipated in Eq. (5.194):

$$f_{xcL}(0) - \left(2 - \frac{2}{d}\right)f_{xcT}(0) = \frac{d^2[n\epsilon_{xc}(n)]}{dn^2}. \qquad (5.258)$$

Similarly, we see that the bulk viscosity $\zeta(\omega)$ vanishes in the limit $\omega \to 0$ by virtue of Eq. (5.196).

The presence of the shear modulus in Eq. (5.245) is (as discussed at the end of Section 5.3.3) essential to obtaining the correct plasmon dispersion, since it takes into account the fact that the local Fermi surface deviates from its quasi-equilibrium shape (i.e., a sphere). More puzzling is the fact that a finite shear modulus persists in the limit of zero frequency. The reason for this strange behavior is that we are taking the $q \to 0$ limit *before* the $\omega \to 0$ limit: because of the infinite wavelength of the disturbance there is not enough time for the electrons to establish local equilibrium, and the system remains "dynamical" down to zero frequency. Even at $q = 0$, however, a complete analysis must take into account the fact that the temperature is never exactly zero. As the frequency becomes smaller than a characteristic relaxation rate $\frac{1}{\tau} \sim \frac{(k_BT)^2}{\hbar\epsilon_F}$ the electron liquid enters the truly static response regime and the shear modulus drops to zero as shown in Fig. 5.31. At the same time the shear viscosity becomes very large i.e., $\eta \sim \mathcal{S}_\infty\tau$, where $\mathcal{S}_\infty$ is the value of $\mathcal{S}(\omega)$ at frequencies large compared to $\frac{1}{\tau}$ but small compared to $\frac{\epsilon_F}{\hbar}$ (this is a consequence of the dispersion relations). It is only at strictly zero temperature that the system remains "dynamical" all the way down to zero frequency.

5.9.3 Spin diffusion

To complete our discussion of the macroscopic dynamics of the electron liquid we want to show that the divergence of the spin local field factor, $G_-(q, \omega)$, at small q and finite ω leads to the appearance of a diffusion pole in the spin–spin response function at finite temperature. We have already encountered a diffusion pole in the density–density response function of an electron liquid in the presence of impurities (see Section 4.6); but the diffusion pole in the spin–spin response function occurs even in the absence of impurities, since it is caused by the interaction between up- and down-spins.

To show this, we go back to our discussion of $G_-(q, \omega)$ at the end Section 5.6.2.2. In the relevant regime of wave vectors $q \ll k_F$ and frequencies $\omega \ll \frac{\epsilon_F}{\hbar}$ we can write (see, however, footnote 36)

$$v_q G_-(q, \omega) \simeq v_q G_-(q, 0) + \frac{i\omega m}{nq^2 \tau_{sd}} , \qquad (5.259)$$

which combines the singular behavior of Eq. (5.191), valid for $\omega \gg qv_F$, with the static limit $v_q G_-(q, 0)$, for $\omega \ll qv_F$. This expression can now be substituted in Eq. (5.110) for the spin response. For $q \ll k_F$ and $\omega \ll qv_F$ we can approximate $\chi_0(q, \omega) \approx -N(0)$. A simple calculation then gives

$$\chi_{S_z S_z}(q, \omega) = \chi(0, 0) \frac{D_S q^2}{-i\omega + D_S q^2}, \qquad (5.260)$$

where

$$\chi(0, 0) = -\lim_{q \to 0} \frac{N(0)}{1 - v_q G_-(q, 0)N(0)} \qquad (5.261)$$

is the static spin susceptibility, and

$$D_S = \frac{n\tau_{sd}}{m|\chi(0, 0)|} \qquad (5.262)$$

is the *spin diffusion constant*. The form of Eq. (5.260) shows that longitudinal spin fluctuations obey a diffusion equation even though there are no impurities in the system, provided the temperature is finite:[46] the diffusive behavior of the spin is intrinsic to the interacting electron liquid.

Exercises

5.1 **Stiffness theorem and dielectric theory in RPA.** Within linear response, the effect of an external charge distribution $en\rho_{\vec{q}} \cos \vec{q} \cdot \vec{r}$ on an otherwise homogeneous three dimensional electron liquid of density n can be described in terms of an effective single particle potential $u \cos \vec{q} \cdot \vec{r}$ acting on each electron.

(a) Determine to leading order in u the increase of the kinetic energy per particle within the non-interacting electron approximation (see also Exercise 1 of Chapter 3).

---

[46] At strictly zero temperature $D_S \to \infty$ for $\omega \to 0$ – see footnote [36].

(b) Show that to linear order in u the electronic density acquires a modulation $n\gamma \cos\vec{q}\cdot\vec{r}$ with $\gamma = \frac{\chi_0(q)u}{n}$, where $\chi_0(q)$ is the static Lindhard function given by Eq. (4.26).

(c) Determine the equilibrium value of u by minimizing with respect to this variable the sum of the (noninteracting) kinetic energy and the electrostatic energy

$$\frac{U(u)}{N} = \frac{\pi n e^2}{q^2}\left(\rho_{\vec{q}} - \gamma\right)^2 .$$

(d) Show that the corresponding total energy change per particle can then be expressed as

$$\frac{\Delta E}{N} = \frac{\pi n e^2 \rho_{\vec{q}}^2}{q^2 \epsilon^{RPA}(q)} ,$$

where we have defined the static RPA dielectric constant as $\epsilon^{RPA}(q) = 1 - v_q \chi_0(q)$.

5.2 **Estimating the screening length.** A seemingly rough guess for the screening length can be obtained by multiplying the typical velocity of propagation of the electrons, v_F, by the typical time scale for charge density variations in the system, i.e. ω_p (see Eq. (5.48)). In 3D except for a numerical factor, this is of course just another way of rewriting Eq. (5.36). Explain why, as obvious from Eq. (5.38), the same type of expression does not apply to the two dimensional case.

5.3 **Measuring the proper compressibility of the electron liquid.** Derive Eq. (5.26) for the variation δE_p of the electric field between the two two-dimensional electron layers in the experimental setup of Fig. 5.3.

5.4 **Equation for the plasmon critical wave vector in 3D.** Show that the critical wave vector q_c for which the plasmon of a three dimensional electron gas enters into the electron–hole continuum in the RPA is determined by the equation

$$\frac{k_3^2}{k_F^2 y_c^2}\left(\frac{1}{2} + \frac{y_c^2 + 2y_c}{y_c}\ln\left|\frac{2+y_c}{y_c}\right|\right) = 1 ,$$

where $y_c = \frac{q_c}{k_F}$.

5.5 **Analytical form of the plasmon dispersion in 2D, I.** Determine the exact RPA dispersion relation for the plasmon of a two dimensional electron gas, Eq. (5.54), and prove that the plasmon mode touches (remaining parallel to) the electron–hole continuum at a critical wave vector q_c determined by the equation $\frac{q_c^2}{\sqrt{2}r_s} + \frac{q_c^3}{4r_s} = 1$.

5.6 **Plasmon oscillator strength.** Making use of Eq. (5.57) and of the analytical expressions for $\epsilon^{RPA}(q,\omega)$, show that the plasmon oscillator strength vanishes at the upper boundary of the electron–hole continuum as $-\frac{1}{\ln|q-q_c|}$ in 3D and as $\sqrt{|q-q_c|}$ in 2D. q_c is the critical wave vector at which the plasmon dispersion enters the electron–hole continuum.

5.7 **Plasmon dispersion from spectral sum rules.** Because the density fluctuation spectrum is dominated, at small wave vectors, by the plasmon, one can write $-\pi^{-1}\Im m\chi_{nn}(q,\omega) \simeq A(q)\delta(\omega - \Omega_p(q))$, where $\Omega_p(q)$ is the plasmon dispersion.

Use the first and third moment sum rules to determine $A(q)$ and $\Omega_p(q)$ to order q^2. Ignoring interaction corrections to the third moment recover the RPA results, i.e., Eqs. (5.52) and (5.53).

5.8 **Correlation energy in the Thomas-Fermi approximation.** Try and estimate the correlation energy of an electron liquid as the difference between the expectation value of the Thomas-Fermi screened interaction in the non-interacting ground-state, and the expectation value of the bare coulomb interaction in the same state. Show that this approach results in a positive, and therefore unphysical, correlation energy and leads to erroneous corrections to the compressibility and the spin susceptibility.

5.9 **Macke's calculation of the correlation energy.** Derive the leading term in the expression for the correlation energy per particle of three dimensional jellium in the $r_s \to 0$ limit by means of the following procedure due to Macke (1950):
  (i) Obtain first an approximate expression for the ground-state wavefunction by correcting the non-interacting Fermi sea to first order within a perturbative expansion in terms of the statically screened Thomas-Fermi potential assumed as an *ad hoc* effective electron-electron interaction.
  (ii) Obtain the correlation energy by calculating the expectation value of the exact hamiltonian in this state.

5.10 **High-density expansion of the correlation energy in 2D.** Following the pattern of the three dimensional calculation obtain the leading term in the high-density expansion of the correlation energy of the two dimensional electron liquid.

5.11 **Low-density behavior of the RPA correlation energy in 2D.** Obtain the the large-r_s form of the RPA correlation energy for the two dimensional electron liquid (see Eq. (5.86) for the result in 3D).

5.12 **Analytical form of the plasmon dispersion in 2D, II.** Neglecting the frequency dependence of the local field factor G_+ obtain an analytic expression similar to Eq. (5.54) for the dispersion of the two dimensional plasmon including the local field correction.

5.13 **Spin-resolved density–density response functions.** Derive the expressions (5.127) for the spin-resolved density–density response functions $\chi_{\sigma\sigma'}(q, \omega)$ of the homogeneous electron liquid.

5.14 **Acoustic plasmon in a bilayer system.** Using the formalism for multicomponent systems described in Section 5.4.5 within the RPA show that a system of two 2D layers separated by a distance d supports two collective modes: the ordinary plasmon, with dispersion $\sim q^{1/2}$, and an acoustic plasmon, with dispersion linear in q for small q. For identical layers, determine the acoustic plasmon velocity and show that the excitation is never Landau damped (Santozo and Giuliani, 1988a). [Hint: the Fourier transform of the coulomb interaction between the layers is $\frac{2\pi e^2}{q}e^{-qd}$.]

5.15 **Effective interactions in the electron–hole liquid.** Derive the electron-electron, hole-hole and electron–hole effective interactions in an electron hole liquid, in which the electrons and the holes have equal densities, opposite charges, and masses m_e and m_h, with $m_h > m_e$. Show that the effective electron-electron interaction can be cast in a form similar to Eq. (5.165), with the holes playing the part of the ions. [Hint: take a look at the paper by Vignale and Singwi (1984).]

5.16 **Vashishta–Singwi theory for the local field factor.** The STLS local field factor introduced by Singwi *et al.* (1968) does not satisfy the compressibility sum rule. The theory of Vashishta and Singwi (1972) (henceforth referred to as VS) corrects this problem while still yielding a pair correlation function (and hence an energy) of comparable quality. The idea of VS is to improve upon the STLS by including the density dependence of the equilibrium pair correlation function. For slowly varying densities this leads to

$$G_+^{VS}(q) = \left(1 + a(n)n\frac{\partial}{\partial n}\right)\left(-\frac{1}{n}\int\frac{d\vec{q}'}{(2\pi)^d}\frac{\vec{q}\cdot\vec{q}'}{q^2}\frac{v_{q'}}{v_q}[S(\vec{q}-\vec{q}')-1]\right), \quad (E5.1)$$

where $a(n)$ is a slowly varying function of the equilibrium density (to be determined) and $S(\vec{q})$ is the structure factor.

(a) Show that the compressibility sum-rule

$$\lim_{q\to 0} v_q G_+^{VS}(q) = \frac{1}{n^2 K_0}\left(1 - \frac{K_0}{K}\right) \quad (E5.2)$$

is satisfied with $a(n) = \frac{1}{2}$ if one ignores the difference between the interacting and non-interacting kinetic energy.

[Hint: recall that according to the quantum mechanical virial theorem, proven in Chapter 1, the compressibility is given by

$$\frac{1}{K} = n\frac{dP}{dn} = \frac{2}{d}n\frac{d[nt(n)]}{dn} + \frac{1}{d}n\frac{d[n\epsilon_{pot}(n)]}{dn}, \quad (E5.3)$$

where $t(n)$ and $\epsilon_{pot}(n)$ are the kinetic and potential energies per electron. Multiplying this by K_0 and expressing the potential energy in terms of the pair correlation function we arrive at

$$\frac{K_0}{K} = \frac{1}{n\epsilon_F}\left\{n\frac{\partial}{\partial n}[nt(n)] + \frac{n^2}{2}\int d\vec{r}\frac{e^2}{r}\right.$$
$$\times\left.\left([g(r,n)-1] + \frac{n}{2}\frac{\partial}{\partial n}[g(r,n)-1]\right)\right\}. \quad (E5.4)$$

Compare this to the small-q limit of Eq. (E5.1).]

(b) When the difference between interacting and non-interacting kinetic energy *is* taken into account, the compressibility sum rule determines the dependence of $a(n)$ on the density. Show that $a(n)$ varies in the range $\frac{1}{2} \le a(n) \le 1$.

5.17 **Large-q expansion of the Lindhard function.** Derive Eq. (5.173) for the large-q expansion of the static Lindhard function.

5.18 **Large-q behavior of the static structure factor from the fluctuation-dissipation theorem.** Derive the large-q behavior of the static structure factor, Eq. (A4.19), from the fluctuation-dissipation theorem (Eq. (5.118)) by making use of the large-q behavior of the local field factor given in footnote.[30] [Hints: for large q expand the density–density response function (5.168) as $\chi_{nn}(q,\omega) \simeq \bar{\chi}_0(q,\omega) + v_q[1 - \bar{G}_+(q,\omega)]\bar{\chi}_0^2(q,\omega)$. In applying the fluctuation-dissipation theorem do the frequency integral along the imaginary frequency axis (as in Eq. (3.159)), and notice that $\bar{\chi}_0(q,i\omega)$ can be replaced by $\chi_0(q,i\omega)$ because the exact occupation number n_k tends to zero, for large k, faster than $\frac{1}{k^{d+2}}$, as required by the finiteness of the kinetic energy.]

5.19 **Regular combination of spin-resolved local field factors.** Due to the conservation of the total momentum in the electron liquid, there is only one linear combination of the quantities $v_q G_{\sigma\sigma'}(q,\omega)$ that remains finite in the limit $q \to 0$ and finite ω. In the paramagnetic case, this is the spin-symmetric sum $v_q G_{\uparrow\uparrow}(q,\omega) + v_q G_{\uparrow\downarrow}(q,\omega)$. What is the form of the regular combination in the general spin-polarized case?

5.20 **High-frequency limit of the exchange-correlation kernels (real part).** Derive Eq. (5.200) for the high-frequency limit of the real parts of the exchange-correlation kernels by taking the small-q limit of the third-moment sum rule.

5.21 **High-frequency limit of the exchange-correlation kernels (imaginary part).** Work out the steps leading from Eq. (5.204) to Eq. (5.209) and then from Eq. (5.209) to the high-frequency form (5.201) of the imaginary part of the exchange-correlation kernel.

5.22 **Vanishing of the bulk viscosity at zero frequency.** Work out in detail the derivation of Eq. (5.211) for the low-frequency limit of the bulk viscosity.

5.23 **Time-dependent Hartree–Fock theory of linear response.** Derive the integral equation (5.226) for the time-dependent Hartree–Fock response functions.

6

The perturbative calculation of linear response functions

6.1 Introduction

No textbook on many-body physics would be complete without a discussion of perturbative methods. In this chapter we focus on the perturbation theoretic approach to the calculation of linear response functions. The problem can be stated as follows: we assume that the (time-independent) hamiltonian has the form

$$\hat{H} = \hat{H}_0 + \hat{H}_1 \tag{6.1}$$

where $\hat{H}_0$ is the hamiltonian of the non-interacting system, and $\hat{H}_1$ represents the interaction. Our goal is to expand the linear response function

$$\chi_{AB}(t) = -\frac{i}{\hbar}\Theta(t)\langle[\hat{A}(t), \hat{B}(0)]\rangle_0 \tag{6.2}$$

in a power series of the dimensionless coupling constant that controls the strength of H_1 relative to H_0.[1] The zero-th order approximation is given by the non-interacting formula (4.8) of Chapter 4. The fundamental assumption on which perturbation theory rests is that such a power series expansion exists, at least in the sense of an *asymptotic series*.[2] In more physical terms, one assumes that the interacting ground-state (or thermal equilibrium state, if we are at finite temperature) can be obtained from the non-interacting one by slowly and continuously "switching on" the interaction.

At first sight, one might be skeptical about the usefulness of a method which is supposed to work well only in the limit of small coupling constant, when most systems of interest have coupling constants of order 1 or larger, and which, moreover, does not seem to apply to broken-symmetry states, as these are not continuously connected to the non-interacting ground-state. It turns out, however, that such a negative view is not entirely justified. One of the great merits of perturbation theory is that it leads to a suggestive graphical representation of physical processes in terms of sequences of free electron propagations and electron-electron interactions. The existence of this graphical representation allows us to identify certain basic quantities such as the "self-energy", the "vertex correction", and the

---

[1] In the case of the electron liquid this coupling constant is r_s.
[2] An asymptotic power series in a coupling constant α is one in which the sum of the first n terms becomes increasingly accurate for $\alpha \to 0$, even though the infinite sum may fail to converge at any given value of α.

"irreducible electron–hole interaction", and to discern exact relationships between them. These relationships typically are cast in the form of integral equations, which can be solved to generate non-perturbative results (including spontaneously broken symmetry) even when their inputs are approximated by low-order perturbation theory. This is the main reason why perturbation theory is useful for strongly interacting systems.

Perturbation theory comes in two versions. The first one is limited to zero temperature and yields an expansion for the response functions in real time or (after Fourier transformation) in real frequency. The second version deals with finite temperature and yields an expansion for the response function in *imaginary time*, or imaginary frequency. The finite temperature results must be analytically continued to real time (or real frequency) in order to obtain the physical response function.

One might ask what is the need for the zero temperature formalism when one can do all the calculations (as well as the experiments) at finite temperature, and take the $T \to 0$ limit only at the very end. The answer is that the analytic continuation from imaginary to real frequencies is a very difficult task when the function to be continued is known only numerically. In such cases the zero temperature formalism provides a more direct access to the physically measurable quantities.

In the following sections we will present both the zero temperature and the finite temperature formalisms, in this order. In both cases the basic strategy will be to express the causal response function (6.2) in terms of a *time-ordered* correlation function, which can be "easily" expanded in powers of the interaction.[3]

6.2 Zero-temperature formalism

6.2.1 Time-ordered correlation function

For reasons that will become clear as we proceed, perturbation theory is more easily formulated for the *time-ordered* correlation function

$$\chi_{AB}^T(t) \equiv -\frac{i}{\hbar} \langle T[\hat{A}(t)\hat{B}] \rangle_0 \qquad (6.3)$$

than for the original causal response function (6.2).[4] We adopt the Heisenberg picture of the time evolution, in which the operators evolve according to the unitary transformation

$$\hat{A}(t) = e^{\frac{i\hat{H}t}{\hbar}} \hat{A} e^{-\frac{i\hat{H}t}{\hbar}} \qquad (6.4)$$

(with $\hat{A}$ being the time-independent operator of the Schrödinger picture), while the states are independent of time. Operators without an explicit time argument, such as $\hat{B}$ in Eq. (6.3), are to be evaluated at $t = 0$, where the Heisenberg and the Schrödinger pictures coincide.

---

[3] It should be noted that, by focusing our discussion on the linear response functions of systems described by a time-independent hamiltonian, we exclude many interesting phenomena that occur far from equilibrium. Although these phenomena are amenable to perturbation theory, the corresponding formalism will not be described here. The interested reader will find an excellent introduction to this subject in the review articles by D. Langreth (1975) and J. Rammer and H. Smith (1986).

[4] As in Chapter 3, we assume that both $\hat{A}$ and $\hat{B}$ have zero average in the ground-state. If this were not the case we would have to work with $\hat{A} - \langle \hat{A} \rangle$ and $\hat{B} - \langle \hat{B} \rangle$.

The time-ordering operator T acting on a product of Heisenberg operators, yields a new product in which the operators are arranged in *chronological order*, i.e., the time decreases from left to right. For example

$$T[\hat{A}(t_1)\hat{B}(t_2)] = \begin{cases} \hat{A}(t_1)\hat{B}(t_2), & t_1 > t_2 \\ \hat{B}(t_2)\hat{A}(t_1), & t_2 > t_1 \end{cases} , \tag{6.5}$$

and, more generally,

$$T[\hat{A}_1(t_1)\hat{A}_2(t_2)\ldots\hat{A}_n(t_n)] = \sum_P \Theta(t_{P_1} - t_{P_2})\Theta(t_{P_2} - t_{P_3})\ldots\Theta(t_{P_{n-1}} - t_{P_n})$$
$$\times \hat{A}_{P_1}(t_{P_1})\hat{A}_{P_2}(t_{P_2})\ldots\hat{A}_{P_n}(t_{P_n}), \tag{6.6}$$

where the sum runs over all the permutations P ($i \to P_i$) of the indices $i = 1, 2 \ldots n$. The above definition holds for operators that can be expressed in terms of an *even number of* fermion or boson creation/destruction operators. *When this is not the case, the T-product must include a minus sign for every interchange of two fermion operators.* Notice that a time-ordered product is invariant under arbitrary permutations of its factors.

A simple relationship exists between the time-ordered correlation function and the causal response function. This is best expressed as a relationship between the Fourier transforms of these two functions. Let us define

$$\chi_{AB}^T(\omega) = \lim_{\eta \to 0^+} \int_{-\infty}^{\infty} \chi_{AB}^T(t)e^{i\omega t - \eta|t|}dt . \tag{6.7}$$

Inserting a sum over a complete set of eigenstates of $\hat{H}$ between $\hat{A}(t)$ and $\hat{B}$ in the expression (6.3) for $\chi_{AB}^T(\omega)$, and then evaluating the integral in Eq. (6.7), we get

$$\chi_{AB}^T(\omega) = \frac{1}{\hbar}\sum_n \left(\frac{A_{0n}B_{n0}}{\omega - \omega_{n0} + i\eta} - \frac{B_{0n}A_{n0}}{\omega + \omega_{n0} - i\eta} \right) , \tag{6.8}$$

where $A_{0n} = \langle\psi_0|\hat{A}|\psi_n\rangle$ is the matrix elements of $\hat{A}$ between the ground-state, $|\psi_0\rangle$, and the n-th excited state, $|\psi_n\rangle$, of $\hat{H}$, and $\hbar\omega_{n0} = E_n - E_0 > 0$ is the corresponding excitation energy.

Eq. (6.8) differs from the expression for the causal response function at zero temperature, i.e.,

$$\chi_{AB}(\omega) = \frac{1}{\hbar}\sum_n \left(\frac{A_{0n}B_{n0}}{\omega - \omega_{n0} + i\eta} - \frac{B_{0n}A_{n0}}{\omega + \omega_{n0} + i\eta} \right) , \tag{6.9}$$

only by the replacement of $+i\eta$ by $-i\eta$ in the denominator of the second term.

The relationship between Eqs. (6.8) and (6.9) takes a particularly simple form in the case $\hat{A} = \hat{B}^\dagger$, because $A_{0n}B_{n0} = |B_{n0}|^2$ is then real. Separating the real and imaginary parts of Eqs. (6.8) and (6.9) we obtain the important result

$$\Re e\chi_{B^\dagger B}(\omega) = \Re e\chi_{B^\dagger B}^T(\omega)$$
$$\Im m\chi_{B^\dagger B}(\omega) = \text{sign}(\omega)\,\Im m\chi_{B^\dagger B}^T(\omega) . \tag{6.10}$$

In the general case $\hat{A} \neq \hat{B}^\dagger$ the desired relationship can be obtained by applying Eq. (6.10) to the response functions $\chi_{\hat{A}+\hat{B}^\dagger,\hat{A}^\dagger+\hat{B}}$ and $\chi_{\hat{A}+i\hat{B}^\dagger,\hat{A}^\dagger-i\hat{B}}$. The detailed construction is left as an exercise (Exercise 1).

Thus, in every case, the calculation of a causal response function can be reduced to the calculation of time-ordered correlation functions.

6.2.2 The adiabatic connection

Perturbation theory begins by establishing a one-to-one correspondence between the states of the system under study and those of the unperturbed system. This is done with the help of the following artifice. Consider the fictitious time-dependent hamiltonian

$$\hat{H}(t) = \hat{H}_0 + \hat{H}_1 e^{-\eta|t|} \qquad (6.11)$$

where the interaction, $\hat{H}_1$, is slowly switched-on and then turned-off again, both processes occurring on a time scale η^{-1} that is much longer than any physical time scale of the system. The two main properties of $\hat{H}(t)$ are (i) it reduces to the non-interacting hamiltonian $\hat{H}_0$ in the limit $t \to \pm\infty$, and (ii) by choosing a sufficiently small η, it provides an arbitrarily accurate approximation to the physical hamiltonian $\hat{H}$ within a "time window" large enough to include the times 0 and t. In other words, $\hat{H}(t)$ is equivalent to $\hat{H}$ during the physical observation time, but is continuously connected to $\hat{H}_0$ in the limits $t \to \pm\infty$.

In the Heisenberg picture of the time evolution the state of the system is constant and can therefore be evaluated at a large negative time, where it coincides with the ground-state of $\hat{H}_0$, which we denote by $|\Phi_0\rangle$. Hence we calculate the time-ordered correlation function as

$$\chi^T_{AB}(t) = -\frac{i}{\hbar} \lim_{\eta \to 0} \langle \Phi_0 | T[\hat{A}(t)\hat{B}] | \Phi_0 \rangle . \qquad (6.12)$$

It is a basic assumption of perturbation theory that the $\eta \to 0$ limit on the right-hand side of this equation exists. From now on, we shall not indicate this limit explicitly in our formulas. Obviously, the advantage of working with Eq. (6.12) rather than with the original Eq. (6.3) is that the average of the operator $\hat{A}(t)\hat{B}$ is evaluated in the non-interacting ground-state $|\Phi_0\rangle$ so that the power of Wick's theorem can be brought to bear.

The Heisenberg-picture operator $\hat{A}(t)$ is related to the Schrödinger-picture operator $\hat{A}$ by the unitary transformation

$$\hat{A}(t) = \hat{U}^{-1}(t,0)\hat{A}\hat{U}(t,0) , \qquad (6.13)$$

where the time-evolution operator $\hat{U}(t,0)$ satisfies the Schrödinger equation

$$i\hbar \frac{\partial}{\partial t}\hat{U}(t,0) = \hat{H}(t)\hat{U}(t,0) \qquad (6.14)$$

with initial condition $\hat{U}(0,0) = \hat{1}$. This operator can be more conveniently expressed as

$$\hat{U}(t,0) = e^{-\frac{i\hat{H}_0 t}{\hbar}} \hat{U}_I(t,0) , \qquad (6.15)$$

where $\hat{U}_I(t,0)$ represents the part of the time evolution that is due to the electron-electron interaction (as opposed to the "trivial" time-evolution, which is due to $\hat{H}_0$). It is easy to verify that $\hat{U}_I(t,0)$ satisfies the equation of motion

$$i\hbar\frac{\partial}{\partial t}\hat{U}_I(t,0) = \hat{H}_{1I}(t)e^{-\eta|t|}\hat{U}_I(t,0) \tag{6.16}$$

with the initial condition $\hat{U}_I(0,0) = \hat{1}$. This equation has the formal solution

$$\hat{U}_I(t,0) = T\exp\left(-\frac{i}{\hbar}\int_0^t \hat{H}_{1I}(t')e^{-\eta|t'|}dt'\right), \tag{6.?}$$

where the time-ordering acts on the power series expansion of the exponential. The operator

$$\hat{H}_{1I}(t) \equiv e^{\frac{i\hat{H}_0 t}{\hbar}}\hat{H}_1 e^{-\frac{i\hat{H}_0 t}{\hbar}} \tag{6.17}$$

is the interaction hamiltonian in the so-called *interaction picture* of the time evolution.[5] We will henceforth attach a subscript I to operators that evolve in time according to this picture; e.g., we set

$$\hat{A}_I(t) \equiv e^{\frac{i\hat{H}_0 t}{\hbar}}\hat{A}e^{-\frac{i\hat{H}_0 t}{\hbar}}. \tag{6.18}$$

Substituting Eqs. (6.15) and (6.13) in Eq. (6.12) we obtain

$$\chi_{AB}^T(t) = -\frac{i}{\hbar}\langle\Phi_0|\hat{T}[\hat{U}_I^{-1}(t,0)\hat{A}_I(t)\hat{U}_I(t,0)\hat{B}]|\Phi_0\rangle. \tag{6.19}$$

The right-hand side of this expression can be greatly simplified with the help of two rather obvious properties of the time-evolution operator:

$$(i)\quad \hat{U}_I(t_2,t_1) = \hat{U}_I^{-1}(t_1,t_2) = \hat{U}_I^\dagger(t_1,t_2) \tag{6.20}$$

(i.e., the inverse of the operator that propagates the state from t_1 to t_2 is the operator that propagates the state from t_2 to t_1), and

$$(ii)\quad \hat{U}_I(t_3,t_2)\hat{U}_I(t_2,t_1) = \hat{U}_I(t_3,t_1). \tag{6.21}$$

Let us focus on the case $t>0$ first. After writing $\hat{U}_I^{-1}(t,0) = \hat{U}_I(0,t) = \hat{U}_I(0,-\infty)\hat{U}_I(-\infty,\infty)\hat{U}_I(\infty,t)$, and noting that the operators that appear in the expansion of

$$U_I(0,-\infty) = T\exp\left(-\frac{i}{\hbar}\int_{-\infty}^0 \hat{H}_{1I}(t')e^{-\eta|t'|}dt'\right) \tag{6.22}$$

are all at times earlier than 0, we see that Eq. (6.19) can be rewritten in the form

$$\chi_{AB}^T(t) = -\frac{i}{\hbar}\langle\Phi_0|\hat{U}_I(-\infty,\infty)\hat{U}_I(\infty,t)\hat{A}_I(t)\hat{U}_I(t,0)\hat{B}_I\hat{U}_I(0,-\infty)|\Phi_0\rangle$$
$$= -\frac{i}{\hbar}\langle\Phi_0|\hat{U}_I(-\infty,\infty)T[\hat{U}_I(\infty,t)\hat{A}_I(t)\hat{U}_I(t,0)\hat{B}_I\hat{U}_I(0,-\infty)]|\Phi_0\rangle. \tag{6.23}$$

At this point we take advantage of the freedom in changing the order of the operators within a time-ordered product to combine the $\hat{U}_I$'s into a single term $\hat{U}_I(\infty,t)\hat{U}_I(t,0)U_I(0,-\infty) =$

[5] The name is justified by the observation that, in this picture, the time-evolution of the *states* is entirely due to the interaction.

$\hat{U}_I(\infty, -\infty)$, so that Eq. (6.23) takes the form

$$\chi^T_{AB}(t) = -\frac{i}{\hbar}\langle\Phi_0|\hat{U}_I(-\infty, \infty)T[\hat{A}_I(t)\hat{B}_I\hat{U}_I(\infty, -\infty)]|\Phi_0\rangle. \tag{6.24}$$

Although this expression has been derived under the assumption $t > 0$, it is easy to see that it must also hold for $t < 0$ since both sides of the equation are invariant upon reversing the order of the operators $\hat{A}_I(t)$ and $\hat{B}_I$ (see Exercise 2).

Finally we observe that the time evolution described by $\hat{U}_I$ is *adiabatic* in the $\eta \to 0$ limit. Therefore, if the system starts from the non-interacting ground-state, it stays in the instantaneous ground-state of $\hat{H}(t)$ at all times, and eventually returns, for $t \to \infty$, to the ground-state of $\hat{H}_0$. From this somewhat dangerous argument we deduce that

$$\hat{U}_I(\infty, -\infty)|\Phi_0\rangle = e^{i\alpha}|\Phi_0\rangle \tag{6.25}$$

where the phase factor $e^{i\alpha}$ (allowed by this physical argument) is given by

$$e^{i\alpha} = \langle\Phi_0|\hat{U}_I(\infty, -\infty)|\Phi_0\rangle . \tag{6.26}$$

Returning to Eq. (6.24), and setting $\langle\Phi_0|\hat{U}_I(\infty, -\infty) = e^{-i\alpha}\langle\Phi_0|$, we arrive at the fundamental result

$$\chi^T_{AB}(t) = -\frac{i}{\hbar}\frac{\langle\Phi_0|T[\hat{A}_I(t)\hat{B}_I\hat{S}]|\Phi_0\rangle}{\langle\Phi_0|\hat{S}|\Phi_0\rangle} , \tag{6.27}$$

where we have defined the *scattering matrix* (also known as *S-matrix*)

$$\hat{S} \equiv \hat{U}_I(\infty, -\infty)$$
$$= T\exp\left(-\frac{i}{\hbar}\int_{-\infty}^{\infty}\hat{H}_{1I}(t')e^{-\eta|t'|}dt'\right) . \tag{6.28}$$

Eqs. (6.27) and (6.28) constitute the starting point for the perturbative treatment of the time-ordered correlation function. All the operators appearing in this formula evolve in time under the non-interacting hamiltonian, and all the averages are evaluated in the non-interacting ground-state.

6.2.3 The non-interacting Green's function

The ground-state averages that appear in the numerator and in the denominator of Eq. (6.27) can be calculated, according to Wick's theorem, by summing over all possible schemes of pairwise time-ordered contractions of creation and destruction operators. The *time-ordered contraction* of two creation/destruction operators $\hat{X}_I(t_1)$ and $\hat{Y}_I(t_2)$ is computed by first bringing the two operators next to each other in chronological order and then carrying out the ordinary contraction defined in Appendix 3.[6] This procedure yields

$$(\hat{X}_I(t_1)\hat{Y}_I(t_2))_T = T[\hat{X}_I(t_1)\hat{Y}_I(t_2)] - : \hat{X}_I(t_1)\hat{Y}_I(t_2) :$$
$$= \langle\Phi_0|T[\hat{X}_I(t_1)\hat{Y}_I(t_2)]|\Phi_0\rangle . \tag{6.29}$$

---

[6] An extra minus sign must be included if, in order to bring the two operators next to each other, we must perform an odd number of interchanges of Fermion operators.

In the rest of this chapter we will refer to time-ordered contractions simply as contractions. It is evident that the contractions of two creation or two destruction operators vanish. On the other hand, the contraction of a destruction operator with a creation operator is, in general, different from zero. This nonvanishing contraction is known as the *non-interacting Green's function*

$$G^{(0)}_{\alpha\beta}(t_1 - t_2) \equiv -i\langle\Phi_0|T[\hat{a}_{\alpha I}(t_1)\hat{a}^\dagger_{\beta I}(t_2)]|\Phi_0\rangle \,, \tag{6.30}$$

and will play a central role in the formulation of perturbation theory.

The physical meaning of the Green's function can be described as follows. For $t_1 > t_2$, $G^{(0)}_{\alpha\beta}(t_1 - t_2)$ is the probability amplitude for an electron added to the non-interacting Fermi sea in state β at time t_2 to be found in state α at time t_1. For $t_2 > t_1$, it is the probability amplitude for a *hole* created within the Fermi sea in state α at time t_1 to be found in state β at the later time t_2. Notice that these amplitudes depend only on the time difference $t = t_1 - t_2$.

In the above definition α and β are arbitrary elements of a complete set of single-particle states. For example, we could choose the "eigenstates" of the position operator, in which case we would have

$$\begin{aligned} G^{(0)}_{\vec{r}_1\sigma_1\,\vec{r}_2\sigma_2}(t_1 - t_2) &\equiv G^{(0)}_{\sigma_1\sigma_2}(\vec{r}_1, \vec{r}_2; t_1 - t_2) \\ &\equiv -i\langle\Phi_0|T[\hat{\Psi}_{\sigma_1,I}(\vec{r}_1, t_1)\hat{\Psi}_{\sigma_2,I}(\vec{r}_2, t_2)]|\Phi_0\rangle \,, \end{aligned} \tag{6.31}$$

where $\hat{\Psi}_\sigma(\vec{r})$ is the the field operator introduced in Section 1.4, and the subscript I specifies that its time evolution is governed by the non-interacting hamiltonian $\hat{H}_0$.

The calculation of $G^{(0)}_{\alpha\beta}(t)$ is particularly simple in the basis that diagonalizes $\hat{H}_0$, i.e., the basis in which

$$\hat{H}_0 = \sum_\alpha \varepsilon_\alpha \hat{a}^\dagger_\alpha \hat{a}_\alpha \,. \tag{6.32}$$

Evaluating the ground-state average we obtain

$$G^{(0)}_{\alpha\beta}(t) = -i\delta_{\alpha\beta}\left[\Theta(t)(1 - n_\alpha) - \Theta(-t)n_\alpha\right]e^{-i\frac{\varepsilon_\alpha t}{\hbar}}e^{-\eta|t|} \,, \tag{6.33}$$

where $n_\alpha = \Theta(\epsilon_F - \varepsilon_\alpha)$ are the usual zero-temperature occupation numbers. Notice that $G^{(0)}_{\alpha\beta}(t)$ is *discontinuous* at $t = 0$, jumping from $-i(1 - n_\alpha)$ for $t = 0^+$ to in_α for $t = 0^-$.

Consider, for example, the homogeneous electron liquid, where the α basis consists of plane waves of wave vector $\vec{k}$ and spin projection σ. Then, Eq. (6.33) becomes

$$G^{(0)}_{\vec{k}\sigma\,\vec{k}'\sigma'}(t) = -i\delta_{\vec{k}\vec{k}'}\delta_{\sigma\sigma'}\left[\Theta(t)(1 - n_{\vec{k}\sigma}) - \Theta(-t)n_{\vec{k}\sigma}\right]e^{-i\frac{\varepsilon_{\vec{k}\sigma}t}{\hbar}}e^{-\eta|t|} \,, \tag{6.34}$$

while its real space representation, Eq. (6.31), takes the form

$$G^{(0)}_{\sigma\sigma'}(\vec{r}_1, \vec{r}_2; t) = \frac{1}{L^d}\sum_{\vec{k}} G^{(0)}_{\vec{k}\sigma\,\vec{k}\sigma}(t)e^{i\vec{k}\cdot(\vec{r}_1 - \vec{r}_2)}\delta_{\sigma\sigma'} \,. \tag{6.35}$$

Exact analytic expressions for $G^{(0)}(r, t)$ are provided in Appendix 12 and are plotted vs time (at $r = 0$) in Fig. 6.1 below.

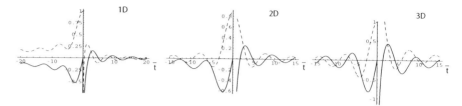

Fig. 6.1. Time dependence of the Green's function $n^{-1}G^{(0)}(r=0,t)$ for noninteracting fermions in d-dimensions. The solid line represents the real part of the Green's function, while the dashed line represents the imaginary part. $\bar{t} = \frac{\varepsilon_F t}{\hbar}$ is a dimensionless time and n is the density. Notice the discontinuity at $t=0$ and the oscillatory behavior at large times.

A much simpler expression is obtained by performing the Fourier transform of the Green's function with respect to time. This gives,

$$G_{\alpha\beta}^{(0)}(\omega) = \int_{-\infty}^{\infty} G_{\alpha\beta}^{(0)}(t)e^{i\omega t}e^{-\eta|t|}dt$$

$$= \delta_{\alpha\beta}\left[\frac{1-n_\alpha}{\omega - \frac{\varepsilon_g}{\hbar} + i\eta} + \frac{n_\alpha}{\omega - \frac{\varepsilon_g}{\hbar} - i\eta}\right]. \tag{6.36}$$

Notice the simple pole structure of this result. The result for the translationally invariant system is obtained by setting $\alpha = \vec{k}\sigma$ and $\beta = \vec{k}'\sigma'$, so that $\vec{k} = \vec{k}'$.

6.2.4 Diagrammatic perturbation theory

Let us return to the problem of evaluating Eq. (6.27). Consider the numerator first. Expanding the exponential in Eq. (6.28) as a power series in the interaction we obtain

$$\sum_{n=0}^{\infty} \frac{1}{n!}\left(-\frac{i}{\hbar}\right)^n \int_{-\infty}^{\infty} dt_1 \dots \int_{-\infty}^{\infty} dt_n \langle\Phi_0|T[\hat{A}_I(t)\hat{B}_I\hat{H}_{1\eta}(t_1)\dots\hat{H}_{1\eta}(t_n)]|\Phi_0\rangle, \tag{6.37}$$

where $\hat{H}_{1\eta}(t) \equiv \hat{H}_{1I}(t)e^{-\eta|t|}$. The denominator has a similar expansion, except that the "external" operators $\hat{A}$ and $\hat{B}$ are absent.

According to Wick's theorem the average is calculated by summing the products of the time-ordered contractions of all the possible pairs of operators, each set of pairs being multiplied by a factor $+1$ or -1 depending on whether the number of interchanges of fermion operators needed to bring the pairs together is even or odd.

Let as assume, for definiteness, that we have a two-body interaction

$$\hat{H}_1 = \frac{1}{2}\sum_{abcd} v_{abcd}\hat{a}_a^\dagger\hat{a}_b^\dagger\hat{a}_c\hat{a}_d \tag{6.38}$$

and that $\hat{A}$ and $\hat{B}$ are both single particle operators, i.e. $\hat{A} = \sum_{\alpha\beta} A_{\alpha\beta}\hat{a}_\alpha^\dagger\hat{a}_\beta$ and $\hat{B} = \sum_{\gamma\delta} B_{\gamma\delta}\hat{a}_\gamma^\dagger\hat{a}_\delta$. The rules for other cases of interest can be easily inferred from this basic example.

A pairing scheme for the n-th order term in Eq. (6.37) can be constructed by listing the $2n + 2$ destruction operators, and as many creation operators in two parallel columns, and choosing any one of the $(2n + 2)!$ ways of pairing an element of the left column with one of the right column.

For example, in the zeroth order ($n = 0$) we have only the four operators contained in $\hat{A}$ and $\hat{B}$

$$\hat{A}(t) \quad \hat{a}_\beta(t) \quad \hat{a}_\alpha^\dagger(t)$$
$$\hat{B} \quad \hat{a}_\delta \quad \hat{a}_\gamma^\dagger \tag{6.39}$$

which can be paired in two different ways: $\hat{a}_\beta(t)$ with $\hat{a}_\alpha^\dagger(t)$ and $\hat{a}_\delta$ with $\hat{a}_\gamma^\dagger$, or $\hat{a}_\beta(t)$ with $\hat{a}_\gamma^\dagger$ and $\hat{a}_\delta$ with $\hat{a}_\alpha^\dagger(t)$.

In the first order ($n = 1$) we already have eight operators

$$\hat{A}(t) \quad \hat{a}_\beta(t) \quad \hat{a}_\alpha^\dagger(t)$$
$$\hat{B} \quad \hat{a}_\delta \quad \hat{a}_\gamma^\dagger$$
$$\hat{H}_1(t_1) \begin{cases} \hat{a}_c(t_1) & \hat{a}_a^\dagger(t_1) \\ \hat{a}_d(t_1) & \hat{a}_b^\dagger(t_1) \end{cases} \tag{6.40}$$

which can be paired in $4! = 24$ ways. Notice that the operators with greek indices arise from $\hat{A}$ and $\hat{B}$ and are therefore called "external", while the operators with latin indices arise from $\hat{H}_1$'s and will be called "internal". Contraction schemes can be classified as either "disconnected" or "connected" according to the following definition. A scheme is said to be "disconnected" if one can find a subset of $\hat{H}_1$'s, whose creation and destruction operators are fully contracted among themselves; it is said to be "disconnected" if no such subset can be found.[7] We now prove that effectively *only the connected schemes contribute to the calculation of the correlation function*.

Consider an arbitrary contraction scheme arising in the evaluation of the n-th order term of Eq. (6.37). In general, this can be written as the product of a connected scheme of order $n_1 \leq n$ times a connected or disconnected scheme of order $n_2 = n - n_1$. There are $\frac{n!}{n_1! n_2!}$ ways of dividing the n interactions in Eq. (6.37) into two such groups. Therefore, we can rewrite Eq. (6.37) as

$$\sum_{n=0}^{\infty} \frac{1}{n!} \left(-\frac{i}{\hbar}\right)^n \sum_{n_1=0}^{\infty} \sum_{n_2=0}^{\infty} \delta_{n_1+n_2,n} \frac{n!}{n_1! n_2!}$$

$$\int_{-\infty}^{\infty} dt_1 \ldots \int_{-\infty}^{\infty} dt_{n_1} \langle \Phi_0 | T[\hat{A}_I(t)\hat{B}_I \hat{H}_1(t_1) \ldots \hat{H}_1(t_{n_1})] | \Phi_0 \rangle_c$$

$$\times \int_{-\infty}^{\infty} dt_{n_1+1} \ldots \int_{-\infty}^{\infty} dt_n \langle \Phi_0 | T[\hat{H}_1(t_{n_1+1}) \ldots \hat{H}_1(t_n)] | \Phi_0 \rangle , \tag{6.41}$$

where the subscript c indicates that only connected terms are retained in the first factor. The sum over n can be done trivially by making use of the constraint $n = n_1 + n_2$: the $n!$

---

[7] This definition of connectedness is weaker than the one we shall finally adopt at the end of this section.

cancels out, leaving us with

$$\sum_{n_1=0}^{\infty} \frac{1}{n_1!} \left(-\frac{i}{\hbar}\right)^{n_1} \int_{-\infty}^{\infty} dt_1 \dots \int_{-\infty}^{\infty} dt_{n_1} \langle \Phi_0 | T[\hat{A}_I(t)\hat{B}_I \hat{H}_1(t_1) \dots \hat{H}_1(t_{n_1})] | \Phi_0 \rangle_c$$

$$\times \sum_{n_2=0}^{\infty} \frac{1}{n_2!} \left(-\frac{i}{\hbar}\right)^{n_2} \int_{-\infty}^{\infty} ds_1 \dots \int_{-\infty}^{\infty} ds_{n_2} \langle \Phi_0 | T[\hat{H}_1(s_1) \dots \hat{H}_1(s_{n_2})] | \Phi_0 \rangle , \quad (6.42)$$

where we have relabelled $t_{n_1+i} = s_i$. But the second factor in Eq. (6.42) is simply the power series expansion of the denominator of Eq. (6.27), and cancels out against the latter. This leads us to the important result,

$$\chi_{AB}^T(t) = -\frac{i}{\hbar} \langle \Phi_0 | T[\hat{A}_I(t)\hat{B}_I \hat{S}] | \Phi_0 \rangle_c , \quad (6.43)$$

i.e. *we can ignore the denominator of Eq. (6.27) provided we include only connected contraction schemes in the evaluation of the numerator.*

A second major simplification occurs when we realize that the connected schemes of contractions can be organized into equivalence classes. Two contraction schemes are equivalent if they differ only by a permutation of the time arguments $t_1 \dots t_n$: this is obvious because the t_i's are just integration variables, which can be arbitrarily renamed. For a two-body interaction of the form (6.38) another equivalence arises from the simultaneous interchange of dummy labels $a \leftrightarrow b$ and $c \leftrightarrow d$ in any of the n interactions. These symmetries suggest that we consider only a single representative pairing scheme for each equivalence class. Two schemes of contractions will be called *topologically distinct* if they are *not* related to each other by a trivial permutation of time or state labels, as described above. Since there are $n!2^n$ connected contraction schemes in each equivalence class we see that we can ignore the factor $1/n!$ from the expansion of the S-matrix as well as the factor $1/2$ in front of the interaction hamiltonian, provided we sum only over *topologically distinct* connected contraction schemes.

The enumeration of topologically distinct and connected contraction schemes leads directly to the introduction of *Feynman diagrams*. The general idea is to draw a solid *oriented* line to represent $G_{\alpha\beta}^{(0)}(t_1 - t_2)$. This line conventionally runs from β, t_2 to α, t_1 as shown in Fig. 6.2. We also introduce a dashed line connecting (a, d) and (b, c) *at equal times* to represent the coulomb interaction matrix element v_{abcd}. A Feynman diagram of order n is constructed by combining n interaction lines and $2n + 2$ oriented solid lines in all possible topologically distinct ways, subject to the constraint that the solid lines cannot reverse their direction at the points of junction (vertices), and that the complete diagram cannot consist of two (or more) disconnected pieces.

Let us return to the zero-order term (see Eq. (6.39)). The direct pairing scheme, $\hat{a}_\beta^\dagger(t)$ with $\hat{a}_\alpha^\dagger(t)$ and $\hat{a}_\delta$ with $\hat{a}_\gamma^\dagger$ gives zero since $\hat{A}$ and $\hat{B}$ have zero average in the ground-state. The other pairing scheme, $\hat{a}_\beta(t)$ with $\hat{a}_\gamma^\dagger$ and $\hat{a}_\delta$ with $\hat{a}_\alpha^\dagger(t)$, gives

$$-\frac{i}{\hbar} \sum_{\alpha\beta\gamma\delta} A_{\alpha\beta} B_{\gamma\delta}(-)\hat{a}_\beta(t)\hat{a}_\gamma^\dagger \hat{a}_\delta \hat{a}_\alpha^\dagger(t) . \quad (6.44)$$

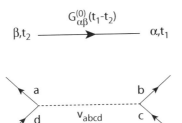

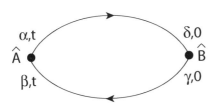

Fig. 6.2. Graphical symbols for the non-interacting Green's function and for the two-body bare interaction.

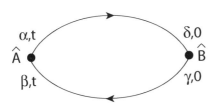

Fig. 6.3. Diagrammatic representation of the pairing scheme of Eq. (6.44). The black dots represents the matrix elements $A_{\alpha\beta}$ and $B_{\gamma\delta}$.

We have deliberately written the contractions with the $\hat{a}$'s preceding the $\hat{a}^\dagger$'s so that the connection with the non-interacting Green's function (6.30) is straightforward. However, in order to accomplish this, we had to interchange fermion operators an odd number of times: hence the minus sign.

The diagrammatic representation of this pairing scheme is shown in Fig. 6.3. The two non-interacting Green's functions are represented by oriented solid lines running in a closed loop from α at time t to δ at time 0 and then back from γ at time 0 to β at time t. There are no interaction lines. The numerical value of Eq. (6.44) is

$$-\frac{i}{\hbar}\sum_{\alpha\beta\gamma\delta} A_{\alpha\beta} B_{\gamma\delta}(-1)[iG^{(0)}_{\beta\gamma}(t)][iG^{(0)}_{\delta\alpha}(-t)] . \qquad (6.45)$$

This value can also be read from the diagram of Fig. 6.3 by attaching a factor $G^{(0)}$ to each solid line. The overall prefactor is the product of $-i/\hbar$ (from the definition of the correlation function) times i^2 (one factor i from the relation between each Green's function and the corresponding contraction) times -1 from the odd number of interchanges needed to arrange the fermion operators in the "convenient" order $(aa^\dagger)(aa^\dagger)$ around the loop.

Let us now consider the first-order term. There are $4! - 2! \times 2! = 20$ connected pairing schemes (see Eq. (6.40)), which, after dividing by the multiplicity factor $2^n n! = 2$, give us 10 topologically distinct and connected diagrams. These diagrams are shown in Fig. 6.4. Notice that some of these "connected" diagrams (the last four), do not look connected at all when drawn on paper! These diagrams represent pairing schemes in which the the

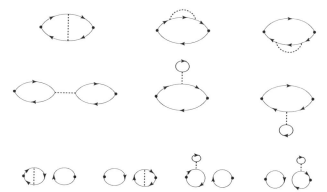

Fig. 6.4. The 10 diagrams arising from the contractions of the eight operators in Eq. (6.40). The black dots represents the matrix elements of the external operators $\hat{A}$ (left) and $\hat{B}$ (right).

operators $\hat{A}$ and $\hat{B}$ are not connected to each other by contractions. We shall call such diagrams *disjoint*.[8] Such disjoint diagrams yield the perturbative expansion of the product $\langle \hat{A} \rangle \langle \hat{B} \rangle$ – a quantity that has been assumed to vanish from the start (see footnote[4]): for this reason they can be neglected. The remaining 6 diagrams are connected in a stronger sense which fortunately happens to coincide with the intuitive notion of connectedness: they are shown on the top two lines of Fig. 6.4.

It is easy at this point to establish the general rules for constructing these and higher-order diagrams. A generic diagram of order n contains $2n + 2$ solid lines and n dashed lines. Solid and dashed lines are joined at "internal" vertices, of which there are $2n$. There are also two "external" vertices at which two solid lines join without an interaction line: these are associated with the external operators and carry labels α, β (at time t) and γ, δ (at time 0) as shown, for example, in Fig. 6.3. The general structure consists of a set of closed Fermion loops linked together by interaction lines. The procedure for evaluating the correlation function at order n is as follows:

(i) Draw all the topologically distinct connected diagrams consisting of two external vertices ($\hat{A}$ and $\hat{B}$), n oriented interaction lines, and $2n + 2$ oriented solid lines.
(ii) Label each solid line with two state labels β, α denoting the outgoing and ingoing state respectively. Assign a time label to each vertex. Assign equal times to vertices joined by an interaction line and times t and 0 to the $\hat{A}$ and $\hat{B}$ vertices respectively.
(iii) Associate a factor $G^{(0)}_{\alpha\beta}(t_1 - t_2)$ to a solid oriented line running from state β at time t_2 to state α at time t_1.
(iv) Associate an interaction amplitude v_{abcd} to a dashed line connecting vertices a, d and b, c at equal times, as in Fig. 6.2.
(v) Associate matrix elements $A_{\alpha\beta}$ and $B_{\gamma\delta}$ to the external vertices, where α, β and γ, δ are the labels of the outgoing and ingoing lines respectively.
(vi) Sum (or integrate) over all the internal state labels and time labels.

---

[8] Notice that a disjoint diagram is still "connected" if it does not contain a subset of interactions that are disjoint from *both* $\hat{A}$ and $\hat{B}$.

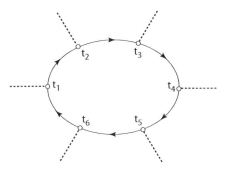

Fig. 6.5. Contractions of fermion operators in a loop carry an overall negative sign.

(vii) Multiply by a factor $\left(-\frac{i}{\hbar}\right)^{n+1}(i)^{2n+2}(-1)^{L} = \left(\frac{i}{\hbar}\right)^{n+1}(-1)^{L}$, where L is the number of Fermion loops.

The prefactor in (vii) arises as follows. One factor $-\frac{i}{\hbar}$ is built into the definition of the response function. A factor $\left(-\frac{i}{\hbar}\right)^{n}$ comes from the expansion of the S-matrix to n-th order in the interaction. A factor i^{2n+2} (an i for each of the $2n+2$ solid lines) arises from the relation (6.30) between the contraction of two operators and the Green's function. Finally, to understand the origin of the factor -1 associated with each Fermion loop consider Fig. 6.5. The loop with p vertices ($p = 6$ in this figure) corresponds to the contraction scheme $[\hat{a}(t_1)\,\hat{a}^{\dagger}(t_2)\,]\ldots[\hat{a}(t_{p-1})\,\hat{a}^{\dagger}(t_p)\,\cdots][\hat{a}(t_p)\,\cdots\hat{a}^{\dagger}(t_1)\,\cdots]$, while the original ordering of the operators in Eq. (6.37) is $[\hat{a}^{\dagger}(t_1)\hat{a}(t_1)]\ldots[\hat{a}^{\dagger}(t_p)\hat{a}(t_p)])$, up to an even permutation. The two orderings differ by an *odd* number of interchanges (i.e., moving the first operator to the last position and re-forming the pairs from left to right): hence the minus sign.

The most remarkable property of the prefactor (vii) is that, if we break up a diagram arbitrarily into two parts of order n_1 and n_2 the total prefactor of the diagram is given by the product of the prefactors associated with the two parts. Thus, Feynman diagrams can be factored in their constituent parts. This is the basis of the skeleton-diagram expansion, discussed later in this chapter.

Some diagrams contain the closed loop $G^{(0)}_{\alpha\beta}(t - t)$, which is ambiguous due to the discontinuity in $G^{(0)}_{\alpha\beta}(t)$ at $t = 0$. The ambiguity is resolved by recalling that pairs of operators at equal time arise from the interaction hamiltonian, in which $\hat{a}^{\dagger}$ precedes $\hat{a}$. This leads to our last rule:

(viii) Interpret the $t \to 0$ limit of $G^{(0)}_{\alpha\beta}(t)$ as $\lim_{t \to 0^{-}} G^{(0)}_{\alpha\beta}(t) = in_{\alpha}\delta_{\alpha\beta}$ (this rule ensures that $\hat{a}^{\dagger}$ precedes $\hat{a}$ in the time-ordered contraction.)

The complexity of the diagrammatic expansion can be appreciated by deriving a formula for the total number of topologically distinct diagrams that contribute to a two-particle response function at a given order n. This number is

$$N_n = C_n - D_n \tag{6.46}$$

where C_n is the number of "connected" diagrams and D_n is the number of connected but disjoint diagrams. It is not too difficult to show (see Exercise 3) that

$$C_n = \frac{C_n^{(2)}}{2^n n!}, \tag{6.47}$$

where $C_n^{(2)}$ is given by the recursion formula

$$C_n^{(2)} = (2n+2)! - \sum_{m=0}^{n-1} \frac{n![(2(n-m)]!}{m!(n-m)!} C_m^{(2)} , \qquad (6.48)$$

with initial value $C_0^{(2)} = 2$, and

$$D_n = \sum_{m=0}^{n} \frac{n!}{m!(n-m)!} \frac{C_m^{(1)}}{2^m m!} \frac{C_{n-m}^{(1)}}{2^{n-m}(n-m)!} , \qquad (6.49)$$

where $C_n^{(1)}$ is given by the recursion relation

$$C_n^{(1)} = (2n+1)! - \sum_{m=0}^{n-1} \frac{n![(2(n-m)]!}{m!(n-m)!} C_m^{(1)} , \qquad (6.50)$$

with initial value $C_0^{(1)} = 1$.[9] Thus, for example, for $n = 1$ and $n = 2$ we calculate $N_1 = 6$ and $N_2 = 46$. The number of diagrams grows *very* rapidly with increasing n. Although many of the diagrams have the same numerical value, it is impossible in practice to perform a diagram-by-diagram calculation beyond the lowest values of n.

6.2.5 Fourier transformation

It is usually the Fourier transform of the time-dependent correlation function that is most directly related the physical properties of a many-particle system. It is possible to formulate the perturbation theory directly for this quantity, thus taking maximum advantage of "translational invariance" in time.

The non-interacting Green's function is expressed in terms of its Fourier components according to the standard formula

$$G_{\alpha\beta}^{(0)}(t) = \int_{-\infty}^{\infty} \frac{d\omega}{2\pi} G_{\alpha\beta}^{(0)}(\omega) e^{-i\omega t} , \qquad (6.51)$$

where $G_{\alpha\beta}^{(0)}(\omega)$ is given by Eq. (6.36) or, more compactly, by

$$G_{\alpha\beta}^{(0)}(\omega) = \frac{\delta_{\alpha\beta}}{\omega - \frac{\varepsilon_\alpha}{\hbar} + i\eta_\alpha} , \qquad (6.52)$$

where $\eta_\alpha = +\eta$ if $\varepsilon_\alpha > \epsilon_F$ and $\eta_\alpha = -\eta$ if $\varepsilon_\alpha < \epsilon_F$.

Use of the Fourier representation for the non-interacting Green's function greatly simplifies the evaluation of the integrals over the internal time labels. Consider a generic Feynman diagram of order n. Each of the n internal time labels, say t_1, appears at two vertices joined by an interaction line. Let $G^{(0)}(t_1 - t_a)$ and $G^{(0)}(t_1 - t_b)$ be the two outgoing Green's function lines and $G^{(0)}(t_c - t_1)$ and $G^{(0)}(t_d - t_1)$ the two incoming Green's function lines, where t_a, t_b, t_c, t_d are time labels different from t_1 that need not be further specified (see Fig. 6.6).

---

[9] $C_n^{(1)}$ is the number of connected diagrams that contribute to the one-particle Green's function at order n.

Fig. 6.6. The dependence of a Feynman diagram on the time variable t_1.

Introducing the Fourier representation these Green's functions, we see that the entire t_1 dependence of the diagram is in the product

$$\int \frac{d\omega_1}{2\pi} \int \frac{d\omega_2}{2\pi} \int \frac{d\omega_3}{2\pi} \int \frac{d\omega_4}{2\pi} G^{(0)}(\omega_1) G^{(0)}(\omega_2) G^{(0)}(\omega_3) G^{(0)}(\omega_4)$$
$$\times e^{-i\omega_1(t_1-t_a)} e^{-i\omega_2(t_1-t_b)} e^{-i\omega_3(t_c-t_1)} e^{-i\omega_3(t_d-t_1)} . \qquad (6.53)$$

Performing the integral over t_1 we obtain a factor $2\pi\delta(\omega_1 + \omega_2 - \omega_3 - \omega_4)$ which ensures that the sum of the incoming frequencies $\omega_1 + \omega_2$ equals the sum of the outgoing frequencies $\omega_3 + \omega_4$. Because the same reasoning can be applied to each pair of internal vertices we conclude that in each interaction process *the sum of the incoming frequencies equals the sum of the outgoing frequencies.*

Notice that the frequency conservation constraint involves two interaction vertices. In order to reduce the constraint to a more manageable form we now assign a direction and a frequency to each interaction line, and require that the sum of the frequencies of the incoming lines *at each vertex* equal the sum of the frequencies of the outgoing lines. It is evident that this condition guarantees frequency conservation in every scattering process: the frequency associated with an interaction line simply represents the *energy transfer* from one particle to another during the scattering process.

As a final step in calculating the Fourier transform, multiply the diagram by $e^{i\omega t}$ (where t is the time label of the external vertex A) and integrate over t. Because the external vertex has only one incoming and one outgoing Green's functions, $G^{(0)}(\omega_{in})$ and $G^{(0)}(\omega_{out})$ respectively, the integral has the form

$$\int dt \int \frac{d\omega_{in}}{2\pi} G^{(0)}(\omega_{in}) \int \frac{d\omega_{out}}{2\pi} G^{(0)}(\omega_{out}) e^{-i\omega_{in}(t-t_a)} e^{-i\omega_{out}(t_b-t)} e^{i\omega t} . \qquad (6.54)$$

Carrying out the integral over t gives us the factor $2\pi\delta(\omega - \omega_{in} + \omega_{out})$ which forces the difference between the incoming and outgoing frequencies at the $\hat{A}$ vertex to be ω. The most effective way to implement this constraint is to associate an outgoing frequency ω with the operator $\hat{A}$ and to require frequency conservation at the $\hat{A}$ vertex.

It is not difficult to verify that the constraints derived above automatically force the difference between the incoming and outgoing frequency at the $\hat{B}$ vertex to be $-\omega$.[10] This

---

[10] One way to see this is to notice that by translational invariance in time, $\langle T[\hat{A}(t)\hat{B}(0)]\rangle = \langle T[\hat{A}(0)\hat{B}(-t)]\rangle$. Therefore, the condition on the incoming and outgoing frequencies at the $\hat{B}$ vertex can be deduced from the analogous condition at the $\hat{A}$ vertex replacing ω by $-\omega$.

condition ensures frequency conservation at the $\hat{B}$ vertex too, provided that we associate an incoming frequency ω with the operator $\hat{B}$.

In summary then, the rules for the diagrammatic calculation of the Fourier transform of a correlation function can be stated as follows:

(i) Draw all the topologically distinct connected diagrams consisting of two external vertices, n oriented interaction lines and $2n + 2$ oriented solid lines.

(ii) Associate a frequency to each solid or interaction line. Conserve frequency at each vertex, including an outgoing frequency ω at the $\hat{A}$ vertex and an incoming frequency ω at the $\hat{B}$ vertex.

(iii) Associate a factor $G_{\alpha\beta}^{(0)}(\omega)$ (given by Eq. (6.52)) to each solid line.

(iv) Associate a factor v_{abcd} to each interaction line connecting vertices a, b and c, d.

(v) Associate matrix elements $A_{\alpha\beta}$ and $B_{\gamma\delta}$ to the external vertices, where α, β and γ,δ are the labels of the outgoing and ingoing lines respectively.

(vi) Integrate over the $n + 1$ independent internal frequencies with a measure $\int \frac{d\omega}{2\pi}$, and sum over the internal state labels.

(vii) Multiply by a factor $(\frac{i}{\hbar})^{n+1}(-1)^L$, where L is the number of closed fermion loops.

(viii) Whenever the integral $\int \frac{d\omega}{2\pi} G_{\alpha\beta}^{(0)}(\omega)$ appears in a diagram, interpret it as $\lim_{\eta\to 0^+} \int \frac{d\omega}{2\pi} G_{\alpha\beta}^{(0)}(\omega)e^{i\eta\omega}$.[11]

6.2.6 Translationally invariant systems

The diagrammatic formalism takes a particularly simple form in translationally invariant systems where the eigenstates of the non-interacting hamiltonian are plane waves labelled by a wave vector and a spin projection. The non-interacting Green's function in the frequency domain is diagonal with respect to the wave vector and spin labels and takes the simple form

$$G_{\sigma}^{(0)}(\vec{k}, \omega) = \frac{1}{\omega - \frac{\varepsilon_{\vec{k}\sigma}}{\hbar} + i\eta_{\vec{k}\sigma}} \, , \tag{6.55}$$

where $\varepsilon_{\vec{k}\sigma} = \frac{\hbar^2 k^2}{2m}$ and $\eta_{\vec{k}\sigma} \equiv \eta \, \text{sign}(k - k_{F\sigma})$. Notice that the Green's function is *spin-independent* in the absence of external magnetic fields that break the spin-rotational symmetry of the non-interacting ground-state.

The fundamental advantage of the wave vector representation is that the matrix elements of the coulomb interaction vanishes unless the sum of the incoming wave vectors equal the sum of the outgoing ones: momentum is conserved. In complete analogy to the case of frequency we take advantage of this conservation law by assigning a wave vector to each interaction line, and imposing wave vector conservation at each vertex. Spin is also conserved along each solid line.

This leads us to the following set of diagrammatic rules for translationally invariant systems in momentum and frequency space:

---

[11] This is the frequency-domain version of rule (viii) of the previous section.

(i) Draw all the topologically distinct connected diagrams consisting of two external vertices, n oriented interaction lines, and $2n + 2$ oriented solid lines.

(ii) Associate a frequency and a wave vector to each solid or interaction line. Conserve frequency at each vertex, including an outgoing frequency ω at the $\hat{A}$ vertex and an incoming frequency ω at the $\hat{B}$ vertex. Conserve wave vector and spin at each internal vertex. In the practically important case in which the operators $\hat{A}$ and $\hat{B}$ carry wave vectors $\vec{q}$ and $-\vec{q}$ (e.g., if $\hat{A} = \hat{n}_{\vec{q}}$ and $\hat{B} = \hat{n}_{-\vec{q}}$) associate an outgoing wave vector $\vec{q}$ at the $\hat{A}$ vertex, an incoming wave vector $\vec{q}$ at the $\hat{B}$ vertex, and conserve wave vector at these vertices.

(iii) Associate a non-interacting Green's function $G_\sigma^{(0)}(\vec{k}, \omega)$ to each solid line.

(iv) Associate a factor $v_{\vec{p}}$ to an interaction line whose wave vector is $\vec{p}$ (physically, this is the momentum transfer in the scattering process).

(v) Associate matrix elements $A_{\vec{k}_1 \vec{k}_2}$ and $B_{\vec{k}_3 \vec{k}_4}$ to the external vertices, where $\vec{k}_1$, $\vec{k}_2$ and $\vec{k}_3$, $\vec{k}_4$ are the wave vectors of the outgoing and incoming lines at those vertices.

(vi) Integrate over the $n + 1$ independent internal wave vectors and frequencies with measures $\int \frac{d\omega}{2\pi}$ and $\int \frac{d\vec{k}}{(2\pi)^d}$ and sum over the internal state labels.

(vii) Multiply by a factor $(\frac{i}{\hbar})^{n+1}(-1)^L$ where L is the number of fermion loops.

(viii) Whenever the integral $\int \frac{d\omega}{2\pi} G_\sigma^{(0)}(\vec{k}, \omega)$ appears in a diagram, interpret it as $\lim_{\eta \to 0^+} \int \frac{d\omega}{2\pi} G_\sigma^{(0)}(\vec{k}, \omega) e^{i\eta\omega}$.

6.2.7 Diagrammatic calculation of the Lindhard function

Evaluating Feynman diagrams is a difficult task. Usually one begins by carrying out the frequency integrals which can in principle be done analytically because of the simple pole structure of the non-interacting Green's function. The remaining wave vector sums however are very hard to do even on a powerful computer and at low order in perturbation theory.

As a warm-up exercise, let us verify that the zero-th order calculation of the density–density response function of a non-interacting electron gas reproduces the Lindhard function of Chapter 4.

According to the rules listed in the previous section, the zero-th order diagram for $\chi_{nn}(q, \omega)$, shown in Fig. 6.7,[12] has the analytical expression

$$\chi_{nn}^{(0)}(q, \omega) = -\frac{i}{\hbar} \sum_\sigma \int \frac{d\vec{k}}{(2\pi)^d} \int_{-\infty}^{\infty} \frac{d\varepsilon}{2\pi} G_\sigma^{(0)}(\vec{k}, \varepsilon) G_\sigma^{(0)}(\vec{k} + \vec{q}, \varepsilon + \omega) \qquad (6.56)$$

The non-interacting Green's function is given by Eq. (6.55), and the frequency integral can be easily evaluated by the method of residues. The integral is zero if both poles are on the same side of the real axis (i.e., if $\vec{k}$ and $\vec{k} + \vec{q}$ are on the same side of the Fermi surface) and nonzero otherwise. A straightforward calculation yields

$$\chi_{nn}^{(0)}(q, \omega) = \sum_\sigma \int \frac{d\vec{k}}{(2\pi)^d} \frac{n_{\vec{k}\sigma} - n_{\vec{k}+\vec{q}\sigma}}{\hbar\omega + \varepsilon_{\vec{k}\sigma} - \varepsilon_{\vec{k}+\vec{q}\sigma} + i\hbar\eta_\omega} , \qquad (6.57)$$

---

[12] Notice that the matrix element of $\hat{n}_{\vec{q}}$ between plane wave states with wave vectors differing by $\vec{q}$ is simply 1, so the black dot at the external vertices has been simply omitted.

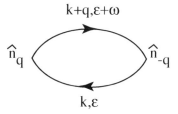

Fig. 6.7. Zero-th order Feynman diagram for the density–density response function $\chi_{nn}(q, \omega)$.

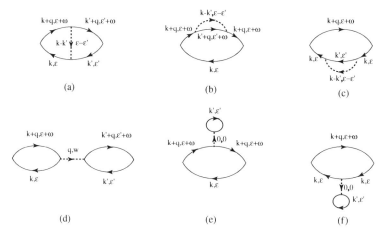

Fig. 6.8. First-order Feynman diagrams for the density–density response function $\chi_{nn}(q, \omega)$.

where $\eta_\omega = \eta \, \text{sign}(\omega)$. Remember that this is an expression for the *time-ordered* response function. To obtain the causal response function, we apply the prescription of Eq. (6.10), which amounts to multiplying the infinitesimal η by $\text{sign}(\omega)$ or, in the present case, replacing η_ω by η. In this way, we precisely recover Eq. (4.10) for the Lindhard function $\chi_0(q, \omega)$.

6.2.8 First-order correction to the density–density response function

Let us now consider the calculation of the first-order correction to the density–density response function $\chi_{nn}(q, \omega)$ of the homogeneous electron liquid. The six first order diagrams are shown in the top two lines of Fig. 6.4. We reproduce these diagrams in Fig. 6.8 with the appropriate wave vector and frequency labels.

The two diagrams (e) and (f) vanish due to the fact that the interaction line carries zero wave vector, and $v_{\vec{q}=0} = 0$ in the jellium-model hamiltonian. Of the remaining diagrams, (d) has a very simple structure: it is the product of two non-interacting response functions times a coulomb interaction with fixed wave vector $\vec{q}$, and can therefore be expressed in terms of the Lindhard function as follows:

$$v_q [\chi_0(q, \omega)]^2. \tag{6.58}$$

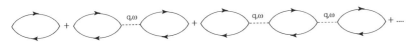

Fig. 6.9. Infinite series of Feynman diagrams corresponding to the RPA for $\chi_{nn}(q, \omega)$.

In fact, this is just the beginning of an infinite series of "bubble diagrams" (shown in Fig. 6.9), which can be summed exactly to yield the RPA response function

$$\chi^{RPA}(q, \omega) = \chi_0(q, \omega) + v_q[\chi_0(q, \omega)]^2 + v_q^2[\chi_0(q, \omega)]^3 + \cdots.$$

$$= \frac{\chi_0(q, \omega)}{1 - v_q \chi_0(q, \omega)}. \tag{6.59}$$

It is already evident, and will be rigorously shown later in Section 6.3.1, that such bubble diagrams do not contribute to the expansion of the "proper" density–density response function, as defined in Section 5.2.2.

The evaluation of the three diagrams (a), (b), and (c) is more difficult. For diagram (b) following the general rules we write

$$\chi^{SE1}(q, \omega) = \frac{1}{\hbar^2} \sum_\sigma \int \frac{d\vec{k}}{(2\pi)^d} \int \frac{d\vec{k}'}{(2\pi)^d} v_{\vec{k}-\vec{k}'} i n_{\vec{k}'+\vec{q}\sigma}$$

$$\times \int \frac{d\varepsilon}{2\pi} [G_\sigma^{(0)}(\vec{k} + \vec{q}, \varepsilon + \omega)]^2 G_\sigma^{(0)}(\vec{k}, \varepsilon), \tag{6.60}$$

where we have used (viii) to express the equal time Green's function in terms of the occupation number: $\int G_\sigma^{(0)}(\vec{k}' + \vec{q}, \varepsilon')e^{i\varepsilon'\eta}\frac{d\varepsilon'}{2\pi} = i n_{\vec{k}'+\vec{q}\sigma}$.

From now on we assume that the system is paramagnetic, so that we can ignore the spin dependence of the various quantities. We substitute Eq. (6.55) for the Green's function in Eq. (6.60) and do the frequency integral by the methods of residues (the integrand vanishes faster than $1/\omega'$ for $\omega' \to \infty$ so that contributions from the semicircle at infinity vanish). This gives

$$\int \frac{d\varepsilon}{2\pi i} \left(\frac{1}{\varepsilon + \omega - \frac{\varepsilon_{\vec{k}+\vec{q}}}{\hbar} + i\eta_{\vec{k}+\vec{q}}} \right)^2 \left(\frac{1}{\varepsilon - \frac{\varepsilon_{\vec{k}}}{\hbar} + i\eta_{\vec{k}}} \right)$$

$$= n_{\vec{k}}(1 - n_{\vec{k}+\vec{q}}) \left(\frac{1}{\omega + \frac{\varepsilon_{\vec{k}} - \varepsilon_{\vec{k}+\vec{q}}}{\hbar} + i\eta} \right)^2$$

$$- n_{\vec{k}+\vec{q}}(1 - n_{\vec{k}}) \left(\frac{1}{\omega + \frac{\varepsilon_{\vec{k}} - \varepsilon_{\vec{k}+\vec{q}}}{\hbar} - i\eta} \right)^2. \tag{6.61}$$

Since we are ultimately interested in calculating the retarded response function we switch the sign of the imaginary part (i.e., the sign of η) in the negative frequency term (the second) and put the integral back into Eq. (6.60) to get

$$\chi^{SE1}(q,\omega) = -\frac{2}{\hbar^2} \int \frac{d\vec{k}}{(2\pi)^d} \int \frac{d\vec{k}'}{(2\pi)^d} \frac{\left(n_{\vec{k}} - n_{\vec{k}+\vec{q}}\right) n_{\vec{k}'+\vec{q}}}{\left(\omega + \frac{\varepsilon_{\vec{k}} - \varepsilon_{\vec{k}+\vec{q}}}{\hbar} + i\eta\right)^2} v_{\vec{k}-\vec{k}'}, \tag{6.62}$$

where the factor 2 comes from the sum over spin. The contribution of diagram (c) leads to a very similar expression, differing only for the replacements $\omega \to -\omega$ and $\vec{q} \to -\vec{q}$ of which the latter is irrelevant. Combining the two terms we therefore get

$$\chi^{SE}(q,\omega) = \frac{2}{\hbar^2} \int \frac{d\vec{k}}{(2\pi)^d} \int \frac{d\vec{k}'}{(2\pi)^d} \frac{(n_{\vec{k}} - n_{\vec{k}+\vec{q}}) v_{\vec{k}-\vec{k}'} (n_{\vec{k}'} - n_{\vec{k}'+\vec{q}})}{\left(\omega + \frac{\varepsilon_{\vec{k}} - \varepsilon_{\vec{k}+\vec{q}}}{\hbar} + i\eta\right)^2}. \tag{6.63}$$

In a similar way we evaluate the diagram (a). According to the general rules this is given by

$$\chi^{EX}(q,\omega) = \frac{1}{\hbar^2} \sum_\sigma \int \frac{d\vec{k}}{(2\pi)^d} \int \frac{d\vec{k}'}{(2\pi)^d} \int \frac{d\varepsilon}{2\pi} \int \frac{d\varepsilon'}{2\pi} G^{(0)}(\vec{k}, \varepsilon)$$
$$\times G^{(0)}(\vec{k}+\vec{q}, \varepsilon+\omega) G^{(0)}(\vec{k}', \varepsilon') G^{(0)}(\vec{k}'+\vec{q}, \varepsilon'+\omega) v_{\vec{k}-\vec{k}'}. \tag{6.64}$$

Performing the two frequency integrals with the method of residues we obtain

$$\chi^{EX}(q,\omega) = -\frac{2}{\hbar^2} \int \frac{d\vec{k}}{(2\pi)^d} \int \frac{d\vec{k}'}{(2\pi)^d} \frac{(n_{\vec{k}} - n_{\vec{k}+\vec{q}}) v_{\vec{k}-\vec{k}'} (n_{\vec{k}'} - n_{\vec{k}'+\vec{q}})}{\left(\omega + \frac{\varepsilon_{\vec{k}} - \varepsilon_{\vec{k}+\vec{q}}}{\hbar} + i\eta\right) \left(\omega + \frac{\varepsilon_{\vec{k}'} - \varepsilon_{\vec{k}'+\vec{q}}}{\hbar} + i\eta\right)}. \tag{6.65}$$

The complete first-order result for the proper density–density response function is obtained by summing Eqs. (6.63) and (6.65), and coincides with the first order term in the expansion of the time-dependent Hartree–Fock expression derived in Chapter 5. Detailed calculations of these difficult integrals have been done by Holas *et al.* (1982) for all frequencies, by Engel and Vosko (1990) for $\omega = 0$, and by Richardson and Ashcroft (1994) for finite imaginary frequency.

6.3 Integral equations in diagrammatic perturbation theory

After the initial phase of enthusiasm for diagrams subsides, a student of many-body theory usually realizes that a systematic sum of diagrams, order by order, cannot be pursued very far. In the absence of a small expansion parameter the problem is hopelessly open-ended. However, one of the most powerful features of the diagrammatic formalism is the possibility of combining infinite series of diagrams into "building blocks" that can be connected to one another in rough analogy to the components of an electrical circuit. The infinite series of diagrams for a given quantity can be expressed recursively in terms of these building blocks, and the problem of summing the series reduces to the solution of an *integral equation*.

In many cases, the building blocks can be successfully approximated by low-order perturbation theory or even just "guessed" without seriously compromising the physics. Or, we may be able to give approximate expressions for the building blocks in terms of response functions, thus arriving at a closed set of self-consistent equations that can be solved without further approximations.

Integral equations are also useful to cure unphysical divergences that often plague the naive order-by-order expansion. The diverging sum is replaced by an integral equation whose kernel is free of divergencies and can therefore be computed by perturbation theory.

The paradoxical outcome of all these procedures is that nonperturbative results can often be obtained from a perturbative formalism. This section is devoted to an exploration of this theme.

6.3.1 Proper response function and screened interaction

Consider the diagrams for the density–density response function of a uniform electron liquid (the following discussion is easily generalized to nonuniform systems). These diagrams fall into two classes: (i) diagrams that can be separated into two parts (each containing one of the external vertices) by the cutting of a *single* interaction line and (ii) diagrams that cannot be separated in this manner. The diagrams of the second class are called "proper" and those of the first "improper". For example, in Fig. 6.9 only the first diagram is proper, while in Fig. 6.8 only (d) is improper.

Let us denote by $\tilde{\chi}_{nn}(q, \omega)$ the sum of all the proper diagrams. It is evident that this group of diagrams can be regarded as a "building block", from which the whole series of diagrams for the density–density response function can be reconstructed. We shall presently show that this building block is nothing but the *proper* density–density response function $\tilde{\chi}_{nn}(q, \omega)$ introduced in Chapter 5.

To see this observe that, by definition, any improper diagram in the expansion of $\chi_{nn}(q, \omega)$ can be written as the product of a proper diagram times an interaction line carrying fixed wave vector $\vec{q}$, times another diagram of $\chi_{nn}(q, \omega)$ that can be either proper or improper. Thus, the diagrammatic expansion of χ_{nn} has the structure illustrated in Fig. 6.10. The white oval represents the sum of all the diagrams contributing to χ_{nn} while the gray oval represents the sum of the proper diagrams only. The resulting relationship between χ_{nn} and $\tilde{\chi}_{nn}$ is

$$\chi_{nn}(q, \omega) = \tilde{\chi}_{nn}(q, \omega) + \tilde{\chi}_{nn}(q, \omega)v_q \chi_{nn}(q, \omega)$$

$$= \frac{\tilde{\chi}_{nn}(q, \omega)}{1 - v_q \tilde{\chi}_{nn}(q, \omega)}, \qquad (6.66)$$

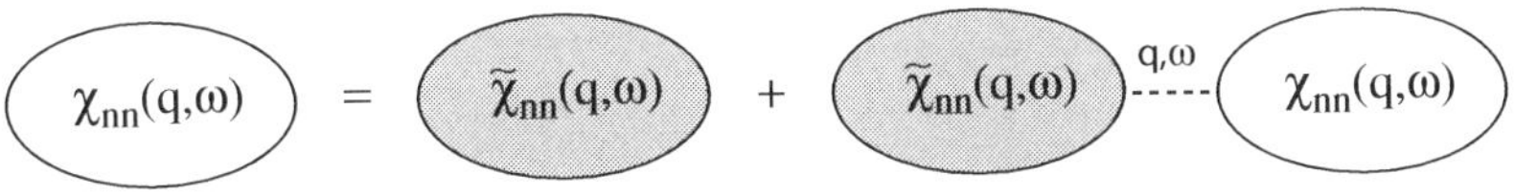

Fig. 6.10. Diagrammatic relationship between the density–density response function $\chi_{nn}(q, \omega)$ and the proper density–density response function $\tilde{\chi}_{nn}(q, \omega)$ defined as the sum of all the proper diagrams.

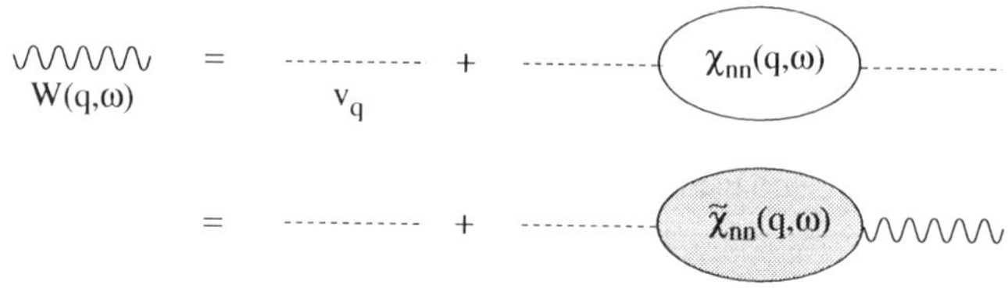

Fig. 6.11. Series of diagrams for the screened interaction between two test particles (wavy line).

which clearly identifies $\tilde{\chi}_{nn}(q, \omega)$ as the proper density–density response function and the quantity

$$\epsilon(q, \omega) = 1 - v_q \tilde{\chi}_{nn}(q, \omega) \tag{6.67}$$

as the dielectric function.

It follows immediately that the *screened interaction* between two test charges

$$W(q, \omega) = \frac{v_q}{1 - v_q \tilde{\chi}_{nn}(q, \omega)} = \frac{v_q}{\epsilon(q, \omega)} \tag{6.68}$$

corresponds to the infinite series of diagrams shown in Fig. 6.11. The screened interaction will be respresented by a wavy line to avoid confusion with the bare interaction, which is represented by a dashed line.

In the above example the improper diagrams are "dangerous" because they contain coulomb interaction lines with fixed wave vector $\vec{q}$, which diverge for $q \to 0$. The proper diagrams are well behaved because the momentum transfer in the interaction lines is not fixed: the small-q divergence of the interaction is rendered harmless by the integration over wave vector.

From Eq. (6.66) we see that any finite-order perturbative approximation for $\tilde{\chi}_{nn}(q, \omega)$ automatically generates an infinite-order approximation for the full response function. For example, we may set $\tilde{\chi}_{nn}(q, \omega) = \chi_0(q, \omega)$: the resulting approximation for $\chi_{nn}(q, \omega)$ is just the random phase approximation.

Notice that in this simple example $\chi_{nn}(q, \omega)$ and $\tilde{\chi}_{nn}(q, \omega)$ are connected by an algebraic equation rather than an integral equation. This feature is due to the translational invariance of the system. The same relation, written for the response functions in real space, would take the form of an integral equation.

A similar resummation can be done for all two-point response functions. Fig. 6.12, for example, shows the resummation of the diagrams for the longitudinal spin–spin response function. Its algebraic form is

$$\chi_{S_z S_z}(q, \omega) = \tilde{\chi}_{S_z S_z}(q, \omega) + v_q^2 \tilde{\chi}_{S_z n}(q, \omega) \chi_{nn}(q, \omega) \tilde{\chi}_{n S_z}(q, \omega). \tag{6.69}$$

In the paramagnetic case the spin–density response function $\tilde{\chi}_{S_z n}$ vanishes by symmetry (see Chapter 3), and therefore $\chi_{S_z S_z}(q, \omega) = \tilde{\chi}_{S_z S_z}(q, \omega)$ in that case.

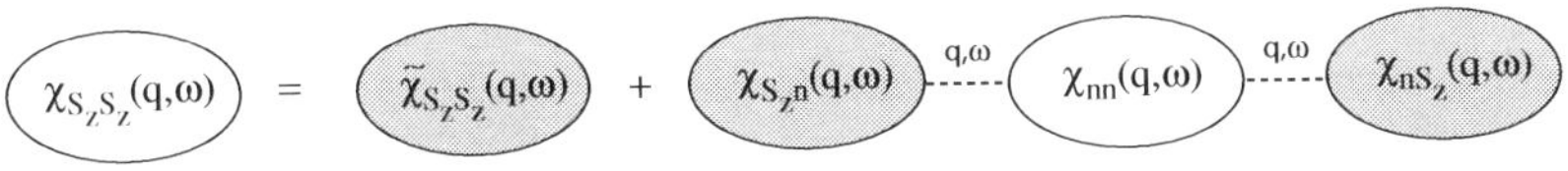

Fig. 6.12. Diagrammatic relationship between the full longitudinal spin–spin response function $\chi_{S_z S_z}(q, \omega)$ and the proper response functions $\tilde{\chi}_{S_z S_z}(q, \omega)$ and $\tilde{\chi}_{n S_z}(q, \omega)$ represented by the gray ovals.

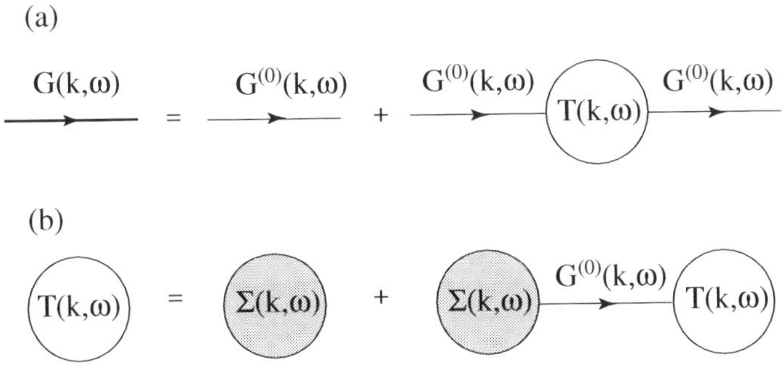

(a)

(b)

Fig. 6.13. Structure of the diagrams for the interacting Green's function.

6.3.2 Green's function and self-energy

Consider the set of diagrams that arise from the "dressing" of a bare Fermion line. The general structure of these diagrams is illustrated in part (a) of Fig. 6.13. One particle comes in and one goes out: processes of arbitrary complexity, represented by the "reducible self-energy", $T(k, \omega)$ can take place in between. It is not difficult to see that the whole set is simply the graphical representation of the perturbation series for the *interacting* Green's function

$$G_{\alpha\beta}(t_1, t_2) = -i \langle \psi_0 | T[\hat{a}_\alpha(t_1) \hat{a}_\alpha^\dagger(t_2)] | \psi_0 \rangle. \tag{6.70}$$

Here (at variance with the non-interacting Green's function of Section 6.2.3) the time dependence of the operators is controlled by the interacting hamiltonian and $|\psi_0\rangle$ is the interacting ground-state. A generic diagram for G is shown in Fig. 6.14: it contains n interaction lines and $2n + 1$ solid lines.

With the exception of the zero-order diagram (the non-interacting Green's function) all the diagrams for G are "dangerous" because they contain at least *two* solid lines with fixed wave vector and frequency $\vec{k}, \omega$ resulting in a pole of at least second order, i.e., a behavior of the form $\left(\frac{1}{\omega - \varepsilon_{\vec{k}}/\hbar + i\eta_{\vec{k}}} \right)^n$ with $n \geq 2$ when ω approaches the free particle energy. These higher order poles are an artifact of perturbation theory: the exact Green's function is expected to have only simple poles.[13]

In order to get around this difficulty we introduce a new "building block", namely, the "irreducible self-energy" $\Sigma(\vec{k}, \omega)$, or self-energy *tout court*. We define $\hbar^{-1} \Sigma(\vec{k}, \omega)$ as the sum of all the diagrams that can be inserted between an incoming and an outgoing solid

---

[13] In a normal Fermi liquid such poles are close to the real axis and define the so-called quasiparticle energies (see Chapter 8).

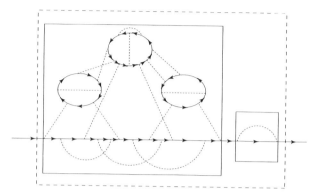

Fig. 6.14. Generic diagram for the Green's function with $n = 18$ interaction lines and $2n + 1 = 37$ solid lines. The reducible self-energy insertion, $T(k, \omega)$, is enclosed in the dashed box. The solid boxes enclose two irreducible self-energy insertions.

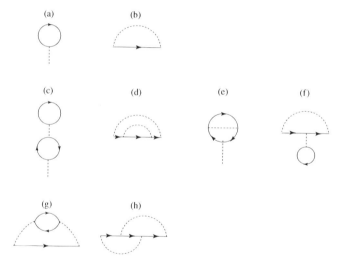

Fig. 6.15. First and second order diagrams for the proper self-energy.

line with wave vector and frequency $\vec{k}$, ω and, furthermore, *cannot be separated into two parts by cutting a single internal line with fixed momentum and frequency* (see Fig. 6.13). Notice that the two external lines are not included in the definition of the self-energy: the diagrams for Σ are shown, up to second order, in Fig. 6.15.

Making use of Σ, the perturbation series for G can be reorganized as follows

$$G(\vec{k}, \omega) = G^{(0)}(\vec{k}, \omega) + [G^{(0)}(\vec{k}, \omega)]^2 \hbar^{-1} \Sigma(\vec{k}, \omega) + [G^{(0)}(\vec{k}, \omega)]^3 \hbar^{-2} [\Sigma(\vec{k}, \omega)]^2 + \cdots$$

(6.71)

where the n-th term on the right-hand side represents the sum of all the diagrams that contain exactly $n - 1$ internal lines with fixed wave vector and frequency.

The sum of the geometric series is equivalent to the solution of the equation

$$G(\vec{k}, \omega) = G^{(0)}(\vec{k}, \omega) + G^{(0)}(\vec{k}, \omega)\hbar^{-1}\Sigma(\vec{k}, \omega)G(\vec{k}, \omega) , \qquad (6.72)$$

which is widely known as the *Dyson equation*.[14] By making use of the expression (6.55) for the non-interacting Green's function we obtain the formal solution

$$G(\vec{k}, \omega) = \frac{1}{\omega - \frac{\varepsilon_{\vec{k}} + \Sigma(\vec{k}, \omega)}{\hbar}} . \qquad (6.73)$$

The physical meaning of the self-energy is clearly revealed by this equation: $\Sigma(k, \omega)$ represents, in a sense, the many-body correction to the free-particle energy $\varepsilon_{\vec{k}}$. The fact that Σ is complex and frequency-dependent simply means that a single particle cannot be described as a closed hamiltonian system. Its dynamical evolution depends on the state of the whole electron liquid or, equivalently, on the previous history of the particle – hence the frequency dependence. The one-particle state itself may decay in time (its time evolution is non-unitary) – hence the imaginary part.

Notice that, in the first order approximation, $\Sigma(k, \omega)$ is given by the sum of the diagrams (a) and (b) in Fig. 6.15. Of these, the first vanishes (because $v_{q=0} = 0$) and the second gives a result independent of frequency, which is easily seen to coincide with the Hartree–Fock uniform exchange energy, $\varepsilon_{\vec{k}}^{(x)}$, calculated in Section 2.4.

It is remarkable that any finite-order approximation for Σ generates, through Eq. (6.73), an approximation for G that has the correct analytical structure. By contrast, any finite-order approximation to G itself would inevitably lead to the appearance of unphysical double or higher order poles at the free particle energy, as discussed above.[15]

We conclude this section by introducing a slightly different form of the Green's function, known as the *retarded Green's function*, which, together with the corresponding retarded self-energy, will play a pivotal role in the study of the single particle properties of a Fermi liquid in Chapter 8. The retarded Green's function in real time is defined as

$$G^{ret}(\alpha, t) = -i\Theta(t)\langle\psi_0|\{\hat{a}_\alpha(t), \hat{a}_\alpha^\dagger\}|\psi_0\rangle , \qquad (6.74)$$

where $\{\hat{a}_\alpha(t), \hat{a}_\alpha^\dagger\}$ is the *anticommutator* of the two Heisenberg-picture operators $\hat{a}_\alpha(t)$ and $\hat{a}_\alpha^\dagger$, the latter being evaluated at $t = 0$. The presence of the Θ-function of time reveals the retarded character of this quantity, which is in many ways analogous to a causal response function, except for the fact that its definition contains an anticommutator rather than a commutator. However, this difference has no influence on the general causality properties, which follow directly from the analyticity of the Fourier transform

$$G^{ret}(\alpha, \omega) = \int_{-\infty}^{+\infty} G^{ret}(\alpha, t)e^{i(\omega+i\eta)t} dt$$

$$= \sum_n \left(\frac{|[\hat{a}_\alpha^\dagger]_{n0}|^2}{\omega - \omega_{n0} + i\eta} + \frac{|[\hat{a}_\alpha]_{n0}|^2}{\omega + \omega_{n0} + i\eta} \right) , \qquad (6.75)$$

---

[14] In an inhomogeneous system the Dyson equation becomes an integral equation for $G(\vec{r}, \vec{r}', \omega)$.
[15] We are now in a position to understand the reason for the singularities in the first order calculation of the local field factor, presented in Section 5.7.1. The cause of the trouble is that diagrams (b) and (c) in Fig. 6.8 include the first order approximation to the Green's function, rather than the first order approximation to the self-energy. This led to the appearance of the double pole in Eq. (5.228) and hence to the divergence of the local field factor at certain frequencies.

in the upper half plane of the complex variable ω (the notation is the same as in Section 3.2.3). In particular, the Kramers–Krönig dispersion relation derived in Section 3.2.7 for causal response functions remains in force:

$$\Re e G^{ret}(\alpha, \omega) = \mathcal{P} \int_{-\infty}^{+\infty} \frac{\Im m G^{ret}(\alpha, \nu)}{\nu - \omega} \frac{d\nu}{\pi} . \tag{6.76}$$

The retarded Green's function can be straightforwardly connected to the time-ordered Green's function $G(\alpha, \omega)$, which appears in perturbation theory. At zero temperature the connection is given by the formulas

$$\Im m G^{ret}(\alpha, \omega) = \Im m G(\alpha, \omega)\mathrm{sign}\left(\omega - \frac{\mu}{\hbar}\right) ,$$

$$\Re e G^{ret}(\alpha, \omega) = \Re e G(\alpha, \omega) . \tag{6.77}$$

The finite-temperature case will be discussed in detail in Section 6.4.

The retarded Green's function is expressed in terms of the retarded self-energy, $\Sigma^{ret}(\alpha, \omega)$, in exactly the same manner as the time-ordered Green's function was expressed in terms of the time-ordered self-energy in Eq. (6.73). In a translationally invariant system, setting $\alpha = \vec{k}$ and ignoring any spin dependence one has

$$G^{ret}(\vec{k}, \omega) = \frac{1}{\omega - \frac{\varepsilon_{\vec{k}} + \Sigma^{ret}(\vec{k}, \omega)}{\hbar}} \tag{6.78}$$

where the retarded self-energy is related to the time-ordered self-energy at $T = 0$ by

$$\Sigma(\vec{k}, \omega) = \Re e \Sigma^{ret}(\vec{k}, \omega) + \mathrm{sign}\left(\omega - \frac{\mu}{\hbar}\right) \Im m \Sigma^{ret}(\vec{k}, \omega) . \tag{6.79}$$

It is not difficult to show that $G^{ret}(\vec{k}, \omega)$ is not only analytic, but also free of zeroes in the upper half of the complex plane (see Exercise 6). It then follows that $\Sigma^{ret}(\vec{k}, \omega)$ is analytic in the upper half plane of ω, and satisfies Kramers–Krönig relations in the form

$$\Re e \Sigma^{ret}(\vec{k}, \omega) = \Re e \Sigma^{ret}(\vec{k}, \infty) + \mathcal{P} \int_{-\infty}^{+\infty} \frac{\Im m \Sigma^{ret}(\vec{k}, \nu)}{\nu - \omega} \frac{d\nu}{\pi} . \tag{6.80}$$

6.3.3 Skeleton diagrams

The definition of the self-energy leads quite naturally to the notion of *skeleton diagrams*. A skeleton diagram is a diagram that does not contain self-energy insertions. For example, in Fig. 6.15, diagrams (a), (b), (g), and (h) are skeleton diagrams, but (c), (d), (e), and (f) are not. A generic Feynman diagram can be uniquely reduced to a skeleton diagram by absorbing all the the self-energy insertion into the solid lines. On the other hand, a given skeleton diagram generates infinitely many diagrams when it is "dressed" with an arbitrary number of self-energies insertions. Accordingly, the self-energy can be formally expressed as the sum of its skeleton diagrams with the proviso that all the lines represent full Green's

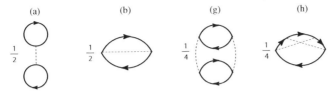

Fig. 6.16. Skeleton diagrams for the generating functional $\Omega[G]$.

functions. This procedure defines the self-energy as a functional, $\Sigma[G]$, of the full Green's function.[16]

The self-energy functional $\Sigma[G]$ has a remarkable property, namely, there exists a "generating functional" of G, denoted by $\Omega[G]$, such that

$$\Sigma_\sigma[G](\vec{k}, \omega) = \frac{\delta\Omega[G]}{\delta G_\sigma(\vec{k}, \omega)} . \tag{6.81}$$

The meaning of this expression is that the variation of $\Omega[G]$ due to an infinitesimal variation of G is given by

$$\delta\Omega[G] = \frac{1}{L^d} \sum_{\vec{k}\sigma} \int \frac{d\omega}{2\pi i} \Sigma_\sigma(\vec{k}, \omega)\delta G_\sigma(\vec{k}, \omega) . \tag{6.82}$$

The Ω functional is constructed as follows: consider the set of all the diagrams obtained by closing a skeleton diagram for Σ with an additional G-line. A few of these *closed skeleton diagrams*, obtained from skeleton diagrams (a), (b), (g), and (h) of Fig. 6.15 are shown in Fig. 6.16. Note that a closed diagram with n interaction lines has $2n$ full G-lines. The numerical value of each diagram is calculated according to the standard Feynman rules *and the result is divided by* $2n$. $\Omega[G]$ is defined as the sum of all the closed skeleton diagrams calculated in this manner.

The proof of Eq. (6.81) is simple. Functional differentiation with respect to G requires that we successively vary each of the $2n$ Green's function in a closed diagram by an infinitesimal amount δG. This procedure amounts to successively removing every G and replacing it by δG. But removing one G-line generates a skeleton diagram for Σ. Furthermore, each skeleton diagram for Σ arises in $2n$ possible ways from closed diagrams containing n interaction lines. This multiplicity exactly cancels the prefactor $1/2n$, leaving us with Eq. (6.82). This result will have an important application in the proof of Luttinger's theorem in Chapter 8. A discussion of the relation between $\Omega[G]$ and the ground-state energy is presented in Appendix 13.

---

[16] A crucial property of skeleton diagrams is that they can be evaluated according to the same set of rules that one uses for ordinary diagrams. The only differences are that to each solid line one must associate a full Green's function rather than a non-interacting one, and that the order of the diagram must be determined only by the number of interaction lines that *explicitly* appear in it. One sees upon reflection that this fortunate fact is a consequence of the simple exponential dependence of the global prefactor of each diagram upon its order.

$$\chi^{T}{}_{AB}= \quad \widehat{A} \bullet \overset{\longrightarrow}{\underset{\longleftarrow}{\bigcirc}} \bullet \widehat{B} \quad + \quad \widehat{A} \bullet \overset{}{\underset{}{\bigcirc \,\Gamma\,}} \bullet \widehat{B}$$

Fig. 6.17. Structure of the skeleton diagrams for the correlation function χ^{T}_{AB}.

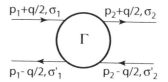

Fig. 6.18. Labelling of the two-body interaction $\Gamma^{\sigma_1\sigma_1',\sigma_2\sigma_2'}_{p_1 p_2}(q)$ in a translationally invariant system.

6.3.4 Irreducible interactions

A central role in the theory of Fermi systems is played by the concept of *two-body scattering amplitude*.

The concept arises naturally in the diagrammatic calculation of correlation functions χ_{AB} where $\hat{A}$ and $\hat{B}$ are single particle operators. The skeleton diagrams for χ_{AB} have the structure shown in Fig. 6.17. The first block of diagrams has a pair of Green's functions running in opposite directions. This can be described as an *electron–hole propagator* because if one line runs forward in time, thus picking the electron component of G, then the other runs backward, thus picking the hole component of G. The electron and the hole are fully dressed by many-body effects (self-energy insertions), and yet do not interact with each other. The second block of diagrams, shown in Fig. 6.17, describes instead processes in which the electron and the hole do interact with each other. The block Γ represents the sum of *all* the diagrams that can be inserted between incoming and outgoing electron–hole pairs (see Fig. 6.18), and its first-order parts are shown in Fig. 6.19. It is evident that Γ, together with the exact Green's function, completely determines the correlation function. While each individual diagram of this kind can be painstakingly evaluated according to standard Feynman rules, it is possible, as we shall see momentarily, to streamline the process by developing special rules for the calculation of diagrams containing whole interaction blocks as well as rules to attach an explicit value to a given block.

The description of Γ is greatly simplified if one assumes that the system is translationally and rotationally invariant.[17] In order to streamline the discussions that follow, it is very convenient to introduce the four-dimensional notation

$$p \equiv (p_0, \vec{p}) \tag{6.83}$$

---

[17] The formalism presented below can be extended with little extra effort to the case of spin-polarized systems that are rotationally invariant about an axis – see Exercise 7.

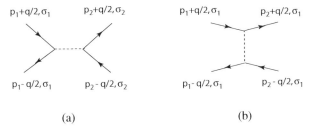

Fig. 6.19. First order diagrams for $\Gamma_{p_1\sigma_1, p_2\sigma_2}(q)$ in a translationally invariant system. The exchange diagram (b) exists only in the case $\sigma_1 = \sigma_2$.

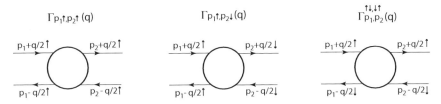

Fig. 6.20. Graphical representation of the spin components of the scattering amplitude defined in Eqs. (6.87) and (6.88).

to denote the frequency (p_0) and the wave vector ($\vec{p}$) of a Fermion line, and the notation

$$\frac{1}{L^d}\sum_p \cdots = \int \frac{d\vec{p}}{(2\pi)^d} \int \frac{dp_0}{2\pi} \cdots \tag{6.84}$$

to denote wave vector and frequency integrals. Let us label by $p_1 + q/2$, σ_1 and $p_2 - q/2$, σ_2' the four-dimensional wave vectors and the spin projections of the incoming particles and by $p_1 - q/2$, σ_1' and $p_2 + q/2$, σ_2 those of the outgoing ones, as shown in Fig. 6.18. The function Γ can then be written as $\Gamma_{p_1,p_2}^{\sigma_1\sigma_1',\sigma_2\sigma_2'}(q)$: the labeling reflects the conservation of wave vector and frequency. Spin conservation imposes the condition

$$\sigma_1 + \sigma_2' = \sigma_1' + \sigma_2. \tag{6.85}$$

The most general form of Γ that is consistent with rotational invariance in spin space is

$$\Gamma_{p_1,p_2}^{\sigma_1\sigma_1',\sigma_2\sigma_2'}(q) = \Gamma_{p_1,p_2}^+(q)\delta_{\sigma_1\sigma_1'}\delta_{\sigma_2\sigma_2'} + \Gamma_{p_1,p_2}^-(q)\vec{\sigma}_{\sigma_1\sigma_1'} \cdot \vec{\sigma}_{\sigma_2\sigma_2'} \tag{6.86}$$

where $\Gamma_{p_1,p_2}^+(q)$ and $\Gamma_{p_1,p_2}^-(q)$ are two *spin-independent* functions, $\hat{\vec{\sigma}}$ is a vector whose components are the Pauli matrices ($\hat{\sigma}_x, \hat{\sigma}_y, \hat{\sigma}_z$), and $\vec{\sigma}_{\sigma_1\sigma_1'} \cdot \vec{\sigma}_{\sigma_2\sigma_2'} = [\sigma_x]_{\sigma_1\sigma_1'}[\sigma_x]_{\sigma_2\sigma_2'} + [\sigma_y]_{\sigma_1\sigma_1'}[\sigma_y]_{\sigma_2\sigma_2'} + [\sigma_z]_{\sigma_1\sigma_1'}[\sigma_z]_{\sigma_2\sigma_2'}$.

All the matrix elements of Γ can be expressed in terms of the scattering amplitudes for parallel and antiparallel spin electrons, defined as follows (see Fig. 6.20)

$$\Gamma_{p_1\uparrow, p_2\uparrow} \equiv \Gamma_{p_1,p_2}^{\uparrow\uparrow,\uparrow\uparrow} = \Gamma_{p_1p_2}^+ + \Gamma_{p_1p_2}^-$$
$$\Gamma_{p_1\uparrow, p_2\downarrow} \equiv \Gamma_{p_1,p_2}^{\uparrow\uparrow,\downarrow\downarrow} = \Gamma_{p_1p_2}^+ - \Gamma_{p_1p_2}^- \tag{6.87}$$

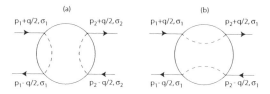

Fig. 6.21. Two general types of diagrams for Γ: (a) direct, (b) exchange.

(we have omitted for brevity the q argument). For example, we have

$$\Gamma^{\uparrow\downarrow,\downarrow\uparrow}_{p_1,p_2} = 2\Gamma^-_{p_1 p_2} = \Gamma_{p_1\uparrow,p_2\uparrow} - \Gamma_{p_1\uparrow,p_2\downarrow} . \qquad (6.88)$$

Henceforth for simplicity's sake we shall often use the notation $\Gamma_{\uparrow\uparrow}$ and $\Gamma_{\uparrow\downarrow}$ in lieu of $\Gamma_{p_1\uparrow,p_2\uparrow}$ and $\Gamma_{p_1\uparrow,p_2\downarrow}$.

We begin our discussion of the diagrammatic rules by explaining how to attach a value to the interaction blocks themselves. The notation will acquire a particularly simple form if we choose to evaluate these diagrams by associating all solid and dashed lines with the corresponding factors of $G(\vec{k}, \omega)$ and v_k (with suitable arguments) and then multiply the result of the sums by the overall numerical factor $\left(\frac{i}{\hbar}\right)^{m-1} (-1)^L (\pm 1)$ where m is the order of the diagram (i.e. the number of interaction lines appearing in it) while L represents the number of its fermion loops. The last factor of ± 1, which can colorfully be referred to as the "loop snapping factor", stems from the way each diagram contributing to Γ influences the number of Fermion loops within the overall diagram in which the interaction block eventually is made use of. Which of the two signs is appropriate is determined by observing that any diagram contributing to Γ can be classified as being one of two distinct types: in the "direct" type, the incoming line $p_1 - q/2$ is continuously connected to the outgoing line $p_1 + q/2$, while in the "exchange" type it is continuously connected to the outgoing line $p_2 - q/2$. This distinction is schematically depicted in Fig. 6.21. It follows from this definition that while all the diagrams for $\Gamma_{\uparrow\downarrow}$ are necessarily of the direct type, the blocks contributing to $\Gamma_{\uparrow\uparrow}$ will contain in general contributions of either type. For example, the first order diagram (a) in Fig. 6.19 is direct, while (b) is of the exchange type. It is clear that substituting an interaction diagram of the exchange type for one of the direct type does quite generally decrease the number of Fermion loops in any Feynman diagram by one. Accordingly we will associate a loop snapping factor of -1 to each exchange diagram.

We can immediately test our rules by using them in the case of the two diagrams of Fig. 6.19. One simply obtains v_q for (a) and $-v_{p_2-p_1}$ for (b).

We establish next the Feynman rules for the evaluation of the skeleton diagrams of a correlation function χ^T_{AB} in which whole interaction blocks appear as represented for instance in Fig. 6.17. It is possible in general for a diagram for Γ to contain both interaction lines and interaction blocks which we will denote with Γ_i.[18] Similarly to the stipulations discussed in Section 6.2.6 we again introduce for the vertices the appropriate matrix elements

---

[18] Notice that even when a diagram does not explicitly contain interaction blocks, it is often convenient to suitably combine elements of the diagram to actually define one or more such devices within it.

of the operators $\hat{A}$ and $\hat{B}$ and associate a factor of $G_\sigma(\vec{k}, \omega)$ to all the solid lines and a factor of $v_{\vec{k}}$ to all the dashed lines. A factor of Γ_i (calculated separately accordingly to the rules given above) will be then associated with each interaction block. Finally the overall factor is given by $\left(\frac{i}{\hbar}\right)^{n+1} (-1)^{L_{MAX}}$ where n is here the putative order of the diagram obtained by adding the total number of explicitly appearing interaction lines to the total number of interaction blocks Γ_i. The integer L_{MAX} represents here the total number of Fermion loops one would identify in the diagram if all the interaction blocks were to be of the direct type (see above). A quick way to determine L_{MAX} is to redraw the diagram substituting diagram (a) of Fig. 6.19 for each of the interaction blocks. The reader is encouraged to verify that these stipulations allow for a correct determination of the overall Fermion loop parity for each of the individual diagrammatic contributions to the interaction blocks.

As an application we make use of our Feynman rules to determine the expression for the density–density response function as represented in Fig. 6.17. In this case $\hat{A} = \hat{n}_{\vec{q}}$ and $\hat{B} = \hat{n}_{-\vec{q}}$, and we readily obtain

$$\chi_{nn}(q) = -\frac{i}{\hbar L^d} \sum_{p\sigma} G_\sigma\left(p - \frac{q}{2}\right) G_\sigma\left(p + \frac{q}{2}\right)$$

$$+ \left(\frac{i}{\hbar L^d}\right)^2 \sum_{p\sigma} G_\sigma\left(p - \frac{q}{2}\right) G_\sigma\left(p + \frac{q}{2}\right)$$

$$\times \sum_{p'\sigma'} \Gamma_{p\sigma, p'\sigma'}(q) G_{\sigma'}\left(p' + \frac{q}{2}\right) G_{\sigma'}\left(p' - \frac{q}{2}\right) .$$

Eq. (6.89) can be written more compactly as

$$\chi_{nn}(q) = -\frac{i}{\hbar L^d} \sum_{p\sigma} G_\sigma\left(p - \frac{q}{2}\right) G_\sigma\left(p + \frac{q}{2}\right) \Lambda_{p\sigma}(q) , \qquad (6.89)$$

where we have defined

$$\Lambda_{p\sigma}(q) \equiv 1 - \frac{i}{\hbar L^d} \sum_{p'\sigma'} \Gamma_{p\sigma, p'\sigma'}(q) G_{\sigma'}\left(p' + \frac{q}{2}\right) G_{\sigma'}\left(p' - \frac{q}{2}\right) . \qquad (6.90)$$

It is evident from Fig. 6.22 that $\Lambda_{p\sigma}(q)$ represents the sum of all the diagrams that can be inserted between an electron–hole propagator $(p - q/2, \sigma; p + q/2, \sigma)$, and an interaction line with four-momentum q, that is, the sum of all the possible "dressings" of the bare interaction vertex. This is the reason why $\Lambda_{p\sigma}(q)$ is called the *vertex function*.

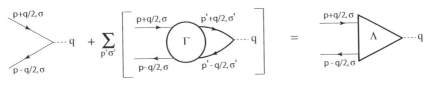

Fig. 6.22. Skeleton diagrams for the vertex function $\Lambda_{p\sigma}(q)$.

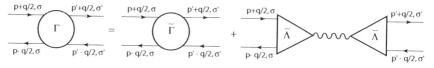

Fig. 6.23. Expressing the full scattering amplitude Γ in terms of the proper scattering amplitude and the proper vertex function.

Similar expressions can easily be written for other linear response functions. For example, the longitudinal spin–spin response function is given by

$$\chi_{S_z S_z}(q) = -\frac{i}{\hbar L^d} \sum_{p\sigma\sigma'} \sigma\sigma' G_\sigma\left(p - \frac{q}{2}\right) G_\sigma\left(p + \frac{q}{2}\right) \Lambda_{p\sigma\sigma'}(q), \qquad (6.91)$$

where the *spin-resolved vertex function*, $\Lambda_{p\sigma\sigma'}(q)$, is defined as

$$\Lambda_{p\sigma\sigma'}(q) \equiv \delta_{\sigma\sigma'} - \frac{i}{\hbar L^d} \sum_{p'} \Gamma_{p\sigma,p'\sigma'}(q) G_{\sigma'}\left(p' + \frac{q}{2}\right) G_{\sigma'}\left(p' - \frac{q}{2}\right). \qquad (6.92)$$

The *proper* density–density response function can also be calculated from Eqs. (6.89) and (6.90), provided the scattering amplitude and the vertex function are replaced by their "proper" versions, described below. The *proper scattering amplitude* $\tilde{\Gamma}_{p\sigma,p'\sigma'}(q)$ is defined as the sum of all the scattering diagrams that cannot be divided into two parts by cutting a single interaction line that carries a fixed four-momentum q. The *proper vertex function* $\tilde{\Lambda}_{p\sigma}(q)$ is constructed from the proper scattering amplitude in the same way that the full vertex function is constructed from the full scattering amplitude. In fact, it is not difficult to see that the graphical identity described in Fig. 6.22 remains valid if one replaces Γ and Λ by $\tilde{\Gamma}$ and $\tilde{\Lambda}$, and the bare interaction line by the wavy screened interaction line. The relation between Λ and $\tilde{\Lambda}$ is simply

$$\Lambda_{p\sigma}(q) = \frac{\tilde{\Lambda}_{p\sigma}(q)}{\epsilon(q)} \qquad (6.93)$$

where $\epsilon(q)$ is the dielectric function. Similarly, in Fig. 6.23 one sees that the full scattering amplitude is related to the proper one by the equation

$$\Gamma_{p\sigma,p'\sigma'}(q) = \tilde{\Gamma}_{p\sigma,p'\sigma'}(q) + \tilde{\Lambda}_{p\sigma}(q)\frac{v_q}{\epsilon(q)}\tilde{\Lambda}_{p'\sigma'}(q). \qquad (6.94)$$

A very useful relation connects the self-energy to the proper vertex function and the screened interaction. It is quite evident that every self-energy diagram is contained once and only once in the skeleton diagram of Fig. 6.24. Therefore, we can write

$$\Sigma_\sigma(p) = \frac{i}{L^d} \sum_q G_\sigma(p+q)W(q)\tilde{\Lambda}_{p\sigma}(q). \qquad (6.95)$$

This is, in a sense, a generalization of the exchange diagram (b) of Fig. 6.15, with the obvious differences that it contains the exact Green's function G_σ instead of the non-interacting one, that the coulomb interaction is screened by the dielectric function, and

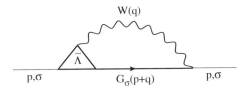

Fig. 6.24. Skeleton diagram for the self-energy.

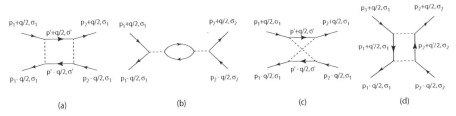

Fig. 6.25. Some second order diagrams for the scattering amplitude. (a) and (b) are reducible in the direct electron–hole channel, while (c) and (d) are irreducible in that channel. However, (d) is reducible in the "cross" electron–hole channel, and (c) is reducible in the electron-electron channel.

that one of the two vertices is "dressed" by the vertex function. The great difficulty of the many-body problem lies in the fact that Eq. (6.95) involves *two-particle* properties such as the dielectric function and the vertex function. Thus, the Dyson equation (6.72) by itself does not lead to a closed set of equations for the Green's function: one must work much harder to find additional equations for Λ and ϵ. In spite of these difficulties, a closed set of equations for G, Λ, and ϵ, *can* be derived, and will be discussed in Section 6.3.5.

Let us now examine in greater detail the diagrams for the two-particle scattering amplitude. The most dangerous diagrams are the ones that contain an intermediate electron–hole propagator with fixed four dimensional wave vector q. Such diagrams can be separated into two parts by cutting a single particle–hole propagator and are therefore said to be *reducible in the direct electron–hole channel*. Some examples are shown in Fig. 6.25.

Diagram (c) and (d) are irreducible in the direct electron–hole channel, while diagrams (a) and (b) are reducible. We will now show that the reducible diagrams are singular in the $q \to 0$ limit and therefore need the most accurate treatment. The irreducible diagrams, on the other hand, can be treated more roughly in the hope that their precise value does not greatly affect the physics at the qualitative level.

It is not difficult to understand the origin of the singularity for $q \to 0$. For example, diagram (b) in Fig. 6.25 contains an intermediate electron–hole bubble that is equivalent to the Lindhard function. As we pointed out in Section 4.4.3, this function is singular for $q \to 0$, because its limiting value depends on the ratio $\nu = \omega/qv_F$ between the frequency and the maximum energy of an electron hole pair. Mathematically, the singularity arises from the coalescence of the poles of the two intermediate Green's functions – the existence of such poles being the defining characteristic of a *Landau Fermi liquid* (see Chapter 8) of which we will anticipate a few ideas here. Let us then assume that $G(p)$ can be written in the form

$$G(p) = G^{(reg)}(p) + \frac{Z_{\vec{p}}}{p_0 - \frac{\epsilon_F + v_F(|\vec{p}| - p_F)}{\hbar} + i\eta_{\vec{p}}}, \tag{6.96}$$

that is, the sum of a regular function, $G^{(reg)}(p)$, and a "quasiparticle pole" of strength $Z_{\vec{p}}$ at energy $\epsilon_F + v_F(p - p_F)$. Physical arguments and detailed calculations, presented in Chapter 8, confirm that this is indeed the correct form of the Green's function in three- and two-dimensional Fermi liquids. Armed with this expression for $G(p)$, it is possible to prove (see Exercise 8) that the electron–hole propagator has the singular limiting form

$$
\begin{aligned}
R(p,q) &\equiv G\left(p + \frac{q}{2}\right) G\left(p - \frac{q}{2}\right) \\
&\stackrel{q \to 0}{\to} 2\pi i Z_{\vec{p}}^2 \frac{\hbar \vec{q} \cdot \vec{v}_{\vec{p}}}{\omega - \vec{q} \cdot \vec{v}_{\vec{p}} + i\eta \, \mathrm{sign}(\omega)} \delta\left(p_0 - \frac{\epsilon_F}{\hbar}\right) \delta(\varepsilon_{\vec{p}} - \epsilon_F) \\
&\quad + \tilde{R}(p,q),
\end{aligned}
\tag{6.97}
$$

where $\vec{v}_{\vec{p}} = \frac{\vec{p}}{m}$, and $\tilde{R}(p,q)$ is a regular function of its arguments. The reader is urged to verify that the last line of this expression, taken in the non-interacting limit (that is, for $Z_{\vec{p}} = 1$ and $\tilde{R} = 0$), and substituted in Eq. (6.56), yields the correct small-q limit of the Lindhard function.

In order to distinguish the diagrams that contain at least one singular electron–hole propagator from those that do not, we introduce the *irreducible electron–hole interaction*, $I_{p\sigma, p'\sigma'}(q)$, defined as the sum of all the two-particle scattering diagrams that are irreducible in the direct electron–hole channel. Notice that the two first-order diagrams of Γ belong to I. The first one, is, in fact, the *only* improper diagram that contributes to I: all the other diagrams in I are easily seen to be proper.[19]

The irreducible electron–hole interaction can be derived in a formal manner by removing (or cutting) a Green's function line at all possible places within the skeleton diagrams for the self-energy (see Fig. 6.26). It is evident that the diagrams generated in this manner are interaction diagrams, and their irreducibility in the particle-hole channel follows from the fact that the original skeleton diagram does not contain self-energy insertions (check this in Exercise 5!). Furthermore, each diagram for I is generated once and only once by this "differentiation" procedure. Thus one arrives at the important relation

$$I_{p\sigma, p'\sigma'}(0) = \frac{\delta \Sigma_\sigma(p)}{\delta G_{\sigma'}(p')}, \tag{6.98}$$

which will play a crucial role in the microscopic derivation of the Landau theory of Fermi liquids in Chapter 8.[20]

---

[19] Improper diagrams have an intermediate interaction line carrying a four momentum q that must necessarily end on an intermediate electron–hole propagator of the same momentum, thus violating the requirement of irreducibility.

[20] Eq. (6.98) lies at the heart of the derivation of the so-called *Ward–Pitaevskii identities*, which correspond to the Ward identities of quantum electrodynamics. These identities relate the vertex function to the self-energy (Noziérés, 1964), and guarantee the satisfaction of the conservation laws. For this reason Eq. (6.98) is sometimes improperly referred to as a Ward identity.

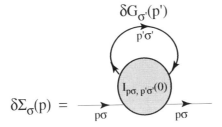

Fig. 6.26. Diagrammatic representation of the identity (6.98).

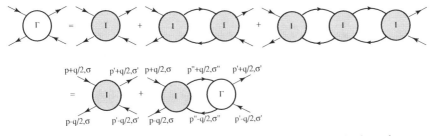

Fig. 6.27. The Bethe–Salpeter equation in the direct electron–hole channel.

Knowledge of the irreducible electron–hole interaction enables one to construct the full scattering amplitude through a standard recursive process illustrated in Fig. 6.27. Basically, diagrams for Γ are classified according to the number of intermediate electron–hole propagators they contain, in much the same way as, in the previous section, diagrams for G were classified according to the number of intermediate electron propagators. The connecting blocks are now irreducible electron–hole interactions, which are presumably regular in the $q \to 0$ limit. The sum of the series is equivalent to the solution of the *Bethe–Salpeter* equation

$$\Gamma_{p\sigma,p'\sigma'}(q) = I_{p\sigma,p'\sigma'}(q)$$
$$+ \frac{1}{\hbar L^d} \sum_{p''\sigma''} I_{p\sigma,p''\sigma''}(q) G_{\sigma''}\left(p'' + \frac{q}{2}\right) G_{\sigma''}\left(p'' - \frac{q}{2}\right) \Gamma_{p''\sigma'',p'\sigma'}(q) . \quad (6.99)$$

Together with Eq. (6.98), the $q \to 0$ limit of this equation will play a central role in the microscopic justification of the Landau theory of Fermi liquids.

In a spin-isotropic system the Bethe–Salpeter equation decouples into two independent equations for the density and spin channels, i.e.

$$\Gamma_{pp'}^+(q) = I_{pp'}^+(q) + \frac{1}{\hbar L^d} \sum_{p''} I_{pp'}^+(q) R(p'', q) \Gamma_{p''p'}^+(q)$$

$$\Gamma_{pp'}^-(q) = I_{pp'}^-(q) + \frac{1}{\hbar L^d} \sum_{p''} I_{pp'}^-(q) R(p'', q) \Gamma_{p''p'}^-(q) , \quad (6.100)$$

where $I_{pp'}^+(q)$ and $I_{pp'}^-(q)$ are defined by analogy to $\Gamma_{pp'}^+(q)$ and $\Gamma_{pp'}^-(q)$ in Eq. (6.86), and $R(p, q)$ is defined in the first line of Eq. (6.97). All these equations remain valid if Γ, I, and Λ are replaced by their proper counterparts.

There are two more channels in which the two-body scattering amplitude can be expanded. These are

(1) The "cross" electron–hole channel, in which the intermediate particle–hole propagator carries a four-momentum $p_1 - p_2$;
(2) The electron-electron channel in which the intermediate pair of Fermion lines run in the same direction and carry a total four-momentum $p_1 + p_2$.

In either case, one defines irreducible diagrams that cannot be decomposed into two parts by cutting a single electron–hole or electron-electron propagator (see also Fig. 6.25). The full interaction is then related to the irreducible interaction by integral equations similar to Eq. (6.99).

The usefulness of the cross electron–hole channel in the Landau theory of Fermi liquids is rather modest, because in a calculation of macroscopic response functions ($q \to 0$) the momentum transfer $p_1 - p_2$ is an integrated variable, which is rarely found within the dangerous region $p_1 - p_2 \sim 0$. The expansion is nevertheless very useful in the calculation of the scattering amplitudes, outlined in Section 6.3.6.

The electron-electron channel is normally not dangerous, unless the effective two-body scattering amplitude happens to be consistently negative ("attractive") for electrons near the Fermi surface. In that case, the channel is not only dangerous, but also extremely interesting, as it drives a transition to the superconducting state of Bardeen, Cooper, and Schrieffer.

For the sake of completeness, we write down the form of the Bethe–Salpeter equation in the electron-electron channel, as can be surmised from Fig. 6.28.

$$\Gamma_{p_1\sigma_1, p_2\sigma_2}(q) = J_{p_1\sigma_1, p_2\sigma_2}(q) + \frac{1}{2\hbar L^d} \sum_{q'} J_{p_1\sigma_1, p_2\sigma_2}\left(q - \frac{q'}{2}\right)$$

$$\times G_{\sigma_1}\left(p_1 - \frac{q'-q}{2}\right) G_{\sigma_2}\left(p_2 + \frac{q'-q}{2}\right) \Gamma_{p_1\sigma_1, p_2\sigma_2}(q' - q). \quad (6.101)$$

Here $J_{p_1\sigma_1, p_2\sigma_2}(q)$ is , of all the diagrams that are irreducible in the particle particle channel, with incoming momenta and spins $(p_1 - q/2, \sigma_1)$, $(p_2 + q/2, \sigma_2)$; and outgoing momenta

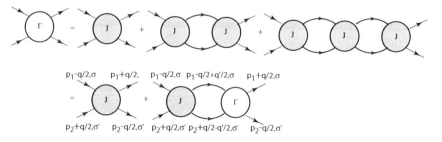

Fig. 6.28. The Bethe–Salpeter equation in the electron-electron channel.

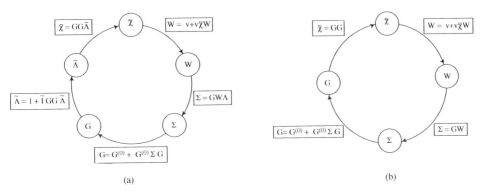

Fig. 6.29. (a) Schematic representation of the self-consistent scheme for the calculation of the self-energy, the screened interaction, and the vertex function. (b) A simplified self-consistent scheme, known as the GW approximation, in which the vertex function is set equal to the bare vertex (adapted from Aulbur *et al.* (2000)).

and spins $(p_1 + q/2, \sigma_1)$, $(p_2 - q/2, \sigma_2)$. The factor $1/2$ in front of the last term corrects for the double-counting due to the equivalence of the two intermediate electron lines (there is no double counting in the electron–hole channel, since the electron and hole lines are distinguished by the opposite direction of their arrows).

6.3.5 Self-consistent equations

The exact relationships between different diagrammatic blocks obtained so far in this section form a closed set of integro-differential equations (Hedin, 1965; Hedin and Lunqvist, 1969) from which the density–density response function, χ_{nn}, the screened interaction, W, the self-energy, Σ, the Green's function, G, and the vertex function, Λ, can be calculated, in principle, exactly. The complete set of equations is summarized in Fig. 6.29(a): notice its circular structure. These equations could be solved by iteration starting, say, from the Hartree approximation for the Green's function, i.e., in the homogeneous electron liquid, $G^{(0)} =$ non-interacting Green's function and $\Sigma^{(0)} = 0$. In the first cycle one would calculate $\tilde{\Lambda}^{(1)}$, $\tilde{\chi}_{nn}^{(1)}$, $W^{(1)}$ and, finally, $\Sigma^{(1)}$ and $G^{(1)}$. The cycle would then be repeated until $G^{(n)}$ and $G^{(n+1)}$ coincide. A major difficulty in the implementation of this program is that the vertex function is defined implicitly through the Bethe–Salpeter equation

$$\Lambda_{p\sigma}(q) = 1 - \frac{i}{\hbar L^d} \sum_{p'\sigma'} I_{p\sigma, p'\sigma'}(q) G_{\sigma'}\left(p' + \frac{q}{2}\right) G_{\sigma'}\left(p' - \frac{q}{2}\right) \Lambda_{p'\sigma'}(q), \quad (6.102)$$

which is obtained by combining Eqs. (6.90) and (6.99). The irreducible electron–hole interaction $I_{p\sigma, p'\sigma'}(q)$ is defined in terms of a functional derivative of the self-energy Σ with respect to the dressed propagator G, which is hard to calculate, since G is connected to Σ by another integral equation (the Dyson equation (6.72)).

In spite of these difficulties, the goal of solving the self-consistent equations has become considerably nearer in recent years, thanks to faster computers and some formal advances.

As an example of the latter, Schindlmayr and Godby (1998) have been able to show that the vertex function at the $n + 1$-th cycle of iteration is related to the vertex function at the *first* cycle of iteration via the equation

$$\Lambda_{p\sigma}^{(n+1)}(q) = 1 - \frac{i}{\hbar L^d} \sum_{p'\sigma'} I_{p\sigma,p'\sigma'}^{(n)}(q) G_{\sigma'}^{(0)}\left(p' + \frac{q}{2}\right) G_{\sigma'}^{(0)}\left(p' - \frac{q}{2}\right) \Lambda_{p'\sigma'}^{(1)}(q) , \quad (6.103)$$

where $I_{p\sigma,p'\sigma'}^{(n)}(q)$ is the irreducible electron–hole interaction obtained by differentiating the self-energy at the n-th cycle of iteration with respect to $G^{(0)}$ (see the original paper for the rather complex derivation of this result).

A very popular approach to the calculation of self-energies and response functions in real materials is the *GW approximation*, whose structure is shown in Fig. 6.29(b). This can be viewed as the first cycle in an iterative solution of Hedin's equation, starting from the Hartree theory. In the GW approximation one sets the vertex function to unity so that the self-energy is just the Green's function times the screened interaction, and the proper density–density response function is a product of two Green's function.

In practice the GW approximation comes in various "flavors". At the simplest level, one forfeits self-consistency by using the non-interacting Green's function $G^{(0)}$ instead of G: the screened interaction is then given by the RPA. At the next level one may seek self-consistency on G while still using the RPA for the screened interaction. Finally, in the full-fledged implementation the self-consistent G is used everywhere, leading to a screened interaction that is no longer of the RPA form. Ironically this last scheme – the fully self-consistent GW – does not work better than the previous ones, due, in part, to the fact that the proper response function constructed from a product of two dressed Green's function fails to satisfy basic requirements such as the f-sum rule (Holm and von Barth, 1998; Schöne and Eguiluz, 1998). On the other hand, the non self-consistent GW leads to results that are usually in qualitatively good agreement with experiment, at least in not too strongly correlated systems (for a recent review, see Aulbur, Jönsson, and Wilkins, 2000).

While a rigorous theoretical justification for the success of the non self-consistent GW approach (also known as G0W) is not available, a heuristic derivation of this scheme based on the existence of a renormalized hamiltonian for low-energy excitations (Yarlagadda and Giuliani, 1989) has been put forth, and will be presented in detail in Section 8.6. There, we will also learn how to take into account some of the effects associated with the existence of vertex corrections through the use of local field factors.

Another important family of approximations, known as *conserving approximations*, stems from the seminal work of Kadanoff and Baym (1962).[21] In this approach one starts from an approximate expression for the generating functional $\Omega[G]$, and derives the self-energy functional $\Sigma[G]$ from functional differentiation, according to Eq. (6.81). Inserting the expression for Σ in Eq. (6.73) one obtains a self-consistent equation for G, which can

---

[21] The main concern, at that time, was to ensure that any approximation to the exact solution of the many-body problem satisfy the local conservation laws for particle density, momentum, and energy. Kadanoff and Baym showed that this requirement is met when the self-energy is derived from a generating functional $\Omega[G]$ according to Eq. (6.81).

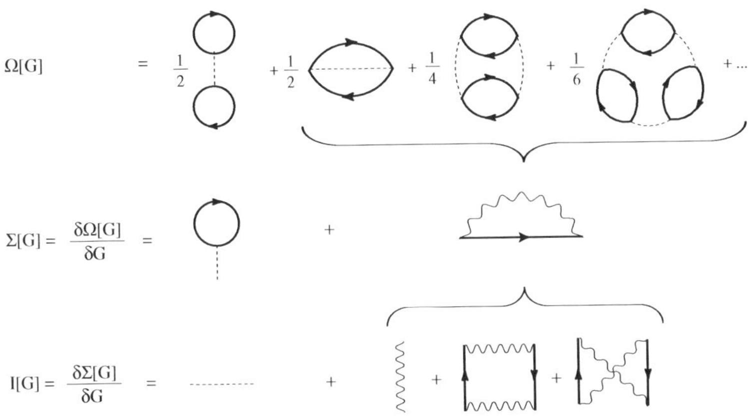

Fig. 6.30. Selected diagrams for the generating functional Ω, the self-energy Σ and the irreducible electron–hole interaction I in a conserving approximation. Notice that the wavy line here represents an RPA-screened interaction.

be solved. Once G is known, one can calculate the irreducible electron–hole interaction $I[G]$ from the functional derivative $\frac{\delta\Sigma[G]}{\delta G}$, insert it in the Bethe–Salpeter equation (6.99), and thus obtain the two-particle scattering amplitude Γ. Finally, once Γ is known, one can calculate two-particle response functions, such as χ_{nn} and $\chi_{S_z S_z}$, from Eqs. (6.89) and (6.91) respectively. A simple example is shown in Fig. 6.30. This approximation is not completely self-consistent, however, because the self-energy computed from Eq. (6.81) does not necessarily satisfy the exact relation (6.95).

Needless to say, the numerical implementation of this program presents formidable difficulties, and only in the last few years computers have become powerful enough to make the task approachable (Bickers, 1989). Conserving approximations for the vertex function have recently been taken to a higher level of sophistication by Takada (Takada, 1995): we refer the interested reader to his original papers.

6.3.6 Two-body effective interaction: the local approximation

To illustrate the use of the diagrammatic formalism as an analytical tool in many-body theory we now employ it to (i) Give a precise definition of the effective interaction between two electrons in the electron liquid and (ii) Extend the calculation of the Kukkonen-Overhauser interaction to cases in which the original derivation presented in Section 5.5.3.1 does not work (Vignale and Singwi, 1984).

First of all, we must carefully explain what we mean by "effective interaction" between two particles. Let v_q be the Fourier transform of the *bare* interaction between the two particles. If the system consisted of only two particles in a vacuum then the diagrams for the scattering amplitude would have the form shown in Fig. 6.31: these are "ladder diagrams" of

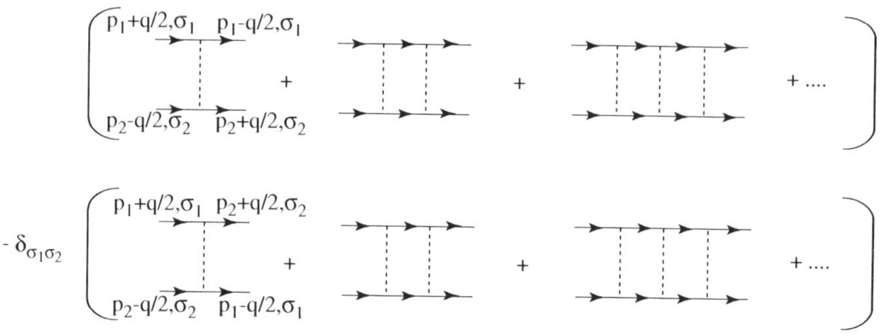

Fig. 6.31. In a system consisting of only two identical particles in a vacuum all the diagrams for the scattering have the "ladder" structure displayed here. The first series represents the direct scattering amplitude and the second series the exchange amplitude.

the direct or exchange type.[22] Notice that the two first-order diagrams, which are *irreducible in the particle-particle channel*, yield the matrix element of the bare interaction between the two-particle states $(p_1 - \frac{q}{2}, \sigma_1; p_2 + \frac{q}{2}, \sigma_2)$ and $(p_1 + \frac{q}{2}, \sigma_1; p_2 - \frac{q}{2}, \sigma_2)$ (direct diagram) and $(p_1 - \frac{q}{2}, \sigma_1; p_2 + \frac{q}{2}, \sigma_2)$ and $(p_2 - \frac{q}{2}, \sigma_2; p_1 + \frac{q}{2}, \sigma_1)$ (exchange diagram).

Taking hint from this simple example, we define the effective interaction as the two-body operator $\hat{W}$ whose matrix elements are

$$\left\langle p_1 - \frac{q}{2}, \sigma_1; p_2 + \frac{q}{2}, \sigma_2 \left| \hat{W} \right| p_1 + \frac{q}{2}, \sigma_1; p_2 - \frac{q}{2}, \sigma_2 \right\rangle = J^D_{p_1\sigma_1, p_2\sigma_2}(q) \,,$$

$$\left\langle p_2 + \frac{q}{2}, \sigma_2; p_1 - \frac{q}{2}, \sigma_1 \left| \hat{W} \right| p_1 + \frac{q}{2}, \sigma_1; p_2 - \frac{q}{2}, \sigma_2 \right\rangle = J^X_{p_1\sigma_1, p_2\sigma_2}(q) \,, \quad (6.104)$$

where $J^D_{p_1\sigma_1, p_2\sigma_2}(q)$ is the sum of all the direct-type diagrams that are irreducible in the electron–electron channel, and $J^X_{p_1\sigma_1, p_2\sigma_2}(q)$ is the sum of all the exchange-type diagrams with the same property. Obviously, with this definition, the "effective interaction" for a system of just two electrons coincides with the bare interaction.[23]

Our next task is to find a simple approximation for $J = J^D - J^X$. We observe that all diagrams for J fall into one of three mutually exclusive classes (see Exercise 10):

(i) Diagrams that are reducible in the direct electron–hole channel;
(ii) Diagrams that are reducible in the cross electron–hole channel;
(iii) Diagrams that are *totally irreducible*, i.e., irreducible in the two electron–hole channels *and* in the electron–electron channel.

In class (iii), we neglect all but the two first-order diagrams, i.e., we set

$$J^{(iii)}_{p_1\uparrow, p_2\uparrow}(q) \simeq v_q - v_{|\vec{p}_1 - \vec{p}_2|}$$

$$J^{(iii)}_{p_1\uparrow, p_2\downarrow}(q) \simeq v_q \,. \quad (6.105)$$

---

[22] The essential characteristic of the ladder diagrams is that all the Fermion lines run forward in time, i.e., they are particle-like. Any process not described by the ladder diagrams would have to involve "holes", but such a process is impossible in a two-particle system since there is no Fermi sea and hence no holes to talk about.

[23] The present definition ignores the renormalization factors associated with the four external lines in the diagrams of Fig. 6.31. The roughness of the approximations that follow does not justify the retention of these fine corrections.

This uncontrolled approximation roughly corresponds to the Kukkonen-Overhauser assumption that the disturbance created by an electron in the electron liquid can be treated by linear response theory.

Consider now the diagrams in class (i) (Fig. 6.27). In order to sum the infinite series we resort to two drastic approximations. First, we approximate the internal electron-hole propagator by *non-interacting* electron–hole propagators. Second, we approximate the irreducible electron–hole interaction as

$$I_{p_1\sigma_1, p_2\sigma_2}(q) = v_q[1 - G_{\sigma_1\sigma_2}(q)] , \qquad (6.106)$$

where $G_{\sigma_1\sigma_2}(q)$ is the dynamical local field factor introduced in Chapter 5 for the calculation of the density and spin response functions.

The first approximation is (roughly) justified by the observation that the Green's function of an interacting electron liquid near the Fermi surface is qualitatively similar to that of a non-interacting Fermi liquid, aside from small renormalization factors. This point will be examined in greater detail in Chapter 8. As for the second approximation, it amounts to replacing the interaction function by some kind of an average calculated with respect to all possible values of p_1 and p_2 near the Fermi surface.[24]

The fact that the irreducible electron–hole interaction, $I(q)$, should be related to the local field factor as in Eq. (6.106) can be seen from the following argument. Consider again the expressions (6.89) and (6.91) for the density–density and spin–spin response functions in terms of the full scattering amplitude Γ. For a paramagnetic system, the density and spin components of Γ (Γ^+ and Γ^-) satisfy independent Bethe–Salpeter equations (6.100) with kernels $I^+(q)$ and $I^-(q)$ respectively. According to Eq. (6.106) these two kernels will be approximated as $I^+(q) \sim v_q[1 - G_+(q)]$ and $I^-(q) \sim -[v_q G_-(q)]$, where G_+ and G_- are the density and spin local field factors introduced in Chapter 5. With this approximation we see that Γ^+ and Γ^- also become independent of p_1 and p_2 and satisfy the algebraic equations

$$\Gamma^+(q) = v_q[1 - G_+(q)][1 + \chi_0(q)\Gamma^+(q)]$$
$$= \frac{v_q[1 - G_+(q)]}{1 - v_q[1 - G_+(q)]\chi_0(q)} , \qquad (6.107)$$

and

$$\Gamma^-(q) = -v_q G_-(q)[1 + \chi_0(q)\Gamma^-(q)]$$
$$= -\frac{v_q G_-(q)}{1 + v_q G_-(q)\chi_0(q)} , \qquad (6.108)$$

where $\chi_0(q)$ is the Lindhard function. Inserting these expressions into Eqs. (6.89) and (6.91) we recover the standard formulas (5.102) and (5.110) for the density–density and spin–spin response function. This shows that our local approximation for I, Eq. (6.106), is consistent with the standard definition of the local field factors.

---

[24] Because the lack of dependence on p_1 and p_2 characterizes the scattering amplitude from a *local* potential, this approximation has been referred to as the "local" approximation.

The problem of calculating the sum of the diagrams in class (i) is now reduced to the summation of a geometric series. We can first switch to the $+,-$ basis (in which the charge and spin contributions are decoupled) and then extract the $\uparrow\uparrow$ and $\uparrow\downarrow$ components. The result, expressed in terms of the density–density and spin–spin response functions, is

$$J^{(i)}_{p_1\uparrow,p_2\uparrow}(q) = \{v_q[1 - G_+(q)]\}^2 \chi_{nn}(q) + [v_q G_-(q)]^2 \chi_{S_z S_z}(q) \,,$$
$$J^{(i)}_{p_1\uparrow,p_2\downarrow}(q) = \{v_q[1 - G_+(q)]\}^2 \chi_{nn}(q) - [v_q G_-(q)]^2 \chi_{S_z S_z}(q) \,. \tag{6.109}$$

Finally we must evaluate the sum of the diagrams in class (ii). The method of evaluation is the same as for the diagrams of class (i), but we must pay attention to the fact that the diagrams that were of the direct type in (i) are of the exchange type in (ii) and vice versa: this gives rise to an extra minus sign

$$J^{(ii)}_{p_1\uparrow,p_2\uparrow}(q) = - \{v_{p_1-p_2}[1 - G_+(p_1 - p_2)]\}^2 \chi_{nn}(p_1 - p_2)$$
$$- [v_{p_1-p_2} G_-(p_1 - p_2)]^2 \chi_{S_z S_z}(p_1 - p_2) \,,$$
$$J^{(ii)}_{p_1\uparrow,p_2\downarrow}(q) = -2[v_{p_1-p_2} G_-(p_1 - p_2)]^2 \chi_{S_z S_z}(p_1 - p_2) \,. \tag{6.110}$$

Combining the three contributions to $J_{p_1\uparrow,p_2\uparrow}$ and $J_{p_1\uparrow,p_2\downarrow}$ after some simple manipulations one recovers the scattering amplitudes $A_{\uparrow\uparrow}$ and $A_{\uparrow\downarrow}$ (Eqs. (5.153) and (5.154)) of the Kukkonen-Overhauser theory.[25] It should be noted that the present derivation of the scattering amplitude, unlike the Kukkonen-Overhauser derivation, can be easily generalized to situations lacking spin-space isotropy or to multicomponent systems. In this manner one derives the general equations (5.155)–(5.158) of Section 5.5.3.2 (Exercise 11).

The scattering amplitude $J^{(ii)}_{p_1\uparrow,p_2\downarrow}$ may be puzzling, at first, since it seems to imply the existence of an exchange process between two particles that are distinguishable by virtue of their opposite spin orientation. The physical process responsible for this strange effect is the production of transverse spin waves (that is, fluctuations in $\langle \hat{S}_{\pm,\vec{q}} \rangle$) whereby the $\uparrow$-electron of four-momentum $p + \frac{q}{2}$ may flip its spin and emerge as the $\downarrow$-electron of four-momentum $p' + \frac{q}{2}$, as shown in Fig. 6.32. The correctness of this interpretation becomes evident after making the approximation $I^{\uparrow\downarrow,\downarrow\uparrow}_{p+p'+q,p+p'-q}(p - p') \simeq -v_{\vec{p}-\vec{p}'} G_{\uparrow\downarrow,\downarrow\uparrow}(p - p')$, where $G_{\uparrow\downarrow,\downarrow\uparrow}(p - p')$ is the transverse spin local field factor discussed after Eq. (5.162). Then the sum of the reducible diagrams of Fig. 6.32 yields an interaction mediated by transverse spin fluctuations of the electron liquid.

6.3.7 Extension to broken symmetry states

One of the most important applications of the diagrammatic integral equation formalism arises in the study of broken-symmetry phases. In Chapters 1 and 2 we have encountered several examples of broken-symmetry phases, such as the Bloch ferromagnetic liquid, Overhauser's charge and/or spin density waves, and the BCS superconducting state. All

---

[25] In order to reproduce the Kukkonen-Overhauser formulas one must make the changes of variables $p_1 \rightarrow p_1 - \frac{q}{2}$ and $p_2 \rightarrow p_2 + \frac{q}{2}$, so that the incoming momenta are p_1 and p_2 and the outgoing ones $p_1 - q$ and $p_2 + q$ as in Chapter 5. The more symmetric labeling adopted in this chapter simplifies the writing of the Bethe–Salpeter equation.

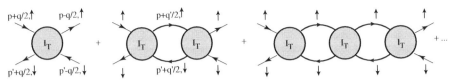

Fig. 6.32. Exchange-like diagrams for the scattering amplitude of opposite spin electrons. I_T is short-hand for $I_{\frac{p+p'+q}{2}, \frac{p+p'-q}{2}}^{\uparrow\downarrow,\downarrow\uparrow}(p-p')$.

such phases are beyond the reach of order-by-order perturbation theory, since the latter is based on the assumption that the interacting system has the same symmetry as the non-interacting one. However, integral equations, which connect infinite series of diagrams, continue to hold even when the Green's function has lower symmetry than the hamiltonian. The reason for this is easy to understand. Imagine to add to the unperturbed hamiltonian, $\hat{H}_0$, an infinitesimal symmetry-breaking external potential, V_{sb}, whose only purpose is to reduce the symmetry of the unperturbed Green's function. For example, to describe a spiral spin density wave state of wave vector $\vec{Q}$, one could introduce an infinitesimal potential that couples $\vec{k} \uparrow$ to $\vec{k} + \vec{Q} \downarrow$ and $\vec{k} + \vec{Q} \uparrow$ to $\vec{k} \downarrow$: the corresponding zero-order Green's function would then contain "anomalous" off-diagonal elements $G_{\vec{k}\uparrow,\vec{k}+\vec{Q}\downarrow}$, and so would all the other diagrammatic blocks such as the self-energy, the vertex function, etc. The diagrammatic formulation of perturbation theory (based on Wick's theorem) remains unchanged, and, in particular, the integral equations depicted in Fig. 6.29 retain their form.

We now solve the self-consistent equations for the Green's function in the presence of a finite V_{sb} and then let V_{sb} tend to zero. Two things can happen. In the first scenario the anomalous components of the Green's function vanish for $V_{sb} \to 0$: this is the case of *unbroken* symmetry. But it may also happen that, in addition to the usual symmetric solution, the self-consistent equations admit a lower symmetry solution, which does not reduce to the symmetric one for $V_{sb} \to 0$. Of course, this second scenario corresponds to the case of broken symmetry. In this case there is usually a critical temperature, T_c, above which the low-symmetry solution disappears. In a continuous transition the anomalous components of the Green's function are infinitesimal just below T_c and this fact allows one to determine T_c by solving a linear eigenvalue problem for these components (for the generalization of the diagrammatic formalism to finite temperature see Section 6.4).

It is not difficult to see that the approach outlined above provides a systematic way to go beyond the Hartree–Fock mean field theory of a broken-symmetry state. Indeed, the HF mean field theory is obtained from the integral equation approach by approximating the self-energy by the two first-order skeleton diagrams of Fig. 6.15, i.e.

$$\Sigma_{\alpha\beta}^{HF} = \sum_{\gamma\delta}(v_{\alpha\gamma\delta\beta} - v_{\alpha\gamma\beta\delta}) \int_{-\infty}^{\infty} G_{\gamma\delta}(\omega)e^{i\eta\omega} \frac{d\omega}{2\pi i} \tag{6.111}$$

and by solving the Dyson equation

$$[G]_{\alpha\beta}^{-1}(\omega) = [G^{(0)}]_{\alpha\beta}^{-1}(\omega) - \Sigma_{\alpha\beta}^{HF} \tag{6.112}$$

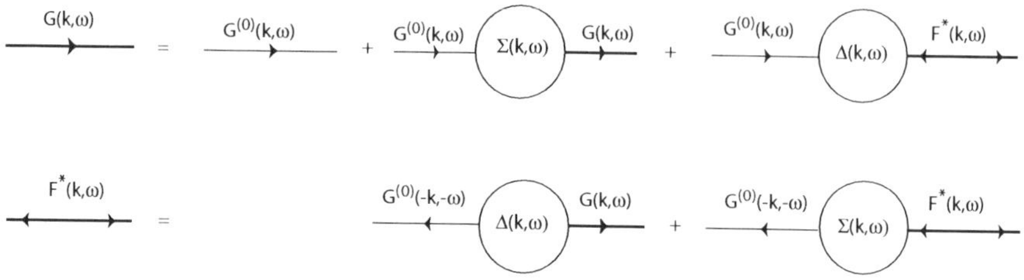

Fig. 6.33. The integral equations for the normal components $G(\vec{k}, \omega)$ and $G(-\vec{k}, -\omega)$ and the anomalous components $F(\vec{k}, \omega)$ and $F^*(\vec{k}, \omega)$ of a superconducting system. $\Sigma(\vec{k}, \omega)$ and $\Delta(\vec{k}, \omega)$ represent the normal and the anomalous self-energy, respectively. Notice the absence of the non-interacting contribution to the anomalous Green's function.

for the broken-symmetry Green's function $G_{\alpha\beta}(\omega)$. Better approximations for Σ will presumably lead to a more accurate description of the broken-symmetry state.

A particularly important and non-trivial example of broken symmetry is provided by the BCS theory of superconductivity, which is briefly described in Section 2.7. In Exercise 9, we make use of the Bogoliubov transformation (2.124), to show that the ordinary Green's function (6.30) associated with the non number-conserving mean field hamiltonian (2.123) is

$$G^{BCS}(\vec{k}, \omega) = \frac{u_{\vec{k}}^2}{\omega - \frac{E_{\vec{k}}}{\hbar} + i\eta} + \frac{v_{\vec{k}}^2}{\omega + \frac{E_{\vec{k}}}{\hbar} - i\eta} , \qquad (6.113)$$

where $u_{\vec{k}}$, $v_{\vec{k}}$, and $E_{\vec{k}}$ were defined in Section 2.7. However, and here is the interesting news, we also have a nonvanishing *anomalous Green's function*,

$$F^{BCS}(\vec{k}, t) = -i\langle \psi_{BCS}| T[\hat{a}_{\vec{k}\uparrow}(t)\hat{a}_{-\vec{k}\downarrow}]|\psi_{BCS}\rangle , \qquad (6.114)$$

which involves the average of two "destruction" operators in the ground-state of $\hat{H}_{BCS}$. Its Fourier transform with respect to time is given by

$$F^{BCS}(\vec{k}, \omega) = \frac{\Delta_{\vec{k}}}{\hbar\omega^2 - \frac{E_{\vec{k}}^2}{\hbar} + i\eta} , \qquad (6.115)$$

where $\Delta_{\vec{k}}$ is the gap function introduced in Section 2.7.

The appearance of the anomalous Green's function F^{BCS} is the clearest indication of the lack of number conservation in the BCS ground-state. From a mathematical point view it means that the contractions between two "destruction" or two "creation" operators, which were zero in the normal state, are in fact different from zero in the superconducting ground-state.[26] It can be proved that Wick's theorem is still valid when the operators are normal-ordered with respect to the BCS ground-state, *provided all the anomalous contractions are included*. For diagrammatic perturbation theory, this means that we must include both normal and anomalous propagators in our diagrams (both shown graphically in Fig. 6.33) making sure that the arrows do not change their direction at the interaction vertices (the interaction *is* number conserving after all!).

---

[26] More precisely, one should say that in the BCS ground-state an operator such as $\hat{a}_{\vec{k}\sigma}$, with $k > k_F$, is no longer a pure destruction operator, but contains also a creation part.

In practice, the integral equation formalism for superconductors is implemented by combining the normal Green's functions $G(\vec{k}, \omega)$ and $G(-\vec{k}, -\omega)$, and the anomalous ones $F(\vec{k}, \omega)$ and $[F(\vec{k}, \omega)]^*$ in a 2×2 *matrix Green's function*, known as the Nambu–Gorkov Green's function. Accordingly, the self-energy also becomes a 2×2 matrix, with normal components $\Sigma(\vec{k}, \omega)$ on the diagonal, and anomalous components $\Delta(\vec{k}, \omega)$ off the diagonal. We will not describe this formalism here, but we will return to this point in Chapter 8, when we briefly discuss the possibility of superconductivity without phonons in the electron liquid.

6.4 Perturbation theory at finite temperature

Let us now turn to the problem of doing perturbation theory in a many-body system at finite temperature. Our specific objective is to set up a perturbative formalism for the calculation of the causal response function

$$\chi_{AB}(t) = -\frac{i}{\hbar}\Theta(t)\langle [\hat{A}(t), \hat{B}]\rangle$$

where the angular brackets denote the quantum average in the grand-canonical ensemble at temperature T and chemical potential μ. For a general observable $\hat{O}$ this average is defined as

$$\langle \hat{O}\rangle = Tr[\hat{\rho}\hat{O}]\,, \tag{6.116}$$

where

$$\hat{\rho} = Z^{-1}e^{-\beta(\hat{H}-\mu\hat{N})} \tag{6.117}$$

is the grand-canonical density matrix,

$$Z = e^{-\beta\Omega} = Tr[e^{-\beta(\hat{H}-\mu\hat{N})}] \tag{6.118}$$

is the partition function, Ω is the grand thermodynamic potential and $\beta = \frac{1}{k_B T}$. Recall that the trace of an operator is defined as the sum of its diagonal matrix elements with respect to an arbitrary basis set, such as the set of the simultaneous eigenstates of $\hat{H}$ and $\hat{N}$. From now on, we will simplify the notation by setting

$$\hat{K} \equiv \hat{H} - \mu\hat{N}\,. \tag{6.119}$$

At first sight one might think that the simplest way to calculate the response function $\chi_{AB}(t)$ is (in analogy to the zero-temperature case) to derive it from the time-ordered correlation function $\chi^T_{AB}(t) = -\frac{i}{\hbar}\langle T[\hat{A}(t)\hat{B}]\rangle$. Indeed, the Fourier transform of $\chi_{AB}(t)$ is related to that of $\chi^T_{AB}(t)$ by a simple generalization of Eq. (6.10), in which sign(ω) is replaced by coth $\frac{\hbar\beta\omega}{2}$ (see Exercise 12). Unfortunately, it turns out that correlation functions of real time are not directly amenable to perturbation theory: however, we will now show that a perturbation theory analogous to the zero-temperature one can be developed for a third object, namely, the time-ordered correlation function in *imaginary* time.

We define the time-ordered correlation function in imaginary time (also known as the *temperature correlation function*) in the following manner:

$$\chi_{AB}^{T}(\tau) \equiv -\frac{1}{\hbar} Tr\{\hat{\rho} T[\hat{A}(\tau)\hat{B}]\}, \tag{6.120}$$

where the evolution of the operators in imaginary time ($t = -i\tau$) is given by

$$\hat{A}(\tau) = e^{\frac{\hat{K}\tau}{\hbar}} \hat{A} e^{-\frac{\hat{K}\tau}{\hbar}}.^{27} \tag{6.121}$$

The time-ordering operator in imaginary time, $T[\ldots]$, arranges the operators in the square bracket in order of decreasing τ from left to right, multiplying the result by a minus sign if an *odd* number of interchanges of fermion operators is needed to effect the rearrangement. In particular, the *temperature Green's function*,

$$\mathcal{G}_{\alpha\beta}(\tau) \equiv -Tr\{\hat{\rho} T[\hat{a}_{\alpha}(\tau)\hat{a}_{\beta}^{\dagger}]\}, \tag{6.122}$$

has the form of Eq. (6.120), and therefore enjoys all the properties that will be shown below to apply to temperature correlation functions.

It is important to appreciate that, because the eigenvalues of $\hat{K}$ can be arbitrarily large, Eq. (6.120) makes sense mathematically only when τ is restricted to the range

$$-\hbar\beta < \tau < \hbar\beta; \tag{6.123}$$

it is only in this range that the exponentially large eigenvalues of $e^{\pm\frac{\hat{K}\tau}{\hbar}}$ are compensated by the exponentially small eigenvalues of the density matrix.

A remarkable identity relates the values of χ_{AB}^{T} at times τ and $\tau \pm \hbar\beta$ within the interval (6.123). The identity reads

$$\chi_{AB}^{T}(\tau) = (-1)^{f} \chi_{AB}^{T}(\tau - \hbar\beta \, \text{sign}(\tau)), \tag{6.124}$$

where f is the number of interchanges of fermion operators that occurs when $\hat{A}$ and $\hat{B}$ are interchanged.[28] Thus $\chi_{AB}^{T}(\tau)$ is *periodic in τ* (with period $\hbar\beta$) if f is even, *antiperiodic* otherwise.

The proof of Eq. (6.124) depends crucially on the form of the density matrix and on the cyclic invariance of the trace (i.e., the fact that $Tr[\hat{A} \ldots \hat{Y}\hat{Z}] = Tr[\hat{Z}\hat{A} \ldots \hat{Y}]$). Consider, for example, the case $\hbar\beta > \tau > 0$, so that $\tau - \hbar\beta < 0$. We have the following chain of identities:

$$\begin{aligned}
\hbar\chi_{AB}^{T}(\tau - \hbar\beta) &= -Tr\{\hat{\rho} T[\hat{A}(\tau - \hbar\beta)\hat{B}]\} \\
&= -(-1)^{f} Z^{-1} Tr\{e^{-\beta\hat{K}} \hat{B} e^{\hat{K}(\frac{\tau}{\hbar} - \beta)} \hat{A} e^{-\hat{K}(\frac{\tau}{\hbar} - \beta)}\} \\
&= -(-1)^{f} Z^{-1} Tr\{e^{-\beta\hat{K}} e^{\hat{K}\frac{\tau}{\hbar}} \hat{A} e^{-\hat{K}\frac{\tau}{\hbar}} \hat{B}\} \\
&= -(-1)^{f} Z^{-1} Tr\{\hat{\rho} T[A(\tau)\hat{B}]\} \\
&= (-1)^{f} \hbar\chi_{AB}^{T}(\tau).
\end{aligned} \tag{6.125}$$

---

[27] As in Chapter 3, operators without an explicit time argument are understood to be evaluated at $\tau = 0$.

[28] This number is $f = f_A f_B$, where f_A and f_B are the numbers of fermion operators contained in $\hat{A}$ and $\hat{B}$ respectively.

The same conclusion is easily reached in the case $\tau < 0$, proving the general validity of Eq. (6.124).

It follows from Eq. (6.124) that the temperature correlation function can be represented as a Fourier series in the following manner:

$$\chi_{AB}^{T}(\tau) = \frac{1}{\hbar\beta} \sum_{k} \chi_{AB}^{T}(i\omega_k) e^{-i\omega_k \tau}, \tag{6.126}$$

where the *Matsubara frequencies* ω_k are even integer multiples of $\frac{1}{\hbar\beta}$ in the periodic case (even f), or odd integer multiples of the same quantity in the antiperiodic case (odd f):

$$\omega_k = \frac{2k}{\hbar\beta} \qquad (even\ f)$$

$$\omega_k = \frac{2k+1}{\hbar\beta} \qquad (odd\ f) \tag{6.127}$$

($k = 0, \pm 1, \pm 2 \ldots$). The coefficients of the series expansion are given by the familiar formula

$$\chi_{AB}^{T}(i\omega_k) = \int_0^{\beta} \chi_{AB}^{T}(\tau) e^{i\omega_k \tau} d\tau. \tag{6.128}$$

Knowledge of the "Matsubara transform" $\chi_{AB}^{T}(i\omega_k)$ enables us to calculate the Fourier transform of the retarded response function through a procedure known as *analytic continuation*. To understand how this works let us introduce the exact eigenstates representation of the temperature correlation function in imaginary time. In the interval $0 \le \tau < \hbar\beta$ we have

$$\chi_{AB}^{T}(\tau) = -\frac{1}{\hbar Z} \sum_{nm} A_{mn} B_{nm} e^{-\omega_{nm}\tau} e^{-\beta \tilde{E}_m}, \tag{6.129}$$

where $\tilde{E}_m = E_m - \mu N_m$ and $\omega_{nm} = \frac{\tilde{E}_n - \tilde{E}_m}{\hbar}$. Inserting this in Eq. (6.128) and performing the integral, we obtain

$$\chi_{AB}^{T}(i\omega_k) = \frac{1}{\hbar Z} \sum_{nm} A_{mn} B_{nm} \frac{P_m - (-1)^f P_n}{i\omega_k - \omega_{nm}}. \tag{6.130}$$

Let us now consider the function $\chi_{AB}^{T}(z)$ obtained by substituting the complex variable z for $i\omega_k$ in Eq. (6.130). The following properties are immediately evident:

(i) $\chi_{AB}^{T}(z)$ is an analytic function of z in both the upper and lower half of the complex plane: all its singularities occur on the real axis in correspondence of the differences of eigenvalues of $\hat{K}$.

(ii) $\chi_{AB}^T(z)$ is *discontinuous* across the real axis and the magnitude of the discontinuity is given by

$$\lim_{\eta \to 0} \left[\chi_{AB}^T(\omega + i\eta) - \chi_{AB}^T(\omega - i\eta) \right] = 2i \Im m \chi_{AB}^T(\omega + i\eta)$$

$$= \frac{2\pi i}{\hbar \mathcal{Z}} \sum_{nm} A_{mn} B_{nm} \left[P_m - (-1)^f P_n \right] \delta \left(\omega - \omega_{nm} \right), \qquad (6.131)$$

where ω is real.

(iii) In the periodic case (even f) $\chi_{AB}^T(\omega + i\eta)$ coincides with the causal response function $\chi_{AB}(\omega)$.

(iv) In the antiperiodic case (odd f) $\chi_{AB}^T(\omega + i\eta)$ is the Fourier transform of the *retarded correlation function*

$$\chi_{AB}^{ret}(t) = -\frac{i}{\hbar} \Theta(t) \langle \{ \hat{A}(t), \hat{B} \} \rangle, \qquad (6.132)$$

which differs from an ordinary response function because it contains an anticommutator instead of a commutator.

Thus, we see that ordinary causal response functions, as well as the retarded fermionic Green's function introduced in Section 6.3.2, can be obtained from the analytic continuation of the appropriate temperature correlation function. It remains to be shown that the latter is amenable to perturbation theory. To do this, let us assume that

$$\hat{K} = \hat{K}_0 + \hat{H}_1, \qquad (6.133)$$

where $\hat{K}_0 = \hat{H}_0 - \mu \hat{N}$ and $\hat{H}_1$ is the interaction hamiltonian. For $\tau > 0$, $\chi_{AB}^T(\tau)$ can be written more explicitly as

$$\chi_{AB}^T(\tau) = -\frac{1}{\hbar} \frac{Tr[e^{-\beta \hat{K}} e^{\frac{\hat{K}\tau}{\hbar}} \hat{A} e^{-\frac{\hat{K}\tau}{\hbar}} \hat{B}]}{Tr[e^{-\beta \hat{K}}]}. \qquad (6.134)$$

The statistical factor $e^{-\beta \hat{K}}$ has the structure of a time-evolution operator in imaginary time, propagating the wave function from time 0 to time $-i\hbar\beta$. We can therefore apply with minor modifications the standard formulas of the real-time evolution:

$$e^{-\beta \hat{K}} = e^{-\beta \hat{K}_0} \hat{U}_I(\hbar\beta, 0), \qquad (6.135)$$

where $\hat{U}_I$ is the imaginary time evolution operator in the interaction picture,

$$\hat{U}_I(\tau_2, \tau_1) = T \exp \left[-\frac{1}{\hbar} \int_{\tau_1}^{\tau_2} \hat{H}_{1I}(\tau') d\tau' \right], \qquad (6.136)$$

and

$$\hat{U}_I(\hbar\beta, 0) = T \exp \left[-\frac{1}{\hbar} \int_0^{\hbar\beta} \hat{H}_{1I}(\tau') d\tau' \right] \equiv \hat{S} \qquad (6.137)$$

is the analogue of the zero-temperature "S-matrix", and is therefore denoted by the same symbol.[29]

Substituting Eq. (6.135) into Eq. (6.134), and making use of the identities $e^{\frac{\hat{K}\tau}{\hbar}} = \hat{U}_I^{-1}(\tau, 0)e^{\frac{\hat{K}_0\tau}{\hbar}}$ and $\hat{U}_I(\hbar\beta, 0)\hat{U}^{-1}(\tau, 0) = \hat{U}_I(\hbar\beta, \tau)$, we obtain

$$\chi_{AB}^T(\tau) = -\frac{1}{\hbar}\frac{Tr[e^{-\beta\hat{K}_0}\hat{U}_I(\hbar\beta, \tau)\hat{A}_I(\tau)\hat{U}_I(\tau, 0)\hat{B}_I]}{Tr[e^{-\beta\hat{K}_0}\hat{U}_I(\hbar\beta, 0)]}, \qquad (6.138)$$

where the operators $\hat{A}_I$ and $\hat{B}_I$ evolve in imaginary time under the unperturbed hamiltonian $\hat{K}_0$. At this point, notice that the operators coming after $e^{-\beta\hat{K}_0}$ in the numerator of Eq. (6.138) are already time-ordered (for $\tau > 0$). We can, therefore, insert a time-ordering operator $\mathcal{T}$ in front of them and *then* use the freedom to change the order of the operators within a $\mathcal{T}$-product to compactify $\hat{U}_I(\hbar\beta, \tau)\hat{U}_I(\tau, 0) = \hat{U}_I(\hbar\beta, 0) = \hat{S}$. The final expression is

$$\chi_{AB}^T(\tau) = -\frac{1}{\hbar}\frac{Tr\{e^{-\beta\hat{K}_0}\mathcal{T}[\hat{A}_I(\tau)\hat{B}_I\hat{S}]\}}{Tr\{e^{-\beta\hat{K}_0}\hat{S}\}}, \qquad (6.139)$$

which closely resembles the corresponding zero-temperature formula, i.e., Eq. (6.27). Thus, we have expressed the correlation function in terms of averages of products of operator whose (imaginary) time-dependence is governed by $\hat{K}_0$. These averages are done in a non-interacting ensemble (with density matrix proportional to $e^{-\beta\hat{K}_0}$) and can therefore be evaluated by means of the finite-temperature Wick's theorem, derived in Appendix 3.

The main ingredient of the perturbative calculation is the non-interacting temperature Green's function, which is defined, in analogy to Eq. (6.30), as the negative of the contraction of the operators $a_\alpha(\tau)$ and $\hat{a}_\beta^\dagger$:

$$\mathcal{G}_{\alpha\beta}^{(0)}(\tau) \equiv -\langle\mathcal{T}[\hat{a}_\alpha(\tau)\hat{a}_\beta^\dagger]\rangle_0, \qquad (6.140)$$

where the time dependence of the operators is governed by $\hat{K}_0$. From our previous discussion it should be clear that $\mathcal{G}_{\alpha\beta}^{(0)}(\tau)$ is an antiperiodic function of τ with period $\hbar\beta$ ($f = 1$). Evaluation of Eq. (6.140) yields

$$\mathcal{G}_{\alpha\beta}^{(0)}(\tau) = -\delta_{\alpha\beta}\left[\Theta(\tau)(1 - n_\alpha)e^{-\tilde{\varepsilon}_\alpha\tau/\hbar} - \Theta(-\tau)n_\alpha e^{-\tilde{\varepsilon}_\alpha\tau/\hbar}\right], \qquad (6.141)$$

where $\tilde{\varepsilon}_\alpha = \varepsilon_\alpha - \mu$ and $n_\alpha = [e^{\beta\tilde{\varepsilon}_\alpha} + 1]^{-1}$ is the Fermi–Dirac thermal occupation factor of single particle state α. The reader is urged to verify that it satisfies Eq. (6.124).

The Matsubara transform of Eq. (6.141), calculated according to Eq. (6.128), has non-vanishing components only at *odd* Matsubara frequencies:

$$\mathcal{G}_{\alpha\beta}^{(0)}(i\omega_k) = \frac{\delta_{\alpha\beta}}{i\omega_k - \frac{\tilde{\varepsilon}_\alpha}{\hbar}}, \qquad \omega_k = \frac{2k + 1}{\hbar\beta}. \qquad (6.142)$$

From this point on we can repeat the analysis of Section 6.2.4, which led to the formulation of perturbation theory in terms of Feynman diagrams. The rules for the construction, labeling, and evaluation of Feynman diagrams are essentially the same we presented

---

[29] Notice that, due to the finiteness of the imaginary time strip ($0 < \tau < \hbar\beta$), we do not need the convergence factor $e^{-\eta|\tau|}$.

in Section 6.2.4: the only difference is that we now assign fermionic (i.e., odd) Matsubara frequencies ($\omega_k = \frac{2k+1}{\hbar\beta}$) to the solid lines, and, correspondingly, bosonic (i.e., even) Matsubara frequencies ($\omega_k = \frac{2k}{\hbar\beta}$) to the interaction lines.[30] In the end, we are left with the following set of rules:

(i) Draw all the topologically distinct connected diagrams consisting of two external vertices, n oriented interaction lines and $2n + 2$ oriented solid lines.

(ii) Associate a fermionic Matsubara frequency to each solid line and a bosonic Matsubara frequency to each interaction line. Conserve frequency at each vertex, including an outgoing frequency $i\omega_k$ (fermionic or bosonic depending on the number of fermion operators in $\hat{A}$) at the $\hat{A}$ vertex.

(iii) Associate a non-interacting temperature Green's function $\mathcal{G}_{\alpha\beta}^{(0)}(i\omega_k)$ (given by Eq. (6.142)) to each solid line.

(iv) Associate a factor v_{abcd} to each interaction line connecting vertices a, b and c, d.

(v) Associate matrix elements $A_{\alpha\beta}$ and $B_{\gamma\delta}$ to the external vertices, where α, β and γ, δ are the labels of the outgoing and ingoing lines respectively.

(vi) Sum over the $n + 1$ internal Matsubara frequencies and over the internal state labels, using the "measure" $\frac{1}{\hbar\beta}\sum_{k=-\infty}^{\infty}\ldots$ for the former.

(vii) Multiply by the factor $(\frac{1}{\hbar})^{n+1}(-1)^L$, where L is the number of closed fermion loops.

(viii) Whenever the sum $\frac{1}{\hbar\beta}\sum_{k=-\infty}^{\infty}\mathcal{G}_{\alpha\beta}^0(i\omega_k)$ appears in a diagram, interpret it as $\lim_{\eta\to 0^+}\frac{1}{\hbar\beta}\sum_{k=-\infty}^{\infty}\mathcal{G}_{\alpha\beta}^0(i\omega_k)e^{i\eta\omega_k}$.

(ix) Finally, obtain the causal response function (or, in the antiperiodic case, the retarded correlation function) through the analytic continuation $i\omega_k \to \omega + i\eta$.

We hasten to say that the last step is highly non-trivial when the dependence of $\chi_{AB}^T(i\omega_k)$ on $i\omega_k$ is not known analytically. The technical difficulty of the analytic continuation is ultimately the main limitation of the finite temperature formalism. Additional information concerning the evaluation of the frequency sums required by rule (vi) via the *spectral representation method* is provided in Appendix 14.

Exercises

6.1 **Connection between causal and time-ordered response functions.** Derive the formulas corresponding to Eq. (6.10) for the real and the imaginary parts of $\chi_{AB}(\omega)$, when $\hat{A} \neq \hat{B}^\dagger$.

6.2 **Derivation of the basic formula of perturbation theory.** Work out the steps leading from Eq. (6.19) to the basic formula of perturbation theory, Eq. (6.24), in the case $t < 0$.

6.3 **Formulas for the number of Feynman diagrams.**

(a) Show that the number of connected diagrams for two-particle response functions χ_{AB} is given by Eqs. (6.47) and (6.48). [Hint: from the total number of contraction schemes subtract the number of schemes in which $n - m$ interaction blocks,

---

[30] The difference of two fermionic Matsubara frequencies is bosonic.

with $0 \leq m < n$, are only contracted among themselves, while the remaining m form a connected diagram.]

(b) Show that the number of connected *and* disjoint diagrams is given by Eqs. (6.49) and (6.50). [Hint: write the disjoint diagram as the product of a connected diagram of order m containing the operator $\hat{A}$, and a connected diagram of order $n - m$ containing the operator $\hat{B}$.]

6.4 **Getting familiar with Feynman diagrams.**

(a) Draw the 10 second-order diagrams for the Green's function.

(b) Draw the 46 second-order diagrams for the density–density response function.

6.5 **Irreducible interaction and self-energy.** Show that all the diagrams obtained by cutting a Green's function line in a skeleton diagram for the self-energy, as mandated by Eq. (6.98), are irreducible in the particle-hole channel with momentum transfer $q = 0$.

6.6 **Analytic properties of the retarded Green's function and the self-energy.**

(a) Making use of the exact eigenstates representation (6.75) show that the spectral function $-\frac{1}{\pi}\Im m G^{ret}(\vec{k}, \omega)$ for real ω is a non-negative quantity.

(b) Show that the retarded Green's function $G^{ret}(\vec{k}, \omega)$ for complex frequency ω in the upper half plane is given by

$$G^{ret}(\vec{k}, \omega) = \int_{-\infty}^{+\infty} \frac{\Im m G^{ret}(\vec{k}, v)}{v - \omega} \frac{dv}{\pi}$$

where the integral over v runs along the real axis. Combine this with the result obtained in (a) to show that $G^{ret}(\vec{k}, \omega)$ has no zeroes in the upper half of the complex ω plane.

(c) Show that $\Sigma^{ret}(\vec{k}, \omega)$ is an analytic function in the upper half of the complex ω plane, and satisfies the dispersion relations (6.80).

6.7 **Two-body scattering amplitude for spin-polarized systems.** Generalize Eq. (6.86) for the two-body scattering amplitude to the case of a system that is spin-polarized along the z-axis.

6.8 **Small-q and ω behavior of the electron–hole propagator in a Fermi liquid.** Derive the long-wavelength and low-frequency behavior of the electron–hole propagator, Eq. (6.97), from the form of the Green's function near the Fermi surface, i.e., Eq. (6.96). [Hint: this problem is discussed in detail in the book by P. Noziéres (1964).]

6.9 **Green's functions in a BCS superconductor.** Starting from the quadratic BCS hamiltonian (2.123) of Section 2.7, and making use of the Bogoliubov–Valatin transformation (2.124), derive the expressions (6.113) and (6.115) for the normal and anomalous Green's function of a BCS superconductor.

6.10 **Diagrams that are reducible in one channel are irreducible in the other two.** Show that if an effective interaction diagram is *reducible* in one of the three channels (direct electron–hole, cross electron–hole, and electron-electron), then it is necessarily irreducible in the other two.

6.11 **Effective interactions in a multicomponent system.** Derive Eqs. (5.155)–(5.158) for the effective interaction between two particles in a multicomponent system. Show that the exchange-like term in A_{ij} arises from the sum of diagrams which are reducible in the cross electron–hole channel and can be approximated in terms of the local field factor $G_{ij,ij}(q)$ defined in Eq. (5.161) of Section 5.5.3.2.

6.12 **Connection between retarded and time-ordered response functions at finite temperature, I.** Show that the Fourier transform of a time-ordered response function (in real time) at finite temperature is related to the Fourier transform of the retarded response function by a simple generalization of Eq. (6.10), in which sign(ω) is replaced by $\coth \frac{\hbar\omega}{2k_B T}$.

6.13 **Connection between retarded and time-ordered response functions at finite temperature, II.** Starting from the exact eigenstates respresentation of the temperature correlation function (Eq. (6.130)) verify that the retarded response function is given by Eq. (6.132).

7

Density functional theory

7.1 Introduction

Density functional theory (DFT) has become in the past few decades one of the most widely used methods for the calculation of the properties of complex electronic systems: molecules, solids, polymers. The basic idea, introduced by Hohenberg, Kohn, and Sham in the 1960s (Hohenberg and Kohn (1964), Kohn and Sham (1965)), is to describe the system in terms of the electronic density (and possibly additional densities such as the spin density, the current density, etc.) without explicit reference to the many-body wave function. At first sight, this seems impossible. How can the subtle correlations encoded in an N-electron wave function be adequately represented by a simple collective variable, such as the density? But Hohenberg and Kohn, in their classic paper, were able to show that the ground-state energy of a quantum system can be determined by minimizing the energy as a *functional* of the density, in much the same way as, in standard quantum mechanics, one can determine the energy by minimizing the expectation value of the hamiltonian with respect to the wave function. Furthermore, the nontrivial part of this functional is *universal*, that is, it has the same form for all physical systems.

The implementation of the Hohenberg–Kohn minimum principle leads to mean-field-like equations, known as the *Kohn–Sham equations*, which are simpler than the Hartree–Fock equations, yet in principle exact as far as the calculation of the ground-state density and energy is concerned. The key advantage of this approach is that the Kohn–Sham equations can be solved on a computer in a time that grows as a power of the number of electrons, whereas an exact solution of the N-electron Schrödinger equation requires a time that grows exponentially with N.

But, such a dramatic simplification comes at a price. First, the effective potential that appears in the Kohn–Sham equations, although local in space (in the sense that it acts on the wave function as a simple multiplication) has a nonlocal dependence on the density, and, for this reason, can only be approximately described. Second, the determinantal wavefunction obtained from the solution of the Kohn–Sham equations, while giving (in principle) the exact ground-state density, is by no means a good approximation to the true ground-state wave function. Finally, the universality of the effective potential, while unifying the treatment of many different systems, tends to obscure the physical peculiarities of each one. In brief, by abandoning the traditional program of approximating the ground-state wave function of a specific system, DFT affords a simpler and broadly applicable formalism for the ground-state energy and density of many systems.

In the following sections we present the main ideas and theorems of DFT, both in its original static version, which addresses only the ground-state properties, and in its more recent dynamical version (Runge and Gross, 1984), which allows the calculation of excitation energies. The density functional theory for thermal ensembles (Mermin, 1965) will be briefly dealt with in Appendix 18. Our presentation freely departs from the historical development of the subject. We make no claim to mathematical rigor, but we try to critically discuss the physical assumptions on which the theory is based. The practically-minded reader may want to skip in a first reading Sections 7.2.6–7.2.9, to get more quickly to the local density approximation and the gradient expansion. Comprehensive treatments of the ground-state DFT can be found in the excellent books by Dreizler and Gross (1990) and Parr and Yang (1989). The time-dependent DFT is reviewed by Gross, Dobson, and Petersilka (1996). Additional material on the recent developments of DFT can be found in the books edited by Gross and Dreizler (1995) and Dobson *et al.* (1998).

7.2 Ground-state formalism

7.2.1 The variational principle for the density

Consider an N-electron system described by the hamiltonian

$$\hat{H} = \hat{T} + \hat{H}_{e-e} + \hat{V} , \tag{7.1}$$

where $\hat{T}$ and $\hat{H}_{e-e}$ are the kinetic energy and the electron-electron interaction, respectively, and $\hat{V}$ is the potential energy associated with an external *local* potential $V(\vec{r})$ (for example the nuclear potential, if the system is an atom). This term can be written as

$$\hat{V} = \int V(\vec{r})\hat{n}(\vec{r})d\vec{r} , \tag{7.2}$$

where

$$\hat{n}(\vec{r}) = \sum_{i=1}^{N} \delta(\vec{r} - \hat{\vec{r}}_i) \tag{7.3}$$

is the density *operator*.

The crucial step in the formulation of DFT is the conversion of the familiar Rayleigh–Ritz variational principle for the quantum mechanical wave function into a variational principle for the density. This conversion is most elegantly accomplished by means of the *constrained search algorithm* introduced by M. Levy (1979), which we now describe.

According to the Rayleigh–Ritz principle the ground-state energy, E, is found by minimizing the expectation value of the hamiltonian with respect to the wave function ψ, i.e.,

$$E = \min_{\psi}\langle\psi|\hat{H}|\psi\rangle , \tag{7.4}$$

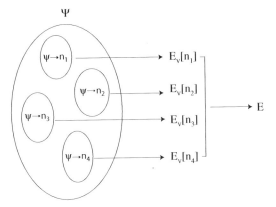

Fig. 7.1. Schematic representation of the constrained search algorithm. The minimization of the energy within the full set of admissible wave functions is executed in two steps: in the first step the energy is minimized within subsets at fixed density, yielding a functional $E_V[n]$; in the second step $E_V[n]$ is minimized with respect to the density.

where ψ must be antisymmetric under interchange of two electron orbital and spin coordinates and satisfy the boundary conditions appropriate to the system under study. The search for the wave function that minimizes (7.4) can be carried out in two steps, as shown in Fig. 7.1.

In the first step we pick a density $n(\vec{r})$ and minimize (7.4) within the subset of wave functions that yield this density.[1] This gives the constrained minimum

$$E_V[n] = \min_{\psi \to n(\vec{r})} \langle \psi | \hat{H} | \psi \rangle$$

$$= F[n] + \int V(\vec{r})n(\vec{r})d\vec{r} , \tag{7.5}$$

where we have defined

$$F[n] = \min_{\psi \to n(\vec{r})} \langle \psi | \hat{T} + \hat{H}_{e-e} | \psi \rangle . \tag{7.6}$$

The notation $\min_{\psi \to n(\vec{r})}$ indicates that the search for the minimum is restricted to antisymmetric wave functions ψ which yield the density $n(\vec{r})$, i.e.,

$$N \sum_{\sigma_1 \dots \sigma_N} \int d\vec{r}_2 \dots d\vec{r}_N |\psi(\vec{r}, \sigma_1; \vec{r}_2, \sigma_2; \dots; \vec{r}_N, \sigma_N)|^2 = n(\vec{r}) . \tag{7.7}$$

It is not difficult to show that one can always find a complete set of antisymmetric wave functions that yield any given reasonable density (see Appendix 15). The minimization problem posed by Eq. (7.6) is therefore guaranteed to have a solution, and this solution defines the functionals $F[n]$ and $E_V[n]$.

---

[1] Any reasonable density must, of course, be positive and continuous, and add up to the total number of electrons in the system, i.e. $\int n(\vec{r})d\vec{r} = N$.

In the second step we minimize $E_V[n]$ with respect to the density. This leads to the desired variational principle for the density:

$$E = \min_{n(\vec{r})} \left\{ F[n] + \int V(\vec{r})n(\vec{r})d\vec{r} \right\} . \qquad (7.8)$$

Looking back at what we have just done we see that the complexity of the original Rayleigh–Ritz variational principle has been absorbed in the definition of the functional $F[n]$, Eq. (7.6). This functional is *universal* in the sense that it is an intrinsic property of the inhomogeneous electron liquid, and thus is the same for all systems, regardless of what the external potential $V(\vec{r})$ is. It is obviously very hard, if not impossible, to calculate $F[n]$ exactly: the study of its rigorous mathematical properties is still an active area of research (see Lieb, 1985).

A particularly important question is that of the existence of the *functional derivative* of $F[n]$ with respect to the density. We say that $D(\vec{r})$ is the functional derivative of $F[n]$ with respect to $n(\vec{r})$ if, for *any* density variation $\delta n(\vec{r})$, and positive number η, we have

$$\lim_{\eta \to 0} \frac{F[n + \eta \delta n] - F[n]}{\eta} = \int D(\vec{r})\delta n(\vec{r})d\vec{r} . \qquad (7.9)$$

The key point is that $D(\vec{r})$ does not depend on the choice of $\delta n(\vec{r})$. If this is the case, we write

$$D(\vec{r}) = \frac{\delta F[n]}{\delta n(\vec{r})} , \qquad (7.10)$$

which is itself a functional of $n(\vec{r})$.

Consider now the variation of the energy functional $E_V[n]$ due to an infinitesimal variation of the density. Assuming that the functional derivative of $F[n]$ exists we write

$$E_V[n + \eta \delta n] - E_V[n] = \eta \int \left[\frac{\delta F[n]}{\delta n(\vec{r})} + V(\vec{r}) \right] \delta n(\vec{r})d\vec{r} + \mathcal{O}(\eta^2) . \qquad (7.11)$$

It follows from the minimum condition (7.8) that the quantity in the square brackets must vanish at the ground-state density, because, if it did not, we could lower the energy by appropriately choosing the sign of η. Therefore, the ground-state density must satisfy the equation

$$\frac{\delta F[n]}{\delta n(\vec{r})} = -V(\vec{r}) , \qquad (7.12)$$

subject to the condition $\int n(\vec{r})d\vec{r} = N$. If Eq. (7.12) has multiple solutions, then the one with the minimum energy must be selected.

A basic difficulty with the above analysis is that a randomly chosen density variation $\delta n(\vec{r})$ will in general change the total particle number. Notice that, up to this point, the functional $F[n]$ has been defined only for densities whose integral over space is an integer number of particles. The extension of $F[n]$ to arbitrary densities will be considered in

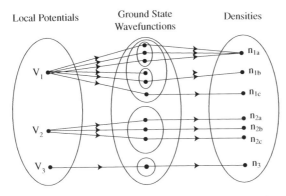

Fig. 7.2. Graphic representation of the Hohenberg–Kohn theorem. Although a local potential $V(\vec{r})$ can have multiple ground-states with different densities, two different potentials can never give the same ground-state density.

Section 7.2.7, and it will then be shown that the extended functional is non-differentiable for variations of the density that change the total particle number. Fortunately, we can still apply the variational principle (7.12), provided we consider only density variations that do not change N, i.e., such that $\int \delta n(\vec{r}) d\vec{r} = 0$. With this restriction, the functional derivative is defined only up to an arbitrary constant, and so is the external potential on the right-hand side of Eq. (7.12).

Are there any other "dangerous" variations of the density? In this chapter we shall informally assume that the answer to this question is "no", i.e. that the functional derivative of $F[n]$ with respect to the density *at constant particle number* can always be safely taken.[2]

7.2.2 The Hohenberg–Kohn theorem

An electronic density $n(\vec{r})$ is said to be *V-representable* if it can occur in the ground-state of $\hat{H}$ with a suitable local potential $V(\vec{r})$.[3] Remarkably, this potential is unique. One cannot find two truly different local potentials (i.e., two potentials that differ by more than a trivial constant) that yield the same ground-state density. A given potential can yield degenerate ground-states and multiple densities associated with them, but two different potentials will never give the same density (see Fig. 7.2). The discovery of this important fact, known as

---

[2] To the best of our knowledge, this has never been rigorously proved. A slight extension of the F functional, in which the constrained minimization is done over a set of *ensembles* (rather than wave functions) yielding the density $n(\vec{r})$, has been shown by Lieb (Lieb, 1985) to be convex, and hence continuous, but not necessarily differentiable.

[3] More accurately one should describe such a density as a *"pure-state V-representable"* density. The nomenclature implies that there are also *non-V-representable densities* or "densities in search of hamiltonians" (Levy, 1982), which cannot arise in any single ground-state. Among these are many of the so-called "ensemble V-representable" densities which arise from an *ensemble* of degenerate ground-states associated with a single local potential. A mathematically rigorous description of the set of V-representable densities is not available to date. On a discrete lattice it can be proved that every physically admissible density is *ensemble V-representable* (Chayes, Chayes, and Ruskai, 1985). In the continuum, it is generally assumed that the set of V-representable densities is sufficiently dense as to be identifiable, for practical purposes, with the set of all admissible densities. We shall adopt this point of view in this chapter. For an example of a non-V-representable density, see Exercise 3.

the *Hohenberg–Kohn* (HK) *theorem*, marked the birth of density functional theory, since it showed that a knowledge of the ground-state density could, at least in principle, determine the hamiltonian, the ground-state wave function, and therefore all the ground-state properties of a many-body system.

The proof of the theorem is by *reductio ad absurdum*. Let us assume, against the thesis of the theorem, that one can find two different potentials $V(\vec{r})$ and $V'(\vec{r})$ differing by more than a constant, such that the two hamiltonians $\hat{H} = \hat{T} + \hat{H}_{e-e} + \hat{V}$ and $\hat{H}' = \hat{T} + \hat{H}_{e-e} + \hat{V}'$ have the same ground-state density $n(\vec{r})$. Let us denote by $|\psi\rangle$ and $|\psi'\rangle$ the (possibly degenerate) ground-states of $\hat{H}$ and $\hat{H}'$, and by E and E' their respective energies. We first observe that $|\psi'\rangle$ cannot be an eigenstate of $\hat{H}$ and, conversely, $|\psi\rangle$ cannot be an eigenstate of $\hat{H}'$. This is because if $|\psi\rangle$ were to be an eigenstate of $\hat{H}$ with eigenvalue E and, at the same time, an eigenstate of $\hat{H}'$ with eigenvalue $\tilde{E}$, then one would have

$$[\hat{V}(\vec{r}) - \hat{V}'(\vec{r})]|\psi\rangle = (E - \tilde{E})|\psi\rangle , \tag{7.13}$$

which would make V and V' differ by a mere constant, against the hypothesis.[4]

It then follows from the Rayleigh–Ritz variational principle that the inequality

$$\begin{aligned} E = \langle\psi|\hat{H}|\psi\rangle &< \langle\psi'|\hat{H}|\psi'\rangle \\ &= \langle\psi'|\hat{H}' + \hat{V} - \hat{V}'|\psi'\rangle \\ &= E' + \int [V(\vec{r}) - V'(\vec{r})]n(\vec{r})d\vec{r} \end{aligned} \tag{7.14}$$

is strictly satisfied. Interchanging the primed and unprimed variables we also obtain

$$E' < E + \int [V'(\vec{r}) - V(\vec{r})]n(\vec{r})d\vec{r} , \tag{7.15}$$

and summing the inequalities (7.14) and (7.15) we arrive at the contradiction

$$E + E' < E' + E , \tag{7.16}$$

which proves the falsehood of the initial assumption.

We conclude that *the ground-state density uniquely determines (up to a constant) the local external potential* $V(\vec{r})$ *that gives rise to it.*[5]

The ground-state wave function, however, is not uniquely determined by the density, because the ground-state of $\hat{H}$ can be degenerate. Fortunately, this ambiguity is of no consequence in the evaluation of $F[n]$. If ψ_1 and ψ_2 are two degenerate ground-states with energy E and density $n(\vec{r})$, then $E = \langle\psi_1|\hat{T} + \hat{H}_{e-e}|\psi_1\rangle + \int V(\vec{r})n(\vec{r})d\vec{r} = \langle\psi_2|\hat{T} + \hat{H}_{e-e}|\psi_2\rangle + \int V(\vec{r})n(\vec{r})d\vec{r}$, showing that $F[n] = \langle\psi|\hat{T} + \hat{H}_{e-e}|\psi\rangle$ has the same value in the two states.

---

[4] The locality of the external potential is essential to the validity of this argument.
[5] An immediate consequence of the HK theorem is that the density–density response function $\chi_{nn}(\vec{r}, \vec{r}')$ of an N-electron system is invertible (up to a constant), i.e., the equation $\int \chi_{nn}(\vec{r}, \vec{r}')V(\vec{r}')d\vec{r}' = 0$ has $V(\vec{r}) = constant$ as the only solution.

7.2.3 The Kohn–Sham equation

The minimum principle (7.8) and its variational version (7.12) offer an elegant approach to the problem of calculating the ground-state density and energy, but we do not know the functional $F[n]$. What are we to do? One possibility is to embark immediately in an effort to find a suitable approximation for $F[n]$. This would not be wise, however, because there is at least one case, that of a non-interacting system, in which the density can be calculated exactly and efficiently from the solution of the Schrödinger equation, without resorting to the density functional formulation. We would like to devise an approach that, regardless of the approximations we will be doing at a later stage, is guaranteed to give the exact result in the case of a non-interacting system.

In 1965 Kohn and Sham (KS) came up with an ingenious strategy to accomplish this. The ground-state density of the interacting system – they argued – can be represented as the ground-state density of a *non-interacting* system in some local external potential $V_{KS}(\vec{r})$.[6] Then, according to the general density-functional formalism, the ground-state density can be obtained by minimizing the non-interacting energy functional

$$E_{V_{KS}}^{(0)}[n] = T_s[n] + \int V_{KS}(\vec{r})n(\vec{r})d\vec{r}\,, \qquad (7.17)$$

where

$$T_s[n] = \min_{\psi \to n(\vec{r})} \langle \psi|\hat{T}|\psi\rangle \qquad (7.18)$$

is the *non-interacting kinetic energy functional*, i.e., the kinetic energy of a non-interacting system whose ground-state density is $n(\vec{r})$.[7] The stationarity condition for the non-interacting energy functional has the form

$$\frac{\delta T_s[n]}{\delta n(\vec{r})} = -V_{KS}(\vec{r})\,, \qquad (7.19)$$

which is completely analogous to Eq. (7.12), and formally defines $V_{KS}(\vec{r})$ as a functional of the density.

Now, following Kohn and Sham, we decompose $F[n]$ as follows:

$$F[n] = T_s[n] + E_H[n] + E_{xc}[n]\,, \qquad (7.20)$$

where $T_s[n]$ is defined by Eq. (7.18), $E_H[n]$ is the familiar Hartree energy functional

$$E_H[n] = \frac{e^2}{2}\int d\vec{r}\int d\vec{r}'\,\frac{n(\vec{r})n(\vec{r}')}{|\vec{r}-\vec{r}'|}\,, \qquad (7.21)$$

---

[6] This is, of course, an unproven assumption. However, the Hohenberg–Kohn theorem guarantees that this potential, if it exists, is unique. Recently, numerical methods have been developed to calculate $V_{KS}(\vec{r})$ from the ground-state density (Zhao, Morrison, and Parr, 1994; Görling, 1992). In the case of one- and two-electron systems the calculation can be done analytically, see Exercises 2 and 5.

[7] It is important to appreciate the difference between the non-interacting kinetic energy $T_s[n]$ and the true kinetic energy functional $T[n]$. The latter is defined as the expectation value of $\hat{T}$ in the state that has density $n(\vec{r})$ and minimizes the expectation value of $\hat{T}+\hat{H}_{e-e}$. It follows from this definition that $T[n] \geq T_s[n]$ for every n. The extra kinetic energy in the interacting system arises from correlations in the ground-state wavefunction.

and $E_{xc}[n]$ is a remainder, known as *exchange-correlation* (xc) energy functional, which is effectively *defined* by Eq. (7.20). Substituting this decomposition into the stationarity condition (7.12) we obtain

$$\frac{\delta T_s[n]}{\delta n(\vec{r})} = -V(\vec{r}) - V_H(\vec{r}) - V_{xc}(\vec{r}), \tag{7.22}$$

where $V_H(\vec{r}) = e^2 \int \frac{n(\vec{r'})}{|\vec{r}-\vec{r'}|} d\vec{r'}$ is the Hartree potential and

$$V_{xc}(\vec{r}) \equiv \frac{\delta E_{xc}[n]}{\delta n(\vec{r})} \tag{7.23}$$

is the *exchange-correlation (xc) potential*. Notice that, by construction, $V_{xc}(\vec{r})$ is a *local* potential, even though its functional dependence on the density is nonlocal. Comparing Eqs. (7.19) and (7.22) we obtain an expression for the Kohn–Sham potential

$$V_{KS}(\vec{r}) = V(\vec{r}) + V_H(\vec{r}) + V_{xc}(\vec{r}). \tag{7.24}$$

We now observe that minimizing $E_{V_{KS}}^{(0)}[n]$ is equivalent to finding the ground-state density of a non-interacting electron system in the presence of the Kohn–Sham potential $V_{KS}(\vec{r})$. All we have to do then, to calculate the ground-state density, is solve the *Kohn–Sham equation*

$$\left[-\frac{\hbar^2}{2m}\nabla^2 + V(\vec{r}) + V_H(\vec{r}) + V_{xc}(\vec{r}) \right] \phi_\alpha(\vec{r}, s) = \varepsilon_\alpha \phi_\alpha(\vec{r}, s), \tag{7.25}$$

for the orbitals $\phi_\alpha(\vec{r})$ with $\alpha = 1, \ldots N$, $\varepsilon_1 < \varepsilon_2 \ldots < \varepsilon_N$. The ground-state density is given by

$$n(\vec{r}) = \sum_{\alpha=1}^{N}\sum_{s} |\phi_\alpha(\vec{r}, s)|^2. \tag{7.26}$$

What has been gained? First of all, the Kohn–Sham equation is guaranteed to yield the right result if the electron-electron interaction is turned off. Second, we no longer need to approximate the whole functional $F[n]$: instead, only its exchange-correlation part $E_{xc}[n]$ and its functional derivative $V_{xc}(\vec{r})$ need to be approximated. Since E_{xc} is usually smaller than E_H, we can hope that errors we make in approximating E_{xc} will have a small impact on the final outcome of a calculation. Last but not least, because the KS potential is local, the Kohn–Sham equation turns out to be simpler than the HF equation, while including correlation effects well beyond the HF approximation.

Given an approximate form for $V_{xc}(\vec{r})$ as a functional of the density, Eq. (7.25) can be solved iteratively. Starting from a reasonable guess for $n(\vec{r})$, which determines $V_{xc}(\vec{r})$, one solves Eq. (7.25) and makes use of Eq. (7.26) to recompute the density. The new density is used to update $V_{xc}(\vec{r})$, and the cycle is repeated until the assumed density and the calculated density coincide to the desired level of accuracy. Once the density is known, the ground-state energy can be computed. Obviously, it is not just the sum of the N lowest-lying eigenvalues of the KS equation. Instead, we notice that, by construction, the non-interacting kinetic energy functional for this density is given by

$$T_s[n] = \sum_{\alpha=1}^{N} \varepsilon_\alpha - \int V_{KS}(\vec{r}) n(\vec{r}) d\vec{r}. \qquad (7.27)$$

Putting this in Eq. (7.20) for $F[n]$ and substituting the resulting expression in Eq. (7.5) we obtain

$$E = \sum_{\alpha=1}^{N} \varepsilon_\alpha - \frac{1}{2} \int d\vec{r} n(\vec{r}) \int d\vec{r}' n(\vec{r}') \frac{e^2}{|\vec{r} - \vec{r}'|} - \int n(\vec{r}) V_{xc}(\vec{r}) d\vec{r} + E_{xc}[n]. \qquad (7.28)$$

7.2.4 Meaning of the Kohn–Sham eigenvalues

It should be clear from the foregoing discussion that neither the KS orbitals ϕ_α, nor the KS eigenvalues ε_α have a compelling physical significance (although they are often used as if they did). In particular, the "Kohn–Sham wave function", that is, the Slater determinant constructed from the N lowest-lying KS orbitals, is by no means a good approximation to the true ground-state wave function: the HF wave function is definitely a better one. Similarly, the KS eigenvalues do not correspond, in general, to physical excitation energies of the system.[8]

There are, however, some significant exceptions. In finite systems, such as atoms and molecules, it can be proved (Levy *et al.*, 1984, Almbladh and von Barth, 1985) that the negative of the highest occupied KS eigenvalue ε_N equals the exact ionization potential when the KS potential vanishes at infinity. The proof makes use of the fact that the KS equation yields the exact ground-state density, whose behavior at large distance from the nucleus is entirely controlled by the eigenvalue of the highest occupied KS orbital: $n(\vec{r}) \propto e^{-\sqrt{2m|\varepsilon_N|}r/\hbar}$. On the other hand, the exact asymptotic behavior of the ground-state density can be independently shown to be $n(\vec{r}) \propto e^{-\sqrt{2mI}r/\hbar}$ where I is the ionization potential. Comparing these two exact results leads to the identification $-\varepsilon_N = I$.

A similar result can be derived for the highest occupied eigenvalue ε_N of an extended system ($N \gg 1$) with no gap at the Fermi level, i.e., a metal. In this case it can be proved rigorously (see Appendix 16) that ε_N gives the true Fermi energy of the metal. Unfortunately, even though the Fermi energy comes out right, the *shape* of the Kohn–Sham Fermi surface in momentum space differs, in general, from the true one (Mearns, 1988).

7.2.5 The exchange-correlation energy functional

Up to this point our definition of the xc energy functional $E_{xc}[n]$ has been purely formal (see Eq. (7.20)). Now we wish to present a more physical definition of this central quantity in terms of the pair correlation function.

---

[8] Differences of KS eigenvalues do provide a kind of zeroth-order approximation to the true excitation energies, when the latter are calculated within the framework of time-dependent DFT (see Section 7.6).

We begin by considering a fictitious system described by the hamiltonian

$$\hat{H}(\lambda) = \hat{T} + \lambda \hat{H}_{e-e} + \int V(\vec{r}, \lambda)\hat{n}(\vec{r})d\vec{r} ,\tag{7.29}$$

where the real parameter λ, which controls the strength of the electron-electron interaction, takes values in the range $0 \leq \lambda \leq 1$. The local potential $V(\vec{r}, \lambda)$ is chosen in such a way that the expectation value of the density in the ground-state of $\hat{H}(\lambda)$ has the prescribed value $n(\vec{r})$ independent of λ. We assume that such a potential can be found for every value of λ in the interval $[0, 1]$: in other words, the density $n(\vec{r})$ is assumed to be V-representable for values of λ in this range. We also assume that the the ground-state $|\psi(\lambda)\rangle$ is a continuous and differentiable function of λ.[9] Although these assumptions are unproven to date, the HK theorem guarantees that the potential $\hat{V}(\vec{r}, \lambda)$, if it exists, is unique.

The physical meaning of $\hat{H}(\lambda)$ is clear. At $\lambda = 1$, $\hat{H}(1)$ is the true hamiltonian of the electronic system with ground-state density $n(\vec{r})$. On the other hand, at $\lambda = 0$, $\hat{H}(0)$ is just the KS hamiltonian, and $V(\vec{r}, 0)$ is the KS potential for the system. For intermediate values of λ, $\hat{H}(\lambda)$ provides a smooth interpolation, known as *adiabatic connection*, between the KS hamiltonian and the true hamiltonian. Now, according to the Hellmann-Feynman theorem, we have

$$\begin{aligned} E(1) - E(0) &= \int_0^1 \langle\psi(\lambda)|\hat{H}_{e-e}|\psi(\lambda)\rangle d\lambda + \int_0^1 d\lambda \int \frac{\partial V(\vec{r}, \lambda)}{\partial \lambda} n(\vec{r})d\vec{r} \\ &= \frac{1}{2}\int_0^1 d\lambda \int d\vec{r}n(\vec{r}) \int d\vec{r}'n(\vec{r}')\frac{e^2}{|\vec{r} - \vec{r}'|}g_\lambda(\vec{r}, \vec{r}') \\ &\quad + \int [V(\vec{r}, 1) - V(\vec{r}, 0)]n(\vec{r})d\vec{r} ,\end{aligned}\tag{7.30}$$

where $g_\lambda(\vec{r}, \vec{r}')$ is the pair correlation function in the ground-state of $\hat{H}(\lambda)$. On the other hand, from the definitions (7.5) and (7.17) we see that

$$E(1) = F[n] + \int V(\vec{r}, 1)n(\vec{r})d\vec{r} ,\tag{7.31}$$

and

$$E(0) = T_s[n] + \int V(\vec{r}, 0)n(\vec{r})d\vec{r} .\tag{7.32}$$

Now, taking the difference of the last two equations and comparing with Eqs. (7.20) and (7.30) we arrive at the following expression for the *xc* energy functional:

$$E_{xc}[n] = \frac{1}{2}\int_0^1 d\lambda \int d\vec{r}n(\vec{r}) \int d\vec{r}'n(\vec{r}')\frac{e^2}{|\vec{r} - \vec{r}'|}[g_\lambda(\vec{r}, \vec{r}') - 1] .\tag{7.33}$$

The physical meaning of this formula is clear: the spatial integral gives the electrostatic energy of interaction between the electron and the "exchange-correlation hole" that surrounds

---

[9] It should be noted that the ground-state wave function of a many-electron system in a fixed external potential is in general not a continuous function of λ. The fact that $V(\vec{r}, \lambda)$ depends on λ is of essence here.

it. The integral over the coupling constant ensures that the correlation-induced changes in kinetic energy are properly taken into account.

Taking the functional derivative of Eq. (7.33) with respect to the density we obtain a formally exact expression for the exchange-correlation potential:

$$
V_{xc}(\vec{r}) = \int_0^1 d\lambda \int d\vec{r}' n(\vec{r}') \frac{e^2}{|\vec{r} - \vec{r}'|} [g_\lambda(\vec{r}, \vec{r}') - 1]
$$
$$
+ \frac{1}{2} \int_0^1 d\lambda \int d\vec{r}_1 n(\vec{r}_1) \int d\vec{r}_2 n(\vec{r}_2) \frac{e^2}{|\vec{r}_1 - \vec{r}_2|} \frac{\delta g_\lambda(\vec{r}_1, \vec{r}_2)}{\delta n(\vec{r})} . \tag{7.34}
$$

The above equations express the xc energy and the xc potential in terms of a third unknown functional – the pair-correlation function of an inhomogeneous electron liquid at density $n(\vec{r})$ and coupling constant λ. For this reason, Eqs. (7.33) and (7.34) are as important theoretically as they are difficult to evaluate in practice.

It is often useful to consider the xc energy functional in the *exchange-only* approximation. This is done by substituting in both Eqs. (7.33) and (7.34) the pair-correlation function $g_0(\vec{r}, \vec{r}')$ of the *non-interacting* system with ground-state density $n(\vec{r})$.[10] Then the coupling constant integral becomes trivial and we get

$$
E_x[n] = \frac{1}{2} \int d\vec{r} \int d\vec{r}' \, n(\vec{r}) n(\vec{r}') \frac{e^2}{|\vec{r} - \vec{r}'|} [g_0(\vec{r}, \vec{r}') - 1] . \tag{7.35}
$$

The constrained-search definition of the exchange-only functional is

$$
E_x[n] = \langle \psi_0[n] | \hat{H}_{e-e} | \psi_0[n] \rangle - E_H[n] , \tag{7.36}
$$

where $\psi_0[n]$ is the wave function that yields $n(\vec{r})$ and minimizes the expectation value of $\hat{T}$. It is not difficult to show that this is just the exchange energy calculated in the Kohn–Sham ground-state:

$$
E_x[n] = -\frac{e^2}{2} \sum_{\alpha=1}^N \sum_{\beta=1}^N \sum_\sigma \int d\vec{r} \int d\vec{r}' \frac{\phi_\alpha^*(\vec{r}, \sigma) \phi_\alpha(\vec{r}', \sigma) \phi_\beta^*(\vec{r}', \sigma) \phi_\beta(\vec{r}, \sigma)}{|\vec{r} - \vec{r}'|} . \tag{7.37}
$$

Thus, $E_x[n]$ is easily written as a functional of the KS orbitals, but it is, at the same time, a very complex implicit functional of the density.

It should also be noted that $E_x[n]$ is quite different from the Hartree–Fock exchange energy (see Chapter 2), which is defined by

$$
E_{x,HF}[n] = \langle \psi_{HF}[n] | \hat{H}_{e-e} | \psi_{HF}[n] \rangle - E_H[n] . \tag{7.38}
$$

Here $\psi_{HF}[n]$ is the single Slater determinant that yields the density $n(\vec{r})$ and minimizes the expectation value of $\hat{T} + \hat{H}_{ee}$, not $\hat{T}$. In practice, the difference is that the Hartree–Fock orbitals satisfy a Schrödinger-like equation which includes a nonlocal potential term (the Fock term). By contrast, the exchange-only potential

$$
V_x(\vec{r}) = \frac{\delta E_x[n]}{\delta n(\vec{r})} \tag{7.39}
$$

---

[10] Because in a non-interacting fermion system the antisymmetry of the wave function is the only source of correlation between particles, this is properly termed the "exchange-only" approximation.

is local and can be shown to coincide with the optimized effective potential introduced in Section 2.8.2 for the physical system at density $n(\vec{r})$ (see Exercise 6).

It is customary to define the *correlation energy functional* as the difference between the exchange-correlation energy functional and the exchange-only functional:

$$E_c[n] = E_{xc}[n] - E_x[n] . \tag{7.40}$$

The constrained search definition of the correlation functional is therefore

$$E_c[n] = \langle \psi[n]|\hat{T} + \hat{H}_{e-e}|\psi[n]\rangle - \langle \psi_0[n]|\hat{T} + \hat{H}_{e-e}|\psi_0[n]\rangle , \tag{7.41}$$

where $\psi[n]$ yields $n(\vec{r})$ and minimizes $\hat{T} + \hat{H}_{e-e}$, while $\psi_0[n]$ yields $n(\vec{r})$ and minimizes $\hat{T}$. It follows immediately that *the correlation energy is always negative.*

All the above definitions can be easily generalized to intermediate values of the coupling constant. Thus, for example, the xc energy functional at a coupling constant γ, $E_{xc,\gamma}[n]$ with $0 \leq \gamma \leq 1$, will be given by Eq. (7.33) with the integral over λ running from 0 to γ, while the exchange-only functional at the same coupling constant will be given by $E_{x,\gamma}[n] = \gamma E_x[n]$.

7.2.6 Exact properties of energy functionals

Some exact properties of the functionals $T_s[n]$, $E_x[n]$, and $E_c[n]$ can be derived from their behavior under the scale transformation

$$n(\vec{r}) \rightarrow n_\gamma(\vec{r}) \equiv \gamma^3 n(\gamma \vec{r}). \tag{7.42}$$

(Levy, 1993). This transformation rescales the coordinate axes by a factor $\gamma > 0$, while keeping the total number of electrons constant:

$$\int n_\gamma(\vec{r})d\vec{r} = \int n(\vec{r})d\vec{r} = N. \tag{7.43}$$

Let us begin by computing $T_s[n_\gamma(\vec{r})]$. What we need is the wave function that yields $n_\gamma(\vec{r})$ and minimizes $\hat{T}$. To determine this, we notice that for any wave function ψ_0 that yields $n(\vec{r})$ there exists a wave function $\psi_{0,\gamma}$ that yields $n_\gamma(\vec{r})$ according to the formula

$$\psi_{0,\gamma}(\vec{r}_1, \ldots, \vec{r}_N) \equiv \gamma^{3N/2}\psi_0(\gamma \vec{r}_1, \ldots, \gamma \vec{r}_N) . \tag{7.44}$$

It is easy to verify that

$$\langle \psi_{0,\gamma}|\hat{T}|\psi_{0,\gamma}\rangle = \gamma^2 \langle \psi_0|\hat{T}|\psi_0\rangle . \tag{7.45}$$

This relation shows that, if ψ_0 is chosen to be the minimizer of $\hat{T}$ at density $n(\vec{r})$, then automatically $\psi_{0,\gamma}$, defined by Eq. (7.44), is the minimizer of $\hat{T}$ at density $n_\gamma(\vec{r})$. It follows that

$$T_s[n_\gamma] = \gamma^2 T_s[n] . \tag{7.46}$$

Let us now consider the behavior of the exchange functional $E_x[n]$ under a scaling transformation. By direct substitution of $n_\gamma(\vec{r})$ in the Hartree energy functional we obtain

$$E_H[n_\gamma] = \gamma E_H[n]. \tag{7.47}$$

Then, from the definition (7.36), together with Eq. (7.44), we see that

$$E_x[n_\gamma] = \gamma E_x[n]. \tag{7.48}$$

The behavior of the correlation energy functional under scaling is more complex. According to Eq. (7.41) what we need to compute is the wave function ψ_γ which yields $n_\gamma(\vec{r})$ and minimizes $\hat{T} + \hat{H}_{e-e}$. To solve this problem let us write $\psi_\gamma(\vec{r}_1, \ldots, \vec{r}_N) = \gamma^{3N/2}\Phi(\gamma\vec{r}_1, \ldots, \gamma\vec{r}_N)$, where $\Phi(\vec{r}_1, \ldots, \vec{r}_N)$ is an undetermined wave function that yields $n(\vec{r})$. Then, after simple transformations, we find

$$\langle \psi_\gamma | \hat{T} + \hat{H}_{e-e} | \psi_\gamma \rangle = \gamma^2 \langle \Phi | \hat{T} + \frac{1}{\gamma}\hat{H}_{e-e} | \Phi \rangle. \tag{7.49}$$

We choose Φ so as to minimize $\hat{T} + \gamma^{-1}\hat{H}_{e-e}$. With this choice we see that

$$
\begin{aligned}
E_c[n_\gamma] &= \langle \psi_\gamma | \hat{T} + \hat{H}_{e-e} | \psi_\gamma \rangle - \langle \psi_{0,\gamma} | \hat{T} + \hat{H}_{e-e} | \psi_{0,\gamma} \rangle \\
&= \gamma^2 \left[\langle \Phi | \hat{T} + \frac{1}{\gamma}\hat{H}_{e-e} | \Phi \rangle - \langle \psi_0 | \hat{T} + \frac{1}{\gamma}\hat{H}_{e-e} | \psi_0 \rangle \right] \\
&= \gamma^2 E_{c,1/\gamma}[n] ,
\end{aligned}
\tag{7.50}
$$

where $E_{c,1/\gamma}[n]$ is the correlation energy functional for the hamiltonian (7.29) with coupling constant $\lambda = 1/\gamma$.

Eq. (7.50) shows that a rescaling of the inhomogeneous density is equivalent (up to a factor) to a rescaling of the coupling constant. In fact, Eqs. (7.46), (7.48), and (7.50), are the generalizations to inhomogeneous systems of the r_s-scaling properties of the different components of the ground-state energy of the uniform jellium model.[11]

The analogy can be pushed further. In the homogeneous electron liquid, we could calculate the ground-state energy by a perturbation expansion in powers of the coupling constant. Similarly, in the inhomogeneous electron liquid, we can expand the ground-state energy of the hamiltonian $\hat{H}(\lambda)$ (Eq. (7.29)) in a power series of λ *at constant density*. This type of perturbation expansion is known as *Görling–Levy (GL) perturbation theory* (Görling, 1994). Its peculiarity lies in the fact that the coupling constant appears both in the interaction term *and* in the external potential $\hat{V}(\lambda)$, which keeps the density independent of λ. The zero-order hamiltonian in the GL perturbation theory is just the ground-state Kohn–Sham hamiltonian, which already includes interaction effects through the Hartree and the KS exchange-correlation potential. It is plausible that, with such an "advanced" zero-order

---

[11] Thus, for example, Eq. (7.50) says that the correlation energy of a homogeneous three-dimensional electron liquid at density $\gamma^3 n$ is equal to γ^2 times the correlation energy of a liquid of fictitious electrons of charge $e\gamma^{-1/2}$ at density n. This is true because the two systems have the same value of $r_s \sim 1/na_B^3 \sim e^6/n$. The factor γ^2 accounts for the different value of the effective Rydberg ($\sim e^4$) in the two systems.

approximation, the expansion will converge more rapidly than the conventional perturbation expansion about the non-interacting ground-state.

The GL perturbation theory allows us to see an interesting property of the correlation energy $E_{c,\lambda}[n]$ of an inhomogeneous system. Recall that, according to Eq. (7.41), the correlation energy functional at coupling constant λ is given by

$$E_{c,\lambda}[n] = \langle \psi_\lambda[n] | \hat{H}(\lambda) | \psi_\lambda[n] \rangle - \langle \psi_0[n] | \hat{H}(\lambda) | \psi_0[n] \rangle , \qquad (7.51)$$

where $\psi_\lambda[n]$ yields $n(\vec{r})$ and minimizes $\hat{T} + \lambda \hat{H}_{e-e}$. Let us write the GL expansion for this functional as

$$E_{c,\lambda}[n] = E_c^{(0)}[n] + \lambda E_c^{(1)}[n] + \lambda^2 E_c^{(2)}[n] + \cdots . \qquad (7.52)$$

The first term on the right side is zero. This follows directly from the definition (7.51) of $E_{c,\lambda}[n]$. Next, by applying the Hellmann–Feynman theorem to Eq. (7.51), we find that

$$\left. \frac{\partial E_{c,\lambda}[n]}{\partial \lambda} \right|_{\lambda=0} = 0 \qquad (7.53)$$

too. Thus, we see that the perturbation expansion for $E_{c,\lambda}[n]$ about $\lambda = 0$ begins with the quadratic term in λ:

$$E_{c,\lambda}[n] = \lambda^2 E_c^{(2)}[n] + \lambda^3 E_c^{(3)}[n] + \cdots . \qquad (7.54)$$

Since, by virtue of Eq. (7.50), $E_{c,\lambda}[n] = \lambda^2 E_c[n_{1/\lambda}]$, we arrive at the interesting result

$$\lim_{\lambda \to 0} E_c[n_{1/\lambda}] = E_c^{(2)}[n]. \qquad (7.55)$$

The interesting point is that while the density $n_{1/\lambda}(\vec{r})$ at the origin of the coordinates tends to infinity for $\lambda \to 0$ (see Eq. (7.42)), the right-hand side of this equation is a finite negative number (an explicit expression for $E_c^{(2)}[n]$ is worked out in Appendix 17). By contrast, the correlation energy of the infinite homogeneous electron gas tends logarithmically to minus infinity as the density tends to infinity (see Section 1.5.6). We shall see in Section 7.3.2 that the popular local density approximation (LDA) for the xc energy functional violates the exact constraint (7.55) and for this reason usually leads to overbinding in small molecular systems.

Finally, we mention that in three dimensions the exchange-correlation energy functional satisfies the *Lieb–Oxford inequality* (Lieb and Oxford, 1981)

$$E_{xc}[n] \geq C \int d\vec{r}\, n^{4/3}(\vec{r}), \qquad (7.56)$$

where C is a negative constant lying in the range $-1.43 \geq C \geq -1.68$.

7.2.7 Systems with variable particle number

Up to this point, we have been considering the energy as a functional of the density at constant particle number. In condensed matter physics, one often needs to consider very large systems ($N \gg 1$) or even, for the sake of mathematical convenience, infinite systems, such as periodic crystals. In such systems it is often convenient to treat N as a continuous variable, which can be tuned through a chemical potential μ. The dependence of the energy functional on the total particle number becomes then a serious issue, with important physical implications.

In order to implement the "variable-number" formalism, one starts from the grand-canonical hamiltonian

$$\hat{K} = \hat{H} - \mu \hat{N}, \qquad (7.57)$$

were $\hat{N}$ is the total number operator and μ is the chemical potential. Because $\hat{N}$ commutes with $\hat{K}$, eigenstates of $\hat{K}$ can be classified according to the number of particles they contain. By appropriately choosing the value of μ we can force the minimum energy state – the ground-state of $\hat{K}$ – to fall into the M-electron sector, where M is any desired positive integer. In this sense μ is the variable conjugated to N in the same way that $v(\vec{r})$ is the variable conjugated to $n(\vec{r})$.

There is, however, an important difference, namely, the number of particles in the ground-state does not uniquely determine the chemical potential. At a mathematical level, this is an obvious consequence of the different nature of the variables μ and N: μ is continuous, while N is quantized, hence it is impossible to have a one-to-one correspondence between the two of them. The non-invertible mapping between μ and N is shown in Fig. 7.4, and its physical origin is explained in Fig. 7.3. This figure shows schematically the spectra of $\hat{K}$ in the sectors $N = M, M + 1$, and $M - 1$ for three different values of μ. In panel (a) the value of the chemical potential is such that the ground-state of $\hat{K}$ has $N = M$ electrons. Increasing the chemical potential lowers the energy levels of the $N = M + 1$ sector relative to those of the $N = M$ sector, but does not change the ground-state until a critical value $\mu_+(M)$ is reached, for which the ground-states of the M- and $M + 1$-electron systems are

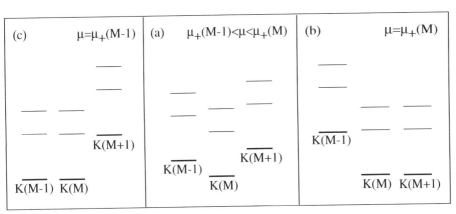

Fig. 7.3. Spectrum of $\hat{K}$ for three different values of the chemical potential μ. (a) In the range $\mu_+(M - 1) < \mu < \mu_+(M)$ the ground-state has $N = M$ and does not depend on μ. (b) At $\mu = \mu_+(M)$ the ground-state jumps into the $M + 1$-particle sector. (c) At $\mu = \mu_+(M - 1)$ the ground-state jumps into the $M - 1$-particle sector. Here $K(N) \equiv E(N) - \mu M$ where $E(M)$ is the ground-state energy of $\hat{H}$ in the M-particle sector. The zero of the energy in each panel is adjusted to keep the lowest eigenvalue constant.

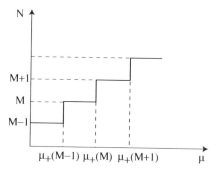

Fig. 7.4. Schematic behavior of the ground-state particle number as a function of the chemical potential. Notice that μ is undefined at integer values of N, but has a well defined value for fractional values of N.

degenerate (Fig. 7.3(b)). At this point the number of electrons in the ground-state jumps from M to $M + 1$. The critical value of μ is given by

$$\mu_+(M) = E(M + 1) - E(M) , \tag{7.58}$$

where $E(M)$ is the energy of the ground-state of $\hat{H}$ (not $\hat{K}$!) in the M-electron sector. The quantity $\mu_+(M)$ is called *addition energy* by condensed matter physicists. Chemists prefer to talk of "electron affinity energy" $A(M) = -\mu_+(M)$ because, if A is positive, then the system lowers its energy by capturing an extra electron.

Similarly, decreasing the value of μ lowers the levels of the $N = M - 1$ sector relative to those of the $N = M$ sector, but does not change the ground-state until the critical value $\mu_+(M - 1) = E(M) - E(M - 1)$ is reached (see Fig. 7.3(c)): at this point the number of electrons jumps from M to $M - 1$. The quantity $I(M) = \mu_+(M - 1)$ is known to chemists as the "first ionization potential" because this is the minimum energy needed to remove an electron from an M-electron ground-state.

An important feature of stable systems is that $\mu_+(M - 1) < \mu_+(M)$, that is, $\mu_+(M)$ increases monotonically with M. This can be deduced from an inspection of Fig. 7.3(a): if $\hat{K}$ has its minimum at $N = M$, then one must have $E(M + 1) - E(M) - \mu > 0$ and $E(M) - E(M - 1) - \mu < 0$: the two conditions are compatible only if the condition $\mu_+(M - 1) < \mu < \mu_+(M)$ is satisfied. Throughout this range of values of μ the ground-state (and hence N) remains unchanged. These "plateaus" are separated by sharp steps where the number of electrons (and hence the ground-state wave function) changes discontinuously. The behaviour of N vs μ is illustrated in Fig. 7.4.

7.2.8 Derivative discontinuities and the band gap problem

The discontinuous behavior of N as a function of μ has some important implications for density functional theory. First of all, to better handle variable particle numbers, we introduce the generalized energy functional

$$E_{V-\mu}[n] = \tilde{F}[n] + \int [V(\vec{r}) - \mu]n(\vec{r})d\vec{r} , \tag{7.59}$$

where $\tilde{F}[n]$ is an extension of $F[n]$ to densities that do not necessarily yield an integer number of particles.[12]

Minimizing $E_{V-\mu}[n]$ with respect to $n(\vec{r})$ should yield both the density and the energy of the ground-state of $\hat{K}$. The problem is that the extended functional $\tilde{F}[n]$ is not differentiable with respect to density variations that change the total number of particles. To see this, suppose that $\tilde{F}[n]$ were differentiable at all densities and in all "directions" in the space of densities. Then, the condition of stationarity for the energy

$$\frac{\delta \tilde{F}[n]}{\delta n(\vec{r})} = \mu - V(\vec{r}), \qquad (7.60)$$

would unambiguously determine $\mu - V(\vec{r})$ as a functional of the ground-state density. But this is impossible, because, as we have seen in the previous section, a sufficiently small constant added to the chemical potential does not change the ground-state of $\hat{K}$: such a constant, therefore, cannot be determined by the ground-state density. We conclude that the derivative of $\tilde{F}[n]$ is *undefined* whenever $\int n(\vec{r}) d\vec{r}$ is an integer.

By the same argument, $\tilde{F}[n]$ can be expected to be differentiable at densities that do *not* yield integer particle numbers, since, in that case, the chemical potential is well defined (see Fig. 7.4). Let us then consider two such densities $n_M^+(\vec{r}) = n_M(\vec{r}) + \eta(\vec{r})$ and $n_M^-(\vec{r}) = n_M(\vec{r}) - \eta(\vec{r})$, where $n_M(\vec{r})$ is the density of the M-particle ground-state, and $\eta(\vec{r})$ is an infinitesimal positive distribution which changes the particle number by the infinitesimal amount $\int \eta(\vec{r}) d\vec{r} = \eta \ll 1$. The chemical potentials corresponding to these two densities are $\mu_+(M)$ and $\mu_+(M-1)$, respectively. Then, according to Eq. (7.60) we have

$$\left. \frac{\delta \tilde{F}[n]}{\delta n(\vec{r})} \right|_{n_M^+(\vec{r})} = \mu_+(M) - V(\vec{r}) \qquad (7.61)$$

and

$$\left. \frac{\delta \tilde{F}[n]}{\delta n(\vec{r})} \right|_{n_M^-(\vec{r})} = \mu_+(M-1) - V(\vec{r}). \qquad (7.62)$$

Taking the difference of these two equations we obtain an exact expression for the derivative discontinuity of the functional $\tilde{F}[n]$ at any density $n_M(\vec{r})$ that yields an integer particle number M:

$$\left. \frac{\delta \tilde{F}[n]}{\delta n(\vec{r})} \right|_{n_M^+(\vec{r})} - \left. \frac{\delta \tilde{F}[n]}{\delta n(\vec{r})} \right|_{n_M^-(\vec{r})} = \mu_+(M) - \mu_+(M-1) \equiv E_g. \qquad (7.63)$$

---

[12] Following Perdew *et al.* (1982), this extended functional can be constructed as follows. Let D be the set of all the density matrices of the form $\hat{\rho} = (1-x)|\psi_N\rangle\langle\psi_N| + x|\psi_{N+1}\rangle\langle\psi_{N+1}|$ obtained by mixing an N-particle state and an $N+1$-particle state with weights $1-x$ and x repectively. The expectation value of $\hat{N}$ for any density matrix in this set is $Tr(\hat{\rho}\hat{N}) = (1-x)N + x(N+1) = N + x$. Define

$$\tilde{F}[n] \equiv \min_{\hat{\rho} \to n(\vec{r})} Tr[\hat{\rho}(\hat{T} + \hat{H}_{e-e})],$$

where the minimum is sought among the density matrices of the set D that yield the density $n(\vec{r})$. It is evident that $\tilde{F}[n]$ reduces to $F[n]$ [Eq. (7.6)] when the particle number is integer. A critical reader might ask whether this definition is unique: for example, why not consider also mixtures of states with particle numbers $N-1$, $N+2$, etc.? The answer is that, due to the convexity of the energy as a function of particle number, the inclusion of such states would not change the outcome of the constrained search for the optimal ensemble, and hence the value of $\tilde{F}[n]$.

The quantity on the far right-hand side of this equation is, by definition, the *gap* in the one-particle excitation spectrum. In chemical language, it is the difference between the ionization energy and the electron affinity energy.[13]

The existence of the derivative discontinuity (7.63) is a potential source of trouble for approximate implementations of DFT. To see why, observe that a derivative discontinuity occurs also in the non-interacting kinetic energy functional $T_s[n]$, once it is suitably extended to accommodate noninteger particle numbers (we call this extended functional $\tilde{T}_s[n]$). This discontinuity is

$$\left.\frac{\delta \tilde{T}_s[n]}{\delta n(\vec{r})}\right|_{n_M^+(\vec{r})} - \left.\frac{\delta \tilde{T}_s[n]}{\delta n(\vec{r})}\right|_{n_M^-(\vec{r})} = E_{g,KS}, \qquad (7.64)$$

where $E_{g,KS}$ is the *Kohn–Sham gap*, defined as the difference of the addition energies $\mu_+(M) - \mu_+(M-1)$ in a non-interacting system with ground-state density $n_M(\vec{r})$. Since $\tilde{F}$ and $\tilde{T}_s$ are two independent functionals of the density, there is no reason to expect their derivative discontinuities E_g and $E_{g,KS}$ to have the same value. Thus, the exact gap is in general different from the KS gap. Since the derivative of the Hartree energy functional is obviously continuous, and $\tilde{E}_{xc}[n] = \tilde{F}[n] - \tilde{T}_s[n] - E_H[n]$, we conclude that the exchange-correlation energy functional must have a derivative discontinuity, whose magnitude is the difference between the true gap and the KS gap:

$$\Delta_{xc} \equiv \left.\frac{\delta \tilde{E}_{xc}[n]}{\delta n(\vec{r})}\right|_{n_M^+(\vec{r})} - \left.\frac{\delta \tilde{E}_{xc}[n]}{\delta n(\vec{r})}\right|_{n_M^-(\vec{r})} = E_g - E_{g,KS}.^{14} \qquad (7.65)$$

The bad news is that some of the most popular approximations for $\tilde{E}_{xc}[n]$ – for instance, the local density approximation (LDA), which will be discussed in detail in Section 7.3.2 – do not present any derivative discontinuity. Functionals of the Kohn–Sham orbitals, such as the exchange-only functional of Eq. (7.37), do naturally lead to derivative discontinuities (for example, in a band insulator the Kohn–Sham orbitals usually change symmetry in going from the valence band to the conduction band), but the inclusion of correlation in this type of functional is more problematic.

Whether the non differentiability of the $\tilde{E}_{xc}$ functional should be a cause of concern or not depends on the system under study. In few-electron systems the difference in density between the ground-states with M and $M+1$ electrons is large, so the functional derivative of E_{xc} is hardly relevant. The situation is completely different in extended systems, because the difference in density between the M and $M+1$-electron ground-states is infinitesimal and it becomes essential to have the correct "slope" of E_{xc} vs density. A key issue, in this case, is whether the differences $E(M \pm 1) - E(M)$ tend to zero in the limit $M \to \infty$, in

---

[13] The one-particle gap differs, in general, from the optical gap, which is determined by the least electron–hole excitation energy.
[14] Walter Kohn (1986) has shown that the derivative discontinuity Δ_{xc} of Eq. (7.65) can be written explicitly in terms of the E_{xc} functional as follows:

$$\Delta_{xc} = \{E_{xc}[n_M + n_c] - E_{xc}[n_M]\} + \{E_{xc}[n_M - n_v] - E_{xc}[n_M]\} - \{E_{xc}[n_M + n_c - n_v] - E_{xc}[n_M]\},$$

where $n_c(\vec{r})$ is the density of a single electron at the bottom of the conduction band and $n_v(\vec{r})$ is the density of a single electron at the top of the valence band (Exercise 12).

Table 7.1. *Band gaps (eV) of selected semiconductors and insulators: LDA vs experiment. (Adapted from Aulbur, Jönsson, and Wilkins, 2000.)*

	Diamond	Si	Ge	LiCl	GaAs
LDA	3.9	0.52	0.07	6.0	0.12
Expt.	5.48	1.17	0.744	9.4	1.52

which case the system is said to be *gapless*, or to a finite constant, independent of M, in which case the system has an energy gap given by

$$E_g = \lim_{M \to \infty} [E(M+1) + E(M-1) - 2E(M)] \ . \qquad (7.66)$$

If one were to do three separate ground-state calculations for $E(M+1)$, $E(M)$, and $E(M-1)$ with an approximate "cusp-free" xc energy functional, one would find $E_g = E_{g,KS}$, where

$$E_{g,KS} = \varepsilon_{M+1} + \varepsilon_{M-1} - 2\varepsilon_M \ , \qquad (7.67)$$

and ε_M, $\varepsilon_{M\pm1}$ are the M-th and $M \pm 1$-th KS eigenvalues calculated in essentially the same xc potential, up to corrections of order $1/M$ (see Appendix 16). Unfortunately, the derivative discontinuity of $\tilde{E}_{xc}[n]$ is, in general, of the same order of magnitude as $E_{g,KS}$, and cannot be neglected.

The difficulty shows up in all its severity when one attempts to calculate the band gap of semiconductors and insulators. As shown in Table 7.1 the standard calculation, based on the Kohn–Sham equation and the LDA for $\tilde{E}_{xc}$, yields a band gap that is typically 40–50% smaller than the observed one. At first, one is tempted the blame this poor performance on the local density approximation. This is, of course, true, but not for the reason most people would have guessed. In fact, Godby, Schlüter, and Sham (1986) showed that even the use of a very accurate Kohn–Sham potential (calculated from the best available ground-state density) did not significantly improve the agreement between theory and experiment in Si. Since the spatial dependence of V_{xc} is virtually exact in their calculation, the remaining 50% error in the band gap must be attributed to the lack of a discontinuous dependence on particle number. If one knew how to add to the LDA (or to any other approximation) the appropriate N-dependent constant as a functional of the ground-state density (or, of the KS orbitals) then one could greatly improve the calculation of the energy gaps. However, no one has been able to accomplish this in a systematic manner so far.

One might hope to circumvent the difficulty with the help of Mermin's finite temperature formalism described in Appendix 18. At finite temperature, the relationship between N and μ is invertible, because the expectation value of $\hat{N}$ in a thermal ensemble varies continuously as a function of μ. This implies that the grand thermodynamic potential $\Omega[n]$ has no derivative discontinuity as a function of N. Unfortunately, this solution is more formal than real, since the zero temperature limit is approached exponentially fast as the temperature drops below the excitation gap. When the temperature is much smaller than the gap, then $\Omega[n]$ is essentially indistinguishable from the ground-state energy functional. Therefore, the thermal smoothening of the cusp does not help to improve the agreement between the Kohn–Sham gap and the true band gap.

Another possible line of attack relies on the time-dependent density functional theory, which is described later in this chapter. It would seem that a theory that is capable of predicting excitation energies should also predict the correct band gap. Even this, however, is not as straightforward as it sounds. We will see that the derivative discontinuity problem resurfaces in time-dependent DFT through the infinite range of the functional derivative $\frac{\delta V_{xc}(\vec{r})}{\delta n(\vec{r}')}$ as a function of $\vec{r} - \vec{r}'$.

In summary, the problem is still unsolved. It is certainly one of the major outstanding problems in contemporary DFT.

7.2.9 Generalized density functional theories

In the forty years following the first formulation of DFT the range of applications of this theory has dramatically broadened. While the original papers merely dealt with the ground-state of electronic systems in an external potential, the work of the past few decades has created a more powerful formalism, which can be applied to systems in magnetic fields, spin-polarized systems, and superconducting systems.

The generalized theories are based on larger sets of density-like variables. For example, in order to treat electronic systems in the presence of a magnetic field, it was natural to enlarge the set of basic variables by including first the spin density and then the orbital current density. This led to the *spin density functional theory* (SDFT), and, later, to the *current density functional theory* (CDFT). Another remarkable extension of DFT (Oliveira, Gross, and Kohn, 1988) allowed the treatment of superconducting systems in terms of the Cooper pair condensate density.

A typical generalized DFT starts with a hamiltonian of the form

$$\hat{H} = \hat{T} + \hat{H}_{e-e} + \sum_i \int \hat{n}_i(\vec{r}) V_i(\vec{r}) d\vec{r} \,, \tag{7.68}$$

where $\hat{n}_i(\vec{r})$ are generalized density operators and $V_i(\vec{r})$ are conjugate fields that couple linearly to them.[15] The prototypical case is that of collinear SDFT in which the two generalized densities are the up- and down-spin densities or, equivalently, the ordinary density and the z-component of the spin density. The corresponding conjugate fields are the ordinary scalar potential, which couples to the density, and the z-component of the magnetic field, which couples to the spin–density.

The formal aspects of the generalization are quite straightforward, *provided one is willing to accept the basic assumption that the generalized densities $n_i(\vec{r})$ can be varied*

---

[15] The current–density functional theory does not quite fit in this scheme, due to the presence of a coupling between the density and the *square* of the vector potential. This theory will be examined separately in Section 7.4.

independently.[16] The universal F functional is defined as

$$F[n_1, n_2, \ldots] = \min_{\psi \to \{n_1(\vec{r}), n_2(\vec{r}), \ldots\}} \langle \psi | \hat{T} + \hat{H}_{e-e} | \psi \rangle , \qquad (7.69)$$

and the Rayleigh–Ritz variational principle leads to the stationarity conditions

$$\frac{\delta F[n_1, n_2, \ldots]}{\delta n_i(\vec{r})} = -V_i(\vec{r}) . \qquad (7.70)$$

Similarly, one defines T_s and E_{xc} to be functionals of several variables. The Kohn–Sham equation follows, and the KS potentials – one for each basic variable – are given by $V_{KS,i}(\vec{r}) = V_i(\vec{r}) + V_{H,i}(\vec{r}) + V_{xc,i}(\vec{r})$, where

$$V_{xc,i}(\vec{r}) = \frac{\delta E_{xc}[n_1, n_2, \ldots]}{\delta n_i(\vec{r})} , \qquad (7.71)$$

and, similarly, $V_{H,i}(\vec{r})$ is the functional derivative of the Hartree energy functional with respect to $n_i(\vec{r})$.

Although the formulation runs on the whole parallel to the single-variable one, a few subtle differences are worth noting. First of all, while it is still true that the generalized densities uniquely determine the ground-state wave function, it is no longer true that they determine the external potentials up to a physically irrelevant constant. For example, the application of a small uniform magnetic field to an atom in a ground-state with total spin $S = 0$ does not change the state of the atom. This implies that, in this particular system, the ground-state spin densities are not sufficient to determine the value of the external magnetic field. A little thought shows that a similar non-uniqueness problem will appear whenever the ground-state of the system is "rigid" with respect to a suitable combination of the external fields (i.e., when the corresponding linear susceptibility vanishes).[17]

The non-uniqueness of the external potentials at certain densities implies that the universal energy functional $F[n_1, n_2 \ldots]$, and hence the E_{xc} functional, has derivative discontinuities at those densities (Capelle and Vignale, 2001). For, if F were differentiable, then the stationarity condition (7.70) would uniquely determine the external potentials $V_i(\vec{r})$. This argument shows that derivative discontinuities can occur, in generalized DFT, even when the particle number is kept constant.

Perhaps the most important thing to be understood in relation to generalized DFTs is the answer to the question: why do we need such theories at all? At first sight, nothing would prevent us from applying the conventional DFT to hamiltonians that include additional external fields. In fact, the ground-state wave function of the general hamiltonian (7.68) can still be proven to be a functional of the ordinary density. We could introduce a *nonuniversal* functional of the density $F[n_1, V_2, V_3 \ldots]$ that depends not only on the density (we assume, for definiteness, $n_1(\vec{r}) \equiv n(\vec{r})$) but also on the external fields $V_2(\vec{r}), V_3(\vec{r}), \ldots$. By contrast, the essence of generalized DFT lies in making use of the *universal* functional $F[n_1, n_2, \ldots]$ defined in Eq. (7.69).

---

[16] In reality, this assumption should be critically examined in each case, since the existence of constraints between the variables is not compatible with the existence of an invertible and differentiable mapping between densities and potentials.

[17] In a typical case, the rigidity of the ground-state follows from the existence of a discrete quantum number that cannot be changed continuously by the application of an external field: for example, the quantization of particle number led in the previous section to a non invertible relation between μ and N. The question of the non-uniqueness of the external potential is further explored in Exercises 9–11.

The fundamental advantage of the generalized DFT is that it allows us to directly compute the densities $n_i(\vec{r})$ even if some of the conjugate fields $V_i(\vec{r})$ happen to be zero. This is absolutely vital when one of the additional densities is the order parameter of a spontaneously broken symmetry. In this case, the presence of the self-consistent exchange-correlation potential $\frac{\delta E_{xc}[n_1, n_2, \ldots]}{\delta n_i(\vec{r})}$ is what allows the existence of a nontrivial solution $n_i(\vec{r}) \neq 0$ even in the absence of the corresponding "driving field".

Even if no symmetry is broken, the inclusion of additional density variables gives us more flexibility in the search for the minimum energy, since the minimization is carried out in a larger functional space. Thus, it is usually better to approximate $F[n_1, n_2, \ldots]$ rather than $F[n_1, V_2, \ldots]$. The clearest example is provided by the SDFT, which has, for all practical purposes, supplanted the original DFT by allowing a more accurate treatment of the local spin density in *non magnetic* systems.

7.3 Approximate functionals

7.3.1 The Thomas-Fermi approximation

It is a curious fact that the first energy functional on record actually predates density functional theory: it is the so-called *Thomas-Fermi* (TF) energy functional (Fermi, 1927; Thomas, 1927), introduced by these authors as the basis of their statistical theory of heavy atoms. In contemporary language one would say that the TF functional is the local density approximation for the non-interacting kinetic energy functional $T_s[n]$.

The TF theory begins with the observation that an inhomogeneous electron liquid of slowly varying density is locally indistinguishable from a homogeneous one, provided that the variation of the density is negligible on the scale of the wavelength of the most energetic (and therefore most numerous) electrons. Mathematically, the condition of slow variation of the density has the form

$$\frac{|\vec{\nabla} n(\vec{r})|}{n(\vec{r})k_F(\vec{r})} \ll 1 , \tag{7.72}$$

where $k_F(\vec{r})$ is the Fermi wave vector of a uniform non-interacting electron gas at the local density $n(\vec{r})$. If this condition is satisfied, then the kinetic energy of a small volume element of the electron liquid can be approximated as the kinetic energy of the homogeneous electron gas of the same density. For a spin-unpolarized system in d dimensions one gets

$$T_s^{TF}[n] = \frac{d}{d+2} \frac{\hbar^2}{2m} \int n(\vec{r}) k_F^2(\vec{r}) d\vec{r} , \tag{7.73}$$

and the complete TF energy functional reads

$$E_V^{TF}[n] = \frac{d}{d+2} \frac{\hbar^2}{2m} \int n(\vec{r}) k_F^2(\vec{r}) d\vec{r} + \frac{e^2}{2} \int d\vec{r} \int d\vec{r}' \frac{n(\vec{r})n(\vec{r}')}{|\vec{r} - \vec{r}'|}$$
$$+ \int V(\vec{r}) n(\vec{r}) d\vec{r} . \tag{7.74}$$

The exchange-correlation contribution is neglected on the assumption that the density is so high that the energy of the electron gas is almost entirely kinetic. For an atom of atomic number Z and nuclear potential $V(\vec{r}) = -\frac{Ze^2}{r}$, the minimization of Eq. (7.74) (with $d = 3$) with respect to the density leads to the equation

$$\frac{\hbar^2}{2m}[3\pi^2 n(\vec{r})]^{2/3} + \int \frac{e^2 n(\vec{r}')}{|\vec{r} - \vec{r}'|}d\vec{r}' - \frac{Ze^2}{r} = \mu , \tag{7.75}$$

where μ is determined by the condition $\int n(\vec{r})d\vec{r} = Z$.[18]

The solution of this equation yields a universal density profile that extends to infinity, but is accurate only in the region $\frac{a_B}{Z} \ll r \ll \frac{a_B}{Z^{1/3}}$, in which the condition (7.72) is satisfied. The upper limit of the region, $\frac{a_B}{Z^{1/3}}$, gives the characteristic size of the Thomas-Fermi atom. Notice that the condition of slow density variation holds throughout the "bulk" of the atom, but fails both in the vicinity of the nucleus, where the coulomb potential diverges, and in the outermost region, where the wave function vanishes exponentially. Other situations in which the TF theory fails include the calculation of surface density profiles of solids, and the binding of atoms to form molecules. In fact, it can be shown rigorously (Teller, 1962) that in the Thomas-Fermi approximation it is impossible to have chemical binding between two atoms.

In spite of these shortcomings the TF approximation is a tool of great practical and theoretical importance. Even the Teller "no-binding" theorem – an apparently negative result – turns out to be the cornerstone of the proof of the stability of fermionic matter by Dyson (1967) and Lieb and Thirring (1975). We refer the reader to the beautiful review article by Larry Spruch (1991) for a complete pedagogical description of the TF theory.

7.3.2 The local density approximation for the exchange-correlation potential

The philosophy of the Thomas-Fermi approximation can be extended to the exchange-correlation energy functional. In Chapter 1 we saw that the exchange-correlation hole $n(\vec{r}')[g(\vec{r}, \vec{r}') - 1]$ of a liquid state is a short-ranged function of the separation $|\vec{r} - \vec{r}'|$, with a characteristic range of the order of the average distance between the electrons. If the density is slowly varying on this scale (which is another way of saying that the condition (7.72) is satisfied) we can ignore the difference between $n(\vec{r})$ and $n(\vec{r}')$ in the integrand of Eq. (7.33). At the same time we can make the approximation

$$g_\lambda(\vec{r}, \vec{r}') \simeq g_\lambda^{(h)}(|\vec{r} - \vec{r}'|; n(\vec{r})) , \tag{7.76}$$

where $g_\lambda^{(h)}(|\vec{r} - \vec{r}'|; n(\vec{r}))$ is the pair correlation function of the *homogeneous* electron liquid at coupling constant λ and density $n(\vec{r})$. With these approximations, Eq. (7.33) becomes

$$E_{xc}^{LDA}[n] \equiv \frac{1}{2} \int_0^1 d\lambda \int d\vec{r} n^2(\vec{r}) \int d\vec{r}' \frac{e^2}{|\vec{r} - \vec{r}'|}[g_\lambda^{(h)}(|\vec{r} - \vec{r}'|; n(\vec{r})) - 1]$$

$$= \int d\vec{r} n(\vec{r})\epsilon_{xc}(n(\vec{r})) , \tag{7.77}$$

---

[18] More precisely, Eq. (7.75) holds only in the regions of space in which $n(\vec{r}) > 0$. In regions where $n(\vec{r}) = 0$ the left-hand side must be larger than μ.

Table 7.2. *Atomic ground-state energies in units of*
$\frac{e^2}{a_B} = 27.2116$ *eV from the Kohn–Sham equation in
the local spin density approximation. Adapted from
Gross and Dreizler (1990), p. 232.*

Atom	E_{HF}	E_{LDA}	$E_{expt.}$
He	-2.862	-2.870	-2.904
Li	-7.433	-7.395	-7.478
Be	-14.573	-14.515	-14.667
Ne	-128.547	-128.410	-128.928
Mg	-199.615	-199.345	-200.043
Ar	-526.817	-526.755	-527.549

where $\epsilon_{xc}(n)$ is the exchange-correlation energy *per particle* of the homogeneous electron liquid at density n. The corresponding exchange-correlation potential is

$$V_{xc}^{LDA}(\vec{r}) = \frac{d[n\epsilon_{xc}(n)]}{dn}\bigg|_{n(\vec{r})}. \tag{7.78}$$

Eqs. (7.77) and (7.78) define the local density approximation for E_{xc} and V_{xc} – better known as the LDA *tout court*.

Perhaps the most remarkable fact about the LDA is that, when used in the Kohn–Sham equation, it works reasonably well even in systems such as atoms, molecules, and metal surfaces, in which the density is definitely *not* slowly varying. In fact, early attempts to improve upon the LDA by adding corrections proportional to the gradients of the density led to functionals that, surprisingly, did not work as well as the plain LDA.

Eventually, it became clear that the reason for the success of the LDA is that the approximation (7.76) respects the most important (from the energetic point of view) property of the exchange-correlation hole, namely, its global strength, which is given by the sum rule (see Appendix 4)

$$\int [g(\vec{r}, \vec{r}') - 1]n(\vec{r}')d\vec{r}' = -1. \tag{7.79}$$

The LDA satisfies this sum rule simply because the xc hole of the homogeneous electron liquid satisfies it in its own right.[19] By contrast, an uncritical inclusion of gradient corrections, while improving the short-range behavior of the xc hole, violates the sum rule (7.79), and may lead to worse results for the energy when the long-range part of the coulomb interaction dominates. We shall see in the next section how this difficulty has been recently overcome by the so-called *generalized gradient approximation* (GGA).

There is, at this time, a whole industry of electronic structure calculations based on the Kohn–Sham equation combined with the LDA. In Tables 7.2–7.4 we report a few examples

[19] Numerical studies have shown that the on-top xc hole $[g(\vec{r}, \vec{r}) - 1]n(\vec{r})$ is given essentially exactly by the LDA.

Table 7.3. *Equilibrium separations in units of a_B and dissociation energies in eV for diatomic molecules from the Kohn–Sham equation in the LDA. Adapted from Gross and Dreizler (1990), p. 240.*

Molecule	HF		LDA		expt.	
H_2	1.39 a_B	3.64 eV	1.44 a_B	4.79 eV	1.40 a_B	4.75 eV
N_2	2.01	5.27	2.16	7.8	2.07	9.91
O_2	2.18	1.43	2.5	4.1	2.28	5.23
F_2	2.50	−1.37	2.91	0.6	2.68	1.66

Table 7.4. *Wigner–Seitz radii in units of a_B and cohesive energies in Ry for bulk metals from the Kohn–Sham equation in the LDA. Adapted from Kohn and Vashishta (1983).*

Metal	LDA		expt.	
Y	3.60 a_B	0.36 Ry	3.75 a_B	0.31 Ry
Zr	3.20	0.50	3.30	0.46
Nb	3.00	0.55	3.05	0.56
Mo	2.90	0.49	2.90	0.50
Ru	2.80	0.56	2.80	0.49
Rh	2.80	0.45	2.80	0.42
Pd	3.05	0.27	3.00	0.28

representative of the performance of this approximation in different classes of systems. Fig. 7.5 shows a comparison between the spherically averaged xc hole calculated by the variational Monte Carlo method and by the LDA at different positions in the unit cell of crystalline silicon.

A serious limitation of the LDA in atomic systems is that the xc potential is tied to the local density and therefore approaches zero exponentially at large distance from the nucleus. The correct behavior, on the other hand, is $-\frac{e^2}{r}$. This is evident from the fact that, far from the atom, an electron should see the coulomb potential of the positively charged ion, namely, $-\frac{e^2}{r}$. Since in this region the nuclear potential is exactly cancelled by the Hartree potential, the $-\frac{e^2}{r}$ must come entirely from the xc potential.

Another characteristic weakness of the LDA is that it tends to overestimate the binding energy of molecular systems. As pointed out at the end of Section 7.2.6, this is partly due to the fact that the high-density limit of E_{xc}^{LDA} has an incorrect logarithmic divergence. This problem is largely corrected by the GGA. It has also been found that the LDA generally does not work well in strongly correlated systems, particularly those in which the Landau Fermi liquid picture (see Chapter 8) is not valid. Finally, a well documented failure of the

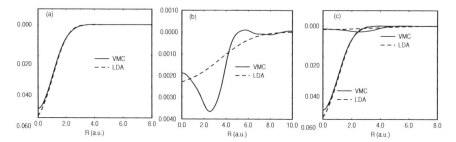

Fig. 7.5. The spherically averaged exchange-correlation hole calculated by the variational Monte Carlo (VMC) method and in LDA at different positions of the fixed electron in the unit cell of crystalline Si: (a) one electron fixed at the Si–Si bond center, (b) one electron fixed at the tetrahedral interstitial site, and (c) plots (a) and (b) superimposed. (Adapted from Hood *et al.* (1998).)

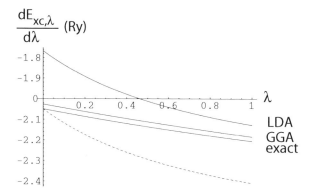

Fig. 7.6. The derivative of the xc energy functional $\frac{dE_{xc,\lambda}}{d\lambda}$ plotted as a function of the coupling constant λ at the ground-state density of a helium atom. The three solid lines show, from top to bottom, the local density approximation (LDA), the generalized gradient approximation (GGA) described in Section 7.3.4, and the exact result. The dashed line is obtained by shifting the LDA curve to make it coincide with the exact result at $\lambda = 0$. From K. Burke, with permission.

LDA is the systematic underestimation of the band-gap of insulators, due to the absence of derivative discontinuities in the LDA functional (see Section 7.2.8).

The LDA xc energy can be neatly separated into exchange and correlation parts: the exchange-only LDA is obtained simply by replacing $\epsilon_{xc}(n)$ with $\epsilon_x(n)$ in Eqs. (7.77) and (7.78), and the correlation-only LDA is the remainder. In view of this fact, one might be tempted to treat the exchange exactly (e.g., by doing a Hartree–Fock or OEP calculation) and add the approximate LDA correlation energy as a correction to the exact exchange. Unfortunately, this reasonable idea often leads to errors that are larger than those associated with the standard LDA for the xc energy. To understand why, consider Fig. 7.6, which shows the derivative of the xc energy functional $E_{xc,\lambda}[n]$ with respect to λ at the ground-state density of a helium atom. The three curves shown in the figure are, starting from the top, the spin-LDA curve, the curve obtained from a generalized gradient approximation (described in Section 7.3.4), and the exact curve. Notice that at $\lambda = 0$ the derivative of the xc energy functional coincides with that of the exchange-only functional, so the large difference between the LDA and the exact result at $\lambda = 0$ reflects the inaccuracy of the LDA exchange. The key point is that the difference between the LDA and the exact result decreases with increasing λ because the xc hole becomes deeper and the LDA is more accurate in the short-range part of the xc hole. Combining the exact exchange with the LDA correlation energy would amount to shifting the whole LDA curve downward to bring it in coincidence with the exact curve at $\lambda = 0$: the shifted curve, shown as a dashed line in Fig. 7.6, clearly results in a poor approximation at the physical value of $\lambda = 1$.

7.3.3 The gradient expansion

The obvious next step beyond the LDA would seem to be the *gradient expansion approximation* (GEA), which was already suggested in the original paper by Hohenberg and Kohn (1964). Consider an electron liquid of density $n(\vec{r}) = \bar{n} + \delta n(\vec{r})$, where $\bar{n}$ is uniform and $\delta n(\vec{r})$ is a density modulation that is both weak *and* slowly varying in space.

According to the stiffness theorem, introduced and discussed in Chapter 3, the F functional has the following expansion to second-order in the amplitude of the density modulation:

$$F[n] = F[\bar{n}] - \frac{1}{2L^d} \sum_{\vec{q}} \delta n(\vec{q}) \chi_{nn}^{-1}(\vec{q}) \delta n(\vec{q}), \tag{7.80}$$

where $\chi_{nn}(q)$ is the static density–density response function, and $\delta n(\vec{q})$ is the Fourier component of $\delta n(\vec{r})$ at wave vector $\vec{q}$. Since, by assumption, $\delta n(\vec{r})$ is slowly varying in space, its Fourier components are appreciable only for wave vectors $q \ll k_F$. In this range the inverse of the static density–density response function can be expanded as

$$\chi_{nn}^{-1}(q) = A(\bar{n}) - B(\bar{n})q^2 + \cdots - v(q), \tag{7.81}$$

where A and B are related to the coefficients of the small-q expansion of the *proper* density–density response function in the following manner:

$$\tilde{\chi}_{nn}(q) = \tilde{\chi}_{nn}(0) + \tilde{\chi}_{nn}^{(2)}(0)q^2 + \cdots,$$

$$A = \frac{1}{\tilde{\chi}_{nn}(0)},$$

$$B = \frac{\tilde{\chi}_{nn}^{(2)}(0)}{\tilde{\chi}_{nn}(0)^2}. \tag{7.82}$$

Substituting Eq. (7.81) into Eq. (7.80) and expressing the result in terms of $\delta n(\vec{r})$ we obtain

$$F[n] = F[\bar{n}] + \frac{e^2}{2} \int d\vec{r} \int d\vec{r}' \frac{\delta n(\vec{r}) \delta n(\vec{r}')}{|\vec{r} - \vec{r}'|} - \frac{1}{2} \int d\vec{r} A(\bar{n})[\delta n(\vec{r})]^2$$

$$+ \frac{1}{2} \int d\vec{r} B(\bar{n}) |\vec{\nabla} n(\vec{r})|^2. \tag{7.83}$$

The first term on the right-hand side is the energy of the uniform electron liquid, the second is the Hartree energy, the third is the second order term in the expansion of the LDA in powers of δn, and, finally, the last term is the desired second-order gradient correction. We generalize Eq. (7.83) to systems of arbitrary slowly varying density (not necessarily close to uniformity) via the replacement $\bar{n} \rightarrow n(\vec{r})$. This identifies the leading gradient correction to the F functional as

$$F^{(2)}[n] = \frac{1}{2} \int d\vec{r} \, B(n(\vec{r})) |\vec{\nabla} n(\vec{r})|^2 \,. \tag{7.84}$$

A similar result is of course obtained for the non-interacting kinetic energy functional

$$T_s^{(2)}[n] = \frac{1}{2} \int d\vec{r} \, B_0(n(\vec{r})) |\vec{\nabla} n(\vec{r})|^2 \,, \tag{7.85}$$

where B_0 is the non-interacting counterpart of B. The reader is urged to verify that the small-q expansion of the Lindhard function yields $B_0(n) = \frac{\hbar^2}{72mn}$ in three dimensions (von Weizsäcker, 1935) and 0 in two dimensions (Exercise 14).

The GEA for the xc energy functional is obtained from the difference of the GEAs for F and T_s. The calculation, first carried out by Sham (1971) in the weak coupling limit, has been refined and extended by several authors.[20] Particularly noteworthy are the expressions obtained by Rasolt and Geldart (1986) for the coefficients $B_{xc,\sigma\sigma'}(n_\uparrow, n_\downarrow)$ of the second-order spin-resolved GEA. These coefficients are defined by the expression

$$E_{xc}^{(2)}[n_\uparrow, n_\downarrow] = \frac{1}{2} \sum_{\sigma\sigma'} \int d\vec{r} \, B_{xc,\sigma\sigma'}(n_\uparrow(\vec{r}), n_\downarrow(\vec{r})) \vec{\nabla} n_\sigma(\vec{r}) \cdot \vec{\nabla} n_{\sigma'}(\vec{r}) \,, \tag{7.87}$$

and their values can be found in the cited reference.

Unfortunately, in applications to real systems, the GEA is often found to be less accurate than the LDA. As already discussed the reason for this disappointing behavior lies in the fact that the gradient expansion for the xc hole fails at large values of $|\vec{r} - \vec{r}'|$, resulting in a violation of the sum rule (7.79). We shall see in the next section how this shortcoming of the naive GEA has been remedied in recent "generalized gradient approximations".

We now turn to an important philosophical question, namely whether a gradient expansion exists or not for a given set of intensive variables. If the answer is affirmative then the variables are probably well-chosen, at least in the sense that they admit the LDA and its systematic refinements. In the opposite case we should probably be looking for a different set of variables!

From the stiffness theorem we immediately see that a necessary condition for the existence of the gradient expansion for a set of densities $n_i(\vec{r})$ is that the matrix inverse of the static

---

[20] The calculation of the exchange contribution to B_{xc} has an interesting history. In 1971 Sham obtained

$$B_x[n] = -\frac{7e^2}{216\pi (3\pi^2)^{1/3}} n^{-4/3} \,, \tag{7.86}$$

assuming an electron-electron interaction of the form $v_q = \frac{4\pi e^2}{q^2 + \eta^2}$ and letting $\eta \rightarrow 0$ at the end of the calculation. It was subsequently shown (Kleinman, 1984) that the correct result for the unscreened coulomb interaction at $T = 0$ is 10/7 times the Sham result. Yet, *at finite temperature*, the result appears to be 24/7 times the Sham result (Geldart *et al.*, 1990).

susceptibility matrix $\chi_{n_i,n_j}(q)$ of the homogeneous electron liquid exists and is Taylor-expandable in powers of q about $q = 0$:

$$[\chi^{-1}]_{n_i,n_j}(q) = [\chi_0^{-1}]_{n_i,n_j}(q) - B_{ij}q^2 + \cdots . \tag{7.88}$$

Let us consider some examples. In the simplest case the variables are just the spin-resolved particle densities. The spin response matrix $\chi_{\sigma\sigma'}(q)$ of the homogeneous electron liquid has a finite macroscopic ($q \to 0$) limit and, therefore the spin density functional theory admits a gradient expansion.

Consider now the set $n(\vec{r})$, $\vec{j}(\vec{r})$ where $\vec{j}(\vec{r})$ is the particle current. Because the current–current response function of the homogeneous electron liquid vanishes for $q \to 0$ (the diamagnetic sum rule, see Section 3.4.2), while the static current–density response function is zero at all q, we see that $[\chi^{-1}]_{jj}(q)$ diverges for $q \to 0$, and the set of variables does not admit a gradient expansion. By contrast, the set $n(\vec{r})$, $\vec{j}_p(\vec{r})$ does admit a gradient expansion, because the current–current response function for the paramagnetic current density tends to the finite value $-\frac{n}{m}$ for $q \to 0$. This is one of the basic reasons for using the paramagnetic current density in the formulation of the static current density functional theory (see Section 7.4).

7.3.4 Generalized gradient approximation

The pioneering work on overcoming the limitations of the naive gradient expansion was done by Langreth and Perdew (1980). The remedies, known as *generalized gradient approximations*, assume the following form for the xc energy functional:

$$E_{xc}^{GGA}[n_\uparrow, n_\downarrow] = \int d\vec{r}\, f(n_\uparrow(\vec{r}), n_\downarrow(\vec{r}), \vec{\nabla}n_\uparrow(\vec{r}), \vec{\nabla}n_\downarrow(\vec{r})) . \tag{7.89}$$

Several forms of the function f have been presented in the literature, but, in the last few years, a consensus has developed about a few that are qualitatively similar for systems of physical interest. These GGAs have been particularly successful in chemical applications, where they tend to reduce the LDA overestimation of the binding energy by about a factor 5.

The main idea of the GGA is that the worst problem of the gradient expansion – the spurious long-range part of the xc hole – can be cured by a real space cutoff procedure, which reinstates the correct sum rules and other exact properties. Although the basic idea is simple, the detailed calculations are quite technical and time-consuming. In this section we outline one of the most recent derivations of the f function due to Perdew, Burke, and Ernzerhof (PBE, 1996).[21] In this derivation the function f is expressed entirely in terms of fundamental constants, which are determined by the theory of the electron liquid. The only empirical step is the real space cutoff procedure, described below.

---

[21] See also the excellent review article by K. Burke, J. P. Perdew, and Yue Wang (1998), which we follow closely.

7.3.4.1 The exchange-correlation hole

The starting point of the GEA is the real-space decomposition (7.33) of the exchange-correlation energy. Define the coupling-constant averaged xc hole

$$n_{xc}(\vec{r}, \vec{r} + \vec{u}) = \int_0^1 d\lambda \, n(\vec{r} + \vec{u})[g_\lambda(\vec{r}, \vec{r} + \vec{u}) - 1] , \qquad (7.90)$$

and its system average

$$\langle n_{xc}(\vec{u}) \rangle = \frac{1}{N} \int d\vec{r} \, n(\vec{r}) n_{xc}(\vec{r}, \vec{r} + \vec{u}) , \qquad (7.91)$$

which depends only on the separation $\vec{u}$. The xc energy (7.33) can then be written as

$$E_{xc} = \frac{N}{2} \int d\vec{u} \, \frac{e^2}{u} \langle n_{xc}(\vec{u}) \rangle . \qquad (7.92)$$

To better analyze the xc hole we now split it into its exchange component, $n_x(\vec{r}, \vec{r} + \vec{u})$, which is obtained by putting the non-interacting pair correlation function g_0 instead of g_λ in Eq. (7.90), and a remainder known as the correlation hole, $n_c(\vec{r}, \vec{r} + \vec{u})$:

$$n_{xc}(\vec{r}, \vec{r} + \vec{u}) = n_x(\vec{r}, \vec{r} + \vec{u}) + n_c(\vec{r}, \vec{r} + \vec{u}) . \qquad (7.93)$$

The exchange hole is a negative quantity (see Appendix 4) whose spatial integral is -1, since the non-interacting pair correlation function satisfies the sum rule (7.79) in its own right. Thus, we have

$$n_x(\vec{r}, \vec{r} + \vec{u}) \le 0 ,$$
$$\int d\vec{u} \, n_x(\vec{r}, \vec{r} + \vec{u}) = -1 ,$$
$$\int d\vec{u} \, n_c(\vec{r}, \vec{r} + \vec{u}) = 0 . \qquad (7.94)$$

These relations are also satisfied by the system-averaged exchange and correlation holes, and play a crucial role in the construction of the GGA.

7.3.4.2 GGA for exchange

The gradient expansion for the exchange hole is known (in three dimensions) up to third order in $\vec{\nabla} n$. The zero-th order term – the only one used in the LDA – gives the exact exchange-hole at $\vec{u} = 0$. The first order term gives the exact slope of the exchange-hole at $\vec{u} = 0$, the second-order term the exact curvature, and so on. Inclusion of each higher order term improves the description of the exchange hole at small u, where the m-th order term varies as u^m, but worsens the description at large u, as the m-th order term has an oscillating component proportional to u^{2m-4}. It is this large-u problem that needs to be corrected.

Because the exchange energy satisfies the spin-scaling relation (see Exercise 13)

$$E_x[n_\uparrow, n_\downarrow] = \frac{1}{2} E_x[2n_\uparrow] + \frac{1}{2} E_x[2n_\downarrow] , \qquad (7.95)$$

where $E_x[n]$ is the exchange energy for a fully spin polarized electron gas, we only need to determine the exchange hole for a fully spin polarized system. The gradient expansion approximation (GEA) for this quantity, or, more precisely, for the closely related quantity $\tilde{n}_x(\vec{r}, \vec{r} + \vec{u})$ discussed in footnote [22], is given by Perdew and Wang (1986) in the form

$$\tilde{n}_x^{GEA}(\vec{r}, \vec{r} + \vec{u}) = -\frac{1}{2}n(\vec{r})\tilde{y}(\vec{r}, \vec{u}), \qquad (7.96)$$

where

$$\tilde{y}(\vec{r}, \vec{u}) = J(z) + \frac{4}{3}L(z)\vec{e}_u \cdot \vec{s} - \frac{16}{27}M(z)(\vec{e}_u \cdot \vec{s})^2 - \frac{16}{3}N(z)s^2,$$

$$\vec{e}_u = \frac{\vec{u}}{u},$$

$$\vec{s}(\vec{r}) = \frac{\vec{\nabla}n(\vec{r})}{2k_F(\vec{r})n(\vec{r})},$$

$$z(\vec{r}, u) = 2k_F(\vec{r})u, \qquad (7.97)$$

and $k_F(\vec{r}) = [3\pi^2 n(\vec{r})]^{1/3}$ is the local Fermi wave vector. The functions J, L, M, N are given in the paper of Perdew and Wang (1986). The function $J(z)$ alone yields the LDA for the exchange hole.[22]

In order to satisfy the requirements (7.94) Perdew and Wang use the "brutal" but unbiased approximation

$$\tilde{n}_x^{GGA}(\vec{r}, \vec{r} + \vec{u}) = -\frac{1}{2}n(\vec{r})\tilde{y}(\vec{r}, \vec{u})\Theta(\tilde{y}(\vec{r}, \vec{u}))\Theta(u_x(\vec{r}) - u), \qquad (7.98)$$

where $\Theta(x) = 1$ for $x > 0$, and 0 otherwise. The first Θ-function enforces the negativity of the exchange hole by simply cutting off the regions of space in which the GEA hole becomes positive. The second Θ-function similarly cuts off the exchange hole beyond a critical distance $u_x(\vec{r})$ chosen in such a way that the integrated strength of the hole equals -1.

Inserting this hole into the real space decomposition (7.92) produces the GGA for exchange, which reads

$$E_x^{GGA}[n] = \int d\vec{r}\, n(\vec{r})\epsilon_x(n(\vec{r}))F_x(s(\vec{r})), \qquad (7.99)$$

where $\epsilon_x(n(\vec{r})) = -\frac{3e^2 k_F}{4\pi}$ yields the exchange-only LDA when $F_x = 1$. $F_x(s)$ – a function of the scaled gradient s only – is the enhancement factor over the LDA exchange. The calculation of $F_x[s]$ is a technically complex, but conceptually straightforward task. The first analytical approximations to F_x where produced by Perdew and Wang in 1986 and 1991. More recently, PBE (1996) have proposed a simpler analytic function which depends

---

[22] There is a difference between the gradient expansion for the true exchange hole $n_x^{GEA}(\vec{r}, \vec{r} + \vec{u})$ and that for the auxiliary quantity $\tilde{n}_x^{GEA}(\vec{r}, \vec{r} + \vec{u})$ introduced in Eq. (7.96). In writing the latter, terms proportional to $\nabla^2 n$ have been transformed to equivalent $|\vec{\nabla}n|^2$ terms, which yield identical results for the system-averaged hole. Thus $n_x^{GEA}(\vec{r}, \vec{r} + \vec{u})$ and $\tilde{n}_x^{GEA}(\vec{r}, \vec{r} + \vec{u})$ are equivalent when used in the integral (7.91), upon which the energy depends, but are different at any given $\vec{r}$. The Perdew–Wang real space cutoff is carried out on $\tilde{n}$, not on n itself.

only on fundamental constants and satisfies all the energetically relevant constraints. This function is

$$F_x(s) = 1 + \kappa - \frac{\kappa}{1 + \mu s^2/\kappa} , \qquad (7.100)$$

where $\mu = 0.21951$ and $\kappa = 0.804$.

7.3.4.3 GGA for correlation

The cutoff procedure for the correlation hole is more straightforward than the one for the exchange hole, as there is no negativity constraint. However, the GEA correlation hole is imperfectly known, especially where its long-range oscillatory behavior is concerned. Correlation introduces a new length scale k_{TF}^{-1} where $k_{TF} = \left(\frac{4k_F}{\pi a_B}\right)^{1/2}$ is the Thomas-Fermi screening wave vector. An additional complication is that the correlation energy does not depend on spin polarization through a simple scaling relation, such as (7.95), and must therefore be calculated for each value of the fractional polarization p. Therefore, after neglecting the small corrections that arise from the gradient of $p(\vec{r})$, the GGA defines a function of *three* variables

$$E_c^{GGA}[n_\uparrow, n_\downarrow] = \int d\vec{r}\, n(\vec{r}) \epsilon_c^{GGA}(r_s(\vec{r}), p(\vec{r}), s(\vec{r})) , \qquad (7.101)$$

where $r_s(\vec{r})$ is the local value of r_s. Because the correlation hole is not required to be negative, one can focus at the outset on its spherical average, which depends on $\vec{r}$ and $u = |\vec{u}|$ and is approximated as

$$\tilde{n}_c(\vec{r}, u) = n_c^{LDA}(r_s(\vec{r}), p(\vec{r}), v(\vec{r}, u)) + t^2(\vec{r})\delta\tilde{n}_c(r_s(\vec{r}), p(\vec{r}), v(\vec{r}, u)) , \qquad (7.102)$$

where

$$t(\vec{r}) = \frac{|\vec{\nabla} n(\vec{r})|}{2k_p(\vec{r}) n(\vec{r})} = \left(\frac{\pi}{4\alpha_3}\right)^{1/2} \frac{s(\vec{r})}{[r_s(\vec{r})]^{1/2}\phi(p(\vec{r}))} \qquad (7.103)$$

is another reduced density gradient, which scales with the Thomas-Fermi wave vector

$$k_p(\vec{r}) = \phi(p(\vec{r}))k_{TF}(\vec{r}) , \qquad (7.104)$$

and

$$v(\vec{r}, u) = k_p(\vec{r})u \qquad (7.105)$$

is another reduced separation expressed on the Thomas-Fermi length scale. The spin scaling factor in the above formulas is

$$\phi(p) = \frac{(1 + p)^{2/3} + (1 - p)^{2/3}}{2} . \qquad (7.106)$$

The relation between $\tilde{n}_c$ and n_c is the same as the relation between $\tilde{n}_x$ and n_x (see footnote[22]).

The LDA correlation hole $n_c^{LDA}(r_s, p, v)$ is accurately given by the Perdew–Wang formula, which is supported by Monte Carlo calculations:

$$n_c^{LDA}(r_s, p, v) = \phi^3(p)k_p^2 A_c(r_s, p, v),\qquad (7.107)$$

where the function $A_c(r_s, p, v)$ can be evaluated exactly from the RPA in the high-density limit ($r_s \to 0$) or in the long-range limit ($v \to \infty$). At larger values of r_s, $A_c(r_s, p, v)$ can be represented as the sum of the exact long-range contribution and an approximate short-range contribution (Perdew and Wang, 1992).

The gradient correction to $\tilde{n}_c$ in Eq. (7.102) is similarly written as

$$\delta\tilde{n}_c(r_s, p, v) = \phi^3(p)k_p^2 B_c(r_s, p, v),\qquad (7.108)$$

where the function $B_c(r_s, p, v)$ is again represented as the sum of long- and short-range contributions (Perdew and Wang, 1992).

With the GEA correlation hole thus fully defined, the spherically-averaged GGA hole is given by

$$\tilde{n}_c^{GGA}(r_s, p, t, v) = \phi^3(p)k_p^2 \left[A_c(r_s, p, v) + t^2 B_c(r_s, p, v) \right]\Theta(v_c - v),\quad (7.109)$$

where v_c is the largest cutoff that satisfies the normalization condition

$$\int_0^{v_c} dv\, 4\pi v^2 \left[A_c(r_s, p, v) + t^2 B_c(r_s, p, v) \right] = 0.\qquad (7.110)$$

Notice that the value of the cutoff v_c decreases with increasing value of the scaled density gradient t, and tends to 0 when the latter tends to infinity, turning off the correlation energy altogether. Inserting Eq. (7.109) into Eq. (7.92) we get the correlation energy per particle in Eq. (7.101) as

$$\epsilon_c^{GGA}(r_s, p, s) = \phi^3(p)\int_0^{v_c} dv\, \frac{4\pi v^2}{2v} \left[A_c(r_s, p, v) + t^2 B_c(r_s, p, v) \right]$$
$$= \epsilon_c(r_s, p) + H(r_s, p, t).\qquad (7.111)$$

A simple analytical representation of the function H that satisfies the most important constraints has been recently proposed by PBE (1996). Its form is

$$H(r_s, p, t) = \frac{e^2}{a_B}\gamma\phi^3(p)\ln\left\{ 1 + \frac{\beta}{\gamma}t^2\left[\frac{1 + At^2}{1 + At^2 + A^2 t^4} \right] \right\},\qquad (7.112)$$

where

$$A = \frac{\beta}{\gamma}\left\{ \exp\left(-\frac{\epsilon_c/\mathrm{Ry}}{2\gamma\phi^3(p)} \right) - 1 \right\}^{-1},\qquad (7.113)$$

$\beta = 0.066725$ and $\gamma = (1 - \ln 2)/\pi^2 \simeq 0.031091$. Notice that H vanishes at $t = 0$ and grows monotonically to the limit $-\epsilon_c$ for $t \to \infty$, so that the correlation energy is suppressed in this limit.

Table 7.5. *Dissociation energies of molecules in eV.*
Adapted from Perdew, Burke, and Ernzerhof (1996).

Molecule	ΔE^{LDA}	ΔE^{GGA}	ΔE^{HF}	ΔE^{exp}
H_2	4.90	4.55	3.64	4.73
LiH	2.60	2.25	1.43	2.51
CH_4	20.0	18.2	14.2	18.2
NH_3	14.6	13.1	8.72	12.9
OH	5.38	4.77	2.95	4.64
H_2O	11.6	10.15	6.72	10.06
N_2	11.6	10.5	4.98	9.93
O_2	7.59	6.24	1.43	5.25
F_2	3.38	2.30	−1.60	1.69

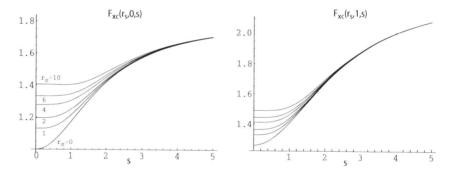

Fig. 7.7. The enhancement factor $F_{xc}(r_s, p, s)$ calculated by Perdew, Burke, and Ernzerhof (1996) at $p = 0$ and $p = 1$ for different values of r_s (courtesy of K. Burke).

7.3.4.4 The exchange-correlation energy

We can now piece together the results of the two previous subsections. To represent the GGA for the full exchange-correlation energy it has become customary to define the enhancement factor F_{xc} over local exchange:

$$E_{xc}^{GGA}[n_\uparrow, n_\downarrow] = \int d\vec{r}\, n(\vec{r}) \epsilon_x(n(\vec{r})) F_{xc}(r_s(\vec{r}), p(\vec{r}), s(\vec{r})) \,. \qquad (7.114)$$

The LDA corresponds to the approximation $F_{xc}(r_s, p, s) \approx F_{xc}(r_s, p, 0)$. Plots of the enhancement factor for $p = 0$ and $p = 1$ are taken from PBE and shown in Fig. 7.7. The overall performance of this approximation for the xc energy is illustrated in Table 7.5, which shows the dissociation energy ΔE of several molecules in LDA, in the PBE version of the GGA, in the HF approximation, and, finally, in experiment. The improvement of the GGA upon the LDA in these calculations is both evident and striking.

7.3.5 Van der Waals functionals

There are situations in which neither the LDA nor the GGA give an adequate description of the physical system. One instance is the "band-gap problem" in insulators. Another instance is provided by the *van der Waals* (vdW) *interaction* between two electrically neutral and spatially separated systems. The typical example is the interaction between two neutral atoms separated by a distance R that is much larger than the atomic size. The mechanism of interaction is qualitatively as follows. A quantum fluctuation produces an instantaneous electric dipole moment of magnitude d_1 on one of the two atoms (by charge conservation, no net charge can be created). The electric dipole is the source of an electric field that scales as $\frac{d_1}{r^3}$, r being the distance from the dipole (we ignore the angular dependence for simplicity). This electric field, acting on the second atom, induces an electric dipole of magnitude $d_2 \sim \frac{\alpha_2 d_1}{R^3}$, where α_2 is the electric polarizability of that atom. Finally, the two dipoles interact with each other, giving a negative interaction energy proportional to $-\frac{d_1^2 \alpha_2}{R^6}$. By the stiffness theorem, the average quadratic value of d_1^2 in the ground-state of an atom is proportional to the electric polarizability α_1. We conclude that the van der Waals interaction energy between two atoms is proportional to $-\frac{\alpha_1 \alpha_2}{R^6}$: the negative sign implies that the interaction is *attractive*. By a similar analysis one can show that a neutral atom placed at a distance R from a metal surface interacts with the latter with a potential proportional to $-\frac{\alpha^2}{R^3}$. On the other hand, the interaction between two parallel metal surfaces is found to scale as $-\frac{1}{R^2}$.

Two things should be stressed about the vdW interaction. The first is that it arises entirely from the *correlation* of density fluctuations in the two atoms, molecules, or surfaces: if the fluctuations were statistically independent, their interaction energy would vanish. The second is that the existence of these correlated fluctuations is a feature of the quantum mechanical ground-state, so the vdW interaction is operative down to the absolute zero of temperature.

Van der Waals interactions play an important role in condensed matter physics, insofar as they are the only mechanism of interaction between neutral objects at large distances. Examples of physical phenomena that are governed by the vdW attraction include adhesion and cohesion of liquid crystals, polymers, and biomolecular objects. Atomic force microscopy allows experimental measurements of the vdW force.

In density functional theory, the vdW interaction energy is part of the correlation energy functional, and must be expressed in terms of the ground-state density. It is not difficult to see why this program cannot be fulfilled by any local density approximation or GGA: the problem is that all such approximations are spatially "short-sighted" and therefore cannot keep track of the correlation that exists between widely separated objects.

A correlation functional that includes vdW effects between widely separated objects can nevertheless be constructed in a relatively straightforward manner following the insights of several authors (Rapcewicz and Ashcroft, 1991; Dobson, 1996; Andersson, 1996). The basic idea is to treat the coulomb interaction between the two objects as a weak perturbation to the zero-th order hamiltonian, which describes the two independent objects. For two atoms of atomic numbers Z_1 and Z_2 this interaction has the form

$$\hat{H}_{12} = \int_{V_1} d\vec{r}_1 \int_{V_2} d\vec{r}_2 \frac{e^2 [\hat{n}_1(\vec{r}_1) - Z_1 \delta(\vec{r}_1)][\hat{n}_2(\vec{r}_2) - Z_2 \delta(\vec{r}_2)]}{|\vec{r}_1 - \vec{r}_2|} , \qquad (7.115)$$

where $\hat{n}_i(\vec{r})$ is the density operator of the i-th atom, and the integrals run over the volumes V_i of the two atoms. Here we have assumed that a sharp distinction can be drawn between electrons that belong to atom 1 and electrons that belong to atom 2: this makes sense only if the two electronic clouds have a negligible overlap. It is an easy exercise to show that the first order correction to the energy of the two isolated atoms vanishes, while the second order correction is given by

$$E^{(2)} = -\frac{\hbar}{2\pi} \int_{V_1} d\vec{r}_1 d\vec{r}_1' \int_{V_2} d\vec{r}_2 d\vec{r}_2' \frac{e^2}{|\vec{r}_1 - \vec{r}_2|} \frac{e^2}{|\vec{r}_1' - \vec{r}_2'|}$$
$$\times \int_0^\infty \chi_{nn,1}(\vec{r}_1, \vec{r}_1', i\omega) \chi_{nn,2}(\vec{r}_2, \vec{r}_2', i\omega) d\omega , \qquad (7.116)$$

where $\chi_{nn,i}(\vec{r}, \vec{r}', i\omega)$ is the density–density response function of the isolated i-th atom.[23]

We now need to express the density–density response functions as functionals of the ground-state density. The simplest way to do this is to apply some kind of local density approximation. As a preliminary step, however, we must make sure that our approximation satisfies charge conservation. This is accomplished by expressing χ_{nn} exactly in terms of the dipole–dipole response function[24] χ_{d_i, d_j}:

$$\chi_{nn}(\vec{r}, \vec{r}', \omega) = -\frac{1}{e^2} \sum_{ij} \nabla_i \nabla_j' \chi_{d_i d_j}(\vec{r}, \vec{r}', \omega) . \qquad (7.117)$$

The presence of the two gradients in Eq. (7.117) ensures that the spatial integral of the density response vanishes, as required by charge conservation, and that no change in density arises from a constant potential, as required by gauge invariance. At this point we make the local density approximation on the dipole–dipole response function, i.e., we assume

$$\chi_{d_i d_j}(\vec{r}, \vec{r}', \omega) \simeq \alpha(\vec{r}, \omega) \delta(\vec{r} - \vec{r}') \delta_{ij} , \qquad (7.118)$$

where $\alpha(\vec{r}, \omega)$ is the macroscopic dipole–dipole response function of a homogeneous electron liquid at density $n(\vec{r})$. The latter is defined as

$$\alpha(\vec{r}, \omega) = - \lim_{q \to 0} \frac{e^2 \chi_{nn}(q, \omega)}{q^2}\bigg|_{n = n(\vec{r})}$$
$$= \frac{e^2 n(\vec{r})}{m(\omega_p^2(\vec{r}) - \omega^2)} , \qquad (7.119)$$

where $\omega_p(\vec{r})$ is the frequency of the homogeneous plasmon at density $n = n(\vec{r})$ and $q = 0$.

---

[23] Notice that this formula is valid not only for two atoms, but for any pair of non-overlapping objects.

[24] The electric dipole operator in a finite system is defined as $\hat{\vec{d}}(\vec{r}) = -e \sum_i \hat{\vec{r}}_i \delta(\vec{r} - \hat{\vec{r}}_i)$.

Substituting Eqs. (7.118) and (7.119) into Eq. (7.116) and integrating by parts we get

$$
E^{(2)}[n] = -\frac{\hbar}{2\pi} \int_{V_1} d\vec{r}_1 \int_{V_2} d\vec{r}_2 \frac{6}{|\vec{r}_1 - \vec{r}_2|^6} \int_0^\infty \alpha_1(\vec{r}_1, i\omega)\alpha_2(\vec{r}_2, i\omega) d\omega
$$

$$
= -\frac{3\hbar e}{2(4\pi)^{3/2} m^{1/2}} \int_{V_1} d\vec{r}_1 \int_{V_2} d\vec{r}_2 \frac{1}{|\vec{r}_1 - \vec{r}_2|^6} \frac{\sqrt{n_1(\vec{r}_1) n_2(\vec{r}_2)}}{\sqrt{n_1(\vec{r}_1)} + \sqrt{n_2(\vec{r}_2)}} , \quad (7.120)
$$

which is the desired functional of the density. In the limit of large separation, the interaction energy reduces to $E^{(2)} = -\frac{C_6}{R^6}$, where

$$
C_6 = \frac{3\hbar}{\pi} \int_0^\infty \bar{\alpha}_1(i\omega)\bar{\alpha}_2(i\omega) d\omega , \quad (7.121)
$$

and $\bar{\alpha}(\omega) = \int d\vec{r} \alpha(\vec{r}, \omega)$ is the total polarizability of the system.

Unfortunately, a straightforward substitution of Eq. (7.119) in Eq. (7.121) usually results in an overestimation of the force constant C_6 between atomic systems, with the largest contribution coming from the low density regions at the outskirts of the atom, where the homogeneous electron gas model for the polarizability is least plausible. This problem was recognized by Rapcewitz and Ashcroft (1991), who proposed as a cure to exclude from the spatial integration the regions in which the local wave vector $k(\vec{r}) = \frac{|\vec{\nabla} n(\vec{r})|}{n(\vec{r})}$ is larger than the local Thomas-Fermi wave vector $\frac{\omega_p(\vec{r})}{v_F(\vec{r})}$ (this criterion is similar to the criterion for the onset of Landau damping on plasma waves). With the help of this somewhat arbitrary cutoff procedure the theory turns out to be remarkably accurate in practice. Fig. 7.8 shows numerical results for C_6 based on Eq. (7.121) plotted vs the corresponding results of time-consuming *ab initio* calculations. The agreement is quite impressive. It must be borne in mind, however, that this simple theory applies only to systems consisting of two well separated parts. Efforts to extend it to systems with overlapping parts are currently underway.

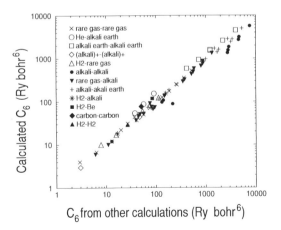

Fig. 7.8. Van der Waals coefficients C_6 in Ry a_B^6 calculated from Eq. (7.121), plotted against corresponding values from first-principles calculations. From Andersson *et al.*, 1996.

7.4 Current density functional theory

In this section we describe the extension of DFT to *nonrelativistic* electronic systems in the presence of strong magnetic fields. This theory is known as *current density functional theory* (CDFT) (Vignale and Rasolt, 1987), and its peculiarity stems from the fact that the external vector potential $\vec{A}(\vec{r})$ (related to the external magnetic field by $\vec{B}(\vec{r}) = \vec{\nabla} \times \vec{A}(\vec{r})$) enters the nonrelativistic hamiltonian with both linear and quadratic terms:[25]

$$\hat{H} = \sum_i \frac{1}{2m}\left[\left(\hat{\vec{p}}_i + \frac{e}{c}\vec{A}(\hat{\vec{r}}_i)\right)^2 + V(\hat{\vec{r}}_i)\right] + \hat{H}_{e-e} \ . \tag{7.122}$$

The nonrelativistic CDFT is particularly suitable for applications to condensed matter systems, since the hamiltonian (7.122) leads directly to Landau levels (see Section 10.2). The theory has proved useful in calculating the properties of disparate systems such as (i) two-dimensional quantum dots in a magnetic field (Ferconi and Vignale, 1994; Steffens, Suhrke, and Rössler, 1998; Reimann and Manninen, 2002), (ii) inhomogeneous quantum Hall liquids (Ferconi, Geller, and Vignale, 1995), (iii) the Wigner crystal at high magnetic field (Vignale, 1994), (iv) ferromagnetic metals (Ebert, Battocletti, and Gross, 1997).

A fundamental property of (7.122) is its gauge-invariance. If we change $\vec{A}(\vec{r})$ to $\vec{A}(\vec{r}) + \vec{\nabla}\Lambda(\vec{r})$, where $\Lambda(\vec{r})$ is an arbitrary smooth function of position, and simultaneously multiply the wave function by the phase factor $\exp[-\frac{ie}{\hbar c}\sum_i \Lambda(\vec{r}_i)]$, then all the *physically observable properties* of the system remain invariant. This is intuitively true, since the transformation does not change the value of the magnetic field.

It is convenient for our purposes to rewrite the hamiltonian as follows:

$$\hat{H} = \hat{T} + \hat{H}_{e-e} + \frac{e}{c}\int \hat{\vec{j}}_p(\vec{r}) \cdot \vec{A}(\vec{r})d\vec{r} + \int \hat{n}(\vec{r})\left[V(\vec{r}) + \frac{e^2}{2mc^2}A^2(\vec{r})\right]d\vec{r} \ , \tag{7.123}$$

where $\hat{T}$ is the "canonical" kinetic energy, defined in terms of the canonical momentum operators, $\hat{p}_i$, as $\hat{T} = \sum_i \frac{\hat{p}_i^2}{2m}$, and $\hat{\vec{j}}_p(\vec{r})$ is the paramagnetic current operator, defined by Eq. (3.166). It must be immediately noted that $\hat{T}$ is *not* the physical kinetic energy – the latter being defined as $\hat{T}_{kin} = \sum_i \frac{m\hat{v}_i^2}{2}$, with $\hat{v}_i = \frac{1}{m}\left(\hat{\vec{p}}_i + \frac{e}{c}\vec{A}(\hat{\vec{r}}_i)\right)$ – and, similarly, $\hat{\vec{j}}_p(\vec{r})$ differs from the physical current density

$$\hat{\vec{j}}(\vec{r}) = \hat{\vec{j}}_p(\vec{r}) + \frac{e}{mc}\hat{n}(\vec{r})\vec{A}(\vec{r}) \ . \tag{7.124}$$

In fact both $\hat{T}$ and $\hat{\vec{j}}_p(\vec{r})$ are non-gauge-invariant quantities: so, while the complete hamiltonian $\hat{H}$ is gauge-invariant, its separate parts are not.

The main idea of CDFT is to use the expectation values of the density and the paramagnetic current density as basic variables. From a purely mathematical point of view the choice of $\vec{j}_p$ rather than $\vec{j}$ is natural, since $\vec{j}_p$ is an intrinsic property of the system, which can be calculated without reference to the external vector potential. Furthermore, one can easily prove the

---

[25] Curiously, the *relativistic* current-DFT (Rajagopal and Callaway, 1973) is more in line with "normal" generalizations of DFT, since in the relativistic hamiltonian the coupling between the four-current density and the external four-potential is linear.

following generalization of the Hohenberg–Kohn theorem: the ground-state values of n and $\vec{j}_p$ uniquely determine the ground-state wave function.[26] There is also a more physical reason for working with $\vec{j}_p$ rather than $\vec{j}$. We have seen that the static current–current response function of the homogeneous electron liquid vanishes at $q = 0$. Thus, the condition for the existence of a gradient expansion discussed at the end of Section 7.3.3 is violated. By contrast, the static response function for the paramagnetic current density has a finite limit for $q \to 0$ and this opens the possibility of a gradient expansion for the ground-state energy as a functional of $\vec{j}_p$.

7.4.1 The vorticity variable

Having established that n and $\vec{j}_p$ are the appropriate variables in CDFT, the challenge is to construct a gauge-invariant theory from non-gauge-invariant variables. Let us define the universal functional $F[n(\vec{r}), \vec{j}_p(\vec{r})]$ as

$$F[n, \vec{j}_p] = \min_{\psi \to \{n(\vec{r}), \vec{j}_p(\vec{r})\}} \langle \psi | \hat{T} + \hat{H}_{e-e} | \psi \rangle , \qquad (7.125)$$

where the minimum is to be found among the wave functions that yield the prescribed densities $n(\vec{r})$ and $\vec{j}_p(\vec{r})$. Minimization of the energy functional

$$E_{V,\vec{A}}[n, \vec{j}_p] = F[n, \vec{j}_p] + \frac{e}{c} \int \vec{j}_p(\vec{r}) \cdot \vec{A}(\vec{r}) d\vec{r}$$
$$+ \int n(\vec{r}) \left[V(\vec{r}) + \frac{e^2}{2mc^2} A^2(\vec{r}) \right] d\vec{r} \qquad (7.126)$$

with respect to $n(\vec{r})$ and $\vec{j}_p(\vec{r})$ must yield ground-state densities $n_0(\vec{r})$ and $\vec{j}_{p0}(\vec{r})$ with the correct gauge-transformation properties. Thus, for $\vec{A}(\vec{r}) \to \vec{A}(\vec{r}) + \vec{\nabla}\Lambda(\vec{r})$ we must have

$$n_0(\vec{r}) \to n_0(\vec{r})$$
$$\vec{j}_{p0}(\vec{r}) \to \vec{j}_{p0}(\vec{r}) - \frac{e}{mc} n_0(\vec{r}) \vec{\nabla}\Lambda(\vec{r}) . \qquad (7.127)$$

The wave function $\psi_0[n(\vec{r}), \vec{j}_p(\vec{r})]$, which yields $n(\vec{r})$ and $\vec{j}_p(\vec{r})$ and minimizes $\hat{T} + \hat{H}_{e-e}$, has the transformation property

$$\psi_0[n, \vec{j}_p - \frac{e}{mc} n\vec{\nabla}\Lambda] = e^{-\frac{ie}{\hbar c}\sum_i \Lambda(\vec{r}_i)} \psi_0[n, \vec{j}_p] , \qquad (7.128)$$

hence we see that

$$F[n, \vec{j}_p - \frac{e}{mc}n\vec{\nabla}\Lambda] = F[n, \vec{j}_p] - \frac{e}{c} \int \vec{j}_p(\vec{r}) \cdot \vec{\nabla}\Lambda(\vec{r}) d\vec{r}$$
$$+ \frac{e^2}{2mc^2} \int n(\vec{r}) |\vec{\nabla}\Lambda(\vec{r})|^2 d\vec{r}. \qquad (7.129)$$

[26] This statement would be false for the variables n and $\vec{j}$, for, if $\psi(\vec{r}_1, \ldots, \vec{r}_N)$ is the ground-state wave function in the presence of a vector potential $\vec{A}$, then $\psi'(\vec{r}_1, \ldots, \vec{r}_N) = \psi(\vec{r}_1, \ldots, \vec{r}_N)\exp[-\frac{ie}{\hbar c}\sum_i \Lambda(\vec{r}_i)]$ is a *different* ground-state wave function (with external vector potential $\vec{A}' = \vec{A} + \vec{\nabla}\Lambda$) which yields the same values of n and $\vec{j}$. This implies that the expectation values of non-gauge-invariant operators, such as $\hat{T}$ and $\hat{\vec{j}}_p$ are not uniquely determined by n and $\vec{j}$.

This property guarantees that the gauge-covariance requirements of Eq. (7.127) are satisfied: it is therefore essential to make sure that any approximation to $F[n, \vec{j}_p]$ respects it.

Let us now examine the implications of Eq. (7.129) for the xc energy functional $E_{xc}[n, \vec{j}_p] \equiv F[n, \vec{j}_p] - T_s[n, \vec{j}_p] - E_H[n]$. First of all, notice that the non-interacting functional $T_s[n(\vec{r}), \vec{j}_p(\vec{r})]$ transforms exactly as $F[n(\vec{r}), \vec{j}_p(\vec{r})]$, since Eq. (7.129) does not involve the electron-electron interaction. Thus, by considering the transformation of $F - T_s - E_H$ we arrive at the important result

$$E_{xc}[n, \vec{j}_p - \frac{e}{mc}n\vec{\nabla}\Lambda] = E_{xc}[n, \vec{j}_p] . \tag{7.130}$$

Since $\Lambda(\vec{r})$ is completely arbitrary, Eq. (7.130) implies that E_{xc} must be a functional of the *vorticity* $\vec{v}(\vec{r})$ – a gauge-invariant variable defined as

$$\vec{v}(\vec{r}) \equiv \vec{\nabla} \times \left(\frac{\vec{j}_p(\vec{r})}{n(\vec{r})} \right) . \tag{7.131}$$

We conclude that

$$E_{xc}[n, \vec{j}_p] = \bar{E}_{xc}[n, \vec{v}] , \tag{7.132}$$

where $\bar{E}_{xc}$ is a functional of the vorticity.

It is not difficult to show that the vorticity of a homogeneous electron liquid subjected to a uniform magnetic field $\vec{B}$ is given by $-\frac{e\vec{B}}{mc}$ (see Section 7.4.4). In an inhomogeneous liquid, Eq. (7.132) tells us that $\vec{v}(\vec{r})$ is the natural variable in terms of which the xc energy functional should be expressed.

7.4.2 The Kohn–Sham equation

The Kohn–Sham equation is obtained, as in ordinary DFT, from the minimization of the energy functional (7.126). The stationarity conditions are

$$\left.\frac{\delta T_s[n, \vec{j}_p]}{\delta n(\vec{r})}\right|_{\vec{j}_p} = -V(\vec{r}) - V_H(\vec{r}) - \frac{e^2}{2mc^2}A^2(\vec{r}) - \left.\frac{\delta \bar{E}_{xc}[n, \vec{v}]}{\delta n(\vec{r})}\right|_{\vec{j}_p} \tag{7.133}$$

and

$$\left.\frac{\delta T_s[n, \vec{j}_p]}{\delta \vec{j}_p(\vec{r})}\right|_{n} = -\frac{e}{c}\vec{A}(\vec{r}) - \left.\frac{\delta \bar{E}_{xc}[n, \vec{v}]}{\delta \vec{j}_p(\vec{r})}\right|_{n} . \tag{7.134}$$

We seek effective potentials $V_{KS}(\vec{r})$ and $\vec{A}_{KS}(\vec{r})$ that yield the correct ground-state density and current via the solution of the single-particle equation

$$\left[\frac{1}{2m}\left(-i\hbar\vec{\nabla} + \frac{e}{c}\vec{A}_{KS}(\vec{r})\right)^2 + V_{KS}(\vec{r}) \right]\phi_\alpha(\vec{r}, \sigma) = \varepsilon_\alpha\phi_\alpha(\vec{r}, \sigma), \tag{7.135}$$

where $n(\vec{r}) = \sum_{\alpha=1,\sigma}^{N} |\phi_\alpha(\vec{r}, \sigma)|^2$ and $\vec{j}_p(\vec{r}) = \frac{\hbar}{m}\Im m \sum_{\alpha=1,\sigma}^{N} \phi_\alpha^*(\vec{r}, \sigma)\vec{\nabla}\phi_\alpha(\vec{r}, \sigma)$.

Eq. (7.135) implies that the KS potentials are related to the functional derivatives of T_s in the following manner:

$$\frac{\delta T_s[n, \vec{j}_p]}{\delta n(\vec{r})}\bigg|_{\vec{j}_p} = -V_{KS}(\vec{r}) - \frac{e^2}{2mc^2}A_{KS}^2(\vec{r}) \tag{7.136}$$

and

$$\frac{\delta T_s[n, \vec{j}_p]}{\delta \vec{j}_p(\vec{r})}\bigg|_{n} = -\frac{e}{c}\vec{A}_{xc}(\vec{r}) . \tag{7.137}$$

The desired expressions for the KS potentials are obtained from the comparison of Eqs. (7.136) and (7.137) with Eqs. (7.133) and (7.134). This yields

$$\vec{A}_{KS}(\vec{r}) = \vec{A}(\vec{r}) + \vec{A}_{xc}(\vec{r}) , \tag{7.138}$$

and

$$V_{KS}(\vec{r}) = V(\vec{r}) + V_H(\vec{r}) + \bar{V}_{xc}(\vec{r})$$
$$- \frac{e}{c}\vec{A}_{xc}(\vec{r}) \cdot \frac{\vec{j}(\vec{r})}{n(\vec{r})} - \frac{e^2}{2mc^2}A_{xc}^2(\vec{r}) , \tag{7.139}$$

where

$$\bar{V}_{xc}(\vec{r}) \equiv \frac{\delta \bar{E}_{xc}[n, \vec{v}]}{\delta n(\vec{r})}\bigg|_{\vec{v}} \tag{7.140}$$

and

$$\frac{e}{c}\vec{A}_{xc}(\vec{r}) = \frac{1}{n(\vec{r})}\vec{\nabla} \times \frac{\delta \bar{E}_{xc}[n, \vec{v}]}{\delta \vec{v}(\vec{r})}\bigg|_{n} . \tag{7.141}$$

Notice that both $\bar{V}_{xc}$ and $\vec{A}_{xc}$ are gauge-invariant functionals of n and $\vec{v}$. The ground-state energy is still given by Eq. (7.28), with $\bar{V}_{xc}$ and $\bar{E}_{xc}$ taking the place of V_{xc} and E_{xc} respectively.

Two properties of Eq. (7.135) are especially pleasing. The first is its gauge-covariance, which follows from the gauge-invariance of $\vec{A}_{xc}$, $\bar{V}_{xc}$ and $\vec{j}$. The second is that its solutions automatically satisfy the static continuity equation $\vec{\nabla} \cdot \vec{j}(\vec{r}) = 0$ by virtue of the identity $\vec{\nabla} \cdot [n(\vec{r})\vec{A}_{xc}(\vec{r})] = 0$, which follows immediately from Eq. (7.141). This is an example of a more general rule: exact conservation laws will be satisfied by the Kohn–Sham equation if and only if the corresponding exact symmetries of the xc energy functional are respected.

7.4.3 Magnetic screening

Thus far, we have ignored the fact that the equilibrium current induced by the external magnetic field is itself the source of an additional magnetic field, which in turn modifies the ground-state current. This effect is obviously absent in a description based on the hamiltonian (7.122), which allows only electrostatic interactions between the electrons. One can treat *magnetic screening* effects exactly by including in the hamiltonian velocity-dependent electron-electron interactions. Or, perhaps more wisely, one can settle for a mean field treatment of the latter, based on simple magnetostatics. In this approach one still works with the hamiltonian (7.122), but the external vector potential $\vec{A}(\vec{r})$ is replaced by the self-consistent vector potential

$$\vec{A}_{sc}(\vec{r}) = \vec{A}(\vec{r}) - \frac{e}{c} \int \frac{\vec{j}(\vec{r}')}{|\vec{r} - \vec{r}'|} d\vec{r}' , \qquad (7.142)$$

where the second term on the right-hand side is generated by the ground-state current and must be determined self-consistently with the latter.[27]

How large is the correction to the external vector potential? To get a feeling for its order of magnitude recall that the magnetization M (magnetic moment per unit volume) induced in the homogeneous electron liquid by an external magnetic field B, is approximately $M = \chi_{orb} B$, where $\chi_{orb} = \frac{\partial^2[n\epsilon(n,B)]}{\partial B^2}$ is the orbital magnetic susceptibility (see Section 3.4.3) – a dimensionless number of the order of $\left(\frac{e^2}{\hbar c}\right)^2 \sim 10^{-4}$ in three dimensions. This suggests that magnetic screening effects are ordinarily quite small. However, interacting systems on the verge of an orbital magnetic instability, or superconductors, have anomalously large χ_{orb}: magnetic screening cannot be neglected in such systems.

7.4.4 The local density approximation

The local density approximation for the CDFT energy functional is constructed, as usual, by calculating the energy of each small volume element according to homogeneous electron gas formulas, and then integrating over space.

Let us apply this idea to the construction of the non-interacting energy functional $T_s[n, \vec{j}_p]$. In order to generate the canonical current $\vec{j}_p$ in a homogeneous electron liquid of density n, we need to subject the liquid to a fictitious vector potential

$$\vec{A}_\nu = -\frac{mc}{e} \frac{\vec{j}_p}{n} . \qquad (7.143)$$

The simplest way to see this is to recall that the current response of the homogeneous electron liquid to a uniform vector potential $\vec{A}$ vanishes (see Chapter 3): therefore, the paramagnetic current is given by $-\frac{ne}{mc}\vec{A}$. We make $\vec{A}_\nu$ a function of position by allowing $\vec{j}_p$ and n on the right-hand side of (7.143) to be slowly varying functions of $\vec{r}$. The curl of $\vec{A}_\nu(\vec{r})$

$$\vec{\nabla} \times \vec{A}_\nu(\vec{r}) \cong \vec{B}_\nu(\vec{r}) = -\frac{mc}{e}\vec{v}(r) \qquad (7.144)$$

is then the fictitious magnetic field, related to the real magnetic field via

$$\vec{B}_\nu(\vec{r}) = \vec{B}(\vec{r}) - \frac{mc}{e}\vec{\nabla} \times \left(\frac{\vec{j}(\vec{r})}{n(\vec{r})}\right) . \qquad (7.145)$$

To construct the local density approximation to the functional $T_s[n, \vec{j}_p]$ we integrate the kinetic energy density $n(\vec{r})\epsilon_0(n(\vec{r}), B_\nu(\vec{r}))$ of the non-interacting electron gas in the fictitious

---

[27] This procedure is analogous to the familiar Hartree approximation for the scalar potential.

magnetic field B_ν and subtract the energy that pertains to the interaction with the fictitious vector potential. This yields

$$T_s[n, \vec{j}_p] \simeq \int n(\vec{r}) \epsilon_0 \left(n(\vec{r}), |\vec{B}_\nu(\vec{r})| \right) d\vec{r} - \frac{e}{c} \int \vec{j}_p(\vec{r}) \cdot \vec{A}_\nu(\vec{r}) d\vec{r}$$
$$- \frac{e^2}{2mc^2} \int n(\vec{r}) A_\nu^2(\vec{r}) d\vec{r} . \tag{7.146}$$

Similarly, the local density approximation to the xc energy functional is constructed as the integral of the xc energy density of the homogeneous electron liquid in the fictitious magnetic field

$$\bar{E}_{xc}[n, \nu] \simeq \int n(\vec{r}) \epsilon_{xc}(n(\vec{r}), B_\nu(\vec{r})) d\vec{r} . \tag{7.147}$$

Collecting all the above, we obtain the local approximation to the energy functional:

$$E_{VA}[n, \vec{j}_p] = \int n(\vec{r}) \epsilon \left(n(\vec{r}), |\vec{B}_\nu(\vec{r})| \right) d\vec{r} + \frac{m}{2} \int \frac{|j(\vec{r})|^2}{n(\vec{r})} d\vec{r}$$
$$+ E_H \left(n(\vec{r}) \right) + \int n(\vec{r}) V(\vec{r}) d\vec{r} , \tag{7.148}$$

where $\epsilon(n, B)$ is the total ground-state energy per particle of the homogeneous electron liquid in a uniform magnetic field B. The second term on the right-hand side represents the kinetic energy associated with the macroscopic motion of the electrons with velocity $\vec{v}(\vec{r}) = \frac{\vec{j}(\vec{r})}{n(\vec{r})}$.

The gauge-invariant functional (7.148) (Vignale, Skudlarski, and Rasolt, 1992) generalizes the Thomas-Fermi approximation of ordinary DFT. When used in the variational principle it yields Thomas-Fermi-like equations for the ground-state density and current.

The problem of calculating the ground-state energy per particle $\epsilon(n, B)$ (or, rather, its xc component $\epsilon_{xc}(n, B)$) as a function of density and magnetic field is discussed in Chapter 10 and in Appendix 22. In the limit of weak magnetic field perturbation theory gives

$$\epsilon(n, B) \simeq \epsilon(n, 0) - \frac{1}{2} \chi_{orb}(n) B^2 , \tag{7.149}$$

where $\chi_{orb}(n)$ is the orbital magnetic susceptibility per particle of the homogeneous electron gas at $B = 0$ and density n. Some values of $\chi_{orb}(n)$ for the interacting three-dimensional electron liquid are listed in Table 7.6.

Table 7.6. *Orbital magnetic susceptibility of the homogeneous electron gas in 3D. (From Vignale, Rasolt, and Geldart (1988).)*

r_s	1	2	3	4	5	6
$\chi_{orb}/\chi_{orb}^{(0)}$	0.981	0.970	0.957	0.942	0.926	0.909

7.5 Time-dependent density functional theory

The success of the density functional theory for ground-state properties has prompted its extension to time-dependent systems. A very attractive feature of the time-dependent density functional theory (TDDFT) is that it allows a first-principles calculation of excitation energies – a target not easily reached within conventional DFT. TDDFT is presently quite an active area of research: in the following sections we will present only the basic ideas and the results that can be considered well established at this time.

7.5.1 The Runge–Gross theorem

The foundation theorem of TDDFT was proved by Runge and Gross (Runge and Gross, 1984). In analogy to the Hohenberg–Kohn theorem of static DFT, this theorem proves the existence of a mapping from time-dependent densities to time-dependent potentials, but there are several differences in the detail.

Let us consider a time-dependent hamiltonian of the form

$$\hat{H}(t) = \hat{T} + \hat{H}_{e-e} + \hat{V}(t), \tag{7.150}$$

where

$$\hat{V}(t) = \int V(\vec{r}, t)\hat{n}(\vec{r})d\vec{r} \tag{7.151}$$

is a time-dependent local potential, and $\hat{T}$ and $\hat{H}_{e-e}$ are, as usual, the kinetic energy and the electron-electron interaction operators. The Runge–Gross theorem can be stated as follows:

Runge–Gross theorem: *For systems described by hamiltonians of the form (7.150), the densities $n(\vec{r}, t)$ and $n'(\vec{r}, t)$ which evolve from a common initial state $|\psi_0\rangle$ under the action of two local potentials $V(\vec{r}, t)$ and $V'(\vec{r}, t)$ that are expandable in Taylor series about the initial time t_0, are necessarily different, provided the potentials differ by more than a purely time-dependent ($\vec{r}$-independent) function, that is*

$$V(\vec{r}, t) \neq V'(\vec{r}, t) + c(t). \tag{7.152}$$

The last condition is easy to understand: if V and V' differed by a purely time-dependent constant, then the corresponding wave functions $|\psi(t)\rangle$ and $|\psi'(t)\rangle$ would differ by a purely time-dependent phase factor, and, consequently, the densities n and n' would be identical. Thus, two truly different potentials must satisfy Eq. (7.152) in much the same way as, in the ground-state formalism, two truly different potentials must differ by more than a constant.

The Runge–Gross theorem depends on two delicate hypotheses, which have no analogue in the ground-state theory. The first one requires that the two densities evolve from the same initial state: this leaves open the possibility that two truly different potentials might yield the same density starting from two different initial states. This possibility can be eliminated by adding the requirement that the system be initially in the ground-state, since, according to the Hohenberg–Kohn theorem, the initial density does then uniquely determine the initial state.[28]

Even more bothersome is the second hypothesis, requiring that the potentials be expandable in Taylor series in a neighborhood of the initial time t_0. This appears to exclude one of the most important cases, namely, the adiabatically switched-on potential, which is the cornerstone of linear response theory: the switching-on function $e^{\eta t}$ cannot be Taylor-expanded about $t = -\infty$! Indeed, it was shown by Oliveira and Kohn (1989) that there are physical systems in which the density–density response function is not invertible at certain frequencies. This implies the existence, for these systems, of infinitely many adiabatically switched-on potentials that yield the same density starting from the same ground-state. Fortunately, the difficulty is more apparent than real. Physically, the long-term response to a periodic perturbation should not depend on the detailed from of the adiabatic switching-on procedure. Thus, we can start the switching-on process at a large but finite negative time t_0 where the hypotheses of the Runge–Gross theorem hold: the only restriction being that the switching-on parameter η must remain much larger than $1/|t_0|$ – not a serious limitation if $|t_0|$ is large. Interestingly, the phenomenon of non-invertibility of the density–density response function, noted by Oliveira and Kohn, disappears if the infinitesimal η is kept finite, or, equivalently, if the system has a continuous, as opposed to discrete, excitation spectrum (see Ng and Singwi, 1987).

Proof of the Runge–Gross theorem:[29] One way to prove the Runge–Gross theorem is to show that, under the stated assumptions, two potentials V and V' that yield the same density $n(\vec{r}, t)$ can differ, at the most, by a time-dependent constant $c(t)$. Because $n(\vec{r}, t)$ arises from the time evolution of both hamiltonians $\hat{H}$ (with potential V) and $\hat{H}'$ (with potential V'), it must simultaneously satify the *two* continuity equations

$$\frac{\partial n(\vec{r}, t)}{\partial t} = -\vec{\nabla} \cdot \vec{j}(\vec{r}, t) \tag{7.153}$$

and

$$\frac{\partial n(\vec{r}, t)}{\partial t} = -\vec{\nabla} \cdot \vec{j}'(\vec{r}, t) , \tag{7.154}$$

where $\vec{j}(\vec{r}, t)$ and $\vec{j}'(\vec{r}, t)$ are the current densities arising from the time evolutions under $\hat{H}$ and $\hat{H}'$ respectively. Taking the difference of these two equations we find that

$$\vec{\nabla} \cdot [\vec{j}'(\vec{r}, t) - \vec{j}(\vec{r}, t)] = 0 , \tag{7.155}$$

implying that the longitudinal components of the current densities in the two systems coincide.

---

[28] An explicit example of two different potentials that produce the same density starting from two different initial states has been worked out by Maitra and Burke (2001), and is presented in Exercise 16.

[29] The proof that follows is due to Robert van Leeuwen (1999).

We next consider the equation of motion for the current density. We have

$$\frac{\partial \vec{j}(\vec{r}, t)}{\partial t} = -\frac{i}{\hbar} \langle \psi(t) | [\hat{\vec{j}}(\vec{r}), \hat{H}(t)] | \psi(t) \rangle \tag{7.156}$$

and

$$\frac{\partial \vec{j}'(\vec{r}, t)}{\partial t} = -\frac{i}{\hbar} \langle \psi'(t) | [\hat{\vec{j}}(\vec{r}), \hat{H}'(t)] | \psi'(t) \rangle , \tag{7.157}$$

where $|\psi(t)\rangle$ and $|\psi'(t)\rangle$ are the states that evolve from the common initial state $|\psi_0\rangle$ under $\hat{H}$ and $\hat{H}'$ respectively. Evaluating the commutator between $\hat{\vec{j}}$ and $\hat{H}$ we obtain

$$[\hat{\vec{j}}(\vec{r}), \hat{H}(t)] = -\hat{n}(\vec{r})\vec{\nabla}V(\vec{r}, t) + \hat{\vec{Q}}(r), \tag{7.158}$$

where the first term on the right-hand side arises from the commutator of $\hat{\vec{j}}(\vec{r})$ and $\hat{V}(t)$, while the second term, which we denote by $\hat{\vec{Q}}(\vec{r})$ (and need not specify further) arises from the commutator of $\hat{\vec{j}}(\vec{r})$ with $\hat{T} + \hat{H}_{e-e}$. A similar expression is obtained for the commutator $[\hat{\vec{j}}(\vec{r}), \hat{H}'(t)]$, the only difference being the replacement of V by V'. Substituting these expressions in Eqs. (7.156)–(7.157) and taking the difference between them, we arrive at

$$\frac{\partial}{\partial t}[\vec{j}'(\vec{r}, t) - \vec{j}(\vec{r}, t)] = -n(\vec{r}, t)\vec{\nabla}[V'(\vec{r}, t) - V(\vec{r}, t)] + \vec{Q}'(\vec{r}, t) - \vec{Q}(\vec{r}, t) , \tag{7.159}$$

where $\vec{Q}(\vec{r}, t)$ and $\vec{Q}'(\vec{r}, t)$ are the expectation values of $\hat{\vec{Q}}(\vec{r})$ in $|\psi(t)\rangle$ and $|\psi'(t)\rangle$ respectively.

The divergence of the left-hand side of this equation vanishes by virtue of the continuity condition (7.155): this implies that

$$\vec{\nabla} \cdot \{n(\vec{r}, t)\vec{\nabla}[V'(\vec{r}, t) - V(\vec{r}, t)]\} = \zeta(\vec{r}, t), \tag{7.160}$$

where $\zeta(\vec{r}, t) \equiv \vec{\nabla} \cdot [\vec{Q}'(\vec{r}, t) - \vec{Q}(\vec{r}, t)]$.

It will now be proved that Eq. (7.160) has the unique solution $V'(\vec{r}, t) - V(\vec{r}, t) = c(t)$ under the assumption that both V and V' are "properly behaved" at infinity (this will be made precise in what follows). The proof is based on the fact that the simpler equation

$$\vec{\nabla} \cdot \{n_0(\vec{r})\vec{\nabla}w(\vec{r})\} = 0 , \tag{7.161}$$

where $n_0(\vec{r})$ is the density at $t = t_0$, has the unique solution $w(\vec{r}) = $ *constant in space*.

Before proving the last assertion, however, let us show how it is used to complete the proof of the Runge–Gross theorem. At $t = t_0$ the right-hand side of Eq. (7.160) vanishes, because by hypothesis $|\psi\rangle = |\psi'\rangle = |\psi_0\rangle$ at the initial time. Then, Eq. (7.161) tells us that $V'(\vec{r}, t_0) - V(\vec{r}, t_0)$ is a constant, independent of $\vec{r}$. We now expand both sides of Eq. (7.160) in a Taylor series about t_0, that is, we write $V(\vec{r}, t) = \sum_{k=0}^{\infty} V_k(\vec{r})(t - t_0)^k$, and similarly expand $n(\vec{r}, t)$ and $\zeta(\vec{r}, t)$. Equating the k-th coefficients ($k \geq 0$) of the expansion on both sides we get

$$\vec{\nabla} \cdot \left\{ \sum_{l=0}^{k} n_{k-l}(\vec{r}) \vec{\nabla}[V_l'(\vec{r}) - V_l(\vec{r})] \right\} = \zeta_k(\vec{r}), \qquad (7.162)$$

where the subscript k denotes the k-th coefficient in the Taylor expansion with respect to time. For $k \geq 1$ let us rewrite Eq. (7.162) as

$$\vec{\nabla} \cdot \{n_0(\vec{r}) \vec{\nabla}[V_k'(\vec{r}) - V_k(\vec{r})]\} = \zeta_k(\vec{r}) - \vec{\nabla} \cdot \left\{ \sum_{l=0}^{k-1} n_{k-l}(\vec{r}) \vec{\nabla}[V_l'(\vec{r}) - V_l(\vec{r})] \right\}, \qquad (7.163)$$

and observe that this equation can be read as a recursion relation connecting $V_k'(\vec{r}) - V_k(\vec{r})$ to the set of $V_l'(\vec{r}) - V_l(\vec{r})$ with $l < k$. The reason why this is so is that the Schrödinger equation $i\hbar \frac{\partial |\psi(t)\rangle}{\partial t} = [\hat{T} + \hat{H}_{e-e} + \hat{V}(t)]|\psi(t)\rangle$ allows us to express the k-th coefficient in the Taylor expansion of $|\psi(t)\rangle$ in terms of coefficients of order $l < k$ in the Taylor expansion of the potential. Hence the quantity $\zeta_k = \vec{\nabla} \cdot [\vec{Q}_k' - \vec{Q}_k]$, with

$$\vec{Q}_k = \frac{1}{k!} \frac{\partial^k}{\partial t^k} \left(\langle \psi(t) | \hat{\vec{Q}}(\vec{r}) | \psi(t) \rangle \right)_{t=t_0}, \qquad (7.164)$$

is completely determined by the differences $V_l'(\vec{r}) - V_l(\vec{r})$ with $l < k$, and vanishes if the latter are constants independent of $\vec{r}$. The difference of the $k = 0$ coefficients $V_0'(\vec{r}) - V_0(\vec{r})$ is, of course, $V'(\vec{r}, t_0) - V(\vec{r}, t_0)$, which we have already established to be a constant independent of $\vec{r}$. Furthermore, it follows from the recursion relation (7.163) that the constancy of the differences $V_l'(\vec{r}) - V_l(\vec{r})$ with $l < k$ implies the constancy of $V_k'(\vec{r}) - V_k(\vec{r})$. By the principle of mathematical induction we can therefore conclude that $V_k'(\vec{r}) - V_k(\vec{r})$ is independent of $\vec{r}$ for all values of k, i.e., $V'(\vec{r}, t) - V(\vec{r}, t)$ is a mere function of time: this is precisely the thesis of the Runge–Gross theorem.

In order to complete the proof, we must still show that the equation

$$\vec{\nabla} \cdot \{n_0(\vec{r}) \vec{\nabla} w(\vec{r})\} = 0, \qquad (7.165)$$

with $n_0(\vec{r}) \geq 0$, has the unique solution $w(\vec{r}) = constant$ subject to appropriate boundary conditions at infinity. To this end, consider the integral

$$W = \int n_0(\vec{r}) |\vec{\nabla} w(\vec{r})|^2 d\vec{r}. \qquad (7.166)$$

Because $n_0(\vec{r})$ is non-negative, W attains the minimum value, zero, when and only when $w(\vec{r})$ is constant in space. On the other hand, W can be rewritten, with the help of Green's theorem, as

$$W = -\int w(\vec{r}) \vec{\nabla} \cdot [n_0(\vec{r}) \vec{\nabla} w(\vec{r})] d\vec{r} + \int_S [n_0(\vec{r}) w(\vec{r}) \vec{\nabla} w(\vec{r})] \cdot d\mathbf{S}, \qquad (7.167)$$

where, in the second term, the integral is done on a surface at infinity. We now impose the following boundary condition: *the behavior of $w(\vec{r})$ at infinity must be such that the surface integral in Eq. (7.167) vanishes.* If $w(\vec{r})$ is a solution of Eq. (7.165) that satisfies this boundary condition, then Eq. (7.167) implies $W = 0$. But we have already seen that $W = 0$ if and only if w is constant: hence, the only admissible solution of Eq. (7.165) is $w = $ constant.

How restrictive is the boundary condition on $w(\vec{r})$? This depends on the nature of the density distribution. In any finite system $n_0(\vec{r})$ vanishes exponentially at infinity: hence the boundary condition is satisfied by every bounded function of $\vec{r}$. Infinite systems are more delicate, but one can verify that the theorem goes through with the standard choice of periodic boundary conditions.[30]

The same methodology can be applied to study more general questions, for which the answer is not so simple. For example, van Leeuwen (1999) has been able to prove an important generalization of the Runge–Gross theorem, which strengthens the foundations of the time-dependent Kohn–Sham theory, described in the next section. The van Leeuwen theorem can be stated as follows:

van Leeuwen theorem: *Let $\hat{H}(t)$ and $\hat{H}'(t)$ be two hamiltonians of the form (7.150) containing not only two different time-dependent local potentials V and V' but also two different interactions $\hat{H}_{e-e}$ and $\hat{H}'_{e-e}$. Let $n(\vec{r}, t)$ be the density that evolves from the initial state $|\psi_0\rangle$ under the hamiltonian $\hat{H}(t)$, and let $|\psi'_0\rangle$ be another initial state with the same density and the same value of $\vec{\nabla} \cdot \vec{j}(\vec{r}, t)$.[31] Then the time-dependent density $n(\vec{r}, t)$ uniquely determines, up to a time-dependent constant, the potential $V'(\vec{r}, t)$ that yields $n(\vec{r}, t)$ starting from $|\psi'_0\rangle$ and evolving under $\hat{H}'$.*

For the details of the proof, which is similar to the proof of the Runge–Gross theorem given above, we refer the reader to the original paper by van Leeuwen (1999).

The Runge–Gross theorem is a special case of van Leeuwen's theorem, and can be easily recovered from the latter by setting $|\psi'_0\rangle = |\psi_0\rangle$ and $\hat{H}_{e-e} = \hat{H}'_{e-e}$. The real value of van Leeuwen's theorem, however, is appreciated when one sets $H'_{e-e} = 0$, for, in that case, the theorem asserts that there is one and only one potential $V'(\vec{r}, t)$ that yields, *in a non-interacting system*, the true time-dependent density of the physical system with interaction $\hat{H}_{e-e}$ and potential $V(\vec{r}, t)$. Another interesting application of van Leeuwen's theorem is presented in Exercise 16.

7.5.2 The time-dependent Kohn–Sham equation

Let us assume that our time-dependent system is initially in its ground-state,[32] and that this ground-state is properly described by the Kohn–Sham equation of Section 7.2.3. Our objective in this section is to generalize the KS equation so that it yields the time evolution of the density.

---

[30] In this case the densities and the potentials become periodic in space, with a period equal to the sample size; the integral over space in Eq. (7.166) is restricted to the volume of the sample, and the surface integral in Eq. (7.167) becomes the integral over the surface of the sample.

[31] $|\psi'_0\rangle$ may coincide with $|\psi_0\rangle$; neither $|\psi_0\rangle$ nor $|\psi'_0\rangle$ needs be the ground-state.

[32] Although not strictly necessary, this assumption will allow us to introduce the time-dependent Kohn–Sham potential as a functional of the density only.

According to van Leeuwen's theorem, one can construct a local potential $V_{KS}(\vec{r}, t)$ that yields, in a non-interacting system, the true density $n(\vec{r}, t)$ of the interacting system. This potential is a unique functional of $n(\vec{r}, t)$ (up to an additive time-dependent constant) and can be written as

$$V_{KS}(\vec{r}, t) = V(\vec{r}, t) + V_H(\vec{r}, t) + V_{xc}(\vec{r}, t) , \qquad (7.168)$$

where all three components (external, Hartree, and exchange-correlation) are functions of time.[33] The time-dependent Kohn–Sham equation then takes the form

$$\left\{ i\hbar \frac{\partial}{\partial t} + \frac{\hbar^2}{2m} \nabla^2 - V_{KS}(\vec{r}, t) \right\} \phi_\alpha(\vec{r}, \sigma, t) = 0 , \qquad (7.169)$$

from which the density can be calculated, as in Eq. (7.26), by adding up the contributions of the N orbitals that are occupied at the initial time. As in the static case, we note that the "Kohn–Sham wave function" – i.e., the Slater determinant constructed from the N selected solutions of Eq. (7.169) – is just a device to produce the correct density and should not be regarded as an approximation to the true many-body wave function. Although both the static and the time-dependent KS potential must be determined self-consistently with the density, there is an important difference between the practical implementation of this self-consistency in the two cases. In the static case, V_{KS} determines, via the KS orbitals, a ground-state density, which is used to recalculate V_{KS} until self-consistency is achieved. In the time-dependent theory, the initial density determines the initial potential which is then used to recalculate the density at an infinitesimally later time, and so on. In other words, the numerical solution of the time-dependent KS equation by incremental steps in time plays the same role as the self-consistency loop of the static theory.

In the static DFT, the xc potential is expressed as the functional derivative of the xc energy functional with respect to the density. Unfortunately, the situation is not nearly as simple in the time-dependent DFT. It was initially hoped that one could write

$$V_{xc}(\vec{r}, t) = \frac{\delta \mathcal{A}_{xc}[n]}{\delta n(\vec{r}, t)} , \qquad (7.170)$$

where $\mathcal{A}_{xc}[n]$ should have been the xc part of an "action functional", defined, in analogy to $F[n]$ as

$$\mathcal{A}[n] = \langle \psi[n] | i\hbar \frac{\partial}{\partial t} - \hat{T} - \hat{H}_{e-e} | \psi[n] \rangle . \qquad (7.171)$$

But, this representation runs into a serious difficulty. It is evident from the physical principle of *causality* that the KS potential at time t can only depend on the density at earlier times $t' < t$: mathematically, this implies that the functional derivative $\frac{\delta V_{xc}(\vec{r}, t)}{\delta n(\vec{r}', t')}$ vanishes for $t' > t$. But, a potential defined through Eq. (7.170) cannot satisfy this condition, since

$$\frac{\delta V_{xc}(\vec{r}, t)}{\delta n(\vec{r}', t')} = \frac{\delta^2 \mathcal{A}_{xc}[n]}{\delta n(\vec{r}, t) \delta n(\vec{r}', t')} \qquad (7.172)$$

---

[33] In a nonrelativistic approximation, the time-dependent Hartree potential is given by the usual static expression, evaluated at the instantaneous density. This neglects retardation effects due to the finite speed of propagation of the electromagnetic field.

is symmetric under the interchange of $\vec{r}, t$ with $\vec{r}', t'$.[34] This disappointing observation indicates that we do not have as yet a complete grasp on the formal properties of the time-dependent xc potential.[35] In what follows, we will assume that van Leeuwen's theorem is a sufficient basis for the existence of the functional $V_{xc}[n](\vec{r}, t)$, and will present approximations for it based on physical considerations.

7.5.3 Adiabatic approximation

The most popular approximation for $V_{xc}(\vec{r}, t)$ is the so-called adiabatic local density approximation (ALDA), which was first introduced by Zangwill and Soven (1980) *before* the advent of formal TDDFT. In this approximation one assumes that the time-dependent xc potential $V_{xc}[n](\vec{r}, t)$ is just the static xc potential evaluated at the instantaneous density $n(\vec{r}, t)$. The static xc potential is then treated within the LDA.[36] Thus, with the same notation of Section 7.3.2, one writes

$$V_{xc}^{ALDA}(\vec{r}, t) = \left(\frac{d[n\epsilon_{xc}(n)]}{dn} \right)_{n=n(\vec{r},t)}. \tag{7.173}$$

The main characteristic of the ALDA is that it is "local in time", as well as in space. Memory effects, whereby the xc potential at a time t might depend on the density at $t' < t$, are completely ignored. Attractive features of the ALDA are the extreme simplicity, the ease of implementation, and the fact that it is not restricted to small deviations from the ground-state density, i.e., to the linear response regime.

In practical applications, the ALDA turns out to be an extremely successful approximation – far more successful than one should expect a priori. Examples of ALDA-based calculations of excitation energies in a broad variety of systems will be presented in the following sections. In the atomic case, the main weakness of the approximation stems from its failure to reproduce the exact long range behavior $-\frac{e^2}{r}$ of the xc potential. This is really a weakness of the ground-state LDA, not of the adiabatic approximation. In fact, it can be remedied by making use of better ground-state potentials, without leaving the framework of the adiabatic approximation.

The success of the ALDA in calculating excitation energies is somewhat paradoxical. Naively, one would expect the adiabatic approximation to work only for frequencies that are much lower than the characteristic excitation frequencies of the system: if this were the case, the approximation would be completely useless as a tool for calculating these very same excitation energies! Before attempting to resolve this paradox, let us take a closer look at the density functional theory of frequency-dependent linear response.

---

[34] The only case in which this contradiction does not arise is when $V_{xc}(\vec{r}, t)$ is approximated as an *instantaneous* functional of the density.

[35] Recently, van Leeuwen (van Leeuwen, 1998, 2001) has found a way out of the impasse by constructing an action functional that depends on the time-evolution of the wave function along a closed path in time – the so called Keldysh contour. A two-time kernel that is time-ordered along the path – and hence symmetric under interchange of the two times along the path – becomes retarded in real time.

[36] In the same spirit, any approximation for the ground-state potential – for instance the generalized gradient approximation of Section 7.3.4 – can be converted into an adiabatic approximation for time-dependent problems.

7.5.4 Frequency-dependent linear response

Consider the response of an electronic system, initially in the ground-state, to a periodic external potential of the form

$$V(\vec{r}, t) = V_1(\vec{r}, \omega)e^{-i\omega t} + c.c. \tag{7.174}$$

(Here and in the following a subscript "1" characterizes all the quantities that are of first order in the perturbing potential, while a subscript "0" characterizes the ground-state quantities.) As a result of the perturbation the density becomes

$$n(\vec{r}, t) = n_0(\vec{r}) + n_1(\vec{r}, \omega)e^{-i\omega t} + c.c. , \tag{7.175}$$

where $n_0(\vec{r})$ is the ground-state density. Both the Hartree and the xc potential can be expanded to first order in n_1 as

$$V_H(\vec{r}, t) = V_{H,0}(\vec{r}) + V_{H,1}(\vec{r}, \omega)e^{-i\omega t} + c.c.$$
$$V_{xc}(\vec{r}, t) = V_{xc,0}(\vec{r}) + V_{xc,1}(\vec{r}, \omega)e^{-i\omega t} + c.c. \tag{7.176}$$

The first-order correction to the ground-state Hartree potential $V_{H,0}$ is

$$V_{H,1}(\vec{r}, \omega) = \int \frac{e^2}{|\vec{r} - \vec{r}'|} n_1(\vec{r}', \omega)d\vec{r}' , \tag{7.177}$$

while the first-order correction to V_{xc} can be written in the form

$$V_{xc,1}(\vec{r}, \omega) = \int f_{xcL}(\vec{r}, \vec{r}', \omega)n_1(\vec{r}', \omega)d\vec{r}', \tag{7.178}$$

which is the most general expression for a linear functional of n_1. The complex function $f_{xcL}(\vec{r}, \vec{r}', \omega)$ – a ground-state property – is actually the generalization to inhomogeneous electron liquids of the *exchange correlation kernel* introduced in Section 5.6.2. It is evident that any approximation for f_{xcL} generates an approximation for V_{xc} and vice versa. For example, in the ALDA one has

$$f_{xcL}^{ALDA}(\vec{r}, \vec{r}', \omega) = \left(\frac{d^2[n\epsilon_{xc}(n)]}{dn^2}\right)_{n=n_0(\vec{r})} \delta(\vec{r} - \vec{r}'), \tag{7.179}$$

which is local in space, real, and frequency-independent. But, in general, $V_{xc}(\vec{r}, t)$ is a retarded functional of the density, i.e., it depends on the density distribution at times earlier than t. By the general principles discussed in Section 3.2.7, this implies that $f_{xcL}(\vec{r}, \vec{r}', \omega)$ is an analytic function of frequency in the upper half plane and satisfies the Kramers–Krönig dispersion relations.

Given the above expressions for the different components of the linearized KS potential it is an easy task to calculate the induced density. According to Eq. (7.169), n_1 is given by

$$n_1(\vec{r}, \omega) = \int \chi_{KS}(\vec{r}, \vec{r}', \omega)\left[V_1(\vec{r}', \omega) + V_{H,1}(\vec{r}', \omega) + V_{xc,1}(\vec{r}', \omega)\right]d\vec{r}' \tag{7.180}$$

where $\chi_{KS}(\vec{r}, \vec{r}', \omega)$ is the density–density response of the *static* Kohn–Sham system, i.e., the non-interacting system subjected to the *static* potential $V_0 + V_{H,0} + V_{xc,0}$. According to the general formulas of Chapter 4, χ_{KS} has the following expression in terms of the ground-state Kohn–Sham orbitals and their eigenvalues:

$$\chi_{KS}(\vec{r}, \vec{r}', \omega) = \sum_{k,j}(f_k - f_j)\frac{\phi_k^{(0)}(\vec{r})^*\phi_j^{(0)}(\vec{r})\phi_j^{(0)}(\vec{r}')^*\phi_k^{(0)}(\vec{r}')}{\omega - \omega_{jk} + i\eta}, \qquad (7.181)$$

where $\omega_{jk} \equiv \frac{\varepsilon_j - \varepsilon_k}{\hbar}$, f_j and f_k are Fermi occupation factors, and η is a positive infinitesimal. Substituting Eqs. (7.177) and (7.178) in Eq. (7.180) we obtain an integral equation for n_1:

$$n_1(\vec{r}, \omega) = \int d\vec{r}' \chi_{KS}(\vec{r}, \vec{r}', \omega)$$
$$\times \left\{ V_1(\vec{r}', \omega) + \int \left[\frac{e^2}{|\vec{r}' - \vec{r}''|} + f_{xcL}(\vec{r}', \vec{r}'', \omega) \right] n_1(\vec{r}'', \omega)d\vec{r}'' \right\}, \quad (7.182)$$

which can be solved numerically, yielding the linear density response. The mathematical relationship between the exact density–density response function $\chi_{nn}(\vec{r}, \vec{r}', \omega)$ and the Kohn–Sham response function, is compactly expressed in term of their inverses:

$$\chi_{nn}^{-1}(\vec{r}, \vec{r}', \omega) = \chi_{KS}^{-1}(\vec{r}, \vec{r}', \omega) - \frac{e^2}{|\vec{r} - \vec{r}'|} - f_{xcL}(\vec{r}, \vec{r}', \omega). \qquad (7.183)$$

Notice that the form of this relationship is characteristic of time-dependent mean field theories (see Chapter 4). In fact, setting $f_{xcL} = 0$, we see that Eq. (7.183) reduces to the RPA for the Kohn–Sham ground-state. However, in contrast to ordinary mean field theories, Eq. (7.183) gives, in principle, an *exact* representation of the response function.

7.6 The calculation of excitation energies

As discussed in Chapter 3, the excitation energies of a *finite* system are found to be in correspondence with the poles of its linear response functions along the real frequency axis. For example, the poles of the density–density response function determine the energies of excited states that are coupled to the ground-state by the density operator.[37] In an *infinite* system the situation is somewhat different, since the poles merge together to form branch cuts that describe a continuous excitation spectrum. The surviving poles (if any) occur in the lower half of the complex frequency plane and correspond to collective modes. Because of this basic difference we discuss the two cases separately.

7.6.1 Finite systems

Let us begin with the calculation of excitation energies in finite systems, such as atoms and molecules. The condition for the occurrence of a pole of the response function $\chi_{nn}(\vec{r}, \vec{r}', \omega)$

---

[37] For generic excitations one may need to consider the poles of different response functions, e.g., the spin–spin response function.

at a frequency $\omega = \omega_\nu$ can be written as

$$\int \chi_{nn}^{-1}(\vec{r}, \vec{r}', \omega_\nu) \Phi(\vec{r}') d\vec{r}' = 0, \qquad (7.184)$$

where $\Phi(\vec{r})$ is an eigenfunction of χ_{nn}^{-1} (the inverse of χ_{nn}, regarded as a matrix with indices $\vec{r}$ and $\vec{r}'$) with eigenvalue zero.

Let us verify the correctness of Eq. (7.184) for the Kohn–Sham response function. The excitation energies of the Kohn–Sham system are given by differences of Kohn–Sham eigenvalues, $\varepsilon_j - \varepsilon_k$, where $\phi_j^{(0)}$ is empty and $\phi_k^{(0)}$ is occupied in the Kohn–Sham ground-state. For ω close to ω_{jk} and far from all the other resonances, we can isolate the singular term in Eq. (7.181) as follows:

$$\chi_{KS}(\vec{r}, \vec{r}', \omega) = \frac{\Phi_{jk}^{(0)}(\vec{r}) \Phi_{jk}^{(0)}(\vec{r}')^*}{\omega - \omega_{jk} + i\eta} + \chi_{KS}^{reg}(\vec{r}, \vec{r}', \omega), \qquad (7.185)$$

where

$$\Phi_{jk}^{(0)}(\vec{r}) \equiv \phi_k^{(0)}(\vec{r})^* \phi_j^{(0)}(\vec{r}), \qquad (7.186)$$

like χ_{KS}^{reg} – the sum of all the regular terms in Eq. (7.181) – is a smooth function of ω in a neighborhood of ω_{jk}.

Eq. (7.185) can be inverted with the help of the *Sherman–Morrison formula* (see Press *et al.*, in *Numerical Recipes*, p. 65), which yields

$$\chi_{KS}^{-1} = [\chi_{KS}^{reg}]^{-1} - \frac{[\chi_{KS}^{reg}]^{-1} |\Phi_{jk}^{(0)}\rangle \langle \Phi_{jk}^{(0)}| [\chi_{KS}^{reg}]^{-1}}{\omega - \omega_{jk} + \langle \Phi_{jk}^{(0)}| [\chi_{KS}^{reg}]^{-1} |\Phi_{jk}^{(0)}\rangle}, \qquad (7.187)$$

where we have introduced the self-explanatory "ket" and "bra" notation

$$[\chi_{KS}^{reg}]^{-1} |\Phi_{jk}^{(0)}\rangle = \int [\chi_{KS}^{reg}]^{-1}(\vec{r}, \vec{r}', \omega) \Phi_{jk}^{(0)}(\vec{r}') d\vec{r}',$$

$$\langle \Phi_{jk}^{(0)}| [\chi_{KS}^{reg}]^{-1} = \int \Phi_{jk}^{(0)}(\vec{r})^* [\chi_{KS}^{reg}]^{-1}(\vec{r}, \vec{r}', \omega) d\vec{r}, \qquad (7.188)$$

and

$$\langle \Phi_{jk}^{(0)}| [\chi_{KS}^{reg}]^{-1} |\Phi_{jk}^{(0)}\rangle = \int d\vec{r} \int d\vec{r}' \Phi_{jk}^{(0)}(\vec{r})^* [\chi_{KS}^{reg}]^{-1}(\vec{r}, \vec{r}', \omega) \Phi_{jk}^{(0)}(\vec{r}'). \qquad (7.189)$$

It is now straightforward to verify that

$$\chi_{KS}^{-1} |\Phi_{jk}^{(0)}\rangle = \frac{\omega - \omega_{jk}}{\omega - \omega_{jk} + \langle \Phi_{jk}^{(0)}| [\chi_{KS}^{reg}]^{-1} |\Phi_{jk}^{(0)}\rangle} [\chi_{KS}^{reg}]^{-1} |\Phi_{jk}^{(0)}\rangle. \qquad (7.190)$$

Because the right-hand side of this equation vanishes for $\omega \to \omega_{jk}$, we see that $\Phi_{jk}^{(0)}(\vec{r})$ is indeed an eigenfunction of $\chi_{KS}^{-1}(\vec{r}, \vec{r}', \omega_{jk})$ with zero eigenvalue.

Let us now return to the problem of solving Eq. (7.184) in a physical system. By virtue of Eq. (7.183) we can write the eigenvalue problem (7.184) in the form

$$\int \left[\chi_{KS}^{-1}(\vec{r}, \vec{r}', \omega) - \frac{e^2}{|\vec{r} - \vec{r}'|} - f_{xcL}(\vec{r}, \vec{r}', \omega) \right] \Phi(\vec{r}') d\vec{r}' = 0 . \qquad (7.191)$$

Due to the presence of the Hartree and xc terms, *the exact excitation energies differ from the excitation energies of the Kohn–Sham system*. The Hartree and xc corrections to the Kohn–Sham excitation energies are not necessarily small. In extreme cases, the presence of these terms gives rise to excitations that could not exist in the Kohn–Sham system. A vivid example of this phenomenon in atomic physics is provided by the *Fano resonance* (Fano, 1961), in which *two* electrons are simultaneously excited from an occupied shell to the ionization continuum.[38]

If, as it is often the case, a certain excitation of the physical system is not too far in energy from a similar excitation of the Kohn–Sham system, then a perturbative approach is possible (Petersilka, Gossmann, and Gross, 1996). Let us expand the excitation energy and the corresponding eigenfunction in Eq. (7.191) as

$$\omega_{jk} = \omega_{jk}^{(0)} + \omega_{jk}^{(1)} \qquad (7.192)$$

and

$$\Phi_{jk} = \Phi_{jk}^{(0)} + \Phi_{jk}^{(1)}, \qquad (7.193)$$

where $\omega_{jk}^{(1)}$ and $\Phi_{jk}^{(1)}$ are first-order shifts due to the Hartree and exchange-correlation terms. Expanding Eq. (7.191) to first order in these shifts we obtain

$$\int \chi_{KS}^{-1}(\vec{r}, \vec{r}', \omega_{jk}^{(0)}) \Phi_{jk}^{(1)}(\vec{r}') d\vec{r}' + \omega_{jk}^{(1)} \int \left. \frac{\partial}{\partial \omega} \chi_{KS}^{-1}(\vec{r}, \vec{r}', \omega) \right|_{\omega=\omega_{jk}^{(0)}} \Phi_{jk}^{(0)}(\vec{r}') d\vec{r}'$$

$$= \int \left[\frac{e^2}{|\vec{r} - \vec{r}'|} + f_{xcL}(\vec{r}, \vec{r}', \omega_{jk}^{(0)}) \right] \Phi_{jk}^{(0)}(\vec{r}') d\vec{r}' . \qquad (7.194)$$

Multiplying both sides of this equation by $\Phi_{jk}^{(0)}(\vec{r})^*$ and integrating over $\vec{r}$, we see that the first term on the left-hand side vanishes, while in the second term, by virtue of Eq. (7.190), we have

$$\int d\vec{r} \int d\vec{r}' \Phi_{jk}^{(0)}(\vec{r})^* \left. \frac{\partial}{\partial \omega} \chi_{KS}^{-1}(\vec{r}, \vec{r}', \omega) \right|_{\omega=\omega_{jk}^{(0)}} \Phi_{jk}^{(0)}(\vec{r}') = 1. \qquad (7.195)$$

Thus, we obtain the extremely useful result

$$\omega_{jk}^{(1)} = \int d\vec{r} \int d\vec{r}' \Phi_{jk}^{(0)}(\vec{r})^* \left[\frac{e^2}{|\vec{r} - \vec{r}'|} + f_{xcL}(\vec{r}, \vec{r}', \omega_{jk}^{(0)}) \right] \Phi_{jk}^{(0)}(\vec{r}') , \qquad (7.196)$$

which expresses the Hartree and xc shifts in the excitation energies in terms of the Kohn–Sham eigenfunction $\Phi_{jk}^{(0)}$ and the Kohn–Sham excitation energy $\omega_{jk}^{(0)}$. Notice that the Hartree

---

[38] Notice that Eq. (7.191) can have solutions in the complex frequency plane. These solutions correspond to resonances (such as the Fano resonance) which decay, after a characteristic lifetime, into the continuum of states they are degenerate with.

Table 7.7. $^1S \rightarrow \ ^1P$ *excitation energies in two-valence-electron atoms. The last two columns on the right contain the KS excitation energies in* Ry *calculated by the static LDA and OEP respectively. (From Petersilka* et al., *1996.)*

Atom	ω_{exp}	ω_{ALDA}	ω_{TDOEP}	$\omega_{LDA}^{(0)}$	$\omega_{OEP}^{(0)}$
He	1.56 Ry	1.552	1.568	–	–
Be	0.388	0.399	0.392	0.257	0.259
Mg	0.319	0.351	0.327	0.249	0.234
Ca	0.216	0.263	0.234	0.176	0.157
Zn	0.426	0.477	0.422	0.352	0.314
Sr	0.198	0.241	0.210	0.163	0.141
Cd	0.398	0.427	0.376	0.303	0.269

term (known to solid state physicists as *depolarization shift*) is always positive, while the xc term is usually negative owing to the negative sign of the static xc kernel (see Eq. (7.179)). A crucial assumption underlying this approach is that $f_{xcL}(\vec{r}, \vec{r}', \omega)$ be a slowly varying function of frequency in a neighborhood including both $\omega_{jk}^{(0)}$ and ω_{jk}. The ALDA obviously satisfies this requirement, since, for better or for worse, it completely ignores the frequency dependence of f_{xcL}.

The application of the perturbative method to atomic systems was pioneered by Petersilka, Gossman, and Gross (1996). In Table 7.7 we report their results for the energy of the $^1S \rightarrow \ ^1P$ transition in a series of atoms with two electrons in the outer s-shell. Both the initial and the final states are spin singlets so that one does not have to worry about the spin density.

The calculations were done in the ALDA and in the time-dependent Optimized Effective Potential (TDOEP) method, another form of the adiabatic approximation based on the static OEP approach (see Section 2.8.2). The latter decays as $-\frac{e^2}{r}$ at large distance from the nucleus, as the exact xc potential should. The form of the xc kernel in OEP theory is (Petersilka *et al.*, 1996)

$$f_{xcL}^{OEP}(\vec{r}, \vec{r}', \omega) = -\frac{|\sum_k f_k \phi_k(\vec{r}) \phi_k^*(\vec{r}')|^2}{|\vec{r} - \vec{r}'| n_0(\vec{r}) n_0(\vec{r}')}, \tag{7.197}$$

where $\phi_k(\vec{r})$ are the ground-state OEP orbitals discussed in Section 2.8.2. Notice that f_{xcL}^{OEP} is still frequency-independent (i.e., instantaneous in time), but, unlike f_{xcL}^{ALDA}, is nonlocal in space.

Inspecting Table 7.7, we see that the OEP performs better than the ALDA, suggesting that the large distance behavior of the potential is important. We also see that the Hartree and exchange-correlation shifts, taken together, *increase* the excitation energy significantly. While this improves the agreement with experiment, it also casts some doubt on the appropriateness of the perturbative treatment, which assumes those shifts to be small.[39]

---

[39] The perturbative approach has been criticized (Vasiliev *et al.*, 1999). According to these authors the ALDA works better than the time-dependent OEP when the excitation energies are calculated by solving the eigenvalue problem (7.191) exactly.

In recent years density functional calculations of excitation energies have been performed on a variety of complex molecular systems and clusters (Casida, 1995; van Gisbergen *et al.*, 1998; Onida *et al.*, 2002), and the spin has been taken into account, so that, for example, the energy difference between the the singlet–singlet $^1S \rightarrow {}^1P$ and the singlet–triplet $^1S \rightarrow {}^3P$ transitions could be resolved (Dobson, 1998). In most cases, the agreement with experiment and with more time-consuming configuration–interaction calculations has been quite satisfactory.

7.6.2 Infinite systems

The formalism described in the previous section is also applicable to infinite systems (systems satisfying periodic boundary conditions in a box of size L tending to infinity) but a more careful discussion is needed to distinguish between single-particle excitations and collective modes.

Continuous spectrum – Single particle excitations in an infinite system give rise to a continuous spectrum, which, in general, consists of several bands. Hartree and exchange-correlation corrections modify not only the spectral density of excitations within each band (intra-band transitions), but also the width and the position of the bands.

The effect of the time-dependent Hartree potential on intra-band transitions is very strong, as we known from the study of the homogeneous electron liquid (see Section 5.3.4), where this potential causes a dramatic shift of spectral density from electron–hole pairs to the plasmon mode. Inter-band transitions in crystalline solids are not so strongly affected, yet dynamical interaction effects cause significant changes, such as the appearance of the two-peak structure seen in the experiments and shown in Fig. 7.9.

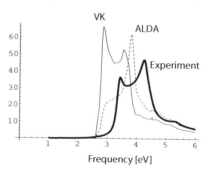

Imaginary part of the dielectric function of Silicon

Fig. 7.9. Imaginary part of the macroscopic dielectric function of Si (proportional to the optical spectrum) calculated in the adiabatic local density approximation (dashed line) and in a simplified version of the current-density functional theory of Vignale and Kohn (VK) described in Section 7.9.2. The experimental spectrum is shown by the thick solid line (courtesy of P. de Boeij).

The calculation of the *optical band gap*, defined, in a crystal, as the lower limit of the electron–hole spectrum at zero Bloch wave vector,[40] poses a challenging problem. Experience shows that the true optical gap of insulators is considerably higher than the gap computed in the ALDA. For example, the difference between the true gap and the calculated one is clearly visible in Fig. 7.9, where the optical spectrum of Si (i.e., the imaginary part of the dielectric function) is plotted as a function of frequency. We know that, in principle, the correct gap could be obtained from Eq. (7.191), if one knew the exact xc kernel. Unfortunately, the features of the exact kernel that are responsible for producing a good optical gap are among the most difficult to capture in an approximate theoretical scheme.

First of all, any approximation that treats f_{xcL} as a real quantity will not produce any change in the KS gap. The reason is that if f_{xcL} is real then, according to Eq. (7.183), the imaginary part of χ^{-1} vanishes if and only if χ_{KS}^{-1} also vanishes. This implies that the two spectral densities $\Im m \chi$ and $\Im m \chi_{KS}$, while numerically different, vanish in the same ranges of frequency, and therefore have the same gaps.

Even endowing f_{xcL} with an imaginary part (and hence with a frequency dependence) is not enough to solve the problem: to get a finite correction to the KS gap it is also necessary that $f_{xcL}(\vec{r}, \vec{r}', \omega)$ be a long-ranged function of $|\vec{r} - \vec{r}'|$ (Tokatly and Pankratov, 2001). To understand why long-rangedness is required, consider again Eq. (7.196) for the exchange-correlation correction to KS excitation energies. Let $\omega^{(0)}$ be the *lowest* electron–hole pair excitation frequency in the KS system and $\Phi^{(0)}$ the corresponding electron–hole wave function (see Eq. (7.186)). Since the KS orbitals are normalized to 1 over the volume of the system, $\Phi^{(0)}$ must scale as $\frac{1}{L^d}$. Thus, if $f_{xcL}(\vec{r}, \vec{r}', \omega)$ is of finite range in $|\vec{r} - \vec{r}'|$, then Eq. (7.196) predicts an energy shift of the order of $\frac{1}{L^d}$, which tends to zero in the relevant limit $L \to \infty$.

What we have just described is an instance of the "ultra-nonlocality problem", which affects several aspects of TDDFT and will be discussed in more detail in Section 7.8.2.

Collective modes – The possibility of collective modes is one of the most striking manifestations of many-body effects in extended systems. These excitations do not exist in the non-interacting Kohn–Sham system, and therefore cannot be obtained by means of perturbation theory.

The best known example is the ordinary plasmon. As discussed in Chapter 5, this mode is primarily sustained by the Hartree potential: exchange-correlation effects are important in determining the correct dispersion and the linewidth of the excitation, but do not introduce qualitatively new features. Therefore exchange-correlation corrections to the plasmon dispersion can be calculated perturbatively, taking the RPA (i.e., the time-dependent Hartree theory) as the "zero-order" approximation. In practice, this means dropping the Hartree term in the square brackets of Eq. (7.196) and using the RPA eigenfunction for $\Phi^{(0)}$.

---

[40] The restriction to zero Bloch wave vector arises from the fact that the electromagnetic field at optical wavelength has essentially zero wave vector. In *indirect-gap* semiconductors, like Ge and Si, the optical gap is much larger than the actual single-particle energy gap.

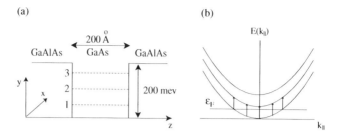

Fig. 7.10. Schematic representation of electron subbands in a two-dimensional quantum well. (a) Quantized energy levels in the direction perpendicular to the plane of the well. (b) Electron energies vs wave vector in the plane of the quantum well. Notice the massive degeneracy of electron–hole pair excitations between the two lowest subbands.

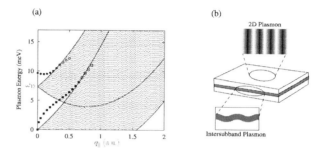

Fig. 7.11. (a) Dispersions of intersubband plasmons (circles) and ordinary 2D plasmons (squares) vs wave vector in the plane of the well. The two shaded regions correspond to the intra- and intersubband electron–hole continua. (b) Character of the density oscillations in the intersubband plasmon *vis-à-vis* the ordinary plasmon (from Ullrich and Vignale, 2002).

A more interesting example is provided by the so-called *intersubband plasmon* in a semiconductor quantum well. In this system the electrons (treated in the effective mass approximation) are free to move in the xy plane, but are strongly confined in the z-direction (see Fig. 7.10(a)). The discrete energy levels associated with motion in the z-direction define the bottoms of the two-dimensional subbands, and only a few subbands are occupied in the ground-state (see Fig. 7.10(b)): thus, the system is atomic-like in the z-direction and electron-liquid like in the xy plane. Electron–hole transitions between the highest occupied and the lowest unoccupied subband of the Kohn–Sham system are massively degenerate at $q_\parallel = 0$ ($q_\parallel$ being the wave vector in the plane) and form a continuum for $q_\parallel \neq 0$, as shown in Fig. 7.11(a). The intersubband plasmon splits off above the intersubband electron–hole continuum, in much the same way as the ordinary plasmon splits-off above the intraband electron–hole continuum (see Fig. 7.11(a)). Physically, however, the intersubband plasmon differs from the ordinary plasmon mainly in that the center of mass of the density distribution oscillates in the z direction rather than in the plane of the well (Fig. 7.11(b)).

Table 7.8 presents the results of ALDA calculation of the intersubband frequency at $q_\parallel = 0$ for several values of the "sheet density" n_s, defined as the the ordinary three-dimensional

Table 7.8. *Intersubband plasmon frequency* ($q_\parallel = 0$) *and linewidth in a GaAs/AlGaAs quantum well. Ullrich and Vignale (2002). Experimental values from Williams* et al. *(2001).*

n_s (cm^{-2})	$\hbar\omega_{exp}$	$\hbar\omega_{ALDA}$	$\hbar\omega_{VUC}$	Γ_{exp}	Γ_{VUC}
5×10^9	9.19 meV	8.69	8.73	0.121	8.2×10^{-3} meV
1×10^{10}	9.23	8.71	8.76	0.128	1.44×10^{-2}
3×10^{10}	9.41	8.91	8.97	0.162	3.19×10^{-2}
5×10^{10}	9.64	9.15	9.21	0.187	4.39×10^{-2}
7×10^{10}	9.83	9.41	9.47	0.201	5.41×10^{-2}
1×10^{11}	10.14	9.79	9.86	0.226	6.82×10^{-2}

density times the width of the quantum well. This table also presents the results of more complex calculations that utilize a non-adiabatic approximation for f_{xcL} (VUC): these will be discussed in Section 7.8. The frequencies obtained from the two theories are quite similar, and compare reasonably well with the experimental data of Williams *et al.* (2001), listed in the second column.

It should be noticed at this point that the ALDA predicts no intrinsic linewidth[41] of the intersubband plasmon, since the mode lies well above the Kohn–Sham electron–hole continuum, in a region where the KS response function is real. Experimentally, the linewidth has been measured, and turns out to be quite an interesting function of the experimental parameters. While part of the linewidth is certainly due to extrinsic effects (impurity scattering, surface roughness) the problem of calculating the intrinsic contribution remains open, and cannot be solved within the framework of the ALDA. We shall return to this question in the following sections, after we have discussed non-adiabatic approximations.

7.7 Reason for the success of the adiabatic LDA

We ended Section 7.5.3 by pointing out a paradox. How can the ALDA be successfully applied to the calculation of excitation energies, when, by construction, it should be expected to work only at frequencies that are much lower than the characteristic frequencies of the system?

The resolution of this paradox is that the scale of the frequency dependence of the exchange-correlation potential is set by the energy of the excitations that are *left out* of the Kohn–Sham spectrum, i.e., correlated multi-particle excitations. Because the energy of such excitations is typically much higher than that of single-particle excitations, it turns out that the adiabatic approximation can be quite reasonable in calculating the latter.

---

[41] Unlike the ordinary plasmon, which is rigorously free of damping in a translationally invariant system at $q = 0$, the intersubband plasmon has, in general, a finite linewidth, even at $q_\parallel = 0$. The only exception occurs when the confining potential well is *parabolic*. In that special case, the generalized Kohn's theorem (also known in the literature as the *harmonic potential theorem* (see Section 7.8) guarantees that the intersubband plasmon at $q_\parallel = 0$ is an exact eigenstate of the system, and that its frequency coincides with the natural frequency of a single electron in the harmonic oscillator potential.

Let us illustrate the idea with two examples.

(1) In a homogeneous electron liquid the scale of the frequency dependence of the xc potential is set by the characteristic energy of multiple electron–hole pair excitations, which is typically of the order of the Fermi energy and much above the cutoff of the electron–hole continuum at small q.
(2) In atoms, the scale is set by the energy of excitations in which two or more electrons are simultaneously promoted to higher orbitals: these are considerably higher than the one-electron excitation energies of interest (e.g., the lowest-lying two-electron excitation in He is the $(1s)^2 \to (2s2p)$ transition, which already lies in the one-electron ionization continuum).

In a sense, electronic correlations produce a kind of "thermal bath", through which the simpler one-electron excitations are created. Due to the highly complex dynamics of the thermal bath, it is reasonable to assume that its coherence time is much shorter than the period of one-electron excitations: hence the approximate validity of the adiabatic approximation.

From the above discussion it follows that:

(i) The adiabatic approximation is generally inadequate to study complex electronic excitations (e.g., multiparticle excitations), which have no counterpart in the independent electron model or in the RPA.
(ii) The adiabatic approximation may also fail quantitatively in the case of excitations that are embedded in a continuum of multiparticle excitations and strongly interact with the latter. In particular, the damping of collective modes due to their interaction with multiple electron–hole pairs is not described by the ALDA.

In all such cases (and in the band-gap problem too) one needs frequency-dependent approximations for V_{xc}. We now proceed to describe some such approximations.

7.8 Beyond the adiabatic approximation

The first attempt to go beyond the ALDA was made by Gross and Kohn (1985) (GK) within the framework of the linear response theory. GK argued that, if the ground-state density $n_0(\vec{r})$ is "sufficiently slowly varying", then the linear xc kernel, $f_{xcL}(\vec{r}, \vec{r}', \omega)$, can be replaced by the xc kernel of a homogeneous electron liquid, $f_{xcL}^h(\vec{r} - \vec{r}', \omega)$, evaluated at the local density $n_0(\vec{r})$. "Sufficiently slowly varying" means that the variation of the density is negligible over a distance of the order of the *range* of f_{xcL}, i.e., the characteristic value of the distance $|\vec{r} - \vec{r}'|$ beyond which $f_{xcL}(\vec{r}, \vec{r}', \omega)$ falls to zero. But, does this range exist?

In the homogeneous electron liquid everything seems to work out. Comparing Eq. (7.183) with the corresponding equation (5.105) of Chapter 5, we see that $f_{xcL}^h(\vec{r} - \vec{r}', \omega)$ is just the Fourier transform of $-v(q)G_+(q, \omega)$:

$$f_{xcL}^h(|\vec{r} - \vec{r}'|, \omega) = -\frac{1}{L^d} \sum_{\vec{q}} v(q) G_+(q, \omega) e^{i\vec{q} \cdot (\vec{r} - \vec{r}')} , \qquad (7.198)$$

where $G_+(q, \omega)$ is the density local field factor. This is a *short-ranged* function of $|\vec{r} - \vec{r}'|$, in the very specific sense that the $q \to 0$ limit of its Fourier transform

$$f_{xcL}(\omega) \equiv - \lim_{q \to 0} [v(q)G_+(q, \omega)] , \qquad (7.199)$$

exists and is finite (see discussion in Section 5.6.2.2 and approximate formulas for $f_{xcL}(\omega)$ in Appendix 11).

Gross and Kohn argued that, if the induced density $n_1(\vec{r}, \omega)$ is also slowly varying, then one can pull n_1 out of the integral formula for $V_{xc,1}$ and get a local relation between $V_{xc,1}$ and n_1:

$$V_{xc,1}(\vec{r}, \omega) \simeq \int f_{xcL}^h(\vec{r} - \vec{r}', \omega)n_1(\vec{r}', \omega)d\vec{r}'$$

$$\simeq \left(\int f_{xcL}^h(|\vec{r} - \vec{r}'|, \omega)d\vec{r}' \right) n_1(\vec{r}, \omega)$$

$$= f_{xcL}(\omega)n_1(\vec{r}, \omega). \qquad (7.200)$$

This chain of approximations is, in effect, equivalent to the single approximation

$$f_{xcL}(\vec{r}, \vec{r}', \omega) \simeq f_{xcL}(\omega)\delta(\vec{r} - \vec{r}') , \qquad (7.201)$$

where $f_{xcL}(\omega)$ is to be evaluated at the local equilibrium density $n_0(\vec{r})$. The above equation appears to provide the desired extension of the local density approximation to finite frequencies, and does, indeed, work rather well in many applications.[42]

Unfortunately, the argument that leads to Eq. (7.201) contains a subtle flaw. It turns out that the xc kernel of an *inhomogeneous* electron liquid at *finite frequency* is *not* a short-ranged function of $|\vec{r} - \vec{r}'|$, even though the xc kernel of the homogeneous electron liquid is. More precisely, it will be shown below that the reciprocal space representation of the xc kernel of an inhomogeneous electron liquid, $f_{xcL}(\vec{k}, \vec{k}', \omega)$, is singular when $k' \to 0$ at constant k, or vice versa. This singularity is "invisible" in the homogeneous case, since, due to translational invariance, $f_{xcL}^h(\vec{k}, \vec{k}', \omega) = 0$ for $\vec{k} \neq \vec{k}'$.

Historically, the first indication of trouble with Eq. (7.201) came from John Dobson's observation (Dobson, 1994) that this approximation fails to satisfy the *harmonic potential theorem*. According to this theorem, the center of mass of a system of electrons confined in a parabolic potential well and subjected to a uniform time-dependent electric field should evolve in time in exactly the same way as a single electron would. But, a calculation based on Eq. (7.201) was found to violate this theorem. Following this observation, several exact properties of f_{xcL} were discovered, and the long-rangedness of f_{xcL} was eventually identified as the basic difficulty (Vignale and Kohn, 1998). This line of investigation resulted in the development of new approximate functionals (Dobson, Bünner, and Gross, 1997; Vignale, Ullrich, and Conti, 1997) which go beyond the adiabatic approximation, satisfy

---

[42] Notice that the $\omega \to 0$ limit of the GK approximation is not quite the same as the ALDA. The difference is due to the fact that (see Section 5.6.2.3), $\lim_{\omega \to 0} \lim_{q \to 0} [v(q)G_+(q, \omega)]$ which appears in Eq. (7.201) is different from $\lim_{q \to 0} \lim_{\omega \to 0} [v(q)G_+(q, \omega)] = -d^2 n \varepsilon_{xc}^h(n)/dn^2$ which appears in the ALDA.

the harmonic potential theorem, and are still local in space. These new functionals, however, are functionals of the *current density*. The story is told in the next few sections.

7.8.1 The zero-force theorem

This theorem can be proved in several ways. We take the most physical approach and start from the requirement that the net force exerted by the coulomb potential on the system must vanish, by virtue of Newton's third law. Since the net force exerted by the Hartree potential

$$-e^2 \int d\vec{r}\, n(\vec{r}, t) \vec{\nabla}_{\vec{r}} \int d\vec{r}' n(\vec{r}', t) \frac{1}{|\vec{r} - \vec{r}'|} \tag{7.202}$$

trivially vanishes (as one can see by interchanging the integration variables $\vec{r}$ and $\vec{r}'$), it follows that the net force exerted by the xc potential must vanish too, i.e.,

$$\int d\vec{r} n(\vec{r}, t) \vec{\nabla}_{\vec{r}} V_{xc}[n](\vec{r}, t) = 0 \,. \tag{7.203}$$

In the linear response approximation at frequency ω this condition takes the form

$$\int d\vec{r} n_0(\vec{r}) \vec{\nabla}_{\vec{r}} V_{xc,1}(\vec{r}, \omega) + \int d\vec{r} n_1(\vec{r}, \omega) \vec{\nabla}_{\vec{r}} V_{xc,0}(\vec{r}) = 0 \,, \tag{7.204}$$

where $n_1(\vec{r}, \omega)$ is an arbitrary induced density. Substituting Eq. (7.178) into this, and integrating by parts, we obtain

$$\int \vec{\nabla}_{\vec{r}} n_0(\vec{r}) f_{xcL}(\vec{r}, \vec{r}', \omega) d\vec{r} = \vec{\nabla}_{\vec{r}'} V_{xc,0}(\vec{r}') \,. \tag{7.205}$$

This equation and its transpose[43]

$$\int f_{xcL}(\vec{r}, \vec{r}', \omega) \vec{\nabla}_{\vec{r}'} n_0(\vec{r}') d\vec{r}' = \vec{\nabla}_{\vec{r}} V_{xc,0}(\vec{r}) \tag{7.206}$$

form the mathematical statement of the zero-force theorem.

Eq. (7.205) establishes a link between the static and the dynamic xc potential. In essence, it says that the ground-state potential $V_{xc,0}$ "rides along" with the density in any rigid time-dependent translation of the latter. It is not difficult to verify (see Exercise 19) that this property guarantees satisfaction of the harmonic potential theorem. We can now see why the approximation (7.201) violates the harmonic potential theorem: upon substituting (7.201) into (7.206) we obtain a frequency-dependent quantity on the left-hand side, which cannot be equal to the frequency-independent quantity on the right-hand side.

7.8.2 The "ultra-nonlocality" problem

We now show that Eq. (7.205) implies spatial long-rangedness of $f_{xcL}(\vec{r}, \vec{r}', \omega)$ and hence the failure of the local density approximation at finite frequency. Consider a uniform

---

[43] The transposition is done with the help of the formula $f_{xcL}(\vec{r}, \vec{r}', \omega) = f_{xcL}(\vec{r}', \vec{r}, \omega)$, which follows from the exact symmetries of the response functions, discussed in Chapter 3.

electron gas modulated by a small sinusoidal potential of wave vector $\vec{q}$. The ground-state density is

$$n_0(\vec{r}) = \bar{n}(1 + \gamma \cos \vec{q} \cdot \vec{r}), \qquad (7.207)$$

where the amplitude of the modulation is $\gamma \ll 1$. Let us expand the xc kernel in powers of γ as follows:

$$f_{xcL}(\vec{r}, \vec{r}'; \omega) = f_{xcL}^h(|\vec{r} - \vec{r}'|, \omega) + \gamma f_{xcL}^{(1)}(\vec{r}, \vec{r}'; \omega) + O(\gamma^2), \qquad (7.208)$$

where the zero-th order term is the homogeneous xc kernel, and let

$$f_{xcL}(\vec{k}, \vec{k}', \omega) = \frac{1}{L^d} \int d\vec{r} \int d\vec{r}' f_{xcL}(\vec{r}, \vec{r}', \omega) e^{-i\vec{k}\cdot\vec{r}} e^{i\vec{k}'\cdot\vec{r}'} \qquad (7.209)$$

be the matrix elements of the xc kernel in reciprocal space. Because of translational periodicity, $f_{xcL}(\vec{k}, \vec{k}', \omega)$ differs from zero only when $\vec{k} - \vec{k}' = m\vec{q}$, where m is an integer. To first order in γ, the diagonal matrix elements $f_{xcL}(\vec{k}, \vec{k}, \omega)$ coincide with those of the underlying homogeneous electron liquid (i.e., $f_{xcL}(\vec{k}, \vec{k}, \omega) = f_{xcL}^h(\vec{k}, \omega) + O(\gamma^2)$), while the matrix elements with $|m| > 1$ are of order γ^2 or higher. Thus, to first order in γ, the matrix elements $f_{xcL}^{(1)}(\vec{k}, \vec{k}', \omega)$ are all zero unless $\vec{k} = \vec{k}' \pm \vec{q}$.

We now expand both sides of the zero-force theorem, Eq. (7.206), to first order in γ. The static xc potential on the right-hand side can be written as (see Eq. (7.178))

$$V_{xc,0}(\vec{r}) = \gamma \bar{n} \int f_{xcL}^h(|\vec{r} - \vec{r}'|, 0) \cos \vec{q} \cdot \vec{r}' d\vec{r}'$$
$$= \gamma \bar{n} f_{xcL}^h(q, 0) \cos \vec{q} \cdot \vec{r}, \qquad (7.210)$$

and, after a few simple rearrangements, Eq. (7.206) takes the form

$$\int f_{xcL}^{(1)}(\vec{r}, \vec{r}'; \omega) \vec{\nabla}_{r'} \bar{n} d\vec{r}' = \bar{n}[f_{xcL}^h(q, 0) - f_{xcL}^h(q, \omega)] \vec{\nabla}_r \cos(\vec{q} \cdot \vec{r}). \qquad (7.211)$$

The right-hand side of this equation is definitely different from zero. The left-hand side, on the other hand, appears to vanish, because the gradient of the homogeneous density $\bar{n}$ is zero. Thus, there is no way to satisfy (7.211) unless $f_{xcL}^{(1)}(\vec{r}, \vec{r}'; \omega)$ is of infinitely long range in $|\vec{r} - \vec{r}'|$, indeed of such a long range that we are forced to take into account the *boundaries* of the system, where $n_0(\vec{r})$ ceases to have the form (7.207).

To understand more clearly the mechanism by which the left-hand side of Eq. (7.211) acquires a nonzero value, let us assume that $\bar{n}$ is constant up to the boundary of a large volume, and zero outside.[44] We do an integration by parts and transform Eq. (7.211) to

$$\bar{n} \int \vec{\nabla}_{r'} f_{xcL}^{(1)}(\vec{r}, \vec{r}'; \omega) d\vec{r}' = -\bar{n}[f_{xcL}^h(q, 0) - f_{xcL}^h(q, \omega)] \vec{\nabla}_r \cos \vec{q} \cdot \vec{r}, \qquad (7.212)$$

---

[44] To avoid mathematical complications at the boundary the transition from constant to zero value can be smoothened over a length scale much smaller than the dimensions of the system.

where the integral runs over the volume of the system. In the limit of infinite volume the integral on the left-hand side of this equation can be written as

$$\lim_{k' \to 0} i\vec{k}' f_{xcL}^{(1)}(\vec{r}, \vec{k}'; \omega) , \qquad (7.213)$$

where $f_{xcL}^{(1)}(\vec{r}, \vec{k}'; \omega)$ denotes the Fourier transform of $f_{xcL}^{(1)}(\vec{r}, \vec{r}'; \omega)$ with respect to $\vec{r}'$, calculated at wave vector $\vec{k}'$. Finally, multiplying both sides of (7.212) by $L^{-d} e^{-i\vec{q} \cdot \vec{r}}$ and integrating over $\vec{r}$ we obtain

$$\lim_{k' \to 0} \vec{k}' f_{xcL}^{(1)}(\vec{q}, \vec{k}'; \omega) = \frac{1}{2} \vec{q} [f_{xcL}^{h}(q, \omega) - f_{xcL}^{h}(q, 0)]. \qquad (7.214)$$

This is the key result we promised. It shows that $f_{xcL}^{(1)}(\vec{q}, \vec{k}'; \omega)$ diverges for $k' \to 0$ provided $\vec{k}'$ has a finite component along the direction of $\vec{q}$. The form of the singularity for $k' \to 0$ is

$$f_{xcL}^{(1)}(\vec{q}, \vec{k}'; \omega) \overset{k' \to 0}{\approx} \frac{\vec{k}' \cdot \vec{q}}{2k'^2} \Delta f_{xcL}^{h}(q, \omega) , \qquad (7.215)$$

where $\Delta f_{xcL}^{h}(q, \omega) \equiv [f_{xcL}^{h}(q, \omega) - f_{xcL}^{h}(q, 0)]$. Notice that the singularity disappears at $\omega = 0$.

Eq. (7.215) implies that for $|\vec{r} - \vec{r}'| \to \infty$, $f_{xcL}(\vec{r}, \vec{r}', \omega)$ behaves as $\gamma \frac{\Delta f_{xcL}^{h}(q, \omega)}{8\pi\bar{n}} \frac{\vec{q}}{q} \cdot \vec{\nabla}_{r'} \frac{1}{|\vec{r}-\vec{r}'|}$.[45] Because of this long-ranged behavior (which is totally unrelated to the range of the electron-electron interaction) a frequency-dependent local density approximation does not exist. This disturbing feature has come to be known as the *ultra-nonlocality* problem of time-dependent DFT.

One should not conclude from the above discussion that the GK approximation is always incorrect. If the magnitude of $\vec{q}$, which characterizes the degree of nonuniformity of the ground-state density, is much smaller than the magnitude of $\vec{k}$, which characterizes the spatial dependence of the time-dependent potential, then the GK approximation is qualitatively correct. It is only in the opposite limit, $k \ll q$, that the GK approximation becomes qualitatively incorrect.

Because the ultra-nonlocality problem is not intrinsic to the physics, but arises from the choice of the density as a basic variable, we may hope to avoid it by the use of a different basic variable. In the next section we show how this is done with the help of the *current density*.

7.9 Current density functional theory and generalized hydrodynamics

We have seen in Section 3.4.1 that a time-dependent scalar potential $V(\vec{r}, t)$ is physically equivalent to (and can be replaced by) a longitudinal vector potential of the form

$$\vec{A}(\vec{r}, t) = -\frac{c}{e} \int_{t_0}^{t} \vec{\nabla} V(\vec{r}, t') dt' . \qquad (7.216)$$

---

[45] The residual dependence of this asymptotic behavior on the direction of the wave vector of the infinitesimal density modulation shows that V_{xc} is not a differentiable functional of the density.

On the other hand, any time-periodic component of the particle density can be extracted from the corresponding component of the current density by means of the continuity equation

$$n(\vec{r}, \omega) = \frac{\vec{\nabla} \cdot \vec{j}(\vec{r}, \omega)}{i\omega}. \tag{7.217}$$

Therefore, the problem of calculating the linear response of the density to a periodic scalar potential is just a special instance of the more general problem of calculating the response of the current to a periodic vector potential

$$\vec{A}_1(\vec{r}, t) = \vec{A}_1(\vec{r}, \omega)e^{-i\omega t} + c.c. \tag{7.218}$$

In view of this, why not try and reformulate the whole time-dependent DFT in terms of the current density?

Efforts in this direction began in the late 1980s. Ghosh and Dhara (1988) proved the relevant generalization of the Runge–Gross theorem, asserting that, for a given initial state, the time-dependent current density determines the vector potential up to a gauge transformation. However, it was only within the framework of the linear response theory that concrete expressions for the current-dependent xc vector potential $(\vec{A}_{xc}[\vec{j}](\vec{r}, t))$ began to emerge (Ng, 1989; Vignale and Kohn, 1996). Vignale and Kohn (VK), in particular, showed that the xc vector potential $\vec{A}_{xc}$ admits a local approximation in terms of the current density: therefore, the nonlocality problem that plagues time-dependent DFT, does not occur in current-density functional theory.

To understand qualitatively how this can happen consider the continuity equation (7.217). This has the same form as the Maxwell equation relating the electric field to the electric charge, and from this analogy we see that the longitudinal component of the current is related to the density by

$$\vec{j}_L(\vec{r}, \omega) = -\frac{i\omega}{4\pi} \int n(\vec{r}', \omega)\vec{\nabla}_r \frac{1}{|\vec{r} - \vec{r}'|} d\vec{r}', \tag{7.219}$$

a highly nonlocal relation. As an extreme example, consider a uniform sphere of charge performing an oscillatory motion whose amplitude is much smaller than the radius of the sphere. In such a motion the periodic component of the density is zero almost everywhere, except in the immediate vicinity of the surface; on the other hand, the periodic component of the current density is finite throughout the the sphere, showing that the current and the density are not locally related.

Similarly, Eq. (7.216) shows that the scalar xc potential V_{xc} of time-dependent DFT will be related in a nonlocal manner (i.e., via a line integral) to the longitudinal part of the xc vector potential $\vec{A}_{xc}$ in the current-DFT. The remarkable fact is that the nonlocality of V_{xc} as a functional of n can be *entirely absorbed* into the nonlocal relations between $\vec{A}_{xc}$ and V_{xc} and n and $\vec{j}$. This is the essence of the VK theory.

It turns out that the form of the local approximation for $\vec{A}_{xc}$ is essentially determined by symmetries and conservation laws. The final expression for $\vec{A}_{xc}$ is local in space and retarded in time, satifies the generalized Kohn's theorem, and allows a consistent calculation of the linewidth of elementary excitations, at least the part of it that arises from intrinsic many-body effects.

7.9.1 The xc vector potential in a homogeneous electron liquid

As a preparation to the study of inhomogeneous systems let us first construct the xc vector potential for a weakly perturbed homogeneous electron liquid of uniform density n. We will use several results from Chapters 4 and 5, where the current–current response functions of the homogeneous electron gas were studied in detail.

First of all, notice that the exact Kohn–Sham vector potential $\vec{A}_{KS}$ for the homogeneous liquid coincides with the effective potential $\vec{A}_{eff}$ introduced in Eqs. (5.130) and (5.132) of Chapter 5: by definition, this is the potential that produces the exact current response when applied to a non-interacting electron gas. As usual, we write

$$\vec{A}_{KS}(\vec{q}, \omega) = \vec{A}_{ext}(\vec{q}, \omega) + \vec{A}_H(\vec{q}, \omega) + \vec{A}_{xc}(\vec{q}, \omega) . \tag{7.220}$$

The Hartree component $\vec{A}_H$ is purely logitudinal (since it is just another way of describing the scalar Hartree potential) and we see from Eq. (5.130) that it is given by

$$\frac{e}{c}\vec{A}_H(\vec{q}, \omega) = \frac{q^2}{\omega^2} v_q j_L(q, \omega)\frac{\vec{q}}{q} . \tag{7.221}$$

The exchange-correlation vector potential, on the other hand, has both longitudinal and transverse components. From Eqs. (5.130) and (5.132) we see that we can write

$$\frac{e}{c}\vec{A}_{xc}(\vec{q}, \omega) = -\frac{q^2}{\omega^2} v_q \left[G_+(q, \omega)j_L(\vec{q}, \omega)\hat{q} + G_{T+}(q, \omega)\vec{j}_T(\vec{q}, \omega) \right] , \tag{7.222}$$

where $\vec{j}_T(\vec{q}, \omega)$ is the Fourier component of the transverse current density, that is, the component of the Fourier-transformed current perpendicular to $\vec{q}$.

In order to derive a local approximation for $\vec{A}_{xc}$ we consider a periodic perturbation that is slowly varying in space, i.e., we work in the limit of $q \to 0$ and finite frequency. In this limit Eq. (7.222) becomes

$$\frac{e}{c}\vec{A}_{xc}(\vec{q}, \omega) \simeq \frac{q^2}{\omega^2} \left[f_{xcL}(\omega)j_L(\vec{q}, \omega)\hat{q} + f_{xcT}(\omega)\vec{j}_T(\vec{q}, \omega) \right] , \tag{7.223}$$

where the functions $f_{xcL(T)}(\omega)$ are defined in Eq. (5.193) of Chapter 5 and their properties are discussed in Section 5.6.2. This equation implies a local relationship between $\vec{A}_{xc}$ and the current density, namely,

$$\frac{e}{c}\vec{A}_{xc}(\vec{r}, \omega) \simeq -\frac{1}{\omega^2} \left\{ f_{xcL}(\omega)\vec{\nabla}[\vec{\nabla} \cdot \vec{j}(\vec{r}, \omega)] + f_{xcT}^h(\omega)\nabla^2 \vec{j}_T(\vec{r}, \omega) \right\} . \tag{7.224}$$

For reasons that will become clear in the next section, it is very convenient at this point to isolate the standard ALDA contribution and to express the remaining "post-ALDA" contributions in terms of the *velocity field*

$$\vec{v}(\vec{r}, \omega) = \frac{\vec{j}(\vec{r}, \omega)}{n} \; . \tag{7.225}$$

For this purpose, the ALDA scalar xc potential of Eq. (7.173) is best rewritten as a current-dependent vector potential:

$$\frac{e}{c}\vec{A}_{xc}^{ALDA}(\vec{q}, \omega) = -\frac{1}{\omega^2}\frac{d^2[n\epsilon_{xc}(n)]}{dn^2}\vec{\nabla}[\vec{\nabla} \cdot \vec{j}(\vec{r}, \omega)] \; , \tag{7.226}$$

Making use of this representation, it is not difficult to verify (Exercise 22) that Eq. (7.224) can be recast in the form

$$\frac{e}{c}A_{xc,i}(\vec{r}, \omega) = \frac{e}{c}A_{xc,i}^{ALDA}(\vec{r}, \omega) - \frac{1}{i\omega n}\sum_j \frac{\partial\sigma_{xc,ij}}{\partial r_j}(\vec{r}, \omega), \tag{7.227}$$

where the visco-elastic "stress tensor" $\sigma_{xc,ij}$ in d dimensions is defined as

$$\sigma_{xc,ij} = \tilde{\eta}_{xc}(n, \omega)\left(\frac{\partial v_i}{\partial r_j} + \frac{\partial v_j}{\partial r_i} - \frac{2}{d}\vec{\nabla} \cdot \vec{v}\delta_{ij}\right) + \tilde{\zeta}_{xc}(n, \omega)\vec{\nabla} \cdot \vec{v}\delta_{ij}, \tag{7.228}$$

and the complex, frequency-dependent viscosities $\tilde{\eta}$ and $\tilde{\zeta}$ are related to f_{xcL} and f_{xcT} in the following manner:

$$\tilde{\eta}_{xc}(n, \omega) = -\frac{n^2}{i\omega}f_{xcT}(\omega) \tag{7.229}$$

and

$$\tilde{\zeta}_{xc}(n, \omega) = -\frac{n^2}{i\omega}\left[f_{xcL}(\omega) - \left(2 - \frac{2}{d}\right)f_{xc,T}(\omega) - \frac{d^2[n\epsilon_{xc}(n)]}{dn^2}\right]. \tag{7.230}$$

The physical significance of Eq. (7.227) is best understood in terms of the classical force exerted by $\vec{A}_{xc}$ on a volume element of the electron liquid. Recall that the "electric field" associated with $\vec{A}_{xc}$ is $\vec{E}_{xc}(\vec{r}, \omega) = \frac{i\omega}{c}\vec{A}_{xc}(\vec{r}, \omega)$, and that the force exerted by this field on an infinitesimal volume element $d\vec{r}$ is $-ne\vec{E}_{xc}(\vec{r}, \omega)d\vec{r}$. According to Eq. (7.227) the force-density is the sum of two pieces:

$$-ne\vec{E}_{xc,i}(\vec{r}, \omega) = -n\nabla_i V_{xc}^{ALDA}(\vec{r}, \omega) + \sum_j \frac{\partial\sigma_{xc,ij}}{\partial r_j}(\vec{r}, \omega) \; . \tag{7.231}$$

The ALDA term corresponds to the hydrostatic force term in hydrodynamics. In fact, it is not difficult to show (Exercise 23) that this term can be rewritten as $-\nabla_i p_{xc}(\vec{r}, \omega)$ where $p_{xc}(\vec{r}, \omega) = -n^2\frac{d\epsilon_{xc}(n)}{dn}$ is the xc contribution to the local equilibrium *pressure* of the electron liquid. The second term, which contains the divergence of the xc stress tensor, is the truly dynamical contribution to the force, arising from the fact that the system is not in local equilibrium (see discussion at the end of Section 5.3.3).

Perhaps the most remarkable feature of Eq. (7.231) is that *it does not reduce to the ALDA in the limit of zero frequency*. In this limit the Fourier amplitude of the velocity field tends to zero in such a way that $-\lim_{\omega\to 0}\frac{\vec{v}(\vec{r},\omega)}{i\omega} = \vec{u}(\vec{r}, 0)$ is a finite displacement field. Now the term proportional to $\tilde{\zeta}(\omega)$ vanishes by virtue of Eqs. (5.257), and so does the term proportional to $\Im m f_{xcT}(\omega)$, but the term proportional to $\Re e f_{xcT}(\omega)$ remains finite. Thus, the zero-frequency limit of the xc force density is

$$-ne\vec{E}_{xc}(\vec{r}, \omega \to 0) = -n\vec{\nabla}V_{xc}^{ALDA}(\vec{r}, 0) + \frac{S_{xc}}{n^2}\left[\nabla^2\vec{u} + \left(1 - \frac{2}{d}\right)\vec{\nabla}(\vec{\nabla}\cdot\vec{u})\right], \quad (7.232)$$

where S_{xc} is the xc contribution to the static shear modulus, and $\vec{u}$ is the static displacement field. As discussed at the end of Section 5.9.2, the difference between the $\omega \to 0$ limit of the dynamical theory and the ALDA arises because, as the wave vector q tends to zero, local equilibrium cannot be attained on any finite time scale. The shear term on the right-hand side of Eq. (7.232) has turned out to be extremely important in a recent application of the time-dependent CDFT to the calculation of the static polarizability of conjugated polymer chains (van Faassen *et al.*, 2002 – see Section 7.9.3).

7.9.2 *The exchange-correlation field in the inhomogeneous electron liquid*

The main result of the previous section, Eq. (7.227), was deliberately written in a form that can be generalized to inhomogeneous systems simply by the substitution $n \to n_0(\vec{r})$, $n_0(\vec{r})$ being the ground-state density of the inhomogeneous system. Of course, f_{xcL} and f_{xcT} must also be evaluated at $n_0(\vec{r})$. This is a nontrivial assertion. Why should the replacement $n \to n_0(\vec{r})$ be done in Eq. (7.227) rather than in the equivalent expression (7.224), which does not contain the velocity field? This is a consequence of general physical principles, which we now discuss.

First of all, from Newton's third law we know that the net force exerted by the xc "electric field" on the system must vanish. At the local level, Newton's third law implies that a volume element of the liquid cannot exert a net force on itself. Accordingly, the net force acting on a volume element must be expressible as the integral of *external* stresses acting on its surface. The mathematical expression of this requirement is that the xc force density must be representable as the divergence of a local stress tensor.

A similar argument holds for the net *torque* acting on a volume element. Again, this must be expressible as a surface integral, and it is not difficult to show that, for this to happen, the stress tensor must be a symmetric 2×2 tensor (see Landau–Lifshitz, *Course of Theoretical Physics*, Vol. 6).

Finally, Galilean invariance requires that the stress tensor vanish identically if the fluid moves as a whole, i.e., if the velocity field is uniform in space. It is for this reason that the stress tensor must depend on derivatives of the velocity field rather than derivatives of the current.

It turns out that the three requirements listed above uniquely determine Eq. (7.227) as the correct starting point for the substitution $n \to n_0(\vec{r})$. This result was originally obtained by Vignale and Kohn (1996, 1998) by a more laborious and apparently quite different path. VK considered a weakly inhomogeneous electron liquid modulated by a charge density wave of small amplitude γ and small wave vector $\vec{q}$ as in Eq. (7.207) of Section 7.8.2. Both k and q were assumed to be small not only relative to the Fermi momentum k_F but also relative to $\frac{\omega}{v_F}$ (v_F being the Fermi velocity). The latter condition assures that the phase velocity of the density disturbance is much faster than the Fermi velocity, so that there can be no static screening. Under these assumptions, all the components of the tensorial kernel $f_{xc,ij}$ connecting $A_{xc,i}$ to j_j could be calculated, up to first order in the amplitude of the charge density wave, and to second order in the wave vectors k and q. The calculation was greatly facilitated by a set of sum rules which are mathematically equivalent to the zero-force and zero-torque requirements discussed above.

An important result of the analysis was that both the diagonal component, $f_{xc,ij}(\vec{k}, \vec{k}, \omega)$, and the off-diagonal component, $f_{xc,ij}(\vec{k} + \vec{q}, \vec{k}, \omega)$, of the tensorial kernel remain finite for k or q tending to zero in any order. This is in marked contrast with the singular behavior of the off-diagonal elements of the scalar xc kernel of TDDFT (see Eq. (7.215)) and is the fundamental reason why the dynamical local density approximation is possible in terms of the current, but not in terms of the density.

The calculated expression for $f_{xc,ij}$ can be translated into a real space expression for $\vec{A}_{xc}(\vec{r}, \omega)$. The derivation is quite lengthy, but, when the dust settles, one is left with a simple and elegant result, given by Eq. (7.227).

7.9.3 The polarizability of insulators

Fig. 7.12 shows an interesting application of TDDFT to the calculation of the electric polarizability of polyacetylene, a long one-dimensional molecule. The problem of calculating the polarizability, and hence the dielectric constant of long polymer chains has

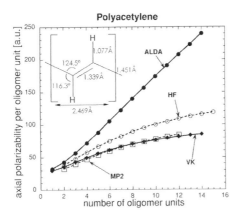

Fig. 7.12. The polarizability of polyacetylene calculated in time-dependent DFT in the ALDA and in time-dependent CDFT with the approximation of Eq. (7.227) (VK). The results of second-order perturbation theory (MP2) are also shown as a benchmark. (Adapted from van Faassen *et al.*, 2002).

Table 7.9. *Dielectric constants of selected insulators: ALDA vs experiment. (Adapted from Aulbur, Jönsson, and Wilkins, 2000.)*

	C	Si	Ge	GaAs
ϵ_{LDA}	5.9	13.5	22	13.7
ϵ_{exp}	5.7	11.4	15.3	10.9

attracted considerable attention since the 1990s (Gonze *et al.*, 1995; Martin and Ortiz, 1997; Vanderbilt, 1997; van Gisbergen *et al.*, 1998). The main difficulty lies in the fact that linear response calculations based on the static LDA systematically overestimate the polarizability and underestimate the band gap (see Table 7.9).[46]

On formal grounds, we know that the band-gap problem (see Section 7.2.8) is caused by the discontinuity of the static xc potential as a function of particle number. This discontinuity implies a singularity in the functional derivative $\frac{\delta V_{xc}(\vec{k})}{\delta n(\vec{k}')} = f_{xcL}(\vec{k}, \vec{k}')$ when the wave vectors $\vec{k}$ and $\vec{k}'$ both tend to zero, and such a singularity is precisely the signature of an ultra-nonlocal relation between V_{xc} and the density.

The question is now whether the ultra-nonlocality of V_{xc} can be avoided by making use of a variable other than the density. Several authors (Gonze *et al.*, 1995; Martin and Ortiz, 1997) have convincingly argued that a more natural (i.e., more local) description of insulators can be given in terms of the *polarization density* $\vec{P}(\vec{r})$ and its associated xc electric field. This suggestion can also be viewed from the vantage point of time-dependent CDFT. Recall that the polarization is the time integral of the current density, while the electric field is the time-derivative of the vector potential. Therefore, the problem of calculating the polarizability of an insulator can be completely and exactly addressed within the framework of time-dependent CDFT. As a first step, it is natural to resort to the local approximation derived in the last two sections. We have already noted that, due to the presence of the transverse term, this approximation does *not* reduce to the ALDA when the frequency tends to zero: this gives us hope that the ensuing correction to the ALDA polarizability might change the result in the right direction. Indeed, recent calculations by van Faassen *et al.*, based on the approximation (7.232), show that the inclusion of the transverse term goes a long way towards resolving the difficulties of the ALDA in this class of systems (see Fig. 7.12). Whether the homogeneous electron liquid-based approximation, which was derived for frequencies higher than the electron–hole continuum, can be legitimately applied to an insulator at zero frequency is, of course, a serious question of principle, and a matter of concern. There is little doubt, however, that the transverse term will survive in the exact theory at zero frequency, since the physical condition for its appearance, namely the inability of the electrons to reach local equilibrium in the

---

[46] The connection between band gap and polarizability fact is a consequence of the stiffness theorem, discussed in Section 3.2.9.

presence of a uniform electric field at a very small yet finite frequency, is satisfied in an insulator.

Another problem in which the the transverse component of the xc field appears to play an important role is the calculation of the optical spectra of semiconductors. The VK spectrum, plotted in Fig. 7.9, was computed in a simplified version of CDFT in which $f_{xcT}(\omega)$ was replaced by its purely real zero-frequency limit. The result of the calculation is striking because the optical spectrum exhibits the experimentally observed two-peak structure, which is absent in the ALDA. The agreement between theory and the experiment would greatly improve if one could shift the calculated spectrum to higher frequency so as to match the experimentally observed gap.

7.9.4 Spin current density functional theory

The analysis of the last two sections can be extended to include spin-current response. An interesting twist arises from the fact that the small q limit of $f_{xc,\sigma\sigma'}(q,\omega)$ diverges as $\frac{1}{q^2}$, as a result of the non conservation of the macroscopic spin current (see discussion in Section 5.6.2.2). This singularity is responsible for the appearance of an additional term in the expression (7.231) for the xc electric field, which now takes the form (Qian *et al.*, 2003)

$$-e\vec{E}_{xc\sigma,i}(\omega) = -\nabla_i V_{xc\sigma}^{LDA} + \frac{1}{n_\sigma}\sum_j \frac{\partial \sigma_{xc\sigma,ij}}{\partial r_j}(\vec{r},\omega)$$

$$+ \frac{in^2 A(\omega)}{4\omega}\sum_{\sigma'}\frac{\sigma\sigma'}{n_\sigma n_{\sigma'}}j_{\sigma',i}(\vec{r},\omega), \tag{7.233}$$

where $A(\omega)$ is defined in Eq. (5.192).[47] The essential feature of the new term is that it produces damping of the spin current proportional to the relative velocity between up- and down-spin electrons. This makes it readily distinguishable from the usual viscous friction (contained in the second term), which is proportional to the *derivatives* of the velocity field. Numerical results for $A(\omega)$ have been recently reported by Qian *et al.* (2003).

7.9.5 Linewidth of collective excitations

One of the signal applications of the time-dependent CDFT is the calculation of the linewidth of collective modes. The simplest way to obtain an expression for the linewidth begins with the eigenvalue equation for the current-density in the collective mode:

$$\sum_j \int \{[\chi_{ALDA}^{-1}]_{ij}(\vec{r},\vec{r}',\omega) - f_{xc,ij}^{dyn}(\vec{r},\vec{r}',\omega)\}j_j(\vec{r}',\omega)d\vec{r}' = 0, \tag{7.234}$$

---

[47] The spin-dependent xc stress tensor in Eq. (7.233) is related to the regular part of the small q expansion of $f_{xc,\sigma\sigma'}^h(q,\omega)$ by straightforward extensions of the formulas for the spin independent case (Qian and Vignale, 2003).

where the first term is the inverse of the current–current response function calculated in ALDA, and the second term is the post-ALDA correction. The post-ALDA xc kernel $f_{xc,ij}^{dyn}(\vec{r}, \vec{r}', \omega)$ is defined by the relation

$$\sum_j \int f_{xc,ij}^{dyn}(\vec{r}, \vec{r}', \omega) j_j(\vec{r}', \omega) d\vec{r}' = -\frac{1}{i\omega n_0(\vec{r})} \sum_j \frac{\partial \sigma_{xc,ij}}{\partial r_j}(\vec{r}, \omega). \tag{7.235}$$

Eq. (7.234) is expected to have a solution at a complex frequency $\omega = \omega^{ALDA} + \omega_1$, where ω^{ALDA} is the ALDA frequency of the collective mode (a purely real quantity), while ω_1 is the post-ALDA correction, which can have both a real and an imaginary part. The linewidth Γ of the collective mode is conventionally defined as twice the imaginary part of ω_1.

Assuming that the correction to the ALDA is small, we calculate ω_1 to first order in $f_{xc,ij}^{dyn}$. Standard manipulations, very similar to the ones described in Section 7.6, lead to

$$\omega_1 = \frac{\sum_{ij} \int d\vec{r} \int d\vec{r}' j_i^*(\vec{r}, \omega) f_{xc,ij}^{dyn}(\vec{r}, \vec{r}', \omega) j_j(\vec{r}', \omega)}{\sum_{ij} \int d\vec{r} \int d\vec{r}' j_i^*(\vec{r}, \omega) \partial_\omega [\chi_{ALDA}^{-1}]_{ij}(\vec{r}, \vec{r}', \omega) j_j(\vec{r}', \omega)}, \tag{7.236}$$

where $\partial_\omega [\chi_{ALDA}^{-1}]$ is the derivative of χ_{ALDA}^{-1} with respect to the frequency, evaluated at $\omega = \omega^{ALDA}$. Because $f_{xc,ij}^{dyn}$ in the numerator already contains two gradients operators we can calculate the denominator to zero-th order in the gradient expansion, i.e., we can set

$$[\chi_{ALDA}^{-1}]_{ij}(\vec{r}, \vec{r}', \omega) \simeq \frac{m}{n_0(\vec{r})} \delta(\vec{r} - \vec{r}') \delta_{ij} - \frac{1}{\omega^2} \nabla_{r,i} \frac{e^2}{|\vec{r} - \vec{r}'|} \nabla_{r',j}. \tag{7.237}$$

The apparently second-order (in ∇) Hartree term is absolutely vital, due to the long range of the coulomb interaction.

From the fact that $[\chi_{ALDA}^{-1}]_{ij}(\vec{r}, \vec{r}', \omega)$ must have a vanishing eigenvalue at $\omega = \omega^{ALDA}$, and from the approximation (7.237), it is easy to deduce that

$$\sum_{ij} \int d\vec{r} \int d\vec{r}' j_i^*(\vec{r}, \omega) \partial_\omega [\chi_{ALDA}^{-1}]_{ij}(\vec{r}, \vec{r}', \omega) j_j(\vec{r}', \omega)$$

$$\approx 2m\omega^{ALDA} \int n_0(\vec{r}) |v(\vec{r})|^2 d\vec{r}. \tag{7.238}$$

Finally, making use of Eq. (7.235) we arrive at the following expression for the linewidth

$$\Gamma = \Re e \frac{\sum_{ij} \int d\vec{r} v_i^*(\vec{r}, \omega) \nabla_{r,j} \sigma_{xc,ij}(\vec{r}, \omega)}{m \int n_0(\vec{r}) |v(\vec{r})|^2 d\vec{r}}. \tag{7.239}$$

The physical interpretation of this result is straightforward. The numerator represents the average energy dissipated by the exchange-correlation force $\sum_j \nabla_{r,j} \sigma_{xc,ij}$ during a cycle of oscillation of the current. The denominator is the total energy stored in the oscillation (twice the kinetic energy of the macroscopic motion).

An application of this theory to intersubband plasmons in quantum wells has been recently worked out by Ullrich and Vignale (2002). Some of their results are listed in Table 7.8. Notice that the intrinsic linewidth of the collective mode is still quite smaller than the experimentally measured one. Additional dissipative effects due to impurities and interfacial roughness must be taken into account in order to get a quantitative agreement between theory and experiment.

7.9.6 Nonlinear extensions

Our discussion of non-adiabatic approximations has been limited to the linear response regime. What if the perturbing field is so strong that the linear approximation is not valid? Since the presence of *two* spatial derivatives of the velocity field in the post-ALDA term is dictated by general principles, Eq. (7.227) is expected to remain a sensible approximation even in the non linear regime, provided v and n are slowly varying. The argument goes as follows. Suppose we tried to extend Eq. (7.227) to the nonlinear regime by including terms of order v^2. Because, by galilean invariance, the stress tensor must depend on derivatives of $v(\vec{r})$ such correction would have to go as $(\nabla v)^2$. But then the expression for the force density, given by the derivative of the stress tensor, would contain at least *three* derivatives. Thus, for sufficently slow spatial variation of the density and velocity fields, the nonlinear terms can be neglected.

Since the ALDA is an intrinsically nonlinear approximation, Vignale, Ullrich, and Conti (1997) proposed that Eq. (7.227), written in the time domain, could provide a reasonable description of both linear and nonlinear response properties. An alternative nonlinear description in terms of a scalar potential V_{xc} has been proposed by Bünner, Dobson, and Gross (1997). These nonlinear extensions are still largely unexplored to date.

Exercises

7.1 **The universal functional *F[n]* as a Legendre transform.** Let $E[V]$ be the ground-state energy of an electronic system, regarded as a functional of the external potential $V(\vec{r})$.

(a) Show that this functional is *concave*, that is, for any pair of potentials $V_1(\vec{r})$ and $V_2(\vec{r})$ and for any α such that $0 \leq \alpha \leq 1$ one has

$$E[V] \geq \alpha E[V_1] + (1-\alpha)E[V_2] \,,$$

where $V(\vec{r}) = \alpha V_1(\vec{r}) + (1-\alpha)V_2(\vec{r})$. Also show that $E[V]$ is "monotonically decreasing," that is, if $V_1(\vec{r}) \leq V_2(\vec{r})$ for all $\vec{r}$, then $E[V_1] \leq E[V_2]$.

(b) Show that the ground-state density $n(\vec{r})$ is given by the functional derivative $\frac{\delta E[V]}{\delta V(\vec{r})}$.

(c) Finally show that the *Legendre transform* of $E[V]$ with respect to $V(\vec{r})$, which exists by virtue of concavity, is just the universal functional $F[n]$.

7.2 **Hohenberg–Kohn potential for a one-electron system.** Given a strictly positive one-electron density $n(\vec{r})$, such that $\int n(\vec{r})d\vec{r} = 1$, find an expression for the one-electron potential $V(\vec{r})$ that yields $n(\vec{r})$ *in the ground-state.*

7.3 **Example of non-V-representable density in a one-electron system.** Show that the one-electron density

$$n(\vec{r}) = \frac{2}{\pi^{3/2}a^3}\left(\frac{r}{a}\right)^2 e^{-\frac{r^2}{a^2}}$$

where a is an arbitrary length scale, cannot be the ground-state density of an electron subjected to a potential that is everywhere finite, but can be realized in an *excited state* of such a potential. Identify this potential. [Hint: do the previous exercise first.]

7.4 **Exchange-correlation potential for a one-electron system.** Show that the exchange-correlation potential associated with a one-electron density such that $\int n(\vec{r})d\vec{r} = 1$ is equal to the negative of the Hartree potential.

7.5 **Exchange-correlation potential for a two-electron system.** Show that the exchange-correlation potential associated with a two-electron density such that $\int n(\vec{r})d\vec{r} = 2$ is given by

$$V_{xc}(\vec{r}) = \frac{\hbar^2}{2m}\frac{\nabla^2 n^{1/2}(\vec{r})}{n^{1/2}(\vec{r})} - \int \frac{e^2}{|\vec{r}-\vec{r}'|}n(\vec{r}')d\vec{r}' - V_{ext}(\vec{r}), \qquad \text{(E7.1)}$$

where $V_{ext}(\vec{r})$ is the external potential that yields $n(\vec{r})$.

7.6 **The optimized effective potential is the exchange-only Kohn–Sham potential.** Show that the exchange-only Kohn–Sham potential $V_{KS,x}(\vec{r}) = V(\vec{r}) + V_H(\vec{r}) + V_x(\vec{r})$, with $V_x(\vec{r})$ defined by Eq. (7.39), coincides with the "optimized effective potential" defined in Section 2.8. That is, the ground-state generated by $V_{KS,x}(\vec{r})$ has the least expectation value of $\hat{T} + \hat{H}_{e-e} + \int V(\vec{r})\hat{n}(\vec{r})d\vec{r}$ among all the Slater determinants that arise from a local potential. [Hint: start from the stationary condition $\frac{\delta T_s[n]}{\delta n(\vec{r})} = -V_{KS,x}(\vec{r})$, and show that this implies the stationarity of $\langle\psi_0[n]||\hat{T} + \hat{H}_{e-e} + \int V(\vec{r})\hat{n}(\vec{r})d\vec{r}|\psi_0[n]\rangle = 0$ with respect to infinitesimal variations of n, and hence of $\psi_0[n]$.]

7.7 **Properties of the correlation energy functional $E_{c,\lambda}[n]$.** With reference to the definition (7.51) of the correlation energy functional at coupling constant λ, prove that $\frac{dE_{c,\lambda}[n]}{d\lambda} < 0$ and $\frac{d^2E_{c,\lambda}[n]}{d\lambda^2} < 0$. [Hint: $\frac{dE_{c,\lambda}[n]}{d\lambda} = \langle\psi_\lambda|\hat{H}_{e-e}|\psi_\lambda\rangle - \langle\psi_0|\hat{H}_{e-e}|\psi_0\rangle < \lambda^{-1}[\langle\psi_\lambda|\hat{T} + \lambda\hat{H}_{e-e}|\psi_\lambda\rangle - \langle\psi_0|\hat{T} + \lambda\hat{H}_{e-e}|\psi_0\rangle] \le 0$. For the second inequality recall that the second-order correction to the ground-state energy of a quantum system is always negative.]

7.8 **Virial theorem in density functional theory.** Show that the quantum virial identity

$$2\langle \hat{T} \rangle = -\sum_{i=1}^{N} \langle \hat{\vec{r}}_i \cdot \vec{\nabla}_{\vec{r}_i} W(\vec{r}_1, \dots, \vec{r}_1) \rangle \, ,$$

where $W(\vec{r}_1, \dots, \vec{r}_N)$ is the *total* potential energy of the system implies the following universal relation between density functionals:

$$2T[n] + U[n] = -\int d\vec{r} \, n(\vec{r}) \, \vec{r} \cdot \vec{\nabla} \frac{\delta F[n]}{\delta n(\vec{r})} \, , \qquad \text{(E7.2)}$$

where $T[n]$ is the kinetic energy functional and $U[n] = \langle \psi[n] | \hat{H}_{e-e} | \psi[n] \rangle$ is the potential energy functional. Show that Eq. (E7.2) reduces to the standard virial theorem if the external potential is of the coulomb form.

7.9 **Non-uniqueness of the external potentials in generalized DFT.** Two different sets of external potentials $\{V_i\}$ and $\{V_i'\}$ in the hamiltonian (7.68) can produce the same ground-state only if the condition

$$\sum_i \int [V_i'(\vec{r}) - V_i(\vec{r})] \hat{n}_i(\vec{r}) d\vec{r} \, |\psi\rangle \; = \; \Delta E |\psi\rangle,$$

where ΔE is a constant, is fulfilled. Show that a solution of this equation can be *systematically* constructed if the hamiltonian $\hat{H}$ admits a constant of the motion $\hat{C}$ that is a linear function of the generalized densities, i.e.,

$$\hat{C} = \sum_i \int c_i(\vec{r}) \hat{n}_i(\vec{r}) d\vec{r} \, ,$$

where $c_i(\vec{r})$ are suitable functions, and $[\hat{C}, \hat{H}] = 0$. The solution is

$$V_i'(\vec{r}) = V_i(\vec{r}) + \Delta E c_i(\vec{r}) \, .$$

7.10 **Example of non-uniqueness of the external potential, I.** In a cylindrically symmetric molecule the z-component of the spin, $\hat{S}_z$, and the z-component of the orbital argument momentum, $\hat{L}_z$, are constants of the motion (spin–orbit coupling is neglected). Show that these two constants of motion, when expressed in terms of the spin density and the current density, respectively, have the form discussed in the previous exercise. Conclude that in spin-DFT and current-DFT the ground-state densities do not uniquely determine the external fields (magnetic field and vector potential).

7.11 **Example of non-uniqueness of the external potential, II.** Let $n(\vec{r})$ and $\vec{j}_p(\vec{r})$ be the ground-state density and paramagnetic current density of a two-electron system subjected to a scalar potential $V(\vec{r})$ and a vector potential $\vec{A}(\vec{r})$. Ignore the Zeeman coupling with the magnetic field. Show that $V(\vec{r})$ and $\vec{A}(\vec{r})$ are given by the following expressions:

$$V(\vec{r}) = \frac{\hbar^2}{2m} \frac{\nabla^2 \sqrt{n(\vec{r})}}{\sqrt{n(\vec{r})}} - \frac{e^2}{2mc^2} \left(\frac{\vec{\nabla} \times \vec{Q}(\vec{r})}{n(\vec{r})} \right)^2 + constant$$

$$\vec{A}(\vec{r}) = -\frac{cm}{e} \frac{j_p(\vec{r})}{n(\vec{r})} + \frac{\vec{\nabla} \times \vec{Q}(\vec{r})}{n(\vec{r})} ,$$

where $\vec{Q}(\vec{r})$ is a small, but otherwise arbitrary vector field.

7.12 **Kohn's formula for the derivative discontinuity of the exchange-correlation potential.** In order to derive Kohn's formula (refer to equation in footnote 14) for the derivative discontinuity of the exchange-correlation potential one begins by writing the "plus" functional derivative in Eq. (7.65) as

$$\left. \frac{\delta E_{xc}[n]}{\delta n(\vec{r})} \right|_{n_m^+} = \lim_{\eta \to 0} \frac{E_{xc}[n + \eta \delta_{\vec{r}}] - E_{xc}[n]}{\eta} ,$$

where $\delta_{\vec{r}}(\vec{r}') \equiv \delta(\vec{r}' - \vec{r})$ is a δ-function density distribution centered at the position $\vec{r}$ at which we want to calculate the functional derivative. This δ-function is then decomposed into two parts: $n_c(\vec{r}')$ – the density of one electron at the bottom of the conduction band, and a remainder $m_{c,\vec{r}}(\vec{r}')$, which leaves the number of electrons unchanged. (For the "minus" derivative one must use $n_v(\vec{r}')$ – the density of an electron at the top of the valence band – and the corresponding remainder $m_{v,\vec{r}}(\vec{r}')$.) Show that this decomposition, combined with the assumption of differentiability of the xc energy functional *at constant particle number*, leads to the equation given in footnote 14.

7.13 **Spin scaling relation for the exchange energy functional.** Show that the exact exchange energy functional $E_x[n_\uparrow, n_\downarrow]$ satisfies the spin-scaling relation (7.95).

7.14 **Gradient correction to non-interacting kinetic energy functional.** Calculate the coefficient B_0 of the leading gradient correction to the non-interacting kinetic energy functional (see Eq. (7.85)) from the small-q expansion of the inverse of the Lindhard function in three and two dimensions.

7.15 **Calculation of van der Waals interaction energy for two neutral atoms.** Derive Eq. (7.116) for the leading correction to the ground-state energy of two widely separated atoms.

7.16 **Different time-dependent potentials can produce the same density.** Two different time-dependent potentials can produce the same density starting from two different initial states, as the following example shows (Maitra and Burke, 2001; a detailed study of this example can be found in Holas and Balawender, 2002). Consider two non-interacting electrons in 1 dimension. A possible initial state is the Slater determinant, $\Phi(x_1, x_2)$, formed from $\phi_1(x) \propto e^{-x^2/2}$ and $\phi_2(x) \propto x e^{-x^2/2}$ – the two lowest energy states of a particle in a harmonic oscillator potential of force constant $k = 1$. Another possible initial state is the Slater determinant, $\tilde{\Phi}(x_1, x_2)$, formed from the orthonormal orbitals

$$\tilde{\phi}_1(x) = \sqrt{1 + f(x)\phi_2^2(x)}\,\phi_1(x)\,,$$

$$\tilde{\phi}_2(x) = \sqrt{1 - f(x)\phi_1^2(x)}\,\phi_2(x)\,,$$

where $f(x) = c(256x^4 - 192x^2 + 12)e^{-2x^2}$, and c is an arbitrary constant with values between -0.172 and 0.147.

(a) Show that these two states have the same density and the same time derivative of the density at the initial time ($t = 0$).

(b) Let $n(\vec{r}, t)$ be the density of the state that evolves from Φ under the time-dependent potential $V(\vec{r}, t)$, and let $\tilde{V}(\vec{r}, t)$ be the potential that generates the same density starting from $\tilde{\Phi}$ (according to van Leeuwen's theorem such a potential exists). Show that $\tilde{V}(\vec{r}, t)$ differs from $V(\vec{r}, t)$ by more than a time-dependent constant.

7.17 **Density and current density determine a one-electron state.**

(a) Show that the ratio of the current density to the density, $\frac{\vec{j}(\vec{r})}{n(\vec{r})}$ in a one-electron system is purely longitudinal (i.e., it is the gradient of a scalar field).

(b) Show that two *one-electron* states that have the same density and the same current density must perforce coincide up to a global time-dependent phase.

7.18 **Time-dependent exchange-correlation potential for a two-electron system.** Calculate the exact time-dependent Kohn–Sham potential for a *time-dependent* 2-electron system that evolves with a density $n(\vec{r}, t)$. (This is the time-dependent version of Exercise 5.)

7.19 **Harmonic potential theorem.** The *harmonic potential theorem* states that the density of a system of electrons subjected to a static harmonic potential $V_0(\vec{r}) = \frac{m}{2}\omega_0^2 r^2$ and to a time-dependent uniform electric field $\vec{E}(t)$ evolves in time as $n(\vec{r} - \vec{x}(t), t)$ where $n(\vec{r}, t)$ is the time evolution of the density in the absence of the electric field and $\vec{x}(t)$ is the solution of the classical equation of motion for a harmonic oscillator of frequency ω_0 starting from rest at $\vec{x} = 0$ under the action of the field $\vec{E}(t)$. Prove that Eqs. (7.205) and (7.206) guarantee that the solution of the time-dependent Kohn–Sham equation satisfies this theorem.

7.20 **Zero-force theorem.** Prove Eq. (7.203) from the Heisenberg equation of motion for the center of mass operator $\hat{\vec{R}}_{cm} = \int \vec{r}\,\hat{n}(\vec{r})\,d\vec{r}$ in a finite system. [Hint: compare the time evolution of $\hat{\vec{R}}$ under the full many-body hamiltonian and the one-particle Kohn–Sham hamiltonian, and recall that the expectation value of $\hat{n}(\vec{r})$ is, by definition, the same in both cases.]

7.21 **Acceleration sum rule for the density–density response function.** Prove that the density–density response function $\chi_{nn}(\vec{r}, \vec{r}', \omega)$ of a finite system of electrons subjected to a static external potential $V_0(\vec{r})$ satisfies the sum rule

$$\int \chi_{nn}(\vec{r}, \vec{r}', \omega) \left[\vec{\nabla}V_0(\vec{r}') - m\omega^2\vec{r}'\right] d\vec{r}' = \vec{\nabla}n_0(\vec{r})\,,$$

where $n_0(\vec{r})$ is the ground-state density (Vignale, 1995). [Hint: look at the system from an accelerated frame of reference that performs small oscillations at frequency ω.]

7.22 **Exchange-correlation stress tensor.** Work out Eq. (7.227) for the xc vector potential in terms of the stress tensor (7.228).

7.23 **Hydrostatic form of the ALDA exchange-correlation force.** Show that the ALDA term in Eq. (7.231) can be written as $-\nabla_i p_{xc}(n(\vec{r}))$ where $p_{xc}(n) = -n^2 \frac{d\epsilon_{xc}(n)}{dn}$ is the exchange-correlation pressure of the homogeneous electron liquid.

8

The normal Fermi liquid

8.1 Introduction and overview of the chapter

This chapter is devoted to the discussion of one of the most useful and sweeping concepts of condensed matter physics: the *Landau Fermi liquid concept*. This paradigm describes not only the low-energy behavior of interacting electrons in metals (our main interest here) but also the low-energy behavior of Fermi systems such as ^{3}He and nuclear matter, irrespective of the complex details that may characterize them.

It had been observed since the first studies of the physical properties of metals that, in spite of their mutual interaction, electrons appear to behave as independent particles, the main observed features being qualitatively understood in terms of the Sommerfeld picture (1928) based on the degenerate ideal Fermi gas model. This model was not easily justified. The average interaction energy per electron can be roughly estimated to be of the order of $e^2 n^{\frac{1}{d}}$. For Na, for which $r_s \simeq 3.93$ for instance, this is gives $\left(\frac{6}{\pi}\right)^{\frac{1}{3}} \frac{Ry}{r_s} \simeq 4.29$ eV, which is already larger than the value $\epsilon_F \simeq 3.24$ eV of the Fermi energy. From a more technical viewpoint, the difficulty stems from the fact that the r_s values appropriate to the historical metallic systems approximately range from 2 to 5, and therefore lie well outside of the high density limit of $r_s \ll 1$ in which the coulomb interaction can be safely considered a perturbation.

It was not until the late fifties that this unresolved state of affairs was finally clarified from a theoretical point of view by L. D. Landau (Landau, 1957a, 1957b, 1959). Although Landau did not provide a rigorous solution of the problem, he provided a firm basis for the understanding of the "normal" low-energy behavior of interacting Fermi systems. The solidity of this basis has been confirmed by subsequent theoretical and experimental work. So strong has been the impact of the Landau theory that the Landau Fermi liquid is often referred to as the Fermi liquid *tout-court*. Nowadays, experiments that report deviations from the normal Landau behavior are considered very exciting, and usually hit the headlines of scientific journals. Some examples of "non-Fermi liquid" behavior in low-dimensional systems will be presented in Chapters 9 and 10.

This chapter is organized as follows.

Section 8.2 provides a general introduction to the Landau Fermi liquid concept. The discussion touches upon most of the relevant ideas and sets the stage for the more specialized treatments of the subsequent sections.

Section 8.3 is devoted to the *macroscopic theory of the Fermi liquids* due to and mostly developed by Landau. The first six subsections provide explicit derivations of the basics results. The section is completed by a short presentation of the *kinetic equation* for the quasiparticle distribution function, and useful discussions of the experimental studies of quasiparticle properties, of the effects of the electron–phonon coupling, and of the idea of dynamic shear modulus.

Section 8.4 presents a simple theory of the *quasiparticle inelastic lifetime* based on the *Fermi golden rule* of elementary quantum mechanics. This section can be read without any knowledge of formal many-body theory.

The *microscopic underpinning* of the Landau theory is presented in Section 8.5. The first subsection contains a formal definition of the quasiparticle concept. The second deals with the Fermi liquid concept from the point of view of field theoretic perturbation theory. The last two subsections deal with more advanced topics, namely, the proof of *Luttinger's theorem* and the construction of the Landau energy functional from the Green's function formalism of Chapter 6.

Section 8.6 contains an in-depth presentation of the *renormalized hamiltonian method*. This powerful and physically appealing technique allows us to obtain explicit formulas for the *quasiparticle energy*, the *quasiparticle interaction* and the *renormalization constant*.[1] This material can be read without any knowledge of field theoretical techniques.

Section 8.7 presents a *practical diagrammatic approach* to the calculation of the properties of a Fermi liquid. In particular, the popular *GW approximation* for the self-energy is discussed in detail and shown to be a special case of the renormalized hamiltonian approach.

Section 8.8 provides a discussion of explicit *numerical results for various quasiparticle properties*.

Finally, Section 8.10 illustrates the peculiar deviations from the standard theory of Fermi liquids that occur in the presence of *disorder*.

8.2 The Landau Fermi liquid

Landau's basic idea was that, under very broad conditions, the low-lying excitations of a system of interacting Fermions with repulsive interactions can be constructed starting from the low-lying states of a non-interacting ideal Fermi liquid by a suitably slow *switching-on* of the interaction between the particles. There are several subtleties in the definition of the switching-on process, beginning with a precise definition of the words "suitably slow" (see footnote[5]). For the time being we will not delve into these subtleties, but simply notice that the switching-on procedure establishes a one-to-one correspondence between

---

[1] The renormalized hamiltonian approach can be seen as a "poor man bosonization" of the electron liquid. This aspect will sharpen as the reader becomes familiar with the ideas developed in Chapters 9 and 10.

the eigenstates of the ideal system and a set of (approximate) eigenstates of the interacting one.[2] Since the eigenstates of the noninteracting system are specified by a set of occupation numbers $\{\mathcal{N}_{\vec{k}\sigma}\}$ of single-particle momentum eigenstates, it follows that the corresponding low-lying excitations of the interacting system can also be described by the same set of occupation numbers. It must be clearly understood that the quantum numbers $\mathcal{N}_{\vec{k}\sigma}$, which specify an excited state of the Fermi liquid, are not the true momentum occupation numbers $n_{\vec{k}\sigma} = \langle \hat{a}^{\dagger}_{\vec{k}\sigma} \hat{a}_{\vec{k}\sigma} \rangle$ for that state. Rather, they are momentum occupation numbers of the ideal system from which the excited state has evolved.

Because in an interacting system the momentum occupation numbers are not constants of the motion, one could have naively expected that any memory of the initial noninteracting momentum distribution $\{\mathcal{N}_{\vec{k}\sigma}\}$ would be completely lost at the end of the switching-on process. The genius of Landau was to recognize that, for states that are weakly excited (i.e., close to the noninteracting Fermi distribution) the occupation numbers change very, very slowly even when particle-particle interactions are strong. The main consequence of this fact is that the $\mathcal{N}_{\vec{k}\sigma}$ retain their validity as *approximate quantum numbers*, which specify an excited state. Thus, low energy elementary excitations of an interacting Fermi liquid can be described in terms of addition or removal of individual *quasiparticles* from a filled Fermi sphere of radius k_F, where k_F is the Fermi momentum of a noninteracting electron gas with the same density.[3] For example, a state of the ideal system containing one particle of momentum $\hbar \vec{k}$ with $k \geq k_F$ outside the full Fermi sphere evolves into an excited state of the interacting system containing one quasiparticle of momentum $\hbar \vec{k}$ outside the (suitably modified) Fermi sphere. Likewise, a state of the noninteracting system containing one empty state (a hole) of momentum $\hbar \vec{k}$ within the Fermi sphere evolves into an excited state of the interacting system containing a *quasihole* of the same momentum. More complex excitations consisting of multiple quasiparticles and quasiholes can be constructed in a similar manner. By definition, the ground-state has $\mathcal{N}_{\vec{k}\sigma} = \Theta(k_F - k)$. In contrast to this, we will show in this chapter that the momentum occupation numbers $n_{\vec{k}\sigma}$ in the ground-state *decrease discontinuously* by an amount Z (with $0 < Z \leq 1$) as $\vec{k}$ crosses the Fermi sphere from inside to outside. The discontinuity, Z, is of course 1 in the noninteracting system and less than 1 in the interacting system. The existence of a discontinuity in the momentum occupation number at $k = k_F$ is one of the distinctive signatures of the normal Fermi liquid.

The physical basis of the Landau theory rests on the surprising ineffectiveness of electron–electron scattering to change the momentum distribution of quasiparticles near the Fermi level. What happens is that most of the states into which a quasiparticle near the Fermi surface might end up after a collision are already occupied by other electrons, and therefore, according to the Pauli exclusion principle, unavailable. Because of this *Pauli blocking* effect, which operates irrespective of the strength of the interaction, the rate at which a quasiparticle is scattered out of a state of momentum $k \simeq k_F$ vanishes for $k \to k_F$.

---

[2] This is often referred to as "adiabatic switching-on". The process is however hardly adiabatic for in view of the large degeneracy of the initial non-interacting excited state, a suitably slow switching-on of the interaction leads to an image state in the interacting electron liquid which is a superposition of many nearly degenerate eigenstates.
[3] This particular statement is referred to as the Luttinger's theorem and is discussed in Section 8.5.4.

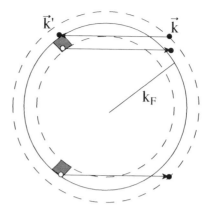

Fig. 8.1. Schematic illustration of two possible decay processes for a quasiparticle near the Fermi surface. The quasiparticle makes a transition from $\vec{k}$ to $\vec{k}'$, producing an electron–hole pair in the process. Momentum and energy conservation restrict the momentum of the hole to the shaded regions.

This result can be obtained from a simple *phase space argument*. Consider a quasiparticle with initial wave vector $\vec{k}$ with $k > k_F$. At zero temperature the empty states into which the quasiparticle can decay lie within a shell of thickness $|k - k_F|$ just above the Fermi surface. The number of states in this region is clearly proportional to $|k - k_F|$ – a result valid in one, two and three dimensions. Now, through the coulomb interaction, the momentum and energy change of the quasiparticle will be offset by the momentum and energy of an electron–hole pair. In two and three dimensions, the state of the hole must lie within a shell of thickness $|k - k_F|$ below the Fermi surface (see Fig. 8.1). This contributes another factor of $|k - k_F|$ to the probability of decay, which, as anticipated, is thus found to be proportional to $(k - k_F)^2$ in three dimensions.[4]

Accordingly in two and three dimensions in the limit $k \to k_F$, the inverse of the scattering rate, i.e. the scattering time $\tau_{\vec{k}}$, becomes long enough to include many cycles of oscillation of an external field that is able to create the quasiparticle excitation out of the ground-state (the frequency of this field being proportional to the excitation energy which is of order $|k - k_F|$). Thus, on a time scale that is *short* compared to τ_k (but still long compared to the inverse excitation frequency) the occupation number $\mathcal{N}_{\vec{k}\sigma}$ can be regarded as a good quantum number for the excited state.[5]

Strictly speaking however the $\mathcal{N}_{\vec{k}\sigma}$ are not exact quantum numbers (for that to be true the scattering rate would have to actually vanish), and if one waits long enough, i.e., up to times

---

[4] In two dimensions, a more accurate calculation shows that the scattering rate vanishes at a somewhat slower rate $-(k - k_F)^2 \ln |k - k_F|$ (see Section 8.4.3). In the presence of disorder the quasiparticle inelastic lifetime is even shorter. This is discussed in Section 8.10.

[5] An alert reader will notice that the phase-space argument is circular: one assumes the existence of quasiparticles to deduce that their lifetime is long. This does not prove the existence of quasiparticles, but shows that one can assume their existence without falling in a contradiction. This discussion does also clarify that a "suitably slow" switching-on must be carried out in a time intermediate between the fast time scale of order $\frac{1}{v_F |k - k_F|}$ related to the resolution of a quasiparticle state and the quasiparticles lifetime which we have shown to be of order $\frac{\varepsilon_F}{\hbar (v_F |k - k_F|)^2}$.

$t \gg \tau_{\vec{k}}$, one will see them change. Thus as anticipated a quasiparticle state is *not* an exact eigenstate of the interacting Fermi liquid. Rather, it must be understood as a superposition of closely spaced exact eigenstates with energies spread over a range of width $\hbar/\tau_{\vec{k}} \ll \varepsilon_{\vec{k}}$ about the median quasi-particle energy $\varepsilon_{\vec{k}}$ (see footnote[2]). Such a state decays with a characteristic lifetime $\tau_{\vec{k}}$, and can be regarded as stationary only for times much shorter than $\tau_{\vec{k}}$. It turns out, and it will be shown in the next section, that these *quasi-eigenstates* completely determine the response of the system to macroscopic perturbations.

The situation is completely different in one-dimensional systems, for in that case the spectral density of electron–hole pairs exhibits a huge peak at small wavevector q and frequency $\omega = v_F q$. All the electron–hole pairs with given momentum $\hbar q$ have essentially the same energy comprised between $\hbar|\omega_-(q)| = \left|\hbar v_F q - \frac{\hbar^2 q^2}{2m}\right|$ and $\hbar\omega_+(q) = \hbar v_F q + \frac{\hbar^2 q^2}{2m}$ (see Fig. 4.2). Due to the massive quasi-degeneracy of the noninteracting spectrum, it turns out that an arbitrarily weak interaction causes a complete reconstruction of the many-body eigenstates, and, in particular, destroys the "Fermi surface" (actually, two points in 1D). Thus, there is no Landau Fermi liquid in one dimension. The correct paradigm in this case is the *Luttinger liquid*, which will be described in detail in Chapter 9. The rest of this chapter will only deal with two and three-dimensional systems.

The main properties of a quasiparticle excitation are embodied in its *effective mass m**, which determines the energy of a single quasiparticle, and in the *Landau interaction function* $f_{\vec{k}\sigma,\vec{k}'\sigma'}$, which tells us how the energy of a quasiparticle is modified by the presence of other quasiparticles. The quasiparticle effective mass differs in general from the bare particle mass m and is related to the interaction function.[6] Because of the interaction between quasiparticles, the response functions of a Fermi liquid differ in general from those of an ideal gas of quasiparticles as the change in the quasiparticle distribution function caused by an external field induces an additional mean field that must be taken into account self-consistently.

The usefulness and beauty of the Landau theory of the Fermi liquid lies in the fact that all the relevant properties characterizing the low energy physical behavior of a metallic system can therefore be described in terms of a few numbers, the *Landau parameters* $F_\ell^{s,a}$ which, as we will show, are simply related to the function $f_{\vec{k}\sigma,\vec{k}'\sigma'}$. In fact even if such a function is unknown the Landau parameters still provide a simple way to connect different macroscopic properties of a Fermi liquid.

The concept of quasiparticle is a truly encompassing one. It applies not only to electrons in metals and doped semiconductors (where the renormalization factor $\frac{m^*}{m}$ and Z remain close to 1) but also to ^{3}He atoms in the liquid phase where $\frac{m^*}{m} \simeq 3$ and to highly correlated *heavy fermion systems* where $\frac{m^*}{m}$ can run in the hundreds. Considering the diversity in coupling constants and physical character of these systems it is quite amazing that they can be subsumed under the same theoretical paradigm. On the other hand, as we have seen, dimensionality is crucial to the existence of a Landau Fermi liquid.

---

[6] In real metals, what we here call the bare particle mass includes the effects of the underlying electronic band structure and the lattice vibrations. The latter will be discussed in Section 8.3.7.

8.3 Macroscopic theory of Fermi liquids

We begin our presentation of the theory of the Fermi liquids by providing an introduction
to the original macroscopic formulation of Landau. This formulation is presently the main
paradigm for the phenomenological description of the low energy properties of metallic
systems in two and three dimensions. The microscopic justification for the main ingredients
of the Landau approach will be provided later, beginning in Section 8.5.

8.3.1 The Landau energy functional

As discussed in the previous section, a set of noninteracting occupation numbers $\mathcal{N}_{k\sigma} = 0$
or 1 defines, by continuity, a quasi-eigenstate of the interacting Fermi liquid. Similarly, a
distribution of *fractional* occupation numbers $0 \leq \mathcal{N}_{k\sigma} \leq 1$ defines an ensemble of quasiparticle states: in such an ensemble the quasiparticle state of momentum $\hbar\vec{k}$ and spin σ
has a probability $\mathcal{N}_{k\sigma}$ to be occupied and $1 - \mathcal{N}_{k\sigma}$ to be empty. At the heart of Landau's
macroscopic theory of the Fermi liquids lies an *Ansatz* for the functional dependence of the
energy of the liquid on the quasiparticle distribution function $\mathcal{N}_{\vec{k}\sigma}$. This functional is in fact
a expansion for the energy to second order in the deviation of the quasiparticle distribution
function from its ground-state value $\mathcal{N}_{\vec{k}\sigma}^{(0)} = \Theta(k_F - k)$:

$$E[\{\mathcal{N}_{\vec{k}\sigma}\}] = E_0 + \sum_{\vec{k}\sigma} \mathcal{E}_{\vec{k}\sigma} \delta\mathcal{N}_{\vec{k}\sigma} + \frac{1}{2} \sum_{\vec{k}\sigma, \vec{k}'\sigma'} f_{\vec{k}\sigma, \vec{k}'\sigma'} \delta\mathcal{N}_{\vec{k}\sigma} \delta\mathcal{N}_{\vec{k}'\sigma'} , \tag{8.1}$$

where E_0 is the ground-state energy, $\mathcal{E}_{\vec{k}\sigma}$ is the isolated quasiparticle energy, $f_{\vec{k}\sigma, \vec{k}'\sigma'}$ is the
Landau interaction function and $\delta\mathcal{N}_{\vec{k}\sigma} = \mathcal{N}_{\vec{k}\sigma} - \mathcal{N}_{\vec{k}\sigma}^{(0)}$ is the deviation of the quasiparticle
distribution function from the ideal Fermi distribution at $T = 0$.[7]

Because the quasiparticles are well defined only in the immediate vicinity of the Fermi
surface, it is evident that this expansion makes sense only when $\delta\mathcal{N}_{\vec{k}\sigma}$ is restricted to a
thin shell of momentum space surrounding the Fermi surface. In addition, since every wave
vector sum introduces a factor L^d, the interaction function $f_{\vec{k}\sigma, \vec{k}'\sigma'}$ must scale as the inverse of
the volume $\frac{1}{L^d}$ in order to keep the energy proportional to the volume in the thermodynamic
limit.

The energy $\mathcal{E}_{\vec{k}\sigma}$ of a single quasiparticle can be formally viewed as the functional derivative of the energy with respect to the quasiparticle distribution function evaluated in the
ground-state:

$$\mathcal{E}_{\vec{k}\sigma} = \left(\frac{\delta E}{\delta\mathcal{N}_{\vec{k}\sigma}} \right)_{\mathcal{N}_{\vec{k}\sigma} = \mathcal{N}_{\vec{k}\sigma}^{(0)}} . \tag{8.2}$$

Since the ground-state of the $N + 1$-particle system is obtained by adding a quasiparticle

---

[7] It is important to keep in mind that, in the case of the electron liquid, Eq. (8.1) does *not* include the electrostatic energy associated
with the space charge $-\frac{1}{L^d} \sum_{\vec{k}\sigma} e\delta\mathcal{N}_{\vec{k}\sigma}$. One can mentally picture Eq. (8.1) as the energy of a locally neutral system, in which
the space charge is perfectly compensated, at no energy cost, by an adjustable background of positive charge.

of wavevector k_F to the ground-state of the N-particle system, it is evident that

$$\mathcal{E}_{k_F\sigma} = \mu \,, \tag{8.3}$$

where μ is the chemical potential.

In an isotropic system, for $|\vec{k}|$ close to k_F, the quasiparticle energy can be expanded as

$$\mathcal{E}_{\vec{k}\sigma} \simeq \mu + \hbar v_F^*(k - k_F) \,, \tag{8.4}$$

where

$$v_F^* = \frac{1}{\hbar}\left|\frac{\partial \mathcal{E}_{\vec{k}\sigma}}{\partial \vec{k}}\right|_{k=k_F} \tag{8.5}$$

defines the *effective Fermi velocity* of a quasiparticle. v_F^* can be conveniently written as

$$v_F^* = \frac{\hbar k_F}{m^*} \,, \tag{8.6}$$

which defines the fundamental notion of *quasiparticle effective mass m^**.

The effective mass determines in turn $N^*(0)$, the quasiparticle density of states (per unit volume) at the Fermi level μ. This can be seen as follows:

$$N^*(0) = \frac{1}{L^d}\sum_{\vec{k}\sigma}\delta(\mathcal{E}_{\vec{k}\sigma} - \mu) = \frac{1}{L^d}\sum_{\vec{k}\sigma}\frac{\delta(k - k_F)}{\hbar v_F^*} = \frac{m^*}{m}N(0) \,, \tag{8.7}$$

where we have made use of Eq. (8.3). Notice that in Eq. (8.7) $N(0)$ is the density of states per unit volume at the Fermi surface for a noninteracting electron gas. Since $N(0)$ is proportional to the electron mass, $N^*(0)$ is in turn given by the noninteracting expression with m simply replaced by m^*.

A fundamental role in the Landau theory is played by the quantity

$$\tilde{\mathcal{E}}_{\vec{k}\sigma} = \frac{\delta E}{\delta \mathcal{N}_{\vec{k}\sigma}} = \mathcal{E}_{\vec{k}\sigma} + \sum_{\vec{k}'\sigma'} f_{\vec{k}\sigma,\vec{k}'\sigma'}\delta \mathcal{N}_{\vec{k}'\sigma'} \,, \tag{8.8}$$

often referred to as the *local quasiparticle energy*. This can be interpreted as the energy of a quasiparticle modified by its interaction with other quasiparticles. From the form of this equation it is clear that, within the Landau theory, this effect is treated by means of a mean field approximation.

We turn next our attention to the quasiparticle interaction function. An inspection of Eq. (8.1) reveals that this quantity can be expressed in terms of functional derivatives of the Landau energy functional with respect to the quasiparticle distribution function as follows:

$$f_{\vec{k}\sigma,\vec{k}'\sigma'} = \frac{\delta^2 E}{\delta \mathcal{N}_{\vec{k}\sigma}\delta \mathcal{N}_{\vec{k}'\sigma'}} = \frac{\delta \tilde{\mathcal{E}}_{\vec{k}\sigma}}{\delta \mathcal{N}_{\vec{k}',\sigma'}} \,, \tag{8.9}$$

where the functional derivatives are evaluated at the ground-state distribution. It is important to appreciate that in order to correctly perform the second derivative appearing in Eq. (8.9), it is necessary to know the energy $E[\{\mathcal{N}_{\vec{k}\sigma}\}]$ as a functional of both $\mathcal{N}_{\vec{k}\uparrow}$ and $\mathcal{N}_{\vec{k}\downarrow}$

up to second order in $\delta\mathcal{N}_{\vec{k}\sigma}$. This implies, for instance, that to derive the expression for the interaction function in a paramagnetic system one needs the knowledge of the energy functional appropriate to an *infinitesimally polarized electron liquid* (Yarlagadda and Giuliani, 1994).[8] This complication clearly does not arise in the case of the quasiparticle energy since its calculation only requires a knowledge of the Landau energy functional up to first order in $\delta\mathcal{N}_{\vec{k}\sigma}$.

Eq. (8.1) is widely used to calculate from a macroscopic point of view the thermal equilibrium properties, the response functions, and the transport properties of an interacting Fermi liquid, and to establish relationships between different such properties. These developments are presented in several textbooks. Particularly authoritative is the discussion presented by Pines and Nozières (1966). We have neither the intention nor the ability to compete with them. For completeness, a brief summary of the main results will be presented in the next few sections.

8.3.2 The heat capacity

As it turns out, the low-temperature specific heat of a Fermi liquid coincides with that of a noninteracting Fermi gas comprised of particles of mass m^*: it is therefore given by

$$c_v(T) = \frac{\pi^2}{3}N^*(0)L^d k_B^2 T , \qquad (8.10)$$

where, as we have seen, $N^*(0)$, the density of quasiparticle energy states per unit volume at the Fermi level in d dimensions, differs from the corresponding quantity in the noninteracting case by the substitution of the bare electronic mass m with m^*. It is remarkable that interaction effects enter only through the effective mass: the Landau interaction function is not explicitly invoked.

To see how Eq. (8.10) comes about observe first that the entropy of the thermal ensemble described by the quasiparticle distribution function $\mathcal{N}_{\vec{k}\sigma}$ is given by

$$S[\{\mathcal{N}_{\vec{k}\sigma}\}] = -k_B \sum_{\vec{k}\sigma}[\mathcal{N}_{\vec{k}\sigma} \ln\mathcal{N}_{\vec{k}\sigma} + (1 - \mathcal{N}_{\vec{k}\sigma})\ln(1 - \mathcal{N}_{\vec{k}\sigma})] , \qquad (8.11)$$

which coincides with entropy of the noninteracting ensemble of origin, and vanishes in the ground-state. According to the general principles of statistical mechanics the average value of the quasiparticle occupation numbers in the state of thermal equilibrium at temperature T and chemical potential μ is determined by minimizing the grand-canonical thermodynamic potential functional

$$\Omega[\{\mathcal{N}_{\vec{k}\sigma}\}] = E[\{\mathcal{N}_{\vec{k}\sigma}\}] - TS[\{\mathcal{N}_{\vec{k}\sigma}\}] - \mu\sum_{\vec{k}\sigma}\mathcal{N}_{\vec{k}\sigma} . \qquad (8.12)$$

---

[8] Some of the earlier theories based on Eq. (8.9) failed to account for this fact and therefore arrived to incorrect expressions for the Landau interaction function.

Carrying out the minimization with the help of the expressions (8.1) and (8.11) yields

$$\mathcal{N}_{\vec{k}\sigma}^{eq}(\mu, T) = \frac{1}{e^{\beta(\tilde{\mathcal{E}}_{\vec{k}\sigma} - \mu)} + 1}, \tag{8.13}$$

where $\beta = 1/k_B T$ and $\tilde{\mathcal{E}}_{\vec{k}\sigma}$ is the local quasiparticle energy. Notice that Eqs. (8.13) and (8.8) should be solved simultaneously since the local quasiparticle energy depends self-consistently on the distribution of thermally excited quasiparticles. Substituting the solution in the functional (8.12) finally yields the grand canonical thermodynamic potential from which all the thermodynamic properties of the system can be derived.

In practice, in an isotropic system at thermal equilibrium, $\delta \mathcal{N}_{\vec{k}'\sigma'}^{eq}$ depends only on the magnitude of $\vec{k}'$ and is restricted to a thin shell of width $k_B T$ about the Fermi surface. In this region the dependence of the Landau interaction function on the magnitudes of $\vec{k}$ and $\vec{k}'$ is negligible: only the dependence on the *angle* between $\vec{k}$ and $\vec{k}'$ matters. Now, since the particle number remains constant up to corrections of order T^2,[9] we have, to a high degree of accuracy, $\sum_{\vec{k}'\sigma'} \delta \mathcal{N}_{\vec{k}'\sigma'}^{eq} = 0$, and hence we see that the second term in Eq. (8.8) vanishes, leaving us with $\tilde{\mathcal{E}}_{\vec{k}\sigma} = \mathcal{E}_{\vec{k}\sigma}$. Thus, the quasiparticle distribution function in the state of thermal equilibrium is that of an ideal Fermi gas of particles of energy $\mathcal{E}_{\vec{k}\sigma}$. Substituting this in Eq. (8.12) we see that the interaction contributions disappear, i.e., *the equilibrium thermodynamic potential is the same as for a system of noninteracting Fermions of energy* $\mathcal{E}_{\vec{k}\sigma}$. The specific heat and other thermodynamic properties simply follow the behavior of the ideal Fermi gas with renormalized mass.

The key feature responsible for the disappearance of the interaction function from the thermodynamic potential is the isotropy of the thermal equilibrium distribution, and the fact that, at low temperature, the number of quasielectrons equals the number of quasiholes. Thus, the thermal excitation of the system does not contribute to the quasiparticle energy. The situation is dramatically different when the excitation is caused by an external field such as pressure or magnetic field. This will be analyzed next.

8.3.3 The Landau Fermi liquid parameters

Before proceeding any further it is useful to introduce at this point the *Landau Fermi liquid parameters*. One starts from the observation that within the dynamically relevant shell where $\delta \mathcal{N}_{\vec{k}'\sigma'}$ is finite, the Landau interaction function depends only on the cosine of the angle θ between $\vec{k}$ and $\vec{k}'$. Accordingly we can set $f_{\vec{k}\sigma, \vec{k}'\sigma'} \simeq f_{\sigma\sigma'}(\cos\theta)$ and introduce the dimensionless quantities $F_\ell^{s,a}$ which are defined in terms of spin symmetric (s) and spin antisymmetric (a) angular averages of $f_{\sigma\sigma'}(\cos\theta)$ as follows:

$$F_\ell^{s,a} = \frac{L^d N^*(0)}{2} \int \frac{d\Omega_d}{\Omega_d} [f_{\uparrow\uparrow}(\cos\theta) \pm f_{\uparrow\downarrow}(\cos\theta)] \begin{cases} P_\ell(\cos\theta), & 3D, \\ \cos\ell\theta, & 2D, \end{cases} \tag{8.14}$$

---

[9] Corrections of order T^2 occur because we are working in the grand-canonical ensemble.

where the $+$ and $-$ signs are associated with s and a respectively, $\Omega_d = 2^{d-1}\pi$ is the solid angle in $d = 3$ or 2 dimensions, and $P_\ell(\cos\theta)$ is the ℓ-th Legendre polynomial.[10]

It must be noted immediately that *this definition differs from the one commonly used in previous texts and in large part of the literature*. The reason we choose to deviate from the norm in this respect is that our notation allows one to express the results of the theory in a form that is the same in two and three dimensions. Nervous readers can readily convert the results of this section into the older notation by simply making the substitution $F_\ell^{s,a} \to \frac{F_\ell^{s,a}}{2\ell+1}$ in three dimensions and $F_\ell^{s,a} \to \frac{F_\ell^{s,a}}{2}(1 + \delta_{\ell 0})$ in two dimensions.

The inverse of Eq. (8.14) is

$$f_{\uparrow\uparrow}(\cos\theta) \pm f_{\uparrow\downarrow}(\cos\theta) = \frac{2}{L^d N^*(0)} \sum_{\ell=0}^{\infty} F_\ell^{s,a} \begin{cases} (2\ell + 1)P_\ell(\cos\theta), & 3D, \\ (2 - \delta_{\ell 0})\cos\ell\theta, & 2D, \end{cases} \tag{8.15}$$

where $+$ is associated with s and $-$ with a. Approximate values of the first few Landau parameters are listed in Tables 8.1 (3D) and 8.5 (2D).

8.3.4 The compressibility

An important property of a charged Fermi liquid is the proper compressibility K, related to the small q limit of the proper density–density response function by the *compressibility sum rule* of Eqs. (5.23) and (5.24). K can in general be obtained from the relation

$$\frac{1}{K} = n^2 \frac{\partial\mu}{\partial n} = \frac{nk_F}{d}\frac{\partial\mu}{\partial k_F}. \tag{8.16}$$

The compressibility determines, among other things, the magnitude of the screening wave vector and the hydrodynamic sound velocity $s = \frac{1}{\sqrt{nmK}}$.

For a noninteracting system one simply has the result

$$K_0 = \frac{N(0)}{n^2}. \tag{8.17}$$

To evaluate the derivative $\frac{\partial\mu}{\partial k_F}$ within the Landau theory of Fermi liquids we first recall that, according to Eq. (8.3), μ is simply the energy of a quasiparticle of any spin at the Fermi surface, i.e., $\mu = \mathcal{E}_{k_F\sigma}$. As the Fermi surface is slightly expanded to accommodate an additional particle density δn the Fermi wave-vector changes from k_F to $k_F + \delta k_F$ (see Fig. 8.2). The change in μ is the sum of two contributions: one is the change in the bare quasiparticle energy when the wave vector varies from k_F to $k_F + \delta k_F$, while the other is the interaction energy with the additional quasiparticles created by the expansion of the Fermi sphere. Thus, by making use of Eq. (8.8) we readily obtain

$$\delta\mu = \tilde{\mathcal{E}}_{k_F+\delta k_F,\sigma} - \tilde{\mathcal{E}}_{k_F,\sigma} = v_F^\star \hbar \delta k_F + \sum_{\vec{k}',|\vec{k}|=k_F,\sigma'} f_{\vec{k}\sigma,\vec{k}'\sigma'} \delta\mathcal{N}_{\vec{k}'\sigma'}, \tag{8.18}$$

where $\delta\mathcal{N}_{\vec{k}\sigma} = 1$ for $k_F < k < k_F + \delta k_F$ and zero otherwise.

---

[10] The Legendre polynomials are defined as $P_\ell(x) = \frac{1}{2^\ell \ell!}\frac{d^\ell}{dx^\ell}(x^2 - 1)^\ell$. The first three polynomials are $P_0(x) = 1$, $P_1(x) = x$, and $P_2(x) = \frac{3x^2-1}{2}$.

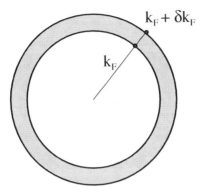

Fig. 8.2. Calculation of the compressibility in the Landau theory of Fermi liquids. The chemical potential, i.e., the energy of a quasiparticle at the Fermi surface, changes due to (i) the variation of the Fermi momentum, (ii) the addition of quasiparticles to the shaded region.

We show next that the quasiparticle effective mass and the Landau parameter F_0^s determine the compressibility of a Fermi liquid. Consider the second term of Eq. (8.18). From our previous discussion this can be rewritten as

$$\sum_{\vec{k}',|\vec{k}|=k_F,\sigma'} f_{\vec{k}\sigma,\vec{k}'\sigma'}\delta\mathcal{N}_{\vec{k}'\sigma'} = \left(\frac{L^d\Omega_d}{(2\pi)^d}\int_{k_F}^{k_F+\delta k_F} dk'\, k'^{d-1}\right)$$

$$\times \left(\int \frac{d\Omega_d}{\Omega_d}\sum_{\sigma'} f_{\sigma\sigma'}(\cos\theta)\right). \qquad (8.19)$$

In this expression the first factor can be seen to coincide with the quantity $\frac{1}{2}\sum_{\vec{k}'\sigma'}\delta\mathcal{N}_{\vec{k}'\sigma'} = \frac{\delta N}{2}$ which, at variance with the thermal excitation case, is here finite and simply represents the number of quasiparticle states within the thin shell of interest divided by two. By definition of δk_F, this number is to linear order given by $\frac{d}{2}N\frac{\delta k_F}{k_F}$, an expression which can be conveniently rewritten as $\frac{1}{2}L^d N^\star(0)v_F^\star\hbar\delta k_F$. The second factor is simply the angular average of $\sum_{\sigma'} f_{\sigma\sigma'}(\cos\theta)$ which, by making use of the definition of Eq. (8.14), is seen to be given by $\frac{2F_0^s}{L^d N^\star(0)}$. Then Eq. (8.18) becomes

$$\delta\mu = \hbar v_F^\star(1 + F_0^s)\delta k_F, \qquad (8.20)$$

which immediately leads to the elegant result

$$\frac{K}{K_0} = \frac{\frac{m^\star}{m}}{1 + F_0^s}. \qquad (8.21)$$

Thus, the electron–electron interaction enters the proper compressibility not only through the effective mass, but also, explicitly, through the spin symmetric spherical average of the Landau interaction function.

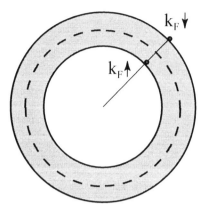

Fig. 8.3. Calculation of the spin susceptibility in the Landau theory of Fermi liquids. In the presence of a magnetic field the up- and down-spin Fermi surfaces split in such a way that the energies of two quasiparticles at the two Fermi surfaces are equal. The additional quasiparticles are down-spin electrons above the unperturbed Fermi surface (dashed line) and up-spin holes below it.

8.3.5 The paramagnetic spin response

The spin susceptibility can be calculated in a completely analogous way. In the presence of an external magnetic field B the hamiltonian is modified by the addition of the Zeeman energy term

$$\hat{H}_Z = \frac{g\mu_B}{2} B \hat{S}_z \, , \tag{8.22}$$

where g is the *bare* g-factor for the electrons and $\hat{S}_z$ is, as usual, the z-component of the spin in units of $\frac{\hbar}{2}$.[11] The energy of an isolated quasiparticle of wave vector $\vec{k}$ and spin σ in the presence of the magnetic field becomes

$$\mathcal{E}_{\vec{k}\sigma}(B) = \mathcal{E}_{\vec{k}\sigma} + \frac{1}{2} g\mu_B B\sigma \, , \tag{8.23}$$

where, it must be noted, the g-factor of the quasiparticle coincides with that of the bare electron, because the many-body state described by the quasiparticle at $\vec{k}\sigma$ is an eigenstate of $\hat{S}_z$ with eigenvalue σ. Because of the Zeeman energy, the Fermi surfaces of up-spin and down-spin electrons shift by equal amounts in opposite directions, i.e., the radius of the down-spin Fermi surface increases by an amount $\delta k_{F\downarrow}$, while the radius of the up-spin Fermi surface decreases by the same amount.[12] In this situation the net magnetization per unit volume is given by

$$M = -\frac{1}{2} g\mu_B(\delta n_\uparrow - \delta n_\downarrow) = \frac{1}{2} g\mu_B N^*(0) v_F^* \hbar \delta k_{F\downarrow} \, . \tag{8.24}$$

---

[11] For free electrons $g \simeq 2$, but this value can be considerably different for electrons in a solid state environment: for example in GaAs one has $g = -0.44$. A further modification of the g-factor associated with many-body effects is introduced in Eq. (8.30).
[12] The fact that $\delta k_{F\uparrow} = -\delta k_{F\downarrow}$ follows of course from the conservation of the total number of particles.

The equilibrium value of $\delta k_{F\downarrow}$ is determined by the condition that the energy of an up-spin quasiparticle at the up-spin Fermi surface be equal to that of a down-spin quasiparticle at the down-spin Fermi surface: if this were not the case one could gain energy by transferring quasiparticles from one Fermi surface to the other. The common value of the energy is, of course, the chemical potential (see Eq. (8.3)). The mathematical form of the equilibrium condition is thus

$$\tilde{\mathcal{E}}_{k_{F\uparrow}\uparrow} + \frac{1}{2}g\mu_B B = \tilde{\mathcal{E}}_{k_{F\downarrow}\downarrow} - \frac{1}{2}g\mu_B B . \tag{8.25}$$

Now, by making use of Eq. (8.8) and following the same basic procedure used for the case of the compressibility, we recast Eq. (8.25) in the form

$$-\hbar v_F^* \delta k_{F\downarrow} - F_0^a \hbar v_F^* \delta k_{F\downarrow} + \frac{1}{2}g\mu_B B = +\hbar v_F^* \delta k_{F\downarrow} + F_0^a \hbar v_F^* \delta k_{F\downarrow} - \frac{1}{2}g\mu_B B , \tag{8.26}$$

which yields

$$\delta k_{F\downarrow} = \frac{\frac{g\mu_B B}{2\hbar v_F^*}}{1 + F_0^a} . \tag{8.27}$$

Substitution in Eq. (8.24) yields the spin susceptibility

$$\chi_S = \left(\frac{g\mu_B}{2}\right)^2 \frac{N^*(0)}{1 + F_0^a} , \tag{8.28}$$

or, equivalently,

$$\frac{\chi_S}{\chi_P} = \frac{\frac{m^*}{m}}{1 + F_0^a} , \tag{8.29}$$

where χ_P is the Pauli spin susceptibility, i.e., the spin susceptibility of the noninteracting electron gas. Notice that the electron–electron interaction enters both through the effective mass in the density of states at the Fermi surface and through the Landau parameter F_0^a. Negative values of F_0^a arising from the exchange interaction enhance the spin susceptibility. Since the static spin susceptibility of a stable paramagnetic system must be positive, Eq. (8.29) implies that the paramagnetic state will be unstable if $F_0^a < -1$.

It is often useful to introduce the notion of *effective g-factor* by considering the energy change associated with the flipping of the spin of a single quasiparticle from down to up in the presence of the spin polarization created by the magnetic field B, *without allowing that polarization to change*, e.g.,

$$g^* \mu_B B \equiv \tilde{\mathcal{E}}_{k_{F\downarrow}\uparrow} - \tilde{\mathcal{E}}_{k_{F\downarrow}\downarrow} . ^{[13]} \tag{8.30}$$

Following the same procedure leading Eq. (8.18) we can now write

$$g^* \mu_B B = g\mu_B B - 2\left(\int \frac{d\Omega_d}{\Omega_d}[f_{\uparrow\uparrow}(\cos\theta) - f_{\uparrow\downarrow}(\cos\theta)]\right)\delta N_\downarrow . \tag{8.31}$$

---

[13] We emphasize that $g^* \mu_B B$ is *not* the spin splitting that is measured in conduction electron spin resonance (CESR). This is because, as discussed in Section 8.3.8, in a CESR experiment the resonance occurs at the *bare* frequency $g\mu_B B/\hbar$.

The number of extra induced down-spin electrons, $\delta N_\downarrow$, can in turn be expressed in terms of the quasiparticle density of states at the Fermi energy and g^*:

$$\delta N_\downarrow = \frac{L^d N^*(0)}{2} \frac{g^* \mu_B B}{2} . \tag{8.32}$$

Noting that the integral in Eq. (8.31) is, by virtue of the definition (8.14), equal to $\frac{2F_0^a}{L^d N^*(0)}$ we immediately arrive at

$$\frac{g^*}{g} = \frac{1}{1 + F_0^a} . \tag{8.33}$$

Notice that this equation, together with Eq. (8.29) leads to the relationship

$$\frac{\chi_S}{\chi_P} = \frac{m^*}{m} \frac{g^*}{g} . \tag{8.34}$$

This relation remains valid even in the presence of more general interactions (e.g. the electron–phonon interaction discussed in Section 8.3.7).

It is evident from Eqs. (8.21) and (8.29) that a connection must exist between the Landau parameters and the many-body local field factors $G_\pm(q,0)$ introduced in Section 5.4, since they both give corrections to the noninteracting compressibility and spin susceptibility. In view of the presence of the effective mass ratio in Eqs. (8.21) and (8.29) however, this connection is by no means simple as the effective mass of a quasiparticle depends in general in a complicated way on both $G_+(q,\omega)$ and $G_-(q,\omega)$.

8.3.6 The effective mass

The effective mass of a quasiparticle and the effective interaction function are not completely independent. They are obtained, respectively, from the first and the second derivative of the energy functional with respect to quasiparticle occupation number, so it is not surprising that a relationship should exist between the two of them. This relationship takes a particularly simple form in a *translationally invariant system*, where one finds

$$\frac{m^*}{m} = 1 + F_1^s , \tag{8.35}$$

as we now proceed to show.[14]

In a translationally invariant system the total momentum $\hat{P}$ is a constant of the motion and the eigenstates of the hamiltonian can therefore be chosen to be eigenstates of $\hat{P}$. Let $\hbar \vec{k}$ be the total momentum of one of these eigenstates, in a reference frame in which the center of mass is at rest, and E its total energy. Let us now look at this state from a reference frame that moves at an infinitesimal velocity $-\vec{v}$ relative to the center of mass frame. By applying the transformation $\vec{p}_i \rightarrow \vec{p}_i + m\vec{v}$ (known as a *Galilean transformation*)

---

[14] Thus, the "normal" ground-state will be unstable if $F_1^s < -1$.

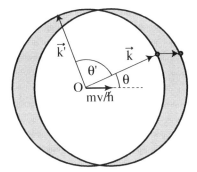

Fig. 8.4. Relation between the effective mass and the Landau parameters. When the original Fermi surface centered at O is viewed from a reference frame moving at speed v it appears to be shifted by an amount $mv/\hbar$. The shift can be described in terms of quasiparticles and quasiholes added to the original Fermi distribution in the shaded crescent-shaped regions.

to the momentum of each electron we see that the energy of our state in the new frame is

$$E' \simeq E + \hbar \vec{v} \cdot \vec{k} , \tag{8.36}$$

where we have neglected on the right-hand side a term $\frac{Nmv^2}{2}$, which is of second order in $\vec{v}$. The change is entirely in the kinetic energy, since the potential energy depends only on the relative positions of the particles, and therefore is not affected by the transformation. Notice that this conclusion does not depend on the physical nature of the state under consideration, only on its total momentum and energy.

Let us now come back to the Landau description of the Fermi liquid, and consider the state that contains a single quasiparticle of momentum $\hbar \vec{k}$ ($k \simeq k_F$) in the center of mass frame. When this state is viewed from the moving frame, both the quasiparticle and the Fermi surface appear to be shifted by $m\vec{v}/\hbar$ in wave vector space (see Fig. 8.4). The change in energy has three contributions:

(i) The change of the bare quasiparticle energy $\mathcal{E}_{\vec{k}\sigma}$ due to the fact that the quasiparticle momentum is shifted from $\hbar \vec{k}$ to $\hbar \vec{k} + m\vec{v}$.

(ii) The energy of interaction between the quasiparticle of wave vector $\vec{k}$ and the additional quasiparticles and quasiholes that appear in the crescent-shaped regions of Fig. 8.4 due to the shift of the Fermi surface.

(iii) The change that arises from the global shift of the states within the Fermi sphere (i.e. the energy to create the quasiholes and the quasiparticles that describe the crescent-shaped regions of Fig. 8.4).

The first contribution is given by

$$(i) \quad \left(\frac{\partial \mathcal{E}_{\vec{k}\sigma}}{\partial \vec{k}} \right)_{k=k_F} \cdot \frac{m\vec{v}}{\hbar} = \frac{m}{m^\star} \hbar \vec{k} \cdot \vec{v} . \tag{8.37}$$

The second one can be computed from Eq. (8.8). To linear order in v we have:

$$\sum_{\vec{k}'\sigma'} f_{\vec{k}+\frac{m\vec{v}}{\hbar}\sigma\vec{k}'\sigma'}[\mathcal{N}^{(0)}_{\vec{k}'-\frac{m\vec{v}}{\hbar}\sigma'} - \mathcal{N}^{(0)}_{\vec{k}'\sigma'}] \simeq \sum_{\vec{k}'\sigma'} f_{\sigma\sigma'}(\cos\theta')\delta(\mathcal{E}_{\vec{k}'\sigma'} - \mu)\frac{m}{m^*}(\hbar\vec{k}' \cdot \vec{v}) , \qquad (8.38)$$

where θ' is the angle between $\vec{k}'$ and $\vec{k}$. By handling this expression as done in Eq. (8.19), we can recast it in the form

$$\left(\frac{1}{2}\sum_{\vec{k}'\sigma'} \delta(\mathcal{E}_{\vec{k}'\sigma'} - \mu)\right)\left(\int \frac{d\Omega'_d}{\Omega_d} \sum_{\sigma'} f_{\sigma\sigma'}(\cos\theta')\frac{m}{m^*}(\hbar\vec{k}' \cdot \vec{v})\right). \qquad (8.39)$$

We next notice that in three dimensions we have

$$\hat{k}' \cdot \hat{v} = \cos\theta'\cos\theta + \sin\theta'\sin\theta\cos(\phi' - \phi) , \quad 3D , \qquad (8.40)$$

where the angles θ', ϕ' and θ, ϕ respectively determine the direction of $\hat{k}'$ and $\hat{v}$ with respect to the z-axis taken along $\vec{k}$. In two dimensions the same expression holds with $\phi = \phi' = 0$. Only the first term on the right-hand side of (8.40) survives angular averaging, and Eq. (8.38) becomes

$$(ii) \quad \frac{m}{m^*}\hbar k_F v \cos\theta \frac{L^d N^*(0)}{2} \int \frac{d\Omega'_d}{\Omega_d} \cos\theta' \sum_{\sigma'} f_{\sigma\sigma'}(\cos\theta') . \qquad (8.41)$$

Finally, contribution (iii) is of order v^2 (in fact, it is precisely $\frac{Nmv^2}{2}$) and can be neglected.

Now, according to Eq. (8.36) the sum of (i) and (ii) must yield $\hbar\vec{v} \cdot \vec{k}$. Thus, we must have

$$\vec{v} \cdot \vec{k} = \frac{m}{m^*}\vec{v} \cdot \vec{k} + \frac{m}{m^*}\vec{v} \cdot \vec{k}\frac{L^d N^*(0)}{2} \int \frac{d\Omega'_d}{\Omega_d} \cos\theta' \sum_{\sigma'} f_{\sigma\sigma'}(\cos\theta') , \qquad (8.42)$$

which finally leads to Eq. (8.35).

The reader is urged to verify (see Exercise 2) that this relation holds true to first-order in the interaction, and that, in the first-order approximation, the effective mass vanishes for long-range coulomb interactions (an anomaly which is of course cured by the inclusion of screening in higher order).

Eq. (8.42) has a simple physical interpretation. First of all notice that in a translationally invariant system momentum conservation implies that the *current* carried by a quasiparticle of momentum $\hbar\vec{k}$ is $\frac{\hbar\vec{k}}{m}$.[15] On the other hand, $\frac{\hbar\vec{k}}{m^*} = \frac{1}{\hbar}\frac{\partial\mathcal{E}_{\vec{k}\sigma}}{\partial\vec{k}}$ is recognized as the *group velocity* of a quasiparticle wave packet with an average wave vector $\vec{k}$. The difference between the current of the plane wave state and that associated with the group velocity of the wave packet can be qualitatively understood in terms of the *backflow current* that arises when plane waves of different wave vectors are superimposed to create the wave packet. Dividing

---

[15] The argument is identical to the one used immediately after Eq. (8.23) to argue that, due to spin conservation, the (bare) g-factor of a quasiparticle is the same as that of the bare particle.

Table 8.1. *Calculated values of Landau Fermi liquid parameters of the three-dimensional electron liquid. The calculation of F_0^s, F_0^a, and F_1^s is described in Section 8.8, while the values of F_1^a, F_2^s, and F_2^a were calculated by Yasuhara and Ousaka (1992).*

r_s	F_0^s	F_0^a	F_1^s	F_1^a	F_2^s	F_2^a
1	−0.21	−0.17	−0.04	−0.0645	−0.0215	−0.0181
2	−0.37	−0.25	−0.03	−0.0825	−0.0168	−0.0126
3	−0.55	−0.32	−0.02	−0.0915	−0.0107	−0.0073
4	−0.74	−0.37	−0.0	−0.0956	−0.0047	−0.0022
5	−0.95	−0.40	−0.03	−0.0965	+0.0009	+0.0023

both sides of Eq. (8.42) by m, we see that the second term on the right-hand side is precisely the negative of the backflow current times $\vec{v}$.[16]

The importance of momentum conservation in the above discussion cannot be overemphasized. Consider, for example, the following question: what is the *spin-current*, $j_\uparrow - j_\downarrow$, carried by a quasiparticle of wave vector $\vec{k}$ and spin $\uparrow$? One might be tempted to answer "$\hbar\vec{k}/m$" on the (false) assumption that there is no down-spin current, but this is incorrect because the difference between the total up- and down-spin momenta $\hat{P}_\uparrow - \hat{P}_\downarrow$ is not a constant of the motion. In fact, the magnitude of the *spin-current* is smaller than $\frac{\hbar\vec{k}}{m}$ (Qian et al., 2004). What happens is that in the process of switching-on the interaction some momentum is transferred from the up- to the down-spin component of the electron liquid. This reduces the spin current without altering the total momentum and spin. The reduction can be expressed in terms of an effective *spin mass* $m_S > m$ such that $j_\uparrow - j_\downarrow = \frac{\hbar\vec{k}}{m_S}$. The relation between m_S and m^* has the same form as the relation (8.35) between the "charge mass", m, and m^*, i.e.

$$\frac{m^*}{m_S} = 1 + F_1^a . \tag{8.43}$$

Numerical values of F_1^a obtained from an approximate microscopic theory (Yasuhara and Ousaka, 1992) are listed in Table 8.1 for several different densities.

8.3.7 The effects of the electron–phonon coupling

We show in this section that while the effective mass and the effective g-factor are renormalized in a simple way by the electron–phonon coupling, the compressibility and the spin susceptibility of a Fermi liquid are to leading order unchanged by this interaction.

As it turns out, field-theoretic calculations are not required to see whether a given quasiparticle property is influenced by the electron–phonon interaction. A careful physical analysis

---

[16] A truly microscopic characterization of the quasiparticle wave packet remains a problematic task (see Heinonen and Kohn, 1987).

will suffice.[17] In order to study the physics of the phenomenon one needs only recall that in leading order, the contribution to the quasiparticle energy due to the electron–phonon coupling is of the form

$$\mathcal{E}_{\vec{k}\sigma}^{(el-ph)} \simeq -\lambda(\mathcal{E}_{\vec{k}\sigma} - \mu), \tag{8.44}$$

where the positive number λ is the *electron–phonon coupling constant* (see for instance Ashcroft and Mermin, 1976, Chapter 26). It is important to observe here that $\mathcal{E}_{\vec{k}\sigma}^{(el-ph)}$ simply "rides along" with the Fermi surface sliding up or down the phonon-free curve for $\mathcal{E}_{\vec{k}\sigma}$.

We begin by noticing that, since, again to leading order in λ, $\mathcal{E}_{\vec{k}\sigma}^{(el-ph)}$ can be approximated by

$$\mathcal{E}_{\vec{k}\sigma}^{(el-ph)} \simeq -\lambda\hbar v_F^*(k - k_{F\sigma}), \tag{8.45}$$

where v_F^* is free of phonon effects, one is left with a straightforward renormalization of the effective mass and of the density of states of the type

$$\frac{m_{el-ph}^*}{m^*} = \frac{N_{el-ph}^*(0)}{N^*(0)} = 1 + \lambda, \tag{8.46}$$

where m^* and $N^*(0)$ are respectively the quasiparticle effective mass and the density of states at the Fermi surface in the absence of electron–phonon coupling.[18]

To understand the consequences of this coupling on the response properties of the system requires a slightly more careful analysis.

Consider first the case of the proper compressibility K. As anticipated, in spite of the structure of Eq. (8.21) and of (8.46), to this level of approximation K is independent of λ. To see this one should return to Eq. (8.18). It is obvious that Eq. (8.44) contributes no explicit new terms to $\delta\mu$. Of the two terms on the right-hand side of (8.18) the one associated with the Landau interaction is unchanged since it can be still manipulated in exactly the same way as done below Eq. (8.19) as to lead to the same result. But also the first term, the change of the bare quasiparticle energy, is unmodified since, as already observed, the energy correction (8.44) does in this case simply slide along the curve $\hbar v_F^*(k - k_F)$, with v_F^* still free of phonon modifications, as k_F is changed. As a consequence K is still given by Eq. (8.21) with the stipulation that all the ingredients are calculated for $\lambda = 0$.

The case of the spin susceptibility χ_S can be analyzed exactly in the same way. In this case the key formula is represented by Eq. (8.26). Here again Eq. (8.44) adds no terms and each of the three terms on both sides of (8.26) are unchanged exactly for the same reasons given for the case of the corresponding terms appearing in Eq. (8.18). Thus also the spin susceptibility is still given by Eq. (8.28) with unspoiled Landau parameters.

Since Eq. (8.34) is valid in general, and as we have shown χ_S is immune to phonons, the g-factor g^* of a Fermi liquid is renormalized by the electron–phonon coupling. It is

---

[17] Early theoretical work on this problem was based on Green's function techniques. See in particular the discussions by Quinn and Ferrell (1961) and Prange and Kadanoff (1964).

[18] Notice that the form of the correction (8.44) to the quasiparticle energy implies that the electron–phonon interaction gives a frequency-dependent but wave-vector-independent contribution to the self-energy (Exercise 8.4):

$$\Sigma^{(el-ph)}(\vec{k}, \omega) \simeq -\lambda(\omega - \mu).$$

immediately deduced that

$$\frac{g^*_{el-ph}}{g^*} = \frac{1}{1+\lambda} . \tag{8.47}$$

The reader is urged to derive this result by making use of Eqs. (8.30) and (8.31) for g^*. This instructive calculation constitutes Exercise 5.

8.3.8 Measuring m^*, K, g^* and χ_S

The effective mass and the proper compressibility, the spin susceptibility and the g-factor of the electron liquid have been the object of many experimental studies.

In three dimensional elemental metals the study of many-body effects is generally hampered by solid-state effects which are hard to identify and eliminate from the analysis, as their relevance differs from metal to metal. In spite of this difficulty valiant attempts have been made to determine a variety of quasiparticle properties and response functions in these systems. A rather formidable albeit dated discussion of the situation can be found in the book by Pines and Noziéres (1966). While a measurement of the quasiparticle effective mass or the electronic compressibility is rather difficult, nuclear magnetic resonance (NMR) and conduction electron spin resonance (CESR) techniques have been successfully employed to determine the spin susceptibility and the g-factor in these systems (see for instance the thorough and elegant discussion by Kushida, Murphy and Hanabusa, 1976). Although the CESR occurs at a frequency determined by the *bare* g-factor,[19] the intensity of the effect, obtained by integrating the absorption curve across the resonance, is proportional to the interacting electronic spin susceptibility χ_S, and therefore does give information about many-body effects. By cleverly taking the ratio of such an intensity to that of the nuclear spin resonance at the same frequency (and therefore at a substantially larger magnetic field) χ_S can be determined in terms of well known nuclear susceptibilities.

An alternative way of measuring the spin susceptibility is provided by the Knight shift, i.e. the shift in the frequency of the nuclear magnetic resonance due to the additional magnetic field ΔH_{Knight} created by the spin polarization of the electron liquid. This can be expressed as

$$\frac{\Delta H_{Knight}}{H} = \frac{8\pi}{3} V P_F \chi_S , \tag{8.48}$$

where P_F is the probability density for an electron to be "riding" the nucleus, i.e., the average over the Fermi surface of the modulus square of the electronic wavefunction at the nuclear position (see for instance Slichter, 1990). In practice, specific values of P_F must be either obtained from a band structure calculation or inferred from a combination of Knight shift and CESR measurements. Knight shift data and measurements of the nuclear relaxation

---

[19] One can understand this phenomenon in terms of a collective Larmor precession of the spins in which the relative orientation of the individual spins and therefore the interaction energy is left unchanged. This is essentially the same physics that leads to the absence of many-body corrections to the cyclotron resonance frequency in a homogeneous system in a magnetic field, as discussed in Section 10.6.4.

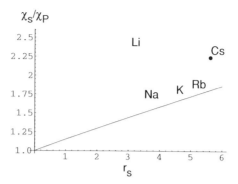

Fig. 8.5. Experimentally determined spin susceptibilities of elemental metals vs theoretical values obtained from the analytical fit (1.128) to the quantum Monte Carlo results.

time, T_1, can also be combined to determine the effective g-factor. This is due to the fact that according to the *Korringa relation* (Slichter, 1990) one has

$$T_1 \left(\frac{\Delta H_{Knight}}{H} \right)^2 = \frac{\hbar}{4\pi k_B T} \frac{\gamma_e^2}{\gamma_n^2} \left[\frac{\chi_S}{\chi_P} \frac{N(0)}{N^*(0)} \right]^2, \tag{8.49}$$

where γ_e and γ_n are the gyromagnetic ratios of the electrons and the nuclei respectively. The quantity in the square brackets is just the ratio $\frac{g^*}{g}$. All in all, the measured spin susceptibility in Na, K, and Rb, compares reasonably well with theoretical calculations (see Fig. 8.5 above), but significant disagreement exists for Li and Cs. This disagreement is largely due to the non-uniformity of the electron liquid, and obligingly disappears when the latter is taken into account (see Vosko, Perdew, and MacDonald, 1975).

As mentioned in Chapter 1, a crucial advantage of two-dimensional systems over three-dimensional ones, is that the electronic density can be changed by electrical means, thus allowing a systematic study of interaction effects. We have already described in Section 5.2.3 the techniques by which the proper compressibility of a 2DEL has been measured. We focus therefore on the measurements of the effective mass and the spin susceptibility. The most accurate measurements so far have been done on systems such as n-type silicon inversion layers (Fang and Stiles, 1968; Pudalov *et al.*, 2002), n-doped GaAs/AlGaAs (Zhu *et al.*, 2003), and p-doped GaAs (Tutuc *et al.*, 2001). While the early experiments were done at densities of the order of 10^{12} cm^{-2}, corresponding to $r_s \sim 3$ in Si, the most recent ones have reached considerably lower densities in both GaAs (1.7×10^9 cm^{-2}, i.e. $r_s = 13.4$) and in Si inversion layers (10^{11} cm^{-2}, i.e. $r_s = 8.4$). At such low densities, the many-body renormalizations (i.e. the Landau parameters) are strong: for example, the spin susceptibility enhancement $\frac{m^* g^*}{m_b g_b}$ (where m_b and g_b are the "bare" mass and g-factor determined by the band structure of the host semiconductors: $g_b \sim 2$ in Si and $g_b \sim 0.44$ in GaAs) can be as large as 5 (Zhu, 2003). This is consistent with the theoretical expectation that the 2DEG should become ferromagnetic at sufficiently low density (see Chapter 1): however, the ferromagnetic transition has not been observed so far.

It is possible in principle to attempt a determination of the quasiparticle effective mass by studying the cyclotron resonance. As it turns out however such an endeavor is by no means straightforward. To start with, while it is well understood that the cyclotron frequency of a homogeneous system is unrenormalized by interactions (see Section 10.6.4), it is not at all clear to what extent the effects of the coulomb interaction are restored by the presence of the crystalline lattice, surfaces and inhomogeneities in general (see Ting, Ying, and Quinn, 1977). Moreover from the experimental point of view residual band structures effects (like for instance a lack of band parabolicity), incipient localization of the electronic orbitals, and other effects can complicate the interpretation of the data.[20] It is for these reasons that, to this date, the best determinations of quasiparticle properties, both in Si inversion layers as well as in GaAs heterostructures, has come from transport studies.

The basic idea of these measurements, contained in the pioneering paper by Fang and Stiles (1968), is to study the longitudinal magneto-resistance of the electron gas layer in the presence of a *tilted magnetic field* that has components both perpendicular and parallel to the plane of the electron liquid. The *perpendicular component* of the magnetic field $B_\perp$ causes the density of states of each spin to split into evenly spaced Landau levels separated by gaps of magnitude $\hbar\omega_c^* \equiv \frac{e\hbar B_\perp}{m^*c}$ (see Chapter 10, Fig. 10.4). The *total magnetic field B*, on the other hand, controls the Zeeman splitting $\Delta E_z = g^*\mu_B B$ between the Landau levels of up- and down-spins. For sufficiently large magnetic fields, the resistance of the electron layer depends on the total density of states at the Fermi level – a quantity that exhibits the *Shubnikov–de Haas* (SdH) *effect*, i.e. characteristic periodic oscillations as a function of density and/or inverse magnetic field.

The left panel of Fig. 8.6, for example, shows the oscillatory behavior of the sheet resistance in a Si sample at $n = 1.06 \times 10^{12}$ cm^{-2} at a temperature of 0.35 K. In the absence of an in-plane magnetic field ($B_\parallel = 0$) the period of the oscillations depends only on the density (upper left panel). In a Fermi liquid the amplitude of the various harmonics of the resistance vs $B_\perp$ depends on the ratio of the cyclotron gap $\hbar\omega_c^*$ to the temperature, and can be used to determine the quasiparticle effective mass. For example, the amplitude of the dominant harmonic component has the temperature dependence

$$\frac{T}{\sinh \frac{2\pi^2 k_B T}{\hbar\omega_c^*}} , \tag{8.50}$$

from which the effective mass can be extracted. The same data allow one to measure the spin susceptibility χ_S by finding a suitable ratio of the two field components that rids the oscillatory behavior of higher harmonics. Once this condition is satisfied, one then infers that the series of energy levels is evenly spaced. This in turn means that $\hbar\omega_c^* = g^*\mu_B B$ thereby leading to a determination of χ_S from Eq. (8.34).

A more generally applicable method for the measurement of χ_S is based the following idea. Including an in-plane magnetic field causes the two sets of Landau levels associated

---

[20] A discussion of the early experimental work in Si inversion layers can be found in Ando, Fowler, and Stern, 1982.

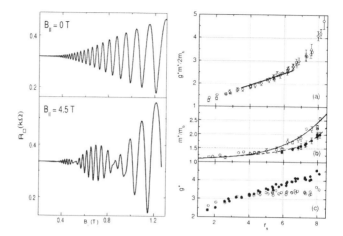

Fig. 8.6. Left side: Oscillatory behavior of the resistance of a two-dimensional electron liquid in a Si inversion layer as a function of $B_\perp$ for $n = 1.06 \times 10^{12}$ cm^{-2} vs $B_\perp$ at a temperature of 0.35 K. The lower panel shows the beats that appear when an in-plane magnetic field $B_\parallel$ is applied. Right side: Parameters $\frac{m^* g^*}{2m_b}$, $\frac{m^*}{m_b}$ and g^* vs r_s. The solid line in (a) shows the data by Okamoto *et al.* (1999). The solid and open dots in (b) and (c) correspond to two different methods of extracting m^* from the data. The solid and the dashed line in (b) are polynomial fits for the two dependences $m^*(r_s)$. Adapted from Pudalov *et al.* (2002).

with up- and down-spins to shift relative to each other: this, in turn, leads to the appearance of "beats", i.e., a modulation in the amplitude of the resistance oscillations, which is shown in the lower left panel of Fig. 8.6. From a theoretical point of view it is quite straightforward to see that in the case of a non-interacting electron gas the leading harmonic of the resistance vs $B_\perp$ is proportional to

$$\cos\left(\frac{2\pi g^* \mu_B B}{\hbar \omega_c^*}\right) \cos\left(\frac{4\pi^2 n\hbar c}{eB_\perp} - \pi\right). \tag{8.51}$$

From the measurement of the beat frequency one can then immediately extract the enhancement of χ_S for the given value of r_s. Notice that there remains some ambiguity in the separation of the $\frac{m^*}{m_b}$ contribution, which enters the susceptibility enhancement via ω_c^*. At any rate, both fundamental ratios $\frac{m^*}{m_b}$ and $\frac{g^*}{g}$ increase with increasing r_s.[21]

Recently, tilted field experiments have allowed the determination of the spin susceptibility of the 2DEL in very narrow AlAs quantum wells in a broad range of densities (Vakili *et al.*, 2004). Fig. 8.7 shows the comparison, obviously very satisfactory, between the measured data and the quantum Monte Carlo values obtained, for example, from the formulas of Section 1.7.2.

---

[21] As noted in Section 8.3.7 it is important to measure g^* at low temperature in order to avoid complications due to the electron–phonon coupling.

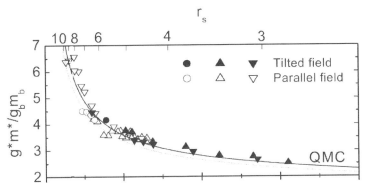

Fig. 8.7. Spin susceptibility enhancement vs r_s for the 2DEL in narrow AlAs quantum wells. Circles and triangles denote experimental data taken in different samples and for different orientations of the magnetic field. The quantum Monte Carlo predictions at zero and full spin polarization are shown by the dotted and solid lines respectively. From Vakili *et al.*, 2004.

8.3.9 The kinetic equation

Perhaps the technical centerpiece of the Landau theory of Fermi liquid is the derivation of the *kinetic equation*, which governs the response of the quasiparticle distribution function to weak and slowly-varying electromagnetic fields. The main idea is to treat the system as an assembly of quasiparticle wave packets characterized by an average position $\vec{r}$ and an average momentum $\hbar\vec{k}$ ($k \simeq k_F$). The quantum mechanical uncertainties in position and energy are negligible on the scale of spatial and temporal variation of the external fields. Obviously, this description makes sense only if the wavevector and frequency of the external field are much smaller than the Fermi wave vector and the Fermi energy respectively. In addition, the thermal energy $k_B T$ must be much smaller than the Fermi energy in order that the notions of Fermi surface and quasiparticles be well defined. Under these assumptions the quasiparticle wave packets can be treated as classical particles, with canonical coordinates and momenta $\vec{r}$ and $\hbar\vec{k}$, described by a "classical" hamiltonian

$$H_{cl}(\vec{r}, \hbar\vec{k}, \sigma) = \mathcal{E}_{\vec{k}\sigma} - e\phi_\sigma(\vec{r}, t) + \sum_{\vec{k}'\sigma'} f_{\vec{k}\sigma, \vec{k}'\sigma'} \delta \mathcal{N}_{\vec{k}'\sigma'}(\vec{r}, t) \tag{8.52}$$

where $\phi_\sigma(\vec{r}, t)$ is a spin-dependent scalar potential. The last term on the right-hand side of Eq. (8.52) describes the effect of the short-range interaction between the quasiparticles.[22] It has the form of a mean effective potential whose strength is controlled by the Landau interaction function. For this reason the theory is often described as a quasi-classical mean field approximation.

---

[22] As usual, the long-range electrostatic potential (Hartree potential) is treated as part of the external field.

The equation of motion for the quasiparticle distribution function follows immediately from Liouville's theorem for a classical flow in phase space

$$\frac{\partial \mathcal{N}_{\vec{k}\sigma}(\vec{r},t)}{\partial t} + \frac{1}{\hbar}\frac{\partial H_{cl}}{\partial \vec{k}} \cdot \frac{\partial \mathcal{N}_{\vec{k}\sigma}(\vec{r},t)}{\partial \vec{r}} - \frac{1}{\hbar}\frac{\partial H_{cl}}{\partial \vec{r}} \cdot \frac{\partial \mathcal{N}_{\vec{k}\sigma}(\vec{r},t)}{\partial \vec{k}} = \left(\frac{\partial \mathcal{N}_{\vec{k}\sigma}(\vec{r},t)}{\partial t}\right)_{coll}. \qquad (8.53)$$

The collisional time derivative on the right-hand side of Eq. (8.53) takes into account the fact that the evolution of the quasiparticle distribution function is affected by collision processes that are not included in the classical mean field hamiltonian. The point of the matter is that the mean field hamiltonian includes quasiparticle interactions only insofar as they give rise to a self-consistent mean field described by the Landau parameters. Actual collisions are not included, and, in fact, cannot be included without going to a more microscopic level of description. As discussed in previous sections, quasiparticle collisions result in a finite lifetime of quasiparticles ($\tau_{\vec{k}\sigma}^{(e)}$) and quasiholes ($\tau_{\vec{k}\sigma}^{(h)}$) near the Fermi surface. We can therefore write

$$\left(\frac{\partial \mathcal{N}_{\vec{k}\sigma}}{\partial t}\right)_{coll} = -\frac{\mathcal{N}_{\vec{k}\sigma}}{\tau_{\vec{k}\sigma}^{(e)}} + \frac{1-\mathcal{N}_{\vec{k}\sigma}}{\tau_{\vec{k}\sigma}^{(h)}}, \qquad (8.54)$$

where the first term represents the rate at which quasiparticles leave the state $\vec{k}\sigma$ and the second is the rate at which they are scattered into it. Microscopic calculations of the lifetimes will be presented in the next section. The total inverse lifetimes will in general include a contribution from electron-impurity scattering, as discussed in Chapter 4.

It is evident that the collisional derivative must vanish when $\mathcal{N}_{\vec{k}\sigma}$ is equal to the thermal equilibrium distribution $\mathcal{N}_{\vec{k}\sigma}^{eq}$, i.e., the Fermi–Dirac distribution with energy $\mathcal{E}_{\vec{k}\sigma}$ (see Eq. (8.13) and the discussion following it). This *principle of detailed balance* leads to an exact relation between quasiparticle and quasihole lifetimes:

$$\frac{\mathcal{N}_{\vec{k}\sigma}^{eq}}{\tau_{\vec{k}\sigma}^{(e)}} = \frac{1-\mathcal{N}_{\vec{k}\sigma}^{eq}}{\tau_{\vec{k}\sigma}^{(h)}}. \qquad (8.55)$$

Expanding Eq. (8.53) to first order in the strength of the external fields we obtain the *linearized kinetic equation* for the deviation of the distribution function from equilibrium. This equation has the form

$$\frac{\partial \delta\mathcal{N}_{\vec{k}\sigma}(\vec{r},t)}{\partial t} + \vec{v}_{\vec{k}\sigma} \cdot \frac{\partial \delta\mathcal{N}_{\vec{k}\sigma}(\vec{r},t)}{\partial \vec{r}} + \vec{v}_{\vec{k}\sigma} \cdot \vec{\mathcal{F}}_{\sigma}(\vec{r},t)\delta(\mathcal{E}_{\vec{k}\sigma}-\mu) = \left(\frac{\partial \delta\mathcal{N}_{\vec{k}\sigma}(\vec{r},t)}{\partial t}\right)_{coll}$$

$$(8.56)$$

where

$$\vec{v}_{\vec{k}\sigma} = \frac{\hbar\vec{k}}{m^*} \qquad (8.57)$$

s the quasiparticle velocity, and

$$\vec{\mathcal{F}}_\sigma(\vec{r},t) = -\vec{\nabla}_{\vec{r}}\left[-e\phi_\sigma(\vec{r},t) + \sum_{\vec{k}'\sigma'} f_{\vec{k}\sigma,\vec{k}'\sigma'}\delta\mathcal{N}_{\vec{k}'\sigma'}(\vec{r},t)\right] \qquad (8.58)$$

is the classical force acting on the quasiparticle. This equation is the starting point for most applications of the Landau theory of Fermi liquids. Among many possibilities, it can be used to study the macroscopic dynamics of the quasiparticle distribution function in the absence of external fields, leading to the prediction of several self-sustained collective modes (i.e., normal modes of oscillation of the Fermi surface) of different symmetries. These collective modes are nothing but the macroscopic ($q \to 0$) limit of the collective modes deduced in Section 3.2.7 from the poles of the microscopic linear response function.[23]

Eq. (8.56) can be generalized to include external vector potentials. For simplicity, we limit our considerations to the case of a weak vector potential $\vec{A}(\vec{r},t)$, which does not depend on spin. To first order in $\vec{A}$, the quasiclassical hamiltonian of Eq. (8.52) becomes

$$H_{cl}(\vec{r},\hbar\vec{k},\sigma) = \mathcal{E}_{\vec{k}\sigma} - e\phi_\sigma(\vec{r},t) + \sum_{\vec{k}'\sigma'} f_{\vec{k}\sigma,\vec{k}'\sigma'}\delta\mathcal{N}_{\vec{k}'\sigma'}(\vec{r},t) + \frac{\hbar\vec{k}}{m}\cdot\frac{e}{c}\vec{A}(\vec{r},t), \quad (8.59)$$

where we have made use of the fact that the coupling with the vector potential is via the actual quasiparticle current $\frac{\hbar\vec{k}}{m}$ and not the group velocity $\frac{\hbar\vec{k}}{m^*}$. The same conclusion can be arrived at by the canonical momentum shift $\vec{k} \to \vec{k} + \frac{e}{\hbar c}\vec{A}$ in Eq. (8.52). Following this route one must however be careful to not only implement the shift in the bare quasiparticle energy (so that $\mathcal{E}_{\vec{k}\sigma} \to \mathcal{E}_{\vec{k}+\frac{e}{\hbar c}\vec{A}\,\sigma}$) but also in the equilibrium distribution (so that $\delta\mathcal{N}_{\vec{k}\sigma} \to \mathcal{N}_{\vec{k}\sigma} - \mathcal{N}^{(0)}_{\vec{k}+\frac{e}{\hbar c}\vec{A}\,\sigma}$). The interaction function $f_{\vec{k}\sigma,\vec{k}'\sigma'}$, on the other hand, is not modified for it represents a velocity-independent interaction. Expanding to first order in $\vec{A}$, and making use of Eq. (8.42) one then easily obtains Eq. (8.59) (see Exercise 7).

The resulting linearized kinetic equation for $\delta\mathcal{N}_{\vec{k}\sigma}$ has still the form of Eq. (8.56), but the expression (8.58) for the force is replaced by the $\vec{k}$-dependent expression

$$\vec{\mathcal{F}}_\sigma(\vec{r},t) = -\vec{\nabla}_{\vec{r}}\left[-e\phi_\sigma(\vec{r},t) + \frac{\hbar\vec{k}}{m}\cdot\frac{e}{c}\vec{A}(\vec{r},t) + \sum_{\vec{k}'\sigma'} f_{\vec{k}\sigma,\vec{k}'\sigma'}\delta\mathcal{N}_{\vec{k}'\sigma'}(\vec{r},t)\right]. \qquad (8.60)$$

The reader is urged to show that Eq. (8.56), together with Eq. (8.60), is gauge-invariant and satisfies the continuity equation (Exercise 8).

8.3.10 The shear modulus

As a specific application of the kinetic equation derived in the previous section, we now use Eqs. (8.56) and (8.60) to derive the formulas for the dynamical shear modulus of a Fermi liquid, which were stated in Section 5.6.2 without proof. The idea is to use Eq. (8.56)

---

[23] The reader is referred to the book of Pines and Noziéres (1966) and the review article by Abrikosov and Khalatnikov (1959) for more details.

to calculate the current response functions for small q and ω. The latter will be however assumed to be large enough to greatly exceed the inverse of the equilibration time so that the system can be safely assumed to be in the collisionless regime. A comparison with Eq. (5.251) of Chapter 5 should then allow us to determine $f_{xcL(T)}(0)$, and hence the bulk and the shear moduli, in terms of Landau parameters.

Following Conti and Vignale (1999) we begin by Fourier transforming Eq. (8.56) with respect to $\vec{r}$ and t. This gives

$$\left(\vec{q} \cdot \vec{v}_{\vec{k}\sigma} - \omega\right) \delta \mathcal{N}_{\vec{k}\sigma}(\vec{q}, \omega) + \vec{q} \cdot \vec{v}_{\vec{k}\sigma} \delta(\mathcal{E}_{\vec{k}\sigma} - \mu)$$
$$\times \left[\sum_{\vec{k}'\sigma'} f_{\vec{k}\sigma, \vec{k}'\sigma'} \delta \mathcal{N}_{\vec{k}'\sigma'}(\vec{q}, \omega) + \frac{\hbar \vec{k}}{m} \cdot \frac{e}{c} \vec{A}(\vec{q}, \omega) \right] = 0 , \qquad (8.61)$$

where, as discussed, we have neglected the collision integral. This equation can be solved for any given value of the ratio $x = \frac{q v_F^*}{\omega}$. After setting

$$\delta \mathcal{N}_{\vec{k}\sigma}(\vec{q}, \omega) = \vec{\Pi}_{\vec{k}\sigma}(x) \cdot \frac{e}{c} \vec{A}(\vec{q}, \omega), \qquad (8.62)$$

we see that $\vec{\Pi}_{\vec{k}\sigma}(x)$ obeys the equation of motion

$$\vec{\Pi}_{\vec{k}\sigma}(x) = R_{\vec{k}\sigma}(x) \left(\frac{\vec{k}}{m} + \sum_{\vec{k}'\sigma'} f_{\vec{k}\sigma, \vec{k}'\sigma'} \vec{\Pi}_{\vec{k}'\sigma'}(x) \right) \qquad (8.63)$$

where

$$R_{\vec{k}\sigma}(x) \equiv -\delta(\mathcal{E}_{\vec{k}\sigma} - \mu) \frac{x \cos \theta}{x \cos \theta - 1}, \qquad (8.64)$$

and θ is the angle between $\vec{k}$ and $\vec{q}$.

In the $x \to 0$ limit we expand $R_{\vec{k}\sigma}(x)$ and $\vec{\Pi}_{\vec{k}\sigma}(x)$ in a power series of x as follows:

$$R_{\vec{k}\sigma}(x) = \delta(\mathcal{E}_{\vec{k}\sigma} - \mu) \sum_{n=1}^{\infty} (x \cos \theta)^n \qquad (8.65)$$

and

$$\vec{\Pi}_{\vec{k}\sigma}(x) = \sum_{n=0}^{\infty} \vec{\Pi}_{\vec{k}\sigma}^{(n)} x^n. \qquad (8.66)$$

Inserting these expansions in Eq. (8.63) we obtain the recursion relation

$$\vec{\Pi}_{\vec{k}\sigma}^{(n)} = \delta(\mathcal{E}_{\vec{k}\sigma} - \mu) \left[\frac{\hbar \vec{k}}{m} \cos^n \theta + \sum_{\vec{k}'\sigma'} f_{\vec{k}\sigma, \vec{k}'\sigma'} \sum_{m=0}^{n-1} \vec{\Pi}_{\vec{k}'\sigma'}^{(m)} \cos^{n-m} \theta \right] , \qquad (8.67)$$

where $\vec{\Pi}_{\vec{k}\sigma}^{(0)} = 0$ and $\vec{\Pi}_{\vec{k}\sigma}^{(1)} = \delta(\mathcal{E}_{\vec{k}\sigma} - \mu) \frac{\hbar \vec{k} \cos \theta}{m}$.

The current response

$$\vec{j}(\vec{q}, \omega) = \frac{n}{m}\frac{e}{c}\vec{A}(\vec{q}, \omega) + \frac{1}{L^d}\sum_{\vec{k}\sigma}\frac{\hbar\vec{k}}{m}\vec{\Pi}_{\vec{k}\sigma}(x) \cdot \frac{e}{c}\vec{A}(\vec{q}, \omega) \tag{8.68}$$

will now be calculated exactly to order $x^2 = \left(\frac{qv_F^*}{\omega}\right)^2$. The relevant term is the one with $n = 2$:

$$\vec{\Pi}_{\vec{k}\sigma}^{(2)} = \delta(\mathcal{E}_{\vec{k}\sigma} - \mu)\cos\theta\left[\frac{\hbar\vec{k}}{m}\cos\theta + \sum_{\vec{k}'\sigma'}f_{\vec{k}\sigma,\vec{k}'\sigma'}\delta(\mathcal{E}_{\vec{k}'\sigma'} - \mu))\frac{\hbar\vec{k}'}{m}\cos\theta'\right]. \tag{8.69}$$

Substituting this in Eq. (8.68) we can easily compute all the components of the current–current response function. In particular, its longitudinal component is given by

$$\chi_L = \frac{n}{m} + \frac{1}{L^d}\sum_{\vec{k}\sigma}\frac{\hbar\hat{q}\cdot\vec{k}}{m}\hat{q}\cdot\vec{\Pi}_{\vec{k}\sigma}^{(2)}(x)x^2 , \tag{8.70}$$

and its trace (sum of the eigenvalues) is

$$\chi_L + (d-1)\chi_T = d\frac{n}{m} + \frac{1}{L^d}\sum_{\vec{k}\sigma}\frac{\hbar\vec{k}}{m}\cdot\vec{\Pi}_{\vec{k}\sigma}^{(2)}(x)x^2. \tag{8.71}$$

The sums in Eqs. (8.70) and (8.71) are evaluated by first expanding, according to Eq. (8.15), the Landau interaction function as a series of Legendre polynomials or cosines, and then repeatedly applying the basic identities

$$\int\frac{d\Omega_3'}{\Omega_3}P_\mu(\hat{k}_1\cdot\hat{k}')P_\nu(\hat{k}'\cdot\hat{k}_2) = \frac{\delta_{\mu\nu}}{2\mu+1}P_\mu(\hat{k}_1\cdot\hat{k}_2) , \tag{8.72}$$

for arbitrary unit vectors $\hat{k}_1$, $\hat{k}_2$ in three dimensions and

$$\int\frac{d\Omega_2'}{\Omega_2}\cos\mu(\theta - \theta')\cos\nu\theta' = \frac{\delta_{\mu\nu}(1 + \delta_{\mu0})}{2}\cos\mu\theta , \tag{8.73}$$

in two dimensions.

The final expressions for χ_L and χ_T are

$$\chi_L(q, \omega) = \frac{n}{m} + \frac{n}{m}\left(\frac{qv_F}{\omega}\right)^2\begin{cases}\frac{3+5F_0^s+4F_2^s}{15(1+F_1^s)} & (3D) \\ \frac{3+2F_0^s+F_2^s}{4(1+F_1^s)} & (2D) ,\end{cases} \tag{8.74}$$

and

$$\chi_T(q, \omega) = \frac{n}{m} + \frac{n}{m}\left(\frac{qv_F}{\omega}\right)^2\begin{cases}\frac{1+F_2^s}{5(1+F_1^s)} & (3D) \\ \frac{1+F_2^s}{4(1+F_1^s)} & (2D) .\end{cases} \tag{8.75}$$

Notice that we have made use of the effective mass equation (8.35) to relate $v_F^* = \frac{\hbar k_F}{m^*}$ to $v_F = \frac{\hbar k_F}{m}$.

At this point a direct comparison between Eqs. (8.74) and (8.75) and the long wavelength expansion of the current–current response functions given by Eq. (5.251) of Chapter 5 leads to explicit expressions for low frequency limit of the longitudinal and transverse local field factors. These expressions are listed in Eqs. (5.194) and (5.195) of Chapter 5, and will not be repeated here.

Once the transverse current response function is known, the shear modulus can be readily obtained from

$$S = n \, \frac{\chi_T(q, \omega) - \frac{n}{m}}{\frac{q^2}{\omega^2}} \, , \tag{8.76}$$

a result that follows immediately from a comparison of Eqs. (5.251) and (5.254). Substituting our result (8.71) in this expression yields

$$\frac{S}{S_0} = \frac{1 + F_2^s}{1 + F_1^s} \, , \tag{8.77}$$

where

$$S_0 = \begin{cases} \frac{2n\epsilon_F}{5} & (3D) \\ \frac{n\epsilon_F}{2} & (2D) \end{cases} \tag{8.78}$$

is the noninteracting shear modulus. Notice that by the same method one can also calculate the bulk modulus $\mathcal{B} = \frac{1}{K}$, which is of course consistent with the compressibility calculated in Section 8.3.4. Numerical values of F_2^s and F_1^s, obtained from approximate microscopic theories can be found in Table 8.1 for several different densities.

8.4 Simple theory of the quasiparticle lifetime

8.4.1 General formulas

The most relevant process contributing to the decay of a quasiparticle state in the electron liquid is *electron–hole pair production*. At zero temperature, given a quasiparticle close to the Fermi surface, there is a certain probability that, because of the coulomb interaction, part of the energy and momentum of the quasiparticle will be transferred to a single electron–hole pair out of the Fermi sea. Upon losing part of its initial energy the quasiparticle makes a transition to an available lower energy state. At finite temperatures the scenario is slightly complicated by the fact that the available final states are neither definitely occupied nor definitely empty, but the basic physical picture remains the same.

In principle the quasiparticle can also decay by exciting *multiple* electron–hole pairs and/or plasmon modes. As it turns out both processes are irrelevant for quasiparticles near the Fermi surface at sufficiently low temperature. Plasmon emission is forbidden by energy and momentum conservation. This is quite obvious in three dimensions, since a quasiparticle near the Fermi surface does not have enough energy to excite a plasmon.

n two dimensions, it is the mismatch between the square-root dispersion of the plasmon frequency $\omega_p(\vec{k}) \sim \sqrt{k}$, and the linear dispersion of the quasiparticle energy (see
Eq. (8.4)) that effectively prevents the emission of plasmons (Giuliani and Quinn, 1982).
As for multiple electron–hole pair excitations, the problem is that their spectral density
vanishes, at low energy, much more rapidly than the spectral density of single electron–hole
pairs.

The simplest way to estimate the electron–hole contribution to the quasiparticle lifetime
is to make use of the *Fermi golden rule* to compute the transition probability between the
initial and the final state of the system. Within this approach (which we will in fact formalize
in Section 8.7 through an explicit approximate calculation of the self-energy) the rate at
which an electron of spin σ and momentum $\hbar\vec{k}$ is scattered by the coulomb interaction into
an empty state of momentum $\hbar(\vec{k} - \vec{q})$, while an electron of spin σ' and momentum $\hbar k'$ is
scattered into an empty state of momentum $\hbar(\vec{k'} + \vec{q})$ is given by

$$\frac{2\pi}{\hbar} \left| \frac{W(\vec{q})}{L^d} \right|^2 \delta(\varepsilon_{\vec{k}-\vec{q}\sigma} + \varepsilon_{\vec{k'}+\vec{q}\sigma'} - \varepsilon_{\vec{k}\sigma} - \varepsilon_{\vec{k'}\sigma'}), \tag{8.79}$$

where $\frac{W(\vec{q})}{L^d}$ is the matrix element of an effective two-particle interaction between the initial
and final plane wave states. The δ-function ensures that the energy is properly conserved
through the coulomb collision.

We hasten to say that Eq. (8.79) is approximate in more than one way. First of all, the
correct two-particle scattering amplitude is a function of $\vec{k}$, $\vec{k'}$, and $\vec{q}$, as well as the relative
spin orientation, not just of the momentum transfer $\vec{q}$.[24] Secondly, Eq. (8.79) violates the
indistinguishability of the electrons, since the scattering amplitude is not antisymmetric
upon interchange of the two final plane wave states of parallel spin (this question will be
discussed in detail in Section 8.4.4 below). Finally, Eq. (8.79) is not entirely self-consistent,
since the quasiparticle energy is approximated by the bare particle energy. Despite these
obvious defects, Eq. (8.79) is still an excellent starting point to begin to understand the
microscopic physics of the Fermi liquid.

The inverse lifetime $1/\tau_{\vec{k}}^{(e)}$ of a plane wave state initially occupied by an electron of
momentum $\hbar\vec{k}$ and spin σ, is given by the sum of the probabilities of all the allowed decay
processes:

$$\frac{1}{\tau_{\vec{k}\sigma}^{(e)}} = \frac{2\pi}{\hbar} \sum_{\vec{q}\vec{k'}\sigma'} \left| \frac{W(\vec{q})}{L^d} \right|^2 n_{\vec{k'}\sigma'}(1 - n_{\vec{k'}+\vec{q}\sigma'})(1 - n_{\vec{k}-\vec{q}\sigma})$$
$$\times \delta(\varepsilon_{\vec{k}-\vec{q}\sigma} + \varepsilon_{\vec{k'}+\vec{q}\sigma'} - \varepsilon_{\vec{k}\sigma} - \varepsilon_{\vec{k'}\sigma'}), \tag{8.80}$$

where the Fermi occupation factors guarantee that the plane wave state $\vec{k'}\sigma'$ is indeed
occupied by an electron, while the final states $\vec{k'} + \vec{q}\ \sigma'$, and $\vec{k} - \vec{q}\ \sigma'$ are empty and

---

[24] Such is, for example, the approximate scattering amplitude derived in Section 5.5.3.1 from the Kukkonen-Overhauser effective
interaction (see Eqs. (5.153) and (5.154)).

therefore available for occupation after the scattering event.[25] A rough estimate of the resul
that ensues from this equation can be quickly arrived at by means of the integral discussec
in Exercise 11.

A nice feature of Eq. (8.80) is that part of the calculation can be carried out withou
specifying the form of the scattering amplitude $W(\vec{q})$. We work, for simplicity, in the
paramagnetic state, and approximate $n_{\vec{k}\sigma}$ by the *noninteracting* occupation numbers $n_{\vec{k}\sigma}^{(0)}$.
Then we make use of the fluctuation-dissipation theorem for the non-interacting electron
gas (see Eq. (3.74)) to evaluate the sum over $\vec{k}'$ and σ':

$$\frac{\pi}{\hbar L^d} \sum_{\vec{k}'\sigma'} n_{\vec{k}'\sigma'}^{(0)} (1 - n_{\vec{k}'+\vec{q}\ \sigma'}^{(0)}) \delta\left(\frac{\varepsilon_{\vec{k}'+\vec{q}\ \sigma'} - \varepsilon_{\vec{k}'\sigma'}}{\hbar} - \omega\right) = -\frac{\Im m \chi_0(q,\omega)}{1 - e^{-\beta\hbar\omega}} . \tag{8.81}$$

Naturally, the appearance of the spectral function $-\Im m \chi_0(q,\omega)$ shows that the spectral
density of electron–hole pairs plays a central role in the process. Eq. (8.80) can now be
rewritten as

$$\frac{1}{\tau_{\vec{k}\sigma}^{(e)}} = -\frac{2}{(2\pi)^d} \int_{-\infty}^{\infty} d\omega \frac{1 - n_F(\varepsilon_{\vec{k}\sigma} - \hbar\omega - \mu)}{1 - e^{-\beta\hbar\omega}}$$

$$\times \int_0^{\infty} dq q^{d-1} |W(\vec{q})|^2 \Im m \chi_0(q,\omega) \int d\Omega_d \delta(\varepsilon_{\vec{k}\sigma} - \varepsilon_{\vec{k}-\vec{q}\sigma} - \hbar\omega) , \tag{8.82}$$

where $n_F(x) = (e^{\beta x} + 1)^{-1}$ is the Fermi–Dirac distribution function at zero chemical poten-
tial, so that $n_F(\varepsilon_{\vec{k}\sigma} - \mu) = [e^{\beta(\varepsilon_{\vec{k}\sigma} - \mu)} + 1]^{-1} = n_{\vec{k}\sigma}^{(0)}$ is the noninteracting occupation number. In obtaining this expression we have introduced the variable $\hbar\omega = \varepsilon_{\vec{k}\sigma} - \varepsilon_{\vec{k}-\vec{q}\sigma}$ through
the introduction of an auxiliary delta function and its corresponding integration. Notice that
the angular integration only involves the delta function.

The reader is invited to verify (Exercise 9) that the corresponding formula for the
lifetime of a quasihole $(\tau_{\vec{k}\sigma}^{(h)})$ is obtained from Eq. (8.82) simply by performing the re-
placements $1 - n_F(\varepsilon_{\vec{k}\sigma} - \hbar\omega - \mu) \to n_F(\varepsilon_{\vec{k}\sigma} - \hbar\omega - \mu)$ and $1 - e^{-\beta\hbar\omega} \to 1 - e^{\beta\hbar\omega}$, and
changing the overall sign. This can be in turn be used to show that the principle of
detailed balance, Eq. (8.55), is satisfied at the appropriate level of accuracy, i.e., we
have

$$\frac{n_{\vec{k}\sigma}}{\tau_{\vec{k}\sigma}^{(e)}} = \frac{1 - n_{\vec{k}\sigma}}{\tau_{\vec{k}\sigma}^{(h)}} . \tag{8.83}$$

We shall henceforth concentrate only on the calculation of the quasiparticle lifetime. The
calculation will be carried out separately for three and two-dimensional systems below.
In one dimension, a calculation of $1/\tau_{\vec{k}\sigma}^{(e)}$ based on Eq. (8.82) would result in a divergent
integral at finite temperatures (see Exercise 12): we can conclude that the Landau Fermi
liquid picture cannot be consistently applied to 1D systems.

---

[25] Again, we must remark that the sum in (8.80) ignores the fact that the state determined by $\vec{k}'\sigma'$ and $\vec{q} = \vec{Q}$ is indistinguishable
from the one determined by $\vec{k}'\sigma'$ and $\vec{q} = \vec{k} - \vec{k}' - \vec{Q}$.

8.4.2 Three-dimensional electron gas

In this case the angular integration is rather simple once the z-axis is taken along the direction of $\vec{q}$. We have:

$$\int_0^{2\pi} d\phi \int_0^\pi d\theta \sin\theta \, \delta(\varepsilon_{\vec{k}\sigma} - \varepsilon_{\vec{k}-\vec{q}\sigma} - \hbar\omega) = 2\pi \int_{-1}^{1} d\eta \, \delta\left(\frac{\hbar^2}{2m}(2kq\eta - q^2) - \hbar\omega\right). \tag{8.84}$$

The integration is trivial and gives:

$$\int d\Omega_3 \delta(\varepsilon_{\vec{k}\sigma} - \varepsilon_{\vec{k}-\vec{q}\sigma} - \hbar\omega) = \frac{2\pi m}{\hbar^2 kq}\, \Theta\left(1 - \left|\frac{q^2 + \frac{2m\omega}{\hbar}}{2kq}\right|\right), \tag{8.85}$$

where $\Theta(x)$ is the familiar Heaviside step function which mandates a precise behavior for the limits of the remaining quadratures. As it turns out however, these limiting conditions are irrelevant, in view of the behavior of the integrand. This can be seen as follows: for positive frequencies the Fermi thermal occupation factor $1 - n_F(\varepsilon_{\vec{k}\sigma} - \hbar\omega - \mu)$ cuts off the integral for ω of the order of $|\varepsilon_{\vec{k}\sigma} - \mu|$ an energy which, by assumption, is much smaller than the Fermi energy ϵ_F. For negative frequencies, on the other hand, it is the thermal occupation factor $(1 - e^{-\beta\hbar\omega})^{-1}$ that cuts off the frequency integral for ω of the order of $k_B T$, an energy scale that we assume to be much smaller than ϵ_F. At very low frequency the most stringent limits on the q-integral come from the factor $\Im m\, \chi_0(q, \omega)$ which contains the structure of the electron–hole continuum as described in Section 4.4.2 and differs from zero only along the segment shown in Fig. 8.8. This sets the lower limit of the q integral at $q \sim \frac{|\omega|}{v_F} \sim 0$ and the upper limit at $q \sim 2k_F + \frac{|\omega|}{v_F} \sim 2k_F$. Notice that the dominant contribution to the integral (for $\omega \to 0$) comes from the region labeled as II in Fig. 8.8. In this region it is

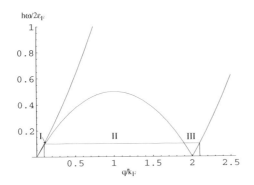

Fig. 8.8. Regions of wave vector integration for the calculation of the low-energy quasiparticle lifetime. For $k \to k_F$ and $\hbar\omega \ll \epsilon_F$ the leading contribution to the wave vector integral comes from region II where $\frac{|\omega|}{v_F} < q < 2k_F - \frac{|\omega|}{v_F}$. Plasmon excitations do not contribute to the lifetime and hence are not shown.

legitimate to approximate $\Im m \chi_0(q, \omega)$ by its zero temperature and low-frequency form

$$\Im m \chi_0(q, \omega) \simeq -\frac{\pi \omega}{2 v_F q} N(0) . \tag{8.86}$$

Accordingly the formula for $\frac{1}{\tau_k^{(e)}}$ in three dimensions is seen to be proportional to the expression

$$\frac{1}{\tau_{\vec{k}\sigma}^{(e)}} \propto \int_{-\infty}^{\infty} \frac{d\omega \, \omega}{(1 + e^{\beta(\hbar\omega - \varepsilon_{\vec{k}\sigma} + \mu)})(1 - e^{-\beta\hbar\omega})} \int_0^{2k_F} dq |W(\vec{q})|^2, \quad 3D. \tag{8.87}$$

The frequency integral can be calculated analytically by means of the exact result

$$\int_{-\infty}^{\infty} dy \frac{x - y}{(1 + e^{-y})(1 - e^{y-x})} = \frac{x^2 + \pi^2}{2(1 + e^{-x})} , \tag{8.88}$$

and is given by

$$\frac{1}{2\hbar^2} \frac{(\varepsilon_{\vec{k}\sigma} - \mu)^2 + (\pi k_B T)^2}{1 + e^{-\beta(\varepsilon_{\vec{k}\sigma} - \mu)}} . \tag{8.89}$$

Deriving Eq. (8.88) provides a useful, entertaining exercise in contour integration.[26] The integral over the wave vector q deserves special attention. It is quite obvious at this point why one cannot make use of the bare coulomb interaction. The integral would emphatically diverge. This is of course a consequence of the long range of the coulomb interaction. It is then quite natural to employ for $W(\vec{q})$ some sort of screened interaction. In this case the q integral presents no problems. A reasonable approximation is provided by the choice $W(\vec{q}) \simeq \frac{v_q}{\varepsilon(q,0)}$ so that

$$\int_0^{2k_F} dq |W(\vec{q})|^2 \simeq \frac{1}{N^2(0)} \int_0^{2k_F} dq \left(\frac{N(0)v_q}{\varepsilon(q, 0)}\right)^2 . \tag{8.90}$$

Further simplification can be achieved by making use of the Thomas-Fermi approximate dielectric function, that is the static long wavelength limit of the RPA dielectric function, given by Eqs. (5.35) and (5.36) of Section 5.3.2. This gives the result

$$\int_0^{2k_F} dq |W(\vec{q})|^2 \simeq \frac{2k_F}{N^2(0)} \xi_3(r_s) , \tag{8.91}$$

where the function $\xi_3(r_s)$ is given by

$$\xi_3(r_s) = \int_0^1 \frac{dy}{\left(1 + \frac{4k_F^2}{\kappa_3^2} y^2\right)^2}$$

$$= \sqrt{\frac{\alpha_3 r_s}{4\pi}} \tan^{-1} \sqrt{\frac{\pi}{\alpha_3 r_s}} + \frac{1}{2\left(1 + \frac{\pi}{\alpha_3 r_s}\right)} . \tag{8.92}$$

For most densities in the metallic range, $\xi_3(r_s) \simeq 1$. Notice that $\xi_3(r_s) \sim \sqrt{r_s}$ as $r_s \to 0$:

---

[26] It can also be looked up in Gradshteyin and Ryzhik, Eq. (3.419.2).

thus, due to the non-perturbative nature of the screening, the quasiparticle decay rate turns out to be proportional to the electron charge e, rather than to e^4, as one could have naively expected.

Collecting the various factors we finally obtain for the inelastic quasiparticle lifetime in three dimensions the following result:

$$\frac{1}{\tau_{\vec{k}\sigma}^{(e)}} \simeq \frac{\pi}{8\hbar\epsilon_F} \frac{(\varepsilon_{\vec{k}\sigma} - \epsilon_F)^2 + (\pi k_B T)^2}{1 + e^{-\beta(\varepsilon_{\vec{k}\sigma} - \epsilon_F)}} \xi_3(r_s), \qquad 3D, \qquad (8.93)$$

where we have approximated μ with ϵ_F and k with k_F.

The inverse lifetime of a quasiparticle at the Fermi surface ($k = k_F$) vanishes as T^2 at small temperatures. On the other hand, at $T = 0$, the inverse lifetime vanishes as $(\varepsilon_{\vec{k}\sigma} - \epsilon_F)^2$. One power of $\varepsilon_{\vec{k}\sigma} - \epsilon_F$ (or T) arises from the phase space restrictions on the scattering process. The second one stems from the linearly vanishing density of electron–hole pairs excitations. The numerical prefactor is simply a Fermi surface average of the statically screened coulomb interaction. This is the expected behavior, an indication that the Landau theory of the electron liquid is consistent with the microscopic perturbative approach.

8.4.3 Two-dimensional electron gas

The two-dimensional case presents a few new twists. The most important difference with the three-dimensional case is the q dependence of the integrand of Eq. (8.82) which, as we shall see, must be handled with special care in the regions $q \simeq 0$ and $q \simeq 2k_F$. This necessitates a more precise treatment of the limits of integration.

We begin by considering the angular integration which in this case gives the interesting result

$$\int_0^{2\pi} d\phi \, \delta(\varepsilon_{\vec{k}\sigma} - \varepsilon_{\vec{k}-\vec{q}\sigma} - \hbar\omega) = \frac{2\Theta\left(1 - \left|\frac{q^2 + \frac{2m\omega}{\hbar}}{2kq}\right|\right)}{\sqrt{\left(\frac{\hbar^2 kq}{m}\right)^2 - \left(\hbar\omega + \frac{\hbar^2 q^2}{2m}\right)^2}}, \qquad (8.94)$$

an expression that features an extra frequency dependence with respect to the three-dimensional case. The other necessary ingredient is the expression for $\Im\chi_0(q,\omega)$ in two dimensions, which, at low frequency and in region II of Fig. 8.8, is approximately given by

$$\Im\chi_0(q,\omega) \simeq -\frac{\omega}{qv_F} \frac{N(0)}{\sqrt{1 - \left(\frac{q}{2k_F}\right)^2}}, \qquad 2D, \qquad (8.95)$$

where $N(0) = \frac{m}{\pi\hbar^2}$. Within the necessary accuracy, we can set $k = k_F$ in the argument of the square root appearing in Eq. (8.94), which can then be rewritten as

$$\frac{\hbar^2}{2m} \sqrt{\left[q^2 - \left(\frac{|\omega|}{v_F}\right)^2\right](4k_F^2 - q^2)}. \qquad (8.96)$$

Accordingly we see that the contribution of region II to the q integral of Eq. (8.82) is

$$8\pi k_F \int_{\frac{|\omega|}{v_F}}^{2k_F - \frac{|\omega|}{v_F}} \frac{|N(0)W(\vec{q})|^2 dq}{\sqrt{q^2 - \left(\frac{|\omega|}{v_F}\right)^2} (4k_F^2 - q^2)} , \qquad 2D. \qquad (8.97)$$

This integral can be evaluated rather easily. Notice that in the limit $\omega \to 0$ it presents a logarithmic divergence originating from the regions $q \simeq 0$ and $q \simeq 2k_F$. To extract the exact coefficient of the logarithmic singularity we set $q = 0$ and $q = 2k_F$ in the regular parts of the integrand, when evaluating the contributions of $q \simeq 0$ and $q \simeq 2k_F$ respectively. Up to corrections that remain finite as $\omega \to 0$ the integral is then found to be equal to

$$\frac{\pi \left(|N(0)W(0)|^2 + \frac{1}{2}|N(0)W(2k_F)|^2\right)}{k_F} \ln \frac{4\epsilon_F}{|\omega|} . \qquad (8.98)$$

It is convenient, at this point, to define the "coupling constant"

$$\xi_2(r_s) \equiv |N(0)W(0)|^2 + \frac{1}{2}|N(0)W(2k_F)|^2 , \qquad (8.99)$$

which, in the Thomas-Fermi approximation (see Eqs. (5.37) and (5.38)), depends on r_s in the following manner:

$$\xi_2(r_s) = 1 + \frac{1}{2} \left(\frac{r_s}{r_s + \sqrt{2}}\right)^2 . \qquad (8.100)$$

Notice that, unlike its three-dimensional counterpart, $\xi_2(r_s)$ tends to a constant, 1, in the high-density limit.[27] Combining Eqs. (8.82), (8.94), (8.95), and (8.97), we find that the quasiparticle lifetime is given by the integral

$$\frac{1}{\tau_{\vec{k}\sigma}^{(e)}} \simeq \frac{\hbar \xi_2(r_s)}{2\pi \epsilon_F} \int_{-\infty}^{\infty} \frac{\omega \ln \frac{4\epsilon_F}{|\omega|}}{(1 + e^{\beta(\hbar\omega - \epsilon_{\vec{k}\sigma} + \mu)})(1 - e^{-\beta\hbar\omega})} d\omega , \qquad 2D. \qquad (8.101)$$

Consider first the "zero-temperature" situation $|\epsilon_{\vec{k}\sigma} - \epsilon_F| \gg k_B T$. In this case it is clear that the main contribution to the integral comes from the region $\omega \simeq \epsilon_{\vec{k}\sigma} - \epsilon_F$. Now, since in this region the logarithm is slowly varying, we can take it out of the integration to give the factor $\ln \frac{4\epsilon_F}{|\epsilon_{\vec{k}\sigma} - \epsilon_F|}$. This leaves us with a frequency integral which coincides with that of Eq. (8.87), which we calculated exactly. The only difference is that in this case we need to take the limit $\frac{k_B T}{|\epsilon_{\vec{k}\sigma} - \epsilon_F|} \to 0$. By making use of Eq. (8.89) we therefore obtain the result

$$\frac{1}{\tau_{\vec{k}\sigma}^{(e)}} \simeq \xi_2(r_s) \frac{(\epsilon_{\vec{k}\sigma} - \epsilon_F)^2}{4\pi\hbar\epsilon_F} \ln \frac{4\epsilon_F}{|\epsilon_{\vec{k}\sigma} - \epsilon_F|} , \qquad k_B T \ll |\epsilon_{\vec{k}\sigma} - \epsilon_F|. \qquad (8.102)$$

The other relevant case is that of $k_B T \gg |\epsilon_{\vec{k}\sigma} - \epsilon_F|$, which corresponds to the case of a quasiparticle lying on the Fermi surface. In this case a direct inspection of Eq. (8.101)

---

[27] The surprising fact that the inverse lifetime fails to vanish in the noninteracting limit $r_s \to 0$ is an artifact due to our asymptotic expansion of the integral (8.97), which requires the limit $k \to k_F$ to be taken *before* the limit $r_s \to 0$. The expansion fails for $r_s < \left|\frac{k}{k_F} - 1\right|$.

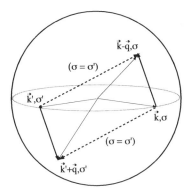

Fig. 8.9. Two indistinguishable "routes" can connect the initial states $\vec{k}$, σ and $\vec{k}'$, σ' to the final states $\vec{k} - \vec{q}$, σ and $\vec{k} + \vec{q}$, σ' on the Fermi sphere. The exchange route is shown by the dashed lines and is allowed only for $\sigma = \sigma'$.

shows that the most relevant contributions to the integral come from a region of the order of $\frac{k_B T}{\hbar}$ centered about the origin. In this situation the logarithm can again be taken out of the integral[28] as to give a factor of $\ln \frac{4\epsilon_F}{k_B T}$. The remaining integral can then again be evaluated by means of Eq. (8.89) in the limit of $\frac{|\varepsilon_{\vec{k}\sigma} - \epsilon_F|}{k_B T} \to 0$. This immediately gives[29]

$$\frac{1}{\tau_{\vec{k}\sigma}^{(e)}} \simeq \xi_2(r_s) \frac{(\pi k_B T)^2}{8\pi \hbar \epsilon_F} \ln \frac{4\epsilon_F}{k_B T} , \qquad |\varepsilon_{\vec{k}\sigma} - \epsilon_F| \ll k_B T . \qquad (8.103)$$

The only significant difference with the three-dimensional case is the appearance here of the logarithmic factors. The basic comments made about the three-dimensional result equally apply.

We emphasize that the above calculation focused on the leading term in the low-energy/low-temperature expansion of the inverse lifetime.[30] The complete calculation of the "subleading" contributions of order $(\varepsilon_{\vec{k}\sigma} - \epsilon_F)^2$ and $(k_B T)^2$ is more tricky: in particular, it can be shown that the regions I and III in the q integral (see Fig. 8.8) do contribute to the result at this order.

8.4.4 Exchange processes

In our calculations of the quasiparticle lifetime we have so far treated the electron quasiparticle as if it were somehow distinguishable from the other electrons in the system. Some effects of indistinguishability can be incorporated in the electron–electron interaction $W(\vec{q})$, as discussed in Section 5.5.3.1. In this section, however, we are concerned with a more explicit effect, which is illustrated in Fig. 8.9. For $\sigma = \sigma'$, the initial states $\vec{k}$, σ and $\vec{k}'$, σ'

---

[28] This is due to the fact that if $f(x)$ is well behaved function in the interval $[-a, a]$, then in the limit $a \to 0$, with logarithmic accuracy, $\int_{-a}^{a} \ln|x| f(x) dx \simeq \ln|a| \int_{-a}^{a} f(x) dx$ as one can readily verify.

[29] The coefficient of Eq. (8.103) can also be quickly inferred from Eq. (8.102) by making use of the general result (8.89).

[30] It is somewhat sobering to remark that a surprisingly vast variety of contradicting results for the coefficients of Eqs. (8.102) and (8.103) have appeared in the literature.

are connected to the final states $\vec{k} - \vec{q}, \sigma$ and $\vec{k} + \vec{q}, \sigma'$, by two indistinguishable routes. According to the general principles of quantum mechanics the probability of the scattering process is given by the squared modulus of the sum of the probability amplitudes of the two routes. Following this prescription, we obtain the following amended expression for the inverse lifetime:

$$\frac{1}{\tau_{\vec{k}\sigma}^{(e)}} = \frac{2\pi}{\hbar L^{2d}} \sum_{\vec{q}\vec{k}'\sigma'} \left(1 - \frac{1}{2}\delta_{\sigma\sigma'}\right) |W(\vec{q}) - \delta_{\sigma\sigma'} W(\vec{k} - \vec{k}' - \vec{q})|^2 n_{\vec{k}'\sigma'}[1 - n_{\vec{k}'+\vec{q}\sigma'}]$$
$$\times [1 - n_{\vec{k}-\vec{q},\sigma}]\delta(\varepsilon_{\vec{k}-\vec{q}\sigma} + \varepsilon_{\vec{k}'+\vec{q}\sigma'} - \varepsilon_{\vec{k}\sigma} - \varepsilon_{\vec{k}'\sigma'}) . \tag{8.104}$$

The factor $\left(1 - \frac{1}{2}\delta_{\sigma\sigma'}\right)$, that is, $\frac{1}{2}$ for $\sigma' = \sigma$ and 1 for $\sigma' = -\sigma$ corrects for the double counting of the final states in the case of parallel spins. The parallel-spin part of the sum (8.104) is invariant under the substitution $\vec{q} \to \vec{k} - \vec{k}' - \vec{q}$ since these two wave vector transfers describe exactly the same physical process. Making use of this symmetry, we see that the inverse lifetime is naturally written as the difference of two terms

$$\frac{1}{\tau_{\vec{k}\sigma}^{(e)}} = \frac{1}{\tau_{d,\vec{k}\sigma}^{(e)}} - \frac{1}{\tau_{x,\vec{k}\sigma}^{(e)}} , \tag{8.105}$$

where the "direct" term is given by Eq. (8.80) and the "exchange" term is given by

$$\frac{1}{\tau_{x,\vec{k}\sigma}^{(e)}} = \frac{2\pi}{\hbar L^{2d}} \sum_{\vec{k}'\vec{q}} W(\vec{q})W(\vec{k} - \vec{k}' - \vec{q})n_{\vec{k}'\sigma}[1 - n_{\vec{k}'+\vec{q}\sigma}]$$
$$\times [1 - n_{\vec{k}-\vec{q},\sigma}]\delta(\varepsilon_{\vec{k}-\vec{q}\sigma} + \varepsilon_{\vec{k}'+\vec{q}\sigma} - \varepsilon_{\vec{k}\sigma} - \varepsilon_{\vec{k}'\sigma}) . \tag{8.106}$$

The calculation of the exchange contribution to the inverse lifetime is an interesting exercise. Taking advantage of the fact that all the wave vectors are close to the Fermi surface one can, through a series of clever transformations, reduce the evaluation of the seemingly complicated sum (8.106) to the calculation of simple angular integrals. The final compact formula in 3D is presented in Exercise 18 together with a few useful hints for its derivation. However, some basic facts about the exchange contribution to the quasiparticle decay rate can be established without detailed calculations.

- Because the direct and the exchange term in Eq. (8.104) have the same dependence on the occupation factors, it follows that they depend on energy and temperature in the same manner (see Exercises 10 and 17 for details).
- Let $\frac{1}{\tau_{\vec{k}\sigma\sigma}^{(e)}}$ denote the part of the inverse lifetime of a σ-spin quasiparticle that arises from collisions with same-spin electrons ($\sigma' = \sigma$), i.e., the term with $\sigma = \sigma'$ in Eq. (8.104). This quantity is definitely positive. If we denote by $\frac{1}{\tau_{d,\vec{k}\sigma\sigma}^{(e)}}$ the contribution of direct scattering processes to this quantity, we then have

$$\frac{1}{\tau_{\vec{k}\sigma\sigma}^{(e)}} = \frac{1}{\tau_{d,\vec{k}\sigma\sigma}^{(e)}} - \frac{1}{\tau_{x,\vec{k}\sigma}^{(e)}} , \tag{8.107}$$

which implies

$$\frac{1}{\tau_{x,\vec{k}\sigma}^{(e)}} \leq \frac{1}{\tau_{d,\vec{k}\sigma\sigma}^{(e)}} , \tag{8.108}$$

where the equality holds only if W is independent of $\vec{q}$ (in which case $\frac{1}{\tau_{\vec{k}\sigma\sigma}^{(e)}}$ vanishes). In the

paramagnetic state we also have $\frac{1}{\tau_{d,\vec{k}\sigma\sigma}^{(e)}} = \frac{1}{2\tau_{d,\vec{k}}^{(e)}}$, independent of σ, and Eq. (8.108) therefore implies

$$\frac{1}{\tau_{x,\vec{k}}^{(e)}} \leq \frac{1}{2\tau_{d,\vec{k}}^{(e)}} . \tag{8.109}$$

Thus, the exchange contribution to the inverse lifetime cannot exceed one half of the direct contribution in the paramagnetic state.

- If $W(\vec{q})$ is positive for all wavevectors (as, for instance, in the Thomas-Fermi approximation) then $\frac{1}{\tau_{x,\vec{k}\sigma}^{(e)}}$ is also positive. In this case exchange processes *decrease* the scattering rate, reinforcing our general conclusions about the validity of Fermi liquid theory.

- In the high-density limit the relative importance of the exchange contribution to the inverse lifetime depends on dimensionality. In three dimensions one finds $\frac{1}{\tau_x} \ll \frac{1}{\tau_d}$, with $\frac{\tau_d}{\tau_x}$ tending to zero for $r_s \to 0$. This can be seen by comparing the magnitudes of $|W(\vec{q})|^2$ (Eq. (8.80)) and $W(\vec{q})W(\vec{k} - \vec{k}' - \vec{q})$ (Eq. (8.106)) in the most important region of integration, where $q \simeq 0$ and $|k - k'| \simeq 2k_F$ (see Fig. 8.9). Using the Thomas-Fermi approximation (which is appropriate for small r_s) we find that the ratio of these two quantities scales as $\frac{k_{TF}^2}{k_F^2}$, which is proportional to r_s. In two dimensions the situation is different because the leading contribution to the decay rate, namely, the logarithmically enhanced quantity on the right-hand side of Eq. (8.102), comes from a region of integration in which both q and $|\vec{k} - \vec{k}'|$ are small relative to k_F. Hence the exchange correction is non-negligible, and significantly reduces the decay rate even as $r_s \to 0$ (Reizer and Wilkins, 1997).

- In the low-density limit $\frac{1}{\tau_x}$ approaches the upper bound of Eq. (8.109), since the difference between $W(\vec{q})$ and $W(\vec{k} - \vec{k}' - \vec{q})$ vanishes as the radius of the Fermi sphere shrinks to zero.

8.5 Microscopic underpinning of the Landau theory

Landau guessed the correct theory of interacting Fermi liquids by the sheer power of physical intuition. Shortly afterwards, ordinary mortals showed that indeed the theory could be "derived" from the microscopic hamiltonian under certain assumptions of continuity and regularity. Nowadays the Landau theory is clearly recognized as a beautiful example of *renormalization*, whereby the exact many-body hamiltonian is transformed, through recursive elimination of fast degrees of freedom, into an effective hamiltonian of weakly interacting quasiparticles (Shankar, 1994). A "poor man" version of this theory, based on seminal work by Hamann and Overhauser (1966), will be presented in Section 8.6.

This and the next section are devoted to achieving an understanding of the Landau theory from a microscopic point of view. In particular, the Landau interaction function, the quasiparticle effective mass, and other relevant Fermi liquid parameters will be expressed in terms of microscopic quantities.

As we have seen, the main features of the Fermi liquid theory are

(i) The idea of a continuous evolution of the low energy spectrum of the non-interacting system into that of the interacting one upon slowly switching-on the interaction;

(ii) The fact that the quasiparticle decay rate vanishes as $|k - k_F|^2$ for $k \to k_F$;

(iii) The discontinuity of the momentum occupation numbers as the wave vector crosses the Fermi surface;

(iv) Luttinger's theorem, i.e., the statement that the volume of the Fermi sphere does not change as the interaction is switched on.

As it turns out, property (ii) translates into the fact that the imaginary part of the electron self-energy behaves as $\Im m \, \Sigma(k_F, \omega) \simeq a(\omega - \mu)^2$ as $\omega \to \mu$, an equation that is often taken as the defining feature of a Fermi liquid. Properties (iii) and (iv) can then be shown to follow from the assumption of continuity, which in the Green's function formalism corresponds to the convergence of the perturbative series. While it is well known that the perturbative series does not converge in the case of attractive interactions,[31] to date no general necessary and sufficient conditions on the strength and form of a repulsive interaction potential have been established for the convergence of the perturbative series. Moreover, as it should be clear from our discussion of the possible states of the electron liquid in Chapter 1, for the coulomb interaction the perturbative approach is only valid in the high density regime while it fails in the low density regime in which the electrons form a Wigner crystal. Where and how the Fermi liquid picture fails is a fundamental and still open question.

8.5.1 The spectral function

A formally rigorous description of single-particle properties in a many-body system can be given in terms of the *spectral function*. At zero temperature this is defined through the following construction. Let us start with the N-particle system in the ground-state, denoted by $|0, N\rangle$, and let us add one particle in the single-particle state $|\psi_\alpha\rangle$ (for example, $|\psi_\alpha\rangle$ can be a plane wave state of wave vector $\vec{k}$ and spin projection σ). This operation is carried out by applying the creation operator $\hat{a}_\alpha^\dagger$ to the ground-state $|0, N\rangle$: the result is a superposition of exact eigenstates of the hamiltonian of the $N + 1$-particle system, which can be written as

$$\hat{a}_\alpha^\dagger |0, N\rangle = \sum_n c_{n,N+1}^\alpha |n, N + 1\rangle . \tag{8.110}$$

The coefficients $c_{n,N+1}^\alpha$ are given by

$$c_{n,N+1}^\alpha = \langle n, N + 1 | \hat{a}_\alpha^\dagger | 0, N \rangle . \tag{8.111}$$

Thus the addition of a particle leaves the the system in a state that does not have a definite energy. However, we can still ask the question: what is the probability of finding the $N + 1$-particle system at the energy $E_{n,N+1}$ of the n-th excited state? Evidently, this is given

---

[31] For attractive interactions the Fermi surfaces collapses due to the onset of superconductivity.

эу

$$\frac{|c_{n,N+1}^{\alpha}|^{2}}{\sum_{n}|c_{n,N+1}^{\alpha}|^{2}} \ , \tag{8.112}$$

where

$$\sum_{n}|c_{n,N+1}^{\alpha}|^{2} = \langle 0, N|\hat{a}_{\alpha}\hat{a}_{\alpha}^{\dagger}|0, N\rangle = 1 - n_{\alpha} \tag{8.113}$$

is the normalization constant.[32] In a large system the eigenstates of the hamiltonian are very closely spaced in energy and a more relevant question is: what is the probability that the system be left in a state with an energy comprised between $E_{0,N} + \hbar\omega$ and $E_{0,N} + \hbar(\omega + d\omega)$? From Eqs. (8.110) and (8.111) we see that this probability is $A_{>}(\alpha, \omega)d\omega$, where

$$A_{>}(\alpha, \omega) = \sum_{n}|\langle n, N+1|\hat{a}_{\alpha}^{\dagger}|0, N\rangle|^{2}\delta\left(\omega - \frac{E_{n,N+1} - E_{0,N}}{\hbar}\right). \tag{8.114}$$

The normalization integral is (see Eq. (8.113))

$$\int_{-\infty}^{\infty} A_{>}(\alpha, \omega)d\omega = 1 - n_{\alpha}. \tag{8.115}$$

In a completely analogous manner we can start from the ground-state $|0, N\rangle$, *remove* a particle from $|\psi_{\alpha}\rangle$, and ask: what is the probability that the system is left in a state with an energy comprised between $E_{0,N} - \hbar\omega$ and $E_{0,N} - \hbar(\omega + d\omega)$? The answer is $A_{<}(\alpha, \omega)d\omega$, where

$$A_{<}(\alpha, \omega) = \sum_{n}|\langle n, N-1|\hat{a}_{\alpha}|0, N\rangle|^{2}\delta\left(\omega + \frac{E_{n,N-1} - E_{0,N}}{\hbar}\right), \tag{8.116}$$

with a normalization integral given by

$$\int_{-\infty}^{\infty} A_{<}(\alpha, \omega)d\omega = n_{\alpha}. \tag{8.117}$$

The complete spectral function is defined as the sum of the probability densities $A_{>}$ and $A_{<}$, i.e.,

$$A(\alpha, \omega) \equiv A_{>}(\alpha, \omega) + A_{<}(\alpha, \omega). \tag{8.118}$$

Physically, this represents the probability density for increasing or decreasing the energy of the N-particle system by an amount comprised between $\hbar\omega$ and $\hbar(\omega + d\omega)$ by adding or removing a single particle in state $|\psi_{\alpha}\rangle$. As befits a probability density, $A(\alpha, \omega)$ is a positive-definite quantity, and satisfies the sum rule

$$\int_{-\infty}^{\infty} A(\alpha, \omega)d\omega = 1 . \tag{8.119}$$

---

[32] Notice that the normalization constant would vanish if the state ψ_{α} were *definitely* occupied in $|0, N\rangle$. A ground-state with this property is annihilated by $\hat{a}_{\alpha}^{\dagger}$ in accordance with Pauli's exclusion principle.

The energy differences in Eqs. (8.114) and (8.116) can be decomposed into two parts, one associated with the mere change in particle number and the other with the excitation energy proper. For example

$$E_{n,N+1} - E_{0,N} = (E_{n,N+1} - E_{0,N+1}) + (E_{0,N+1} - E_{0,N}) . \qquad (8.120)$$

The term in the first bracket is an excitation energy of the $N + 1$-particle system and therefore is, by definition, positive. The term in the second bracket is the *addition energy*, i.e., the difference between the ground-state energies of the $N + 1$- and N-particle systems: following the notation of Chapter 7 we denote this difference by $\mu_+(N)$. Similarly,

$$E_{n,N-1} - E_{0,N} = (E_{n,N-1} - E_{0,N-1}) + (E_{0,N-1} - E_{0,N}) , \qquad (8.121)$$

where the term in the first bracket is positive and the second is $-\mu_+(N - 1)$. Thus we can write

$$A_>(\alpha, \omega) = \sum_n |\langle n, N + 1|\hat{a}_\alpha^\dagger|0, N\rangle|^2 \delta\left(\omega - \frac{\mu_+(N)}{\hbar} - \omega_{n0}(N + 1)\right) ,$$

$$A_<(\alpha, \omega) = \sum_n |\langle n, N - 1|\hat{a}_\alpha|0, N\rangle|^2 \delta\left(\omega - \frac{\mu_+(N - 1)}{\hbar} + \omega_{n0}(N - 1)\right) ,$$

$$(8.122)$$

where $\omega_{n0}(N) \equiv \frac{E_{n,N} - E_{0,N}}{\hbar}$ denotes the n-th excitation frequency of an N-particle system. These formulas show that, at zero temperature, $A_>(\alpha, \omega)$ differs from zero only for $\omega > \mu_+(N)$ and $A_<(\alpha, \omega)$ only for $\omega < \mu_+(N - 1)$. In a large *gapless* system the addition energies $\mu_+(N)$ and $\mu_+(N - 1)$ coincide in the thermodynamic limit and are both equal to the chemical potential $\mu(N) = \frac{\partial E_0(N)}{\partial N}$ up to corrections of the order of $1/N$. Combining this observation with the sum rule (8.117) and (8.118) we see that the zero temperature occupation number can be calculated from

$$n_\alpha = \int_{-\infty}^{\frac{\mu}{\hbar}} A(\alpha, \omega)d\omega , \qquad T = 0 . \qquad (8.123)$$

This formula also holds for gapped systems (e.g. insulators) provided μ lies within the gap between $\mu_+(N)$ and $\mu_+(N - 1)$.

8.5.1.1 Noninteracting electron gas

Before proceeding with the formalism let us pause to calculate the spectral function for a uniform noninteracting electron gas. Let $\alpha = (\vec{k}\sigma)$ be the label of a plane wave state of wave vector $\vec{k}$ and spin projection σ. It is evident from the character of the noninteracting ground-state that one can add a particle in a state of wave vector $\vec{k}$ only if $k > k_F$. Similarly, one can remove a particle from a state of wave vector $\vec{k}$ only if $k < k_F$. In the former case the energy added to the system is $\varepsilon_{\vec{k}} = \frac{\hbar^2 k^2}{2m}$, larger than the chemical potential potential $\mu = \frac{\hbar^2 k_F^2}{2m}$. In the latter case the energy subtracted from the system is also $\varepsilon_{\vec{k}}$, now smaller than the chemical potential. Thus we see that

$$A^{(0)}_{>\sigma}(\vec{k}, \omega) = (1 - n^{(0)}_{\vec{k}\sigma})\delta\left(\omega - \frac{\varepsilon_{\vec{k}}}{\hbar}\right),$$

$$A^{(0)}_{<\sigma}(\vec{k}, \omega) = n^{(0)}_{\vec{k}\sigma}\delta\left(\omega - \frac{\varepsilon_{\vec{k}}}{\hbar}\right), \qquad (8.124)$$

and

$$A^{(0)}_{\sigma}(\vec{k}, \omega) = \delta\left(\omega - \frac{\varepsilon_{\vec{k}}}{\hbar}\right). \qquad (8.125)$$

(The (0) superscript denotes noninteracting properties everywhere.) Note that the sum rules (8.115), (8.117), and (8.119) are satisfied. Integrating $A^{(0)}$ over frequency from $-\infty$ to $\mu = \frac{\hbar^2 k_F^2}{2m}$ we obtain $n^{(0)}_{\vec{k}\sigma} = \Theta(k_F - k)$, the zero-temperature Fermi distribution.

8.5.1.2 Interacting electron gas

The fact that the noninteracting spectral function is a single δ-function of frequency means that the addition or removal of a particle in a plane wave state generates an exact eigenstate of the noninteracting system. There is nothing surprising about this, but what happens when the interactions are turned on? As a first guess, upon assuming continuity, one might expect the δ-function to continuously evolve into some kind of Lorentzian peak with a width proportional to the interaction strength. This expectation is not entirely incorrect, but misses a most essential feature, namely, that for k near k_F the width of the Lorentzian peak vanishes as $(k - k_F)^2$, *no matter how strong the interaction is*. This behavior is enforced by kinematic constraints on the quasiparticle decay process, described in Section 8.4, and is what characterizes the normal Fermi liquid phase. In the limit $k \to k_F$ the spectral function has a sharp quasiparticle peak at $\omega \simeq v_F^*(k - k_F)$ with a width $\frac{1}{\tau_{\vec{k}\sigma}} \sim (k - k_F)^2$. This behavior is qualitatively shown in Fig. 8.10. The quasiparticle peak absorbs a fraction Z $(0 < Z \leq 1)$

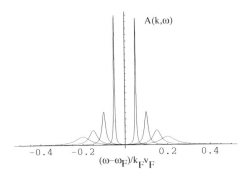

Fig. 8.10. Qualitative sketch of the zero temperature spectral function of an interacting Fermi liquid for different values of k/k_F near the Fermi surface. Here $\omega_F = \epsilon_F/\hbar$. The eight curves, from left to right, correspond to $k/k_F = 0.8, 0.85, 0.9, 0.95, 1.05, 1.1, 1.15, 1.2$. The peak at $k = k_F$ is infinitely sharp and for this reason has not been plotted.

of the total spectral weight, while the remaining $1 - Z$ is distributed over a structureless background. Thus, the notions of quasiparticle energy and lifetime, introduced in a somewhat hand-waving way in the previous sections, acquire a precise mathematical meaning as properties of the spectral function.

Looking back at the definitions of of $A_>$ and $A_<$ in Eqs. (8.122), we see that the renormalization constant Z is just the squared modulus of the overlap between the state obtained by adding (or removing) an electron of wave vector k_F to the N-particle ground-state, and the ground-state of the $N \pm 1$ particle systems: $|\langle 0, N + 1|\hat{a}_{k\sigma}^{\dagger}|0, N\rangle|^2_{|\vec{k}|=k_F} = |\langle 0, N - 1|\hat{a}_{\vec{k}\sigma}|0, N\rangle|^2_{|\vec{k}|=k_F} = Z_{k_F\sigma}$, for $N \to \infty$. This implies that the ground-state of the $N \pm 1$-particle system gives the spectral function a contribution comparable to that of all the other eigenstates taken together. Z represents the probability of adding or removing an electron at $k = k_F$ *without creating excitations*. The fact that one can inject or remove a particle without essentially disturbing the system is thus seen to be a fundamental property of the normal Fermi liquid state.

8.5.1.3 Finite temperature

The definition of the spectral function can be easily generalized to finite temperature. The main difference from the zero-temperature case is that, instead of starting from the ground-state of the N-particle system, one needs to start from the canonical equilibrium ensemble in which the state $|n, N\rangle$ is occupied with probability $P_{n,N} = Z^{-1}e^{-E_{n,N}/k_B T}$. The general definition (8.118) still applies, i.e. the spectral function is still the sum of two parts $A_>$ and $A_<$ now given by

$$A_>(\alpha, \omega) = \sum_{n,m} P_{m,N}|\langle n, N + 1|\hat{a}_{\alpha}^{\dagger}|m, N\rangle|^2 \delta\left(\omega - \frac{E_{n,N+1} - E_{m,N}}{\hbar}\right),$$

$$A_<(\alpha, \omega) = \sum_{n,m} P_{m,N}|\langle n, N - 1|\hat{a}_{\alpha}|m, N\rangle|^2 \delta\left(\omega + \frac{E_{n,N-1} - E_{m,N}}{\hbar}\right). \quad (8.126)$$

The sum rules (8.115), (8.117), and (8.119) remain unchanged, provided that the n_α's are interpreted as temperature-dependent occupation factors.

With the help of these definitions it is not difficult to verify that the formulas (8.124), (8.125) remain valid for a noninteracting Fermi system at finite temperature, with $n_k^{(0)}$ given by the Fermi–Dirac distribution function. In particular, we see that the full spectral spectral function $A(k, \omega)$ of a noninteracting Fermi gas is independent of temperature. This remarkable property does not hold in the presence of interactions.

8.5.1.4 Green's function and self-energy

It would be quite difficult in practice to calculate the spectral function directly from the exact eigenstates representation of Eq. (8.122). Fortunately, we can take advantage of the powerful machinery of diagrammatic many-body theory and link the spectral function to the imaginary part of the retarded Green's function introduced in Section 6.3.2. The relevant relationship is

$$A(\alpha, \omega) = -\frac{1}{\pi}\Im m G^{ret}(\alpha, \omega) ,$$ (8.127)

where

$$G^{ret}(\alpha, t) = -i\Theta(t)\langle\{\hat{a}_\alpha(t), \hat{a}_\alpha^\dagger(0)\}\rangle$$ (8.128)

($\langle ..\rangle$ is the thermal average and $\{., .\}$ is the *anti*commutator) and

$$G^{ret}(\alpha, \omega) = \int_{-\infty}^{\infty} G^{ret}(\alpha, t)e^{i\omega t}dt .$$ (8.129)

The proof of Eq. (8.127) is achieved by expanding Eq. (8.128) in a complete set of exact eigenstates, Fourier-transforming with respect to time, and taking the imaginary part: the resulting expression for the right-hand side of Eq. (8.127) coincides with the spectral function $A = A_> + A_<$.

The retarded Green's function can be straightforwardly computed from the time-ordered Green's function, $G(\alpha, \omega)$, according to the formulas (6.77) of Section 6.3.2. Taking the imaginary part of Eq. (6.78) we immediately obtain

$$A_\sigma(\vec{k}, \omega) = -\frac{\hbar}{\pi} \frac{\Im m \Sigma_\sigma^{ret}(\vec{k}, \omega)}{[\hbar\omega - \varepsilon_{\vec{k}\sigma} - \Re e \Sigma_\sigma^{ret}(\vec{k}, \omega)]^2 + [\Im m \Sigma_\sigma^{ret}(\vec{k}, \omega)]^2}.$$ (8.130)

Thus we see that the calculation of the spectral function requires the knowledge of both the real and the imaginary parts of the retarded self-energy. Given $A_\sigma(\vec{k}, \omega)$ the spectral function for any other single particle state $|\psi_\alpha\rangle$ is obtained via the transformation

$$A(\alpha, \omega) = \sum_{\vec{k}\sigma} |\langle\psi_\alpha|\vec{k}\sigma\rangle|^2 A_\sigma(\vec{k}, \omega) .$$ (8.131)

8.5.1.5 Measuring the spectral function

Thanks to the great improvements in the manufacture of high-quality quantum well systems it has recently been possible to directly measure, by means of precise and elegant tunneling experiments between parallel identical quantum wells, the low-energy spectral function of a two-dimensional electron liquid (Murphy *et al.*, 1995).

The basic idea of the experiment is shown in Fig. 8.11. The two identical parabolas separated by an energy eV represent the energy vs wave vector relation of the quasiparticle states in the two quantum wells, and the shaded regions show the energy spread of these states due to finite lifetime of a quasiparticle at the Fermi surface (we are, of course, at finite temperature). Here $k_\parallel$ is the two-dimensional in-plane wave vector, V is the electric potential difference between the wells, and Γ is the width of the spectral function at $k_\parallel = k_F$. Under the assumption that electron–impurity and electron–phonon scattering are negligible the two-dimensional wave vector of the tunneling electrons is conserved and overall energy conservation causes the tunneling probability to decrease sharply when eV exceeds Γ. More precisely, the tunneling conductance vs voltage is approximately a Lorentzian centered at

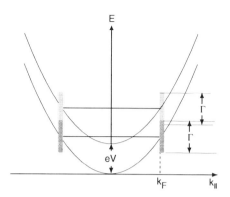

Fig. 8.11. Momentum-conserving tunneling between two identical free-electron bands separated by a potential difference eV. Because of energy conservation, the tunneling probability decreases rapidly when eV exceeds the spectral width Γ of the single-particle states in each band.

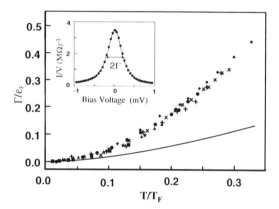

Fig. 8.12. Inset: Lorentzian lineshape of the current–voltage (I–V) relation in tunneling between two-dimensional GaAs quantum wells. Main figure: a plot of the half-width Γ, identified with the inverse of the quasiparticle lifetime, vs temperature for systems of different density (a residual $T = 0$ contribution, attributed to disorder, has been subtracted). The solid line is the theoretical prediction from Eq. (8.103) which is only applicable at very low temperature. Adapted from S. Murphy, 2004.

zero voltage with full width at half maximum equal to 2Γ, as shown in the inset of Fig. 8.12. From this quantity the width of the spectral function in each of the layers and hence the quasiparticle lifetime can be inferred.

In practice, the interpretation of the experimental data is complicated by the presence of disorder, which leads to imperfect momentum conservation and a finite linewidth even in the limit of zero temperature. This extrinsic contribution, however, is expected to be nearly independent of temperature, and when one subtracts it from the data one obtains values that are in reasonably good agreement with the theory presented in this section (see Fig. 8.12).

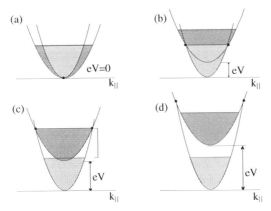

Fig. 8.13. Tunneling between two dissimilar two-dimensional parabolic quantum wells probes the quasiparticle spectral function in a range of energies above the Fermi level. The two wells in this figure have different Fermi wave vectors and different effective masses. Tunneling that conserves both the energy and the in-plane momentum $k_\parallel$ can occur only at the points of intersection between the two parabolas (solid dots), and only if at these points one band is occupied while the other is empty. When the heavy-electron band is raised by a bias energy eV relative to the light-electron one, tunneling becomes possible within a range of voltages intermediate between the situation depicted in part (b) and the situation depicted in part (c). The region of energies that can thus be probed is shown by the square bracket in part (c) (Marinescu, Quinn, and Giuliani, 2002).

The above experiments give information on the spectral function at the Fermi energy. In order to be able to access the value of the spectral function at finite excitation energies it is necessary that tunneling between the two layers involve at least one state far from the Fermi surface. It has been recently proposed that a way to realize such a process is to study tunneling between dissimilar quantum wells (Marinescu, Quinn, and Giuliani, 2002). A brief description of this idea is provided in the caption of Fig. 8.13.

8.5.2 The momentum occupation number

The behavior of the average occupation number of plane wave states $n_{\vec{k}}$ can easily be inferred from Eq. (8.123). It is the presence of an infinitely sharp quasiparticle peak at $k = k_F$ that causes the occupation number to jump discontinuously – by an amount Z – at $k = k_F$. Naturally, the discontinuity is superimposed to a smooth background that arises from the structureless background of the spectral function. This behavior is shown in Fig. 8.14. The very fact that $n_{\vec{k}\sigma}$ deviates from the ideal Fermi gas expression for wave vectors near k_F indicates that the actual many-body ground-state wave function cannot be simply expressed as a single Slater determinant of plane wave states. In particular for homogeneous systems this deviation is a direct evidence of correlations effects beyond Hartree–Fock.

Approximate analytical calculations of $n_{\vec{k}\sigma}$ have been performed quite some time ago within the RPA (see for instance the elegant work of Daniel and Vosko (1960) and Lam (1971)). More recently, numerical determinations of $n_{\vec{k}\sigma}$ have become possible thanks to

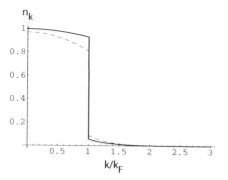

Fig. 8.14. Behavior of the plane wave states average occupation number in three dimensions from Eq. (8.132) at $r_s = 2$ (solid line) and $r_s = 5$ (dashed line).

the quantum Monte Carlo method. It is useful to summarize the results of these studies by means of relatively simple interpolation formulas designed to satisfy the normalization sum rule and to allow one to correctly reproduce the value of the interacting kinetic energy, that is, $\sum_{\vec{k}\sigma} n_{\vec{k}\sigma} \frac{\hbar^2 k^2}{2m}$. In three dimensions we have (Senatore, Moroni, and Ceperley, 1996):

$$n_{\vec{k},\sigma} \simeq \begin{cases} a_1 + a_2 x^2 & (x < 1) \\ \frac{a_3}{x^8} + a_4 e^{-a_5 x^2} & (x > 1), \end{cases} \qquad (8.132)$$

where $x = \frac{k}{k_F}$ and the parameters a_i are given in Table 8.2. There we also give the values of the renormalization constant Z as determined by Monte Carlo (Ortiz and Ballone, 1997).[33]

For the two-dimensional case the corresponding interpolation formula is instead given by (Conti, 1997):

$$n_{\vec{k},\sigma} \simeq \begin{cases} a_0 + a_1 x + a_2 x^2 + a_3 x^3 + a_4 x^4 + a_{10} x^5 & (x < 1) \\ \frac{4g(0)r_s^2}{x^6} + \left(a_7 + a_8 x + a_9 x^2\right) \exp\left\{ -\frac{x-1}{a_5} - \frac{(x-1)^2}{a_6^2} \right\} & (x > 1). \end{cases} \qquad (8.133)$$

The corresponding parameters are given in Table 8.3 alongside the Monte Carlo values for Z. The necessary numerical values for $g(0)$ can be found in Appendix 4.

8.5.3 Quasiparticle energy, renormalization constant, and effective mass

The essential property of a "normal Fermi liquid" can now be restated as follows: there exists a Fermi surface at $k = k_F$ such that for all wave vectors close to it the spectral function has a sharp peak at the quasiparticle energy $\mathcal{E}_{\vec{k}\sigma} = \mu + \mathcal{O}(k - k_F)$, and the width of the

---

[33] The large-k behavior of $n_{\vec{k}\sigma} \sim a_3 \left(\frac{k_F}{k}\right)^8$ is believed to be exact, with the coefficient a_3 calculated by Yasuhara and Kawazoe (1976): $a_3 = \frac{8}{9} \left(\frac{\alpha_3 r_s}{\pi}\right)^2 g(0)$. What makes the calculation doable at large wave vectors is the fact that only two-body collisions matter in this regime – a fact already noted in the calculation of the large-q behavior of the structure factor.

Table 8.2. *Fit parameters for Eq. (8.132) and momentum distribution discontinuity Z for 3D fermions. (Adapted from Senatore, Moroni, and Ceperley, 1996; the values of Z are from Ortiz and Ballone, 1997.)*

r_s	1	2	3	5	8	10
a_1	1.0017	0.9933	0.9838	0.9658	0.9018	0.9007
a_2	−0.0283	−0.0677	−0.0959	−0.1591	−0.1698	−0.2198
a_3	0.0067	0.0154	0.0203	0.0142	0.0079	0.0059
a_4	0.0165	0.1119	0.1372	0.2437	0.3270	0.2970
a_5	1.0494	1.3988	1.2539	1.2019	1.1122	0.9967
Z	0.952	0.889	0.842	0.725	0.651	0.593

Table 8.3. *Fit parameters for Eq. (8.133) and momentum distribution discontinuity Z for 2D fermions. (Adapted from Conti, 1997.)*

r_s	1	2	5	10	20	40
a_0	1.958	1.9	1.67	1.435	1.143	0.8292
a_1	−0.01297	0.00528	0.009721	0.02612	0.007975	0.008649
a_2	0.02773	0.004692	−0.003434	0.108	0.01505	0.00832
a_3	−0.01206	−0.06888	−0.05649	−0.3253	−0.09478	−0.06632
a_4	−0.03796	−0.06144	−0.1666	−0.008001	−0.1839	−0.1415
a_5	8.714	1.811	1.059	1.169	1.163	1.346
a_6	1.745	1.888	1.914	3.66	2.01	2.119
a_7	0.1115	0.4622	0.9791	1.039	0.9777	0.7707
a_8	−0.07	−0.2755	−0.5134	−0.404	−0.3819	−0.26
a_9	0.01101	0.04319	0.08415	0.04138	0.08323	0.07823
a_{10}	0.01039	0.04699	0.09488	0.06176	0.1111	0.08985
Z	1.754	1.553	0.9657	0.5349	0.2059	0.0533

peak tends to zero as $(k - k_F)^2$.[34] Notice that, up to this point, we have not said that k_F coincides with the Fermi wave vector of the noninteracting electron gas at the same density. However, it turns out that if a special wavevector k_F exists, such that the above properties are satisfied, then this wave vector *must* coincide with the normal Fermi wave vector. This important result is the content of *Luttinger's theorem*, which will be proved in the next section.

Inspecting Eq. (8.130) for the spectral function, we see that the validity of the Fermi liquid scenario requires that

$$\mathcal{E}_{\vec{k}\sigma} = \varepsilon_{\vec{k}\sigma} + \Re e \Sigma_\sigma^{ret}(\vec{k}, \mathcal{E}_{\vec{k}\sigma}) \tag{8.134}$$

[34] In fact any peak-width that tends to zero faster than $k - k_F$ will define a Landau Fermi liquid, albeit an unconventional one.

and

$$\Im m \Sigma_\sigma^{ret}(\vec{k}, \mathcal{E}_{\vec{k}\sigma}) \sim -a(k - k_F)^2 \,, \tag{8.135}$$

where a is a (positive) constant. The first equation amounts to a microscopic definition of the quasiparticle/quasihole energy (see discussion below to determine which is which) and the two equations together guarantee that the spectral function has a Lorentzian peak of vanishing width at $\omega = \frac{\mathcal{E}_{\vec{k}\sigma}}{\hbar}$. To see this more clearly, let us expand the two terms in the denominator of Eq. (8.130) to the lowest nonvanishing order in $\hbar\omega - \mathcal{E}_{\vec{k}\sigma}$. For the first term we have

$$\hbar\omega - \mathcal{E}_{\vec{k}\sigma} - \Re e \Sigma_\sigma^{ret}(\vec{k}, \omega) \approx \frac{(\hbar\omega - \mathcal{E}_{\vec{k}\sigma})}{Z_{\vec{k}\sigma}} \,, \tag{8.136}$$

where

$$Z_{\vec{k}\sigma} \equiv \left(1 - \frac{1}{\hbar}\frac{\partial}{\partial\omega} \Re e \Sigma_\sigma^{ret}(k, \omega) \bigg|_{\hbar\omega = \mathcal{E}_{\vec{k}\sigma}} \right)^{-1} \,, \tag{8.137}$$

while in the second term we can simply put $\omega = \frac{\mathcal{E}_{\vec{k}\sigma}}{\hbar}$. If we now define

$$\frac{\hbar}{2\tau_{\vec{k}\sigma}} \equiv Z_{\vec{k}\sigma} |\Im m \Sigma_\sigma^{ret}(k, \mathcal{E}_{\vec{k}\sigma})| \,, \tag{8.138}$$

then the spectral function can be cast in the more incisive form

$$A_\sigma(k, \omega) \simeq \frac{Z_{\vec{k}\sigma}}{\pi} \frac{\frac{1}{2\tau_{\vec{k}\sigma}}}{\left(\omega - \frac{\mathcal{E}_{\vec{k}\sigma}}{\hbar}\right)^2 + \left(\frac{1}{2\tau_{\vec{k}\sigma}}\right)^2} \,, \tag{8.139}$$

that is, the equation of a Lorentzian of strength $Z_{\vec{k}\sigma}$ and half-width $\frac{1}{2\tau_{\vec{k}\sigma}}$ centered at the quasiparticle energy.

The strength of the Lorentzian peak at $k = k_F$ is nothing but the renormalization constant introduced in the previous section and must therefore be a positive number between 0 and 1. This means that the consistency of the Landau Fermi liquid scenario requires that the inequality

$$\frac{\partial}{\partial\omega} \Re e \Sigma_\sigma^{ret}(k_F, \omega) \bigg|_{\hbar\omega = \mu} < 0 \tag{8.140}$$

be satisfied. Another fundamental condition for the consistency of the Landau Fermi liquid scenario emerges when one considers the Kramers–Krönig dispersion relation (6.80) for the retarded self-energy. As shown in Exercise 27, this relation implies that if $\Im m \Sigma_\sigma^{ret}(k_F, \omega) \to \left(\omega - \frac{\mu}{\hbar}\right)^\alpha$, with $\alpha < 1$, for $\omega \to \frac{\mu}{\hbar}$, then $\Re e[\Sigma_\sigma^{ret}(k_F, \omega) - \Sigma_\sigma^{ret}(k_F, \frac{\mu}{\hbar})]$ also varies, in the same limit, as $\left(\omega - \frac{\mu}{\hbar}\right)^\alpha$, implying that the renormalization constant vanishes. The case $\alpha = 1$ is marginal, as the derivative of the real part of the self-energy diverges only logarithmically for $\omega \to \frac{\mu}{\hbar}$. *Thus, it is essential to the validity of the Landau theory of*

Fermi liquids that the imaginary part of the self-energy vanish for $\omega \to \frac{\mu}{\hbar}$ more rapidly than $\omega - \frac{\mu}{\hbar}$ itself.

The width of the Lorentzian peak in the spectral function should be interpreted as the inverse lifetime of the corresponding plane wave state. This can be inferred from the fact that the retarded Green's function has a pole in the lower half of the complex frequency plane at $z = \frac{\mathcal{E}_{\vec{k}\sigma}}{\hbar} - \frac{i}{2\tau_{\vec{k}\sigma}}$. The imaginary part of the frequency implies an exponential decay of the squared amplitude of a plane wave state with a characteristic time scale $\tau_{\vec{k}\sigma}$ given by Eq. (8.138). The above considerations imply that the *time-ordered* Green's function, as anticipated in Eq. (6.96), has the form

$$G_\sigma(\vec{k}, \omega) = G_\sigma^{(reg)}(\vec{k}, \omega) + \frac{Z_{\vec{k}\sigma}}{\omega - \frac{\mathcal{E}_{\vec{k}\sigma}}{\hbar} - \frac{i}{2\tau_{\vec{k}\sigma}}\text{sgn}(k - k_F)}, \qquad (8.141)$$

where $G_\sigma^{(reg)}(\vec{k}, \omega)$ is a regular function of $\vec{k}$ and ω.

The lifetime of a plane wave state is closely related to, but not identical with the lifetime of a quasiparticle ($\tau_{\vec{k}\sigma}^{(e)}$) or a quasihole ($\tau_{\vec{k}\sigma}^{(h)}$) discussed in Section 8.4. The tricky point is that, at finite temperature, a plane wave state is sometimes empty and sometimes occupied, and in the former case it can host a quasiparticle, while in the latter it can only host a quasihole. Consequently, the inverse lifetimes of the quasiparticle and the quasihole are related to the inverse lifetime of the plane wave state as follows:

$$\frac{1}{\tau_{\vec{k}\sigma}^{(e)}} = \frac{1 - n_{\vec{k}\sigma}}{\tau_{\vec{k}\sigma}},$$

$$\frac{1}{\tau_{\vec{k}\sigma}^{(h)}} = \frac{n_{\vec{k}\sigma}}{\tau_{\vec{k}\sigma}}. \qquad (8.142)$$

Notice that the principle of detailed balance, Eq. (8.83), is satisfied. Only when $k_B T \ll \mathcal{E}_{\vec{k}\sigma} - \mu \ll \epsilon_F$, one can definitely associate a quasiparticle to a plane wave state with $k > k_F$ and a quasihole to a plane wave state with $k < k_F$. In the opposite limit of $\mathcal{E}_{\vec{k}\sigma} - \mu \ll k_B T \ll \epsilon_F$ the quasiparticle and quasihole contributions cannot be resolved and $1/\tau_{\vec{k}\sigma}$, differs from both $1/\tau_{\vec{k}\sigma}^{(e)}$ and $1/\tau_{\vec{k}\sigma}^{(h)}$ and is in fact the sum of the two.

Finally, let us examine the dispersion of the quasiparticle energy as k moves away from k_F. We assume that the system is isotropic, so that the quasiparticle energy depends only on the magnitude of $\vec{k}$. Expanding $\mathcal{E}_{\vec{k}\sigma}$ to first order in $k - k_F$ and using the fact that $\mathcal{E}_{k_F\sigma} = \mu$, we can write

$$\mathcal{E}_{k\sigma} \simeq \mu + \frac{\hbar^2 k_F(k - k_F)}{m^*}, \qquad (8.143)$$

which effectively defines the quasiparticle *effective mass* m^* as

$$\frac{\hbar^2 k_F}{m^*} = \frac{d\mathcal{E}_{k\sigma}}{dk}\bigg|_{k=k_F}. \qquad (8.144)$$

Differentiating both sides of Eq. (8.134) with respect to k we get

$$\frac{d\mathcal{E}_{k\sigma}}{dk}\bigg|_{k=k_F} = \frac{\hbar^2 k_F}{m} + \frac{\partial \Re e \Sigma_\sigma^{ret}(k,\mu)}{\partial k}\bigg|_{k=k_F} + \frac{1}{\hbar}\frac{\partial \Re e \Sigma_\sigma^{ret}(k_F,\omega)}{\partial \omega}\bigg|_{\omega=\mu}\frac{d\mathcal{E}_{k\sigma}}{dk}\bigg|_{k=k_F},$$

(8.145)

which is readily solved, yielding the effective mass

$$\frac{m^*}{m} = \frac{1}{Z_{k_F\sigma}\left[1 + \frac{m}{\hbar^2 k_F}\frac{\partial}{\partial k}\Re e\Sigma_\sigma^{ret}(k,\mu)\bigg|_{k=k_F}\right]}.$$

(8.146)

The value of $\frac{m^*}{m}$ is determined by the renormalization constant $Z_{k_F\sigma}$, which is always smaller than 1, and by the derivative of $\Re e \Sigma_\sigma^{ret}(k,\omega)$ with respect to k. Because the latter can have either sign, the effective mass can be larger or smaller than the bare mass. In a weakly interacting electron liquid ($r_s \ll 1$), we expect the self-energy to be dominated by exchange: since the exchange self-energy $\Sigma_\sigma^{HF}(k,\omega) = \varepsilon_x(k)$ is a frequency-independent, monotonically increasing function of k, we expect, in this limit, $Z_{k_F\sigma} = 1$ and $\frac{\partial}{\partial k}\Re e\Sigma_\sigma^{ret}(k,\mu) > 0$, resulting in an effective mass that is *smaller* than the bare mass.[35]

The situation is quite different at lower density. As r_s increases correlations become sufficiently strong to make $Z_{k_F\sigma} < 1$ and $\frac{\partial}{\partial k}\Re e\Sigma_\sigma^{ret}(k,\mu) < 0$: this gives an effective mass that is *larger than the bare mass*, as one can see in Tables 8.4 and 8.6. It is still an unsettled question to date whether $m^* > m$ or $m^* < m$ in the *metallic density range*. The most recent calculations suggest that the second scenario is correct.

8.5.4 Luttinger's theorem

We are now ready to finally provide a formal proof for the statement that the Fermi wave vector of a normal interacting paramagnetic Fermi liquid coincides with that of a noninteracting Fermi liquid of the same density. We will, for simplicity, confine ourselves to the case of the spin-unpolarized (paramagnetic) Fermi liquid, and drop the spin indices accordingly. The proof cannot be given without the help of a few basic assumptions whose validity is equivalent to the validity the normal Fermi liquid concept. These assumptions are as follows:

(i) The imaginary part of the self-energy vanishes at $\omega = \mu$, that is, $\Im m\Sigma(\vec{k},\mu) = 0$.
(ii) The inverse of the Green's function

$$G^{-1}(\vec{k},\mu) = \mu - \varepsilon_{\vec{k}} - \Sigma(\vec{k},\mu)$$

(8.147)

is positive for $k > k_F$, negative for $k < k_F$, and vanishes for $k = k_F$. This condition provides a

---

[35] Recall that in the Hartree–Fock approximation, one would have $m^* = 0$! The small-r_s behavior of the effective mass is worked out in Exercise 22.

precise definition of the Fermi wave vector as the solution of the equation

$$\varepsilon_{k_F} + \Sigma(k_F, \mu) = \mu ,$$ (8.148)

and tells us that the Green's function has a pole at $k = k_F$ and $\omega = \mu$.

(iii) The renormalization constant $Z = \left(1 - \frac{\partial \mathfrak{Re} \Sigma(k_F, \omega)}{\hbar \partial \omega} \Big|_{\hbar\omega = \mu} \right)^{-1}$ is positive and less than 1.

(iv) The self-energy is a differentiable functional of the Green's function G, and can be written as the functional derivative with respect to G of a generating functional $\Omega[G]$, as described in Section 6.3.3:

$$\Sigma[G](\vec{k}, \omega) = \frac{\delta \Omega[G]}{\delta G(\vec{k}, \omega)} .$$ (8.149)

The reader will notice that conditions (ii) and (iii) taken together guarantee that the plane wave occupation number jumps by Z at $k = k_F$ and that the effective mass of a quasiparticle near the Fermi surface, Eq. (8.146), is a positive quantity – two highly desirable properties. However, we have not yet assigned any value to k_F. As for condition (iv), recall that the existence of the differentiable functional $\Omega[G]$ is provable only within diagrammatic perturbation theory. Therefore this condition mandates that the state of the interacting liquid be connected by a perturbative expansion to the noninteracting ground-state.

We now proceed to counting the number of plane wave states per unit volume for which $k < k_F$ (or, equivalently, $G(\vec{k}, \mu) > 0$) and show that it equals the electronic density. We shall see that the proof of this fact is essentially topological. The electronic density can be expressed (see Eq. (8.123)) as the integral of the spectral function over frequencies $\omega < \frac{\mu}{\hbar}$

$$n = 2 \int \frac{d\vec{k}}{(2\pi)^d} \int_{-\infty}^{\frac{\mu}{\hbar}} d\omega A(\vec{k}, \omega) ,$$ (8.150)

where the factor 2 accounts for spin multiplicity. We have already noted (see Eq. (6.131)) that the spectral function can be viewed as $-(1/2\pi i)$ times the discontinuity of the complex frequency Green's function $G(\vec{k}, z)$ across the real axis:

$$A(\vec{k}, \omega) = -\frac{1}{2\pi i}[G(\vec{k}, \omega + i0^+) - G(\vec{k}, \omega - i0^+)].$$ (8.151)

The complex frequency Green's function $G(\vec{k}, z)$ is completely specified by the following properties: (i) it is analytic in both the upper and lower half planes of the complex variable z and tends to zero as z^{-1} for $z \to \infty$ in any direction, (ii) it has a branch cut discontinuity across the real axis, such that $G(\vec{k}, \omega + i0^+)$ coincides with the retarded Green's function for real frequency ω and $G(\vec{k}, \omega - i0^+) = G^*(\vec{k}, \omega + i0^+)$ (the latter quantity is known as the "advanced" Green's function).[36]

In addition to these general properties, the Green's function of a Fermi liquid is also real, and therefore free of discontinuity, at $\hbar\omega = \mu$. This is due to the vanishing of the imaginary part of Σ and it is what allows us to rewrite Eq. (8.150) as a contour integral in the complex

---

[36] At $T = 0$ the time-ordered Green's function coincides with the retarded one for $\hbar\omega > \mu$ and with the advanced one for $\hbar\omega < \mu$.

z plane along a path that encircles the $\omega < \frac{\mu}{\hbar}$ portion of the real axis counterclockwise starting and ending at $-\infty$ as shown in Fig. 8.15.

$$n = 2 \int \frac{d\vec{k}}{(2\pi)^d} \int_C \frac{dz}{2\pi i} G(\vec{k}, z) . \tag{8.152}$$

Let us now change, for fixed $\vec{k}$, the variable of integration from z to

$$G(\vec{k}, z) = \frac{\hbar}{\hbar z - \varepsilon_{\vec{k}} - \Sigma(\vec{k}, z)} . \tag{8.153}$$

Making use of the transformation

$$dz = -\frac{dG}{G^2} + \frac{d\Sigma}{\hbar} , \tag{8.154}$$

we can write

$$n = \frac{2}{2\pi i} \int \frac{d\vec{k}}{(2\pi)^d} \left(-\int_\Gamma \frac{dG}{G} + \frac{1}{\hbar} \int_\Gamma G d\Sigma \right) . \tag{8.155}$$

As z runs around the contour C, G describes a *closed* path Γ in the complex plane starting and ending at $G = 0^-$ – the image of $z = -\infty$ (see Fig. 8.15). The important minus superscript reminds us that $G = 0$ is approached from the negative side since z tends to infinity along the *negative* real axis. The precise form of the path Γ is quite complicated, but one crucial property can be discerned without calculation. Because the imaginary part of $G(\vec{k}, z)$ vanishes only for $|z| = \infty$ and $z = \frac{\mu}{\hbar}$ the path Γ intersects the real axis only at the two points $G = 0^-$ and $G = G(\vec{k}, \mu)$. The value of $G = G(\vec{k}, \mu)$ is positive or negative depending on whether $k < k_F$ or $k > k_F$. Thus, the integration path encloses the point $G = 0$ for $k < k_F$ and excludes it for $k > k_F$. Further inspection of the map $z \to G$ reveals that Γ is traversed clockwise when $k < k_F$ and counterclockwise when $k > k_F$ (see Fig. 8.15).

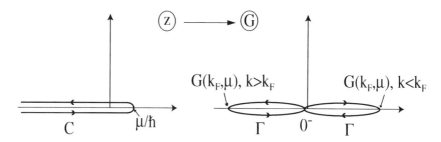

Fig. 8.15. Mapping of the integration contour for Eq. (8.152) from C in the complex z-plane to Γ in the complex G-plane.

The second contour integral on the right-hand side of Eq. (8.155) can be transformed as follows:

$$\int d\vec{k} \int_\Gamma G d\Sigma = -\int d\vec{k} \int_\Gamma \Sigma dG$$

$$= -\int d\vec{k} \int_\Gamma \frac{\delta\Omega[G]}{\delta G} dG, \tag{8.156}$$

where we have made use of $\int_\Gamma d(\Sigma G) = 0$, together with our Fermi liquid assumption (iv). This term vanishes because it is the integral of an exact differential form around a closed loop in the complex G plane. On the other hand, the value of the first contour integral is determined by the residue at the pole at $G = 0$ and vanishes for $k > k_F$ since, in that case, the pole falls outside the integration contour:

$$-\frac{1}{2\pi i}\int_\Gamma \frac{dG}{G} = \Theta(k_F - k). \tag{8.157}$$

Substituting this in Eq. (8.155) we obtain

$$n = 2\int \frac{d\vec{k}}{(2\pi)^d}\Theta(k_F - k), \tag{8.158}$$

i.e. the number of plane wave states with $k < k_F$ equals the number of electrons. This completes the proof of Luttinger's theorem.

8.5.5 The Landau energy functional

Eq. (8.134) provides a microscopic definition of the quasiparticles energy, but we still have to prove that this definition is indeed equivalent to the Landau definition in terms of the functional derivative of the total energy with respect to the quasiparticle distribution function (see Eq. (8.2)). Furthermore, we would like to identify the microscopic counterpart of the Landau interaction function. These two issues are addressed below.

The key step is the construction of a precise microscopic expression for the ground-state energy as a functional of the quasiparticle distribution function. We start from the exact expression for the ground-state energy of a homogeneous Fermi system derived in Appendix 13:

$$E = E_0 - \hbar \sum_{\vec{k}\sigma} \int_{-\infty}^{\infty} \frac{d\omega}{2\pi i}\left\{\ln\left[1 - \hbar^{-1}G_\sigma^{(0)}(\vec{k},\omega)\Sigma_\sigma(\vec{k},\omega)\right]\right.$$

$$\left. + \hbar^{-1}G_\sigma(\vec{k},\omega)\Sigma_\sigma(\vec{k},\omega)\right\} + L^d\Omega[G], \tag{8.159}$$

where G, $G^{(0)}$, and Σ are respectively the exact Green's function, the noninteracting Green's function, and the self-energy. $\Omega[G]$ is the generating function of the self-energy (see Eq. (8.149)). Assuming that perturbation theory converges, all these quantities can ultimately be expressed as power series in the noninteracting time-ordered Green's function

$$G_\sigma^{(0)}(\vec{k}, \omega) = \frac{n_{\vec{k}\sigma}^{(0)}}{\omega - \frac{\varepsilon_{\vec{k}\sigma}}{\hbar} - i\eta} + \frac{1 - n_{\vec{k}\sigma}^{(0)}}{\omega - \frac{\varepsilon_{\vec{k}\sigma}}{\hbar} + i\eta}, \qquad (8.160)$$

where $n_{\vec{k}\sigma}^{(0)} = \Theta(k_F - k)$ are the plane wave occupation numbers of the noninteracting ground-state. Thus, the ground-state energy is ultimately determined by the $n_{\vec{k}\sigma}^{(0)}$'s.

We now define the Landau energy functional $E[\{\mathcal{N}_{\vec{k}\sigma}\}]$ as the energy obtained from Eq. (8.159) after performing the replacement

$$n_{\vec{k}\sigma}^{(0)} \to \mathcal{N}_{\vec{k}\sigma} \qquad (8.161)$$

in Eq. (8.160). The quasiparticle distribution function $\mathcal{N}_{\vec{k}\sigma}$ differs from the ground-state distribution $n_{\vec{k}\sigma}$ within a thin shell of wave vectors near the Fermi surface.

To verify the correctness of this definition, let us consider the effect on the energy of a small variation in $\mathcal{N}_{\vec{k}\sigma}$. Because the expression (8.159) (regarded as a functional of G) is stationary with respect to a small variation of G (see Appendix 13) one sees that the first-order variation of the energy stems solely from the variation of the noninteracting Green's function $G^{(0)}$. Thus we have

$$\delta E = \delta E_0 - \hbar \sum_{\vec{k}\sigma} \int_{-\infty}^{\infty} \frac{d\omega}{2\pi i} \delta \ln \left[1 - G_\sigma^{(0)}(\vec{k}, \omega) \frac{\Sigma_\sigma(\vec{k}, \omega)}{\hbar} \right], \qquad (8.162)$$

where Σ is not affected by the variation. Let us now pick a wavevector $\vec{k}$ with $k > k_F$ and change $\mathcal{N}_{\vec{k}\sigma}$ from 0 to 1 leaving the other occupation numbers unchanged. From Eq. (8.160) we see that this amounts to changing $G_\sigma^{(0)}(\vec{k}, \omega)$ from $\frac{1}{\hbar\omega - \varepsilon_{\vec{k}\sigma} + i\eta}$ to $\frac{1}{\hbar\omega - \varepsilon_{\vec{k}\sigma} - i\eta}$. Then, according to Eq. (8.162) the variation in energy is

$$\delta E = \varepsilon_{\vec{k}\sigma} - \hbar \int_{-\infty}^{\infty} \frac{d\omega}{\pi} \Im m \left\{ \ln \left(\frac{\hbar\omega - \varepsilon_{\vec{k}\sigma} - \Sigma_\sigma(\vec{k}, \omega) - i\eta}{\hbar\omega - \varepsilon_{\vec{k}\sigma} - i\eta} \right) \right\}. \qquad (8.163)$$

Notice, and this is crucial, that the self-energy $\Sigma_\sigma(\vec{k}, \omega)$ has been treated as a purely real quantity because only values of $\hbar\omega$ close to μ contribute to the integral. The complex logarithms appearing in the above formula are evaluated with the help of the formula

$$\Im m \ln(x - i\eta) = -\pi \Theta(-x) \qquad (8.164)$$

for x real and $\eta \to 0^+$. Then we obtain

$$\delta E = \varepsilon_{\vec{k}\sigma} + \hbar \int_{-\infty}^{\infty} d\omega \left\{ \Theta \left[\varepsilon_{\vec{k}\sigma} + \Sigma_\sigma(\vec{k}, \omega) - \hbar\omega \right] - \Theta \left[\varepsilon_{\vec{k}\sigma} - \hbar\omega \right] \right\}. \qquad (8.165)$$

The first Θ-function restricts the integral to frequencies such that $\hbar\omega < \mathcal{E}_{\vec{k}\sigma}$, where $\mathcal{E}_{\vec{k}\sigma}$ is the quasiparticle energy defined in Eq. (8.134). Similarly, the second Θ-function restricts the integral to frequencies such that $\hbar\omega < \varepsilon_{\vec{k}\sigma}$. The frequency integral is therefore trivial and yields $\mathcal{E}_{\vec{k}\sigma} - \varepsilon_{\vec{k}\sigma}$. Putting this in Eq. (8.165) we finally get

$$\delta E = \mathcal{E}_{\vec{k}\sigma}. \qquad (8.166)$$

Since we have varied $\mathcal{N}_{\vec{k}\sigma}$ by $+1$, this is exactly the functional derivative of the energy functional with respect to $\mathcal{N}_{\vec{k}\sigma}$.

Next let us consider the Landau interaction function defined in Eq. (8.9). Let $\delta\Sigma_\sigma(\vec{k}, \omega)$ be the variation of the self-energy, *at fixed values of its arguments*, caused by a change in the quasiparticle distribution. According to Eq. (8.134), the corresponding variation in the quasiparticle energy satisfies the relation

$$\mathcal{E}_{\vec{k}\sigma} + \delta\mathcal{E}_{\vec{k}\sigma} = \varepsilon_{\vec{k}\sigma} + \Sigma_\sigma(\vec{k}, \mathcal{E}_{\vec{k}\sigma} + \delta\mathcal{E}_{\vec{k}\sigma}) + \delta\Sigma_\sigma(\vec{k}, \mathcal{E}_{\vec{k}\sigma}). \tag{8.167}$$

Expanding to first order in $\delta\mathcal{E}_{\vec{k}\sigma}$ and making use of the definition (8.137) of the renormalization constant we see that

$$\delta\mathcal{E}_{\vec{k}\sigma} = Z_{k_F\sigma}\,\delta\Sigma_\sigma(\vec{k}, \mathcal{E}_{\vec{k}\sigma}) \tag{8.168}$$

leading us to

$$f_{\vec{k}\sigma,\vec{k}'\sigma'} = Z_{k_F\sigma} \frac{\delta\Sigma_\sigma(\vec{k}, \mathcal{E}_{\vec{k}\sigma})}{\delta\mathcal{N}_{\vec{k}'\sigma'}}. \tag{8.169}$$

Let us now calculate the variation of the self-energy at fixed values of its arguments. Since Σ is a functional of the exact Green's function G it is evident that the variation arises entirely from the change in G. We can therefore make use of the identity (6.98), discussed in Section 6.3.4, to write

$$\delta\Sigma_\sigma(\vec{k}, \omega) = -\frac{1}{L^d}\sum_{\vec{k}''\sigma''}\int\frac{d\omega''}{2\pi i}\, I_{\vec{k}\omega\sigma;\vec{k}''\omega''\sigma''}(0)\,\delta G_{\sigma''}(\vec{k}'', \omega''), \tag{8.170}$$

where $I_{\vec{k}\omega\sigma;\vec{k}''\omega''\sigma''}(0)$ is the irreducible electron–hole interaction in the limit of vanishing $d+1$-momentum transfer $q = (\vec{q}, \Omega)$ (see Chapter 6 for the definition and discussion of this quantity). The $d+1$-momenta of the incoming electron and hole are $p + q/2$ and $p - q/2$ respectively, where $p \equiv (\vec{k}, \omega)$. Similarly, the $d+1$-momenta of the outgoing electron and hole are $p'' + q/2$ and $p'' - q/2$, where $p'' \equiv (\vec{k}'', \omega'')$. Irreducibility in the electron–hole channel means that, in computing the scattering amplitude, one excludes processes in which an intermediate electron–hole pair of $d+1$-momentum q is produced. It is the exclusion of these processes that guarantees the existence of a well defined limit for $q \to 0$.[37]

The variation of the exact Green's function

$$G_\sigma(\vec{k}, \omega) = \frac{\mathcal{N}_{\vec{k}\sigma}}{\omega - \frac{\varepsilon_{\vec{k}\sigma} + \Sigma_\sigma(\vec{k},\omega)}{\hbar} - i\eta} + \frac{1 - \mathcal{N}_{\vec{k}\sigma}}{\omega - \frac{\varepsilon_{\vec{k}\sigma} + \Sigma_\sigma(\vec{k},\omega)}{\hbar} + i\eta} \tag{8.171}$$

arises both explicitly from the variation of the occupation numbers $\mathcal{N}_{\vec{k}\sigma}$ and implicitly though the variation of the self-energy.

---

[37] By contrast, the full electron–hole scattering amplitude is a singular function of $q = (\vec{q}, \Omega)$: its $q \to 0$ limit depends on the ratio $\frac{q}{\Omega}$.

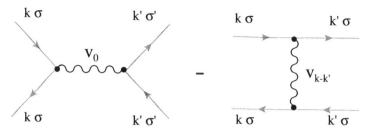

Fig. 8.16. The two diagrams that contribute to the $q \to 0$ limit of the irreducible electron–hole interaction to first order in the coulomb interaction.

Changing only the occupation number of the quasiparticle state of wave vector $\vec{k}'$ and spin σ' from 0 to 1 we obtain

$$\delta G_{\sigma''}(\vec{k}'', \omega) = -2\pi i Z_{k_F \sigma'} \delta\left(\omega - \frac{\mathcal{E}_{\vec{k}'\sigma'}}{\hbar}\right) \delta_{\vec{k}'\vec{k}''} \delta_{\sigma'\sigma''}$$
$$+ \hbar^{-1} G_{\sigma''}^2(\vec{k}'', \omega) \delta \Sigma_{\sigma''}(\vec{k}'', \omega) \,, \qquad (8.172)$$

where, once again, we have made use of the reality of the self-energy for $\hbar\omega \simeq \mu$. Substituting (8.172) in Eq. (8.170) we see that the functional derivative of the self-energy with respect to the quasiparticle distribution satisfies the integral equation

$$\frac{\delta \Sigma_{\sigma}(\vec{k}, \omega)}{\delta \mathcal{N}_{\vec{k}'\sigma'}} = \frac{1}{L^d} I_{\vec{k}\omega\sigma;\vec{k}'\mathcal{E}_{\vec{k}'\sigma'}\sigma'}(0) Z_{k_F \sigma'}$$
$$- \frac{1}{\hbar L^d} \sum_{\vec{k}''\sigma''} \int \frac{d\omega''}{2\pi i} I_{\vec{k}\omega\sigma;\vec{k}''\omega''\sigma''}(0) G_{\sigma''}^2(\vec{k}'', \omega'') \frac{\delta \Sigma_{\sigma''}(\vec{k}'', \omega'')}{\delta \mathcal{N}_{\vec{k}'\sigma'}} \,. \qquad (8.173)$$

Combining this with Eq. (8.169) we see that the Landau interaction function can be written as

$$f_{\vec{k}\sigma,\vec{k}'\sigma'} = \frac{1}{L^d} Z_{k_F \sigma} Z_{k_F \sigma'} I_{\vec{k}E_{\vec{k}\sigma}\sigma;\vec{k}'E_{\vec{k}'\sigma'},\sigma'}^f(0) \,, \qquad (8.174)$$

where $I_{\vec{k}\omega\sigma;\vec{k}'\omega'\sigma'}^f(0)$ is the solution of the integral equation

$$I_{\vec{k}\omega\sigma;\vec{k}'\omega'\sigma'}^f(0) = I_{\vec{k}\omega\sigma;\vec{k}'\omega'\sigma'}(0)$$
$$- \frac{1}{L^d} \sum_{\vec{k}''\sigma''} \int \frac{d\omega''}{2\pi i} I_{\vec{k}\omega\sigma;\vec{k}''\omega''\sigma''}(0) G_{\sigma''}^2(\vec{k}'', \omega'') I_{\vec{k}''\omega''\sigma'';\vec{k}'\omega'\sigma'}^f(0) \,. \qquad (8.175)$$

Thus, we have expressed the Landau interaction function in terms of quantities that have a well defined microscopic meaning and can be calculated (in principle, at least) from diagrammatic perturbation theory.

As a simple application of this rather forbidding formalism consider the calculation of $f_{\vec{k}\sigma,\vec{k}'\sigma'}$ to first-order in the coulomb interaction. In this case the self-energy is just the Hartree–Fock self-energy, which is independent of frequency, so $Z_{\vec{k}\sigma} = 1$. The quasiparticle nergy is then

$$\mathcal{E}_{\vec{k}\sigma} = \varepsilon_{\vec{k}\sigma} + \Sigma^{HF}(\vec{k}\sigma) \qquad (8.176)$$

and the effective mass tends to zero logarithmically.[38] The two diagrams contributing to I to first order in the interaction are shown in Fig. 8.16. The direct diagram vanishes because $v_{q=0} = 0$ when the positively charged background is taken into account (see Section 1.3.2). The "exchange" diagram gives

$$I_{\vec{k}\omega\sigma;\vec{k}'\omega'\sigma'}(0) = -v_{\vec{k}-\vec{k}'}\delta_{\sigma\sigma'} , \qquad (8.177)$$

independent of frequencies. Because in this first-order approximation I coincides with I^f we obtain

$$f_{\vec{k}\sigma,\vec{k}'\sigma'} = --\frac{1}{L^d}v_{\vec{k}-\vec{k}'}\delta_{\sigma\sigma'} . \qquad (8.178)$$

It is quite tempting to try and improve upon this rather crude approximation by simply screening the interaction, while maintaining the spin structure unaltered. An example of Landau interaction function of such form for the three dimensional case was suggested by Quinn and Ferrell (1961)

$$f_{\vec{k}\sigma,\vec{k}'\sigma'} = -\frac{1}{L^d}\frac{4\pi e^2}{|\vec{k}-\vec{k}'|^2 + \kappa_3^2}\delta_{\sigma\sigma'} , \qquad (8.179)$$

where κ_3 is the three-dimensional Fermi-Thomas screening wave vector defined in Eq. (5.36). While this expression results in a microscopic theory in which some of the ills of the HF approximation are cured (e.g. the effective mass and therefore the heat capacity are well behaved), it still leads to unphysical results for the compressibility and the spin susceptibility (see Exercise 8).

8.6 The renormalized hamiltonian approach

We now turn to a physically intuitive derivation of the *renormalized hamiltonian* for the low-energy degrees of freedom of the electron liquid. The procedure, originally devised by D. Hamann and A. W. Overhauser (1966) and recently revisited and extended by Yarlagadda and Giuliani (1989, 1994) provides an explicit construction of the Landau quasiparticles. Through this approach we will be able to clarify the physical significance of the GW approximation and its generalizations, which are described in detail in Section 8.7 and are regularly used in numerical calculations of quasiparticle properties. For example, what is known in the literature as the "coulomb hole" contribution to the self-energy will be shown to be the analogue of the energy shift of an electron interacting with the phonons of a crystal (the so-called "polaron shift"). We also show that the common practice of ignoring

---

[38] Strictly speaking, this means that the first-order approximation violates the assumptions of the Landau theory of Fermi liquids. This difficulty is avoided by cutting off the divergence of the coulomb interaction at some characteristic screening wavevector. The effective mass is then finite.

self-consistency in the GW approximation for weak coupling can be given a justification within the framework of the renormalized hamiltonian approach. More generally, the present approach provides a link between the quasiparticle description of the normal Fermi liquid and the collective description of low-dimensional electron liquids presented in Chapters 9 and 10.

Having thus listed the merits of the renormalized hamiltonian approach we must at this point acknowledge that it is still based on an uncontrolled second-order approximation with respect to the strength of the interaction between the quasiparticles and the density and spin fluctuations of the electron liquid. For this reason the approach becomes increasingly unreliable as the strength of this coupling increases.

8.6.1 Separation of slow and fast degrees of freedom

Let us begin by separating the Hilbert space into "slow" and "fast" sectors: the first one contains only plane wave states with wavevectors $\vec{k}$ close to the Fermi surface, i.e., such that $|k - k_F| < \Lambda$, where Λ is an arbitrarily small cutoff, and the second contains all the other states. The existence of the Fermi surface at $k = k_F$ is of course assumed.

Let us denote by $\hat{\alpha}_{\vec{k}\sigma}$ the destruction operators for the states close to the Fermi surface, and by $\hat{\beta}_{\vec{k}\sigma}$ those for the states far from the Fermi surface:

$$\hat{a}_{\vec{k}\sigma} = \begin{cases} \hat{\alpha}_{\vec{k}\sigma}, & |k - k_F| < \Lambda, \\ \hat{\beta}_{\vec{k}\sigma}, & |k - k_F| > \Lambda. \end{cases} \tag{8.180}$$

Following the folklore stemming from the original literature we shall refer to the $\hat{\alpha}$ operators as the *test particle* operators. Our objective is to derive an effective hamiltonian for these test particles in the slow sector of the Hilbert space. To accomplish this we begin by rewriting the exact many-electron hamiltonian in terms of the operators $\hat{\alpha}$ and $\hat{\beta}$ and their hermitian conjugates. The result is expressed as the sum of three terms

$$\hat{H} = \hat{H}_0 + \hat{H}_M + \hat{H}_I. \tag{8.181}$$

The first term contains only $\hat{\alpha}$ operators:

$$\hat{H}_0 = \sum_{\vec{k}\sigma} \tilde{\varepsilon}_{\vec{k}} \hat{\alpha}_{\vec{k}\sigma}^\dagger \hat{\alpha}_{\vec{k}\sigma} + \frac{1}{2L^d} \sum_{\vec{q}\neq 0} v_q \sum_{\vec{k}_1\sigma_1 \vec{k}_2\sigma_2} \hat{\alpha}_{\vec{k}_1+\vec{q}\sigma_1}^\dagger \hat{\alpha}_{\vec{k}_2-\vec{q}\sigma_2}^\dagger \hat{\alpha}_{\vec{k}_2\sigma_2} \hat{\alpha}_{\vec{k}_1\sigma_1} , \tag{8.182}$$

where $\vec{k}_1, \vec{k}_2, \vec{k}_1 + \vec{q}$, and $\vec{k}_2 - \vec{q}$ all lie within a shell of thickness Λ about k_F and $\tilde{\varepsilon}_{\vec{k}}$ is the free-particle energy relative to the chemical potential.

The second term, $\hat{H}_M$, contains only $\hat{\beta}$ operators. Since these degrees of freedom are of interest only insofar as they affect the behavior of the test particles we shall regard them as a kind of *medium* and refer to the hamiltonian $\hat{H}_M$ as the hamiltonian of this medium:

$$\hat{H}_M = \sum_{\vec{k}\sigma} \tilde{\varepsilon}_{\vec{k}} \hat{\beta}_{\vec{k}\sigma}^\dagger \hat{\beta}_{\vec{k}\sigma} + \frac{1}{2L^d} \sum_{\vec{q}\neq 0} v_q \sum_{\vec{k}_1\sigma_1 \vec{k}_2\sigma_2} \hat{\beta}_{\vec{k}_1+\vec{q}\sigma_1}^\dagger \hat{\beta}_{\vec{k}_2-\vec{q}\sigma_2}^\dagger \hat{\beta}_{\vec{k}_2\sigma_2} \hat{\beta}_{\vec{k}_1\sigma_1} , \tag{8.183}$$

where all the wave vectors lie outside a shell of width Λ about k_F. Notice, and this is crucial, that $\hat{H}_M$ reduces to the full electron liquid hamiltonian in the limit $\Lambda \to 0$. We will repeatedly make use of this fact in what follows.

The third term, $\hat{H}_I$, describes the interaction between the test particles and the medium. This interaction is the sum of fourteen different terms, but, guided by physical intuition, we only keep the ones that *separately* conserve the number of test particles and medium particles. These are terms of the form $\hat{\alpha}^\dagger \hat{\alpha} \hat{\beta}^\dagger \hat{\beta}$, while, for example, terms of the form $\hat{\alpha}^\dagger \hat{\alpha}^\dagger \hat{\beta} \hat{\beta}$ are excluded. The rationale for this approximation is that the neglected terms typically create high-energy virtual excitations, which may be absorbed in a renormalization of the low-energy effective hamiltonian. So in the end we are left with

$$\hat{H}_I \simeq \frac{1}{L^d} \sum_{\vec{q} \neq 0} v_q \sum_{\vec{k}_1 \sigma_1 \vec{k}_2 \sigma_2} \hat{\alpha}^\dagger_{\vec{k}_1 + \vec{q} \sigma_1} \hat{\beta}^\dagger_{\vec{k}_2 - \vec{q} \sigma_2} \hat{\beta}_{\vec{k}_2 \sigma_2} \hat{\alpha}_{\vec{k}_1 \sigma_1}$$

$$- \frac{1}{L^d} \sum_{\vec{q} \neq 0} v_{\vec{k}_2 - \vec{k}_1 - \vec{q}} \sum_{\vec{k}_1 \sigma_1 \vec{k}_2 \sigma_2} \hat{\alpha}^\dagger_{\vec{k}_1 + \vec{q} \sigma_1} \hat{\beta}^\dagger_{\vec{k}_2 - \vec{q} \sigma_2} \hat{\beta}_{\vec{k}_2 \sigma_1} \hat{\alpha}_{\vec{k}_1 \sigma_2} \,, \tag{8.184}$$

where $\vec{k}_1$ and $\vec{k}_1 + \vec{q}$ are within the Λ shell, while $\vec{k}_2$ and $\vec{k}_2 - \vec{q}$ are outside. The first term can be expressed exactly as an interaction between *density fluctuations* of two different types of particles. The second term represents an exchange process in which a particle from the medium replaces a test particle and vice versa. Notice that the two particles can have opposite spins ($\sigma_1 = -\sigma_2$), and, when this is this case, the second term of $\hat{H}_I$ transfers spin angular momentum between the test particles and the medium. Therefore any attempt to express the second term of $\hat{H}_I$ in terms of collective variables must necessarily involve an interaction between the *spin fluctuations* of the test particles and the medium.

This observation leads us to our second and crucial approximation: we assume that the interaction between the test particles and the medium can, in an average sense, be expressed as an interaction between density and spin fluctuations of the test particles and the medium. Thus we write

$$\hat{H}_I \simeq \frac{1}{L^d} \sum_{\vec{q} \neq 0} v_n(q) \sum_{\vec{k} \sigma_1} \hat{n}_{-\vec{q}} \, \hat{\alpha}^\dagger_{\vec{k} - \vec{q} \sigma_1} \hat{\alpha}_{\vec{k} \sigma_1}$$

$$+ \frac{1}{L^d} \sum_{\vec{q} \neq 0} v_s(\vec{q}) \sum_{\vec{k} \sigma_1 \sigma_2} \hat{\vec{S}}_{-\vec{q}} \cdot \left(\hat{\alpha}^\dagger_{\vec{k} - \vec{q} \sigma_1} \vec{\sigma}_{\sigma_1 \sigma_2} \hat{\alpha}_{\vec{k} \sigma_2} \right) \,, \tag{8.185}$$

where $\hat{n}_q = \sum_{\vec{k}\sigma} \hat{\beta}^\dagger_{\vec{k} - \vec{q} \sigma} \hat{\beta}_{\vec{k} \sigma}$ and $\hat{\vec{S}}_q = \sum_{\vec{k} \sigma_1 \sigma_2} \hat{\beta}^\dagger_{\vec{k} - \vec{q} \sigma_1} \vec{\sigma}_{\sigma_1 \sigma_2} \hat{\beta}_{\vec{k} \sigma_2}$ are, respectively, the density and spin fluctuation operators for the medium. The effective interaction potentials between density and spin fluctuations $v_n(q)$ and $v_s(q)$ must approximately include both the exchange effect (the second term of Eq. (8.184)) and the correlation (all the high energy processes excluded from $\hat{H}_I$). This rather tall order can be approximately fulfilled with the help of the many-body local field factors introduced in Chapter 5. Thus, we write

$$v_n(q) \equiv v_q [1 - G_+(q)] \,, \tag{8.186}$$

and

$$v_s(q) \equiv -v_q G_-(q) , \qquad (8.187)$$

where $G_+(q)$ and $G_-(q)$ are the static local field factors for density and spin fluctuations respectively.[39] We will see that this approximation is consistent, at least in the sense of leading to an effective electron–electron interaction that coincides with the Kukkonen-Overhauser interaction derived in Section 5.5.3.1.

8.6.2 Elimination of the fast degrees of freedom

The approximate hamiltonian (8.181) presents some distinct advantages over the exact one. First of all, when the width Λ of the low-energy shell tends to zero, the direct interaction between test particles, which contains *two* wave vector sums tends to zero more rapidly than the kinetic energy, which contains only one. We are therefore justified in treating this interaction by first order perturbation theory. As for the interaction between the test particles and the medium, our strategy is as follows. We seek (and find) a unitary transformation that eliminates this interaction up to second order in its strength. We then average the transformed hamiltonian over the ground-state (or thermal equilibrium ensemble) of the medium. In this manner we arrive at a *renormalized hamiltonian* that is expressed solely in terms of the $\hat{\alpha}$ operators and contains no terms higher than quartic in these operators. Finally, we treat the quartic terms by first order perturbation theory, higher order corrections being negligible in the limit $\Lambda \to 0$.

A unitary transformation maps the original hamiltonian $\hat{H}$ to a new hamiltonian $\hat{\tilde{H}}$

$$\hat{\tilde{H}} = e^{i\hat{F}} \hat{H} e^{-i\hat{F}} , \qquad (8.188)$$

which has the same eigenvalues, but transformed eigenfunctions. In order that the transformation be unitary, its generator $\hat{F}$ must be hermitian. Now, Eq. (8.188) can be readily expanded as follows:

$$\hat{\tilde{H}} = \hat{H} + i[\hat{F}, \hat{H}] - \frac{1}{2!}[\hat{F}, [\hat{F}, \hat{H}]] - \frac{i}{3!}[\hat{F}, [\hat{F}, [\hat{F}, \hat{H}]]] + \cdots .^{[40]} \qquad (8.189)$$

Substituting Eq. (8.181) for $\hat{H}$ we obtain

$$\hat{\tilde{H}} \simeq \hat{H}_0 + \hat{H}_M + \hat{H}_I + i[\hat{F}, \hat{H}_0 + \hat{H}_M] + i[\hat{F}, \hat{H}_I]$$
$$- \frac{1}{2}\{[\hat{F}, [\hat{F}, \hat{H}_0 + \hat{H}_M]] , \qquad (8.190)$$

where all the neglected terms contain at least three powers of the test particle–medium interaction. This can be seen as follows. First of all, in order to eliminate the interaction term $\hat{H}_I$ we must choose $\hat{F}$ as the solution of the operator equation

$$i[\hat{F}, \hat{H}_0 + \hat{H}_M] = -\hat{H}_I . \qquad (8.191)$$

---

[39] Strictly speaking the local field factors should be evaluated at a frequency corresponding to the energy transfer in the scattering process. This is, however, a negligible difference for particles in the immediate vicinity of the Fermi surface.

[40] Defining $\hat{H}(\lambda) \equiv e^{i\lambda\hat{F}} \hat{H} e^{-i\lambda\hat{F}}$, we see that $\hat{H} = \hat{H}(0)$, $\hat{\tilde{H}} = \hat{H}(1)$, and $i[\hat{F}, \hat{H}(0)] = \left(\frac{d\hat{H}(\lambda)}{d\lambda}\right)_{\lambda=0}$, so that Eq. (8.189) is nothing but the Taylor expansion of $\hat{H}(1)$ about $\lambda = 0$.

This characterizes $\hat{F}$ as quantity of first order in the test particle–medium interaction, so it is evident that the omitted term $[\hat{F}, [\hat{F}, \hat{H}_I]]$ is of third order in that interaction. Similarly, all the omitted terms with three or more $\hat{F}$ operators are at least of third order in the test charge–medium interaction. Thus, we are left with the "second-order" hamiltonian

$$\hat{H} \simeq \hat{H}_0 + \hat{H}_M + \frac{i}{2}[\hat{F}, \hat{H}_I] . \tag{8.192}$$

Finally, to complete the elimination of the fast degrees of freedom we average $\hat{H}$ in the ground-state $|0\rangle$ of the medium (or in the thermal equilibrium ensemble) and discard any constant term. The result is the renormalized hamiltonian

$$\hat{H}_{qp} = \hat{H}_0 + \frac{i}{2} \langle 0|[\hat{F}, \hat{H}_I]|0\rangle , \tag{8.193}$$

which we now proceed to evaluate.

8.6.3 The quasiparticle hamiltonian

As a first step, let us solve Eq. (8.191) for the generator $\hat{F}$ of the unitary transformation (8.188). We seek a solution in the form of a one-particle operator in the space of test particles, i.e.,

$$\hat{F} = \frac{1}{L^d} \sum_{\vec{q} \neq 0} \sum_{\vec{k}\sigma\sigma'} \hat{\mathcal{F}}_{\sigma\sigma'}(\vec{k}, \vec{q}) \hat{\alpha}^{\dagger}_{\vec{k}-\vec{q}\sigma} \hat{\alpha}_{\vec{k}\sigma'} , \tag{8.194}$$

where $\hat{\mathcal{F}}_{\sigma\sigma'}(\vec{k}, \vec{q})$ is an operator in the Hilbert space of the medium.[41] At first sight this seems inconsistent, because the commutator of such a $\hat{F}$ with the interaction part of $\hat{H}_0$ contains four $\hat{\alpha}$ operators, whereas $\hat{H}_I$ contains only two. However, we will see shortly that this quartic term is inconsequential, because it vanishes upon averaging in the ground-state of the medium. Thus, evaluating only the quadratic part of the commutator in Eq. (8.191) we arrive at the equation

$$i \left\{ (\tilde{\varepsilon}_{\vec{k}} - \tilde{\varepsilon}_{\vec{k}-\vec{q}}) \hat{\mathcal{F}}_{\sigma\sigma'}(\vec{k}, \vec{q}) + [\hat{\mathcal{F}}_{\sigma\sigma'}(\vec{k}, \vec{q}), \hat{H}_M] \right\}$$
$$= -v_n(\vec{q}) \hat{n}_{-\vec{q}} \delta_{\sigma\sigma'} - v_s(\vec{q}) \hat{\vec{S}}_{-\vec{q}} \cdot \vec{\sigma}_{\sigma\sigma'} . \tag{8.195}$$

Taking the matrix element of both sides of this equation between exact eigenstates $\langle m|$ and $|n\rangle$ of the medium (with eigenvalues E_m and E_n respectively) gives the solution for $\hat{\mathcal{F}}_{\sigma\sigma'}(\vec{k}, \vec{q})$ in the form

$$i\langle m|\hat{\mathcal{F}}_{\sigma\sigma'}(\vec{k}, \vec{q})|n\rangle = -v_n(\vec{q}) \frac{\langle m|\hat{n}_{-\vec{q}}|n\rangle \delta_{\sigma\sigma'}}{\tilde{\varepsilon}_{\vec{k}} - \tilde{\varepsilon}_{\vec{k}-\vec{q}} + E_n - E_m}$$
$$- v_s(\vec{q}) \frac{\langle m|\hat{\vec{S}}_{-\vec{q}}|n\rangle \cdot \vec{\sigma}_{\sigma\sigma'}}{\tilde{\varepsilon}_{\vec{k}} - \tilde{\varepsilon}_{\vec{k}-\vec{q}} + E_n - E_m} . \tag{8.196}$$

---

[41] The wave vector sums are of course restricted to the shell $|k - k_F| < \Lambda$.

Notice that the expectation value of $\hat{\mathcal{F}}_{\sigma\sigma'}(\vec{k},\vec{q})$ in the ground-state of the medium vanishes (for $\vec{q} \neq 0$): this justifies, *a posteriori*, our neglect of the commutator between $\hat{F}$ and the interaction part of $\hat{H}_0$.

We are now ready to evaluate Eq. (8.193). First we rewrite it as

$$\hat{H}_{qp} = \hat{H}_0 + \frac{i}{2}\sum_n \left\{ \langle 0|\hat{F}|n\rangle\langle n|\hat{H}_I|0\rangle - \langle 0|\hat{H}_I|n\rangle\langle n|\hat{F}|0\rangle \right\} \tag{8.197}$$

and substitute in this expression Eqs. (8.185), (8.194), and (8.196). A straightforward calculation (which, however, makes use of the translational and spin rotational invariance of the medium) gives the following result:

$$\begin{aligned}
\hat{H}_{qp} = &\sum_{\vec{k}\sigma} \tilde{\varepsilon}_{\vec{k}}\hat{\alpha}^\dagger_{\vec{k}\sigma}\hat{\alpha}_{\vec{k}\sigma} + \frac{1}{2L^d}\sum_{\vec{q}\neq 0} v_q \sum_{\vec{k}_1\sigma_1\vec{k}_2\sigma_2} \hat{\alpha}^\dagger_{\vec{k}_1+\vec{q}\sigma_1}\hat{\alpha}^\dagger_{\vec{k}_2-\vec{q}\sigma_2}\hat{\alpha}_{\vec{k}_2\sigma_2}\hat{\alpha}_{\vec{k}_1\sigma_1} \\
&+ \frac{1}{2L^d}\sum_{\vec{q}\neq 0} v_n^2(q) \sum_{\vec{k}_1\vec{k}_2\sigma_1\sigma_2} M_n(\vec{k}_1,\vec{k}_2,\vec{q})\hat{\alpha}^\dagger_{\vec{k}_1+\vec{q}\sigma_1}\hat{\alpha}_{\vec{k}_1\sigma_1}\hat{\alpha}^\dagger_{\vec{k}_2-\vec{q}\sigma_2}\hat{\alpha}_{\vec{k}_2\sigma_2} \\
&+ \frac{1}{2L^d}\sum_{\vec{q}\neq 0} v_s^2(q) \sum_{\vec{k}_1\vec{k}_2\sigma_1\sigma_2} M_s(\vec{k}_1,\vec{k}_2,\vec{q})\hat{\alpha}^\dagger_{\vec{k}_1+\vec{q}\sigma_1}\vec{\sigma}_{\sigma_1\sigma_2}\hat{\alpha}_{\vec{k}_1\sigma_2} \\
&\cdot \sum_{\tau_1\tau_2} \hat{\alpha}^\dagger_{\vec{k}_2-\vec{q}\tau_1}\vec{\sigma}_{\tau_1\tau_2}\hat{\alpha}_{\vec{k}_2\tau_2},
\end{aligned} \tag{8.198}$$

where we have defined

$$M_n(\vec{k}_1,\vec{k}_2,\vec{q}) \equiv \frac{1}{L^d}\sum_n \left\{ \frac{|\langle n|\hat{n}_{\vec{q}}|0\rangle|^2}{\tilde{\varepsilon}_{\vec{k}_2} - \tilde{\varepsilon}_{\vec{k}_2-\vec{q}} + E_0 - E_n} - \frac{|\langle n|\hat{n}_{\vec{q}}|0\rangle|^2}{\tilde{\varepsilon}_{\vec{k}_1} - \tilde{\varepsilon}_{\vec{k}_1+\vec{q}} + E_n - E_0} \right\}, \tag{8.199}$$

and

$$M_s(\vec{k}_1,\vec{k}_2,\vec{q}) \equiv \frac{1}{L^d}\sum_n \left\{ \frac{|\langle n|\hat{S}_{z,\vec{q}}|0\rangle|^2}{\tilde{\varepsilon}_{\vec{k}_2} - \tilde{\varepsilon}_{\vec{k}_2-\vec{q}} + E_0 - E_n} - \frac{|\langle n|\hat{S}_{z,\vec{q}}|0\rangle|^2}{\tilde{\varepsilon}_{\vec{k}_1} - \tilde{\varepsilon}_{\vec{k}_1+\vec{q}} + E_n - E_0} \right\}. \tag{8.200}$$

The last two terms in $\hat{H}_{qp}$ describe an effective interaction between the test particles that is mediated by density and spin fluctuations respectively. Interestingly, these terms also include a *self-interaction* of each test particle with itself. Physically, this is due to the fact that a test particle polarizes the medium which in turn acts back on the test particle. Mathematically, the presence of the self-interaction is evident from the fact that the interaction operators in Eq. (8.198) have the form $\hat{\alpha}^\dagger\hat{\alpha}\hat{\alpha}^\dagger\hat{\alpha}$ rather than the normal-ordered form $\hat{\alpha}^\dagger\hat{\alpha}^\dagger\hat{\alpha}\hat{\alpha}$.[42]

In order to extract the self-interaction we rearrange the operators according to the formula

$$\begin{aligned}
\hat{\alpha}^\dagger_{\vec{k}_1+\vec{q}\sigma_1}\hat{\alpha}_{\vec{k}_1\sigma_2}\hat{\alpha}^\dagger_{\vec{k}_2-\vec{q}\tau_1}\hat{\alpha}_{\vec{k}_2\tau_2} = &\,\hat{\alpha}^\dagger_{\vec{k}_1+\vec{q}\sigma_1}\hat{\alpha}^\dagger_{\vec{k}_2-\vec{q}\tau_1}\hat{\alpha}_{\vec{k}_2\tau_2}\hat{\alpha}_{\vec{k}_1\sigma_2} \\
&+ \hat{\alpha}^\dagger_{\vec{k}_2\sigma_2}\hat{\alpha}_{\vec{k}_2\tau_2}\delta_{\vec{k}_2,\vec{k}_1+\vec{q}}\delta_{\sigma_2\tau_1}.
\end{aligned} \tag{8.201}$$

---

[42] Recall that according to the rules of second quantization (see Chapter 1) any two-body operator *that is free of self-interaction* has a representation of the form $\hat{a}^\dagger\hat{a}^\dagger\hat{a}\hat{a}$, i.e., it is normal ordered with respect to the vacuum.

The last term on the right-hand side is a one-particle operator, which can be combined with the one-body part of $\hat{H}_0$, i.e. the kinetic energy. The complete expression for the *one-body part* of H_{qp} is then

$$
\hat{H}_{qp,1} = \sum_{\vec{k}\sigma} \left\{ \tilde{\varepsilon}_{\vec{k}\sigma} + \frac{1}{L^d} \sum_{\vec{q}\neq 0} v_n^2(q) \frac{1}{L^d} \sum_n \frac{|\langle n|\hat{n}_{\vec{q}}|0\rangle|^2}{\tilde{\varepsilon}_{\vec{k}} - \tilde{\varepsilon}_{\vec{k}-\vec{q}} + E_0 - E_n} \right.
$$
$$
\left. + 3\frac{1}{L^d} \sum_{\vec{q}\neq 0} v_s^2(q) \frac{1}{L^d} \sum_n \frac{|\langle n|\hat{S}_{z,\vec{q}}|0\rangle|^2}{\tilde{\varepsilon}_{\vec{k}} - \tilde{\varepsilon}_{\vec{k}-\vec{q}} + E_0 - E_n} \right\} \hat{\alpha}^\dagger_{\vec{k}\sigma} \hat{\alpha}_{\vec{k}\sigma} , \qquad (8.202)
$$

where the factor 3 in front of the spin term arises from the equivalence of the three spin directions in a paramagnetic system.[43] This expression implies a shift in the single particle energy – a shift that will soon be shown to coincide with the "coulomb hole" part of standard self-energy calculations. We will thus arrive at a physical interpretation of the coulomb hole self-energy as the shift in the energy of a test particle due to its interaction with the polarization it induces in the medium.

Notice that the sum over exact eigenstates of the medium in Eq. (8.202) can be written as

$$
\frac{1}{L^d} \sum_n \frac{|\langle n|\hat{n}_{\vec{q}}|0\rangle|^2}{\tilde{\varepsilon}_{\vec{k}} - \tilde{\varepsilon}_{\vec{k}-\vec{q}} + E_0 - E_n} = -\frac{1}{\pi}\mathcal{P}\int_0^\infty d\omega \frac{\Im m\, \chi_{nn}(q,\omega)}{\frac{\tilde{\varepsilon}_{\vec{k}}-\tilde{\varepsilon}_{\vec{k}-\vec{q}}}{\hbar} - \omega} , \qquad (8.203)
$$

where $\chi_{nn}(q,\omega)$ is the density–density response function of the medium, which coincides, for $\Lambda \to 0$, with the density–density response function of the electron liquid itself. A similar expression is obtained for the spin-dependent term. This identity will play a crucial role in establishing the connection between the present results and the GW approximation in the next section.

As for the two-body part of Eq. (8.201), we observe that it is symmetric under the variable changes $\vec{k}_1 \leftrightarrow \vec{k}_2$, $\sigma_1 \leftrightarrow \sigma_2$, $\tau_1 \leftrightarrow \tau_2$, and $\vec{q} \to -\vec{q}$. Making use of these symmetries, together with the standard expressions for the density–density and spin–spin response functions of the medium, it is easy to see that the *two-body part* of $\hat{H}_{qp}$ has the form

$$
\hat{H}_{qp,2} = \frac{1}{2L^d} \sum_{\vec{q}\neq 0} \left[v_q + v_n^2(q)\Re e\,\chi_{nn}\left(q, \frac{\varepsilon_{\vec{k}} - \varepsilon_{\vec{k}-\vec{q}}}{\hbar}\right) \right]
$$
$$
\times \sum_{\vec{k}_1\sigma_1\vec{k}_2\sigma_2} \hat{\alpha}^\dagger_{\vec{k}_1+\vec{q}\sigma_1} \hat{\alpha}^\dagger_{\vec{k}_2-\vec{q}\sigma_2} \hat{\alpha}_{\vec{k}_2\sigma_2} \hat{\alpha}_{\vec{k}_1\sigma_1}
$$
$$
+ \frac{1}{2L^d} \sum_{\vec{q}\neq 0} v_s^2(q)\Re e\,\chi_{S_zS_z}\left(q, \frac{\varepsilon_{\vec{k}} - \varepsilon_{\vec{k}-\vec{q}}}{\hbar}\right) \sum_{\sigma_1\sigma_2\tau_1\tau_2} \vec{\sigma}_{\sigma_1\sigma_2} \cdot \vec{\sigma}_{\tau_1\tau_2}
$$
$$
\times \sum_{\vec{k}_1\vec{k}_2} \hat{\alpha}^\dagger_{\vec{k}_1+\vec{q}\sigma_1} \hat{\alpha}^\dagger_{\vec{k}_2-\vec{q}\tau_1} \hat{\alpha}_{\vec{k}_2\tau_2} \hat{\alpha}_{\vec{k}_1\sigma_2} . \qquad (8.204)
$$

---

[43] The reader is encouraged to try and generalize this formula to spin polarized systems (Exercise 20).

The form of this two-body interaction is very interesting. It coincides with the Kukkonen-Overhauser effective interaction we derived heuristically in Section 5.5.3.1 and from diagrammatic perturbation theory in Section 6.3.6, provided the interactions $v_n(\vec{q})$ and $v_s(q)$ are indeed given by Eqs. (8.186) and (8.187) respectively. This fact provides *a posteriori* confirmation for the validity of our choice of v_n and v_s. Interestingly, the present derivation also suggests that, in calculating quasiparticle properties, one should not attempt to use the Kukkonen-Overhauser interaction beyond the first-order approximation.

8.6.4 The quasiparticle energy

The renormalized quasiparticle hamiltonian

$$\hat{H}_{qp} = \hat{H}_{qp,1} + \hat{H}_{qp,2} \tag{8.205}$$

allows us to calculate the quasiparticle energy in a very simple manner. Recall that, in the limit $\Lambda \to 0$, the quartic term $\hat{H}_{qp,2}$ is small, and can be treated by first-order perturbation theory. Doing this, we obtain a *screened exchange* correction to the quasiparticle-energy – a correction that should be combined with the coulomb-hole shift built in $\hat{H}_{1,qp}$ to give the complete quasiparticle energy.[44] Thus, the quasiparticle energy, relative to the chemical potential, is the sum of three contributions

$$\mathcal{E}_{\vec{k}} - \mu = \tilde{\varepsilon}_{\vec{k}} + \Sigma_{ch,\vec{k}} + \Sigma_{sx,\vec{k}}, \tag{8.206}$$

where $\Sigma_{ch,\vec{k}}$ is the coulomb-hole self-energy, and $\Sigma_{sx,\vec{k}}$ is the screened-exchange self-energy. The coulomb-hole self-energy, Eq. (8.202), is rewritten with the help of Eq. (8.203) in the form

$$\Sigma_{ch,\vec{k}} = -\frac{\hbar}{\pi} \int \frac{d\vec{q}}{(2\pi)^d} \mathcal{P} \int_0^\infty d\omega \frac{v_n^2(q)\Im m \chi_{nn}(q,\omega) + 3v_s^2(q)\Im m \chi_{S_z S_z}(q,\omega)}{\varepsilon_{\vec{k}} - \varepsilon_{\vec{k}-\vec{q}} - \hbar\omega}. \tag{8.207}$$

The screened exchange self-energy

$$\Sigma_{sx,\vec{k}} = -\int \frac{d\vec{q}}{(2\pi)^d} \left[v_q + v_n^2(q)\Re e \chi_{nn}\left(q, \frac{\varepsilon_{\vec{k}} - \varepsilon_{\vec{k}-\vec{q}}}{\hbar}\right) \right.$$
$$\left. + 3v_s^2(q)\Re e \chi_{S_z S_z}\left(q, \frac{\varepsilon_{\vec{k}} - \varepsilon_{\vec{k}-\vec{q}}}{\hbar}\right) \right] n_{\vec{k}-\vec{q}}^{(0)} \tag{8.208}$$

is formally just an ordinary exchange self-energy, which is calculated, however, with the Kukkonen-Overhauser interaction (evaluated at the frequency $\omega = \frac{\varepsilon_{\vec{k}} - \varepsilon_{\vec{k}-\vec{q}}}{\hbar}$), rather than with the bare coulomb interaction.[45]

---

[44] The screened exchange contribution can also be described as the one-body term that arises from the normal-ordering of the two-body hamiltonian *with respect to the Fermi sea* (see Section 2.3).

[45] Strictly speaking the wave vector sum in Eq. (8.208) should be restricted to the region $|\vec{k} - \vec{q}| - k_F < \Lambda$. This restriction is irrelevant if one is interested only in the behavior of the *difference* $\mathcal{E}_{\vec{k}} - \mathcal{E}_{k_F}$ for k close to k_F. To see this, notice that the derivative of $\Sigma_{sx,\vec{k}}$ with respect to $\vec{k}$ can be calculated under the integral sign in Eq. (8.208). Since the real parts of the response functions $\chi_{nn}(q,\omega)$ and $\chi_{S_z S_z}(q,\omega)$ have zero derivative with respect to ω at $\omega = 0$, we see that the derivative with respect to $\vec{k}$ acts only on $n_{\vec{k}-\vec{q}}^{(0)}$, yielding a δ-function of $|\vec{k} - \vec{q}| - k_F$. However, this discussion points to the fact that the renormalized hamiltonian approach does not determine the absolute self-energy, or, equivalently, the chemical potential $\mu = \mathcal{E}_{k_F}$.

Eq. (8.206) for the quasiparticle energy can also be derived in a much more formal manner by starting from the identity (6.98), and performing a "local approximation" on the irreducible electron–hole interaction I (Ng, 1984). The resulting differential relation can be integrated, yielding Eq. (8.206) up to an integration constant. This derivation will be presented in Section 8.7.2.

8.6.5 Physical significance of the renormalized hamiltonian

The renormalized hamiltonian $\hat{H}_{qp}$ has been constructed by applying the unitary transformation (8.188) to the original model hamiltonian (8.181) of test particles interacting with a medium. After the eigenstates of $\hat{H}_{qp}$ have been obtained, one must still undo the transformation (8.188) in order to get the eigenstates of the original hamiltonian, which describes the physical system.

So, while the operator $\hat{a}^{\dagger}_{\vec{k}\sigma}$ creates a plane wave state of wave vector $\vec{k}$ and spin σ, the application of the inverse transformation (8.188) transforms this plane wave state into a quasiparticle state with the same quantum numbers. The corresponding operator, meant to be applied to the original interacting ground-state of $\hat{H}$ is therefore given by

$$\hat{\tilde{a}}_{\vec{k}\sigma} = e^{i\hat{F}} \hat{a}_{\vec{k}\sigma} e^{-i\hat{F}}. \tag{8.209}$$

It is in this sense that the operators $\hat{a}^{\dagger}_{\vec{k}\sigma}$ appearing in the renormalized hamiltonian can now be *reinterpreted* as quasiparticle creation operators.

Since the renormalized hamiltonian is already expressed in terms of quasiparticle operators it should not be surprising that the renormalization constant Z does not explicitly enter the formulation. On the other hand, we know that Z measures the strength of the quasiparticle peak in the spectral function and is, in a sense, the "order parameter" of the normal Fermi liquid state. It is therefore conceptually important to show that the renormalized hamiltonian can be used to predict not only the quasiparticle energies, but also the renormalization constant. This section is devoted to this task.

We begin by noting that the Green's function for plane wave states close to the Fermi surface is given by

$$G_{\sigma}(\vec{k}, t) = -i \langle \Psi_0 | T \hat{a}_{\vec{k}\sigma}(t) \hat{a}^{\dagger}_{\vec{k}\sigma} | \Psi_0 \rangle , \tag{8.210}$$

where both the ground-state $|\Psi_0\rangle$ and the time-evolution of the operators are determined by the original hamiltonian $\hat{H} = \hat{H}_0 + \hat{H}_M + \hat{H}_I$. In order to take advantage of the simplicity of the transformed hamiltonian $\hat{\tilde{H}}$ of Eq. (8.192) we now apply the unitary transformation (8.188) to both $\hat{\alpha}$ and $|\Psi_0\rangle$. This gives

$$G_{\sigma}(\vec{k}, t) = -i \langle \bar{\Psi}_0 | T \hat{\tilde{a}}_{\vec{k}\sigma}(t) \hat{\tilde{a}}^{\dagger}_{\vec{k}\sigma} | \bar{\Psi}_0 \rangle , \tag{8.211}$$

where $|\bar{\Psi}_0\rangle$ is the ground-state of $\hat{\tilde{H}}$. The advantage of Eq. (8.211) is that the ground-state of $\hat{\tilde{H}}$ can be written as the direct product of the ground-state for the quasiparticles and the ground-state for the medium. Similarly, the time evolution under $\hat{\tilde{H}}$ is the product of the time evolutions under $\hat{H}_{qp}$ and $\hat{H}_M$ respectively. There is one complication, however: the transformed operators are no longer simple. To calculate these transformed operators let us note that, according to Eq. (8.194), the commutator of $\hat{F}$ with $\hat{\alpha}_{\vec{k}\sigma}$ is

$$[\hat{F}, \hat{\alpha}_{\vec{k}\sigma}] = -\frac{1}{L^d} \sum_{\vec{k}'\sigma'} \hat{\mathcal{F}}_{\sigma,\sigma'}(\vec{k}, \vec{k} - \vec{k}') \hat{\alpha}_{\vec{k}'\sigma'}, \tag{8.212}$$

a linear function of $\hat{\alpha}$. Then, by iterating the commutator, according to the general expansion (8.189), we obtain

$$\hat{\alpha}_{\vec{k}\sigma} = \sum_{\vec{k}',\sigma'} \left(e^{-i\hat{M}} \right)_{\vec{k}\sigma,\vec{k}'\sigma'} \hat{\alpha}_{\vec{k}'\sigma'}, \tag{8.213}$$

where

$$\hat{M}_{\vec{k}\sigma,\vec{k}'\sigma'} \equiv \frac{1}{L^d} \hat{\mathcal{F}}_{\sigma,\sigma'}(\vec{k}, \vec{k} - \vec{k}') \tag{8.214}$$

is a *matrix* with indices $\vec{k}\sigma$ and $\vec{k}'\sigma'$, and must be exponentiated as such on the right-hand side of Eq. (8.213). Then we see that the Green's function takes the form

$$G_\sigma(\vec{k}, t) = \sum_{\vec{k}'\sigma'} \left\langle \left(e^{-i\hat{M}(t)} \right)_{\vec{k}\sigma,\vec{k}'\sigma'} \left(e^{i\hat{M}} \right)_{\vec{k}'\sigma',\vec{k}\sigma} \right\rangle_M G_{qp,\sigma'}(\vec{k}', t), \tag{8.215}$$

where the time evolution of $\hat{M}$ and the average $\langle \ldots \rangle_M$ are solely determined by the hamiltonian of the medium and

$$G_{qp,\sigma}(\vec{k}, t) = -i e^{-\frac{i}{\hbar}(\mathcal{E}_{\vec{k}\sigma} - \mu)t} [\Theta(t)(1 - n_{\vec{k}\sigma}^{(0)}) - \Theta(-t)n_{\vec{k}\sigma}^{(0)}] \tag{8.216}$$

is the usual free-fermion Green's function generated by the hamiltonian $\hat{H}_{qp}$, with the quartic term appropriately neglected. Due to the presence of the factor $\left\langle \left(e^{-\hat{M}(t)} \right)_{\vec{k}\sigma,\vec{k}'\sigma'} \left(e^{-\hat{M}^\dagger} \right)_{\vec{k}'\sigma',\vec{k}\sigma} \right\rangle_M$ the form of the Green's function is considerably more complicated than one could have naively expected from the simple form of the quasiparticle hamiltonian.

Nevertheless, a closer inspection of Eq. (8.215) reveals that the Green's function retains a long-term coherence that is characteristic of the Landau Fermi liquid state. To see how this comes about, we observe that in the long-time limit ($t \to \infty$) the average of the product of the two exponentials reduces to the product of two independent averages, each independent of time. In addition, due to the translational and rotational invariance of the medium, only the terms with $\vec{k} = \vec{k}'$ and $\sigma = \sigma'$ survive the averaging, i.e. we have

$$\left\langle \left(e^{-i\hat{M}(t)} \right)_{\vec{k}\sigma,\vec{k}'\sigma'} \left(e^{i\hat{M}} \right)_{\vec{k}'\sigma',\vec{k}\sigma} \right\rangle_M \xrightarrow{t \to \infty} \left| \left\langle \left(e^{-i\hat{M}} \right)_{\vec{k}\sigma,\vec{k}\sigma} \right\rangle_M \right|^2 \delta_{\vec{k},\vec{k}'} \delta_{\sigma\sigma'}. \tag{8.217}$$

Notice that the right-hand side of this expression does not depend on the direction of $\vec{k}$. We can therefore define the (positive) constant

$$Z_\sigma \equiv \left| \left\langle \left(e^{-i\hat{M}} \right)_{k_F\sigma, k_F\sigma} \right\rangle_M \right|^2 . \tag{8.218}$$

Substituting in Eq. (8.215) we obtain the long time behavior of the Green's function in the form

$$G_\sigma(\vec{k}, t) \to Z_\sigma G_{qp,\sigma}(\vec{k}, t) . \tag{8.219}$$

This shows that the Green's function retains a particle-like long-term coherence, only with a reduced strength Z_σ.[46] Fourier transformation of Eq. (8.219) yields a quasiparticle peak in the spectral function at $\omega = \mathcal{E}_{\vec{k}\sigma}$ with strength Z_σ. The latter can therefore be identified as the quasiparticle renormalization constant introduced in Section 8.5.3.[47]

It remains to be shown that Eq. (8.218) is equivalent to the previously given expression (8.137). Consistent with our procedure we need only consider the expansion of Eq. (8.218) to second order in the particle–medium interaction:[48] this gives

$$Z_\sigma = 1 + \frac{1}{L^{2d}} \sum_{\vec{k}'\sigma'} \sum_n \langle 0|\hat{\mathcal{F}}_{\vec{k}\sigma,\vec{k}'\sigma'}|n\rangle \langle n|\hat{\mathcal{F}}_{\vec{k}'\sigma',\vec{k}\sigma}|0\rangle , \tag{8.220}$$

where $k = k_F$. Substituting Eq. (8.196) in this expression and making use of Eqs. (8.203) and (8.206)–(8.208) one can show that

$$Z_\sigma = 1 + \frac{1}{\hbar} \left. \frac{\partial \Re e \Sigma_\sigma(k_F, \omega)}{\partial \omega} \right|_{\hbar\omega=\mu} , \tag{8.221}$$

where $\Sigma_\sigma(k_F, \omega)$ is the self-energy. This is equivalent to Eq. (8.137), up to first-order in the self-energy. The details of the calculation are left as an exercise.

8.7 Approximate calculations of the self-energy

In the previous section we have derived explicit formulas for the energy of a quasiparticle and for the effective interaction between quasiparticles. Before proceeding to a numerical evaluation of those formulas, we want to show that equivalent results can be obtained by a more formal (but by no means more physically justifiable) calculation of the self-energy. This approach has been extensively applied to the study of quasiparticle properties in solids: the reader will find more information in the recent review article by Aulbur, Jönsson, and Wilkins (2000).

---

[46] The condition $Z_\sigma < 1$ follows from the unitarity of the matrix $\left(e^{-i\hat{M}} \right)_{\vec{k}\sigma,\vec{k}'\sigma'}$, which, in turn, is implied by Eq. (8.213).

[47] Eq. (8.218) shows that Z_σ is analogous to the *Debye–Waller factor*, i.e. the factor that measures the reduction in the intensity of a Bragg diffraction peak due to the thermal vibrations of a periodic lattice. In the theory of diffraction, the Debye–Waller factor gives the probability of scattering without exciting quantized lattice vibrations (phonons). Similarly, the renormalization constant is the probability of injecting (or extracting) a particle at the Fermi level without exciting the medium.

[48] The first order terms vanish by virtue of the condition $\hat{M}_{\vec{k}\sigma,\vec{k}\sigma} = 0$.

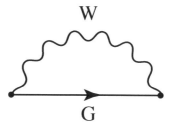

Fig. 8.17. Feynman diagram for the self-energy in the GW approximation. The wavy line denotes a dynamically screened coulomb interaction.

8.7.1 The GW approximation

The exact expression for the self-energy in terms of the Green's function, G, the dielectric function, ϵ, and the vertex function, Λ, was given in Eq. (6.95) of Chapter 6. If both the vertex correction and the screening of the interaction are neglected by setting $\Lambda = 1$ and $\epsilon = 1$, one simply recovers the Hartree–Fock self-energy $\Sigma_\sigma(\vec{k}) = \epsilon_{x\sigma}(\vec{k})$, with all the shortcomings discussed in Section 2.4.1. Notice that, after setting $\Lambda = 1$ and $\epsilon = 1$ in Eq. (6.95), it does not make any difference whether one uses the self-consistent G or the noninteracting $G = G^{(0)}$. The reason is that a frequency-independent self-energy, such as the HF self-energy, can be eliminated by the change of variable $\hbar\omega \to \hbar\omega + \Sigma_\sigma(\vec{k})$ in the frequency integral.

To go beyond the HF approximation the first step is to screen the coulomb interaction through the introduction of the frequency-dependent dielectric function. If one continues to ignore the vertex correction – a major approximation that will be critically examined later on – one arrives at what is known in the literature as the *GW approximation* (see Fig. 8.17). In the finite temperature formalism we have

$$\Sigma_\sigma(\vec{k}, i\omega_n) = -\frac{1}{\hbar\beta} \int \frac{d\vec{q}}{(2\pi)^d} \sum_{m=-\infty}^{\infty} \mathcal{G}_\sigma(\vec{k} - \vec{q}, i\omega_n - i\Omega_m) W(q, i\Omega_m), \quad (8.222)$$

where $\mathcal{G}$ is the temperature Green's function and

$$W(q, i\omega) = \frac{v_q}{\epsilon(q, i\omega)} = v_q + v_q^2 \chi_{nn}(q, i\omega) \quad (8.223)$$

is the screened coulomb interaction. According to the general rules described in Chapter 6, $\omega_n = \frac{(2n+1)\pi}{\hbar\beta}$ is a fermionic Matsubara frequency and the sum runs over all the bosonic Matsubara frequencies $\Omega_m = \frac{2m\pi}{\hbar\beta}$. Notice that the Green's function still depends on Σ in the following manner:

$$\mathcal{G}_\sigma(\vec{k}, i\omega_n) = \frac{1}{i\omega_n - \frac{\tilde{\varepsilon}_{\vec{k}\sigma} + \Sigma_\sigma(\vec{k}, i\omega_n)}{\hbar}}, \quad (8.224)$$

where as usual $\tilde{\varepsilon}_{\vec{k}\sigma} \equiv \varepsilon_{\vec{k}\sigma} - \mu$. Thus Eqs. (8.222) and (8.224) form a set of coupled equations for $\Sigma_\sigma(\vec{k}, i\omega_n)$. Only after this has been solved we can compute the retarded self-energy via he analytic continuation

$$\Sigma_\sigma^{ret}(\vec{k}, \omega) = \Sigma_\sigma\left(\vec{k}, i\omega_n \to \omega - \frac{\mu}{\hbar} + i\eta\right).^{49} \tag{8.225}$$

The coupled equations (8.222) and (8.224) constitute a set that can only be solved numerically. In order to pursue the subject analytically we must give up self-consistency, and replace the exact Green's function by the noninteracting one $\mathcal{G}^{(0)}$. The combination of these two approximations – neglect of vertex correction, and replacement of $\mathcal{G}$ by $\mathcal{G}^{(0)}$ – leads to the simpler G0W self-energy

$$\Sigma_\sigma(\vec{k}, i\omega_n) = -\frac{1}{\hbar\beta} \int \frac{d\vec{q}}{(2\pi)^d} \sum_{m=-\infty}^{\infty} \mathcal{G}_\sigma^{(0)}(\vec{k} - \vec{q}, i\omega_n - i\Omega_m) W(q, i\Omega_m). \tag{8.226}$$

It is very convenient at this point to introduce the *spectral representation* of the effective interaction (see Appendix 14):

$$W(q, i\Omega_m) = v_q - v_q^2 \int_{-\infty}^{\infty} \frac{\Im m \chi_{nn}(q, \omega') \, d\omega'}{i\Omega_m - \omega'} \frac{}{\pi}. \tag{8.227}$$

Inserting this and the analytic expression for $\mathcal{G}^{(0)}$ in Eq. (8.226), and recalling that the bare interaction, v_q, generates the HF self-energy, we obtain

$$\Sigma_\sigma(\vec{k}, i\omega_n) = \varepsilon_{x\sigma}(\vec{k}) + \int \frac{d\vec{q}}{(2\pi)^d} v_q^2 \int_{-\infty}^{\infty} \frac{d\omega'}{\pi} \Im m \chi_{nn}(q, \omega')$$

$$\times \frac{1}{\beta} \sum_{m=-\infty}^{\infty} \frac{1}{\hbar(i\omega_n - i\Omega_m) - \tilde{\varepsilon}_{\vec{k}-\vec{q}\sigma}} \frac{1}{i\Omega_m - \omega'}. \tag{8.228}$$

The sum over bosonic Matsubara frequencies, carried out with the help of the technique described in Chapter 6, gives

$$\frac{1}{\beta} \sum_{m=-\infty}^{\infty} \frac{1}{\hbar(i\omega_n - i\Omega_m) - \tilde{\varepsilon}_{\vec{k}-\vec{q}\sigma}} \frac{1}{i\Omega_m - \omega'} = -\frac{n_F(-\tilde{\varepsilon}_{\vec{k}-\vec{q}\sigma}) + n_B(\hbar\omega')}{i\omega_n - \omega' - \frac{\tilde{\varepsilon}_{\vec{k}-\vec{q}\sigma}}{\hbar}}, \tag{8.229}$$

where $n_F(x) = \frac{1}{e^{\beta x}+1}$ and $n_B(x) = \frac{1}{e^{\beta x}-1}$ are the Fermi–Dirac and Bose–Einstein distributions at zero chemical potential and energy x. Substituting this in Eq. (8.228) and performing the analytic continuation (8.225) to real frequencies, we finally obtain the GW expression for the retarded self-energy

$$\Sigma_\sigma^{ret}(\vec{k}, \omega) = \varepsilon_{x\sigma}(\vec{k}) - \int \frac{d\vec{q}}{(2\pi)^d} v_q^2 \int_{-\infty}^{\infty} \frac{d\omega'}{\pi} \Im m \chi_{nn}(q, \omega')$$

$$\times \frac{n_F(-\tilde{\varepsilon}_{\vec{k}-\vec{q}\sigma}) + n_B(\hbar\omega')}{\omega - \omega' - \frac{\varepsilon_{\vec{k}-\vec{q}\sigma}}{\hbar} + i\eta}. \tag{8.230}$$

---

[49] The shift by $-\frac{\mu}{\hbar}$ is required because the temperature Green's function is defined with reference to the hamiltonian $\hat{K} = \hat{H} - \mu\hat{N}$ (see Eq. (6.119)); however, after performing the continuation to the real frequency axis, we revert to the "absolute" hamiltonian $\hat{H}$.

In the weak coupling limit, one can approximate $\chi_{nn}(q, \omega)$ by the RPA response function function $\frac{\chi_0(q,\omega)}{1 - v_q \chi_0(q,\omega)}$. The density fluctuation spectrum then takes the form

$$\Im m\,\chi_{nn}(q, \omega) \simeq \frac{\Im m\,\chi_0(q, \omega)}{|\epsilon^{RPA}(q, \omega)|^2} . \tag{8.231}$$

Substituting this into Eq. (8.230), one can easily show (see Exercise 16) that the absolute value of the imaginary part of the self-energy, evaluated at $\omega = \frac{\varepsilon_{k\sigma}}{\hbar}$, reduces to the Fermi golden rule expression for the sum of the inverse lifetimes of a quasiparticle and a quasihole at wave vector k (see Section 8.4).

We now want to show that the real part of Eq. (8.230), when evaluated "on the energy shell", that is, at $\omega = \frac{\varepsilon_{k\sigma}}{\hbar}$, yields the same quasiparticle energy as the renormalized hamiltonian of Section 8.6 [i.e. Eqs. (8.206)–(8.208)], provided $W(q, \omega)$ is replaced by the Kukkonen-Overhauser effective interaction

$$W(q, \omega) = v_q + v_n^2(q)\chi_{nn}(q, \omega) + 3v_s^2(q)\chi_{S_z S_z}(q, \omega) . \tag{8.232}$$

This is very interesting because it suggests that, despite its lack of self-consistency, the G0W approximation has a better physical justification than the fully self-consistent GW approximation itself. Apparently, it is preferable to use the G0W formula with an accurate effective interaction than try to achieve self-consistency between the Green's function and the self-energy. Recent numerical solutions of the self-consistent GW equations (Holm and von Barth, 1998) have confirmed this fact.

In order to establish the equivalence between the G0W formula (8.230) (on the energy shell) and Eqs. (8.206)–(8.208) we must identify the terms that correspond to the "screened exchange" and "coulomb hole" parts of the self-energy. This decomposition is based on the fact that Eq. (8.230) can be naturally divided into parts proportional to the Bose distribution (the coulomb hole) and the Fermi distribution (the screened exchange) respectively. We begin by writing

$$n_F(-\tilde{\varepsilon}_{k-q}) + n_B(x) = -n_{k-q} - n_B(-x) , \tag{8.233}$$

which follows from the relations $n_F(x) + n_F(-x) = 1$, and $n_B(x) + n_B(-x) = -1$. We then note that, by virtue of the dispersion relations, one has

$$-\frac{1}{\pi} \int_{-\infty}^{\infty} d\omega' \frac{v_q^2 \Im m\,\chi_{nn}(q, \omega')}{\omega - \omega' - \frac{\varepsilon_{k-q\sigma}}{\hbar} + i\eta} = v_q^2 \chi_{nn}\left(q, \omega - \varepsilon_{k-q}\right) . \tag{8.234}$$

Combining this with the Hartree–Fock self-energy, and using the relation $\frac{1}{\epsilon(q,\omega)} = 1 + v_q \chi_{nn}(q, \omega)$ we get

$$\Sigma_\sigma(k, \omega) = \Sigma_{sx,\sigma}(\vec{k}, \omega) + \Sigma_{ch,\sigma}(\vec{k}, \omega) , \tag{8.235}$$

where

$$\Sigma_{sx,\sigma}(\vec{k}, \omega) = - \int \frac{d\vec{q}}{(2\pi)^d} \frac{v_q}{\epsilon\left(q, \omega - \frac{\varepsilon_{k-q\sigma}}{\hbar}\right)} n_{k-q\sigma}^{(0)} , \tag{8.236}$$

ind

$$\Sigma_{ch,\sigma}(\vec{k}, \omega) = \int \frac{d\vec{q}}{(2\pi)^d} v_q^2 \int_{-\infty}^{\infty} \frac{d\omega'}{\pi} \frac{\Im m\, \chi_{nn}(q, \omega') n_B(-\hbar\omega')}{\omega - \omega' - \frac{\varepsilon_{\vec{k}-\vec{q}\sigma}}{\hbar} + i\eta} . \tag{8.237}$$

We see that the real parts of Eqs. (8.236) and (8.237), evaluated at $\omega = \frac{\varepsilon_{\vec{k}\sigma}}{\hbar}$, correspond, respectively, to the "screened exchange" (Eq. (8.208)) and coulomb hole (Eq. (8.207)) parts of the quasiparticle energy in the renormalized hamiltonian approach of Section 8.6. The correspondence becomes perfect if we use the effective interaction (8.232) in Eq. (8.226).

Thus, the renormalized hamiltonian approach provides a physical justification for the common practice of using the on-shell G0W approximation in calculations of quasiparticle properties. Furthermore, through the inclusion of the local field factors in the effective interaction, the renormalized hamiltonian approach provides a reasonable, albeit uncontrolled, way to include vertex corrections without losing the mathematical simplicity of the G0W approximation.

8.7.2 Diagrammatic derivation of the generalized GW self-energy

Following up on the discussion of the previous section we present here an alternative derivation of the generalized GW self-energy (i.e., the GW self-energy with W given by the Kukkonen–Overhauser formula (8.232)) based on diagrammatic perturbation theory (Ng and Singwi, 1986). The starting point is the identity (6.98) of Section 6.3.4, which relates the self-energy to the irreducible electron–hole interaction at zero momentum transfer:

$$\delta\Sigma_\sigma(p) = i \sum_{\sigma'} \int I_{p\sigma, p'\sigma'}(0) \delta G_{\sigma'}(p') \frac{d^{d+1}p'}{(2\pi)^{d+1}} , \tag{8.238}$$

where $p = (\vec{p}, \omega)$ is the $d + 1$-momentum (see Fig. 6.26). This differential relation cannot be integrated as it stands because I is a complicated functional of G. The idea of Ng and Singwi was to use an approximate form of $I_{p\sigma, p'\sigma'}(0)$ that does not depend on G, so that Eq. (8.238) can be integrated. The "local" approximation introduced in Section 6.3.6 is ideal for this purpose since it yields an expression of the form

$$I_{p\sigma, p'\sigma'}(0) \simeq V_{\sigma,\sigma'}^{eff}(p - p') , \tag{8.239}$$

where $V_{\sigma\sigma'}^{eff}$ is just a function of $p - p'$. Thus, the main characteristic of the Ng–Singwi approach is that the *the key approximation, Eq. (8.239), is done on the irreducible electron–hole interaction rather than on the self-energy itself.* With his approximation we can integrate Eq. (8.238) and obtain, up to an integration constant,

$$\Sigma_\sigma(p) = i \sum_{\sigma'} \int V_{\sigma\sigma'}^{eff}(p - p') G_{\sigma'}(p') \frac{d^{d+1}p'}{(2\pi)^{d+1}}. \tag{8.240}$$

After making the change of variable $p - p' = q$ we see that this expression has the form

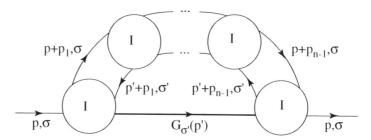

Fig. 8.18. Structure of the diagrams included in the approximate evaluation of the self-energy from Eq. (8.240). Within the local approximation the value of each block I depends only on the electron–hole momentum $p - p'$ and on the orientations of the incoming and outgoing spins, according to Eqs. (8.241) and (8.242). The integral over the $n - 1$ internal four-momenta $p_1, \ldots, p_{n-1}$ (with $n \geq 2$) generates a factor proportional to $[\chi_0(p - p')]^{n-1}$, where χ_0 is the Lindhard function. Notice that these diagrams are irreducible (as they should) in the electron–hole channel defined by the two external legs, which carries zero momentum.

of the GW self-energy (Eq. (8.222)), except for two crucial differences:

- The effective interaction, $V^{eff}_{\sigma\sigma'}(p - p')$, includes vertex corrections and is therefore more general than the screened test charge–test charge interaction $W(q, \omega)$ that appears in the standard GW approximation (8.222).
- The expression (8.240) contains an arbitrary integration constant, which must be fixed by independent means, for example by requiring that $\Sigma(k_F, \mu)$ yield the correct value of the chemical potential, determined from the derivative of the quantum Monte Carlo energy with respect to particle number.

In both respects, Eq. (8.240) is completely consistent with the outcome of the pseudo-hamiltonian analysis developed in Section 8.6.

The derivation of Eq. (8.239) closely follows the pattern of our earlier derivation of the Kukkonen–Overhauser interaction in Section 6.3.6. With reference to Fig. 6.26, the diagrams that contribute to $I_{p\sigma, p'\sigma'}(0)$ are sorted into three groups:

(i) Diagrams that are reducible in the "crossed" electron–hole channel, which carries four-momentum $p - p'$;[50]
(ii) Diagrams that are reducible in the electron–electron channel;
(iii) Diagrams that are "totally irreducible", i.e., irreducible in the electron–electron channel and in both electron–hole channels.

For an approximate evaluation of the diagrams in class (i), shown in Fig. 8.18, we resort to the local approximation, which treats each irreducible electron–hole interaction block I as a function of the electron–hole four-momentum $p - p'$. Borrowing from Section 6.3.6 we write

$$I^{\sigma\sigma,\sigma'\sigma'} \simeq v_{\vec{p}-\vec{p}'}[1 - G_{\sigma\sigma'}(p - p')], \qquad (8.241)$$

---

[50] To avoid confusion, keep in mind that the validity of Eq. (8.238) requires the interaction I to be irreducible in the "direct" particle-hole channel, which carries zero four momentum, but not in the "crossed" one (see Fig. 8.18).

nd

$$I_{p,p'}^{\uparrow\downarrow,\downarrow\uparrow} \simeq -v_{\vec{p}-\vec{p}'}G^T(p-p') , \tag{8.242}$$

where the labeling of the irreducible blocks is graphically shown in Fig. 6.18, and $G^T = G^{\uparrow\downarrow,\downarrow\uparrow}$ is the transverse local field factor. If, in addition, the intermediate Green's functions in Fig. 8.18 are replaced by noninteracting Green's functions, then the integral over the internal four-momenta $p_1, \ldots, p_{n-1}$ can be done analytically, giving a result proportional to $[\chi_0(p-p')]^{n-1}$ where χ_0 is the Lindhard function, and the series of diagrams can be summed algebraically yielding

$$\begin{aligned}
V_{\sigma\sigma'}^{eff,(i)}(p-p') &= [v_{\vec{p}-\vec{p}'}G_+(p-p')]^2\chi_{nn}(p-p')\delta_{\sigma\sigma'} \\
&\quad + [v_{\vec{p}-\vec{p}'}G_-(p-p')]^2\chi_{S_zS_z}(p-p')\delta_{\sigma\sigma'} \\
&\quad + [v_{\vec{p}-\vec{p}'}G_T(p-p')]^2\frac{1}{2}\chi_{S_\perp S_\perp}(p-p')\delta_{\sigma,-\sigma'} .
\end{aligned} \tag{8.243}$$

The subscripts $+$, $-$, and T denote as usual the charge, the longitudinal spin, and the transverse spin channels; $S_\perp$ denotes any transverse component, x or y, of the spin. The density–density and spin–spin response functions are expressed in terms of the Lindhard function and many-body local field factors according to the formulas of Section 5.4.1.

The evaluation of the remaining terms, i.e., the electron–electron ladder diagrams and the totally irreducible diagrams, is far more complex. The only simple diagram is the bare exchange interaction with momentum transfer $\vec{p} - \vec{p}'$. This yields the Hartree–Fock self-energy. All the other terms are ignored in the probably naive hope that they are smaller than the terms retained in (8.243). Keeping only the bare exchange interaction and combining it with Eq. (8.243) we arrive at an effective interaction of the form

$$\begin{aligned}
V_{\uparrow\uparrow}^{eff}(q) &= v_q + \{v_q[1 - G_+(q)]\}^2\chi_{nn}(q) + \{v_qG_-(q)\}^2\chi_{S_z,S_z}(q) , \\
V_{\uparrow\downarrow}^{eff}(q) &= \{v_qG_T(q)\}^2\frac{1}{4}\chi_{S_+S_-}(q) ,
\end{aligned} \tag{8.244}$$

where, in the paramagnetic state, we have $G_T = 2G_-$ and $\chi_{S_+S_-} = 2\chi_{S_zS_z}$ (see Section 5.4.5 for the handling of the transverse spin term). Putting this in Eq. (8.240) and repeating the analytical transformations described in Section 8.7.1, we finally recover the expression (8.202) for the self-energy.

The above derivation has been criticized (Mahan, 1994) because it *appears* to imply that

$$\Sigma_\sigma(p) = i\sum_{\sigma'}\int I_{p\sigma,p'\sigma'}(0)G_{\sigma'}(p')\frac{d^{d+1}p'}{(2\pi)^{d+1}} , \quad (wrong!) , \tag{8.245}$$

a manifestly incorrect relation, which double-counts diagrams.

Before passing judgment, however, one must appreciate that Eq. (8.245) gives the self-energy only *after* $I_{p\sigma,p'\sigma'}(0)$ has been approximated by Eq. (8.239), and then only up to an integration constant, which is fixed *a posteriori* by imposing the correct value of the chemical potential. Since all the quasiparticle properties depend on relative variations of the self-energy, i.e. on $\delta\Sigma$ rather than on Σ, and since $\delta\Sigma$ is correctly given by Eq. (8.238) it seems very likely that our procedure will yield accurate results for the quasiparticle properties. At the same time, the absolute error in Σ, associated with the double counting of diagrams in Eq. (8.245), will be largely corrected by the imposition of the true chemical potential. The crucial test of reasonableness, from this point of view, is whether the local approximation *for I* double-counts diagrams or not. It should be evident from the foregoing discussion that no diagram for $I_{p\sigma,p'\sigma'}(0)$ has been double-counted, but many diagrams have been neglected. The approximation appears therefore to be free of the double-counting problem.

On the other hand, let us see what would have happened if we had applied the local approximation directly to the self-energy, starting from the exact expression

$$\Sigma_\sigma(p) = i \int \frac{v_{\bar{p}-\bar{p}'}}{\varepsilon(p-p')} \tilde{\Lambda}_{p\sigma}(p-p')G_\sigma(p')\frac{d^{d+1}p'}{(2\pi)^{d+1}} . \tag{8.246}$$

The diagrammatic expansion for the proper vertex part is shown in Fig. 6.22. Within the local approximation one finds (Rahman and Vignale, 1984; Sernelius and Mahan, 1989)

$$\tilde{\Lambda}_{p\sigma}(q) = \frac{1}{1 + v_q G_+(q)\chi_0(q)} , \tag{8.247}$$

which depends only on G_+. Now although this expression is definitely free of double-counting problems, we see that it is unfortunately also free of the contribution of spin fluctuations. The root of the difficulty lies in the fact that the local approximation for Λ is not good enough to capture the contribution of spin–density fluctuations. On the other hand the dependence of I on spin fluctuations is manifest in the terms proportional to G_-^2 and G_T^2 in the local approximation (8.243). This is the main physical reason why it is better to apply the local approximation to the differential relation (8.238) than to the integral relation (8.246).

8.8 Calculation of quasiparticle properties

Numerical calculations of the self-energy, $\Sigma(k, \omega)$, aimed at computing the quasiparticle properties and the Landau parameters, date back to the 1960s and have been repeated over the years by several authors,[51] using different forms of the effective interaction. Given an approximation for $\Sigma(k, \omega)$, the quasiparticle energy can be calculated either by solving the Dyson equation (8.134), or by making the on-shell approximation

$$\mathcal{E}_{\bar{k}\sigma} \approx \varepsilon_{\bar{k}\sigma} + \mathfrak{Re}\Sigma^{ret}\left(\vec{k}\sigma, \frac{\varepsilon_{\bar{k}\sigma}}{\hbar}\right) . \tag{8.248}$$

The first approach would be rigorous, if only one knew the exact self-energy. With an approximate Σ it is not so clear which of the two approaches works better. The pseudo-hamiltonian approach of Section 8.6 suggests that the on-shell approximation is more natural

---

[51] A partial list of references includes Rice, 1965; Hedin and Lundqvist, 1969; Rahman and Vignale, 1984; Yarlagadda and Giuliani, 1989; Santoro and Giuliani, 1989; Mahan and Sernelius, 1989; Zhu and Overhauser, 1986; Ng and Singwi, 1986; Yasuhara and Ousaka, 1992; Holm and von Barth, 1998; Simion and Giuliani, 2003.

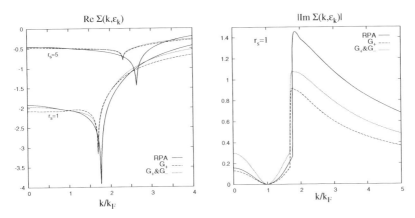

Fig. 8.19. Real and imaginary parts of the on-shell self-energy $\Sigma\left(k, \frac{\varepsilon_k}{\hbar}\right)$ (in Rydberg) in the G0W approximation in 2D. The calculations are done in RPA (solid line), including the density local field factor G_+ (thick dashes) and including both the density and the spin local field factors G_+ and G_-. The local field factors are the ones calculated by Davoudi *et al.*, 2001. For each r_s the curves have been "renormalized" (see footnote[53]) so that at $k = k_F$ they all reduce to the "exact" exchange-correlation chemical potential given by quantum Monte Carlo calculations. From Asgari *et al.*, 2004.

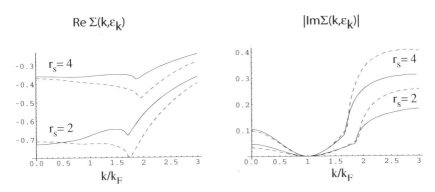

Fig. 8.20. Real and imaginary parts of the on-shell self-energy $\Sigma\left(k, \frac{\varepsilon_k}{\hbar}\right)$ (in Rydberg) calculated in the G0W approximation in 3D. The dashed line shows the absolute self-energy obtained in the RPA, without "renormalization". The solid line shows the result obtained by including the local field factors $G_+(q)$ (given by Eq. (A11.5)) and $G_-(q)$ (given by Eq. (A11.6)), using the renormalization procedure described in footnote[53] to fix the chemical potential to the quantum Monte Carlo value. From Simion and Giuliani, 2005.

when Σ is obtained from the G0W approximation. Accordingly, many authors have used Eq. (8.248) as the basis for the calculation of quasiparticle energies.[52]

It turns out that the gross qualitative behavior of the self-energy is independent of the details of the approximations: Figs. 8.19 and 8.20 show the behavior of the on-shell self-energy

---

[52] Early arguments in favor of the on-shell approximation, based on the alleged cancellation of higher-order corrections (Rice, 1965), are not expected to apply in a regime in which the many-body corrections to the effective mass are not small.

in 2D and 3D respectively. The imaginary part of the self-energy, which is proportional to the inverse of the quasiparticle lifetime, vanishes, of course, at $k = k_F$, and exhibits a sharp rise at $k = k_p$ – the smallest value of k for which a quasiparticle can decay into plasmons (see Exercise 24). The real part of the self-energy, on the other hand, is weakly k-dependent in the region $0 < k < k_F$. The near constancy of $\Re e \Sigma \left(k, \frac{\varepsilon_k}{\hbar}\right)$ over this range implies that the self-energy is a short-ranged operator in real space and provides some justification for the common practice of describing many-body effects by means of local effective potentials. This behavior should be contrasted with that of the exchange self-energy, which varies by 50% in the same wave vector range. That strong wave vector dependence corresponds to the strongly nonlocal character of the exchange potential.[53]

Unfortunately, quasiparticle properties, such as the effective mass, are determined by the *derivatives* of the self-energy with respect to k and ω, quantities that in turn depend significantly on the chosen form of the approximation for the effective interaction and the local field factors. For example, in the weak coupling limit the effective mass can be calculated either from the exact expression (8.146) or from the on-shell approximation (8.248), and in the latter case we have (see Exercise 23)

$$\frac{m}{m^*} = 1 + \frac{m}{\hbar^2 k_F} \frac{\partial}{\partial k} \Re e \Sigma_\sigma^{ret} \left(k, \frac{\mu}{\hbar}\right)\bigg|_{k=k_F} + \frac{1}{\hbar} \frac{\partial}{\partial \omega} \Re e \Sigma_\sigma^{ret}(k_F, \omega)\bigg|_{\omega=\mu/\hbar} . \qquad (8.249)$$

Now, since in the weak coupling limit the exchange self-energy, which is a monotonically increasing function of k, dominates the total self-energy, it is not too surprising, in view of Eq. (8.249), to find that the effective mass is *reduced* by the interaction: $\frac{m^*}{m} < 1$.[54] However, at larger values of r_s the correlation contribution makes $\Sigma \left(k, \frac{\varepsilon_k}{\hbar}\right)$ a weakly decreasing function of k at $k = k_F$, and hence $m^* > m$ (see Figs. 8.19 and 8.20). Furthermore, the difference between the on-shell effective mass and the effective mass determined from Eq. (8.146) grows with increasing r_s: the on-shell effective mass may even diverge (unphysically) if the negative quantity $\frac{\partial \Sigma^{ret}(k_F \sigma, \omega)}{\partial \omega}$ is large enough to cancel the other two terms of Eq. (8.249) (see footnote[52]).

In Figs. 8.21 and 8.22 we present the results of different calculations of $\frac{m^*}{m}$ in two and three dimensions. The uncertainty in the value and even in the sign of $\frac{m^*}{m} - 1$ (which, incidentally, is just the Landau parameter F_1^s) is considerable. One might think that the uncertainty could easily be cleared by performing a quantum Monte Carlo calculation of the quasiparticle energy. But this is not as easy as it sounds, since it requires an accurate determination of the wave function of an excited state. A first step in this direction was taken by Kwon, Ceperley, and Martin (1994), for the two dimensional electron liquid. Actually, these authors computed the energy of a quasiparticle *and* a quasihole, rather than the energy of a single quasiparticle or quasihole. The two approaches should be equivalent in the thermodynamic limit, since

---

[53] The calculation of the generalized G0W self-energy including the local field factors presents a technical problem, namely, the integral over q in Eq. (8.207) for the coulomb-hole self-energy diverges at large q due to the large-q divergence of the static local field factors. This difficulty is solved by noting that both in the renormalized hamiltonian approach, and in the diagrammatic derivation of Section 8.7.2, the generalized G0W self-energy is determined only in relation to the chemical potential. Taking the latter from quantum Monte Carlo calculation one therefore needs to calculate only the difference $\Sigma(k, \omega) - \Sigma \left(k_F, \frac{\mu}{\hbar}\right)$ which is given by a convergent integral. Also, the derivatives of $\Sigma(k, \omega)$ with respect to k and ω are given by convergent integrals.

[54] The exact analytic behavior of the effective mass in the limit $r_s \to 0$ is worked out in Exercise 22.

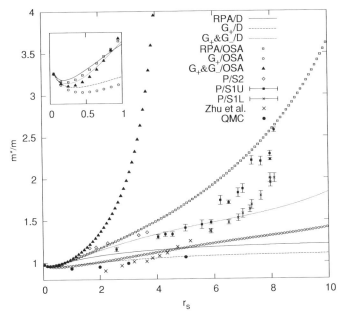

Fig. 8.21. Effective mass enhancement in a strictly two-dimensional electron liquid calculated in the G0W approximation with the following effective interactions: RPA, KO including only the density local field factor G_+, and KO including both the density and the spin local field factors G_+ and G_-. The calculation is done both in the on-shell approximation (OSA), Eq. (8.249), and making use of the formally exact Dyson equation (8.146) (D). The local field factors are given by Eq. (A11.8) in Appendix 11. The curve labeled QMC shows the results of the quantum Monte Carlo calculation by Kwon *et al.* (1994). The three sets of symbols with error bars (P/S1U, P/S1L, and PS2) represent data from different Si inversion layers samples. (Pudalov *et al.*, 2002). The crosses represent measured values of the effective mass in 2-dimensional GaAs samples (Zhu *et al.*, 2003). The inset shows an enlargement of the results for $0 \leq r_s \leq 1$. From Asgari *et al.*, 2004.

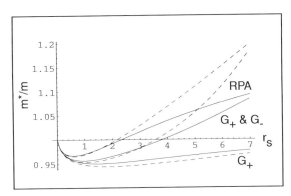

Fig. 8.22. Effective mass enhancements for the three-dimensional electron liquid. Effective masses calculated from Eq. (8.146) are shown as solid lines, while the ones obtained from the on-shell approximation of Eq. (8.249) are shown as dashed lines. The local field factors are given by Eqs. (A11.5) (for G_+) and (A11.1–A11.4) (for G_-). From Simion and Guiliani, 2005.

Table 8.4. *Effective mass enhancement and renormalization constant in a strictly two-dimensional electron liquid calculated in the G0W approximation with the following forms of the effective interaction: RPA (second column), KO interaction with local field factor G_+ only (third column), and KO interaction with local field factors G_+ and G_- (fourth column). In each of these columns two values of the mass enhancement are reported: the first one is obtained from the exact formula, Eq. (8.146), while the one in parentheses is calculated in the on-shell approximation of Eq. (8.249). The local field factors were computed according to Eq. (A11.8) of Appendix 11. Also listed in the fifth column is the mass enhancement obtained from the quantum Monte Carlo calculation of Kwon et al. (1994). (Adapted from Asgari et al. (2004).)*

r_s	$(m^*/m)_{RPA}$	$(m^*/m)_{G_+}$	$(m^*/m)_{G_+ \& G_-}$	$(m^*/m)_{QMC}$	Z
1	1.03 (1.04)	0.98 (0.98)	1.03 (1.05)	0.93(1)	0.67
2	1.09 (1.17)	1.02 (1.04)	1.15 (1.36)	0.95(1)	0.50
3	1.12 (1.32)	1.06 (1.11)	1.26 (2.01)	0.99(1)	0.41
5	1.17 (1.65)	1.10 (1.27)	1.43 (–)	1.06(1)	0.30
8	1.20 (2.32)	1.13 (1.51)	1.65 (–)	–	0.22

the energy of interaction between the quasiparticle and the quasihole, $-\frac{1}{L^d} f_{\vec{p}\sigma, \vec{p}'\sigma'}$, tends to 0 for $L \to \infty$: in a finite-size system, however, the presence of the interaction energy can be a serious source of error. The results of Kwon *et al.* for the effective mass enhancement are listed in the fifth column of Table 8.4 and are plotted in Fig. 8.21. Notice that these values lie below all the G0W values and, perhaps more importantly, are farther removed from the experimental data shown in Fig. 8.6.[55]

As we saw in Chapter 1, quantum Monte Carlo provides accurate values of the ground-state energy, from which the compressibility and the spin susceptibility can be reliably calculated. Thus, if we choose a fiducial value of $\frac{m^*}{m}$ from one of the available microscopic theories, for example the one obtained from Eq. (8.146) making use of both G_+ and G_-, we can automatically calculate the values of the Landau parameters F_0^s and F_0^a. We have done this in Tables 8.5 and 8.7. Another important quantity is the renormalization constant $Z = Z_{k_F}$, which is tabulated, as a function of density, in the rightmost column of Tables 8.4 and 8.6. As expected, Z decreases monotonically with increasing r_s, but remains finite, confirming the existence of the Fermi liquid, up to rather large values of r_s.

What about experiments? Measurements of $\frac{m^*}{m}$ and $\frac{g^*}{g}$ in high purity low-density 2DELs are in qualitative agreement with generalized GW calculations of these quantities, provided one includes both G_+ and G_- and the density is not too low.[56] The agreement breaks down quite dramatically at lower densities, with the theoretically predicted mass enhancement

---

[55] The comparison between theory and experiment is complicated by the fact that the theoretical calculations presented here have been done for a strictly 2D electron liquid, while the experiments are done on layers of finite thickness.

[56] Particularly poor results are obtained if one includes only G_+ and not G_-. As Figs. 8.22 and 8.21 show, this is worse than not including any G's, i.e., doing the RPA.

Table 8.5. *Calculated values of some of the relevant Landau Fermi liquid parameters of the two-dimensional electron liquid. The values of the effective mass are taken from the fourth column of Table 8.4. Compressibility and spin stiffness are calculated from the formulas of Section 1.8. Notice that F_0^s becoming more negative than -1 is not a sign of instability, but simply reflects the change in sign of the proper compressibility of the jellium model.*

r_s	K_0/K	χ_P/χ_S	F_0^s	F_0^a	F_1^s
1	0.533	0.691	−0.45	−0.29	0.03
2	0.018	0.525	−0.98	−0.40	0.15
3	−0.538	0.421	−1.68	−0.47	0.26
5	−1.737	0.296	−3.48	−0.58	0.43
8	−3.657	0.196	−7.03	−0.68	0.65

Table 8.6. *Same as Table 8.4 for the three-dimensional electron liquid. The local field factors $G_+(q)$ and $G_-(q)$ are taken from Eqs. (A11.5) and (A11.1) of Appendix 11, with the parameter β_- determined self-consistently to match the value of the spin susceptibility enhancement. (Adapted from Simion et al. (unpublished).)*

r_s	$(m^*/m)_{RPA}$	$(m^*/m)_{G_+}$	$(m^*/m)_{G_+\&G_-}$	Z
1	0.97 (0.97)	0.95 (0.95)	0.96 (0.95)	0.88
2	0.99 (0.99)	0.95 (0.94)	0.97 (0.96)	0.80
3	1.02 (1.03)	0.96 (0.95)	0.98 (0.98)	0.73
4	1.04 (1.07)	0.96 (0.95)	1.00 (1.01)	0.67
5	1.06 (1.11)	0.97 (0.96)	1.03 (1.06)	0.62
6	1.08 (1.15)	0.97 (0.96)	1.06 (1.11)	0.58

Table 8.7. *Calculated values of some of the relevant Landau Fermi liquid parameters of the three-dimensional electron liquid. The values of the effective mass are taken from the fourth column of Table 8.6. Compressibility and spin stiffness are calculated from the formulas of Section 1.8.*

r_s	K_0/K	χ_P/χ_S	F_0^s	F_0^a	F_1^s
1	0.827	0.867	−0.21	−0.17	−0.04
2	0.645	0.770	−0.37	−0.25	−0.03
3	0.454	0.693	−0.55	−0.32	−0.02
4	0.256	0.631	−0.74	−0.37	0.0
5	0.052	0.580	−0.95	−0.40	0.03
6	−0.157	0.537	−1.17	−0.43	0.06

growing much more rapidly than the experiment suggests. In 3D metals (where, of course band structure effects cannot be ignored) a somewhat indirect, yet important clue came in the mid-1980s from angle-resolved photoemission experiments (Jensen and Plummer 1985; Lyo and Plummer, 1988). These experiments measured the width of the quasiparticle band, that is, the difference between the energy of a quasiparticle at $k = k_F$ and $k = 0$ in the conduction band of Na. A word of caution should be spent here: these experiments involve lots of hard-to-control surface effects which complicate the interpretation of the data. Furthermore the notion of quasiparticle is well defined only in the immediate vicinity of the Fermi surface, and may not make much sense near the bottom of the conduction band. These experiments indicate a significant reduction of the bandwidth from the band structure value of 3.14 eV, to an observed value of ~ 2.5 eV – a result implying that $\Sigma\left(k_F, \frac{\epsilon_F}{\hbar}\right) < \Sigma(0, 0)$. While this conclusion is roughly consistent with the behavior of the RPA self-energy, inclusion of the local field factors leads to a self-energy that is a much "flatter" function of k in the metallic density range (see Fig. 8.20). In both cases, the difference is not large enough to account for the narrowing observed in the experiment. The important point is that the exchange self-energy alone would go in the opposite direction, predicting a bandwidth substantially larger than the free-electron theory. This demonstrates the importance of correlations in neutralizing the effect of the strongly non-local exchange potential.

8.9 Superconductivity without phonons?

Superconductivity in metals is understood to be a consequence of an effective attractive interaction between electrons, mediated by lattice vibrations (phonons). It is precisely this interaction that lowers the energy of the BCS wave function below that of the normal Fermi liquid state.

A more subtle question is whether superconductivity may occur *without phonons*, e.g. in the rigid jellium model, even though the fundamental interaction between the particles is purely repulsive. This question was first answered, in the affirmative, by Kohn and Luttinger (Kohn and Luttinger, 1965). They observed that, due to the sharpness of the Fermi surface, the statically screened interaction between two electrons has a long-range oscillatory part – which is, of course, the Friedel oscillation discussed in Section 5.4.3. By taking advantage of the attractive parts of the effective interaction Cooper pairs can form, thus giving rise to superconductivity below some (very small) critical temperature.

Since this conclusion was reached within the framework of the BCS mean field theory it was not clear at all that it would survive the inclusion of quantum fluctuations. For this reason, in recent years efforts have focused on going beyond the mean field approximation by making use of the techniques introduced in Section 6.3.7. The main idea has been to generalize the GW theory by allowing for the possible presence of an anomalous Green's function, which, as we saw in Section 6.3.7, is the signature of superconductivity.

The diagrammatic representation of the GW theory for a superconductor near the critical temperature, T_c, is shown in Fig. 8.23. These diagrams are obtained from the more general

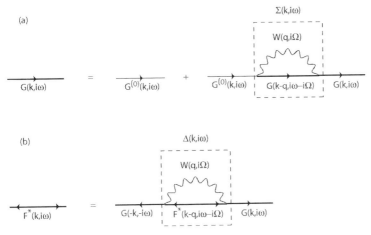

Fig. 8.23. GW approximation for superconducting systems near the transition temperature: (a) gives the normal-state GW self-energy without reference to the anomalous propagator $\mathcal{F}^*$; (b) has the structure of an eigenvalue problem for $\mathcal{F}^*(\vec{k}, i\omega)$, from which T_c can be computed.

ones of Fig. 6.33 by discarding terms of second and higher order in $\mathcal{F}^*(k, \omega)$ (a procedure justified near T_c) and substituting the GW expressions for the normal and the anomalous components of the self-energies, Σ and Δ respectively (see Exercise 25).

Notice that, at variance with the Kohn–Luttinger analysis, which considered only the statically screened interaction between the electrons, here one takes into account the full frequency-dependent effective electron–electron interaction $W(q, \omega)$. Accordingly, the anomalous self-energy is also frequency-dependent, and adjusts its form to take maximal advantage of the features of $W(q, \omega)$.

The frequency-dependent electron–electron interaction is formally very similar to a phonon-mediated interaction, the role of the electron–phonon propagator being taken up by the density–density and spin–spin response functions (see Section 5.5.3.3). The essential difference between the two interactions lies in the energy scale of the intermediate excitations: in the phonon-mediated case this is the Debye energy $\hbar\omega_D$, while in the present case it is of the order of, or even larger than the Fermi energy, ϵ_F. In the phonon-mediated case, the existence of the small parameter $\hbar\omega_D/\epsilon_F$ leads to major simplifications: for example, it can be proved (Migdal's theorem) that vertex corrections are of the order of $\hbar\omega_D/\epsilon_F$, and hence negligible. In the purely electronic case, on the other hand, no such simplification is possible, and for this reason the GW approach, with its neglect of vertex corrections, is an uncontrolled approximation.

Be as it may, the diagrams shown in Fig. 8.23 tell us that the anomalous Green's function will first appear at a critical temperature T_c determined by the solution of the equation

$$\mathcal{F}^*(\vec{k}, i\omega) = -\mathcal{G}(\vec{k}, i\omega)\mathcal{G}(-\vec{k}, -i\omega)\frac{1}{\beta L^d}\sum_{q,m} W(q, i\Omega_m)\mathcal{F}^*(\vec{k} - \vec{q}, i\omega - i\Omega_m), \quad (8.250)$$

where $\mathcal{G}(\vec{k}, i\omega)$ is the Green's function calculated by the GW approximation in the normal state ($\mathcal{F} = 0$), $i\omega$ is a fermionic Matsubara frequency and $i\Omega_m$ are bosonic ones. This equation has the structure of an eigenvalue problem in which $\mathcal{F}^*(\vec{k}, i\omega)$ is an eigenvector with eigenvalue 1. Requiring that the kernel has such an eigenvalue yields the critical temperature.

Early numerical calculations in which Eq. (8.250) was solved numerically using the RPA effective interaction found embarrassingly high transition temperatures at the density of Na (Rietschel and Sham, 1983). Later calculations, which included approximate vertex corrections (Grabowski and Sham, 1984) considerably reduced these initial estimates. More recently Takada (Takada, 1993) has examined the problem by making use of the Kukkonen-Overhauser interaction (see Section 5.5.3) with suitable local field factors. His conclusion is that the ground-state of the electron gas exhibits superconductivity starting at $r_s = 4$. Unfortunately, in view of the large uncertainties in the approximation of the vertex corrections, and of the sensitive dependence of T_c on these corrections, none of the above results can be considered truly definitive. Quantum Monte Carlo studies of the superconducting state will be most helpful in settling this delicate question.

8.10 The disordered electron liquid

Thus far in this chapter we have only considered translationally invariant electron liquids. But real electron liquids are invariably subjected to potentials that break translational invariance, such as the periodic potential of a crystal lattice, or the potential from randomly distributed impurities. It would certainly be very disappointing if the Landau Fermi liquid model turned out to be inapplicable to such systems. Fortunately it turns out that the concept is still applicable; yet, there are some important differences between homogeneous and inhomogeneous electron liquids.

A general approach to inhomogeneous interacting Fermi liquids begins with a consideration of the exact eigenstates $|\phi_\alpha\rangle$ of a single particle – an electron in our case – in the external potential. These are Bloch waves in a perfectly periodic crystal lattice, but have no definite symmetry in the presence of a random impurity potential. In any case, however, the ground-state of the noninteracting system is obtained by singly occupying the eigenstates with the N lowest energies ε_α. The highest occupied eigenvalue defines a Fermi energy $\epsilon_F = \varepsilon_N$, but not a Fermi surface. Excited states are obtained by promoting some electrons from below the Fermi level to above the Fermi level. All these states are described by a set of occupation numbers $\mathcal{N}_\alpha$.

We now start from one of these states, and slowly turn on the electron–electron interaction. By the reasoning that worked so well in the homogeneous case we expect to generate long-lived interacting states characterized by a quasiparticle distribution $\mathcal{N}_\alpha$. But here comes an important difference. While in the homogeneous case the interaction between quasiparticles is made ineffective by the Pauli exclusion principle *and* by the conservation of momentum and energy, in an inhomogeneous system the law of momentum conservation does not

∎old. It is therefore expected that the quasiparticles will interact more strongly and decay more rapidly than in the homogeneous case. How much more rapidly? This is an essential question, since the very existence of the Fermi liquid requires that the decay rate of a quasiparticle of energy ε tend to zero more rapidly than the $\varepsilon - \epsilon_F$ when $\varepsilon \to \epsilon_F$ (see Exercise 27).

The answer depends on whether the system is periodic or disordered. In periodic systems crystal momentum (the Bloch wave vector) is conserved up to reciprocal lattice vectors. While the occurrence of "umklapp" processes can alter the numerical value of the lifetime it does not lead to qualitative departures from the homogeneous picture.[57]

In disordered systems the lack of momentum conservation has more serious consequences. First of all, if disorder is sufficiently strong, it can lead to *localization* of the quasiparticle states and hence change the electrical properties of the system from metal to insulator.[58] Even in the weak disorder regime, i.e., for $k_F \ell \gg 1$, where ℓ is the electron mean free path, the combined effects of disorder and interactions can be significant.[59] As we saw in Chapter 4, the dynamics of density fluctuations in this regime is governed, at long wavelengths, by the diffusion equation. Because density fluctuations relax at a slower pace than in a perfect crystal the electrons within them stay together for a longer time and hence interact more strongly: this *electron loitering* leads to an enhanced quasiparticle decay rate. In three dimensions the decay rate goes as $\frac{1}{\tau^{(e)}} \sim (\varepsilon - \epsilon_F)^{3/2}$ and $\frac{1}{\tau^{(e)}} \sim (k_B T)^{3/2}$ in the limits of $k_B T \ll \varepsilon - \epsilon_F$ and $k_B T \gg \varepsilon - \epsilon_F$ respectively (Schmid, 1974). This is considerably larger than $(\varepsilon - \epsilon_F)^2$ and $(k_B T)^2$ yet still small compared to $\varepsilon - \epsilon_F$. In two dimension, on the other hand, one finds (Abrahams *et al.*, 1981; Schmid, 1974) $\frac{1}{\tau^{(e)}} \sim (\varepsilon - \epsilon_F)$ at $T = 0$, and $\frac{1}{\tau^{(e)}} \sim (k_B T) \ln(k_B T)$ at finite temperature, which implies that the conventional Fermi liquid picture is, at best, *marginally valid*. A derivation of these results will be presented in Section 8.10.1.

In concomitance with the enhancement of the quasiparticle decay rate, the single-particle density of states, $N(\varepsilon)$, is *reduced* in the vicinity of the Fermi level. In Section 8.10.2 we will show that the many-body correction to the density of states goes as $(\varepsilon - \epsilon_F)^{1/2}$ in three dimensions, and $|\ln(\varepsilon - \epsilon_F)|$ in two dimensions. These results are based on perturbation theory, and therefore cannot be trusted when the correction to the unperturbed density of states becomes too large (e.g., in two dimensions, for $\varepsilon \to \epsilon_F$). Notice, however, that a suppression of the density of states at the Fermi level is also predicted in the limit of strong disorder (this is the classical "coulomb gap" discussed in Section 2.2.4), even though the physics appears to be quite different in that regime. Other thermodynamic and transport properties, such as the heat capacity and the conductivity, show non-analytic behavior, which

[57] This is not to say that *all* homogeneous Fermi liquid properties will carry over to crystals. For example, the effective mass equation (8.35), which depends on momentum conservation, does not hold.

[58] In fact, according to the *scaling theory of localization* (Abrahams, Anderson, Licciardello, and Ramakrishnan, 1979) *all* the one-electron states are localized in a two-dimensional disordered system, although the localization length becomes exponentially large as the strength of disorder tends to zero.

[59] Besides the combined effects of interaction and disorder, there are also quantum interference effects, due to disorder, which do not explicitly depend on the interaction: they will be briefly discussed in Section 8.10.3.

can be calculated more or less straightforwardly from perturbation theory (Altshuler and Aronov, 1985).[60] Remarkably, there is no non-analytic correction to the compressibility.[61]

Going beyond perturbation theory, in the genuine spirit of the Landau theory of Fermi liquid, has turned out to be an extremely difficult task. The most complete treatment of the problem to date is based on a mapping to the "nonlinear sigma model" (Finkelstein, 1983) and the renormalization group analysis (Castellani *et al.*, 1984 and 1987) inspired by it. Exposing the details of the above theories is well beyond the scope of this section. However, the essential ideas can be summarized as follows. The system is described in terms of three "running variables", whose values change as we focus on lower and lower energies or larger and larger length scales. These variables are (i) the inverse of the dimensionless conductance, $t = \frac{e^2}{2\hbar\sigma L^{d-2}}$, and (ii)–(iii) the proper isotropic scattering amplitudes γ^s and γ^a for the density and the spin channel respectively.[62] The fact that only two isotropic scattering amplitudes are relevant (as opposed to an infinite set of Landau parameters in the "clean" Fermi liquid) is certainly a simplification, but it is unfortunately offset by the fact that these scattering amplitudes are now scale-dependent. In terms of observable properties one has

$$\overline{\frac{\partial n}{\partial \mu}} = N(0)\frac{\overline{c_v}}{c_v}(1 - \gamma^s), \tag{8.251}$$

and

$$\overline{\chi_s} = N(0)\frac{\overline{c_v}}{c_v}(1 - \gamma^a), \tag{8.252}$$

where $\overline{\frac{\partial n}{\partial \mu}}$, $\overline{c_v}$, and $\overline{\chi_s}$ are, respectively, the proper compressibility, the heat capacity, and the spin susceptibility of the disordered system, while $N(0)$ and c_v are the quasiparticle density of states at the Fermi level and the heat capacity of the clean system. Remarkably, it turns out that $\overline{\frac{\partial n}{\partial \mu}}$ is scale-independent, and its value is essentially the same as in the clean electron liquid. Therefore, γ^s and γ^a are seen to determine the scale-dependent heat capacity and spin susceptibility of the disordered electron liquid.

The scale-dependence of γ^s, γ^a, and g is controlled by a set of first-order differential equations – the renormalization-group equations – which can been derived from diagrammatic calculations to leading order in $\frac{1}{\epsilon_F \tau}$ (Castellani *et al.*, 1984) . We will not report these equations here, but simply note the most striking features of their solution. In two dimensions, the spin-antisymmetric amplitude, γ^a, is found to increase monotonically with the length scale L_s, and to diverge at a finite value of L_s. The divergence of one of the running variables at a finite length scale means that the renormalization group approach

---

[60] To avoid misunderstandings we emphasize that the specific heat is related to the quasiparticle density of states, which, in a disordered Fermi liquid, can be different from the single particle density of states. The latter includes contributions from the incoherent part of the Green's functions, and is measured in dynamical experiments, such as tunneling.

[61] It should mentioned, for completeness, that additional non-analytic corrections to the density of states and the specific heat arise when the transverse electromagnetic interaction between the electrons is taken into account. See Reizer (1989) for details.

[62] In a clean electron liquid the scattering amplitudes γ^s and γ^a are given by $\frac{F_0^s}{1+F_0^s}$ and $\frac{F_0^a}{1+F_0^a}$, respectively.

ultimately fails (at least to leading order in $\frac{\hbar}{\epsilon_F \tau}$) to produce an effective hamiltonian in the low-energy sector. Nevertheless, the manner in which the theory breaks down is very suggestive. The divergence of γ^a implies a diverging spin susceptibility associated with the formation of ferromagnetic domains at the length scale of the divergence. At the same time the inverse conductance g is seen to initially increase, as if the system were going towards an insulating state, but it then reaches a maximum and begins to decrease as the divergence of γ^a takes over. These findings show very clearly that the noninteracting theory of localization, which is based on a single scaling variable g (see footnote [58]), needs serious revision in the presence of electron–electron interactions. In particular, the possibility of a metallic state in the disordered 2D electron liquid cannot be discounted.

In three dimensions one has a richer scenario, depending on the values of the running variables at the microscopic length scale. In the first scenario, associated with weak disorder, γ^a can have any value, while g tends to zero (infinite conductance in the thermodynamic limit). In the other scenario, associated with strong disorder, γ^a diverges at a finite length scale, and the system behaves qualitatively as in the two-dimensional case discussed above, leaving again open the possibility of a ferromagnetic instability. Unfortunately, the runaway behavior of γ^a has prevented, thus far, a proper treatment of the metal-insulator transition in 3D. The problem remains open (for a recent review, see Di Castro and Raimondi, 2004).

8.10.1 The quasiparticle lifetime

According to the Fermi golden rule the inverse lifetime of an electron initially occupying the exact one-electron eigenstate state $|\phi_\alpha\rangle$ is given by

$$\frac{1}{\tau_\alpha^{(e)}} = \frac{2\pi}{\hbar} \sum_{\beta\gamma\delta} \left| \frac{W_{\gamma\delta\beta\alpha}}{L^d} \right|^2 n_\beta (1 - n_\delta)(1 - n_\gamma)\delta(\varepsilon_\alpha + \varepsilon_\beta - \varepsilon_\gamma - \varepsilon_\delta), \qquad (8.253)$$

where

$$W_{\gamma\delta\beta\alpha} = \sum_{\vec{q}} W(\vec{q})(\hat{n}_{\vec{q}})_{\gamma\alpha}(\hat{n}_{-\vec{q}})_{\delta\beta} \qquad (8.254)$$

is the matrix element of the effective coulomb interaction between exact eigenstates $|\phi_\alpha\rangle$, $|\phi_\beta\rangle$ (initially occupied) and $|\phi_\gamma\rangle$, $|\phi_\delta\rangle$ (initially empty). Notice that Eq. (8.253) is the natural generalization of Eq. (8.80): as in Section 8.4 the occupation numbers n_α will be approximated by the noninteracting value $n_\alpha^{(0)}$ (i.e., $\Theta(\epsilon_F - \varepsilon_\alpha)$ at $T = 0$) and exchange processes will be ignored in a first approximation.

Due to the randomness of the system we are not so much interested in the lifetime of one particular state as in the *average decay rate* of all the states at a given energy $\varepsilon > \epsilon_F$. Thus, the relevant quantity is

$$\frac{1}{\tau_\varepsilon^{(e)}} = \frac{1}{N(0)L^d} \overline{\sum_\alpha \frac{1}{\tau_\alpha^{(e)}} \delta(\varepsilon - \varepsilon_\alpha)}, \qquad (8.255)$$

where $N(0)$ is the average density of states at the Fermi level, and the over bar denotes an average over realizations of disorder. Substituting Eq. (8.253) into Eq. (8.255) we come to the seemingly complicated expression

$$
\frac{1}{\tau_\varepsilon^{(e)}} = \frac{2\pi}{\hbar} \frac{1}{N(0)L^{3d}} \sum_{\vec{q}\vec{q}'} W(\vec{q})W(-\vec{q}')
$$

$$
\times \overline{\sum_{\alpha\gamma}(1 - n_\gamma^{(0)})(\hat{n}_{\vec{q}})_{\gamma\alpha}(\hat{n}_{-\vec{q}'})_{\alpha\gamma}\delta(\varepsilon - \varepsilon_\alpha)}
$$

$$
\times \overline{\sum_{\beta\delta} n_\beta^{(0)}(1 - n_\delta^{(0)})(\hat{n}_{-\vec{q}})_{\delta\beta}(\hat{n}_{\vec{q}'})_{\beta\delta}\delta(\varepsilon - \varepsilon_\gamma - \varepsilon_\delta + \varepsilon_\beta)}. \tag{8.256}
$$

A first major simplification occurs when we realize that the last two factors in this expression can be averaged independently, since they are related to the imaginary part of the density–density response function, which is a *self-averaging* quantity, that is, a quantity whose value does not depend on the particular realization of disorder (see footnote[30] in Chapter 3). The complete argument goes as follows. We first notice that the sum on the last line of Eq. (8.256) can be written as

$$
\frac{1}{L^d} \sum_{\beta\delta} n_\beta^{(0)}(1 - n_\delta^{(0)})(\hat{n}_{-\vec{q}})_{\delta\beta}(\hat{n}_{\vec{q}'})_{\beta\delta}\delta(\varepsilon - \varepsilon_\gamma - \varepsilon_\delta + \varepsilon_\beta)
$$

$$
= -\frac{1}{\pi}\Im m \chi_0^{imp}\left(q, \frac{\varepsilon - \varepsilon_\gamma}{\hbar}\right)\Theta(\varepsilon - \varepsilon_\gamma)\delta_{\vec{q}\vec{q}'}, \tag{8.257}
$$

where in the range $v_F q\tau \ll 1, \omega\tau \ll 1$

$$
\chi_0^{imp}(q, \Omega) \simeq -N(0)\frac{Dq^2}{-i\Omega + Dq^2}, \tag{8.258}
$$

(see Eq. (4.53)) is the density–density response function of the noninteracting system in the presence of disorder. Notice that there is no need for a bar over the left-hand side of Eq. (8.257), because we have assumed that the density–density response function is self-averaging. Furthermore, terms with $\vec{q} \neq \vec{q}'$ have been assumed to vanish since averaging over disorder restores translational invariance.

We now put Eq. (8.257) into Eq. (8.256) and notice that, due to the presence of the occupation factor $1 - n_\gamma^{(0)}$ and $\Theta(\varepsilon - \varepsilon_\gamma)$, the admissible values of the energy ε_γ range from ϵ_F to ε. Setting $\varepsilon - \varepsilon_\gamma = \hbar\Omega$ we write

$$
\frac{1}{\tau_\varepsilon^{(e)}} = -\frac{2}{\hbar L^{2d}} \frac{1}{N(0)} \sum_{\vec{q}} |W(\vec{q})|^2 \int_0^{\frac{\varepsilon - \epsilon_F}{\hbar}} d\Omega \Im m \chi_0^{imp}(q, \Omega)
$$

$$
\times \overline{\sum_{\alpha\gamma} |(\hat{n}_{\vec{q}})_{\gamma\alpha}|^2\delta(\varepsilon - \varepsilon_\alpha)\delta(\varepsilon - \hbar\Omega - \varepsilon_\gamma)}, \tag{8.259}
$$

where the second δ-function guarantees that $\varepsilon - \varepsilon_\gamma = \hbar\Omega$. In the final step of the calculation we evaluate the disorder average on the second line of the above equation. This seems a hopeless task until we realize that the expression under study is once again closely related

o $\Im m \chi_0^{imp}(q, \Omega)$. To see this rewrite Eq. (8.257) for $\vec{q} = \vec{q}'$ and $\varepsilon - \varepsilon_\gamma = \hbar\Omega > 0$ in the
orm

$$\frac{1}{L^d} \int_{\epsilon_F}^{\epsilon_F + \hbar\Omega} d\varepsilon \sum_{\beta\delta} |(\hat{n}_{\vec{q}})_{\beta\delta}|^2 \delta(\varepsilon - \varepsilon_\delta)\delta(\hbar\Omega - \varepsilon + \varepsilon_\beta) = -\frac{1}{\pi}\Im m\chi_0^{imp}(q, \Omega), \quad (8.260)$$

where the limits of integration ensure that $\varepsilon_\beta < \epsilon_F < \varepsilon_\delta$. In the limit of small Ω this tells
us that

$$\lim_{\varepsilon \to \epsilon_F} \sum_{\alpha\gamma} |(\hat{n}_{\vec{q}})_{\gamma\alpha}|^2 \delta(\varepsilon - \varepsilon_\alpha)\delta(\varepsilon - \hbar\Omega - \varepsilon_\gamma) \simeq -\frac{L^d}{\pi}\frac{\Im m\chi_0^{imp}(q, \Omega)}{\Omega}$$

$$\simeq \frac{N(0)L^d}{\pi}\frac{Dq^2}{\Omega^2 + (Dq^2)^2}$$

for $\epsilon_F < \varepsilon < \epsilon_F + \hbar\Omega$. Thus, we see that the seemingly innocent assumption of diffusion
implies a strong correlation between eigenstates with energies on opposite sides of the
Fermi level. Substituting Eq. (8.261) into Eq. (8.259) and making use of Eq. (8.257) we get

$$\frac{\hbar}{\tau_\varepsilon^{(e)}} = \frac{2}{N(0)\pi} \int \frac{d\vec{q}}{(2\pi)^d}|N(0)W(\vec{q})|^2 \int_0^{\frac{\varepsilon-\epsilon_F}{\hbar}} d\Omega\Omega \left(\frac{Dq^2}{\Omega^2 + (Dq^2)^2}\right)^2. \quad (8.261)$$

For a quick estimate of the lifetime let us use for $W(\vec{q})$ the statically screened coulomb
interaction $\frac{v_q}{\varepsilon(q,0)}$. For $\varepsilon \to \epsilon_F$ the leading contribution to the integral comes from the $q \simeq 0$
region where we can approximate $N(0)W(\vec{q}) \simeq 1$. The integrals in Eq. (8.261) are best
done with the help of the substitution $q = \sqrt{\frac{\Omega}{D}}x$ and one easily finds

$$\frac{\hbar}{\tau_\varepsilon^{(e)}} \simeq \frac{2(\varepsilon - \epsilon_F)^{\frac{d}{2}}}{d\pi^d(\hbar D)^{\frac{d}{2}}N(0)} \int_0^\infty dx\frac{x^{d+3}}{(1 + x^4)^2}, \quad d = 2, 3. \quad (8.262)$$

The value of the integral is $\frac{3\pi}{8\sqrt{2}}$ in three dimensions, and $\frac{\pi}{8}$ in two dimensions.[63]

The $(\varepsilon - \epsilon_F)^{\frac{d}{2}}$ behavior of the inverse lifetime is valid at zero temperature and finite
disorder.[64] At finite temperature an additional logarithmic singularity appears in two
dimensions, leading to $\frac{\hbar}{\tau^{(e)}} \sim (k_B T)|\ln T|$ (see Abrahams, 1981). The devoted reader may
want to attempt this calculation using the method outlined in this section.

8.10.2 *The density of states*

To calculate the density of states $N(\varepsilon)$ in the weak coupling approximation we start from
the expression

$$N(\varepsilon) = \frac{1}{L^d} \sum_\alpha \delta(\varepsilon - \varepsilon_\alpha - \Sigma_{\alpha\alpha}), \quad (8.263)$$

---

[63] A more careful treatment of the *dynamical* screening, using the frequency-dependent effective interaction $W(\vec{q}, \Omega) = \frac{v_q}{1 - v_q \chi_0^{imp}(q,\Omega)}$ gives slightly different values of the numerical factors, namely $\frac{\pi}{2\sqrt{2}}$ in three dimensions and $\frac{\pi}{4}$ in two dimensions (Exercise 26).

[64] Note that the coefficient of the anomalous lifetime vanishes in a clean system as the diffusion constant $D \to \infty$.

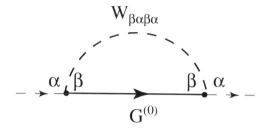

Fig. 8.24. Diagrammatic representation of the diagonal screened-exchange self-energy $\Sigma_{\alpha\alpha}$ for an electron in exact eigenstate $|\phi_\alpha\rangle$.

where $\Sigma_{\alpha\alpha}$ is the screened-exchange contribution to the self-energy of a particle in state $|\phi_\alpha\rangle$ (see Fig. 8.24):

$$\Sigma_{\alpha\alpha} = -\frac{1}{L^d} \sum_\beta n_\beta^{(0)} W_{\beta\alpha\beta\alpha}$$

$$= -\frac{1}{L^d} \sum_{\vec{q}} W(\vec{q}) \sum_\beta n_\beta^{(0)} |(\hat{n}_{\vec{q}})_{\alpha\beta}|^2 \,, \tag{8.264}$$

with $W(\vec{q}) = \frac{v_q}{\epsilon(q,0)}$. Expanding $N(\varepsilon)$ to first order in $\Sigma_{\alpha\alpha}$ we get

$$N(\varepsilon) \simeq N(0)\left[1 - \frac{\partial \Sigma_\varepsilon}{\partial \varepsilon}\right], \tag{8.265}$$

where Σ_ε – the average value of the exchange energy shift for states at energy ε – is defined as

$$\Sigma_\varepsilon = -\frac{1}{L^d} \sum_{\vec{q}} W(\vec{q}) \frac{1}{N(0)L^d} \overline{\sum_{\alpha\beta} n_\beta^{(0)} |(\hat{n}_{\vec{q}})_{\alpha\beta}|^2 \delta(\varepsilon - \varepsilon_\alpha)}$$

$$= -\frac{1}{L^d} \sum_{\vec{q}} W(\vec{q}) \frac{1}{N(0)L^d} \int_{-\infty}^{\epsilon_F} d\varepsilon' \overline{\sum_{\alpha\beta} n_\beta |(\hat{n}_{\vec{q}})_{\alpha\beta}|^2 \delta(\varepsilon - \varepsilon_\alpha)\delta(\varepsilon' - \varepsilon_\beta)} \,. \tag{8.266}$$

At this point we make use of Eq. (8.261) to write the average energy shift as

$$\Sigma_\varepsilon = -\frac{1}{L^d} \sum_{\vec{q}} W(\vec{q}) \int_{-\infty}^{\epsilon_F} \frac{d\varepsilon'}{\pi} \frac{\hbar^2 D q^2}{(\varepsilon - \varepsilon')^2 + (\hbar D q^2)^2} \,. \tag{8.267}$$

Taking the derivative with respect to ε and using Eq. (8.265) we get

$$\frac{\delta N(\varepsilon)}{N(0)} = -\int \frac{d\vec{q}}{(2\pi)^d} W(\vec{q}) \frac{\hbar^2 D q^2}{(\varepsilon - \epsilon_F)^2 + (\hbar D q^2)^2} \,. \tag{8.268}$$

Simple power counting shows that the correction to the density of states, which is negative, has a $\sqrt{|\varepsilon - \epsilon_F|}$ dependence in three dimensions, and diverges as $\ln|\varepsilon - \epsilon_F|$ in two dimensions (Altshuler and Aronov, 1985).[65] The logarithmic divergence is, of course, an

---

[65] A more careful treatment, including the dynamical screening of the interaction (i.e., the "coulomb-hole" contribution to the self-energy) leads to a stronger divergence, proportional to $\ln^2 |\varepsilon - \epsilon_F|$, in two dimensions.

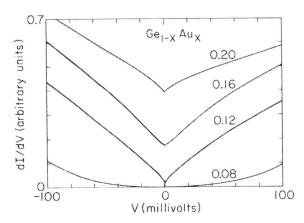

Fig. 8.25. Tunneling conductance $\frac{dI}{dV}$ of a three-dimensional $Ge_{1-x}Au_x$ alloys vs bias voltage for different values of x. The tunneling conductance, which is proportional to the density of states at $\varepsilon - \epsilon_F = eV$, exhibits a downward cusp at the Fermi level. From McMillan and Mochel (1981).

indication that the weak coupling approximation for the self-energy fails when $\varepsilon \to \epsilon_F$. This result is consistent with the already noticed "marginal stability" of the Landau Fermi liquid in two dimensions (see discussion after Eq. (8.140) and Exercise 27).

The anomalous behavior of the density of states in a three-dimensional disordered metal is clearly visible in Fig. 8.25 which shows measurements of the tunneling density of states in conducting germanium–gold alloys.

8.10.3 Coulomb lifetimes and weak localization in two-dimensional metals

Weakly disordered two-dimensional metals at very low temperature exhibit a decrease in conductivity with decreasing temperature due to a quantum interference effect that enhances the probability of a particle to retrace its own path. This phenomenon, which in itself has nothing to do with coulomb interactions, is known as *weak localization*. How strong the effect is depends entirely on the "effective size" of the system, namely, how far an electron can travel before losing phase coherence due to interactions with the environment and, in particular, interactions with other electrons. This effective size is given by $L_\phi = \sqrt{D\tau_\phi}$, the average distance an electron can diffuse during a "dephasing time" τ_ϕ. The conductivity is predicted to be given by

$$\sigma = \sigma_{cl} - \frac{e^2}{\pi^2 \hbar} \ln \frac{L_\phi}{\ell} , \tag{8.269}$$

where σ_{cl} is the Drude conductivity, and ℓ is a microscopic length scale of the order of the elastic mean free path. At high temperature the dephasing length approaches ℓ and the weak localization corrections are correspondingly small. The dephasing length increases, however, with decreasing temperature, and is theoretically expected to become infinite at $T = 0$. The weak localization effect grows accordingly causing the zero temperature conductivity to vanish for an infinite system.

The coulomb lifetime clearly offers an upper bound to the actual inelastic lifetime, which, in turn, is an upper bound to the dephasing time.[66] Since at low temperature the coulomb interaction is the primary dephasing mechanism, Eq (8.269) predicts that the conductivity should vary as $\ln T$, which is reasonably well confirmed by experiments. Notice that for a long time, (i.e., before the advent of the tunneling experiments, which directly probe the spectral function) Eq. (8.269) has been the only way to obtain information about quasiparticle lifetimes.

Exercises

8.1 **Compressibility, spin susceptibility and Landau interaction.** By making use of the results of Section 8.3 show that the ratios $\frac{K}{K_0}$ and $\frac{\chi_S}{\chi_P}$ can be expressed solely in terms of the Landau interaction functions as follows:

$$\frac{1}{1 + \frac{L^d N(0)}{2} \left(\langle f_{\uparrow\uparrow} \pm f_{\uparrow\downarrow} \rangle_{\ell=0} - \langle f_{\uparrow\uparrow} + f_{\uparrow\downarrow} \rangle_{\ell=1} \right)},$$

where the plus sign refers to the compressibility while the minus sign applies to the spin susceptibility. The notation $\langle \ldots \rangle_{\ell=n}$ stands for the $\ell = n$ angular average, the notation being the same as in Eqs. (8.14) and (8.15). $N(0)$ is the density of states of bare particles at the Fermi surface.

8.2 **Landau parameters and effective mass to first order.** Consider a system of fermions interacting via the pair potential $v(r) = e^2 \frac{e^{-\lambda r}}{r}$ (Fourier transform $v_{\vec{q}} = \frac{4\pi e^2}{q^2 + \lambda^2}$). The energy of the state that arises from the non-interacting state with momentum occupation numbers $\mathcal{N}_{\vec{k}\sigma}$ to first-order in the interaction strength is given by

$$E[\mathcal{N}_{\vec{k}\sigma}] = \sum_{\vec{k}\sigma} \frac{\hbar^2 k^2}{2m} \mathcal{N}_{\vec{k}\sigma} + \frac{1}{2V} \sum_{\vec{k}\sigma\vec{k}'\sigma'} [v_{\vec{0}} - v_{\vec{k}-\vec{k}'} \delta_{\sigma\sigma'}] \mathcal{N}_{\vec{k}\sigma} \mathcal{N}_{\vec{k}'\sigma'}. \tag{E8.1}$$

(a) Substitute $\mathcal{N}_{\vec{k}\sigma} = \mathcal{N}_{\vec{k}\sigma}^{(0)} + \delta\mathcal{N}_{\vec{k}\sigma}$ (where $\mathcal{N}_{\vec{k}\sigma}^0 = \Theta(k_F - k)$ are the ground-state occupation numbers) to obtain the Landau energy functional. Give explicit expressions for the quasiparticle energy and the Landau interaction function.

(b) Calculate the Landau parameter F_1^s and the effective mass of the quasiparticle. What happens for $\lambda \to 0$?

(c) Calculate the chemical potential μ from the energy of a quasi-particle at k_F.

(d) Calculate the Landau parameter F_0^s and the compressibility.

(e) Calculate the Landau parameter value F_0^a and the spin susceptibility. In the limit $\lambda \to 0$, at what value of the density do we predict a ferromagnetic instability (infinite spin susceptibility)?

---

[66] This arises from the fact that some elastic or quasi-elastic processes can lead to loss of coherence without contributing to the inelastic lifetime.

8.3 **General relation between g^*, m^* and χ_S.** Make use of the basic definition of the spin susceptibility and of Eq. (8.32) to arrive at the general result Eq. (8.34).

8.4 **Electron–phonon renormalizations for noninteracting electrons.** Make use of the electron–phonon self-energy, given in footnote[19] and the Dyson equation (8.134) to determine the phonon renormalizations for the mass and the g-factor in the case of noninteracting electrons.

8.5 **Electron–phonon renormalization of the anomalous g-factor.** Make use of Eqs. (8.31) and (8.32) to derive the result of Eq. (8.47).
[Hint: notice that, at variance with what happens in Eqs. (8.18) and (8.31), in this case $\mathcal{E}^{(el-ph)}_{\vec{k}\sigma}$ (see Eq. (8.44)) does contribute the extra term $-\lambda(g^*\mu_B B)$ to the energy.]

8.6 **A simple model for zero-sound.** Consider the linearized transport Eq. (8.56) with the classical force term (8.58). Assuming that the Landau interaction function has the form $f_{\vec{k}\sigma,\vec{k}'\sigma'} = f_0$, independent of wave vectors and spin orientations, that there is no external field, and that quasiparticle collisions are negligible, show that the kinetic equation always admits the zero-sound solution

$$\delta\mathcal{N}_{\vec{k}\sigma}(\vec{q},\omega) \propto \delta(\mathcal{E}_{\vec{k}\sigma} - \mu)\frac{f_0}{L^d}\frac{\cos\theta}{\frac{\omega}{v_{Fq}} - \cos\theta}$$

where θ is the angle between $\vec{k}$ and $\vec{q}$, and $\frac{\omega}{v_{Fq}} \equiv x$ is the solution of the equation

$$\frac{x}{2}\ln\frac{x+1}{x-1} - 1 = \frac{1}{N(0)f_0},$$

with $x > 1$.

8.7 **Quasiclassical quasiparticle hamiltonian.** Verify the form of the quasiclassical quasiparticle hamiltonian of Eq. (8.59).

8.8 **Transport equation, gauge-invariance and number conservation.** Prove that the linearized kinetic equation (8.56) with the force given by Eq. (8.60) is gauge-invariant and satisfies the continuity equation (7.217), with the current density given by

$$\vec{j}(\vec{r},t) = \sum_{\vec{k}\sigma}\frac{\hbar\vec{k}}{m}\mathcal{N}_{\vec{k}\sigma}(\vec{r},t) + \frac{ne}{mc}\vec{A}(\vec{r},t).$$

8.9 **Quasihole lifetime.** Verify that the quasihole lifetime is obtained from the quasiparticle lifetime through a change of sign and the replacements $1 - n_F(\varepsilon_{\vec{k}\sigma} - \hbar\omega - \mu) \to n_F(\varepsilon_{\vec{k}\sigma} - \hbar\omega - \mu)$ and $1 - e^{-\beta\hbar\omega} \to 1 - e^{\beta\hbar\omega}$.

8.10 **Calculation of the quasiparticle lifetime.** Fig. 8.26 shows an ingenious scheme to simplify the calculation of the six-dimensional integral (8.80) for the quasiparticle lifetime in 3D (Abrikosov and Khalatnikov, 1959).

The integral over $\vec{k}'$ is done in polar coordinates with polar axis along the direction of $\vec{k}$:

$$\frac{1}{L^3}\sum_{\vec{k}'}\cdots = \frac{1}{(2\pi)^3}\int dk'k'^2\sin\theta d\theta d\phi'\cdots$$

The integral over $\vec{q}$ is transformed into an integral over $\vec{k}_f \equiv \vec{k} - \vec{q}$ and is done in cylindrical coordinates with the cylinder axis along the direction of the conserved

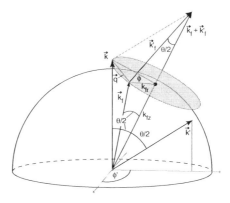

Fig. 8.26. Calculation of the quasiparticle lifetime in 3D.

total wave vector $\vec{k} + \vec{k}'$:

$$\frac{1}{L^3} \sum_{\vec{k}_f} \cdots = \frac{1}{(2\pi)^3} \int dk_{fz} dk_{fr} k_{fr} d\phi \,,$$

where k_{fz} and k_{fr} are, respectively, the "axial" and the "radial" components of $\vec{k}_f$ (see Fig. 8.26). Finally, the integrals over the *magnitude* of $\vec{k}'$, k_{fz}, and k_{fr} are replaced by integrals over the energies ε', ε_f, and ε'_f, subject to the constraint of energy conservation.

(a) Taking advantage of the fact that all the wave vectors are close in magnitude to the Fermi wave vector k_F show that

(i) $q \simeq 2k_F \sin \frac{\theta}{2} \sin \frac{\phi}{2}$;

(ii) $k_{fr} \simeq k_F \sin \frac{\theta}{2}$;

(iii) $\begin{cases} d\varepsilon' = \frac{\hbar^2 k_F}{m} dk' \\ d\varepsilon_f = \frac{\hbar^2 k_F}{m} \left(dk_{fz} \cos \frac{\theta}{2} + dk_{fr} \sin \frac{\theta}{2} \right) \\ d\varepsilon'_f = \frac{\hbar^2 k_F}{m} \left(-dk_{fz} \cos \frac{\theta}{2} + dk_{fr} \sin \frac{\theta}{2} \right) \end{cases}$.

(b) Making use of (i)–(iii) show that the integral for the quasiparticle lifetime can be rewritten as the product of a term that depends only on the angular average of the interaction, and a term that depends on the energy of the quasiparticle, ε, and the temperature T:

$$\frac{1}{\tau_{\vec{k}\sigma}^{(e)}} = \frac{1}{16\hbar \epsilon_F} \int_0^\pi d\theta \sin \frac{\theta}{2} \int_0^{2\pi} d\phi \left| \bar{W} \left(2k_F \sin \frac{\theta}{2} \sin \frac{\phi}{2} \right) \right|^2$$
$$\times I(\varepsilon, T) \,,$$

where $\bar{W} \equiv N(0)W$ and

$$I(\varepsilon, T) \equiv \int_{-\infty}^\infty d\varepsilon_1 \int_{-\infty}^\infty d\varepsilon_2 \int_{-\infty}^\infty d\varepsilon_3 n_F(\varepsilon_1) n_F(\varepsilon_2) n_F(\varepsilon_3)$$
$$\times \delta(\varepsilon + \varepsilon_1 + \varepsilon_2 + \varepsilon_3) \,.$$

[Hint: use the identity $n_F(\varepsilon) = 1 - n_F(-\varepsilon)$.]

.11 **A beautiful integral.** With reference to the previous exercise show that

$$I(\varepsilon, T) = \frac{(\varepsilon - \mu)^2 + (\pi k_B T)^2}{2(1 + e^{-\frac{\varepsilon - \mu}{k_B T}})}.$$

[Hint: write the δ-function in the form $\delta(x) = \frac{1}{2\pi} \int_{-\infty}^{\infty} dt\, e^{itx}$ and use complex plane integration to show that $\int_{-\infty}^{\infty} \frac{dx}{2\pi i} \frac{e^{ix(t-i\eta)}}{e^x + 1} = -\frac{1}{2 \sinh \pi (t - i\eta)}.$]

This result shows that the temperature dependence of the quasiparticle lifetime can be deduced from its energy dependence, and vice versa. Discuss why this simple and elegant result does not hold in two and one dimensions.

8.12 **Quasiparticle lifetime in 1D.** Making use of Eq. (8.82) and of the expression (4.45) for the imaginary part of the Lindhard function in one dimension, show that the inverse lifetime of a putative electron quasiparticle in a strictly one-dimensional system at $k = k_F$ and finite temperature would be proportional to

$$\int_0^{\infty} \frac{dq}{q} |W(q)|^2 \left(\frac{1}{\cosh \beta \hbar \omega_-(q)} + \frac{1}{\cosh \beta \hbar \omega_+(q)} \right),$$

where $\omega_{\pm}(q)$ are the boundaries of the electron–hole continuum defined in Eq. (4.34). This integral diverges at the lower limit.

8.13 **Coulomb drag.** For the bilayer system of Exercise 4 calculate the rate of momentum transfer between the layers, assuming that one layer carries a current $I = \frac{e\hbar k_D}{mL}$ and is described by the non-equilibrium distribution function $\Theta(k_F - |\vec{k} + \vec{k}_D|)$, while the other carries no current and is described by the equilibrium distribution $\Theta(k_F - k)$ (work to first order in k_D). Also determine the magnitude of the electric field that must be present in the layer that carries no current in order to keep the electrons from drifting. This phenomenon is known as *coulomb drag*.

8.14 **Spectral function and retarded Green's function.** Verify the correctness of Eq. (8.127) for the spectral function.

8.15 **On the renormalization constant.** Show that the renormalization constant Z can be interpreted as the square of the overlap between the $N + 1$-electron ground-state and the state obtained by adding an electron with $k = k_F$ to the N-electron ground-state.

8.16 **Weak coupling limit and GW self-energy.** Prove that in the weak coupling limit the imaginary part of the GW self-energy (Eq. (8.230)) reduces to

$$|\Im m \, \Sigma^{ret}(\vec{k}\sigma, \varepsilon_{\vec{k}\sigma})| \simeq \frac{\pi}{\hbar^2} \int \frac{d\vec{q}}{(2\pi)^d} |W(\vec{q})|^2$$

$$\times \left[\left(1 - n_{\vec{k}\sigma}^{(0)} \right) S_0 \left(\vec{q}, \frac{\varepsilon_{\vec{k}\sigma} - \varepsilon_{\vec{k}-\vec{q}\sigma}}{\hbar} \right) + n_{\vec{k}\sigma}^{(0)} S_0 \left(-\vec{q}, \frac{\varepsilon_{\vec{k}-\vec{q}\sigma} - \varepsilon_{\vec{k}\sigma}}{\hbar} \right) \right], \qquad \text{(E8.2)}$$

where

$$S_0(\vec{q}, \omega) = \frac{1}{L^d} \sum_{\vec{k}'\sigma'} n_{\vec{k}'\sigma'}^{(0)} \left(1 - n_{\vec{k}'+\vec{q}\sigma'}^{(0)} \right) \delta \left(\omega - \frac{\varepsilon_{\vec{k}'+\vec{q}\sigma'} + \varepsilon_{\vec{k}'\sigma'}}{\hbar} \right) \qquad \text{(E8.3)}$$

is the dynamic structure factor of the noninteracting electron gas, and $W(\vec{q})$ is the RPA screened interaction. Then show that the two terms in the square brackets of

Eq. (E8.2) yield the Fermi golden rule expressions for the inverse lifetimes of
quasielectron and a quasihole respectively (see Section 8.4).
[Hint: use the fluctuation dissipation theorem $-\frac{\Im m \chi_0(\vec{q},\omega)}{1-e^{-\beta\hbar\omega}} = \frac{\pi}{\hbar} S_0(\vec{q},\omega)$. Note the
identity: $n_F(-\varepsilon) + n_B(\omega) = \frac{1-n_F(\varepsilon)}{1-e^{-\beta\hbar\omega}} - \frac{n_F(\varepsilon)}{1-e^{\beta\hbar\omega}}$.]

8.17 **Exchange contribution to the quasiparticle lifetime in 3D, I.** Show that the exchange contribution to the quasiparticle decay rate can be obtained from a simple
modification of Eq. (E8.2): what is it? (Notice that this implies that the direct and
exchange contributions to the decay rate have the same energy and temperature
dependence.)

8.18 **Exchange contribution to the quasiparticle lifetime in 3D, II.** An alternative
approach to the calculation of this quantity in 3D is the following one, due to Penn
(1980).

(a) With reference to Fig. 8.9, the sums over $\vec{k}'$ and $\vec{q}$ in Eq. (8.106) are replaced by
sums over the final wave vectors $\vec{k}_f \equiv \vec{k} - \vec{q}$ and $\vec{k}'_f \equiv \vec{k}' + \vec{q}$. Making use of the fact
that all the wave vectors are close to the Fermi surface, the sums over $\vec{k}_f$ and $\vec{k}'_f$ can
be converted into integrals over the final energies ε_{k_f} and $\varepsilon_{k'_f}$ (both running from ϵ_F
to ε_k) and over the directions of $\vec{k}_f$ and $\vec{k}'_f$. To this end, write the momentum transfers
$\vec{q}$ and $\vec{k} - \vec{k}' - \vec{q}$ approximately as $\sqrt{2(1 - \cos\theta_f)}k_F$ and $\sqrt{2(1 - \cos\theta'_f)}k_F$, where
θ_f is the angle between $\vec{k}_f$ and $\vec{k}$, and θ'_f is the angle between $\vec{k}'_f$ and $\vec{k}$. Show that
the δ-function for energy conservation, i.e. $\delta\left(\varepsilon_{k_f} + \varepsilon_{k'_f} - \varepsilon_k - \varepsilon_{|\vec{k}_f + \vec{k}'_f - \vec{k}|}\right)$, takes the
form

$$\frac{1}{2\epsilon_F}\delta\left([1 - \cos\theta_f][1 - \cos\theta'_f] + \sin\theta_f \sin\theta'_f \cos\phi\right) ,$$

where ϕ is the azimuthal angle between the directions of $\vec{k}_f$ and $\vec{k}'_f$. Perform analytically the integral over ϕ, and those over ε_{k_f} and $\varepsilon_{k'_f}$ to arrive at

$$\frac{1}{\tau^{(e)}_{x,\vec{k}\sigma}} = \frac{\left(\varepsilon_{\vec{k}} - \epsilon_F\right)^2}{8\sqrt{2}\hbar\epsilon_F} \int_{-1}^{+1} d(\cos\theta_f) \int_{\cos\theta_f}^{+1} d(\cos\theta'_f)$$

$$\times \frac{\bar{W}\left(\sqrt{2(1 - \cos\theta_f)}k_F\right)\bar{W}\left(\sqrt{2(1 - \cos\theta'_f)}k_F\right)}{\sqrt{(1 - \cos\theta_f)(1 - \cos\theta'_f)}|\cos\theta_f + \cos theta'|} ,$$

where $\bar{W}(q) \equiv N(0)W(q)$.

(b) Make the change of variables $y = \sqrt{\frac{1-\cos\theta_f}{2}}$, $\sin\gamma = \sqrt{\frac{1-\cos\theta'_f}{1+\cos\theta_f}}$ to reduce the
above expression to the simpler form

$$\frac{1}{\tau^{(e)}_{x,\vec{k}}} \approx \frac{\left(\varepsilon_{\vec{k}} - \epsilon_F\right)^2}{2\hbar\epsilon_F} \int_0^1 dy \int_0^{\pi/2} d\gamma \, \bar{W}(2k_F y) \, \bar{W}\left(2k_F\sqrt{1 - y^2}\sin\gamma\right) . \quad \text{(E8.4)}$$

(c) What happens if one attempts to apply this approach to the two-dimensional case?

8.19 **Line and residue decomposition.** An interesting alternative to the screened exchange–coulomb hole decomposition of the G0W self-energy is provided by the so-called *line and residue decomposition* (Quinn and Ferrell, 1958). The idea is to write the G0W self-energy, Eq. (8.226), as the sum of two terms

$$\Sigma_{GOW}(\vec{k}, \omega) = \Sigma_{line}(\vec{k}, \omega) + \Sigma_{res}(\vec{k}, \omega),$$ (E8.5)

where Σ_{line} is obtained by performing the analytic continuation $i\omega_n \to \omega - \frac{\mu}{\hbar} + i\eta$ *before* carrying out the Matsubara sum in Eq. (8.226), and the remainder, Σ_{res}, corrects for the non-commutativity of these two operations (see Appendix 14).

Show that Σ_{line} and Σ_{res} are given by

$$\Sigma_{line}(\vec{k}, \omega) = -\int \frac{d\vec{q}}{(2\pi)^d} \frac{1}{\beta} \sum_m \frac{W(q, i\Omega_m)}{\omega - i\Omega_m - \frac{\tilde{\varepsilon}_{\vec{k}-\vec{q}\sigma}}{\hbar}},$$ (E8.6)

and

$$\Sigma_{res}(\vec{k}, \omega) = -\int \frac{d\vec{q}}{(2\pi)^d} W\left(q, \omega - \frac{\tilde{\varepsilon}_{\vec{k}-\vec{q}\sigma}}{\hbar}\right)$$
$$\times \left[n_F(-\tilde{\varepsilon}_{\vec{k}-\vec{q}\sigma}) + n_B(\hbar\omega - \tilde{\varepsilon}_{\vec{k}-\vec{q}\sigma})\right],$$ (E8.7)

and prove that $\Sigma_{line}(\vec{k}, \omega)$ is purely real.

8.20 **Pseudo-hamiltonian for a spin-polarized electron liquid.** Generalize the pseudo-hamiltonian approach of Section 8.6 to the case of a spin-polarized system.

8.21 **Fröhlich interaction.** Consider the hamiltonian

$$\hat{H} = \hat{a}^{\dagger}_{\vec{k},\sigma} \hat{a}_{\vec{k}\sigma} + \sum_{\vec{q}} \hat{b}^{\dagger}_{\vec{q}} \hat{b}_{\vec{q}} (\hbar\omega_{\vec{q}} + \frac{1}{2})$$
$$- i \sum_{\vec{k},\vec{q},\sigma} A_{\vec{q}} \left(\hat{b}_{\vec{q}} + \hat{b}^{\dagger}_{-\vec{q}}\right) \hat{a}^{\dagger}_{\vec{k}+\vec{q}\sigma} \hat{a}_{\vec{k}\sigma},$$ (E8.8)

where the $\hat{b}$'s are phonon destruction operators. $\hat{H}$ describes non-interacting electrons coupled to a phonon bath. Following the procedure employed in Section 8.6.2, derive an effective electronic hamiltonian to second order in the electron–phonon coupling $A_{\vec{q}}$ (the phonon degrees of freedom having been averaged out by taking the expectation value of the suitably transformed hamiltonian over the phonon ground-state). The result is the famed phonon-induced effective electron–electron interaction due to Fröhlich.

8.22 **Effective mass in the Thomas-Fermi approximation.** The behavior of the effective mass in the high density limit can be extracted from a simple calculation of the exchange energy, in which the bare coulomb interaction is replaced by the

Thomas-Fermi screened interaction. Show that this leads to the formula

$$\frac{m}{m^*} = 1 + \frac{1}{2\pi^{3-d}} \int \frac{d\Omega_d}{\Omega_d} \bar{W}\left(\sqrt{2(1-\cos\theta)}k_F\right),$$

where $\bar{W}(q)$ is the Thomas-Fermi screened interaction, given by Eqs. (5.35) and (5.37), times the density of states at the Fermi surface, and Ω_d is the solid angle in $d = 3$ and 2 dimensions. Verify that for $r_s \ll 1$ one has

$$\frac{m}{m^*} \simeq 1 - c_d \frac{\alpha_d r_s}{\pi} \ln r_s + \mathcal{O}(r_s),$$

where α_d is defined in Eq. (1.79) and $c_3 = \frac{1}{2}$, $c_2 = \frac{1}{\pi}$.

8.23 **On-shell approximation.** Verify that the on-shell effective mass derived from Eq. (8.248) is given by Eq. (8.249). Also show that the on-shell effective mass and the Dyson effective mass (Eq. (8.146)) coincide in the weak coupling limit.

8.24 **Self-energy and plasmon emission.** Calculate the value of the wave vector k_p at which $\Im m \, \Sigma(k, \varepsilon_k)$ has an upward kink in the three-dimensional electron liquid. [Hint: this is the smallest value of k for which a quasiparticle can make a transition to the Fermi surface by emitting a plasmon.]

8.25 **Superconductivity in the electron liquid.** Show that the superconducting transition temperature of an electron liquid is determined by the solution of the self-consistent equations shown diagrammatically in Fig. 8.23. [Hint: start from the diagrams of Fig. 6.33 and recall that the anomalous self-energy is infinitesimal just below T_c.]

8.26 **On the coulomb lifetime in a disordered electron liquid.** Repeat the calculation of the lifetime of a quasiparticle in a disordered electron liquid (Section 8.10.1) with the dynamically screened interaction of footnote[63] and show that the result has the form of Eq. (8.262), but with different numerical factors.

8.27 **Admissible behavior of the self-energy in a Landau Fermi liquid.** (a) Show that if, for $\omega \to \frac{\mu}{\hbar}$, $\Im m \, \Sigma^{ret}(k_F, \omega) \simeq \left(\omega - \frac{\mu}{\hbar}\right)^\alpha$, with $\alpha < 1$, then also $\Re e[\Sigma^{ret}(k_F, \omega) - \Sigma^{ret}(k_F, \frac{\mu}{\hbar})] \simeq \left(\omega - \frac{\mu}{\hbar}\right)^\alpha$ in the same limit. [Hint: make use of the dispersion relation (6.80) for the retarded self-energy.] (b) Show that, under the conditions stated in (a), the renormalization constant Z vanishes: hence one cannot have a Landau Fermi liquid in which the inverse of the quasiparticle lifetime vanishes, for $\mathcal{E} \to \mu$, more slowly than the quasiparticle energy $\mathcal{E} - \mu$. (c) Show that the case $\alpha = 1$ represents a marginal situation since Z vanishes logarithmically for $\omega \to \frac{\mu}{\hbar}$. (d) Assuming that $\alpha > 1$ and that the imaginary part of the self-energy vanishes at high frequency, what is the behavior of $\Re e[\Sigma^{ret}(k_F, \omega) - \Sigma^{ret}(k_F, \frac{\mu}{\hbar})]$ for $\omega \to \frac{\mu}{\hbar}$?

9

Electrons in one dimension and the Luttinger liquid

9.1 Non-Fermi liquid behavior

In this chapter we begin the study of electronic systems that are not Landau Fermi liquids. These systems are like totalitarian societies in which the behavior of the individual is subordinated to the needs of the organization: their low-energy excitations are collective, rather than single-particle-like. A strongly collective behavior is not at all unusual in condensed matter. For example, the low energy excitations of a crystal lattice are acoustic phonons, which are collective oscillations of the atoms about their equilibrium positions. However, such examples are usually associated with a broken symmetry (translational symmetry in this case): when it comes to homogeneous electron liquids, the familiar picture of Landau's quasiparticles is so ingrained that we tend to regard any departure from it as a surprising phenomenon. Nevertheless, departures from the normal Fermi liquid pattern can and do occur in two typical scenarios:

(1) In three- and two-dimensional systems in which the electron liquid is *strongly correlated*, i.e., when the order of magnitude of the coulomb interaction greatly exceeds the kinetic bandwidth,

(2) In quasi-one-dimensional systems, for any strength of the interaction.

A trivial example of the first scenario is offered by the three-dimensional electron liquid at very low-density. In this case the electrons form a Wigner crystal and their collective behavior arises immediately from the loss of translational symmetry.[1] A much more complex example is offered by the two-dimensional electron liquid at high magnetic field. Because the kinetic energy is effectively suppressed by the magnetic field, the structure of the ground-state and the low-lying excitations is entirely controlled by the coulomb interaction. As a result, this system presents the most radical departures from Landau Fermi liquid theory so far encountered in any condensed matter system. Its unusual properties will be discussed in detail in Chapter 10.

The second scenario is realized in quasi-one dimensional metallic systems such as Bechgaard salts, TTF-TCNQ, and carbon nanotubes (see Chapter 1). The reduced effective dimensionality of these systems hinders single particle motion to the point that the electrons must be regarded as strongly correlated even when their interactions are weak. The

---

[1] A fascinating question is in this context whether the collective behavior sets in *before* crystallization, when the system is still in the homogeneous phase. The answer is not known at this time.

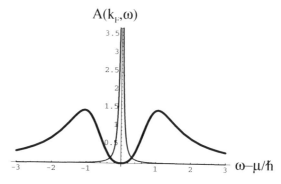

Fig. 9.1. Characteristic behavior of the zero-temperature spectral function $A(k_F, \omega)$ for a Luttinger liquid in the weak coupling regime (thin line) and in the strong coupling regime (thick line). In both cases $A(k_F, \omega) \sim \frac{1}{|\omega - \frac{\mu}{\hbar}|^\nu}$ for $\omega \to \frac{\mu}{\hbar}$, with the exponent ν tending to 1 for weak interactions and becoming negative for strong interactions. Notice the absence of the quasiparticle δ-function peak at $\omega = \frac{\mu}{\hbar}$.

ensuing highly collective state of the electron liquid is known as the *Luttinger liquid* and the study of its peculiar properties is the main object of this chapter.

An interesting phenomenon, known by the colorful name of *orthogonality catastrophe*, lies at the heart of the distinction between Fermi liquids and non Fermi liquids. Imagine injecting an extra electron into the ground-state of a strongly correlated N-electron system. Since the new electron lacks the appropriate correlations with the pre-existing electrons, the state of the $N + 1$-electron system after the injection is essentially orthogonal to the ground-state. In mathematical terms, the orthogonality catastrophe implies that the renormalization constant Z, defined in Section 8.5.1 as the square of the overlap between the state of the system immediately after the injection of an electron and the ground-state of that system, vanishes in the thermodynamic limit. We take this to be the defining feature of a non-Fermi liquid state.

An immediate consequence of this situation is the disappearance of the quasiparticle δ-function peak in the spectral function $A(k_F, \omega)$ at the chemical potential: there are no single-electron quasiparticles. For weak interactions the δ-function peak is replaced by a power-law divergence for $\omega \to \frac{\mu}{\hbar}$. With increasing coupling strength a sort of energy gap develops, whereby $A(k_F, \omega)$ vanishes with a power law for $\omega \to \frac{\mu}{\hbar}$ as shown in Fig. 9.1 (this situation is reminiscent of the "coulomb gap" discussed in Section 2.2.4). Later on, we will see that the position of the lateral maxima in the spectral function is a rough measure of the energy of the disturbance created by the injection of the new electron in the liquid, while the "width" of these maxima is inversely proportional to the time needed for the many-electron system to adjust to the presence of the new electron.[2] Another consequence of the vanishing of Z is that the plane wave occupation number n_k is no longer discontinuous at $k = k_F$, even though a singularity persists in its derivatives with respect to k.

---

[2] The process is analogous to the reconstruction of a crystal after the introduction of an interstitial defect.

Non-Fermi liquid states often exhibits anomalous transport properties. For example, the electrical conductivity of an interacting quasi-one dimensional system is expected to vanish at zero temperature. One might find this unsurprising since it is known that in a one dimensional system any amount of disorder causes localization of the one-electron states, and hence a vanishing conductivity at $T = 0$. But, in the Luttinger liquid any perturbation that breaks translational invariance, e.g., *even a single impurity*, leads to an insulating state at $T = 0$. The physical reason for this effect is that the perfectly clean system is on the verge of spontaneously forming a charge density wave (CDW) of wave vector $2k_F$. Under these conditions even a single impurity can pin down an insulating CDW state. This phenomenon will be described in detail in Section 9.11.

Disentangling the Luttinger liquid behavior from disorder-related localization effects is an extremely difficult task, which requires very high levels of purity and control of the system geometry. These requirements are just beginning to be met: for example, it is now possible to experimentally study the conductance of a *quantum point contact*, e.g., a single scattering center placed in the middle of perfectly clean one-dimensional channel (see Section 9.11). Yet, to date the best evidence in support of the Luttinger liquid paradigm has come from experiments on quasi-one-dimensional conduction channels that are formed at the edge of a two-dimensional electron liquid at high magnetic field. In these *edge Luttinger liquids* disorder is rendered irrelevant by the chirality of the electron motion: the relevant experiments will be discussed in Chapter 10.

In this chapter we introduce and solve a simple model, known as the *Luttinger model*, which exhibits all the interesting features of the general Luttinger liquid paradigm. In fact, one may say that the Luttinger model is to the Luttinger liquid concept what the non-interacting Fermi liquid is to the normal Landau Fermi liquid. These two exactly solvable models serve as archetypes for two qualitatively different classes of systems – Fermi liquids and Luttinger liquids: a perturbative connection exists only between two systems in the same class, but not between two systems in different classes.

9.2 The Luttinger model

The Luttinger model (Luttinger, 1963) was introduced, building on previous work by Tomonaga (Tomonaga, 1950), to describe the behavior of an interacting Fermi liquid in one dimension. The correct solution of the model was first obtained by Mattis and Lieb (1965) and further developed by Luther and Peschel (1974). Finally, the concept of Luttinger liquid was codified by Haldane (Haldane, 1981) in a seminal paper, on which our presentation is largely based.[3]

To arrive at the Luttinger model let us start with the ground-state of a non-interacting system of ordinary fermions with parabolic dispersion $\epsilon_k = \frac{\hbar^2 k^2}{2m}$. Since we are in one dimension the wave vector k is now just a number that can be positive or negative. In the non-interacting ground-state the plane wave states with $|k| < k_F$ are occupied while those

---

[3] For recent reviews see Schulz, Cuniberti, and Pieri, 1998 and Voit, 1994.

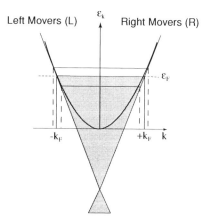

Fig. 9.2. The Luttinger liquid model is obtained by linearizing the one-electron dispersion in a region of wave vectors about $\pm k_F$. The shaded region indicates the states that are occupied in the "vacuum". The dashed lines define the "low-energy sector" of the Hilbert space.

with $|k| > k_F$ are empty. We shall, for the time being, ignore the (considerable) complication of the spin and assume that the fermions are spinless so that the relation between the Fermi wave vector k_F and the density is $n = \frac{k_F}{\pi}$. Let us now slowly turn on an interaction between the Fermions. At this stage we do not want to specify the nature of the interaction (that is, short-range vs long range, etc.) but we do assume that it is translationally invariant. As a result of the interaction several electron-hole pairs (with vanishing total momentum) get mixed into the ground-state. From perturbation theory we see that the most strongly affected states are the ones that lie within an energy V of the Fermi level, where V is the order of magnitude of the interaction. Assuming that $V \ll \epsilon_F$ we see that within the important region $|\epsilon_k - \epsilon_F| < V \ll \epsilon_F$ the parabola of the single particle energies can be approximated by two straight lines tangent to it, as shown in Fig. 9.2. The two linear dispersion relations $\epsilon_{k,R} = \epsilon_0 + \hbar v_F k$ and $\epsilon_{k,L} = \epsilon_0 - \hbar v_F k$ (with $\epsilon_0 = \epsilon_F - \hbar v_F k_F$) accurately describe the energies of the single particle states near the one-dimensional "Fermi surface", which consists of just the two points $k = \pm k_F$. States in the "R" branch have positive velocity and are therefore called *right movers*, while states in the "L" branch have negative velocity and are called *left movers*. It is clear that this model provides an accurate description of the one dimensional Fermi liquid only in a narrow band of wave vectors $||k| - k_F| < \Lambda$ about the Fermi surface. The arbitrary ultraviolet cutoff $\Lambda \ll k_F$ defines what will be called the *low-energy sector* of the Hilbert space.

Due to the presence of infinitely many states with arbitrarily large negative energies, the model, as it now stands, does not even admit a ground-state. Since there are no negative energy states in the physical system, one must find a way to exclude them from the model too. This is accomplished by an artifice first employed by Dirac (1928) to cure the problem of the negative energy solutions of the wave equation for a relativistic electron. We imagine that the "*vacuum state*" consists of an infinite Fermi sea in which all the right movers with $k < k_F$ and the left movers with $k > -k_F$ are occupied, while the other states are empty.

Thus the momentum occupation numbers of the "vacuum" are

$$n_{k,\tau}^{(0)} = \Theta(k_F - \tau k) , \qquad (9.1)$$

where $\tau = +1$ for right movers (also denoted by "R"), and $\tau = -1$ for left movers ("L").

It is evident that the so-called vacuum state actually contains an infinite number of particles, but this does not pose any problem, since we are only interested in *excitations* from the vacuum, and these excitations only involve a finite number of particles. Excited states are constructed by transferring an arbitrary number of particles from states lying below the Fermi energy to states lying above it. It is also possible to construct states with $N + n$ particles (where n is a positive integer) by putting $n + m$ particles above the Fermi level and m holes below the Fermi level, where m is a non negative integer. For example, the ground-state of the non-interacting $N + n$-particle system is obtained by putting n particles into the n lowest energy states above the Fermi level and no holes below it. Similarly, states of the $N - n$ particle system are generated by adding m electrons and $n + m$ holes. The key point is that all the low-lying states – the states that are most strongly affected by interactions – involve only particles in the immediate vicinity of the Fermi level. But this is precisely the region of momentum space in which our linearization of the original parabolic dispersion is accurate.[4]

Let us now take a closer look at the structure of the Hilbert space and the form of the main observables. To this end, it is convenient to introduce the *normal-ordered* number operators

$$\hat{N}_{k,\tau} \equiv\, : \hat{a}_{k,\tau}^{\dagger} \hat{a}_{k,\tau} :\, , \qquad (9.2)$$

which measure the number of particles *relative to the vacuum*. The creation and destruction operators for right and left movers, $\hat{a}_{k,\tau}^{\dagger}$ and $\hat{a}_{k,\tau}$, satisfy the usual fermionic anticommutation relations

$$\{\hat{a}_{k,\tau}, \hat{a}_{k',\tau'}\} = \{\hat{a}_{k,\tau}^{\dagger}, \hat{a}_{k',\tau'}^{\dagger}\} = 0 ,$$
$$\{\hat{a}_{k,\tau}, \hat{a}_{k',\tau'}^{\dagger}\} = \delta_{kk'}\delta_{\tau\tau'} , \qquad (9.3)$$

and the normal ordering (see Appendix 3) is done with respect to the "vacuum" defined above. The eigenvalues of $\hat{N}_{k,\tau}$ are the *relative* occupation numbers

$$N_{k,\tau} = n_{k,\tau} - n_{k,\tau}^{(0)} , \qquad (9.4)$$

and can take values $+1, 0$ (for electrons) or $-1, 0$ (for holes).

With these definitions, it is easy to see that the normal-ordered non-interacting hamiltonian can be written as

$$\hat{H}_0 = \hbar v_F \sum_{k,\tau} (\tau k - k_F)\hat{N}_{k,\tau} , \qquad (9.5)$$

---

[4] Notice that by introducing two types of operators for each value of k we have apparently *doubled* the number of degrees of freedom of the original physical problem. For example, at a given $k > 0$ we have both a right mover and a left mover whereas the physical parabolic dispersion admits only a right mover. Under closer inspection, however, we see that only right movers with $k \sim k_F$ and left movers with $k \sim -k_F$ are close to the Fermi level. Thus, there are no unphysical degrees of freedom in the low-energy sector of the Hilbert space.

where the single-particle energies $\hbar v_F(\tau k - k_F)$ have been defined so as to vanish at $k = \tau k_F$. Eigenstates of $\hat{H}_0$ can be written as $|\{N_{k,\tau}\}\rangle$, where $\{N_{k,\tau}\}$ denotes a set of relative occupation numbers of free-particle states, and the corresponding eigenvalues are

$$E_0[\{N_{k\tau}\}] = \hbar v_F \sum_{k,\tau}(\tau k - k_F)N_{k,\tau} \ . \tag{9.6}$$

Notice that these eigenvalues are all non-negative since $\tau k - k_F$ is necessarily negative when $N_{k,\tau} = -1$ and positive when $N_{k,\tau} = +1$: the lowest eigenvalue, zero, is attained in the non-interacting N-particle ground-state, which coincides with the "vacuum".

Let us now introduce the interaction between the particles. The fundamental interaction hamiltonian is

$$\hat{H}_{int} = \frac{1}{2L}\sum_{q \neq 0}v_q\hat{n}_{-q}\hat{n}_q \ , \tag{9.7}$$

where $\hat{n}_q$ is the electron density fluctuation operator, v_q is the Fourier transform of the interaction potential, and the $q = 0$ term is excluded from the sum.[5] It is important to appreciate that the electron density fluctuation operator $\hat{n}_q$ is not just the sum of the density fluctuation operators for right movers and left movers, even if we confine ourselves to the low-energy sector of the Hilbert space. The key point is that a physical electron is neither a right mover nor a left mover: rather, it is a right mover for $k > 0$ and a left mover for $k < 0$. Accordingly, its destruction operator is

$$\hat{a}_{k,e} = \Theta(k)\hat{a}_{k,R} + \Theta(-k)\hat{a}_{k,L} \tag{9.8}$$

and its density fluctuation operator is given by

$$\begin{aligned}\hat{n}_q &= \sum_{ke}\hat{a}^\dagger_{k-q,e}\hat{a}_{k,e} \\ &= \sum_k \Big[\Theta(k-q)\Theta(k)\hat{a}^\dagger_{k-q,R}\hat{a}_{k,R} + \Theta(-k+q)\Theta(-k)\hat{a}^\dagger_{k-q,L}\hat{a}_{k,L} \\ &\quad + \Theta(k-q)\Theta(-k)\hat{a}^\dagger_{k-q,R}\hat{a}_{k,L} + \Theta(-k+q)\Theta(k)\hat{a}^\dagger_{k-q,L}\hat{a}_{k,R}\Big] \ . \end{aligned} \tag{9.9}$$

Recall that in the low energy sector of the Hilbert space both k and $k - q$ are restricted to be in the vicinity of the "Fermi surface" i.e., the two points $\pm k_F$. This leaves us with only two possibilities for q: either $|q| \sim 0$ (on the scale of k_F), in which case k and $k - q$ have the same sign, or $|q| \sim 2k_F$, in which case k and $k - q$ have opposite signs. In the former case only the terms on the second line of Eq. (9.9) differ from zero, and we see that the electron density fluctuation is indeed the sum of the density fluctuations of right and left movers. For $|q| \sim 2k_F$, however, it is only the terms on the third line of Eq. (9.9) that differ from zero, and these terms transfer an electron from the right-moving branch with $k \sim k_F$

---

[5] Thus, we are assuming the presence of a neutralizing background of charge. For completeness, we should also subtract from the right-hand side of Eq. (9.7) the self-interaction term $\frac{1}{2L}\sum_q v_q\hat{N}$, where $\hat{N}$ is the total number of physical particles. This term, however, has no effect other than shifting the chemical potential, and will therefore disappear when we finally set the zero of the single particle energy at the interacting chemical potential.

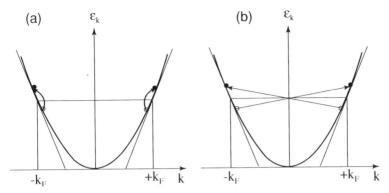

Fig. 9.3. The two types of processes, (a) forward and (b) back scattering that contribute to the interaction hamiltonian in the low energy sector.

to the left moving branch with $k \sim -k_F$, or viceversa (see Fig. 9.3). Thus, in summary, we have

$$\hat{n}_q \simeq \begin{cases} \sum_k \left(\hat{a}^\dagger_{k-q,R} \hat{a}_{k,R} + \hat{a}^\dagger_{k-q,L} \hat{a}_{k,L} \right), & |q| \sim 0, \\ \sum_k \hat{a}^\dagger_{k-q,L} \hat{a}_{k,R}, & q \sim 2k_F, \\ \sum_k \hat{a}^\dagger_{k-q,R} \hat{a}_{k,L}, & q \sim -2k_F. \end{cases} \quad (9.10)$$

Substituting this expression into Eq. (9.7) we see that the electron-electron interaction has two distinct contributions, which are illustrated in Fig. 9.3. The first one, coming from $q \sim 0$ is responsible for *forward scattering* processes in which the direction of motion of the particles does not change:

$$\hat{H}_{int,1} = \frac{1}{2L} \sum_{|q|<\Lambda} v_q \left(\hat{n}_{-q,R} + \hat{n}_{-q,L} \right) \left(\hat{n}_{q,R} + \hat{n}_{q,L} \right), \quad (9.11)$$

where

$$\hat{n}_{q,\tau} \equiv \sum_k \hat{a}^\dagger_{k-q,\tau} \hat{a}_{k,\tau} \quad (9.12)$$

are the density fluctuation operators for right and left movers. The second one, coming from large momentum transfers $|q| \sim 2k_F$, is responsible for *back-scattering* processes in which particles in the proximity of the Fermi surface are scattered from one branch of movers to the other. This contribution has the form

$$\hat{H}_{int,2} = \frac{1}{2L} \sum_{||q|-2k_F|<\Lambda} v_q \sum_{k_1 k_2} \left(\hat{a}^\dagger_{k_1+q,R} \hat{a}_{k_1,L} \hat{a}^\dagger_{k_2-q,L} \hat{a}_{k_2,R} + L \leftrightarrow R \right). \quad (9.13)$$

Now, since we are working in the low energy sector, we can set $k_2 = k_F$, $k_1 = -k_F$ and $q = 2k_F + \tilde{q}$ in the first term in the brackets, and $k_2 = -k_F$, $k_1 = k_F$ and $q = -2k_F + \tilde{q}$

in the second one. Then, after reordering the operators, we get

$$\hat{H}_{int,2} \simeq -\frac{1}{2L} \sum_{|\tilde{q}|<\Lambda} v_{2k_F} \left(\hat{a}^\dagger_{k_F+\tilde{q},R} \hat{a}_{k_F,R} \hat{a}^\dagger_{-k_F-\tilde{q},L} \hat{a}_{-k_F,L} \right.$$
$$\left. + \hat{a}^\dagger_{-k_F+\tilde{q},L} \hat{a}_{-k_F,L} \hat{a}^\dagger_{k_F-\tilde{q},R} \hat{a}_{k_F,R} \right) , \tag{9.14}$$

which is equivalent to

$$\hat{H}_{int,2} \simeq -\frac{1}{2L} \sum_{|\tilde{q}|<\Lambda} v_{2k_F} \left(\hat{n}_{-q,R} \hat{n}_{q,L} + \hat{n}_{-q,L} \hat{n}_{q,R} \right) . \tag{9.15}$$

Combining Eqs. (9.5), (9.11), and (9.15), and sending the cutoff Λ to infinity we finally write the complete *Luttinger model hamiltonian* as the sum of a kinetic, a forward scattering, and a back-scattering term:

$$\hat{H}_{LM} = \hbar v_F \sum_{k,\tau} (\tau k - k_F) \hat{N}_{k,\tau}$$
$$+ \frac{1}{2L} \sum_{q \neq 0} V_1(q) \left(\hat{n}_{-q,R} \hat{n}_{q,R} + \hat{n}_{-q,L} \hat{n}_{q,L} \right)$$
$$+ \frac{1}{2L} \sum_{q \neq 0} V_2(q) \left(\hat{n}_{-q,R} \hat{n}_{q,L} + \hat{n}_{-q,L} \hat{n}_{q,R} \right) , \tag{9.16}$$

where

$$V_1(q) = v_q$$
$$V_2(q) = v_q - v_{2k_F} . \tag{9.17}$$

It is generally believed that this form accurately mimics the physical one-dimensional fermion hamiltonian, within the low-energy sector of the Hilbert space.[6]

Although the expressions (9.17) for $V_2(q)$ and $V_1(q)$ have been physically motivated, it is customary in much of the literature to treat $V_1(q)$ and $V_2(q)$ as two independent functions of $|q|$,[7] subject to the following *stability conditions*:

$$\text{(i)} \quad 2\pi\hbar v_F + V_1(q) > |V_2(q)| > 0 ,$$
$$\text{(ii)} \quad \lim_{q \to \infty} \frac{q^{\frac{1}{2}} V_2(q)}{2\pi v_F + V_1(q)} = 0 . \tag{9.18}$$

The reason for these conditions will be explained in Section 9.5.1.[8]

We have deliberately ignored, so far, the spin of the particles. One might think that the spin is straightforwardly included by adding an index σ to all Fermion and density fluctuation

---

[6] Notice that in writing the hamiltonian (9.16) we have dropped the restriction $|q| < \Lambda$ and thus created a model that is defined at all length scales. However, the bothersome *ultraviolet cutoff* Λ will resurface in Section 9.6.1.

[7] In the specialized literature (see, for example, Schulz *et al.* (1998)) the following notation is adopted for the "coupling constants" of the Luttinger model: $V_1(q)$ and $V_2(q)$ are called $g_4(q)$ and $g_2(q)$ respectively, while the coupling constant associated with the spin-flip term (see Eq. (9.19)) is called $g_1(q)$.

[8] The conditions (9.18) are satisfied, in particular, by the screened coulomb interaction $v_q = -e^2 e^{(q^2 + \lambda^{-2})a^2} \text{Ei}(-(q^2 + \lambda^{-2})a^2)$, where a is the radius of the one dimensional wire and λ is the screening length (see Appendix 1).

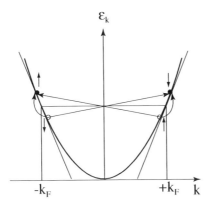

Fig. 9.4. A back-scattering process between antiparallel spin electrons can be mimicked by a forward *spin-flip* scattering process.

operators. This works perfectly well for the interaction between parallel spin electrons, but fails for antiparallel spin electrons. Consider the back-scattering of an up-spin electron from the right to the left moving branch, with a down-spin electron going from the left to the right-moving branch as shown in the figure. For parallel spins such a process can be mimicked by an interaction of the form $-V(2k_F)\hat{n}_{-q\sigma,R}\hat{n}_{q\sigma,R}$, which is the obvious extension of what we did in the spinless case. The same mimicking in the case of antiparallel spins would require a term of the form

$$\sum_{|\bar{q}|<\Lambda} V_1(2k_F)\hat{a}^{\dagger}_{k_F+\bar{q},R,\downarrow}\hat{a}_{k_F,R,\uparrow}\hat{a}^{\dagger}_{-k_F-\bar{q},L,\uparrow}\hat{a}_{-k_F,L,\downarrow} . \tag{9.19}$$

Unfortunately, this term cannot be expressed in terms of spin density fluctuation operators: instead, it involves the "spin-flip" operators $\sum_k \hat{a}^{\dagger}_{k-q,\tau,\uparrow}\hat{a}_{k,\tau,\downarrow}$, as shown in Fig. 9.4. The appearance of such *spin-flip processes* leads to a considerably more complicated hamiltonian, which can no longer be solved exactly.

9.3 The anomalous commutator

The solution of the Luttinger model has an interesting history. Luttinger (1963), who invented the model, fell victim to a subtle paradox and therefore did not achieve a correct solution of the problem he himself had posed. The paradox lies in the failure of the commutator of two density fluctuation operators to vanish when common sense and our experience based on finite-density systems tell us it should vanish.

It is instructive to fall once into the Luttinger trap. Luttinger's idea was to "gauge away" the interaction between the fermions by means of a unitary transformation of the form

$$e^{i\hat{F}}\hat{H}_{LM}e^{-i\hat{F}}, \tag{9.20}$$

where the operator $\hat{F}$ is chosen to satisfy the operator equations

$$i[\hat{H}_0, \hat{F}] = -\hat{H}_{int} , \tag{9.21}$$

and

$$i[\hat{H}_{int}, \hat{F}] = 0 , \qquad (9.22)$$

with $\hat{H}_0$ and $\hat{H}_{int}$ being respectively the kinetic and the interaction parts of the Luttinger hamiltonian. If an operator $\hat{F}$ that satisfies these two properties exists, then one can easily verify (see Eqs. (8.188) and (8.189) and footnote[41]) that

$$e^{i\hat{F}} \hat{H}_{LM} e^{-i\hat{F}} = \hat{H}_0 - \hat{H}_{int} + \hat{H}_{int}$$
$$= \hat{H}_0 . \qquad (9.23)$$

Then $\hat{H}_{LM}$ would be unitarily related to $\hat{H}_0$, implying that the eigenvalue spectrum of $\hat{H}_{LM}$ coincides with that of $\hat{H}_0$, and leading to an elegant but disappointingly trivial solution of the problem.

Noting that $\hat{H}_{int}$ can be written in the compact form

$$\hat{H}_{int} = \frac{1}{2L} \sum_{q\neq 0} \sum_{\tau\tau'} V_{\tau\tau'}(q)\hat{n}_{-q\tau}\hat{n}_{q\tau'} , \qquad (9.24)$$

with $V_{\tau\tau}(q) = V_1(q)$ and $V_{\tau-\tau}(q) = V_2(q)$, one could then too hastily think to have found in the expression

$$i\hat{F} = \frac{1}{4L} \sum_{q\neq 0} \sum_{\tau\tau'} \frac{V_{\tau\tau'}(q)}{q} \hat{n}_{-q\tau}\hat{n}_{q\tau'} \qquad (9.25)$$

an operator satisfying the conditions (9.21) and (9.22). The reader is urged to explicitly verify that the commutators of $i\hat{F}$ with $\hat{H}_0$ and $\hat{H}_{int}$, calculated according to the standard rules of second quantization – in which two density fluctuation operators commute with each another – do satisfy Eqs. (9.21) and (9.22). Thus, it would appear that the Luttinger model is just a non-interacting model in disguise, which would hardly justify the attention devoted to it in this book.

Fortunately for the fate of the Luttinger model, this conclusion is completely wrong. One can see that something must have gone seriously wrong from the fact that simple perturbation theory predicts a dependence of the ground-state energy on the magnitude of the interaction coupling constant, while, according to Eq. (9.23) such dependence should be absent. The problem is that within our special Hilbert space, which accommodates infinitely many particles in the vacuum, the commutator of two right or left density fluctuation operators fails to vanish, even though, in first quantization, these operators depend only on the coordinates of the particles.

Fig. 9.5 shows in detail how the non-commutativity arises. We first apply $\hat{n}_{-q,R}$, with $q = \frac{4\pi}{L}$, to the vacuum state $|0\rangle$, and end up in a superposition of two excited states. At this point, the application of $\hat{n}_{q,R}$ reduces each term of the superposition to the vacuum yielding

$$\hat{n}_{q,R}\hat{n}_{-q,R}|0\rangle = 2|0\rangle = \frac{qL}{2\pi}|0\rangle. \qquad (9.26)$$

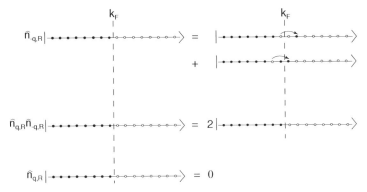

Fig. 9.5. Demonstration of the non-commutativity of density fluctuations for right movers.

Had we applied the two density fluctuation operators in reverse order we would have obtained 0 because, due to the Pauli exclusion principle, it is impossible, starting from the vacuum, to decrease the momentum of the right movers. Thus, for this, as for any other positive q, one has

$$\hat{n}_{-q,R}\hat{n}_{q,R}|0\rangle = 0 \ . \tag{9.27}$$

Eqs. (9.26) and (9.27) show that density fluctuation operators with opposite values of q do not commute.

To calculate the commutator in general we note that the standard fermionic anticommutation relations (9.3) give

$$[\hat{n}_{q,\tau}, \hat{n}_{-q,\tau}] = \sum_k \left(\hat{a}^\dagger_{k-q,\tau}\hat{a}_{k-q,\tau} - \hat{a}^\dagger_{k,\tau}\hat{a}_{k,\tau} \right) \ . \tag{9.28}$$

In the *ordinary* Hilbert space (one with a finite density of particles) one could at this point make the change of variable $k \to k+q$ in the first term of the sum, and obtain the familiar result $[\hat{n}_{q,\tau}, \hat{n}_{-q,\tau}] = 0$. But in the present case both terms in the sum are unbounded operators, (i.e., operators with arbitrarily large eigenvalues) and the difference of two infinities is an undetermined quantity. The correct way to proceed is to write

$$\hat{a}^\dagger_{k,\tau}\hat{a}_{k,\tau} = \hat{N}_{k,\tau} + n^{(0)}_{k,\tau} \tag{9.29}$$

where the first term is normal-ordered (see Eq. (9.2)) and the second term is the ground-state occupation number given by Eq. (9.1). We now substitute Eq. (9.29) in (9.28). Because the normal-ordered occupation number is a bounded operator one can cancel it out by the usual trick, after which one is left with

$$[\hat{n}_{q,\tau}, \hat{n}_{-q,\tau}] = \sum_k \left(n^{(0)}_{k-q,\tau} - n^{(0)}_{k,\tau} \right) \ . \tag{9.30}$$

There remains the task of calculating the sum on the right-hand side. Of course we cannot apply the change of variables trick here, since both terms in the sum are infinite. However, knowing the explicit expression of $n_{k,\tau}^{(0)}$, Eq. (9.1), we see that the summand vanishes unless $k - q$ and k are on opposite sides of k_F. When the summand does not vanish, it equals $+1$ for $q > 0$ and $\tau = R$ or $q < 0$ and $\tau = L$ and -1 in the remaining cases. Thus, in summary

$$\sum_k \left(n_{k-q,\tau}^{(0)} - n_{k,\tau}^{(0)} \right) = \frac{Lq\tau}{2\pi}. \tag{9.31}$$

The same technique can easily be applied to the calculation of the commutator between two density fluctuations with wave vectors q and q' of different magnitude and/or different τ's. In both cases, the result is zero. We can summarize these results in the compact formula

$$[\hat{n}_{q,\tau}, \hat{n}_{-q',\tau'}] = \frac{qL\tau}{2\pi} \delta_{qq'} \delta_{\tau\tau'} . \tag{9.32}$$

Of course, this result is essential to both the definition and the solution of the Luttinger model.

9.4 Introducing the bosons

The results of the previous section suggest that we define, for each $q \neq 0$, a boson creation/destruction operator pair $(\hat{b}_q^\dagger, \hat{b}_q)$ in the following manner:

$$\hat{b}_q^\dagger = \sqrt{\frac{2\pi}{L|q|}} \left(\Theta(q)\hat{n}_{q,R}^\dagger + \Theta(-q)\hat{n}_{q,L}^\dagger \right),$$

$$\hat{b}_q = \sqrt{\frac{2\pi}{L|q|}} \left(\Theta(q)\hat{n}_{q,R} + \Theta(-q)\hat{n}_{q,L} \right). \tag{9.33}$$

With these definitions, it is straightforward to verify that the $\hat{b}$'s are proper boson operators, i.e. they satisfy

$$[\hat{b}_q, \hat{b}_{q'}^\dagger] = \delta_{q,q'} ,$$

$$[\hat{b}_q, \hat{b}_{q'}] = [\hat{b}_q^\dagger, \hat{b}_{q'}^\dagger] = 0. \tag{9.34}$$

The inverse relations, expressing the density fluctuations in terms of boson operators, are:

$$\hat{n}_{q,R} = \sqrt{\frac{L|q|}{2\pi}} \left(\Theta(q)\hat{b}_q + \Theta(-q)\hat{b}_{-q}^\dagger \right),$$

$$\hat{n}_{q,L} = \sqrt{\frac{L|q|}{2\pi}} \left(\Theta(-q)\hat{b}_q + \Theta(q)\hat{b}_{-q}^\dagger \right). \tag{9.35}$$

Notice that $\hat{n}_{q,\tau} = \hat{n}_{-q,\tau}^\dagger$.

The fundamental property of $\hat{b}_q^\dagger$ is that it acts as a rising operator for $\hat{H}_0$, i.e., it transforms an eigenstate of $\hat{H}_0$ with eigenvalue E_0 into another eigenstate of $\hat{H}_0$ with eigenvalue $E_0 + \hbar v_F |q|$. Similarly, $\hat{b}_q$ is a lowering operator, which decreases the energy by $\hbar v_F |q|$.

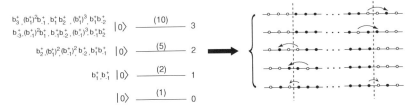

Fig. 9.6. Equivalence of the bosonic and fermionic representations for the first few energy levels of the non-interacting Luttinger model hamiltonian. Energies are in units of $\hbar v_F \frac{2\pi}{L}$, with 0 being the energy of the ground-state. Degeneracies are indicated by the numbers in the round brackets. The bosonic representations of the states in each multiplet are shown on the left-hand side (wave vectors are in units of $\frac{2\pi}{L}$). The fermionic representation of the 5-fold degenerate multiplet of states in the second excited energy level is shown on the right-hand side of the figure.

These assertions follow immediately from the commutation relations

$$[\hat{H}_0, \hat{b}_q^\dagger] = \hbar v_F |q| \hat{b}_q^\dagger ,$$
$$[\hat{H}_0, \hat{b}_q] = -\hbar v_F |q| \hat{b}_q , \qquad (9.36)$$

which the reader can easily verify, following the definitions (9.5) and (9.33). Notice that $\hat{b}_q$ and $\hat{b}_q^\dagger$ commute with the total number operators $\hat{N}_\tau$ and therefore do not change the total number of left or right movers. Thus, starting from the ground-state $|0\rangle$ we can construct a whole family of eigenstates of $\hat{H}_0$ by repeated application of $\hat{b}_q^\dagger$. At the same time, $\hat{b}_q|0\rangle = 0$ because $|0\rangle$ is the minimum energy state for given N.

Let us now consider a set of non-negative integers $\mathcal{N}_q$, one for each q, and denote by

$$|\{\mathcal{N}_q\}\rangle = \prod_{q\neq 0} \frac{(\hat{b}_q^\dagger)^{\mathcal{N}_q}}{\sqrt{\mathcal{N}_q!}} |0\rangle \qquad (9.37)$$

the state that contains $\mathcal{N}_q$ bosons in the mode of wave vector q. Clearly this is an eigenstate of $\hat{H}_0$ with eigenvalue

$$E[\{\mathcal{N}_q\}] = \sum_{q\neq 0} \hbar v_F |q| \mathcal{N}_q , \qquad (9.38)$$

and it is easy to see that *all* the eigenvalues of $\hat{H}_0$ are given by this formula.[9] The real difficulty lies in proving that the eigenvalues (9.38) have the correct degeneracies, so that the set described by Eq. (9.37) is equivalent to the complete set of eigenstates of $\hat{H}_0$. Fig. 9.6 shows the degeneracies of the first few energy levels, together with their bosonic representations. For example, the second excited level has degeneracy 5, and can be obtained, in the fermionic basis, either as an electron-hole pair of wave vector $\frac{4\pi}{L}$ (this can be done in two distinct ways, as shown in the figure), or as an electron-hole pair of wave vector $-\frac{4\pi}{L}$

---

[9] Every eigenstate of $\hat{H}_0$ can be viewed as a collection of electron-hole pairs carrying momenta $q_1 = \frac{2\pi n_1}{L}$, $q_2 = \frac{2\pi n_2}{L}$, etc., where $n_1, n_2 \ldots$ are integers. The energy spectrum is therefore of the form $\hbar \sum_i v_F \frac{2\pi}{L} |n_i|$, which coincides with the spectrum described by Eq. (9.38).

(two more states), or, finally, by combining an electron-hole pair of wave vector $\frac{2\pi}{L}$ and one of wave vector $-\frac{2\pi}{L}$ (one more state).

A very ingenious and elegant proof of the equivalence between the fermionic and the bosonic representation was given by F. D. M. Haldane (1981). The idea of the proof is to calculate the partition function $\mathcal{Z} = Tr\{e^{-\beta \hat{H}_0}\}$ first in the fermionic basis set $|\{N_{k,\tau}\}$ defined just before Eq. (9.6), and then in the bosonic basis set $|\{\mathcal{N}_q\}\rangle$ of Eq. (9.37). In either case

$$\mathcal{Z} = \sum_n D_n e^{-\beta E_n} , \tag{9.39}$$

where the sum runs over the eigenstates of $\hat{H}_0$, E_n is the energy of the n-th eigenstate, and D_n is its degeneracy. Because $\mathcal{Z}$ is a sum of positive terms, any difference between the degeneracies of the two basis sets would lead to different values of $\mathcal{Z}$. Since the results of the two calculations are in fact identical at all temperatures (see Appendix 19), we conclude that the two sets have identical degeneracies.

An immediate consequence of the existence of the bosonic basis set (9.37) is that $\hat{H}_0$ can be written as

$$\hat{H}_0 = \sum_{q \neq 0} \hbar v_F |q| \hat{b}_q^\dagger \hat{b}_q . \tag{9.40}$$

This is evident from the fact that the operators on the two sides of the equality have same eigenvalues and eigenvectors given by Eqs. (9.38) and (9.37). More generally, the existence of the bosonic basis set ensures that *every* operator, *including the Fermion creation and destruction operators*, can be represented in terms of boson operators combined with operators that raise or lower the *total* numbers of particles. The implications of this fact are far-reaching.

9.5 Solution of the Luttinger model

With the help of Eqs. (9.40) and (9.35) we can now write the spinless Luttinger model Hamiltonian, Eq. (9.16), in terms of $\hat{b}^\dagger$ and $\hat{b}$. After some straightforward manipulations we obtain

$$\hat{H}_{LM} = \sum_{q \neq 0} \left[\left(\hbar v_F + \frac{V_1(q)}{2\pi} \right) |q| \hat{b}_q^\dagger \hat{b}_q + \frac{V_2(q)}{4\pi} |q| \left(\hat{b}_q^\dagger \hat{b}_{-q}^\dagger + \hat{b}_{-q} \hat{b}_q \right) \right] . \tag{9.41}$$

This hamiltonian is quadratic in the boson operators, and can therefore be solved exactly.[10]

---

[10] It is interesting to remark that in arriving to Eq. (9.41) we have pursued a route opposite to the one commonly followed when tackling a many-body problem by mean field theory. In the mean field approach the effects of the two-body interactions are mimicked by an effective one-body potential. Here on the other hand the strategy of the exact solution has been to express the kinetic energy (a one-body operator) as a quadratic form of the bosonic operators, i.e. to attribute to it the same operatorial structure of the two-body interaction term.

9.5.1 Exact diagonalization

f the interaction $V_2(q)$ between right and the left movers were absent, then the hamiltonian (9.41) would be already diagonal in the bosonic representation: in this special case, the ground-state of $\hat{H}_{LM}$ coincides with the ground-state of $\hat{H}_0$. This makes perfect sense, since this is the state with the minimum (maximum) possible value of the total momentum for right (left) movers and, of course, the interaction $V_1(q)$ cannot change the value of the total momentum. Higher eigenstates of $\hat{H}_{LM}$ are also eigenstates of $\hat{H}_0$, but the converse is not true unless $V_1(q)$ is independent of q.

In the general case in which both V_1 and V_2 are present, we diagonalize (9.41) by means of the unitary transformation

$$\hat{b}_q^\dagger = \cosh\varphi_q \hat{B}_q^\dagger - \sinh\varphi_q \hat{B}_{-q} ,$$
$$\hat{b}_{-q} = \cosh\varphi_q \hat{B}_{-q} - \sinh\varphi_q \hat{B}_q^\dagger , \tag{9.42}$$

where φ_q is a "rotation angle", chosen is such a way as to eliminate off-diagonal terms in the transformed hamiltonian (i.e., terms proportional to $\hat{B}_q^\dagger \hat{B}_q^\dagger$). The reader can easily verify that this is accomplished by the choice

$$\tanh 2\varphi_q = \frac{\frac{V_2(q)}{2\pi\hbar}}{v_F + \frac{V_1(q)}{2\pi\hbar}}, \tag{9.43}$$

which always admits a solution for φ_q if the first stability condition (9.18) is satisfied.[11]

Substituting Eq. (9.42) into Eq. (9.41), making use of Eq. (9.43), and normal-ordering the resulting expression, we obtain

$$\hat{H}_{LM} = \sum_{q\neq 0} \hbar c_q |q| \hat{\beta}_q^\dagger \hat{\beta}_q, \tag{9.44}$$

where the $\hat{\beta}$ operators satisfy the standard bosonic commutation relations, and the "sound velocity" c_q is given by

$$c_q = \sqrt{\left|v_F + \frac{V_1(q)}{2\pi\hbar}\right|^2 - \left|\frac{V_2(q)}{2\pi\hbar}\right|^2}. \tag{9.45}$$

From the form of the hamiltonian (9.44) we see that the Luttinger model is equivalent to a system of independent massless bosons with dispersion $\omega_q = c_q|q|$ – the "quanta" of a collective density fluctuation field. Notice also that c_q is real only if the first stability condition (9.18) is satisfied.

The ground-state $|\psi_0\rangle$ of the interacting Hamiltonian is annihilated by the destruction operators $\hat{\beta}_q$ just as the non-interacting ground-state is annihilated by the $\hat{b}_q$'s. Since $\hat{\beta}_q = \cosh\varphi_q \hat{b}_q + \sinh\varphi_q \hat{b}_{-q}^\dagger$ we see that $|\psi_0\rangle$ must satisfy the equation

$$\hat{b}_q |\psi_0\rangle = -\tanh\varphi_q \hat{b}_{-q}^\dagger |\psi_0\rangle, \tag{9.46}$$

---

[11] Note that the rotation angle $\varphi_q = \frac{1}{4} \ln \frac{2\pi\hbar v_F + V_1(q) + V_2(q)}{2\pi\hbar v_F + V_1(q) - V_2(q)}$ tends to infinity for $q \to 0$ in the case of long range interactions $V_1(q) \to \infty$, $V_1(q) - V_2(q) = v_{2k_F}$.

for all $q \neq 0$. Making use of the commutation relation

$$[\hat{b}_q, e^{-\tanh \varphi_q \hat{b}_q^\dagger \hat{b}_{-q}^\dagger}] = -\tanh \varphi_q \hat{b}_{-q}^\dagger e^{-\tanh \varphi_q \hat{b}_q^\dagger \hat{b}_{-q}^\dagger}, \tag{9.47}$$

we realize that the ground-state of the Luttinger model has the explicit form

$$|\psi_0\rangle = A \exp\left(-\sum_{q \neq 0} \tanh \varphi_q \hat{b}_q^\dagger \hat{b}_{-q}^\dagger\right) |0\rangle. \tag{9.48}$$

The excited states are then generated by repeatedly applying the creation operators $\hat{\beta}_q^\dagger$ to the ground-state.

The normalization constant A in Eq. (9.48) is given by

$$\begin{aligned}
A^{-2} &= \langle 0| e^{-\sum_{q \neq 0} \tanh \varphi_q \hat{b}_q \hat{b}_{-q}} e^{-\sum_{q \neq 0} \tanh \varphi_q \hat{b}_q^\dagger \hat{b}_{-q}^\dagger} |0\rangle \\
&= \int \prod_{q \neq 0} \frac{d\Re e \phi_q \, d\Im m \phi_q}{\pi} e^{-\sum_{q \neq 0} [|\phi_q|^2 + (\phi_q \phi_{-q} + \phi_q^* \phi_{-q}^*) \tanh \phi_q]} \\
&= \prod_{q > 0} \cosh^2 \varphi_q, \tag{9.49}
\end{aligned}$$

where we have made use of the coherent state representation described in Appendix 21. This result is expressed more compactly as

$$A = e^{-\frac{L}{2\pi} \int_0^\infty dq \, \ln \cosh \varphi_q}, \tag{9.50}$$

from which we see that the normalization exists only if the rotation angle φ_q vanishes, for large q, faster than $|q|^{-\frac{1}{2}}$. This explains the origin of the second *stability condition* (9.18).

To gain more insight into the structure of the many-body correlations encoded in the deceptively simple-looking state (9.48), the reader is urged to work out the expression for the ground-state wave function of the Luttinger model in the *coordinate representation*. The calculation is outlined in Exercise 5 and leads to the following interesting result:

$$\psi_0(x_1, \ldots, x_N) \propto \prod_{i < j} \sin \frac{\pi(x_i - x_j)}{L} \left| \sin \frac{\pi(x_i - x_j)}{L} \right|^{\frac{1}{g} - 1}, \tag{9.51}$$

where $g \equiv \lim_{q \to 0} e^{-2\varphi_q}$. The first factor of this expression is (up to a proportionality constant) the ground-state wave function of non-interacting fermions in one dimension, i.e., a Slater determinant of plane waves with wave vectors ranging from $-k_F$ to k_F (check this in Exercise 3!). The last factor introduces the correlation between the electrons: notice that this factor reduces to a trivial 1 if the interaction is absent, i.e., if $g = 1$.

In summary, the energy eigenvalues of the Luttinger model are simply those of a system of non-interacting bosons with dispersion $\omega_q = c_q |q|$, but the eigenfunctions have a rather complicated and highly correlated structure, which could not have been obtained from any simple-minded perturbative approach.

9.5.2 Physical properties

Physical properties that depend only on low-energy/long-wavelength excitations can be easily calculated from the hamiltonian (9.44). For example, the low-temperature heat capacity $c_v(T)$ is readily inferred from the equilibrium energy

$$\frac{E}{L} \simeq \int_0^\infty \frac{dq}{2\pi} \frac{\hbar cq}{e^{\beta\hbar cq} - 1} = \frac{\pi}{12} \frac{(k_B T)^2}{\hbar c} , \tag{9.52}$$

where c is the $q \to 0$ limit of the sound velocity. This gives

$$c_v(T) = L \frac{\pi}{6} \frac{k_B^2 T}{\hbar c} , \tag{9.53}$$

which looks superficially like that of a normal Fermi liquid with an "effective mass" $\frac{m^*}{m} = \frac{c_v}{c_{v0}} = \frac{v_F}{c}$. However, the notion of effective mass is meaningless here, since the elementary excitations are massless bosons. This result highlights the great difficulty of distinguishing a Luttinger liquid from a normal Fermi liquid by means of thermodynamic measurements.

Another interesting property is the compressibility. We calculate it from the small-q limit of the density–density response function, which, as discussed in Section 3.3, is generally given by

$$\chi_{nn}(q, \omega) = -\frac{i}{\hbar L} \int_0^\infty \langle [\hat{n}_q(t), \hat{n}_{-q}] \rangle e^{i(\omega + i\eta)t} dt . \tag{9.54}$$

At small q the density fluctuation operator reduces to the sum of the right and left density fluctuation operators, i.e., $\hat{n}_q \simeq \hat{n}_{q,R} + \hat{n}_{q,L}$ (see Eq. (9.10)). Then, making use of Eqs. (9.35) and (9.42), we get

$$\hat{n}_q \simeq \sqrt{\frac{L|q|}{2\pi}} \left(\hat{b}_q + \hat{b}_{-q}^\dagger \right)$$

$$= \sqrt{\frac{L|q|}{2\pi}} e^{-\varphi} \left(\hat{\beta}_q + \hat{\beta}_{-q}^\dagger \right) , \tag{9.55}$$

where φ is the $q \to 0$ limit of φ_q, and $\simeq$ is a reminder that the first equality is valid only in the limit of small wave vector.

The time evolution of the $\hat{\beta}$ operators is simple ($\hat{\beta}_q(t) = \hat{\beta}_q e^{-ic|q|t}$, etc.), and so are their commutation relations. Thus, we can easily evaluate Eq. (9.54) to get

$$\chi_{nn}(q, \omega) = \frac{1}{\pi \hbar v_N} \frac{(cq)^2}{(\omega + i\eta)^2 - (cq)^2} , \tag{9.56}$$

where we have defined

$$v_N \equiv c e^{2\varphi} = v_F + \lim_{q \to 0} \frac{V_1(q) + V_2(q)}{2\pi\hbar} . \tag{9.57}$$

The compressibility K is then obtained as

$$K = -\frac{1}{n^2} \lim_{q \to 0} \chi_{nn}(q, 0) = \frac{1}{\pi \hbar n^2 v_N} . \tag{9.58}$$

Notice that the ratio of the interacting compressibility to the non-interacting one is

$$\frac{K}{K_0} = \frac{v_F}{v_N} = e^{-2\varphi}\frac{c_v}{c_{v0}} . \tag{9.59}$$

Comparing this with the normal Fermi liquid results (8.10) and (8.21) we see that $g \equiv e^{-2\varphi}$ plays the role of the Landau renormalization factor $\frac{1}{1+F_0^s}$, and provides a dimensionless measure of the strength of the interaction.

The calculation of the *proper* density–density response function is more delicate. Because the interaction V_2 between right and left movers is different from the interaction V_1 between two right movers or two left movers, the system must be treated as a two-component liquid. Applying the multi-component formalism of Section 5.4.5, we introduce a matrix of density–density response functions

$$\chi(q,\omega) = \begin{pmatrix} \chi_{RR}(q,\omega) & \chi_{RL}(q,\omega) \\ \chi_{LR}(q,\omega) & \chi_{LL}(q,\omega) \end{pmatrix} , \tag{9.60}$$

where $\chi_{\tau\tau'}(q,\omega) \equiv -\frac{i}{\hbar L}\int_0^\infty \langle [\hat{n}_{q\tau}(t), \hat{n}_{-q\tau'}]\rangle e^{i(\omega+i\eta)t}dt$ describes the density response of τ movers to a potential that couples only to τ' movers. Using the representation (9.35) for the density operators, and the Bogoliubov transformation (9.42), one can easily show (Exercise 7) that

$$\chi_{\tau\tau}(q,\omega) = \frac{\tau q}{2\pi\hbar}\frac{\omega + \tau cq\cosh 2\varphi}{\omega^2 - c^2q^2} ,$$

$$\chi_{RL}(q,\omega) = -\frac{q}{2\pi\hbar}\frac{cq\sinh 2\varphi}{\omega^2 - c^2q^2} = \chi_{LR}(q,\omega) , \tag{9.61}$$

where $\tau = 1$ for right movers and $\tau = -1$ for left movers. The full density–density response function $\chi_{nn}(q,\omega)$ of Eq. (9.56) is obtained by summing $\chi_{\tau\tau'}(q,\omega)$ over τ and τ'. The matrix of the *proper* density–density response functions is then given by the relation

$$\tilde{\chi}^{-1}(q,\omega) = \chi^{-1}(q,\omega) + \begin{pmatrix} V_1 & V_2 \\ V_2 & V_1 \end{pmatrix} . \tag{9.62}$$

Only after performing the matrix inversion can we calculate the full proper density–density response function from the definition $\tilde{\chi}_{nn}(q,\omega) = \sum_{\tau\tau'}\tilde{\chi}_{\tau\tau'}(q,\omega)$. The result of this calculation is very interesting. One finds (see Exercise 8) that $\tilde{\chi}$ coincides with the noninteracting density–density response function, i.e., with Eq. (9.61) evaluated for $c = v_F$ and $\varphi = 0$. This means that, *in the $q \to 0$ limit*, the random phase approximation gives the exact density–density response functions of the Luttinger liquid![12]

Finally, consider the current–density fluctuation operator $\hat{j}_q$, whose fundamental definition is provided by the continuity equation

$$\frac{\partial\hat{n}_q(t)}{\partial t} = -iq\hat{j}_q(t). \tag{9.63}$$

---

[12] This is most easily verified in the case $V_1 = V_2$, for, in this case, one can use the one-component formalism and calculate $\tilde{\chi}$ directly from Eqs. (5.21) and (5.22) of Chapter 5.

Substituting Eq. (9.55) for the density operator, and calculating the time derivative, we obtain

$$\hat{j}_q = \sqrt{\frac{L|q|}{2\pi}} c e^{-\varphi} \text{sign}(q) \left(\hat{\beta}_q - \hat{\beta}_{-q}^\dagger \right)$$
$$= v_J \left(\hat{n}_{q,R} - \hat{n}_{q,L} \right) , \tag{9.64}$$

where we have defined

$$v_J \equiv c e^{-2\varphi} = v_F + \lim_{q \to 0} \frac{V_1(q) - V_2(q)}{2\pi\hbar} . \tag{9.65}$$

This result leads to the physical interpretation of v_J as a renormalized Fermi velocity, which controls the current carried by the fermions.

9.6 Bosonization of the fermions

One of the most striking consequences of the existence of the bosonic basis set $|\{\mathcal{N}_q\}\rangle$ [see Eq. (9.37)] is the possibility of expressing the original fermion operators $\hat{a}_{k,\tau}$ and their associated fields

$$\hat{\psi}_\tau^\dagger(x) = \frac{1}{\sqrt{L}} \sum_k \hat{a}_{k,\tau}^\dagger e^{-ikx} \tag{9.66}$$

in terms of boson creation and destruction operators. Even aside from the fun of it, this construction is needed for calculating some of the most interesting physical properties of the Luttinger model, such as the Green's function, the momentum occupation number, and the density–density response function for $q \sim 2k_F$.

9.6.1 Construction of the fermion fields

Since the fermion operators change the particle number it will be necessary in this section to explicity keep track of the latter. To this end, we denote by $|N_R, N_L, \{\mathcal{N}_q\}\rangle$ the state of the system which contains $\mathcal{N}_q$ bosons in the collective mode of wave vector q, N_R right movers and N_L left movers *in excess* of the number of right and left movers in the "vacuum". The admissible values of N_L and N_R range over all the integers from $-\infty$ to $+\infty$.

In order to construct the fermion operators we introduce two global number-rising operators $\hat{U}_R$ and $\hat{U}_L$, which increment the values of N_L and N_R according to the equation

$$\hat{U}_R |N_L, N_R, \{\mathcal{N}_q\}\rangle = (-1)^{+N_L/2} |N_L, N_R + 1, \{\mathcal{N}_q\}\rangle$$
$$\hat{U}_L |N_L, N_R, \{\mathcal{N}_q\}\rangle = (-1)^{-N_R/2} |N_L + 1, N_R, \{\mathcal{N}_q\}\rangle , \tag{9.67}$$

for any choice of $|N_L, N_R, \{\mathcal{N}_q\}\rangle$. The choice of the phase factor guarantees that the two operators *anticommute* with each other, i.e. $\hat{U}_L \hat{U}_R = -\hat{U}_R \hat{U}_L$ – a property that will prove handy in constructing the fermion fields. From Eq. (9.67) one can also deduce that the $\hat{U}_\tau$ are unitary operators, i.e. $\hat{U}_\tau^\dagger \hat{U}_\tau = \hat{U}_\tau \hat{U}_\tau^\dagger = \hat{1}$.

Our strategy is to express $\hat{U}_\tau$ in terms of fermion fields and boson operators: this will implicitly define the fermion fields in terms of $\hat{U}_\tau$ and boson operators. To do this we observe that

(i) In order to increase the fermion numbers by 1, $\hat{U}_\tau$ must be a *linear* function of the fermion field operator $\hat{\psi}_\tau^\dagger(x)$.

(ii) In order to leave the boson occupation numbers $\{\mathcal{N}_q\}$ unchanged $\hat{U}_\tau$ must *commute* with all the boson operators.

(iii) When operating on the non-interacting ground-state $|N_L, N_R, \{0\}\rangle$, $\hat{U}_\tau$ must be equivalent to $\hat{a}_{k_{F\tau},\tau}^\dagger$, where $k_{F\tau}$ is the wave vector of the lowest unoccupied plane wave state of τ-movers: $k_{FR} = k_F + \frac{2\pi}{L} N_R$ and $k_{FL} = -k_F - \frac{2\pi}{L} N_L$.

The main difficulty is that the fermion fields do not commute with the bosons. For example, the naive choice

$$\hat{U}_\tau = \hat{a}_{\tau k_F,\tau}^\dagger = \frac{1}{\sqrt{L}} \int_0^L e^{ik_{F\tau}x} \hat{\psi}_\tau^\dagger(x)dx \quad \text{(wrong)}, \tag{9.68}$$

which obviously satisfies conditions (i) and (iii), fails miserably on condition (ii). In fact, making use of the definition (9.66) together with Eqs. (9.33) and (9.12), it is straightforward to show that (Exercise 10)

$$\left[\hat{\psi}_\tau^\dagger(x), \hat{b}_q\right] = -\sqrt{\frac{2\pi}{L|q|}} \Theta(\tau q) e^{-iqx} \hat{\psi}_\tau^\dagger(x),$$

$$\left[\hat{\psi}_\tau^\dagger(x), \hat{b}_q^\dagger\right] = -\sqrt{\frac{2\pi}{L|q|}} \Theta(\tau q) e^{+iqx} \hat{\psi}_\tau^\dagger(x), \tag{9.69}$$

which is, in a sense, a restatement of the fact that $\hat{\psi}_\tau^\dagger(x)$ creates an additional τ-mover at point x.[13]

What we need is a linear combination of the $\hat{\psi}_\tau$'s that commute with the bosons. With this in mind let us introduce the bosonic field

$$\hat{\phi}_\tau(x) = \sum_{\tau q>0} \sqrt{\frac{2\pi}{L|q|}} e^{+iqx} \hat{b}_q \tag{9.70}$$

and observe that the commutator of $e^{\hat{\phi}_\tau(x)}$ with $\hat{b}_q^\dagger$ has the same form as the commutator of

---

[13] From Eq. (9.35) one sees that he density of right or left movers (minus the uniform density of the "vacuum") can be written as

$$\hat{n}_\tau(x) = \frac{\hat{N}_\tau}{L} + \frac{1}{L} \sum_{q\neq 0} \Theta(\tau q)\sqrt{\frac{L|q|}{2\pi}} \left(\hat{b}_q e^{+iqx} + \hat{b}_q^\dagger e^{-iqx}\right).$$

Combining this with Eq. (9.69) we get

$$\left[\hat{\psi}_\tau^\dagger(x), \hat{n}_{\tau'}(x')\right] = -\hat{\psi}_\tau^\dagger(x)\delta_{\tau,\tau'}\delta(x - x'),$$

which shows that $\hat{\psi}_\tau^\dagger(x)$ increases the density by $\delta(x' - x)$ – the density of the additional particle.

$\hat{\psi}_\tau^\dagger(x)$ with $\hat{b}_q^\dagger$, but the opposite sign:

$$\left[e^{\hat{\phi}_\tau(x)}, \hat{b}_q^\dagger\right] = \sqrt{\frac{2\pi}{L|q|}} \Theta(\tau q) e^{iqx} e^{\hat{\phi}_\tau(x)} . \tag{9.71}$$

The commutator of $e^{-\hat{\phi}_\tau^\dagger(x)}$ with $\hat{b}_q$ has the same form as the commutator of $\hat{\psi}_\tau^\dagger(x)$ with $\hat{b}_q$, but the opposite sign:

$$\left[e^{-\hat{\phi}_\tau^\dagger(x)}, \hat{b}_q\right] = \sqrt{\frac{2\pi}{L|q|}} \Theta(\tau q) e^{-iqx} e^{-\hat{\phi}_\tau^\dagger(x)} . \tag{9.72}$$

These formulas imply that the combination

$$e^{-\hat{\phi}_\tau^\dagger(x)} \hat{\psi}_\tau^\dagger(x) e^{\hat{\phi}_\tau(x)} , \tag{9.73}$$

commutes with both $\hat{b}_q$ and $\hat{b}_q^\dagger$, suggesting that the solution of our problem is

$$\hat{U}_\tau = \frac{1}{\sqrt{L}} \int_0^L e^{ik_{F\tau}x} e^{-\hat{\phi}_\tau^\dagger(x)} \hat{\psi}_\tau^\dagger(x) e^{\hat{\phi}_\tau(x)} dx . \tag{9.74}$$

Indeed, it is evident that this form satisfies conditions (i) and (ii): we need to show that it also satisfies (iii). Let us consider, for definiteness, the case of right movers, $\tau = +1$. Let $\hat{U}_R$, defined by Eq. (9.74), operate on the non-interacting ground-state $|N_L, N_R, \{0\}\rangle$. The rightmost exponential $e^{\hat{\phi}_R(x)}$ acts as the identity operator because $\hat{b}_q|N_L, N_R, \{0\}\rangle = 0$. The other exponential $e^{-\hat{\phi}_R^\dagger(x)}$ can be expanded as $\sum_{n=0}^\infty \frac{\left(-\hat{\phi}_R^\dagger(x)\right)^n}{n!}$. Since only positive wave vectors enter the definition of $\hat{\phi}_R^\dagger$ (the hermitian conjugate of Eq. (9.70)) we realize that the x dependence of every term in this expansion is of the form e^{-iqx}, where $q > 0$, except for the $n = 0$ term, which is just the identity. Carrying out the integral over x we see that every term beyond $n = 0$ in the expansion of the exponential gives a contribution proportional to

$$\frac{1}{L} \int_0^L e^{i(k_{FR}-q)x} \hat{\psi}_R^\dagger(x) dx = \hat{a}_{k_{FR}-q,R}^\dagger \tag{9.75}$$

which kills the non-interacting ground-state for $q > 0$. Thus, only the $n = 0$ term survives, and our $\hat{U}_R$ reduces to $\hat{a}_{k_{FR},R}^\dagger$, as desired.

The bosonic form of the Fermion field operator is now trivially obtained by inverting Eq. (9.74) and making use of the fact that $\hat{U}_\tau$ commutes with all the bosons. This gives

$$\hat{\psi}_\tau^\dagger(x) = \frac{1}{\sqrt{L}} e^{-i\tau(k_F + \frac{2\pi}{L}\hat{N}_\tau)x} e^{\hat{\phi}_\tau^\dagger(x)} e^{-\hat{\phi}_\tau(x)} \hat{U}_\tau , \tag{9.76}$$

which is the central result of this section.

In algebraic manipulations involving $\hat{\psi}_\tau$ one frequently needs the commutator between the boson field operators that appear in the exponent of Eq. (9.76). From the definition (9.70) and the canonical commutation relations we immediately get

$$[\hat{\phi}_\tau(x), \hat{\phi}_\tau^\dagger(x')] = \sum_{q\tau>0} \frac{2\pi}{L|q|} e^{iq(x-x')} . \tag{9.77}$$

Unfortunately, the sum over q fails to converge at $x = x'$. This should not come as a complete surprise, since we know that the Luttinger model breaks down at length scales shorter than Λ^{-1}, where Λ is the ultraviolet cutoff introduced in Section 9.2. The difficulty is solved by multiplying the summand by the convergence factor $e^{-|q|/\Lambda}$ and letting Λ tend to infinity only at the end of the calculation.[14] This gives

$$[\hat{\phi}_\tau(x), \hat{\phi}_\tau^\dagger(x')] = \sum_{n=1}^{\infty} \frac{1}{n} \left(e^{\frac{2\pi i \tau(x-x')}{L}} e^{-\frac{2\pi}{L\Lambda}} \right)^n$$

$$= -\ln\left(1 - e^{\frac{2\pi i \tau(x-x'+i\tau/\Lambda)}{L}} \right), \tag{9.78}$$

where we have set $q = \frac{2\pi n}{L}$ (n integer) and made use of the power series expansion of the logarithm: $\sum_{n=1}^{\infty} \frac{x^n}{n} = -\ln(1-x)$.

9.6.2 Commutation relations

Let us now verify that Eq. (9.76) satisfies the canonical anticommutation relations

$$\{\hat{\psi}_\tau(x), \hat{\psi}_{\tau'}^\dagger(x')\} = \delta(x - x')\delta_{\tau,\tau'},$$
$$\{\hat{\psi}_\tau(x), \hat{\psi}_{\tau'}(x')\} = \{\hat{\psi}_\tau^\dagger(x), \hat{\psi}_{\tau'}^\dagger(x')\} = 0. \tag{9.79}$$

The case $\tau \neq \tau'$ is trivial since $\hat{\phi}_R(x)$ commutes with $\hat{\phi}_L(x')$, $\hat{U}_\tau$ commutes with $\hat{N}_{-\tau}$, and $\hat{U}_R$ anticommutes with $\hat{U}_L$ as noted after Eq. (9.67).

To treat the case $\tau = \tau'$ we must become proficient at handling products of exponentials of boson operators. This is done with the help of the *disentanglement lemma*, which is stated below and proved in Appendix 20.

The disentanglement lemma. *If two operators $\hat{A}$ and $\hat{B}$ have the property that $[\hat{A}, \hat{B}]$ commutes with both $\hat{A}$ and $\hat{B}$ then*

$$e^{\hat{A}+\hat{B}} = e^{-\frac{1}{2}[\hat{A},\hat{B}]} e^{\hat{A}} e^{\hat{B}}. \tag{9.80}$$

An immediate and useful consequence of the lemma is that

$$e^{\hat{A}} e^{\hat{B}} = e^{\hat{B}} e^{\hat{A}} e^{[\hat{A},\hat{B}]}. \tag{9.81}$$

Let us apply the lemma to the calculation of the anticommutators (9.79). By repeated application of (9.80) and (9.81) we rewrite the anticommutator of two fermion fields as

$$\{\hat{\psi}_\tau^\dagger(x), \hat{\psi}_\tau^\dagger(x')\} = \frac{e^{-i\tau[k_F + \frac{2\pi}{L}\hat{N}_\tau](x+x')}}{L} e^{(\phi_\tau^\dagger(x)+\phi_\tau^\dagger(x'))} \hat{U}_\tau^2 e^{-(\phi_\tau(x)+\phi_\tau(x'))}$$

$$\times \left(e^{-\frac{2\pi i \tau}{L}x} e^{-[\phi_\tau(x),\phi_\tau^\dagger(x')]} + e^{-\frac{2\pi i \tau}{L}x'} e^{-[\phi_\tau(x'),\phi_\tau^\dagger(x)]} \right). \tag{9.82}$$

The exponential factors $e^{-\frac{2\pi i \tau}{L}x}$ in the last line arise from the commutation rule

$$e^{-\frac{2\pi i \tau}{L}\hat{N}_\tau x} \hat{U}_\tau = e^{-\frac{2\pi i \tau}{L}x} \hat{U}_\tau e^{-\frac{2\pi i \tau}{L}\hat{N}_\tau x}. \tag{9.83}$$

[14] For a more physical perspective on the choice of Λ see footnote[18].

Notice that the application of the disentanglement lemma to the writing of Eq. (9.82) is justified only because the commutator $[\hat{\phi}_\tau(x), \hat{\phi}_\tau^\dagger(x')]$ is a number. Substituting Eq. (9.78) in the second line of Eq. (9.82) and taking the limit $\Lambda \to \infty$ we get

$$e^{-\frac{2\pi i \tau x}{L}}\left(1 - e^{\frac{2\pi i \tau(x-x')}{L}}\right) + e^{-\frac{2\pi i \tau x'}{L}}\left(1 - e^{\frac{2\pi i \tau(x'-x)}{L}}\right) = 0 \qquad (9.84)$$

showing that the anticommutator is null, as expected.

In a similar way we find that

$$\{\hat{\psi}_\tau(x), \hat{\psi}_\tau^\dagger(x')\} = \frac{e^{-i\tau[k_F + \frac{2\pi}{L}\hat{N}_\tau](x'-x)}}{L}e^{(\phi_\tau^\dagger(x')-\phi_\tau^\dagger(x))}e^{-(\phi_\tau(x')-\phi_\tau(x))}$$

$$\times \left(e^{\frac{2\pi i \tau}{L}(x-x')}e^{[\phi_\tau(x),\phi_\tau^\dagger(x')]} + e^{[\phi_\tau(x'),\phi_\tau^\dagger(x)]}\right). \qquad (9.85)$$

Making use of the commutator (9.78), we rewrite the contents of the round bracket on the second line of this equation as

$$\frac{1}{e^{\frac{2\pi i \tau}{L}(x'-x)} - e^{-\frac{2\pi}{L\Lambda}}} + \frac{1}{1 - e^{\frac{2\pi i \tau}{L}(x'-x+i\tau/\Lambda)}}. \qquad (9.86)$$

In the limit $\Lambda \to \infty$, this expression vanishes for $x \neq x'$ but diverges for $x = x'$. In fact, for x close to x', expanding the exponentials to first order in $x - x'$ we see that Eq. (9.86) reduces to

$$\frac{iL\tau}{2\pi}\left(\frac{1}{x'-x+i\tau/\Lambda} - \frac{1}{x'-x-i\tau/\Lambda}\right) \overset{\Lambda\to\infty}{\to} L\delta(x'-x). \qquad (9.87)$$

Replacing the round brackets of Eq. (9.85) with this expression yields the canonical anti-commutator.

9.6.3 Construction of observables

The beautiful algebra derived in the previous sections can be used to directly construct physical observables, such as the density, the kinetic energy etc.

To begin with, consider the construction of the density operator $\hat{n}_\tau(x) = \hat{\psi}_\tau^\dagger(x)\hat{\psi}_\tau(x)$. Using the bosonic representation of the fermion field (9.76) and the disentanglement lemma we rewrite this as

$$\hat{n}_\tau(x) = \frac{1}{L}e^{\hat{\phi}_\tau^\dagger(x)}e^{-\hat{\phi}_\tau(x)}e^{-\hat{\phi}_\tau^\dagger(x)}e^{\hat{\phi}_\tau(x)}$$

$$= \frac{1}{L}e^{[\hat{\phi}_\tau(x),\hat{\phi}_\tau^\dagger(x)]}$$

$$= \frac{1}{L}\frac{1}{1 - e^{-\frac{2\pi}{L\Lambda}}}. \qquad (9.88)$$

This is just a number that does not depend on x and diverges for $\Lambda \to \infty$! The fact does not come as a surprise, since in setting up the model we postulated an infinite homogeneous density of fermions filling the space: the physically relevant part of the density – an x-dependent operator – remains invisible against this infinite background. What we really want to calculate is the normal-ordered density operator, which measures density fluctuations relative to the vacuum density. To do this we resort to a cute mathematical trick known as *point splitting*. The idea is to put the two fermion field operators at two slightly different points in space, and consider the limit in which the two points coincide. In other words, we define

$$\hat{n}_\tau(x) = \lim_{a \to 0} \hat{\psi}_\tau^\dagger \left(x + \frac{a}{2} \right) \hat{\psi}_\tau \left(x - \frac{a}{2} \right) , \tag{9.89}$$

where a is real. The right-hand side of this equation can be straightforwardly expressed in terms of boson operators. All we need is the disentanglement lemma, the identity $\hat{U}_\tau \hat{U}_\tau^\dagger = 1$, and the commutator

$$\left[\hat{\phi}_\tau \left(x + \frac{a}{2} \right), \hat{\phi}_\tau^\dagger \left(x - \frac{a}{2} \right) \right] = - \ln \left(1 - e^{\frac{2\pi i \tau a}{L}} \right) \tag{9.90}$$

where, since a is finite, there is no need for any convergence factor. The result is

$$\hat{\psi}_\tau^\dagger \left(x + \frac{a}{2} \right) \hat{\psi}_\tau \left(x - \frac{a}{2} \right) = \frac{1}{L} \frac{e^{-i\tau k_F a}}{1 - e^{\frac{2\pi i \tau a}{L}}} e^{-i\frac{2\pi \tau}{L} \hat{N}_\tau a} e^{\hat{D}_\tau^\dagger(x)} e^{-\hat{D}_\tau(x)} , \tag{9.91}$$

where we have defined

$$\hat{D}_\tau(x) \equiv \hat{\phi}_\tau \left(x + \frac{a}{2} \right) - \hat{\phi}_\tau \left(x - \frac{a}{2} \right) . \tag{9.92}$$

Of course this expression still diverges in the limit $a \to 0$: however, we are now in a position to regularize it by simply subtracting the vacuum expectation value

$$\left\langle 0 \left| \hat{\psi}_\tau^\dagger \left(x + \frac{a}{2} \right) \hat{\psi}_\tau \left(x - \frac{a}{2} \right) \right| 0 \right\rangle = \frac{1}{L} \frac{e^{-i\tau k_F a}}{1 - e^{\frac{2\pi i \tau a}{L}}} . \tag{9.93}$$

This procedure leads to the following expression for the *normal-ordered* density operator:

$$: \hat{n}_\tau(x) := \frac{1}{L} \lim_{a \to 0} \frac{e^{-i\tau k_F a}}{1 - e^{\frac{2\pi i \tau a}{L}}} \left(e^{-\frac{2\pi i \tau}{L} \hat{N}_\tau a} e^{\hat{D}_\tau^\dagger(x)} e^{-\hat{D}_\tau(x)} - 1 \right) . \tag{9.94}$$

The $a \to 0$ limit is well behaved and can be easily worked out to be

$$\begin{aligned} : \hat{n}_\tau(x) : &= \frac{\hat{N}_\tau}{L} + i\frac{\tau}{2\pi} \left(\frac{\partial \hat{\phi}_\tau^\dagger(x)}{\partial x} - \frac{\partial \hat{\phi}_\tau(x)}{\partial x} \right) \\ &= \frac{\hat{N}_\tau}{L} + \frac{1}{L} \sum_{\tau q > 0} \sqrt{\frac{L|q|}{2\pi}} \left(\hat{b}_q e^{iqx} + \hat{b}_q^\dagger e^{-iqx} \right), \end{aligned} \tag{9.95}$$

which is consistent with Eq. (9.35) (see also footnote[13]).

To calculate the kinetic energy observe that

$$\begin{aligned} \sum_k e^{-ik\tau a} : \hat{a}_{k,\tau}^\dagger \hat{a}_{k,\tau} : &= \int_0^L : \hat{\psi}_\tau^\dagger \left(x + \tau\frac{a}{2} \right) \hat{\psi}_\tau \left(x - \tau\frac{a}{2} \right) : dx \\ &= \frac{e^{-i\tau k_F a}}{1 - e^{\frac{2\pi i \tau a}{L}}} \frac{1}{L} \int_0^L \left(e^{-\frac{2\pi i \tau}{L} \hat{N}_\tau a} e^{\hat{D}_\tau^\dagger(x)} e^{-\hat{D}_\tau(x)} - 1 \right) dx . \end{aligned} \tag{9.96}$$

Expanding both sides of this equation to first order in a we obtain

$$\sum_k \tau k : \hat{a}^\dagger_{k,\tau} \hat{a}_{k,\tau} := k_F \hat{N}_\tau + \frac{\pi}{L} \hat{N}^2_\tau + \sum_{\tau q > 0} |q| \hat{b}^\dagger_q \hat{b}_q . \tag{9.97}$$

The quantity on the left-hand side is precisely the kinetic energy (divided by $\hbar v_F$), and this equation reduces to Eq. (9.40) when one sets $N_\tau = 0$ and sums over τ (Exercise 12).

Finally, let us obtain the expression for the density fluctuation operator at $q = 2k_F$. Combining Eqs. (9.10) and (9.76) and carrying out the sum over k we obtain

$$\hat{n}_{2k_F} = \frac{1}{L} \int_0^L \hat{\psi}^\dagger_L(x) \hat{\psi}_R(x) e^{-2ik_F x} dx . \tag{9.98}$$

This expression will be needed for the calculation of the "Friedel oscillations" and the transmission probability across a barrier in Sections 9.10 and 9.11, respectively.

9.7 The Green's function

9.7.1 Analytical formulation

In order to explore the single-particle properties of the Luttinger model we now begin the study of the Green's function

$$G_\tau(x, t) = \Theta(t) G_{\tau,>}(x, t) + \Theta(-t) G_{\tau,<}(x, t) , \tag{9.99}$$

where

$$G_{\tau,>}(x, t) = -i \langle \hat{\psi}_\tau(x, t) \hat{\psi}^\dagger_\tau(0, 0) \rangle \tag{9.100}$$

and

$$G_{\tau,<}(x, t) = +i \langle \hat{\psi}^\dagger_\tau(0, 0) \hat{\psi}_\tau(x, t) \rangle \tag{9.101}$$

are the Green's functions for positive and negative times respectively. The ground work has already been laid out. We set, for simplicity, $N_R = N_L = 0$. First, the fermion field operator (9.76) is rewritten in the form

$$\hat{\psi}^\dagger_\tau(x) = \frac{e^{-i\tau k_F x} e^{-\hat{\theta}_\tau(x)}}{\sqrt{L(1 - e^{-\frac{2\pi}{L\Lambda}})}} \hat{U}_\tau , \tag{9.102}$$

where

$$\hat{\theta}_\tau(x) \equiv \hat{\phi}_\tau(x) - \hat{\phi}^\dagger_\tau(x)$$
$$= \sum_{\tau q > 0} \sqrt{\frac{2\pi}{L|q|}} \left(e^{+iqx} \hat{b}_q - e^{-iqx} \hat{b}^\dagger_q \right) . \tag{9.103}$$

Putting this in Eq. (9.99) we see that the Green's function for $t > 0$ is given by

$$G_{\tau,>}(x, t) = -i \frac{e^{i\tau k_F x} \left\langle e^{\hat{\theta}_\tau(x,t)} e^{-\hat{\theta}_\tau(0,0)} \hat{U}^\dagger_\tau(t) \hat{U}_\tau \right\rangle}{L(1 - e^{-\frac{2\pi}{L\Lambda}})} . \tag{9.104}$$

The time dependence of $\hat{\theta}_\tau$ and $\hat{U}_\tau$ is controlled by the hamiltonian $\hat{H}_{LM}$ of Eq. (9.41)
$\hat{U}_\tau$ commutes with $\hat{H}_{LM}$ and therefore $\hat{U}_\tau^\dagger(t)\hat{U}_\tau = \hat{U}_\tau^\dagger\hat{U}_\tau = 1$.[15] As for $\hat{\theta}_\tau$, after making
the transformation (9.42) to express the boson operators $\hat{b}_q$ and $\hat{b}_q^\dagger$ in terms of the boson
operators $\hat{\beta}_q$ and $\hat{\beta}_q^\dagger$ that diagonalize $\hat{H}_{LM}$ we get

$$\hat{\theta}_\tau(x,t) = \sum_{\tau q > 0} \sqrt{\frac{2\pi}{|q|L}} \cosh \varphi_q \left(e^{-i(\omega_q t - qx)} \hat{\beta}_q - e^{i(\omega_q t - qx)} \hat{\beta}_q^\dagger \right)$$

$$+ \sum_{\tau q < 0} \sqrt{\frac{2\pi}{L|q|}} \sinh \varphi_q \left(e^{-i(\omega_q t - qx)} \hat{\beta}_q - e^{i(\omega_q t - qx)} \hat{\beta}_q^\dagger \right)$$

$$= \hat{\theta}_{\tau,+}(x,t) + \hat{\theta}_{\tau,-}(x,t) . \qquad (9.105)$$

Notice that $\hat{\theta}_\tau(x,t)$ is the sum of two mutually commuting terms $\hat{\theta}_{\tau,+}$ and $\hat{\theta}_{\tau,-}$ containing
boson operators with $\tau q > 0$ and $\tau q < 0$ respectively. Therefore, the exponentials of $\hat{\theta}_{\tau,+}$
and $\hat{\theta}_{\tau,-}$ can be averaged independently and the Green's function takes the form

$$G_{\tau,>}(x,t) = -i \frac{e^{i\tau k_F x}}{L(1 - e^{-\frac{2\pi}{L\Lambda}})} \left\langle e^{\hat{\theta}_{\tau,+}(x,t)} e^{-\hat{\theta}_{\tau,+}(0,0)} \right\rangle \left\langle e^{\hat{\theta}_{\tau,-}(x,t)} e^{-\hat{\theta}_{\tau,-}(0,0)} \right\rangle . \qquad (9.106)$$

9.7.2 Evaluation of the averages

The equilibrium averages appearing in Eq. (9.106) can be evaluated with the help of a
powerful theorem, which we state here and prove in Appendix 21.

The independent boson theorem. *Let $\{\hat{b}_i\}$ denote a set of boson operators and let*

$$\langle \hat{A} \rangle = \frac{Tr e^{-\beta \hat{H}} \hat{A}}{Tr e^{-\beta \hat{H}}}$$

*denote the equilibrium average of an operator $\hat{A}$ in the thermal ensemble with hamiltonian
$\hat{H} = \sum_i \hbar\omega_i \hat{\beta}_i^\dagger \hat{\beta}_i$, where $\{\hat{\beta}_i\}$ denotes set of boson operators that are connected to the $\hat{b}_i$'s
and $\hat{b}_i^\dagger$'s by a linear unitary transformation.[16] Then*

$$\left\langle e^{\sum_i \left(x_i \hat{b}_i^\dagger + y_i \hat{b}_i \right)} \right\rangle = e^{\frac{1}{2} \sum_i \left\langle \left(x_i \hat{b}_i^\dagger + y_i \hat{b}_i \right)^2 \right\rangle} , \qquad (9.107)$$

where $\{x_i\},\{y_i\}$ are two arbitrary sets of complex numbers.

To apply this theorem to the evaluation of Eq. (9.106) we first use once again the disentanglement lemma to write

$$e^{\hat{\theta}_{\tau,\pm}(x,t)} e^{-\hat{\theta}_{\tau,\pm}(0,0)} = e^{\{\hat{\theta}_{\tau,\pm}(x,t) - \hat{\theta}_{\tau,\pm}(0,0) - \frac{1}{2}[\hat{\theta}_{\tau,\pm}(x,t),\hat{\theta}_{\tau,\pm}(0,0)]\}} , \qquad (9.108)$$

---

[15] By setting $N_\tau = 0$ we have implicitly set the chemical potential μ at zero energy. In the general case one has $\hat{U}_\tau^\dagger(t) = \hat{U}_\tau^\dagger e^{-i\mu\tau t/\hbar}$.
[16] As a special case, one can have $\hat{\beta}_i = \hat{b}_i$.

where the commutator that appears in the exponent is just a number. The independent boson theorem then tells us that

$$\left\langle e^{\hat{\theta}_{\tau,\pm}(x,t)} e^{-\hat{\theta}_{\tau,\pm}(0,0)} \right\rangle = e^{F_{\tau,\pm}(x,t)} , \qquad (9.109)$$

where

$$F_{\tau,\pm}(x,t) = \left\{ \frac{1}{2} \left\langle \left(\hat{\theta}_{\tau,\pm}(x,t) - \hat{\theta}_{\tau,\pm}(0,0)\right)^2 \right\rangle - \frac{1}{2}[\hat{\theta}_{\tau,\pm}(x,t), \hat{\theta}_{\tau,\pm}(0,0)] \right\} . \qquad (9.110)$$

The calculation of $F_{\tau,\pm}(x,t)$ is straightforward. For example, from the first line of Eq. (9.105) we get

$$\frac{1}{2} \left\langle \left(\hat{\theta}_{\tau,+}(x,t) - \hat{\theta}_{\tau,+}(0,0)\right)^2 \right\rangle$$

$$= -\sum_{\tau q > 0} \frac{2\pi}{|q|L} \cosh^2 \varphi_q \left| e^{i(\omega_q t - qx)} - 1 \right|^2 \left(n_B(\hbar\omega_q) + \frac{1}{2} \right) , \qquad (9.111)$$

where $n_B(\hbar\omega_q) = \langle \hat{\beta}_q^\dagger \hat{\beta}_q \rangle = (e^{\beta\hbar\omega_q} - 1)^{-1}$ is the thermal occupation number for the mode of wave vector q. We also have

$$\frac{1}{2}[\hat{\theta}_{\tau,+}(x,t), \hat{\theta}_{\tau,+}(0,0)] = \frac{1}{2} \sum_{q \neq 0} \Theta(\tau q) \frac{2\pi}{|q|L} \cosh^2 \varphi_q \left(e^{i(\omega_q t - qx)} - e^{-i(\omega_q t - qx)}\right) . \qquad (9.112)$$

Combining these two equations we get

$$F_{\tau,+}(x,t) = \sum_{\tau q > 0} \frac{2\pi}{|q|L} \cosh^2 \varphi_q \left(e^{-i(\omega_q t - qx)} - 1\right) [1 + n_B(\hbar\omega_q)]$$

$$+ \sum_{\tau q > 0} \frac{2\pi}{|q|L} \cosh^2 \varphi_q \left(e^{i(\omega_q t - qx)} - 1\right) n_B(\hbar\omega_q) . \qquad (9.113)$$

Notice that in Eq. (9.113) we can write $|q|$ as τq and $\omega_q = c\tau q$.[17] Since $1 + n_B(\hbar c\tau q) = -n_B(-\hbar c\tau q)$ (while φ_q is an even function of q) we make the change of variable $q \to -q$ in the first line and combine it with the second line to get the compact expression

$$F_{\tau,+}(x,t) = \sum_{q \neq 0} \frac{2\pi}{qL} \cosh^2 \varphi_q n_B(\hbar c q) \left(e^{iq(ct - \tau x)} - 1\right) e^{q/\Lambda}$$

$$\equiv F_+(ct - \tau x) . \qquad (9.114)$$

The factor $e^{q/\Lambda}$, where Λ is the ultraviolet cutoff discussed in Section 9.2, has been introduced to ensure the convergence of the integral at the $q \to -\infty$ limit. Convergence for $q \to \infty$ is guaranteed by the Bose occupation factor, provided $k_B T < \hbar c \Lambda$. Physically, this action is justified by the observation that the Luttinger model is meaningful only for wave vectors and temperatures that are smaller than Λ and $\hbar c \Lambda$ respectively.

---

[17] Strictly speaking, the sound velocity c should be a function of q. We shall ignore this complication since it has no effect on the qualitative behavior of the Green's function. The case of coulomb interactions, in which both c and φ_q diverge for $q \to 0$ will be analyzed separately.

Essentially the same expression is obtained for $F_{\tau,-}(x,t)$, except for the fact that $\cosh^2 \varphi$ is replaced by $\sinh^2 \varphi$ and x by $-x$. Thus

$$F_{\tau,-}(x,t) = \sum_{q\neq 0} \frac{2\pi}{qL} \sinh^2 \varphi_q n_B(\hbar c_q q) \left(e^{iq(c_q t + \tau x)} - 1\right) e^{q/\Lambda}$$

$$\equiv F_-(ct + \tau x) . \tag{9.115}$$

Then, upon combining Eqs. (9.106) and (9.109) we come to the following expression for $G_{\tau,>}$:

$$G_{\tau,>}(x,t) = -i\frac{e^{i\tau k_F x}}{L(1 - e^{-\frac{2\pi}{L\Lambda}})} e^{F_+(ct - \tau x)} e^{F_-(ct + \tau x)}, \tag{9.116}$$

where F_+ and F_- are defined by Eqs. (9.114) and (9.115) respectively.

The Green's function for negative times, $G_\tau(x,t) = G_{\tau,<}(x,t)$, is obtained by interchanging the order of the exponentials in Eq. (9.104). It is easy to see that this amounts to replacing (x,t) by $(-x,-t)$ in the two F-functions. Thus, without repeating the calculation, we find

$$G_{\tau,<}(x,t) = i\frac{e^{i\tau k_F x}}{L(1 - e^{-\frac{2\pi}{L\Lambda}})} e^{F_+(-ct + \tau x)} e^{F_-(-ct - \tau x)} . \tag{9.117}$$

This completes the analytical calculation of the Green's function for both positive and negative times.

A remarkable feature of the result is the appearance of left-propagating terms (functions of $ct + x$) in the Green's function for right movers and vice versa. The physical reason for this mixing is clear. When one adds a right mover to the system one perturbs, via the coulomb interaction, the density of left movers as well as that of right movers. Similarly to casting a pebble into a calm pond the injection of the extra particle is accompanied by trains of collective waves propagating in both directions. This is described by the two exponential factors in Eq. (9.116). The mixing disappears in the absence of interaction, since $\sinh \varphi_q = 0$ in that case.

9.7.3 Non-interacting Green's function

The Green's function can be written in a perhaps more revealing form as the product of the Green's function of a "non-interacting" system (in which, however, we set $v_F = c$) and a correction factor. To accomplish this, notice that

$$F_+(x) = F_+^{(0)}(x) + F_-(x) , \tag{9.118}$$

where

$$F_+^{(0)}(x) = \sum_{q\neq 0} \frac{2\pi}{qL} n_B(\hbar cq) \left(e^{iqx} - 1\right) e^{q/\Lambda} . \tag{9.119}$$

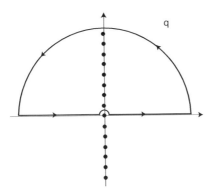

Fig. 9.7. Integration contour in the complex q-plane for the evaluation of Eq. (9.122).

Substituting this in Eq. (9.116) we get

$$G_{>,\tau}(x, t) = G^{(0)}_{>,\tau}(x, t)e^{F_-(ct-\tau x)}e^{F_-(ct+\tau x)} ,$$ (9.120)

where

$$G^{(0)}_{>,\tau}(x, t) = -i \frac{e^{i\tau k_F x}}{L(1 - e^{-\frac{2\pi}{L\Lambda}})}e^{F^{(0)}_+(ct-\tau x)} .$$ (9.121)

To calculate the key function $F^{(0)}_+(x)$ we convert the sum over q in Eq. (9.119) into an integral:

$$F^{(0)}_+(x) = \mathcal{P} \int_{-\infty}^{+\infty} \frac{dq}{q} \frac{e^{iqx} - 1}{e^{\beta \hbar v_F q} - 1}e^{q/\Lambda} ,$$ (9.122)

where the principal part $\mathcal{P}$ excludes the contribution from $q = 0$. The integral is easily done by the method of residues using the fact that the Bose distribution function has poles with residue $\frac{1}{\beta \hbar c}$ at $q = i\frac{2\pi n}{\beta \hbar c}$ where n is an integer. The path of integration is shown in Fig. 9.7 and the result of the calculation is (Exercise 13)

$$F^{(0)}_+(x) = -\ln \frac{1 - e^{-\frac{2\pi}{\beta \hbar c}}(x - i/\Lambda)}{1 - e^{i\frac{2\pi}{\beta \hbar c \Lambda}}} - \frac{\pi}{\beta \hbar v_F}(x - i/\Lambda) ,$$ (9.123)

where the second term on the right-hand side subtracts the contribution of the small semi-circle around $q = 0$.

Substituting the expression (9.123) into Eq. (9.116) and going to the thermodynamic limit $L \to \infty$ we get

$$G^{(0)}_{\tau,>}(x, t) = \frac{e^{i\tau k_F x}}{2\pi} \frac{\frac{\pi}{\beta \hbar c}}{\sinh\left[\frac{\pi}{\beta \hbar c}(\tau x - ct + i/\Lambda)\right]} .$$ (9.124)

This is the same result one would obtain by directly calculating

$$G^{(0)}_{\tau,>}(x, t) = -\frac{i}{L}\sum_k (1 - n^{(0)}_{k,\tau})e^{ikx}e^{-ic(\tau k - k_F)t}e^{-k/\Lambda} ,$$ (9.125)

as the reader can verify by converting the sum over k into an integral and making use of the formula

$$\int_{-\infty}^{\infty} dx \, \frac{e^{izx}}{e^x + 1} = -i \frac{\pi}{\sinh \pi z} \, , \qquad (-1 < \Im m \, z < 0) \, . \tag{9.126}$$

Notice that the true non-interacting Green's function is obtained by putting $c = v_F$ in Eq. (9.124).

9.7.4 Asymptotic behavior

In this Section we study in detail the long-time behavior of the Green's function. In a noninteracting system at *zero temperature* the Green's function (9.124) reduces to

$$G_{\tau,>}^{(0)}(x,t) = \frac{e^{i\tau k_F x}}{2\pi} \frac{1}{\tau x - ct + i/\Lambda} \, , \tag{9.127}$$

which decays rather slowly, as t^{-1}, for $t \to \infty$. This slow decay indicates the presence of well defined quasiparticle excitations at the Fermi level. On the other hand, in the presence of interaction, the Green's function vanishes more rapidly than $\frac{1}{|t|}$, reflecting the absence of fermionic quasiparticles.

To calculate the behavior of the interacting Green's function at large times we go back to Eq. (9.120) and notice that the zero-temperature limit of F_- is

$$F_-(ct \pm \tau x) = \sum_{q>0} \frac{2\pi}{qL} \sinh^2 \varphi_q \left(e^{-iq(ct\pm\tau x)} - 1 \right) \, . \tag{9.128}$$

Unfortunately, the presence of the q-dependent rotation angle φ_q prevents us from calculating this sum analytically. But, in the limit in which $ct + \tau x$ and $ct - \tau x$ are both large, an analytic evaluation becomes possible.

The scale on which $ct \pm \tau x$ is large is set by the inverse of the scale of variation of φ_q as a function of q, which we identify with the screening length λ. For $|ct \pm \tau x| \gg \lambda$ we can replace φ_q by its $q \to 0$ limit φ, provided the wave vector sum in Eq. (9.128) is cutoff at $q \sim \frac{1}{\lambda}$, for example by multiplying the summand by $e^{-q\lambda}$.[18]

At *zero temperature* this procedure gives

$$\begin{aligned}
F_-(ct \pm \tau x) &= \sinh^2 \varphi \sum_{q>0} \frac{2\pi}{qL} \left(e^{-iq(ct\pm\tau x)} - 1 \right) e^{-q\lambda} \\
&= -\sinh^2 \varphi \ln \frac{1 - e^{-\frac{2\pi i}{L}(ct\pm\tau x - i\lambda)}}{1 - e^{-\frac{2\pi}{L}\lambda}} \\
&\overset{L\to\infty}{\to} -\sinh^2 \varphi \ln \left(i \frac{ct \pm \tau x - i\lambda}{\lambda} \right) \, ,
\end{aligned} \tag{9.129}$$

---

[18] In practice, this procedure corresponds to replacing the formal ultraviolet cutoff Λ by the physical screening wave vector λ^{-1}. The results of calculations done in this manner will be reliable only at length scales larger than λ, which, in any case, is roughly the length scale below which the Luttinger liquid model breaks down.

which, substituted into Eq. (9.120), yields

$$G_{\tau,>}(x,t) \overset{|x \pm ct| \gg \lambda}{\longrightarrow} \frac{e^{i\tau k_F x}}{2\pi \, (\tau x - ct)} \left(\frac{\lambda}{\tau x - ct} \right)^{\sinh^2 \varphi} \times \left(\frac{\lambda}{\tau x + ct} \right)^{\sinh^2 \varphi} . \qquad (9.130)$$

The Green's function tends to zero for large times as $|t|^{-(1+2\sinh^2 \varphi)}$, which is faster than $|t|^{-1}$, confirming the absence of quasiparticles at the Fermi level.

At finite temperature, the asymptotic behavior (9.130) holds only up to a temperature-dependent cutoff length L_T of the order of $\frac{\hbar c}{\pi k_B T}$, i.e. for $\lambda \ll |\tau x \pm ct| \ll L_T$. For larger values of $|\tau x \pm ct|$ the Green's function tends to zero exponentially. The finite temperature expression can be obtained from the zero-temperature one via the substitution

$$\frac{1}{\tau x \pm ct} \to \frac{1}{L_T \sinh\left[(\tau x \pm ct)/L_T\right]} , \qquad (9.131)$$

where

$$L_T \equiv \frac{\hbar c}{\pi k_B T} \qquad (9.132)$$

is the thermal cutoff length. Notice that this gives the correct result for the non-interacting Green's function (see Eq. (9.124)).

9.8 The spectral function

Once the Green's function is known, the spectral function can be calculated as the Fourier transform of $G_{\tau,>}(k,t)$:

$$A_{\tau,>}(k,\omega) = -\frac{1}{2\pi i} \int_{-\infty}^{+\infty} dt \int_{-\infty}^{+\infty} dx \, G_{\tau,>}(x,t) e^{i(\omega t - kx)} , \qquad (9.133)$$

a fact that can be immediately verified from the formal definitions of these quantities. $A_{\tau,<}(k,\omega)$ is similarly related to $G_{\tau,<}(x,t)$. The two components of the spectral function are the mirror image of each other, in the sense that $A_{\tau,<}(\tilde{k},\omega) = A_{\tau,>}(-\tilde{k},-\omega)$, where $\tilde{k} = k - \tau k_F$. Furthermore, at $T = 0$ $A_>$ vanishes for $\omega < 0$ and $A_<$ for $\omega > 0$.

Zero temperature – Since we are interested in the behavior of the spectral function for $\tilde{k} \sim 0$ and $\omega \sim 0$ we can work with the asymptotic form of the Green's function, Eq. (9.130). What we need is the convolution product of the Fourier transforms of the three factors appearing in Eq. (9.130), namely, $\left(\tau x - ct + i\frac{\eta L}{2\pi} \right)^{-1}$, $(\tau x - ct + i\lambda)^{-\alpha}$, and $(\tau x + ct - i\lambda)^{-\alpha}$, where $\alpha = \sinh^2 \varphi$.[19] The first Fourier transform gives just the non-interacting spectral

---

[19] The Fourier transform of a product equals the convolution of the Fourier transforms:

$$\int dx \int dt f(x,t) g(x,t) e^{i(\omega t - kx)} = \int \frac{dk'}{2\pi} \int \frac{d\omega'}{2\pi} f(k',\omega') g(k-k',\omega-\omega') ,$$

where $f(k,\omega)$ and $g(k,\omega)$ are the Fourier transforms of f and g.

function

$$A^{(0)}_{\tau,>}(k,\omega) = \Theta(\omega)\delta(\omega - \tau c\tilde{k}) .$$ (9.134)

For the other two, the relevant integrals have the form

$$\int_{-\infty}^{+\infty} dt \int_{-\infty}^{+\infty} dx \frac{\lambda^\alpha e^{i(\omega t - \tilde{k}x)}}{(\tau x \pm ct \mp i\lambda)^\alpha} .$$ (9.135)

Making the change of variables $(t,x) \to (z,x)$ where $z = \tau x \pm ct$ and integrating over x this becomes

$$2\pi \delta(\omega \pm \tau c\tilde{k}) \int_{-\infty}^{+\infty} dz \frac{\lambda^\alpha e^{\pm i \frac{\omega}{c} z}}{(z \mp i\lambda)^\alpha} .$$ (9.136)

Notice that the integral vanishes if ω is negative. For positive ω it can be calculated analytically (see Gradshteyn and Ryzhik, 1965, 3.382.6) with the following result:

$$\int_{-\infty}^{+\infty} dz \frac{\lambda^\alpha e^{\pm i \frac{\omega}{c} z}}{(z \mp i\lambda)^\alpha} = (\pm i)^\alpha I_\alpha(\omega) ,$$ (9.137)

where we have defined the function

$$I_\alpha(\omega) \equiv \frac{2\pi\lambda}{\Gamma(\alpha)} \Theta(\omega) \left(\frac{\lambda\omega}{c} \right)^{\alpha-1} e^{-\frac{\lambda\omega}{c}} .$$ (9.138)

It is now straightforward, if somewhat tedious, to calculate the convolution product of the three Fourier transforms. From the convolution of the last two factors we get

$$\int d\tilde{k}' \int d\omega' \delta(\omega' - \tau c\tilde{k}')\delta(\omega - \omega' + \tau c\tilde{k} - \tau c\tilde{k}')I_\alpha(\omega')I_\alpha(\omega - \omega')$$
$$= \frac{1}{2c} I_\alpha \left(\frac{\omega + \tau c\tilde{k}}{2} \right) I_\alpha \left(\frac{\omega - \tau c\tilde{k}}{2} \right) .$$ (9.139)

A second convolution with the factor $\Theta(\omega)\delta(\omega - \tau c\tilde{k})$ yields

$$A_{\tau,>}(k,\omega) = \frac{1}{8\pi^2 c^2} I_\alpha \left(\frac{\omega - \tau c\tilde{k}}{2} \right) \int_0^{\frac{\omega + \tau c\tilde{k}}{2}} I_\alpha(\omega')d\omega' .$$ (9.140)

The frequency integral is elementary, and the final result evaluated at the Fermi wave vector ($\tilde{k} = 0$) is

$$A_{\tau,>}(k_F,\omega) = \frac{\lambda}{c} \frac{1}{2\Gamma^2(\alpha)} \left(\frac{\lambda\omega}{2c} \right)^{\alpha-1} e^{-\frac{\lambda\omega}{2c}} \gamma \left(\alpha, \frac{\lambda\omega}{2c} \right) \Theta(\omega) ,$$ (9.141)

where $\gamma(\alpha, u) \equiv \int_0^u dx e^{-x} x^{\alpha-1}$ is the incomplete γ-function. Notice that this expression varies as $\omega^{2\alpha-1} = \omega^{2\sinh^2 \varphi - 1}$ for $\omega \to 0$.

Thus we see that in the presence of interactions, no matter how weak, the behavior of the spectral function is changed from a δ-function to a power law. The exponent is negative for weak coupling, but becomes positive for $\varphi > \sinh^{-1} \left(\frac{1}{\sqrt{2}} \right)$. Physically, this is

manifestation of the orthogonality catastrophe, whereby the state obtained by adding an
lectron to the highly correlated ground-state results in a state that is essentially orthogonal
⊃ the ground-state of the $N + 1$-electron system.

From a mathematical point of view, the vanishing of the spectral function implies the
existence of a pole singularity in the self-energy at the chemical potential: it is evident
hat such a pole could not have been obtained by doing perturbation theory about the
non-interacting ground-state.

Finite temperature – Let us now consider the calculation of the spectral function at finite
temperature. We continue to focus on $A_>(k, \omega)$ since, as we have seen, $A_<(k, \omega)$ can be
obtained from it simply by a mirror reflection about $\tilde{k} = 0$ and $\omega = 0$. The main difference
from the zero temperature case is that the system needs not initially be in the ground-
state, and therefore $A_>(\omega)$ has a finite value for negative frequencies. In practice, only
excited states within a characteristic energy $E \sim k_B T$ above the ground-state occur with
significant probability in the equilibrium distribution. Therefore $A_>(\omega)$ still has *de facto*
an onset frequency at $\omega \simeq -k_B T$. Assuming that $k_B T$ is small compared to the "bandwidth" $\frac{\hbar c}{\lambda}$ we see that the effect of the finite temperature is roughly equivalent to shifting
the $T = 0$-spectral function by $-k_B T$ along the frequency axis. This description implies
that $A_>(k_F, 0) \sim T^{2\alpha-1}$ if, at zero temperature, $A_>(k_F, \omega) \sim \omega^{2\alpha-1}$. At higher frequencies
($\omega \gg k_B T$) the spectral function is still described by the zero-temperature formula.

A quick way to mathematically derive the correct behavior is to notice that in the calculations of Section 9.8 the asymptotic behavior of the Green's function is cutoff exponentially
at distances $|\tau x \pm ct| \sim L_T = \frac{\hbar c}{\pi k_B T}$. This can be taken qualitatively into account by re-
stricting the integral in Eq. (9.137) to the region $|z| < L_T$. The calculation can then be
carried out analytically. Choosing the upper sign in Eq. (9.137) we get

$$\int_{-L_T}^{+L_T} dz \frac{\lambda^\alpha e^{i\frac{\omega}{c}z}}{(z - i\lambda)^\alpha} = \lambda\lambda_T^{\alpha-1}\int_{-1}^{+1} dx \frac{e^{i\omega_T x}}{(x - i\lambda_T)^\alpha}, \qquad (9.142)$$

where $\omega_T = \frac{\omega L_T}{c} = \frac{\hbar\omega}{\pi k_B T}$ and $\lambda_T = \frac{\lambda}{L_T}$. The integral on the right-hand side of the above
equation tends to a constant $\frac{1-e^{-i\pi(1-\alpha)}}{1-\alpha}$ in the relevant limit $\hbar\omega \ll k_B T$ and $\lambda_T \ll 1$. Thus, we
see that the function $I_\alpha(\omega)$, defined by Eq. (9.137), scales as $T^{\alpha-1}f_\alpha\left(\frac{\hbar\omega}{\pi k_B T}\right)$, where $f_\alpha(\omega_T)$
is a function that tends to a finite limit for $\omega_T \to 0$. In particular, in the zero-frequency
limit, $I_\alpha(0)$ is found to scale as $T^{\alpha-1}$.

We can now repeat all the steps that led us to Eq. (9.140) in Section 9.8. The first
convolution goes through unchanged, leading to Eq. (9.139). In the second convolution,
the only difference is that the factor $\Theta(\omega)$ is replaced by by the Fermi-Dirac distribution
function $n_F(-\omega)$. Thus, Eq. (9.140) takes the form

$$A_{\tau,>}(k_F, 0) = \frac{1}{8\pi^2 c^2} I_\alpha(0) \int_{-\infty}^{\infty} n_F(-\omega')I_\alpha(-\omega')d\omega', \qquad (9.143)$$

where the temperature dependence of I_α is discussed in the previous paragraph. Since bot $n_F(-\omega')$ and $I_\alpha(-\omega')$ are functions of $\frac{\hbar\omega'}{k_B T}$, the temperature dependence of the integral ca be made explicit through the change of variable $\frac{\hbar\omega'}{\pi k_B T} = w$. This gives an extra factor 7 between $d\omega'$ and dw, leaving us with $A_{\tau,>}(k_F, 0) \propto T^{2\alpha-1}$ as anticipated.

9.9 The momentum occupation number

One of the best-known signatures of a normal Fermi liquid is the sharp discontinuity in the momentum occupation number, which, at zero temperature, drops by an amount Z when k crosses k_F. As we have anticipated in our preliminary discussion, this scenario does no apply to the Luttinger liquid. To see explicitly how this comes about let us go back to the basic relations

$$1 - n_{k,\tau} = \int d\omega A_{\tau,>}(k, \omega),$$

$$n_{k,\tau} = \int d\omega A_{\tau,<}(k, \omega) . \tag{9.144}$$

The momentum distribution has the same form for right and left movers, so we consider in detail only the case of right movers ($\tau = +1$). From the fact that $A_{\tau,>}(k, \omega)$ is the mirror image of $A_{\tau,<}(k, \omega)$ about $k = k_F$ and $\omega = 0$ it immediately follows that

$$n_{k_F+\tilde{k},R} = 1 - n_{k_F-\tilde{k},R} . \tag{9.145}$$

In particular, setting $\tilde{k} = 0$ we get

$$n_{k_F,R} = 1 - n_{k_F,R} = \frac{1}{2} . \tag{9.146}$$

For $k > k_F$, integrating Eq. (9.140) with respect to frequency we obtain

$$1 - n_{k,R} = \frac{1}{4\pi^2 c^2} \int_0^\infty d\omega I_\alpha(\omega) \int_0^{\omega+c\tilde{k}} d\omega' I_\alpha(\omega') , \tag{9.147}$$

which gives

$$n_{k,R} = 1 - \frac{1}{\Gamma^2(\alpha)} \int_0^\infty dx\, x^{\alpha-1} e^{-x} \int_0^{x+\tilde{k}\lambda} dy\, y^{\alpha-1} e^{-y} , \quad (\tilde{k} > 0) . \tag{9.148}$$

The results of a numerical evaluation of this formula are plotted in Fig. 9.8 for three different values of $\alpha = 0.1, 1$, and 2. Notice that the function is now continuous at $k = k_F$ even though a singularity (not visible in the figure) persists in its derivatives (Exercise 14).

9.10 Density response to a short-range impurity

Let us now consider the effect of a single impurity on the density of the Luttinger liquid. For simplicity we assume that the electron–impurity potential is a δ-function of dimensionless

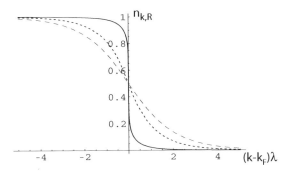

Fig. 9.8. The momentum distribution of right movers in the Luttinger model for $\alpha = \sinh^2 \varphi = 0.1$ (solid line), 1.0 (short-dashed line), and 2.0 (long-dashed line).

strength u located at $x = 0$, i.e., $V_{e-i}(x) = \hbar v_F u \delta(x)$. The electron density operator in real space is given by the Fourier transform of Eq. (9.10), which yields

$$\hat{n}(x) \simeq \hat{n}_0(x) + \hat{n}_{2k_F}(x) \tag{9.149}$$

where

$$\hat{n}_0(x) = \hat{\psi}_R^\dagger(x)\hat{\psi}_R(x) + \hat{\psi}_L^\dagger(x)\hat{\psi}_L(x) \tag{9.150}$$

is the part associated with zero-momentum transfer (forward scattering) and

$$\hat{n}_{2k_F}(x) = \hat{\psi}_R^\dagger(x)\hat{\psi}_L(x) + \hat{\psi}_L^\dagger(x)\hat{\psi}_R(x) \tag{9.151}$$

is the part associated with momentum transfer $\pm 2k_F$ (back-scattering). The electron impurity interaction operator is thus

$$\hat{V}_{e-i} = \hbar v_F u \left[\hat{n}_0(0) + \hat{n}_{2k_F}(0) \right] . \tag{9.152}$$

It is intuitively clear, and mathematically easy to show (see Exercise 14), that the forward scattering term simply multiplies the one-electron wave functions by a position-dependent phase factor, with no observable consequences. Hence this term will be disregarded from now on. On the other hand, the back-scattering interaction $\hbar v_F u \hat{n}_{2k_F}(0)$ has important and unexpected consequences, which we now examine in detail.

Our objective is to calculate the $2k_F$ component of the static density induced by the interaction

$$\hat{V}_{e-i} \simeq \hbar v_F u \left(\hat{\psi}_L^\dagger(0)\hat{\psi}_R(0) + \hat{\psi}_R^\dagger(0)\hat{\psi}_L(0) \right) . \tag{9.153}$$

It turns out that this problem can be solved exactly (Egger and Grabert, 1995), but the solution is not simple and lies well beyond the scope of our introductory treatment. Perturbation theory (to first order in the electron–impurity interaction) provides a much simpler approach, even though, unfortunately, it fails at large distance from the impurity, as we now show.

According to linear response theory the $2k_F$ component of the induced density is given to first order in u, by

$$n_{2k_F}(x) = \chi_{2k_F}(x)\hbar v_F u , \qquad (9.154$$

where

$$\chi_{2k_F}(x) = -\frac{i}{\hbar} \lim_{\omega \to 0} \int_0^\infty \langle [\hat{n}_{2k_F}(x,t), \hat{n}_{2k_F}(0,0)] \rangle e^{i\omega t} dt \qquad (9.155)$$

is the $2k_F$ component of the density–density response function.

To calculate this function we first make use of Eq. (9.102) to rewrite the operator $\hat{\psi}_L^\dagger(x)\hat{\psi}_R(x)$ as

$$\hat{\psi}_L^\dagger(x)\hat{\psi}_R(x) = \frac{e^{2ik_F x} e^{-[\hat{\theta}_L(x) - \hat{\theta}_R(x)]}}{L(1 - e^{-\frac{2\pi}{L\Lambda}})} \hat{U}_L \hat{U}_R^\dagger . \qquad (9.156)$$

The fields $\hat{\theta}_L$ and $\hat{\theta}_R$ commute with each other, and the combination

$$\hat{\theta}_L(x,t) - \hat{\theta}_R(x,t) = \sum_{q<0} \sqrt{\frac{2\pi}{L|q|}} e^{-\varphi_q} \left[e^{-i(\omega_q t - qx)} \hat{\beta}_q - e^{i(\omega_q t - qx)} \hat{\beta}_q^\dagger \right]$$

$$- \sum_{q>0} \sqrt{\frac{2\pi}{L|q|}} e^{-\varphi_q} \left[e^{-i(\omega_q t - qx)} \hat{\beta}_q - e^{i(\omega_q t - qx)} \hat{\beta}_q^\dagger \right] \qquad (9.157)$$

looks like a single θ-field (see Eq. (9.105)), except for the fact that the factors $\cosh \varphi_q$ and $\sinh \varphi_q$ are both replaced by $e^{-\varphi_q}$. We can therefore avail ourselves of the results previously obtained for the one-particle Green's function to calculate the more complicated two-particle Green's function

$$\chi_>(x,t) = -\frac{i}{\hbar} \left\langle \hat{\psi}_L^\dagger(x,t)\hat{\psi}_R(x,t)\hat{\psi}_R^\dagger(0,0)\hat{\psi}_L(0,0) \right\rangle + L \leftrightarrow R . \qquad (9.158)$$

Once $\chi_>(x,t)$ is known, the linear response function is obtained from the integral

$$\chi_{2k_F}(x) = \lim_{\omega \to 0} \int_0^\infty \left[\chi_>(x,t) - \chi_>(-x,-t) \right] e^{i\omega t} dt . \qquad (9.159)$$

Repeating the steps of Section 9.7.2 we obtain

$$\chi_>(x,t) = -\frac{i}{\hbar} \frac{2\cos 2k_F x}{[L(1 - e^{-\frac{2\pi}{L\Lambda}})]^2} e^{F(ct-x)} e^{F(ct+x)} , \qquad (9.160)$$

where

$$F(z) = \sum_{q\neq 0} \frac{2\pi}{qL} e^{-2\varphi_q} n_B(\hbar cq) \left(e^{iqz} - 1 \right) e^{q/\Lambda} . \qquad (9.161)$$

As discussed in Section 9.7.4 we now separate $e^{-2\varphi_q}$ into a "one" that can be calculated exactly and yields the non-interacting response, and a residue $e^{-2\varphi_q} - 1$, which is calculated with "asymptotic accuracy" by setting $\varphi_q = \varphi$ up to a cutoff of order $1/\lambda$. At zero

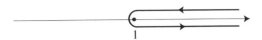

Fig. 9.9. Contour in the complex plane for the evaluation of the integral in Eq. (9.164).

...emperature this yields the following result:

$$\chi_>(x,t) = -i\frac{\cos 2k_F x}{2\pi^2\hbar}\frac{1}{(x-ct+i/\Lambda)(x+ct-i/\Lambda)}$$

$$\times\left(\frac{\lambda}{x-ct+i\lambda}\right)^{g-1}\left(\frac{\lambda}{x+ct-i\lambda}\right)^{g-1}, \qquad (9.162)$$

where $g = e^{-2\varphi}$. Substituting this expression in Eq. (9.159) and making the change of variable $z = \frac{ct}{|x|}$ one can easily show that, for large $|x|$ ($|x| \gg \lambda$) and $\Lambda \to \infty$,

$$\chi_{2k_F}(x) \simeq \frac{1}{\pi^2\hbar c}\frac{\cos 2k_F x}{|x|}\left(\frac{\lambda}{|x|}\right)^{2g-2}\frac{1}{2i}\int_C\frac{dz}{(1-z^2)^g}, \qquad (9.163)$$

where the contour integral is done along a path C that encircles the $x > 1$ portion of the real axis as shown in Fig. 9.9 (Exercise 15). This integral can be evaluated analytically, yielding $-i\sqrt{\pi}\frac{\Gamma(g-1/2)}{\Gamma(g)}$. Therefore the final expression for the density induced at large distance from the impurity is

$$n(x) \simeq \frac{u v_F}{2\pi c}\frac{\cos 2k_F x}{|x|}\times\frac{\Gamma(g-1/2)}{\sqrt{\pi}\Gamma(g)}\left(\frac{\lambda}{|x|}\right)^{2g-2}. \qquad (9.164)$$

The first factor on the right-hand side of this expression is the familiar result for the Friedel oscillations in a non-interacting one-dimensional Fermi liquid. The second factor, which reduces to 1 for $g \to 1$, contains the effect of the interaction. The anomalously slow decay of the density oscillations for repulsive interaction ($g < 1$) indicates a strong tendency towards the formation of a charge-density wave: in the normal Fermi liquid, the exponent would be 1 irrespective of the strength of the interaction.

We stated earlier that the calculation based on linear response theory fails at large distance from the impurity. We can now be more precise. First of all, notice that the factor $\Gamma(g-1/2)$ in Eq. (9.164) diverges at $g = 1/2$, indicating a breakdown of the linear response theory. In fact, for $g < 1/2$ Eq. (9.164) predicts an induced density that grows for $x \to \infty$ – a patently absurd prediction. Even in the "reasonable" regime $1/2 < g < 1$, Eq. (9.164) cannot be trusted at distances larger than $x_c \simeq \frac{\lambda c}{u v_F}$. The physical origin of this restriction lies in the fact that the perturbation theory cannot be applied to collective modes whose energy $\hbar c q$ is lower than the electron–impurity energy scale $\frac{\hbar v_F u}{\lambda}$. But these modes, having wavevectors $q < q_c = \frac{u v_F}{\lambda c}$, are precisely the ones that control the behavior of the density at distances larger than $q_c^{-1} = x_c$.

It can be shown from the exact solution (Egger and Grabert, 1995) that the amplitude of the density oscillations for $x \gg x_c$ scales as $\left(\frac{\lambda u}{x}\right)^g$ (notice the nonanalytic dependence on

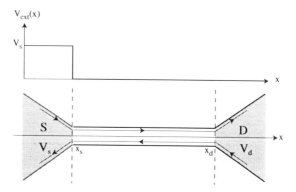

Fig. 9.10. A translationally invariant Luttinger liquid is placed between two electrodes, a "source" and a "drain". The source electrode sets the chemical potential of the right movers to V_s, while the drain sets that of the left movers to V_d ($= 0$ in this example). As a result, a net current flows in the system.

u). However, for weak impurity scattering (i.e., for $u \ll 1$), there is still a wide range of distances $\lambda \ll x \ll x_c$ in which the linear response formula (9.164) applies.

9.11 The conductance of a Luttinger liquid

Some of the most interesting features of a Luttinger liquid emerge when one considers the behavior of its conductance. To define the latter precisely we put the one-dimensional electron liquid between two electrodes, a source and a drain, held at different potentials, as shown in Fig. 9.10. Our objective is to calculate the current that flows in this 1D channel as a function of the potential difference $V_{sd} \equiv V_s - V_d$.

Let us first do the calculation without impurities. We set $V_d = 0$ and consider the current induced by a small time-dependent potential $V_s \cos \omega t$ applied to the source electrode. With reference to Fig. 9.10, the applied potential is assumed to be spatially uniform and equal to $V_s \cos \omega t$ in the region $x < x_s$, and to vanish for $x > x_s$. The potential drops discontinuously to zero at $x = x_s$. This model is justified for a uniform electron liquid since there is no induced electronic charge, and hence no electric field in the interior of the wire. According to this model, the potential drop (and hence the electric field) is confined to the region of contact between the left reservoir and the wire.

It is tempting, but, unfortunately, incorrect, to calculate the current response by means of the linear response formula

$$I = -e \lim_{\omega \to 0} \int_{-\infty}^{x_s} \chi_{jn}(x - x', \omega) V_s \, dx' \quad \text{(wrong!)} , \qquad (9.165)$$

where $\chi_{jn}(x - x', \omega)$ is the current–density response function.

The reason why this is wrong is that V_s is not a truly external potential, but one that arises, in part, from charges induced in the electron liquid by the external potential. The correct xpression is

$$I = -e \lim_{\omega \to 0} \int_{-\infty}^{x_s} \tilde{\chi}_{jn}(x - x', \omega) V_s \, dx' , \qquad (9.166)$$

where $\tilde{\chi}_{jn}(x - x', \omega)$ is the *proper* current–density response function. But, in the Luttinger model the proper current–density response function coincides with the non-interacting current–density response function.[20] Therefore we can immediately conclude, without further calculations, that the conductance of the homogeneous Luttinger liquid (without impurities) is the same as that of the non-interacting one-dimensional electron gas.

It is nevertheless useful to do the calculation according to the wrong formula (9.165), since we will later encounter a physical realization of the Luttinger model – the edge of a quantum Hall liquid – in which the screening is absent and therefore Eq. (9.165) gives the right result.

The calculation of

$$\chi_{jn}(x - x', \omega) = -\frac{i}{\hbar L} \int_{-\infty}^{+\infty} \frac{dq}{2\pi} e^{iq(x-x')} \int_0^\infty \langle [\hat{j}_q(t), \hat{n}_{-q}] \rangle e^{i(\omega+i\eta)t} dt \qquad (9.167)$$

is easily done with the help of the bosonic expressions (9.55) and (9.64) for the density and the current, and we get

$$\chi_{jn}(x - x', \omega) = \frac{g\omega}{\pi\hbar c} \int_{-\infty}^{+\infty} \frac{dq}{2\pi} \frac{q e^{iq(x-x')}}{\left(\frac{\omega+i\eta}{c}\right)^2 - q^2}$$

$$= -i\frac{g\omega}{\hbar c} e^{-i\frac{\omega+i\eta}{c}|x-x'|} , \qquad (9.168)$$

where $g = e^{-2\varphi}$, and the integral over q has been done by the method of residues. Putting this into Eq. (9.166) we obtain

$$I = ie\frac{g\omega}{\hbar c} \lim_{\omega \to 0} \lim_{\eta \to 0} \int_{-\infty}^{x_s} e^{i\frac{\omega+i\eta}{c}(x-x')}(-eV_s)dx'$$

$$= g\frac{e^2}{h} V_s , \qquad (9.169)$$

independent of x. This result indicates that the "improper" conductance of the one-dimensional electron liquid is renormalized to $g\frac{e^2}{h}$. The *proper* conductance, however, is obtained, as discussed above, by turning off the interactions, i.e., by setting $g = 1$ in the above expression. The final result is therefore $\frac{e^2}{h} \approx 4 \times 10^{-5} S$, or twice as much if the spin is included. More details are given in Exercise 9.

Let us now study what happens when a single δ-function impurity is introduced at $x = 0$. A possible approach to the problem is to use perturbation theory to calculate the effect of the back-scattering interaction (9.153) on the current–density response function. It is simpler, however, to compute the *back-scattering current*, I_B, defined as the net current that the

---

[20] The quickest way to see this is to recall that the proper density–density response function coincides with the non-interacting one (see Section 9.5.2 and Exercise 8), and that the density response determines the current response via the continuity equation.

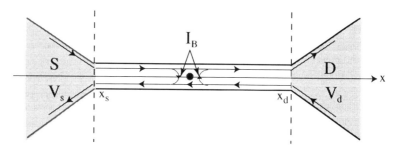

Fig. 9.11. Same as Fig. 9.10, but now with an impurity at $x = 0$. The impurity scatters particles from the right- to the left-moving branch and vice versa. As a result, the net current is reduced by an amount I_B – the back-scattering current. Although the chemical potentials of the right and left movers are no longer uniform along the system, their difference stays approximately constant and equal to $-e(V_s - V_d)$.

impurity causes to flow *across* the wire at $x = 0$, i.e., the rate of transfer of right movers to the left-moving branch minus the rate of transfer of left movers to the right-moving branch. As shown in Fig. 9.11 the back-scattering current *reduces* the magnitude of the net current that flows along the wire (i.e., between the source and the drain) at a given bias voltage. The conductance of the channel is reduced accordingly to

$$ G = \frac{e^2}{h} - \frac{dI_B}{dV_{sd}} . \tag{9.170} $$

The operator of the back-scattering current $\hat{I}_B$ is obtained from the time derivative of the number of right movers:

$$ \hat{I}_B = -i\frac{e}{\hbar}[\hat{V}_{e-i}, \hat{N}_R] $$

$$ = \frac{ieuv_F}{\hbar}\left(\hat{\psi}_L^\dagger(0)\hat{\psi}_R(0) - \hat{\psi}_R^\dagger(0)\hat{\psi}_L(0) \right) , \tag{9.171} $$

where we have used the fact that only the electron–impurity interaction term in the hamiltonian fails to commute with $\hat{N}_R$. According to linear response theory, the back-scattering current is then given by

$$ I_B = -\frac{i}{\hbar}\lim_{\omega \to 0}\int_0^\infty \langle[\hat{I}_B(t), \hat{V}_{e-i}]\rangle e^{i(\omega+i\eta)t} . \tag{9.172} $$

The calculation of I_B is very similar to the one we did in the previous section for the Friedel oscillations, but there is an important new twist: the system must be driven out of equilibrium in order to obtain a non-zero current. Let us examine this point in detail. From Eqs. (9.171) and (9.153) we see that

$$ I_B = \lim_{\omega \to 0}\int_0^\infty [K_>(t) + K_>(-t)] e^{i(\omega+i\eta)t} dt , \tag{9.173} $$

where

$$K_>(t) = ieu^2 v_F \left(-\frac{i}{\hbar}\right) \left\{ \langle \hat{\psi}_L^\dagger(0,t)\hat{\psi}_R(0,t)\hat{\psi}_R^\dagger(0,0)\hat{\psi}_L(0,0)\rangle - L \leftrightarrow R \right\}. \quad (9.174)$$

To drive the system out of equilibrium we assume that right and left movers have slightly different chemical potentials μ_R and μ_L, where $\mu_R - \mu_L = -e(V_s - V_d)$. The physical idea is that the right movers are in equilibrium with the source, and the left movers with the drain: this makes sense, since anything that moves to the right must have come directly from the source and knows nothing about the drain, while the converse is true for left movers.[21] The two different chemical potentials affect the time evolution of the number rising operators $\hat{U}_L$ and $\hat{U}_R^\dagger$ in different ways. We now have

$$\hat{U}_L(t) = \hat{U}_L e^{-\frac{i}{\hbar}\mu_L t} \quad \text{and} \quad \hat{U}_R^\dagger(t) = \hat{U}_R^\dagger e^{\frac{i}{\hbar}\mu_R t}, \quad (9.175)$$

so the time dependences of the two operators no longer cancel out, but combine to give a factor $e^{-i\frac{eV_{sd}}{\hbar}t}$ where $V_{sd} \equiv V_s - V_d$. Substituting Eq. (9.156) into Eq. (9.174) we get

$$\begin{aligned} K_>(t) &= ieu^2 v_F^2 \frac{2\sin\frac{eV_{sd}t}{\hbar}}{[L(1-e^{-\frac{2\pi}{L\Lambda}})]^2} e^{2F(ct)} \\ &= -i\frac{eu^2 v_F^2}{2\pi^2} \frac{\sin\frac{eV_{sd}t}{\hbar}}{(-ct+i/\Lambda)(ct-i/\Lambda)}\left(\frac{\lambda}{-ct+i\lambda}\right)^{g-1}\left(\frac{\lambda}{ct-i\lambda}\right)^{g-1}, \quad (9.176)\end{aligned}$$

where $F(z)$ is still defined by Eq. (9.161). At sufficiently low voltages, i.e., for $|V_{sd}| \ll \frac{\hbar c}{e\lambda}$, only the large time behavior matters, and we can ignore the microscopic length scale λ in the denominators of Eq. (9.176). Putting the resulting expression into Eq. (9.173), changing integration variable from t to $z = \frac{e|V_{sd}|t}{\hbar}$, and making use of the integral

$$\int_0^\infty dz \frac{\sin z}{z^{2g}} = \frac{\pi}{2\sin\pi g}\frac{1}{\Gamma(2g)}, \quad (g<1), \quad (9.177)$$

we finally obtain

$$I_B = \frac{u^2}{2\pi\Gamma(2g)}\left(\frac{v_F}{c}\right)^2 \frac{e^2 V_{sd}}{\hbar}\left(\frac{\hbar c}{e|V_{sd}|\lambda}\right)^{2-2g}, \quad \left(|V_{sd}| \ll \frac{\hbar c}{e\lambda}\right). \quad (9.178)$$

In the noninteracting liquid ($g = 1$) the back-scattering current is simply proportional to the source-drain voltage. This implies a finite negative correction to the conductance of the pure liquid, as expected for an ordinary ohmic conductor. The surprise is that this behavior changes qualitatively in the presence of interactions, no matter how weak. The back-scattering current scales as $|V_{sd}|^{2g-1}$, which is much larger than $|V_{sd}|$ for $V_{sd} \to 0$: this implies a *diverging negative correction to the conductance* (see Eq. (9.170)). Of course

[21] We ignore in this argument the effect of the weak back-scattering caused by the impurity: this is sufficient for a calculation to first-order in the strength of the impurity potential.

our perturbative treatment ceases to be valid as soon as the back-scattering current become comparable to the source-drain current. However, the failure of perturbation theory tells u that something quite dramatic is happening, namely, the system becomes an insulator in th limit of vanishing source-drain voltage.

A physical way to understand this surprising behavior is to recall that a Luttinger liquid is on the verge of a charge density wave instability. A single impurity precipitates the instability by inducing long range density oscillations at wave vector $2k_F$. The phase of the ensuing charge-density wave system is pinned by the impurity, and it takes a finite bias to overcome the pinning: hence the conductance vanishes at zero voltage.

The above calculations can be extended to finite temperature by applying the transformation (9.131) to the zero-temperature propagator (9.176). All the steps are the same as in the zero-temperature calculation, except that at the very end the integral (9.177) is replaced by

$$
\int_0^\infty dz \, \frac{\sin z}{\left[\bar{V} \sinh \left(\frac{z}{\bar{V}}\right)\right]^{2g}} = \frac{\pi}{2 \sin \pi g} \frac{1}{\Gamma(2g)}
$$

$$
\times \frac{1}{\pi} \left(\frac{2}{\bar{V}}\right)^{2g-1} \sinh \left(\frac{\pi \bar{V}}{2}\right) \left|\Gamma \left(g - i\frac{\bar{V}}{2}\right)\right|^2, \quad (9.179)
$$

where $\bar{V} = \frac{e|V_{sd}|}{\pi k_B T}$.

Comparing Eq. (9.179) to (9.177) we see that the back-scattering current at finite temperature is given by

$$
I_B = I_0 \frac{\bar{u}^2}{\pi \Gamma(2g)} \left(\frac{T}{T_0}\right)^{2g-1} \left|\Gamma \left(g - i\frac{e|V_{sd}|}{2\pi k_B T}\right)\right|^2 \sinh \left(\frac{e|V_{sd}|}{2k_B T}\right), \quad (9.180)
$$

where $k_B T_0 = eV_0 \equiv \frac{\hbar c}{2\pi \lambda}$ is a microscopic energy scale, $I_0 \equiv \frac{e^2 V_0}{\hbar}$, and $\bar{u} \equiv \frac{u v_F}{c}$. Thus, we see that a finite temperature restores the normal "ohmic" behavior, but only for bias voltages so small that $e|V_{sd}| \ll k_B T$: the system then remains a conductor for vanishingly small voltage, but its conductance has a power-law dependence on temperature and quickly vanishes as T tends to zero. This behavior is illustrated in Fig. 9.12, where the back-scattering current and the differential conductance dI_B/dV_{sd} are plotted (in arbitrary, but consistent units) for different temperatures. We will return to these results in Section 10.20.5 of Chapter 10, when we discuss recent experiments on tunneling between the edges of a quantum Hall liquid.

9.12 Spin–charge separation

Up to this point we have ignored the spin of the electron. Let us try and remedy this omission. In order to do so, we again inspect Eq. (9.16). The kinetic energy term is trivially generalized by attaching a spin index $\sigma = \uparrow$ or $\downarrow$ to the number operators, i.e., replacing $\hat{N}_{k,\tau}$ by $\hat{N}_{k\sigma,\tau}$ and summing over σ. The V_1 interaction term needs no change insofar as the density

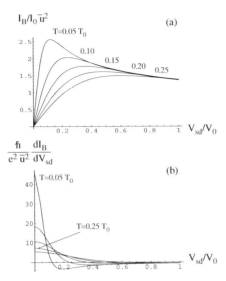

Fig. 9.12. (a) Back-scattering current calculated from Eq. (9.180) at $g = 0.33$ for different values of T. The scales $k_B T_0 (= eV_0)$ and I_0, and the dimensionless coupling constant $\bar{u}$, are defined after Eq. (9.180). Notice that, for $T \to 0$ the limiting curve $I_B \propto V_{sd}^{2g-1}$ is approached. (b) The dimensionless differential conductance $\frac{\hbar}{e^2 \bar{u}^2} \frac{dI_B}{dV_{sd}}$ for the same g and Ts as in (a). The zero-bias conductance diverges for $T \to 0$.

fluctuation operators $\hat{n}_{q\tau}$ are redefined as sums of up-spin and down-spin components:

$$\hat{n}_{q,\tau} \equiv \hat{n}_{q\uparrow,\tau} + \hat{n}_{q\downarrow,\tau}. \tag{9.181}$$

In the V_2 interaction term, write $V_2(q) = V_1(q) + (V_2(q) - V_1(q))$. The piece associated with $V_1(q)$ needs no change. The piece associated with $V_2(q) - V_1(q)$, however, arises from scattering between the right- and left-moving branches (see Eq. (9.17)), and can be expressed in terms of density fluctuations only when the spins of the interacting electrons are parallel. When they are antiparallel, instead, one gets an interaction of the form (9.19), which *cannot* be expressed in terms of density fluctuations. Ignoring, for the time being, this troublesome term, we see that the hamiltonian of the "spinful" Luttinger model takes the form

$$\hat{H} = \hbar v_F \sum_{k\sigma,\tau} (\tau k - k_F) \hat{N}_{k\sigma,\tau}$$

$$+ \frac{1}{2L} \sum_{q \neq 0} V_1(q) \left(\hat{n}_{-q,R} + \hat{n}_{-q,L} \right) \left(\hat{n}_{q,R} + \hat{n}_{q,L} \right)$$

$$+ \frac{1}{2L} \sum_{q \neq 0,\sigma} [V_2(q) - V_1(q)] \left(\hat{n}_{-q\sigma,R} \hat{n}_{q\sigma,L} + \hat{n}_{-q\sigma,L} \hat{n}_{q\sigma,R} \right). \tag{9.182}$$

To "bosonize" the hamiltonian we define the spin-dependent boson operators $\hat{b}_{q\sigma}$ starting from Eqs. (9.33) and (9.35) and attaching a spin index σ to the operators on both sides.

544   *Electrons in one dimension and the Luttinger liquid*

Operators with different spin indices commute with each other. With these definitions it is straightforward to verify that Eq. (9.182) takes the form

$$\hat{H} = \sum_{q\neq0,\sigma} \hbar v_F |q| \hat{b}^\dagger_{q\sigma} \hat{b}_{q\sigma}$$

$$+ \sum_{q\neq0,\sigma\sigma'} \frac{V_1(q)}{2\pi} |q| \left[\hat{b}^\dagger_{q\sigma} \hat{b}_{q\sigma'} + \frac{1}{2}\left(\hat{b}^\dagger_{q\sigma} \hat{b}^\dagger_{-q\sigma'} + \hat{b}_{-q\sigma} \hat{b}_{q\sigma'} \right) \right]$$

$$+ \sum_{q\neq0,\sigma} \frac{V_2(q)-V_1(q)}{4\pi} |q| \left(\hat{b}^\dagger_{q\sigma} \hat{b}^\dagger_{-q\sigma} + \hat{b}_{-q\sigma} \hat{b}_{q\sigma} \right) . \tag{9.183}$$

The coupling between up-spin and down-spin operators can be eliminated by the introduction of the spin-symmetric and spin-antisymmetric combinations

$$\hat{b}^C_q = \frac{1}{\sqrt{2}}(\hat{b}_{q\uparrow} + \hat{b}_{q\downarrow}),$$

$$\hat{b}^S_q = \frac{1}{\sqrt{2}}(\hat{b}_{q\uparrow} - \hat{b}_{q\downarrow}), \tag{9.184}$$

where "C" and "S" stand for the charge channel and the spin channel respectively. Notice that the charge and spin-fluctuation operators commute with each other: $[\hat{b}^{C\dagger}_q, \hat{b}^S_q] = 0$. Armed with these definitions and with some patience, one can then show that the complete Hamiltonian separates into two mutually commuting parts, one for the charge, and one for the spin degree of freedom:

$$\hat{H} = \hat{H}_C + \hat{H}_S , \tag{9.185}$$

where

$$\hat{H}_C = \sum_{q\neq0} \left[\left(\hbar v_F + \frac{V_1(q)}{\pi} \right) |q| \hat{b}^{C\dagger}_q \hat{b}^C_q + \frac{V_1(q)+V_2(q)}{4\pi} |q| \left(\hat{b}^{C\dagger}_q \hat{b}^{C\dagger}_{-q} + \hat{b}^C_{-q} \hat{b}^C_q \right) \right]$$

$$\tag{9.186}$$

and

$$\hat{H}_S = \sum_{q\neq0} \left[\hbar v_F |q| \hat{b}^{S\dagger}_q \hat{b}^S_q + \frac{V_2(q)-V_1(q)}{4\pi} |q| \left(\hat{b}^{S\dagger}_q \hat{b}^{S\dagger}_{-q} + \hat{b}^S_{-q} \hat{b}^S_q \right) \right] . \tag{9.187}$$

The fact that the hamiltonian separates into two independent pieces is remarkable. Each piece can be diagonalized exactly by means of a Bogoliubov transformation similar to the one used to diagonalize (9.41): we do not present the details of the calculation (which is straightforward) but summarize the main results. Equations (9.185)–(9.187) show that the system supports two independent branches of excitations: density waves, which propagate with velocity

$$v^C_q = \sqrt{ \left(v_F + \frac{V_1(q)}{\pi\hbar} \right)^2 - \left(\frac{V_1(q)+V_2(q)}{2\pi\hbar} \right)^2 } , \tag{9.188}$$

ıd spin waves, which propagate with velocity

$$v_q^S = \sqrt{v_F^2 - \left(\frac{V_1(q) - V_2(q)}{2\pi\hbar}\right)^2}.$$ (9.189)

'he model has a stable ground-state only if the expressions under the square roots in the above two equations are positive. The density–density and spin–spin response function ꭓave the form of Eq. (9.56) with c given by Eqs. (9.188) and (9.189) and v_N given by ᵥF $+ \lim_{q\to 0} \frac{3V_1(q)+V_2(q)}{2\pi\hbar}$ for the charge channel and by $v_F + \lim_{q\to 0} \frac{V_2(q)-V_1(q)}{2\pi\hbar}$ for the spin ꞔhannel.[22]

One of the most striking features of the spinful Luttinger model is the phenomenon ᴋnown as *spin-charge separation*. Consider the expression (9.76) for the creation operator ɔf a right or left-moving particle. This is generalized to the spinful case simply by attaching ᴛhe appropriate spin index to all the operators. The spin-dependent boson operators are then ᴇxpressed in terms of charge and spin operators $\hat{b}^C$ and $\hat{b}^S$ according to Eq. (9.184). After ᴅdoing this, we see that the creation operator of the particle, initially carrying a charge and ᴀ spin, factors into the product of two terms: one creates charge fluctuations but no spin fluctuation, and the other creates spin fluctuations but no charge fluctuations. The key point is that these two factors evolve in time according to what are effectively two different and independent hamiltonians $\hat{H}_C$ and $\hat{H}_S$. Since the charge and spin velocities are different, the charge and the spin of an electron injected into the system at one and the same position will eventually separate in space. This behavior should be contrasted to that of a normal Fermi liquid, in which the charge and the spin of the electron quasiparticle always travel together.

A few words on the role of the back-scattering term, Eq. (9.19), which has so far been neglected, are now in order. We have already pointed out that this term cannot be expressed in terms of charge and spin operators. In fact, one can make use of the bosonization formalism to express it as an exponential function of the $\hat{b}$'s. Adding this term to the hamiltonian (9.185) results in a model that can no longer be analytically solved. It is still possible, however, to treat the new term perturbatively. The calculation is rather technical, and beyond the scope of our presentation, but its outcome is easy to grasp. Basically the perturbative term does not change the qualitative features of the solution (i.e. the existence of two independent sets of modes propagating with different velocities), but *renormalizes* the values of parameters such as the mode velocities, and induces an interaction between previously independent modes. This is reminiscent of what happens in a normal Fermi liquid, where the electron-electron interaction does not change the qualitative features of the non-interacting liquid, but renormalizes the quasiparticle energies and introduces an interaction between the quasiparticles. The *robustness* of the qualitative features of the ideal Fermi gas with respect to electron-electron interactions is the essence of the Fermi liquid concept. In much the same way, the robustness of the Luttinger model hamiltonian (9.185) with respect to perturbations such as (9.19) is the essence of the Luttinger liquid concept and the ultimate

---

[22] One must also include a factor 2 to take into account the doubled density of electrons.

reason for its usefulness. These hand-waving considerations can be rigorously formulated within the framework of the *renormalization group theory* (Schulz, Cuniberti, and Pier 1998). It can be shown that the perturbative term is *irrelevant* in the technical sense of the word, i.e. it ultimately disappears in the effective hamiltonian that describes the long wavelength and low-energy properties of the model.

9.13 Long-range interactions

In most of this section we have assumed that the electron-electron interaction is of finite range, i.e., the $q \to 0$ limit of $V_1(q)$ and $V_2(q)$ is finite. This is not unreasonable in many situations, since the one-dimensional system is usually embedded within a larger three-dimensional structure which eventually screens the coulomb interaction at sufficiently large distance. It is of interest, however, to see what happens in the ideal case of the bare coulomb interaction. We have seen in Chapter 1 that the Fourier transform of the coulomb interaction in one dimension has the form $V_1(q) = v_q = -2e^2 \ln qa$ where a is a length scale related to the diameter of the one-dimensional channel. At the same time, $V_2(q) = v_q - v_{2k_F}$. We then have

$$\left| v_F + \frac{v_q}{2\pi\hbar} \right|^2 - \left| \frac{v_q - v_{2k_F}}{2\pi\hbar} \right|^2 \overset{q\to 0}{\to} -\frac{4e^2}{h}\left(v_F + \frac{v_{2k_F}}{2\pi\hbar} \right) \ln qa . \tag{9.190}$$

Substituting this in Eq. (9.45) we see that the frequency of the collective mode at long wavelength is

$$\omega_q = 2\sqrt{\frac{e^2}{2\pi\hbar v_F}\left(1 + \frac{v_{2k_F}}{2\pi\hbar v_F} \right) v_F |q| |\ln qa|^{1/2}} . \tag{9.191}$$

This differs from the RPA result by the presence of the "local field correction" $\left(1 + \frac{v_{2k_F}}{2\pi\hbar v_F} \right)$.

The long range of the interaction has an interesting effect on the exponents that characterize the long time/long distance behavior of the Green's function, the density–density response function, and the physical properties these quantities are associated with. Basically, the Bogoliubov rotation angle φ_q diverges (although very slowly) for $q \to 0$:

$$g = e^{-2\varphi_q} \to \frac{v_{2k_F} + 2\pi\hbar v_F}{4e^2 |\ln qa|} . \tag{9.192}$$

When this asymptotic form is substituted in the expressions for the functions F (defined in Section 9.7.2) one sees that the power law decay of these functions is replaced by a much slower decay of the form $e^{-c\sqrt{\ln(x/a)}}$, where c is a nonuniversal constant. Interestingly, this behavior coincides with that of a classical Wigner crystal of regularly spaced electrons along a line. We refer the reader to the literature (Schulz, 1993) for further discussion of this point.

Exercises

9.1 **Eigenvalues of the non-interacting Luttinger model hamiltonian.** Verify that the eigenstates of the non-interacting Luttinger model hamiltonian defined in the opening paragraph of Appendix 19 have energies of the form (A19.1).

9.2 **The ground-state of the Luttinger model is a liquid.** Show that the ground-state of the Luttinger model, given by Eq. (9.48) is a homogeneous liquid, i.e., its density is uniform.

9.3 **Ground-state wave function of a non-interacting 1D Fermi liquid.** Show that the first factor of Eq. (9.51) is the ground-state wave function of non-interacting fermions in one dimension, i.e., a Slater determinant of plane waves with wave vectors ranging from $-k_F$ to k_F. [Hint: show that the Slater determinant can be written in the form

$$
e^{-i\frac{\pi}{L}\sum_{i\neq j}(x_i+x_j)}\det
\begin{pmatrix}
1 & z_1 & z_1^2 & . & . & . & z_1^{N-1} \\
1 & z_2 & z_2^2 & . & . & . & z_2^{N-1} \\
. & . & . & . & . & . & . \\
. & . & . & . & . & . & . \\
. & . & . & . & . & . & . \\
1 & z_N & z_N^2 & . & . & . & z_N^{N-1}
\end{pmatrix}
$$

where $z_i \equiv e^{i\frac{2\pi}{L}x_i}$. This is an example of a *Vandermonde determinant* (see also Section 10.7.1).]

9.4 **Oscillator representation of the Luttinger model hamiltonian.**

    (a) Show that the Luttinger model hamiltonian, Eq. (9.44), can be rewritten (up to a constant) in the harmonic oscillator form

$$
\hat{H}_{LM} = \frac{1}{2}\sum_{q\neq 0}\hbar c_q|q|\left(\hat{P}_{-q}\hat{P}_q + \hat{X}_{-q}\hat{X}_q\right),
$$

where the dimensionless "momentum" and "position" operators are defined as

$$
\hat{P}_q \equiv \frac{\hat{\beta}_{-q}^{\dagger} + \hat{\beta}_q}{\sqrt{2}}, \quad \hat{X}_q \equiv \frac{\hat{\beta}_{-q}^{\dagger} - \hat{\beta}_q}{\sqrt{2}i},
$$

and satisfy the fundamental commutation relations $[\hat{P}_q, \hat{X}_{-q'}] = -i\delta_{qq'}$; $[\hat{P}_q, \hat{P}_{-q'}] = [\hat{X}_q, \hat{X}_{-q'}] = 0$.

    (b) Verify that the ground-state wave function of the Luttinger model in the "momentum" representation has the form

$$
\psi_0\left(\{P_q\}\right) \sim e^{-\frac{1}{2}\sum_{q\neq 0}P_{-q}P_q},
$$

where P_q are the eigenvalues of $\hat{P}_q$.

9.5 **Ground-state wave function of the Luttinger model.** Use the results of the previous exercise to show that the ground-state wave function of the Luttinger model in the coordinate representation is given by Eq. (9.51) when the distances $|x_i - x_j|$ are larger than the ultraviolet cutoff length Λ^{-1}, but still much smaller than L. [Hint:

make use of Eqs. (9.35) and (9.55) to show that

$$\hat{P}_{-q}\hat{P}_q = \frac{1}{g}\frac{\pi}{L|q|}\hat{n}_{-q}\hat{n}_q\,, \quad \text{for } q < \Lambda\,,$$

and hence

$$\sum_{q\neq0}P_{-q}P_q \simeq \frac{1}{g}\frac{\pi}{L}\sum_{i,j}\sum_{q\neq0}\frac{e^{iq(x_i-x_j)}}{|q|}\,.$$

Finally, use Eq. (9.78) to evaluate the sum over q. Don't forget to antisymmetrize the resulting wave function.]

9.6 **Normalizing the ground-state of the Luttinger model.** Starting from the second-quantized representation (9.48) for the ground-state of the Luttinger model, derive Eq. (9.50) for the normalization constant, and show that it is finite only if the stability condition 9.18(ii) is satisfied.

9.7 **Density–density response functions of the Luttinger model.** Calculate the partial density–density response functions $\chi_{\tau\tau'}(q,\omega)$ of the Luttinger model, and show that they are given by Eq. (9.61).

9.8 **The density–density response functions of the Luttinger model at long wavelength are correctly given by the RPA.** Starting from Eq. (9.62) show that the proper density–density response functions of the Luttinger model coincide with the non-interacting density–density response functions. Conclude that the RPA yields the exact density–density response functions in the limit $q \to 0$.

9.9 **The proper conductance of a ballistic 1D channel is unaffected by interactions.** Explicitly verify the cancellation of g in the proper current–density response function of the Luttinger liquid in the special case that $V_1 = V_2$. [Hint: recall that $\tilde{\chi}_{jn}(q,\omega) = \epsilon(q,\omega)\chi_{jn}(q,\omega)$ and $\epsilon(q,\omega) = \frac{\omega^2-c^2q^2}{\omega^2-v_F^2q^2}$ in this special case.]

9.10 **Algebra of fermion field operators.** Verify the commutation rules (9.69) between the field operators of right and left movers and the bosons.

9.11 **Expression of fermion field operators in terms of displacement and phase fields.** Although Eq. (9.76) is sufficient for all calculations on the Luttinger liquid, it has become customary in the literature (see Glazman and Fischer, 1996) to express the fermion fields operator in terms of a *phonon-displacement field*

$$\hat{\theta}(x) = \frac{1}{2i}\sum_{q\neq0}\sqrt{\frac{2\pi}{L|q|}}\mathrm{sign}(q)\left(\hat{b}_q e^{iqx} - \hat{b}_q^\dagger e^{-iqx}\right)\,, \tag{E9.1}$$

and a *phase field*

$$\hat{\phi}(x) = \frac{1}{2i}\sum_{q\neq0}\sqrt{\frac{2\pi}{L|q|}}\left(\hat{b}_q e^{iqx} - \hat{b}_q^\dagger e^{-iqx}\right)\,. \tag{E9.2}$$

(a) Prove that

$$[\hat{\phi}(x),\hat{\theta}(x')] = i\frac{\pi}{2}\mathrm{sign}(x-x')\,. \tag{E9.3}$$

(b) Show that the right and left density operators are given by

$$\hat{n}_\tau(x) = \frac{\hat{N}_\tau}{L} + \frac{1}{2\pi}\frac{\partial}{\partial x}\left[\hat{\phi}(x) + \tau\hat{\theta}(x)\right] . \tag{E9.4}$$

(c) Show that the fermion field operator (9.76) has the form

$$\hat{\psi}_\tau^\dagger(x) = \frac{e^{-i\left[\tau k_F x + \tau\hat{\theta}(x) + \hat{\phi}(x)\right]}}{\sqrt{L(1 - e^{-\eta})}} . \tag{E9.5}$$

(d) Prove that the hamiltonian of the Luttinger model takes the form

$$\hat{H}_{LM} = \frac{\hbar v_F}{2\pi}\int_0^L\left[g\left(\frac{\partial\hat{\phi}(x)}{\partial x}\right)^2 + \frac{1}{g}\left(\frac{\partial\hat{\theta}(x)}{\partial x}\right)^2\right]dx , \tag{E9.6}$$

where

$$g = e^{-2\varphi} = \sqrt{\frac{2\pi\hbar v_F + V_1 - V_2}{2\pi\hbar v_F + V_1 + V_2}} . \tag{E9.7}$$

9.12 **Calculation of the kinetic energy in the bosonization formalism.** Work out the steps leading from the generating function of Eq. (9.96) to the bosonized expression (9.97) for the kinetic energy.

9.13 **Calculation of the non-interacting Green's function.** Explicitly carry out the integral leading to Eq. (9.123) for the exponent of the non-interacting Green's function.

9.14 **Irrelevancy of forward scattering in the Luttinger model.** Show that scattering from a local potential in a system of right- or left-movers, can be eliminated by multiplying the one-electron states by a position-dependent phase factor.

9.15 $2k_F$ **density oscillations in the Luttinger model.** Starting from Eqs. (9.159) and (9.162) for the $2k_F$ density–density response function, derive Eq. (9.164) for the $2k_F$ oscillations induced by a short-range impurity in the Luttinger model.

10

The two-dimensional electron liquid at high magnetic field

10.1 Introduction and overview

Thinking of the two-dimensional electron liquid (2DEL) at high magnetic field is to many the same as thinking about the quantum Hall effect. Indeed, the discovery of the incredibly accurate quantization of the Hall conductance of a 2DEL at integral and special fractional multiples of

$$\frac{e^2}{h} = \frac{1}{25,812.807572 \ \Omega} \tag{10.1}$$

(von Klitzing, Dorda, and Pepper, 1980; Tsui, Störmer, and Gossard, 1982) (see Fig. 10.7) is one of the most beautiful discoveries of the twentieth century.[1]

However, the physics of the 2DEL at high magnetic field goes well beyond the quantum Hall effects: the presence of a strong magnetic field in the 1–10 T range at densities between 10^{10} and 10^{12} cm^{-2} creates the possibility of quantum phase transitions to broken-symmetry states, which would be inaccessible under ordinary conditions. The full scenario is still far from being completely understood, and it seems fair to say that Robert Laughlin only scratched the surface when he introduced his celebrated wave function for the fractional quantum Hall effect (Laughlin, 1983). In brief, the 2DEL at high magnetic field is a marvellous physics laboratory to study interaction effects in electron liquids, and it is precisely this aspect that we wish to emphasize in this chapter.

The basic reason for the rich behavior advertised in the previous paragraph is the *quenching of the kinetic energy* at high magnetic field. This effect occurs in classical as well as in quantum mechanics.

In classical mechanics, a magnetic field $\vec{B}$ forces each electron to move on a circular orbit at the cyclotron frequency

$$\omega_c = \frac{eB}{m_b c}, \tag{10.2}$$

where m_b is the electron band mass ($m_b = 0.067m$ in GaAs). Additional forces, due to external potentials, interactions with other electrons etc., cause the center of the circular

---

[1] The discovery has led to two Nobel prizes awarded to Klaus von Klitzing (1985) for the integral effect, and to R. B. Laughlin, H. L Störmer, and D. C. Tsui (1998) for the fractional effect. The quantum Hall effects are reviewed in several excellent books (Prange and Girvin, 1987; Chakraborty and Pietiläinen, 1988; Das Sarma and Pinczuk, 1990; Heinonen, 1998).

▬rbit to "drift" with a velocity $v_D = \frac{cF}{eB}$, in a direction perpendicular to both $\vec{B}$ and the ▬dditional force $\vec{F}$. As the magnitude of the magnetic field increases, the frequency of the ▬otational motion of the electron goes up while the drift velocity tends to zero. In the limit ▬f infinite magnetic field the slow drift motion of the center of the orbit becomes the only ▬hysically meaningful motion, that is, the only one that survives averaging over a "time ▬indow" fixed by finite experimental resolution. External forces affect only the slow motion ▬f the center of the orbit, while the large kinetic energy associated with the fast rotational ▬motion remains unchanged. An energy that cannot be changed is as good as no energy at all: it is in this sense that one talks of a "quenching" of the kinetic energy in an electronic ▬system at high magnetic field.[2]

In quantum mechanics, the kinetic energy of an electron in a magnetic field is quantized in units of $\hbar\omega_c$ so that its admissible values are

$$\epsilon_n = \left(n + \frac{1}{2}\right)\hbar\omega_c \qquad (10.3)$$

where n is a non-negative integer. These energy levels are known as *Landau levels*.[3] The number of degenerate states in each Landau level is proportional to the magnetic field and to the area of the system, and is given by

$$N_L = \frac{eB\mathcal{A}}{hc} . \qquad (10.4)$$

The macroscopically large degeneracy corresponds to the fact that the center of a classical circular orbit can be located anywhere in the plane. The gap between two consecutive Landau levels $\hbar\omega_c$ increases with increasing magnetic field, and so does the "capacity", N_L, of each Landau level. Thus, in the limit of infinite magnetic field all the electrons fall in the lowest Landau level, with negligible admixture of higher Landau levels. Once again, we can conclude that the kinetic energy, in this limit, is essentially a constant equal to the number of electrons times $\frac{\hbar\omega_c}{2}$.[4]

The quenching of the kinetic energy in a 2DEL at high magnetic field has far-reaching consequences. The key parameter that controls the properties of the system is the *filling factor*

$$\nu = \frac{N}{N_L} . \qquad (10.5)$$

Its integer part, $[\nu]$, (the largest integer smaller than ν) is the number of completely filled Landau levels in the noninteracting ground-state, while the remainder, $\nu - [\nu]$, determines

---

[2] The dynamics of an electron at high magnetic field is analogous to the dynamics of a fast top under the action of the gravitational field. The precessional motion of the axis of the top corresponds to the drift motion of the center of the electron's orbit, while the fast spinning motion of the top is the analogue of the cyclotron motion of the electron. Equivalently, one may say that the electron obeys a "massless dynamics", governed by a lagrangian that does not contain the kinetic energy term: this implies that the net classical force acting on the electron (including the Lorentz force) must vanish.

[3] The "Landau levels" (Landau, 1930) were actually first derived by Fock (1928) who solved the more general problem of an electron subjected to both a uniform magnetic field *and* a harmonic oscillator potential.

[4] In reality, the condition for the quenching of the kinetic energy is $\hbar\omega_c \gg V$ where V is the energy scale of Landau-level-mixing interactions such as external potentials, electron-electron interactions, etc. This condition does not require that all the electrons reside in the lowest Landau level.

the population of the highest occupied Landau level, which is only partially filled by electrons.

Consider first the generic situation in which ν is *not* an integer. In this case the noninteracting ground-state is highly degenerate because any rearrangement of the electrons within the highest occupied Landau level costs no kinetic energy. The degeneracy is removed, at least in part, by the interaction. Broken symmetry phases, which, in the absence of a magnetic field would be preempted by the kinetic energy cost, become energetically possible. Consider, for example, the Wigner crystal. In Section 1.6 we saw that the gain in potential energy with respect to the homogeneous liquid phase is partially offset by the higher kinetic energy, which scales as $r_s^{-\frac{3}{2}}$ at large r_s (as opposed to r_s^{-2} in the homogeneous phase). Obviously this does not happen in a partially filled Landau level, since the kinetic energy within the Landau level remains constant. For this reason we expect Wigner crystallization to occur at higher density than in the zero field case. Indeed, there is experimental evidence that Wigner crystallization occurs at filling factors $\nu < \nu_c \simeq 0.18$, which corresponds to $r_s > 2.7$ at $B = 10$ T: compare this with the condition $r_s > 34$ for the appearance of the Wigner crystal in the 2DEL at zero magnetic field!

Once it is realized that the nature of the ground-state is entirely controlled by interactions, it is natural to look for strongly correlated states. The *incompressible quantum Hall liquid*, introduced by Laughlin (1983) to explain the quantum Hall effect at filling factors of the form $\nu = \frac{1}{2k+1}$, where k is a positive integer, is the oldest and best known example. The many-body wave function that describes it has multiple zeroes (of order $2k + 1$) on the hypersurfaces of configuration space on which two electrons come in contact: in this sense, the zeroes of the wave function are "bound" to the particles.[5] These multiple zeroes are far more powerful than the simple zeroes required by the Pauli exclusion principle, and are ultimately responsible for giving the Laughlin state a particularly low interaction energy. More general incompressible liquid states, which explain the quantum Hall effect at filling factors of the form $\nu = \frac{n}{2kn+1}$, where n is a non-zero integer, have been constructed by Jain, based on the beautiful idea of *composite fermions* (Jain, 1989, 2000), about which more will be said below. The wave functions and the physical properties of these incompressible states will be described in detail in this chapter.

Strong correlation effects also have an interesting effect on the evolution of the spin polarization as a function of density. In a Landau level of a given spin orientation one would naively expect the ground-state to have the maximum spin polarization allowed by the Pauli exclusion principle, because in such a state the exchange energy attains its most negative value at no extra cost in kinetic energy.[6] This is, indeed, the case for filling factor $\nu = 1$ and for many other filling factors, including those of the Laughlin fractional quantum Hall states. However, there is strong experimental evidence (Barrett, 1995) that, as the filling factor is

---

[5] In general, as was noted in footnote [46] of Chapter 1, the nodal hypersurfaces on which a many-body wavefunction vanishes do not coincide with the hypersurfaces on which two particles are in contact. The Laughlin wave function is quite exceptional in that the two types of hypersurfaces coincide.

[6] The reader will notice that this is analogous to Hund's first rule for partially filled atomic shells. The spontaneous polarization occurs even if the g-factor is tuned to be nearly zero (which can be done by clever band-structure engineering) so that the Zeeman coupling to the external magnetic field is absent.

lightly reduced from 1 the spin polarization decreases very rapidly, whereas, according to the naive exchange-energy argument, it should remain constant. These observations indicate that the loss of exchange energy of the partially spin-polarized state relative to the fully spin-polarized one is more than compensated by a gain in correlation energy.

The "quenching of the kinetic energy", which normally enhances the importance of correlations, can, under certain circumstances, result in a "suppression of correlations". In such cases the ground-state turns out to be essentially uncorrelated, i.e., a single Slater determinant. To appreciate this, consider a completely full Landau level, filled with electrons of both spins (the Zeeman coupling splits the Landau level into two levels, separated by an energy gap, but both levels are full). In analogy to a closed-shell atomic configuration, this state has total spin zero. Notice that, by conservation of angular momentum, only states with the same value of the total spin can be mixed by the coulomb interaction. Thus, if $\hbar\omega_c$ is much larger than the coulomb interaction energy, then the mixing of excited states is negligible and the noninteracting closed-shell configuration is essentially the exact ground-state.

The dichotomy between strong correlation in partially filled Landau levels and weak correlation in full Landau levels is turned to one's advantage in the composite fermion approach (for reviews, see Kamilla and Jain, 1997 and Heinonen, 1998). In this approach one introduces composite particles consisting of an electron attached to an infinitely thin flux tube carrying an even number $2k$ of magnetic flux quanta Φ_0.[7] These composite particles, like the original electrons, are subject to the external magnetic field B and interact with each other via the coulomb interaction. The vector potential produced by the magnetic flux tubes exerts no force on the particles, and therefore does not contribute to the particle-particle interaction. When two composite particles are adiabatically interchanged along a path that does not enclose other particles the wave function is multiplied by an Aharonov–Bohm (AB) phase factor $e^{i\pi(2k+1)} = -1$: this indicates that the composite particles are fermions and the problem of interacting composite particles is mathematically identical to the original problem of interacting electrons.

The real advantage of this description appears only when one introduces the mean field approximation. To this end, the fictitious magnetic flux attached to the particles is replaced by an average magnetic flux proportional to the particle density. This average magnetic flux combines with the external physical flux to yield an effective flux

$$B^*\mathcal{A} = B\mathcal{A} - 2k\Phi_0 N \qquad (10.6)$$

where the flux tubes have been assumed to be *antiparallel* to the external field for $k > 0$ and parallel for $k < 0$. The composite fermions, in this approximation, feel the effective field B^* and their effective filling factor v^* is therefore given by (see Eqs. (10.4), (10.5), and (10.6))

$$\frac{1}{v^*} = \frac{1}{v} - 2k . \qquad (10.7)$$

---

[7] The magnetic flux quantum is defined as

$$\Phi_0 \equiv \frac{hc}{e} \simeq 1.2 \times 10^{-6} \text{ T/m}^2 .$$

Notice that the degeneracy of a Landau level N_L is the total magnetic flux $B\mathcal{A}$ expressed in units of Φ_0.

Thus, by cleverly choosing the values of $2k$ and v, it is possible to convert a strongly correlated problem of electrons at *fractional* filling factor v into a weakly correlated problem of composite fermions at *integral* filling factor v^*. For example if $v = \frac{1}{3}$ and $2k = 2$ then we have $v^* = 1$, which corresponds to a full Landau level of composite fermions. Although this conversion is inspired by a questionable mean field approximation, it provides a conceptual framework for constructing correlated wave functions of excellent quality *for electrons* starting from uncorrelated wave functions for composite fermions. The composite fermion picture also establishes a beautiful connection between the integral and the fractional quantum Hall effects. According to this picture, the fractional quantum Hall effect of electrons is nothing but the integral quantum Hall effect of composite fermions (see Fig. 10.23).

The special case of a half-filled Landau level ($v = \frac{1}{2}$) provides an extreme example of the power of the composite fermion idea. In this case, Eq. (10.7), with $2k = 2$, yields $v^* = \infty$, implying that the composite fermions experience no magnetic field on the average.[8] If the mean field approximation makes sense one would then expect to see here some of the characteristic signatures of a two-dimensional Fermi liquid, such as a two-dimensional Fermi surface (Halperin, Lee, and Read, 1993). In fact, this has been quite convincingly demonstrated in surface acoustic wave propagation experiments (Willett, 1993). However, one must keep in mind that the surprising "Fermi liquid" behavior is limited to density and current response properties: it certainly does not apply to single-particle properties. For example, tunneling experiments (Eisenstein *et al.*, 1992) show a pseudogap in the spectral density of one-electron excitations and no sign of the quasiparticle peak characteristic of a normal Fermi liquid. The reasons for this puzzling state of affairs will be discussed in detail.

The *edge* of a 2DEL at high magnetic field provides yet another example of novel physics beyond the quantum Hall effect. It will be shown that the collective oscillations of the density in such an edge are dynamically equivalent to the collective oscillations of a *chiral Luttinger liquid* (χLL), (Wen, 1990) i.e., a Luttinger liquid in which, after diagonalizing the electron-electron interaction, only one half of the bosons – those propagating to the right *or* those propagating to the left – are retained. Unlike previous putative realizations of the Luttinger liquid paradigm, the edges of the 2DEL are essentially free of disorder effects and, more importantly, the Luttinger liquid coupling constant $g = e^{-2\phi}$ (see Section 9.11) coincides with the bulk filling factor v. This has allowed a rather detailed experimental verification of the non-universal exponents in the power-law decay of the correlation functions of the Luttinger liquid (Milliken *et al.*, 1995; Chang *et al.*, 1996; Roddaro *et al.*, 2003). Studies of the tunneling current between edges in the fractional quantum Hall regime (Goldman and Su, 1995; Saminadayar *et al.*, 1997; De Picciotto *et al.*, 1997) have also provided the first convincing evidence of fractionally charged excitations in condensed matter systems.

---

[8] Density fluctuations, however, produce fluctuations of the effective magnetic field, which must be taken into account not only to going beyond the mean field approximation, but even to calculate the response to an external field within the framework of the time-dependent mean field theory.

10.2 One-electron states in a magnetic field

The hamiltonian of a free electron in a uniform magnetic field parallel to the z-axis $\vec{B} = -B\vec{e}_z$ coincides with the kinetic energy operator

$$\hat{H}_0 = \frac{m_b \hat{v}^2}{2} \tag{10.8}$$

where the velocity operator $\hat{v}$ is expressed in terms of the canonical momentum operator $\hat{\vec{p}} = -i\hbar\vec{\nabla}$ as

$$\hat{v} = \frac{1}{m_b}\left(\hat{\vec{p}} + \frac{e}{c}\vec{A}(\hat{\vec{r}})\right) \tag{10.9}$$

and $\vec{A}(\vec{r})$ is the vector potential, defined by the relation

$$\vec{\nabla} \times \vec{A}(\vec{r}) = \vec{B}. \tag{10.10}$$

In this section we ignore the Zeeman coupling $\hat{H}_Z = \frac{g\mu_B}{2}B$.[9]

The vector potential is determined by Eq. (10.10) only up to a *gauge transformation*

$$\vec{A}(\vec{r}) \to \vec{A}(\vec{r}) + \vec{\nabla}\Lambda(\vec{r}), \tag{10.11}$$

where $\Lambda(\vec{r})$ is a regular function of $\vec{r}$. Such a transformation does not change $\vec{B}$. Two widely used choices for $\vec{A}$ are the *Landau gauge*

$$\vec{A}(\vec{r}) = -Bx\vec{e}_y, \tag{10.12}$$

and the *symmetric gauge*

$$\vec{A}(\vec{r}) = \frac{1}{2}\vec{B} \times \vec{r}. \tag{10.13}$$

The choice of a particular gauge may offer technical advantages in a specific calculation. For example, the Landau gauge is translationally invariant in the y-direction, which makes it particularly suitable for the study of translationally invariant systems. The symmetric gauge, being rotationally invariant about the z-axis, is the gauge of choice for circularly symmetric systems, such as quantum dots. Despite these technical differences it must be borne in mind that all the physical results are completely independent of the arbitrary choice of the gauge.

An immediate consequence of Eq. (10.9) is that the x- and y-components of the velocity operator do not commute with each other. One can easily verify that

$$[\hat{v}_x, \hat{v}_y] = i\frac{\hbar\omega_c}{m_b}, \tag{10.14}$$

where ω_c is the cyclotron frequency defined in Eq. (10.2). This important result is at the root

---

[9] The choice of the negative sign of $\vec{B}$ is a matter of convenience. It will allow us to express the wavefunctions in the lowest Landau level in terms of the complex variable $z = x + iy$ rather than its complex conjugate z^*.

of the complex behavior of electrons in a magnetic field. On the other hand, $\hat{v}_z$ continues to commute with $\hat{v}_x$ and $\hat{v}_y$, so the motion in the z-direction is separable.

10.2.1 Energy spectrum

To calculate the eigenvalues of $\hat{H}_0$ let us define the dimensionless operators

$$\hat{\Pi} \equiv \frac{m_b \ell}{\hbar} \frac{\hat{v}_x + i\hat{v}_y}{\sqrt{2}},$$

$$\hat{\Pi}^\dagger \equiv \frac{m_b \ell}{\hbar} \frac{\hat{v}_x - i\hat{v}_y}{\sqrt{2}}, \tag{10.15}$$

where ℓ is the *magnetic length* defined as follows:

$$\ell \equiv \sqrt{\frac{\hbar c}{eB}}. \tag{10.16}$$

Notice that unlike the cyclotron frequency the magnetic length is independent of material parameters.[10]

According to Eq. (10.14), these operators satisfy the bosonic commutation rule

$$[\hat{\Pi}, \hat{\Pi}^\dagger] = 1 \tag{10.17}$$

as can be easily verified. Substituting the expressions for $\hat{v}_x$ and $\hat{v}_y$ in terms of $\hat{\Pi}$ and $\hat{\Pi}^\dagger$ in Eq. (10.8) we obtain

$$\hat{H}_0 = \hbar\omega_c \left(\hat{\Pi}^\dagger \hat{\Pi} + \frac{1}{2} \right) + \frac{\hat{p}_z^2}{2m}, \tag{10.18}$$

where we have assumed, for simplicity, that $\vec{A}$ lies in the x–y plane. From Eqs. (10.18) and (10.17) we see at a glance that the energy spectrum of $\hat{H}_0$ is that of a harmonic oscillator of frequency ω_c plus that of a free particle moving along the z-axis. The energy levels are therefore given by

$$\epsilon_n(k_z) = \left(n + \frac{1}{2} \right) \hbar\omega_c + \frac{\hbar^2 k_z^2}{2m_b}, \tag{10.19}$$

where $\hbar k_z$ is the z-component of the linear momentum and n is a non-negative integer. For electrons confined to the x–y plane – the case of primary interest in this chapter – the last term on the right-hand side of Eq. (10.19) can be set to zero[11] and we recover Eq. (10.3).

Let us now examine the *degeneracy* of the energy levels. In quantum mechanics, degeneracy is caused by the existence of two non-commuting constants of the motion. In the

---

[10] It may be useful to remember that $\ell \simeq 250/\sqrt{B}\,\text{Å}$ where B is the value of the magnetic field in Tesla. ℓ is the radius of a disk that encloses *one half* of a magnetic flux quantum $\Phi_0 = \frac{hc}{e}$.
[11] Strictly speaking, it should be set equal to the energy of the lowest state of quantized motion along the z-axis.

resent case, the existence of two such constants of the motion is deduced from the solution of the classical mechanical problem, which can be written in the form

$$z(t) = Z + R_c e^{-i\omega_c t},$$
$$v(t) = -i\omega_c R_c e^{-i\omega_c t}, \qquad (10.20)$$

where the complex variables $z = x + iy$ and $v = v_x + iv_y$ describe the position and the velocity of the particle in the x–y plane, while $Z = X + iY$ denotes the position of the center of a circular orbit of radius R_c. Eliminating R_c from these two equations we obtain the expressions for the coordinates of the center of the orbit in terms of the particle position and velocity:

$$X = x + \frac{v_y}{\omega_c},$$
$$Y = y - \frac{v_x}{\omega_c}. \qquad (10.21)$$

Since X and Y are constants of the classical motion, we expect that the corresponding quantum mechanical operators $\hat{X}, \hat{Y}$ constructed in terms of $\hat{v}$ and $\hat{r}$ according to Eq. (10.21) should be constants of the motion too. This guess is correct, and one can easily verify it by noting that the commutators of $\hat{X}$ and $\hat{Y}$ with both $\hat{v}_x$ and $\hat{v}_y$ vanish, so that $[\hat{X}, \hat{H}_0] = [\hat{Y}, \hat{H}_0] = 0$. Notice, however, that $\hat{X}$ and $\hat{Y}$ do not commute with each other:

$$[\hat{X}, \hat{Y}] = -i\ell^2. \qquad (10.22)$$

To determine the "amount of degeneracy" we need more insight into the quantum mechanical significance of $\hat{X}$ and $\hat{Y}$. First of all let us define the operators

$$\hat{T}_x \equiv -m_b \omega_c \hat{Y} = m_b(\hat{v}_x - \omega_c \hat{y}),$$
$$\hat{T}_y \equiv m_b \omega_c \hat{X} = m_b(\hat{v}_y + \omega_c \hat{x}), \qquad (10.23)$$

which are just $\hat{Y}$ and $\hat{X}$ (in this order) rescaled by $\mp m\omega_c$. Observe that these operators, on one hand, commute with the velocity, on the other hand satisfy the commutation relations

$$[\hat{T}_x, \hat{x}] = [\hat{T}_y, \hat{y}] = -i\hbar,$$
$$[\hat{T}_x, \hat{y}] = [\hat{T}_y, \hat{x}] = 0 . \qquad (10.24)$$

Because these relations are of the same form as the commutation relations between the canonical momentum and the position operator, we see that $\hat{\vec{T}} = (\hat{T}_x, \hat{T}_y)$ must be the generator of the translations in the presence of the magnetic field, in the sense that $\exp(i\hat{\vec{T}} \cdot \vec{u}/\hbar)$ translates the position operator by $\vec{u}$ without affecting the velocity operator. Notice that the familiar nonmagnetic translation operator $\exp(i\hat{\vec{p}} \cdot \vec{u}/\hbar)$ would not work in this case, because it fails to commute with the velocity operator. Both $\hat{T}_x$ and $\hat{T}_y$ are constants of the motion, but since

$$[\hat{T}_x, \hat{T}_y] = -\frac{i\hbar^2}{\ell^2} \qquad (10.25)$$

we see that the algebra of *magnetic translations*, i.e., translations generated by $\hat{\vec{T}}$, is quite different from that of ordinary translations. In particular, $\hat{T}_x$ and $\hat{T}_y$ cannot be simultaneously assigned a definite value in a given state. We can, however, classify the eigenstates of $\hat{H}$ according to the eigenvalues of $\hat{T}_y$, which we denote by $\hbar k_y$:

$$\hat{T}_y |\epsilon_n, k_y\rangle = \hbar k_y |\epsilon_n, k_y\rangle. \tag{10.26}$$

We assume that the particle is confined within a large rectangular box of dimensions L_x and L_y, and impose periodic boundary conditions in the y-direction: this means that $\exp(i\hat{T}_y L_y/\hbar) = 1$, and therefore k_y must be quantized in units of $2\pi/L_y$:

$$k_y = \frac{2\pi}{L_y} k , \tag{10.27}$$

where k is an integer. Now observe that, by virtue of Eq. (10.24), the operator $\hat{R} = \exp(2\pi i \ell^2 \hat{T}_x/\hbar L_y)$ acting on $|\epsilon_n, k_y\rangle$ increases the value of k_y by $\frac{2\pi}{L_y}$ without changing the energy: $\hat{R}|\epsilon_n, \frac{2\pi}{L_y}k\rangle = |\epsilon_n, \frac{2\pi}{L_y}(k+1)\rangle$. Finally, recognize that $\hat{R}$ is the operator that translates the $\hat{x}$ coordinate by an amount $\Delta x = \frac{2\pi \ell^2}{L_y}$. Since the quantization box has a finite extent L_x in the x-direction such a translation can be repeated up to a maximum of $N_L = \frac{L_x}{\Delta x} = \frac{L_x L_y}{2\pi \ell^2}$ times. N_L is therefore, the number of different values of k_y that are allowed in a given Landau level in a finite box. Setting $L_x L_y = \mathcal{A}$ (the area of the quantization box) and $2\pi \ell^2 = \frac{hc}{eB}$ we see that the degeneracy of the Landau level is given by $N_L = \frac{eBA}{hc}$ as stated in Eq. (10.4).[12]

10.2.2 One-electron wave functions

Due to the massive degeneracy of the Landau levels the form of the eigenfunctions of $\hat{H}_0$ depends strongly on the choice of the gauge.

Landau gauge – Making use of Eq. (10.12), we write the Schrödinger equation in the form

$$\left[-\frac{\hbar^2}{2m_b} \frac{\partial^2}{\partial x^2} + \frac{1}{2m_b} \left(-i\hbar \frac{\partial}{\partial y} - \frac{eB}{c}x \right)^2 \right] \psi_n(x, y) = \epsilon_n \psi_n(x, y), \tag{10.28}$$

where n is the Landau level index. Because the Hamiltonian does not depend on y, we can seek the solution in the form

$$\psi_n(x, y) = \frac{1}{\sqrt{L_y}} e^{ik_y y} \psi_{n,k_y}(x), \tag{10.29}$$

where $\psi_{n,k_y}(x)$ is the solution of the one-dimensional equation

$$\left[-\frac{\hbar^2}{2m_b} \frac{d^2}{dx^2} + \frac{m_b \omega_c^2}{2} \left(x - k_y \ell^2 \right)^2 \right] \psi_{n,k_y}(x) = \epsilon_n \psi_{n,k_y}(x). \tag{10.30}$$

---

[12] It is interesting to notice that, due to the non-commutativity of magnetic translations in the x- and y-directions, periodic boundary conditions can be imposed along *one* direction but not along both.

This is the familiar Schrödinger equation for a one-dimensional harmonic oscillator of frequency ω_c with equilibrium position at $X_{k_y} \equiv k_y \ell^2$. Its normalized solution is

$$\psi_{n,k_y}(x) = \frac{1}{\sqrt{2^n n! \pi^{\frac{1}{2}} \ell}} e^{-\frac{(x-k_y \ell^2)^2}{2\ell^2}} H_n \left(\frac{x - k_y \ell^2}{\ell} \right), \tag{10.31}$$

where $H_n(x)$ is the Hermite polynomial of order n. Thus the solution is plane-wave-like, with wave vector k_y, in the y-direction, and is exponentially localized in the x-direction with center at $k_y \ell^2$ and localization length of the order of ℓ. Note that, in this gauge, $\hbar k_y$ is the eigenvalue of the magnetic translation generator $\hat{T}_y = -i\hbar \nabla_y$. The allowed values of k_y are given by Eq. (10.27) with $0 \leq k < N_L - 1$, ensuring that the center of the state falls in the range $0 < x < L_x$. The separation between the centers of two consecutive states is $\Delta x = 2\pi \ell^2 / L_y$ which is much smaller than ℓ: thus, *there is a macroscopically large number of overlapping states within a single magnetic length.*

Symmetric gauge – In this gauge the hamiltonian has the form

$$\hat{H}_0 = -\frac{\hbar^2}{2m_b} \nabla^2 + \frac{m_b}{2} \left(\frac{\omega_c}{2} \right)^2 r^2 - \frac{\omega_c}{2} \hat{L}_z, \tag{10.32}$$

where $\hat{L}_z = (\hat{r} \times \hat{p})_z$ is the z-component of the canonical angular momentum, i.e., the generator of rotations about the z-axis.

The first two terms in $\hat{H}_0$ form the hamiltonian of an isotropic two-dimensional harmonic oscillator of frequency $\frac{\omega_c}{2}$. The eigenfunctions of this system in polar coordinates r and ϕ are

$$\psi_{n_r,m}(r, \phi) = \frac{1}{\ell} \sqrt{\frac{n_r!}{(n_r + |m|)!}} \left(\frac{r}{\ell \sqrt{2}} \right)^{|m|} e^{-\frac{r^2}{4\ell^2}} L_{n_r}^{|m|} \left(\frac{r^2}{2\ell^2} \right) e^{im\phi}, \tag{10.33}$$

where $L_n^{|m|}(r)$ are the associated Laguerre polynomials,[13] the integer m, which runs from $-\infty$ to $+\infty$, is the angular momentum quantum number, and $n_r(\geq 0)$ is a "radial quantum number", which counts the number of zeroes in the radial part of the wave function. The corresponding eigenvalues $(|m| + 2n_r + 1) \frac{\hbar\omega_c}{2}$ are just the energy levels of the two-dimensional harmonic oscillator.

Since $\psi_{n_r,m}$ is an eigenfunction of $\hat{L}_z$, it follows that it is automatically an eigenfunctions of the complete hamiltonian $\hat{H}_0$ with eigenvalues

$$\epsilon_{n_r,m} = (|m| + 2n_r + 1) \frac{\hbar\omega_c}{2} - m \frac{\hbar\omega_c}{2}, \tag{10.34}$$

where the first term on the right-hand side describes the energy levels of the harmonic

---

[13] The definition of the Laguerre polynomials is (see Gradshteyn–Ryzhik 8.97)

$$L_n^\alpha(x) = \frac{1}{n!} e^x x^{-\alpha} \frac{d^n}{dx^n} \left(e^{-x} x^{n+\alpha} \right) = \sum_{m=0}^{n} (-1)^m \binom{n+\alpha}{n-m} \frac{x^m}{m!}.$$

oscillator of frequency $\frac{\omega_c}{2}$. Eq. (10.34) can be rewritten as

$$\epsilon_{n_r,m} = \left(n_r + \frac{|m| - m}{2} + \frac{1}{2}\right)\hbar\omega_c , \qquad (10.35)$$

which is immediately seen to be equivalent to Eq. (10.3) since $|m| - m$ is an even, non-negative integer. The comparison between the two equations, however, reveals the existence of a restriction on the admissible values of the angular momentum within a given Landau level: because $n = n_r + \frac{|m|-m}{2}$, where both n and n_r are non-negative integers, we must have

$$m \geq -n . \qquad (10.36)$$

Thus, only non-negative values of m are permitted in the lowest Landau level, only values greater than or equal to -1 in the first Landau level, and so on. The asymmetry in the range of allowed values of m is a manifestation of the broken time-reversal symmetry of the hamiltonian.

Taking a closer look at Eq. (10.33) one sees that the symmetric-gauge wave-functions are extended in the azimuthal direction, and localized in the radial direction, their shape being roughly that of a ring of width ℓ and radius $r_{|m|} \simeq \sqrt{2m}\ell$ (for large m), so that exactly $|m|$ quanta of magnetic flux are enclosed. For $m \gg 1$ the radial distance between two consecutive rings scales as $\ell/\sqrt{m}$ and thus is much smaller than the width of each ring.

10.2.3 Fock–Darwin levels

As we pointed out in footnote[3], the Landau levels are actually a special case of the energy levels of an electron subjected to both a magnetic field *and* a harmonic oscillator potential of the form $\frac{1}{2}m_b\omega_0^2 r^2$. This interesting problem was first solved by Fock (1928) and Darwin (1930), and the resulting energy levels are known as the Fock–Darwin levels. The Fock–Darwin (FD) hamiltonian in the symmetric gauge has the form

$$\hat{H}_{FD} = -\frac{\hbar^2}{2m_b}\nabla^2 + \frac{m_b}{2}\Omega^2 r^2 - \frac{\omega_c}{2}\hat{L}_z , \qquad (10.37)$$

where

$$\Omega = \sqrt{\frac{\omega_c^2}{4} + \omega_0^2} . \qquad (10.38)$$

A quick comparison with Eq. (10.32) shows that the eigenfunctions of this hamiltonian are still given by Eq. (10.33), with the proviso that the magnetic length ℓ be replaced by $\ell_\Omega = \sqrt{\frac{\hbar}{2m_b\Omega}}$. The Fock–Darwin levels are then given by

$$\epsilon_{n_r,m} = (|m| + 2n_r + 1)\hbar\Omega - m\frac{\hbar\omega_c}{2} , \qquad (10.39)$$

and notice that they reduce to the Landau levels in the limit $B \to \infty$.

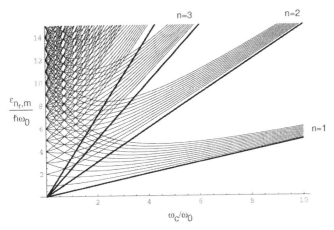

Fig. 10.1. A plot of the Fock–Darwin levels (10.39) vs. $\frac{\omega_c}{\omega_0}$, for $0 \le n_r \le 10$ and $0 \le |m| \le 10$, exhibits an intricate structure of *level crossings*. Notice that, for $\omega_c \to \infty$ the levels bunch up into Landau levels, shown by the straight lines.

The Fock–Darwin level structure (see Fig. 10.1) has recently been observed in mesoscopic systems known as *quantum dots* in which a variable number of electrons can be trapped in an approximately parabolic potential. The agreement between theory and experiment is impressive. Notice that the degeneracy of the Landau levels is lifted by the confinement potential. But, one can see in Fig. 10.1 how the degenerate structure of the Landau levels emerges as a bunching of Fock–Darwin levels with different values of the angular momentum in the limit of large magnetic field.

10.2.4 Lowest Landau level

The symmetric-gauge wave functions in the lowest Landau level (LLL) have a striking property of *analyticity*. Setting $n = 0$ and $m \ge 0$ in Eq. (10.35) we see that $n_r = 0$ in the LLL. Putting this in Eq. (10.33) and noting that $L_0^{|m|}(x) = 1$ for all values of m and x we get

$$\psi_{0,m}(r, \phi) \equiv \psi_m(z) = \frac{1}{\sqrt{2\pi \ell^2 2^m m!}} z^m e^{-|z|^2/4\ell^2} \tag{10.40}$$

where $m \ge 0$ and $z = x + iy$. Any wavefunction that lies entirely in the LLL can be written as a superposition (finite or infinite) of wave functions of the form (10.40), i.e., integer powers of z times the exponential factor $e^{-|z|^2/4\ell^2}$. This implies that any such wave function will have the form

$$f(z)e^{-|z|^2/4\ell^2} \tag{10.41}$$

where $f(z)$ is an *analytic* function of z. An N-electron wave function that similarly lies entirely in the LLL will have the form

$$f(z_1, \ldots, z_N) \prod_{i=1}^{N} e^{-|z_i|^2/4\ell^2} \tag{10.42}$$

where f is an analytic function of the N variables $z_1, \ldots, z_N$.

This property is unique to the LLL – it is easy to see that wave functions in higher Landau levels inevitably contain powers of z^* and therefore cannot be written in the form (10.41). We shall often make use of this property in the following sections.

10.2.5 Coherent states

Another useful set of one-electron states in the lowest Landau level is provided by the so-called "coherent states". These states are exponentially localized about a center, and will be used in Section 10.19 to construct variational wave functions for the Wigner crystal. The simplest coherent state coincides with the $m = 0$ state in the symmetric gauge, that is,

$$\phi_0(z) = \frac{1}{\sqrt{2\pi \ell^2}} e^{-|z|^2/4\ell^2} . \tag{10.43}$$

Other coherent states are constructed by applying to ϕ_0 the magnetic translation operator $e^{\frac{i}{\hbar} \hat{T} \cdot \vec{R}}$ where $\vec{R} = (X, Y)$ is an arbitrary two-dimensional vector. This gives

$$\phi_{\vec{R}}(\vec{r}) = \frac{1}{\sqrt{2\pi \ell^2}} e^{-|\vec{r}-\vec{R}|^2/4\ell^2} e^{i(xY-yX)/2\ell^2} . \tag{10.44}$$

It must be noted at this point that coherent states are not mutually orthogonal but form a vastly overcomplete set of states in the LLL. An interesting question is whether one can extract a complete set of linearly independent localized states by keeping only the coherent states that are centered on the sites of a lattice of density $N_L/\mathcal{A}$. For a square lattice this amounts to selecting the set of coherent states centered at $\vec{R} = n_1 a \vec{e}_x + n_2 a \vec{e}_y$, where n_1 and n_2 are integers and $a = \sqrt{2\pi \ell^2}$ is the lattice constant. If such a set were complete, then the electron liquid in a fully occupied Landau level would be no different from an ordinary band insulator, in which the valence band is completely occupied, and the Hall effect would be absent.[14] However, it can be shown that the states in this set are not completely independent due to the existence of the linear relationship (Thouless, 1984)

$$\sum_{n_1, n_2} (-1)^{n_1+n_2+n_1 n_2} \phi_{n_1 a \vec{e}_x + n_2 a \vec{e}_y}(\vec{r}) = 0 , \tag{10.45}$$

which holds for every $\vec{r}$ (when $a = \sqrt{2\pi \ell^2}$). In more physical language this identity implies the existence of an extended state which is orthogonal to *all* the coherent states in the set. It

---

[14] Technically, one could then construct a complete set of *Wannier functions* localized on the sites of the lattice in complete analogy to the Wannier functions of a crystal lattice.

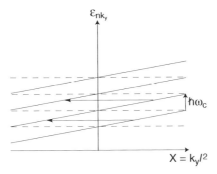

Fig. 10.2. Landau levels in the presence of an electric field $\mathcal{E}$ (solid lines) and without the electric field (dashed lines). The kinetic energy $\left(n + \frac{1}{2}\right)\hbar\omega_c$ increases in jumps of $\hbar\omega_c$ in vertical transitions, while the potential energy decreases as one moves to the left in a Landau level. Energy-conserving transitions between different Landau levels, indicated by horizontal arrows, can occur via tunneling between states that are separated in space by a large distance $\hbar\omega_c \gg e\mathcal{E}$.

is the presence of this extended state that makes the difference between a full Landau level and an insulator, and allows the Hall effect. An explicit construction of the extended state is worked out in Exercise 2.

10.2.6 Effect of an electric field

Ordinarily, a free electron under the action of an electric field gains kinetic energy in a continuous manner at the expenses of the potential energy. But, in the presence of a strong magnetic field, the kinetic energy can only increase in steps of $\hbar\omega_c$. As shown in Fig. 10.2, the "jump" to a higher Landau level is really a tunneling process between two states that are separated in space by a distance $\hbar\omega_c/e\mathcal{E}$. This distance is typically much larger than the characteristic width ℓ of a one-electron orbital: for example, at $B = 10$ T it would take an electric field of $\mathcal{E} = 10$ V/μm to make $\hbar\omega_c/e\mathcal{E}$ comparable to $\ell \approx 80$ Å. Thus, for moderate electric fields, the tunneling amplitude, which is proportional to the overlap of the initial and final states, is negligible, and the kinetic energy does not change. This, in turn, implies that the electron moves along a trajectory of constant potential energy, i.e., one that is everywhere orthogonal to the direction of the electric field. The velocity of this motion will turn out to be directly proportional to the magnitude of the electric field, and inversely proportional to the magnitude of the magnetic field.

Let us see in detail how this happens in the simple case of a uniform electric field. Without loss of generality we orient the x-axis along the direction of the electric field: the potential energy is thus

$$V(x) = e\mathcal{E}x. \qquad (10.46)$$

Working in the Landau gauge, we see that it is still possible to represent the solutions of the Schrödinger equation in the form of Eq. (10.29), where $\psi_{n,k_y}(x)$ is the solution of the one-dimensional equation

$$\left[-\frac{\hbar^2}{2m_b}\frac{d^2}{dx^2} + \frac{m_b}{2}\omega_c^2\left(x - k_y\ell^2\right)^2 + e\mathcal{E}x\right]\psi_{n,k_y}(x) = E_{n,k_y}\psi_{n,k_y}(x). \quad (10.47)$$

This can be rewritten as

$$\left[-\frac{\hbar^2}{2m}\frac{d^2}{dx^2} + \frac{m}{2}\omega_c^2\left(x - X_{k_y}\right)^2 + e\mathcal{E}k_y\ell^2 - \frac{m}{2}v_d^2\right]\psi_{n,k_y}(x) = E_{n,k_y}\psi_{n,k_y}(x), \quad (10.48)$$

where

$$X_{k_y} = k_y\ell^2 - \frac{v_d}{\omega_c}, \quad (10.49)$$

and

$$v_d = c\frac{\mathcal{E}}{B}. \quad (10.50)$$

From this we can immediately read the eigenvalues, which are given by

$$E_{n,k_y} = \left(n + \frac{1}{2}\right)\hbar\omega_c + e\mathcal{E}X_{k_y} + \frac{m}{2}v_d^2, \quad (10.51)$$

and the eigenfunctions, which are still given by Eq. (10.31), but now centered about $k_y\ell^2 - \frac{v_d}{\omega_c}$. We see that the effect of the electric field is simply to displace the "center" of the wavefunction. Because of this displacement, the quantum state $|n, k_y\rangle$, which would have carried zero current in the absence of an electric field, acquires a finite current in the y-direction. This current can be calculated from the derivative of the energy with respect to $\hbar k_y$ (Exercise 1) with the following result:

$$I_{n,k_y} = -e\frac{\partial E_{n,k_y}}{\hbar\partial k_y} = -\frac{e^2\mathcal{E}\ell^2}{\hbar} = -ev_d. \quad (10.52)$$

It is for this reason that one says that the electric field induces a "drift" motion of the electron with velocity $\vec{v}_d$ perpendicular to $\vec{\mathcal{E}}$.

Eq. (10.51) for the eigenvalues can be interpreted as follows: the first term is the energy of the unperturbed Landau level; the second term is the electrostatic potential evaluated at the center of the wavefunction, and the last term is the kinetic energy of the drift motion. The dependence of these energies on k_y is linear, as shown in Fig. 10.2. Strictly speaking, there is no lowest energy, and hence no ground-state. However, once an electron is trapped in a state with given k_y transitions to lower energy states are prevented by translational invariance.

10.2.7 Slowly varying potentials and edge states

The results of the previous subsection are easily generalizable to the case of potentials that have an arbitrary dependence on x, but vary *slowly* on the scale of the magnetic length.

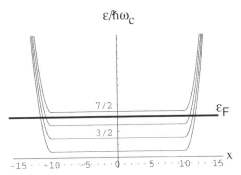

Fig. 10.3. Model for the behavior of the Landau levels in the bulk and at the edges of a two-dimensional strip. The potential is assumed to depend only on the coordinate (x) perpendicular to the edges of the strip, which, in this figure, are located at $x = \pm 10$. The Fermi level (thick line) separates the occupied states from the empty states.

Specifically, if we assume that the local electric field $\mathcal{E}(x)$ satisfies the inequality $e\mathcal{E}(x)\ell \ll \hbar\omega_c$, then the variation of the electric potential is small over the region in which a given Landau level orbital is appreciably different from zero, and therefore perturbation theory is applicable even if the potential itself is large. To first order in $\alpha \equiv \frac{e\mathcal{E}(x)\ell}{\hbar\omega_c}$ the positions of the centers of the orbitals are shifted from $k_y\ell^2$ to $k_y\ell^2 + \alpha\ell$ and the corresponding energies from ϵ_n to

$$E_{n,k_y} = \epsilon_n + V(k_y\ell^2). \tag{10.53}$$

The above approximation is commonly employed to describe the electronic states in the vicinity of the *edges* of the two-dimensional system, where the translational invariance in the direction perpendicular to the edge is broken. It is assumed (see Fig. 10.3) that the electrons are confined within the geometrical boundaries of the system by a potential that rises sharply at the edge of the system. If the potential is slowly varying on the scale of ℓ then the energy levels simply follow its rise as shown in the Fig. 10.3.[15]

Potentials that depend on both x and y pose a considerably more complicated problem. Of particular interest is the structure of the energy levels in a *random* but slowly varying potential landscape. Because the kinetic energy is "absent", we expect that the electronic states will closely follow the lines of constant potential energy. In the example of Fig. 10.4 the equipotential lines are immediately seen to be of two types: closed lines, which encircle regions of low or high potential (the light and the dark regions, respectively) and open lines, which connect two opposite sides of the system – the left and the right in this example.[16]

---

[15] In reality, due to electron-electron interactions, the confinement potential must be determined self-consistently with the electronic density, for example by solving the Kohn–Sham equations of density functional theory (DFT). In this context we note that Fig. 10.3 applies to *sharp edges*, i.e. edges in which the electronic density drops to zero over a distance of the order of the magnetic length. The qualitatively different behavior of *smooth edges* will be discussed in Section 10.20.2.

[16] Equipotential lines that connect adjacent sides of the system are also considered closed for our purposes.

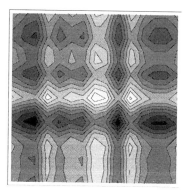

Fig. 10.4. Equipotential lines of a random, but slowly varying potential in the x–y plane. The gray scale indicates the magnitude of the potential, with darker regions being at higher potential. Notice the extended states that run from left to right about the center of the figure.

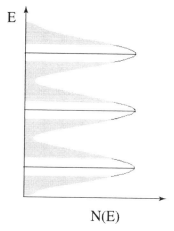

Fig. 10.5. Schematic density of states for the disordered two-dimensional electron gas in a strong magnetic field. Extended states exist near the center of the original Landau level. Regions of localized states are shaded in gray.

It is intuitively evident that only states associated with open lines, i.e. *extended states*, contribute to the electrical conductivity. Furthermore, it follows from topological considerations that, for a bounded potential, there is always at least one extended state at about the center of the potential energy band. The situation is depicted in Fig. 10.5, which shows disordered-broadened Landau levels, with extended states at intermediate energies and localized states in the tails.[17] This situation is profoundly different from the one in effect

---

[17] The problem of determining whether a *finite fraction* of the total number of states is extended is very delicate. It is believed that this fraction tends to zero in the thermodynamic limit in the absence of a uniform electric field $\mathcal{E}$, but becomes finite when $\mathcal{E}$ is non zero (Trugman, 1983).

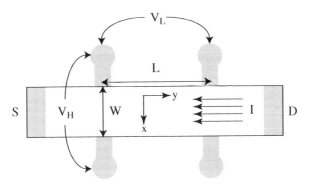

Fig. 10.6. Standard quantum Hall bar device for the measurement of the Hall resistance and the longitudinal resistance. The magnetic field is in the $-\vec{e}_z$ direction, perpendicular to the plane of the electron gas.

at $B = 0$, where, in the presence of disorder, *all* the states are expected to be localized (Abrahams *et al.*, 1979). The mechanism by which the extended states leave the Hilbert space with decreasing magnetic field is known as *levitation*, and is still a subject of active research.

10.3 The integral quantum Hall effect

10.3.1 Phenomenology

The phenomenology of the quantum Hall effects (both integral and fractional) is summarized in Figs. 10.6 and 10.7. Figure 10.6 presents the schematic set-up of a four-terminal quantum Hall measurement. A current I is driven through the two-dimensional electron liquid from a "source" (S) into a "drain" (D). The magnetic field is perpendicular to the plane of the figure. The Hall voltage V_H is measured as the potential difference between two electrodes on opposite sides of the current path, while the longitudinal voltage V_L is the potential difference between two electrodes on the same side of the current path. The longitudinal resistance R_L and the Hall resistance R_H are defined as the ratios V_L/I and V_H/I respectively.

In a classical (i.e. unquantized) Hall measurement the values of R_L and R_H depend on extrinsic properties (such as the geometric arrangement of the measuring electrodes and the concentration of impurities in the device) beside the electron density. For example, a classical calculation of the Drude *resistivity*, similar to what we did in Section 4.6.1, but including the Lorentz force, would give (Ashcroft and Mermin, 1976)

$$\rho_{xx} = \frac{m_b}{ne^2\tau} = \rho_{yy} \, ,$$

$$\rho_{xy} = \frac{B}{nec} = -\rho_{yx} \, . \tag{10.54}$$

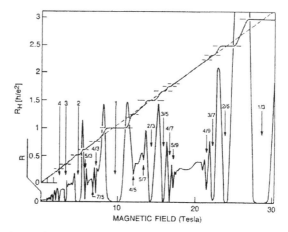

Fig. 10.7. Both the integral and the fractional quantum Hall effects are characterized by well-developed plateaus at quantized values of the Hall resistance R_H, and vanishing values of the longitudinal resistance R_L in correspondence to the plateaus. Here the dashed diagonal line represents the classical Hall resistance, and the solid line the experimental results (Stormer, 1992).

The two resistances R_L and R_H are related to the diagonal and off-diagonal *resistivities* by the familiar formulas

$$R_L = \rho_{yy} \left(\frac{L}{W} \right),$$

$$R_H = \rho_{xy} \left(\frac{W}{W} \right) = \rho_{xy}, \qquad (10.55)$$

where L and W are the length and the width of the device shown in Fig. 10.6. Clearly, these quantities depend on the electronic density n, on the momentum relaxation time τ (which is inversely proportional to the concentration of impurities), on the magnetic field B, and on the ratio $\frac{L}{W}$.

Not so in the quantum Hall effect: Fig. 10.7 shows broad ranges of values of densities and magnetic fields near an integral filling factor ν in which the Hall resistance R_H has *precisely* the value $\frac{h}{\nu e^2} \approx \frac{25812.807572}{\nu}$ Ω. The accuracy of this quantized value is spectacular: better than 1 part in 10^{10}. In the same range of filling factors the longitudinal resistivity is found to vanish, implying that there is no dissipation. A similar phenomenology is observed in the vicinity of some fractional filling factors ($\nu = \frac{1}{3}$, for instance), but we shall focus only on the integral effect in this section.

The persistence of the quantized Hall resistance over a range of densities and magnetic fields, and its complete insensitivity to details such as sample shape and impurity concentration are astounding facts, which call for an explanation. It is the very nature of the Hall effect that makes it independent of the geometrical arrangement of the electrodes. For, if the longitudinal resistance vanishes, then the electric potential must be constant along each of the two edges of Fig. 10.6, implying that the Hall voltage cannot depend on where the contacts are placed.[18] Observe that the value of the quantized Hall resistance $R_H = \frac{h}{ke^2}$ at integer filling factor $\nu = k$ is correctly given by the naive Drude theory (Eq. (10.54)) for an electron gas *at the same filling factor*. What the naive theory cannot explain is why the longitudinal resistivity vanishes in the experiment, and why the value of R_H remains constant even when the filling factor is significantly different from k. Fig. 10.7 contrasts the observed values of R_H vs. B with the naive prediction, which is just a straight line through the origin. The obvious difference is the appearance of *plateaus* in R_H near integer filling factors: this suggests the existence of incompressible states, which successfully "resist" a change in the values of the external parameters.

After the experimental discovery by von Klitzing, Dorda, and Pepper in 1980, there has been no lack of theoretical explanations of the integral quantum Hall effect. The theories vary in their level of mathematical sophistication and emphasize different aspects of the physics. Ultimately, the effect is shown to be topological in nature, and this is the reason for its extraordinary robustness. We now present three theoretical "derivations" of the integral Hall effect, and discuss their strengths and weaknesses.

10.3.2 The "edge state" approach

Let us start with the somewhat unrealistic case of a smooth random potential $V(x)$ that depends only on the x coordinate across the Hall bar. The energy levels look more or less like the ones in Fig. 10.3 with the addition of small random oscillations in the flat parts. The states are still labelled by the wave vector k_y, which determines the location of their centers at $k_y l^2$. The energy is $\epsilon_n + V(k_y l^2)$, which rises sharply at the edge due to the presence of the confining potential. Let us assume that the chemical potential μ falls in between two Landau levels in the bulk of the sample, so that there are precisely ν occupied Landau levels in the bulk. In the present model, this would be a very rare occurrence, since there are very few states (all at the edge of the system) to "pin" the Fermi level at such an energy. But, the introduction of physical disorder will change the situation dramatically, creating a large number of localized states in between the bulk Landau levels. In fact, it is known from detailed calculations that the density of states in the presence of disorder looks like a superposition of broadened Landau levels, with the extended states residing near the centers of the levels (see Fig. 10.5) and localized states in tails. Thus, there is no lack of localized states to pin the Fermi level at the assumed position.[19]

Since the electrons are noninteracting, the total current is the sum of the currents carried by all the occupied states. Making use of Eqs. (10.52) and (10.53) we obtain

$$I = -\frac{e}{\hbar L} \sum_{n=0}^{\nu-1} \sum_{k_y} \frac{\partial V(k_y \ell^2)}{\partial k_y}, \tag{10.56}$$

[18] It is worth pointing out that the matrix elements of the conductivity are related to those of the resistivity by a matrix inversion, such that $\sigma_{xx} = \rho_{xx}/(\rho_{xx}^2 + \rho_{xy}^2)$ and $\sigma_{xy} = -\rho_{xy}/(\rho_{xx}^2 + \rho_{xy}^2)$. Thus both σ_{xx} and ρ_{xx} vanish in the quantum Hall effect, and $\sigma_{xy} = 1/R_H$.

[19] The density of states of the LLL in the presence of impurities has been calculated analytically by Wegner, 1983.

where the sum over $k_y = \frac{2\pi k}{L_y}$ includes all the states that lie below the Fermi energy i— the Landau levels with $n = 0, 1, \ldots, \nu - 1$. In a macroscopic system, we can convert th— discrete sum over k_y to an integral over the variable $X = k_y \ell^2$, which runs, in each Landa— level, from $X_{n<}$ to $X_{n>}$ – the two solutions of $\epsilon_n + V(X) = \mu$. The precise correspondenc— is $\frac{1}{L} \sum_{k_y} \cdots \rightarrow \int_{X_{n<}}^{X_{n>}} \frac{dX}{\ell^2} \cdots$. Then, Eq. (10.56) becomes

$$I = -\frac{e}{\hbar \ell^2} \sum_{n=0}^{\nu-1} [V(X_{n>}) - V(X_{n<})] = 0 , \qquad (10.57$$

which vanishes, as expected, when the system is in equilibrium at a single chemical poten- tial.[20]

Suppose now a potential difference V_H is applied between the two edges of the system. Because the edges are separated by a macroscopic distance, it can be safely assumed that they are in equilibrium at two *different* chemical potentials μ_L and μ_R such that $\mu_L - \mu_R = -eV_H$. The current can be calculated as before, but the limits of integration for X are determined by the equations $V(X_{n<}) = \mu_L$ and $V(X_{n>}) = \mu_R$. Putting this in Eq. (10.57) we obtain

$$I = \nu \frac{e^2}{\hbar} V_H , \qquad (10.58)$$

which completes our first derivation of the formula for the integrally quantized Hall conduc- tance. The longitudinal resistance is zero, because each edge is in equilibrium at a constant chemical potential: no voltage drop is measured between two probes attached to the same edge.

Of course, there are several weaknesses in the argument. We have invoked disorder to provide the much needed reservoir of localized states that pins the chemical potential between Landau levels. Will this not invalidate our calculation of the current, which was explicitly based on the assumption of translational invariance in the y-direction? Actually, it does not, and for a beautiful reason. Going back to the calculation of the current we can see that, although all the states contribute, most contributions cancel out, except the ones coming from relatively small number of states centered on the two edges. These are the states that are most readily populated or depleted by a change in the Hall voltage.[21] These crucial *edge states* have the remarkable property that they flow in one and only one direction along the edge of the sample, and cannot be back-scattered by random impurities. If an impurity is present, the edge state will simply flow around it, keeping on an equipotential line. The only way the direction of motion could be reversed is scattering to the opposite edge, where the states flow in the opposite direction. But the two edges are macroscopically separated, and the probability of such a tunneling event is extremely small. From this point of view,

---

[20] In a *mesoscopic system* this is no longer true. The difference between the sum in Eq. (10.56) and the integral in Eq. (10.57) can result in a finite *persistent current* of the order of $\frac{1}{L}$.
[21] Notice that we are not saying that there is no current in the bulk of the system. But, the bulk current vanishes when integrated over the entire sample.

the absence of scattering between the two edges is the fundamental reason for the robust quantization of the integral Hall effect.

There are additional difficulties. According to the picture presented above the Hall resistance is determined by the number of Landau levels below the chemical potential. The calculation, however, appears to break down when the eV_H is larger than $\hbar\omega_c$, for, in that case, the number of occupied Landau levels on the left and/or the right edge differs from the number of Landau levels occupied in the bulk. In reality, the quantum Hall effect is known to hold up to potential differences much larger than $\hbar\omega_c/e$. How can this be?

The problem is that the above "derivation" is, strictly speaking, valid only for noninteracting electrons. Electrostatically defined Hall devices usually have smooth edge regions, in which the density decays to zero on the scale of a screening length, much larger than the magnetic length. A detailed discussion of this will be presented in Section 10.20.2, but, for the time being, it is sufficient to note that the screened confining potential is slowly varying, as shown in Fig. 10.27, rather than sharply rising as in Fig. 10.3. Because the applied Hall voltage is distributed over a large spatial region, the number of occupied Landau levels remains constant throughout the system, provided the condition $eE_H\ell \ll \hbar\omega_c$ is satisfied. Since the Hall electric field is $E_H = \frac{V_H}{L_x}$, we see that the condition $eV_H \ll \hbar\omega_c$ is replaced by the much weaker condition $eV_H \ll \frac{L_x}{\ell}\hbar\omega_c$, which allows the Hall effect to persist up to voltages much larger than $\frac{\hbar\omega_c}{e}$.

10.3.3 Strěda formula

An elegant derivation of the quantized Hall conductance, which relies only on general thermodynamic relations, is provided by the *Strěda formula* (Strěda, 1982). According to this

$$\sigma_H = ec\left(\frac{\partial n}{\partial B}\right)_\mu, \tag{10.59}$$

where $\sigma_H = \frac{1}{R_H}$ is the inverse of the Hall resistance, n is the bulk electronic density, and its derivative with respect to B is taken at constant chemical potential.

Consider, for example, the two-dimensional electron gas with the chemical potential falling in a region of localized states between Landau levels. Upon changing B the number of states below the chemical potential changes by $\frac{\nu e \delta B A}{hc}$, where ν is the number of full Landau levels, leading to $\sigma_H = ec\left(\frac{\partial n}{\partial B}\right)_\mu = \frac{\nu e^2}{h}$. Thus, the conditions for the quantization of the Hall conductance are (i) that the chemical potential fall in a gap of the energy spectrum, and (ii) that the number of occupied states below the chemical potential be proportional to the number of magnetic flux quanta. The second condition guarantees that the Hall conductivity of a band insulator is zero, even though the chemical potential lies in the gap between the valence and the conduction band: in this case, the total number of states in the

valence band is fixed by the size of the unit cell of the lattice and does not depend on th
strength of the magnetic field.[22]

The Strĕda formula (10.59) can be derived in the following manner (MacDonald, 1989)
For a system in a slowly varying potential, the charge current density $-e\vec{j}(\vec{r})$ is the curl o
the *local* orbital magnetization density

$$\vec{M}(\vec{r}) = -\frac{\partial \Omega(\mu(\vec{r}), B)}{\partial B}\vec{e}_z, \tag{10.60}$$

where $\Omega(\mu(\vec{r}), B)$ is the grand-canonical potential $\Omega(\mu, B)$ of the homogeneous electron
liquid evaluated at the local chemical potential $\mu(\vec{r})$, which is in turn determined by the
local density via the relation $n = \frac{\partial \Omega(\mu,B)}{\partial \mu}$. Thus, we can write

$$-e\vec{j}(\vec{r}) = c\vec{\nabla} \times \vec{M}(\vec{r}). \tag{10.61}$$

The magnetization density has a constant value in the bulk of the system, where the density
$n(\vec{r})$ is constant, and vanishes outside. This implies that the current density differs from
zero only in the relatively narrow region in which the density drops from the bulk value to
zero. The net current I that flows along an edge of the system is obtained by integrating
the current density (10.61) along a direction perpendicular to that edge. Making use of the
identity $\vec{\nabla} \times \vec{M}(\vec{r}) = \vec{\nabla}M(\vec{r}) \times \vec{e}_z$, and the fact that M vanishes outside the system, this is
readily found to be

$$I = cM(\mu, B), \tag{10.62}$$

where $M(\mu, B)$ is the value of the magnetization density, evaluated at the chemical potential of the bulk immediately adjacent to the edge under consideration. Up to this point
we have considered only the equilibrium situation. To apply a Hall voltage we shift the
chemical potentials of the right and left edges by amounts $-eV_R$ and $+eV_L$ respectively,
with V_R and V_L such that $V_R - V_L = V_H$. Because the separation between the two edges is
macroscopically large, we can assume that each one is in thermodynamic equilibrium with
its own chemical potential, $\mu_R = \mu - eV_R$ and $\mu_L = \mu + eV_L$. Then the application of the
Hall voltage causes the edge currents to change by amounts

$$\delta I_R = c\frac{\partial M(\mu, B)}{\partial \mu}(-eV_R),$$

$$\delta I_L = c\frac{\partial M(\mu, B)}{\partial \mu}(eV_L). \tag{10.63}$$

Since, from elementary thermodynamics,

$$\frac{\partial M(\mu, B)}{\partial \mu} = -\frac{\partial^2 \Omega(\mu, B)}{\partial \mu \partial B} = -\left(\frac{\partial n}{\partial B}\right)_\mu, \tag{10.64}$$

---

[22] For a deeper analysis of the quantum Hall effect in the presence of a periodic crystal potential see the article by Thouless in the
book edited by Prange and Girvin (1987).

e see that

$$I_H = \delta I_R - \delta I_L = ec \left(\frac{\partial n}{\partial B} \right)_\mu V_H . \tag{10.65}$$

10.3.4 The Laughlin argument

Here is yet another "derivation" of the integral quantum Hall effect, which was first suggested by Laughlin (1981) and later refined by several authors (Halperin, 1982; Giuliani *et al.*, 1983).

The quantum Hall bar is bent into a two-dimensional annulus, as shown in Fig. 10.8. In addition to the uniform magnetic field, $\vec{B}$, we now also have a magnetic flux Φ through the central hole – a flux that does not physically touch the electrons, but affects their wave functions through the azimuthal vector potential

$$\vec{A}_\Phi(\vec{r}) = \frac{\Phi}{2\pi r} \vec{e}_\phi . \tag{10.66}$$

A potential difference V_H is maintained between the inner and outer edge of the annulus. As a result, a Hall current I flows around the annulus, and we want to establish the relation between I and V_H. To do this, we make use of the fact that the current is the derivative of the ground-state energy (or, at finite temperature, of the thermodynamic potential) with

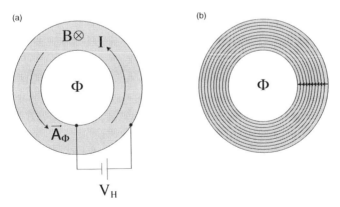

Fig. 10.8. (a) Corbino-disk geometry for a variant of the Laughlin argument. In this version of the argument the magnetic field $\vec{B}$ is perpendicular to the plane of the annulus and the two edges of the annulus are kept at different potentials so that a Hall current I flows. The central magnetic flux drives an azimuthal electric field, which causes a transfer of charge between the edges. (b) Schematic picture of the one-electron energy levels. The arrows indicate the direction of displacement of the charge due to a varying magnetic flux.

respect to the central magnetic flux Φ:

$$I = -c\frac{\partial E}{\partial \Phi} .^{23} \tag{10.67}$$

The physical content of this equation is that a variation of the central flux induces, by Faraday's law, an azimuthal electric field, which performs a work on the current, thus changing the energy: the change in energy is proportional to the magnitude of the azimuthal current. Imagine now to slowly change the central flux from zero to $\Phi_0 = \frac{hc}{e}$. At the end of the process the spectrum of the hamiltonian (including disorder and interactions) is unchanged, while its eigenfunctions have been multiplied by the phase factor $e^{i\sum_{j=1}^{N}\phi_j}$ where ϕ_j are the angular coordinates of the electrons.[24] This does not imply, however, that the energy of the system is unchanged: an eigenstate of the energy may have evolved into another eigenstate with a different energy.

Let us then ask what the final state can be. We cannot have excitations in the bulk of the electron liquid because of the large gap ($\sim\hbar\omega_c$) in the spectrum of extended states (see Fig. 10.5). At first sight, it seems that one might induce transitions between localized states, which are not "gapped". But this is false, because the localized states, being not linked to the magnetic flux, are not affected by its variation. The wave function of a localized state vanishes exponentially as one moves away from the localization center. Since the variation of the angle ϕ over the extent of the wavefunction is negligible (we are assuming that the circumference of the annulus is much larger than the localization length) the multiplication by the phase factor $e^{(i\frac{e\Phi}{\hbar c})\phi}$ reduces to multiplication by a constant, which does not change the state.

The only remaining possibility is that each extended state is slightly modified resulting in a net transfer of electrons between the edges of the annulus. In other words, the slow addition of flux may remove k electrons from the Fermi level at one edge and inject them at the Fermi level of the other edge, acting as a pump. The corresponding change in energy is $\Delta E = -keV_H$, and, putting this in Eq. (10.67), we get

$$I = -c\frac{\Delta E}{\Phi_0} = k\frac{e^2}{h}V_H . \tag{10.68}$$

Notice that this argument does not tell us what the actual number of electrons transferred between the edges of the annulus is, only that it is an integer k, possibly equal to zero in the case of an insulator. The main source of error in the argument is the finite temperature, which allows excitations across the gap in the bulk of the system. The probability of such

---

[23] The mathematical proof of this equation begins with the observation that the current–density operator is the derivative of the hamiltonian with respect to the vector potential: $\frac{e}{c}\hat{j}(\vec{r}) = \frac{\partial \hat{H}}{\partial \vec{A}(\vec{r})}$. Making use of the Hellman–Feynman theorem and of the expression (10.66) for the vector potential we obtain

$$-c\frac{\partial E}{\partial \Phi} = -e\int \frac{j_\phi(\vec{r})}{2\pi r}d^2r = \frac{1}{2\pi}\int_0^{2\pi}d\phi\int_{r_{min}}^{r_{max}}dr(-e)j_\phi(\vec{r}),$$

where $j_\phi(\vec{r})$ is the azimuthal component of the current density. Because the integral of $-ej_\phi$ over r is just the electric current I, which is independent of ϕ, the rightmost term of this equation is equal to I.

[24] We note that this *gauge transformation* is legitimate only when Φ varies by an integer multiple of Φ_0. Otherwise, the phase factor is not single-valued.

excitations, however, is exponentially small at low temperature. Thus we see that the absence of dissipation – and hence the vanishing of the longitudinal resistivity – is vital to the exactness of the quantum Hall effect.

The general argument can be explicitly verified in the case of the noninteracting electron gas, for which we know the wave functions. Upon inserting a quantum of central magnetic flux, the orbital of angular momentum m (in the symmetric gauge) evolves adiabatically into an orbital of the same angular momentum, but with the radial wavefunction that previously corresponded to angular momentum $m - 1$. To see this observe that the state $e^{im\phi}\psi_{n_r,m}(r)$ at $\Phi = 0$ corresponds to the state $e^{i(m+1)\phi}\psi_{n_r,m}(r)$ at $\Phi = \Phi_0$, in the sense that these two states have the same energy. Similarly, the state $e^{i(m-1)\phi}\psi_{n_r,m-1}(r)$ at $\Phi = 0$ corresponds to the state $e^{im\phi}\psi_{n_r,m-1}(r)$ at $\Phi = \Phi_0$. Consider now the actual *evolution* of the state $e^{im\phi}\psi_{n_r,m}(r)$ upon adiabatic insertion of a flux quantum. It is evident that such an evolution cannot change the value of a discrete quantum number such as the angular momentum: therefore the state $e^{im\phi}\psi_{n_r,m}(r)$ must necessarily evolve into the state whose wave function is $e^{im\phi}\psi_{n_r,m-1}(r)$. This state has the same angular dependence as the initial one, but its radius is reduced – the electron has been slightly "pulled in" and what used to be its position is now occupied by the electron with angular momentum $m + 1$, which in turn has been pulled in from its previous position. The process terminates at the edges of the annulus, with v electrons (one per occupied Landau level) being added to the inner edge and v electrons being removed from the outer edge (see Fig. 10.8). The net effect is to transfer v electrons from the outermost to the innermost edge of the annulus, as expected on the basis of the general argument.

There is a tacit assumption in this argument, namely, that the charge transferred between the edges must be an integral multiple of the electron charge. Natural as it sounds, this assumption is not forced on us by the fundamental laws of physics. In a highly correlated state it is conceivable that collective excitations might transfer a fractional amount of charge from one edge to the other. We will see that this is precisely what happens in the fractional QHE.

10.4 Electrons in full Landau levels: energetics

In this section we begin the study of the *interacting* 2DEL at high magnetic field. We consider the simplest situation – an integer number of full Landau levels of each spin. The density of σ-spin electrons is $n_\sigma = \frac{v_\sigma}{2\pi \ell^2}$. As discussed in the introduction, this state is very well described by a single Slater determinant: correlations are associated with virtual transitions to unoccupied Landau levels, and are insignificant if $\frac{e^2}{\epsilon_b \ell} \ll \hbar\omega_c$.[25] Since the high magnetic field regime defined by this inequality is experimentally attainable, we have here an excellent opportunity to test some Hartree–Fock physics in the lab. Let us assume, then, that there are $v_\uparrow$ and $v_\downarrow$ fully occupied levels of spin up and spin down respectively. In the following we calculate the kinetic energy, the exchange energy, and the RPA correlation energy for such a system, and use the results to predict an exchange-driven transitions in a tilted magnetic field.

---

[25] Here and in the following ϵ_b is the dielectric constant of the host material, $\epsilon_b \approx 12$ in GaAs.

10.4.1 Noninteracting kinetic energy

The noninteracting kinetic energy is readily calculated from the energy levels (10.3) and their degeneracies (10.4):

$$T = N_L \sum_\sigma \sum_{j=0}^{\nu_\sigma - 1} \left(j + \frac{1}{2} \right) \hbar \omega_c$$

$$= \frac{\hbar \omega_c}{2} \left(N_\uparrow \nu_\uparrow + N_\downarrow \nu_\downarrow \right) , \qquad (10.69)$$

where $N_\sigma = N_L \nu_\sigma$ is the number of σ-spin electrons. Notice that this can be rewritten as

$$T = \frac{1}{2} \left(N_\uparrow \frac{\hbar^2 k_{F\uparrow}^2}{2m_b} + N_\downarrow \frac{\hbar^2 k_{F\downarrow}^2}{2m_b} \right) , \qquad (10.70)$$

where $k_{F\sigma}$ are the usual Fermi wave vectors in two dimensions. Thus, the kinetic energy of a state with an integer number of full Landau levels coincides with the kinetic energy of the same number of electrons in zero magnetic field.[26]

10.4.2 Density matrix

The one-particle density matrix of the state with ν_σ full Landau levels of spin σ is given by

$$\rho(\vec{r}, \vec{r}') = \sum_\sigma \sum_{j=0}^{\nu_\sigma - 1} \rho_j(\vec{r}, \vec{r}'), \qquad (10.71)$$

where

$$\rho_j(\vec{r}, \vec{r}') = \sum_{k_y} \psi_{j,k_y}^*(\vec{r}) \psi_{j,k_y}(\vec{r}') \qquad (10.72)$$

is the density matrix of the j-th Landau level, calculated in the Landau gauge. The calculation of $\rho_j(\vec{r}, \vec{r}')$ is straightforward. First, replace the sum over k_y by an integral over $X = k_y \ell^2$ according to $\frac{1}{L_y} \sum_{k_y} \dots \to \int_{-\infty}^{\infty} \frac{dX}{2\pi \ell^2} \dots$. Then, make the change of variable $X \to X + \frac{x'+x}{2} + i \frac{y'-y}{2}$. Finally, make use of the integral (Gradshteyn and Ryzhik, 1965, 7.377)

$$\int_{-\infty}^{\infty} e^{-x^2} H_j(a - x) H_j(b - x) dx = 2^j \pi^{1/2} j! L_j^0(-2ab) . \qquad (10.73)$$

The final result is

$$\rho_j(\vec{r}, \vec{r}') = \frac{1}{2\pi \ell^2} e^{i \frac{(x'+x)(y'-y)}{2\ell^2}} e^{-\frac{|\vec{r}'-\vec{r}|^2}{4\ell^2}} L_j^0 \left(\frac{|\vec{r}' - \vec{r}|^2}{2\ell^2} \right) . \qquad (10.74)$$

Notice that ρ_j is not invariant under ordinary translations, but is invariant under magnetic translations.

---

[26] This remarkable result does not hold at fractional filling factors: in that case the kinetic energy in the presence of a magnetic field is higher than at zero field: the noninteracting electron gas is diamagnetic.

10.4.3 Pair correlation function

In Chapter 2, we saw that the pair correlation function of a single Slater determinant is related to the density matrix by Eq. (2.137). In the present case we have

$$g(\vec{r}, \vec{r}') = 1 - \frac{1}{n^2} \sum_\sigma \left| \sum_{j=0}^{\nu_\sigma - 1} \rho_j(\vec{r}, \vec{r}') \right|^2 , \qquad (10.75)$$

where the squared quantity is the σ-spin component of the full density matrix. Putting Eq. (10.74) in Eq. (10.75) we obtain

$$g(\vec{r}, \vec{r}') = 1 - \frac{1}{\nu^2} e^{-\frac{|\vec{r}' - \vec{r}|^2}{2\ell^2}} \sum_\sigma \left[\sum_{j=0}^{\nu_\sigma - 1} L_j^0 \left(\frac{|\vec{r}' - \vec{r}|^2}{2\ell^2} \right) \right]^2 , \qquad (10.76)$$

where $\nu = \nu_\uparrow + \nu_\downarrow$ is the total filling factor. Note that this expression is gauge-invariant as well as translationally and rotationally invariant. Setting $\vec{r}' = 0$ and making use of the identity (GR 8.974.3) $\sum_{j=0}^{\nu-1} L_j^0(x) = L_{\nu-1}^1(x)$ we arrive at the more compact expression

$$g(r) = 1 - \frac{1}{\nu^2} e^{-\frac{r^2}{2\ell^2}} \sum_\sigma \left[L_{\nu_\sigma - 1}^1 \left(\frac{r^2}{2\ell^2} \right) \right]^2 . \qquad (10.77)$$

10.4.4 Exchange energy

As discussed in Section 1.5.5, the exchange energy per particle is the energy of an electron in the electric potential created by its own exchange hole (divided by two to eliminate double counting):

$$\frac{E_x}{N} = \frac{n}{2} \int \frac{e^2}{\epsilon_b r} [g(r) - 1] d^2 r$$

$$= -\frac{e^2}{\epsilon_b \ell} \frac{1}{2\nu} \sum_\sigma \int_0^\infty e^{-x^2/2} \left[L_{\nu_\sigma - 1}^1 \left(\frac{x^2}{2} \right) \right]^2 dx . \qquad (10.78)$$

An interesting property of $\frac{E_x}{N}$ is that, once expressed in units of $\frac{e^2}{\epsilon_b \ell}$, it depends only on the filling factors ν_σ, and not on the actual value of the electron density, or, equivalently, of the magnetic field.

In Table 10.1 we present the values of $\frac{E_x}{N}$ obtained from Eq. (10.78) for the first few integer values of ν of a given spin.[27] These energies lie sligthly below those of the homogeneous unpolarized 2DEL of the same density in zero magnetic field. The zero field limit is approached very rapidly with increasing ν.

It must be borne in mind that a set of full Landau levels is *not* an exact solution of the Hartree–Fock equation, even though it approaches one in the limit of high magnetic field.

---

[27] Notice that the integral in Eq. (10.78) can be done analytically. Since exchange couples only electrons with parallel spins, the exchange energy can be calculated, in the general case, by adding the independent contributions of up and down-spins.

Table 10.1. *Exchange energy per particle, ϵ_x, for an integer number of full Landau levels. ϵ_x^e and ϵ_x^h are the exchange components of the energy of, respectively, an electron in the lowest unoccupied Landau level and a hole in the highest occupied one. The meaning of β_x will be explained in Section 10.4.8. All energies are in units of $\frac{e^2}{\epsilon_b \ell}$.*

Conversion to Rydberg is accomplished by multiplying these numbers by $\frac{2a_B^*}{\ell} = \frac{2}{r_s}\sqrt{\frac{2}{\nu}}$.

ν	0	1	2	3	4	5	6
$-\epsilon_x$	0.0	0.6266	0.8616	1.0476	1.2061	1.3465	1.4737
$-\epsilon_x^e$	0.0	0.6266	1.0183	1.3218	1.5770	1.8009	2.0025
ϵ_x^h	0.0	1.2533	1.5666	1.8212	2.0415	2.2386	2.4187
$-\beta_x$	0.6266	0.4700	0.4014	0.3598	0.3308	0.3089	0.2916

This is because the exchange potential, which, according to Eq. (2.23) is given by

$$V_x(\vec{r}, \vec{r}') = -\frac{e^2}{\epsilon_b |\vec{r} - \vec{r}'|} \sum_{j=0}^{\nu_\sigma - 1} \rho_j(\vec{r}, \vec{r}'), \tag{10.79}$$

mixes different Landau levels, with a mixing amplitude of the order of $\frac{e^2}{\epsilon_b \ell \hbar \omega_c}$. On the other hand, k_y remains a good quantum number. Thus, in the limit of high magnetic field, the HF orbitals approach the free-electron orbitals $\psi_{j,k_y}(\vec{r})$, and the HF eigenvalues are approximately given by

$$\epsilon_{n,k_y}^{HF} \simeq \left(n + \frac{1}{2}\right)\hbar\omega_c + \int d^2r \int d^2r' \psi_{n,k_y}^*(\vec{r}) V_x(\vec{r}, \vec{r}') \psi_{n,k_y}(\vec{r}')$$

$$= \left(n + \frac{1}{2}\right)\hbar\omega_c - \frac{e^2}{\epsilon_b \ell} \int_0^\infty e^{-x^2/2} L_n^0\left(\frac{x^2}{2}\right) L_{\nu_\sigma - 1}^1\left(\frac{x^2}{2}\right) dx . \tag{10.80}$$

Not surprisingly, the Hartree–Fock eigenvalues are still independent of k_y: this is expected, since the degeneracy follows from the invariance of the HF hamiltonian under magnetic translations.

In Table 10.1 we list the exchange components of the energy of an electron in the lowest unoccupied Landau level, $\epsilon_x^e \equiv \epsilon_\nu^{HF} - \left(\nu + \frac{1}{2}\right)\hbar\omega_c$, and that of a hole in the highest fully occupied one, $\epsilon_x^h \equiv -\left[\epsilon_{\nu-1}^{HF} - \left(\nu - \frac{1}{2}\right)\hbar\omega_c\right]$, for the first few values of ν.

10.4.5 The "Lindhard" function

According to the formulas of Chapter 4 the density–density response function of the non-interacting 2DEL at filling factors ν_σ is given by

$$\chi_0(q, \omega) = \frac{1}{\mathcal{A}} \sum_{\sigma, j, j', k_y, k_y'} \frac{f(\epsilon_{j\sigma}) - f(\epsilon_{j'\sigma})}{\hbar\omega + \epsilon_j - \epsilon_{j'} + i\eta} |\langle j', k_y' | \hat{n}_q | j, k_y \rangle|^2, \tag{10.81}$$

where $f(\epsilon_{j\sigma})$ is the occupation number for states of spin σ in the j-th Landau level, and $\mathcal{A}$

the area of the system. The matrix element of the density fluctuation operator $\hat{n}_{\vec{q}}$ between Landau-gauge wavefunctions (Eq. (10.31)) is (Exercise 5)

$$\langle j', k'_y | \hat{n}_{\vec{q}} | j, k_y \rangle = e^{-i\frac{q_x(k_y+k'_y)\ell^2}{2}} F_{j'j}(\vec{q}) \delta_{k_y - k'_y, q_y} \qquad (10.82)$$

where

$$F_{j'j}(\vec{q}) = \sqrt{\frac{j!}{j'!}} \left[\frac{(q_y - iq_x)\ell}{\sqrt{2}} \right]^{j'-j} e^{-q^2\ell^2/4} L_j^{j'-j}\left(\frac{q^2\ell^2}{2}\right) \quad (j' \geq j), \qquad (10.83)$$

and $F_{j'j}(\vec{q}) = [F_{jj'}(-\vec{q})]^*$ for $j' < j$. The Krönecker δ in Eq. (10.82) ensures that only the term with $k'_y = k_y - q_y$ contributes to the sum in (10.81), and from Eq. (10.82) we see that the squared modulus of the matrix element does not depend on k_y. So the sum over k_y reduces to a multiplication by $N_L = \frac{A}{2\pi\ell^2}$, and Eq. (10.81) takes the form

$$\chi_0(q, \omega) = \frac{1}{2\pi\ell^2} \sum_{\sigma, j, j'} \frac{f(\epsilon_{j\sigma}) - f(\epsilon_{j'\sigma})}{\hbar\omega + (j - j')\hbar\omega_c + i\eta} |F_{j'j}(\vec{q})|^2. \qquad (10.84)$$

Setting $j' - j = k$ this expression can be rewritten in the form

$$\chi_0(q, \omega) = \frac{e^{-\frac{q^2\ell^2}{2}}}{2\pi\hbar\ell^2} \sum_{\sigma, k=1}^{\infty} {\sum_j}' \frac{j!}{(j+k)!} \left(\frac{q^2\ell^2}{2}\right)^k \left[L_j^k\left(\frac{q^2\ell^2}{2}\right) \right]^2$$

$$\times \left[\frac{1}{\omega - k\omega_c + i\eta} - \frac{1}{\omega + k\omega_c + i\eta} \right], \qquad (10.85)$$

where the primed sum runs over the range $\max(0, \nu_\sigma - k) \leq j < \nu_\sigma$.

From the above expression one can immediately read the noninteracting dynamical structure factor $S(q, \omega) = -\frac{\hbar}{\pi n}\Im m \chi_0(q, \omega)$, which is simply a sum of delta-functions centered at $\omega = k\omega_c$ $(k \geq 1)$. The noninteracting static structure factor is

$$S(\vec{q}) = \int_0^{\infty} S(\vec{q}, \omega)d\omega$$

$$= 1 - \frac{1}{\nu} \sum_{\sigma} \sum_{j, j'=0}^{\nu_\sigma - 1} |F_{j'j}(\vec{q})|^2. \qquad (10.86)$$

The proof of this result is left as an exercise. Notice that at filling factor $\nu = 1$ we have $S(\vec{q}) = 1 - e^{-q^2\ell^2/2}$.

10.4.6 Static screening

The behavior of the static density–density response function $\chi_0(q, 0)$ for different values of ν is compared in Fig. 10.9 with that of the Lindhard function at the same density and zero magnetic field.[28]

---

[28] At $\nu = 1$ the infinite sum of Eq. (10.85) can be evaluated analytically, yielding

$$\chi_0(q, 0) = -\frac{1}{\pi\ell^2\hbar\omega_c} \left[\text{Ei}\left(\frac{q^2\ell^2}{2}\right) - \ln\left(\frac{q^2\ell^2}{2}\right) - \gamma \right] e^{-\frac{q^2\ell^2}{2}}, \qquad (\nu = 1),$$

where $\text{Ei}(x)$ is the exponential integral function and $\gamma = 0.577216$ is the Euler–Mascheroni constant.

$$-\frac{\chi_0(q,0)}{\hbar\omega_c \ell^2}$$

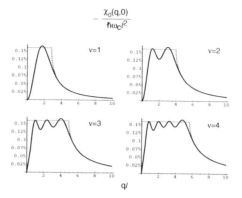

Fig. 10.9. Behavior of the static density–density response function of a 2DEL with $\nu = 1 - 4$ completely filled Landau levels of a given spin orientation. In each case the dashed line is the Lindhard function at the same density and zero magnetic field. Notice that $k_{F_o}\ell = \sqrt{2\nu}$ and $\frac{N_o(0)}{\hbar\omega_c \ell^2} = \frac{1}{2\pi} \approx 0.16$.

The most important difference occurs in the $q \to 0$ limit, where the finite-ν response function vanishes as q^2, due to the absence of low-energy excitations: by contrast, the Lindhard function tends to a constant in the same limit. Mathematically, this is due to the fact that the sum over k in Eq. (10.85) is dominated by the $k = 1$ term, i.e. by transitions between the highest occupied and the lowest unoccupied Landau level. Fig. 10.9 also shows oscillations, at finite ν, in the intermediate wave vector range: these oscillations, however, quickly subside as ν increases, and the ordinary Lindhard function is recovered in the limit of large ν.

Noting that $L_j^1(0) = j + 1$ we can easily calculate the small-q limit of the static RPA dielectric function, which turns out to be

$$\epsilon_{RPA}(q, 0) \overset{q \to 0}{\to} 1 + 2(\nu + 1)\frac{\ell^2}{a_B}q \,, \quad \left(q \ll \frac{1}{\ell}\right) . \tag{10.87}$$

Of particular interest is the case of weak magnetic field ($\nu \gg 1$) in which the *cyclotron radius* $R_c = k_F\ell^2 = \sqrt{2\nu}\ell$ is much larger than the magnetic length ℓ. In the range $\frac{1}{\ell} \ll q \ll \frac{R_c}{\ell^2}$ the static RPA dielectric function is well described by the Thomas–Fermi approximation

$$\epsilon_{RPA}(q, 0) \approx 1 + \frac{2}{a_B q} . \tag{10.88}$$

The Fourier transform of $\frac{v_q}{\epsilon_{RPA}(q,0)}$ determines the density distribution around a static impurity, as well as the effective interaction, $V_{tt}(r)$, between two static test charges embedded in the 2DEL.

Concerning the first question, we note that the presence of the gap suppresses the Friedel oscillations at large distances from the impurity (Simion and Giuliani, 2005). However, if the magnetic field is sufficiently weak, oscillations may still occur at a finite distance from the impurity, i.e., in a range $\frac{\ell^2}{R_c} \ll r \ll \ell$.

As for the effective interaction, the limiting forms (10.87) and (10.88) imply that $V_{tt}(r)$ as the following properties:

(i) It reduces to the bare coulomb interaction $\frac{e^2}{\epsilon_b r}$ for $r \gg \frac{R_c^2}{a_B}$;

(ii) It has the Thomas-Fermi form $\frac{e^2 a_B^2}{\epsilon_b r^3}$ for $a_B \ll r \ll R_c$;

(iii) In the intermediate region, $R_c \ll r \ll \frac{R_c^2}{a_B}$, it varies as

$$V_{tt}(r) \simeq \frac{\hbar \omega_c}{2\nu} \ln\left(\frac{R_c^2}{a_B r}\right), \quad \left(R_c \ll r \ll \frac{R_c^2}{a_B}\right), \quad (10.89)$$

which is nearly independent of r (Aleiner and Glazman, 1995 – see also Exercise 8).

We will return to the last property in the discussion of the "stripe phase" in Section 10.19.

10.4.7 Correlation energy – the random phase approximation

We can use the RPA dielectric function to calculate the correlation energy, E_c, for an integer number of full Landau levels in the random phase approximation. The basic formula, derived in Chapter 5, is

$$E_c = -\frac{N}{n} \int \frac{d^2q}{(2\pi)^2} \int_0^\infty \frac{d\omega}{2\pi} \{Q_0(q, i\omega) - \ln[1 + Q_0(q, i\omega)]\}, \quad (10.90)$$

where $Q_0(q, i\omega) \equiv -v_q \chi_0(q, i\omega)$. An explicit expression for Q_0 is

$$Q_0(q, i\omega) = \frac{1}{q a_B} \sum_{\sigma, j, j'} \frac{f(\epsilon_{j\sigma})|F_{j'j}(\vec{q})|^2 (j' - j)}{(\omega/\omega_c)^2 + (j' - j)^2}. \quad (10.91)$$

Unlike the exchange energy, the correlation energy depends on two parameters, the filling factor ν and the electronic density – the latter being described, as usual, by r_s. In fact, it is easy to see that the dimensionless function Q_0 is proportional to r_s, so the correlation energy vanishes for $r_s \to 0$. This is, of course, expected, but the interesting fact is that, if one fixes ν at some integer value (say $\nu = 1$), then, for B tending to infinity, the density becomes very large, and r_s tends to zero. The conclusion is that the high magnetic field limit of a full Landau level corresponds to $r_s = 0$ and is completely free of correlation effects.

Table 10.2 presents the values of the correlation energy computed from Eq. (10.90) for various integer values of ν (1 – 6) and r_s (1 – 2) in a fully spin polarized 2DEL. These values lie slightly above those of an electron liquid of the same density in zero magnetic field.

10.4.8 Fractional filling factors

The simple energy calculations presented in this section fail at fractional filling factors, due to the massive degeneracy of the noninteracting ground-state. However, there is still a sense in which a perturbative calculation of the interaction effects can be done. Assume

Table 10.2. *Correlation energy per electron in fully spin-polarized Landau levels in units of* $\frac{e^2}{\epsilon_b \ell}$. *Conversion to Rydberg is accomplished by multiplying these numbers by* $\frac{2a_B^*}{\ell} = \frac{2}{r_s}\sqrt{\frac{2}{\nu}}$.

ν	1	2	3	4	5	6
$-\epsilon_c(r_s = 1)$	0.075	0.114	0.144	0.168	0.188	0.207
$-\epsilon_c(r_s = 2)$	0.136	0.205	0.256	0.297	0.332	0.364

that the system is not exactly at zero temperature, but at some finite temperature T. Then the noninteracting equilibrium *ensemble* is well defined, and attributes the same fractional occupation number $p = \nu - [\nu] < 1$ to all the states in a partially occupied Landau level. One can do finite-temperature perturbation theory starting from this ensemble, and let T tend to zero at the end of the calculation. This approach can be used, in particular, to calculate the noninteracting kinetic energy (zero-th order perturbation theory) and the exchange energy (first order). The result is that the kinetic energy depends *linearly* on particle number in each interval $jN_L \leq N \leq (j+1)N_L$, with j a non negative integer:

$$T = N_L \hbar \omega_c \left\{ \frac{j^2}{2} + (\nu - j)\left(j + \frac{1}{2}\right)\right\}, \quad j \leq \nu \leq j+1. \tag{10.92}$$

Thus, the graph of the noninteracting kinetic energy vs. density consists of segments of straight lines, with cusps at integral filling factors.

In the same interval $jN_L \leq N \leq (j+1)N_L$ the exchange energy can be written as the sum of three terms:

(i) The exchange energy of j full Landau levels (which we have already calculated);
(ii) The exchange energy of the electrons in the partially occupied j-th Landau level;
(iii) The exchange energy due to the interaction between the electrons in the j-th level and those in the lower levels.

It is evident that (iii) depends linearly on the occupation of the j-th level, while (ii) depends quadratically on it. Thus, the graph of the exchange-energy vs density consists of arcs of parabola, with cusps at integral filling factors:

$$E_x = N_L \frac{e^2}{\epsilon_b \ell} \left\{ j\epsilon_x(j) + \epsilon_x^e(j)(\nu - j) + \beta_x(j)(\nu - j)^2 \right\}, \quad j \leq \nu \leq j+1. \tag{10.93}$$

Here $\epsilon_x(j)$ is the exchange energy per particle for j full Landau levels, and is tabulated together with the energies $\epsilon_x^e(j)$ and $\beta_x(j)$ in Table 10.1.

At higher order in perturbation theory there is no reason to expect a smooth dependence of the correlation energy on filling factor at non-integer values: in fact nonperturbative theories, described later in this chapter, predict that the energy vs density graph should

xhibit cusps at the filling factors of the fractional quantum Hall effect. Yet there is no
evidence to date that a perturbative approach can produce such cusps.

10.5 Exchange-driven transitions in tilted field

The absence of correlation effects in full Landau levels at high magnetic fields leads to the
possibility of exchange-driven phase transitions, similar in spirit to the Bloch ferromagnetic
transition predicted in Hartree–Fock theory and discussed in Section 2.4.2 (Giuliani and
Quinn, 1985). Consider the paramagnetic ground-state of the 2DEL at $v_\uparrow = v_\downarrow = 1$ in
a system with Zeeman splitting E_Z comparable to the cyclotron gap. The first excited
state in a Hartree–Fock sense is obtained by promoting one electron from the down-spin
component of the lowest Landau level to the up-spin component of the next Landau level at
a kinetic+Zeeman energy cost $\hbar\omega_c - E_Z$ (see Fig. 10.10). We now gradually tilt the sample,
and, at the same time, increase the magnitude of the magnetic field in such a way that $B_\perp$,
the component of $\vec{B}$ perpendicular to the plane of the 2DEL, remains constant. The filling
factor and the cyclotron energy gap, which are proportional to $B_\perp$ are thus unaffected by
the tilt, while the Zeeman splitting, which is proportional to the magnitude of $\vec{B}$, increases.
Then, at some critical value of the tilt angle the energy cost for promoting electrons from
the $0 \downarrow$ to the $1 \uparrow$ level will be exactly balanced by the gain in exchange energy arising
from having all the spins point in the same direction. At this point the ground-state turns
from paramagnetic to ferromagnetic Fig. 10.10. It should be noted that this transition would
occur even in a noninteracting system, but then only when $\hbar\omega_c = E_Z$. The effect of the
exchange energy is to cause the transition to occur at a smaller value of the tilt angle, that is,
when $\hbar\omega_c$ is still larger than E_Z. If E_Z is sufficiently large (in fact, larger than the coulomb
interaction energy scale) then the transition will occur in a regime in which the uncorrelated
description of the ground-state is essentially exact.

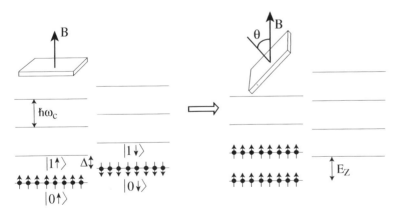

Fig. 10.10. Schematic description of the paramagnetic to ferromagnetic transition in a 2DEL in tilted
magnetic field. Notice that the cyclotron gap and the filling factor remain constant while the Zeeman
splitting increases. In this figure the g-factor is assumed to be negative (as appropriate for GaAs,
where $g = -0.44$), so the up-spin states are lower in energy than their down-spin counterparts.

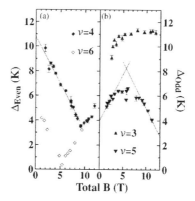

Fig. 10.11. Experimental measurement of the transport gap vs. magnetic field in the tilted field geometry in a two-dimensional *hole* gas in GaAs. (a) Even filling factors, (b) odd filling factors. The sharp minimum in Δ in (a) is taken as evidence of the exchange-driven transition discussed in the text. From Daneshvar *et al.*, 1997.

Following its theoretical prediction, the transition described above has been experimentally observed (Daneshvar *et al.*, 1997). The gap, Δ, which separates the highest occupied state from the lowest empty one, is determined by measuring the behavior of the longitudinal resistivity as a function of temperature, $\rho_{xx}(T) \sim e^{-\frac{\Delta}{2k_B T}}$, and for this reason is known as the *transport gap*. The behavior of Δ vs B depends critically on whether the filling factor is even or odd. At even filling factor the situation is shown in Fig. 10.11(a). With increasing B, Δ at first decreases to a minimum positive value and then begins to increase as soon as the electrons are transferred from the highest occupied $\downarrow$-spin Landau level to the lowest empty $\uparrow$-spin one. The fact that Δ does not vanish at the transition is consistent with our qualitative description of the effect of the exchange energy. In the case of odd filling factor ($\nu \geq 3$) no such transition occurs. The observed behavior of Δ vs. B in this case is shown in Fig. 10.11(b). Δ increases as long as $E_Z < \hbar\omega_c$, since the highest occupied Landau level has spin $\uparrow$, parallel to $\vec{B}$, and begins to decrease as soon as E_Z exceeds $\hbar\omega_c$, since the spin of the highest occupied Landau level becomes then antiparallel to $\vec{B}$. No transfer of electrons occurs between different Landau levels. These experiments provide one of the most elegant and vivid demonstrations of the complex Hartree–Fock scenario of the electron liquid.

10.6 Electrons in full Landau levels: dynamics

In this section we describe the dynamics of the interacting electron gas at integer filling factors. We show that the density fluctuation spectrum is dominated by a single collective mode – the magneto-exciton – whose frequency tends, in the long wavelength limit, to the bare cyclotron frequency ω_c. The dispersion of the magneto-exciton will be calculated in the RPA and in the time-dependent Hartree–Fock approximation.

10.6.1 Classification of neutral excitations

Due to the non-commutativity of the magnetic translation operators (see Section 10.2.1) it is impossible, in general, to classify the excitations of the 2DEL in a magnetic field in terms of conserved wave vector $\vec{q} = (q_x, q_y)$. There is, however, a very important exception: if the excitation is *neutral*, i.e., if it consists of equal numbers of positively and negatively charged particles (e.g. an electron–hole pair) then, for such excitations, the magnetic translations do commute, and a wave vector can be specified. In particular, collective density and spin excitations are labelled by a conserved wave vector $\vec{q}$. The proof is very simple. Consider a system of n positively charged and n negatively charged particles in a magnetic field. The magnetic translation operator for such a system is constructed by adding the translation operators for each particle. Evaluating the commutator between $\hat{T}_x$ and $\hat{T}_y$ we see that each negatively charged particle contributes $-i\frac{\hbar^2}{\ell^2}$ (see Eq. (10.25)) while each positively charged particle contributes $+i\frac{\hbar^2}{\ell^2}$. Thus, the commutator vanishes and the two components of $\vec{q}$ can be simultaneously specified. In the next two sections we will take advantage of this fact to study the collective excitations of the 2DEL at high magnetic field.

10.6.2 Collective modes

The frequency of collective modes in an electron liquid with an integer number of full Landau levels is determined, as usual, by finding the zeroes of the dielectric function, $1 - v_q \chi_0(q, \omega)$ in RPA. The calculation can be done analytically in the limit of high magnetic field, i.e. for $\hbar\omega_c \gg \frac{e^2}{\epsilon_b \ell}$. We find an infinite set of modes whose frequencies tend to $m\omega_c$ (with m an integer ≥ 1) for $q \to 0$ and disperse upward as $(q\ell)^{2m-1}$ for small q. The complete analytic expression is easily obtained from Eq. (10.85), noting that for ω close to $m\omega_c$ only the term with $k = m$ is relevant. Thus, we get

$$\omega_m^{RPA}(q) \simeq m\omega_c + \frac{e^2}{\hbar\epsilon_b\ell} e^{-\frac{q^2\ell^2}{2}} \frac{(q\ell)^{2m-1}}{2^m} \sum_\sigma \sum_j {}' \frac{j!}{(j+m)!} \left[L_j^m \left(\frac{q^2\ell^2}{2} \right) \right]^2, \quad (10.94)$$

where the primed sum runs over the range $\max(0, \nu_\sigma - m) \leq j < \nu_\sigma$.

It should be noted that, for $q \to 0$, the spectral weight is entirely concentrated in the $m = 1$ mode, in agreement with *Kohn's theorem*, which will be discussed in Section 10.6.4. The positive shift in the frequency of the collective mode at finite q is an example of the "depolarization shift", first mentioned in Section 7.6: this shift is caused by the electrostatic field created by the density fluctuation. Additional corrections to the dispersion appear when one goes beyond the RPA. There is, in particular, a negative "excitonic shift", which is due to the exchange field, and can be calculated within the time-dependent Hartree–Fock approximation.

10.6.3 Time-dependent Hartree–Fock theory

In Section 5.7.1 we derived the time-dependent Hartree–Fock equation (5.216) for the "mother of all response functions", namely, the density matrix-density matrix response function $\chi_{\alpha\beta,\gamma\delta}(\omega)$. This equation can be easily put to work in the limit of high magnetic field. The main trick is to introduce the reduced response function

$$\chi_{jm}(\vec{q}, \omega) = \sum_{k_y} e^{iq_x k_y \ell^2} \chi_{j+m\ k_y+q_y/2\ \sigma\ j\ k_y-q_y/2\ \sigma,j\ k_y-q_y/2\sigma_{j+m}\ k_y+q_y/2\ \sigma}(\omega) , \quad (10.95$$

where j is the label of one of the fully occupied Landau levels and $j + m$ is that of an empty one. In this representation, the TDHF equation (5.216) reduces to a set of algebraic equations because the wave vector $\vec{q}$ is a rigorous quantum number, and m is a good quantum number in the limit of high magnetic field (the eigenfrequencies $\omega_m(\vec{q})$ are close to the unperturbed values $m\hbar\omega_c$).

However, one obtains the same result and learns more physics by explicitly constructing the collective states within the HF theory (Lerner and Lozovik, 1978). To do this we note that a state with energy close to $m\hbar\omega_c$ and wave vector $\vec{q}$ must necessarily be a superposition of states of the form

$$|\psi_j\rangle = \frac{1}{\sqrt{N_L}} \sum_{k_y} e^{iq_x k_y \ell^2} \hat{a}^\dagger_{j+m\ k_y+q_y/2\ \sigma} \hat{a}_{j\ k_y-q_y/2\ \sigma} |\nu_\uparrow, \nu_\downarrow\rangle , \quad (10.96)$$

where $|\nu_\uparrow, \nu_\downarrow\rangle$ is the state with ν_σ fully occupied Landau levels of spin σ, j is the label of an occupied Landau level, and $j + m$ is that of an empty Landau level. Eq. (10.96) describes a superposition of electron–hole pairs with center of mass momentum $\hbar\vec{q}$ in Landau levels $j + m$ and j. Unlike two electrons, which, as we will show later, circle around each other in the magnetic field, an electron and a hole travel together in a straight line, their mutual attraction being compensated by the Lorentz force. Such a state is commonly referred to as a *magnetic exciton*. We consider, for simplicity, only singlet excitons (the electron and the hole have the same spin), but it is clear that the theory can be easily generalized to triplet (spin-flip) excitons.

To calculate the energies of the collective modes we diagonalize the hamiltonian in the exciton basis (10.96): that is, we diagonalize the matrix of the amplitudes $H_{jj'} = \langle\psi_j|\hat{H}|\psi_{j'}\rangle$. In most cases of interest, this is quite a small matrix: for $m = 1$ it reduces to just one number.

The calculation of $H_{jj'}$ is a nice exercise in second quantization. First of all, as in Section 2.3, we rewrite $\hat{H}$ as the sum of a constant plus one- and two-body operators that are normal-ordered terms with respect to the "vacuum" $|\nu_\uparrow, \nu_\downarrow\rangle$. The constant is just the ground-state energy and is of no interest here. The one-body term is diagonal in the exciton basis and its expectation value is just the sum of the HF energy of an electron in level $j + m$ and the HF energy of a hole in level j:

$$[\hat{H}_1]_{jj'} = \left(m\hbar\omega_c + \epsilon^e_{x,j+m} + \epsilon^h_{x,j}\right)\delta_{jj'} . \quad (10.97)$$

$\epsilon^e_{x,j+m}$ and $\epsilon^h_{x,j}$ are the exchange energies of electrons and holes in Landau levels $j + m$ and j respectively, and can be extracted from Eq. (10.80) (for $j = \nu - 1$ and $m = 1$ these are tabulated in Table 10.1).

Finally, the relevant part of the normal-ordered two-body hamiltonian is written as

$$\hat{H}_2 = \frac{1}{2}\sum_{m=1}^{\infty}\sum_{j,j'}{}'\sum_{k_y\sigma k_y'\sigma'} v_{j+m\ k_y+q_y/2,\ j'\ k_y'-q_y/2,\ j'+m\ k_y'+q_y/2,\ j\ k_y-q_y/2}$$

$$\times \hat{a}^\dagger_{j+m\ k_y+q_y/2\ \sigma}\hat{a}_{j\ k_y-q_y/2\ \sigma}\hat{a}^\dagger_{j'\ k_y'-q_y/2\ \sigma'}\hat{a}_{j'+m\ k_y'+q_y/2\ \sigma'}, \tag{10.98}$$

where j and j' denote occupied Landau levels and $j+m$, $j'+m$ empty ones. The matrix element of $\hat{H}_2$ between exciton states is the sum of a "direct" and an "exchange" term:

$$[\hat{H}_2]_{jj'} = [\hat{H}_2]^d_{jj'} + [\hat{H}_2]^x_{jj'}, \tag{10.99}$$

where

$$[\hat{H}_2]^d_{jj'} = \frac{1}{N_L}\sum_{k_y k_y'} e^{-iq_x(k_y-k_y')\ell^2} v_{j+m\ k_y+q_y/2,\ j'\ k_y'-q_y/2,\ j'+m\ k_y'+q_y/2,\ j\ k_y-q_y/2}, \tag{10.100}$$

and

$$[\hat{H}_2]^x_{jj'} = -\frac{1}{N_L}\sum_{k_y k_y'} e^{-iq_x(k_y-k_y')\ell^2} v_{j+m\ k_y+q_y/2,\ j'\ k_y'-q_y/2,\ j\ k_y-q_y/2,\ j'+m\ k_y'+q_y/2}. \tag{10.101}$$

Both terms will be calculated analytically. For the direct term (which actually corresponds to an exchange interaction between the electron and the hole) we make use of the Fourier representation $\frac{e^2}{r} = \frac{1}{A}\sum_{\vec{k}}\frac{2\pi e^2}{k}e^{i\vec{k}\cdot\vec{r}}$ together with Eqs. (10.82) and (10.83) to find

$$[\hat{H}_2]^d_{jj'} = \frac{e^2}{\epsilon_b\ell}e^{-\frac{q^2\ell^2}{2}}\frac{(q\ell)^{2m-1}}{2^m}\sqrt{\frac{j!j'!}{(j+m)!(j'+m)!}}L_j^m\left(\frac{q^2\ell^2}{2}\right)L_{j'}^m\left(\frac{q^2\ell^2}{2}\right). \tag{10.102}$$

Because this matrix has the separable form $u_j u_{j'}$, its only nonvanishing eigenvalue is $\sum_j u_j^2$, which in fact yields the RPA "depolarization shift" obtained in the previous section by a different method (see Eq. (10.94)).

The calculation of the exchange term (which actually corresponds to the direct interaction between the electron and the hole) is somewhat more laborious, but equally straightforward in principle. We leave to the reader the pleasure of the derivation (Exercise 11), and present here only the final result expressed in terms of a two-dimensional integral that can be easily evaluated numerically:

$$[\hat{H}_2]^x_{jj'} = -\frac{e^2}{\epsilon_b\ell}\sqrt{\frac{j!j'!}{(j+m)!(j'+m)!}}$$

$$\times \int\frac{d\vec{w}}{2\pi}\frac{e^{-\frac{w^2}{2}}}{|\vec{w}+\vec{q}\ell\times\vec{e}_z|}\left(\frac{w^2}{2}\right)^m L_j^m\left(\frac{w^2}{2}\right)L_{j'}^m\left(\frac{w^2}{2}\right). \tag{10.103}$$

As a simple application, let us calculate the dispersion of the $m=1$ magneto-exciton for the simplest case of $v_\uparrow = v_\downarrow = v$. In this case there is only one admissible value of $j = j' = v - 1$ (the index of the highest occupied Landau level) and the matrix $[\hat{H}_2]_{jj'}$

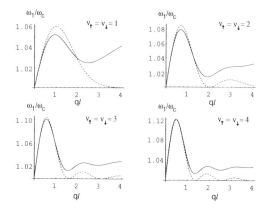

Fig. 10.12. The dispersion of the $m = 1$ magnetic exciton for an integer number of full Landau levels $\nu_\uparrow = \nu_\downarrow = 1 - 4$. The solid line is calculated from Eq. (10.104), while the dashed line is the result of the RPA. For illustrative purposes we have set $\frac{e^2}{\hbar\epsilon_b \ell \omega_c} = \frac{\ell}{a_B^*} = 0.1$ – a very large magnetic field indeed!

reduces to a single number. Combining this with the expressions for $[\hat{H}_1]_{jj'}$, $\epsilon^e_{x,\nu}$ and $\epsilon^h_{x,\nu-1}$ (Eq. (10.81)) we obtain:

$$
\omega_1^{TDHF}(q) = \omega_c + \frac{e^2}{\hbar\epsilon_b \ell} \frac{q\ell}{\nu} e^{-\frac{q^2 \ell^2}{2}} \left[L^1_{\nu-1}\left(\frac{q^2 \ell^2}{2}\right) \right]^2
$$

$$
+ \frac{e^2}{\hbar\epsilon_b \ell} \int_0^\infty dx\, e^{-\frac{x^2}{2}} L^1_{\nu-1}\left(\frac{x^2}{2}\right) \left[L^0_{\nu-1}\left(\frac{x^2}{2}\right) - L^0_\nu\left(\frac{x^2}{2}\right) \right]
$$

$$
- \frac{e^2}{\hbar\epsilon_b \ell} \frac{1}{\nu} \int \frac{d\vec{w}}{2\pi} \frac{e^{-\frac{w^2}{2}}}{|\vec{w} + \vec{q}\ell \times \vec{e}_z|} \left(\frac{w^2}{2}\right) \left[L^1_{\nu-1}\left(\frac{w^2}{2}\right) \right]^2 . \quad (10.104)
$$

This formula shows that the difference $\omega_1(q) - \omega_c$ consists of three parts: the RPA depolarization shift (first line), the exchange shifts in the energies of the electron and the hole (second line), and the excitonic shift due to the electron–hole attraction (last line). It is interesting to notice that the second and the third line cancel against each other at $q = 0$ by virtue of the identity $x L^1_{n-1}(x) = n[L^0_{n-1}(x) - L^0_n(x)]$ (see Abramowitz and Stegun (1964), Eq. 22.7.31). Since the RPA correction also vanishes for $q \to 0$ we conclude that the energy of the uniform $(q = 0)$ $m = 1$ mode remains unchanged at $\hbar\omega_c$. This is a manifestation of Kohn's theorem – which holds at all orders in the interaction strength and will be discussed in detail in the next section.

The last term of Eq. (10.104) depends on $\vec{q}$ via the combination $\vec{q}\ell \times \vec{e}_z$. This quantity is just the average separation between the electron and the hole in the exciton bound state, expressed in units of ℓ (this can be directly verified by writing the exciton wave function (10.96) in real space: see Exercise 12). Thus, for large q the electron and the hole are far apart and the energy of the exciton should reduce to the sum of the Hartree–Fock energies of the electron and the hole (independent of q) plus the electron–hole interaction energy $-\frac{e^2}{q\ell^2}$. This is exactly what Eq. (10.104) predicts for $q\ell \gg 1$. In this limit, the velocity of the enter of mass of the exciton is $\vec{v}(\vec{q}) = \vec{\nabla}_{\vec{q}}\omega_1(q) = \frac{e^2\vec{q}}{\hbar q^3 \epsilon_b \ell^2}$. Thus, we see that the Lorentz orce experienced by each particle in the pair ($\frac{e}{c}\vec{v}(\vec{q}) \times \vec{B}$ for the hole) exactly cancels the oulomb force experienced by the same particle ($\frac{e^2}{\epsilon_b q^3 \ell^4}\vec{q} \times \vec{e}_z$ for the hole), as demanded y the classical equation of motion in the limit of high magnetic field (see footnote[2]).

The other case in which one expects a single mode is $\nu_\uparrow = \nu$, $\nu_\downarrow = 0$, which differs from ne above only by having the RPA shift on the first line reduced by a factor 2 (Exercise 13). Aore general spin-polarized situations will lead to two $m = 1$ modes associated with the wo inequivalent orientations of the spin (for details see Kallin and Halperin, 1984).

10.6.4 Kohn's theorem

We have seen in the previous section that the $q \to 0$ limit of the magnetic exciton energy has he noninteracting value $\hbar\omega_c$, independent of the electronic density. This is a special case of *Kohn's theorem* (Kohn, 1961), which asserts that the center of mass of a translationally invariant 2DEL in a uniform magnetic field of arbitrary strength responds to a uniform time-dependent electric field like a single quantum particle of mass $M = Nm_b$ and charge $-Ne$. The cyclotron resonance thus occurs at $\omega = \omega_c$ regardless of interactions.

The proof of the theorem is very direct. Let us choose a gauge in which the static magnetic field is described by the vector potential $\vec{A}(\vec{r}) = \frac{1}{2}\vec{B} \times \vec{r}$, and the uniform electric field is described by a *time-dependent* vector potential $\vec{A}_{ext}(t) = \vec{A}_0 e^{-i\omega t} + c.c.$ The hamiltonian

$$\hat{H}(t) = \frac{1}{2m_b}\sum_{i-1}^{N}\left[\hat{\vec{p}}_i + \frac{e}{2c}\vec{B} \times \hat{\vec{r}}_i + \frac{e}{c}\vec{A}_{ext}(t)\right]^2 + \frac{1}{2}\sum_{i \neq j}\frac{e^2}{\epsilon_b|\hat{\vec{r}}_i - \hat{\vec{r}}_j|} \quad (10.105)$$

can be written as the sum of two mutually commuting terms,

$$\hat{H}(t) = \hat{H}_{cm}(t) + \hat{H}_{rel} \,, \quad (10.106)$$

where

$$\hat{H}_{cm}(t) = \frac{1}{2Nm_b}\left[\hat{\vec{P}}_{cm} + \frac{Ne}{2c}\vec{B} \times \hat{\vec{R}}_{cm} + \frac{Ne}{c}\vec{A}_{ext}(t)\right]^2 \quad (10.107)$$

is the hamiltonian of the center of mass, while

$$\hat{H}_{rel} = \frac{1}{2m_e}\sum_{i=1}^{N}\left[\hat{\vec{p}}'_i + \frac{e}{c}\vec{B} \times \hat{\vec{r}}'_i\right]^2 + \frac{1}{2}\sum_{i \neq j}\frac{e^2}{\epsilon_b|\hat{\vec{r}}'_i - \hat{\vec{r}}'_j|} \quad (10.108)$$

governs the relative motion.

In the above equations $\hat{\vec{R}}_{cm} = \frac{1}{N}\sum_{i=1}^{N}\hat{\vec{r}}_i$ and $\hat{\vec{P}}_{cm} = \sum_{i=1}^{N}\hat{\vec{p}}_i$ are, respectively, the position and the canonical momentum of the center of mass; while $\hat{\vec{r}}'_i = \hat{\vec{r}}_i - \hat{\vec{R}}_{cm}$ and $\hat{\vec{p}}'_i = \hat{\vec{p}}_i - \frac{\hat{\vec{P}}_{cm}}{N}$ are, respectively, the position and the momenta of the individual electrons relative to the center of mass. It turns out that $\hat{H}_{rel}$ can be expressed exactly in terms of $N-1$ internal coordinates (the so-called Jacobi coordinates) and their associated momenta.[29] Therefore, the full N-electron wave function is the product of a center of mass wave function and an "internal" wave function that depends only on the Jacobi coordinates. Since the time-dependent external field enters only $\hat{H}_{cm}(t)$ we see that the internal state of the electron liquid is not affected by the electric field, while its center of mass evolves like a single particle of mass Nm_b and charge $-Ne$ subjected to both the electric and the magnetic field. This remarkable result remains valid even in the presence of a static *harmonic* potential, such as the potential experienced by the electrons in a quantum dot (Exercise 15).

Because $\hat{H}_{cm}(t)$ is a simple quadratic hamiltonian, the time evolution of the center of mass wave function can be calculated analytically. We will not do this exercise here, but simply notice that the average values of $\hat{\vec{R}}_{cm}$ and $\hat{\vec{P}}_{cm}$ are determined by the classical equations of motion for a charged particle in the presence of electric and magnetic fields. In Exercise 14 we show that the complete N-electron wave function has the form

$$\Psi(\vec{r}_1, \vec{r}_2 \ldots, \vec{r}_N; t) = e^{-\frac{i}{\hbar} S(t)} \prod_{i=1}^{N} e^{\frac{i}{\hbar} \vec{P}_{cm} \cdot \frac{\vec{r}_i}{N}} e^{-\frac{1}{\hbar} E_i t}$$

$$\times \Psi_i \left(\vec{r}_1 + \vec{R}_{cm}(t), \vec{r}_2 + \vec{R}_{cm}(t), \ldots \vec{r}_N + \vec{R}_{cm}(t) \right) , \quad (10.109)$$

where Ψ_i is the wave function of the initial state, assumed to be a stationary state of the zero electric field hamiltonian with energy E_i, $\vec{R}_{cm}(t)$ and $\vec{P}_{cm}(t)$ are the solutions of the classical equations of motion with initial conditions $\vec{R}_{cm}(0) = 0$ and $\vec{P}_{cm}(0) = 0$, and

$$S(t) = N \int_0^t \left\{ \frac{m_b V_{cm}^2(t)}{2} - m_b \omega_c \vec{V}_{cm}(t) \cdot [\vec{R}_{cm}(t) \times \vec{e}_z] \right\} dt , \quad (10.110)$$

where $\vec{V}_{cm} \equiv \vec{P}_{cm}/M$. An immediate application of Kohn's theorem is the calculation of the current–current response function of the 2DEL at $\vec{q} = 0$ and finite frequency. In the presence of an external vector potential, $\vec{A}_{ext}(t)$, the macroscopic current is given by

$$\hat{\vec{j}}_{\vec{0}} \equiv \lim_{q \to 0} \hat{\vec{j}}_{\vec{q}} = \frac{1}{m_b} \left[\hat{\vec{P}}_{cm} + \frac{Ne}{2c} \vec{B} \times \hat{\vec{R}}_{cm} + \frac{Ne}{c} \vec{A}_{ext}(t) \right] . \quad (10.111)$$

The equation of motion for the expectation value of $\hat{\vec{j}}_{\vec{0}}$ is easily worked out to be

$$\frac{\partial \vec{j}_{\vec{0}}(t)}{\partial t} = \frac{Ne}{m_b c} \dot{\vec{A}}_{ext}(t) + \omega_c \vec{j}_{\vec{0}}(t) \times \vec{e}_z . \quad (10.112)$$

Solving this linear equation for the Fourier transform $\hat{\vec{j}}_{\vec{0}}(\omega)$ we obtain

$$\begin{pmatrix} j_{\vec{0},x}(\omega) \\ j_{\vec{0},y}(\omega) \end{pmatrix} = \frac{Ne}{m_b c} \frac{\omega^2}{\omega^2 - \omega_c^2} \begin{pmatrix} 1, & i\frac{\omega_c}{\omega} \\ -i\frac{\omega_c}{\omega}, & 1 \end{pmatrix} \cdot \begin{pmatrix} A_{ext,x}(\omega) \\ A_{ext,y}(\omega) \end{pmatrix} . \quad (10.113)$$

Notice that the linear response function exhibits a pole at $\omega = \omega_c$ independent of the strength of the interaction, and reduces to the correct limit in zero magnetic field. With little extra

---

[29] The Jacobi coordinates $\vec{q}_1, \ldots, \vec{q}_{N-1}$ are defined as follows: $\vec{q}_1 = \vec{r}_1 - \vec{r}_2, \vec{q}_2 = \frac{\vec{r}_1 + \vec{r}_2}{2} - \vec{r}_3, \ldots, \vec{q}_{N-1} = \frac{\vec{r}_1 + \vec{r}_2 + \cdots + \vec{r}_{N-1}}{N-1} - \vec{r}_N$. They are invariant under simultaneous translation of all the $\vec{r}_i$'s by the same vector $\vec{a}$, and for this reason they are sufficient to express any translationally invariant function of the $\vec{r}_i$'s.

effort, we can include a static harmonic potential. In this case the response function has poles at $\omega = \left| \frac{\omega_c}{2} \pm \sqrt{\frac{\omega_c^2}{4} + \omega_0^2} \right|$, where ω_0 is the frequency of the harmonic oscillator (Exercise 16).

10.7 Electrons in the lowest Landau level

In this section we begin the study of electronic correlations in the partially filled lowest Landau level (filling factors $\nu < 1$). Due to the quenching of the kinetic energy and the consequent degeneracy of the noninteracting ground-state, we must abandon all familiar ideas about Landau Fermi liquids and take a fresh look at the problem starting with highly correlated wave functions. As stated in the introduction, there have been major advances in the solution of this problem thanks to the insights afforded by the Laughlin wave function and by the composite fermion theory.

We begin our discussion by assuming that the ground-state and the low-lying excited states of the system are adequately represented by wave functions that lie entirely within the Hilbert space spanned by the LLL orbitals. This is another example of the "renormalization principle" on which the Landau Fermi liquid paradigm and the Luttinger liquid paradigm are also based. Strictly speaking, the projection in the LLL is justified only at very high magnetic field, i.e., when $\frac{e^2}{\epsilon_b \ell \hbar \omega_c} \ll 1$, but it turns out that the model obtained by excluding higher Landau level components of the wave function, is qualitatively correct even when $\frac{e^2}{\epsilon_b \ell \hbar \omega_c}$ is of the order of unity, as often is the case.

A major advantage of working in the LLL is that one can exploit the powerful property of analyticity of the wave function (see Section 10.2.4) to establish a connection between the radial and the angular dependence of the wave function for each pair of particles. To see this, let us begin with some warm-up exercises.

10.7.1 One full Landau level

This state is constructed by singly occupying each orbital in the LLL. Working in the symmetric gauge we sequentially occupy orbitals with $m = 0, 1, 2, \ldots, N-1$ where N is the total number of electrons. In the limit $N \to \infty$ this describes a uniform state with density $\frac{1}{2\pi \ell^2}$.[30] The wave function is then given by the single Slater determinant

$$\psi_1(z_1, \ldots, z_N) = \det \begin{pmatrix} 1 & z_1 & z_1^2 & \cdot & \cdot & z_1^{N-1} \\ 1 & z_2 & z_2^2 & \cdot & \cdot & z_2^{N-1} \\ \cdot & \cdot & \cdot & \cdot & \cdot & \cdot \\ \cdot & \cdot & \cdot & \cdot & \cdot & \cdot \\ \cdot & \cdot & \cdot & \cdot & \cdot & \cdot \\ 1 & z_N & z_N^2 & \cdot & \cdot & z_N^{N-1} \end{pmatrix} \prod_{i=1}^{N} e^{-\frac{|z_i|^2}{4\ell^2}}. \tag{10.114}$$

---

[30] Due to the fact that symmetric gauge orbitals are localized in the radial direction, the occupation of additional orbitals with $m \geq N$ does not affect the property of the system at any finite distance from the origin, when $N \to \infty$.

Although the main properties of this wave function (e.g., the density matrix and the pair correlation function) have already been calculated in Section 10.4, it is useful to take a second look at it from a different point of view. The determinant of the matrix in Eq. (10.114) is the well-known *Vandermonde determinant* – a homogeneous polynomial of degree $\frac{N(N-1)}{2}$ that is antisymmetric with respect to the interchange of any pair of variables (z_i, z_j), and has simple zeroes at $z_i = z_j$. Since a polynomial is completely determined by the position and order of its zeroes we see that Eq. (10.114) can be written in the more incisive form (Bychkov, Iordanskii, and Eliashberg, 1981)

$$\psi_1(z_1, \ldots, z_N) = \prod_{i<j}^{N}(z_i - z_j) \prod_{i=1}^{N} e^{-\frac{|z_i|^2}{4\ell^2}}. \tag{10.115}$$

This representation highlights the role of Pauli correlations (the factor $z_i - z_j$) and is most suitable for later generalizations to fractional filling factors.

10.7.2 Two-particle states: Haldane's pseudopotentials

Let us consider another extreme case: only two electrons in an otherwise empty LLL. This problem can be solved exactly owing to the fact that, as we saw in Section 10.6.4, the hamiltonian separates into two parts: one that governs the motion of the center of mass and one for the relative motion. Accordingly, the wave function of the two-particle system can be written as the product of two factors, which depend on the coordinate of the center of mass

$$Z \equiv \frac{z_1 + z_2}{2} \tag{10.116}$$

and the relative coordinate

$$z \equiv z_1 - z_2. \tag{10.117}$$

Explicitly, we have

$$\psi(z_1, z_2) = F(Z) e^{-\frac{|Z|^2}{2\ell^2}} f(z) e^{-\frac{|z|^2}{8\ell^2}}, \tag{10.118}$$

where both F and f are *analytic* functions of their arguments. Notice that the product of the two exponentials in Eq. (10.118) is $e^{-(|z_1|^2 + |z_2|^2)/4\ell^2}$ in agreement with the general structure of wave functions in the LLL.

To determine the form of the functions F and f, we make use of the fact that the angular momentum of the center of mass $\mathcal{M}$ and the relative angular momentum m are separately conserved. The center of mass wave function must therefore have an angular dependence of the form $\exp(i\mathcal{M}\Phi)$ where Φ is the angle formed by Z with the x-axis. By analyticity, this gives us immediately

$$F(Z) = Z^{\mathcal{M}}. \tag{10.119}$$

The relative wave function, on the other hand, must have an angular dependence of the form $\exp(im\phi)$, where ϕ is the angle formed by z with the x-axis. This gives

$$f(z) = z^m. \tag{10.120}$$

Putting the two pieces together, and introducing the appropriate normalization factors, we get

$$\psi_{\mathcal{M},m}(z_1, z_2) = \frac{Z^{\mathcal{M}} e^{-\frac{|Z|^2}{2\ell^2}} z^m e^{-\frac{|z|^2}{8\ell^2}}}{2\pi \ell^2 \sqrt{2^{\mathcal{M}+m} \mathcal{M}! m!}}. \tag{10.121}$$

Because this is the *only* wave function that lies entirely in the LLL and has the desired values of the angular momenta, it follows that it *must* be an eigenfunction of the two-particle hamiltonian. Of course, if the electrons have parallel spins, only the states with *odd* values of m are allowed by the antisymmetry of the wave function.

Notice that we have determined the wave function without making use of any specific property of the interaction other than the general rotational and translational invariance. The former ensures conservation of angular momentum, while the latter allows separation into center of mass and relative coordinates.

The eigenfunctions (10.121) represent *bound states*, the characteristic distance d between the electrons being fixed by the relative angular momentum at $d \sim \ell \sqrt{m}$. How can two-electrons, which by the coulomb interaction ought to repel each other, form a bound state? The answer, in classical language, is that the electrons orbit around each other in such a way that the repulsive radial force is overcompensated by the Lorentz force, resulting in a net centripetal force. The rotational motion of the electrons is in fact nothing but their classical drift under the action of crossed electric and magnetic fields.

The energies of the two-electron bound states depend on the relative angular momentum, but not of $\mathcal{M}$. These energies, denoted by Δ_m, can be calculated directly from the expectation value of $V(r)$ in the wave function (10.121):

$$\Delta_m = \frac{\int_0^\infty r^{2m+1} e^{-\frac{r^2}{4\ell^2}} V(r) dr}{\int_0^\infty r^{2m+1} e^{-\frac{r^2}{4\ell^2}}}. \tag{10.122}$$

Substituting $V(r) = \frac{e^2}{\epsilon_b r}$ for the coulomb interaction potential we obtain $\Delta_m = \frac{e^2}{\epsilon_b \ell} \frac{\Gamma(m+\frac{1}{2})}{2m!}$. The first few values of Δ_m for this case are listed in Table 10.3.

Table 10.3. *Haldane pseudopotentials (in units of $\frac{e^2}{\epsilon_b \ell}$) for the coulomb interaction.*

m	0	1	2	3	4	5
Δ_m	0.8862	0.4431	0.3323	0.2769	0.2423	0.2181

The Δ_m's provide a complete and unambiguous description of *any* isotropic interaction potential in the LLL. Recall that in second-quantization the electron-electron interaction is completely described by its matrix elements between two-particle states. But two-particle states can always be represented in the form of Eq. (10.121), and any rotationally invariant interaction will be diagonal in this basis, with eigenvalues given by (10.122).[31] We can then turn the whole machinery backward, introducing an arbitrary set of Δ_m's to parametrize an arbitrary interaction potential. This procedure was first suggested by Haldane, who proposed to study an interaction potential characterized by $\Delta_m = \Delta$ for $m = 1$, and 0 otherwise (more about this in the next section). For this reason the Δ_m's are commonly referred to as the "Haldane pseudopotentials".

10.8 The Laughlin wave function

10.8.1 A most elegant educated guess

With his highly original proposal of a wave function to explain the fractional quantum Hall effect, Laughlin set the stage for one of the most beautiful developments in the physics of the twentieth century. Let us consider a uniform 2DEL at filling factor $\nu = \frac{1}{3}$ – the most prominent fraction of the FQHE. What is its ground-state wave function? Assuming that the state is homogeneous and isotropic, we expect the electrons to uniformly fill a disk of area $\mathcal{A}$ such that $\frac{N}{\mathcal{A}} = \frac{\nu}{2\pi\ell^2}$. The radius of this disk is approximately $\sqrt{\frac{2N}{\nu}}\ell$, implying that only single-particle orbitals up to a maximum angular momentum $\frac{N}{\nu}$ ($N \gg 1$) are significantly occupied in the ground-state. This, in turn, implies that each variable z_i in the polynomial part of the many-body wave function is raised to a maximum power $\frac{N}{\nu}$. Taking into account the indistinguishability and antisymmetry requirements we are then led to the following guess for the wavefunction:

$$\psi_M(z_1, \ldots, z_N) = \prod_{i<j}^N (z_i - z_j)^M \prod_{i=1}^N e^{-\frac{|z_i|^2}{4\ell^2}}, \qquad (10.123)$$

where $M \equiv \frac{1}{\nu}$ is an odd integer ($M = 3$ in the specific case under discussion). This is the famous Laughlin wavefunction.[32] Like the wave function of a full Landau level, to which it reduces for $M = 1$, it has zeroes at the positions of the particles. But, these zeroes are far more "powerful", being of order $M = 3$, and leading to a pair correlation function that vanishes as $r^{2M} = r^6$ as the distance r between two particles tends to zero. Because of the higher exponent of $z_i - z_j$, Eq. (10.123) cannot be written as a single Slater determinant

---

[31] The reader might wonder why, if the interaction is diagonal in the basis (10.121), one cannot solve the many-particle problem as easily as the two-particle problem. The reason is that the values of the quantum numbers $\mathcal{M}, m$ cannot be simultaneously defined for more than one pair of electrons. The projectors upon states with definite values of $\mathcal{M}$ and m do not commute for different pairs.

[32] In retrospect one sees that the Laughlin wave function is amazingly similar to the ground-state wave function of the Luttinger liquid model, given by Eq. (9.51). Like that wave function, it can be written as the product of a single Slater determinant, $\prod_{i<j}^N (z_i - z_j) \prod_{i=1}^N e^{-\frac{|z_i|^2}{4\ell^2}}$, and a correlation factor, $\prod_{i<j}^N (z_i - z_j)^{M-1}$. The complex coordinate z_i corresponds to $e^{i\frac{2\pi}{L}x_i}$, and M corresponds to $\frac{1}{g}$ (see also Exercise 3).

single-particle orbitals: indeed, the polynomial factor embodies correlations far stronger than those due to the Pauli exclusion principle. Notice, however, that M *must* be odd in order to respect the antisymmetry of the wave function.

It is not difficult to see that Eq. (10.123) is an eigenstate of the canonical angular momentum with eigenvalue $\frac{N(N-1)M}{2}$. To see this just operate on ψ_M with the rotation operator $\exp\left(\frac{i}{\hbar}\hat{L}_z\phi\right)$. This rotates the complex coordinate of each electron by an angle ϕ, i.e., it multiplies z_i by $e^{i\phi}$ with the net result of multiplying the whole wavefunction by $e^{i\frac{N(N-1)M\phi}{2}}$, where $\frac{N(N-1)}{2}$ is the number of distinct pairs of electrons. The fact that the angular momentum is M time larger than in the full Landau level reflects the fact that the electrons are spread over a disk whose area is M times larger. Finally, the form of the wave function shows that each pair of electrons, while definitely *not* in an eigenstate of the pair's relative angular momentum, cannot have a relative angular momentum smaller than M, due to the presence of the factor $(z_i - z_j)^M$.

Calculating the expectation values of observables in the Laughlin wave function is a difficult task. Even the proof that the density is uniform is not easy if approached too directly.[33] Fortunately, there is an extremely clever method, first pointed out by Laughlin himself, that allows one to calculate the most important physical properties of this wave function in a relatively painless way. This is called the *classical plasma analogy*, which we now describe.

10.8.2 The classical plasma analogy

Consider the squared modulus of the Laughlin wave function

$$|\psi_M(z_1,\ldots,z_N)|^2 = \prod_{i<j}^{N} |z_i - z_j|^{2M} \prod_{i=1}^{N} e^{-\frac{|z_i|^2}{2\ell^2}}, \qquad (10.124)$$

which gives the relative probability of finding the electrons at the positions $z_1,\ldots,z_N$. Clearly, the expectation value of any operator that depends only on the electron coordinates (for instance the coulomb interaction energy) can be expressed as an integral over this probability.

Consider now the probability distribution of a system of *classical* particles interacting with a potential energy $U(z_1,\ldots,z_N)$, in two dimensions. According to classical statistical mechanics the probability of a configuration $z_1,\ldots,z_N$ is proportional to

$$e^{-\beta U(z_1\ldots z_n)} \qquad (10.125)$$

where $\beta = 1/k_B T$ is the inverse temperature. We now ask, for what form of the dimensionless potential energy βU does the classical probability distribution (10.125) coincide

---

[33] Part of the difficulty lies in the fact that Eq. (10.123) is the wave function of a *finite* system, which is therefore not uniform in the vicinity of its physical boundary. Uniformity is attained only in the limit $N \to \infty$ and sufficiently far from the boundary.

with the quantum probability distribution function (10.124)? The answer is

$$\beta U(z_1, \ldots, z_N) = -2M \sum_{i<j} \ln|z_i - z_j| + \sum_i \frac{|z_i|^2}{2\ell^2}. \qquad (10.126)$$

The value of β is completely arbitrary, and we exploit this freedom by setting $\beta = \frac{2}{M}$ in some arbitrary units of energy, so that

$$U(z_1, \ldots, z_N) = -M^2 \sum_{i<j} \ln|z_i - z_j| + M \sum_i \frac{|z_i|^2}{4\ell^2} \qquad (10.127)$$

in those units.[34]

Eq. (10.127) has a simple physical interpretation. The first term is the potential energy of a system of N point particles of "charge" M, interacting with one another via the two-dimensional coulomb potential $\Phi_2(|z_i - z_j|) = \ln|z_i - z_j|$. The logarithmic form of the coulomb potential arises from the solution of the *two-dimensional* Poisson equation

$$\nabla^2 \Phi(\vec{r}) = -2\pi M \delta(\vec{r}), \qquad (10.128)$$

which determines the potential created at the point $\vec{r} = (x, y)$ by a particle of charge M placed at the origin of the coordinates.[35]

The second term in Eq. (10.127) is the potential energy of N particles of charge M interacting with a uniform background of *charge density* $-\frac{1}{2\pi \ell^2}$. To see this, calculate the electric potential created by a uniform charge distribution according to Eq. (10.128). The calculation is done most efficiently using the fact that the electric field ($\vec{E} = -\vec{\nabla}\Phi$) obeys the two-dimensional version of Gauss' theorem, according to which the flux of electric field through a closed contour equals 2π times the charge enclosed by the contour. We then obtain (Exercise 18)

$$\Phi_{background}(r) = \frac{r^2}{4\ell^2}, \qquad (10.129)$$

i.e., a harmonic potential that tends to keep the particles *inside* the system, as expected. We have now shown that all the information about positional correlations in the Laughlin wave function can be obtained from the study of a classical one component plasma (OCP) of charged particles interacting with potential energy (10.127). This is a well studied problem in classical statistical mechanics (Baus and Hansen, 1980). Its solution depends on a single dimensionless coupling constant $\Gamma = 2M$, which measures the strength of the interaction (M^2) relative to the temperature ($k_B T = \frac{M}{2}$). For not too large values of Γ the equilibrium state is known to be uniform and electrically neutral, implying that the average particle density adjusts itself to cancel the electric charge of the background.[36] Since the classical

---

[34] The arbitrary choice of the energy scale does not affect the probability distribution, which depends only on the dimensionless quantity βU.

[35] The reader will recognize this as the potential created by a long wire of linear charge density M running parallel to the z-axis and piercing the x–y plane at the origin.

[36] Nonuniform equilibrium states can occur at very large values of Γ. For example, Monte Carlo studies indicate a freezing transition into a Wigner crystal at $\Gamma \simeq 130$. Such large values of Γ are of no interest here.

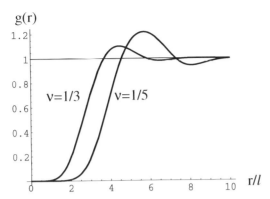

Fig. 10.13. Pair correlation functions in the Laughlin states at $M = 3$ and 5 calculated from Eq. (10.130), with coefficients c_m fitted to numerical Monte Carlo results.

particles have charge M, they must assume a uniform density $n = \frac{1}{2\pi\ell^2 M}$ in order to neutralize the background whose charge density is $-\frac{1}{2\pi\ell^2}$. And this is also the electronic density of the state described by the Laughlin wave function.

Not only the density, but every property that depends only on positional correlations – the pair correlation function, the static structure factor, the ground-state energy – can be calculated with the help of the plasma analogy. We shall not dwell on how these calculations are done but simply reproduce the essential results.[37]

The pair correlation function was calculated numerically by Levesque et al. (1984), using the classical Monte Carlo method. Girvin (1984) showed that the pair correlation function of any rotationally and translationally invariant state in the LLL can be expressed in the form

$$g(r) = 1 - e^{-\frac{r^2}{2\ell^2}} + 2 \sum_{k=0}^{\infty} \frac{c_{2k+1}}{(2k+1)!} \left(\frac{r^2}{4\ell^2} \right)^{2k+1} e^{-\frac{r^2}{4\ell^2}} . \qquad (10.130)$$

Combining the above equation with the Monte Carlo results, the values of the coefficients c_{2k+1} could be determined (see Girvin et al. (1986) for a complete table of values of c_{2k+1} up to $2k + 1 = 27$ for the states with $M = 3$ and $M = 5$). The resulting accurate analytical formula for $g(r)$ is plotted in Fig. 10.13, for $M = 3$ and $M = 5$. Notice how small $g(r)$ is for small values of r. This is due to the multiple zeroes of the wavefunction: the analytic behavior is $g(r) \sim r^{2M}$ for $r \to 0$. The deep correlation hole is the basic reason for the low energy of the Laughlin wave function. Competing states, such as the quantum Wigner crystal, have pair correlation functions that vanish only as r^2, in accordance with Pauli's principle. In the case of the Wigner crystal, however, the coefficient of r^2 decreases

---

[37] As a historical curiosity, we mention that, years before the discovery of the Laughlin wave function, the classical plasma with logarithmic interactions had been found to be analytically solvable at $\beta = 2$. The reason for this surprising occurrence is clearly understood, in retrospect, in terms of the plasma analogy, according to which the positional correlations of the classical plasma at $\beta = 2$ coincide with those of quantum mechanical electrons in a full Landau level ($M = 1$), which can be calculated analytically.

Table 10.4. *Energy per electron in the Laughlin state (Levesque et al., 1984) and in the Wigner crystal (Lam and Girvin, 1984) in units of $\frac{e^2}{\epsilon_b \ell}$.*

M	3	5	7
$-\epsilon_L$	0.4100	0.3277	0.2810
$-\epsilon_{WC}$	0.3948	0.3258	0.2816

exponentially with decreasing density: it is for this reason that the Wigner crystal eventually "beats" the Laughlin liquid state at sufficiently low filling factors. Numerical values of the energy per electron in the Laughlin state are listed in Table 10.4 together with the energies of the Wigner crystal calculated by Lam and Girvin (1984) (see Section 10.19).

What about properties that do not depend only on positional correlations but involve, say, the one and two-particle density matrices? It turns out that these too can be computed from the density and the pair correlation function owing to a remarkable theorem that allows us to express the complete one- and two-particle density matrices in terms of their diagonal elements, i.e., precisely, the density and the pair correlation function. This theorem, which follows from the analyticity of the LLL wavefunctions is proved, for the case of the one-particle density matrix, in Appendix 23.

It must be borne in mind that, despite its low energy and its striking properties, the Laughlin wavefunction is *not* the exact ground-state wave function of the system. In fact the only known case in which it is demonstrably exact is that of a fictitious system in which all the Haldane pseudopotentials are zero for $m \geq 3$: the ground-state energy is then zero (Exercise 19). The real importance of the Laughlin wave function lies in the fact that it is the simplest realization of a new *paradigm* of correlated behavior in the two-dimensional electron liquid. The central feature of this paradigm is that the electrons capture magnetic flux quanta and turn themselves into a new type of objects known as *composite fermions* (CF) (Jain, 1989), which experience a reduced magnetic field. The Laughlin state turns out to be the state that arises from a full Landau level of such composite fermions, while other fractional quantum Hall states arise from CF states with integral filling factors larger than one. The composite fermion theory provides a unified description of integral and fractional quantum Hall states, and will be discussed in detail in Section 10.16.

10.8.3 Structure factor and sum rules

It is well known that the structure factor of *any* plasma (classical or quantal) must vanish for $q \to 0$ simply because (see Appendix 4)

$$S(q) = 1 + n \int [g(r) - 1] e^{-i\vec{q}\cdot\vec{r}} d^2 r \qquad (10.131)$$

nd

$$n \int [g(r) - 1] d^2r = -1 . \tag{10.132}$$

The latter is just a statement of the conservation of the particle number: if one particle is definitely located at $\vec{r} = 0$, then it is not somewhere else: there must be one particle missing from the rest of the system.

The behavior of $S(q)$ for small q gives information about the amplitude of long-wavelength density fluctuations in the liquid. To find out more, we make use of the fact that in a classical plasma the structure factor is related to the static density–density response function by the classical version of the fluctuation dissipation theorem (see Eq. (3.212) in Exercise 4)

$$S(q) = -\frac{k_B T}{n} \chi_{nn}(q, 0) . \tag{10.133}$$

Recall that, according to the formalism of Chapter 5 (which applies to classical systems as well),

$$\chi_{nn}(q, 0) = \frac{\tilde{\chi}_{nn}(q, 0)}{1 - v_{2D}(q)\tilde{\chi}_{nn}(q, 0)} , \tag{10.134}$$

where $\tilde{\chi}_{nn}$ is the proper density–density response function and

$$v_{2D}(q) = \frac{2\pi M^2}{q^2} \tag{10.135}$$

is the Fourier transform of the interaction potential in the classical plasma, in units such that $k_B T = \frac{M}{2}$.[38] The small-q limit of the proper density–density response function is related to the compressibility of the neutral plasma $K = \frac{1}{n^2}\frac{\partial n}{\partial \mu}$ by the compressibility sum rule

$$\lim_{q \to 0} \tilde{\chi}(q, 0) = -n^2 K . \tag{10.136}$$

Putting all this together in Eq. (10.133) with $n = \frac{1}{2\pi \ell^2 M}$, we see that the long wavelength expansion of the structure factor up to order $(q\ell)^4$ is

$$S_M(q) \sim \frac{(q\ell)^2}{2} \left[1 - \frac{(q\ell)^2}{nMK} \right] , \tag{10.137}$$

where the compressibility is expressed in units in which $k_B T = \frac{M}{2}$.

The leading term in the small-q expansion of $S(q)$ is independent of M. This "universal" behavior is actually dictated by Kohn's theorem (see Exercise 20), and must therefore hold in any translationally invariant state, with or without interactions. An immediate implication of the small-q expansion (10.137) is the sum rule

$$n \int \frac{r^2}{2\ell^2}[g(r) - 1]d^2r = -1 \tag{10.138}$$

[38] The form of $v_{2D}(q)$ follows immediately from the fact that the potential satisfies the two dimensional Poisson equation (10.128).

(see Exercise 21). In the present context, this sum rule can be seen as a manifestation of the perfect screening of density fluctuations in the equivalent classical plasma: for this reason it is also referred to as the *perfect screening sum rule*.

The coefficient of $(q\ell)^4$ in Eq. (10.137) is controlled by the *proper compressibility* of the classical plasma, and does depend on the value of M. The proper compressibility can be calculated analytically owing to the special form taken by the virial theorem. In a two-dimensional plasma with a logarithmic interaction (Baus and Hansen, 1980) the theorem states that

$$2\langle \hat{T} \rangle - \frac{1}{2} N M^2 = 2 P \mathcal{A} \qquad (10.139)$$

where $\langle \hat{T} \rangle = N k_B T = \frac{NM}{2}$ is the expectation value of the kinetic energy, P is the pressure, and $\mathcal{A}$ is the area. Notice the extremely simple form of the potential energy term (the second term on the left-hand side), which is a simple quadratic function of the coupling constant. Eq. (10.139) tells us that the pressure, P, is given by

$$P = \frac{nM}{2}\left(1 - \frac{M}{2}\right) \qquad (10.140)$$

and the inverse of the proper compressibility, K, has exactly the same value:

$$\frac{1}{K} = n\frac{\partial P}{\partial n} = P . \qquad (10.141)$$

Substituting Eqs. (10.140) and (10.141) into Eq. (10.137) we obtain

$$S_M(q) \sim \frac{1}{2}(q\ell)^2 + \left(\frac{M}{8} - \frac{1}{4}\right)(q\ell)^4 . \qquad (10.142)$$

This result will play an important role in the theory of the dynamics of the quantum Hall liquid, described in Section 10.13.1.

10.8.4 Interpolation formula for the energy

We conclude this section by reporting an interpolation formula due to Levesque, Weis, and MacDonald (1984), which is useful to approximate the ground-state energy per particle of the 2DEL as a function of filling factor in the LLL, i.e., in the range $0 < \nu \leq 1$. This formula reads

$$\epsilon_{LWM}(\nu) = -\left[0.782133\sqrt{\nu}\left(1 - 0.211\nu^{0.74} + 0.012\nu^{1.7}\right)\right]\frac{e^2}{\epsilon_b\ell} \qquad (10.143)$$

and has the following desirable features: (i) it reproduces the energy of the Laughlin wave function at $\nu = \frac{1}{3}, \frac{1}{5}$ and $\frac{1}{7}$; (ii) it reduces to the energy of the classical Wigner crystal, $\epsilon_{WC} = -0.782133\sqrt{\nu}\frac{e^2}{\epsilon_b\ell}$, for $\nu \to 0$. However, unlike the true energy, which is expected to exhibit a rich and complex structure of *derivative discontinuities* (see Fig. 10.14) this

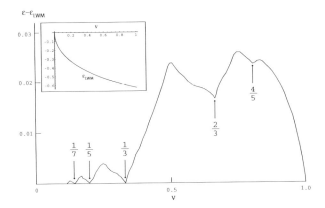

Fig. 10.14. Expected behavior of the difference $\epsilon - \epsilon_{LWM}$ (in units of $\frac{e^2}{\epsilon_b \ell}$), where ϵ is the ground-state energy and ϵ_{LWM} is the smooth interpolation (10.143): only a few major cusps are shown (adapted from Halperin, 1984). Inset: $\epsilon_{LWM}(\nu)$ (in units of $\frac{e^2}{\epsilon_b \ell}$) vs ν.

approximate energy is a smooth function of ν and therefore does not give any information about the cusps in the energy vs ν relation. See Exercise 23 for additional comments on Eq. (10.143).

10.9 Fractionally charged quasiparticles

One of the most striking properties of the Laughlin liquid state is the existence of fractionally charged excitations. In a normal Fermi liquid we constructed the low-lying quasiparticle excitations by adding an electron to the noninteracting ground-state and then adiabatically turning on the interaction. We will again make use of an adiabatic continuation here, but its nature is completely different. We start from the *interacting* ground-state described by the Laughlin wave function and adiabatically insert one quantum of magnetic flux through an infinitely thin solenoid perpendicular to the plane of the 2DEL, piercing the latter at a point z_0. At the end of this process the hamiltonian is equivalent, up to a gauge transformation, to the original one, and its eigenvalue spectrum is therefore unchanged. However, the original ground-state wave function will have evolved into the wave function that corresponds, after the gauge transformation, to an excited state of the original hamiltonian. It is this excited state that we interpret as a "quasiparticle" state. Notice that such a state is naturally labelled by the position z_0 at which the flux was inserted, suggesting that the quasiparticle is somehow localized in space.

To calculate the effect of the adiabatically inserted flux, let us first consider the situation at large distance from the insertion point. The many-body wave function can be expanded as a sum of products of single-particle orbitals in the symmetric gauge with origin at z_0. At large distance from z_0 only orbitals with large angular momentum m (relative to z_0) contribute significantly to the expansion. Such orbitals have the form $\psi_m = (z - z_0)^m e^{-|z-z_0|^2/4\ell^2}$. Because the inserted flux does not physically touch these distant orbitals, we are effectively

in the situation described in Section 10.3.4 *a propos* the Laughlin argument for the integr.
quantum Hall effect. If the inserted flux has the same sign as the pre-existing flux, we se
that the orbital $(z - z_0)^m e^{-|z-z_0|^2/4\ell^2}$ evolves into the orbital $(z - z_0)^{m+1} e^{-i\phi} e^{-|z-z_0|^2/4}$
which has the same angular momentum but is slightly displaced radially in the *outwar*
direction. Removing the extra flux by a gauge transformation shows that the new state i
equivalent to $(z - z_0)^{m+1} e^{-|z-z_0|^2/4\ell^2}$: thus, the effect of the flux insertion, when viewed i
the original gauge, is simply to transform the single-particle orbital $\psi_m(z)$ into $\psi_{m+1}(z$
through a multiplication by the factor $z - z_0$. Since the many-body wave function is mad
up of products of such orbitals we conclude that, when all the z_i's are far from z_0, the stat
that evolves from the Laughlin wave function has the form

$$\psi_M^{h,z_0}(z_1, \dots, z_N) = \prod_i (z_i - z_0)\psi_M(z_1, \dots, z_N). \tag{10.144}$$

This wave function was first introduced by Laughlin in his seminal 1983 paper as a variational approximation for an excited state.

We now show, in two different ways, that this state contains a fractionally charged
"quasihole" of charge $+\nu e$ localized in a region of characteristic size ℓ about z_0.[39] The
first approach is to compute the amount of charge displaced outwardly as a result of the flux
insertion. Because of the continuity equation, this can be calculated as the amount of charge
displaced across a very large circle centered at z_0. As the orbital ψ_m adiabatically moves over
to the orbital ψ_{m+1}, its average radial distance from the center increases from $r_m = \ell\sqrt{2m}$
to $r_{m+1} = \ell\sqrt{2m + 2}$, and the total electric charge transported outwardly equals the charge
contained in the annulus $r_m < r < r_{m+1}$, that is, $\frac{-e\nu\pi(r_{m+1}^2 - r_m^2)}{2\pi\ell^2} = \frac{-eB\nu\pi(r_{m+1}^2 - r_m^2)}{\Phi_0}$. Since, by
definition, $\frac{B\pi r_m^2}{\Phi_0} = m$, we see that the net charge transported outwardly is $-\nu e$. In the bulk
of the system – many magnetic lengths away from both z_0 and the outer boundary – the
transformation amounts to a rigid outward displacement of the ground-state – no charge
being accumulated. This means that the net effect of the process is to remove a charge $-\nu e$
from a region of size $\sim\ell$ near z_0 and relocate it at the outer boundary of the system. Letting
the size of the system go to infinity we are left with just one "quasihole" of fractional
charge $+\nu e$ localized about z_0. The beauty of the argument lies in the fact that the total
charge of the hole depends only on the behavior of the wave function (10.144) far from z_0,
which is clearly understood. The detailed shape of the charge distribution, the localization
length, the excitation energy etc., are only roughly predicted by this wave function, which
is reasonable, but certainly not exact, near z_0.

The second approach employs the classical plasma analogy to compute the properties
of the wave function (10.144). The effective classical potential energy, obtained from the
logarithm of the square modulus of this wave function, differs from the one studied in the
previous section only by the addition of a term $M \sum_i \ln |z_i - z_0|$, which can be interpreted
as the potential energy of an impurity particle of charge $+1$ located at z_0. The classical

---

[39] To avoid misunderstandings, we emphasize that "localization" in this context means that the charge of the quasiparticle is
concentrated in a finite region of space about z_0. In the absence of impurities, the quasiparticle is free to move, and therefore
not localized in the sense of transport theory.

plasma screens this impurity by creating a charge deficiency within a Debye screening length of z_0. The "screening cloud" contains exactly $\frac{1}{M}$ classical particles of charge M and decays exponentially as a function of the distance from the impurity. Because there is a one-to-one correspondence between the classical particles of the equivalent plasma and the electrons of the original problem, we see that the total number of electrons excluded from the neighborhood of z_0 is $\frac{1}{M}$, leaving behind a charge $\frac{e}{M} = \nu e$.

In order to construct fractionally charged "quasielectrons", we simply reverse the direction of the inserted flux. This operation "pulls in" single-particle orbitals far from z_0 and thus accumulates a net negative charge $-\frac{e}{3}$ in a neighborhood of z_0. The mathematics is slightly more tricky, however, because one cannot simply divide the Laughlin wave function by $\prod_i(z_i - z_0)$: that operation would create a singularity at $z = z_0$. A better procedure is to operate on ψ_m with the hermitian conjugate of $\prod_i(z_i - z_0)$, regarded as an operator in the lowest Landau level. For, if $z_i - z_0$ is the operator that raises the angular momentum (relative to z_0) by one unit, then its hermitian conjugate lowers it by the same amount. It is not difficult to convince oneself that the hermitian conjugate of z_i, *within the LLL*, is $2\ell^2\frac{\partial}{\partial z_i}$, where it is understood that *this differential operator acts only on the polynomial part of the wave function* (Girvin and Jach, 1984):[40]

$$z^\dagger \overset{LLL}{\to} 2\ell^2 \frac{\partial}{\partial z}. \tag{10.146}$$

The "quasielectron" wave function is therefore written as

$$\psi_M^{e,z_0}(z_1,\ldots,z_N) = \prod_{i=1}^N e^{-\frac{|z_i|^2}{4\ell^2}} \left(2\frac{\partial}{\partial z_i} - \frac{z_0^*}{\ell^2}\right) \prod_{i<j}^N (z_i - z_j)^m. \tag{10.147}$$

The energies of quasiparticle excitations based on the trial wavefunctions (10.147) and (10.144) have been calculated by several authors (Laughlin, 1983; Chakraborty, 1985; Morf and Halperin, 1986). At $\nu = \frac{1}{3}$ Morf and Halperin report $\varepsilon^e = (0.073 \pm 0.008)\frac{e^2}{\ell}$ and $\varepsilon^h = (0.0268 \pm 0.0033)\frac{e^2}{\ell}$. Since then better wave functions, based on the composite fermion theory, have become available. A small compilation of numerical results is presented in Table 10.5.

Fig. 10.15 shows the electronic density associated with the Laughlin quasihole state at $\nu = \frac{1}{3}$, calculated numerically for systems containing 30, 42, and 72 electrons. The figure clearly shows that the fractional charge of the quasi-hole becomes sharply defined only in the thermodynamic limit, that is, when the size of the system L greatly exceeds the size of the quasiparticle "core" ($\sim\ell$). Equivalently, we can say that the fractional quantization

---

[40] This follows from the chain of identities

$$\int f^*(z)e^{-|z|^2/4\ell^2}zg(z)e^{-|z|^2/4\ell^2}dz = -\int f^*(z)g(z)2\ell^2\frac{\partial}{\partial z^*}e^{-|z|^2/2\ell^2}dz$$

$$= \int \left[2\ell^2\frac{\partial}{\partial z^*}f(z^*)\right]g(z)e^{-|z|^2/2\ell^2}dz$$

$$= \int \left[2\ell^2\frac{\partial f(z)}{\partial z}\right]^* e^{-|z|^2/4\ell^2}g(z)e^{-|z|^2/4\ell^2}dz, \tag{10.145}$$

where $g(z)e^{-|z|^2/4\ell^2}$ and $f(z)e^{-|z|^2/4\ell^2}$ are any two single-particle wave functions in the LLL.

Table 10.5. *Energies of fractionally
charged quasielectrons and
quasiholes, in units of* $\frac{e^2}{\epsilon_b \ell}$ *(from
Kamilla and Jain, 1997).*

ν	$\frac{1}{3}$	$\frac{1}{5}$
ε^e	0.077	0.017
ε^h	0.0264	0.007

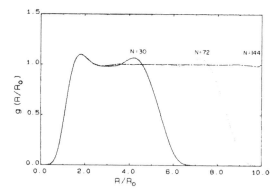

Fig. 10.15. Electronic density distribution in the Laughlin quasihole state at $\nu = \frac{1}{3}$. Here $R_0 = \sqrt{6}\ell$ (from Morf and Halperin, 1986).

of the charge is accurate up to corrections of order $e^{-L/\ell}$, where L is the length scale of the observations.[41] Thus, the occurrence of fractionally charged quasiparticles is truly an "emerging property" of the macroscopic limit, and beautifully exemplifies P. W. Anderson's dictum *"More is different"*.

Fig. 10.15 also helps to resolve the following paradox. Imagine to capture a fractionally charged quasiparticle within a large box of size L. Because the box encloses an integer number of electrons, there is no doubt that the total charge within it is an integral multiple of e. What happened to the fractional charge? The answer is that the the box causes an accumulation/depletion of charge in the vicinity of the walls. The space integral of the extra charge created near the walls exactly compensates for the fractional charge of the quasiparticle at the center of the box.[42] Again, it is the separation of length scales $L \gg \ell$ that allows us to divide the integral of the charge density into a quasiparticle contribution and a boundary contribution.

---

[41] Similarly, fluctuations in the fractional charge occur on a time scale shorter than $\frac{\hbar}{\Delta}$ where Δ is the energy gap for a neutral excitation of the system. Such fluctuations do not affect the low-energy physics.

[42] The situation is somewhat reminiscent of the order parameter of a superconductor which appears to vanish if the particle number is fixed by enclosing the system in a box.

Fractionally charged excitations have another remarkable property, which is known as *fractional statistics*. This is most easily seen in the case of Laughlin's quasiholes. If one slowly transports a quasihole around a closed path that encircles another quasi-hole, then the quantum state of the system is not restored to the initial value, but acquires a phase factor $e^{i\gamma}$, where $\gamma = \frac{2\pi}{3}$ at $\nu = \frac{1}{3}$ (Arovas, Schrieffer, and Wilczek, 1984). This phase is independent of how the path is traversed provided the rate of change in the position of the quasi-hole remains everywhere much smaller than $\frac{\Delta}{\hbar}$, where Δ is the minimum excitation energy (see Section 10.12 for a precise definition of the latter). To verify this, we make use of the quasi-hole wave function (10.144). As the position of the quasi-hole z_0 changes adiabatically in time, the evolution of the wave function can be written as

$$\psi(t) = \psi^{h,z_0(t)} e^{-i\frac{\epsilon h_t}{\hbar} t} e^{i\gamma(t)}, \tag{10.148}$$

where $\gamma(t)$ is, for the time being, an undetermined phase. Substituting (10.148) into the time-dependent Schrödinger equation we see that $\gamma(t)$ obeys the equation of motion

$$\frac{d\gamma(t)}{dt} = i\left\langle \psi^{h,z_0(t)} \left| \frac{d\psi^{h,z_0(t)}}{dt} \right. \right\rangle. \tag{10.149}$$

The total phase γ accumulated as z_0 is transported around a closed path is therefore

$$\gamma = i \oint dz_0 \left\langle \psi^{h,z_0} \left| \frac{d\psi^{h,z_0}}{dz_0} \right. \right\rangle, \tag{10.150}$$

which is equal to 2π times the average number of electrons enclosed by the path (see Exercise 24). If the loop encloses a charge-$\frac{1}{3}$ quasi-hole, then this average number must be $N - \frac{1}{3}$, where N is an integer. Thus, the phase factor is $e^{2i\pi/3}$.

What does this have to do with statistics? Notice that one can interchange two quasi-holes by means of a $180°$ rotation about their center of mass. Two such interchanges applied in succession restore the initial state, but in the process one particle has gone a complete $360°$ loop about the other: hence the wave function is multiplied by $e^{i\gamma}$. Now if a double interchange multiplies the wave function by $e^{i\gamma}$, then a single interchange must multiply it by $e^{i\gamma/2}$. For particles obeying ordinary Fermi or Bose statistics this factor must be ± 1, so $e^{i\gamma} = 1$ and $\gamma = 0$. In the present case, however, the nontrivial value of $e^{i\gamma/2} = e^{i\pi/3}$ (at $\nu = \frac{1}{3}$) suggests that the Laughlin quasiholes obey a new kind of statistics, in which the wave function is multiplied by $e^{i\pi/3}$ when two quasiholes are interchanged.

It must be borne in mind that the wave function of the system is completely antisymmetric upon interchange of the coordinates of two particles, z_1 and z_2, as required by the fundamental principles of quantum mechanics. The fractional statistics appears only when one quasiparticle is adiabatically transported around the other along a closed loop. The physical process of transporting two quasiparticles around each other is quite different from the mathematical interchange of two electron coordinates in the wave function. In the former case all the electronic coordinates are affected, since the quasihole coordinate, z_0, is a global label of the wave function: it is only at the level of the collective coordinate z that a nontrivial phase can arise.[43]

10.10 The fractional quantum Hall effect

The most striking property of the state described by the Laughlin wave function is, of course, the fractional quantum Hall effect. According to Laughlin, this can be viewed as a manifestation of the fractional charge of the quasiparticles. We can repeat, point by point the adiabatic transport argument presented in Section 10.3.4 for the IQHE. We still have a gap to prevent dissipation as the magnetic flux is threaded through the center of the annulus The only difference is that, after the insertion of one flux quantum, the net amount of charge transferred between the edges of the annulus equals ve – the quasiparticle charge. The fractionally quantized Hall conductance follows at once. Notice that impurities are still needed to account for the finite width of the FQHE plateaus. As in the integral case, they create a reservoir of localized quasiparticle states within the gap.

Although the above argument is very suggestive, it must be kept in mind that the FQHE can also be explained without explicit reference to fractionally charged quasiparticles. For example, one may say that the fractionally quantized Hall conductance follows from the Středa formula of Section 10.3.3, and from the fact that the system has a gap at a fractional filling factor.

10.11 Observation of the fractional charge

It is perfectly legitimate at this point to ask if fractionally charged quasiparticles can ever be directly observed in an experiment. Very recently, this question has been answered in the affirmative by three experimental groups (Goldman and Su, 1995; Saminadayar *et al.*, 1997; De Picciotto *et al.*, 1997). The basic idea of one of these experiments (De Picciotto *et al.*, 1997) is to induce tunneling of quasiparticles between the edges of a quantum Hall bar by "squeezing" it between two electrodes as shown, for example, in Fig. 10.32. The size of random fluctuations in the tunneling current is proportional to the size of the "grains of charge", in much the same way as the size of fluctuations in the noise produced by rain on the roof of a log cabin gives a measure of the average size of the rain drops. The measured values of the current noise spectrum, shown in Fig. 10.16, are consistent with a fractional charge $\frac{e}{3}$ at $v = \frac{1}{3}$.

10.12 Incompressibility of the quantum Hall liquid

One often hears the statement that the Laughlin wavefunction describes an *incompressible quantum liquid*. What does this mean? From a thermodynamic point of view it means that

---

[43] Fractional statistics can only be defined in two dimensions, because this is the only case in which one can draw a sharp topological distinction between the paths that enclose a quasiparticle and those that do not.

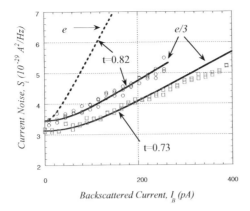

Fig. 10.16. Quantum noise measurement of the fractional charge at $v = \frac{1}{3}$. The data are consistent with the notion that the quasiparticles that tunnel between the edges of the Hall bar carry a fractional charge $\frac{e}{3}$ (solid lines) and inconsistent with charge e (dashed line). Notice that the conductance of the device, given by $\frac{te^2}{h}$, is *not* quantized due to tunneling between the edges. From De Picciotto *et al.*, 1997. (Courtesy of R. Reznikov.)

the ground-state energy vs density exhibits downward cusps at the filling factors of the FQHE. A discontinuity in the first derivative of the energy (i.e., the chemical potential) signals the existence of a *gap* in the one-particle excitation spectrum. This can be roughly understood if one thinks of the Laughlin state as a kind of commensurate state, in which the electronic density is tied to the magnetic flux by a rational relation (e.g., one electron per three flux quanta) in much the same way as in an ordinary crystal the density is tied to the area of the unit cell. Adding or removing an electron creates a defect in this commensurate structure, and the defect costs a finite energy.[44]

The fundamental difficulty in verifying these ideas is that, up to this point, we know nothing about the form of the ground-state wave function for filling factors different from $\frac{1}{M}$. What we want to calculate is the *thermodynamic gap*, defined as

$$E_g = \left(\frac{\partial E}{\partial N}\right)_{v^+} - \left(\frac{\partial E}{\partial N}\right)_{v^-}$$

$$= [E_0(N+1, B) - E_0(N, B)] - [E_0(N, B) - E_0(N-1, B)], \quad (10.151)$$

where $E_0(N, B)$ is the ground-state energy as a function of electron number and magnetic field. What we have discussed so far is the energy of quasi-hole and quasi-electron states (ε^e and ε^h), obtained by inserting or removing one quantum of magnetic flux at constant particle number. Fortunately, it turns out that this information is sufficient: the thermodynamic energy gap at filling factor $\frac{1}{M}$ is just the energy needed to create M well separated

---

[44] It is essential, in this and in the following arguments, that the extra electron be added/removed in the *bulk* of the system. By contrast, edge excitations are gapless (see Section 10.20).

(i.e., noninteracting) quasi-electrons and quasi-holes:

$$E_g = M(\varepsilon^h + \varepsilon^e).$$ (10.152)

The proof of this assertion rests on the fact that the ground-state energy per electron, when expressed in units of $\frac{e^2}{\epsilon_b \ell}$, is a function of the filling factor only:

$$E_0(N, B) = N\epsilon(\nu)\frac{e^2}{\epsilon_b \ell}.$$ (10.153)

Now, increasing the magnetic field from B to $B + \frac{B}{N}$ and, at the same time, the number of electrons from N to $N + 1$, introduces $\frac{B \mathcal{A}}{N \Phi_0} = M$ additional flux quanta while leaving the filling factor unchanged. Since the magnetic length scales as $B^{-1/2}$ we have

$$E_0\left(N + 1, B + \frac{B}{N}\right) - E_0(N, B) = \left[\left(\frac{N+1}{N}\right)^{\frac{3}{2}} - 1\right] E_0(N, B)$$

$$\simeq \frac{3}{2}\epsilon(\nu)\frac{e^2}{\epsilon_b \ell}.$$ (10.154)

Compare the above result to the change in energy that occurs when B is changed at constant N. In this case we have

$$E_0\left(N, B + \frac{B}{N}\right) - E_0(N, B) = \frac{1}{2}\epsilon(\nu)\frac{e^2}{\epsilon_b \ell} + M\varepsilon^h,$$ (10.155)

where the first term on the right-hand side is due to the variation of the global energy scale, $\frac{e^2}{\epsilon_b \ell}$, with B, and the second term arises from the fact that we must create M well-separated quasi-holes in order to accommodate the M additional units of magnetic flux (the quasi-holes are put infinitely far from one another so as not to incur an electrostatic energy cost).

The difference between Eq. (10.154) and Eq. (10.155) is the energy variation due to a change in N at constant B. Thus, we obtain

$$E_0(N + 1, B) - E_0(N, B) = \epsilon(\nu)\frac{e^2}{\epsilon_b \ell} + M\varepsilon^h.$$ (10.156)

By a completely parallel argument we also arrive at

$$E_0(N - 1, B) - E_0(N, B) = -\epsilon(\nu)\frac{e^2}{\epsilon_b \ell} + M\varepsilon^e.$$ (10.157)

The result for the thermodynamic gap, Eq. (10.152), follows immediately from the substitution of the last two equations in Eq. (10.151).

The thermodynamic gap is not directly accessible to experiments, but it is generally believed that the energy of a well separated quasi-electron/quasi-hole pair, $\frac{E_g}{M}$, is what controls the temperature dependence of the longitudinal resistivity ρ_{xx}. At low temperature, and within the quantum Hall plateaus, ρ_{xx} is expected to vary as $e^{-\frac{\Delta}{2k_B T}}$ where $\Delta = \varepsilon^e + \varepsilon^h$. Measuring the "transport gap" Δ should therefore yield the thermodynamic gap E_g via Eq. (10.152). Unfortunately, the presence of disorder and the finite thickness of the samples

ave prevented so far a "clean" comparison between theory and experiment: the activa-
on energy appears to be significantly smaller (by about a factor 2) than the theoretically
redicted value. There are also more fundamental effects, such as the interaction between
ne quasi-electron and the quasi-hole, which could, in principle, modify the value of the
activation energy.

The thermodynamic gap should not be confused with the *tunneling gap*, which measures
ne energies of an electron and a hole injected into the system before thermal equilibrium is
eached (the precise mathematical definition is given in terms of the one-electron spectral
unction in Section 10.14). Since thermal equilibration can only reduce the energy, it is
evident that the tunneling gap is always larger than the thermodynamic gap. In fact, we will
see in Section 10.14.2 that a finite tunneling gap (or, more precisely, a "pseudo-gap") is
present even in states that are thermodynamically gapless.

10.13 Neutral excitations

The low-lying *neutral* excitations of an incompressible quantum Hall liquid are intra-Landau
level versions of the magnetic excitons we examined in Section 10.6.3. They lie within the
lowest Landau level, and their energy is therefore entirely determined by the coulomb
interaction. A key difference between the intra-LLL exciton and the Kohn cyclotron mode
is that the latter has a dipolar character (i.e., it couples directly to the electric field), while
the former is quadrupolar, and thus couples to the spatial derivatives of the electric field.
This fact makes the experimental observation of intra-LLL excitons an extremely difficult
task, since their electromagnetic cross-section is reduced by a factor $(q\ell)^2$ relative to the
main cyclotron resonance.

From a theoretical point of view the main difficulty in calculating the energies of intra-LL
excitons is that neither the RPA nor the TDHFA can be applied. Instead, two different lines
of attack have been developed:

(1) The single-mode approximation of Girvin, MacDonald, and Platzman (1986), in which the wave
    function of the intra-LLL exciton is constructed from the ground-state wave function through a
    generalization of Feynman's theory of collective excitations in superfluid ^{4}He. Interestingly this
    approach can be interpreted in terms of an effective elasticity theory (Conti and Vignale, 1998).
(2) The *composite fermion approach*, in which the wave function of the intra-LL exciton is derived
    from that of the inter-LLL exciton of "composite fermions".

We first present the single mode approximation and its interpretation in terms of elasticity
theory. The composite fermions approach will be discussed in Section 10.16.

10.13.1 The single mode approximation

Shortly after the introduction of the Laughlin wave function, Girvin, MacDonald, and
Platzman (GMP) proposed the following ingenious *Ansatz* for the intra-LLL exciton state.

$$|\psi_{\vec{q}}\rangle = \hat{\bar{n}}_{\vec{q}}|\psi_0\rangle, \tag{10.158}$$

where $|\psi_0\rangle$ is the ground-state and $\hat{n}_{\vec{q}}$ denotes the density fluctuation operator *projected i* the LLL. In a second-quantized representation based on the Landau gauge we have

$$\hat{n}_{\vec{q}} = \sum_{n,n',k_y,k'_y} \langle n', k'_y | e^{-i\vec{q}\cdot\vec{r}} | n, k_y\rangle \hat{a}^\dagger_{n',k'_y} \hat{a}_{n,k_y} \qquad (10.159)$$

and

$$\hat{\tilde{n}}_{\vec{q}} = \sum_{k_y,k'_y} \langle 0, k'_y | e^{-i\vec{q}\cdot\vec{r}} | 0, k_y\rangle \hat{a}^\dagger_{0,k'_y} \hat{a}_{0,k_y}. \qquad (10.160)$$

Notice that $|\psi_{\vec{q}}\rangle$ is *orthogonal* to $|\psi_0\rangle$ because $\langle\psi_0|\hat{n}_{\vec{q}}|\psi_0\rangle = 0$ for $\vec{q} \neq 0$.

The *Ansatz* (10.158) was inspired, in part, by the Feynman theory of ^{4}He in which the low-lying excited state of wave vector $\vec{q}$ is written as $\hat{n}_{\vec{q}}|\psi_0\rangle$ (Feynman, 1953). The rationale for this choice is that all the favorable correlations responsible for the low energy of the ground-state are still contained in $|\psi_0\rangle$, while the application of the operator $\hat{n}_{\vec{q}}$ assigns higher probability to configurations in which the particle density oscillates in space with wave vector $\vec{q}$. The average density remains, nevertheless, uniform, because the overall phase of the periodic density modulation is completely undetermined. Eq. (10.158) is essentially the Feynman state projected in the LLL. The projection removes higher Landau level components of the wave function and is essential to get a low energy at high magnetic field.

We gain additional insight into the meaning of Eq. (10.158) by considering the first quantized form of the projected density fluctuation operator. According to the general technique described in Appendix 24 the projection of an operator in the LLL is done by first writing that operator in terms of z_i and z_i^*, then ordering the expression in such a way that the z_i^*'s precede the z_i's, and finally making the replacement $z_i^* \to 2\ell^2\frac{\partial}{\partial z_i}$. In the case under consideration this procedure yields

$$\hat{\tilde{n}}_{\vec{q}} = \overline{\sum_{j=1}^N e^{-i\vec{q}\cdot\hat{\vec{r}}_j}}$$

$$= \sum_{j=1}^N \overline{e^{-iqz_j^*/2}e^{-iq^*z_j/2}}$$

$$= \sum_{j=1}^N e^{-iq\ell^2\frac{\partial}{\partial z_j}} e^{-iq^*z_j/2}$$

$$= e^{-q^2\ell^2/2} \sum_{j=1}^N e^{-iq^*z_j/2}e^{-iq\ell^2\frac{\partial}{\partial z_j}}, \qquad (10.161)$$

where we have introduced the complex quantity $q \equiv q_x + iq_y$. Thus we have

$$\hat{\tilde{n}}_{\vec{q}}\psi_M(z_1,\ldots,z_N) = \sum_{j=1}^N e^{-\frac{q^2\ell^2}{2}} e^{-i\frac{q^*z_j}{2}}$$

$$\times \psi_M(z_1,\ldots,z_{j-1},z_j - iq\ell^2, z_{j+1}\ldots,z_N). \qquad (10.162)$$

t is evident from this expression that $\hat{n}_{\vec{q}}$ acts in part as a translation operator, shifting the position of each particle by a distance q in the direction perpendicular to $\vec{q}$, and superimposing he resulting N configurations with weights $e^{-iq^*z_j/2}$. Recalling that the zeroes of the Laughlin wave function coincide with the positions of the electrons, we see that the GMP wavefunction has some of its zeroes displaced a distance q from the particles. This accounts for the increase in energy. The coherent displacement of electrons from z_i to $z_i - iq\ell^2$ is what replaces the coherent superposition of electron–hole pairs in an inter-Landau level exciton.

We now proceed to the calculation of the energy of the GMP wave function

$$E_{\vec{q}} = \frac{\langle \psi_{\vec{q}} | \hat{H} | \psi_{\vec{q}} \rangle}{\langle \psi_{\vec{q}} | \psi_{\vec{q}} \rangle} . \tag{10.163}$$

We note that the numerator can be rewritten as

$$E_0 \langle \psi_0 | \hat{n}_{\vec{q}}^\dagger \hat{n}_{\vec{q}} | \psi_0 \rangle + \frac{1}{2} \langle \psi_0 | [\hat{n}_{\vec{q}}^\dagger, [\hat{H}, \hat{n}_{\vec{q}}]] | \psi_0 \rangle , \tag{10.164}$$

where $E_0 = \langle \psi_0 | \hat{H} | \psi_0 \rangle$ is the ground-state energy. (To derive this result we have made use of the inversion symmetry of the ground-state and of the identity $\hat{n}_{-\vec{q}} = \hat{n}_{\vec{q}}^\dagger$.) Substituting Eq. (10.164) in Eq. (10.163) we get

$$E_{\vec{q}} = E_0 + \frac{\bar{f}(\vec{q})}{\bar{S}(\vec{q})} \tag{10.165}$$

where we have defined

$$\bar{f}(q) \equiv \frac{1}{2N} \langle \psi_0 | [\hat{n}_{\vec{q}}^\dagger, [\hat{H}, \hat{n}_{\vec{q}}]] | \psi_0 \rangle \tag{10.166}$$

and

$$\bar{S}(q) \equiv \frac{1}{N} \langle \psi_0 | \hat{n}_{\vec{q}}^\dagger \hat{n}_{\vec{q}} | \psi_0 \rangle . \tag{10.167}$$

The *projected structure factor* $\bar{S}(q)$ differs from the ordinary structure factor,

$$S(q) = \frac{1}{N} \langle \psi_0 | \hat{n}_{\vec{q}}^\dagger \hat{n}_{\vec{q}} | \psi_0 \rangle , \tag{10.168}$$

in that the unprojected density fluctuation operator $\hat{n}_{\vec{q}}$ couples different Landau levels. In Appendix 24 we show, by the application of formal projection techniques, that

$$\bar{S}(q) = S(q) - S_1(q), \tag{10.169}$$

where $S_1(\vec{q}) = 1 - e^{-q^2\ell^2/2}$ is the structure factor of the full LLL (see Section 10.4.5).[45]

Up to this point we have not specified the ground-state, and, in fact, any reasonable choice that lies entirely within the LLL would be admissible. Let us now assume that $\nu = \frac{1}{M}$ and

---

[45] This result can be quickly obtained by noting that the difference between $S(q)$ and $\bar{S}(q)$ is entirely due to the contribution of virtual transitions from the lowest to the higher Landau levels. But these are the *only* transitions that contribute to $S(q)$, when the Landau level is full, and must therefore add up to $S_1(q)$.

that the ground-state is described by the Laughlin wave function ψ_M. The structure factor of this state is calculated by combining Eq. (10.131) with Eq. (10.130). It then follows from Eqs. (10.142) and (10.169) that, for small q,

$$\bar{S}(q) \overset{q \to 0}{\to} \frac{M-1}{8}(q\ell)^4 , \qquad (10.170)$$

while, for large q, one can easily show that (see Exercise 22)

$$\bar{S}(q) \overset{q \to \infty}{\to} \frac{M-1}{M} e^{-\frac{q^2\ell^2}{4}} . \qquad (10.171)$$

Let us now turn to the calculation of $\bar{f}(q)$. The corresponding *unprojected* quantity, $f(q)$, is closely related to the f-sum rule (see Eqs. (3.91) and (3.141)), and quickly works out to be $\frac{\hbar^2 q^2}{2m_b}$. Since, according to Kohn's theorem, the f-sum rule is exhausted (to order q^2) by the cyclotron mode, we expect that a projection in the LLL will leave us with a quantity that tends to zero faster than q^2. Indeed, we can show that $\bar{f}(q)$ vanishes as q^4 so that, by Eqs. (10.165) and (10.170), the excitation energy $E_{\vec{q}} - E_0$ is seen to approach a finite limit for $q \to 0$. The calculation is done as follows. First, we notice that, for the present purposes, the full hamiltonian $\hat{H}$ can be replaced by the projected hamiltonian

$$\begin{aligned}
\bar{H} &= \frac{1}{2\mathcal{A}} \sum_{\vec{q}} v_q \left(\overline{\hat{n}_{\vec{q}}^\dagger \hat{n}_{\vec{q}}} - N \right) \\
&= \frac{1}{2\mathcal{A}} \sum_{\vec{q}} v_q \left(\hat{\bar{n}}_{\vec{q}}^\dagger \hat{\bar{n}}_{\vec{q}} - N e^{-q^2\ell^2/2} \right) ,
\end{aligned} \qquad (10.172)$$

where we have made use of Eq (10.169) in the last line. Next, we calculate the double commutator of $\bar{H}$ with $\hat{\bar{n}}_{\vec{q}}^\dagger$ and $\hat{\bar{n}}_{\vec{q}}$. This is done by repeated application of the fundamental commutator

$$[\hat{\bar{n}}_{\vec{k}}, \hat{\bar{n}}_{\vec{q}}] = \left(e^{\frac{k^* q \ell^2}{2}} - e^{\frac{k q^* \ell^2}{2}} \right) \hat{\bar{n}}_{\vec{k}+\vec{q}} , \qquad (10.173)$$

which can be derived directly from Eq. (10.161). The final result (obtained after a considerable amount of algebra) is

$$\bar{f}(q) = 2 \frac{e^{-\frac{q^2\ell^2}{2}}}{\mathcal{A}} \sum_{\vec{k}} \sin^2 \left(\frac{(\vec{q} \times \vec{k}\ell^2) \cdot \vec{e}_z}{2} \right) \bar{S}(k) \left[v_{|\vec{k}-\vec{q}|} e^{\vec{k} \cdot \vec{q}\ell^2} e^{-\frac{q^2\ell^2}{2}} - v_k \right] . \qquad (10.174)$$

It is evident from this expression that $\bar{f}(q) \sim q^4$ at small q. In fact, with little more effort one can show that

$$\bar{f}(q) \overset{q \to 0}{\to} \frac{e^2}{32\ell} \left[\int_0^\infty dx \, \bar{S}\left(\frac{x}{\ell}\right)(x^4 - 2x^2 - 1) \right] \cdot (q\ell)^4 . \qquad (10.175)$$

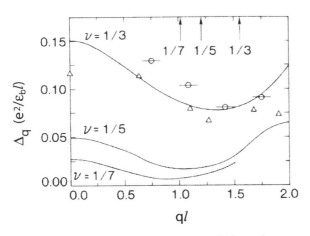

Fig. 10.17. Exciton dispersion curves $\Delta_{\vec{q}} \equiv E_{\vec{q}} - E_M$ at $v = \frac{1}{3}, \frac{1}{5}$ and $\frac{1}{7}$ calculated from the GMP theory. Circles (with error bars) and triangles are the results of exact diagonalizations for 7- and 6-particle systems. (Haldane and Rezayi, 1985) at $v = \frac{1}{3}$. The vertical arrows indicate the magnitude of the primitive reciprocal lattice vector of the corresponding Wigner crystal. From Girvin, MacDonald, and Platzman, 1986.

In the opposite limit of large wave vector, use of Eqs. (10.171) and (10.169) yields

$$\bar{f}(q) \overset{q \to \infty}{\to} \frac{1}{A} \sum_{\vec{k}} v_{\vec{k}} e^{-\frac{k^2 \ell^2}{2}} \{[1 - S(k)] - v[1 - S_1(k)]\}$$

$$= -2e^{-\frac{q^2 \ell^2}{2}} \frac{E_M - v E_1}{N}, \tag{10.176}$$

where E_M is the ground-state energy at filling factor $v = \frac{1}{M}$.

The exciton dispersion curves evaluated from Eqs. (10.165), (10.169) and (10.174) at $v = \frac{1}{3}, \frac{1}{5}$ and $\frac{1}{7}$ are plotted in Fig. 10.17. The essential features of these curves are (i) the existence of an energy gap, Δ_0, at $q \to 0$,[46] (ii) the minimum in the dispersion at $q \sim \frac{1}{\ell}$, and (iii) the constant limit for $q \to \infty$. We now discuss these features.

Concerning (i), it must be kept in mind that the GMP theory, being based on a variational *Ansatz*, provides only an *upper bound* to the energy of the lowest excited state. Strictly speaking, one cannot infer the existence of a gap from such a bound. In fact, from our previous discussion of the behavior of $\bar{S}(q)$, it should be clear that the GMP theory would predict a gap for *any* liquid ground-state, in contrast with both theory and expectation.

What really *is* rigorous in the GMP theory is the statement that $E_{\vec{q}} - E_0$ is the *average* excitation energy at wave vector $\vec{q}$:

$$E_{\vec{q}} - E_0 = \frac{\hbar \int_0^\infty \omega \bar{S}(q, \omega) d\omega}{\int_0^\infty \bar{S}(q, \omega) d\omega} \tag{10.177}$$

---

[46] To avoid potential confusion, keep in mind that $q \to 0$ here means $q \ll 1/\ell$, but $q \gg 1/L$, where L is the size of the system. This condition excludes gapless edge excitations.

where $\bar{S}(q, \omega)$ is the projected dynamical structure factor, defined as

$$\bar{S}(q, \omega) = \sum_\alpha |\langle \psi_\alpha | \hat{\bar{n}}_{\vec{q}} | \psi_0 \rangle|^2 \delta(\omega - \omega_\alpha), \tag{10.178}$$

and the sum runs over all the excited states, labeled by α (Exercise 17). Thus, the GMP theory would be exact (assuming, of course, that the ground-state is known) if all the spectral weight of neutral excitations at wave vector $\vec{q}$ were concentrated in a single collective mode. Such behavior is characteristic of strongly correlated elastic networks, as well as bosonic superfluids: in both cases there are no single particle excitations, and the dynamics of the system is completely described in terms of collective modes. What makes the quantum Hall liquid truly unique is the fact that the single mode, which results from the collectivization of the electrons, is "gapped" for $q \to 0$. We shall return to this point in the next section. Notice that the estimated gap at $q = 0$ and $v = \frac{1}{3}$ is about $0.15 \frac{e^2}{\epsilon_b \ell}$, which is considerably higher than the energy needed to create well separated quasi-electron and quasi-hole excitations $\left[(0.073 + 0.027) \frac{e^2}{\epsilon_b \ell} \simeq 0.10 \frac{e^2}{\epsilon_b \ell} \right]$, but still about one half the thermodynamic energy gap $\left(0.30 \frac{e^2}{\epsilon_b \ell} \right)$.

The second important feature of the results plotted in Fig. 10.17 is the minimum in $E_{\vec{q}}$ at $q = q_{min} \approx \frac{1}{\ell}$. This is very similar to the "roton minimum" in the phonon spectrum of liquid ^{4}He. The depth of the minimum increases with decreasing density, and its energy approaches zero, signaling an instability of the ground-state. This is consistent with the expected transition from electron liquid to Wigner crystal at sufficiently low density. Indeed, the GMP theory relates the minimum in the exciton dispersion to a peak in the static structure factor, which is usually interpreted as a signature of incipient crystallization.

Finally a few comments on the large-q behavior of the exciton dispersion are in order. This is easily obtained by combining Eqs. (10.165), (10.171), and (10.176). The result is

$$\lim_{q \to \infty} (E_{\vec{q}} - E_0) = \frac{2|E_0 - v E_1|}{N(1 - v)}, \tag{10.179}$$

where E_1 is the ground-state energy at filling factor 1. There is however no reason to expect the single mode approximation to be accurate for large q. A more resonable approach in this limit would be to construct the exciton directly as a bound state of a quasi-electron and a quasi-hole in close analogy to what was done in the inter-Landau level case. The characteristic separation, Δr, between the quasi-electron and the quasi-hole is now related to exciton wave vector by $\Delta r = q\ell^2 M$ (see discussion at the end of Section 10.6.3). The energy of the exciton is therefore expected to go as

$$E_{\vec{q}} - E_M = \varepsilon^e + \varepsilon^h - \frac{1}{M^3 q \ell} \frac{e^2}{\epsilon_b \ell}, \tag{10.180}$$

for large q. This formula, evaluated at $q = q_{min}$, is in reasonable agreement with the energy of the roton minimum, that is, about $0.09 \frac{e^2}{\epsilon_b \ell}$ at $v = \frac{1}{3}$.

The above simple picture breaks down at smaller values of q, as indicated by the change in the sign of the derivative of the dispersion curve at $q = q_{min}$. Small-q excitations can be ualitatively described as "bi-exciton" complexes, consisting of two quasi-electron/quasi-ole pairs which are bound together in such a way that the net dipole moment vanishes. This icture is supported by the fact that the optical absorption cross section for small q goes s $(q\ell)^4$ – a behavior characteristic of quadrupolar excitations. The smallness of the cross ection makes the observation of intra-LLL excitons a challenging task. In order to zero-in on the quadrupole moment one must resort to techniques such as inelastic light scattering. This has been done by Pinczuk *et al.* (1993, 1998): the measured excitation energies agree quite well with the GMP theory (after allowing for the finite thickness of the experimental amples) and even better with the more recent composite fermion theory. All in all, the success of the GMP theory is superior to that of the Feynman theory for ^{4}He.[47]

10.13.2 Effective elasticity theory

In Section 5.9 we showed that the dispersion of the plasmon mode at zero magnetic field could be correctly calculated, to order q^2, by treating the electron liquid as a classical elastic medium endowed with a non-vanishing dynamic shear modulus. In the present section we apply the same approach to the description of intra-Landau level collective modes. The main motivation for doing this is that any set of classical linear equations of motion, when quantized, leads to the appearance of Bose particles, in much the same way as the equations of classical lattice dynamics lead to phonons. This opens the way to the application of powerful bosonization techniques for the calculation of single-particle properties.

Let us then treat our two-dimensional electron liquid as a homogeneous elastic medium of uniform equilibrium density n_0, subjected to a perpendicular magnetic field $-B\vec{e}_z$. Deviations from equilibrium are described by a two-dimensional displacement field $\vec{u}(\vec{r}, t)$ in the following manner: a small volume element that, at equilibrium, was located at $\vec{r}$, will be located at $\vec{r} + \vec{u}(\vec{r}, t)$ in the state described by the displacement field $\vec{u}(\vec{r}, t)$. The particle current density in this state is given by

$$\vec{j}(\vec{r}, t) = n_0 \frac{\partial \vec{u}(\vec{r}, t)}{\partial t} \tag{10.181}$$

and use of the continuity equation immediately shows that the density fluctuation is given by

$$n(\vec{r}, t) - n_0 = -n_0 \vec{\nabla} \cdot \vec{u}(\vec{r}, t) . \tag{10.182}$$

According to classical two-dimensional elasticity theory [see Section 5.9.1, Eq. (5.233)] the equation of motion for $\vec{u}(\vec{r}, t)$ is

$$m_b n_0 \frac{\partial^2 \vec{u}(\vec{r}, t)}{\partial t^2} = B\vec{\nabla}\left[\vec{\nabla} \cdot \vec{u}(\vec{r}, t)\right] + S\nabla^2 \vec{u}(\vec{r}, t)$$

$$- \frac{eB}{c} n_0 \vec{e}_z \times \frac{\partial \vec{u}(\vec{r}, t)}{\partial t} + n_0^2 e^2 \vec{\nabla} \int \frac{\vec{\nabla}' \cdot \vec{u}(\vec{r}', t)}{|\vec{r} - \vec{r}'|} d\vec{r}' , \tag{10.183}$$

---

[47] GMP attribute this success in part to the fact that the current density associated with a stationary exciton wave-packet in the LLL is automatically divergence-free, and therefore does not violate the continuity equation.

where $\mathcal{B}$ is the bulk modulus and $\mathcal{S}$ is the shear modulus. The last two terms on the right hand side of this equation represent, respectively, the Lorentz force and the force due to the electric field generated by the charge density. Notice that there are no dissipative terms – an approximation that corresponds to the single-mode approximation in the GMP theory.

Fourier transforming Eq. (10.183) with respect to $\vec{r}$ and t and resolving the displacement field into its longitudinal (u_L) and transverse (u_T) components with respect to the direction of $\vec{q}$, we arrive at the hermitian eigenvalue problem

$$\begin{pmatrix} -m_b n_0 \omega^2 + \left(n_0^2 v_q + \mathcal{B} + \mathcal{S}\right)q^2 & im_b n_0 \omega \omega_c \\ -im_b n_0 \omega \omega_c & -m_b n_0 \omega^2 + \mathcal{S}q^2 \end{pmatrix} \cdot \begin{pmatrix} u_L \\ u_T \end{pmatrix} = 0 , \quad (10.184)$$

which has non-trivial solutions only when the determinant of the 2×2 matrix is zero. The ensuing algebraic equation is readily solved and yields two distinct collective modes with frequencies

$$\omega_\pm(q) = \frac{1}{2}\left\{ \sqrt{[\omega_L(q) + \omega_T(q)]^2 + \omega_c^2} \pm \sqrt{[\omega_L(q) - \omega_T(q)]^2 + \omega_c^2} \right\}, \quad (10.185)$$

where

$$\omega_L(q) = \sqrt{\frac{n_0^2 v_q + \mathcal{B} + \mathcal{S}}{m_b n_0}}\, q \qquad (10.186)$$

and

$$\omega_T(q) = \sqrt{\frac{\mathcal{S}}{m_b n_0}}\, q . \qquad (10.187)$$

Here $\omega_L(q)$ is the frequency of a purely longitudinal wave in the absence of the magnetic field, while $\omega_T(q)$ is the frequency of a purely transverse wave under the same conditions. The magnetic field mixes the longitudinal and transverse modes, leading to hybrid eigenmodes whose frequencies are given by Eq. (10.185). Notice that $\omega_+(q)$ tends to ω_c for $q \to 0$, in agreement with Kohn's theorem, and its dispersion is linear in q for small q, in agreement with the calculations of Section 10.6.3.[48]

Our main interest is in the low-frequency solution, $\omega_-(q)$, which tends to zero in the limit of high magnetic field. Letting $\omega_c \to \infty$ in Eq. (10.185) – a step that corresponds to the projection in the lowest Landau level in the GMP theory – we obtain

$$\omega_-(q) = \frac{\omega_L(q)\omega_T(q)}{\omega_c} = \sqrt{\mathcal{S}\left[v_q + \frac{\mathcal{B} + \mathcal{S}}{n_0^2}\right]\frac{(q\ell)^2}{\hbar}} , \qquad (10.188)$$

which is *independent of the electron mass*, and, inspite of appearance, of $\hbar$ too.[49] The

---

[48] We assume that both $\mathcal{B}$ and $\mathcal{S}$ have finite values at a frequency $\omega \sim \omega_c$.

orresponding eigenfunction is

$$\vec{u}_-(\vec{q}) = \begin{pmatrix} -iq\ell\sqrt{\dfrac{S}{2\hbar\omega_-(q)n_0}} \\ \dfrac{1}{q\ell}\sqrt{\dfrac{\hbar\omega_-(q)n_0}{2S}} \end{pmatrix}, \tag{10.189}$$

where the upper and lower components of the column vector are the projections of $\vec{u}_-(\vec{q})$ along $\vec{q}$ and $\vec{e}_z \times \vec{q}$ respectively.

Because it does not depend on the electron mass, Eq. (10.188) is an excellent candidate or the frequency of the intra-Landau level exciton. There is a serious difficulty, however. According to Eq. (10.188) the energy of the exciton would vanish as $q^{3/2}$ for $q \to 0$, instead of tending to a constant as it should. The origin of the trouble is that we have been too cavalier in assuming that the bulk modulus is independent of q in the frequency range of intra-Landau level excitations. After all, we know that the Laughlin liquid is *incompressible*, and this implies that its static bulk modulus tends to infinity for q tending to zero. How strong is this divergence? Eq. (10.188) shows that $\mathcal{B}$ must diverge as $\frac{1}{q^4}$ if the exciton frequency is to remain finite for $q \to 0$. Let us then set

$$\mathcal{B}(q) = \frac{\alpha n}{(q\ell)^4} \tag{10.190}$$

where α is an as yet undetermined positive constant, with the dimensions of an energy. Substituting (10.190) in (10.188) and taking the $q \to 0$ limit we obtain the following relation between α, the $q = 0$ gap Δ_0 (defined below), and the shear modulus:

$$\Delta_0 \equiv \lim_{q\to 0} \hbar\omega_-(q) = \sqrt{\frac{\alpha S}{n_0}}. \tag{10.191}$$

Notice that the v_q term in Eq. (10.188) has become irrelevant, being dominated, in the $q \to 0$ limit, by the more strongly diverging bulk modulus.

To verify the consistency of the chosen form for $\mathcal{B}(q)$, let us calculate the proper density–density response function $\tilde{\chi}_{nn}(q, \omega)$ by solving the equation of motion for the displacement field in the presence of an external potential $V(q, \omega)$:

$$\begin{pmatrix} -m_b n_0 \omega^2 + \left(n_0^2 v_q + \mathcal{B} + S\right)q^2 & im_b n_0 \omega\omega_c \\ -im_b n_0 \omega\omega_c & -m_b n_0\omega^2 + Sq^2 \end{pmatrix}\cdot\begin{pmatrix} u_L \\ u_T \end{pmatrix} = -iqn_0\begin{pmatrix} V \\ 0 \end{pmatrix}. \tag{10.192}$$

The density response is given by $\delta n(\vec{q}, \omega) = -iqn_0 u_L(q, \omega)$. If all goes well we should find that the compressibility sum rule, Eq. (5.23), is satisfied, i.e.,

$$\lim_{q\to 0} \tilde{\chi}_{nn}(q, 0) = -\frac{n_0^2}{\mathcal{B}(q)}.$$

[49] In fact, we could have obtained this result directly by setting $m_b = 0$ in Eq. (10.183).

Indeed, solving Eq. (10.192) in the limit $\omega_c \to \infty$ (or $m_b \to 0$) we obtain

$$\delta n(\vec{q}, \omega) = \tilde{\chi}_{nn}(q, \omega) V(\vec{q}, \omega) \qquad (10.19?)$$

with

$$\tilde{\chi}_{nn}(q, \omega) = \frac{S(ql)^4}{\hbar^2 \left[(\omega + i\eta)^2 - \omega_-^2(q)\right]}. \qquad (10.194)$$

Setting $\omega = 0$ first, and then letting q tend to zero we get

$$\tilde{\chi}_{nn}(q, 0) \overset{q \to 0}{\to} \frac{S(q\ell)^4}{\Delta_0^2}, \qquad (10.195)$$

which coincides with Eq. (10.190) if Δ_0 is related to α by Eq. (10.191).

Next, we apply the fluctuation-dissipation theorem

$$\bar{S}(q, \omega) = -\frac{\hbar}{\pi n_0} \Im m\, \tilde{\chi}_{nn}(q, \omega)$$

$$= \frac{S(q\ell)^4}{2n_0\hbar\omega_-(q)} \delta\left(\omega - \omega_-(q)\right) \qquad (10.196)$$

to calculate the static structure factor

$$\bar{S}(q) = \int_0^\infty \bar{S}(q, \omega) d\omega$$

$$= \frac{S(q\ell)^4}{2n_0\hbar\omega_-(q)}. \qquad (10.197)$$

It should be noted that the spectral function $\Im m\, \tilde{\chi}_{nn}(q, \omega)$ does not include contributions from higher Landau levels, which have been sent to infinite energy by our limiting procedure: hence $\bar{S}(q)$ must be identified with the projected structure factor of the GMP theory. Substituting in the above expression the small-q limit of $\bar{S}(q)$, given by Eq. (10.170) with $M = \frac{1}{\nu}$, we obtain a direct relationship between the $q = 0$ gap and the shear modulus:

$$\frac{\bar{S}(q)}{(q\ell)^4} = \frac{1 - \nu}{8\nu} = \frac{S}{2n_0\Delta_0}. \qquad (10.198)$$

In fact, this can be rewritten in a form that is very similar to Eq. (10.165) of the GMP theory, namely,

$$\Delta_0 = \lim_{q \to 0} \frac{S(q\ell)^4}{2n_0\bar{S}(q)}. \qquad (10.199)$$

Thus, we see that the projected f-sum $\bar{f}(q)$ of the GMP theory corresponds to the quantity $\frac{S(q\ell)^4}{2n_0}$ in the effective elasticity theory.

The above discussion shows that the dynamics of an incompressible quantum Hall liquid is equivalent to that of an unconventional elastic medium in which the bulk modulus diverges as $\frac{1}{q^4}$ for $q \to 0$ while the shear modulus remains finite. The consistency of this description rests squarely on the existence of a gap in the density fluctuation excitation spectrum at

$= 0$. It is the presence of the gap that allows us to neglect dissipation (since there are no excitations for the collective mode to decay into) and, at the same time, to treat the liquid as an elastic solid characterized by a finite shear modulus.

It is interesting to observe that the eigenfunction (10.189) describes a motion that is almost purely transversal for $q \to 0$: the transverse component of $\vec{u}_-(\vec{q})$ diverges as $\frac{1}{q}$ while the longitudinal component vanishes. The trajectory of a small volume element is therefore a very elongated ellipse perpendicular to $\vec{q}$. A localized excitation can be constructed by superimposing plane-wave solutions with wave vectors $q < \ell^{-1}$: we see that the displacement field for such an excitation is given by $\int d\vec{q} \, \frac{\vec{e}_z \times \vec{q}}{q^2} e^{i\vec{q}\cdot\vec{r}} \sim \frac{\vec{e}_z \times \vec{r}}{2\pi r^2}$, which is the well known shape of a *vortex*.

Eq. (10.199) can be taken as a starting point for approximate evaluations of the gap (Conti and Vignale, 1998). For example, approximating S as the shear modulus of a classical two-dimensional Wigner crystal of average density n_0, namely, $S = 0.09775 n_0 \frac{e^2}{\epsilon_b \ell} \nu^{1/2}$ (Bonsall and Maradudin, 1977), one obtains

$$\Delta_0(\nu) = 0.391 \frac{\nu^{\frac{3}{2}}}{1 - \nu} \frac{e^2}{\epsilon_b \ell} , \tag{10.200}$$

which gives $\Delta_0(\frac{1}{3}) = 0.1129$, $\Delta_0(\frac{1}{5}) = 0.0437$, and $\Delta_0(\frac{1}{7}) = 0.0246$ (always in units of $\frac{e^2}{\epsilon_b \ell}$) in excellent agreement with exact diagonalizations.

The theory outlined in this section is obviously applicable to the long-wavelength dynamics of the classical Wigner crystal, where S and B are finite and B is negative: the dispersion of the low-lying magneto-phonon mode is proportional to $q^{3/2}$. The theory becomes problematic in compressible liquid states, which *do not* exhibit a gap in the excitation spectrum (an important example will be discussed in Section 10.17). In this case, a naive application of the theory with finite values of B and S, would incorrectly predict a collective mode with a $q^{3/2}$ dispersion and a projected structure factor $\bar{S}(q) \sim q^{5/2}$, in conflict with the correct behavior $\bar{S}(q) \sim q^4$. The problem is that S cannot be assumed to be independent of q in these states. Rather, as $q \to 0$, S becomes imaginary, leading to an overdamped diffusive mode which has the correct spectral strength.

10.13.3 Bosonization

Our discussion of elasticity theory has been purely classical so far, but it is easy to see that the classical displacement field can be quantized in terms of Bose particles analogous to phonons in a crystal (Conti and Vignale, 1998). First of all we note that the classical equation of motion (10.184) for $\vec{u}(\vec{r}, t)$ can be derived from the lagrangian

$$L[\vec{u}, \partial\vec{u}/\partial t] = n_0 \int d\vec{r} \left\{ \frac{m_b}{2} \left(\frac{\partial u(\vec{r}, t)}{\partial t} \right)^2 + \frac{eB}{c} u_x(\vec{r}, t) \frac{\partial u_y(\vec{r}, t)}{\partial t} - E_p(\vec{u}, \partial\vec{u}) \right\}$$

$$\tag{10.201}$$

where $E_p(\vec{u}, \partial\vec{u})$ is the elastic potential energy density – a function of the displacement field and its derivatives. The limit of large magnetic field is obtained by setting $m_b = 0$. In this limit, the canonical momentum conjugate to $u_y(\vec{r}, t)$ is

$$\frac{\delta L}{\delta(\partial\vec{u}_y(\vec{r}, t)/\partial t)} = n_0 \frac{\hbar}{\ell^2} u_x(\vec{r}, t) . \tag{10.202}$$

The theory is quantized by imposing the commutation relation

$$[\hat{u}_x(\vec{r}, t), \hat{u}_y(\vec{r}', t)] = -i\frac{\ell^2}{n_0}\delta(\vec{r} - \vec{r}') \tag{10.203}$$

between the components of the displacement field *operator*. Notice that, in view of Eq. (10.182), this equation is consistent with the long-wavelength limit of the commutator between projected density fluctuations, Eq. (10.173).

Not surprisingly, the hamiltonian constructed from the above lagrangian via the canonical procedure contains only the elastic potential energy. Its explicit form is

$$\hat{H} = \frac{1}{2\mathcal{A}}\sum_{\vec{q}} q^2 \left\{ \hat{u}_{-\vec{q},L} \left[n_0^2 v_q + \mathcal{B}(q) + \mathcal{S} \right] \hat{u}_{\vec{q},L} + \hat{u}_{-\vec{q},T}\mathcal{S}\hat{u}_{\vec{q},T} \right\} , \tag{10.204}$$

where $\hat{u}_{\vec{q},L}$ and $\hat{u}_{\vec{q},T}$ are, respectively, the longitudinal and the transverse component of the Fourier-transformed displacement field $\hat{\vec{u}}(\vec{q})$. Since $\hat{H}$ is a quadratic function of $\hat{\vec{u}}$, and the commutator (10.203) is just a number, we see that the quantum equation of motion for $\hat{\vec{u}}$ is linear, and coincides in form with the classical one.

The commutation relation (10.203) implies that the displacement field operator can be represented in terms of standard boson operators $\hat{b}_{\vec{q}}$ and $\hat{b}_{\vec{q}}^\dagger$, which satisfy the commutation relation $[\hat{b}_{\vec{q}}, \hat{b}_{\vec{q}'}^\dagger] = \delta_{\vec{q},\vec{q}'}$. Indeed, it is not difficult to show that

$$\hat{\vec{u}}(\vec{r}) = \frac{\ell}{\sqrt{n_0\mathcal{A}}}\sum_{\vec{q}} \left(\hat{b}_{\vec{q}}\, \vec{u}_-(\vec{q})e^{i\vec{q}\cdot\vec{r}} + \hat{b}_{\vec{q}}^\dagger\, \vec{u}_-^*(\vec{q})e^{-i\vec{q}\cdot\vec{r}} \right) , \tag{10.205}$$

satisfies the commutation relations (10.203), if $\vec{u}_-(\vec{q})$ are the classical eigenfunctions normalized as in Eq. (10.189). One can also show, by making use of the completeness properties of the eigenfunctions (10.189), that the hamiltonian (10.204) can be rewritten as

$$\hat{H} = \sum_{\vec{q}} \hbar\omega_-(q) \left(\hat{b}_{\vec{q}}^\dagger\hat{b}_{\vec{q}} + \frac{1}{2} \right) , \tag{10.206}$$

where $\omega_-(q)$ is the frequency of the intra-LLL exciton. The interested reader can explore this construction in detail in Exercise 27.

We have thus accomplished the desired "bosonization" of the hamiltonian. It is now an easy matter to calculate the correlation functions of operators that can be expressed solely terms of the $\hat{b}$'s. For example, the projected density fluctuation operator is

$$\hat{n}_{\vec{q}} = -iqn_0\hat{u}_L(\vec{q})$$

$$= -q^2\ell^2\sqrt{\frac{S\mathcal{A}}{2\hbar\omega_-(q)}}\left(\hat{b}_{\vec{q}} + \hat{b}^\dagger_{-\vec{q}}\right). \qquad (10.207)$$

The reader may want to verify that the quantum mechanical density–density response function coincides with the classical one, which is given by Eq. (10.194).

Expressing the electron field operators in terms of bosonic displacement field is an outstanding challenge. In the next section, we will present an approximate construction of the field operator, inspired by the theory of the one-dimensional Luttinger liquid, where such a construction can be done without approximations. We will then apply it to a heuristic treatment of the one-electron spectral function.

10.14 The spectral function

In this section we examine the spectral function for addition or removal of one electron in the LLL orbital of angular momentum 0.[50] In keeping with the notation of Chapter 8 we split the spectral function into two parts,

$$A_>(\omega) = \sum_n |\langle n, N+1|\hat{a}_0^\dagger|0, N\rangle|^2\delta\left(\omega - \omega_{n0} - \frac{\mu_+}{\hbar}\right) \qquad (10.208)$$

for particle addition and

$$A_<(\omega) = \sum_n |\langle n, N-1|\hat{a}_0|0, N\rangle|^2\delta\left(\omega + \omega_{n0} - \frac{\mu_-}{\hbar}\right) \qquad (10.209)$$

for particle removal. Here $|0, N\rangle$ is the ground-state of the N-particle system and the sums run over the excited states $|n, N \pm 1\rangle$ with excitation energies $\hbar\omega_{n0} \pm \mu_\pm$. μ_+ and μ_- are the chemical potentials for adding or removing an electron.

The calculation of the spectral function in a partially filled Landau level is particularly difficult, due to the failure of standard perturbation theory. In this section we first present an exact calculation of the *average* one-electron and one-hole excitation energies (Haussman, Mori, and MacDonald, 1996), and then study the *shape* of the spectrum within the framework of the bosonization approach of the preceding section (Conti and Vignale, 1998).

10.14.1 An exact sum rule

The average one-electron removal energy is given by

$$\epsilon_- = -\frac{\int_{-\infty}^{\infty} \hbar\omega A_<(\omega)d\omega}{\int_{-\infty}^{\infty} A_<(\omega)d\omega}$$

$$= -\mu_- + \frac{\langle 0, N|\hat{a}_0^\dagger[\hat{H}, \hat{a}_0]|0, N\rangle}{\nu}, \qquad (10.210)$$

---

[50] This is a gaussian (coherent state) centered at the origin of the coordinates.

where v is the filling factor. Similarly, the average addition energy is given by

$$\epsilon_+ = \frac{\int_{-\infty}^{\infty} \hbar\omega A_>(\omega)d\omega}{\int_{-\infty}^{\infty} A_>(\omega)d\omega}$$

$$= \mu_+ + \frac{\langle 0, N | \hat{a}_0[\hat{H}, \hat{a}_0^\dagger]|0, N \rangle}{1 - v} . \tag{10.21}$$

The calculation of the averages is simple at high magnetic field, because only the intra Landau level part of the Hamiltonian needs to be considered:

$$\hat{H} \simeq \tilde{H} = \frac{1}{2} \sum_{m_1,m_2,m_3,m_4} V_{m_1,m_2,m_3,m_4} \hat{a}_{m_1}^\dagger \hat{a}_{m_2}^\dagger \hat{a}_{m_3} \hat{a}_{m_4} , \tag{10.212}$$

where V_{m_1,m_2,m_3,m_4} is the matrix elements of the coulomb interaction between LLL orbital in the symmetric gauge. Making use of this expression, it is straightforward to show that

$$\hat{a}_0[\hat{H}, \hat{a}_0] = -\sum_{m_2,m_3,m_4} V_{0,m_2,m_3,m_4} \hat{a}_0^\dagger \hat{a}_{m_2}^\dagger \hat{a}_{m_3} \hat{a}_{m_4} . \tag{10.213}$$

Now notice that if the ground-state is translationally invariant then the label 0 of the "removal site" is immaterial, i.e., it can be replaced by any other value m_1 without affecting the outcome of the calculation. We take advantage of this to replace the right-hand side of Eq. (10.213) by

$$-\frac{1}{N_L} \sum_{m_1,m_2,m_3,m_4} V_{m_1,m_2,m_3,m_4} \hat{a}_{m_1}^\dagger \hat{a}_{m_2}^\dagger \hat{a}_{m_3} \hat{a}_{m_4} , \tag{10.214}$$

whose expectation value is proportional to the ground-state energy E_0 (a negative quantity). Substituting this in (10.210) we obtain

$$\epsilon_- = \mu_- - 2\frac{E_0}{N} , \tag{10.215}$$

and, following a similar procedure,

$$\epsilon_+ = \mu_+ + \frac{2v}{1-v}\frac{E_0 - E_1}{N} , \tag{10.216}$$

where E_1 is the energy of the full Landau level ($v = 1$).

10.14.2 Independent boson theory

Let us now investigate in detail the shape of the spectral function. Our approach is in essence the same we used in Chapter 8 to study the properties of the normal Fermi liquid. We consider the effect of the electrostatic interaction between a single electron localized in the $m = 0$ orbital at $\vec{r} = 0$ and the density fluctuations of the 2DEL. This interaction has the form

$$\hat{H}_{int} = \int \frac{e^2}{\epsilon_b r'} \delta\hat{n}(\vec{r}')d\vec{r}' , \tag{10.217}$$

here we have neglected the finite extent of the $m = 0$ orbital. The dynamics of the density fluctuations is adequately described by the bosonic hamiltonian (10.206). Making use of the bosonic representation (10.207) for the density fluctuations we arrive at an effective hamiltonian of the form

$$
\hat{H} = \frac{\hbar\omega_c}{2}\hat{a}_0^\dagger\hat{a}_0 + \sum_{\bar{q}}\hbar\omega_q\hat{b}_{\bar{q}}^\dagger\hat{b}_{\bar{q}}
$$

$$
- \frac{2\pi e^2}{\epsilon_b\sqrt{A}}\sum_{\bar{q}}\sqrt{\frac{S}{2\hbar\omega_q}}\,q\ell^2\left(\hat{b}_{\bar{q}} + \hat{b}_{-\bar{q}}^\dagger\right)\hat{a}_0^\dagger\hat{a}_0 , \tag{10.218}
$$

which describes the coupling between a single fermion and a "bath" of phonons of frequencies $\omega_q \equiv \omega_-(q)$. The idea is to use this hamiltonian to calculate the correlation function $\langle\hat{a}_0(t)\hat{a}_0^\dagger\rangle$ and then obtain the spectral function from the integral

$$
A_>(\omega) = \int_{-\infty}^{\infty}\frac{dt}{2\pi}\left\langle\hat{a}_0(t)\hat{a}_0^\dagger\right\rangle e^{i\omega t} . \tag{10.219}
$$

(The reader may want to verify that this indeed reproduces the definition (10.208) of the spectral function for particle addition.)

The status of the effective hamiltonian (10.218) is analogous to that of the Hamann–Overhauser hamiltonian of Section 8.6 *before* applying the unitary transformation that eliminates the coupling between the "test particles" and the medium to second order in the strength of their interaction.[51] In the present case, however, due to the absence of the kinetic energy, the unitary transformation eliminates the electron–phonon coupling *to all orders*, and allows an exact solution of the model.[52]

The main steps in the solution are presented in Appendix 25. The final result *at zero temperature* is

$$
A_>(\omega) = \int_{-\infty}^{\infty}\frac{dt}{2\pi}e^{i\left(\omega - \frac{\tilde{\mu}_+}{\hbar}\right)t + \sum_{\bar{q}}\left|\frac{M_q}{\hbar\omega_q}\right|^2(e^{-i\omega_q t} - 1)} , \tag{10.220}
$$

where

$$
M_q \equiv \frac{2\pi e^2 q\ell}{\epsilon_b\sqrt{A}}\sqrt{\frac{S}{2\hbar\omega_q}} \tag{10.221}
$$

is the "electron–phonon" matrix element, and $\tilde{\mu}_+$ is the minimum energy required for injecting one electron with $m = 0$ into the system.[53]

---

[51] At variance with the Hamann–Overhauser theory we have neglected local field corrections, which are at this time unknown for the 2DEL in the lowest Landau level.

[52] The hamiltonian (10.218) is one of the best known exactly solved models in condensed matter physics: it describes, for example, the optical absorption from localized electronic states interacting with phonons in polar semiconductors.

[53] Unfortunately, the calculation of the injection energy $\tilde{\mu}_+$ lies beyond the power of the macroscopic elasticity theory since it includes, besides the negative "polaron shift" (see Appendix 25), a positive core energy from the local disruption of the electron liquid. In general $\tilde{\mu}_+$ is expected to be larger than the regular addition energy μ_+.

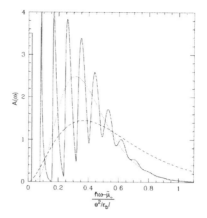

Fig. 10.18. The spectral function of a 2DEL calculated from the independent-boson model in an incompressible state at $\nu = \frac{1}{3}$ (full curve) and in a compressible state at $\nu = 0.3$ (dotted curve, coulomb interaction; dashed curve, short-range interaction). The marked differences between these three curves reflect the different nature of the collective mode dispersion, which is gapped in the first case, and vanishes as $q^{3/2}$ and q^2 for $q \to 0$ in the latter two. In the incompressible case, notice the δ-function peak at $\hbar\omega = \tilde{\mu}_+$ (pictured, for clarity, as a narrow rectangle), where $\tilde{\mu}_+$ is the minimum energy for injection of an electron with $m = 0$ (from Conti and Vignale, 1998).

Notice that $A_>(\omega)$ vanishes for $\omega < \frac{\tilde{\mu}_+}{\hbar}$, while $A_<(\omega)$ vanishes for $\omega > \frac{\tilde{\mu}_-}{\hbar}$: within this model the electron and the hole spectral functions are just mirror images of each other about the midgap energy $\frac{\tilde{\mu}_+ + \tilde{\mu}_-}{2}$.

Fig. 10.18 shows the plot of $A_>(\omega)$ for the incompressible quantum Hall liquid at $\nu = \frac{1}{3}$ – the case in which the bosonization procedure rests on firmer ground. Because of the nearly dispersionless character of the magnetoplasmon, the spectral function consists of a series of sharp peaks: the lowest energy peak is the "zero-phonon line" corresponding to the injection of an electron without production of collective excitations. The spectral weight under this peak is easily seen to be $Z = e^{-\sum_{\vec{q}} \left| \frac{M_q}{\hbar \omega_q} \right|^2}$, which can be interpreted as the "renormalization constant" for an electron quasiparticle in the LLL.[54] The second peak corresponds to injection with emission of one phonon, the third to injection with emission of two phonons and so on. These peaks are yet to be seen in tunneling experiments: their observation would constitute new evidence for the existence of gapped collective modes.

Fig. 10.18 also shows the results for a compressible electron liquid at a generic filling factor $\nu < 1$. The validity of the elasticity theory in this case is uncertain, but it has been argued (Johansson and Kinaret, 1994) that, for the purpose of calculating a localized disturbance, one can approximate the collective excitations of the liquid by the phonons of a classical Wigner crystal at the same density. The most prominent feature of the plotted curve is the "pseudogap" in the spectral function near the chemical potential. This is in

---

[54] The presence of a gap in ω_q is vital to the existence of the zero-phonon peak: a gapless phonon dispersion $\omega_q \propto cq$ for $q \to 0$ would lead to a logarithmically diverging $\sum_{\vec{q}} \left| \frac{M_q}{\hbar \omega_q} \right|^2$ and hence a vanishing Z.

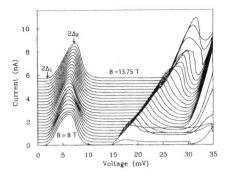

Fig. 10.19. Tunneling $I - V$ characteristics at $T = 0.6$ K for magnetic fields B from 8 T ($\nu = 0.83$) to 13.75 T ($\nu = 0.48$) in steps of 0.25 T. The low-voltage peak corresponds to the coulomb energy $\frac{e^2}{\epsilon_b r_s a_B}$ and is nearly independent of B. The high-voltage peak is associated with the cyclotron energy and increases linearly with B (adapted from Eisenstein *et al.*, 1992b).

excellent qualitative agreement with experimental observations (Eisenstein *et al.*, 1992b), which show a strong suppression of the tunneling density of states near zero bias voltage (see Fig. 10.19).

The physical reason for the pseudogap is that the injected electron goes at first into an excited state that contains a large number of phonons. The peak in the spectral function occurs precisely at the average energy of this excited state, while the width of the peak is proportional to the inverse lifetime of the excess charge density. The reader will recognize the physics of the "orthogonality catastrophe", discussed at length in Chapter 9. It appears that a pseudogap is a generic feature of systems that do not support electron quasi-particles.

10.15 Chern–Simons theory

The Laughlin theory of the fractional quantum Hall liquid is based on an insightful choice of wave functions at special filling factors, rather than on a systematic method of solution. So it is not surprising that one runs into difficulties when one tries to extend the theory to general filling factors, or even just to refine it in the cases in which it works. It would be clearly desirable to have some kind of mean-field approximation that captures the physics of the Laughlin wavefunction, but can also be used as a starting point for systematic improvements and generalizations. The *Chern–Simons (CS) theory* appears to provide such a starting point. This elegant and inspiring approach maps the 2DEL problem into an equivalent many-body problem of interacting composite particles, for which a mean-field approximation is meaningful. Unfortunately, the practical application of the method is still plagued by formidable technical difficulties. This section presents an elementary introduction to this complex and fascinating area of research.[55]

---

[55] For a recent review see Murthy and Shankhar (2003).

10.15.1 Formulation and mean field theory

Let $\psi(z_1, \ldots, z_N)$ be an eigenfunction of the 2DEL hamiltonian, and let us consider th transformed wave function

$$\psi_{CS}(z_1, \ldots, z_N) \equiv \psi(z_1, \ldots, z_N) \prod_{i<j} \left(\frac{z_i^* - z_j^*}{|z_i - z_j|} \right)^p , \qquad (10.222$$

where p is an integer. The factor on the right-hand side can be written in the form

$$\prod_{i<j} e^{-ip\mathrm{Arg}(z_i - z_j)} , \qquad (10.223)$$

where $\mathrm{Arg}(z)$ is the angle formed by $z = x + iy$ with the x-axis in the complex plane. Notice that if ψ obeys Fermi statistics, i.e. if it is antisymmetric under interchange of any pair of variables, then ψ_{CS} will have the same property provided p *is even*. On the other hand, if p is odd, then ψ_{CS} will be a symmetric function of its arguments, and thus obey Bose statistics. In either case, Eq. (10.222) implies that, if ψ is an eigenfunction of the many-electron hamiltonian with vector potential $\vec{A}(\vec{r})$, then ψ_{CS} is an eigenfunction of the *Chern–Simons hamiltonian*

$$\hat{H}_{CS} = \frac{1}{2m_b} \sum_{i=1}^{N} \left\{ \hat{\vec{p}}_i + \frac{e}{c} \left[\vec{A}(\hat{\vec{r}}_i) + \sum_{j \neq i} \vec{A}_{CS}(\hat{\vec{r}}_{ij}) \right] \right\}^2 + \frac{1}{2} \sum_{i \neq j} \frac{e^2}{|\epsilon_b \hat{\vec{r}}_{ij}|} , \qquad (10.224)$$

where $\hat{\vec{r}}_{ij} \equiv \hat{\vec{r}}_i - \hat{\vec{r}}_j$ and $\hat{\vec{A}}_{CS}(\hat{\vec{r}}_{ij})$ is the *Chern–Simons (CS) vector potential* defined as

$$\vec{A}_{CS}(\hat{\vec{r}}_{ij}) = \frac{p\Phi_0}{2\pi} \frac{\vec{e}_z \times \vec{r}_{ij}}{|\vec{r}_{ij}|^2} . \qquad (10.225)$$

These observations, in themselves quite trivial and easy to verify, have far-reaching consequences. The CS vector potential can be interpreted as the vector potential created by infinitely thin flux tubes attached to the electrons: each tube carries p quanta of magnetic flux. Eq. (10.224) tells us that the original N-electron problem is equivalent to the problem of N *composite particles* consisting of an electron plus an infinitely thin flux tube attached to it. The composite particles are fermions for even p and bosons for odd p. They interact with each other not only through the electric charge, but also via the vector potential that they create.

An *exact* solution of the CS hamiltonian is even more difficult than the solution of the original problem, due to the presence of the three-body operator $\sum_i \sum_{j,k \neq i} \vec{A}_{CS}(\hat{\vec{r}}_{ij}) \cdot \vec{A}_{CS}(\hat{\vec{r}}_{ik})$, which arises from the square of the CS vector potential. The real advantage of the formulation becomes apparent only when the CS potential is treated in a mean field approximation quite akin to the Hartree approximation. It follows from Eq. (10.225) that ﬔe CS magnetic field experienced by the i-th electron is

$$\vec{B}_{CS}\left(\hat{\vec{r}}_i, \{\hat{\vec{r}}_j\}\right) = \vec{\nabla}_{\vec{r}_i} \times \sum_{j \neq i} \vec{A}_{CS}(\hat{\vec{r}}_{ij})$$

$$= p\Phi_0 \sum_{j \neq i} \delta(\hat{\vec{r}}_i - \hat{\vec{r}}_j)\vec{e}_z . \tag{10.226}$$

ﾌotice that the CS magnetic field acting on one particle differs from zero only at the ﾖositions of the other particles. In the mean field approximation we replace the operator ﬁeld $\vec{B}_{CS}\left(\hat{\vec{r}}_i, \{\hat{\vec{r}}_j\}\right)$ by the average magnetic field created by all the composite particles at ﾖosition $\vec{r}$, namely

$$\vec{B}_{CS}(\vec{r}) = p\Phi_0 n_0(\vec{r})\hat{z} , \tag{10.227}$$

ﾰhere $n_0(\vec{r})$ is the *average* ground-state density. Thus, the mean CS magnetic field is not ﬁed to the particles, but to their probability distribution in space. Although the physics of ﬄhe problem may have been dangerously altered in this step, it is impossible to resist the ﬔemptation to go ahead and see what happens.

In a homogeneous system the mean CS magnetic field can be combined with the external ﾰmagnetic field to produce a uniform effective field

$$\vec{B}^* = (-B + p\Phi_0 n_0)\vec{e}_z$$

$$= -B(1 - p\nu)\vec{e}_z. \tag{10.228}$$

The corresponding mean field Chern-Simons hamiltonian is

$$\hat{H}_{CS}^{mf} = \frac{1}{2m_b} \sum_{i=1}^{N} \left[\hat{\vec{p}}_i + \frac{e}{c}\vec{A}^*(\hat{\vec{r}}_i)\right]^2 + \frac{1}{2} \sum_{i \neq j} \frac{e^2}{\epsilon_b |\hat{\vec{r}}_{ij}|}, \tag{10.229}$$

where $\vec{A}^*(\vec{r}) = \frac{\vec{B}^* \times \vec{r}}{2}$ is the vector potential associated with the effective magnetic field $\vec{B}^*$ in the symmetric gauge. Notice that this is still a many-body hamiltonian as far as the electron-electron interaction is concerned. However, by suitably choosing p we can either cancel the external magnetic field, or reduce it to a more convenient value. In either case, the original problem will be mapped into a new problem which may be more easily treated by standard many-body theory.

As a first example, consider the case $\nu = \frac{1}{M}$ (M odd) and choose $p = M$, i.e., attach M flux quanta to each electron. Then, we have $B^* = 0$ and $\hat{H}_{CS}^{mf}$ describes a 2D liquid of *charged bosons* in zero magnetic field. This is a very well studied system. The ground-state wave function is real and nodeless and can be effectively computed by quantum Monte Carlo methods.

To be totally unsophisticated, let us neglect the coulomb interaction and write the wave function of the non-interacting composite bosons as

$$\psi_{CS}(z_1, \ldots, z_N) = 1 . \tag{10.230}$$

Then, the wave function of the original electrons, obtained via the transformation (10.222) is

$$\psi(z_1, \dots z_N) = \prod_{i<j} \left(\frac{z_i - z_j}{|z_i - z_j|} \right)^M .$$

(10.231)

Now this is a *very* bad wave function: it does not even vanish when $z_i = z_j$ and certainly does not lie within the LLL. However, its *projection* into the LLL is probably not too different from the Laughlin wave function $\prod_{i<j} (z_i - z_j)^M \prod_i e^{-\frac{|z_i|^2}{4\ell^2}}$. We will return to this point later, when we discuss the far more powerful composite fermion approach.

The mapping of the $\frac{1}{M}$ Laughlin state to charged bosons has been a source of interesting analogies.[56] For example, the incompressibility of the Laughlin state corresponds to the Meissner effect,[57] and the quasihole excitation is the analogue of a vortex in the superfluid Bose system.

10.15.2 Electromagnetic response of composite particles

In Section 4.7 we described a general procedure for transforming a static mean field theory into a time-dependent one. Here we apply that procedure to the CS mean field theory and use it to calculate the electromagnetic response of the 2DEL.

Let $\vec{A}^{ext}(\vec{q}, \omega)$ be the Fourier component of the external time-dependent vector potential at wave vector $\vec{q}$ and frequency ω. We write the linear current response (see Section 3.4) as

$$j_\alpha(\vec{q}, \omega) = \frac{e}{c} \sum_\beta \chi_{\alpha\beta}^{J*}(\vec{q}, \omega) \left[A_\beta^{ext}(\vec{q}, \omega) + \delta A_{CS,\beta}(\vec{q}, \omega) \right],$$

(10.232)

where $\chi_{\alpha\beta}^{J*}(\vec{q}, \omega)$ is the current response function of a system of composite particles subjected to the mean magnetic field $\vec{B}^*$, and $\delta A_{CS,\beta}(\vec{q}, \omega)$ is the first-order variation of the CS mean field due to the induced density and current.[58]

To calculate $\delta \vec{A}_{CS}(\vec{q}, \omega)$ observe that a change $\delta n(\vec{q}, \omega)$ in the density causes a change

$$\delta \vec{B}_{CS}(\vec{q}, \omega) = p\Phi_0 \delta n(\vec{q}, \omega) \vec{e}_z$$

(10.233)

in the CS mean *magnetic field*. The corresponding change in the transverse part of the vector potential is

$$\delta \vec{A}_{CS,T}(\vec{q}, \omega) = ip\Phi_0 \frac{\vec{e}_z \times \vec{q}}{q^2} \delta n(\vec{q}, \omega)$$

$$= i\frac{p\Phi_0}{\omega} \vec{e}_z \times \vec{j}_L(\vec{q}, \omega),$$

(10.234)

where we have made use of the continuity equation $\delta n(\vec{q}, \omega) = \frac{\vec{q} \cdot \vec{j}_L(\vec{q}, \omega)}{\omega}$, to relate $\delta n(\vec{q}, \omega)$ to the longitudinal part of the particle current.

---

[56] The reader is referred to the review paper by S.-C. Zhang (1992) for a detailed discussion of this mapping.
[57] Indeed, any global change in the average density of the liquid would create a uniform CS magnetic field, which is forbidden by the Meissner effect: hence the incompressibility.
[58] Thus $\chi^{J*}(q, \omega)$ includes the effects of the coulomb interaction between the composite particles. In practice, $\chi^{J*}(q, \omega)$ itself needs to be approximated, typically through another time-dependent mean field approximation, such as the RPA.

According to Eq. (10.234) a longitudinal current produces a transverse CS vector poten-
al – a completely predictable effect, given that the CS magnetic field is proportional to the
verage electronic density. Far less obvious is the fact that *a transverse current causes a
hange in the longitudinal part of the CS vector potential*, so that the complete relationship
etween $\delta \vec{A}_{CS}$ and $\vec{j}$ takes the form

$$\delta \vec{A}_{CS}(\vec{q}, \omega) = i \frac{p\Phi_0}{\omega} \, \vec{e}_z \times \vec{j}(\vec{q}, \omega) . \tag{10.235}$$

On a formal level this can be viewed as a consequence of rotational invariance: Eq. (10.235) is
the only rotationally invariant linear relationship between $\delta \vec{A}_{CS}$ and $\vec{j}$ that is consistent with
Eq. (10.234). The physical origin of the longitudinal vector potential can be understood by
considering more carefully the operator nature of the CS vector potential. In Eq. (10.227) we
have been quick at replacing the CS magnetic field (and the corresponding vector potential)
by a numerical ground-state average, which is determined by the ground-state density. In
the linear response theory, however, one needs to consider density fluctuations about the
ground-state value, and this requires that we continue to treat the density, and hence the CS
vector potential, as operators. We therefore write

$$\hat{\vec{A}}_{CS}(\vec{r}) = \frac{p\Phi_0}{2\pi} \int d\vec{r}' \, \frac{\vec{e}_z \times (\vec{r} - \vec{r}')}{|\vec{r} - \vec{r}'|^2} \hat{n}(\vec{r}') , \tag{10.236}$$

and consider more carefully the coupling between this operator and the electronic current.
The coupling has the form

$$\begin{aligned}
\hat{H}_{j-A} &= \frac{e}{m_b c} \int d\vec{r} \, \hat{\vec{j}}(\vec{r}) \cdot \hat{\vec{A}}_{CS}(\vec{r}) \\
&= \frac{e}{m_b c} \frac{p\Phi_0}{2\pi} \int d\vec{r} \int d\vec{r}' \, \hat{\vec{j}}(\vec{r}) \cdot \frac{\vec{e}_z \times (\vec{r} - \vec{r}')}{|\vec{r} - \vec{r}'|^2} \hat{n}(\vec{r}') \\
&= i \frac{e p\Phi_0}{4\pi^2 m_b c} \int d\vec{q} \, \hat{\vec{j}}_{-\vec{q}} \cdot \frac{\vec{e}_z \times \vec{q}}{q^2} \hat{n}_{\vec{q}} .
\end{aligned} \tag{10.237}$$

Now, in a mean-field "factorization" (see Section 2.3) this coupling is approximated as

$$\hat{H}_{j-A} \approx i \frac{e p\Phi_0}{4\pi^2 m_b c} \int d\vec{q} \, \left(\frac{\vec{q} \times \hat{\vec{j}}_{-\vec{q}}}{q^2} \cdot \vec{e}_z \langle \hat{n}_{\vec{q}} \rangle + \frac{\vec{q} \times \langle \hat{\vec{j}}_{-\vec{q}} \rangle}{q^2} \cdot \vec{e}_z \, \hat{n}_{\vec{q}} \right) , \tag{10.238}$$

where $\hat{\vec{j}}_{\vec{q}}$ is the Fourier transform of the current operator. From this expression we see
that the average transverse current produces a scalar potential that couples to the density,
in much the same way as the average density produces a transverse vector potential that
couples to the transverse current.

Substituting Eq. (10.235) into Eq. (10.232) and solving for $\vec{A}^{ext}$ we obtain

$$A_\alpha^{ext}(\vec{q}, \omega) = \sum_\beta \left(\left[\frac{e}{c} \chi^{J*}(\vec{q}, \omega) \right]_{\alpha\beta}^{-1} + i \frac{pe\Phi_0}{\omega c} \epsilon_{\alpha\beta} \right) j_\beta(\vec{q}, \omega) , \tag{10.239}$$

from which we deduce the following RPA-like expression for the electromagnetic response function:

$$[\chi^J(\vec{q}, \omega)]^{-1}_{\alpha,\beta} = [\chi^{J*}(\vec{q}, \omega)]^{-1}_{\alpha\beta} + i \frac{pe\Phi_0}{\omega c}\epsilon_{\alpha\beta} , \qquad (10.240)$$

or, in matrix form:

$$[\chi^J(\vec{q}, \omega)]^{-1} = \begin{pmatrix} [\chi^{J*}(\vec{q}, \omega)]^{-1}_{LL} & [\chi^{J*}(\vec{q}, \omega)]^{-1}_{LT} + i \frac{pe\Phi_0}{\omega c} \\ [\chi^{J*}(\vec{q}, \omega)]^{-1}_{TL} - i \frac{pe\Phi_0}{\omega c} & [\chi^{J*}(\vec{q}, \omega)]^{-1}_{TT} \end{pmatrix} , \qquad (10.241)$$

where L and T denote, as usual, the longitudinal and transverse components relative to the direction of $\vec{q}$.

The above equations provide the basis for several interesting applications of the Chern–Simons mean field theory. Consider, for example, the choice $p = \frac{1}{\nu}$ so that p is an odd integer. This maps the original fermion problem into a problem of composite bosons in zero magnetic field. Since $B^* = 0$, the off-diagonal components $\chi^{J*}_{LT}(q, \omega)$, $\chi^{J*}_{TL}(q, \omega)$ vanish identically while, at small q, $\chi^{J*}_{LL}(q, \omega) = \chi^{J*}_{TT}(q, \omega) = \frac{n}{m_b} + O(q^2)$. The inverse of the electromagnetic response tensor at $q = 0$ and finite frequency takes the form

$$[\chi^J(0, \omega)]^{-1} = \frac{m_b}{n} \begin{pmatrix} 1 & +i\frac{\omega_c}{\omega} \\ -i\frac{\omega_c}{\omega} & 1 \end{pmatrix} , \qquad (10.242)$$

where $\omega_c = \frac{eB}{m_b c}$ is the cyclotron frequency associated with the original magnetic field. Inverting the matrix we see that the $q = 0$ response function has a single pole at $\omega = \omega_c$, and Kohn's theorem is satisfied, *even though the composite bosons see no average magnetic field in the ground-state*! Furthermore, one can easily verify that the Hall conductivity, which, according to Eq. (3.185), is related to the current–current response function by

$$\sigma_{xy} = \lim_{\omega \to 0} \frac{ie^2 \chi^J_{LT}(0, \omega)}{\omega} , \qquad (10.243)$$

is given by $\frac{\nu e^2}{h}$. This answers an important and potentially disturbing question, namely: if the composite bosons experience, on the average, no magnetic field, and hence no classical Lorentz force, how can they exhibit the Hall effect? The answer is that the magnetic flux current, which inevitably accompanies the composite boson current, produces a fictitious Hall electric field, which must be exactly cancelled by a real Hall electric field (the field measured by the voltmeter) so that the composite bosons experience no force.

Considering carefully the flow of the foregoing argument one sees that no use has been made of specific properties of the composite bosons. The $q = 0$ form of the current–current response tensor of a translationally invariant system of charged particles in an effective magnetic field B^* is entirely determined by Kohn's theorem, which does not depend on the statistics of the particles. Thus, for any statistics one has

$$[\chi^{J*}(0, \omega)]^{-1} = \frac{m_b}{n} \begin{pmatrix} 1 & +i\frac{\omega_c^*}{\omega} \\ -i\frac{\omega_c^*}{\omega} & 1 \end{pmatrix} , \qquad (10.244)$$

where $\omega_c^* = \frac{eB^*}{m_b c}$ is the fictitious cyclotron frequency associated with the effective Chern–Simons field B^*. The CS term on the right-hand side of Eq. (10.240) changes B^* to B in the off-diagonal terms of the response tensor: this guarantees the existence of the pole at $v = \omega_c$.

In spite of the fact that it satisfies Kohn's theorem, Eq. (10.241) exposes the fundamental weakness of the CS mean field theory. At high magnetic field the response function should exhibit a pole at the frequency of the intra-LLL exciton, which scales as $\frac{e^2}{\epsilon_b \ell}$ and is completely independent of the band mass. Unfortunately, Eq. (10.241) does not do this: the energy scale of the diagonal response functions χ_{LL}^{J*} and χ_{TT}^{J*} is still set by the band mass. In order to replace the kinetic energy scale by a coulomb energy scale, one would need to include strong interaction effects in the calculation of χ_{LL}^{J*} and χ_{TT}^{J*}. For example, in an incompressible quantum Hall state, we would expect χ^{J*} to be the response function of an elastic medium with diverging bulk modulus and finite shear modulus as discussed in Section 10.13.2. While this is fairly obvious at the phenomenological level, it has been very difficult to make progress in this direction, so the promise of the CS theory remains largely unfulfilled at this time.[59]

10.16 Composite fermions

From the time of its introduction in the late 1980s the composite fermion (CF) approach (Jain, 1989) has gradually been established as the most accurate method for calculating the physical properties of the 2DEL at high magnetic field. The composite-fermion wavefunctions yield ground-state energies, transport gaps, and exciton energies of quality superior to any other approach, and novel applications keep appearing. Here we briefly review the main points of the theory: for more technical detail the reader is referred to recent review articles (see, for example, the books edited by Das Sarma and Pinczuk (1996) and Heinonen (1998)).

We take a hint from the Chern–Simons transformation (10.222) and attach to each electron an even number $p = 2k$ flux quanta, where k is an integer ≥ 1. This transformation maps the problem of interacting electrons at filling factor v to a problem of interacting *composite fermions* at a different filling factor v^*, which is related to v by

$$\frac{1}{v^*} = \frac{1}{v} - 2k . \tag{10.245}$$

We expect that the transformed problem will be "simple" when the composite fermions occupy an integer number of Landau levels in the effective magnetic field $B^* = B(1 - pv)$ corresponding to v^*. In this case, the mean field state has a gap – a fact that should dramatically reduce the importance of correlations, as discussed extensively in Section 10.4. Setting $v^* = n$ in Eq. (10.245), where n is a positive or negative integer (a negative n simply meaning that the direction of the effective field B^* is opposite to the direction of the original

---

[59] Recently some progress in the direction of constructing a truly microscopic hamiltonian formulation based on the CS theory has been made by Murthy and Shankhar (2003).

field B) we find the following set of "magic" filling factors:

$$\nu = \frac{n}{2kn+1} \,, \tag{10.246}$$

with $n = \pm 1, \pm 2$, etc. Remarkably, the sequence with $k = 1$, known as the "Jain sequence" includes some of the most prominent fractional quantum Hall plateaus at $\nu = \frac{1}{3}, \frac{2}{5}, \frac{3}{7} \ldots$ for $n = 1, 2, 3 \ldots$, and $\nu = \frac{2}{3}, \frac{3}{5}, \frac{4}{7} \ldots$ for $n = -2, -3, -4 \ldots$, both sequences converging to $\nu = \frac{1}{2}$ when n tends to infinity. It appears that the fractional QHE of electrons is nothing but an integral QHE of composite fermions (see Fig. 10.23). The limiting state at $\nu = \frac{1}{2}$ shows no quantum Hall effect – consistent with the fact that, at this filling factor, the original system is mapped into a Fermi liquid of composite fermions in zero magnetic field.[60] Similarly, for $k = 2$ we have the sequences $\frac{1}{5}, \frac{2}{9}, \frac{3}{13} \ldots$ for $n = 1, 2, 3 \ldots$, and $\nu = \frac{2}{7}, \frac{3}{11}, \frac{4}{15} \ldots$ for $n = -2, -3, -4 \ldots$, both converging to $\frac{1}{4}$ for $n \to \infty$. Notice that incompressible states of composite fermions carrying $2k$ flux quanta (denoted by ^{2k}CF, where ^0CF are ordinary electrons) have filling factors at least as large as $\frac{1}{2k+1}$: therefore low-ν quantum Hall states must necessarily involve high-k composite fermions.

It should be noted, at this point, that the spin of the composite fermion is $\frac{1}{2}$ and couples to the physical magnetic field B (not to B^*!) via the magnetic moment $\frac{g\mu_B}{2}$. Then, depending on the magnitude of B and the value of the g-factor (which can be quite different from the band-structure value) several occupation schemes of composite fermion Landau levels are possible, leading to states of different spin polarization.[61] For example at $\nu^* = 1$ ($\nu = \frac{1}{3}$) one expects a fully polarized ground-state, but at $\nu^* = 2$ or -2 ($\nu = \frac{2}{5}$ or $\nu = \frac{2}{3}$) one would ordinarily expect a spin-unpolarized ground-state, that is, unless the Zeeman splitting is so large that it becomes energetically convenient for the composite Fermions to occupy two Landau levels of the same spin orientation.[62] Indeed, some of these different spin polarization scenarios have been experimentally observed and are described in section 10.8 of the book edited by DasSarma and Pinczuk (1996)).

A major advantage of the CF description is that it allows the construction of accurate wave functions through a relatively simple procedure. The ground-state CF wave function $\Phi_{\nu^*}(z_1, \ldots, z_N)$ is a single Slater determinant, with $|n|$ Landau levels completely filled in an effective magnetic field B^*. The ground-state wave function of the corresponding fractional quantum Hall state is constructed by multiplying this Slater determinant by the factor

$$\Phi_1^{2k}(z_1, \ldots, z_N) = \prod_{i<j}(z_i - z_j)^{2k} \prod_i \left(e^{-\frac{|z_i|^2}{4\ell_1^2}} \right)^{2k}, \tag{10.247}$$

and projecting the resulting wave function in the LLL *of the original fermions*:

$$\Psi_\nu(z_1, \ldots, z_N) = \hat{P}_{LLL}\Phi_1^{2k}(z_1, \ldots, z_N)\Phi_{\nu^*}(z_1, \ldots, z_N) \,. \tag{10.248}$$

---

[60] Recently, the quantum Hall effect has been observed at some fractions that do not belong to the sequence (10.246) (Pan *et al.*, 2003), e.g. $\nu = \frac{4}{11}$ and $\frac{5}{13}$: this has been interpreted as the fractional quantum Hall effect of composite fermions.

[61] A direct measurement of the g-factor of composite fermions has been reported by Schulze-Wischeler *et al.* (2004).

[62] The alert reader will recognize here the physics of the exchange-driven transitions in a tilted magnetic field, discussed in Section 10.5.

Eq. (10.247) $\ell_1^2 = \frac{\hbar c}{evB}$ is the magnetic length associated with a fictitious magnetic field $B_1 = \nu B$ that would cause the system to have filling factor 1. Thus, $\Phi_1(z_1, \ldots, z_N)$ is just the wave function of the fully occupied LLL in a magnetic field B_1.

To understand the rationale behind this construction, notice that the $\Phi_1^{2k}(z_1, \ldots, z_N)$ is the product of three factors:

(i) The Chern-Simons phase factor $\left(\prod_{i<j} \frac{z_i - z_j}{|z_i - z_j|}\right)^{2k}$;

(ii) The radial correlation factor $|z_i - z_j|^{2k}$;

(iii) The exponential factor $\left(\prod_i e^{-\frac{|z_i|^2}{4\ell_1^2}}\right)^{2k}$.

The first factor is required by the CS transformation from the original fermions to composite fermions with $2k$ flux quanta attached to each. The second factor introduces the "correlation hole" i.e., $2k$ additional zeroes of the wavefunction at the location of each particle.[63] Finally, the third factor combines with the factor $\left(\prod_i e^{-\frac{|z_i|^2}{4\ell*^2}}\right)^{2k}$ (which is part of the definition of $\Phi_{\nu*}(z_1, \ldots, z_N)$) to produce the correct exponential factor $\left(\prod_i e^{-\frac{|z_i|^2}{4\ell^2}}\right)^{2k}$, where the magnetic length ℓ now has the physical value $\left(\frac{\hbar c}{eB}\right)^{1/2}$. The truth of the last assertion is a consequence of the identity $\frac{2k}{\ell_1^2} + \frac{1}{\ell*^2} = \frac{1}{\ell^2}$ which is just a restatement of the relation (10.228) between $B*$ and B. Terms (ii) and (iii) together increase the average interparticle distance and thus change the filling factor from $\nu*$ to ν as required.

It should be clear that the construction (10.248), while inspired by the CS transformation, goes well beyond the framework of the CS theory. Furthermore, the construction is not limited to the ground-state but can be applied to any chosen state of the noninteracting composite fermion system, e.g., to the inter-Landau level exciton discussed in Section 10.6.3.

Kamilla and Jain (1997) have shown that Ψ_ν can be efficiently computed by replacing each single particle orbital $\phi_{n,m}(z_i)$ (n is the Landau level index and m is the angular momentum) that appears in the Slater determinant $\Phi_{\nu*}$ by a function $\phi_{i,n,m}^{CF}(z_1, \ldots, z_N)$ defined in the following manner:

$$\phi_{i,n,m}^{CF}(z_1, \ldots, z_N) = e^{-\frac{|z|^2}{4\ell^2}} z_i^{n+m} \prod_{j\neq i}(z_i - z_j)^{k-1} \frac{\partial^n}{\partial z_i^n} \prod_{j\neq i}(z_i - z_j). \qquad (10.249)$$

The wave functions obtained from this construction, both for the ground-state and for the excited states, have been shown to have nearly 100% overlap with the exact wave functions computed by exact diagonalization in few-electron system. The energies are also accurate to better than 0.1%. Some representative results (calculated for the strictly two-dimensional interaction potential) are reported in Table 10.6.

The fractional charge of the elementary excitations in the quantum Hall liquid state can be easily understood within the framework of the CF mean field theory. First of all, observe that

---

[63] The appearance of the zeroes can be understood if one imagines inserting the flux quanta adiabatically so as to induce a Faraday-law electric field that circulates around the point of insertion and pushes other electrons away from it. One may say that each composite fermion has $2k$ *vortices* rather than just $2k$ flux tubes attached to it.

Table 10.6. *Ground-state energies and gaps (in units of $\frac{e^2}{\epsilon_b \ell}$) calculated in the CF approach and extrapolated to the thermodynamic limit for a strictly two-dimensional system. The numbers in italics are for large but finite systems ($N = 50 - 60$). Adapted from Kamilla and Jain, 1997 and Scarola, Park, and Jain, 2000.*

ν	Ground-State Energy	Transport Gap $\epsilon^e + \epsilon^h$	Exciton $q = q_{min}$	Exciton $q = 0$
$\frac{1}{3}$	-0.409828	0.106	0.066	0.15
$\frac{2}{5}$	-0.432804	0.058	0.037	0.087
$\frac{3}{7}$	-0.442281	0.047	0.027	0.068
$\frac{4}{9}$	-0.447442	0.035	–	–
$\frac{5}{11}$	-0.450797	0.023	–	–
$\frac{1}{2}$	-0.4653	0.0	–	–
$\frac{1}{5}$	-0.327499	0.025	0.0095	–
$\frac{2}{9}$	-0.342782	0.016	*0.010*	*0.033*
$\frac{3}{13}$	-0.348349	0.014	*0.010*	–
$\frac{4}{17}$	-0.351189	0.011	–	–

adding a composite fermion to the system means adding a charge $-e$ *and* a flux $\Phi = 2k\Phi_0$ in the positive z-direction. The adiabatic insertion of the flux creates, by Faraday's law, an azimuthal electric field $E \sim \frac{\dot{\Phi}}{c}$ proportional to the rate of change of the flux ($\dot{\Phi}$). This electric field drives a *radial* current $j = \frac{\nu e^2 E}{h}$ which, integrated over time, gives the total outwardly displaced charge, $2k\nu e$. Substituting $\nu = \frac{n}{2kn+1}$ in this expression we see that the net charge created in the immediate vicinity of the insertion point is $e^* = -e + e\frac{2kn}{2kn+1} = -\frac{e}{2kn+1}$. Notice that this is in perfect agreement with the charge of the Laughlin quasi-electron at $\nu = \frac{1}{3}$ ($n = 1, k = 1, e^* = -\frac{e}{3}$), and its conjugate quasi-hole excitation at $\nu = \frac{2}{3}$ ($n = -2$, $k = 1, e^* = \frac{e}{3}$).[64] Let us now examine more closely the values of the transport gap, defined as the energy needed to create a quasi-electron and a quasi-hole infinitely far apart. Neglecting interactions between composite fermions, one is immediately led to identifying this gap with

[64] Interestingly, this construction predicts $e^* = 0$ at $\nu = \frac{1}{2}$. In this case, the composite fermion appears to have a *dipole moment* but no net charge.

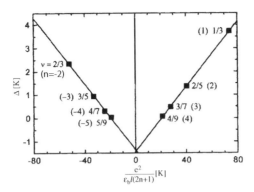

Fig. 10.20. Observed scaling of the quasiparticle excitation gaps of the fractional quantum Hall states at filling factors $\nu = \frac{n}{2kn+1}$, with $k = 1$ and $n = 1, \pm 2, \pm 3, \pm 4$, and -5 (value of n in parentheses). Adapted from Manoharan et al. (1994).

the cyclotron gap of the composite fermions in the effective magnetic field B^*, namely, $\hbar\omega_c^* = \frac{eB^*}{m^*c} = \frac{e\hbar B}{m^*|2kn+1|c}$, where m^* is, by definition, the *effective mass* of the composite fermion. Now, since the energy of every state in the LLL – and therefore also the energy gap – scales as $\frac{e^2}{\epsilon_b\ell} \sim \sqrt{B}$ at constant filling factor, we see from the above equation that m^* must also scale as $\sqrt{B}$. This observation led Halperin et al. (1993) to suggest that the mass of the composite fermion is given by

$$m^* = C^{-1}\frac{\hbar^2\epsilon_b}{e^2\ell}, \qquad (10.250)$$

where C is a constant to be determined by microscopic calculations, or, perhaps, by experiment (notice that m^* bears no relation to the band mass m_b). Substituting Eq. (10.250) in the formula for the cyclotron frequency leads to the following expression for the transport gap:

$$\Delta = \hbar\omega_c^* = \frac{C}{|2kn+1|}\frac{e^2}{\epsilon_b\ell}. \qquad (10.251)$$

Thus, we see that, if the above ideas is correct, the transport gap should tend to zero as $\frac{1}{2n+1}$ as $n \to \infty$ along the Jain sequence. At the same time, the thermodynamic gap $E_g = |2kn+1|\Delta$ (see Section 10.12) should be nearly independent of n.

These expectations are reasonably well confirmed by numerical calculations of the energy gap, based on composite fermion wave function (Kamilla and Jain, 1997). The value of C is about 0.31, and depends weakly on density and filling factor.[65] Experimental measurements of the gap, shown in Fig. 10.20, also confirm the overall picture, even though, as pointed out earlier, a truly quantitative comparison between theory and experiment is hampered by disorder and by the finite thickness of the electron layer.

Notice that the mass of the composite fermion is *defined* in terms of the transport gap, rather than being derived from a fundamental definition, and therefore it is slightly different

---

[65] Thus, we have $m^* \simeq 4m_b$ in GaAs ($\epsilon_b \simeq 12$) at 10 T.

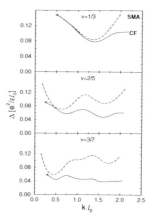

Fig. 10.21. The dispersions of the CF magnetoexciton at $v = \frac{1}{3}$, $\frac{2}{5}$, and $\frac{3}{7}$ for the pure coulomb interaction. The solid curves are the dispersion curves obtained from discrete points of finite systems, with the typical Monte Carlo uncertainty shown by the error bar at the beginning of each curve, while the dashed curve is the dispersion of the GMP mode obtained in the single-mode approximation. Adapted from Scarola, Park, and Jain (2000).

for different filling factors, even when the number of flux quanta k attached to each particle is kept constant. Calculations of the CF effective mass within the framework of the CS theory suggest that Eq. (10.250) is not valid in the immediate vicinity of the half-filled Landau level. Indeed, for $n \to \infty$ and $B^* \to 0$ the composite fermion effective mass appears to exhibit a logarithmic divergence, which ultimately results in the closing of the thermodynamic gap at $v \to \frac{1}{2}$.[66]

The composite fermion theory has also been applied to the calculation of the energy of intra-LLL excitons. One can proceed in two ways: (i) apply the single mode approximation of Section 10.13.1 with the static structure factor calculated from the CF wave function, or (ii) construct the exciton wave function directly from Eq. (10.248) using an inter-Landau-level exciton wavefunction in lieu of the single Slater determinant Φ_{v^*}. Both possibilities have been explored, and a comparison between them is shown in Fig. 10.21, for the case of a strictly two-dimensional electron gas. While the single mode approximation works reasonably well at $v = \frac{1}{3}$ and small q, the differences between it and the wave function method increase ad different filling factors and large q.

The energies obtained from the wave function approach at $q = 0$ and at the minimum of the dispersion are listed in the last two columns of Table 9.13. However, to make a quantitative comparison with the experiments (Pinczuk *et al.*, 1993, 1998) one must still take into account the finite thickness of the electron layer, which modifies the form of the effective interaction (see for instance Scarola, Park, and Jain, 2000).

---

[66] This logarithmic singularity is reminiscent of the one identified by Reizer (1989) in the normal electron liquid when transverse electromagnetic interactions are taken into account (see footnote [62]).

10.17 The half-filled state

The $\nu = \frac{1}{2}$ state has a special status in the theory of the 2DEL at high magnetic field. It is *not* a fractional quantum Hall state, yet it is the limit of the Jain sequence of FQHE states. According to the CF mean field theory the effective magnetic field vanishes at this filling factor, and therefore one should be left with just a two-dimensional Fermi liquid of composite fermions in zero magnetic field. If the Landau description of this Fermi liquid is correct, the response function χ^{J*}, and hence the full response function χ^{J} of Eq. (10.240), should depend on the Fermi wave vector of the (spin-polarized) composite fermions, $k_F = \sqrt{4\pi n}$.[67] Experiments that probe the response of the system at finite wave vectors and frequencies should be sensitive to the value of k_F. Indeed, such experiments have been done, and have apparently confirmed the existence of the CF Fermi surface.

The theory of the half-filled state has been discussed at length by Halperin, Lee, and Read (1993). Here we present a somewhat naive treatment based on the assumption that the composite fermions can be assigned a finite effective mass m^*.[68]

We begin by quoting some of the results obtained in Chapter 4 for the current response functions of the noninteracting two-dimensional Fermi liquid. Since, as we shall see momentarily, the interesting regime is $\omega \ll qv_F$, we can write the various components of the current–current response function of non-interacting composite fermions in the magnetic field $B^* = 0$ in the following manner:

$$\chi_{LL}^{J*(0)}(q, \omega) \sim \frac{\omega^2}{q^2} \frac{m^*}{2\pi\hbar^2} \left[-1 - 2i \frac{\omega}{qv_F} \right], \tag{10.252}$$

$$\chi_{TT}^{J*(0)}(q, \omega) \sim \frac{n}{m^*} \left[\frac{q^2}{12\pi n} - 2i \frac{\omega}{qv_F} \right], \tag{10.253}$$

and

$$\chi_{LT}^{J*(0)}(q, \omega) = \chi_{TL}^{J*(0)}(q, \omega) = 0, \tag{10.254}$$

where in the first two equations we have kept only the terms of leading order in $\frac{\omega}{qv_F}$, and the superscript (0) stands, as usual, for *noninteracting*.[69] Notice that $\chi_{LL}^{J*(0)}(q, \omega)$ vanishes for $\omega \to 0$ as required by gauge invariance. On the other hand, the coefficient of q^2 in $\chi_{TT}^{J*(0)}(q, 0)$ is related to the Landau diamagnetic susceptibility of the free Fermi gas, as discussed in Section 4.5. Finally, the imaginary parts of $\chi_{LL}^{J*(0)}(q, \omega)$ and $\chi_{TT}^{J*(0)}(q, \omega)$ are related to the spectral density of noninteracting electron–hole pairs. Notice that, to first order in $\frac{\omega}{qv_F}$, the excitation spectrum is entirely transversal – a fact that was given a geometrical interpretation in the caption of Fig. 4.3.

---

[67] The fact that the composite fermion liquid at $\nu = \frac{1}{2}$ is fully spin-polarized demonstrates an additional point, namely that the spin of the composite fermions responds to the physical magnetic field B, not to B^*. It is the Zeeman coupling to B that drives the polarization of the CF liquid.

[68] While the uncertainty in the value and even the qualitative behavior of m^* remains a serious weakness of the CS theory, it must be pointed out that m^* does not enter the main result derived in this section and probed by the experiment.

[69] The longitudinal and transverse components of the current–current response function, which were denoted simply by χ_L and χ_T in Section 3.4, are now assigned *two* indices to allow for the appearance of off-diagonal elements χ_{LT}^{J} in the full response function.

The coulomb interaction between composite fermions, treated in the RPA, modifies only the longitudinal term, which is given by

$$\frac{1}{\chi_{LL}^{J*}(q,\omega)} = \frac{1}{\chi_{LL}^{J*(0)}(q,\omega)} - \frac{q^2 v_q}{\omega^2}$$

$$\simeq -\frac{cq^2}{e\omega^2}\left[v_q + \frac{2\pi\hbar^2}{m^*}\right]$$

$$\simeq -\frac{q^2 v_q}{\omega^2}. \tag{10.255}$$

The inverse of the current–current response tensor is obtained by substituting Eq. (10.255) in Eq. (10.241) with $p = 2$:

$$[\chi^J(\vec{q},\omega)]^{-1} = \begin{pmatrix} -v_q\frac{q^2}{\omega^2}, & i\frac{4\pi\hbar}{\omega} \\ -i\frac{4\pi\hbar}{\omega}, & \left[\frac{q^2}{12\pi m^*} - 2i\frac{n\omega}{m^* q v_F}\right]^{-1} \end{pmatrix}. \tag{10.256}$$

The collective excitation frequencies are readily obtained from the zeroes of the determinant of this matrix. The resulting equation is

$$\frac{v_q q^2}{-\frac{q^2}{12\pi m^*} + 2i\frac{n\omega}{m^* q v_F}} = 16\pi^2\hbar^2, \tag{10.257}$$

whose solution is

$$\omega = -i\frac{\hbar k_F q^3}{16\pi n m^*}\left(\frac{m^* v_q}{2\pi\hbar^2} + \frac{2}{3}\right). \tag{10.258}$$

For small q, the solution tends to zero as q^2, and satisfies the condition $\omega \ll q v_F$, justifying *a posteriori* our assumption that this is the relevant frequency range.

The fact that the zero of the determinant lies on the negative imaginary axis means that what we have found is no ordinary propagating mode. The system does not oscillate about an equilibrium state, rather it relaxes exponentially towards it, like a classical overdamped oscillator. We have already encountered an overdamped mode in our discussion of the density–density response function of a disordered electron gas with diffusion constant D: in that case, the pole occurred at $\omega = -iDq^2$, implying that macroscopic density fluctuations obey the classical diffusion equation. The present result is formally very similar, except that *the diffusive behavior arises entirely from the coulomb interaction*. More precisely, putting $v_q = \frac{2\pi e^2}{\epsilon_b q}$ in Eq. (10.258) and neglecting the 2/3 at small q, we get

$$\omega = -iD_{eff}q^2 \tag{10.259}$$

where the "diffusion constant" is

$$D_{eff} = \frac{e^2}{4\epsilon_b\hbar k_F}. \tag{10.260}$$

This result is extremely interesting because it establishes a direct relation between an observable quantity, D_{eff}, and the Fermi wave vector of the composite fermions. Notice that this relation does not depend on the uncertain value of the effective mass m^*.

A very similar expression is obtained for the static longitudinal conductivity $\sigma_{xx}(q, 0)$. According to the standard linear response formalism of Section 3.4.4 we have

$$\Re e\sigma_{xx}(q, \omega) = -e^2 \frac{\Im m \tilde{\chi}^J_{LL}(q, \omega)}{\omega}, \tag{10.261}$$

where $\tilde{\chi}^J_{LL}(q, \omega)$ denotes the LL component of the *proper* response matrix $\tilde{\chi}^J$. The inverse of $\tilde{\chi}^J$ is obtained from the inverse of χ^J by simply setting $v_q = 0$ in Eq. (10.256). Neglecting terms of order q^2 we get

$$[\tilde{\chi}^J(\vec{q}, \omega)]^{-1} \sim \begin{pmatrix} 0, & i\frac{4\pi\hbar}{\omega} \\ -i\frac{4\pi\hbar}{\omega}, & 2\pi i\frac{\hbar q}{k_F\omega} \end{pmatrix}, \tag{10.262}$$

and carrying out the inversion, we finally obtain[70]

$$\lim_{\omega \to 0} \sigma_{xx}(q, \omega) = \frac{e^2 q}{8\pi\hbar k_F}. \tag{10.263}$$

The q-dependent conductivity $\sigma_{xx}(q)$ can be experimentally determined by measuring the q-dependence of the velocity of a surface acoustic wave (SAW) which is coupled to the 2DEL via the longitudinal electric field due to the piezoelectric effect (Willett *et al.*, 1993). The formula for the fractional change in sound velocity is

$$\frac{\Delta v_s}{v_s} = \left(\frac{\alpha^2}{2}\right) \frac{1}{1 + \frac{\sigma^2_{xx}(q)}{\sigma^2_m}}, \tag{10.264}$$

where α is a constant proportional to the piezoelectric coefficient and $\sigma_m = \frac{v_s \epsilon_b}{2\pi}$. By measuring Δv_s vs q one can thus determine the longitudinal conductivity of the 2DEL. The results of such an experiment, done on GaAs, are shown in Fig. 10.22, and are consistent with the existence of a Fermi surface of composite fermions at $v = \frac{1}{2}$ (and $v = \frac{1}{4}$ as well).

10.18 The reality of composite fermions

Let us summarize what has been accomplished so far by the composite fermion paradigm.

- The fractional and the integral quantum Hall effect have been described in a formally unified manner (see Fig. 10.23).
- All the observed values of the quantized Hall resistance have been explained in terms of a simple construction.
- The long-standing puzzle of why there is no quantum Hall effect at even denominator filling factors has been solved.
- The spin structure of several quantum Hall states has been elucidated.

---

[70] Strictly speaking this result is valid only for $\omega \gg \frac{\hbar q^2}{2m^*}$. The coefficient of q is expected to change when one goes beyond the RPA.

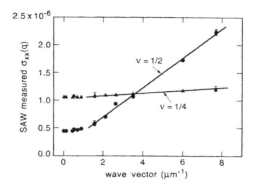

Fig. 10.22. Longitudinal conductivity vs wave vector measured from changes in the surface acoustic wave velocity at $\nu = \frac{1}{2}$ and $\nu = \frac{1}{4}$. Adapted from Willett *et al.*, 1993b.

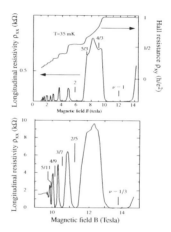

Fig. 10.23. The striking similarity between the fractional and the integral quantum Hall effect is naturally explained by the composite fermion concept. (Courtesy of J. Jain.)

• An appealing picture of the unquantized $\nu = \frac{1}{2}$ state has been developed.
• Accurate wave functions have been produced for all the fractional quantum Hall states and their excitations, without adjustable parameters.

Composite fermions are observed through their most basic property, namely the effective magnetic field $B^* = B(1 - p\nu)$ which they experience. In particular, near $\nu = \frac{1}{2}$ a $p = 2$ composite fermion at the Fermi surface follows a circular semiclassical orbit of radius $R_c^* = k_F \ell^{*2}$, where ℓ^* is the magnetic length for the field B^* and $k_F = \sqrt{4\pi n}$ is the Fermi wave vector. Several experiments are sensitive to the cyclotron radius. In one such experiment (Goldman, 2000) the composite fermions are injected into one constriction and collected into another, as shown in Fig. 10.24. When the distance between the constrictions is a multiple

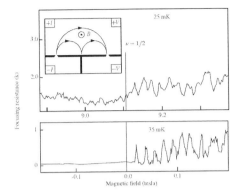

Fig. 10.24. One can measure the effective magnetic field B^* felt by composite fermions by magneti- cally focusing them, i.e., injecting them into one constriction and collecting them from another. The lower panel shows the focusing peaks for *electrons* in a weak magnetic field (B small) while the upper panel shows the corresponding peaks for *composite fermions* near $\nu = \frac{1}{2}$ (B^* small). The two sets of peaks align when one scales B^* by a factor $\sqrt{2}$ to account for the difference between the Fermi wave vector of electrons ($k_F = \sqrt{2\pi n}$) and that of composite fermions ($k_F = \sqrt{4\pi n}$). (Adapted from Goldman, 2000, courtesy of J. Jain.)

of the classical cyclotron orbit diameter, $2R_c^*$, an enhanced arrival rate is detected. By measuring the positions of the maxima in arrival rate vs magnetic field one can identify k_F.

Additional evidence is provided by the surface acoustic wave experiments (briefly de- scribed in the previous section), which show the presence of geometric resonances in $\sigma_{xx}(q)$ at wave vectors q such that $q R_c^* = X_n$, where X_n is the position of the n-th zero of the Bessel function J_1. For a review of these experiments, see the article by Willett (1997).

Taken together, these observations provide compelling evidence that composite fermions are the true quasiparticles of the lowest Landau level.

10.19 Wigner crystal and the stripe phase

To complete our discussion of the 2DEL at high magnetic field we must say something about broken symmetry states that under certain conditions may be lower in energy than the uniform liquid state. The Wigner crystal state is the best known of them. At sufficiently low density it is almost certainly the winner since the classical electrostatic energy dominates in this limit. A nice variational wave function for the Wigner crystal was proposed by Lam and Girvin (1984): its form is

$$\Psi_{WC}(z_1, \ldots, z_N) = e^{\sum_{ij} \frac{(z_i - R_i)B_{ij}(z_j - R_j)}{4\ell^2}} \det\left[\phi_{R_i}(z_j)\right], \qquad (10.265)$$

where

$$\phi_{R_i}(z_i) = \frac{1}{\sqrt{2\pi \ell^2}} e^{-\frac{|z_i - R_i|^2}{4\ell^2}} e^{\frac{(z_i^* R_i - z_i R_i^*)}{4\ell^2}} \qquad (10.266)$$

are coherent states (introduced in Section 10.2.5) centered at the sites $\vec{R}_i = (X_i, Y_i)$ ($i =$ $1, \ldots, N$) of a classical Wigner lattice. The complex variable R_i is defined as $R_i \equiv X_i + i Y_i$ ($i = 1, \ldots, N$), and the coefficients B_{ij} are variational parameters. Notice that this wave function lies in the LLL since the exponential factor in Eq. (10.265) is an analytic function of the z_i's. The Slater determinant of coherent-state orbitals contains no correlations other than those implied by the antisymmetry of the wave function. The role of the exponential factor is to build physical correlations between density fluctuations on different lattice sites.

An elegant way to derive the Lam–Girvin wave function is to start from the hamiltonian of N distinguishable electrons localized near lattice sites $\vec{R}_i$:

$$\hat{H}_{WC} = \frac{1}{2m_b} \sum_{i=1}^{N} \left(\hat{\vec{p}}_i + \frac{e}{c} \vec{A}(\vec{R}_i + \hat{\vec{u}}_i) \right)^2 + \frac{1}{2} \sum_{i,j} \hat{\vec{u}}_i \; \overleftrightarrow{V}_{ij} \; \hat{\vec{u}}_j \,, \qquad (10.267)$$

where $\hat{\vec{u}}_i = \hat{\vec{r}}_i - \vec{R}_i$ are displacement operators, $\overleftrightarrow{V}_{ij}$ is the tensor obtained by expanding the coulomb potential energy to second-order in the displacements (see Chapter 1), and $\vec{A}(\vec{r})$ is the vector potential in the symmetric gauge. This hamiltonian is quadratic in the electron displacements and can therefore be diagonalized exactly. The actual solution is somewhat laborious, as it involves a transformation to normal modes labeled by a wave vector $\vec{q}$ and an additional gauge-transformation for each $\vec{q}$ (Chui, Hakim, and Ma, 1986). In the limit of high magnetic field the ground-state wave function of $\hat{H}_{WC}$ is found to have the form of Eq. (10.265), with the Slater determinant replaced by the simple product $\prod_{i=1}^{N} \phi_{R_i}(z_i)$. The explicit form for B_{ij} is

$$B_{ij} = \sum_{\vec{q}} \frac{\omega_{\vec{q}L} - \omega_{\vec{q}T}}{\omega_{\vec{q}L} + \omega_{\vec{q}T}} e^{i\theta_{\vec{q}}} e^{i\vec{q} \cdot (\vec{r}_i - \vec{R}_j)} \,, \qquad (10.268)$$

where $\omega_{\vec{q}L}$ and $\omega_{\vec{q}T}$ are the longitudinal and transverse phonon frequencies of the classical Wigner crystal at zero magnetic field and θ_q is a rotation angle that describes the mixing of longitudinal and transverse modes in a magnetic field.

By comparing the energy of the Lam–Girvin wave function to that of the uniform liquid one finds that the Wigner crystal becomes stable below about $\nu = 0.14$. Experimentally, there are several indications that crystallization occurs at a considerably higher value of ν (~ 0.2), possibly exhibiting re-entrant behavior (Goldman, 1990). More complex calculations have led to higher values of the transition density, in better agreement with experiment (Esfarjani and Chui, 1990; see also Vignale, 1993).

The tendency of the electron liquid to break translational symmetry takes a peculiar form in partially occupied Landau levels of higher index n. Studies based on the Hartree–Fock approximation indicate that near half-integer filling factors $\nu > 2$ the homogeneous liquid phase is unstable to the formation of a *stripe phase* – a state in which Landau-gauge orbitals centered at $X = k_y \ell^2$ are occupied according to a one-dimensional periodic pattern (Koulakov, Fogler, and Shklovskii, 1996). The existence of a stripe phase has recently received experimental support from the observation of a marked anisotropy of the

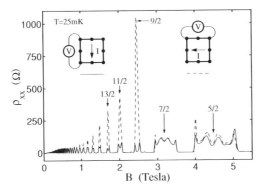

Fig. 10.25. The anisotropy of ρ_{xx} provides evidence for the stripe phase of a 2DEL in higher Landau levels. The two traces result from simply changing the direction of the current through the sample. At $\nu = \frac{9}{2}$ the ratio of the resistances in the two directions is close to 100. Adapted from Lilly *et al.*, 1999.

longitudinal resistivity at $\nu = \frac{9}{2}$ and $\nu = \frac{11}{2}$ (Lilly *et al.*, 1999). Some of the results are displayed in Fig. 10.25.

We will not present the details of the Hartree–Fock analysis (see Exercise 30 for more discussion of this point) but simply explain the physical reason why the stripe phase is favored in higher Landau levels, but not in the lowest one. At first sight, the stripe phase appears to be very much like a charge density wave, which, as we saw in Chapter 2, is favored by the exchange energy and yet, ordinarily, suppressed by its high electrostatic energy cost. But, in a high Landau level, one can modulate the density of the centers of the orbitals (hereafter referred to as *guiding centers*) without creating a charge density modulation and, therefore, without incurring the electrostatic energy cost. To see how this can happen, consider a stripe phase of wave vector $\vec{q}$ in the partially occupied n-th Landau level, with $n \geq 1$. The Hartree–Fock potential created by the stripe will mix the state ψ_{n,k_y} with the states $\psi_{n,k_y \pm q_y}, \psi_{n,k_y \pm 2q_y} \ldots$, and so on, but will not affect the wave function of the n fully occupied Landau levels, which just provide a kind of inert background (as usual, we neglect mixing of orbitals from different Landau levels). Hence, we will only focus on the part of the wave function that lies in the n-th Landau level.

To arrive at a proper definition of "stripe order" we notice that the expectation value of the density operator $\hat{n}_{\vec{q}}$ for $\vec{q} \neq 0$ can be written as

$$\langle \hat{n}_{\vec{q}} \rangle = F_{nn}(\vec{q}) \Delta(\vec{q}) , \tag{10.269}$$

where

$$F_{nn}(\vec{q}) = e^{-\frac{q^2 \ell^2}{4}} L_n^0 \left(\frac{q^2 \ell^2}{2} \right) \tag{10.270}$$

is the form factor of the n-th Landau level, and

$$\Delta(\vec{q}) = \sum_{k_y} e^{-iq_x k_y \ell^2} \left\langle \hat{a}^\dagger_{j,k_y+q_y/2} \hat{a}_{j,k_y-q_y/2} \right\rangle \tag{10.271}$$

is (by definition) the Fourier transform of the density of guiding centers.[71]

Because the Laguerre polynomial $L_n^0\left(\frac{q^2\ell^2}{2}\right)$ with $n \geq 1$ has a zero at $q = q_0 = \frac{c_n}{R_c}$, (where $R_c = \ell\sqrt{2n}$ is the cyclotron radius and $c_n \approx 2.4$ for $n \gg 1$) we see that by choosing the density of guiding center in such a way that $\Delta(\vec{q})$ differs from zero only in a neighborhood of q_0 we can suppress the density modulation $\langle \hat{n}_{\vec{q}} \rangle$, and hence the electrostatic energy

$$E_H = \frac{1}{2} \sum_{\vec{q}} \frac{v_q |F_{nn}(\vec{q})|^2}{\epsilon_n(q,0)} |\Delta(\vec{q})|^2 \ . \tag{10.272}$$

(Here $\epsilon_n(q,0)$ is the dielectric function arising from the n fully occupied Landau levels – see Section 10.4.6.) The interesting fact is that although the guiding centers are distributed according to a periodic pattern, the density remains uniform, because the period of the pattern coincides with the distance between the center of an orbital and its first zero. The exchange energy, on the other hand, is always negative: this means that the translationally invariant state is unstable with respect to the appearance of such an order.

There remains the question of whether the nearly "monocromatic" $\Delta(q)$ discussed in the previous paragraph can be realized in a self-consistent HF ground-state. Detailed calculations show that the actual solution of the HF equation produces a pattern of alternating high-density and low-density stripes, each stripe having a width of the order of $\sim R_c$. Although this segregation does produce a finite electrostatic energy, it is nevertheless advantageous for the following reason. We noted in Section 10.4.8 that in the presence of n full Landau levels the screened interaction is essentially independent of distance in the range $R_c < r < \frac{R_c^2}{a_B}$. Due to this special feature, while the gain in exchange energy is considerable, there is not much electrostatic energy penalty for compactifying the guiding centers into stripes.

10.20 Edge states and dynamics

10.20.1 Sharp edges vs smooth edges

The picture of edge states presented in Section 10.2.7 was based on the non-interacting electron model. The characteristic feature of that model is the sharp transition from occupied states to empty states depicted in Fig. 10.3: the electronic density drops from its bulk value to zero within a distance of the order of the cyclotron radius of the highest occupied Landau level.

This "sharp edge" picture is not always supported by fact. It certainly breaks down in the case of fractional occupation of the bulk Landau level. Even in the case of integral

---

[71] Eq. (10.269) is easily verified with the help of Eqs. (10.82) and (10.83).

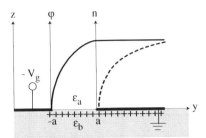

Fig. 10.26. A simple model for the cross-section of one of the edges of a 2DEL. (a) The thick lines represent conducting regions, i.e. the gate, kept at a negative potential $-V_g$, and the 2DEL connected to the ground at zero potential. The plusses represent the uniform positive background due to the donors. The dashed and the solid line represent the electronic density (n) and the electrostatic potential (φ) respectively. The spaces below and above the conducting regions are occupied by semiconductors with dielectric constants ϵ_b and ϵ_a, with $\epsilon_b \gg \epsilon_a$. A positively charged *depletion region* of width $2a$ is formed between the two conducting regions. (Adapted from Chklovskii, Shklovskii, and Glazman, 1992).

occupation, it strictly applies only to extremely sharp edges, such as the ones that are obtained by cleaving a crystal with atomic precision. Edges of quantum Hall liquids are often defined electrostatically, by applying voltages to *gates*, and in this case the characteristic width of the edge is much larger than the magnetic length.

In the next subsection we describe the *static* properties of an electrostatically defined edge, including the effect of the electron-electron interaction. We will see that the most natural description of the non-homogeneous electron liquid under these circumstances is a collective one, even though an effective single-particle description is still possible. This is no longer the case when one considers the *dynamics* of the edge. At sufficiently high magnetic field no vestige of independent-electron behavior survives: edges, sharp or smooth, are intrinsically collective systems, described by the paradigm of the *chiral Luttinger liquid*.

10.20.2 Electrostatics of edge channels

Consider the model depicted in Fig. 10.26 for the edge of a two-dimensional electron liquid. The liquid initially lies on a rigid background of positive charge in the half-plane $z = 0$, $y > -a$, and is connected to an external reservoir of charge, the ground, which fixes its electric potential to zero. The system is translationally invariant in the x direction. A semiconductor substrate of large dielectric constant ϵ_b fills the half-space $z < 0$. An external gate is placed on the left of the rigid background and is connected to the negative pole of a battery at a voltage $-V_g$. This electrode repels the two-dimensional electron liquid: as a consequence, a *depletion region* of width $2a$, in which the electronic density is essentially zero, appears in the region adjacent to the gate.

To develop a physical understanding of this system, let us begin by making use of the *perfect screening* rule of classical electrostatics, i.e., the fact that the electrons in a metal

adjust their density in such a way as to completely eliminate any component of the electr
field in the $z = 0$ plane. Of course, we know that this is not absolutely accurate, for a sma
electric field can be balanced by a gradient in the chemical potential of the electrons
quantum mechanical effect). As we show below however, such a residual electric field
of the order of $\sqrt{\frac{a_B}{W}}$ where a_B is the Bohr radius and W is the length scale over which th
density varies significantly.

Due to the assumed translational invariance in the x-direction, the electrostatic potentia
φ, depends only on y and z and satisfies the Laplace equation $\left(\frac{\partial^2}{\partial y^2} + \frac{\partial^2}{\partial z^2}\right)\varphi(y,z) = $
for $z > 0$, along with the boundary conditions $\varphi(y,0) = -V_g$ for $y < -a$, $\varphi(y,0) = $
for $y > a$, and $\left.\frac{\partial \varphi(y,z)}{\partial z}\right|_{z \to 0^{\pm}} = \mp \frac{4\pi e n_b}{\epsilon_b}$ for $|y| < a$ (n_b being the background density).[72] I
mechanically stable solution of these equations, i.e., a solution in which the y-componen
of the electric field vanishes at the edge of the 2DEL ($y = a$), was found by Chklovski
et al. (1992) to exist only for

$$a = \frac{\epsilon_b V_g}{4\pi^2 n_b e}. \tag{10.273}$$

Notice that, for typical values of $V_g \approx 1\,\mathrm{V}$, $n_b = 10^{11}\,\mathrm{cm}^{-2}$ and $\epsilon_b \approx 12$, one obtains
$a \approx 2000\,\text{Å}$, i.e. the width of the depletion region is much larger than any characteristic
microscopic length scale. The electrostatic potential in this solution is given by

$$\varphi(y,z) = \frac{V_g}{\pi} \Re \left[\sin^{-1}\left(\frac{\zeta}{a}\right) - \frac{\pi}{2} - |z| + \sqrt{1 - \left(\frac{\zeta}{a}\right)^2} \right], \tag{10.274}$$

where $\zeta \equiv y - i|z|$.[73] The behavior of this function along the real axis ($z = 0$) is shown as a
solid line in Fig. 10.26. At the same time, the electronic density in the $z = 0$ plane is given by

$$n(y) = \begin{cases} n_b \sqrt{\frac{y-a}{y+a}}, & y \geq a, \\ 0, & |y| < a, \\ n_b \left(\sqrt{\frac{y-a}{y+a}} - 1\right), & y \leq -a. \end{cases} \tag{10.275}$$

Thus, the length scale over which the electronic density varies significantly coincides with
the width of the depletion region. The reader is urged to verify that Eqs. (10.274) and
(10.275) are truly a solution of the posed problem (Exercise 29).

In order to estimate the quantum mechanical corrections to the above calculation, we
resort to density functional theory. Because the density is slowly varying on the scale of
the magnetic length we may use the local-density approximation for both the kinetic and
the exchange-correlation energy functionals (see Sections 7.3.1 and 7.4.4). Minimization

---

[72] The last condition follows from Gauss' theorem combined with the fact that any surface charge density $-ne$ supplied by
extrinsic donors is partially screened by polarization charges in the semiconductors: the effective surface charge density is
$-\frac{en}{\frac{\epsilon_a + \epsilon_b}{2}} \approx -\frac{2en}{\epsilon_b}$ for $\epsilon_b \gg \epsilon_a$ (see Exercise 28).

[73] The inverse sine function of complex argument is given by $\sin^{-1}(\zeta) = -i \ln\left(i\zeta + \sqrt{1 - \zeta^2}\right)$.

f the total energy then leads to the condition

$$-e\varphi(y, 0) + \mu_k(n(y)) + \mu_{xc}(n(y)) = 0, \quad (y > a), \tag{10.276}$$

where $\mu_k(n) = \frac{d[n\epsilon_k(n)]}{dn}$ and $\mu_{xc}(n) = \frac{d[n\epsilon_{xc}(n)]}{dn}$ are, respectively, the non-interacting and the exchange-correlation parts of the chemical potential of a homogeneous electron liquid of density n in the presence of a magnetic field B. An analytic solution of Eq. (10.276) is not easily found, since the electrostatic potential $\varphi(y, z)$ must still satisfy the boundary conditions described above. However, the qualitative features of the solution can be understood without detailed calculations.

Let us ignore, at first, the exchange-correlation potential. The non-interacting chemical potential at $T = 0$ is given by the formula (see Eq. (10.92))

$$\mu_k(n) = \hbar\omega_c \left([v] + \frac{1}{2} \right), \tag{10.277}$$

where $[v]$ denotes the integer part of v. This is constant in every interval between two consecutive integral filling factors, but can assume any value between $\hbar\omega_c \left(v - \frac{1}{2} \right)$ and $\hbar\omega_c \left(v + \frac{1}{2} \right)$ when v is exactly an integer. Substituting Eq. (10.277) into Eq. (10.276) and setting $\mu_{xc} = 0$, we see that the electrostatic potential can vary only in the regions in which v is integer and must be constant wherever v is non-integer. This curious behavior leads to the formation of regions of constant density, known as *incompressible strips* at $v = 1, 2, 3, \ldots$. Notice that these incompressible regions support a varying electrostatic potential, i.e. an electric field across them. To understand the formation of an incompressible strip at, say, $v = 1$, imagine following the classical density profile of Fig. 10.27 from the outer edge ($v = 0$) towards the interior of the system. When the density exceeds the value $v = 1$ some electrons must necessarily populate the first excited Landau level. But this costs an extra kinetic energy $\hbar\omega_c$ per electron, so the system prefers to keep its density constant at $v = 1$

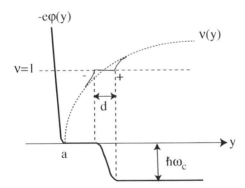

Fig. 10.27. Formation of an incompressible strip at $v = 1$ in a 2DEL with bulk filling factor $v_b = 1.5$. The dashed line is the filling factor calculated in the classical electrostatic approach. The difference between the actual density and the classical one forms an electric dipole. Notice that the electrostatic potential (thick solid line) is constant where the density varies and viceversa. (Adapted from Chklovskii et al., 1992.)

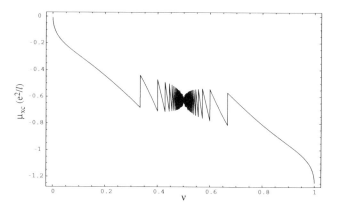

Fig. 10.28. A qualitative picture of the exchange-correlation potential as a function of filling factor in the lowest Landau level (courtesy of A. H. MacDonald (unpublished), adapted from Ferconi *et al.*, 1995).

even though the electrostatic energy thus becomes higher. The width of the constant density region grows up to a maximum value d until it finally becomes energetically convenient to populate the higher Landau level. How large is d? From Eq. (10.276) we see that the variation of the electrostatic potential across the strip must exactly cancel the variation of the chemical potential μ_k, that is, we must have $\Delta\varphi = \frac{\hbar\omega_c}{e}$: hence the electric field within the strip is $E \approx \frac{\hbar\omega_c}{ed}$. The structure of the excess charge within the strip (measured with reference to the classical electrostatic solution) is shown in Fig. 10.27: the "dipole" consists roughly of two oppositely charged wires of linear charge densities $\pm\frac{en'd^2}{8}$ placed a distance d apart (n' is the derivative of the classical density $n(y)$ with respect to y, evaluated at the center of the strip). The average electric field between the wires is $E \approx \frac{en'd}{4\epsilon_b}$, and equating this to $\frac{\hbar\omega_c}{ed}$ we obtain the desired estimate for the width of the incompressible strip:

$$d = 2\sqrt{\frac{\epsilon_b\hbar\omega_c}{e^2 n'}} .^{74}$$

(10.278)

It is instructive to compare d with the width a of the non uniform density region. Making the approximation $n' \simeq \frac{\nu_b}{2\pi\ell^2 a}$ (where ν_b is the bulk filling factor) we see that

$$\frac{d}{a} \simeq 2\sqrt{\frac{a_B}{\nu_b a}} .$$

(10.279)

Thus, both the width of the incompressible region and the associated electric field are very small if $a_B \ll a$.

A similar analysis can be carried out for incompressible strips that arise from discontinuities in the xc potential. As previously discussed, such discontinuities are expected to occur at the filling factors of the fractional quantum Hall effect. A schematic plot of V_{xc} as a function of ν for $0 < \nu < 1$ is shown in Fig. 10.28. In this case $\hbar\omega_c$ is replaced by the

---

[74] Notice that this vanishes in the classical limit.

~ermodynamic gap of the quantum Hall liquid: apart from this modification, the estimate ~f the width is unchanged.[75]

There are, of course, several problems with this quasi-classical description. For example, ~e sharp termination of the electronic density at $y = a$ is certainly incorrect: the physical ~ensity is expected to vanish exponentially on a length scale of order ℓ, but such a behavior is ~eyond the accuracy of the approach. By treating the noninteracting kinetic energy exactly, ~e Kohn–Sham equation of density functional theory offers, in principle, a natural way ~o go beyond the quasi-classical approximation. There is, however, a conceptual difficulty: ~ccording to Section 10.2.7, the one-electron eigenfunctions in a slowly varying potential ~re not too different from Landau level orbitals with shifted centers. If the Kohn–Sham Slater ~round-state is constructed by occupying the N eigenfunctions with energies lower than the ~ermi level, we are inevitably led to a density distribution that drops sharply from the bulk ~value to zero over a distance of order ℓ. How can we then obtain, within this scheme, the smoothly varying density profile predicted by classical electrostatics? The answer is that the standard Kohn–Sham scheme must be modified by allowing for *fractional occupation* of the Kohn–Sham orbitals. The simplest way to allow for fractional occupation is to implement the Kohn–Sham scheme at finite temperature. If our physical picture is correct, in the $T \to 0$ limit the self-consistent Kohn–Sham potential must become essentially constant (up to variations of order $k_B T$) over finite regions of space, in order to allow for fractional occupation of orbitals at the Fermi level. The Kohn–Sham potential that produces a density profile $v(y)$ near the edge is given by

$$V_{KS}(y) = \mu + k_B T \ln\left(\frac{1 - v(y)}{v(y)}\right), \tag{10.280}$$

which tends to a constant for $T \to 0$.

The correctness of these ideas has been verified in some detail. Self-consistent, finite temperature Kohn–Sham calculations of the edge structure in the LLL have been carried out for a model in which the density of the positive background dropped from the bulk value to zero linearly over a region of width W (Ferconi, Geller, and Vignale, 1995). The xc potential employed in these calculations has the form depicted in Fig. 10.28. Tiny incompressible strips were indeed observed, but only for the smoothest density profiles. They were, in all cases, small corrections to the overall density profile, which is largely determined by classical electrostatics.

10.20.3 Collective modes at the edge

In addition to bulk collective modes a 2DEL also supports *edge collective modes*: their distinctive feature is that the density disturbance is confined to the edge region and vanishes rapidly in the interior of the system.

---

[75] It should be noted that, in the presence of the xc contribution, or in any case at finite temperature, the electrostatic potential will not be strictly constant in the region occupied by the electrons. As a consequence, while it will be still much weaker than in the incompressible strips, the electric field will not be perfectly screened in the compressible regions.

All collective modes of a many-body system can be obtained from poles of the appropri ate response function (the density–density response function in the case of modes involving a density fluctuation) and the edge modes are no exception. However, a microscopic calcu lation of the density–density response function in the edge region is usually very difficul Fortunately, as we have seen in Chapter 5, the long wavelength response of the electro liquid is governed by simple hydrodynamic equations of motion, and this provides a muc simpler line of attack to the problem.[76]

For simplicity, we consider the geometry illustrated in Fig. 10.26. The system is transla tionally invariant in the x-direction and its equilibrium density $n_0(y)$ grows smoothly from zero to the bulk value $n_b = \frac{v_b}{2\pi \ell^2}$ over a distance of the order of a. Incompressible strip. are neglected, because their width is much smaller than a. Our treatment is restricted to collective modes whose spatial variation occurs on a length scale comparable to a in the y direction, and much longer than ℓ in the x-direction. In this regime the density fluctua tion, $\delta n(\vec{r}, t) = n(\vec{r}, t) - n_0(y)$, obeys hydrodynamic equations of motion, which may be regarded as the macroscopic limit of the RPA:

$$
\begin{cases}
\frac{\partial \delta n(\vec{r},t)}{\partial t} + \vec{\nabla} \cdot \left[n_0(y)\vec{v}(\vec{r}, t) \right] = 0 \\
\qquad \text{(continuity equation)} \\
\frac{\partial \vec{v}(\vec{r},t)}{\partial t} + \omega_c \vec{e}_z \times \vec{v}(\vec{r}, t) + \frac{1}{m}\vec{\nabla}_{\vec{r}} \int \frac{e^2 \delta n(\vec{r}',t)}{\epsilon_b |\vec{r}-\vec{r}'|} d\vec{r}' = 0 \\
\qquad \text{(Euler equation)}
\end{cases}
\tag{10.281}
$$

where $v(\vec{r}, t) = \frac{\vec{j}(\vec{r},t)}{n_0(y)}$ is the velocity field. Let us seek solutions in the form

$$
\delta n(\vec{r}, t) = \Re e[\delta n(y)e^{i(qx-\omega t)}],
$$
$$
\vec{v}(\vec{r}, t) = \Re e[\vec{v}(y)e^{i(qx-\omega t)}] ,
\tag{10.282}
$$

which have a simple periodic variation with wave vector q in the x-direction. Solving the Euler equation for v_x and v_y in terms of δn, and substituting the solution in the continuity equation, we obtain

$$
(\omega_c^2 - \omega^2) \delta n(y) - \frac{1}{m_b} \left\{ q^2 n_0(y) - n_0(y)\frac{d^2}{dy^2} \right.
$$
$$
\left. - n_0'(y)\frac{d}{dy} + \frac{q}{\omega}\omega_c n_0'(y) \right\} \int_0^\infty v(q, y - y')\delta n(y')dy' = 0 ,
\tag{10.283}
$$

where

$$
v(q, y - y') \equiv \int_{-\infty}^\infty \frac{e^2}{\epsilon_b |\vec{r} - \vec{r}'|} e^{iq(y-y')} dy
$$
$$
= 2\frac{e^2}{\epsilon_b} K_0[|q(y - y')|] .
\tag{10.284}
$$

---

[76] The following discussion is based on the paper by Aleiner and Glazman (1994).

Notice that the Bessel function $K_0(z)$ diverges as $-\ell n\, z$ for $z \to 0$, so the effective one-dimensional interaction is logarithmically divergent at small q.

Up to this point we have made no assumption about the strength of the magnetic field. In the large field limit ($\omega_c \to \infty$) we can neglect all but the first and the last term on the left-hand side of Eq. (10.283) and we readily arrive at the eigenvalue problem

$$\delta n(y) = \frac{q\, n_0'(y)}{\omega\, m_b \omega_c} \int_0^\infty v(q, y - y')\delta n(y')dy' , \qquad (10.285)$$

whence the mass and Planck's constant have disappeared. It is instructive to compare this equation with the eigenvalue problem (10.184) for the elastic displacement field in the homogeneous 2DEL. In the homogeneous case the frequency of the collective mode vanishes unless the shear modulus S differs from zero (see Eq. (10.188)). In the present case the shear modulus is assumed to be zero, yet a finite frequency arises from the non uniformity of the equilibrium density. Eq. (10.285) shows clearly that the density fluctuation in an edge mode *differs from zero only in the region in which the density is nonuniform*, that is, where $n_0'(y) \neq 0$.

It is not difficult to show (Conti and Vignale, 1996) that the eigenvalue problem (10.285) is hermitian with respect to the scalar product

$$(f, g) \equiv \int_0^\infty f(y)g(y)\frac{n_0'(y)}{n_b}dy . \qquad (10.286)$$

For every value of q there is a complete set of orthonormal solutions $f_{jq}(y)$, $j = 0, 1, 2 \dots$, such that

$$(f_{iq}, f_{jq}) = \delta_{ij} \qquad (10.287)$$

and

$$\sum_{j=0}^\infty f_{jq}(y)f_{jq}(y') = \frac{n_b}{n_0'(y)}\delta(y - y') . \qquad (10.288)$$

According to standard theorems, the $j = 0$ eigenfunction is nodeless in the interval $0 < y < a$. This mode is proportional to $n_0'(y)$ if the interaction is of finite range, i.e., if the Fourier transform $v_{q,y-y'}$ has a finite limit for $q \to 0$. The integral $e\int_0^a \delta n(y)dy$ of the displaced charge density across the edge is maximal in this mode. All other eigenfunctions have nodes – j of them in the j-th eigenfunction – so that the integrated charge density is very small: these modes are approximately neutral. It must be borne in mind that the hydrodynamic description breaks down when the scale of variation of the eigenfunctions, a/j, becomes comparable to the magnetic length. Thus, although the mathematics allows for an infinite number of solutions, only the solutions with $j \ll a/\ell$ are physically meaningful. An analytic solution of Eq. (10.285) was obtained by Aleiner and Glazman (1994) for the special equilibrium density profile

$$n_0(y) = \frac{2n_b}{\pi}\tan^{-1}\sqrt{\frac{y}{a}}, \quad (0 < y < a), \qquad (10.289)$$

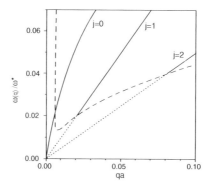

Fig. 10.29. The first three branches of edge excitations as calculated by Aleiner and Glazman (1993) The frequency scale of the plot is $\omega^* = \frac{2n_b e^2}{\epsilon_b m \omega_c a}$, where a is the width of the edge. The dashed line separates the regions of strong damping (below the line) from those of weak damping (above the line). In the latter region the acoustic mode becomes observable. Adapted from Aleiner and Glazman (1995).

which is representative of a typical smooth density profile. We will not give the details of their solution, but simply show, in Fig. 10.29, the frequencies of the first few collective modes as functions of q. At long wavelength, the charge mode has a frequency

$$\omega_0(q) = \frac{2n_b e^2 q}{\epsilon_b m_b \omega_c} \ln\left(\frac{e^{-\gamma}}{2|qa|}\right) \tag{10.290}$$

which tends to zero as $q \ln qa$ for small q ($\gamma = 0.577216$ is here the Euler constant). This behavior is characteristic of the plasmon mode in a one-dimensional system. The positive value of the phase velocity ω/q indicates that the mode has a *definite chirality*, i.e. that it can only propagate in one direction (the positive x-direction in this case) determined by the sign of the derivative $n_0'(y)$ (positive in this case) and by the direction of the magnetic field. The neutral modes are all *lower* in frequency, and their dispersion is linear in q:

$$\omega_j(q) = \frac{2n_b e^2}{\epsilon m_b \omega_c j} q \ , \tag{10.291}$$

for $j = 1, 2 \ldots \ll a/\ell$.

One can get a good feeling for the solutions of the eigenvalue problem by setting $n_0'(y) = n_0'$ independent of y (a linear density profile) and making the variational *Ansatz*

$$\delta n_j(y) \propto \frac{1}{\sqrt{|n_0'(y)|}} \cos\left(\frac{\pi j y}{a}\right) , \tag{10.292}$$

which satisfies the orthogonality condition (10.287) and has j nodes in the interval $0 < y < a$. An upper bound to the eigenfrequencies is then given by

$$\omega_j(q) \simeq \frac{q n_0'}{m_b \omega_c} \frac{\int_0^a dy \int_0^a dy' \cos\frac{\pi j y}{a} v(q, y - y') \cos\frac{\pi j y'}{a}}{\int_0^a dy \cos^2\frac{\pi j y}{a}} . \tag{10.293}$$

10.20.4 The chiral Luttinger liquid

The edge of a quantum Hall liquid bears an intriguing relation to the Luttinger model of the one-dimensional electron liquid. More precisely, it was pointed out by X. G. Wen in the early 1990s that the density fluctuations at the quantum Hall edge can be described in terms of a *chiral* version of the Luttinger model. To understand clearly the concept, recall that the standard Luttinger model is characterized by *two* sets of electronic density fluctuation operators, $\hat{n}_{q,R}$ and $\hat{n}_{q,L}$ (R and L standing for right and left movers, respectively), which satisfy the commutation relations $[\hat{n}_{q,\tau}, \hat{n}_{-q',\tau'}] = \frac{Lq\tau}{2\pi}\delta_{qq'}\delta_{\tau\tau'}$, and can therefore be expressed in terms of boson operators $\hat{b}_q$ and $\hat{b}_q^\dagger$ according to Eqs. (9.32). The interaction between right and left movers can be eliminated by a Bogoliubov transformation – see Eq. (9.42) – leading to a noninteracting hamiltonian of the form

$$\hat{H}_{LM} = \sum_{q \neq 0} \hbar c |q| \hat{\beta}_q^\dagger \hat{\beta}_q , \qquad (10.294)$$

where $c|q|$ is the frequency of the mode of wave vector q, and $\hat{\beta}_q$ are the transformed bosons. In order to arrive at the *chiral Luttinger model* (χLM) one discards all the modes with a given sign of q, say $q < 0$, and keeps only the ones with $q > 0$, i.e., the right-moving *bosons*.[77] Thus, the hamiltonian takes the form

$$\hat{H}_{\chi LM} = \sum_{q > 0} \hbar c q \hat{\beta}_q^\dagger \hat{\beta}_q . \qquad (10.295)$$

To completely define the model, however, we still need to know how the bosons appearing in this hamiltonian are related to electronic density fluctuations. In the *non-chiral* Luttinger model the long-wavelength density fluctuation operator was given by $\hat{n}_q = \hat{n}_{q,R} + \hat{n}_{q,L} = \sqrt{\frac{L|q|}{2\pi}} e^{-\phi}\left(\hat{\beta}_q + \hat{\beta}_{-q}^\dagger\right)$, where ϕ is the $q \to 0$ limit of the Bogoliubov rotation angle ϕ_q (see Eq. (10.207)). To obtain the corresponding expression for the *chiral* Luttinger model, we again discard all the boson operators with $q < 0$. The result is

$$\hat{n}_q = e^{-\phi}\sqrt{\frac{L|q|}{2\pi}}\left(\Theta(q)\hat{\beta}_q + \Theta(-q)\hat{\beta}_{-q}^\dagger\right) , \qquad (10.296)$$

which satisfies the commutation relation

$$[\hat{n}_q, \hat{n}_{-q'}] = e^{-2\phi}\frac{L|q|}{2\pi}\delta_{qq'} . \qquad (10.297)$$

The hamiltonian (10.295), together with Eq. (10.296), constitutes the proper specification of the χLM. What makes this model different from a non-interacting liquid of right-moving particles in one dimension is the presence of the scale factor $e^{-\phi}$ in the relation between the density fluctuations and the bosons.

We will now show that the edge dynamics of a 2DEL at long wavelengths and high magnetic field is described by a set of independent χLMs (one for each hydrodynamic

---

[77] Care should be taken not to confuse these right-moving bosons with the right-moving fermions of the original Luttinger model.

mode), *regardless of whether the system exhibits the quantum Hall effect or not.* We sta▪
from the LLL-projected hamiltonian

$$\hat{H} = \frac{e^2}{2\epsilon_b} \int d\vec{r} \int d\vec{r}' \frac{\hat{n}(\vec{r})\hat{n}(\vec{r}')}{|\vec{r} - \vec{r}'|} + \int d\vec{r} V(y)\hat{n}(\vec{r}) , \qquad (10.298▪)$$

where $V(y)$ is the confining potential, which we assume to depend only on the coordina▪
y perpendicular to the edge, and $\hat{n}(\vec{r})$ is the density operator projected in the LLL. In th▪
latter, we separate out the equilibrium density, $n_0(y)$, and write

$$\hat{n}(\vec{r}) = n_0(y) + \delta\hat{n}(\vec{r}) , \qquad (10.299▪)$$

where $n_0(y)$ is obtained from the minimization of the classical potential energy and depend▪
only on y. Upon substituting this into Eq. (10.298), we see that the term linear in $\delta\hat{n}(\vec{r}$
vanishes by virtue of the classical equilibrium condition[78] and we are left with the simple
quadratic hamiltonian

$$\hat{H} = \frac{e^2}{2\epsilon_b} \int d\vec{r} \int d\vec{r}' \frac{\delta\hat{n}(\vec{r})\delta\hat{n}(\vec{r}')}{|\vec{r} - \vec{r}'|} . \qquad (10.300▪)$$

The projected density fluctuation operators satisfy the commutation relations (10.173).
Since we are only interested in long-wavelength fluctuations we expand the right-hand side
of Eq. (10.173) to leading order in $k\ell$ and $q\ell$ and transform to real space to get

$$[\delta\hat{n}(\vec{r}), \delta\hat{n}(\vec{r}')] = i\ell^2 n_0'(y)\frac{\partial}{\partial x}\delta(\vec{r} - \vec{r}') . \qquad (10.301)$$

Strictly speaking, the right-hand side of this equation should have contained the derivative
of the density operator rather than the derivative of the equilibrium density. However, the
replacement of $\hat{n}'(\vec{r})$ by its equilibrium expectation value $n_0'(y)$ is justified when studying
the dynamics in the linear response regime.

The remarkable feature of Eq. (10.301) is that the commutator is proportional to the
derivative of the equilibrium density. This implies that hydrodynamic density fluctuations
are bound to regions in which the equilibrium density is non-constant, i.e., in practice, to
the edges of the system.

Let us now define "normal mode" operators $\delta\hat{n}_{jq}$ in the following manner:

$$\delta\hat{n}_{jq} = \int_0^L dx e^{-iqx} \int_0^\infty dy f_{jq}(y)/\delta\hat{n}(x, y), \qquad (10.302)$$

where $f_{jq}(y)$ are the eigenfunctions of the hydrodynamic eigenvalue problem (10.285), and
L is the length of the sample in the x direction. The commutator of two of these operators is
readily deduced from the algebra (10.301) and Eqs. (10.286)–(10.287) and works out to be

$$[\delta\hat{n}_{jq}, \delta\hat{n}_{j'-q'}] = v_b\frac{Lq}{2\pi}\delta_{qq'}\delta_{jj'} , \qquad (10.303)$$

---

[78] On a sufficiently large length scale one can ignore the difference between the classical equilibrium density and the true equilibrium density.

or an edge with $n_0'(y) < 0$. Comparing this with Eq. (10.297) we see immediately that each normal mode obeys the algebra of a χLL with $e^{-2\phi} = v_b$. Accordingly, we can express the edge density fluctuations in terms of boson operators $\hat{\beta}_q$ as follows:

$$\delta \hat{n}_{jq} = \sqrt{v_b} \sqrt{\frac{L|q|}{2\pi}} \left(\Theta(q)\hat{\beta}_{jq} + \Theta(-q)\hat{\beta}^\dagger_{j-q} \right) . \tag{10.304}$$

It remains to be shown that the hamiltonian (10.300) can be recast in the form

$$\hat{H} = \sum_{j,q>0} \hbar\omega_j(q)\hat{\beta}^\dagger_{jq}\hat{\beta}_{jq} , \tag{10.305}$$

where $\omega_j(q)$ are the frequencies of the classical collective modes, given by Eq. (10.285). This follows from the completeness properties of the solutions of the classical eigenvalue problem (see Eqs. (10.287) and (10.288)), and can be readily verified by substituting in Eq. (10.300) the expression

$$\delta \hat{n}(x, y) = \frac{1}{L} \sum_{jq} \delta \hat{n}_{jq} f_{jq}(y) \frac{n_0'(y)}{n_b} e^{iqx} \tag{*}$$

obtained by inverting Eq. (10.302) and making use of Eq. (10.304).

As anticipated, the hamiltonian (10.305) is the sum of independent chiral Luttinger models with different "sound velocities" but a single "mixing angle" ϕ such that $e^{-2\phi} = v_b$. It is important to appreciate that, at variance with the ordinary Luttinger model, here ϕ has absolutely nothing to do with a physical interaction between the two opposite edges of the system: such an interaction is utterly negligible when the two edges are separated by a macroscopically large distance. The origin of ϕ in the present model is far more fundamental, as it lies in the peculiar algebra of density fluctuations projected first in the LLL and then on the classical edge-mode solutions.

The index j in Eq. (10.305) is restricted to values such that the transverse wavelength a/j is much larger than ℓ. Large values of j are therefore excluded. This restriction becomes more and more severe as the edge becomes sharper. In the limit of sharp edge only the $j = 0$ mode, that is, the charge mode, survives. Thus, a sharp edge is equivalent to a chiral Luttinger liquid.[79]

10.20.5 Tunneling and transport

In the previous section we have shown that density fluctuations at the edge of the 2DEL at high magnetic field are described by the chiral Luttinger model. Can this theory be tested experimentally? The answer is yes. To date, two crucial experiments have confirmed the validity of the χLM. One is the measurement of the tunneling of electrons from a metal into the edge of the 2DEL (Grayson *et al.*, 1998). The second is the measurement of the tunneling of fractionally charged quasiparticles between the edges of a quantum Hall liquid (Roddaro *et al.*, 2003).

---

[79] Although our discussion of edge modes has been based on the smooth edge model, it turns out that the $j = 0$ mode, being a global (x-dependent) displacement of the edge, has the same behavior irrespective of whether the edge is sharp or smooth.

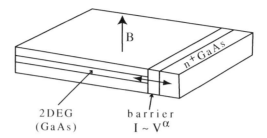

2DEG   b a r r i e r
(GaAs) $I \sim V^{\alpha}$

Fig. 10.30. Schematics of the Grayson–Chang experiment. Electrons tunnel from a heavily n-doped GaAs electrode (the "metal") into the edge of the 2DEL. The tunneling current I, indicated by the double arrow, is found to follow a power law of the form V^{α}, where V is the potential difference between the metal and the 2DEL.

Let us begin with the former. A sharp edge is realized by cleaving a GaAs quantum well along a plane perpendicular to the plane of the 2DEL. An Al–Ga–As insulating barrier is grown on one side of the quantum well, as shown in Fig. 10.30 and is capped by a heavily n-doped GaAs, which plays the role of a metal electrode. Electrons tunnel through the insulating barrier into the edge of the 2DEL. The tunneling current, plotted as a function of the bias voltage V (the potential difference between the electrode and the 2DEL) is found to follow a power law $I \sim V^{\alpha}$ for $k_B T \ll eV$. As eV drops below $k_B T$ an ordinary linear dependence $I = GV$ is recovered, but the conductance G has a nonlinear dependence on temperature: $G \sim T^{\alpha}$. The exponent α is smaller than 1, indicating a *suppression* of tunneling. More importantly, the value of α is found to be very close to $\frac{1}{\nu_b}$, in a wide range of fractional filling factors $\frac{1}{4} < \nu_b < 1$ (see Fig. 10.31).

The power-law suppression of tunneling at low voltages is reminiscent of the power-law suppression of the spectral function in a normal Luttinger liquid – a phenomenon that was explained in Chapter 9 in terms of the "orthogonality catastrophe" concept. The same concept was invoked earlier in this chapter to explain the *exponential suppression* of the tunneling density of states in the bulk of the 2DEL in the fractional quantum Hall regime (see Fig. 10.19). The result of the Grayson–Chang experiment can also be understood in terms of the orthogonality catastrophe, with the added bonus that the exponent of the power law behavior of the current is cleanly related to the bulk density, thus providing a clear-cut verification of the algebra of the density fluctuation operators.

Electrons that tunnel into the edge of the 2DEL create a linear superposition of edge waves. Because the edge is very sharp, we assume that only the charged edge mode – the $j = 0$ mode – is relevant. We also assume that the range of the electron-electron interaction is effectively reduced to a finite value by external screening, due to gates. Then the properly normalized eigenfunction of the $j = 0$ mode is

$$f_{0q}(y) = 1 ,$$ (10.306)

and the density fluctuation operator (Eq. (∗)) takes the form

$$\delta\hat{n}(x, y) = \frac{\sqrt{\nu_b}}{L}\frac{n_0'(y)}{n_b}\sum_{q>0}\sqrt{\frac{Lq}{2\pi}}\left[e^{iqx}\hat{\beta}_q + e^{-iqx}\hat{\beta}_q^{\dagger}\right] ,$$ (10.307)

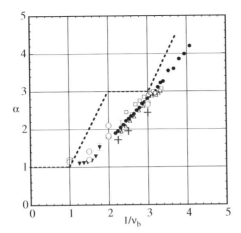

Fig. 10.31. Plot of the experimentally determined exponents α of the I vs. V relation vs. ν_b in the range $\frac{1}{4} < \nu_b < 1$ (adapted from Grayson et al., 1998). The dashed line, with its striking plateau for $\frac{1}{3} < \nu_b < \frac{1}{2}$, is the prediction of the fermionic Chern-Simons theory (Shytov et al., 1992). Non-interacting composite fermion theory also predicts a constant exponent $\alpha = 3$ at filling factors of the Jain sequence $\nu_b = \frac{n}{2n+1}$ (all these states correspond to integral filling of composite fermion Landau levels). The crosses at $\nu_b = \frac{1}{3}, \frac{2}{5}$, and $\frac{3}{7}$ are the results of recent Monte Carlo calculations by Mandal and Jain (2002), which include *interactions* between the composite fermions. These results are dramatically different from those of the non-interacting theory, and are seen to be in good agreement with the experiment and with the simple hydrodynamic theory outlined in this section.

where $\hat{\beta}_q(\equiv \hat{\beta}_{0q})$ is the operator associated with the $j = 0$ mode of wave vector q. Notice that this mode corresponds to a global oscillation of the edge with an x-dependent amplitude. Integrating over the transverse coordinate y we obtain the *edge density fluctuation* of our chiral Luttinger model in the following form:

$$\delta \hat{n}(x) = \int_0^W \delta \hat{n}(x, y) dy$$

$$= \frac{\sqrt{\nu_b}}{L} \sum_{q>0} \sqrt{\frac{Lq}{2\pi}} \left[e^{iqx} \hat{\beta}_q + e^{-iqx} \hat{\beta}_q^\dagger \right] . \qquad (10.308)$$

As expected, this is consistent with Eq. (10.296) of the chiral Luttinger model after making the identification $e^{-\phi} = \sqrt{\nu_b}$.

In order to describe tunneling into the edge we now introduce an "electron tunneling operator" $\hat{\psi}^\dagger(x)$, which adds one unit of charge at point x. This is accomplished by requiring that $\hat{\psi}^\dagger(x)$ satisfy the commutation relation

$$[\hat{\psi}^\dagger(\vec{x}), \delta \hat{n}(\vec{x}')] = -\hat{\psi}^\dagger(x)\delta(x - x') . \qquad (10.309)$$

Notice that the extra charge created by $\psi^\dagger(x)$ is uniformly distributed across the edge, so this is different from an electron field operator, which creates charge at just one point in space.

The commutation relation (10.309) has precisely the same form as the commutation relation between the electron field operator and the density in the ordinary Luttinger model. The
only difference is the extra factor $\sqrt{v_b}$ in the definition (10.308) of the density fluctuation.
then follows from the work of Chapter 9 that a possible solution for the tunneling operator
has the form

$$\hat{\psi}^\dagger(x) = \frac{1}{\sqrt{L(1 - e^{-\frac{2\pi}{L\Lambda}})}} e^{-\hat{\theta}(x)}\hat{U} \ , \tag{10.310}$$

where

$$\hat{\theta}(x) = \sum_{q>0} \sqrt{\frac{2\pi}{Lqv_b}} \left[e^{iqx}\hat{B}_q - e^{-iqx}\hat{B}_q^\dagger \right] \tag{10.311}$$

$\hat{U}$ is an operator that commutes with all the bosons and increases the total number of
electrons on the edge by 1, and Λ is the usual ultraviolet wave vector cutoff.

Notice that this expression differs from the corresponding Eq. (9.103) for right movers in
the Luttinger model only by the the presence of the factor $\frac{1}{\sqrt{v_b}}$. Because of this extra factor
$\psi^\dagger(x)$ fails, in general, to satisfy fermionic anticommutation relations at different points.[80]
This is not necesarily a fatal flaw, since $\hat{\psi}^\dagger(x)$, as mentioned earlier, is not a true electron
field operator.

From this point on our work closely parallels the work done in Chapter 9 for the Green's
function of the Luttinger model. The tunneling conductance G is proportional to the local
density of states $A_>(\omega)$ evaluated at $\omega = \frac{eV}{\hbar}$, where V (>0) is the bias voltage. The local
density of states is given by

$$A_>(\omega) = -\frac{1}{2\pi i} \int_{-\infty}^{\infty} G_>(t)e^{i\omega t}\,dt \ , \tag{10.312}$$

where

$$G_>(t) = -\frac{i}{\hbar} \langle \hat{\psi}(0, t)\hat{\psi}^\dagger(0, 0) \rangle$$

$$= -\frac{i}{\hbar} \frac{e^{F(t)}}{L(1 - e^{-\eta})} \ , \tag{10.313}$$

and

$$F(t) = \sum_{q>0} \frac{2\pi}{qLv_b} n_B(\hbar cq) \left(e^{-icqt} - 1 \right) e^{\eta \frac{qL}{2\pi}} \ , \tag{10.314}$$

with c the speed of the $j = 0$ mode. At $v_b = 1$ the above expressions yields the Green's
function of a one-dimensional Fermi liquid. This goes as $\frac{1}{t}$ for large times at the absolute
zero of temperature, implying that $F(t)$ must be proportional to $-\ln t$ in this limit. For
$v_b < 1$, Eq. (10.314) shows that $F(t) \sim -\frac{1}{v_b}\ln t$, and therefore $G_>(t) \sim \frac{1}{t^{\frac{1}{v_b}}}$. Taking the

---

[80] An important exception occurs when v_b is the inverse of an odd integer.

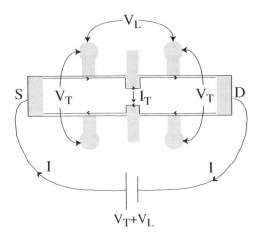

Fig. 10.32. Set-up of the experiment of Roddaro *et al.* (2003). Quasiparticle tunneling between the top and bottom edges occurs at the constriction induced by the split gate.

Fourier transform with respect to time, we immediately find $A_>(\omega) \sim \omega^{\frac{1}{\nu_b}-1}$ for $\omega \to 0$. This implies that the conductance vanishes for $V \to 0$, and the tunneling current varies as $V^{\frac{1}{\nu_b}}$ in beautiful agreement with the experimental results. Physically the suppression of the conductance is due to the fact that the incoming electron disrupts the pre-existing correlations of the χLL and creates a state that is essentially orthogonal to the ground-state. This analysis is easily extended to finite temperatures, with the predictable result that the zero-bias conductance vanishes, at low temperature, as $T^{\frac{1}{\nu_b}-1}$.[81]

Let us now turn to the recent experiments by Roddaro *et al.* (2003). The experimental set-up is shown in Fig. 10.32. A split metallic gate grown on top of a quantum Hall bar brings the edges of the quantum Hall liquid close enough to allow tunneling (remember that in the *ideal* quantum Hall effect there is no tunneling between the edges: thus, the purpose of the experimental setup is to induce small deviations from the ideal quantum Hall effect and allow their measurement). Due to the highly correlated nature of the quantum Hall liquid, the objects that tunnel between the edges are not bare electrons but quasiparticle excitations with an effective charge e^* ($e^* = -\frac{e}{3}$ at $\nu = \frac{1}{3}$). The tunneling conductance $G_T = \frac{dI_T}{dV_T}$ is measured as follows: one begins by driving a current I from source to drain, so that the potential difference V_T between the edges of the Hall bar is approximately given by $\frac{hI}{\nu_b e^2}$. Due to the presence of a weak inter-edge tunneling current, I_T, a potential difference $V_L \approx \frac{hI_T}{\nu_b e^2}$ appears across the constriction: the differential longitudinal resistance $R_L \equiv \frac{dV_L}{dI}$ is then measured as a function of I (or, equivalently, V_T), and, finally, the differential tunneling

---

[81] An interesting situation occurs when the edge is smooth, i.e., when its width greatly exceeds the magnetic length (Conti and Vignale, 1996). One finds that not only the charged mode, but each additional branch of edge waves gives a positive contribution to the tunneling exponent. Thus the exponent increases monotonically with increasing width of the edge and diverges when the latter tends to ∞. This result is understandable. In the limit of infinite width the electron liquid in the edge region becomes indistinguishable from a uniform electron liquid with unquantized fractional density. But, as discussed in Section 10.14, the tunneling density of states of such a system vanishes exponentially at low energy, hence faster than any power law.

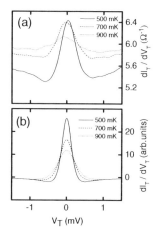

Fig. 10.33. (a) The measured tunneling conductance in Ω^{-1} at $v_b = \frac{1}{3}$ for three different temperatures, 900 mK, 700 mK, and 500 mK. (b) Theoretical prediction for the same quantity, based on Eq. (9.180). Adapted from Roddaro *et al.*, 2003.

conductance is obtained via the relation[82]

$$G_T = \frac{dI_T}{dV_T} = \left(\frac{v_b e^2}{\hbar}\right)^2 \frac{dV_L}{dI}. \tag{10.315}$$

The results of the experiment are striking: rather than being suppressed, as in the Grayson–Chang experiment, the tunneling conductance now *grows* as the potential difference V_T decreases for temperatures T such that $k_B T \ll eV$. Fig. 10.33 shows G_T having a peak at zero bias at $v = \frac{1}{3}$. It appears that instead of an orthogonality catastrophe one is now observing an "overlap catastrophe" in which electronic correlations encourage the tunneling process.

To understand the reason for this strange behavior, observe that the edges of the system can be modeled as two independent chiral Luttinger liquids: the collective excitations travel to the right in the top edge and to the left in the bottom edge.[83] Even though the two edges are physically decoupled (aside from the weak tunneling coupling, which is treated perturbatively) we can still formally regard them as the right and left halves of a non-chiral Luttinger model. More precisely, the right-propagating bosons of the top χLM are identified with the $q > 0$ bosons of the non-chiral LM, while the left-propagating bosons of the bottom χLM are identified with the $q < 0$ modes of the nonchiral LM. *The two independent chiral edges, taken together, are equivalent to a single non-chiral LM with "coupling constant"* $e^{-2\phi} \equiv g = v_b$.

---

[82] In practice, the differential longitudinal resistance is determined from the ratio $\frac{\Delta V_L}{\Delta I}$, where ΔI is the amplitude of a low-frequency modulation of the steady current I, and ΔV_L is the amplitude of the corresponding modulation in V_L.

[83] We assume that the screening length is short enough to suppress any coulomb coupling between density fluctuations on opposite edges.

We saw in Section 9.11 that the ballistic conductance of a nonchiral Luttinger model
(i.e., the source-drain conductance in the absence of impurities) would be $\frac{ge^2}{h}$ if electrostatic
screening could be neglected (it can't). But this is precisely what one is allowed to do in
the present case, due to a combination of two facts: (i) the bulk of the quantum Hall liquid
is incompressible, hence it does not screen the external field, and (ii) the coupling constant
of the chiral LM at the edge does not reflect a physical interaction between the electrons:
electron-electron interactions along the edge have already been "used" in the construction
of the collective modes. Thus, the calculation presented in Section 9.11, now applied to a
chiral Luttinger model, leads to the conclusion that the conductance of the quantum Hall
edge is indeed $v_b \frac{e^2}{h}$, as experimentally found.

In Section 9.11 we also discussed in detail how the back-scattering caused by an impurity
in the nonchiral Luttinger model (see Fig. 9.11) is amplified by interaction effects. The back-
scattering operator $\hat{\psi}_L^\dagger(x)\hat{\psi}_R(x)$ was found to be proportional to $e^{-[\hat{\theta}_L(x)-\hat{\theta}_R(x)]}$ where (see
Eq. (9.157))

$$\hat{\theta}_R(x) = \sum_{q>0} \sqrt{\frac{2\pi}{L|q|}} e^{-\phi_q} \left[e^{iqx}\hat{\beta}_q - e^{-iqx}\hat{\beta}_q^\dagger \right] , \qquad (10.316)$$

and $\hat{\theta}_L(x)$ is given by the same expression with the sum running over negative wavevectors.
On the other hand, the $\hat{\theta}$ fields for our two chiral Luttinger liquids are given by
Eq. (10.311):

$$\hat{\theta}_t(x) = \sum_{q>0} \sqrt{\frac{2\pi}{L|q|v_b}} \left[e^{iqx}\hat{\beta}_q - e^{-iqx}\hat{\beta}_q^\dagger \right] , \qquad (10.317)$$

for the operator of the "top" (t) edge, and by the same expression, summed over negative
wavevectors, for the operator $\hat{\theta}_b(x)$ of the "bottom" (b) edge. Making use of the correspondence $e^{-\phi} = \sqrt{v_b}$ we see that the relations between the two sets of operators is

$$\hat{\theta}_{R(L)}(x) = v_b\hat{\theta}_{t(b)}(x) , \qquad (10.318)$$

showing that the back-scattering operator studied in Section 9.11 can be expressed as
$e^{-v_b[\hat{\theta}_b(x)-\hat{\theta}_t(x)]}$. Notice that this operator transfers a fractional charge $e^* = v_b e$ between the
edges of the quantum Hall liquid. We have thus arrived at a very important result: *the tunnel-
ing of fractionally charged quasiparticles between the edges of a quantum Hall liquid corresponds to the back-scattering of electrons in a non-chiral Luttinger model and vice versa.*

We can now understand the enhanced tunneling of quasiparticles at a constriction between quantum Hall edges as a manifestation of the enhanced back-scattering of electrons
by an impurity in a non-chiral Luttinger model. We saw in Chapter 9 that an arbitrarily
weak impurity potential in the Luttinger model is sufficient to cause complete reflection of
electrons, turning the system into an insulator at sufficiently low voltage or temperature. In
the regime in which perturbation theory is applicable we found that the correction to the
ballistic conductance $\frac{ge^2}{h}$ is negative and scales as $|V|^{2g-2}$ for $e|V| \gg k_B T$, or as T^{2g-2} for
$e|V| \ll k_B T$. By sheer power of analogy, we can now argue that the tunneling conductance

of fractionally charged quasiparticles between the edges of a quantum Hall liquid mu
grow as $|V_T|^{2\nu_b-2}$ for $e|V_T| \gg k_BT$, or as $T^{2\nu_b-2}$ for $e|V_T| \ll k_BT$. The tunneling curre
will be given by Eq. (9.180), with g replaced by ν_b and eV_{sd} replaced by e^*V_T Numerica
results for the peak in the G_T vs V_T curve obtained from this formula at $\nu_b = \frac{1}{3}$ are plotte
in part (b) of Fig. 10.33: notice the qualitative agreement with the experimental data. Th
full source-to-drain conductance is given by $G = \frac{\nu_b e^2}{h} - G_T$.

Perturbation theory is valid only as long as $G_T \ll \frac{\nu_b e^2}{h}$ and will therefore break down a
sufficiently low temperature and/or voltage. In that limit the system is better described by
a different model, in which the potential barrier created by the split gate effectively break
the electron liquid into *two* parts separated by a very low density region. Accordingly, we
now expect ordinary *electrons*, and not fractionally charged quasiparticles, to tunnel across
the depletion region that separates the two parts of the system. This can be viewed as a
generalization of the Grayson–Chang experiment, differing from the latter only for the fac
that tunneling occurs between two χLM's, rather than between a χLM and a normal metal.
The linear tunneling conductance G_T, which in this case coincides with the source-drain
conductance G, is easily shown to vanish as $T^{2\left(\frac{1}{\nu_b}-1\right)}$ – the square of the Chang–Grayson
result. Notice that the expression for the tunneling conductance of electrons is related to
the corresponding expression for quasiparticles by the simple transformation $\nu_b \to \frac{1}{\nu_b}$: this
transformation connects the strong and the weak back-scattering regimes.

In spite of the good qualitative agreement between the data shown in Fig. 10.33(a) and
the predictions of the chiral Luttinger model, many aspect of these experiments, including
their dependence on filling factor and temperature remain puzzling. It can only be hoped
that this and other fascinating puzzles that continuing experimentation will undoubtedly
confront us with, will be successfully resolved by the thoughtful reader of this heavy tome.

Exercises

10.1 **Current carried by Landau level orbitals.** Apply the Hellman–Feynman theorem
to the eigenfunctions of Eq. (10.47) to show that the electric current carried by a
Landau-gauge eigenstate $\frac{1}{\sqrt{L}}e^{ik_y y}\psi_{n,k_y}(x)$ in the y direction is given by $-e\frac{\partial E_{n,k_y}}{\hbar \partial k_y}$.

10.2 **Coherent states in the lowest Landau level.**
(a) Show that the coherent states defined in Section 10.2.5, when written in the
Landau gauge, have the form

$$\phi_{\vec{R}}(\vec{r}) = \frac{1}{\sqrt{2\pi \ell^2}}e^{-\frac{|\vec{r}-\vec{R}|^2}{4\ell^2}}e^{i\frac{(x+X)(y-Y)}{2\ell^2}}, \tag{E10.1}$$

where $\vec{R} = (X, Y)$.
(b) Show that the overlap between two such states is

$$\langle \phi_{\vec{R}'}|\phi_{\vec{R}}\rangle = e^{-\frac{|\vec{R}-\vec{R}'|^2}{4\ell^2}}e^{i\frac{(X-X')(Y+Y')}{2\ell^2}}. \tag{E10.2}$$

(c) Making use of the identity (10.45) show that the extended state

$$\psi_0(\vec{r}) = \sum_{n_1,n_2} \phi_{(n_1+\frac{1}{2})a\vec{e}_x+(n_2+\frac{1}{2})a\vec{e}_y}(\vec{r}) , \qquad (E10.3)$$

where n_1 and n_2 are integers of any sign and $a = \sqrt{2\pi \ell^2}$ is orthogonal to all the coherent states $\phi_{m_1 a\vec{e}_x+m_2 a\vec{e}_y}(\vec{r})$. Hence, the set of the coherent states centered at the lattice sites is not complete.

   (d) Generalize the form of Eq. (E10.3) to obtain a set of "Bloch wavefunctions" $\psi_{\vec{k}}(\vec{r})$ that satisfy the boundary conditions $\psi_{\vec{k}}(x, y + Na) = \psi_{\vec{k}}(x, y)$ and $\psi_{\vec{k}}(x + Na, y) = e^{2\pi i Ny} \psi_{\vec{k}}(x, y)$, where N is an integer.

10.3 **Non-interacting kinetic energy at integer filling factor.** Show that the non-interacting kinetic energy of a 2D electron liquid occupying an integer number of Landau levels coincides with the non-interacting kinetic energy of an electron liquid of the same density in the absence of a magnetic field. [Hint: make use of Eq. (10.70).]

10.4 **One-particle density matrix of a full Landau level.** Show that it is given by Eq. (10.74).

10.5 **Matrix elements of the density fluctuation operator between Landau-gauge orbitals in the same or in different Landau levels.** Verify that they are given by Eq. (10.82).

10.6 **Static structure factor for non-interacting electrons at integer filling factor** Show that it is given by Eq. (10.86).

10.7 **Density–density response function of the non-interacting electron liquid at filling factor $\nu = 1$.** Derive the closed form expression reported in footnote [28].

10.8 **Effective interaction between charged particles in a weak magnetic field.** Making use of the limiting form (10.87) show that the static effective interaction between two test particles, computed as the Fourier transform of $\frac{v_q}{\epsilon_{RPA}(q,0)}$ in a weak magnetic field ($\nu \gg 1$) is approximately given by Eq. (10.89) in the range $R_c \ll r \ll \frac{R_c^2}{a_B}$, where $R_c = k_F \ell^2 = \sqrt{2\nu}\ell$ is the cyclotron radius.

10.9 **Ensemble-averaged kinetic and exchange energy for fractionally filled Landau levels.** Verify that the average of the kinetic energy of a 2D electron liquid in a magnetic field in the non-interacting thermal ensemble in the limit of zero temperature is given by Eq. (10.92), and that the same average for the potential energy is given by Eq. (10.93).

10.10 **Hartree–Fock potential in the lowest Landau level.** Making use of Eq. (A23.6) show that the Hartree–Fock potential *within the lowest Landau level* has the form

$$V_{HF}(\vec{r}) = V_{ext}(\vec{r}) + \int d\vec{r}' \left[\frac{e^2}{\epsilon_b |\vec{r} - \vec{r}'|} + V_x(\vec{r} - \vec{r}') \right] n(\vec{r}') , \qquad (E10.4)$$

where $n(\vec{r})$ is the density and the two-dimensional Fourier transform of $V_x(\vec{r})$ is given by

$$V_x(\vec{q}) = -\frac{2\pi e^2 \ell}{\epsilon_b} \sqrt{\frac{\pi}{2}} e^{\frac{q^2 \ell^2}{4}} I_0\left(\frac{q^2 \ell^2}{4}\right).$$
(E10.5)

10.11 **Derivation of the magneto-exciton hamiltonian.** Derive Eqs. (10.102) and (10.103) for the direct and the exchange terms of the effective magneto-exciton hamiltonian.

10.12 **Magneto-exciton wave function.** Using the explicit forms of the Landau level orbitals work out the real-space expression for the magneto-exciton wave function (10.96).

10.13 **Magneto-exciton dispersion.** Calculate the dispersion of the magneto-exciton for $v_\uparrow = v, v_\downarrow = 0$.

10.14 **Kohn's theorem for the many-electron wave function.** Show that the wavefunction of a homogeneous many-electron system in a uniform magnetic field that evolves under the influence of a time-dependent and uniform electric field is given by Eq. (10.109), where $\vec{R}_{cm}(t)$ and $\vec{P}_{cm}(t)$ are the solutions of the classical equations of motion for the center of mass in the presence of the magnetic and the electric field, with initial conditions $\vec{R}_{cm}(0) = \vec{P}_{cm}(0) = 0$.

10.15 **Generalized Kohn's theorem for the wave function.** Show that Kohn's theorem continues to hold if the electrons are also subjected to a static harmonic potential. [Hint: a harmonic potential admits the separation

$$\frac{m_b}{2}\omega_0^2 \sum_{i=1}^{N} \hat{r}_i^2 = \frac{m_b}{2}\omega_0^2 \left[N\hat{R}_{cm}^2 + \sum_{i=1}^{N} \hat{r'}_i^2 \right],$$

where $\vec{r'}_i = \vec{r}_i - \vec{R}_{cm}$.]

10.16 **Macroscopic current–current response function in the presence of a static parabolic potential.** Generalize Eq. (10.113) for the macroscopic ($q = 0$) current–current response function to include a static parabolic potential with oscillator frequency ω_0 acting on the electrons.

10.17 **The single-mode approximation yields the correct first moment of the spectral function.** This is the content of Eq. (10.177): verify it.

10.18 **Strictly two-dimensional jellium model potential.** Verify that Eq. (10.129) is the correct expression for the potential created by a uniform background of charge in the strictly two-dimensional electrostatics defined by Eq. (10.128).

10.19 **Model for which the Laughlin wave function is the exact ground-state.** Show that the Laughlin wave function is the exact ground-state wave function (in the lowest Landau level) of an interacting system in which all the Haldane pseudopotentials with $m \geq 3$ are zero. Show that the ground-state energy is zero.

10.20 **Small-q behavior of the static structure factor from Kohn's theorem.** Show that the leading term in the small-q expansion of the static structure factor of a uniform electron liquid in a magnetic field is constrained by Kohn's theorem to have the form reported in Eq. (10.137). [Hint: start from the current–current response functions given in Eq. (10.113), derive from these the small-q form of the density–density response function, and combine the latter with the fluctuation-dissipation theorem to obtain the small-q behavior of $S(q)$.]

0.21 **Perfect screening sum rule.**
(a) Show that the form of the small-q expansion of $S(q)$ in Eq. (10.137) implies the sum rule (10.138).
(b) What does this have to do with "perfect screening"?

0.22 **Large-q behavior of the projected structure factor.** Show that the large-q behavior of the LLL-projected structure factor (10.169) is given by Eq. (10.171). [Hint: combine Eqs. (10.131) for the structure factor with Eq. (10.130) for the pair correlation function, and observe that the Fourier transform of $g(r) - 1$ is dominated, for large q, by the contribution of the first two terms on the right-hand side of Eq. (10.130).]

10.23 **Electron–hole symmetry in the lowest Landau level.** Assuming that the ground-state wave function lies entirely within the lowest Landau level show that the energy per particle of the 2DEL satisfies the electron–hole symmetry relation

$$v\,[\epsilon(v) - \epsilon(1)] = (1 - v)\,[\epsilon(1 - v) - \epsilon(1)] \ . \tag{E10.6}$$

[Hint: start from the second-quantized form of the hamiltonian in the LLL and apply the electron–hole transformation $\hat{a}_{e,k_y} \to \hat{a}^\dagger_{h,-k_y}$ where $\hat{a}^\dagger_{e,k_y}$ creates an electron of momentum k_y in the vacuum and $\hat{a}^\dagger_{h,-k_y}$ creates a hole of momentum $-k_y$ in the full Landau level.] Notice that Eq. (10.143) for the energy does not satisfy electron–hole symmetry. A similar expression that does satisfy electron–hole symmetry was proposed by Fano and Ortolani (1988) and reads

$$\epsilon(v) = -\frac{e^2}{\epsilon_b \ell} \left[\sqrt{\frac{\pi}{8}}\, v + 0.782133\, v^{\frac{1}{2}}(1 - v)^{\frac{3}{2}} \right.$$
$$\left. -0.55\, v(1 - v)^2 + 0.463\, v^{\frac{3}{2}}(1 - v)^{\frac{5}{2}} \right] \ . \tag{E10.7}$$

10.24 **Berry phase for fractionally charged excitations.** Show that the phase γ accumulated by a Laughlin quasihole that is transported around a closed loop according to Eq. (10.150) equals the average number of particles enclosed by the integration contour.

10.25 **Commutation relations between projected density fluctuations in real space.** Expand the commutation relation (10.173) between the lowest Landau level-projected density fluctuation to leading order in $k\ell$ and $q\ell$ and transform to real space to get the commutation relations (10.301).

10.26 **Bosonization of the elastic displacement field.** Verify that the expression (10.205) for the elastic displacement field satisfies the canonical commutation relation (10.203).

10.27 **Bosonization of the two-dimensional electron liquid in the lowest Landau level.** Replace the bosonized expression for the Fourier components of the elastic displacement field

$$\hat{u}_{\vec{q}} = \hat{u}^{\dagger}_{-\vec{q}} = \frac{\vec{u}_{-}(\vec{q})}{\sqrt{n_0}} \, \hat{b}_{\vec{q}}$$

in the hamiltonian (10.204) of the effective elasticity theory, and make use of the explicit form of the eigenfunctions $\vec{u}_{-}(\vec{q})$, given in Eq. (10.189), to show that the hamiltonian reduces to a sum of independent harmonic oscillators, i.e., to Eq. (10.206)

10.28 **Screening charge at the interface of two dielectric media.** Use Gauss' theorem to show that an electric charge density $-ne$ placed at the interface between two media of dielectric constants ϵ_a and ϵ_b is reduced by polarization charges to the effective value $-\frac{en}{\frac{\epsilon_a+\epsilon_b}{2}}$ (see also footnote[72]).

10.29 **Electrostatics of edge channels.** Verify that the electric potential (10.274) together with the electronic density distribution (10.275) provide a solution to the electrostatic problem defined in Section 10.20.2.

10.30 **Hartree–Fock energy of the stripe phase.** Express the Hartree–Fock energy of the stripe phase in the n-th Landau level in terms of the order parameter $\Delta(\vec{q})$, defined by Eq. (10.271).

Appendix 1

Fourier transform of the coulomb interaction in low dimensional systems

In this appendix we develop the general procedure for determining the appropriate form of the interaction potential in systems of reduced dimensionality. To be specific, we examine the representation of the Yukawa interaction

$$v(r, \kappa) = e^2 \frac{e^{-\kappa r}}{r} \qquad (A1.1)$$

in an infinite cylindrical wire of radius a (Fig. A1.1(a)) and in an infinite slab of thickness a (Fig. A1.1(b)).

In both cases a is assumed to be so small that all the electrons reside in the lowest subband, ϕ_0, of transverse motion, and higher subbands are so high in energy as to be practically irrelevant. Thus, the only relevant one-electron wave functions have the form

$$\phi_{\vec{k}_\parallel \sigma}(\vec{r}_\parallel, \vec{r}_\perp, s) = \phi_0(\vec{r}_\perp) \frac{e^{i\vec{k}_\parallel \cdot \vec{r}_\parallel}}{\sqrt{L^d}} \delta_{s\sigma} , \qquad (A1.2)$$

where $\vec{r}_\parallel$ and $\vec{r}_\perp$ (and, similarly, $\vec{k}_\parallel$ and $\vec{k}_\perp$) are the components of $\vec{r}$ (and $\vec{k}$) parallel and perpendicular to the structure (e.g. $\vec{r}_\perp = (x, y)$ and $\vec{r}_\parallel = z$ in the 1D wire). Notice that these states are labelled by quantum numbers $\vec{k}_\parallel$ and σ, just as the states of a literally d-dimensional system, and will be denoted, in what follows, by $|\vec{k}_\parallel \sigma\rangle$.

The objective of our exercise is to construct the electron-electron interaction within the subspace spanned by the states of the form (A1.2). To this end, we consider the matrix element of the interaction between two two-electron states formed from the product of wave functions in the lowest subband:

$$v_{q_\parallel}(\kappa) \equiv e^2 \langle i, \vec{k}_{\parallel,1} + \vec{q}_\parallel \, \sigma_1 | \langle j, \vec{k}_{\parallel,2} - \vec{q}_\parallel \, \sigma_2 | \frac{e^{-\kappa r_{ij}}}{r_{ij}} | j, \vec{k}_{\parallel,2} \sigma_2 \rangle | i, \vec{k}_{\parallel,1} \sigma_1 \rangle , \qquad (A1.3)$$

where $r_{ij} \equiv |\vec{r}_i - \vec{r}_j|$. Making use of the three dimensional Fourier representation of $v_\kappa(r)$,

$$e^2 \frac{e^{-\kappa r_{ij}}}{r_{ij}} = \int \frac{d\vec{q}}{(2\pi)^3} \frac{4\pi e^2}{q^2 + \kappa^2} e^{i\vec{q} \cdot (\vec{r}_i - \vec{r}_j)}, \qquad (A1.4)$$

we can rewrite $v_{q_\parallel}(\kappa)$ more compactly as

$$v_{q_\parallel}(\kappa) = \int \frac{d\vec{q}_\perp}{(2\pi)^{3-d}} \frac{4\pi e^2}{q_\perp^2 + q_\parallel^2 + \kappa^2} |F(\vec{q}_\perp)|^2 , \qquad (A1.5)$$

(a)   (b)

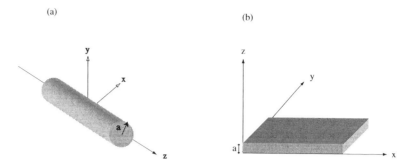

Fig. A1.1. Two standard geometries for the electron liquid in low dimension: thin 1D wire and flat 2D slab.

where

$$F(\vec{q}_\perp) = \int d\vec{r}_\perp |\phi_0(\vec{r}_\perp)|^2 e^{-i\vec{q}_\perp \cdot \vec{r}_\perp} . \qquad (A1.6)$$

The quantity $v_{q_\parallel}(\kappa)$ defined above – a function of the magnitude of $\vec{q}_\parallel$ – constitutes our definition of the Fourier transform of the interaction in a system of reduced dimensionality. The quantity $|F(\vec{q}_\perp)|^2$ is the *form factor* associated with the transverse density distribution $|\phi_0(\vec{r}_\perp)|^2$.

To be more explicit, we need to specify the form of the transverse wavefunction. Let us assume, for the sake of definiteness, that ϕ_0 has the gaussian form

$$\phi_0(\vec{r}_\perp) = \left(\frac{2}{\pi a^2}\right)^{\frac{3-d}{4}} e^{-r_\perp^2/a^2} , \qquad (A1.7)$$

where d is the dimensionality of the system, i.e., $d = 1$ for the wire and $d = 2$ for the slab. Then Eq. (A1.6) gives

$$F(\vec{q}_\perp) = e^{-q_\perp^2 a^2/2} \qquad (A1.8)$$

for both values of d. Substituting this in Eq. (A1.5) we obtain for a 1D wire ($\vec{q}_\perp = (q_x, q_y)$)

$$\begin{aligned}
v_{q_\parallel}(\kappa) &= 4\pi e^2 \int \frac{d\vec{q}_\perp}{(2\pi)^2} \frac{e^{-q_\perp^2 a^2}}{q_\perp^2 + q_\parallel^2 + \kappa^2} \\
&= e^2 \int_0^\infty dt \frac{e^{-t}}{t + (q_\parallel^2 + \kappa^2) a^2} \\
&= -e^2 e^{(q_\parallel^2 + \kappa^2) a^2} \text{Ei}\left(-(q_\parallel^2 + \kappa^2) a^2\right) \quad , \quad 1D \text{ wire} , \qquad (A1.9)
\end{aligned}$$

where $\text{Ei}(x)$ is the *exponential-integral function* (see Gradshteyn and Rhyzhik, 1965, 8.21). The above expression coincides with the one given in Eq. (1.18) with $q_\parallel = q$.

As immediately determined from its definition, $\text{Ei}(x)$ diverges logarithmically for $x \to 0$. We then see that the $a \to 0$ limit of $v_{q_\parallel}(\kappa)$ does not exist. Thus, the use of a finite a in one dimensional wires, is not only a physically realistic feature, but also a mathematical necessity. From a physical view point the need for a finite a stems from the obvious fact at in literally one dimensional space electrons cannot avoid each other, and therefore ce the full brunt of their mutual Coulomb interaction, which diverges as $\frac{1}{r}$ for $r \to 0$. On the other hand, for a two-dimensional slab ($\vec{q}_\perp = q_z$), we have

$$
v_{q_\parallel}(\kappa) = 4\pi e^2 \int \frac{dq_z}{2\pi} \frac{e^{-q_z^2 a^2}}{q_z^2 + q_\parallel^2 + \kappa^2}
$$

$$
= 4 e^2 a \int_0^\infty dt \, \frac{e^{-t^2}}{t^2 + (q_\parallel^2 + \kappa^2) a^2}
$$

$$
= \frac{2\pi e^2}{\sqrt{q_\parallel^2 + \kappa^2}} \left[1 - \mathrm{erf}\left(a\sqrt{q_\parallel^2 + \kappa^2} \right) \right] e^{a\sqrt{q_\parallel^2 + \kappa^2}} , \quad 2D \text{ slab}, \quad (A1.10)
$$

where $\mathrm{erf}(x) \equiv \frac{2}{\sqrt{\pi}} \int_0^x e^{-t^2} dt$ is the *error function* (see Gradshteyn and Rhyzhik, 1965, 3.466). At variance with the 1D case, the $a \to 0$ limit of this expression exists, and yields the interaction quoted in Eq. (1.17), with $q_\parallel = q$.

Appendix 2

Second-quantized representation of some useful operators

In this appendix we derive for easy reference the second-quantized representations of a number of operators of frequent use.

- **Spin resolved densities**

We define the density operators for $\uparrow$ and $\downarrow$ spins in the following manner:

$$\hat{n}_{\uparrow(\downarrow)}(\vec{r}) = \sum_i \frac{1 \pm \hat{\sigma}_{z,i}}{2} \delta(\vec{r} - \hat{\vec{r}}_i), \tag{A2.1}$$

where $\hat{\sigma}_{z,i}$ is the operator of the z-component of the spin of the i-th electron, expressed in units of $\frac{\hbar}{2}$, and the sum runs over all the electrons in the system, $i = 1 \ldots N$. The upper and the lower signs apply to the $\uparrow$ and $\downarrow$ components of the spin respectively.

Throughout this book we use the standard representation of the components of the spin operator (in units of $\frac{\hbar}{2}$) as 2×2 matrices, known as the *Pauli matrices*:

$$\hat{\sigma}_x = \begin{pmatrix} 0 & 1 \\ 1 & 0 \end{pmatrix}, \quad \hat{\sigma}_y = \begin{pmatrix} 0 & -i \\ i & 0 \end{pmatrix}, \quad \hat{\sigma}_z = \begin{pmatrix} 1 & 0 \\ 0 & -1 \end{pmatrix}. \tag{A2.2}$$

These matrices have the following properties:

$$\hat{\sigma}_\alpha^2 = \hat{1},$$
$$\hat{\sigma}_\alpha \hat{\sigma}_\beta = -\hat{\sigma}_\beta \hat{\sigma}_\alpha = i \sum_\gamma \epsilon_{\alpha\beta\gamma} \hat{\sigma}_\gamma \quad (\alpha \neq \beta),$$
$$\left[\hat{\sigma}_\alpha, \hat{\sigma}_\beta \right] = 2i \sum_\gamma \epsilon_{\alpha\beta\gamma} \hat{\sigma}_\gamma, \tag{A2.3}$$

where the greek indices denote cartesian components (x, y, or z) and $\epsilon_{\alpha\beta\gamma}$ is the Levi–Civita antisymmetric tensor.

Because $\hat{\sigma}_{z,i}$ has eigenvalues ± 1, it is clear that the sum in Eq. (A2.1) is effectively restricted to the electrons of a given spin orientation. Notice that

$$\hat{n}(\vec{r}) = \hat{n}_\uparrow(\vec{r}) + \hat{n}_\downarrow(\vec{r}) \tag{A2.4}$$

is the complete number density operators, while

$$\hat{S}_z(\vec{r}) = \hat{n}_\uparrow(\vec{r}) - \hat{n}_\downarrow(\vec{r}) \tag{A2.5}$$

is the z-component of the spin density operator in units of $\frac{\hbar}{2}$.

The Fourier transformed density operators are given by

$$\hat{n}_{\vec{q},\uparrow(\downarrow)} = \int \hat{n}_{\uparrow(\downarrow)}(\vec{r}) \, e^{-i\vec{q}\cdot\vec{r}} \, d\vec{r} = \sum_i \frac{1 \pm \hat{\sigma}_{z,i}}{2} e^{-i\vec{q}\cdot\hat{\vec{r}}_i}. \tag{A2.6}$$

These are operators of the general form (1.57), so their second-quantized representation in terms of plane waves is

$$\hat{n}_{\vec{q},\uparrow(\downarrow)} = \sum_{\vec{k}\sigma,\vec{k}'\sigma'} \langle \vec{k}\sigma| \frac{1 \pm \hat{\sigma}_z}{2} e^{-i\vec{q}\cdot\hat{\vec{r}}} |\vec{k}'\sigma'\rangle \, \hat{a}^\dagger_{\vec{k}\sigma} \hat{a}_{\vec{k}'\sigma'}$$

$$= \sum_{\vec{k}\sigma} \hat{a}^\dagger_{\vec{k}-\vec{q}\,\uparrow(\downarrow)} \hat{a}_{\vec{k}\uparrow(\downarrow)}. \tag{A2.7}$$

• **Spin density**

The spin density operator $\hat{\vec{S}}(\vec{r})$ (in units of $\frac{\hbar}{2}$) is defined as follows:

$$\hat{\vec{S}}(\vec{r}) = \sum_i \hat{\vec{\sigma}}_i \, \delta(\vec{r} - \hat{\vec{r}}_i), \tag{A2.8}$$

where the spin operators $\hat{\vec{\sigma}}_i$ act in the spin space of the i-th electron and are defined by Eq. (A2.2).

The corresponding expression in Fourier space is

$$\hat{\vec{S}}_{\vec{q}} = \int \hat{\vec{S}}(\vec{r}) \, e^{-i\vec{q}\cdot\vec{r}} \, d\vec{r} = \sum_i \hat{\vec{\sigma}}_i \, e^{-i\vec{q}\cdot\hat{\vec{r}}_i}, \tag{A2.9}$$

and its Fourier transform has the cartesian components

$$\hat{S}_{\vec{q},\alpha} = \sum_{\vec{k}\mu\nu} (\hat{\sigma}_\alpha)_{\mu\nu} \, \hat{a}^\dagger_{\vec{k}-\vec{q}\,\mu} \hat{a}_{\vec{k}\nu}. \tag{A2.10}$$

Thus, for example, we have

$$\hat{S}_{\vec{q},x} = \sum_{\vec{k}} \left(\hat{a}^\dagger_{\vec{k}-\vec{q}\,\uparrow} \hat{a}_{\vec{k}\downarrow} + \hat{a}^\dagger_{\vec{k}-\vec{q}\,\downarrow} \hat{a}_{\vec{k}\uparrow} \right). \tag{A2.11}$$

• **Current density**

The *current density operator* $\hat{\vec{j}}(\vec{r})$ is defined in terms of the velocity and position operators $\hat{\vec{v}}_i$ and $\hat{\vec{r}}_i$ of the i-th electron as follows:

$$\hat{\vec{j}}(\vec{r}, t) = \frac{1}{2} \sum_i [\hat{\vec{v}}_i \delta(\vec{r} - \hat{\vec{r}}_i) + \delta(\vec{r} - \hat{\vec{r}}_i) \hat{\vec{v}}_i]. \tag{A2.12}$$

The velocity is in turn related to the canonical momentum operator by the relation

$$\hat{\vec{v}}_i = \frac{1}{m} \left[\hat{\vec{p}}_i + \frac{e}{c} \vec{A}(\hat{\vec{r}}_i) \right], \tag{A2.13}$$

where $\vec{A}(\vec{r})$ is the electromagnetic vector potential.[1]

---

[1] See Feynman's *Lectures on Physics* for an engaging discussion of the physical origin of the velocity–momentum relation.

It follows that the current density can be written as

$$\hat{\vec{j}}(\vec{r}) = \hat{\vec{j}}_p(\vec{r}) + \frac{e}{mc}\hat{n}(\vec{r})\vec{A}(\vec{r}),\qquad\qquad\text{(A2.14)}$$

where we have defined the *paramagnetic current density operator* $\hat{\vec{j}}_p(\vec{r})$ as

$$\hat{\vec{j}}_p(\vec{r}) = \frac{1}{2m}\sum_i[\hat{\vec{p}}_i\delta(\vec{r}-\hat{\vec{r}}_i)+\delta(\vec{r}-\hat{\vec{r}}_i)\hat{\vec{p}}_i].\qquad\text{(A2.15)}$$

The second term on the right-hand side of Eq. (A2.14) is known as the *diamagnetic current density*.

The Fourier components of $\hat{\vec{j}}_p(\vec{r})$ are given by

$$
\begin{aligned}
\hat{\vec{j}}_{p,\vec{q}} &= \int \hat{\vec{j}}_p(\vec{r})\, e^{-i\vec{q}\cdot\vec{r}}\, d\vec{r}\\
&= \frac{1}{2m}\sum_i\left[\hat{\vec{p}}_i e^{-i\vec{q}\cdot\hat{\vec{r}}_i}+e^{-i\vec{q}\cdot\hat{\vec{r}}_i}\hat{\vec{p}}_i\right]\\
&= \frac{1}{m}\sum_i\left(\hat{\vec{p}}_i+\frac{\hbar\vec{q}}{2}\right)e^{-i\vec{q}\cdot\hat{\vec{r}}_i},
\end{aligned}\qquad\text{(A2.16)}
$$

and those of $\hat{\vec{j}}(\vec{r})$ by

$$\hat{\vec{j}}_{\vec{q}} = \hat{\vec{j}}_{p,\vec{q}} + \frac{e}{mcL^d}\sum_{\vec{k}}\vec{A}_{\vec{q}-\vec{k}}\hat{n}_{\vec{k}},\qquad\text{(A2.17)}$$

where $\vec{A}_{\vec{q}}$ is the Fourier transform of $\vec{A}(\vec{r})$.

The second-quantized representation of the paramagnetic current operator follows immediately from the matrix element (Exercise 18)

$$\langle\vec{k}\sigma|\hat{\vec{j}}_{p,\vec{q}}|\vec{k}'\sigma'\rangle = \frac{\hbar}{m}\left(\vec{k}+\frac{\vec{q}}{2}\right)\delta_{\vec{k},\vec{k}'-\vec{q}}\delta_{\sigma\sigma'}.\qquad\text{(A2.18)}$$

According to Eq. (1.57) we then have

$$\hat{\vec{j}}_{p,\vec{q}} = \frac{\hbar}{m}\sum_{\vec{k}\sigma}\left(\vec{k}+\frac{\vec{q}}{2}\right)\hat{a}^{\dagger}_{\vec{k}-\vec{q}\,\sigma}\hat{a}_{\vec{k}\sigma}.\qquad\text{(A2.19)}$$

• **Spin current density**

In modern applications of electron liquid theory it is often necessary to consider, in addition to the ordinary particle current density, the current of spin angular momentum or *spin current*. In the absence of spin–orbit coupling this is best defined as the tensor operator

$$[\hat{\mathbf{J}}(\vec{r})]_{\alpha\beta} = \frac{1}{2}\sum_i\hat{\sigma}_{\alpha,i}[\hat{v}_{\beta,i}\delta(\vec{r}-\hat{\vec{r}}_i)+\delta(\vec{r}-\hat{\vec{r}}_i)\hat{v}_{\beta,i}],^{*}\qquad\text{(A2.20)}$$

---

* If spin–orbit interaction is present, the product $\hat{\sigma}_{\alpha,i}\hat{v}_{\beta,i}$ must be replaced by $\frac{1}{2}(\hat{\sigma}_{\alpha,i}\hat{v}_{\beta,i}+\hat{v}_{\beta,i}\hat{\sigma}_{\alpha,i})$.

which gives the β-component of the current density for the α-component of the spin (in units of $\frac{\hbar}{2}$). The Fourier transform of this operator has the second-quantized representation

$$[\hat{\mathbf{J}}_{\vec{q}}]_{\alpha\beta} = \frac{\hbar}{m} \sum_{\vec{k}\mu\nu} [\hat{\sigma}_\alpha]_{\mu\nu} \left(k_\beta + \frac{q_\beta}{2} \right) \hat{a}^\dagger_{\vec{k}-\vec{q}\,\mu} \hat{a}_{\vec{k}\nu} + \frac{e}{mcL^d} \sum_{\vec{k}} A_{\vec{q}-\vec{k},\beta} \hat{S}_{\vec{k},\alpha} \, . \tag{A2.21}$$

Continuity equations

A fundamental relationship exists between the particle density operator and the current density operator of particles whose number is locally conserved. It is known as the *continuity equation* and is expressed as

$$\frac{\partial}{\partial t} \hat{n}(\vec{r}) = -\vec{\nabla} \cdot \hat{\vec{j}}(\vec{r}) \, , \tag{A2.22}$$

where the time derivative of $\hat{n}(\vec{r})$ – as well as the time derivative of any quantum mechanical operator in the Heisenberg picture of the time evolution – is defined by

$$\frac{\partial}{\partial t} \hat{n}(\vec{r}) \equiv -\frac{i}{\hbar} [\hat{n}(\vec{r}), \hat{H}] \, . \tag{A2.23}$$

The continuity equation can be verified via a direct calculation of the commutator of the density operator with the full many-body hamiltonian (see Exercise 19). Actually, only the kinetic energy operator needs to be considered, since the potential energy commutes with $\hat{n}_{\vec{q}}$ (see however Section 9.3 for an interesting exception to this rule).

By Fourier transforming in space both sides of Eq. (A2.22) we arrive at

$$\frac{\partial}{\partial t} \hat{n}_{\vec{q}} = -i\vec{q} \cdot \hat{\vec{j}}_{\vec{q}} \, . \tag{A2.24}$$

Notice that this equation relates the density operator to the *longitudinal component* of the current operator, but says nothing about the transverse component of the current operator.

It is also possible to derive a kind of continuity equation for the vector spin density operator (A2.8), even though this quantity is not conserved when an external magnetic field is present. Assuming that the interaction between the magnetic field and the spin density operator has the standard Zeeman form

$$\hat{H}_Z = \frac{g\mu_B}{2} \int \hat{\vec{S}}(\vec{r}) \cdot \vec{B}(\vec{r}) d\vec{r} \tag{A2.25}$$

we find that the evaluation of the commutator of $\hat{S}_z(\vec{r})$ with the hamiltonian (i.e., in practice, with the kinetic energy and with the Zeeman energy) leads to the generalized continuity equation

$$\frac{\partial}{\partial t} \hat{\vec{S}}(\vec{r}) = -\vec{\nabla} \cdot \hat{\vec{\mathbf{J}}}(\vec{r}) + \frac{g\mu_B}{\hbar} \vec{B}(\vec{r}) \times \hat{\vec{S}}(\vec{r}) \, . \tag{A2.26}$$

The second term on the right-hand side represents the non-conservation of spin angular momentum due to the torque exerted by the magnetic field on the local magnetic moment. Notice that the divergence of the spin current tensor in the above expression is taken with respect to the *second index*, i.e. $[\vec{\nabla} \cdot \hat{\vec{\mathbf{J}}}(\vec{r})]_\alpha = \sum_\beta \nabla_{r_\beta} [\hat{\vec{\mathbf{J}}}(\vec{r})]_{\alpha\beta}$.

Appendix 3

Normal ordering and Wick's theorem

In practical applications of the second-quantization formalism one often needs to calculate the average of long products of creation and destruction operators in the vacuum state $|0\rangle$. This daunting task is greatly simplified by the use of a theorem due to Giancarlo Wick and based on the notion of *normal ordering*, which we now describe (Wick, 1950). For definiteness we focus on the fermion case, but we will state all the theorems in a way that is valid for bosons too.

A3.1 Normal ordering with respect to the vacuum

A product of creation and destruction operators is said to be "normal-ordered with respect to the vacuum" if all the destruction operators (i.e., the operators that annihilate the vacuum) are placed on the right of all the creation operators. Let us denote by $\hat{A}$, $\hat{B}$, $\hat{C}$ ... some arbitrary creation or destruction operators. The operation that implements normal order on the product $\hat{A}\hat{B}\hat{C}$ ... is called *normal ordering* and is denoted by *colon brackets*. Thus, we define

$$: \hat{A}\hat{B}\hat{C}\ldots : \equiv (-1)^P\, \hat{B}\hat{C}\hat{D}\ldots \tag{A3.1}$$

where the operators on the right-hand side are normal-ordered, and P is the number of interchanges of fermion operators that are needed to go from the initial ordering $\hat{A}\hat{B}\hat{C}$ ... to the normal ordering $\hat{B}\hat{C}\hat{D}$ .... A little thought leads to the conclusion that this definition is completely unambiguous, even though the final ordering of the operators is not unique. For example

$$: \hat{a}_2^\dagger \hat{a}_3 \hat{a}_4^\dagger \hat{a}_2 \hat{a}_1^\dagger : = -\hat{a}_2^\dagger \hat{a}_4^\dagger \hat{a}_1^\dagger \hat{a}_3 \hat{a}_2. \tag{A3.2}$$

The minus sign arises because of the three interchanges needed to implement normal ordering. Had we decided to put $\hat{a}_4^\dagger$ before $\hat{a}_2^\dagger$ (which is still a normal-ordered arrangement), we would have had one more interchange, and therefore a plus sign, consistent with the fact that $\hat{a}_4^\dagger \hat{a}_2^\dagger = -\hat{a}_2^\dagger \hat{a}_4^\dagger$. It should be clear from the above definition that the operators within colon brackets can be freely permuted, provided one includes a factor -1 for each interchange of fermion operators.

The most important property of a normal-ordered product of operators is also the easiest to see: its vacuum average is zero, because there is always either a destruction perator on the right or a creation operator on the left to kill the vacuum state:

$$\langle 0| : \hat{A}\hat{B}\hat{C}\ldots : |0\rangle = 0. \tag{A3.3}$$

A somewhat more subtle notion is that of *contraction* of two operators $\hat{A}$ and $\hat{B}$. This is efined as the difference between the regular product and the normal-ordered product of he two operators, and is denoted by equal numbers of dots affixed to the operators to be ontracted together, e.g.

$$\hat{A}^{\cdot}\hat{B}^{\cdot} \equiv \hat{A}\hat{B} - : \hat{A}\hat{B} : \tag{A3.4}$$

Because the anticommutator of $\hat{A}$ and $\hat{B}$ is a c-number it is easy to see that the contraction of $\hat{A}$ and $\hat{B}$ must be a c-number too. Indeed : $\hat{A}\hat{B}$: is either $\hat{A}\hat{B}$ or $-\hat{B}\hat{A}$, implying that he contraction of the two operators is either 0, or $\{\hat{A}, \hat{B}\}$. Since the average of a c-number s the c-number itself, taking the vacuum average of both sides of Eq. (A3.4) we obtain an alternative definition of the contraction

$$\hat{A}^{\cdot}\hat{B}^{\cdot} = \langle 0|\hat{A}\hat{B}|0\rangle. \tag{A3.5}$$

Yet a third way to think about the contraction $\hat{A}^{\cdot}\hat{B}^{\cdot}$ is to notice that it vanishes if either $\hat{B}$ is a destruction operator or $\hat{A}$ is a creation operator, and equals $\{\hat{A}, \hat{B}\}$ otherwise.

Eqs. (A3.4) and (A3.5) are just the simplest instance of a much more general theorem according to which any product of N creation and/or destruction operators operators can be expanded into a sum of normal ordered products of $N, N-2, N-4\ldots$ operators times $0, 1, 2\ldots$ contractions respectively. Wick's theorem is precisely stated and proved next.

A3.2 Wick's theorem

Let $\hat{A}\hat{B}\hat{C}\ldots\hat{W}\hat{X}\hat{Y}$ be a product of N creation and/or destruction operators in arbitrary order. How is this related to the normal-ordered product of the same operators? To go from the original ordering to the normal ordering we must interchange operators that are not already in normal order. Every time we do this, we pick a minus sign plus the difference between the original and the normal ordered product of the two operators, that is, their contraction. The remaining $N-2$ operators must still be put in normal order, generating additional contractions. This line of reasoning suggests that the difference between the original and the normal ordered product can be expanded in a sum of terms containing $1, 2, 3\ldots$ contractions (up to a maximum of $\left[\frac{N}{2}\right]$ – the integral part of $\frac{N}{2}$). The precise statement of this idea is the content of *Wick's theorem*:

$$\hat{A}\hat{B}\hat{C}\ldots\hat{X}\hat{W}\hat{Y} = \; : \hat{A}\hat{B}\hat{C}\ldots\hat{W}\hat{X}\hat{Y} :$$
$$+ : \hat{A}^{\cdot}\hat{B}^{\cdot}\hat{C}\ldots\hat{W}\hat{X}\hat{Y} : + : \hat{A}^{\cdot}\hat{B}\hat{C}^{\cdot}\ldots\hat{W}\hat{X}\hat{Y} : + \ldots + : \hat{A}\hat{B}\hat{C}\ldots\hat{W}\hat{X}^{\cdot}\hat{Y}^{\cdot} :$$
$$+ : \hat{A}^{\cdot}\hat{B}^{\cdot}\hat{C}^{\cdot\cdot}\ldots\hat{W}^{\cdot\cdot}\hat{X}\hat{Y} : + \ldots + : \hat{A}^{\cdot}\hat{B}^{\cdot}\hat{C}\ldots\hat{W}\hat{X}^{\cdot\cdot}\hat{Y}^{\cdot\cdot} : + : \hat{A}\hat{B}\hat{C}\ldots\hat{W}^{\cdot}\hat{X}^{\cdot\cdot}\hat{Y}^{\cdot\cdot} :$$
$$+ \ldots$$
$$+ \hat{A}^{\cdot}\hat{B}^{\cdot}\hat{C}^{\cdot\cdot}\ldots\hat{W}^{\cdot\cdot}\hat{X}^{\cdots}\hat{Y}^{\cdots} + \ldots \; . \tag{A3.6}$$

The second line of this equation is meant to represent the sum of all the terms that contain only one pair of contracted operators. Similarly, the second line represents the sum of all terms that contain two pairs of contracted operators, and so on until the last line, which represents all the terms containing the maximum possible number $\left[\frac{N}{2}\right]$ of contractions.

Remember that the contractions, being c-numbers, can always be pulled out of the colon brackets. Also, in writing Eq. (A3.6) we have stipulated that *the contraction of two operators within a product of many operators is executed by bringing the two operators next to each other and multiplying by a factor −1 for each interchange of fermion operators that is necessary to do this.* Since the contraction of two operators that are already in normal order vanishes, we see that effectively only the pairs of operators that were not initially in normal order contribute to the expansion, in agreement with our preliminary discussion.

We now proceed to the proof of the theorem. The theorem is trivially true for $N = 1$ and 2 (the latter case being nothing else than the definition of contraction), and will now be proved by induction for $N > 2$. That is, we assume that the theorem holds true for N arbitrary operators (such as the ones shown in Eq. (A3.6)) and deduce that it must also be true for $N + 1$ operators.

Let

$$\hat{A}\hat{B}\hat{C}\ldots\hat{W}\hat{X}\hat{Y}\hat{Z} \tag{A3.7}$$

be an arbitrary product of $N + 1$ operators. The last operator $\hat{Z}$ is either a destruction operator or a creation operator. If it is a destruction operator, then we simply multiply it by both sides of Eq. (A3.6). Because it is a destruction operator we can safely bring it inside the colon brackets, provided we keep it in the rightmost position in every term. In this way, we obtain a true identity that looks very much like Wick's theorem for $N + 1$ operators, except for the fact that the terms in which $\hat{Z}$ is contracted with other operators are missing. However, these terms are zero, because the contraction $\hat{A}^{\cdot}\hat{Z}^{\cdot}$ vanishes when $\hat{Z}$ is a destruction operator. Thus, the absence of those contractions is harmless, and we have proved that Wick's theorem holds for a product of $N + 1$ operators *if the last one is a destruction operator*. In a completely analogous way we can prove that Wick's theorem holds for a product of $N + 1$ operators *if the first one is a creation operator*.

Consider now the case that $\hat{Z}$ in Eq. (A3.7) is a creation operator. We begin by moving it to the first position on the left. To do this, it is necessary to interchange $\hat{Z}$ successively with each of the preceding operators. These interchanges are done with the help of the formula $\hat{A}\hat{Z} = -\hat{Z}\hat{A} + \{\hat{A}, \hat{Z}\}$. It is essential here to notice that $\hat{A}^{\cdot}\hat{Z}^{\cdot} = \{\hat{A}, \hat{Z}\}$ when $\hat{Z}$ is a creation operator. Thus, after N interchanges we obtain

$$\hat{A}\hat{B}\hat{C}\ldots\hat{W}\hat{X}\hat{Y}\hat{Z} = (-1)^N \hat{Z}\hat{A}\hat{B}\hat{C}\ldots\hat{W}\hat{X}\hat{Y}$$
$$+ \hat{A}\hat{B}\hat{C}\ldots\hat{W}\hat{X}\hat{Y}^{\cdot}\hat{Z}^{\cdot} + \hat{A}\hat{B}\hat{C}\ldots\hat{W}\hat{X}^{\cdot}\hat{Y}\hat{Z}^{\cdot} + \ldots\hat{A}^{\cdot}\hat{B}\hat{C}\ldots\hat{W}\hat{X}\hat{Y}\hat{Z}^{\cdot}. \tag{A3.8}$$

The first term on the right-hand side of this equation has a creation operator in the first position. It is therefore possible to expand it in normal-ordered products according to the restricted form of Wick's theorem proved above. All the contractions involving $\hat{Z}$ vanish, because $\hat{Z}$ is the first operator on the left. Only after the expansion is complete are we free to move $\hat{Z}$ back to the last position on the right *within the normal product*: this operation "absorbs" the factor $(-1)^N$. We are now very close to the desired result, but the terms in which the operators $\hat{A}\ldots\hat{Y}$ are contracted with $\hat{Z}$ are still missing. However, these terms are supplied by the remaining products on the right-hand side of Eq. (A3.8). These are products of $N - 1$ operators and can therefore be expanded, by the inductive hypothesis, according to the scheme of Eq. (A3.6). When this is done, we see that they generate all and only the missing terms containing the contractions of $\hat{A}\ldots\hat{Y}$ with $\hat{Z}$. This completes the proof of Wick's theorem.

We emphasize that Wick's theorem establishes the exact identity between operators without any reference to averages. It applies to creation and destruction operators in *any* single-particle basis. Its greatest value however lies in the simplicity it brings to the problem of calculating vacuum averages. For, when we average both sides of Eq. (A3.6) in the vacuum, all the terms containing normal products vanish, and only the fully contracted ones survive (assuming, of course, that the product contains an even number of operators). Since the contraction of two operators has the same value as the vacuum average of their product we arrive at the following fundamental result:

The vacuum average of a product of an even number of creation and destruction operators equals the sum of the products of the vacuum averages of all the possible pairs of operators, each term in the sum being multiplied by a factor +1 or −1 depending on whether the number of interchanges of fermion operators needed to bring the pairs together is even or odd.

A3.3 Normal ordering with respect to a "Fermi sea"

In the study of Fermi systems one typically needs to calculate the average of a product of creation and/or destruction operators in the noninteracting ground-state

$$|\Psi_0\rangle = \prod_{i \leq N} \hat{a}^\dagger_{\alpha_i} |0\rangle \,, \tag{A3.9}$$

where α_i are the labels of the occupied states. Such a state is pictorially described as a "Fermi sea" of occupied states. In principle, one could calculate averages in $|\Psi_0\rangle$ by a direct application of Wick's theorem for vacuum averages. In practice, this would be a rather clumsy approach, due to the large number of operators appearing in Eq. (A3.9). The calculation can be done in a much more elegant way, thanks to the fact that the Fermi sea itself is, because of the exclusion principle, a kind of vacuum state.

To see this, observe that the Fermi sea is annihilated either by the destruction operator of an empty state $\hat{a}_{\alpha_i}$ with $i > N$, *or* by the creation operator of an occupied state $\hat{a}^\dagger_{\alpha_i}$ with $i \leq N$. Hence both types of operators can be regarded as *destruction operators with respect to the Fermi sea*. In the same sense the corresponding hermitian conjugate operators $\hat{a}^\dagger_{\alpha_i}$ with $i > N$, and $\hat{a}_{\alpha_i}$ with $i \leq N$ can be regarded as *creation operators with respect to the Fermi sea* since they create either an additional particle in an initially empty state, or the lack of a particle (a *hole*) in one of the originally occupied states.

These considerations lead us to *generalize* the notion of creation and destruction operators in the following manner: a generalized destruction operator is an operator that annihilates the Fermi sea, while a generalized creation operator is the hermitian conjugate of a generalized destruction operator.

The notions of normal-ordered product can then be similarly generalized. A product of creation and/or destruction operators is said to be *normal-ordered with respect to the Fermi sea* if all the generalized destruction operators are placed to the right of all the generalized creation operators. For example, if the Fermi sea consists of three occupied states with labels 1, 2 and 3, while the states with labels 4,5 ... are all empty then

$$: \hat{a}^\dagger_2 \hat{a}^\dagger_4 \hat{a}_2 \hat{a}^\dagger_1 : = \hat{a}^\dagger_4 \hat{a}_2 \hat{a}^\dagger_2 \hat{a}^\dagger_1. \tag{A3.10}$$

Notice that, by construction, the average in the Fermi sea of a generalized normal-ordered product vanishes. In this sense the Fermi sea plays the role of a *vacuum*.

From this point on we can repeat all the formal steps that led in the previous section to the proof of Wick's theorem. In particular, we can define the contraction of two generalized creation or destruction operators as the difference between their regular product and their (generalized) normal ordered product. With this definition, Eq. (A3.6) carries over without changes. Again, the value of a contraction can be obtained by taking the average of the product of the two operators involved in the Fermi sea.

To gain practice with the use of the theorem, the reader is encouraged to work out the following important example:

$$\hat{a}_\alpha^\dagger \hat{a}_\beta^\dagger \hat{a}_\gamma \hat{a}_\delta = \langle \hat{a}_\alpha^\dagger \hat{a}_\delta \rangle \langle \hat{a}_\beta^\dagger \hat{a}_\gamma \rangle - \langle \hat{a}_\alpha^\dagger \hat{a}_\gamma \rangle \langle \hat{a}_\beta^\dagger \hat{a}_\delta \rangle$$

$$+ \langle \hat{a}_\alpha^\dagger \hat{a}_\delta \rangle : \hat{a}_\beta^\dagger \hat{a}_\gamma : + \langle \hat{a}_\beta^\dagger \hat{a}_\gamma \rangle : \hat{a}_\alpha^\dagger \hat{a}_\delta :$$

$$- \langle \hat{a}_\alpha^\dagger \hat{a}_\gamma \rangle : \hat{a}_\beta^\dagger \hat{a}_\delta : - \langle \hat{a}_\beta^\dagger \hat{a}_\delta \rangle : \hat{a}_\alpha^\dagger \hat{a}_\gamma :$$

$$+ : \hat{a}_\alpha^\dagger \hat{a}_\beta^\dagger \hat{a}_\gamma \hat{a}_\delta : \, , \qquad\qquad\qquad\text{(A3.11)}$$

where $\langle \dots \rangle$ now represents the average in the Fermi sea. Notice that no averages of the type $\langle \hat{a}_\alpha^\dagger \hat{a}_\beta^\dagger \rangle$ or $\langle \hat{a}_\alpha \hat{a}_\beta \rangle$ appear in the expression as the contractions $\hat{a}_\alpha^\dagger \hat{a}_\beta^\dagger$ and $\hat{a}_\alpha \hat{a}_\beta$ always vanish.

As in the previous version of the theorem, the most important aspect of this result is the simplicity it brings to the problem of calculating averages in a Fermi sea. In this case the final rule can be stated as follows:

The average in the Fermi sea of a product of an even number of (generalized) creation or destruction operators equals the sum of the products of the averages of all the possible pairs of operators, each term in the sum being multiplied by a factor $+1$ or -1 depending on whether the number of interchanges of fermion operators needed to bring the pairs together, is even or odd.

For example, in the case of Eq. (A3.11) we get

$$\langle \hat{a}_\alpha^\dagger \hat{a}_\beta^\dagger \hat{a}_\gamma \hat{a}_\delta \rangle = \langle \hat{a}_\alpha^\dagger \hat{a}_\delta \rangle \langle \hat{a}_\beta^\dagger \hat{a}_\gamma \rangle - \langle \hat{a}_\alpha^\dagger \hat{a}_\gamma \rangle \langle \hat{a}_\beta^\dagger \hat{a}_\delta \rangle \, . \qquad\qquad\text{(A3.12)}$$

A wonderful feature of the generalized Wick's theorem is that it can be used to calculate averages in any *single* determinantal state, even when the latter is constructed from a set of states that are different from the ones with respect to which the creation and destruction operators are defined. The reason is that creation or destruction operators in a given basis can always be expressed as linear combinations of creation or destruction operators in any other basis.[1]

A3.4 Wick's theorem at finite temperature

Let us now consider the problem of calculating the thermal average of a product of creation and/or destruction operators in a *noninteracting thermal ensemble*.

At first sight, one might think that Wick's theorem is of very limited value in this case. To be sure, the operator identity (A3.6) is still valid, but the thermal average of the normal-ordered products does not vanish and no major simplification seems to be

---

[1] Wick's theorem for averages directly applies also to linear superpositions of generalized creation and destruction operators. A relevant example of such an application is its use with the vacuum averages of the particle field operators $\hat{\psi}_\sigma^\dagger(\vec{r})$ and $\hat{\psi}_\sigma(\vec{r})$. The reader is encouraged to verify this useful property (see Exercise 15).

~ossible. Fortunately, the situation is much better than this superficial analysis would
~uggest. It turns out that, thanks to a sort of small algebraic miracle, the thermal average
f a product of creation and/or destruction operators in a noninteracting equilibrium
~nsemble can still be calculated as a sum of products of averages of pairs, in precisely the
~ame way as the noninteracting ground-state average. We shall prove this below working
~n the *grand canonical* ensemble, which is presumably equivalent to any other ensemble
~n the thermodynamic limit.

We first recall a few basic facts about quantum statistical mechanics. The thermal
~verage of an operator $\hat{A}$ in the grand-canonical ensemble associated with a
~oninteracting hamiltonian $\hat{H}_0 = \sum_\alpha \varepsilon_\alpha \hat{a}_\alpha^\dagger \hat{a}_\alpha$ is given by

$$\langle \hat{A} \rangle = \text{Tr}[\hat{\rho}_0 \hat{A}] , \tag{A3.13}$$

~where the normalized grand-canonical density matrix is given by

$$\hat{\rho}_0 = e^{\beta(\Omega - \hat{H}_0 - \mu \hat{N})} , \tag{A3.14}$$

~the symbol "Tr" denotes the trace of an operator in the Hilbert space (i.e., the sum of its
~diagonal elements in an arbitrary representation), $\beta = 1/k_B T$ is the inverse temperature,
μ is the chemical potential, $\hat{N} = \sum_\alpha \hat{a}_\alpha^\dagger \hat{a}_\alpha$ is the number operator, and Ω is the grand
thermodynamic potential, defined by the normalization condition

$$e^{-\beta\Omega} = \text{Tr}[e^{-\beta(\hat{H}_0 - \mu \hat{N})}]. \tag{A3.15}$$

For example, the average of an occupation number operator $\hat{N}_\alpha = \hat{a}_\alpha^\dagger \hat{a}_\alpha$ is given (for
fermions) by

$$\begin{aligned}
\langle \hat{N}_\alpha \rangle &= \text{Tr}[\hat{\rho}_0 \hat{a}_\alpha^\dagger \hat{a}_\alpha] \\
&= -\frac{1}{\beta} \frac{\partial}{\partial \varepsilon_\alpha} \ln \text{Tr}[e^{-\beta(\hat{H}_0 - \mu \hat{N})}] \\
&= \frac{1}{e^{\beta(\varepsilon_\alpha - \mu)} + 1} \equiv n_\alpha .
\end{aligned} \tag{A3.16}$$

This is, of course, nothing but the well known Fermi-Dirac distribution for noninteracting
fermions. For noninteracting bosons, a similar calculation would have yielded the
Bose–Einstein distribution $n_\alpha = \frac{1}{e^{\beta(\varepsilon_\alpha - \mu)} - 1}$.

Let us now consider the thermal average of an *even* number of Fermion operators[2]

$$\text{Tr}[\hat{\rho}_0 \hat{A} \hat{B} \hat{C} \ldots \hat{X} \hat{Y} \hat{Z}] , \tag{A3.17}$$

where $\hat{A}$, $\hat{B}$, etc. are creation and/or destruction operators $\hat{a}_\alpha$ or $\hat{a}_\alpha^\dagger$ in the basis that
diagonalizes $\hat{H}_0$. In this section creation and destruction operators are defined with respect
to the vacuum.

We begin by moving $\hat{A}$ to the right in successive steps:

$$\begin{aligned}
\text{Tr}[\hat{\rho}_0 \hat{A} \hat{B} \hat{C} \ldots \hat{X} \hat{Y} \hat{Z}] &= \text{Tr}[\hat{\rho}_0 \{\hat{A}, \hat{B}\} \hat{C} \ldots \hat{X} \hat{Y} \hat{Z}] \\
&- \text{Tr}[\hat{\rho}_0 \hat{B} \{\hat{A}, \hat{C}\} \ldots \hat{X} \hat{Y} \hat{Z}] + \ldots + \text{Tr}[\hat{\rho}_0 \hat{B} \hat{C} \ldots \hat{X} \hat{Y} \{\hat{A}, \hat{Z}\}] \\
&- \text{Tr}[\hat{\rho}_0 \hat{B} \hat{C} \ldots \hat{X} \hat{Y} \hat{Z} \hat{A}] .
\end{aligned} \tag{A3.18}$$

---

[2] The average of the product of an odd number of creation and/or destruction operators is obviously zero.

Note that the last sign is a minus because $\hat{A}$ has been commuted $N - 1$ times, and N is even.

The last term on the right-hand side is actually proportional to the one on the left-hand side. This result follows from combining the cyclic invariance property of the trace (according to which the trace of a product of operators is invariant under a cyclic permutation of the operators in the product) with the commutation rule

$$\hat{A}\hat{\rho}_0 = e^{\lambda_A \beta(\varepsilon_A - \mu)}\hat{\rho}_0 \hat{A} . \tag{A3.19}$$

Here ε_A is the energy of the single-particle state to which $\hat{A}$ refers, and $\lambda_A = 1$, if $\hat{A}$ is a creation operator or $\lambda_A = -1$ if $\hat{A}$ is a destruction operator. Thus, we obtain

$$\text{Tr}[\hat{\rho}_0 \hat{B}\hat{C} \ldots \hat{X}\hat{Y}\hat{Z}\hat{A}] = e^{\lambda_A \beta(\varepsilon_A - \mu)}\text{Tr}[\hat{\rho}_0 \hat{A}\hat{B}\hat{C} \ldots \hat{X}\hat{Y}\hat{Z}] . \tag{A3.20}$$

Substituting this in Eq. (A3.18), we get

$$\text{Tr}[\hat{\rho}_0 \hat{A}\hat{B}\hat{C} \ldots \hat{X}\hat{Y}\hat{Z}] = \frac{\{\hat{A}, \hat{B}\}}{e^{\lambda_A \beta(\varepsilon_A - \mu)} + 1}\text{Tr}[\hat{\rho}_0 \hat{C} \ldots \hat{X}\hat{Y}\hat{Z}]$$
$$- \frac{\{\hat{A}, \hat{C}\}}{e^{\lambda_A \beta(\varepsilon_A - \mu)} + 1}\text{Tr}[\hat{\rho}_0 \hat{B} \ldots \hat{X}\hat{Y}\hat{Z}] + \ldots + \frac{\{\hat{A}, \hat{Z}\}}{e^{\lambda_A \beta(\varepsilon_A - \mu)} + 1}\text{Tr}[\hat{\rho}_0 \hat{B}\hat{C} \ldots \hat{X}\hat{Y}] , \tag{A3.21}$$

where we have used the fact that here the anticommutators are numbers. We now come to the crucial observation, namely, we note that

$$\frac{\{\hat{A}, \hat{B}\}}{e^{\lambda_A \beta(\varepsilon_A - \mu)} + 1} = \langle \hat{A}\hat{B} \rangle . \tag{A3.22}$$

This follows from the fact that both the anticommutator on the left and the thermal average on the right vanish unless $\hat{A}$ and $\hat{B}$ pertain to the same single particle state α and are not both destruction or both creation operators. Thus, the only nonvanishing cases are $\hat{A} = \hat{a}_\alpha^\dagger$, $\hat{B} = \hat{a}_\alpha$ and $\hat{A} = \hat{a}_\alpha$, $\hat{B} = \hat{a}_\alpha^\dagger$. Substituting in Eq. (A3.22) we obtain, in the first case

$$\frac{1}{e^{\beta(\varepsilon_\alpha - \mu)} + 1} = \langle \hat{a}_\alpha^\dagger \hat{a}_\alpha \rangle , \tag{A3.23}$$

and, in the second case,

$$\frac{1}{e^{-\beta(\varepsilon_\alpha - \mu)} + 1} = \langle \hat{a}_\alpha \hat{a}_\alpha^\dagger \rangle . \tag{A3.24}$$

Both equations are correct, and therefore Eq. (A3.22) is correct in general.

We can then rewrite Eq. (A3.21) in the following form:

$$\langle \hat{A}\hat{B}\hat{C} \ldots \hat{X}\hat{Y}\hat{Z} \rangle = \langle \hat{A}\hat{B} \rangle \langle \hat{C} \ldots \hat{X}\hat{Y}\hat{Z} \rangle$$
$$- \langle \hat{A}\hat{C} \rangle \langle \hat{B} \ldots \hat{X}\hat{Y}\hat{Z} \rangle + \ldots + \langle \hat{A}\hat{Z} \rangle \langle \hat{B}\hat{C} \ldots \hat{X}\hat{Y} \rangle. \tag{A3.25}$$

(Notice the sign, which always has the parity of the number of interchanges of fermion operators necessary to bring the two paired operators together.)

Thus, we have succeeded in "factoring out" the averages of all the pairs of operators that contain $\hat{A}$. We can apply the same trick to each of the remaining averages of $N - 2$ operators. The process can be continued until the initial average is completely expressed a sum of products of averages of pairs formed in all possible ways, the sign of each term being determined by the number of fermion interchanges. Thus, Wick's theorem for the finite temperature average has precisely the same form as for the ground-state average. Also in this version the theorem applies not only to products of creation and/or destruction operators that appear in $\hat{H}_0$, but also to any of their linear combinations.

Appendix 4
The pair correlation function and the structure factor

In this appendix we summarize the basic definitions and properties of the *pair correlation function* and its Fourier transform, the static structure factor.

A4.1 The pair correlation function

The pair correlation function $g(\vec{r}_2, \vec{r}_1)$ is defined as the normalized probability of simultaneously finding an electron at $\vec{r}_1$ *and* one at $\vec{r}_2$:

$$g(\vec{r}_2, \vec{r}_1) \equiv \frac{1}{n(\vec{r}_2)n(\vec{r}_1)} \left\langle \sum_{i \neq j} \delta(\vec{r}_1 - \vec{r}_i)\delta(\vec{r}_2 - \vec{r}_j) \right\rangle . \tag{A4.1}$$

This expression can also be rewritten in terms of the density operator as follows:

$$g(\vec{r}_2, \vec{r}_1) = \frac{\langle \hat{n}(\vec{r}_2)\hat{n}(\vec{r}_1) \rangle}{n(\vec{r}_2)n(\vec{r}_1)} - \frac{\delta(\vec{r}_2 - \vec{r}_1)}{n(\vec{r}_1)}, \tag{A4.2}$$

where the second term excludes the trivial correlation of a particle with itself. In the limit of large separation $|\vec{r}_1 - \vec{r}_2|$ the correlation between $\hat{n}(\vec{r}_1)$ and $\hat{n}(\vec{r}_2)$ is expected to vanish and, therefore, the average $\langle \hat{n}(\vec{r}_2)\hat{n}(\vec{r}_1) \rangle$ should reduce to the product of the two independent averages, i.e. $n(\vec{r}_2)n(\vec{r}_1)$. Thus we have

$$\lim_{|\vec{r}_1 - \vec{r}_2| \to \infty} g(\vec{r}_2, \vec{r}_1) = 1 . \tag{A4.3}$$

In a homogeneous and isotropic system $g(\vec{r}_2, \vec{r}_1)$ is a function of the magnitude of $\vec{r} = |\vec{r}_2 - \vec{r}_1|$ and in that case we write $g(\vec{r}, 0) \equiv g(r)$.

The quantity

$$h(\vec{r}_2, \vec{r}_1) \equiv n(\vec{r}_2)[g(\vec{r}_2, \vec{r}_1) - 1] \tag{A4.4}$$

describes the modification of the average electronic density at $\vec{r}_2$ due to the presence of an electron at $\vec{r}_1$. This quantity, integrated over all space, must give -1 because, if an electron is present at $\vec{r}_1$ then it is not somewhere else, implying that there is a net deficiency of one electron in the rest of the system. Thus we have the important sum rule

$$\int h(\vec{r}_2, \vec{r}_1)d\vec{r}_2 = -1 , \tag{A4.5}$$

682

hich ought to explain why $h(\vec{r}_2, \vec{r}_1)$ is commonly referred to as the *exchange-correlation* le for an electron at $\vec{r}_1$.

In a *noninteracting* Fermi system, the antisymmetry of the wave function ensures that $(\vec{r}_2, \vec{r}_1)$ is everywhere negative, i.e., the presence of an electron at $\vec{r}_1$ causes a reduction in ectron density everywhere else. This depletion effect is known as the *exchange hole* and an be easily quantified starting from the second-quantized expression for $g(\vec{r}_2, \vec{r}_1)$ of q. (1.100), i.e.,

$$g(\vec{r}_2, \vec{r}_1) = \sum_{\sigma_1 \sigma_2} \frac{\left\langle \hat{\Psi}^\dagger_{\sigma_2}(\vec{r}_2) \hat{\Psi}^\dagger_{\sigma_1}(\vec{r}_1) \hat{\Psi}_{\sigma_1}(\vec{r}_1) \hat{\Psi}_{\sigma_2}(\vec{r}_2) \right\rangle}{n(\vec{r}_2)n(\vec{r}_1)}, \tag{A4.6}$$

where the average is taken in the ground-state of the non-interacting system. This average s evaluated with the help of Wick's theorem (see Appendix 3), by decomposing it into the um of products of averages of pairs of operators. This leads us to the following general xpression for the pair correlation function of a non interacting Fermi system:

$$g^{(0)}(\vec{r}_2, \vec{r}_1) = 1 - \sum_{\sigma_1 \sigma_2} \frac{\left| \left\langle \hat{\Psi}^\dagger_{\sigma_1}(\vec{r}_1) \hat{\Psi}_{\sigma_2}(\vec{r}_2) \right\rangle \right|^2}{n(\vec{r}_2)n(\vec{r}_1)}, \tag{A4.7}$$

which is manifestly smaller than 1, thus proving the negativity of the exchange hole.

In general, however, the pair correlation function of an interacting Fermi system can be larger than 1 (see Fig. A4.1), and the full exchange-correlation hole $h(\vec{r}_2, \vec{r}_1)$ can be positive in some ranges of separations, even though it satisfies the global constraint (A4.5).

The pair correlation function $g(\vec{r}_2, \vec{r}_1)$ as defined so far is blind to the spin of the electrons. For a finer description of positional *and* spin correlations we introduce the

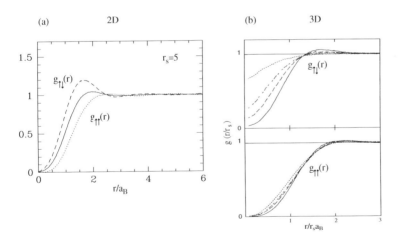

Fig. A4.1. (a) The pair correlation functions $g_{\uparrow\downarrow}$ and $g_{\uparrow\uparrow}$ for a paramagnetic 2D electron liquid at $r_s = 5$ (solid line represents the average of the two functions). (From Rapisarda and Senatore, 1996, courtesy of G. Senatoze.) (b) The same functions for a paramagnetic 3D electron liquid computed by QMC for $r_s = 1$ (dotted line), $r_s = 3$ (dash-dotted line), $r_s = 5$ (dashed line), and $r_s = 10$ (full line). (From Ortiz and Ballone, 1994.)

spin-resolved pair correlation function

$$g_{\sigma_2\sigma_1}(\vec{r}_2, \vec{r}_1) = \frac{\langle \hat{n}_{\sigma_2}(\vec{r}_2)\hat{n}_{\sigma_1}(\vec{r}_1)\rangle}{n_{\sigma_2}(\vec{r}_2)n_{\sigma_1}(\vec{r}_1)} - \delta_{\sigma_1\sigma_2}\frac{\delta(\vec{r}_2 - \vec{r}_1)}{n_{\sigma_1}(\vec{r}_1)}$$

$$= \frac{\langle \hat{\Psi}^{\dagger}_{\sigma_2}(\vec{r}_2)\hat{\Psi}^{\dagger}_{\sigma_1}(\vec{r}_1)\hat{\Psi}_{\sigma_1}(\vec{r}_1)\hat{\Psi}_{\sigma_2}(\vec{r}_2)\rangle}{n_{\sigma_2}(\vec{r}_2)n_{\sigma_1}(\vec{r}_1)} . \qquad (A4.8)$$

which gives the probability of simultaneously finding an electron at $\vec{r}_1$ with spin projection σ_1 and one at $\vec{r}_2$ with spin projection σ_2. The normalization is chosen in such a way that we still have

$$\lim_{|\vec{r}_1 - \vec{r}_2| \to \infty} g_{\sigma\sigma'}(\vec{r}_2, \vec{r}_1) = 1 . \qquad (A4.9)$$

It should be evident that the full pair correlation function is obtained from the spin-resolved one in the following manner:

$$g(\vec{r}_2, \vec{r}_1) = \sum_{\sigma_1\sigma_2} \frac{g_{\sigma_2\sigma_1}(\vec{r}_2, \vec{r}_1)n_{\sigma_2}(\vec{r}_2)n_{\sigma_1}(\vec{r}_1)}{n(\vec{r}_1)n(\vec{r}_2)} . \qquad (A4.10)$$

Fig. A4.1 shows the values of $g_{\uparrow\downarrow}$ and $g_{\uparrow\uparrow}$ computed by QMC in a paramagnetic electron liquid at various values of r_s. Notice that $g_{\uparrow\uparrow}(0) = 0$, consistent with Pauli's exclusion principle. Furthermore, both functions become larger than 1 in certain ranges of separations. Analytical expressions for $g_{\sigma_2\sigma_1}(\vec{r}_2, \vec{r}_1)$ in the homogeneous electron liquid (where this function only depends on the distance $|\vec{r}_2 - \vec{r}_1|$) can be found in the recent papers by Gori-Giorgi *et al.* (2001, 2004).

A quantity of particular interest is $g(\vec{r}, \vec{r})$, i.e., the probability of finding two electrons on top of each other. In a classical electron liquid this would be zero because the electrons cannot penetrate the infinitely repulsive coulomb barrier. But, quantum mechanical tunneling through the coulomb barrier allows for a finite value of this quantity. Following an original idea by Overhauser (1995) it has been possible to derive a simple analytic expression for $g(\vec{r}, \vec{r})$ in a homogeneous electron liquid (Gori-Giorgi and Perdew, 2001). This is usually written as $g(0)$, and its approximate expression is

$$g(0) = \frac{1 - p^2}{2}g_{\uparrow\downarrow}(0)$$

$$g_{\uparrow\downarrow}(0) = \left(1 + Ar_s + Br_s^2 + Cr_s^3 + Dr_s^4\right)e^{-Er_s} , \qquad (A4.11)$$

where p is the fractional polarization defined in Eq. (1.84), and the coefficients $A - E$ are listed in Table A4.1.[1] It turns out that even in a non-homogeneous system one has approximately $g(\vec{r}, \vec{r}) = g(0)$, provided $g(0)$ is evaluated at the local density $n(\vec{r})$.

It is often useful to express the pair correlation function $g_{\sigma\sigma'}(\vec{r}_2, \vec{r}_1)$ in terms of the relative coordinate $\vec{r} \equiv \vec{r}_2 - \vec{r}_1$ and the center of mass coordinate $\vec{R} \equiv \frac{\vec{r}_1 + \vec{r}_2}{2}$. The resulting function $g_{\sigma\sigma'}(\vec{r}, \vec{R})$ satisfies the so-called *cusp conditions* (Kimball, 1973; Rajagopal,

---

[1] In three dimensions, a more accurate description of the dependence of $g(0)$ on the degree of spin polarization p is achieved by evaluating Eq. (A4.11) at a rescaled r_s value: $r_s \to \frac{2r_s}{(1+p)^{\frac{1}{3}} + (1-p)^{\frac{1}{3}}}$ (Gori-Giorgi and Perdew, 2001).

Table A4.1. *Parameters of the fit (A4.11) for the pair correlation function at zero separation in d dimensions. From Gori-Giorgi et al. (2001, 2004).*

	A	B	C	D	E
$d = 2$	0.088	0.258	0.00037	0.0	1.46
$d = 3$	0.0207	0.08193	−0.01277	0.001859	0.7524

Kimball, and Banerjee, 1978), according to which

$$\lim_{r \to 0} \frac{\partial g_{\uparrow\uparrow}(\vec{r}, \vec{R})}{\partial \vec{r}}\bigg|_{\vec{r}=0} = 0 , \tag{A4.12}$$

and

$$\lim_{r \to 0} \frac{\partial g_{\uparrow\downarrow}(\vec{r}, \vec{R})}{\partial \vec{r}}\bigg|_{\vec{r}=0} = \begin{cases} 2\frac{g_{\uparrow\downarrow}(\vec{0},\vec{R})}{a_B} , & (2D) , \\[2mm] \frac{g_{\uparrow\downarrow}(\vec{0},\vec{R})}{a_B} , & (3D) , \end{cases} \tag{A4.13}$$

These relations follow from the fact that the short-range behavior of the pair correlation function is controlled by the coulomb repulsion between just two electrons (see Exercise 17).

We conclude this brief summary of properties of the pair correlation function by presenting the results for $g_{\sigma\sigma'}(r) \equiv g_{\sigma\sigma'}(\vec{r}, 0)$ in a homogeneous noninteracting electron gas. In this case $g_{\uparrow\downarrow}(r) = g_{\downarrow\uparrow}(r) = 1$ at all distances because there are no Pauli correlations between noninteracting electrons of opposite spin. Then Eqs. (A4.7) and (A4.10) easily lead us to

$$g_{\sigma\sigma'}^{(0)}(\vec{r}) = 1 - \delta_{\sigma\sigma'} \left| \frac{1}{N_\sigma} \sum_{|\vec{k}| \le k_{F\sigma}} e^{-i\vec{k}\cdot\vec{r}} \right|^2 . \tag{A4.14}$$

The sum over wave vectors can be evaluated analytically with the following results:

$$\frac{1}{N_\sigma} \sum_{|\vec{k}| \le k_{F\sigma}} e^{-i\vec{k}\cdot\vec{r}} = \begin{cases} 3\frac{\sin k_{F\sigma} r - k_{F\sigma} r \cos k_{F\sigma} r}{(k_{F\sigma} r)^3} & 3D , \\[2mm] 2\frac{J_1(k_{F\sigma} r)}{k_{F\sigma} r} & 2D , \\[2mm] \frac{\sin(k_{F\sigma} r)}{k_{F\sigma} r} & 1D . \end{cases} \tag{A4.15}$$

where $J_1(x)$ is the Bessel function. Notice that in a paramagnetic liquid $g_{\uparrow\uparrow}^{(0)}(\vec{r})$ is a universal (i.e., r_s-independent) function of $k_F r = \frac{r}{\alpha_d r_s a_B}$ (see Eq. (1.78)) that vanishes at $r = 0$: this function is plotted in Fig. 1.10. The full pair correlation function is obtained by substituting Eq. (A4.15) in Eq. (A4.10).

A4.2 The static structure factor

The static structure factor $S(q)$ of a translationally invariant liquid is defined in terms of the pair correlation function as follows:

$$S(q) = 1 + n \int [g(r) - 1] e^{-i\vec{q}\cdot\vec{r}} d\vec{r} . \tag{A4.16}$$

The inverse of this relation is

$$g(r) - 1 = \frac{1}{n} \int [S(q) - 1] e^{i\vec{q}\cdot\vec{r}} \frac{d\vec{q}}{(2\pi)^d} . \tag{A4.17}$$

The sum rule (A4.5) ensures that:[2]

$$\lim_{q \to 0} S(q) = 0. \tag{A4.18}$$

In the opposite limit of large q, Eq. (A4.16) implies that $\lim_{q \to \infty} S(\vec{q}) = 1$. More precisely, it can be shown that the cusp conditions (A4.12) and (A4.13), imply the following limiting behavior in d dimensions:

$$\lim_{q \to \infty} q^{d+1} [S(q) - 1] = -\frac{2^d \pi n}{a_B} g(0) . \tag{A4.19}$$

For a derivation of this result see Exercise 20.

Substituting in Eq. (A4.16) the form (1.99) of $g(r)$ we obtain an expression for $S(q)$ in terms of the Fourier components of the density fluctuation operator

$$S(q) = \sum_{i,j} \frac{\langle e^{-i\vec{q}\cdot(\hat{\vec{r}}_i - \hat{\vec{r}}_j)} \rangle}{N} = \frac{\langle \hat{n}_{-\vec{q}} \hat{n}_{\vec{q}} \rangle}{N} . \tag{A4.20}$$

Thus $S(q)$ is a measure of the average squared amplitude of density fluctuations of wave vector $\vec{q}$.

The spin-resolved version of the static structure factor is given by

$$S_{\sigma\sigma'}(\vec{q}) \equiv \frac{\langle \hat{n}_{-\vec{q},\sigma} \hat{n}_{\vec{q},\sigma'} \rangle}{N}$$

$$= \frac{n_\sigma}{n} \delta_{\sigma\sigma'} + \frac{n_\sigma n_{\sigma'}}{n} \int [g_{\sigma\sigma'}(r) - 1] e^{-i\vec{q}\cdot\vec{r}} d\vec{r} , \tag{A4.21}$$

where $S(\vec{q}) = \sum_{\sigma\sigma'} S_{\sigma\sigma'}(\vec{q})$. The inverse relation for the spin-resolved pair correlation function is

$$g_{\sigma\sigma'}(r) - 1 = \frac{n}{n_\sigma n_{\sigma'}} \frac{1}{L^d} \sum_{\vec{q} \neq 0} \left[S_{\sigma\sigma'}(\vec{q}) - \frac{n_\sigma}{n} \delta_{\sigma\sigma'} \right] e^{i\vec{q}\cdot\vec{r}} . \tag{A4.22}$$

The structure factor of the noninteracting electron gas at zero temperature can be calculated directly from the definition (A4.20). The results in three, two, and one

---

[2] The limit $q \to 0$ here is taken after the thermodynamic limit $L \to \infty$, where L is the size of the system. In practice, this means that $q \ll k_F$, but $q \gg 1/L$. At strictly $q = 0$ one has $S(0) = N$ (see Eq. (A4.20) below).

dimensions are

$$
S_{\sigma\sigma'}^{(0)}(\vec{q}) = \delta_{\sigma\sigma'}
\begin{cases}
\frac{3}{8}\frac{q}{k_{F\sigma}} - \frac{1}{32}\frac{q^3}{k_{F\sigma}^3}, & 3D, \\[2mm]
\frac{1}{\pi}\sin^{-1}\frac{q}{2k_{F\sigma}} + \frac{1}{2\pi}\frac{q}{k_{F\sigma}}\sqrt{1 - \left(\frac{q}{2k_{F\sigma}}\right)^2}, & 2D, \\[2mm]
\frac{q}{4k_{F\sigma}}, & 1D,
\end{cases}
\tag{A4.23}
$$

for $|q| \le 2k_{F\sigma}$, and

$$
S_{\sigma\sigma'}^{(0)}(\vec{q}) = \frac{\delta_{\sigma\sigma'}}{2},
\tag{A4.24}
$$

for $|q| > 2k_{F\sigma}$ in all dimensions. These formulas are used in Chapter 1 to calculate the exchange energy of the uniform electron liquid (see Eq. (1.93)).

Appendix 5

Calculation of the energy of a Wigner crystal via the Ewald method

In this appendix we supply the steps leading to Eq. (1.118) for the electrostatic energy of a Wigner crystal. The basic idea of the procedure is originally due to Ewald (1921). The starting point is Eq. (1.108). Quite similarly to the procedure for regularizing the coulomb interaction, we replace the pure coulomb interaction by the Yukawa form $e^2 \frac{e^{-\kappa r}}{r}$, where κ is a screening wave vector that will be sent to zero at the end of the calculation. Then, in the thermodynamic limit, the second and the third term on the right-hand side of Eq. (1.108) can be combined to yield

$$-\frac{nNe^2}{2} \int d\vec{r}\, \frac{e^{-\kappa r}}{r} = -\frac{nN}{2} v_{\vec{q}=0}(\kappa), \tag{A5.1}$$

where $v_{\vec{q}}(\kappa)$, given by Eq. (1.17), is the Fourier transform of the Yukawa interaction.

To evaluate the first term on the right-hand side of Eq. (1.108) we make use of the identity

$$\frac{1}{R} = \left(\frac{\eta}{\pi}\right)^{\frac{1}{2}} \int_0^\infty dt\, t^{-\frac{1}{2}} e^{-\eta t R^2}, \tag{A5.2}$$

which holds for any positive value of η. This allows us to write

$$\frac{1}{2} \sum_{i \neq j} \frac{e^2}{R_{ij}} e^{-\kappa R_{ij}} = \frac{Ne^2}{2} \left(\frac{\eta}{\pi}\right)^{\frac{1}{2}} \int_0^\infty dt\, t^{-\frac{1}{2}} \sum_{\vec{R} \neq 0} e^{-\eta t R^2 - \kappa R}, \tag{A5.3}$$

where $\vec{R}_i$ and $\vec{R}$ are vectors of a simple Bravais lattice, and $R_{ij} \equiv |\vec{R}_i - \vec{R}_j|$.

Next we split the integration over t into two parts: $0 < t < 1$ and $t > 1$. The large-t part of the integral can be immediately expressed in terms of the Misra function, $\phi_\nu(z) \equiv \int_1^\infty dt\, e^{-zt} t^\nu$, as follows:

$$\frac{Ne^2}{2} \left(\frac{\eta}{\pi}\right)^{\frac{1}{2}} \int_1^\infty dt\, t^{-\frac{1}{2}} \sum_{\vec{R} \neq 0} e^{-\eta t R^2 - \kappa R} = \frac{Ne^2}{2} \left(\frac{\eta}{\pi}\right)^{\frac{1}{2}} \sum_{\vec{R} \neq 0} \phi_{-\frac{1}{2}}(\eta R^2)\, e^{-\kappa R}. \tag{A5.4}$$

As for the small-t part of the integral, we begin by rewriting it as

$$\frac{Ne^2}{2} \left(\frac{\eta}{\pi}\right)^{\frac{1}{2}} \left(\int_0^1 dt\, t^{-\frac{1}{2}} \sum_{\vec{R}} e^{-\eta t R^2 - \kappa R} - \int_0^1 dt\, t^{-\frac{1}{2}} \right), \tag{A5.5}$$

where the sum now runs over *all* the lattice vectors $\vec{R}$, *including* $\vec{R} = 0$. The second integral in the brackets is equal to 2. In the first one, we convert the sum over lattice vectors $\vec{R}$) into a sum over reciprocal lattice vectors ($\vec{G}$), with the help of the general identity

$$\sum_{\vec{R}} f(\vec{R}) = \frac{1}{\Omega} \sum_{\vec{G}} \tilde{f}(\vec{G}), \qquad (A5.6)$$

where $\Omega = \frac{1}{n}$ is the volume of the unit cell of the Bravais lattice, $f(\vec{R})$ is an arbitrary regular function, and $\tilde{f}(\vec{G}) \equiv \int d\vec{R} \, f(\vec{R}) \exp(-i\vec{G} \cdot \vec{R})$ is its Fourier transform at wave vector $\vec{G}$. Setting

$$f(\vec{R}) = e^{-\eta t R^2 - \kappa R} \qquad (A5.7)$$

we rewrite the first integral in the brackets of Eq. (A5.5) as

$$\frac{1}{\Omega} \int_0^\infty dt \, t^{-\frac{1}{2}} \tilde{f}(\vec{0}) - \frac{1}{\Omega} \int_1^\infty dt \, t^{-\frac{1}{2}} \tilde{f}(\vec{0}) + \frac{1}{\Omega} \sum_{\vec{G} \neq 0} \int_0^1 dt \, t^{-\frac{1}{2}} \tilde{f}(\vec{G}), \qquad (A5.8)$$

where the first two terms take care of the contribution of $\vec{G} = 0$, and the last one, with

$$\tilde{f}(\vec{G}) = \int d\vec{R} \, e^{-\eta t R^2 - \kappa R} e^{-i\vec{G} \cdot \vec{R}}$$

$$\overset{\kappa \to 0}{\to} \left(\frac{\pi}{\eta t} \right)^{\frac{d}{2}} e^{-\frac{G^2}{4\eta t}}, \qquad (A5.9)$$

contains the sum over all the remaining $\vec{G}$.

The first term in Eq. (A5.8) is easily seen, with the help of Eq. (A5.2), to be equal to $\left(\frac{\pi}{\eta} \right)^{\frac{1}{2}} \frac{v_{\vec{q}=0}(\kappa)}{\Omega}$. The remaining two terms are regular for $\kappa \to 0$ and will therefore be evaluated in precisely that limit. In particular, making use of the second line of Eq. (A5.9), we see that the second term in Eq. (A5.8) reduces to

$$-\frac{1}{\Omega} \left(\frac{\pi}{\eta} \right)^{\frac{d}{2}} \int_1^\infty \frac{dt}{t^{\frac{d+1}{2}}} = -\frac{1}{\Omega} \left(\frac{\pi}{\eta} \right)^{\frac{d}{2}} \frac{2}{d-1}. \qquad (A5.10)$$

Finally, the third term in Eq. (A5.8), reduces, upon the change of variable $t \to 1/t$, to an integral from 1 to ∞, which is again expressible in terms of a Misra function:

$$\frac{1}{\Omega} \left(\frac{\pi}{\eta} \right)^{\frac{d}{2}} \int_1^\infty dt \, t^{\frac{d-3}{2}} \sum_{\vec{G} \neq 0} e^{-\frac{G^2 t}{4\eta}} = \frac{1}{\Omega} \left(\frac{\pi}{\eta} \right)^{\frac{d}{2}} \sum_{\vec{G} \neq 0} \phi_{\frac{d-3}{2}} \left(\frac{G^2}{4\eta} \right). \qquad (A5.11)$$

Collecting the various terms we see that the singular part proportional to $v_{\vec{q}=0}(\kappa)$ cancels out, as expected, while the remaining regular terms combine to yield Eq. (1.118) of the main text.

Appendix 6

Exact lower bound on the ground-state energy of the jellium model

In this appendix we present the proof, due to Lieb and Narnhofer (1975), of the fact that the three-dimensional jellium model is stable, in the sense that its ground-state energy E_G satisfies the inequality

$$E_G > \left[\frac{3}{5} \left(\frac{9\pi}{4} \right)^{\frac{2}{3}} \frac{1}{r_s^2} - \frac{9}{5 r_s} \right] N \, Ry \,, \tag{A6.1}$$

where N is the number of electrons.[1]

The ground-state energy of the jellium model is given by

$$E_G = \langle \Phi_0 | \hat{H} | \Phi_0 \rangle = \langle \Phi_0 | \hat{T} | \Phi_0 \rangle + \langle \Phi_0 | \hat{U} | \Phi_0 \rangle, \tag{A6.2}$$

where $|\Phi_0\rangle$ is the ground-state of the jellium hamiltonian.

An obvious lower bound for the kinetic energy is provided by the kinetic energy of the non interacting system, which we have calculated in Section 1.5.1. Thus we have

$$\langle \Phi_0 | \hat{T} | \Phi_0 \rangle \geq \frac{3}{5} \left(\frac{9\pi}{4} \right)^{\frac{2}{3}} \frac{1}{r_s^2} N \, Ry \,. \tag{A6.3}$$

As for the potential energy, we notice that the "eigenstates" of the potential energy operator $\hat{U}(\hat{\vec{r}}_1, \ldots, \hat{\vec{r}}_N)$ are products of δ-functions centered at positions $\vec{R}_1, \ldots, \vec{R}_N$, with eigenvalues $U(\vec{R}_1, \ldots, \vec{R}_N)$ – the classical electrostatic energy of a system of N classical point particles of charge $-e$ located at positions $\vec{R}_1, \ldots, \vec{R}_N$ on a neutralizing background of charge density $\frac{Ne}{L^d} = ne$. It follows immediately that the expectation value of $\hat{U}$ in the ground-state $|\Phi_0\rangle$ is necessarily larger than U_{min} – the minimum electrostatic energy of N point particles on the uniform background. Thus, we have

$$E_G > \frac{3}{5} \left(\frac{9\pi}{4} \right)^{\frac{2}{3}} \frac{1}{r_s^2} N \, Ry + U_{min} \,. \tag{A6.4}$$

We now prove that

$$U_{min} \geq -\frac{9}{5} \frac{N}{r_s} Ry \,. \tag{A6.5}$$

---

[1] This result may be viewed as a special case of a general theorem of stability of fermionic matter, which establishes a lower bound of the form $E_G > -AN$, with A a positive constant, for a system of N electrons and protons interacting via the coulomb potential. See Dyson (1967), Dyson and Lenard (1967), Lenard and Dyson (1968).

The electrostatic energy of a system of N electrons localized at points $\vec{R}_1, \ldots, \vec{R}_N$ on a uniform background of charges ne is given by

$$U(\vec{R}_1 \ldots \vec{R}_N) = E_{B-B} + \sum_i U_i + \sum_{i<j} U_{ij}, \qquad (A6.6)$$

where E_{B-B} is the self-energy of the background, and we have introduced the following notation:

- U_i – the interaction energy of the electron at $\vec{R}_i$ with the background;
- U_{ij} – the interaction between two electrons at $\vec{R}_i$ and $\vec{R}_j$.

We also define

- $\tilde{U}_{ij}$ – the interaction energy (or twice the self-energy, if $i = j$) of uniform spheres of radius a and total charge $-e$ centered at $\vec{R}_i$ and $\vec{R}_j$;
- $\tilde{U}_i$ – the interaction energy of a uniform sphere of charge $-e$ and radius a centered at $\vec{R}_i$ with the background.

The potential energy of the configuration can then be rewritten as

$$\begin{aligned}
U = \; & E_{B-B} + \sum_i \tilde{U}_i + \sum_{i<j} \tilde{U}_{ij} + \tfrac{1}{2} \sum_i \tilde{U}_{ii} \\
& + \sum_i (U_i - \tilde{U}_i) - \tfrac{1}{2} \sum_i \tilde{U}_{ii} \\
& + \sum_{i<j} (U_{ij} - \tilde{U}_{ij}).
\end{aligned} \qquad (A6.7)$$

At this point we notice the following:

(1) The sum of the terms on the first line is positive, being the total electrostatic energy of the background charge and the negatively charged spheres.

(2) In the last line $U_{ij} - \tilde{U}_{ij} \geq 0$. Indeed, this quantity vanishes in three dimensions if the two spheres do not overlap (Gauss' theorem). If they do overlap, then $\tilde{U}_{ij}$ is smaller than $U_{ij} = \frac{e^2}{|\vec{R}_i - \vec{R}_j|}$, because the potential *inside* a uniformly charged sphere is *smaller* than $\frac{e}{r}$ (see Eq. (1.110).[2]

Thus, in three dimensions, we obtain

$$U \geq \sum_i (U_i - \tilde{U}_i) - \frac{1}{2} \sum_i \tilde{U}_{ii}. \qquad (A6.8)$$

The right-hand side of this inequality can now be evaluated exactly. Because the combined potentials of the electron and the sphere cancel outside the sphere, we must only consider the interaction of these two objects with the positive background within the sphere. This gives $U_i = -2\pi ne^2 a^2$ and $\tilde{U}_i = -\frac{8\pi}{5} ne^2 a^2$ so that

$$U_i - \tilde{U}_i = -\frac{2\pi}{5} ne^2 a^2 = -\frac{3e^2 a^2}{10 r_s^3 a_B^3}, \qquad (A6.9)$$

---

[2] It should be noted that this argument fails in two dimensions, since $U_{ij} - \tilde{U}_{ij} < 0$ for widely separated disks.

where we have made use of Eq. (1.22). The self-energy of a sphere of radius a is $\frac{\tilde{U}_{ii}}{2} = \frac{3e^2}{5a}$. Combining these results we find that U, and hence U_{min}, satisfies the inequalit

$$U \geq -N \left(\frac{3e^2 a^2}{10 r_s^3 a_B^3} + \frac{3e^2}{5a} \right). \qquad \text{(A6.10)}$$

It should be mentioned at this point that the value of the radius a of the spheres is still arbitrary. We optimize our lower bound by choosing the value of a that maximizes Eq. (A6.10). This occurs when $a = r_s a_B$. Substituting this value of a in Eq. (A6.10) we obtain the inequality (A6.5) and hence Eq. (A6.1).

Appendix 7

The density–density response function in a crystal

In a crystal, as in any inhomogeneous system, the response function $\chi_{nn}(\vec{r}, \vec{r}', \omega)$ depends separately on $\vec{r}$ and $\vec{r}'$. On the other hand because of the periodicity of the lattice the following relationship must hold true:

$$\chi_{nn}(\vec{r} + \vec{R}_i, \vec{r}' + \vec{R}_i, \omega) = \chi_{nn}(\vec{r}, \vec{r}', \omega),\tag{A7.1}$$

where $\vec{R}_i$ is any of the lattice vectors. This condition leads to a considerable simplification of the problem.

In order to make explicit use of Eq. (A7.1) we sum the above relationship over all the N sites $\vec{R}_i$ of the lattice and immediately find:

$$\frac{1}{N} \sum_{i=1}^{N} \chi_{nn}(\vec{r} + \vec{R}_i, \vec{r}' + \vec{R}_i, \omega) = \chi_{nn}(\vec{r}, \vec{r}', \omega).$$

We then make use of the double spatial Fourier transform for χ_{nn} with respect to $\vec{r}$ and $\vec{r}'$ to write

$$\chi_{nn}(\vec{r}, \vec{r}', \omega) = \frac{1}{L^{2d}} \sum_{\vec{q}_1, \vec{q}_2} e^{i\vec{q}_1 \cdot \vec{r}} e^{-i\vec{q}_2 \cdot \vec{r}'} \left(\frac{1}{N} \sum_{i=1}^{N} e^{i(\vec{q}_1 - \vec{q}_2) \cdot \vec{R}_i} \right) \chi_{nn}(\vec{q}_1, \vec{q}_2, \omega).$$

At this point we recall that:

$$\sum_{i=1}^{N} e^{i\vec{q} \cdot \vec{R}_i} = N \sum_{\vec{G}} \delta_{\vec{q}, \vec{G}},$$

where the sum in the right-hand side runs over all *reciprocal lattice vectors* $\vec{G}$. The expression for χ_{nn} can then be rewritten as:

$$\chi_{nn}(\vec{r}, \vec{r}', \omega) = \frac{1}{L^{2d}} \sum_{\vec{q} \in BZ, \vec{G}, \vec{G}'} e^{i(\vec{q}+\vec{G}) \cdot \vec{r}} e^{-i(\vec{q}+\vec{G}') \cdot \vec{r}'} \chi_{nn}(\vec{q} + \vec{G}, \vec{q} + \vec{G}', \omega),\tag{A7.2}$$

where the sum over $\vec{q}$ is now limited to the first *Brillouin zone*.

Consider next the spatial Fourier transform of the linearly induced density $n_1(\vec{q} + \vec{G}, \omega)$. After substituting the previous expression for $\chi_{nn}(\vec{r}, \vec{r}', \omega)$ in Eq. (3.122) of the text we readily obtain:

$$n_1(\vec{q} + \vec{G}, \omega) = \sum_{\vec{G}'} \chi_{nn}(\vec{q} + \vec{G}, \vec{q} + \vec{G}', \omega) V_{ext}(\vec{q} + \vec{G}', \omega), \qquad (A7.\!)$$

which is the sought matrix equation.

In view of the ubiquity of the reduced vector $\vec{q}$ we can rewrite this result by means of the following more transparent notation:

$$n_{1,\vec{G}}(\vec{q}, \omega) = \sum_{\vec{G}'} \chi_{\vec{G},\vec{G}'}(\vec{q}, \omega) V_{ext,\vec{G}'}(\vec{q}, \omega),$$

where $n_{1,\vec{G}}(\vec{q}, \omega) \equiv n_1(\vec{q} + \vec{G}, \omega)$, $\chi_{\vec{G},\vec{G}'}(\vec{q}, \omega) \equiv \chi_{nn}(\vec{q} + \vec{G}, \vec{q} + \vec{G}', \omega)$, and so on.

In this situation also the inverse of the scalar dielectric constant (see Chapter 5) has a similar matrix structure and, with the same notation, it is given by:

$$\left[\epsilon^{-1}\right]_{\vec{G},\vec{G}'}(\vec{q}, \omega) = \delta_{\vec{G},\vec{G}'} + v_{\vec{G}}(\vec{q}) \chi_{\vec{G},\vec{G}'}(\vec{q}, \omega). \qquad (A7.4)$$

As an exercise, the reader is invited to show that the *macroscopic dielectric constant* of an insulating crystal, defined as the ratio of the average electric dispacement field $\vec{D}$ to the average electric field $\vec{E}$, is given by $\left\{\left[\epsilon^{-1}\right]_{\vec{0},\vec{0}}(0, 0)\right\}^{-1}$, which is, in general, different from $\epsilon_{\vec{0},\vec{0}}(0, 0)$ (see Exercise 3.14).

The condition for the existence of collective modes in a crystal is obtained by applying the inverse of the matrix $\chi_{\vec{G},\vec{G}'}(\vec{q}, \omega)$ to the equation for $n_{1,\vec{G}}(\vec{q}, \omega)$

$$\sum_{\vec{G}'} \chi^{-1}_{\vec{G},\vec{G}'} n_{1\vec{G}'} = V_{ext,\vec{G}},$$

and then requiring that a non trivial solution for $n_{1\vec{G}}$ exists even when $V_{ext,\vec{G}} = 0$. This gives the condition

$$\det[\chi^{-1}_{\vec{G},\vec{G}'}(\vec{q}, \omega)] = 0.$$

The dispersion of collective density waves in solids (including phonons) can be calculated with this formalism.

Appendix 8

Example in which the isothermal and adiabatic responses differ

Consider the electron gas in the presence of a homogeneous, slowly time-dependent magnetic field $B(t)$ directed along the z axis. The hamiltonian is

$$\hat{H}_B(t) = \hat{H} + \bar{B}(t)\hat{S}_z \qquad (A8.1)$$

where $\bar{B}(t) = \frac{g\mu_B}{2}B(t)$, $\mu_B = \frac{e\hbar}{2mc}$ is the Bohr magneton, and $\hat{S}_z$ is the operator of the z-component of the total spin in units of $\frac{\hbar}{2}$. Our objective is to calculate the induced magnetic moment $-\frac{g\mu_B}{2}\langle\hat{S}_z\rangle_1$ in the linear response approximation.

Because $\hat{S}_z$ is a constant of the unperturbed motion, it is evident (see footnote[12] in Chapter 3) that the adiabatic response function (3.47) vanishes at all frequencies. This follows from the fact that the eigenfunctions of $\hat{H}$ (with eigenvalue E_n) can be chosen to be eigenfunctions of $\hat{S}_z$ (with eigenvalue S_z), so that their time evolution amounts to a simple phase factor

$$|\psi_n(t)\rangle = e^{-\frac{i}{\hbar}\left[E_n t + S_z \int_0^t \bar{B}(t')dt'\right]}|\psi_n(0)\rangle . \qquad (A8.2)$$

This evolution has no effect on the expectation value of $\hat{S}_z$, which therefore remains undisturbed.

Let us now consider the isothermal response. Because the system reaches equilibrium in the presence of an external magnetic field we expect a nonvanishing induced magnetization. This can be directly calculated from Eq. (3.119),

$$\chi_S^{iso} = \left(\frac{g\mu_B}{2}\right)^2 \beta\left[\langle\hat{S}_z^2\rangle_0 - \langle\hat{S}_z\rangle_0^2\right]. \qquad (A8.3)$$

In the paramagnetic state the equilibrium average $\langle\hat{S}_z\rangle_0$ is zero, but $\langle\hat{S}_z^2\rangle_0$ is not, and Eq. (A8.3) gives a nonvanishing susceptibility.

Of particular interest for the electron gas is the limit of zero temperature. The fluctuations in S_z tend to zero in this limit, and the static isothermal susceptibility

$$\chi_S^{iso} = \left(\frac{g\mu_B}{2}\right)^2 \frac{\langle\hat{S}_z^2\rangle_0}{k_B T} \qquad (A8.4)$$

can therefore reach a finite limit. To calculate this, we note that, in the $T \to 0$ limit, the partition function is dominated by the states of lowest energy. This allows us to write

$$\langle \hat{S}_z^2 \rangle_0 = \frac{\sum_{S_z=-N/2}^{N/2} S_z^2 e^{-\beta E(S_z)}}{\sum_{S_z=-N/2}^{N/2} e^{-\beta E(S_z)}} , \qquad (A8.5$$

where $E(S_z)$ is the energy of the *lowest-lying state* of spin S_z. Because N is large in the thermodynamic limit, the sum over S_z can be converted into an integral from $-\infty$ to ∞. Furthermore, for $T \to 0$ only small values of $S_z (\ll N)$ contribute significantly to the sum and we can use the second order expansion

$$E(S_z) \simeq E(0) + \frac{1}{2} \alpha_S S_z^2 , \qquad (A8.6)$$

where E_0 is the energy of the ground-state and

$$\alpha_S = \frac{\partial^2 E(S_z)}{\partial S_z^2} \bigg|_{S_z=0} \qquad (A8.7)$$

is the spin stiffness. Putting (A8.6) into (A8.5) and performing the gaussian integral over S_z we arrive at

$$\langle \hat{S}_z^2 \rangle_0 = \frac{\int_{-\infty}^{\infty} S_z^2 e^{-\frac{\beta \alpha_S S_z^2}{2}} dS_z}{\int_{-\infty}^{\infty} e^{-\frac{\beta \alpha_S S_z^2}{2}} dS_z} = \frac{k_B T}{\alpha_S} . \qquad (A8.8)$$

Thus, the isothermal spin susceptibility (A8.3) is simply proportional the inverse of the spin stiffness, in perfect agreement with the general stiffness theorem of Section (3.2.9). In a noninteracting electron gas the energy $E(S_z)$ of the spin-polarized ground-state can be directly read from Eq. (1.87) of Chapter 1. From this, we calculate

$$\chi_S^{iso} = N \left(\frac{g\mu_B}{2} \right)^2 \frac{d}{2\epsilon_F} ,$$

where d is the number of spatial dimensions. In Section 4.4.1, we will recover this result by taking the $q \to 0$ limit of the q-dependent static spin–spin response function.

Appendix 9

Lattice screening effects on the effective electron–electron interaction

The theory of the electron liquid based on the ideas of linear response and many-body local field corrections, allows a very elegant treatment of the effects of lattice screening on the electron-electron effective interaction. In this appendix we develop such a treatment and derive the expression (5.163) for the effective interaction. The discussion is based on the original paper of Kukkonen and Overhauser (1979).

As a first step we determine the effective interaction between an electron or an ion and a test-charge embedded in the system. This instructive exercise will also give us, as a byproduct, an expression for the phonon frequency. We make use of Eq. (5.98) with $V_{ext\uparrow} = V_{ext\downarrow} = v_q\rho$ and write for the (spin independent) effective potential "felt" by an electron:

$$V_{eff} = v_q\rho + v_q(1 - G_+)n_1 - V_{e-i}n_1^{(ion)} , \qquad (A9.1)$$

where ρ is the Fourier transform of the density of the test particle, n_1 and $n_1^{(ion)}$ are, respectively, the Fourier transforms of the induced electronic density and the induced ionic density, and V_{e-i} is the Fourier transform of the bare electron–ion interaction (the arguments q and ω are omitted for simplicity). The corresponding expression for the effective potential felt by an ion is

$$V_{ion} = -V_{e-i}(\rho + n_1) + V_{i-i}n_1^{(ion)} , \qquad (A9.2)$$

where V_{i-i} is the Fourier transform of the bare ion-ion interaction.[1]

At this point we need to establish the relationship between $n_1^{(ion)}$ and V_{ion}. This is done by writing the classical equations for the motion of the ions. We describe the ionic lattice as a deformable continuum medium and let $\vec{u}(\vec{r}, t)$ be the displacement from equilibrium of the ion located at position $\vec{r}$. The equation of motion for $\vec{u}(\vec{r}, t)$ is

$$M\ddot{\vec{u}}(\vec{r}, t) = -\vec{\nabla}V_{ion}(\vec{r}, t) , \qquad (A9.3)$$

where M is the mass of the ions. In Fourier transform this immediately gives:

$$\vec{u} = i\frac{V_{ion}}{M\omega^2}\vec{q} . \qquad (A9.4)$$

The induced ionic density is expressed in terms of the displacement field as

$$n_1^{(ion)}(\vec{r}, t) = -n_{ion}\vec{\nabla} \cdot \vec{u}(\vec{r}, t) , \qquad (A9.5)$$

---

[1] Notice that we have defined V_{e-i} to be a positive quantity. We have also made the reasonable assumption that the ions interact with the test charge with the potential V_{e-i}.

where n_{ion} is the equilibrium ionic density, or equivalently

$$n_1^{(ion)} = -i\, n_{ion}\, \vec{q} \cdot \vec{u} \,, \qquad \text{(A9.}\bullet$$

i.e.

$$n_1^{(ion)} = \frac{n_{ion} q^2}{M\omega^2} V_{ion} \,, \qquad \text{(A9.}\,\check{}$$

which is the sought relationship.

We are now in the position to solve the coupled equations (A9.1) and (A9.2) for V_{eff} and V_{ion}. The solution for V_{ion}, in particular, can be cast in the form

$$V_{ion} = -V_{e-i}(1 + v_q \chi_{nn})\rho + \frac{n_{ion} q^2 \chi_{nn} V_{e-i}^2}{M\omega^2} V_{ion} + \frac{n_{ion} q^2 V_{i-i}}{M\omega^2} V_{ion}, \qquad \text{(A9.8}\bullet$$

where we have used the relation $n_1 = \chi_{nn}(v_q \rho - V_{e-i} n_1^{(ion)})$. At this point the phonon frequency can be obtained by seeking a finite self-consistent solution of this equation in the absence of an external perturbation, i.e. for $\rho \to 0$. This immediately gives:

$$\omega_q^2 = \frac{n_{ion}}{M}\left(V_{i-i} + \chi_{nn} V_{e-i}^2\right) q^2 \,, \qquad \text{(A9.9}\rangle$$

which for convenience we rewrite as:

$$\omega_q^2 = \omega_0^2 + \frac{n_{ion} V_{e-i}^2}{v_q M \epsilon} q^2 \,, \qquad \text{(A9.10)}$$

where we have defined

$$\omega_0^2 = \frac{n_{ion}}{M}\left(V_{i-i} - \frac{V_{e-i}^2}{v_q}\right) q^2 \,. \qquad \text{(A9.11)}$$

Here ω_0^2 represents the contributions to the square of the phonon frequency from non-coulombic effects: this quantity vanishes if all the interactions are assumed to have the coulomb form (i.e. for $V_{e-i} = V_{i-i} = v_q$).

Having established the phonon spectrum we can go on to determine the effective electron-electron interaction. The calculation proceeds exactly as in Sections 5.4.1 and 5.5.1. The effective potentials experienced by the electrons and the ions due to a spin up electron are given by:

$$V_{eff,\uparrow} = v_q(1 - 2G_{\uparrow\uparrow})(\rho_\uparrow + n_{1\uparrow}) + v_q(1 - 2G_{\uparrow\downarrow})n_{1\downarrow} - V_{e-i} n_1^{(ion)} \,, \qquad \text{(A9.12)}$$

$$V_{eff,\downarrow} = v_q(1 - 2G_{\uparrow\downarrow})(\rho_\uparrow + n_{1\uparrow}) + v_q(1 - 2G_{\uparrow\uparrow})n_{1\downarrow} - V_{e-i} n_1^{(ion)} \,, \qquad \text{(A9.13)}$$

and

$$V_{ion} = -V_{e-i}(\rho_\uparrow + n_{1\uparrow} + n_{1\downarrow}) + V_{i-i} n_1^{(ion)} \,, \qquad \text{(A9.14)}$$

where now the electronic effective potentials also contain a contribution proportional to $n_1^{(ion)}$.

This system of coupled equations is handled as usual by introducing the relations $n_{1\sigma} = \chi_{0\sigma} V_{eff,\sigma}$ and $n_1^{(ion)} = \frac{n_{ion} q^2}{M\omega^2} V_{ion}$ (see Eq. (A9.7)).

We immediately notice that the difference $V_{eff,\uparrow} - V_{eff,\downarrow}$ satisfies an equation in which no ionic contribution appears. As expected on physical grounds, this implies that the spin dependent part of Eq. (5.152) is unchanged by lattice screening.[2]

At this point, in order to obtain the ionic term that appears in the spin independent part of the effective interaction W_+ (as done in Section 5.5.1) we add Eqs. (A9.12) and (A9.13). The algebra is a bit tedious but it can be simplified by isolating all the terms proportional to V_{e-i}^2, since all the other contributions reproduce the result appropriate to a rigid lattice. The final result is given by Eq. (5.163) of the main text.

This reflects the fact that within the present model there is no coupling between the induced electronic magnetization and the ionic lattice.

Appendix 10

Construction of the STLS exchange-correlation field

In this appendix we briefly summarize the procedure used by Singwi, Tosi, Land, and Sjölander (STLS) in their classic 1968 paper to construct an approximate and highly successful form of the density local field factor.

The equation of motion for the classical space and momentum distribution function $f(\vec{r}, \vec{p}; t)$ of a many-body system is

$$\frac{\partial}{\partial t} f(\vec{r}, \vec{p}; t) + \vec{v} \cdot \vec{\nabla}_r f(\vec{r}, \vec{p}; t) - \vec{\nabla}_r V_{ext}(x, t) \cdot \vec{\nabla}_p f(\vec{r}, \vec{p}; t)$$

$$- \int \vec{\nabla}_r v(|\vec{r} - \vec{r}'|) \cdot \vec{\nabla}_p f(\vec{r}, \vec{p}; \vec{r}', \vec{p}'; t) dx' dp' = 0, \qquad (A10.1)$$

where $f(\vec{r}, \vec{p}; \vec{r}', \vec{p}'; t)$ is the two-particle distribution function, which gives the joint probability of finding a pair of particles with momenta $\vec{p}$ and $\vec{p}'$ at positions $\vec{r}$ and $\vec{r}'$. To obtain a closed equation of motion for $f(\vec{r}, \vec{p}; t)$ STLS made the Ansatz

$$f(\vec{r}, \vec{p}; \vec{r}', \vec{p}'; t) = f(\vec{r}, \vec{p}; t) f(\vec{r}', \vec{p}'; t) g(|\vec{r} - \vec{r}'|), \qquad (A10.2)$$

where $g(|\vec{r} - \vec{r}'|)$ is the *equilibrium* pair correlation function. Substituting Eq. (A10.2) in (A10.1), and writing

$$f(\vec{r}, \vec{p}; t) = f_0(p) + f_1(\vec{r}, \vec{p}; t) \qquad (A10.3)$$

where $f_0(p)$ is the equilibrium distribution of the homogeneous system, one obtains, to first order in V_{ext}, the following equation for f_1:

$$\left[\frac{\partial}{\partial t} + \vec{v} \cdot \vec{\nabla}_r \right] f_1(\vec{r}, \vec{p}; t)$$

$$- \left(\vec{\nabla}_r V_{ext}(r, t) + \int \vec{\nabla}_r \psi(|\vec{r} - \vec{r}'|) f_1(\vec{r}', \vec{p}'; t) dx' dp' \right) \cdot \vec{\nabla}_p f_0(p) = 0, \quad (A10.4)$$

where

$$\vec{\nabla}_r \psi(r) = g(r) \vec{\nabla}_r v(r) . \qquad (A10.5)$$

It is evident that Eq. (A10.4) has the form of the transport equation for a non-interacting system with an effective force field of the form

$$\vec{E}_{eff} = -\vec{\nabla}_r V_{ext}(x, t) - \int \vec{\nabla}_r v(|\vec{r} - \vec{r}'|) f_1(\vec{r}', \vec{p}'; t) d\vec{r}' d\vec{p}'$$

$$- \int \left[g(|\vec{r} - \vec{r}'|) - 1 \right] \vec{\nabla}_r v(|\vec{r} - \vec{r}'|) f_1(\vec{r}', \vec{p}'; t) d\vec{r}' d\vec{p}' . \qquad (A10.6)$$

Since $\int f_1(\vec{r}', \vec{p}'; t) dp'$ is the induced density, we see that the second and third term on the right-hand side of this equation correspond respectively to the Hartree field (RPA) and to the STLS exchange-correlation field given by Eq. (5.119).

Appendix 11

Interpolation formulas for the local field factors

The quest for simple practical formulas for the many-body local field factors is as old as the concept of local field itself. The idea is to construct analytical formulas that satisfy the known constraints at small and large q and/or ω, while providing reasonable interpolations at intermediate values of these variables. We present below some of the best known interpolation formulas that have been developed for the paramagnetic state in recent years.

A11.1 Wave vector dependence

One possible and rather popular solution is to borrow the analytical form of the original Hubbard local field factor and modify it in such a way that the compressibility sum rule is exactly satisfied and the overall scale of variation with q is approximately correct. This leads to the formulas

$$G_\pm(q,0) \simeq \bar{G}_\pm^{(\infty)} \frac{q^2}{q^2 + \bar{G}_\pm^{(\infty)} \beta_\pm \kappa_3^2} \, , \qquad 3D \, , \qquad (A11.1)$$

and

$$G_\pm(q,0) \simeq \bar{G}_\pm^{(\infty)} \frac{q}{q + \bar{G}_\pm^{(\infty)} \beta_\pm \kappa_2} \, , \qquad 2D \, , \qquad (A11.2)$$

where

$$\beta_\pm^{-1} = 1 - \frac{K_0}{K} \, , \qquad (A11.3)$$

and

$$\beta_\pm^{-1} = 1 - \frac{\chi_P}{\chi_S} \, . \qquad (A11.4)$$

In the above equations, κ_d is the Thomas–Fermi wave vector in d dimensions and $\bar{G}_\pm^{(\infty)}$ is given by the expressions provided in Table 5.1.[1] The compressibility and spin susceptibility enhancements $\frac{K}{K_0}$ and $\frac{\chi_S}{\chi_P}$ are calculated from the second derivatives of the

---

[1] An approximation similar to Eqs. (A11.1) and (A11.2) was proposed by Iwamoto and Pines (1984) for $G_{\uparrow\uparrow}$ and $G_{\uparrow\downarrow}$ separately.

xchange-correlation energy with respect to density and spin polarization respectively. ince the correlation energy can in turn be calculated from the knowledge of the local eld factors, via the density–density response function (see Section 3.3.5) we see that the bove equations (A11.1) and (A11.2) can be used to calculate $\beta_\pm$ self-consistently.[2]

The main weakness of this approach is, of course, the total arbitrariness of the chosen unctional form of $G_\pm(q, 0)$. In particular, this form tends to a finite limit for $q \to \infty$ nstead of diverging, as it should.[3] As it turns out, recent Monte Carlo calculations of the tatic local field factor $G_+(q, 0)$ (Moroni *et al.*, 1992, 1995) indicate that the value of this unction is completely dominated by the compressibility sum rule up to wave vectors of he order of k_F. This can be appreciated by a direct inspection of Figs. 5.19 and 5.20. One nust notice moreover that the local field factors enter many formulas in such a way that only their values roughly between $q = 0$ and $q \simeq 2k_F$ are relevant. This leads to the conclusion that the incorrect behavior of Eqs. (A11.1) and (A11.2) at large q is not a serious problem in most application.

For the three dimensional case a more accurate rendition of the QMC results for $G_+(q, 0)$ in the range $2 < r_s < 10$, can be achieved by means of the interpolation formula originally suggested by Moroni *et al.* (1995) or the more tractable version (Corradini *et al.*, 1998)

$$G_+(q, 0) = C\bar{q}^2 + \frac{B\bar{q}^2}{g + \bar{q}^2} + a\bar{q}^4 e^{-b\bar{q}^2} , \qquad (A11.5)$$

where $\bar{q} = \frac{q}{k_F}$. The parameters that appear in this expression are defined in Table A11.1. The comparison between Eq. (A11.5) and the original Monte Carlo data is shown in Fig. 5.19.

There is, at present, no comparable QMC-based expression for $G_-(q)$ in three dimension, but, recently, Simion and Giuliani have proposed

$$G_-(q, 0) = C_1\bar{q}^2 + \frac{B_1 q^2}{g_1 + \bar{q}^2} , \qquad (A11.6)$$

where $C_1 = C$, $B_1 = B - 1 + 2g(0)$, and

$$g_1 = \frac{4\alpha_3 r_s B_1}{\pi \left(1 - \frac{x_P}{x_S}\right) - 4\alpha_3 r_s C} . \qquad (A11.7)$$

This local field factor has been used in Section 8.8 to calculate the quasiparticle properties of a 3D electron liquid (Simion *et al.*, 2005).

Accurate expressions for the local field factors of the two-dimensional electron gas have recently been proposed by Davoudi *et al.* (2005). These expressions read

$$G_\pm(q, 0) = A\bar{q} \left[\frac{\mathcal{K}}{\sqrt{1 + [A\mathcal{K}\bar{q}/B]^2}} + [1 - \mathcal{K}] e^{-\bar{q}^2/4} \right]$$
$$+ C\bar{q} \left(1 - e^{-B\bar{q}^2}\right) + P(\bar{q}) e^{-D\bar{q}^2} , \qquad (A11.8)$$

---

[2] It is understood that in order to calculate the correlation energy of the spin-polarized electron liquid one must resort to the multicomponent formalism of Section 5.4.5, which involves all the spin components of the local field factor (see Eq. (5.127) in particular).

[3] One may say that what we have here is an approximation for $\bar{G}(q, 0)$ defined in Section 5.6, while $G_n(q, 0)$ is neglected.

Table A11.1. *Parameters of the analytic*
representation (A11.5) for the density local field
factor in three dimensions. The correlation
energy per electron, $\epsilon_c(r_s)$ is expressed in Ry
and is accurately given by the expressions
provided in Section 1.7.2. ϵ_c' and ϵ_c'' denote the
first and the second derivative of ϵ_c with respect
to r_s. Here we have $a_1 = 2.15$, $a_2 = 0.435$,
$b_1 = 1.57$, $b_2 = 0.409$, and $\alpha_3 = (4/9\pi)^{1/3}$.

A	$\frac{1}{4} + \frac{\pi\alpha_3 r_s^2}{8}\left(\frac{2}{3}\epsilon_c' - r_s\epsilon_c''\right)$
B	$\frac{1 + a_1 r_s^{1/2} + a_2 r_s^{3/2}}{3 + b_1 r_s^{1/2} + b_2 r_s^{3/2}}$
C	$-\frac{\pi\alpha_3 r_s}{4}\left(\epsilon_c + r_s\epsilon_c'\right)$
g	$\frac{B}{A-C}$
a	$\frac{1.5A}{r_s^{1/4} g B}$
b	$\frac{1.2}{g B}$

where $\bar{q} = \frac{q}{k_F}$, $\bar{r}_s \equiv \frac{r_s}{10}$, and $P(\bar{q})$ is the polynomial $P(\bar{q}) = g_2\bar{q}^2 + g_4\bar{q}^4 + g_6\bar{q}^6 + g_8\bar{q}^8$.
The parameters appearing in the above expression are defined in the Table A11.2 for both
the density $(+)$ and the spin $(-)$ cases. The comparison with the original Monte Carlo data
is shown in Fig. 5.20.

A11.2 Frequency dependence

The frequency dependent local field factor has both a real and an imaginary part.
However, given an approximate formula for the imaginary part, the real part can be easily
obtained from the dispersion relation (5.183) together with the exact high frequency limit
of the real part. We will therefore focus on approximating the imaginary part of $G_+(q, \omega)$
in this section.

Perhaps the most popular expression for the frequency-dependence of the local field
factor is the the formula for $\Im m f_{xcL}(\omega) \equiv -\lim_{q\to 0} v_q \Im m G_+(q, \omega))$ proposed by Gross
and Kohn (GK) (1985). In three dimensions the GK formula reads

$$\Im m \bar{f}_{xcL}(\omega) = \frac{a(n)\bar{\omega}}{\left[1 + b(n)\bar{\omega}^2\right]^{\frac{5}{4}}}, \tag{A11.9}$$

where $\bar{f}_{xcL} \equiv \frac{f_{xcL}}{e^2 a_B^2}$, $\bar{\omega} = \frac{\hbar\omega}{2\,Ry}$, and $a(n)$ and $b(n)$ are fixed so that the high frequency limit
is given by Eq. (5.201), and the sum rule

$$\bar{f}_{xcL}(0) - \bar{f}_{xcL}(\infty) = P \int_{-\infty}^{\infty} \frac{d\omega}{\pi} \frac{\Im m \bar{f}_{xcL}(\omega)}{\omega} \tag{A11.10}$$

Table A11.2. *Parameters of the analytic representation (A11.8) for the density and spin local field factors in two dimensions.*

	G_+	G_-
A	$\frac{1}{r_s\sqrt{2}}\left[\frac{\sqrt{2}}{\pi}r_s - \frac{r_s^3}{8}\left(r_s\epsilon_c'' - \epsilon_c'\right)\right]$	$\frac{1}{r_s\sqrt{2}}\left[\frac{\sqrt{2}}{\pi}r_s - \frac{r_s^2}{2}\left.\frac{\partial^2\epsilon_c(r_s,p)}{\partial p^2}\right|_{p=0}\right]$
$\mathcal{K}$	$e^{\bar{r}_s}$	$\sqrt{1 + 0.0082r_s^2}$
B	$1 - \frac{1/2}{1+1.372r_s+0.0830r_s^2}$	$\frac{1/2}{1+1.372r_s+0.0830r_s^2}$
C	$-\frac{r_s}{2\sqrt{2}}\left(\epsilon_c + r_s\epsilon_c'\right)$	$-\frac{r_s}{2\sqrt{2}}\left(\epsilon_c + r_s\epsilon_c'\right)$
D	$\frac{0.1598+0.8931\bar{r}_s^{0.9218}}{1+0.8793\bar{r}_s^{0.9218}}$	$\exp\left[-\frac{0.2231+81.2115\bar{r}_s}{1+54.6665\bar{r}_s+50.7534\bar{r}_s^2}\right]$
β	1	$0.8089 - 0.4025\bar{r}_s^3 - 0.0941\bar{r}_s^{1/2}$
g_2	$0.5824\bar{r}_s^2 - 0.4272\bar{r}_s$	$\exp\left[-\frac{12.6262+20.9673\bar{r}_s}{1+12.4002\bar{r}_s}\right]$
g_4	$0.2960\bar{r}_s - 1.003\bar{r}_s^{5/2} + 0.9466\bar{r}_s^3$	$0.0531\left(1 - e^{-2.154\bar{r}_s^2}\right) - 0.4984\bar{r}_s^{3/2} + 0.4021\bar{r}_s^2$
g_6	$-0.0585\bar{r}_s^2$	$-0.0076\left(1 - e^{-\bar{r}_s}\right) + 0.0977\bar{r}_s^{3/2} - 0.0726\bar{r}_s^2$
g_8	$0.0131\bar{r}_s^2$	$-0.0027\bar{r}_s$

is satisfied.[4] These two conditions lead to the following expressions for a and b:

$$a(n) = -c\left(\frac{\gamma}{c}\right)^{\frac{5}{3}}\left[\bar{f}_{xcL}(\infty) - \bar{f}_{xcL}(0)\right]^{\frac{5}{3}},$$

$$b(n) = \left(\frac{\gamma}{c}\right)^{\frac{4}{3}}\left[\bar{f}_{xcL}(\infty) - \bar{f}_{xcL}(0)\right]^{\frac{4}{3}},$$

$$c = \frac{23\pi}{15}, \quad \gamma = \frac{\left[\Gamma(\frac{1}{4})\right]^2}{4\sqrt{2\pi}}, \tag{A11.11}$$

where $\bar{f}_{xcL}(0)$ and $\bar{f}_{xcL}(\infty)$ are obtained from Eqs. (5.194) and (5.200) respectively.
    The analogous formula for the two-dimensional case reads (Holas and Singwi, 1989)

$$\Im m\,\bar{f}_{xcL}(\omega) = \frac{a(n)\bar{\omega}}{b^2(n) + \bar{\omega}^2}, \tag{A11.12}$$

where $\bar{f}_{xcL} \equiv \frac{f_{xcL}}{e^2 a_B}$, $a(n) = -\frac{11\pi^2}{32}$ and $b(n) = \frac{a(n)}{\bar{f}_{xcL}(0)-\bar{f}_{xcL}(\infty)}$.
    The GK interpolation formula can be extended to finite wave vectors (Dabrowski, 1986) under the assumption that the high frequency limit of $-v_q \Im m G_+(q, \omega)$ at finite q is still given by the Glick and Long formula (5.201). The formula reads like Eq. (A11.9), but the coefficients $a(n)$ and $b(n)$ become functions of q and n.
    Several aspects of the GK approximation are unsatisfactory. First of all, the low frequency behavior of $\Im m f_{xcL}$ is determined by global sum rule arguments, instead of

---

[4] This sum rule is an obvious consequence of the Kramers-Krönig relation (5.183)

being directly related to the low frequency excitation spectrum of the electron gas. A perturbative calculation of $\Im m f_{xcL}(\omega)$ (Qian and Vignale, 2002) shows that the GK formula overestimates the value of this function at low frequency. Second, the approximate formula does not take into account the possibility of specific spectral structures associated, for instance, with two-plasmon excitations. A comparison between the GK interpolation formula and the mode-decoupling calculation described in Section 5.7.3 is particularly revealing (see Fig. 5.24). The mode-decoupling calculation exhibits a strong peak just above twice the plasmon frequency, but there is no trace of this structure in the GK formula. Moreover, the GK parameters a and b were calculated under the incorrect assumption that $f_{xcL}(0)$ is equal to the bulk modulus (see Eqs. (5.194)).

An improved interpolation formula that takes into account the above criticism has been very recently proposed (Qian and Vignale, 2002). The main idea is to correct the low-frequency behavior of the GK formula by shifting some of the low-frequency spectral weight to a peak region located about $2\omega_p$ (in three dimensions) as suggested by the mode-decoupling theory. The redistribution of the spectral weight is accomplished by the formula (covering now both longitudinal and transverse cases)

$$\Im m f_{xcL(T)}(\omega) = -\tilde{\omega}\left[\frac{a_3^{L(T)}}{(1+b_3^{L(T)}\tilde{\omega}^2)^{\frac{5}{4}}} + \tilde{\omega}^2 e^{-\frac{\left(|\tilde{\omega}|-\Omega_3^{L(T)}\right)^2}{\Gamma_3^{L(T)}}}\right]\left(\frac{2\omega_p}{n}\right), \quad (3D), \quad (A11.13)$$

where $\tilde{\omega} = \frac{\omega}{2\omega_p}$ and $\Omega_3^{L(T)} = 1 - \frac{3}{2}\Gamma_3^{L(T)}$. The relation between Ω_3 and Γ_3 is chosen so that the second term in the square bracket peaks at $\omega = 2\omega_p$. The remaining free parameter Γ_3 is then fixed by requiring that the slope of $\Im m f_{xcL}(\omega)$ vs ω at small ω agrees with the perturbative calculation. The corresponding proposal in two dimensions has the form

$$\Im m f_{xcL(T)}(\omega) = -\tilde{\omega}\left[\frac{a_2^{L(T)}}{1+b_2^{L(T)}\tilde{\omega}^2} + \tilde{\omega}^2 e^{-\frac{\left(|\tilde{\omega}|-\Omega_2^{L(T)}\right)^2}{\Gamma_2^{L(T)}}}\right]\left(\frac{Ry}{n}\right), \quad (2D), \quad (A11.14)$$

where now $\tilde{\omega} = \frac{\hbar\omega}{(2^7 r_s^2)^{1/3}\epsilon_F}$ is the frequency in units of an effective "two-plasmon" frequency, and $\Omega_2^{L(T)} = 1 - \frac{3}{2}\Gamma_2^{L(T)}$. The calculation of the parameters appearing in the above expressions is slightly technical. The interested reader will find a description of the calculation and a tabulation of the parameters in the paper by Qian and Vignale (2002).

Appendix 12

Real space-time form of the noninteracting Green's function

The Green's function of a spinless, non-interacting, and translationally invariant Fermi liquid has the form

$$G^{(0)}(r, t) = \Theta(t) G_>^{(0)}(r, t) + \Theta(-t) G_<^{(0)}(r, t),$$

(A12.1)

where, in d dimensions,

$$G_>^{(0)}(r, t) = -i \int \frac{d\vec{k}}{(2\pi)^d} \left(1 - n_{\vec{k}}\right) e^{-i\varepsilon_{\vec{k}} t/\hbar} e^{i\vec{k}\cdot\vec{r}},$$

(A12.2)

and

$$G_<^{(0)}(r, t) = i \int \frac{d\vec{k}}{(2\pi)^d} n_{\vec{k}} e^{-i\varepsilon_{\vec{k}} t/\hbar} e^{i\vec{k}\cdot\vec{r}}.$$

(A12.3)

Let $\bar{t} \equiv \frac{\varepsilon_F t}{\hbar}$ and $\bar{r} \equiv k_F r$. Then, at the absolute zero of temperature, after performing the angular integrals in (A12.2) and (A12.3), we have

$$G_>^{(0)}(\bar{r}, \bar{t}) = -ind \int_1^\infty e^{-i\bar{t}x^2} x^{d-1} F_d(x\bar{r}) dx$$

(A12.4)

and

$$G_<^{(0)}(\bar{r}, \bar{t}) = +ind \int_0^1 e^{-i\bar{t}x^2} x^{d-1} F_d(x\bar{r}) dx,$$

(A12.5)

where n is the electron density, and

$$F_d(x) = \begin{cases} \cos x, & d = 1, \\ J_0(x), & d = 2, \\ \frac{\sin x}{x}, & d = 3. \end{cases}$$

(A12.6)

The two functions $G_<^{(0)}$ and $G_>^{(0)}$ are related in the following manner:

$$G_>^{(0)}(\bar{r}, \bar{t}) = G_<^{(0)}(\bar{r}, \bar{t}) - ind \int_0^\infty e^{-i\bar{t}x^2} x^{d-1} F_d(x\bar{r}) dx,$$

(A12.7)

where

$$
\int_0^\infty e^{-i\bar{t}x^2} x^{d-1} F_d(x\bar{r}) dx = e^{i\bar{r}^2/4\bar{t}}
\begin{cases}
\frac{1}{2}\sqrt{\frac{\pi}{i\bar{t}}}, & d = 1, \\
\frac{1}{2i\bar{t}}, & d = 2, \\
-\frac{\sqrt{i\pi}}{4\bar{t}}, & d = 3.
\end{cases}
\tag{A12.8}
$$

The integral for $G_<^{(0)}(\bar{r}, \bar{t})$ can be done analytically. In one dimension we get

$$
G_<^{(0),1D}(\bar{r}, \bar{t}) = \pm \frac{n}{4}\sqrt{\frac{i\pi}{\bar{t}}} e^{i\bar{r}^2/4\bar{t}} \left[\mathrm{erf}\left(\sqrt{i\bar{t}} - \frac{\bar{r}}{2}\sqrt{\frac{i}{\bar{t}}} \right) + \mathrm{erf}\left(\sqrt{i\bar{t}} + \frac{\bar{r}}{2}\sqrt{\frac{i}{\bar{t}}} \right) \right], \tag{A12.9}
$$

where $\mathrm{erf}(z)$ is the *error function* of complex argument (see Abramowitz–Stegun, 1964) and the upper sign applies for negative times, the lower sign for positive ones. Similarly, in three dimensions we find

$$
G_<^{(0),3D}(\bar{r}, \bar{t}) = -\frac{3}{\bar{r}}\frac{\partial}{\partial \bar{r}} G_<^{(0),1D}(\bar{r}, \bar{t})
$$

$$
= -\frac{3i}{2\bar{t}} G_<^{(0),1D}(\bar{r}, \bar{t}) - \frac{3ne^{-i\bar{t}} \sin \bar{r}}{2\bar{r}\bar{t}}, \tag{A12.10}
$$

where use has been made of the identity $\frac{d}{dz}\mathrm{erf}(z) = \frac{2e^{-z^2}}{\sqrt{\pi}}$. The calculation in the two-dimensional case is left as an exercise for the dedicated reader.

Representative plots of the Green's functions are shown in Fig. 6.1. Notice that in one dimension $G^{(0)}$ decreases as $\frac{1}{\sqrt{\bar{t}}}$ for $\bar{t} \to -\infty$ and more rapidly, as $\frac{1}{\bar{t}}$, for $\bar{t} \to +\infty$. In two and three dimensions, on the other hand, the asymptotic behavior is $\frac{1}{\bar{t}}$ for both positive and negative times.

Appendix 13

Calculation of the ground-state energy and thermodynamic potential

We saw in Section 3.3.5 that the ground-state energy (or the thermodynamic potential) of a many-body system can be calculated from the density–density response function. In this appendix we show two alternative methods for calculating these quantities: in the first one we express the energy in terms of the Green's function and the self-energy, in the second a direct diagrammatic expansion for the energy is provided. For definiteness we concentrate on the calculation of the ground-state energy: the resulting formulas are straightforwardly generalized to the case of the thermodynamic potential.

A13.1 Expression in terms of the self-energy

Consider a system of particles interacting via the scaled two-body interaction λv_q, with $0 < \lambda < 1$. For $\lambda = 1$ this coincides with the physical many-body system, while for $\lambda = 0$ it is a non-interacting system. According to the Hellman–Feynman theorem, the difference between the ground-state energy of the physical system and that of the non-interacting system is given by $\int_0^1 \frac{U(\lambda)}{\lambda} d\lambda$, where $U(\lambda)$ – the average potential energy of the system at coupling constant λ – is the key quantity to be computed.

Although the average potential energy is a two-particle property it can still be expressed in terms of the Green's function with the help of the following trick. Observe that

$$\lim_{t \to 0^-} \frac{\partial G_\sigma(\vec{k}, t)}{\partial t} = \frac{1}{\hbar} \langle \hat{a}_{\vec{k},\sigma}^\dagger [\hat{a}_{\vec{k},\sigma}, \hat{H}] \rangle \tag{A13.1}$$

where $\hat{H} = \hat{T} + \hat{U}$ is the hamiltonian. For a system with two-body interactions it is easy to verify that

$$\sum_{\vec{k},\sigma} \hat{a}_{\vec{k},\sigma}^\dagger [\hat{a}_{\vec{k},\sigma}, \hat{H}] = \hat{T} + 2\hat{U}.^1 \tag{A13.2}$$

Thus, we have

$$\langle \hat{U} \rangle = \frac{1}{2} \sum_{\vec{k},\sigma} \left(\hbar \lim_{t \to 0^-} \frac{\partial G_\sigma(\vec{k}, t)}{\partial t} - \langle \hat{T} \rangle \right). \tag{A13.3}$$

---

[1] This is most readily seen from the analogy between taking the commutator of $\hat{a}_{\vec{k},\sigma}$ with $\hat{H}$ and "differentiating" $\hat{H}$ with respect to $\hat{a}_{\vec{k},\sigma}^\dagger$.

The average kinetic energy, $\langle \hat{T} \rangle$, can be immediately expressed in terms of the Green's function as follows:

$$\langle \hat{T} \rangle = -i \sum_{\vec{k},\sigma} \frac{\hbar^2 k^2}{2m} \lim_{t \to 0^-} G_\sigma(\vec{k}, t) .$$ (A13.4)

Switching from the time to the frequency domain we get

$$\langle \hat{T} \rangle = \sum_{\vec{k},\sigma} \int_{-\infty}^{\infty} \frac{d\omega}{2\pi i} \frac{\hbar^2 k^2}{2m} G_\sigma(\vec{k}, \omega) e^{i\omega\eta} ,$$ (A13.5)

and putting this into Eq. (A13.3) we arrive at

$$\langle \hat{U} \rangle = \frac{1}{2} \sum_{\vec{k},\sigma} \int_{-\infty}^{\infty} \frac{d\omega}{2\pi i} \left(\hbar\omega - \frac{\hbar^2 k^2}{2m} \right) G_\sigma(\vec{k}, \omega) e^{i\omega\eta}$$ (A13.6)

with $\eta \to 0^+$. In the limit of zero interaction, the potential energy must vanish. This can be formally verified substituting $G = G^{(0)}$ in Eq. (A13.6), and noting that

$$\int_{-\infty}^{\infty} \frac{d\omega}{2\pi i} \frac{\hbar\omega - \frac{\hbar^2 k^2}{2m}}{\omega - \frac{\hbar k^2}{2m} + i\eta \, \text{sign}(k - k_F)} = 0$$ (A13.7)

for $\eta \to 0^+$. Thus, subtracting the left-hand side of Eq. (A13.7) from the right-hand side of Eq. (A13.6), and making use of the expression (6.73) for the Green's function in terms of the self-energy, we see that

$$\langle \hat{U} \rangle = \frac{1}{2} \sum_{\vec{k},\sigma} \int_{-\infty}^{\infty} \frac{d\omega}{2\pi i} G_\sigma(\vec{k}, \omega) \Sigma_\sigma(\vec{k}, \omega).$$ (A13.8)

Finally, integrating over the coupling constant, we obtain

$$E = E_0 + \frac{1}{2} \int_0^1 \frac{d\lambda}{\lambda} \sum_{\vec{k},\sigma} \int_{-\infty}^{\infty} \frac{d\omega}{2\pi i} G_\sigma^\lambda(\vec{k}, \omega) \Sigma_\sigma^\lambda(\vec{k}, \omega) ,$$ (A13.9)

where $G_\sigma^\lambda(\vec{k}, \omega)$ and $\Sigma_\sigma^\lambda(\vec{k}, \omega)$ are the Green's function and the self-energy at coupling constant λ. This result is straightforwardly extended to finite temperature provided that E and E_0 are replaced by the interacting and non-interacting thermodynamic potential respectively, and the frequency integral by a sum over Matsubara frequencies.

A13.2 Linked-cluster expansion

There exists another representation of the ground-state energy that does not require an integration over the coupling constant and will prove useful in the microscopic derivation of the Landau theory of Fermi liquids in Chapter 8. The alternative expression reads

$$E = E_0 - \hbar \sum_{k\sigma} \int \frac{d\omega}{2\pi i} \left\{ \ln\left[1 - \hbar^{-1} G_\sigma^{(0)}(k, \omega) \Sigma_\sigma(k, \omega) \right] \right.$$

$$\left. + \hbar^{-1} G_\sigma(k, \omega) \Sigma_\sigma(k, \omega) \right\} + L^d \Omega[G] ,$$ (A13.10)

where $\Omega[G]$ is the generating functional for the self-energy, introduced and discussed in Section 6.3.3.

Here is the proof: We begin by noting that the whole expression on the right-hand side is stationary with respect to an infinitesimal variation of the self-energy. This is a direct consequence of the identity (6.81), i.e., $\frac{\delta\Omega[G]}{\delta G} = \Sigma[G]$, as the reader can easily verify. Consider now the derivative of the right-hand side of Eq. (A13.10) with respect to the scale factor λ that multiplies the interaction. Because this expression is stationary with respect to a variation of Σ, a change in λ affects the right-hand side of Eq. (A13.10) only insofar as it changes the numerical value the interaction lines in the skeleton diagrams for $\Omega[G]$. Thus Eq. (A13.10) implies that

$$\frac{\partial E(\lambda)}{\partial \lambda} = L^d \frac{\partial\Omega^\lambda[G]}{\partial \lambda}, \qquad (A13.11)$$

where the partial derivative is carried out only with respect to the *explicit* λ-dependence of the functional, i.e., at constant G. All that needs to be proved now is that

$$L^d \frac{\partial\Omega^\lambda[G]}{\partial \lambda} = \frac{1}{2\lambda} \sum_{\vec{k},\sigma} \int_{-\infty}^{\infty} \frac{d\omega}{2\pi i} G_\sigma^\lambda(\vec{k},\omega)\Sigma_\sigma^\lambda(\vec{k},\omega). \qquad (A13.12)$$

If this holds true, then the equivalence of Eqs. (A13.10) and (A13.9) follows from the fact that their right-hand sides have the same derivative with respect to λ and both vanish at $\lambda = 0$. To show that this is indeed the case, notice that the diagrams in Ω contain two Green's functions for each interaction line: therefore rescaling each interaction line by a factor λ at constant G can also be viewed as rescaling each Green's function by a factor $\sqrt{\lambda}$ at constant interaction strength:

$$\Omega^\lambda[G] = \Omega^1[\sqrt{\lambda}G]. \qquad (A13.13)$$

With the help of this scaling relation it is not difficult to see that

$$\frac{\partial\Omega^\lambda[G]}{\partial \lambda} = \frac{1}{L^d} \sum_{\vec{k}\sigma} \int \frac{d\omega}{2\pi i} \frac{\delta\Omega^1[\sqrt{\lambda}G]}{\delta\sqrt{\lambda}G_\sigma(\vec{k},\omega)} \frac{\partial\sqrt{\lambda}G_\sigma(\vec{k},\omega)}{\partial \lambda}$$

$$= \frac{1}{2\lambda L^d} \sum_{\vec{k}\sigma} \int \frac{d\omega}{2\pi i} \Sigma_\sigma^\lambda(\vec{k},\omega)G_\sigma(\vec{k},\omega), \qquad (A13.14)$$

which proves the point.

The complicated formula (A13.10) for the ground-state energy has a simple interpretation in terms of diagrams: it turns out that the interaction contribution is the sum of *all* the closed and connected diagrams (skeleton diagrams or not) evaluated according to the usual Feynman rules, except for the fact that a diagram of order n must be multiplied by an extra factor $1/n$. Perhaps the simplest way to convince oneself of this fact is to first prove it for the grand thermodynamic potential at finite temperature, and then extend the result to the ground-state energy by taking the zero-temperature limit. From Eqs. (6.118), (6.135), and (6.136) of Section 6.4 we can easily infer that the grand thermodynamic potential is given by

$$\Omega = \Omega_0 - k_B T \ln\langle\hat{S}\rangle_0, \qquad (A13.15)$$

where the S-matrix is defined in Eq. (6.137) and Ω_0 is the non-interacting thermodynamic potential. The diagrammatic expansion for $\langle \hat{S} \rangle_0$ consists of diagrams with no free ends that can be either connected or consist of different unconnected parts. It can be shown that every closed and *connected* diagram, i.e., every *linked* diagram, can be evaluated by the usual Feynman rules with an extra factor $\frac{1}{n}$, where n is the number of interaction lines.[2] The general diagram contributing to $\langle \hat{S} \rangle_0$ contains p_1 linked diagrams with 1 interaction line, p_2 linked diagrams with 2 interaction lines, and so on. The total value of the diagram is obtained simply by multiplying the values of its linked parts. Denoting by Q_n the sum of all the linked diagrams with n interaction lines ($1/n$ factor included), it follows that $\langle \hat{S} \rangle_0$ is given by

$$\langle \hat{S} \rangle_0 = \sum_{p_1, p_2, \dots p_n \dots} \frac{Q_1^{p_1}}{p_1!} \frac{Q_2^{p_2}}{p_2!} \dots \frac{Q_3^{p_n}}{p_n!}$$

$$= \exp\left(Q_1 + Q_2 + \dots + Q_n + \dots\right). \tag{A13.16}$$

The factors $p_i!$ in the denominator ensure that each diagram contributing to the S-matrix is counted once and only once. Taking the logarithm of Eq. (A13.16) we see that

$$\Omega - \Omega_0 = -k_B T \sum_{i=1}^{\infty} Q_i , \tag{A13.17}$$

which means that the interaction correction to the thermodynamic potential is given by the sum of all the linked diagrams containing at least one interaction line. This elegant result is known as the "linked cluster theorem". Its usefulness stems from the fact that it eliminates the need of calculating of $\hat{S}$ – a quantity with a rather unwieldy diagrammatic expansion. Its main weakness comes from the need of including the troublesome $\frac{1}{n}$ factor in the calculation of the diagrams.

---

[2] The factor $1/n$ for a linked diagram arises from the fact that cyclic permutations of the n time labels are immaterial (not only they do not generate a different diagram, but not even a different scheme of contractions). Thus there are only $2^n(n-1)!$ different schemes of contraction for each diagram. This multiplicity, divided by the standard factor $\frac{1}{2^n n!}$, which arises from the expansion of the the exponential in Eq. (6.28), yields the $\frac{1}{n}$ factor.

Appendix 14
Spectral representation and frequency summations

In finite-temperature perturbation theory one often needs to evaluate infinite frequency sums of the form:

$$f(0^-) = \frac{1}{\hbar\beta} \sum_k f(i\omega_k)e^{i\omega_k\eta} , \qquad (A14.1)$$

or

$$\int_0^\beta f(\tau)g(-\tau)e^{i\omega_n\tau}d\tau = \frac{1}{\hbar\beta} \sum_k f(i\omega_n + i\omega_k)g(i\omega_k) , \qquad (A14.2)$$

where $f(z)$ and $g(z)$ are temperature correlation functions (e.g. the temperature Green's function $\mathcal{G}$) that vanish at infinity and are analytic in the complex plane except for a discontinuity across the real axis (see Eq. (6.130)). According to Eq. (6.131), the discontinuity is

$$f(\omega + i\eta) - f(\omega - i\eta) = 2i \,\Im m f(\omega + i\eta) , \qquad (A14.3)$$

where ω is real and $\eta \to 0^+$. A direct consequence of the above properties is that $f(i\omega_k)$ has the *spectral representation*

$$f(i\omega_k) = -\frac{1}{\pi} \int_{-\infty}^{\infty} dx \frac{\Im m f(x+i\eta)}{i\omega_k - x} , \qquad (A14.4)$$

where the integral runs along the real axis. This extremely useful formula allows us to systematically reduce the evaluation of infinite sums of the type (A14.1) and (A14.2) to the calculation of integrals along the real axis. Consider, for example, the sum in Eq. (A14.1). According to Eq. (A14.4) it can be rewritten as

$$f(0^-) = -\frac{1}{\pi} \int_{-\infty}^{\infty} dx \,\Im m f(x+i\eta) \frac{1}{\hbar\beta} \sum_k \frac{e^{i\omega_k\eta}}{i\omega_k - x} . \qquad (A14.5)$$

The sum over Matsubara frequencies is just the Green's function evaluated at $\tau = 0^-$, and can therefore be expressed in terms of thermal occupation factors as follows:

$$\lim_{\eta\to0^+} \frac{1}{\hbar\beta} \sum_k \frac{e^{i\omega_k\eta}}{i\omega_k - x} = \pm n_\pm(x) , \qquad (A14.6)$$

713

where $n_\pm(x) = \frac{1}{e^{\beta x}\pm 1}$, the upper sign holds for fermions, and the lower one for bosons. Thus, we have

$$f(0^-) = \mp\frac{1}{\pi}\int_{-\infty}^{\infty} dx\,\Im mf(x+i\eta)n_\pm(x)\,. \qquad (A14.7)$$

Consider now the calculation of the sum in Eq. (A14.2). First, its right-hand side is rewritten as

$$\frac{1}{\pi^2}\int_{-\infty}^{\infty} dx\,\Im mf(x+i\eta)\int_{-\infty}^{\infty} dy\,\Im mg(y+i\eta)$$
$$\times\frac{1}{\hbar\beta}\sum_k\frac{1}{i\omega_n+i\omega_k-x}\frac{1}{i\omega_k-y}\,. \qquad (A14.8)$$

Making use of the identity

$$\frac{1}{i\omega_n+i\omega_k-x}\frac{1}{i\omega_k-y} = \frac{1}{i\omega_n-x+y}\left(\frac{1}{i\omega_k-y}-\frac{1}{i\omega_n+i\omega_k-x}\right)\,, \qquad (A14.9)$$

we reduce the evaluation of the frequency sum to repeated applications of Eq. (A14.6). The outcome of the calculation depends on whether the frequencies $i\omega_k$ and $i\omega_n$ are fermionic or bosonic. For example, in the case that both functions f and g are fermionic, one sees that both $i\omega_k$ and $i\omega_n+i\omega_k$ must be fermionic, and therefore $i\omega_n$ is bosonic. So the sum of the first term in the brackets of Eq. (A14.9) is $n_+(y)$, while the sum of the second term is $-n_-(x-i\omega_k) = n_+(x)$. Combining the two sums we get

$$\frac{1}{\hbar\beta}\sum_k\frac{1}{i\omega_n+i\omega_k-x}\frac{1}{i\omega_k-y} = \frac{n_+(y)-n_+(x)}{i\omega_n-x+y} \qquad (A14.10)$$

(n even, k odd). The analytic continuation of this expression to real frequency is done by substituting $i\omega_n \to \omega+i\eta$ on the right-hand side. *It is imperative that the substitution be done only after performing the sum over k, since interchanging the order of these two operations leads in general to a different, and incorrect result.*

Other cases can be worked out in a similar way. Notice that the above formulas are generalizable, by recursion, to products of more than two functions.

Appendix 15

Construction of a complete set of wavefunctions with a given density

Let $n(\vec{r}) > 0$, with $\int n(\vec{r})\,d\vec{r} = N$, be a positive density for an N-electron system. We want to show that it is possible to construct a complete set of orthonormal Slater determinants, all of which yield the density $n(\vec{r})$. To this end, notice that a well-behaved density distribution can be interpreted as the Jacobian of a suitable coordinate transformation, i.e., we can find coordinates $\vec{\xi}(\vec{r})$ such that

$$d\vec{\xi}(\vec{r}) = n(\vec{r})d\vec{r} \; . \tag{A15.1}$$

This assertion is certainly plausible on physical grounds, since every nonuniform density can relax to a uniform density by means of a dynamical flow that displaces a volume element from $\vec{r}$ to $\vec{\xi}(\vec{r})$. Such a flow establishes $\vec{\xi}(\vec{r})$ as the coordinate system associated with $n(\vec{r})$.[1]

Consider now the one-electron wavefunctions

$$\phi_{\vec{k}}(\vec{r}) = \sqrt{\frac{n(\vec{r})}{N}}\,e^{i\vec{k}\cdot\vec{\xi}(\vec{r})} \; , \tag{A15.2}$$

where $\vec{k}$ is an arbitrary wave vector. The density associated with each of these wave functions is $\frac{n(\vec{r})}{N}$, and the normalization integral is 1. These wave functions are mutually orthogonal because

$$\int \phi_{\vec{k}'}^*(\vec{r})\phi_{\vec{k}}(\vec{r})d\vec{r} = \int e^{i(\vec{k}-\vec{k}')\cdot\vec{\xi}(\vec{r})}\frac{n(\vec{r})}{N}d\vec{r}$$

$$= \frac{1}{N}\int e^{i(\vec{k}-\vec{k}')\cdot\vec{\xi}}d\vec{\xi} = \frac{(2\pi)^3}{N}\delta^3(\vec{k}-\vec{k}') \; , \tag{A15.3}$$

and, furthermore, they form a complete set since

$$\int \phi_{\vec{k}}^*(\vec{r})\phi_{\vec{k}}(\vec{r}')\frac{d\vec{k}}{(2\pi)^3} = \frac{\sqrt{n(\vec{r})n(\vec{r}')}}{N}\delta^3(\vec{\xi}(\vec{r})-\vec{\xi}(\vec{r}'))$$

$$= \frac{\delta^3(\vec{r}-\vec{r}')}{N} \tag{A15.4}$$

by virtue of Eq. (A15.1).

---

[1] This fact has been pointed out to the authors by Qian Niu.

Starting from this complete set of orthonormal single particle orbitals one can easily construct a complete set of orthogonal N-particle Slater determinants

$$\frac{1}{\sqrt{N!}} det[\phi_{\vec{k}_i}(\vec{r}_j)] \, , \tag{A15.5}$$

where $\vec{k}_1 \ldots \vec{k}_N$ are N different wave vectors, and $\vec{r}_1 \ldots \vec{r}_N$ are the coordinates of the N particles. The density of this wavefunction is the sum of the densities of the N occupied orbitals, i.e. $n(\vec{r})$, for any choice of $\vec{k}_1 \ldots \vec{k}_N$.

Appendix 16

Meaning of the highest occupied Kohn–Sham eigenvalue in metals

Let us consider the variation of the ground-state energy of a system upon addition (or removal) of a single electron. We focus, for definiteness, on the case of addition. Let

$$\delta n(\vec{r}) = n_N(\vec{r}) - n_{N-1}(\vec{r}) \tag{A16.1}$$

be the difference between the ground-state density of the N electron system and that of the $N-1$ electron system. According to Eq. (7.28), the difference between the corresponding ground-state energies is given by

$$
\begin{aligned}
E_N - E_{N-1} = \varepsilon_N &+ \sum_{\alpha=1}^{N-1}(\varepsilon_{\alpha,N} - \varepsilon_{\alpha,N-1}) \\
&- (E_H[n_N] - E_H[n_{N-1}]) + (E_{xc}[n_N] - E_{xc}[n_{N-1}]) \\
&- \int [n_N(\vec{r})V_{xc,N}(\vec{r}) - n_{N-1}(\vec{r})V_{xc,N-1}(\vec{r})]d\vec{r} ,
\end{aligned}
\tag{A16.2}
$$

where $V_{xc,N}(\vec{r})$ denotes the xc potential associated with the ground-state density of the N-electron system.

We now assume that $N \gg 1$, so that $\delta n(\vec{r}) \ll n_N(\vec{r})$. The finite differences in Eq. (A16.2) can then be approximated by "Taylor expansions", e.g.,

$$E_H[n_N] - E_H[n_{N-1}] \sim \int V_{H,N-1}(\vec{r})\delta n(\vec{r})d\vec{r} ,$$

$$E_{xc}[n_N] - E_{xc}[n_{N-1}] \sim \int V_{xc,N-1}(\vec{r})\delta n(\vec{r})d\vec{r} , \tag{A16.3}$$

where we have used the fact that $\frac{\delta E_{xc}[n]}{\delta n(\vec{r})} = V_{xc}(\vec{r})$ and $\frac{\delta E_H[n]}{\delta n(\vec{r})} = V_H(\vec{r})$.

After some simple algebra, we arrive at

$$
\begin{aligned}
E_N - E_{N-1} = \varepsilon_N &+ \sum_{\alpha=1}^{N-1} \int \frac{\delta\varepsilon_\alpha}{\delta n(\vec{r})}\delta n(\vec{r})d\vec{r} \\
&- \int d\vec{r} \int d\vec{r}' \left[\frac{\delta V_{xc}(\vec{r}')}{\delta n(\vec{r})} + \frac{\delta V_H(\vec{r}')}{\delta n(\vec{r})}\right]n(\vec{r}')\delta n(\vec{r}) .
\end{aligned}
\tag{A16.4}
$$

It must be emphasized that the validity of this equation depends on the existence of the functional derivative of the exchange-correlation energy functional – an assumption that breaks down (see Section 7.2.8) in systems with a gap at the Fermi level.

The functional derivative of the Kohn–Sham eigenvalues ε_α with respect to the density is calculated with the help of the Hellman–Feynman theorem

$$\frac{\delta \varepsilon_\alpha}{\delta n(\vec{r})} = \langle \phi_\alpha | \frac{\delta \hat{H}_{KS}}{\delta n(\vec{r})} | \phi_\alpha \rangle$$

$$= \int \left[\frac{\delta V_{xc}(\vec{r}')}{\delta n(\vec{r})} + \frac{\delta V_H(\vec{r}')}{\delta n(\vec{r})} \right] |\phi_\alpha(\vec{r}')|^2 d\vec{r}'. \qquad (A16.5)$$

Since

$$n(\vec{r}') = \sum_{\alpha=1}^{N-1} |\phi_\alpha(\vec{r}')|^2 , \qquad (A16.6)$$

the last two terms on the right-hand side of Eq. (A16.4) cancel out, and we are left with the simple result

$$E_N - E_{N-1} = \varepsilon_N \quad (N \gg 1, \text{ gapless systems}) . \qquad (A16.7)$$

Appendix 17

Density functional perturbation theory

In this appendix, we supply a few more details about the Görling–Levy (GL) perturbation theory (Görling and Levy, 1994), i.e., the perturbation theory at constant density. In particular, we provide the explicit expression for the second-order correlation energy $E_c^{(2)}[n]$, defined in Eq. (7.52).

To set up the GL perturbation theory, observe that the external potential $V_\lambda(\vec{r})$, which yields a given density $n(\vec{r})$ at coupling constant λ, can be written as

$$V_\lambda(\vec{r}) = -\frac{\delta F_\lambda[n]}{\delta n(\vec{r})} . \tag{A17.1}$$

The $\lambda = 0$ potential, $V_0(\vec{r})$, yields $n(\vec{r})$ in the non-interacting system, and must therefore be identified with the Kohn–Sham potential: $V_0(\vec{r}) \equiv V_{KS}(\vec{r})$.

We now partition the full hamiltonian $\hat{H}(\lambda)$ of Eq. (7.29) in the following manner:

$$\hat{H}(\lambda) = \hat{H}(0) + \lambda \hat{H}_{e-e} - \sum_{i=1}^{N} \left(\frac{\delta F_\lambda[n]}{\delta n(\vec{r}_i)} - \frac{\delta F_0[n]}{\delta n(\vec{r}_i)} \right)$$

$$= \hat{H}(0) + \lambda \hat{H}_{e-e} - \sum_{i=1}^{N} \left[\lambda V_H(\vec{r}_i) + \lambda V_x(\vec{r}_i) + V_{c,\lambda}(\vec{r}_i) \right] , \tag{A17.2}$$

where $\hat{H}(0)$ is just the Kohn–Sham hamiltonian, and $V_{c,\lambda}(\vec{r}) = \frac{\delta E_{c,\lambda}[n]}{\delta n(\vec{r})}$ is the correlation potential at density n and coupling constant λ. We see that, because the "zero-order" hamiltonian already includes some interaction effects through the Kohn–Sham potential, the "perturbation" consists of the electron-electron interaction *minus* a sum of one-body counterterms. Now from Eq. (7.54) we have

$$V_{c,\lambda}(\vec{r}) = \lambda^2 \frac{\delta E_c^{(2)}[n]}{\delta n(\vec{r})} + \lambda^3 \frac{\delta E_c^{(3)}[n]}{\delta n(\vec{r})} + \dots . \tag{A17.3}$$

This enables us to express $\hat{H}(\lambda)$ in a form that is convenient for perturbation theory, namely

$$\hat{H}(\lambda) = \hat{H}(0) + \lambda \hat{H}^{(1)} + \lambda^2 \hat{H}^{(2)} + \dots \tag{A17.4}$$

719

where

$$\hat{H}^{(1)} = \hat{H}_{e-e} - \sum_{i=1}^{N} \left[V_H(\vec{r}_i) + V_x(\vec{r}_i) \right] ,$$ (A17.5)

and

$$\hat{H}^{(j)} = -\sum_{i=1}^{N} \frac{\delta E_c^{(j)}[n]}{\delta n(\vec{r}_i)} , \qquad (j \geq 2) .$$ (A17.6)

Thus, denoting by $|\psi_k(\lambda)\rangle$ the k-th eigenstate of $\hat{H}(\lambda)$ and by $E_k(\lambda)$ its eigenvalue (with $k = 0$ for the ground-state), we can immediately write down the familiar perturbation expansion for $|\psi_0(\lambda)\rangle$ and $E_0(\lambda)$ in powers of λ:

$$|\psi_0(\lambda)\rangle = |\psi_0(0)\rangle + \lambda \sum_{k \neq 0} \frac{\langle \psi_k(0)| \hat{H}^{(1)} |\psi_0(0)\rangle}{E_0(0) - E_k(0)} |\psi_k(0)\rangle + \dots$$ (A17.7)

and

$$E_0(\lambda) = E_0(0) + \lambda^2 \left[\sum_{k \neq 0} \frac{|\langle \psi_k(0)| \hat{H}^{(1)} |\psi_0(0)\rangle|^2}{E_0(0) - E_k(0)} + \langle \psi_0(0)| \hat{H}^{(2)} |\psi_0(0)\rangle \right] + \dots .$$ (A17.8)

All the eigenfunctions and eigenvalues on the right-hand side of these expressions are obtained from the solution of the Kohn–Sham equation for the ground-state density. Notice that the first-order term in the expansion of $E_0(\lambda)$ is absent because $\langle \psi_0(0)| \hat{H}^{(1)} |\psi_0(0)\rangle = 0$.

To calculate the second-order term in the perturbative expansion of the correlation energy functional (Eq. (7.54)), i.e., $\lambda^2 E_c^{(2)}$, we differentiate both sides of Eq. (7.51) with respect to λ. Making use of the Hellmann–Feynman theorem we obtain, for small λ,

$$2\lambda E_c^{(2)} \simeq \langle \psi_0(\lambda)| \hat{H}^{(1)} |\psi_0(\lambda)\rangle - \langle \psi_0(0)| \hat{H}^{(1)} |\psi_0(0)\rangle .$$ (A17.9)

Keeping only the first-order term in the expansion (A17.7) of $\psi_0(\lambda)$ we arrive at

$$E_c^{(2)} = \sum_{k \neq 0} \frac{|\langle \psi_k(0)| \hat{H}^{(1)} |\psi_0(0)\rangle|^2}{E_0(0) - E_k(0)} ,$$ (A17.10)

which is always negative. The usefulness of this result lies in the fact that in small size systems (e.g., atoms) the electronic density is usually very high, so the leading approximation to the correlation energy is often sufficient. For example, Fig. 7.6 shows very clearly that $\frac{dE_{xc,\lambda}}{d\lambda}$ at the ground-state density of a helium atom is an approximately linear function of λ, implying that $E_{c,\lambda}$ is approximately quadratic. Notice, however that the GL expansion of the correlation energy is to be used in conjunction with the DFT definition of exchange (Eq. (7.35)) rather than with the Hartree–Fock exchange. A perturbative expression for the correlation energy, which can be added to the Hartree–Fock exchange is also available and can be found in the paper by Görling and Levy (1994).

Appendix 18

Density functional theory at finite temperature

The extension of DFT to finite temperature ensembles is originally due to Mermin (1965). Here we follow a slightly different path, which parallels our presentation of the zero-temperature formalism.

First of all, we work in the grand-canonical ensemble, where the hamiltonian $\hat{H}$ of Eq. (7.1) is replaced by

$$
\begin{aligned}
\hat{K} &= \hat{H} - \mu \hat{N} \\
&= \hat{T} + \hat{H}_{e-e} + \hat{V} - \mu \hat{N} .
\end{aligned}
\tag{A18.1}
$$

The grand thermodynamic potential Ω of this system is given by

$$
\Omega = \mathrm{Tr} \left\{ \hat{\rho}_{eq} [\hat{K} + k_B T \ln \hat{\rho}_{eq}] \right\} ,
\tag{A18.2}
$$

where Tr denotes the trace of an operator, and the equilibrium density matrix is

$$
\hat{\rho}_{eq} = \frac{e^{-\beta \hat{K}}}{\mathrm{Tr}[e^{-\beta \hat{K}}]} .
\tag{A18.3}
$$

The equilibrium density matrix has the following variational property: the functional

$$
\Omega_{V-\mu}[\hat{\rho}] \equiv \mathrm{Tr} \left\{ \hat{\rho}[\hat{K} + k_B T \ln \hat{\rho}] \right\}
\tag{A18.4}
$$

is minimum when $\hat{\rho} = \hat{\rho}_{eq}$. The minimization with respect to $\hat{\rho}$ can be carried out in two steps: first, we find the minimum among the $\hat{\rho}$'s that yield a given density $n(\vec{r}) = \frac{\mathrm{Tr}[\hat{n}(\vec{r})\hat{\rho}]}{\mathrm{Tr}\hat{\rho}}$; second, we minimize with respect to $n(\vec{r})$. In analogy to Eq. (7.6) we define the universal functional of the density

$$
F[n] = \min_{\hat{\rho} \to n(\vec{r})} \mathrm{Tr} \left\{ \hat{\rho}[\hat{T} + \hat{H}_{e-e} + k_B T \ln \hat{\rho}] \right\} .
\tag{A18.5}
$$

The true grand thermodynamic potential is then given by the minimum, with respect to $n(\vec{r})$, of the functional

$$
\Omega_{V-\mu}[n] = F[n] + \int [V(\vec{r}) - \mu] n(\vec{r}) d\vec{r} :
\tag{A18.6}
$$

721

this is the finite-temperature generalization of Eq. (7.5). Assuming differentiability of the functional $F[n]$, the true equilibrium density is determined by the stationarity condition

$$\frac{\delta F[n]}{\delta n(\vec{r})} = \mu - V(\vec{r}).$$

(A18.7)

If the functional derivative exists for every "reasonable" density, then the above equation proves the existence of a unique external potential $V(\vec{r}) - \mu$ for every such density. At variance with the zero temperature case, the potential $V(\vec{r})$ is completely determined – relative to the chemical potential μ, without the possibility of adding an arbitrary constant. Furthermore, all the examples given in Section 7.2.9, of different external potentials that yield against the same ground-state, disappear at finite temperature: a thermal ensemble can never be completely "rigid" the application of an external field. Unfortunately, the resolution of these difficulties is more formal than real, as discussed in Section 7.2.8.

The finite-temperature version of the Hohenberg–Kohn theorem, namely that two different potentials $V(\vec{r}) - \mu$ and $V'(\vec{r}) - \mu'$ cannot give the same equilibrium density follows from a straightforward extension of the proof presented in Section 7.2.2. Let us assume, against the thesis of the theorem, that the two potentials give the same density n. Denoting by $\hat{\rho}$ and $\hat{\rho}'$ the two (necessarily different) equilibrium density matrices associated with the two potentials we have

$$\Omega_{V'-\mu'}[\hat{\rho}'] = Tr\left\{\hat{\rho}'[\hat{T} + \hat{H}_{e-e} + \hat{V}' - \hat{\mu}'\hat{N} + k_B T \ln \hat{\rho}']\right\}$$

$$= \int \left\{[V'(\vec{r}) - \mu'] - [V(\vec{r}) - \mu]\right\} n(\vec{r}) + \Omega_{V-\mu}[\hat{\rho}']$$

$$> \int \left\{[V'(\vec{r}) - \mu'] - [V(\vec{r}) - \mu]\right\} n(\vec{r}) + \Omega_{V-\mu}[\hat{\rho}].$$

(A18.8)

Interchanging the primed and unprimed variables, and adding the two resulting inequalities we obtain the contradiction

$$\Omega_{V-\mu} + \Omega_{V'-\mu'} > \Omega_{V-\mu} + \Omega_{V'-\mu'},$$

(A18.9)

proving the falsehood of the initial assumption.

The Kohn–Sham equation can also be straightforwardly generalized to finite temperatures. To this end we simply write

$$F[n] = F_0[n] + E_H[n] + E_{xc}[n],$$

(A18.10)

where $F_0[n]$ is the non-interacting version of the F functional.[1] Then, the stationarity condition (A18.7) takes the form

$$\frac{\delta F_0[n]}{\delta n(\vec{r})} = \mu - V(\vec{r}) - \frac{\delta E_H[n]}{\delta n(\vec{r})} - \frac{\delta E_{xc}[n]}{\delta n(\vec{r})}$$

$$= \mu - V(\vec{r}) - V_H(\vec{r}) - V_{xc}(\vec{r})$$

$$\equiv \mu - V_{KS}(\vec{r}).$$

(A18.11)

The last line of this equation defines the finite-temperature Kohn–Sham potential as the potential that produces the true equilibrium density in a non-interacting system at the same temperature and chemical potential.

[1] Note that $F_0[n]$ now differs from the non-interacting kinetic energy $T_s[n]$, due to the presence of the non-interacting entropy term.

The single particle representation of the variational problem (A18.11) leads to Kohn–Sham equations in the standard form (7.25). The only differences are (i) the expression of the density in terms of Kohn–Sham orbitals becomes

$$n(\vec{r}) = \sum_{\alpha,s} f(\varepsilon_\alpha - \mu) |\phi_\alpha(\vec{r}, s)|^2, \qquad\qquad (A18.12)$$

where $f(\varepsilon_\alpha - \mu)$ is the Fermi-Dirac distribution, and (ii) the xc potential is explicitly temperature-dependent. The local density approximation retains the standard form (7.77), except that the xc energy per particle of the homogeneous electron liquid must be replaced by the temperature-dependent Gibbs free energy per particle. This has been calculated, for example, by Tanaka and Ichimaru (1989). Explicit expressions for the exchange-correlation potential at finite temperature have been given by Perrot (1982) and by Perrot and Dharma-Wardana (1984).

Appendix 19

Completeness of the bosonic basis set for the Luttinger model

In this appendix we prove the completeness of the *extended* bosonic basis set $|N_L, N_R, \{\mathcal{N}_q\}\rangle$, where N_L and N_R are the numbers of left and right movers counted from the reference "vacuum" in which all the right-mover states are occupied for $k < k_F$ and all the left-mover states are occupied for $k > -k_F$. Thus N_L and N_R can be positive or negative, with a positive value indicating an excess of particles (relative to the "vacuum") and a negative value indicating a deficiency of particles. The basis set (9.37) defined in Section 9.4 is just a special case with $N_L = N_R = 0$. The energies of these states (relative to the energy of the "vacuum") are

$$E[N_L, N_R, \{\mathcal{N}_q\}] = \hbar v_F \left[\frac{\pi}{L} \left(N_L^2 + N_R^2 \right) + \sum_{q \neq 0} |q| \mathcal{N}_q \right], \qquad (A19.1)$$

where the first term $\hbar v_F \frac{\pi}{L} \left(N_L^2 + N_R^2 \right)$ represents the excess ground-state energy of the non-interacting system with N_τ extra particles and no boson excitations (Exercise 1).[1]

The grand-canonical partition function[2] $\mathcal{Z} = Tr\{e^{-\beta \hat{H}_0}\}$ can be calculated in both the bosonic and the fermionic basis sets. The sum over N_τ, being rapidly convergent, can be safely extended to $-\infty$. In the bosonic basis set we have

$$\mathcal{Z}_B = \sum_{N_R = -\infty}^{\infty} e^{-\beta \hbar \pi v_F \frac{N_R^2}{L}} \times \sum_{N_L = -\infty}^{\infty} e^{-\beta \hbar \pi v_F \frac{N_L^2}{L}}$$

$$\times \prod_{q \neq 0} \sum_{\mathcal{N}_q = 0}^{\infty} e^{-\beta \hbar v_F |q| \mathcal{N}_q} . \qquad (A19.2)$$

Setting $w \equiv e^{-\frac{\beta \hbar \pi v_F}{L}}$ and carrying out the sum of the geometric series on the last line of Eq. (A19.2) we get

$$\mathcal{Z}_B = \left(\sum_{n=-\infty}^{\infty} w^{n^2} \right)^2 \times \prod_{q \neq 0} \frac{1}{1 - w^{\frac{qL}{\pi}}}$$

$$= \left(\sum_{n=-\infty}^{\infty} w^{n^2} \right)^2 \times \left(\prod_{n=1}^{\infty} \frac{1}{1 - w^{2n}} \right)^2 , \qquad (A19.3)$$

---

[1] The reason for the quadratic behavior is that the energy of an extra particle vanishes at the Fermi level.
[2] Notice that the chemical potential term is built into our definition of $\hat{H}_0$.

here we have used the fact that $q = \frac{2\pi}{L} n$, with n a nonzero integer.

On the other hand, in the fermionic basis set, we can write

$$\mathcal{Z}_F = \prod_k \left(\sum_{n_{k,R}=0}^{1} e^{-\beta \hbar v_F (k - k_F)(n_{k,R} - n_{k,R}^{(0)})} \times \sum_{n_{k,L}=0}^{1} e^{+\beta \hbar v_F (k + k_F)(n_{k,L} - n_{k,L}^{(0)})} \right), \quad \text{(A19.4)}$$

with $n_{k,\tau}^{(0)}$ defined in Eq. (9.1). Carrying out the sums over fermionic occupation numbers
(0 or 1) we get

$$\mathcal{Z}_F = \prod_{k>k_F} \left(1 + w^{\frac{(k-k_F)L}{\pi}} \right) \times \prod_{k<k_F} \left(1 + w^{-\frac{(k-k_F)L}{\pi}} \right)$$

$$\times \prod_{k<-k_F} \left(1 + w^{\frac{(k+k_F)L}{\pi}} \right) \times \prod_{k>-k_F} \left(1 + w^{-\frac{(k+k_F)L}{\pi}} \right). \quad \text{(A19.5)}$$

Now setting $k \pm k_F = \frac{2\pi}{L} n$ we rewrite the above equation as

$$\mathcal{Z}_F = \left(\prod_{n=1}^{\infty} (1 + w^{2n-1})^2 \right)^2. \quad \text{(A19.6)}$$

To show that Eqs. (A19.3) and (A19.6) are indeed two different representations of the same
quantity we refer the reader to Eqs. (8.180.3) and (8.181.2) of Gradshteyn–Ryzhik (1965,
p. 921), which give two alternative representations of the Jacobi elliptic theta-function

$$\theta_3(0, w) = \sum_{n=-\infty}^{\infty} w^{n^2}$$

$$= \prod_{n=1}^{\infty} (1 + w^{2n-1})^2 (1 - w^{2n}), \quad (|w| < 1). \quad \text{(A19.7)}$$

Then we see that $\mathcal{Z}_B = \mathcal{Z}_F$ and by the argument of Section 9.4 we conclude that the
fermionic and bosonic basis sets are equivalent.

Appendix 20

Proof of the disentanglement lemma

In this appendix we present a proof of the crucial identity (9.80), i.e.

$$e^{\hat{A}+\hat{B}} = e^{-\frac{1}{2}[\hat{A},\hat{B}]}e^{\hat{A}}e^{\hat{B}} ,$$ (A20.1)

under the assumption that $[\hat{A}, \hat{B}]$ commutes with $\hat{A}$ and $\hat{B}$.

Let us define

$$\hat{B}(\lambda) \equiv e^{-\lambda\hat{A}}\hat{B}e^{\lambda\hat{A}} ,$$ (A20.2)

where λ is a real number in the range $0 \leq \lambda \leq 1$. $\hat{B}(\lambda)$ satisfies the "equation of motion"

$$\frac{d\hat{B}(\lambda)}{d\lambda} = -e^{-\lambda\hat{A}}[\hat{A}, \hat{B}]e^{\lambda\hat{A}} = -[\hat{A}, \hat{B}] ,$$ (A20.3)

where we have used the fact that $[\hat{A}, \hat{B}]$ commutes with $\hat{A}$. Notice that the right–hand side of this equation is independent of λ: thus, integrating over λ, and noting that $\hat{B}(0) \equiv \hat{B}$, we arrive at the important result

$$\hat{B}(\lambda) = \hat{B} - \lambda[\hat{A}, \hat{B}] .$$ (A20.4)

As a consequence the two operators $\hat{B}(\lambda)$ and $\hat{B}(\lambda')$ commute with each other for arbitrary values of λ and λ'. Next consider the operator

$$\hat{U}(\lambda) \equiv e^{-\lambda\hat{A}}e^{\lambda(\hat{A}+\hat{B})} .$$ (A20.5)

It is easy to verify that $\hat{U}(\lambda)$ satisfies the equation of motion

$$\frac{d\hat{U}(\lambda)}{d\lambda} = \hat{B}(\lambda)\hat{U}(\lambda) .$$ (A20.6)

The formal solution of this equation is

$$\hat{U}(\lambda) = T_\lambda e^{\int_0^\lambda \hat{B}(\lambda')d\lambda'} ,$$ (A20.7)

where T_λ arranges the operators $\hat{B}(\lambda)$ in order of decreasing values of λ (see Section 6.2.2, where a similar equation is solved to obtain the time-evolution operator). In this case, the T_λ operator is however unnecessary, for, as we have just seen, the operators $\hat{B}(\lambda)$ with different values of λ always commute with each other. Thus, the solution for $\hat{U}(\lambda)$, evaluated at $\lambda = 1$, is simply given by

$$\hat{U}(1) = e^{\int_0^1 (\hat{B} - \lambda'[\hat{A},\hat{B}])d\lambda'}$$
$$= e^{\hat{B}} \, e^{-\frac{1}{2}[\hat{A},\hat{B}]} \,, \tag{A20.8}$$

where we have used Eq. (A20.4) and the fact that $[\hat{A}, \hat{B}]$ commutes with $\hat{B}$. Notice that, by the definition of $\hat{U}(\lambda)$ in Eq. (A20.5), we also have

$$\hat{U}(1) = e^{-\hat{A}} \, e^{\hat{A}+\hat{B}} \,. \tag{A20.9}$$

Thus, comparing Eqs. (A20.8) and (A20.9), we arrive at Eq. (A20.1).

Appendix 21

The independent boson theorem

In this appendix we provide a proof for the important formula (9.107). As a first step we prove that

$$
\left\langle e^{\sum_i x_i \hat{b}_i^\dagger} e^{\sum_i y_i \hat{b}_i} \right\rangle = e^{\sum_i x_i y_i \langle \hat{b}_i^\dagger \hat{b}_i \rangle} , \tag{A21.1}
$$

where x_i and y_i are c-numbers and the thermal average is taken over a set of independent bosons with hamiltonian $\hat{H} = \sum_i \hbar \omega_i b_i^\dagger \hat{b}_i$.

For simplicity, we consider here explicitly the case of a single boson mode. The generalization to many modes is trivial since different boson operators commute with each other. A possible way of carrying out the calculation of the above average takes advantage of the idea of *boson coherent states*. Accordingly, before proceeding any further we review the definition and some of the most relevant properties of these states. In the process we will also introduce some devices which will prove useful in performing some of the calculations in Chapter 10.[1]

One defines the coherent boson state $|\phi\rangle$ as follows:

$$
|\phi\rangle \equiv e^{\phi \hat{b}^\dagger} |0\rangle , \tag{A21.2}
$$

where ϕ is a complex number and $|0\rangle$ is the boson vacuum.

As it is readily verified, the coherent states are eigenstates of the destruction operator with eigenvalue ϕ, i.e. they satisfy the conjugate relations

$$
\hat{b}|\phi\rangle = \phi|\phi\rangle , \tag{A21.3}
$$

and

$$
\langle \phi|\hat{b}^\dagger = \langle \phi|\phi^* . \tag{A21.4}
$$

Evidently, they also satisfy the equations

$$
\hat{b}^\dagger |\phi\rangle = \frac{\partial}{\partial \phi} |\phi\rangle , \tag{A21.5}
$$

and

$$
\langle \phi|\hat{b} = \frac{\partial}{\partial \phi^*} \langle \phi| . \tag{A21.6}
$$

---

[1] For more details on the idea of coherent states the reader is referred to the book by Negele and Orland, 1988.

Since $\hat{b}$ is non-hermitian the following useful relation can be obtained using the identity (9.81):

$$e^{\alpha\hat{b}}|\phi\rangle = e^{\alpha\hat{b}}e^{\phi\hat{b}^\dagger}|0\rangle = e^{\alpha\phi}e^{\phi\hat{b}^\dagger}e^{\alpha\hat{b}}|0\rangle = e^{\alpha\phi}|\phi\rangle .\qquad (A21.7)$$

Clearly the $|\phi\rangle$'s are not mutually orthogonal and their overlap is immediately obtained from Eq. (A21.7):

$$\langle\phi|\phi'\rangle = e^{\phi^*\phi'} .\qquad (A21.8)$$

Even though non-orthogonal, the coherent states enjoy an important completeness relation that allows every state to be expressed as a linear combination of them. This completeness (or, in fact, overcompleteness) relation reads

$$\int_{-\infty}^{+\infty} \frac{d\phi_1 d\phi_2}{\pi} e^{-|\phi|^2}|\phi\rangle\langle\phi| = \hat{1} ,\qquad (A21.9)$$

where we have introduced the notation $\phi = \phi_1 + i\phi_2$, i.e. ϕ_1 and ϕ_2 are, respectively, the real and the imaginary parts of ϕ. Here $\hat{1}$ denotes the identity operator in the Hilbert space.

Eq. (A21.9) can be proved by showing that its left-hand side commutes with both $\hat{b}$ and $\hat{b}^\dagger$ and therefore must be proportional to the identity operator. The proof of this is simple and relies on the fact that, according to Eqs. (A21.3)–(A21.6), the commutator of (say) $\hat{b}$ with $|\phi\rangle\langle\phi|$ is given by

$$[\hat{b}, |\phi\rangle\langle\phi|] \equiv \hat{b}|\phi\rangle\langle\phi| - |\phi\rangle\langle\phi|\hat{b} = \left(\phi - \frac{\partial}{\partial\phi^*}\right)|\phi\rangle\langle\phi| ,\qquad (A21.10)$$

where we have also used the fact that ϕ and ϕ^* are independent and therefore the partial derivative with respect to ϕ^* can be commuted with $|\phi\rangle$. Then the integral on the left-hand side of Eq. (A21.9) is seen to commute with $\hat{b}$ since, upon integrating by parts with respect to both ϕ_1 and ϕ_2, one is left with an expression proportional to $\left(\phi + \frac{\partial}{\partial\phi^*}\right)e^{-|\phi|^2}$ which in turn clearly vanishes. The same procedure can be used for the case of the creation operator. The correctness of Eq. (A21.9) can be finally verified by taking the expectation value of its left-hand side in the vacuum state $|0\rangle$, and showing that it equals 1. Thus, any state $|\psi\rangle$ can be written as a superposition of the states $|\phi\rangle$'s according to the equation

$$|\psi\rangle = \int_{-\infty}^{+\infty} \frac{d\phi_1 d\phi_2}{\pi} e^{-|\phi|^2}|\phi\rangle\langle\phi|\psi\rangle .\qquad (A21.11)$$

After this brief excursus into the properties of coherent states we are now ready for the proof of the independent boson theorem. The sought average is explicitly written as

$$\langle e^{\alpha\hat{b}^\dagger}e^{\beta\hat{b}}\rangle = \frac{\sum_{n=0}^{\infty}\langle n|e^{\alpha\hat{b}^\dagger}e^{\beta\hat{b}}e^{-\frac{\hbar\omega}{k_B T}\hat{b}^\dagger\hat{b}}|n\rangle}{\sum_{n=0}^{\infty}\langle n|e^{-\frac{\hbar\omega}{k_B T}\hat{b}^\dagger\hat{b}}|n\rangle} .\qquad (A21.12)$$

We begin by making use of Eq. (A21.11) to rewrite the numerator as

$$\sum_{n=0}^{\infty}\langle n|e^{\alpha\hat{b}^\dagger}e^{\beta\hat{b}}e^{-\frac{\hbar\omega}{k_B T}\hat{b}^\dagger\hat{b}}|n\rangle = \int\frac{d\phi_1 d\phi_2}{\pi}e^{-|\phi|^2}\sum_{n=0}^{\infty}\langle n|\phi\rangle\langle\phi|e^{\alpha\hat{b}^\dagger}e^{\beta\hat{b}}e^{-\frac{\hbar\omega}{k_B T}\hat{b}^\dagger\hat{b}}|n\rangle .\quad (A21.13)$$

The sum over n can now be carried out exactly in view of the completeness relation $|\phi\rangle = \sum_{n=0}^{\infty} \langle n|\phi\rangle |n\rangle$. We are therefore left with a trace over coherent states:

$$\int \frac{d\phi_1 d\phi_2}{\pi} e^{-|\phi|^2} \langle \phi| e^{\alpha \hat{b}^\dagger} e^{\beta \hat{b}} e^{-\frac{\hbar\omega}{k_B T} \hat{b}^\dagger \hat{b}} |\phi\rangle \, . \tag{A21.14}$$

Next we observe that, with the notation $z = e^{-\frac{\hbar\omega}{k_B T}}$, we can write

$$e^{-\frac{\hbar\omega}{k_B T} \hat{b}^\dagger \hat{b}} |\phi\rangle = e^{-\frac{\hbar\omega}{k_B T} \hat{b}^\dagger \hat{b}} e^{\phi \hat{b}^\dagger} |0\rangle = e^{-\frac{\hbar\omega}{k_B T} \hat{b}^\dagger \hat{b}} e^{\phi \hat{b}^\dagger} e^{\frac{\hbar\omega}{k_B T} \hat{b}^\dagger \hat{b}} |0\rangle = e^{z\phi \hat{b}^\dagger} |0\rangle = |z\phi\rangle \, , \quad (A21.15)$$

where we have made use of the fact that $e^{\frac{\hbar\omega}{k_B T} \hat{b}^\dagger \hat{b}} |0\rangle = |0\rangle$, and have recognized the formal similarity between the operator product in the third expression from the left, and a "time evolution" with hamiltonian $\hat{H} = \hbar\omega \hat{b}^\dagger \hat{b}$ over an imaginary time interval $-i\frac{\hbar}{k_B T}$. Thus, finally, making use of Eqs. (A21.14), (A21.7), and Eq. (A21.8), the numerator of Eq. (A21.12) can be recast in the form

$$\int \frac{d\phi_1 d\phi_2}{\pi} e^{-|\phi|^2} e^{\beta \phi z} \langle \phi| \alpha + z\phi \rangle = \int \frac{d\phi_1 d\phi_2}{\pi} e^{-|\phi|^2(1-z)} e^{\alpha \phi^*} e^{\beta \phi z} \, . \tag{A21.16}$$

Proceeding in exactly the same manner, the denominator of that equation can be rewritten as

$$\sum_{n=0}^{\infty} \langle n| e^{-\frac{\hbar\omega}{k_B T} \hat{b}^\dagger \hat{b}} |n\rangle = \int \frac{d\phi_1 d\phi_2}{\pi} e^{-|\phi|^2(1-z)} . \tag{A21.17}$$

The integrals appearing in the last two equations are standard gaussian integrals, which can be done analytically. Accordingly, for the numerator one obtains the expression

$$(1-z)^{-1} e^{\frac{z\alpha\beta}{1-z}} \, , \tag{A21.18}$$

while for the denominator one gets the well known expression for the partition function of non-interacting bosons, i.e.,

$$\frac{1}{1-z} = \frac{1}{1 - e^{-\frac{\hbar\omega}{k_B T}}} \, . \tag{A21.19}$$

By taking the ratio of the last two equations, and noting that $\frac{z}{1-z} = n_B(\hbar\omega) = \langle \hat{b}^\dagger \hat{b} \rangle$, we arrive at Eq. (A21.1).

Finally, the disentanglement lemma allows us to combine the two exponentials on the left-hand side of Eq. (A21.1) into a single one to get the following identity:

$$\left\langle e^{\sum_i (x_i \hat{b}_i^\dagger + y_i \hat{b}_i)} \right\rangle = e^{\sum_i x_i y_i \left(\hat{b}_i^\dagger \hat{b}_i - \frac{1}{2}[\hat{b}_i^\dagger, \hat{b}_i] \right)} \, . \tag{A21.20}$$

Since $x_i y_i \left(\hat{b}_i^\dagger \hat{b}_i - \frac{1}{2}[\hat{b}_i^\dagger, \hat{b}_i] \right) = \frac{1}{2} \left\langle (x_i \hat{b}_i^\dagger + y_i \hat{b}_i)^2 \right\rangle$ (the average of $\hat{b}_i^2$ and $(\hat{b}_i^\dagger)^2$ being zero), we see that

$$\left\langle e^{\sum_i (x_i \hat{b}_i^\dagger + y_i \hat{b}_i)} \right\rangle = e^{\frac{1}{2} \sum_i \left\langle (x_i \hat{b}_i^\dagger + y_i \hat{b}_i)^2 \right\rangle} \, , \tag{A21.21}$$

and Eq. (9.107) is proved.

It is easy to see that the theorem is valid even if the average is done in the thermal ensemble of the hamiltonian $\sum_i \hbar\omega_i \hat{\beta}_i^\dagger \hat{\beta}_i$, where the $\hat{\beta}_i$'s are related to $\hat{b}_i$ and $\hat{b}_i^\dagger$ by a linear unitary transformation. Indeed, expressing $\hat{b}$ and the $\hat{b}^\dagger$ in terms of $\hat{\beta}$ and $\hat{\beta}^\dagger$ amounts to a redefinition of the coefficients x_i and y_i on both sides of Eq. (A21.21), with no impact on the validity of that equation.

Appendix 22

The three-dimensional electron gas at high magnetic field

In this appendix we summarize some essential facts pertaining to the energetics of the *three-dimensional* electron liquid at high magnetic field. The one-electron energy subbands, given by Eq. (10.19), are plotted vs k_z in Fig. A22.1(a). We have introduced a Zeeman splitting $\Delta = \frac{g}{2}\hbar\omega_c$ between up-spin ($\sigma = +1$) and down-spin ($\sigma = -1$) states. The zero of the energy is set at the bottom of the lowest down-spin subband. Two dimensionless parameters characterize this system, namely, the usual electron gas parameter r_s and the ratio of the magnetic length ℓ to the Bohr radius a_B:

$$\frac{\ell}{a_B} = \sqrt{\frac{2}{\hbar\omega_c/\text{Ry}}} .$$

$$(A22.1)$$

A22.1 Noninteracting kinetic energy

Let $\bar{\epsilon}_F \equiv \frac{\epsilon_F}{\hbar\omega_c}$ be the Fermi energy in units of the cyclotron energy. The j-th occupied subband crosses the Fermi level at wave vectors $k_z = \pm k_{j\sigma}$, where

$$k_{j\sigma}\ell = \sqrt{2\left(\bar{\epsilon}_F - j - g\frac{\sigma+1}{4}\right)} .$$

$$(A22.2)$$

It is convenient to set $k_{j\sigma} = 0$ for the unoccupied subbands, so that the three-dimensional electron density n is simply given by

$$n = \frac{1}{2\pi^2\ell^2} \sum_{j\sigma} k_{j\sigma} .$$

$$(A22.3)$$

Notice that, even though the sum over j is formally unrestricted, our definition of $k_{j\sigma}$ guarantees that only the occupied subbands contribute to the density. Eq. (A22.3) can be inverted numerically (see Fig. A22.1(b)), yielding the Fermi energy as a function of n. Once $\bar{\epsilon}_F$ is known, the non-interacting kinetic energy per particle, ϵ_0, is obtained from

$$\epsilon_0(n, B) = \frac{\hbar\omega_c}{6\pi^2 n} \sum_{j\sigma} k_{j\sigma}^3 .$$

$$(A22.4)$$

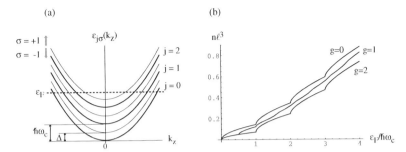

Fig. A22.1. (a) Landau level subbands for a three-dimensional electron gas with $g = 1$ in a magnetic field. Thick and thin lines correspond to down-spin and up-spin dispersions respectively. (b) Electronic density vs Fermi energy at three different values of the g-factor.

A22.2 Density–density response function

The calculation of the noninteracting density–density response function presented in Section 10.4.5 is easily extended to three dimensions, yielding

$$\chi_0(\vec{q}, \omega) = -\frac{1}{4\pi^2 \hbar \omega_c \ell^3} \sum_{jj'\sigma} \frac{|F_{j'j}(\vec{q}_\perp)|^2}{q_z \ell} \ln \left\{ \frac{\omega^2 - \left[\omega_{jj'} - \frac{\hbar q_z^2}{2m_b} - \frac{\hbar k_{j\sigma} q_z}{m_b} \right]^2}{\omega^2 - \left[\omega_{jj'} - \frac{\hbar q_z^2}{2m_b} + \frac{\hbar k_{j\sigma} q_z}{m_b} \right]^2} \right\} , \quad (A22.5)$$

where $\omega_{jj'} \equiv (j - j')\omega_c$ and $\vec{q}_\perp$ is the component of $\vec{q}$ in the plane perpendicular to the magnetic field. The form factors $F_{j'j}(\vec{q}_\perp)$ are defined in Eq. (10.84).

A22.3 Noninteracting structure factor

The noninteracting static structure factor can be calculated directly from the fluctuation-dissipation theorem as follows:

$$S_0(\vec{q}) = -\frac{\hbar}{\pi n} \int_0^\infty \chi_0(\vec{q}, i\omega) d\omega$$

$$= \frac{1}{4\pi^2 n \ell^3} \sum_{jj'\sigma} \frac{|F_{j'j}(\vec{q}_\perp)|^2}{q_z \ell} \left[\left| \frac{q_z^2 \ell^2}{2} + k_{j\sigma} q_z \ell^2 + j' - j \right| \right.$$

$$\left. - \left| \frac{q_z^2 \ell^2}{2} - k_{j\sigma} q_z \ell^2 + j' - j \right| \right] , \quad (A22.6)$$

where use has been made of the integral $\int_0^\infty \ln \frac{\omega^2 + a^2}{\omega^2 + b^2} d\omega = \pi(|a| - |b|)$. Because $S_0(\vec{q})$ is even under inversion of q_z there is no loss of generality in assuming $q_z > 0$. Evaluating the difference of the absolute values in Eq. (A22.6) leads to three different results, depending on the relative values of q_z, j, and j': the three regimes are

$$|\ldots| - |\ldots| = \begin{cases} \text{(i)} & 2k_{j\sigma} q_z \ell^2 , & q_z \geq k_{j\sigma} + k_{j'\sigma} , \\ \text{(ii)} & q_z^2 \ell^2 + 2(j' - j) , & |k_{j\sigma} - k_{j'\sigma}| < q_z < k_{j\sigma} + k_{j'\sigma}, \\ \text{(iii)} & 2k_{j\sigma} q_z \ell^2 \, \text{sign}(j' - j) , & 0 < q_z \leq |k_{j\sigma} - k_{j'\sigma}| . \end{cases} \quad (A22.7)$$

Notice that for $q_z > 2k_{0\downarrow}$ we are always in case (i) and it is easy to verify that $S_0(\vec{q}) = 1$. We can now take advantage of the fact that the sum over j and j' in Eq. (A22.6) is formally unrestricted to further simplify the summand: the term $2(j' - j)$ in case (ii) vanishes upon summation over j and j'; in (i) $2k_{j\sigma}$ can be replaced by $k_{j\sigma} + k_{j'\sigma}$, and in (iii) $2k_{j\sigma}\,\mathrm{sign}(j' - j)$ is replaced by $(k_{j\sigma} - k_{j'\sigma})\mathrm{sign}(j' - j) = |k_{j\sigma} - k_{j'\sigma}|$. Thus, after a few simple manipulations one arrives at

$$S_0(\vec{q}) - 1 \;=\; \frac{1}{4\pi^2 n\ell^2}\sum_{jj'\sigma}|F_{j'j}(\vec{q}_\perp)|^2\,\Theta_{j'j_\sigma}(|q_z|) \tag{A22.8}$$

where

$$\Theta_{j'j_\sigma}(|q_z|) = \begin{cases} 0\,, & |q_z| \ge k_{j\sigma} + k_{j'\sigma}\,, \\ |q_z| - k_{j\sigma} - k_{j'\sigma}\,, & |k_{j\sigma} - k_{j'\sigma}| < |q_z| < k_{j\sigma} + k_{j'\sigma}\,, \\ |k_{j\sigma} - k_{j'\sigma}| - k_{j\sigma} - k_{j'\sigma}\,, & |q_z| \le |k_{j\sigma} - k_{j'\sigma}|\,. \end{cases} \tag{A22.9}$$

Notice that only the occupied subbands contribute to the static structure factor.

A22.4 Exchange energy

The exchange energy per particle is given by

$$\epsilon_x(n, B) \;=\; \frac{1}{2}\int \frac{d^3q}{(2\pi)^3}\,\frac{4\pi e^2}{q^2}\,\big[S_0(\vec{q}) - 1\big]\,. \tag{A22.10}$$

Substituting in this expression the non-interacting structure factor $S_0(\vec{q})$, and performing the integral over q_z analytically we arrive at a one-dimensional integral, which can be evaluated numerically:

$$\epsilon_x(n, B) = \frac{e^2}{4\pi^2 n\ell^2}\sum_{jj'\sigma}\int_0^{\infty} dq_\perp\, q_\perp |F_{j'j}(\vec{q}_\perp)|^2\,H_{j'j_\sigma}(q_\perp)\,, \tag{A22.11}$$

where $H_{j'j_\sigma}(q_\perp) = \int_0^\infty dq_z\,\frac{\Theta_{j'j_\sigma}(q_z)}{q_\perp^2 + q_z^2}$.

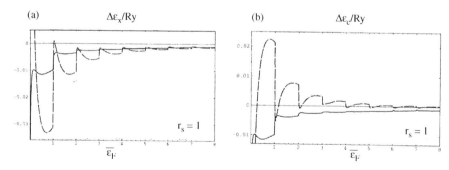

Fig. A22.2. (a) Dashed line: the difference $\Delta\epsilon_x(n, B) \equiv \epsilon_x(n, B) - \epsilon_x(n, 0)$ vs $\bar{\epsilon}_F = \frac{\epsilon_F}{\hbar\omega_c}$ at constant density ($r_s = 1$), for vanishing Zeeman splitting ($g = 0$). (b) Dashed line: the difference $\Delta\epsilon_c(n, B) \equiv \epsilon_c(n, B) - \epsilon_c(n, 0)$ vs $\bar{\epsilon}_F$ at constant density ($r_s = 1$) for $g = 0$. In both panels the solid line represents the difference $\Delta\epsilon_{xc}(n, B) \equiv \epsilon_{xc}(n, B) - \epsilon_{xc}(n, 0)$, where $\epsilon_{xc}(n, B)$ is the exchange-correlation energy. (Adapted from Skudlarski and Vignale, 1993.)

In Fig. A22.2(a) we show how the exchange energy evolves as a function of magnetic field at constant density. Notice that the exchange energy is generally more negative than that of the electron liquid at the same density in zero magnetic field.

A22.5 Correlation energy

The correlation energy in RPA is straightforwardly calculated from Eq. (10.91) (Skudlarski and Vignale, 1993): its evolution as a function of magnetic field is plotted in Fig. A22.2(b). Unlike the exchange energy, the RPA correlation energy is generally less negative than in the electron liquid at the same density in zero magnetic field. The full exchange-correlation energy, however, is always rendered more negative by the magnetic field. These qualitative features have been recently confirmed by Takada and Goto (1998). These authors have calculated the correlation energy in the STLS approximation (see Section 5.4.4) and have also provided plots of the STLS local field factor $G_+(q)$ in the presence of a magnetic field. Analytical fits for the exchange-correlation energy vs density and magnetic field can be found in the cited papers.

Appendix 23

Density matrices in the lowest Landau level

In this appendix we show that the one-particle density matrix of a many-body state in the lowest Landau level can be explicitly computed from the electronic density of that state (MacDonald and Girvin, 1988). The reason for this surprising result lies in the special analytic structure of the wave functions in the lowest Landau level. Recall that, according to Eq. (10.41), the one-particle density matrix

$$\rho(\vec{r}, \vec{r}') = \int \psi^*(\vec{r}, \vec{r}_2 \ldots \vec{r}_N) \psi(\vec{r}', \vec{r}_2 \ldots \vec{r}_N) d\vec{r}_2 \ldots d\vec{r}_N \tag{A23.1}$$

can be written in the form

$$\rho(z, z') = e^{-\frac{|z|^2 + |z'|^2}{4\ell^2}} F(z^*, z') , \tag{A23.2}$$

where z and z' are the complex variables associated with the two-dimensional vectors $\vec{r}$ and $\vec{r}'$, and the function F is analytic in each of its two arguments. The density, in this representation, is given by

$$n(\vec{r}) = \rho(\vec{r}, \vec{r}) = e^{-\frac{z^* z}{2\ell^2}} F(z^*, z) . \tag{A23.3}$$

Now observe that an arbitrary function of position $n(\vec{r}) = n(x, y)$ can be represented as a function of z and z^* simply by substituting for the coordinates x and y their expressions $x = \frac{z + z^*}{2}$ and $y = \frac{z - z^*}{2i}$.

Comparing this with Eq. (A23.3) we immediately see that

$$F(z^*, z) = e^{\frac{z^* z}{2\ell^2}} n \left(\frac{z + z^*}{2}, \frac{z - z^*}{2i} \right) . \tag{A23.4}$$

The key point is that this equation completely determines the analytic dependence of $F(z^*, z)$ on each of its two arguments. By varying these arguments independently we arrive at the identification

$$F(z^*, z') = e^{\frac{z^* z'}{2\ell^2}} n \left(\frac{z' + z^*}{2}, \frac{z' - z^*}{2i} \right) . \tag{A23.5}$$

Finally, by substituting this formula in Eq. (A23.2) we obtain the one-particle density matrix

$$\rho(z, z') = e^{-\frac{|z|^2 + |z'|^2 - 2z^* z'}{4\ell^2}} n \left(\frac{z' + z^*}{2}, \frac{z' - z^*}{2i} \right) . \tag{A23.6}$$

In the special case of a state of uniform density, Eq. (A23.6) tells us that the density matrix reduces to the simple form

$$\rho(z, z') = n e^{-\frac{|z-z'|^2}{4\ell^2}} e^{\frac{z^* z' - z z'^*}{4\ell^2}} , \quad \text{(uniform density)} , \quad (A23.7)$$

which tends to zero exponentially for $|z - z'| \to \infty$.

Eq. (A23.6) is used in Exercise 10 to obtain the explicit expression of the Hartree–Fock potential in the LLL in terms of the projected density operator. A similar analysis allows the construction of the two-particle density matrix in terms of the pair correlation function.

Appendix 24

Projection in the lowest Landau level

Every one-electron wave function $\psi(z)$ in the LLL has the form

$$\psi(z) = f(z)e^{-\frac{|z|^2}{4\ell^2}} \,, \tag{A24.1}$$

where $f(z)$ is an analytic function of z. Conversely, every analytic function $f(z)$ determines a one-electron wave function in the LLL. Operators that depend on the coordinates x and y can be expressed in terms of z and z^*. It is evident that $z\psi(z)$ belongs to the LLL if $\psi(z)$ does, but what about $z^*\psi(z)$? To answer this question consider the "matrix element" of z^* between two states in the LLL:

$$\int g(z^*)z^* f(z)e^{-\frac{|z|^2}{2\ell^2}} dxdy \,. \tag{A24.2}$$

Writing $z^*e^{-\frac{|z|^2}{2\ell^2}}$ as $2\ell^2\frac{\partial}{\partial z}e^{-\frac{|z|^2}{2\ell^2}}$ and integrating by parts we see that this integral can be rewritten as

$$\int g(z^*)\left[2\ell^2\frac{\partial}{\partial z}f(z)\right]e^{-\frac{|z|^2}{2\ell^2}} dxdy \,. \tag{A24.3}$$

Thus the action of z^* within the LLL is equivalent to that of $2\ell^2\frac{\partial}{\partial z}$ *acting only on the analytic part of the wave function* ($f(z)$ in this case). This suggests that operators that are functions of z and z^* are projected in the LLL simply by the replacement $z^* \to 2\ell^2\frac{\partial}{\partial z}$. However, because z and the projected z^* do not commute with each other, one must still pay attention to their relative order. By considering, for example, the matrix element of zz^* between two LLL wave functions, we see that

$$\int g(z^*)zz^* f(z)e^{-\frac{|z|^2}{2\ell^2}} dxdy = \int g(z^*)\left[2\ell^2\frac{\partial}{\partial z}zf(z)\right]e^{-\frac{|z|^2}{2\ell^2}} dxdy \,. \tag{A24.4}$$

Thus, the operator $\frac{\partial}{\partial z}$ must be placed on the left of all the z's. This leads us to the final form of the prescription (Girvin and Jach, 1984): to project in the LLL an operator that is a function of z and z^*

(i) Expand the operator in a power series of z and z^*;
(ii) Put all the z^*'s on the left of all the z's;
(iii) Replace z^* by $2\ell^2\frac{\partial}{\partial z}$;

iv) Remember that $\frac{\partial}{\partial z}$ operates only on the polynomial part of the wave function (*not* on the exponential part).

As a simple application, let us work out the relationship (10.170) between the projected structure factor and the full structure factor. Using the prescriptions given above, together with the disentanglement lemma (9.80) we see that

$$\hat{\bar{n}}_{\vec{q}}\hat{\bar{n}}_{-\vec{q}} = \sum_{j} e^{-iq\ell^2 \frac{\partial}{\partial z_j}} e^{-i\frac{q^* z_j}{2}} \sum_{k} e^{+iq\ell^2 \frac{\partial}{\partial z_k}} e^{+i\frac{q^* z_k}{2}}$$

$$= \sum_{j,k} e^{-iq\ell^2 \left[\frac{\partial}{\partial z_j} - \frac{\partial}{\partial z_k}\right]} e^{-iq^* \left[\frac{z_j - z_k}{2}\right]} e^{-\frac{q^2 \ell^2}{2} \left[\frac{\partial}{\partial z_k}, z_j\right]}$$

$$= \overline{\hat{n}_{\vec{q}}\hat{n}_{-\vec{q}}} + N \left[e^{-\frac{q^2 \ell^2}{2}} - 1 \right]. \tag{A24.5}$$

Eq. (10.170) follows immediately from this and the fact that $\langle \psi_0 | \hat{n}_{\vec{q}}^{\dagger} \hat{n}_{\vec{q}} | \psi_0 \rangle = \langle \psi_0 | \hat{n}_{\vec{q}}^{\dagger} \hat{n}_{\vec{q}} | \psi_0 \rangle$. This is a fine example of the general property that the LLL projection of the product of two operators differs from the product of the individual projections.

Appendix 25

Solution of the independent boson model

The independent boson model is defined by the hamiltonian

$$\hat{H} = \epsilon_0 \hat{a}_0^\dagger \hat{a}_0 + \sum_{\vec{q}} \hbar \omega_{\vec{q}} \hat{b}_{\vec{q}}^\dagger \hat{b}_{\vec{q}} + \sum_{\vec{q}} M_{\vec{q}} (\hat{b}_{\vec{q}} + \hat{b}_{-\vec{q}}^\dagger) \hat{a}_0^\dagger \hat{a}_0 , \qquad (A25.1)$$

where $\hat{a}_0^\dagger$ is the creation operator for a fermion in a localized state, $\hat{b}_{\vec{q}}$ are boson operators and $M_{-\vec{q}} = M_{\vec{q}}^*$ is the fermion-boson coupling. This model is solved by performing a unitary transformation

$$\hat{\tilde{H}} = e^{\hat{S}} \hat{H} e^{-\hat{S}} , \qquad (A25.2)$$

that eliminates the fermion-boson coupling. The generator $\hat{S}$ that accomplishes this is

$$\hat{S} = \sum_{\vec{q}} \frac{M_{\vec{q}}}{\hbar \omega_{\vec{q}}} (\hat{b}_{-\vec{q}}^\dagger - \hat{b}_{\vec{q}}) \hat{a}_0^\dagger \hat{a}_0 . \qquad (A25.3)$$

Indeed, by applying the transformation (A25.2) to the individual fermion and boson operators we find

$$\hat{\tilde{b}}_{\vec{q}} \equiv e^{\hat{S}} \hat{b}_{\vec{q}} e^{-\hat{S}} = \hat{b}_{\vec{q}} - \frac{M_{\vec{q}}^*}{\hbar \omega_{\vec{q}}} \hat{a}_0^\dagger \hat{a}_0 , \qquad (A25.4)$$

$$\hat{\tilde{a}}_0 \equiv e^{\hat{S}} \hat{a}_0 e^{-\hat{S}} = e^{-\sum_{\vec{q}} \frac{M_{\vec{q}}}{\hbar \omega_{\vec{q}}} \left(\hat{b}_{-\vec{q}}^\dagger - \hat{b}_{\vec{q}} \right)} \hat{a}_0 , \qquad (A25.5)$$

and

$$\hat{\tilde{H}} = \left(\epsilon_0 - \sum_{\vec{q}} \frac{|M_{\vec{q}}|^2}{\hbar \omega_{\vec{q}}} \right) \hat{a}_0^\dagger \hat{a}_0 + \sum_{\vec{q}} \hbar \omega_{\vec{q}} \hat{b}_{\vec{q}}^\dagger \hat{b}_{\vec{q}} . \qquad (A25.6)$$

Notice that Eq. (A25.6) expresses the transformed hamiltonian in terms of the *original fermion and boson operators*, not the transformed ones. The eigenvalues and the eigenvectors of $\hat{\tilde{H}}$ are trivially obtained: the former coincide with the eigenvalues of $\hat{H}$, while the latter are related to them by the unitary transformation $e^{\hat{S}}$.

In what follows we focus on the calculation of the spectral function, Eq. (10.220). Applying the unitary transformation (A25.2) to $\hat{a}_0$ and $\hat{H}$ we see that

$$\left\langle \hat{a}_0(t) \hat{a}_0^\dagger \right\rangle_{\hat{H}} = \left\langle \hat{\tilde{a}}_0(t) \hat{\tilde{a}}_0^\dagger \right\rangle_{\hat{\tilde{H}}} , \qquad (A25.7)$$

where $\langle \ldots \rangle_{\hat{H}}$ denotes the average in the thermal ensemble of $\hat{H}$, with the time-dependence of the operators governed by $\hat{H}$ according to the Heisenberg picture of time evolution.

Since there is no fermion-boson coupling in $\hat{H}$ the time evolution of $\hat{a}_0$ and $\hat{b}_{\bar{q}}$ is completely trivial. Substituting Eq. (A25.5) for $\hat{a}_0$ on the right-hand side we get

$$\left\langle \hat{a}_0(t)\hat{a}_0^{\dagger} \right\rangle_{\hat{H}} = [1 - n_F(\bar{\epsilon}_0)]e^{-i\frac{\bar{\epsilon}_0 t}{\hbar}} \left\langle e^{-\sum_{\bar{q}} \frac{M_{\bar{q}}}{\hbar \omega_{\bar{q}}}\left[\hat{b}_{-\bar{q}}^{\dagger}(t) - \hat{b}_{\bar{q}}(t)\right]} e^{\sum_{\bar{q}} \frac{M_{\bar{q}}}{\hbar \omega_{\bar{q}}}\left[\hat{b}_{\bar{q}}^{\dagger} - \hat{b}_{-\bar{q}}\right]} \right\rangle_{\hat{H}}, \quad \text{(A25.8)}$$

where $\bar{\epsilon}_0 = \epsilon_0 - \sum_{\bar{q}} \frac{|M_{\bar{q}}|^2}{\hbar \omega_{\bar{q}}} - \mu$ is the "polaron-shifted" energy of the localized state, and $n_F(\bar{\epsilon}_0)$ is the corresponding thermal occupation factor.

The average of boson operators in Eq. (A25.8) is calculated with the help of the disentanglement lemma and the independent boson theorem, proved in Appendices 20 and 21 respectively. The final result is

$$\left\langle \hat{a}_0(t)\hat{a}_0^{\dagger} \right\rangle_{\hat{H}} = [1 - n_F(\bar{\epsilon}_0)]e^{-i\frac{\bar{\epsilon}_0 t}{\hbar}} e^{\sum_{\bar{q}} \Phi_{\bar{q}}(t)} e^{\sum_{\bar{q}}\left(\Phi_{\bar{q}}(t) + \Phi_{\bar{q}}^*(t)\right)n_B(\hbar \omega_{\bar{q}})}, \quad \text{(A25.9)}$$

where we have defined

$$\Phi_{\bar{q}}(t) \equiv \left|\frac{M_{\bar{q}}}{\hbar \omega_{\bar{q}}}\right|^2 (e^{-i\omega_{\bar{q}} t} - 1). \quad \text{(A25.10)}$$

Putting this into Eq. (10.220) for the spectral function and setting $T = 0$ we obtain the result (10.221) reported in Section 10.14.2.

References

Abrahams, E., Anderson, P. W., Licciardello, D. C., and Ramakrishnan, T. V. (1979). Scaling Theory of Localization: Absence of Quantum Diffusion in Two Dimensions, *Physical Review Letters* **42**, 673–676.

Abrahams, E., Anderson, P. W., Lee, P. A., and Ramakrishnan, T. V. (1981). Quasiparticle lifetime in disordered two-dimensional metals, *Physical Review B* **24**, 6783–6789.

Abramowitz, M. and Stegun, I. A., editors, *Handbook of mathematical functions*, (Dover, 1964).

Abrikosov, A. A. and Khalatnikov, I. M. (1959). The theory of a Fermi liquid, *Reports on Progress in Physics, The Physical Society of London* **22**, 329–367.

Abrikosov, A. A., Gorkov, L. P., and Dzyaloshinski, I. E., *Methods of Quantum Field Theory in Statistical Physics* (Dover Publications Inc., New York, 1963).

Aleiner, I. L. and Glazman, L. I. (1994). Novel edge excitations of two-dimensional electron liquid in a magnetic field, *Physical Review Letters* **72**, 2935–2938.

(1995). Two-dimensional electron liquid in a weak magnetic field, *Physical Review B* **52**, 11296–11312.

Allen, S. J., Jr., Tsui, D. C., and Logan, R. A. (1977). Observation of the two-dimensional plasmon in silicon inversion layers, *Physical Review Letters* **38**, 980–983.

Almbladh, C.-O. and von Barth, U. (1985). Exact results for the spin densities, exchange-correlation potentials, and density-functional eigenvalues, *Physical Review B* **31**, 3231–3244.

Altshuler, B. L. and Aronov, A. G. (1985). Electron-electron interactions in disordered conductors, in *Electron-electron interactions in disordered systems*, edited by A. L. Efros and M. Pollak (North Holland, Amsterdam), p. 1–153.

Anderson, P. W. (1958). Random-Phase Approximation in the Theory of Superconductivity, *Physical Review* **112**, 1900–1916.

Basic Notions of Condensed Matter Physics (Benjamin/Cummings, Menlo Park, 1984).

Andersson, Y., Langreth, D. C., and Lunqvist, B. I. (1996). van der Waals Interactions in Density-Functional Theory, *Physical Review Letters* **76**, 102–105.

Ando, T., Fowler, A. B., and Stern, F. (1982). Electronic properties of two-dimensional systems, *Review of Modern Physics* **54**, 437–672.

Arovas, D., Schrieffer, J. R., and Wilczek, F. (1984). Fractional statistics and the quantum Hall effect, *Physical Review Letters* **53**, 722–723.

Asgari, R., Davoudi, B., Polini, M., Tosi, M. P., Giuliani, G. F., and Vignale, G. (2004). Comparative study of many-body theories of quasiparticle properties in a two-dimensional electron gas, *unpublished*.

Ashcroft, N. and Wilkins, J. (1965). Low temperature specific heat of simple metals, *Physics Letters* **14**, 285–287.

Ashcroft, N. W. and Mermin, N. D., *Solid State Physics*, Saunders College, Philadelphia (1976).

Attaccalite, C., Moroni, S., Gori-Giorgi, P., and Bachelet, G. (2002). Correlation energy and spin polarization in the 2D electron gas, *Physical Review Letters* **88**, 256601.

Atwal, G. S., Khalil, I. G., and Ashcroft, N. W. (2003). Dynamical local-field factors and effective interactions in the two-dimensional electron liquid, *Physical Review B* **67**, 115107.

Aulbur, W. G., Jonsson, L. W., and Wilkins, J. W. (2000). Quasiparticle calculations in solids, *Solid State Physics*, edited by H. Ehrenreich and F. Spaepen (Academic Press, New York) **54**, 1–218.

Bach, V., Lieb, E. H., Loss, M., and Solovej, J. P. (1994). There are no unfilled shells in unrestricted Hartree-Fock theory, *Physical Review Letters* **72**, 2981–2984.

Bardeen, J. (1951). Electron-Vibration Interactions and Superconductivity, *Review of Modern Physics* **23**, 261–270.

Bardeen, J., Cooper, L. N., and Schrieffer, J. R. (1957). Theory of Superconductivity, *Physical Review* **108**, 1175–1204.

Barrett, S. E., Dabbagh, G., Pfeiffer, L. N., West, K. W., and Tycko, R. (1995). Optically Pumped NMR Evidence for Finite-Size Skyrmions in GaAs Quantum Wells near Landau Level Filling $\nu = 1$, *Physical Review Letters* **74**, 5112–5115.

Baus, M. and Hansen, J.-P. (1980). Statistical Mechanics of Simple Coulomb Systems, *Physics Reports* **59**, 1–94.

Baym, G. and Kadanoff, L. P., *Quantum Statistical Mechanics* (W. A. Benjamin, New York, 1962).

Bernu, B., Candido, L., and Ceperley, D. M. (2001). Exchange frequencies in the 2D Wigner crystal, *Physical Review Letters* **86**, 870–873.

Bickers, N. E., Scalapino, D. J., and White, S. R. (1989). Conserving Approximations for Strongly Correlated Electron Systems: Bethe-Salpeter Equation and Dynamics for the Two-Dimensional Hubbard Model, *Physical Review Letters* **62**, 961–964.

Bloch, F. (1929). The electron theory of ferromagnetism and electrical conductivity, Z. Physik **57**, 545–555.

Bohm, D. and Pines, D. (1953). A Collective Description of Electron Interactions: III. Coulomb Interactions in a Degenerate Electron Gas, *Physical Review* **92**, 609–625.

Böhm, H., Conti, S., and Tosi, M. P. (1996). Plasmon dispersion and dynamic exchange-correlation potentials from two-pair excitations in degenerate plasmas, *Journal of Physics: Condensed Matter* **8**, 781–797.

Bonsall, L. and Maradudin, A. A. (1977). Some static and dynamical properties of a two-dimensional Wigner crystal, *Physical Review B* **15**, 1959–1973.

Brooks, H. (1953). Cohesive Energy of Alkali Metals, *Physical Review* **91**, 1027–1028.
(1958). Quantum theory of cohesion, *Nuovo Cimento Supplement* **7**, 165–244.

Brosens, F., Devreese, J. T., and Lemmens, L. F. (1980). Dielectric function of the electron gas with dynamical-exchange decoupling. I and II, *Physical Review B* **21**, 1349–1379.

Brout, R., *Phase Transitions*, W. A. Benjamin, New York (1965).

Brückner, K. A. and Sawada, K. (1958). Magnetic Susceptibility of an Electron Gas at High Density, *Physical Review* **112**, 328–329.

Burke, K., Perdew, J. P., and Wang, Y. (1998) in *Electronic density functional Theory: recent progress and new directions* ed. Dobson, J. F., Vignale, G., and Das, M. P. (Plenum Press, New York, 1998), p. 81–111.

Bychkov, Yu. A., Iordanskii, S. V., and Eliashberg, G. M. (1981). Two-dimensional electrons in a strong magnetic field, *Soviet Physics JETP Letters* **33**, 143–146.

Callen, H. B. and Welton, T. R. (1951). Irreversibility and Generalized Noise, *Physical Review* **83**, 34–40.

Capelle, K. and Vignale, G. (2001). Nonuniqueness of the potential of spin-density functional theory, *Physical Review Letters* **86**, 5546–5549.

Carr, W. J., Jr. (1961). Energy, Specific Heat, and Magnetic Properties of the Low-Density Electron Gas, *Physical Review* **122**, 1437–1446.

Casida, M. E. (1995) in *Recent Advances in Density-Functional Methods*, ed. D. P. Chong (World Scientific, Singapore), Part I, Chapter 5.

Castellani, C., Di Castro, C., Lee, P. A., and Ma, M. (1984). Interaction-driven metal-insulator transitions in disordered fermion systems, *Physical Review B* **30**, 527–543.

Castellani, C., Kotliar, G., and Lee, P. A. (1986). Fermi-liquid theory of interacting disordered systems and the scaling theory of the metal-insulator transition, *Physical Review Letters* **59**, 323–326.

Celli, V. and Mermin, N. D. (1965). Ground State of an Electron Gas in a Magnetic Field, *Physical Review* **140**, A839–A853.

Ceperley, D. M. and Alder, B. J. (1980). Ground State of the Electron Gas by a Stochastic Method, *Physical Review Letters* **45**, 566–569.

Ceperley, D. M. (2004). Introduction to quantum Monte Carlo methods applied to the electron gas, to appear in *The electron liquid paradigm in condensed matter physics*, Proceedings of the International School of Physics "Enrico Fermi", Course CLVII, edited by G. F. Giuliani and G. Vignale.

Chakraborty, T. (1985). Elementary excitations in the fractional quantum Hall effect, *Physical Review B* **31**, 4026–4028.

Chakraborty, T. and Pietiläinen, P. (1988), *The Fractional Quantum Hall Effect: Properties of an incompressible Quantum Fluid* (Springer-Verlag, New York).

Chang, A. M., Pfeiffer, L. N., and West, K. W. (1996). Observation of Chiral Luttinger Behavior in Electron Tunneling into Fractional Quantum Hall Edges, *Physical Review Letters* **77**, 2538–2541.

Chang, A. M. (2003). Chiral Luttinger liquids at the fractional Quantum Hall edge, *Reviews of Modern Physics* **75**, 1449–1505.

Chaplik, A. V. (1971). Energy spectrum and electron scattering processes in inversion layers, *Soviet Physics JETP* **33**, 997–1000.

Chayes, J. T., Chayes, L., and Ruskai, M. B. (1985). Density functional approach to quantum lattice systems, *Journal of Statistical Physics* **38**, 497–518.

Chklovskii, D. B., Shklovskii, B. I., and Glazman, L. I. (1992). Electrostatics of edge channels, *Physical Review B* **46**, 4026–4034; *ibidem* **46**, 15606 (E).

Chow, W. W., Koch, S. W., and Sargent, M. (1994), *Semiconductor Laser Physics* (Springer-Verlag, Berlin), Chapter 6.

Chui, S. T., Hakim, T. M., and Ma, K. B. (1986). Solid versus fluid, and the interplay between fluctuations, correlations, and exchange in the fractional quantized Hall effect, *Physical Review B* **33**, 7110–7121.

Conti, S. and Vignale, G. (1996). Collective modes and electronic spectral function in smooth edges of quantum Hall liquids, *Physical Review B* **54**, 14309–14312.

Conti, S. (1997). *Ground state properties and excitation spectrum of correlated electron systems*, Ph. D. thesis, Scuola Normale Superiore, Pisa, Italy, July 1997.

Conti, S. and Vignale, G. (1998). Dynamics of the two-dimensional electron gas in the lowest Landau level: a continuum elasticity approach, *J. Cond. Mat. Phys* **10**, L779–786.

—— (1999). Elasticity of an electron liquid, *Physical Review B* **60**, 7966–7980.

Corradini, M., Del Sole, R., Onida, G., and Palummo, M. (1998). Analytical expressions for the local-field factor $G(q)$ and the exchange-correlation kernel $K_{xc}(r)$ of the homogeneous electron gas, *Physical Review B* **57**, 14569–14571.

Coulter, P. G. and Datars, W. R. (1980). Open Orbits in Potassium, *Physical Review Letters* **45**, 1021–1024.

Crommie, M. F., Lutz, C. P., and Eigler, D. M. (1993). Imaging standing waves in a two-dimensional electron gas, *Nature* **363**, 524–527.

Czachor, A., Holas, A., Sarma, S. R., and Singwi, K. S. (1982). Dynamical correlations in a two-dimensional electron gas: First-order perturbation theory, *Physical Review B* **25**, 2144–2159.

Dabrowski, B. (1986). Dynamical local-field factor in the response function of an electron gas, *Physical Review B* **34**, 4989–4995.

D'Amico, I. and Vignale, G. (2000). Theory of spin Coulomb drag in spin-polarized transport, *Physical Review B* **62**, 4853–4857.

Daneshvar, A. J. *et al.* (1997). Magnetization instability in a two-dimensional system, *Physical Review Letters* **79**, 4449–4452.

Daniel, E. and Vosko, S. H. (1960). Momentum Distribution of an Interacting Electron Gas, *Physical Review* **120**, 2041–2044.

Darwin, C. G. (1930). The diamagnetism of the free electron, *Proceedings of the Cambridge Philosophical Society* **27**, 86–90.

Das Sarma, S. and Pinczuk, R., editors (1990). *Perspectives in Quantum Hall Effect* (Wiley, New York).

Davisson, C. and Germer, L. H. (1927). Diffraction of electrons by a nickel crystal, *Physical Review* **30**, 705–740.

Davoudi, P., Polini, M., Giuliani, G. F., and Tosi, M. P. (2001a). Analytical expressions for the charge-charge local-field factor and the exchange-correlation kernel of a two-dimensional electron gas, *Physical Review B* **64**, 153101.

—— (2001b). Analytical expressions for the spin-spin local-field factor and the spin-antisymmetric exchange-correlation kernel of a two-dimensional electron gas, *Physical Review B* **64**, 233110.

Decker, P. L., Mapother, D. E., and Shaw, R. W. (1958). Critical Field Measurements on Superconducting Lead Isotopes, *Physical Review* **112**, 1888–1898.

De Picciotto, R., Reznikov, M., Heiblum, M., Umansky, V., Bunin, G., and Mahalu, D. (1997). Direct observation of a fractional charge, *Nature* **389**, 162–164.

Di Castro, C. and Raimondi, R. (2004). Disordered electron systems, to appear in *The electron liquid paradigm in condensed matter physics*, Proceedings of the International School of Physics "Enrico Fermi", Course CLVII , edited by G. F. Giuliani and G. Vignale.

Dirac, P. A. M. (1928). Quantum theory of the electron, *Proceedings of the Royal Society, London* **117**, 610–624.

—— (1929). Quantum mechanics of many-electron systems, *Proceedings of the Royal Society, London* **123**, 714–733.

—— (1930). Exchange phenomena in the Thomas atom, *Proceedings of the Cambridge Philosophical Society* **26**, 367–385.

Dobson, J. F. (1994). Harmonic-Potential Theorem: Implications for Approximate Many-Body Theories, *Physical Review Letters* **73**, 2244–2247.

Dobson, J. F., Dinte, B. P., and Wang, J. (1996). Van der Waals functionals via local approximations for susceptibilities in *Electronic Density Functional Theory: Recent Progress and New Directions*, ed. by J. F. Dobson, G. Vignale, and M. P. Das (Plenum, 1998), p. 261–284.

Dobson, J. F., Bünner, M., and Gross, E. K. U. (1997). Time-Dependent Density Functional Theory beyond Linear Response: An Exchange-Correlation Potential with Memory, *Physical Review Letters* **79**, 1905–1908.

Dobson, J. F., Vignale, G., and Das, M. P., editors (1998). *Electronic Density Functional Theory: Recent Progress and New Directions* (Plenum Press, New York).

Dolgov, O. V., Kirzhnits, D. A., and Maksimov, E. G. (1981). On an Admissible Sign of the Static Dielectric Function of Matter, *Reviews of Modern Physics* **53**, 81–93.

Doniach, S. and Sondheimer, E. H., *Green's functions for solid state physicists* (Benjamin/Cummings, Reading, 1974).

Dreizler, R. M. and Gross, E. K. U. (1990). *Density Functional Theory* (Springer-Verlag, Heidelberg).

Dresselhaus, M. S., Dresselhaus, G., and Eklund, P. C., *Science of fullerenes and carbon nanotubes* (Academic Press, 1996).

Drummond, N. D., Radnai, Z., Trail, J. R., Towler, M. D., and Needs, J. R. (2004). Diffusion quantum Monte Carlo study of three-dimensional Wigner crystals, *Physical Review B* **69**, 085116.

DuBois, D. F. (1959). Electron interactions. I. Field theory of a degenerate electron gas, *Annals of Physics* **7**, 174–237.

Dyson, F. J. and Lenard, A. J. (1967). Stability of matter I, *Journal of Mathematical Physics* **8**, 423–434.

Dyson, F. J. (1967). Ground-state energy of a finite system of charged particles, *Journal of Mathematical Physics* **8**, 1538–1545.

Ebert, H., Battocletti, M., and Gross, E. K. U. (1997). Current density functional theory of spontaneously magnetized solids, *Europhysics Letters* **40**, 545–550.

Efros, A. L. and Shklovskii, B. I. (1975). Coulomb gap and low temperature conductivity of disordered systems, *Journal of Physics C* **8**, L49–L51.

Egger, R. and Grabert, H. (1995). Friedel oscillations for interacting fermions in one dimension, *Physical Review Letters*, **75**, 3505–3508.

Ehrenreich, H. and Cohen, M. (1959). Self-Consistent Field Approach to the Many-Electron Problem, *Physical Review* **115**, 786–790.

Eisenstein, J. P., Pfeiffer, L. N., and West, K. W. (1992). Coulomb barrier to tunneling between parallel two-dimensional electron systems, *Physical Review Letters* **69**, 3804–3807.

(1994). Negative Compressibility of Interacting Two dimensional Electron and Quasiparticle Gases, *Physical Review Letters* **68**, 674–677.

Engel, E. and Vosko, S. H. (1990). Wave-vector dependence of the exchange contribution to the electron-gas response functions: An analytic derivation, *Physical Review B* **42**, 4940–4953.

Esfarjani, K. and Chui, S. T. (1990). Solidification of the two-dimensional electron gas in high magnetic fields, *Physical Review B* **42**, 10758–10760.

Ewald, P. P. (1921). Die Berechnung optischer und elektrostatischer Gitterpotentiale, *Annalen der Physik (Leipzig)* **64**, 253–287.

Fang, F. F. and Stiles, P. J. (1968). Effect of a tilted magnetic field on a two-dimensional electron gas, *Physical Review* **174**, 823–828.

Fano, G. and Ortolani, F. (1988). Interpolation formula for the energy of a two-dimensional electron gas in the lowest Landau level, *Physical Review B* **37**, 8179–8181.

Fano, U. (1961). Effects of Configuration Interaction on Intensities and Phase Shifts, *Physical Review* **124**, 1866–1878.

Farid, B., Heine, V., Engel, G. E., and Robertson, I. J. (1993). Extremal properties of the Harris-Foulkes functional and an improved screening calculation for the electron gas, *Physical Review B* **48**, 11602–11621.

Fedders, A. and Martin, P. C. (1966). Itinerant Antiferromagnetism, *Physical Review* **143**, 245–259 (1966).

Ferconi, M. and Vignale, G. (1994). Current-density-functional theory of quantum dots in a magnetic field, *Physical Review B* **50**, 14722–14725.

Ferconi, M., Geller, M. R., and Vignale, G. (1995). Edge structure of fractional quantum Hall liquids from density functional theory, *Physical Review B* **52**, 16357–16360.

Fermi, E. (1925). Quantization of the monatomic perfect gas, *Rendiconti dell'Accademia Nazionale dei Lincei* **3**, 145–149.

(1927). Application of statistical gas methods to electronic systems, *Rendiconti dell'Accademia Nazionale dei Lincei* **6**, 602–607.

Ferrell, R. A. (1957). Characteristic Energy Loss of Electrons Passing through Metal Foils. II. Dispersion Relation and Short Wavelength Cutoff for Plasma Oscillations, *Physical Review* **107**, 450–462.

(1958). Rigorous Validity Criterion for Testing Approximations to the Electron Gas Correlation Energy, *Physical Review Letters* **1**, 443–445.

Fetter, A. L. and Walecka, J. D., *Quantum Theory of Many-Particle Systems* (McGraw-Hill, New York, 1971).

Feynman, R. P. (1953). Atomic theory of the lambda transition in helium, *Physical Review* **91**, 1291–1301. See also *Statistical Mechanics* (W. A. Benjamin, New York, 1972).

Finkelstein, A. M. (1983). Influence of Coulomb interaction on the properties of disordered metals, *Soviet Physics JETP* **57**, 97–108.

Fock, V. (1928). Quantising a harmonic oscillator in a magnetic field, *Zeitschrift für Physik* **47**, 446–448.

(1930). Approximate method of solution of the problem of many bodies in quantum mechanics, *Zeitschrift für Physik* **61**, 126–148.

Foulkes, W. M. C., Mitas, L., Needs, R. J., and Rajagopal, G. (2001). Quantum Monte Carlo simulations of solids, *Rev. Mod. Phys.* **73**, 33–83.

Friedel, J. (1958). Metallic alloys, *Nuovo Cimento Suppl.* **7**, 287–311.

Fröhlich, H. (1950). Theory of the Superconducting State. I. The Ground State at the Absolute Zero of Temperature, *Physical Review* **79**, 845–856.

(1952). Interaction of electrons with lattice vibrations, *Proc. Roy. Soc. (London)* **A215**, 291–298.

Gell-Mann, M. (1957). Specific Heat of a Degenerate Electron Gas at High Density, *Physical Review* **106**, 369–372.

Gell-Mann, M. and Brueckner, K. A. (1957). Correlation Energy of an Electron Gas at High Density, *Physical Review* **106**, 364–368.

Ghosh, S. K. and Dhara, A. K. (1988). Density-functional theory of many-electron systems subjected to time-dependent electric and magnetic fields, *Physical Review A* **38**, 1149–1158.

Girvin, S. M. (1984). Anomalous quantum Hall effect and two-dimensional classical plasmas: Analytic approximations for correlation functions and ground-state energies, *Physical Review B* **30**, 558–560.

Girvin, S. M. and Jach, T. (1984). Formalism for the quantum Hall effect: Hilbert space of analytic functions, *Physical Review B* **29**, 5616–5625.

Girvin, S. M., MacDonald, A. H., and Platzman, P. M. (1986). Magneto-roton theory of collective excitations in the fractional quantum Hall effect, *Physical Review B* **33**, 2481–2494.

Giuliani, G. F. and Quinn, J. J. (1982). Lifetime of a quasiparticle in a two-dimensional electron gas, *Physical Review B* **26**, 4421–4428.

(1985a). Spin-polarization instability in a tilted magnetic field of a two-dimensional electron gas with filled Landau levels, *Physical Review B* **31**, 6228–6232.

(1985b). Breakdown of the random phase approximation in the fractional quantum Hall effect regime, *Physical Review B* **31**, 3451–3455.

Giuliani, F. G., Quinn, J. J., and Ying, S. C. (1983). Quantization of the Hall conductance in a two-dimensional electron gas, *Physical Review B* **28**, 2969–2978.

Glazman, L. I. and Fischer, M. P. A. (1996). Transport in a one-dimensional Luttinger liquid, in *Mesoscopic electron transport*, edited by L. Sohn, L. P. Kowenhoven, and G. Schön, (NATO ASI Series No. 345, Kluwer, 1997).

Glick, A. and Ferrell, R. A. (1959). Single particle excitations of a degenerate electron gas, *Annals of Physics* **11**, 359–376.

Glick, A. J. and Long, W. F. (1971). High-Frequency Damping in a Degenerate Electron Gas, *Physical Review B* **4**, 3455–3460.

Godby, R. W., Sham, L. J., and Schlüter, M. (1986). Accurate Exchange-Correlation Potential for Silicon and Its Discontinuity on Addition of an Electron, *Physical Review Letters* **56**, 2415–2418.

Goldman, V. J., Santos, M., Shayegan, M., and Cunningham, J. E. (1990). Evidence for two-dimentional quantum Wigner crystal, *Physical Review Letters* **65**, 2189–2192.

Goldman, V. J. and Su, B. (1995). Resonant tunneling in the quantum Hall regime: measurement of fractional charge, *Science* **267**, 1010–1012.

Goldman, V. J. (2000). Quantum Hall effect today, *Physica B* **280**, 372–377.

Goldstone, J. and Gottfried, K. (1959). Collective excitations of Fermi gases, *Nuovo Cimento* **13**, 849–852.

Gonze, X., Ghosez, Ph., and Godby, R. W. (1995). Density-Polarization Functional Theory of the Response of a Periodic Insulating Solid to an Electric Field, *Physical Review Letters* **74**, 4035–4038.

Goodman, B. and Sjölander, A. (1973). Application of the Third Moment to the Electric and Magnetic Response Function, *Physical Review B* **8**, 200–214.

Gori-Giorgi, P. and Perdew, J. P. (2001). Short-range correlation in the uniform electron gas: Extended Overhauser model, *Physical Review B* **64**, 155102.

Gori-Giorgi, P., Moroni, S., and Bachelet, G. B. (2004). Pair-distribution function of the two-dimensional electron gas, cond-mat/0403050.

Görling, A. (1992). Kohn-Sham potentials and wave functions from electron densities, *Physical Review A* **46**, 3753–3757.

Görling, A. and Levy, M. (1994). Exact Kohn-Sham scheme based on perturbation theory, *Physical Review A* **50**, 196–204.

Goudsmit, S. and Uhlenbeck, G. E. (1925). Coupling possibilities of the quantum vectors in the atom, *Zeitschrift für Physik* **35**, 618–625.

Grabowski, M. and Sham, L. J. (1984). Superconductivity from non-phonon interactions, *Physical Review B* **29**, 6132–6142.

Gradshteyn, I. S. and Ryzhik, I. M., *Table of Integrals, Series, and Products*, 4th Edition, Academic Press, San Francisco (1965).

Grayson, M., Tsui, D. C., Pfeiffer, L. N., West, K. W., and Chang, A. M. (1998). Continuum of Chiral Luttinger Liquids at the Fractional Quantum Hall Edge, *Physical Review Letters* **80**, 1062–1065.

Greene, M. P., Lee, H. J., Quinn, J. J. and Rodriguez, S. (1969). Linear response theory for a degenerate electron gas in a strong magnetic field, *Physical Review B* **177**, 1019–1036.

Gross, E. K. U. and Kohn, W. (1985). Local density-functional theory of frequency-dependent linear response, *Physical Review Letters* **55**, 2850–2853; **57**, 923(E).

Gross, E. K. U. and Dreizler, R. M., editors (1995), *Density Functional Theory*, NATO ASI Series, Vol. 137 (Plenum Press, New York).

Gross, E. K. U., Dobson, J. F., Petersilka, M. in *Density Functional Theory II*, ed. R. F. Nalewajski, Vol. 181 of Topics in Current Chemistry, (Springer, Berlin, 1996), p. 81.

Haldane, F. D. M. (1981). Luttinger liquid theory' of one-dimensional quantum fluids:I. Properties of the Luttinger model and their extension to general 1D interacting spinless Fermi gas, *Journal of Physics C: Solid State Physics* **14**, 2585–2609.

Halperin, B. I. (1982). Quantized Hall conductance, current-carrying edge states, and the existence of extended states in a two-dimensional disordered potential, *Physical Review B* **25**, 2185–2190.

(1984). Statistics of Quasiparticles and the Hierarchy of Fractional Quantized Hall States, *Physical Review Letters* **52**, 1583–1586, 2390(E).

Halperin, B. I., Lee, P. A., and Read, N. (1993). Theory of the half-filled Landau level, *Physical Review B* **47**, 7312–7343.

Ham, F. (1962). Energy Bands of Alkali Metals. II. Fermi Surface, *Physical Review* **128**, 2524–2541.

Hamann, D. R. and Overhauser, A. W. (1966). Electron-Gas Spin Susceptibility, *Physical Review* **143**, 183–197.

Hameeuw, K. J., Brosens, F., and Devreese, T. J. (2003). Dynamical exchange corrections to the dielectric function in a two-dimensional electron gas, *Solid State Communications* **126**, 695–698.

Hammond, B. L., Lester, Jr., W. A., and Reynolds, P. J. (1994). *Monte Carlo methods in ab initio quantum chemistry*, (World Scientific, Singapore, 1994).

Hartree, D. R. (1928). Wave mechanics of an atom with non Coulomb central field. Part I. Theory and methods, *Cambridge Philosophical Society Proceedings* **24**, 89–110.

Hasegawa, M. and Watabe, J. (1969). Theory of plasmon damping in metals. I. General formulation and application to an electron gas, *Journal of the Physical Society of Japan* **27**, 1393–1414.

Hasegawa, T. and Shimizu, M. (1975). Electron correlations at metallic densities, II. Quantum mechanical expression of dielectric function with Wigner distribution function, *Journal of the Physical Society of Japan* **38**, 965.

Hatsugai, Y., Bares, P.-A., and Wen, X. G. (1993). Electron spectral function of an interacting two dimensional electron gas in a strong magnetic field, *Physical Review Letters* **71**, 424–427.

Haussmann, R., Mori, H., and MacDonald, A. H. (1996). Correlation Energy and Tunneling Density of States in the Fractional Quantum Hall Regime, *Physical Review Letters* **76**, 979–982.

He, S., Platzman, P. M., and Halperin, B. I. (1993). Tunneling into a two-dimensional electron system in a strong magnetic field, *Physical Review Letters* **71**, 777–780.

Hedin, L. (1965). New Method for Calculating the One-Particle Green's Function with Application to the Electron-Gas Problem, *Physical Review* **139**, A796–A823.

Hedin, L. and Lunqvist, S. (1969), *Solid State Physics* **23**, eds. F. Seitz, D. Turnbull, and H. Ehrenreich (Academic, New York), p. 1.

Hedin, L. and Lunqvist, B. I. (1971). Explicit local exchange-correlation potentials, *Journal of Physics C* **4**, 2064–2083.

Heinonen, O. and Kohn, W. (1987). Internal structure of a Landau quasiparticle wave packet, *Physical Review B* **36**, 3565–3576.

Heinonen, O., editor (1998), *Composite Fermions* (World Scientific, Singapore).

Hirjibehedin, C. F., Pinczuk, A., Dennis, B. S., Pfeiffer, L. N., and West, K. W. (2002). Evidence of electron correlation in plasmon dispersion of ultralow density two-dimensional electron systems, *Physical Review B* **65**, 161309.

Hohenberg, P. and Kohn, W. (1964). Inhomogeneous electron gas, *Physical Review* **136**, B864–871.

Holas, A., Aravind, P. K., and Singwi, K. S. (1979). Dynamic correlations in an electron gas. I. First-order perturbation theory, *Physical Review B* **20**, 4912–4934.

(1982). Dynamic correlations in an electron gas. II. Kinetic-equation approach, *Physical Review B* **25**, 561–578.

Holas, A. and Rahman, S. (1987). Dynamic local-field factor of an electron liquid in the quantum versions of the Singwi-Tosi-Land-Sjölander and Vashishta-Singwi theories, *Physical Review B* **35**, 2720–2731.

Holas, A. and Singwi, K. S. (1989). High-frequency damping of collective excitations in fermion systems. I. Plasmon damping and frequency-dependent local-field factor in a two-dimensional electron gas, *Physical Review B* **40**, 158–166.

Holas, A. (1992) in *Strongly Coupled Plasma Physics*, edited by F. J. Rogers and H. Dewitt, NATO Advanced Study Institute Series B: Physics (Plenum, New York, 1986), Vol. **154**, 463–482.

Holas, A. and Balawender, R. (2002). Maitra-Burke example of initial state dependence in time-dependent density functional theory, *Physical Review A* **65**, 034502.

Holm, B. and von Barth, U. (1998). Fully self-consistent GW self-energy of the electron gas, *Physical Review B* **57**, 2108–2117.

Hood, R. Q., Chou, M. Y., Williamson, A. J., Rajagopal, G., and Needs, R. J. (1998). Exchange and correlation in silicon, *Physical Review B* **57**, 8972–8982.

Hubbard, J. (1957). The description of collective motions in terms of many-body perturbation theory: II. The correlation energy of a free electron gas, *Proceedings of the Royal Society (London)* **A243**, 336–352.

Iijima, S. (1991). Helical microtubules of graphitic carbon, *Nature* **354**, 56.

Iwamoto, N. and Pines, D. (1984). Theory of electron liquids. I. Electron-hole pseudopotential, *Physical Review B* **29**, 3924–3935.

Jain, J. K. (1989). Composite-fermion approach for the fractional quantum Hall effect, *Physical Review Letters* **63**, 199–202.

Jain, J. K. (2000). The composite fermion: a quantum particle and its quantum fluids, *Physics Today* **53**, Issue 4, 39–45.

ensen, E. and Plummer, E. W. (1985). Experimental Band Structure of Na, *Physical Review Letters* **55**, 1912–1915.

ohansson, P. and Kinaret, J. M. (1994). Tunneling between two two-dimensional electron systems in a strong magnetic field, *Physical Review B* **50**, 4671–4686.

ones, M. D. and Ceperley, D. M. (1996). Crystallization of the One-Component Plasma at Finite Temperature, *Physical Review Letters* **76**, 4572–4575.

Kadanoff, L. P. and Baym, G. (1962), *Quantum Statistical mechanics* (The Benjamin/Cummings Publishing Company, Reading, Massachussetts).

Kallin, C. and Halperin, B. I. (1984). Excitations from a filled Landau level in the two dimensional electron gas, *Physical Review B* **30**, 5655–5668.

Kamilla, R. K. and Jain, J. K. (1997). Composite Fermions in the Hilbert Space of the Lowest Electronic Landau Level, *International Journal of Modern Physics B* **11**, 2621–2660.

Kasuya, T. (1956). Electrical resistance of ferromagnetic metals, *Progress of Theoretical Physics* **16**, 58–63.

Kim, Y.-H. and Görling, A. (2002). Excitonic Optical Spectrum of Semiconductors Obtained by Time-Dependent Density-Functional Theory with the Exact-Exchange Kernel, *Physical Review Letters* **89**, 096402.

Kimball, J. C. (1973). Short-Range Correlations and Electron-Gas Response Functions, *Physical Review A* **7**, 1648–1652.

Kleinman, L. (1984). Exchange density-functional gradient expansion, *Physical Review B* **30**, 2223–2225.

Klimontovitch, Y. and Silin, V. P. (1952). On the spectra of systems of interacting particles, *Zh. Exp. Teor. Fiz.* **23**, 151–160.

Kohn, W. and Vosko, S. H. (1960). Theory of Nuclear Resonance Intensity in Dilute Alloys, *Physical Review* **119**, 912–918.

Kohn, W. (1961). Cyclotron Resonance and de Haas-van Alphen Oscillations of an Interacting Electron Gas, *Physical Review* **123**, 1242–1244 (1961).

(1986). Discontinuity of the exchange-correlation potential from a density-functional viewpoint, *Physical Review B* **33**, 4331–4333.

Kohn, W. and Luttinger, J. M. (1965). New Mechanism for Superconductivity, *Physical Review Letters* **15**, 524–526.

Kohn, W. and Sham, L. J. (1965). Self-Consistent Equations Including Exchange and Correlation Effects, *Physical Review* **140**, A1133–A1138.

Kohn, W. and Vashishta, P. (1983). General density functional theory, in *Theory of the inhomogeneous electron gas*, edited by S. Lundqvist and N.H. March (New York : Plenum Press).

Koulakov, A. A., Fogler, M. M., and Shklovskii, B. I. (1996). Charge Density Wave in Two-Dimensional Electron Liquid in Weak Magnetic Field, *Physical Review Letters* **76**, 499–502.

Krieger, J. B., Li, Y., and Iafrate, G. J. (1992). Construction and application of an accurate local spin-polarized Kohn-Sham potential with integer discontinuity: Exchange-only theory, *Physical Review A* **45**, 101–126.

Ku, W. and Eguiluz, A. G. (1999). Plasmon Lifetime in K: A Case Study of Correlated Electrons in Solids Amenable to Ab Initio Theory, *Physical Review Letters* **82**, 2350–2353.

Kubo, R. (1957). Statistical Mechanical Theory of Irreversible Processes I, *Journal of the Physical Society of Japan* **12**, 570–586.

Kukkonen, C. A. and Overhauser, A. W. (1979). Electron-electron interaction in simple metals, *Physical Review B* **20**, 550–557.

Kushida, T., Murphy, J. C., and Hanabusa, M. (1976). Volume dependence of the Pauli susceptibility and the amplitude of the wave functions for Li and Na, *Physical Review* **B 13**, 5136–5153.

Kwon, Y., Ceperley, D. M., and Martin, R. M. (1994). Quantum Monte Carlo calculation of the Fermi-liquid parameters in the two-dimensional electron gas, *Physical Review B* **50**, 1684–1694.

Lam, J. (1971). Momentum Distribution and Pair Correlation of the Electron Gas at Metallic Densities, *Physical Review* **B3**, 3243–3248.

Lam, P. K. and Girvin, S. M. (1984). Liquid-solid transition and the fractional quantum Hall effect, *Physical Review B* **30**, 473–475; **31**, 613 (1985) (E).

Lamelas, F. J., Werner, S. A., Shapiro, S. M., and Mydosh, J. A. (1995). Intrinsic spin-density-wave magnetism in Cu-Mn alloys, *Physical Review B* **51**, 621–624.

Landau, L. D. (1930). Paramagnetism of metals, *Zeitschrift für Physik*, **64**, 629–637.
 (1957a). Theory of the Fermi liquid, *Soviet Physics JETP* **3**, 920–925.
 (1957b). Oscillations in a Fermi liquid, *Sov. Phys. JETP* **5**, 101–108.
 (1959). On the theory of the Fermi liquid, *Sov. Phys. JETP* **8**, 70–74.

Langer, J. S. and Vosko, S. H. (1959). The shielding of a fixed charge in a high-density electron gas, *Journal of Physics and Chemistry of Solids* **12**, 196–205.

Langreth, D. (1975). Linear and nonlinear response theory with applications, in *Linear and Nonlinear transport in Solids*, ed. J. T. Devreese and V. E. van Doren (Plenum Press) p. 3–32.

Langreth, D. C. and Perdew, J. P. (1980). Theory of nonuniform electronic systems. I. Analysis of the gradient approximation and a generalization that works, *Physical Review B* **21**, 5469–5493.

Larson, B. C. *et al.* (1996). Inelastic X-ray scattering as a probe of the many-body local-field factor in metals, *Physical Review Letters* **77**, 1346–1349.

Laughlin, R. B. (1981). Quantized Hall conductivity in two dimensions, *Physical Review B* **23**, 5632–5633.
 (1983). Anomalous Quantum Hall Effect: An Incompressible Quantum Fluid with Fractionally Charged Excitations, *Physical Review Letters* **50**, 1395–1398.
 (1999). Nobel lecture: Fractional quantization, *Reviews of Modern Physics* **71**, 863–874.

Leiderer, P. (1997). Nonlinear Effects – The Multielectron Dimple, in *Two-Dimensional Electron Systems*, edited by E. Andrei (Kluwer Academic Publisher, 1997), p. 317–339.

Lein, M., Gross, E. K. U., and Perdew, J. P. (2000). Electron correlation energies from scaled exchange-correlation kernels: Importance of spatial versus temporal nonlocality, *Physical Review B* **61**, 13431–13437.

Lenard, A. and Dyson, F. J. (1968). Stability of Matter. II, *Journal of Mathematical Physics* **9**, 698–711.

Lerner, I. V. and Lozovik, Yu. E. (1978). Mott exciton in a quasi-two-dimensional semiconductor in a strong magnetic field, *Soviet Physics JETP* **51**, 588–592 (1980).

Levesque, D., Weis, J. J., and MacDonald, A. H. (1984). Crystallization of the incompressible quantum-fluid state of a two-dimensional electron gas in a strong magnetic field, *Physical Review B* **30**, 1056–1058.

Levy, M. (1979). Universal variational functionals of electron densities, first-order density matrices, and natural spin-orbitals and solution of the v-representability problem *Proceedings of the National Academy of Science USA* **76**, 6062–6065.
 (1982). Electron densities in search of Hamiltonians, *Physical Review A* **26**, 1200–1208.

evy, M., Perdew, J. P., and Sahni, V. (1984). Exact differential equation for the density and ionization energy of a many-particle system, *Physical Review A* **30**, 2745–2748.

Lieb, E. H. and Narnhofer, H. (1975). The thermodynamic limit for jellium, *Journal of Statistical Physics* **12**, 291–310.

Lieb, E. H. and Thirring, W. E. (1975). Bound for the Kinetic Energy of Fermions Which Proves the Stability of Matter, *Physical Review Letters* **35**, 687–689; **35**, 1116 (E).

Lieb, E. H. and Oxford, S. (1981). Improved lower bound on the indirect Coulomb energy, *International Journal of Quantum Chemistry*, **19**, 427–439.

Lieb, E. H. (1985). Density functionals for Coulomb systems, *Density Functional Methods in Physics, NATO ASI Series* **B123**, ed. by Dreizler, R. M. and Da Providencia, J. (Plenum Press, New York), p. 31–80.

Lien, W. H. and Phillips, N. E. (1964). Low-Temperature Heat Capacities of Potassium, Rubidium, and Cesium, *Physical Review* **133**, A1370–A1377.

Lilly, M. P. *et al.* (1999). Evidence for an anisotropic state of two dimensional electrons in high Landau levels, *Physical Review Letters* **82**, 394–397.

Lindhard, J. (1954). On the properties of a gas of charged particles, *Det Kongelige Danske Videnskabernes Selskab, Matematisk-fysike Meddlelser* **28**, No. 8.

Liu, K. L. and Vosko, S. H. (1989). A time-dependent spin density functional theory for the dynamical spin susceptibility, *Canadian Journal of Physics* **67**, 1015–1021.

Liu, K. L. (1991). Exchange-correlation effects on the dynamical spin susceptibility of a homogeneous electron gas, *Canadian Journal of Physics* **69**, 573–580.

Lopez, A. and Fradkin, E. (1991). Fractional quantum Hall effect and Chern-Simons gauge theories, *Physical Review B* **44**, 5246–5262.

Luther, A. and Peschel, I. (1974). Single-particle states, Kohn anomaly, and pairing fluctuations in one dimensions, *Physical Review B*, **9**, 2911–2919.

Luttinger, J. M. and Kohn, W. (1955). Motion of Electrons and Holes in Perturbed Periodic Fields, *Physical Review*, **97**, 869–883.

Luttinger, J. M. (1963). An exactly soluble model of a many-fermion system, *Journal of Mathematical Physics*, **4**, 1154–1162.

Lyo, In-Whan and Plummer, E. W. (1988). Quasiparticle band structure of Na and simple metals, *Physical Review Letters* **60**, 1558–1561.

Maasilta, I. J. and Goldman, V. I. (1997). Line shape of resonant tunneling between fractional quantum Hall edges, *Physical Review B* **55**, 4081–4084.

MacDonald, A. H. and Girvin, S. M. (1988). Density matrices for states in the lowest Landau level of a two-dimensional electron gas, *Physical Review B* **38**, 6295–6297.

MacDonald, A. H. (editor) *Quantum Hall effect: a perspective* (Kluwer, Boston, 1989).

Macke, W. von (1950). Über die Wechselwirkungen im Fermi-Gas, *Zeitschrift für Naturforschung* **5a**, 192–208.

Mahan, G. D., *Many-Particle Physics* (Plenum Press, New York, 1981).

(1994). GW approximations, *Comments in Condensed Matter Physics* **16**, 333–354.

Mahan, G. D. and Sernelius, B. E. (1989). Electron-electron interactions and the bandwidth of metals, *Physical Review Letters* **62**, 2718–2720.

Maitra, N. and Burke, K. (2001). Demonstration of initial-state dependence in time-dependent density-functional theory, *Physical Review A* **63**, 042501; *ibidem* **64**, 039901 (E).

Maldague, P. F. (1978). Many-body corrections to the polarizability of the two dimensional electron gas, *Surface Science* **73**, 296–302.

Mandal, S. S. and Jain, J. K. (2002). Relevance of Inter-Composite-Fermion Interaction to the Edge Tomonaga-Luttinger Liquid, *Physical Review Letters* **89**, 096801.

Manoharan, H. C., Shayegan, M., and Klepper, S. J. (1994). Signatures of a novel Fermi liquid in a two-dimensional composite metal, *Physical Review Letters* **73**, 3270–3273.

Marinescu, D. C., Quinn, J. J., and Giuliani, G. F. (2002). Tunneling between dissimilar quantum wells: A probe of the energy-dependent quasiparticle lifetime, *Physical Review B* **65**, 045325.

Martin, D. L. (1961). Specific Heat of Sodium at Low Temperatures *Physical Review* **124**, 438–441.

Martin, B. D. and Heer, C. V. (1964). Atomic heat of Cesium from 0.3° K to 2° K, *Bulletin of the American Physical Society* **9**, 250.

Martin, R. M. and Ortiz, G. (1997). Functional theory of extended Coulomb systems, *Physical Review B* **56**, 1124–1140.

Mattis, D. C. and Lieb, E. H. (1965). Exact solution of a many-fermion system and its associated boson field, *Journal of Mathematical Physics*, **6**, 304–312.

Mayer, M. and El Naby, M. H. (1963). Die optischen Konstanten des Kaliums und die Theorie von Drude, *Zeitschrift für Physik* **174**, 280–288.

McMillan, W. L. and Mochel, J. (1981). Electron Tunneling Experiments on Amorphous $Ge_{1-x}Au_x$, *Physical Review Letters* **46**, 556–557.

Mearns, D. (1988). Inequivalence of the physical and Kohn-Sham Fermi surfaces, *Physical Review B* **38**, 5906–5912.

Mermin, N. D. (1965). Thermal Properties of the Inhomogeneous Electron Gas, *Physical Review* **137**, A1441–1443.

(1970). Lindhard Dielectric Function in the Relaxation-Time Approximation, *Physical Review B* **1**, 2362–2363.

Mermin, N. D. and Wagner, H. (1966). Absence of Ferromagnetism or Antiferromagnetism in One- or Two-Dimensional Isotropic Heisenberg Models, *Physical Review Letters* **17**, 1133–1136.

Millikan, R. A. (1911). On the elementary electrical charge and the Avogadro constant, *Physical Review* **32**, 349.

Milliken, F. P., Umbach. C. P., and Webb, R. (1995). Indications of a Luttinger liquid in the fractional quantum Hall regime, *Solid State Communications* **97**, 309–313.

Morf, R. and Halperin, B. I. (1986). Monte Carlo evaluation of trial wave functions for the fractional quantized Hall effect: Disk geometry, *Physical Review B* **33**, 2221–2246.

(1992). Static response from Quantum Monte Carlo calculations, *Physical Review Letters* **69**, 1837–1840.

Moroni, S., Ceperley, D. M., and Senatore, G. (1995). Static Response and Local Field Factor of the Electron Gas, *Physical Review Letters* **75**, 689–692.

Mukhopadhyay, G., Kalia, R. K., and Singwi, K. S. (1975). Dynamic Structure Factor of an Electron Liquid, *Physical Review Letters* **34**, 950–952.

Murphy, S. Q., Eisenstein, J. P., Pfeiffer, L. N., and West, K. W. (1995). Lifetime of two-dimensional electrons measured by tunneling spectroscopy, *Physical Review B* **52**, 14825–14828.

Murphy, S. (2004). A Brief Guide to Electronic Lifetimes in 2D, to appear in *The electron liquid paradigm in condensed matter physics*, Proceedings of the International School of Physics "Enrico Fermi", Course CLVII , edited by G. F. Giuliani and G. Vignale.

Murthy, G. and Shankar, R. (2003). Hamiltonian theories of the fractional quantum Hall effect, *Reviews of Modern Physics* **75**, 1101–1158.

Nagao, T., Hildenbrandt, T., Henzler, M., and Hasegawa, S. (2001). Dispersion and damping of a two-dimensional plasmon in a metallic surface-state band, *Physical Review Letters* **86**, 5747–5750.

Negele, J. W. and Orland, H., *Quantum Many-Particle Systems* (Addison-Wesley, Redwood City, 1988).

Ng, T. K. and Singwi, K. S. (1986). Effective interactions for self-energy. I. Theory, *Physical Review B* **34**, 7738–7742; Effective interactions for self-energy. II. Application to electron and electron-hole liquids, *ibidem*, 7743–7747.

(1987). Time-Dependent Density-Functional Theory in the Linear-Response Regime, *Physical Review Letters* **59**, 2627–2630.

Ng, T. K. (1989). Transport properties and a current-functional theory in the linear-response regime, *Physical Review Letters* **62**, 2417–2420.

Nifosí R., Conti, S., and Tosi, M. P. (1998). Dynamic exchange-correlation potentials for the electron gas in dimensionality $D = 3$ and $D = 2$, *Physical Review B* **58**, 12758–12769.

Niklasson, G. (1974). Dielectric function of the uniform electron gas for large frequencies or wave vectors, *Physical Review B* **10**, 3052–3061.

Nozières, P. and Pines, D. (1958a). Correlation Energy of a Free Electron Gas, *Physical Review* **111**, 442–454.

(1958b). A dielectric formulation of the many-body problem: Application to the free electron gas, *Nuovo Cimento* **9**, 470–490.

Nozières, P., *Theory of interacting Fermi systems* (W. A. Benjamin, New York, 1964).

Okamoto, T., Osoya, K., Kawaji, S., and Yagi, A. (1999). Spin Degree of Freedom in a Two-Dimensional Electron Liquid, *Physical Review Letters* **82**, 3875–3878.

Oliveira, L. N., Gross, E. K. U., and Kohn, W. (1988). Density-Functional Theory for Superconductors, *Physical Review Letters* **60**, 2430–2433.

Oliveira, L. N., Mearns, D., and Kohn, W. (1988). Zeros of the Frequency-Dependent Linear Density Response, *Physical Review Letters* **61**, 1518.

Onida, G., Reining, L., and Rubio, A. (2002). Electronic excitations: density-functional versus many-body Green's-function approaches, *Reviews of Modern Physics* **74**, 601–659.

Onsager, L., Mittag, L., and Stephen, M. J. (1966). Integrals in the theory of electron correlations, *Annalen der Physik* **7**, 71–77.

Ortiz, G. and Ballone, P. (1994). Correlation energy, structure factor, radial distribution function, and momentum distribution function of the spin-polarized uniform electron gas, *Physical Review B* **50**, 1391–1405; *ibidem* **56**, 9970 (E).

Ortiz, G., Harris, M., and Ballone, P. (1999). Zero temperature phases of the electron gas, *Physical Review Letters* **82**, 5317–5320.

Overhauser, A. W. (1960a). Giant Spin Density Waves, *Physical Review Letters* **4**, 462–465.

(1960b). Mechanism of antiferromagnetism in dilute alloys, *Journal of Physics and Chemistry of Solids* **13**, 71–80.

(1962). Spin Density Waves in an Electron Gas, *Physical Review* **128**, 1437–1452.

(1968). Exchange and Correlation Instabilities of Simple Metals, *Physical Review* **167**, 691–698.

(1971). Simplified theory of electron correlations in metals, *Physical Review B* **3**, 1888–1898.

(1985). Broken symmetry in simple metals, *Highlights of Condensed Matter Theory*, edited by F. Bassani, F. Fumi, and M. P. Tosi (North Holland, 1985).

(1995). Pair correlation function of an electron gas, *Canadian Journal of Physics* **73**, 683–686.

Pan, W. *et al.* (2003). Fractional Quantum Hall Effect of Composite Fermions, *Physical Review Letters* **90**, 016801.

Parks, R. D., editor, *Superconductivity*, M. Dekker Publisher, New York (1969).

Parr, R. G. and Yang, W. (1989). *Density Functional Theory of Atoms and Molecules* (Oxford University Press, New York).

Pauli, W. (1925). Relation between the closing-in of electronic groups in the atom and the structure of complexes in the spectrum, *Zeitschrift für Physik* **31**, 765–783.

(1927). Gas degeneration and paramagnetism, *Zeitschrift für Physik* **41**, 81–102.

Peierls, R. E. (1955), *Quantum theory of solids* (Clarendon, Oxford).

Pellegrini, V., Pinczuk, A., Dennis, B. S., Plaut, A. S., Pfeiffer, L. N., and West, K. W. (1998). Evidence of Soft-Mode Quantum Phase Transitions in Electron Double Layers, *Science* **281**, 799–802.

Penn, D. R. (1980). Mean free paths of very-low-energy electrons: the effects of exchange and correlation, *Physical Review B* **22**, 2677–2682.

Perdew, J. P. and Zunger, A. (1981). Self-interaction correction to density-functional approximations for many-electron systems, *Physical Review B* **23**, 5048–5079.

Perdew, J. P., Parr, R. G., Levy, M., and Balduz, J. L. (1982). Density-Functional Theory for Fractional Particle Number: Derivative Discontinuities of the Energy, *Physical Review Letters* **49**, 1691–1694.

Perdew, J. P. and Wang Y. (1986). Accurate and simple density functional for the electronic exchange energy: Generalized gradient approximation, *Physical Review B* **33**, 8800–8802; *ibidem* **40**, 3399 (1989) (E).

(1992a). Accurate and simple analytic representation of the electron-gas correlation energy, *Physical Review B* **45** 13244–13249.

(1992b). Pair-distribution function and its coupling-constant average for the spin-polarized electron gas, *Physical Review B* **46**, 12947–12954; *ibidem* **56**, 7018 (1997) (E).

Perdew, J. P., Burke, K., and Ernzerhof, M. (1996). Generalized Gradient Approximation Made Simple. *Physical Review Letters* **77**, 3865–3868.

Perrot, F. (1982). Temperature-dependent nonlinear screening of a proton in an electron gas, *Physical Review A* **25**, 489–495.

Perrot, F. and Dharma Wardana, M. W. C. (1984). Exchange and correlation potentials for electron-ion systems at finite temperatures, *Physical Review A* **30**, 2619–2626.

Petersilka, M., Gossmann, U. J., and Gross, E. K. U. (1996). Excitation Energies from Time-Dependent Density-Functional Theory, *Physical Review Letters* **76**, 1212–1215.

Pinczuk, A., Schmitt-Rink, S., Danan, G., Valladares, J. P., Pfeiffer, L. N., and West, K. W. (1989). Large exchange interactions in the electron gas of GaAs quantum wells, *Physical Review Letters* **63**, 1633–1636.

Pinczuk, A., Dennis, B. S., Pfeiffer, L. N., and West, K. W. (1993). Observation of collective excitations in the fractional quantum Hall effect, *Physical Review Letters* **70**, 3983–3986.

(1998). Light scattering by collective excitations in the fractional quantum hall regime, *Physica B* **249**, 40–43.

ines, D. and Bohm, D. (1952). A Collective Description of Electron Interactions: II. Collective vs Individual Particle Aspects of the Interactions, *Physical Review* **85**, 338–353.

ines, D. (1953). A Collective Description of Electron Interactions: IV. Electron Interaction in Metals *Physical Review* **92**, 626–636.

(1955). Electron interaction in metals, *Solid State Physics* (F. Seitz and D. Turnbull, editors) **1**, 368–450.

(1960). Plasma oscillations of electron gases, *Physica* **26**, S103-S123.

(1962). *The Many Body Problem*, W.A. Benjamin, New York (1962).

(1963). *Elementary Excitations in Solids*, W.A. Benjamin, New York (1963).

Pines, D. and Nozières, P. (1966). *The Theory of Quantum Liquids* (W. A. Benjamin, Inc., New York).

Platzman, P. M. *et al.* (1992). X-ray-scattering determination of the dynamic structure factor of Al metal, *Physical Review B* **46**, 12943–12946.

Prange, R. and Kadanoff, L. P. (1964). Transport Theory for Electron-Phonon Interactions in Metals, *Physical Review* **134**, A566–A580.

Prange, R. E. and Girvin, S. M., editors (1987), *The Quantum Hall Effect* (Springer-Verlag, New York).

Press, W. H., Teukolsky, S. A., Vetterling, W. T., and Flannery, B. P., *Numerical Recipes*, (Cambridge University Press, Cambridge, 1992).

Pudalov, V. M., D'Iorio, M., Kravchenko, S. V., and Campbell, J. W. (1993). Zero-magnetic-field collective insulator phase in a dilute 2D electron system, *Physical Review Letters* **70**, 1866–1869.

Pudalov, V. M. *et al.* (2002). Low density spin susceptibility and effective mass of mobile electrons in Si inversion layers, *Physical Review Letters* **88**, 196404.

Puff, R. D. (1965). Application of Sum Rules to the Low-Temperature Interacting Boson System, *Physical Review* **137**, A406–A416.

Qian, Z. and Vignale, G. (2002). Dynamical exchange-correlation potentials for an electron liquid, *Physical Review B* **65**, 235121.

(2003). Dynamical exchange-correlation potentials for the electron liquid in the spin channel, *Physical Review B* **68**, 195113.

Qian, Z., Constantinescu, A., and Vignale, G. (2003). Solving the ultra-nonlocality problem in time dependent spin density functional theory, *Physical Review Letters* **90**, 066402.

Qian, Z., Vignale, G., and Marinescu, D. C. (2004). The spin-mass of an electron liquid, *Physical Review Letters* **93**, 106601.

Qian, Z. (2004). Long wavelength behavior of the dynamical spin-resolved local field factor in a two-dimensional electron liquid, cond-mat/0404325.

Quinn, J. J. and Ferrell, R. A. (1958). Electron Self-Energy Approach to Correlation in a Degenerate Electron Gas, *Physical Review* **112**, 812–827.

(1961). Quasi-particle approach to interaction in an idealized metal, *Journal of Nuclear Energy, Part C*, **2**, 18–24.

Quong, A. A. and Eguiluz, A. G. (1993). First-principles evaluation of dynamical response and plasmon dispersion in metals, *Physical Review Letters* **70**, 3955–3958.

Rahman, S. and Vignale, G. (1984). Fine structure in the dynamic form factor of an electron liquid, *Physical Review B* **30**, 6951–6959.

Rajagopal, A. K. and Callaway, J. (1973). Inhomogeneous Electron Gas, *Physical Review B* **7**, 1912–1919.

Rajagopal, A. K. and Kimball J. C. (1977). Short-ranged correlations and the ferromagnetic electron gas, *Physical Review B* **18**, 2819–2825.

Rajagopal, A. K., Kimball J. C., and Banerjee, M. (1978). Correlations in a two-dimensional electron system, *Physical Review B* **15**, 2339–2345.

Rammer, J. and Smith, H. (1986). Quantum field theoretical methods in transport theory of metals, *Review of Modern Physics* **58**, 323–359.

Rapcewicz, K. and Ashcroft, N. W. (1991). Fluctuation attraction in condensed matter: A nonlocal functional approach, *Physical Review B* **44**, 4032–4035.

Rapisarda, F. and Senatore, G. (1996). Diffusion Monte Carlo study of electrons in two-dimensional layers, *Australian Journal of Physics*, **49**, 161–182.

Rasolt, M. and Geldart, D. J. W. (1986). Exchange and correlation energy in a nonuniform fermion fluid, *Physical Review B* **34**, 1325–1328.

Reimann, S. M. and Manninen, M. (2002). Electronic structure of quantum dots, *Reviews of Modern Physics* **74**, 1283–1342.

Reizer, M. Yu. (1989). Relativistic effects in the electron density of states, specific heat, and the electron spectrum of normal metals, *Physical Review B* **40**, 11571–11575.

Reizer, M. and Wilkins, J. W. (1997). Electron-electron relaxation in heterostructures, *Physical Review B* **55**, R7363–7366.

Rice, T. M. (1965). The effects of electron-electron interactions on the properties of metals, *Annals of Physics* **31**, 100–129.

Richardson, C. F. and Ashcroft, N. W. (1994). Dynamical local-field factors and effective interactions in the three-dimensional electron liquid, *Physical Review B* **50**, 8170–8181.

Ritchie, R. N. (1959). Interaction of Charged Particles with a Degenerate Fermi-Dirac Electron Gas, *Physical Review* **114**, 644–654.

Rimberg, A. J. and Westerwelt, R. M. (1989). Electron energy levels for a dense electron gas in parabolic GaAs/Al_xGa_{1-x}As quantum wells, *Physical Review B* **40**, 3970–3974.

Roddaro, S. *et al.* (2003). Nonlinear Quasiparticle Tunneling between Fractional Quantum Hall Edges, *Physical Review Letters* **90**, 046805.

Roukes, M. L. *et al.* (1987). Quenching of the Hall Effect in a One-Dimensional Wire, *Physical Review Letters* **59**, 3011–3014.

Rowland, T. (1960). Nuclear Magnetic Resonance in Copper Alloys. Electron Distribution Around Solute Atoms, *Physical Review* **119**, 900–912.

Rudermann, M. A. and Kittel, C. (1954). Indirect Exchange Coupling of Nuclear Magnetic Moments by Conduction Electrons, *Physical Review* **96**, 99–102.

Runge, E. and Gross, E. K. U. (1984). Density-Functional Theory for Time-Dependent Systems, *Physical Review Letters* **52**, 997–1000.

Sakurai, J. J. (1985). *Modern Quantum Mechanics* (Addison-Wesley Publishing Company, Inc., Redwood City).

Saminadayar, L., Glattli, D. C., Jin, Y., and Etienne, B. (1997). Observation of the e/3 Fractionally Charged Laughlin Quasiparticle, *Physical Review Letters* **79**, 2526–2529.

Santoro, G. and Giuliani, G. F. (1988a). Acoustic plasmons in a conducting double layer, *Physical Review B* **37**, 937–940.

(1988b). Exact limits of the many-body local fields in a two-dimensional electron gas, *Physical Review B* **37**, 4813–4815.

(1989). Electron self-energy in two dimensions, *Physical Review B* **39**, 12818–12827.

Sawada, K., Brueckner, K. A., Fukuda, N., and Brout, R. (1957). Correlation Energy of an Electron Gas at High Density: Plasma Oscillations, *Physical Review* **108**, 507–514.

Scarola, V. W., Park, K., and Jain, J. K. (2000). Rotons of composite fermions: Comparison between theory and experiment, *Physical Review B* **61**, 13064–13072.

Schindlmayr, A. and Godby, R. W. (1998). Systematic Vertex Corrections through Iterative Solution of Hedin's Equations Beyond the GW Approximation, *Physical Review Letters* **80**, 1702–1705.

Schmid, A. (1974). On the dynamics of electrons in an impure metal, *Zeitschrift für Physik* **271**, 251–256.

Schöne, W.-D. and Eguiluz, A. G. (1998). Self-consistent calculation of quasiparticle states in metals and semiconductors, *Physical Review Letters* **81**, 1662–1665.

Schülke, W., Schulte-Schrepping, H., and Schmitz, J. R. (1993). Dynamic structure of electrons in Al metal studied by inelastic x-ray scattering, *Physical Review B* **47**, 12426–12436.

Schulz, H. J. (1993). Wigner crystal in one dimension, *Physical Review Letters* **71**, 1864–1867.

Schulz, H. J., Cuniberti, G., and Pieri, P. (1995). Fermi liquids and Luttinger liquids, in *Field Theories for Low-Dimensional Condensed Matter Systems*, edited by G. Morandi *et al.* (Springer, Berlin, 2000) (cond-mat/9807366).

Schulze-Wischeler, F., Mariani, E., Hohls, F., and Haug, R. J. (2004). Direct measurement of the g-factor of composite fermions, *Physical Review Letters* **92**, 156401.

Schumacher, R. T., Carver, T., and Slichter, C. P. (1954). Measurement of the Spin Paramagnetism of Conduction Electrons, *Physical Review* **95**, 1089–1090.

Schumacher, R. T. and Slichter, C. P. (1956). Electron Spin Paramagnetism of Lithium and Sodium, *Physical Review* **101**, 58–65.

Schumacher, R. T. and Vehse, W. E. (1960). Paramagnetic susceptibility of Sodium metal at $4°$ K, *Bulletin of the American Physical Society* **4**, 296.

(1963). The paramagnetic susceptibility of sodium metal, *Journal of Physics and Chemistry of Solids* **24**, 297–307.

Seitz, F., *Modern Theory of Solids*, McGraw-Hill, New York (1940).

Senatore, G., Moroni, S., and Ceperley, D. M. (1996). The Local Field of the Electron Gas, *Proceedings of the International Conference on the Physics of Strongly Coupled Plasmas*, edited by W. D. Kraeft and M. Schlanges (World Scientific, Singapore), p. 429–434.

(1999). Static response of homogeneous quantum fluids by diffusion Monte Carlo, *Quantum Monte Carlo methods in physics and chemistry*, edited by M.P. Nightingale and C. J. Umrigar (Kluwer, Dordrecht).

Sham, L. J. (1971). In *Computational methods in band theory*, ed. by Marcus, P. J., Janak, F. and Williams, A. R. (Plenum, New York), p. 458.

Shankar, R. (1994). Renormalization-group approach to interacting fermions, *Reviews of Modern Physics* **66**, 129–192.

Sharp, R. T. and Horton, G. K. (1953). A Variational Approach to the Unipotential Many-Electron Problem, *Physical Review* **90**, 317.

Sholl, C. A. (1967). The calculation of electrostatic energies in metals by plane-wise summation, *Proceedings of the Physical Society, London* **92**, 434–445.

Shytov, A. V., Levitov, L. S., and Halperin, B. I. (1998). Tunneling into the Edge of a Compressible Quantum Hall State, *Physical Review Letters* **80**, 141–144.

Silverstein, S. D. (1962). Influence of Electron Interactions on Metallic Properties. I. Specific Heat, *Physical Review* **128**, 631–637.

(1963). Effects of Electron Correlations on the Properties of Alkali Metals, *Physical Review* **130**, 912–913.

Simkin, D., Ph.D. Thesis, Univ. Illinois, unpublished (1963).

Singwi, K. S., Tosi, M. P., Land, R. H., and Sjölander, A. (1968). Electron Correlations at Metallic Densities, *Physical Review* **176**, 589–599.

Singwi, K. S. and Tosi, M. P. (1981). Correlations in electron liquids *Solid State Physics* **36**, edited by H. Ehrenreich, F. Seitz, and D. Turnbull (Academic, New York), 177–266.

Skudlarski, P. and Vignale, G. (1993). Exchange-correlation energy of a three-dimensional electron gas in a magnetic field, *Physical Review B* **48**, 8547–8559.

Slater, J. C. (1951). A Simplification of the Hartree-Fock Method, *Physical Review* **81**, 385–390.

Slichter, C. P., *Principles of Magnetic Resonance, 3rd edition* (Springer-Verlag, Berlin, 1990), Chapter 4.

Sommerfeld, A. (1928). Electron theory of metals on the basis of Fermi statistics, *Zeitschrift für Physik* **47**, 1–32 and 43–60.

Spruch, L. (1991). Pedagogic notes on Thomas-Fermi theory (and on some improvements): atoms, stars, and the stability of bulk matter, *Review of Modern Physics* **63**, 151–209.

Steffens, O., Suhrke, M., and Rössler, U. (1998). Spontaneously broken time-reversal symmetry in quantum dots, *Europhysics Letters* **44**, 222–228.

Stern, F. (1967). Polarizability of a Two-Dimensional Electron Gas, *Physical Review Letters* **18**, 546–548.

Stormer, H. L. (1992). Two-dimensional electron correlation in high magnetic fields, *Physica B* **177**, 401–408.

Středa, P. (1982). Theory of quantised Hall conductivity in two dimensions, *J. Phys. C* **15**, L717–L722.

Sturm, A. and Gusarov, K. (2000). Dynamical correlations in the electron gas, *Physical Review B* **62**, 16474–16491.

Suhl, H. and Werthamer, R. N. (1961). Higher Random-Phase Approximations in the Many-Body Problem, *Physical Review* **122**, 359–366.

(1962). Renormalization of Many-Fermion Momentum-Space Distributions in Higher Random Phase Approximations, *Physical Review* **125**, 1402–1406.

Takada, Y. (1993). s- and p-wave pairings in the dilute electron gas: Superconductivity mediated by the Coulomb hole in the vicinity of the Wigner-crystal phase, *Physical Review B* **47**, 5202–5211.

(1995). Exact self-energy of the many-body problem from conserving approximations, *Physical Review B* **52**, 12708–12719.

Takada, Y. and Goto, H. (1998). Exchange and correlation effects in the three-dimensional electron gas in strong magnetic fields and application to graphite, *Journal of Physics: Condensed Matter* **10**, 11315–11325.

Tamm, I. (1932). Possible type of electron binding on crystal surfaces, *Zeitschrift für Physik* **76**, 849–850.

Tanaka, S. and Ichimaru, S. (1989). Spin-dependent correlations and thermodynamic functions for electron liquids at arbitrary degeneracy and spin polarization, *Physical Review B* **39**, 1036–1051.

anatar, B. and Ceperley, D. M. (1989). Ground state of the two-dimensional electron gas, *Physical Review B* **39**, 5005–5016.

Teller, E. (1962). On the Stability of Molecules in the Thomas-Fermi Theory, *Review of Modern Physics* **34**, 627–631.

Tesanovic, Z., Axel, F., and Halperin, B. I. (1988). "Hall crystal" versus Wigner crystal, *Physical Review B* **39**, 8525–8551.

Theis, T. N. (1980). Plasmons in inversion layers, *Surface Science* **98**, 515–532.

Thomas, L. H. (1927). Calculation of atomic fields, *Proceedings of the Cambridge Philosophical Society* **33**, 542–548.

Thomson, J. J. (1897). Cathode rays, *Philosophical Magazine* **44**, 293–316.

Thouless, D. J. (1984). Wannier functions for magnetic subbands, *J. Phys. C: Solid State Physics* **17**, L325–L327.

Ting, C. S., Ying, S. C., and Quinn, J. J. (1977). Theory of cyclotron resonance of interacting electrons in a semiconducting surface inversion layer, *Physical Review* **B 16**, 5394–5404.

Toigo, F. and Woodruff, T. O. (1971). Calculation of the Dielectric Function for a Degenerate Electron Gas with Interaction. II. Frequency Dependence, *Physical Review B* **4**, 4312–4315.

Tokatly, I. V. and Pankratov, O. (2001). Many-Body Diagrammatic Expansion in a Kohn-Sham Basis: Implications for Time-Dependent Density Functional Theory of Excited States, *Physical Review Letters* **86**, 2078–2081.

Tomonaga, S. (1950). Remarks on Bloch's method of sound waves applied to many-fermion problems, *Progress of Theoretical Physics* **5**, 544–569.

Trugman, S. A. (1983). Localization, percolation, and the quantum Hall effect, *Physical Review B* **27**, 7539–7546.

Tsui, D. C., Störmer, H. L., and Gossard, A. C. (1982). Two-Dimensional Magnetotransport in the Extreme Quantum Limit, *Physical Review Letters* **48**, 1559–1562.

Tutuc, E., De Poortere, E. P., Papadakis, S. J., and Shayegan, M. (2001). In-plane magnetic field-induced spin polarization and transition to insulating behavior in two-dimensional hole system, *Physical Review Letters* **86**, 2858–2861.

Ueda, S. (1961). The pair correlation function of an imperfect electron gas in high densities, *Progress of Theoretical Physics* **26**, 45–50.

Ullrich, C. A. and Vignale, G. (2002). Time-dependent current density functional theory for the linear response of weakly disordered systems, *Physical Review B* **65**, 245102.

Vakili, K., Shkolnikov, Y. P., Tutuc, E., De Poortere, E. P., and Shayegan, M. (2004). Spin susceptibility of two-dimensional electrons in narrow AlAs quantum wells, *Physical Review Letters* **92**, 226401.

Vanderbilt, D. (1997). Nonlocality of Kohn-Sham Exchange-Correlation Fields in Dielectrics, *Physical Review Letters* **79**, 3966–3999.

van Faassen, M., de Boeij, P. L., van Leeuwen, R., Berger, J. A., and Snijders, J. G. (2002). Ultranonlocality in Time-Dependent Current-Density-Functional Theory: Application to Conjugated Polymers, *Physical Review Letters* **88**, 186401.

van Gisbergen, S. J. A., Kootstra, F., Schipper, P. R. T., Gritsenko, O. V., Snijders, J. G., and Baerends, E. J. (1998). Density-functional-theory response-property calculations with accurate exchange-correlation potentials, *Physical Review A* **57**, 2556–2571.

van Gisbergen, S. J. A., Schipper, P. R. T., Gritsenko, O. V., Baerends, E. J., Snijders, J. G., Champagne, B., and Kirtman, B. (1999). Electric Field Dependence of the

Exchange-Correlation Potential in Molecular Chains, *Physical Review Letters* **83**, 694–697.

van Leeuwen, R. (1998). Causality and Symmetry in Time-Dependent Density-Functional Theory, *Physical Review Letters* **80**, 1280–1283.

(1999). Mapping from Densities to Potentials in Time-Dependent Density-Functional Theory, *Physical Review Letters* **82**, 3863–3866.

(2001). Key concepts in time-dependent density functional theory, *International Journal of Modern Physics* **15**, 1969–2023.

van Wees, B. J., van Houten, H., Beenakker, C. W. J., Williamson, J. G., Kouwenhoven, L. P., van der Marel, D., Foxon, C. T., and Harris, J. J. (1988). Quantized conductance of point contacts in a two-dimensional electron gas, *Physical Review Letters* **60**, 848–850.

Vashishta, P. and Singwi, K. S. (1972). Electron Correlations at Metallic Densities. V, *Physical Review B* **6**, 875–887.

Vasiliev, I., Öğüt, S., and Chelikowsky, J. R. (1999). Ab Initio Excitation Spectra and Collective Electronic Response in Atoms and Clusters, *Physical Review Letters* **82**, 1919–1922.

Vignale, G. and Singwi, K. S. (1984). Possibility of superconductivity in the electron-hole liquid, *Physical Review B* **31**, 2729–2749.

(1985). Effective two-body interaction in Coulomb Fermi liquids, *Physical Review B* **32**, 2156–2166.

Vignale, G. and Rasolt, M. (1987). Density functional theory in strong magnetic fields, *Physical Review Letters* **59**, 2360–2363.

Vignale, G., Rasolt, M., and Geldart, D. J. W. (1988). Diamagnetic susceptibility of a dense electron gas, *Physical Review B* **37**, 2502–2507.

Vignale, G. (1988). Exact behavior of the density and spin susceptibilities of a Fermi liquid for large wave vectors: Derivation from diagrammatic many-body theory, *Physical Review B* **38**, 6445–6451.

(1993). Current-density-functional theory of the two-dimensional Wigner crystal in a strong magnetic field, *Physical Review B* **47**,10105–10111.

(1995). Sum rule for the linear density response of a driven electronic system, *Physics Letters A* **209**, 206–210.

Vignale, G., Skudlarski, P., and Rasolt, M. (1992). Current-density functional theory of surface properties of electron-hole droplets at high magnetic fields, *Physical Review B* **45**, 8494–8497.

Vignale, G. and Kohn, W. (1996). Current dependent exchange-correlation potential for dynamical linear response theory, *Physical Review Letters* **77**, 2037–2040.

(1998). Current-density functional theory of linear response to time-dependent electromagnetic fields, in *Electronic Density Functional Theory*, ed. by J. F. Dobson, G. Vignale, and M. P. Das (Plenum, 1998), p. 199–216.

Vignale, G., Ullrich, C. A., and Conti, S. (1997). Time-dependent density functional theory beyond the adiabatic local density approximation, *Physical Review Letters* **79**, 4878–4881.

Voit, J. (1994). One dimensional Fermi liquids, *Reports of Progress in Physics*, **58**, 977–1116.

vom Felde, A., Sprösser-Prou, J., and Fink, J. (1989). Valence-electron excitations in the alkali metals, *Physical Review B* **40**, 10181–10193.

on Barth, U. and Hedin, L. (1972). A local exchange-correlation potential for the spin polarized case. *Journal of Physics C: Solid State Physics* **5**, 1629–1642.

on Klitzing, K., Dorda, G., and Pepper, M. (1980). New Method for High-Accuracy Determination of the Fine-Structure Constant Based on Quantized Hall Resistance, *Physical Review Letters* **45**, 494–497.

Vosko, S. H., Perdew, J. P., and MacDonald, A. H. (1975). Ab initio calculation of the spin susceptibility for the alkali metals using the density-functional formalism, *Physical Review Letters* **35**, 1725–1728.

Vosko, S. H., Wilk, L., and Nuisair, M. (1980). Accurate spin-dependent electron liquid correlation energies for local spin density calculations: a critical analysis, *Canadian Journal of Physics* **58**, 1200.

Watabe, M. (1962). Transport coefficients of a quantum plasma, *Progress of Theoretical Physics* **28**, 265–282.

Wegner, F. (1983). Exact density of states for lowest Landau level in white noise potential, *Zeitschrift für Physik B* **51**, 279–285.

Weizsäcker, C. F. von (1935).Theory of nuclear masses *Zeitschrift für Physik* **96**, 431–458.

Wen, X. G. (1990). Chiral Luttinger liquid and the edge excitations in the fractional quantum Hall states, *Physical Review B* **41**,12838–12844.

Wick, G. C. (1950). The Evaluation of the Collision Matrix, *Physical Review* **80**, 268–272.

Wigner, E. P. (1934). On the Interaction of Electrons in Metals, *Physical Review* **46**, 1002–1011.

(1938). Effects of the electron-electron interactions on the energy levels of electrons in metals, *Transactions of the Faraday Society* **34**, 678–685.

Wigner, E. P. and Seitz, F. (1933). On the Constitution of Metallic Sodium, *Physical Review* **43**, 804–810.

(1934). On the Constitution of Metallic Sodium. II, *Physical Review* **46**, 509–524.

Willett, R. L., Ruel, R. R., West, K. W., and Pfeiffer, L. N. (1993a). Experimental demonstration of a Fermi surface at one-half filling of the lowest Landau level, *Physical Review Letters* **71**, 3846–3849.

Willett, R. L., Ruel, R. R., Paalanen, M. A., West, K. W., and Pfeiffer, L. N. (1993b). Enhanced finite-wave-vector conductivity at multiple even-denominator filling factors in two-dimensional electron systems, *Physical Review B* **47**, 7344–7347.

Willett, R. L. (1997). Experimental evidence for composite fermions *Advances in Physics* **46**, 447–544.

Williams, P. F. and Bloch, A. (1974). Self-consistent dielectric response of a quasi-one-dimensional metal at high frequencies, *Physical Review B* **10**, 1097–1108.

Williams, J. B., Sherwin, M. S., Maranowski, K. D., and Gossard, A. C. (2001). Dissipation of Intersubband Plasmons in Wide Quantum Wells, *Physical Review Letters* **87**, 037401.

Wolfe, J. P. *et al.* (1975). Photograph of an electron-hole drop in Germanium, *Physical Review Letters* **34**, 1292–1293.

Yarlagadda, S. and Giuliani, G. F. (1989). Spin susceptibility in a two-dimensional electron gas, *Physical Review B* **40**, 5432–5440.

(1994a). Quasiparticle pseudo-Hamiltonian of an infinitesimally polarized Fermi liquid, *Physical Review B* **49**, 7887–7897.

(1994b). Landau theory of Fermi liquids and the integration-over-the-coupling-constant algorithm, *Physical Review B* **49**, 14172–14178.

(1994c). Many-body local fields and Fermi-liquid parameters in a quasi-two-dimensional electron liquid, *Physical Review B* **49**, 14188–14196.

Yasuhara, H. and Kawazoe, Y. (1976). A note on the momentum distribution function for an electron gas, *Physica A* **85**, 416–424.

Yasuhara, H. and Ousaka, Y. (1992). Effective mass, Landau interaction function and self-energy in an electron liquid, *International Journal of Modern Physics B* **6**, 3089–3145.

Yoon, J., Li, C. C., Shahar, D., Tsui, D. C., and Shayegan, M. (1999). Wigner Crystallization and Metal-Insulator Transition of Two-Dimensional Holes in GaAs at $B = 40$, *Physical Review Letters* **82**, 1744–1747.

Yosida, K. (1957). Magnetic Properties of Cu-Mn Alloys, *Physical Review* **106**, 893–898.

Zangwill, A. and Soven, P. (1980). Resonant Photoemission in Barium and Cerium, *Physical Review Letters* **45**, 204–207.

Zhang, S.-C. (1992). The Chern-Simons-Landau-Ginzburg theory of the fractional quantum Hall effect, *International Journal of Modern Physics B* **6**, 25–58.

Zhao, Q., Morrison, R. C., and Parr, R. G. (1994). From electron densities to Kohn-Sham kinetic energies, orbital energies, exchange-correlation potentials, and exchange-correlation energies, *Physical Review A* **50**, 2138–2142.

Zheng, L. and Das Sarma, S. (1996). Coulomb scattering lifetime of a two-dimensional electron gas, *Physical Review B* **53**, 9964–9967.

Zhu, X. and Overhauser, A. W. (1984). Exact exchange and correlation corrections for large wave vectors, *Physical Review B* **30**, 3158–3163.

(1986). Plasmon-pole and paramagnon-pole model of an electron liquid, *Physical Review* **33**, 925–936.

Zhu, J., Stormer, H. L., Pfeiffer, L. N., Baldwin, K. W., and West, K. W. (2003). Spin susceptibility of ultra-low density two-dimensional electron system, *Physical Review Letters* **90**, 056805.

Zong, F. H., Lin, C., and Ceperley, D. M. (2002). Spin polarization of the low-density three-dimensional electron gas, *Physical Review E* **66**, 036703.

Index

Made in the USA
Lexington, KY
03 January 2013